T0213360

Handbook of
Environmental and
Ecological Statistics

Handbook of Environmental and Ecological Statistics

Alan Gelfand, PhD
Montse Fuentes, PhD
Jennifer A. Hoeting, PhD
Richard L. Smith, PhD

CRC Press
Taylor & Francis Group
Boca Raton London New York

CRC Press is an imprint of the
Taylor & Francis Group, an **informa** business

A CHAPMAN & HALL BOOK

CRC Press
Taylor & Francis Group
6000 Broken Sound Parkway NW, Suite 300
Boca Raton, FL 33487-2742

First issued in paperback 2020

ISBN-13: 978-1-4987-5202-2 (hbk)
ISBN-13: 978-0-367-73178-6 (pbk)

Library of Congress Cataloging-in-Publication Data

Names: Gelfand, Alan E., 1945- author.
Title: Handbook of environmental and ecological statistics / Alan E. Gelfand, Montserrat Fuentes, Jennifer A. Hoesting, Richard Smith.
Description: Boca Raton : Taylor & Francis, 2018. | Includes bibliographical references.
Identifiers: LCCN 2018013123| ISBN 9781498752022 (hardback : alk. paper) | ISBN 9781351648547 (ebook)
Subjects: LCSH: Environmental sciences--Statistical methods. | Ecology--Statistical methods.
Classification: LCC GE45.S73 G45 2018 | DDC 557.072/7--dc23
LC record available at https://lccn.loc.gov/2018013123

Visit the Taylor & Francis Web site at
http://www.taylorandfrancis.com

and the CRC Press Web site at
http://www.crcpress.com

Contents

9 Environmental Sampling Design   181

Dale L. Zimmerman and Stephen T. Buckland

10 Accommodating so many zeros: univariate and multivariate data   211

James S. Clark and Alan E. Gelfand

Preface

We are delighted to contribute this Handbook to the growing Chapman & Hall/CRC series of Statistics handbooks. This particular volume focuses on the enormous literature applying statistical methodology and modeling to environmental and ecological processes. The 21st century statistics community has become increasingly interdisciplinary, bringing a large collection of modern tools to all areas of application in the environmental processes. In addition, the environmental community has substantially increased its scope of data collection including, e.g., observational data, satellite-derived data, and computer model output. The resultant impact in this latter community has been substantial; no longer are simple regression and analysis of variance methods adequate. The contribution of this handbook is to assemble, in 35 chapters, a state-of-the-art view of this interface.

As with the previous volumes in the Handbook series, we envision this book as a resource for researchers. Our intended audience is statisticians who engage in inter-disciplinary research applied, as the title suggests, to challenges in the environmental and ecological sciences. By providing a current view of these areas, for researchers looking to explore problems in these fields, it offers an overview of and facilitate access to active problem areas. A reciprocal audience will be environmental scientists and ecologists seeking a current view of statistical methodology and modelling in their fields of interest. The four editors have international reputations in this area, bringing collectively more than 80 years of experience. We think we have recognized the need and the scope for such a volume and think it has come together well.

The book is written at the level of a Masters in Statistics, making it accessible to a broad range of quantitative researchers. Attractively, it will be partitioned into 35 chapters commencing, after an Introduction, with 10 chapters of foundational material followed by 8 chapters on ecological processes, then 7 chapters on environmental exposure, and concluding with 9 chapters on climatology.

The editors thank the chapter authors for taking their valuable to time to provide strong contributions to this volume. We also thank David Grubbs, the senior editor at Taylor & Francis for overseeing this handbook and helping to shepherd it through to completion. In addition, we thank Robin Lloyd-Starkes, the project editor at Taylor & Francis for handling the production of a volume which has turned out to be much bigger than the editors anticipated when they started.

Alan E Gelfand, Duke University
Montserrat Fuentes, Virginia Commonwealth University
Jennifer Hoeting, Colorado State University
Richard L. Smith, University of North Carolina, Chapel Hill
2018

1

Introduction

Alan E Gelfand
Duke University

Montserrat Fuentes
Virginia Commonwealth University

Jennifer Hoeting
Colorado State University

Richard L. Smith
University of North Carolina Chapel Hill

The papers in Section I focus on methodology. The intent is to present a collection of approaches that span a wide area of statistical thinking but also a broad range of methodology needed for the challenges that arise in working with environmental and ecological data. Each of these chapters is fairly formal, somewhat technical, elaborating properties and behaviors but also suggestive and, in some cases, inclusive of application. The level of the papers varies technically but an effort has been made to make them quite accessible and focused.

Chapter 2 (Gelfand) is a modeling chapter, discussing the basics of modeling for environmental processes, broadly reviewing stochastic modeling, hierarchical modeling, capturing uncertainty, model adequacy, and model comparison. Chapter 3 (Craigmile) takes up time series methodology, in particular, classical methods, long-range dependence and change points. Chapter 4 (Schmidt and Lopes) turns to dynamic models, opening up state space ideas, stochastic partial differential equations, integro-difference equations, projection ideas, along with discretization for implementation.

Next, we move to spatial approaches. Chapter 5 (Banerjee) takes up geostatistical modeling for environmental processes, reviewing the basic tools, e.g., variograms, covariance functions, kriging, particularly working with Gaussian processes. Chapter 6 (Illian) considers point patterns, offering basic theory and inference beginning with one dimensional models. However, primarily, it focuses on point patterns in two dimensions, again with associated inference. Chapter 7 (Berrocal) reviews data fusion, summarizing the literature from climatology. It then considers fusing environmental data sources, in particular, combining deterministic models with observations, using, e.g., melding and downscaling.

Chapter 8 (Cooley, Hunter, and Smith) turns to the analysis of extremes. Here, we start with basic theory but then move to multivariate data and spatially referenced extremes, citing illustrations for climate and for environmental exposure. Chapter 9 (Zimmerman and Buckland) focuses on techniques for environmental sampling. Topics include sampling for locations, e.g., for monitoring sites, geometric/space filling strategies, model-based strategies, sampling for field sites and transects with distance sampling. Chapter 10 (Clark and Gelfand) takes up a frequent challenge with ecological data, the problem of zeros. Zeros require care in terms of accommodating so-called zero-inflation settings but, more generally, need attention in the context of multinomial trials and joint species distribution modeling. Chapter 11 (Palmer), the last chapter of this section, takes up ordination methods, also referred to as gradient analysis. Tools like principal components analysis (PCA), canonical correlation analysis (CCA), and factor analysis are the bread and butter for clustering and dimension reduction approaches in ecological data analysis.

Turning to Section II, by definition, ecology is the study of how organisms interact with one another and with their surroundings. Ecologists seek to understand factors that impact population size, the distribution of populations, and interactions with other species and the environment. All of these factors can vary over space and time. There is a long history of statisticians developing new methodology to address challenging problems in ecology. Likewise, ecologists often use sophisticated statistical methods. The chapters in this section provide a comprehensive overview of the most important and currently relevant topics in ecological statistics.

There are many reasons for the close relationship between statisticians and ecologists. Data collected to study ecological processes are often observational, messy, and can involve very large data sets. Thus basic statistical approaches like ANOVA models are typically inadequate. In addition, the field of theoretical ecology is closely linked to mathematical biology–a field in which applied mathematical tools are used to address problems in biology. Statistical models for ecological processes are often inspired by or directly incorporate these theoretical models allowing for statistical inference on the parameters of the theoretical models. Altogether, in order to develop new and useful statistical methods, a statistician working on ecological problems needs to bring substantial understanding of the ecological process under study as well as a broad knowledge of statistical methodology.

Species distribution models are a primary focus in the study of animal and plant populations in order to develop a better understanding of where species live and why. Chapter 12 (Ovaskainen) presents an overview including models for individual species as well as models for more than one species. The focus in this chapter is on species-distribution models for presence-absence data via generalized mixed models.

A related problem is population size estimation. Chapter 13 (Barker) surveys the vast field of capture-recapture and distance sampling methodology to estimate population abundance. Knowledge of the population under study as well as study design influence the choice of model for abundance estimation. There are separate models for open and closed populations and for different study designs such as mark-recapture and tag-recovery designs. In addition, sampling challenges must be addressed in the modeling such as imperfect detection of animals. Furthermore, these models allow for estimation of other key parameters of interest such as birth and death rates, probability of survival, and population growth rates.

Technological advances have made it less expensive to collect fine-scale temporal data on geo-referenced animal locations. In response, a rich set of statistical models for telemetry data have been proposed to better understand movement of individual animals in space. Chapter 14 (Hooten and Johnson) surveys parametric spatio-temporal animal movement models including point-process as well as discrete- and continuous time-models. Models of animal movement can be used improve understanding of the impact of human-caused changes in habitat size and availability for animal populations or to better understand migration patterns for bird, insect or mammal populations.

Understanding the population size and age structure of animal populations can aid in basic understanding or improve efforts aimed at managing fishery and wildlife populations. Chapter 15 (Newman) presents traditional as well state-of-the-art statistical methodology for modeling the structure and dynamics of biological populations. Key models used in animal demography are described including population dynamics, matrix projection, integral projection, individual-based, and state-space models. This chapter also examines the various statistical approaches used to estimate parameters for this wide-ranging set of models.

In traits-based analyses, organisms are categorized by their biological attributes. For example, in order to understand the impact of stream disturbances, stream macroinvertebrate species may be categorized based on pollution sensitivity or tolerance to water temperature changes due to drought. Traits-based analyses allow ecologists to move beyond a narrow focus on one or two species, particularly useful when many species live in a given habitat type. Trait-based analyses are used by ecologists to better understand patterns of species occurrence, predict changes in community composition, and understand ecosystem function. Chapter 16 (Aiello-Lammens and Silander) reviews statistical methods for trait-based ecological modeling. The authors describe data structures, exploratory data analyses methods as well as algorithmic approaches and statistical model-based inference for traits data.

The section on models for ecological processes closes with two chapters focusing on statistical models to address specific ecological problems. Chapter 17 (Pereira and Turkman) provides an overview of statistical models for vegetation fires like grass and forest fires. With the changing climate, forest fires have already or are expected to increase in size and intensity in many regions of the world, with the recent large fires in Canada and Australia as two examples. Careful modeling of spatial, temporal and spatio-temporal patterns of vegetation fires can help government planners to predict and manage future fires. The authors provide a comprehensive overview of models for fire size, fire incidence, and fire risk maps which have been used by ecologists, fire scientists and others to advance the field of fire science.

Similarly, the changing climate has impacted precipitation patterns and timing which in turn impacts stream flow patterns. Chapter 18 (Ver Hoef, Peterson, and Isaak) provides an in-depth study of models for stream networks. Stream data requires unique models because streams work differently from other physical processes. Stream data are spatially-referenced but come from physically-connected processes. It is not appropriate to use Euclidean distance to measure stream distance because two locations that are nearby in physical space may not be part of the same stream network. Using hydrological distance in a standard geostatistical model has been proposed as a solution to this problem, but naïve use of hydrological distance in geostatistical models can produce non-positive covariance matrices and other mathematical inconsistencies. Thus neither standard time-series nor standard spatial statistical models can be applied to data collected on stream networks. The authors describe models specifically developed to address stream-network construction. They show that such models can produce accurate predictions that are consistent with the physical processes of streams.

For Section III of this handbook we include eight chapters offering a comprehensive coverage of topics in environmental exposure. We include standard and state-of-art methods and models for environmental exposure, while illustrating the methods in the context of different case studies.

Chapter 19 (Fuentes, Reich and Huang) is an introductory chapter that defines exposure and introduces methods for mapping across space and time of ground data, going from inverse distance weighting to Bayesian interpolation. It then follows with approaches for data fusion to combine ground data, satellite data, numerical model output and other sources of relevant information while characterizing errors and uncertainty in all sources of data.

Chapter 20 (Bell and Warren) introduces two alternative models for estimating environmental exposure: land use regression and a stochastic model that simulates human exposure. This chapter discusses limitations and opportunities for enhancement of theses alternative methods. Chapter 21 (Diggle and Giorgi) illuminates limitations in our exposure assessment due to sampling bias.

Chapter 22 (Zidek and Zimmerman) introduces a very comprehensive review of the principled statistical methods and approaches for monitoring network design. Many air pollution and environmental stressors are associated to different sources. Source apportionment aims to decompose the observed mixture of environmental stressors, as represented by measurements of individual constituents, into information about the sources that contribute to the mixture. Chapter 23 (Krall and Chang) describes statistical methods and the challenges for source apportionment. The concept of confounding is critical in providing estimates of health effects that are unbiased and precise. Chapter 24 (Dominici and Wilson) presents a high level overview of the important concept of confounding in environmental health studies.

Chapter 25 (Szpiro) presents an overview of statistical methods applied to exposure assessment, describing statistical theory on the consequences for health effect inferences of measurement error resulting from using predicted rather than true exposures, while presenting methods to optimize health effects inference. Chapter 26 (Sheppard), the final chapter in the environmental exposure section, focuses on study designs for environmental epidemiology, as a fundamental aspect of exposure assessment.

The chapters in Section IV of this Handbook concern statistical methods in climate science. Climate science has long been concerned with large datasets databases of historical weather data now cover tens of thousands of stations that in many cases contain daily data from the mid nineteenth century onwards. Added to that, there are many new sources of remote sensing data such as NASA's OCO2 satellite, launched in 2014, which measures near-surface carbon dioxide data at resolutions down to a kilometer. On the modeling side, climate model datasets have grown in little more than a decade from 40 terabytes (the CMIP3 dataset, completed in 2006) to 2 petabytes (CMIP5, 2012) to an estimated 10-30 petabytes for the CMIP6 dataset that is projected to become available in 2019. These datasets are compilations of climate model simulations from centers all over the world and play a major role in the successive reports of the Intergovernmental Panel on Climate Change. These large datasets pose major challenges for statistical analysis.

Chapter 27 by Craigmile and Guttorp addresses one of the oldest but still very critical questions in climate science: how does one measure and determine the size of trends? This chapter focuses on classical methods based on time series analysis but gives a particular focus to the problem of measurement error in the observed series. Chapter 28 by Stephenson is a review of climate models and the kind of datasets they produce, and includes a short review of the multi-model ensemble problem, which is about combining the output of different climate models for prediction purposes.

Chapter 29 by Nychka and Wikle discusses spatial statistics given that most climate datasets are indexed in space as well as time, this is a key statistical methodology for anyone hoping to do statistical research in this field. Chapter 30, by Budhiraja and colleagues, is about data assimilation, which is the process of combining climate and weather models with observational data so that the unobserved state of the system is continuously updated. Classical applied mathematics techniques such as 3D-VAR and 4D-VAR are compared with more statistically oriented approaches such as particle filtering. These methods have long been used for numerical weather forecasting but are finding increasing application for long-term climate prediction as well.

Chapter 31, by Davison and colleagues, is a continuation of Chapter 8 on extremes, in this case, focused on spatial extremes. This chapter describes the modern class of stochastic process models known as max-stable processes, including the estimation methods of composite likelihood and exact likelihood analysis, and concludes with two examples applied to large spatiotemporal datasets. Chapter 32 by Wikle discusses the use of statistics in oceanography, covering both time series and spatial statistics methods and illustrating how Bayesian hierarchical models may be used for predicting processes such as El Niño. Chapter 33, by Craigmile and colleagues, also demonstrates the huge power of Bayesian hierarchical models to combine data from different sources, here applied to the problems of paleoclimatology, which is about the use of proxy datasets such as tree rings and ice cores to reconstruct historical temperature records before modern measurement techniques were available.

Chapter 34 by Hammerling and colleagues is about detection and attribution, the technique that exists for deciding whether trends in climate data are directly associated with anthropogenic signals (the detection problem) and, if so, how the trend may be apportioned among the different anthropogenic and natural sources of variation (attribution). They also touch upon problems of extreme event attribution, which is concerned with how much specific extreme events (such as the successive hurricanes that devastated parts of Central and North America during the summer of 2017) are associated with anthropogenic sources of climate change. Finally in Chapter 35, Ebi and co-authors have provided an expert review of many different aspects of climate and health, covering both vector-borne diseases such as malaria and the direct health effects of extreme weather such as heatwaves, and address many of the interdisciplinary challenges that this field raises.

Part I

Methodology for Statistical Analysis of Environmental Processes

2

Modeling for environmental and ecological processes

Alan E. Gelfand

Department of Statistical Science, Duke University, Durham, NC

CONTENTS

Abstract

The 21st century is witnessing dramatic changes in the way that datasets are being analyzed. Data collection is more and more focused on observational data rather than designed experiments. Datasets are growing increasingly larger. This is particularly true for environmental and ecological data settings. This chapter suggests that a valuable way to infer from such datasets is through formal stochastic model specification. That is, in order to understand and explain the behavior of a complex environmental or ecological process, it can be most effective to build process-based models. Such models introduce simplification but attempt to capture key features of the process and link the model generatively to the data. This chapter attempts to provide a general overview of such modeling enterprise. Many of the subsequent chapters develop sophisticated examples, fabricated according to their context.

2.1   Introduction

In a handbook on environmental and ecological statistics, the swath of statistical work to review is enormous. Statistical analyses span the spectrum from purely descriptive presentation to sophisticated optimization strategies to multi-level modeling. Over the ensuing chapters, the full range of techniques will be presented for learning about complex environmental and ecological processes from associated data. However, this chapter suggests that a contemporary perspective adopts a fully model based approach.

Arguing that the goal of a statistical analysis is formal inference, we devote this chapter to stochastic modeling, hierarchical modeling (usually handled within a Bayesian framework), identifying sources of modeling error, incorporating uncertainty appropriately, propagating uncertainty, and model assessment - adequacy and comparison. All of these issues will be considered in the subsequent chapters; the goal here is present an overview. Moreover, the amount of material is far more than this chapter can accommodate; each topic could fill more than a chapter. As a result, additional references will be supplied.

Perhaps the most important message to take away is the encouragement of *coherent* modeling. We prefer to work with models that are coherent in the sense that they are generative; the specification could have yielded the data that were observed. While all models are simplifications of the process under investigation, within environmental science and ecology, when we observe realizations from a process it seems appropriate to ask that our simplified explanation of the process could yield them. This does not assert that there is only one model specification to consider. There can be many plausible generative specifications. This is why we need to assess model adequacy and model choice. Moreover, when model specifications become too complex to enable tractable fitting, we will readily accede to suitable approximation. However, this does not deny an objective of a coherent model.

The format of this chapter is as follows. Section 2 provides a general perspective on stochastic modeling. Section 3 focuses on the Bayesian paradigm. Section 4 opens up the door to hierarchical modeling, the primary theme of the chapter. Section 5 leads us to latent variables, a catchall for many of the unknowns introduced in specifying hierarchical models. Section 6 briefly looks at mixture models while Section 7 takes us back to random effects with linkages to other chapters in this Handbook. Section 8 offers some words on dynamic models, connecting to a subsequent chapter with fuller development. Sections 9 and 10 consider model assessment, adequacy in the former, comparison in the latter. We conclude in Section 11 with a brief summary.

2.2   Stochastic modeling

Again, here we work with probability model specifications. In particular, such a model provides a distributional specification for observed data $\mathbf{Y}$ given unknown *parameters*, $\boldsymbol{\theta}$, which we write as $f(\mathbf{y}|\boldsymbol{\theta})$. This expression is almost too general to be useful. What is $\mathbf{Y}$? For instance, what sorts of variables is $\mathbf{Y}$ composed of - continuous, categorical, ordinal, nominal? Do we have vectors of observations? Do we have observations that are independent or are they dependent? Are they indexed temporally, spatially, or both?

What is $\boldsymbol{\theta}$? If we envision a process that generates $\mathbf{Y}$, then $\boldsymbol{\theta}$ can be envisioned as the process. Practically, the process is typically specified through a finite collection of parameters which we can takes as $\boldsymbol{\theta}$. Again, this is almost too general to be useful. Moreover, it raises the question of whether parameters are "real"? Since they are features of a model to explain $\mathbf{Y}$ and since the specified model is essentially never the true model, rather an approximation, these parameters need not be real. This does not suggest that it is not useful to infer about these parameters. Such inference can facilitate how covariates/predictors affect responses.

As a result, before continuing discussion of inference for $\boldsymbol{\theta}$ under $f(\mathbf{Y}|\boldsymbol{\theta})$, we digress to remark upon parametric inference, usually called estimation vs. predictive inference. The point here is that estimation takes place in the $\boldsymbol{\theta}$ space while prediction takes place in the $\mathbf{Y}$ space. Inference about $\boldsymbol{\theta}$ can not be compared with any *truth*. On the other hand, inference about $\mathbf{Y}$, perhaps a new $\mathbf{Y}_0$, takes place in the space of the data and can be compared with observed data, perhaps data held out from fitting the model. Predictive inference has a long history in the Statistics literature [25]. It is playing an increasingly important role in current statistical work with observational, rather than designed, data collection.

In any event, the implication is that, while a model may be used for both estimation and prediction, assessment of model performance, i.e., model adequacy and model comparison, should be carried out in the data space. We will elaborate this theme further in Sections 9 and 10 below.

Model selection is a critical issue in developing explanations for environmental and ecological processes. There is a huge literature on model selection. It is impossible to make a case for any particular criterion as being "best." Even if one is prepared to specify the utility for a model, there still will not be a mutually agreed upon criterion. We elaborate this a bit further in Section 10 but defer fuller discussion to the literature [13, 17].

A specific model selection challenge is variable selection. When models are developed for explanation, that is, regression models, then, according to application, we can have a rich range of covariates to consider. With regard to environmental and ecological processes, the list can be substantial. For instance, with regard to explaining biodiversity, we have factors like history, habitat heterogeneity, competition, climate and climate variability, productivity, and disturbance. Each of these factors can be captured with a variety of regressors. And, these regressors can interact with each other producing more regressors. Evidently, explanation can be very complicated. Of course, only a subset will be available in a given application but still the selection of variables to include in the model can become an issue.

The need for variable selection arises because with, for example k predictors, we have 2^k models. With increasing k, we very quickly exceed the possibility of exploring all of them. Usually, variable selection has, as a general goal, parsimony; we seek simpler explanations. If a model is essentially as complicated as the process, then it may not readily facilitate our understanding of the process. Apart from the size of the model space, variable selection is further complicated by the multicollinearity between variables. In different words, different subsets of variables can perform equally well with regard to explanation. In a high dimensional model space, we can find many regions where model performance is indistinguishable. And, since none of the models are true, it can be difficult to prefer one to another. We often fall back to an initial selection of variables that, from a process perspective, are agreed to be *important* and then investigate the addition of further variables. Variable selection criteria can also provide a measure of importance for each variable. See Section 9.

Further complications are the following. We can have different distributional specifications for models, different ways to capture stochasticity. This takes us back to model comparison and may overarch variable selection. We can have multilevel models where explanation can enter at different levels (see Sections 4-8 below). We can have nonlinearities in explanation. That is, usual modeling attempts to explain the mean, perhaps on a transformed scale, and introduces regressors linearly, a generalized linear model [13, 48]. However, with environmental and ecological processes, most relationships between variables are nonlinear. The nature of the nonlinearity, e.g., exponential, asymptotic, sigmoidal, convex, concave, etc. has been gleaned to some extent from various historical experiments. Nonlinear models are much more challenging to fit and have received little attention with regard to variable selection.

Altogether, there is too much to say about variable selection here. We offer some very brief discussion below but, since this issue is not an important focus of this chapter, we are only able to supply some useful references to the literature. Not surprisingly, this literature is, by now, enormous [13, 17], and is an acknowledged challenge across many fields.

Random effects play a critical role in the ensuing modeling ideas. They have a long history in the literature [see, e.g., 56] but in the past twenty years have emerged with a new purpose. That is, traditional random effects modeling was introduced primarily to recognize that in some experiments, individuals selected for treatments or in fact, treatments themselves, e.g., plots, arose randomly from populations. Hence, the effects that they introduced were random rather than fixed in the design and that it was vital to incorporate this source of randomness into the modeling in order to better assess significance of effects. Much of this work played out in designed experiments, typically in analysis of variance or covariance settings.

Again, the past twenty years has seen a dramatic shift toward the analysis of observational data. In this context, the role of random effects has changed dramatically. Rather than being a nuisance because effects could not comfortably assumed fixed, now they have become a valuable modeling tool. They serve as devices for model enrichment, for clarifying uncertainty.

To elaborate a bit, they are often introduced as surrogates for unobservable regressors in a process specification. That is, they flexibly soak up a portion of residual variability from a model with *fixed* regressors. They can capture dependence, e.g., if we obtain a vector of measurements for a set of individuals, assigning a common random effect to each individual in the modeling makes the measurements associated with that individual dependent. They can capture structured dependence in time, e.g., autoregressive behavior. They can capture dependence in space to reflect stronger *association* for observations closer to each other in space. In fact, they can capture space-time dependence, allowing space-time interaction in dependence. All of these examples suggest hierarchical structure. That is, the random effects may be introduced at the first stage of the modeling but their distributional specifications will be supplied at a second level of modeling. We illustrate some of these ideas below but they are fleshed out much more fully in the ensuing chapters.

The likelihood is the starting place for essentially all of the modeling in this handbook. Some chapters will take the likelihood and add a prior, leading to Bayesian modeling and inference, the focus of much of the remainder of this chapter. But, at the least, we would want a likelihood which, viewed as a density for the data, becomes a generating mechanism given parameters. It can be anticipated that likelihood-based inference will be a benchmark in formal statistical analysis into the foreseeable future. In this regard, the Likelihood Principle, in its simplest form, asserts that inference should only depend upon the data observed but not on the sampling mechanism used to obtain the data [10]. The process is what it is and does not change because of the way we collect the data[1]. However, in practice,

---

[1] The Likelihood Principle is not universally supported from a philosophical perspective. See, e.g., [47]

fully specifying a stochastic data generation mechanism and then employing the associated likelihood will usually provide inference with good properties.

The likelihood takes the density for the data, say $f(\mathbf{y}|\boldsymbol{\theta})$ and views it as a function of $\boldsymbol{\theta}$ for fixed, observed data $\mathbf{y}$. The customary point estimate for $\boldsymbol{\theta}$ is the maximum likelihood estimate (MLE), the value of $\boldsymbol{\theta}$ which maximizes the likelihood. The literature on MLE's is enormous [see, e.g., 41], and, though the concept is older, it has been popular in the Statistics community for more than 100 years, dating to the work of Fisher [4]. Optimality properties for MLE's under fairly general conditions are well-established. Asymptotic uncertainty is routinely obtained using Fisher information matrices. The EM algorithm [18] is very widely used, extending likelihood inference to missing and incomplete data settings.

However, working with the likelihood in the 21st century brings challenges. Maximization of a function for a high dimensional $\boldsymbol{\theta}$ is demanding; local optima are common. Inference is limited to a point estimate and an asymptotic covariance matrix, typically adding asymptotic normality for interval estimation. Though we may have rates of convergence, obtained technically, we often can't employ them usefully in practice. Perhaps, most importantly, likelihood inference is not a good fit with general hierarchical modeling; in fact, the likelihood itself is not uniquely determined (see Section 4). The Bayesian framework much more easily accommodates such models, as we argue for the remainder of this chapter.

2.3   Basics of Bayesian inference

We envision that multi-level models will be needed to build effective stochastic models for complex processes. Inference and fitting for such models is most directly handled within a Bayesian framework. So, here, we briefly remind the reader of the basics of Bayesian inference.

To start, how did Bayes' Theorem, an elementary result in probability, become an inference paradigm and, in fact, a controversial one? Recall the theorem in its simplest form, with two events, A and B, i.e., $P(A|B) = \frac{P(B|A)P(A)}{P(B)}$. Suppose we move to events associated with random variables. We obtain $P(X \in A|Y \in B) = \frac{P(Y \in B|X \in A)}{P(Y \in B)}$. Then, we take the simple step to densities, yielding $f(x|y) = \frac{f(y|x)f(x)}{f(y)}$.

Finally, letting $\mathbf{Y}$ denote what you observe, i.e., the data in the previous section and replacing X with $\boldsymbol{\theta}$ denoting the parameters, as in the previous section, i.e., what you don't know (didn't observe), we arrive at

$$f(\boldsymbol{\theta}|\mathbf{Y}) = \frac{f(\mathbf{Y}|\boldsymbol{\theta})\pi(\boldsymbol{\theta})}{f(\mathbf{Y})} \tag{2.1}$$

or

$$f(\boldsymbol{\theta}|\mathbf{Y}) \propto f(\mathbf{Y}|\boldsymbol{\theta})\pi(\boldsymbol{\theta}). \tag{2.2}$$

So, we have an intuitively appealing inference paradigm using the density on the left side: infer about what you don't know given what you have seen. This contrasts with the classical inference approach, using sampling distributions, $T(\mathbf{Y})$ given $\boldsymbol{\theta}$ which asks you to imagine what you might see given what you don't know.

In fact, we are really just thinking about two ways of writing a joint distribution:

$$f(\mathbf{Y}, \boldsymbol{\theta}) = f(\mathbf{Y}|\boldsymbol{\theta})\pi(\boldsymbol{\theta}) = f(\boldsymbol{\theta}|\mathbf{Y})f(\mathbf{Y}). \tag{2.3}$$

In this specification, we see our friend from the previous section, the model for $\mathbf{Y}$ given

$\boldsymbol{\theta}$, the likelihood. The specification adds a distribution for $\boldsymbol{\theta}$, the prior. The result is a generative specification: First a value of $\boldsymbol{\theta}$ is drawn from $\pi(\boldsymbol{\theta})$ and then a realization of the data is achieved by using this $\boldsymbol{\theta}$ in $f(\mathbf{y}|\boldsymbol{\theta})$ to obtain a $\mathbf{Y}$.

Conditioning in the reverse order provides so-called posterior inference, again inferring about $\boldsymbol{\theta}$ given the observed $\mathbf{Y}$. Particularly attractive is that with $f(\boldsymbol{\theta}|\mathbf{Y})$, we obtain an entire distribution to use for inference about $\boldsymbol{\theta}$. Apart from familiar measures of centrality and spread, we can obtain quantiles and probabilities of events for $\boldsymbol{\theta}$. This would seem to be much more satisfying than say the results from a likelihood analysis, i.e., a point estimate, an asymptotic variance, and a confidence interval, typically justified by an appropriate central limit theorem.

Furthermore, prediction is an immediate probabilistic consequence of the specification. That is, $f(Y_0|\mathbf{Y})$ arises from

$$f(Y_0|\mathbf{Y}) = \int f(Y_0|\mathbf{Y}, \boldsymbol{\theta}) f(\boldsymbol{\theta}|\mathbf{Y}) d\boldsymbol{\theta}. \tag{2.4}$$

The first term under the integral becomes $f(Y_0|\boldsymbol{\theta})$ if Y_0 is independent of $\mathbf{Y}$. If not, it would be obtained from the joint distribution, $f(\mathbf{Y}, Y_0|\boldsymbol{\theta})$ which would be available from the model specification.

2.3.1   Priors

The controversial ingredient in the above is $\pi(\boldsymbol{\theta})$, the prior. There are two issues here. Should $\boldsymbol{\theta}$ be viewed as fixed or random? Then, if it is viewed as random, where does $\pi(\boldsymbol{\theta})$ come from? The case for assuming $\boldsymbol{\theta}$ to be random is based upon the idea that, if the model is not true, then there is no *true* $\boldsymbol{\theta}$. If we allow $\boldsymbol{\theta}$ to be random, we incorporate the fact that there is no fixed $\boldsymbol{\theta}$ into the model. We can more appropriately propagate uncertainty in inference regarding $\boldsymbol{\theta}$ as well as in prediction of data. We could simply say, naively, that if what we observe is random, shouldn't what we can not observe also be taken as random.

However, once we assume $\boldsymbol{\theta}$ to be random, then we have to specify a distribution for it. This is where the controversy lies. If you choose one prior and I choose another, whose inference should be adopted? Is there a role for subjectivity in statistical inference? The argument can be made that the prior enables you to incorporate any information you have about the unknowns in the model specification. The case can be made that we always know something about unknowns, e.g., we always have some idea of what the magnitude of regression coefficients can be because, when applied to regressors, we can see if they provide predictions outside the range of what is observable. Similarly, we can argue that we always know something about variances. If uncertainty is allowed to be too large, again, realizations beyond what is observable will occur. Furthermore, we can (and should) engage in sensitivity analysis to the prior. Typically, the data contribution regarding $\boldsymbol{\theta}$ will be much stronger than that of the prior so there will be little prior sensitivity.

What does the prior do mathematically? If we look at the product, $f(\mathbf{Y}|\boldsymbol{\theta})\pi(\boldsymbol{\theta})$, we see that the prior rescales the likelihood, changing its shape, perhaps its mode, perhaps its spread. What does this mean practically? The prior and the likelihood combine multiplicatively to provide our information about $\boldsymbol{\theta}$. Intuitively, we arrive at an inferential compromise between the two functions, which becomes the posterior distribution, to provide inference about $\boldsymbol{\theta}$. In practice, we often use what we call weak, vague, or noninformative priors. These terms themselves are *vague* and, at times, their impacts are not properly appreciated [40].

In the Bayesian literature there is substantial discussion of objective priors, so-called automatic specifications to remove potential subjectivity. These priors go under various names, e.g., Jeffreys priors, reference priors, g-priors, etc. [8]. For instance, for location parameters, they are just constant over R^1, for scale parameters, they are the reciprocal of the parameter or the square of the reciprocal. Typically these priors are improper (i.e., not integrable) raising the possibility that, with an improper joint distribution, we could have an improper posterior [9]. There is also a substantial literature on elicitation [see, e.g., 52, and references therein]. In some settings this can be a valuable exercise but, for most of the hierarchical modeling that is currently employed, there is not much of a role for elicitation. That is, elicitation makes sense for parameters that are, in some sense *real*. In much of hierarchical modeling, we introduce many parameters for which elicitation would not be feasible.

The fundamental issue here is whether we view the prior as an inference device as opposed to a component of a model specification to provide a generative specification. As suggested above, if you define a parameter, you will almost always know something about it so you can always propose a sensible proper prior and maintain a generative model. In any event, to avoid potentially improper posteriors, perhaps it is best to play it safe and adopt proper priors. A last word here, returning to weak or vague priors, if they are *nearly* improper, then the posterior may be nearly improper. With usual Markov chain Monte Carlo or Gibbs sampling used for model fitting (see Section 3.3), it may be difficult to see nearly improper posteriors.

2.3.2 Posterior inference

Armed with an entire posterior distribution for an unknown, inference becomes nothing more than presenting features of the distribution. We can provide customary measures of centrality like posterior means or medians. We can provide customary measures of uncertainty like posterior variances, standard deviations, ranges. We can provide quantiles and probability statements. In particular, if we provide a predictive interval for an unknown, it is a probability statement for that unknown, a so-called credible interval. It is not a confidence interval created by pivoting some sampling distribution. For a full development of Bayesian inference, see [23]. We defer discussion of Bayesian hypothesis testing, equivalently model comparison, to Section 10.1.

2.3.3 Bayesian computation

The foregoing elegance and clarity attached to Bayesian inference ignores the significant computational challenges associated with implementing the inference. The explicit challenge is integration. That is, (2.2) expresses the posterior up to normalizing constant. This constant is needed to calculate probabilities with the this distribution. Further integration is required to calculate any moments of the distribution. In practice, none of these integrations are available explicitly. When θ is high dimension, it is not feasible to handle these integrations numerically. The breakthrough that enables feasible Bayesian computation for the rich classes of models we are interested in replaces integration by sampling; to learn about a high dimensional posterior distribution, we just sample it as much as we wish. These days, sampling high dimensional distributions is accomplished, fairly straightforwardly, using Markov chain Monte Carlo (MCMC) and Gibbs sampling [58]. These techniques have become the workhorses of Bayesian computation in the 21st century. See [23] or [58] for full discussions. The remaining material in this chapter as well as the subsequent chapters that fit hierarchical models will assume familiarity with MCMC and Gibbs sampling for model fitting.

2.4   Hierarchical modeling

As noted in the abstract, moving into the second decade of the 21st century, we are witnessing a dramatic paradigm shift in the way that statisticians collaborate with researchers from other disciplines. Disappearing are the days when the statistician was called in at the end of a project to provide some routine data analysis and some summary displays. Now the statistician is an integral player in a research team, helping to formulate hypotheses, identify data needs, develop suitable stochastic models, and implement fitting of the resulting challenging models.

As part of this shift, there is increasing attention paid to bigger picture science, to looking at complex processes with an integrative perspective, to bringing a range of knowledge and expertise to this effort. Increasingly, we find researchers working with observational data, less with designed experiments, recognizing that the latter can help inform about the former but the gathering of such experiments provides only one source of data for learning about the complex process. Other information sources, empirical, theoretical, physical, etc. should also be included in the synthesis.

The primary result of all of this is the development of a multi-level stochastic model which attempts to incorporate the foregoing knowledge, inserting it at various levels of the modeling, as appropriate. Following the vision of Mark Berliner [11], we imagine a general three stage hierarchical specification:

$$\text{First stage} : [data|process, parameters]$$
$$\text{Second stage} : [process|parameters]$$
$$\text{Third stage} : [(hyper)parameters].$$

$$(2.5)$$

The simple form of this specification belies its breadth. The process component can include multiple levels. It can be dynamic, it can be spatial. The data can be conditioned on whatever aspects of the process are appropriate. The stochastic forms can be multivariate, perhaps infinite dimensional with parametric and/or nonparametric specifications. Moreover, while the focus here is on applications in environment and ecology, the range of applications to which this generic specification has been applied runs the scientific gamut, e.g., biomedical and health sciences, economics and finance, engineering and natural science, political and social science.

In view of the above, hierarchical modeling has taken over the landscape in contemporary stochastic modeling. Though analysis of such modeling can be attempted through nonBayesian approaches [57], working within the Bayesian paradigm enables exact inference and proper uncertainty assessment (see below) within the given specification.

We can immediately extend the Bayesian framework of the previous section to the hierarchical setting. Specifically, we add one further level, parametrically. Now, we would have a prior distribution $\pi(\boldsymbol{\theta}|\boldsymbol{\lambda})$, where $\boldsymbol{\lambda}$ is a vector of hyperparameters. Connecting with (2.5), we can think of $\boldsymbol{\theta}$ even more generally as the "process" of interest with some parts known and some parts unknown. Then, we can write $f(\mathbf{y}|\text{process}, \boldsymbol{\theta})f(\text{process}, \boldsymbol{\theta}|\boldsymbol{\lambda})\pi(\boldsymbol{\theta}|\boldsymbol{\lambda})\pi(\boldsymbol{\lambda})$, evidently, a *hierarchical* specification. If $\boldsymbol{\lambda}$ is known, the posterior distribution for $\boldsymbol{\theta}$ is given by

$$p(\boldsymbol{\theta}|\mathbf{y},\boldsymbol{\lambda}) \quad = \quad \frac{p(\mathbf{y},\boldsymbol{\theta}|\boldsymbol{\lambda})}{p(\mathbf{y}|\boldsymbol{\lambda})} = \frac{p(\mathbf{y},\boldsymbol{\theta}|\boldsymbol{\lambda})}{\int p(\mathbf{y},\boldsymbol{\theta}|\boldsymbol{\lambda})\, d\boldsymbol{\theta}}$$

$$= \quad \frac{f(\mathbf{y}|\boldsymbol{\theta})\pi(\boldsymbol{\theta}|\boldsymbol{\lambda})}{\int f(\mathbf{y}|\boldsymbol{\theta})\pi(\boldsymbol{\theta}|\boldsymbol{\lambda})\, d\boldsymbol{\theta}} = \frac{f(\mathbf{y}|\boldsymbol{\theta})\pi(\boldsymbol{\theta}|\boldsymbol{\lambda})}{m(\mathbf{y}|\boldsymbol{\lambda})}.$$

In practice, $\boldsymbol{\lambda}$ will not be known. A second stage hyperprior distribution $h(\boldsymbol{\lambda})$ will be required, as proposed in (2.5), so that

$$p(\boldsymbol{\theta}|\mathbf{y}) = \frac{p(\mathbf{y},\boldsymbol{\theta})}{p(\mathbf{y})} = \frac{\int f(\mathbf{y}|\boldsymbol{\theta})\pi(\boldsymbol{\theta}|\boldsymbol{\lambda})h(\boldsymbol{\lambda})\, d\boldsymbol{\lambda}}{\int f(\mathbf{y}|\boldsymbol{\theta})\pi(\boldsymbol{\theta}|\boldsymbol{\lambda})h(\boldsymbol{\lambda})\, d\boldsymbol{\theta}d\boldsymbol{\lambda}}. \tag{2.6}$$

This is the path that is usually pursued. However, alternatively, we might replace $\boldsymbol{\lambda}$ in $p(\boldsymbol{\theta}|\mathbf{y},\boldsymbol{\lambda})$ by an estimate $\hat{\boldsymbol{\lambda}}$. This is called empirical Bayes analysis [14]. Because such estimation removes the uncertainty associated with $\hat{\boldsymbol{\lambda}}$, uncertainty in overall inference will tend to be underestimated. However, in practice, with models of sufficient complexity, we often implement a bit of empirical Bayes inference, using a bit of data-based *fixing* of some parameters.

An issue with likelihood based inference for hierarchical models emerges. What is the likelihood? From the multi-level model, we can marginalize over $\boldsymbol{\lambda}$ to obtain $p(\mathbf{y}|\boldsymbol{\theta})\pi(\boldsymbol{\theta})$ where $\pi(\boldsymbol{\theta}) = \int_{\boldsymbol{\lambda}} \pi(\boldsymbol{\theta}|\boldsymbol{\lambda})h(\boldsymbol{\lambda})d\boldsymbol{\lambda}$, interpreting $p(\mathbf{y}|\boldsymbol{\theta})$ as the likelihood. Alternatively, we can marginalize over $\boldsymbol{\theta}$ obtaining $p(\mathbf{y}|\boldsymbol{\lambda})h(\boldsymbol{\lambda})$ where $p(\mathbf{y}|\boldsymbol{\lambda}) = \int_{\boldsymbol{\theta}} p(\mathbf{y}|\boldsymbol{\theta})\pi(\boldsymbol{\theta}|\boldsymbol{\lambda})d\boldsymbol{\theta}$. Now the likelihood would become $p(\mathbf{y}|\boldsymbol{\lambda})$. Which likelihood should we work with? One might argue that we should *focus* on the parameters of interest. However, it may be more satisfying to fit the full hierarchical model, enabling posterior inference regarding all of $\boldsymbol{\theta}$ and $\boldsymbol{\lambda}$.

2.4.1 Introducing uncertainty

The generality reflected in [11] enables rich model specification. In fact, as hierarchical modeling has taken over a substantial portion of the statistical modeling landscape, we are finding increasingly challenging examples. With such richness comes the recognition of the opportunity for introducing uncertainty and how to do it appropriately. Again, the adopted stochastic models are only approximations to the complex process so error will always be introduced. The benefit of the transparency implicit in building multi-level models is that it allows us to determine where and how to introduce error. And, once we do, the fact that we are working in a fully probabilistic setting, enables the uncertainty to be properly propagated to the posterior distribution.

What are the various types of uncertainty to consider? Possibilities include: stochastic uncertainty reflecting not just the noise in realizations but also the randomness with regard to where sampling units are; measurement uncertainty (experimental uncertainty), reflecting measurement error in the devices recording the observations; parameter uncertainty, evidently captured in some fashion through prior specifications; model uncertainty, reflecting missing or unobservable information which are part of the process but not accessible; uncertainty in model dimension or functional uncertainty, reflecting the fact that functional forms employed in relating variables in the specifications are only approximate; and multiple model uncertainty, reflecting the interplay between the modeling levels, how specification at one level influences uncertainty at another level. The implications of introducing uncertainty in various ways to achieve various objectives will not be assessable analytically. Rather, a substantial amount of model fitting and model comparison will be required to make such assessments.

2.4.2   Random effects and missing data

Arguably, the utilization of hierarchical models initially blossomed in the context of handling random effects and missing data, using the E-M algorithm [18] for likelihood analysis and Gibbs sampling [30] for fully Bayesian analysis. In this subsection, we offer some elementary remarks, first on random effects, then on missing data.

With regard to random effects, both classical and frequentist modeling specify a stochastic model for these effects, usually assumed to be a normal distribution with an associated variance component. These effects can be introduced at different levels of the modeling but, regardless, in much of the literature, they are assumed to be exchangeable, in fact i.i.d. More recently, we are seeing random effects with structured dependence in, e.g., dynamic, spatial and spatio-temporal models (see Chapters 4 and 5).

A typical linear version with i.i.d. effects takes the following form. At the first stage:

$$Y_{ij} = X_{ij}^T \beta + \phi_i + \epsilon_{ij}.$$

At the second stage, β has a Gaussian prior while the ϕ_i are i.i.d. $\sim N(0, \sigma_\phi^2)$. The ϵ_{ij} are i.i.d. $\sim N(0, \sigma_\epsilon^2)$. The variance components become the third stage hyperparameters, i.e., we require prior specifications for σ_ϕ^2, σ_ϵ^2. As has been learned over recent years, care is required in these specifications. Recalling our remarks in the previous section on improper priors, they can lead to improper posteriors [9]. The frequently-employed inverse gamma priors, $IG(\epsilon, \epsilon)$ for small ϵ are *nearly* improper and result in *nearly* improper posteriors as well as badly behaved MCMC in practice. A protective recommendation is an $IG(1, b)$ or $IG(2, b)$. Both are far from improper; the former has no integer moments, the latter has a mean but no variance. Evidently, we can revise the model to have a nonGaussian first stage. Again, care is needed with prior specifications as well as in model fitting.

In collecting information on, e.g., individuals, we often have vectors of data with one or more components of the components missing. It is unattractive to confine ourselves to analyzing only the complete data cases. This may discard too much data and possibly introduce bias with regard to the ones retained. To use the individuals with missing data, we must *complete* them, so-called imputation. There is by now a very substantial literature on imputation [see, e.g., 42]. However, to do a fully model-based imputation in the Bayesian setting results in latent variables (Section 5) and looping with Gibbs samplers, updating parameters given missing data, then updating missing data given parameters. In this sense, the Gibbs sampler extends the EM algorithm to provide full posterior inference rather than an MLE with an asymptotic variance.

As a simple example, consider multivariate normal data, $\mathbf{Y}_i \sim N(\boldsymbol{\mu}_i, \Sigma)$ where the components of $\boldsymbol{\mu}_i$ may have regression forms in suitable covariates. Some components of some of the $\mathbf{Y}_i$'s are missing. In order to perform the imputation, we do basic Gibbs sampling: we update $(\boldsymbol{\mu}_i, \Sigma)$ given values for the missing Y's, then we update the missing Y's given values for the $(\boldsymbol{\mu}_i, \Sigma)$. Another standard example considers missing categorical counts within a multinomial model where the multinomial cell probabilities might be modeled using some sort of multivariate logit model [1]. For instance, some categories are aggregated/collapsed so counts for the disaggregated categories are missing. Again, we can envision a looping of the Gibbs sampler: update the parameters given values for all the counts, update the missing counts given values for the parameters.

2.5 Latent variables

Latent variables are at the heart of most hierarchical modeling. Here, we provide examples which suggest they can be envisioned beyond random effects or missing data.[2] Latent variable models customarily result in a hierarchical specification of the form $f(\mathbf{Y}|\mathbf{Z})f(\mathbf{Z}|\boldsymbol{\theta})\pi(\boldsymbol{\theta})$. Here, the Y's are observed, the Z's are latent and the "regression" modeling is shifted to the second stage.

An elementary version of a latent variable model arises with binary data models. In particular, the usual binary response model adopts a logit or probit link function. Illustrating with the probit, suppose $Y_i \sim \text{Bernoulli}(p(\mathbf{X}_i))$ (more generally, we can have $Y_i \sim Bi(n_i, p(\mathbf{X}_i))$). Specifically, let $\Phi^{-1}(p(\mathbf{X}_i)) = \mathbf{X}_i\boldsymbol{\beta}$ with a prior on $\boldsymbol{\beta}$. In fitting this model using MCMC computation, it is awkward to sample $\boldsymbol{\beta}$ using the likelihood in this form. If we introduce $Z_i \sim N(\mathbf{X}_i\boldsymbol{\beta}, 1)$ then, immediately, $P(Y_i = 1) = \Phi(\mathbf{X}_i\boldsymbol{\beta}) = 1 - \Phi(-\mathbf{X}_i\boldsymbol{\beta}) = P(Z_i \geq 0)$. Once we bring in these Z_i's, we achieve a routine Gibbs sampler: update the Z's given $\boldsymbol{\beta}, \mathbf{y}$ (this requires sampling from a truncated normal), update $\boldsymbol{\beta}$ given the Z's and $\mathbf{y}$ (this is the usual, typically conjugate normal updating). This approach was first articulated in the literature by [3].

It is clear that this approach can readily extend to general ordinal categorical data settings [1]. In particular, for each i, the Bernoulli trial is replaced with a multinomial trial. There is still a latent Z_i, still following say a Gaussian linear regression. Now the multinomial outcomes are created by introducing cut points along the real line; the intervals determined by the cut points allocate probabilities to each of the multinomial outcomes. The cut points will be random as well, noting that, in order to identify the intercept in the regression, the smallest cut point can be taken to be 0, without loss of generality.

Another routine generalization of this approach accommodates censored or truncated data models. As a simple illustration, suppose we observe a variable with a point mass at 0 and the remainder of its mass spread over R^+ or perhaps $(0, 1]$. In the first case, such data arise when studying, for instance, daily precipitation at a location; in the second case when considering, for instance, the proportion of a particular land use classification over a region. Here, with observed Y_i's, we can introduce Z_i's such that $Y_i = g(Z_i), Z_i > 0, Y_i = 0, Z_i \leq 0$, with $g(\cdot)$ a link function from R^+ to R^+ or perhaps to $(0, 1]$. Then, we could model the Z_i's using say a usual Gaussian linear regression.

A further setting for latent variables is change point problems [7]. Frequently, we observe a process over the course of time during which we would like to assess whether some sort of change in regime has occurred. Practically speaking, this requires the notion of a "least" significant change. That is, we may be able to identify even a very small change with enough data but we will find challenging the case where the support for the change has "no change" as a boundary point.

In the change point setting two sampling scenarios can be envisioned. In the first, we have a full set of data. Then we look, retrospectively, to try to find if change(s) occurred. In the second, we look at the data sequentially and we try to identify change(s) as the data collection proceeds. We illustrate with a simple version of the first scenario. Let $f_1(y|\theta_1)$ be the density for i.i.d. observations before the change point, $f_2(y|\theta_2)$ the density after the change point. With data $Y_i, i = 1, 2, ..., n$, let K be the change point indicator, i.e., $K \in \{1, 2, ..., n\}$ where $K = k$ means change at observation $k + 1$; $k = n$ means "no change." Then, the model is:

---

[2]While, in a sense, all variables we can not observe are "missing," in the previous subsection we took missing to mean some components of the data while here they will be variables different from the data.

$$L(\theta_1, \theta_2, k; \mathbf{y}) = \Pi_{i=1}^{k} f_1(y_i|\theta_1)\Pi_{i=k+1}^{n} f_2(y_i|\theta_2). \tag{2.7}$$

Again, we have a hierarchical model, $[\mathbf{y}|k, \theta_1, \theta_2][K = k][\theta_1, \theta_2]$. (Note that we do not include any parameters in the prior for K; with only one change point, we could not hope to learn about such parameters.) With a prior on θ_1, θ_2, K, we have a full model specification. Again, a simple Gibbs sampler emerges for model fitting: update θ's given $k, \mathbf{y}$ (so, we know exactly which observations are assigned to which density); update k given θ's and $\mathbf{y}$ (this is just a discrete distribution, easily sampled). Obvious generalizations would allow the y's to be dependent, to have order restrictions on θ's, to imagine multiple change points. Also, in this version, time is discretized to the set of times when the measurements were collected. Extension to continuous time and multiple change point settings is available using point process models [22].

Errors in variables models [24] offer another latent variables setting. Loosely stated, the objective is to learn about the relationship between say Y and X. Unfortunately, X is not observed. Rather, we observe say W instead of X. In some cases, W will be a version of X, subject to measurement error, i.e., W may be X_{obs} while X may be X_{true}. In other cases W may be a variable (variables) that play the role of a surrogate for X. In any event, if we envision a joint distribution for W and X, we may condition in either direction. If we specify a model for $W|X$ we refer to this as a measurement error model [15, 24], imagining W to vary around the true or desired X; if we specify a model for $X|W$ we refer to this as a Berkson model [15]. In fact, we can imagine a further errors in variables component - perhaps we observe Z, a surrogate for Y. Altogether we have a hierarchical model with latent X's, possibly Y's. In particular, for the measurement error case, with independent observations, we have:

$$\Pi_i f(Z_i|Y_i, \gamma)f(Y_i|X_i, \beta)f(W_i|X_i, \delta)f(X_i|\alpha) \tag{2.8}$$

while for the Berkson case we have:

$$\Pi_i f(Z_i|Y_i, \gamma)f(Y_i|X_i, \beta)f(X_i|W_i, \delta) \tag{2.9}$$

Typically, we will also have some *validation* data to inform about the components of the specification. We might have some X, Y pairs or perhaps some X, W pairs. It is noteworthy that the measurement error version requires a model (prior) for X while the Berkson model does not. In many applications the former will be more natural. However, in some contexts, model fitting is only feasible with the latter [6]. In any event, what is most remarkable is that, within this hierarchical framework, using a full Bayesian specification, we can learn about the relationship between Y and X without ever observing X (and, possibly, without observing Y as well). This reveals the power of hierarchical modeling but, evidently, what we can learn depends upon the form of what we specify and the data we have.

2.6   Mixture models

Mixture models have now become a staple of modern stochastic modeling [50, 62]. This has arisen on at least two accounts: (i) their flexibility to model unknown distributional shapes and (ii) their intuition in representing a population in terms of groups/clusters that may exist but are unidentified. Mixture models come in several flavors - parametric or nonparametric, incorporating finite, countable or uncountable mixing. In this regard, they are sometimes referred to as classification problems or discriminant analysis, reflecting a goal of assigning an individual to a population or assessing whether individuals belong to the same population.

The most rudimentary finite mixture version takes the form:

$$\mathbf{Y} \sim \sum_{l=1}^{L} p_l f_l(\mathbf{Y}|\boldsymbol{\theta}_l). \tag{2.10}$$

Often the f_l are normal densities, whence we obtain a normal mixture. If we assume L is specified and we observe $\mathbf{Y}_i, i = 1, 2, ..., n$, then what is latent is a *label* for each $\mathbf{Y}_i$, i.e., an indicator of which component of the mixture Y_i was drawn from. These latent labels would be such that if $L_i = l$, then $\mathbf{Y}_i \sim f_l(\mathbf{Y}|\boldsymbol{\theta}_l)$. Upon introducing these labeling variables, the resulting hierarchical model becomes:

$$\Pi_i f(\mathbf{Y}_i|L_i, \boldsymbol{\theta})\Pi_i f(L_i|\{p_l\})\pi(\boldsymbol{\theta})\pi(\{p_l\}). \tag{2.11}$$

Here, $\boldsymbol{\theta}$ denotes the collection of $\boldsymbol{\theta}_l$. Once again, Gibbs sampling is routine to implement. We create a loop that updates $\boldsymbol{\theta}, \{p_l\}$ given the L's and the data. With observations assigned to components, this becomes the equivalent of an analysis of variance problem to learn about the $\boldsymbol{\theta}_l$. To update the L_i's given $\boldsymbol{\theta}, \{p_l\}$ and the data requires sampling from an L-valued discrete distribution where, for a given i, the mass on the l's is determined by the relative likelihood for the observed $\mathbf{Y}_i$ as well as the prior on the p_l's (often a uniform). Hence, we obtain individual assignment probabilities as well as global assignment weights. Richer versions introduce covariates into the $\boldsymbol{\theta}_l$'s and, possibly, into the p_l's [46]. Further challenge is added if L is unknown with a prior specification. Now, since the dimension of the model changes with L we may attempt reversible jump MCMC [70] to learn about L. A simpler alternative might be to carry out model choice across a set of L's.

It is evident that such mixture models are not identifiable, i.e., the subscripts can be permuted and the same mixture distribution results. This has led to discussion in the literature with regard to introducing identifiability constraints or the possibility of fitting the MCMC, allowing multi-modality in the posterior in the absence of identifiability [46]. The former path seems to be the most widely used, with order constraints on the means, imposed in some fashion, being the most common choice for achieving identifiability.

Next, we recall that many familiar distributional models can be developed through continuous mixing, i.e., closed forms are achieved by virtue of conjugacy between the mixed model and the mixing distribution. Well-known examples include scale mixing of normals to obtain t-distributions, as well as Poisson-Gamma (equivalently negative binomial) and beta-binomial models. In some modeling situations we may seek individual level mixing variables, whence we would introduce such variables as latent quantities. Again, a hierarchical model arises. An illustration is in the case of outlier detection through the use of suitable individual level gamma mixing of normals. Here outliers are "detected" through the magnitudes of their associated mixing variables, i.e., the heavier the posterior tails, the more we are inclined to classify the observation as an outlier. See, e.g., [27, 63].

2.7 Random effects

Returning to the random effects setting, let us first consider individual level longitudinal data with interest in explanation through growth curves. A natural specification would model individual level curves centered around a population level curve. We would need the population level curve to see *average* behavior of the process; we need individual level curves in order, for example, to prescribe *individual* level treatment. With parameters at each level and a third stage of hyperparameters, we again see a hierarchical form.

More precisely, if Y_{ij} is jth measurement for ith individual, let

$$Y_{ij} = g(\mathbf{X}_{ij}, \mathbf{Z}_i, \boldsymbol{\beta}_i) + \epsilon_{ij}$$

where $\epsilon_{ij} \sim N(0, \sigma_i^2)$. The form for g depends upon the application. It is often linear but need not be. We set $\boldsymbol{\beta}_i = \boldsymbol{\beta} + \boldsymbol{\eta}_i$ where the $\boldsymbol{\eta}_i$ have mean $\mathbf{0}$ (or perhaps replace $\boldsymbol{\beta}$ with a regression in the $\mathbf{Z}_i$). Then the $\boldsymbol{\beta}_i$ (or the $\boldsymbol{\eta}_i$) are the random effects. They provide the individual curves with $\boldsymbol{\beta}$ providing the global curve. Learning with regard to any individual curve will borrow strength from the information about the other curves.

Customarily, random effects are modeled using normality. However, they need not be i.i.d. That is, if say the scalar ω_i is associated with individual i, we need not insist that the vector, $\boldsymbol{\omega}$, of ω_i's, be distributed as say, $\boldsymbol{\omega} \sim N(\mathbf{0}, \sigma^2 I)$. We can replace $\sigma^2 I$ with $\Sigma(\boldsymbol{\theta})$ where $\Sigma(\boldsymbol{\theta})$ has structured dependence. That is, with say n individuals, we could not learn about an arbitrary positive definite $n \times n$ matrix Σ but we could learn about Σ defined as a function of only a few parameters. Structured dependence is at the heart of time series, spatial and spatio-temporal modeling (see Chapters 3 and 5) and is frequently specified through a Gaussian process (GP). For example, in the spatial setting, we envision data in the form $Y(s_i), i = 1, 2, ..., n$, i.e., n observations at n different spatial locations s_i so $s_i \in R^2$. We could imagine three dimensional locations and, more generally, replacing geographic space with say covariate space. This takes us into the growing world of computer models [44, 45, 49]. In any event, with a GP, we need only specify finite dimensional (e.g., n) joint distributions with the joint dependence determined by a valid covariance function. Customarily, the covariance function assigns stronger association to variables that are closer to each other in geographic space. A common example is the exponential, $\text{cov}(Y(s), Y(s')) = \sigma^2 \exp(-\phi||s - s'||)$. With n locations, $\Sigma(\boldsymbol{\theta})_{ij} = \sigma^2 \exp(-\phi||s_i - s_j||)$.

Data fusion presents another important area for hierarchical modeling (see Chapter 7). The context is frequently spatial or spatio-temporal. For instance, data on environmental exposure can be available from different sources, e.g., monitoring stations, computer models, and satellites. The fusion objective is to use all data sources to infer about the true exposure. We confront misalignment of scales in space, in time. To reconcile the misalignment we may elect to scale up or scale down.

2.8   Dynamic models

Dynamic models have now become a standard formulation for a wide variety of processes, including financial and environmental applications. Alternate names for them in the literature include Kalman filters, state space models and hidden Markov models [33, 51, 59]. They introduce a first stage (or observational model) and then a second stage (or transition model), with third stage hyperparameters. Again, the first stage provides the data model while the second stage provides a latent dynamic process model. See, e.g., [64] for a full development. In particular, there is substantial modeling opportunity at the second stage, allowing evolution of process variables or process parameters, in either case, driven by covariate information. Chapter 4 presents a full development. Here, we just note the hierarchical structure.

Specifically, the basic dynamic model takes the form:

$$\mathbf{Y}_t = g(\mathbf{X}_t, \boldsymbol{\theta}) + \boldsymbol{\epsilon}_t, \quad \text{the observation equation} \qquad (2.12)$$

with

$$\mathbf{X}_t = h(\mathbf{X}_{t-1}; \boldsymbol{\theta}_2) + \boldsymbol{\eta}_t, \quad \text{the transition equation.} \qquad (2.13)$$

Evidently, time t is discrete and we are putting the dynamics in the mean.

We illustrate with a dynamic space-time model. Consider:

Stage 1: Measurement equation

$$Y(s,t) = \mu(s,t) + \epsilon(s,t); \ \epsilon(s,t) \overset{ind}{\sim} N\left(0, \sigma_\epsilon^2\right).$$
$$\mu(s,t) = \mathbf{x}^T(s,t) \tilde{\boldsymbol{\beta}}(s,t).$$
$$\tilde{\boldsymbol{\beta}}(s,t) = \boldsymbol{\beta}_t + \boldsymbol{\beta}(s,t)$$

with

Stage 2: Transition equation

$$\boldsymbol{\beta}_t = \boldsymbol{\beta}_{t-1} + \boldsymbol{\eta}_t, \ \boldsymbol{\eta}_t \overset{ind}{\sim} N_p\left(\mathbf{0}, \Sigma_{\boldsymbol{\eta}}\right).$$
$$\boldsymbol{\beta}(s,t) = \boldsymbol{\beta}(s,t-1) + \mathbf{w}(s,t)$$

where the $\mathbf{w}(s,t)$ are independent (over t) innovations of a spatial process [see, e.g., 28]. As noted above, this specification can be connected to a linear Kalman filter [39]. Thus, Bayesian model fitting using the forward filter, backward sample (ffbs) algorithm [16, 23] becomes the customary approach. Again, Chapter 4 explores such models in considerable detail.

Furthermore, Wikle and colleagues [38, 65] have adapted these dynamic forms to the fitting of models motivated by stochastic partial differential equations (SPDE's). The approach is to discretize time. In particular, there are many interesting ecological diffusions (characterized by SPDE's) which can be applied to study the behavior over time (and, perhaps space) of: (i) emerging diseases such as avian flu or H1N1 flu; (ii) exotic organisms, e.g., invasive plants and animals; (iii) the evolution of the distribution of size or age of a species; (iv) the dynamics explaining phenomena such as transformation of landscape, deforestation, land use classifications, and urban growth. Our objective for such processes is to forecast likely spread in space and time with associated uncertainty. We anticipate that the evolution will be nonlinear and nonhomogeneous in space and time, driven by explanatory covariates.

2.9 Model adequacy

In a world filled with rich modeling opportunity, it becomes necessary to investigate model adequacy and model comparison. Two brief comments are appropriate here. First, model adequacy is usually viewed as an *absolute* issue. If we develop a criterion, it is usually difficult to calibrate performance under this criterion. How adequate is adequate? The difference between model adequacy and model comparison is therefore clear. The latter is relative; with a criterion, we can make comparison using this criterion. Second, when we propose complex multi-level models, they are often too big for the data. They often overfit but would rarely be inadequate.

As a result, we tend to treat model adequacy informally, e.g., in the realm of exploration to discard inadequate models, to retain adequate ones for comparison. In this spirit, there is a rich array of diagnostic tools for studying model failures of different sorts. These include investigation of residuals (looking for departure from distributional assumptions), comparison of predicted vs. observed (looking for under or overfitting), empirical Q-Q plots, etc. In fact, more formal goodness of fit tests are well known and have a long history in the literature [35]. However, they are rarely relevant for hierarchical model adequacy.

Within the Bayesian framework, the marginal density ordinate for the observed data is a historical criterion [12]. That is, we plug the data into the marginal density, $f(\mathbf{Y})$. This criterion is difficult to calculate since it requires marginalizing over θ and is also difficult to calibrate for high dimensional data. It is not used with hierarchical models. Variants, such as conditional predictive ordinates [54] have been proposed. These are "leave-one-out" criteria, data deletion approaches which also can be very computationally demanding with large amounts of data.

More common in the current literature are posterior predictive checks, as advocated by [32], and prior predictive checks, as in [19]. Both of these ideas approaches employ discrepancies to capture differences between what is observed and what is predicted. The former uses posterior predictive distributions based upon the observed data and then checks the observations against these predictive distributions. Use of the data twice in this fashion makes this approach insufficiently critical of models. The latter makes comparison using prior predictive distributions, obtained under the model but without using the data. Hence, they are more critical of the actual model specification. They can be most easily implemented using Monte Carlo tests [21], i.e., generating realizations of a feature of the data under the model and comparing with the observed value of the feature.

It can be argued that prior predictive distributions are based upon priors which are generally weak, often not proper, and are only introduced to complete model specifications. Posterior predictive distributions are obtained using posteriors and posteriors are the distributions that will be used for inference. The counterargument is that, if you want to work with generative models (proper priors) and want to criticize such a model, you should criticize what is specified, not what is realized upon applying it to the data.

Finally, we suggest the use of empirical vs. nominal coverage as a convenient model checking strategy. The idea is based upon specifying a hold-out or validation dataset, using the remainder for model fitting. The general approach is discussed further in the next section. Here, we would simply suggest to use the fitted model to obtain posterior predictive credible intervals for for a held out observation and then compare this interval with the actual observation. Doing this for the set of held out points enables comparison between the nominal coverage associated with the interval vs. the observed empirical coverage across the set of hold out observations. Empirical undercoverage suggests that the model is not capturing uncertainty well enough. Empirical overcoverage implies uncertainty is greater than it should be, perhaps the model is overfitting. Again, implementation of this model checking strategy should be done out-of-sample. In-sample checking again suffers from using the data twice.

2.10 Model comparison

As we have argued above, model comparison is often done in the parameter space, leading to penalized likelihood criteria such as AIC, BIC and, with hierarchical models, DIC. AIC and BIC have been widely used in the biology/ecology community [13]. DIC [61] has been advocated within the Bayesian community as a criterion which, attractively, can be implemented directly in the BUGS software for Bayesian model fitting [43].

We suggest that it may be preferable to make comparison in the predictive space using (posterior) predictive distributions. In predictive space, having observations, we can examine quantities like predictive mean square error, predictive coverage, and continuous ranked probability scores. In Section 10.1 we review model comparison in the parameter space while in Section 10.2 we review model selection in predictive space.

2.10.1 Bayesian model comparison

Historically, Bayesian model comparison has been addressed through Bayes factors. We start with a formal application of Bayes' theorem to this problem. Suppose we have a set of k models under consideration and, a priori, we assume p_j is the prior probability that model j is true, $j = 1, 2, ..., k$. Then, using the theorem, with data $\mathbf{y}$ and denoting the jth model by M_j,

$$P(M_j|\mathbf{y}) = \frac{P(\mathbf{y}|M_j)p_j}{\sum_{j=1}^{k} P(\mathbf{y}|M_j)p_j}. \tag{2.14}$$

We would then choose the model with the highest posterior probability. On the surface, this idea is attractive. However, calculating $P(\mathbf{y}|M_j) = \int P(\mathbf{y}|\boldsymbol{\theta}_j; M_j)\pi(\boldsymbol{\theta}_j|M_j)d\boldsymbol{\theta}_j$ for each j when $\boldsymbol{\theta}_j$ is large can be challenging. Furthermore, where do the p_j's come from? We have repeatedly asserted that none of these models is "true." Would we assume a prior that they are equally likely? This seems inappropriate since, in application, we would want to reward parsimony or sparseness.

Briefly, we return to the variable selection problem discussed in Section 2, i.e., variable selection in a regression model where parsimony in model dimension, in number of variables, is to be rewarded. Hence, we might place a $Po(\lambda)$ distribution on the number of variables, adding a prior on λ. In fact, a more useful idea is to consider p, a prior inclusion probability for a variable, with exchangeable Bernoulli selection. That is, a priori, P(variable X_k is selected) is p, i.e., $P(I_k = 1) = p$, adding a prior on p. Then, under a binomial model for the number of variables selected, we seek the posterior probability of $I_k = 1$. Suppose we reparameterize the regression coefficient for X_k to $\theta_k = I_k\beta_k$. Then, we introduce a "spike and slab" prior for θ_k using the auxiliary variable I_k. That is, when $I_k = 1$, X_k is in the model with coefficient β_k; when $I_k = 0$, X_k is not in the model. There is a growing literature on variable selection starting from these ideas. See, e.g., [53] and further references therein.

Returning to the model selection criterion above, can we avoid specifying the p_j's? Suppose we look at models in pairs (so, really only suitable for a small number of models). For a pair of models M_0 and M_1, define the Bayes Factor for M_0 vs M_1 as $B_{01} = \frac{P(\mathbf{Y}|M_0)}{P(\mathbf{Y}|M_1)}$. This ratio is customarily interpreted as providing a weight of evidence for model M_0 relative to M_1. The ratio also offers another interpretation. With two models, we have

$$\frac{P(M_0|\mathbf{Y})}{P(M_1|\mathbf{Y})} = \frac{P(\mathbf{Y}|M_0)}{P(\mathbf{Y}|M_1)} \times \frac{p_o}{p_1}$$

where $p_1 = 1 - p_0$. That is, the Bayes factor provides the conversion from the prior odds to the posterior odds without having to provide the prior odds. In practice, the Bayes factor reduces model comparison to, essentially, a hypothesis testing problem. We have two actions with what can be expressed as, essentially, a $0 - 1$ loss function.

The Bayes factor requires proper priors for both models since it is arises as a ratio of density ordinates which can only be interpreted when priors are proper. In fact, this reveals a problem with *nearly* improper priors. Suppose $Y \sim N(\theta, 1)$. $M_0 \equiv H_0 : \theta = 0$, $M_1 \equiv H_A : \theta \neq 0$, i.e., $M_0 \subset M_1$. Then, with a Uniform$(-L, L)$ prior on θ, $B_{01} = 2L\phi(Y)/(\Phi(L - Y) - \Phi(-L - Y))$. The support for M_0 tends to ∞ as L increases.

Finally, as we commented with the general model selection criterion, calculation of the Bayes factor can be challenging, Marginalization over $\boldsymbol{\theta}$ is required for both the numerator and denominator. In fact, calculation of Bayes factors attracted a fair bit of attention in the literature at the turn of the 21st century, exploring Monte Carlo calculation ideas, bridge and path sampling approaches, etc. [58]. However, Bayes factors are seldom used with hierarchical models.

We mention one last feature of Bayes factors which enables connection to penalized likelihood and therefore to AIC and BIC type model selection criteria. This is the Lindley paradox. Suppose $Y_1, Y_2, ..., Y_n$ $i.i.d. N(\theta, 1)$ with prior $\pi(\theta) = N(0, 1)$. Consider $H_0 : \theta = 0$ vs $H_A : \theta \neq 0$. Then, $B_{01} = \frac{N(\bar{Y}|0,\frac{1}{n})}{N(\bar{Y}|0,1+\frac{1}{n})}$. The usual test statistic is $Z = \sqrt{n}\bar{Y}$ in which case

$$B_{01} = \sqrt{n+1}|Z|e^{-\frac{n}{n+1}Z^2/2}.$$

So, regardless of $|Z|$, as $n \to \infty$, $B_{01} \to \infty$, i.e., we choose M_0. This implies that the Bayes factor is "too large."

This leads to the issue of *penalized* likelihood. What do we mean by this? Consider the usual likelihood ratio test which, for $H_0 : \theta \in \Theta_0$ vs. $H_A : \theta \in \Theta - \Theta_0$, takes the form: Reject if

$$\lambda(\mathbf{Y}) = \frac{\sup_{\theta \in \Theta_0} L(\theta; \mathbf{Y})}{\sup_{\theta \in \Theta} L(\theta; \mathbf{Y})} < c.$$

It can be written more generally for models M_1 and M_2 such that $M_1 \subset M_2$.

Under weak conditions, when M_1 is true, $-2\log\lambda \approx \chi^2_{p_2-p_1}$ where p_2 is the dimension of model M_2 and p_1 is the dimension of model M_1. So, $P(\lambda < c|M_1) = P(-2\log\lambda > c') \approx P(\chi^2_{p_2-p_1} > c') > 0$. We see that $P(\text{reject } M_1|M_1 \text{ true}) \nrightarrow 1$ as $n \to \infty$; λ, equivalently $\log\lambda$ is too small. This leads to proposing a penalty function on $-2\log L(\hat{\theta}_{M_j}; \mathbf{Y}), j = 1, 2$. Here, we have nearly fifty years of literature dating to [2] and [60]. Forms for the penalty include functions of only p_j, e.g., cp_j. This form of penalty leads to the Akaike Information Criterion (AIC). Numerous choices of c have been proposed in the literature; $c = 2$ is commonly used. Another choice considers functions of n and p_j, e.g., $\log np_j$. These are called Bayesian Information Criteria (BIC) in the literature. See [13] for full details. This choice results in the criterion $-2\log\lambda_n + (p_2 - p_1)\log n$. Above, we saw that the Bayes factor was too large and $\log\lambda$ is too small. Is there a relationship between them? [26] show that

$$\log BF = \log\lambda_n + \log n\frac{p_2 - p_1}{2} + O(1). \tag{2.15}$$

We note that these criteria really only apply to the case of "fixed" effects modeling where model dimension is clear. With random effects, latent variables, etc., it is no longer clear what model dimension is. So, we can assert that these criteria play a small role in today's modeling landscape. In this regard, the Deviance Information Criterion (DIC) [61] has been proposed as a version suitable for hierarchical models. Computation is easy, following on from MCMC model fitting. However, examples of negative model dimension, of sensitivity to choice of parametrization, of selecting models marginalized to focus on different parameters, have raised issues regarding its effectiveness as a criterion. Perhaps the real issue is that the criterion is being applied over parameter space while model comparison with hierarchical models may be better done in predictive space.

2.10.2 Model comparison in predictive space

We have suggested above that model comparison may be more appropriately implemented in the space of the data rather than in the space of the parameters. This leads to comparison of the posterior predictive distribution for an observation with the actual observed value of the realization. Expression (4) shows how the predictive distribution for an observation Y_0 arises. MCMC model fitting provides posterior samples of θ's. If we take a posterior draw θ^* and then draw a Y_0^* from $f(Y_0|\mathbf{Y}, \theta^*)$, so-called composition sampling [5], then $Y_0^* \sim f(Y_0|\mathbf{Y})$. So, we can draw posterior predictive samples one-for-one with posterior parameter samples.

A critical question emerges. Shall we do model comparison in-sample or out-of-sample? In-sample comparison is attractive in the sense that we fit each model under consideration to all of the data. Then, we can choose which observations we want to use in the model comparison. The disadvantage, as with model adequacy, is, in using the data twice, in-sample validation tends to show better model performance than would be seen in practice. Out-of-sample validation [55]), also called cross-validation, with randomly selected hold out data, seems preferred. We see how well the models do with data that was not used to fit them. The challenge here is how much data to hold out. We create a training/fitting set and a test/validation set. What proportion of the data should we allocate to each. Customary practice employs $10 - 20\%$ hold out according to the overall size of the dataset. Note that hold out is only used for model comparison. Once a model is selected, the inference results will be presented based upon the full dataset.

[29] proposed a model comparison criterion in the spirit of DIC but in predictive space. It is viewed as an in-sample criterion. Adopting a balanced loss function, they obtained a two component criterion. One component rewards fidelity of prediction relative to the actual observation. The other rewards small posterior predictive uncertainty.

More commonly used these days are criteria like predictive mean square error or predictive mean absolute error, applied out-of-sample. Hold out observations are compared with the associated posterior predictive mean or median, respectively, averaged over the hold out set. A possibly more attractive criterion is the ranked or continuous ranked probability score [34], according to whether the predictive distribution has discrete support (like a species count) or continuous support (like a temperature or ozone level). The objective is to compare the entire predictive distribution to the held out value. Intuitively, the more concentrated the predictive distribution is around the hold out value, the better the model performance.

The proposed measure is the squared integrated distance between the predictive distribution and the degenerate distribution at the observed value,

$$CRPS(F, y) = \int_{-\infty}^{\infty} (F(u) - 1(u \geq y))^2 du \tag{2.16}$$

where F is the predictive distribution and y is the observed value. For us, Y_0 is the observation and F is the posterior predictive distribution for Y_0. With a collection of such hold out observations and associated predictive distributions, we would sum the CRPS over these observations to create the model comparison criterion. Recall that, under MCMC model fitting, we will not have F explicitly but, rather, a sample from F. Fortunately, a convenient alternative computational form, provided F has a first moment, is

$$CRPS(F, y) = \frac{1}{2} E_F |Y - Y'| + E_F |Y - y| \qquad (2.17)$$

where Y and Y' are independent replicates from F. With samples from F, we have immediate Monte Carlo integrations to compute (2.17).

As some last thoughts here, one may question the comfort of taking a very complex multi-level model and reducing it to a single number for model comparison. Perhaps, we might be interested in the *local* performance of a model. Perhaps, we could choose different models in different parts of the predictive space. Perhaps we could average over models (Bayesian model averaging, [83]). The overarching question asks what is the utility for the model; different models would be preferred under different utilities.

2.11   Summary

We have argued that hierarchical models provide the stochastic framework within which to develop integrative process models. We have shown, with a variety of examples, that these models typically share a common structure. There is a first stage data model, there is a second stage process model that is latent, i.e., it is endowed with a full model specification but it is unobserved, and a third stage which incorporates prior specifications for all of the remaining parameters in the model. We have noted that, in order to get the uncertainty right, these models should be fitted within the Bayesian framework. We have also noted that fitting of these models typically introduces familiar looping in Gibbs sampling. Hence, it is straightforward to envision how the MCMC model fitting should be implemented. However, according to the size of the dataset and the complexity of the specifications, such model fitting can be very challenging, perhaps infeasible. Indeed, this limitation will become more of a constraint as we continue to seek models which stretch the limits of our computing capabilities. Hence, we imagine a computing future built around simulation based model fitting of these hierarchical forms but incorporating suitable approximation. Thus, the "art" will encompass both specification (with comparison) and approximate fitting (to enable inference and comparison).

Bibliography

[1] Agresti, A. (2013). *Categorical data analysis*. John Wiley & Sons, Hoboken, NJ, third edition.

[2] Akaike, H. (1974). A new look at the statistical model identification. *IEEE Transactions on Automatic Control*, 19(6):716–723.

[3] Albert, J. H. and Chib, S. (1993). Bayesian analysis of binary and polychotomous response data. *Journal of the American Statistical Association*, 88(422):669–679.

[4] Aldrich, J. (1997). RA fisher and the making of maximum likelihood 1912-1922. *Statistical Science*, pages 162–176.

[5] Banerjee, S., Carlin, B. P., and Gelfand, A. E. (2014). *Hierarchical modeling and analysis for spatial data*. Chapman & Hall/CRC Press, Boca Raton, FL, second edition.

[6] Barber, J. J., Gelfand, A. E., and Silander, J. A. (2006). Modelling map positional error to infer true feature location. *Canadian Journal of Statistics*, 34(4):659–676.

[7] Barry, D. and Hartigan, J. A. (1993). A Bayesian analysis for change point problems. *Journal of the American Statistical Association*, 88(421):309–319.

[8] Berger, J. (2006). The case for objective Bayesian analysis. *Bayesian Analysis*, 1(3):385–402.

[9] Berger, J. O., Strawderman, W., and Tang, D. (2005). Posterior propriety and admissibility of hyperpriors in normal hierarchical models. *The Annals of Statistics*, 33(2):606–646.

[10] Berger, J. O., Wolpert, R. L., Bayarri, M., DeGroot, M., Hill, B. M., Lane, D. A., and LeCam, L. (1988). The likelihood principle. *Lecture Notes-Monograph Series*, 6:iii–199.

[11] Berliner, L. M. (1996). Hierarchical Bayesian time series models. In *Maximum entropy and Bayesian methods*, pages 15–22. Kluwer Academic Publishers.

[12] Box, G. E. and Tiao, G. C. (1973). *Bayesian inference in statistical analysis*. Addison-Wesley, Massachusetts.

[13] Burnham, K. P. and Anderson, D. R. (2002). *Model selection and multimodel inference: a practical information-theoretic approach*. Springer-Verlag, New York.

[14] Carlin, B. P. and Louis, T. A. (2008). *Bayesian methods for data analysis*. Chapman and Hall/CRC Press, Boca Raton, FL, third edition.

[15] Carroll, R. J., Ruppert, D., Crainiceanu, C. M., and Stefanski, L. A. (1995). *Measurement error in nonlinear models*. Chapman and Hall/CRC, Boca Raton, FL.

[16] Carter, C. K. and Kohn, R. (1994). On gibbs sampling for state space models. *Biometrika*, 81(3):541–553.

[17] Claeskens, G. and Hjort, N. L. (2008). *Model selection and model averaging*. Cambridge University Press, Cambridge.

[18] Dempster, A. P., Laird, N. M., and Rubin, D. B. (1977). Maximum likelihood from incomplete data via the EM algorithm. *Journal of the Royal Statistical Society: Series B (statistical methodology)*, pages 1–38.

[19] Dey, D. K., Gelfand, A. E., Swartz, T. B., and Vlachos, P. K. (1998). A simulation-intensive approach for checking hierarchical models. *TEST*, 7(2):325–346.

[13] Dobson, A. J. and Barnett, A. (2008). *An introduction to generalized linear models*. Chapman & Hall/CRC press, Boca Raton, FL, third edition.

[21] Dufour, J.-M. (2006). Monte Carlo tests with nuisance parameters: A general approach to finite-sample inference and nonstandard asymptotics. *Journal of Econometrics*, 133(2):443–477.

[22] Fearnhead, P. and Liu, Z. (2011). Efficient Bayesian analysis of multiple changepoint models with dependence across segments. *Statistics and Computing*, 21(2):217–229.

[23] Frühwirth-Schnatter, S. (1994). Data augmentation and dynamic linear models. *Journal of Time Series Analysis*, 15(2):183–202.

[24] Fuller, W. A. (1987). *Measurement error models*. John Wiley & Sons, New York.

[25] Geisser, S. (1993). *Predictive Inference: An Introduction*, volume 55. Chapman & Hall, New York, NY.

[26] Gelfand, A. E. and Dey, D. K. (1994). Bayesian model choice: asymptotics and exact calculations. *Journal of the Royal Statistical Society: Series B (statistical methodology)*, pages 501–514.

[27] Gelfand, A. E., Dey, D. K., and Chang, H. (1992). Model determination using predictive distributions with implementation via sampling-based method. *Bayesian Statistics*, 4:147–167.

[28] Gelfand, A. E., Diggle, P., Guttorp, P., and Fuentes, M. (2010). *Handbook of Spatial Statistics*. Chapman & Hall/CRC press, Boca Raton, FL.

[29] Gelfand, A. E. and Ghosh, S. K. (1998). Model choice: a minimum posterior predictive loss approach. *Biometrika*, 85(1):1–11.

[30] Gelfand, A. E. and Smith, A. F. (1990). Sampling-based approaches to calculating marginal densities. *Journal of the American Statistical Association*, 85(410):398–409.

[23] Gelman, A., Carlin, J. B., Stern, H. S., Dunson, D. B., Vehtari, A., and Rubin, D. B. (2013). *Bayesian Data Analysis*. Chapman & Hall/CRC Press, Boca Raton, FL, third edition.

[32] Gelman, A., Meng, X.-L., and Stern, H. (1996). Posterior predictive assessment of model fitness via realized discrepancies. *Statistica Sinica*, pages 733–760.

[33] Girón, F. and Rojano, J. (1994). Bayesian Kalman filtering with elliptically contoured errors. *Biometrika*, 81(2):390–395.

[34] Gneiting, T. and Raftery, A. E. (2007). Strictly proper scoring rules, prediction, and estimation. *Journal of the American Statistical Association*, 102(477):359–378.

[35] González-Manteiga, W. and Crujeiras, R. M. (2013). An updated review of goodness-of-fit tests for regression models. *TEST*, 22(3):361–411.

[70] Green, P. J. (1995). Reversible jump Markov chain Monte Carlo computation and Bayesian model determination. *Biometrika*, 82(4):711–732.

[83] Hoeting, J. A., Madigan, D., Raftery, A. E., and Volinsky, C. T. (1999). Bayesian model averaging: a tutorial. *Statistical Science*, 14(4):382–401.

[38] Hooten, M. B. and Wikle, C. K. (2008). A hierarchical Bayesian non-linear spatio-temporal model for the spread of invasive species with application to the Eurasian Collared-Dove. *Environmental and Ecological Statistics*, 15(1):59–70.

[39] Kent, J. and Mardia, K. (2002). Modelling strategies for spatial-temporal data. In *Spatial Cluster Modelling*, pages 213–226. Chapman & Hall/CRC press, Boca Raton, FL.

[40] Lambert, P. C., Sutton, A. J., Burton, P. R., Abrams, K. R., and Jones, D. R. (2005). How vague is vague? a simulation study of the impact of the use of vague prior distributions in MCMC using winbugs. *Statistics in Medicine*, 24(15):2401–2428.

[41] Lehmann, E. L. and Casella, G. (2006). *Theory of point estimation.* Springer, New York, second edition.

[42] Little, R. J. and Rubin, D. B. (2014). *Statistical analysis with missing data,* volume 333. John Wiley & Sons.

[43] Lunn, D., Spiegelhalter, D., Thomas, A., and Best, N. (2009). The bugs project: evolution, critique and future directions. *Statistics in Medicine,* 28(25):3049–3067.

[44] Lynch, P. (2006). *The emergence of numerical weather prediction: Richardson's dream.* Cambridge University Press, Cambridge.

[45] Lynch, P. (2008). The origins of computer weather prediction and climate modeling. *Journal of Computational Physics,* 227(7):3431–3444.

[46] Marin, J.-M., Mengersen, K., and Robert, C. P. (2005). Bayesian modelling and inference on mixtures of distributions. In *Handbook of Statistics,* volume 25, pages 459–507. Elsevier.

[47] Mayo, D. G. (2014). On the Birnbaum argument for the strong likelihood principle. *Statistical Science,* pages 227–239.

[48] McCullagh, P. and Nelder, J. A. (1989). *Generalized linear models,* volume 37. Chapman & Hall/CRC press, Boca Raton, FL, second edition.

[49] McGuffie, K. and Henderson-Sellers, A. (2005). *A climate modelling primer.* John Wiley & Sons, Chichester, third edition.

[50] McLachlan, G. and Peel, D. (2004). *Finite mixture models.* John Wiley & Sons, New York.

[51] Minka, T. (2002). Bayesian inference in dynamic models: an overview. Technical report, Carnegie Mellon University.

[52] Oakley, J. E. and O'hagan, A. (2007). Uncertainty in prior elicitations: a nonparametric approach. *Biometrika,* 94(2):427–441.

[53] O'Hara, R. B., Sillanpää, M. J., et al. (2009). A review of bayesian variable selection methods: what, how and which. *Bayesian Analysis,* 4(1):85–117.

[54] Pettit, L. (1990). The conditional predictive ordinate for the normal distribution. *Journal of the Royal Statistical Society: Series B (statistical methodology),* 52(1):175–184.

[55] Picard, R. R. and Cook, R. D. (1984). Cross-validation of regression models. *Journal of the American Statistical Association,* 79(387):575–583.

[56] Raudenbush, S. W. (1994). Random effects models. In *The Handbook of Research Synthesis,* volume 421. Russell Sage Foundation, New York.

[57] Raudenbush, S. W. and Bryk, A. S. (2002). *Hierarchical linear models: Applications and data analysis methods.* Sage, Newbury Park, CA, second edition.

[58] Robert, C. P. and Casella, G. (2004). *Monte Carlo Statistical Methods.* Springer-Verlag, New York, second edition.

[59] Robert, C. P., Ryden, T., and Titterington, D. M. (2000). Bayesian inference in hidden Markov models through the reversible jump Markov chain Monte Carlo method. *Journal of the Royal Statistical Society: Series B (statistical methodology)*, 62(1):57–75.

[60] Schwarz, G. et al. (1978). Estimating the dimension of a model. *The Annals of Statistics*, 6(2):461–464.

[61] Spiegelhalter, D. J., Best, N. G., Carlin, B. P., and Van Der Linde, A. (2002). Bayesian measures of model complexity and fit. *Journal of the Royal Statistical Society: Series B (statistical methodology)*, 64(4):583–639.

[62] Titterington, D. M., Smith, A. F., and Makov, U. E. (1985). *Statistical analysis of finite mixture distributions*. Wiley, New York.

[63] Verdinelli, I. and Wasserman, L. (1991). Bayesian analysis of outlier problems using the Gibbs sampler. *Statistics and Computing*, 1(2):105–117.

[64] West, M. and Harrison, J. (1997). *Bayesian forecasting and dynamic models*. Springer, New York, second edition.

[65] Wikle, C. K. (2003). Hierarchical Bayesian models for predicting the spread of ecological processes. *Ecology*, 84(6):1382–1394.

3

Time series methodology

Peter F. Craigmile

Department of Statistics, The Ohio State University, Columbus, Ohio, USA.

CONTENTS

3.1 Introduction

In environmental science and ecology we often make statistical inferences based on data collected sequentially in time. To perform accurate inferences we need to build *time series models or processes* to precisely represent the dependencies observed in *time series data*. Key to this modeling exercise is being able to specify different model components that account for all the effects that we observe in the data over time. For example, we may choose to model a time series in terms of the long term changes in the mean (the trend), regular periodic components (the seasonality), and the noise. This is called the classical decomposition (see Section 3.5.1). Typically for most *time series analyses* the noise cannot

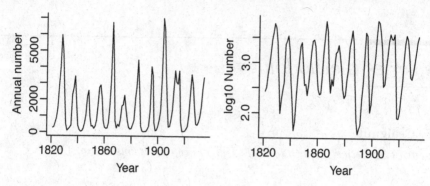

FIGURE 3.1

Original (left) and log-transformed (right) annual number of Canadian lynx trapped around the MacKenzie River between 1821 and 1934.

be assumed to be independent and identically distributed, and we need to find a model to describe the dependence that we observe in the noise process.

One application of this idea is to regression modeling. When the response variable is a time series, the noise, once we account for covariates, is unlikely to be independent over time. We specify a time series model for the noise, to perform accurate inference for the regression parameters. In other situations, we could also fit a time series model to remove certain features from the data. For example, we may want to account for, and remove seasonality in the data that is not of direct interest (we *deseasonalize*). We can also *detrend* to remove long-term mean effects.

An important application of time series analysis is to the *prediction* of future, past, or missing values. When predicting into the future we call this *forecasting*; when predicting the past, *hindcasting* (Hindcasting is useful, for example in paleoclimate when the interest is in studying past climate). We are interested not just in obtaining a point prediction, but in producing measures of prediction uncertainty.

There has also been attention to using time series models as simulators for physical phenomena. For example, time series models can be used as *statistical emulators* of turbulence [16, 40, 41] or hydrologic systems [32].

In this chapter we introduce useful classes of time series processes and models, and outline their theoretical properties. We take care to differentiate between so-called stationary and nonstationary processes. We then provide commonly used methods of statistical analysis for time series data. First, we introduce a motivating time series that we will explore as we progress through the chapter. [22] studied an ecological time series of the annual number of Canadian lynx trapped around the MacKenzie River between 1821 and 1934 (See [13] for an early review of statistical methods of analysis for the Canadian lynx dataset). Time series plots of the original and log-transformed series (Figure 3.1) indicate that the data have a seasonal pattern with a period of around 10 years, but that the amplitude of the seasonality are not constant in time. The log transformation makes the series more symmetrically-distributed, but the seasonal patterns are still not constant in time. We will demonstrate that these *quasi-periodicities* can be modeled using time series processes.

3.2 Time series processes

Formally a *time series process* is a stochastic process, a family of random variables (RVs) indexed in time,

$$\{X_t : t \in T\},$$

defined on a probability space. When $T \subset \mathbb{Z}$ we have a *discrete-time* time series process; when $T \subset \mathbb{R}$, a *continuous-time* time series process. In this chapter we focus on discrete-time processes. (See [10] for a review of continuous-time processes, including an exposition of continuous autoregressive moving average (CARMA) processes.) If X_t is a univariate RV for all t then $\{X_t : t \in T\}$ is a *univariate time series process*, and when X_t is a random vector, we have a *multivariate time series process*. We focus here on univariate processes, although the methods extend straightforwardly to multivariate series [11, 44].

The functions $\{X_t(\omega) : \omega \in \Omega, t \in T\}$ are known as the *realizations* or *sample paths* of the process $\{X_t\}$. For this chapter we informally call these realizations the *time series data* or just the *time series*, denoted $\{x_t\}$. Typically we want to perform inference upon the time series process on the basis of a finite collection of $n > 1$ observations, $\{x_t : t = 1, \ldots, n\}$.

Mathematically, Kolmogorov's theorem (e.g., [11], Theorem 1.2.1) can be used to check for existence of a time series process by verifying conditions on the finite dimensional distributions of the RVs over time. More commonly we demonstrate existence by building a time series process from a measurable function of another time series process; for example, in Section 3.3.1 we generate processes by filtering. An interesting subclass are the *Gaussian processes*: time series processes, for which all finite-dimensional distributions are multivariate normal.

3.3 Stationary processes

Stationary processes are time series processes for which some characteristic of the distribution of the series does not depend on the time points, only on the distance between the points. (A distance in time is often called the *time lag* or just the *lag*.) While stationarity may only be an approximation in practice, the assumption that a process is stationary (or the noise in a regression model is stationary, for example) permits a parsimonious representation of dependence. This allows, for example, for efficient estimation and prediction with time series.

We contrast between strictly and weakly stationary processes. For strictly stationary processes the joint distribution of the process at any collection of time points are unaffected by time shifts: formally $\{X_t : t \in T\}$ is *strictly stationary* (also called *narrow-sense*) when $(X_{t_1}, \ldots, X_{t_n})$ is equal in distribution to $(X_{t_1+h}, \ldots, X_{t_n+h})$, for all integers h and $n \geq 1$, and time points $\{t_k \in T\}$.

On the other hand, weakly stationary processes are defined only in terms of the mean and covariance of the process. A time series process $\{X_t : t \in T\}$ is *(weakly) stationary* if

1. $E(X_t) = \mu_X$ for some constant μ_X which does not depend on t; and

2. $\mathrm{cov}(X_t, X_{t+h}) = \gamma_X(h)$, a finite constant that can depend on the lag h but not on t.

Other terms for weakly stationary include *second-order, covariance, wide-sense*. In the above definition $\gamma_X(\cdot)$ is called the *autocovariance function (ACVF)* of the stationary process. It can also be useful to work with the unitless *autocorrelation function (ACF)* of the stationary process,

$$\rho_X(h) = \text{corr}(X_t, X_{t+h}) = \frac{\gamma_X(h)}{\gamma_X(0)}, \quad h \in \mathbb{Z}.$$

A strictly stationary process $\{X_t\}$ is also weakly stationary as long as $E(X_t)$ and $\text{var}(X_t)$ are finite for all t. Weak stationarity does not imply strict stationarity. However, a Gaussian weakly stationary process is strictly stationary.

Any ACVF of a stationary process has the following properties:

1. The ACVF is even: $\gamma_X(h) = \gamma_X(-h)$ for all h;
2. The ACVF is *nonnegative definite (nnd)*: $\sum_{j=1}^{n} \sum_{k=1}^{n} a_j \gamma_X(j-k) a_k \geq 0$ for all positive integers n and real-valued constants $\{a_j : j = 1, \ldots, n\}$.

More formally, Bochner's theorem [5, 6] states that a real-valued function $\gamma_X(\cdot)$ defined on the integers is the ACVF of a stationary process if and only if it is even and nnd. Typically we do not use this theorem to construct ACVF functions for modeling stationary time series from even and nnd functions. (There are exceptions: e.g., [24] shows that the Cauchy function is even and nnd, and demonstrates the utility of using this ACVF to model long-range dependence – for a definition of long-range dependence, see Section 3.6). Instead we typically create new processes from measurable functions of other processes. The most popular way to create stationary processes is via linear filtering.

3.3.1 Filtering preserves stationarity

In the context of discrete-time processes, a *linear time invariant (LTI) filter* is an operation that takes a time series process $\{W_t : t \in T\}$ and creates a new process $\{X_t\}$ via

$$X_t = \sum_{j=-\infty}^{\infty} \psi_j W_{t-j}, \quad t \in T,$$

where $\{\psi_j : j \in \mathbb{Z}\}$ is a sequence of *filter coefficients* that do not depend on time and that are absolutely summable: $\sum_j |\psi_j| < \infty$.

Then, if $\{W_t\}$ is a mean zero stationary process with ACVF $\gamma_W(h)$, $\{X_t\}$ is a mean zero stationary process with ACVF

$$\gamma_X(h) = \sum_{j=-\infty}^{\infty} \sum_{k=-\infty}^{\infty} \psi_j \, \psi_k \, \gamma_W(j-k+h), \quad h \in \mathbb{Z}.$$

This is the *linear filtering preserves stationarity* result. We now use this result to define useful classes of stationary processes.

3.3.2 Classes of stationary processes

Here follows some popular classes of stationary processes that are often used for modeling time series. (We also discuss other useful classes of time series processes, some of which are stationary, later in the chapter.)

3.3.2.1 IID noise and white noise

The simplest strictly stationary processes is *IID noise*: $\{X_t : t \in T\}$ is a set of independent and identically distributed RVs. There are no dependencies across time for this process.

The analogous "no correlation" model that is weakly stationary is *white noise*: $\{X_t\}$ in this case satisfies $E(X_t) = \mu$, a constant independent of time for all t, with

$$
\gamma_X(h) = \begin{cases} \sigma^2, & h = 0; \\ 0 & \text{otherwise.} \end{cases}
$$

We write $\mathrm{WN}(\mu, \sigma^2)$ for a white noise process with mean μ and variance $\sigma^2 > 0$.

3.3.2.2 Linear processes

Suppose that $\{\psi_j : j \in \mathbb{Z}\}$ is an absolutely summable sequence that is independent of time and $\{Z_t : t \in T\}$ is a $\mathrm{WN}(0, \sigma^2)$ process. Then by the filtering preserves stationary result, the *linear process* $\{X_t : t \in T\}$, defined by a filtering of the white noise process,

$$
X_t = \sum_{j=-\infty}^{\infty} \psi_j Z_{t-j}, \quad t \in T,
$$

is a stationary process with zero mean and an ACVF given by

$$
\gamma_X(h) = \sigma^2 \sum_{j=-\infty}^{\infty} \psi_j \psi_{j+h}, \quad h \in \mathbb{Z}.
$$

This class of models is useful for theoretical studies of linear stationary processes, but in practice for statistical modeling we use linear filters of finite time support (i.e., $\psi_j \neq 0$ for a finite collection of indexes j) as we now describe.

3.3.2.3 Autoregressive moving average processes

Autoregressive moving average processes, arguably the most famous time series model, are defined in terms of two filtering operations – one for the process $\{X_t : t \in T\}$ itself, and one for the white noise $\{Z_t : t \in T\}$. These two filtering operations produce a flexible class for modeling linear dependence.

Suppose $\{Z_t\}$ is a $\mathrm{WN}(0, \sigma^2)$ process throughout this section. We say that $\{X_t\}$ is an *autoregressive moving average process of autoregressive order p and moving average order q*, or an *ARMA(p, q) process* for short, if there exists a stationary solution to the following equation:

$$
X_t - \sum_{j=1}^{p} \phi_j X_{t-j} = Z_t + \sum_{k=1}^{q} \theta_k Z_{t-k}, \quad t \in T,
$$

assuming that $\phi_p \neq 0$ and $\theta_q \neq 0$. By "there exists a stationary solution" we mean that we can represent $\{X_t\}$ as a (stationary) linear process, as defined by Section 3.3.2.2. The above ARMA process is a mean zero process; more generally $\{X_t\}$ is an ARMA(p, q) process with mean μ if

$$
(X_t - \mu) - \sum_{j=1}^{p} \phi_j (X_{t-j} - \mu) = Z_t + \sum_{k=1}^{q} \theta_k Z_{t-k}, \quad t \in T.
$$

There are two important simplifications of ARMA models. When $q = 0$ we obtain the *autoregressive process of order p* (the AR(p) process; [65, 66])

$$X_t - \sum_{j=1}^{p} \phi_j X_{t-j} = Z_t, \quad t \in T. \tag{3.1}$$

With $p = 0$, the *moving average process of order q (the MA(q) process)* satisfies

$$X_t = Z_t + \sum_{k=1}^{q} \theta_k Z_{t-k}, \quad t \in T. \tag{3.2}$$

For ARMA processes the *AR coefficients* $\{\phi_j : j = 1, \ldots, p\}$, the *MA coefficients* $\{\theta_j : j = 1, \ldots, q\}$, and the *innovation variance* σ^2 determine the statistical properties of the process $\{X_t\}$, such as the ACVF. Further, examining the AR coefficients allow us to test for stationarity and causality. (For a *causal* process X_t can be written as an MA process of infinite order; i.e., we set $q = \infty$ in (3.2); an example of an AR processes that is not causal is a linear process for which $\psi_{-1} \neq 0$. Assuming causality is useful for prediction and calculating the ACVF.) The MA coefficients determine if we can represent X_t as an AR process of infinite order (such *invertible* processes allow us to find unique parameter estimates for ARMA processes). See e.g., [11] and [64] for properties of ARMA processes.

We now provide simple examples of ARMA processes. For any given application there is no a priori reason that these models will fit adequately, but they do illustrate simple models of time series dependence. The MA(1) process $\{X_t : t \in T\}$, defined by

$$X_t = Z_t + \theta Z_{t-1}, \quad t \in T,$$

is a mean zero stationary process for all values of θ. The ACVF is

$$\gamma_X(h) \;=\; \begin{cases} \sigma^2(1 + \theta^2), & h = 0; \\ \sigma^2\theta, & h = \pm 1; \\ 0, & \text{otherwise,} \end{cases}$$

and the ACF is

$$\rho_X(h) \;=\; \begin{cases} 1 & h = 0; \\ \theta/(1 + \theta^2), & h = \pm 1; \\ 0, & \text{otherwise.} \end{cases}$$

The MA(1) process is a model for any stationary time series in which we only have a lag-one correlation (this is also called a 1-correlated process). As this type of dependence is uncommon, in many applications an AR(1) process is more reasonable than an MA(1) process. The AR(1) process, defined by

$$X_t = \phi X_{t-1} + Z_t, \quad t \in T, \tag{3.3}$$

is stationary, has mean zero, and is causal for $-1 < \phi < 1$. The ACVF,

$$\gamma_X(h) = \sigma^2 \frac{\phi^{|h|}}{1 - \phi^2}, \quad h \in \mathbb{Z}$$

and ACF $\rho_X(h) = \phi^{|h|}$, both decay exponentially in magnitude to zero. In many applications, we might imagine that we observe an AR(1) process with independent additive measurement error; this can be represented as a special case of an ARMA(1,1) process:

$$X_t \;=\; \phi X_{t-1} + Z_t + \theta Z_{t-1}, \quad t \in T.$$

This process is again stationary, mean zero, and causal for $-1 < \phi < 1$. The ACVF,

$$
\gamma_X(h) \;=\;
\begin{cases}
\sigma^2 \left[1 + \dfrac{(\theta + \phi)^2}{1 - \phi^2} \right], & h = 0; \\[2ex]
\sigma^2 \left[\theta + \phi + \dfrac{(\theta + \phi)^2 \phi}{1 - \phi^2} \right], & h = \pm 1; \\[2ex]
\phi^{|h|-1} \gamma_X(1) & \text{otherwise,}
\end{cases}
$$

exhibits features in common with both the MA(1) and AR(1) processes.

3.4 Statistical inference for stationary series

Suppose throughout this section that $\{X_t : t \in T\}$ is a stationary process with mean μ_X and ACVF $\gamma_X(h)$. Let $\rho_X(h)$ denote the associated ACF. We now outline methods of estimation for the mean, ACVF, and ACF based on time series data $x = (x_1, \ldots, x_n)^T$; let $X = (X_1, \ldots, X_n)^T$ denote the associated random vector. For a stationary process we note that the $n \times n$ covariance matrix of X, Σ, is a *symmetric Toeplitz* matrix with (j, k) element given by $\Sigma_{jk} = \gamma_X(j - k)$.

3.4.1 Estimating the process mean

The most commonly used estimator of the process mean μ_X is the sample mean,

$$
\overline{X}_n = \frac{1}{n} \sum_{t=1}^{n} X_t.
$$

Except when $\{X_t\}$ is WN or IID noise, this unbiased estimator does not have the smallest variance among all linear unbiased estimators of μ_X (i.e., it is not the best linear unbiased estimator; BLUE). In its favor, it does not require knowledge of the ACVF to calculate it.

The BLUE estimator of μ_X, $\widehat{\mu}_n$, does require knowledge of the ACVF:

$$
\widehat{\mu}_n \;=\; [\mathbf{1}^T \Sigma^{-1} \mathbf{1}]^{-1} \mathbf{1}^T \Sigma^{-1} X,
$$

where $\mathbf{1}$ is a vector of ones.

The sampling distribution of $\overline{X}_n$ and $\widehat{\mu}_n$ both depend on the (typically unknown) ACVF. With the sample mean, for example, $\operatorname{var}(\overline{X}_n) = v_n/n$, where

$$
v_n \;=\; \sum_{h=-(n-1)}^{(n-1)} \left(1 - \frac{|h|}{n} \right) \gamma_X(h).
$$

When $\{X_t\}$ is a stationary Gaussian process,

$$
\frac{\overline{X}_n - \mu_X}{\sqrt{v_n/n}}
$$

follows a standard normal distribution, which forms the basis for statistical inference for μ_X. As for the independent case, there are central limit theorems for stationary process (e.g., [11], Section 7.1; [59], Section 1.3). These theorems allow us to carry out approximate inference on the process mean for certain non-Gaussian stationary processes. Often a large sample approximation is made for v_n.

To calculate v_n we need to estimate the ACVF from the data (see Section 3.4.2 – typically we truncate the number of lags to include in the estimator for v_n; see e.g., [12], p.59), or we can propose a time series model for our data and calculate v_n using the ACVF from the estimated model (Section 3.4.5). For this reason we postpone inference for the mean in the lynx example until we have a time series model for the data.

3.4.2   Estimating the ACVF and ACF

Since the ACVF for a stationary process is even and nnd, we look for estimators of the ACVF and ACF that are even and nnd. Thus, to estimate

$$\gamma_X(h) \quad = \quad \text{cov}(X_t, X_{t+h}) \quad = \quad E[(X_t - \mu_X)(X_{t+h} - \mu_X)],$$

we use the *sample ACVF*, defined by

$$\widehat{\gamma}_X(h) \quad = \quad \frac{1}{n} \sum_{t=1}^{n-|h|} (X_t - \overline{X})(X_{t+|h|} - \overline{X}), \quad |h| < n.$$

The *sample ACF* is

$$\widehat{\rho}_X(h) \quad = \quad \frac{\widehat{\gamma}_X(h)}{\widehat{\gamma}_X(0)}, \quad |h| < n.$$

Clearly, given the definition of the sample ACVF and ACF, these estimators are more untrustworthy for larger lags h, relative to the sample size n (since there is less information in the estimator). [8] suggests using these estimators to select a time series model only when $n \geq 50$ which is too limiting in practice (they suggest using "experience and past information to derive a preliminary model" when $n < 50$). As for how many lags to interpret, [8] suggests $h \leq n/4$. For univariate series, R ([53]) displays estimates at lags $|h| \leq 10\log_{10}n$, although this does not suggest how many lags to actually interpret. For complicated dependence structures we may need to investigate the ACVF and ACF at longer lags.

Both the sample ACVF and ACF are biased estimators at finite sample sizes n. At a fixed lag they become less biased as n increases. Bartlett [2] provides the large sample distribution of the sample ACF for certain stationary processes, which can be useful to test the fit of time series models. In particular Bartlett's result can be used as a basis for testing whether or not a time series process is IID. When $\{X_t\}$ is an IID process with finite variance, the sample ACFs $\{\widehat{\rho}_X(h) : h = 1, 2, \ldots\}$ are approximately IID normal RVs each with mean 0 and variance $1/n$.

As a demonstration, Figure 3.2(a) displays a plot of the sample ACF versus the lag for the log-transformed Canadian lynx series (we discuss panel (b) after we define the PACF below). According to Bartlett's result, if the process is IID, then approximately 95% of the non-zero lags of the ACF should be within the bands $\pm 1.96/\sqrt{n}$ (denoted by the dashed horizontal lines in the figure). The lynx series is clearly not a sample of IID noise; there is a strong seasonal pattern to the sample ACF, that possibly decays in amplitude at longer lags. The sample ACF at lags h and $h+10$ are similar, which suggests a periodicity in the log lynx numbers (as claimed by [22]).

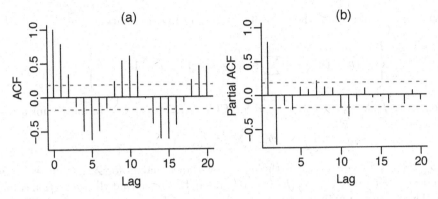

FIGURE 3.2
(a) A plot of the sample ACF for the log-transformed Canadian lynx series; (b) A plot of the sample PACF.

Interpreting an ACF plot with pointwise bands leads to multiple testing concerns. This can be alleviated by building a test based on sums of squares of the ACF. The Box-Pierce test [7] tests the null hypothesis that $\{X_t\}$ is IID (versus not IID). Under the null hypothesis the test statistic

$$Q = n \sum_{h=1}^{k} \widehat{\rho}_X^2(h)$$

is approximately chi-squared on k degrees of freedom. [66] suggested a revised statistic with improved power, although the advice is mixed on how many lags k to base any IID noise test on. For the log lynx series using 20 lags (the same number as presented in Figure 3.2), the Box-Pierce test statistic is 459.4 on 20 degrees of freedom – the corresponding P-value is very close to zero, which is strong evidence against the time series being generated from an IID process. This hypothesis test is not particularly useful here since a glance of the time series plot for the Canadian lynx series indicates that the data are not IID, and that a time series analysis is necessary. IID noise tests are more commonly used to check the fit of various models for time series – see Section 3.4.6 below. Also consult, e.g., [12] and [64] for other IID and WN tests.

3.4.3 Prediction and forecasting

Earlier we pointed out that we are often interested in predicting or forecasting values, along with some measure of uncertainty. (Similar methods apply to predicting missing values or values in the past.) In time series analysis, prediction is also a key ingredient in identifying, fitting, diagnosing, and comparing models.

Again suppose that $\{X_t : t \in T\}$ is our stationary process with mean μ_X and ACVF $\gamma_X(h)$. Based on $\boldsymbol{X} = (X_1, \ldots, X_n)^T$ we want to forecast X_{n+h} at some integer lag in the future, $h > 0$. The *best linear unbiased predictor (BLUP)* of X_{n+h} based on a linear combination of $X_1, X_2, \ldots, X_n$ is given by the following conditional expectation:

$$E(X_{n+h}|X_n, \ldots, X_1) = \mu_X + \sum_{j=1}^{n} \phi_{n,j}(X_{n+1-j} - \mu_X),$$

where $\phi_{n,j}$ is the solution of the following system of linear equations:

$$\left\{ \gamma_X(h+1-k) = \sum_{j=1}^{n} \phi_{n,j} \gamma_X(k-j) \ : \ k = 1, \ldots, n \right\}. \tag{3.4}$$

We often write $P_n X_{n+h}$ as a shorthand for $E(X_{n+h}|X_n, \ldots, X_1)$. In matrix form we have

$$\phi_n = \Sigma^{-1} \gamma_{n,h}, \tag{3.5}$$

where $\phi_n = (\phi_{n,1}, \ldots, \phi_{n,n})^T$, Σ is the $n \times n$ covariance matrix for X introduced in Section 3.4.1, and $\gamma_{n,h} = (\gamma_X(h), \ldots, \gamma_X(h+n-1))^T$. It can be shown that the prediction coefficients ϕ_n only depend on the ACF for the stationary process. In practice when the ACF or ACVF is unknown, we often substitute their estimates into (3.5).

Formally the predictor $P_n X_{n+h}$ minimizes the *mean squared prediction error* given by

$$\text{MSPE} \ = \ E\left[(X_{n+h} - P_n X_{n+h})^2\right] \ = \ E\left[U_{n+h}^2\right],$$

where $U_{n+h} = X_{n+h} - P_n X_{n+h}$ is called the *prediction error* for X_{n+h}. In practice we solve (3.4) recursively using the Levinson-Durbin algorithm or the Innovation Algorithm (see, e.g., [11] for a full description of both algorithms). The Levinson-Durbin algorithm recursively calculates the coefficients of $P_n X_{n+1}$ given the coefficients of the previous prediction equation for $P_{n-1} X_n$, whereas the Innovations Algorithm calculates recursively the coefficients for calculating the next prediction error U_{n+1} on the basis of coefficients from past prediction errors. The minimum value of the MSPE is also calculated recursively in each algorithm.

This minimum value of the MSPE, given by

$$\nu_{n,h}^2 \ = \ \gamma_X(0) - \phi_n^T \gamma_{n,h}, \tag{3.6}$$

provides an estimate of the variance for the predictor $P_n X_{n+h}$. For a stationary Gaussian process $\{X_t\}$, a $100(1-\alpha)\%$ prediction interval for X_{n+h} is

$$P_n X_{n+h} \pm z_{1-\alpha/2} \sqrt{\nu_{n,h}^2},$$

where z_q is the qth quantile of the standard normal distribution. This interval is approximate for non-Gaussian processes.

3.4.4   Using measures of correlation for ARMA model identification

A traditional method of identifying plausible time series models is to examine the ACF and the *partial ACF (PACF)*. The ACF looks at the correlation a certain number of lags apart, whereas the PACF looks at the correlation at a certain lag apart, while ignoring the effect of variables at intervening lags.

More formally the PACF for lag $h \geq 1$ is the correlation between the prediction error after predicting X_{h+1} using $X_2, \ldots, X_h$, and the prediction error after predicting X_1 using $X_2, \ldots, X_h$ (For $h = 1$ we interpret this to mean simply the correlation between X_2 and X_1). We calculate the lag h PACF, $\alpha_X(h)$, from the prediction equations (3.5):

$$\alpha_X(h) \ = \ \phi_{h,h}, \quad h = 1, 2 \ldots.$$

To calculate the sample PACF we calculate the prediction coefficients using the sample ACF.

We can now summarize how to identify ARMA processes:

1. The PACF is ideally suited to the identification of AR(p) processes, because we always have that the theoretical PACF is zero at all lags greater than p. Moreover Quenouille [52] proved that the sample PACF are approximately IID normal with zero mean and variance $1/n$, at all lags greater than h. For AR(p) processes, the ACF decays exponentially in magnitude after lag p.

2. For MA(q) processes, the ACF is zero after lag q, but the PACF decays exponentially in magnitude.

3. For ARMA(p, q) processes neither the ACF and PACF is zero at longer lags; instead, at longer lags, they both decay exponentially in magnitude.

The sample ACF and PACF will be noisy versions of this result. Often there can be several models that are plausible based on these guidelines; we should be cautious not to choose models that are too complicated (i.e., we should not choose p and q to be too large – we discuss the issue of overfitting in Section 3.4.6 below).

Returning to Figure 3.2, we examine the sample ACF and PACF plots produced using R. The PACF follows the pattern of an AR(2) process: there is a non-zero lag 2 PACF followed by smaller PACF values after lag 2 – in reference to the Quenouille-based confidence intervals, these longer lags of the PACF may not be significantly different from zero; also, the ACF exponentially decays in magnitude. On the other hand, we could argue that the larger values at longer lags of the PACF require us to consider more involved AR or ARMA process alternatives. The pattern in the ACF rules out an MA process.

3.4.5 Parameter estimation

Once we have selected a time series model for our data $\boldsymbol{x} = (x_1, \ldots, x_n)^T$, we need to estimate the model parameters $\boldsymbol{\omega}$ (which we will assume is an m-dimensional vector). When $\{X_t : t \in T\}$ is stationary, for example, we estimate the mean μ_X and the model parameters that determine the ACVF $\gamma_X(h)$.

There are numerous estimation methods for time series models (see, e.g., [77], [12], and [64]). Here we discuss two commonly used methods: (i) the method of moments Yule-Walker scheme for AR processes; and (ii) the maximum likelihood scheme which applies more generally, but requires us to specify the joint distribution of $\boldsymbol{X} = (X_1, \ldots, X_n)^T$.

For AR(p) processes as defined by (3.1), we estimate $\boldsymbol{\omega} = (\boldsymbol{\phi}_p^T, \sigma^2)^T$, where $\boldsymbol{\phi}_p = (\phi_1, \ldots, \phi_p)^T$ are the p unknown AR coefficients. The *Yule-Walker (Y-W) method* sets the sample ACVF equal to the theoretical ACVF, and solves for the unknown parameter values (under the assumption that the AR process is causal). It can be shown that the Y-W estimates for $\boldsymbol{\phi}_p$ exactly correspond to solving the prediction equations given in Section 3.4.3, but substituting in the sample ACVF for the theoretical ACVF. The estimate for σ^2 is the minimum value of the MSPE given by (3.6), substituting in the sample ACVF for the theoretical ACVF. Reusing notation from that section, with hat notation to denote sample-based quantities, the Y-W estimates for an AR(p) model are

$$\widehat{\boldsymbol{\phi}}_p = \widehat{\boldsymbol{\Sigma}}_p^{-1} \widehat{\boldsymbol{\gamma}}_{n,p}, \quad \text{with} \quad \widehat{\sigma}^2 = \widehat{\gamma}_X(0) - \widehat{\boldsymbol{\phi}}_p^T \widehat{\boldsymbol{\gamma}}_{n,p}.$$

In practice we use the Levinson-Durbin algorithm to estimate AR models recursively: first we obtain the estimates for the AR(1) model, and then for each $k = 2, \ldots, p$ we find the AR(k) model estimates in terms of the AR($k-1$) estimates. This recursive estimation method is computationally efficient, and can help speed up our calculations when we try to choose the order p of the AR model that fits well to the data (See Section 3.4.6 below). In terms of statistical inference for the AR model parameters, under certain conditions the Y-W estimators have the same sampling distribution as the maximum likelihood estimates [11], which we now discuss.

Maximum likelihood for time series models in its most basic form is the same as likelihood estimation for other statistical models. We write down the *likelihood* $L_n(\omega)$, the joint distribution of the data x, evaluated at the model parameters ω. A Gaussian process is the most commonly made assumption for estimation in time series analysis (of course this assumption needs to be verified in practice; in other situations the process may not be Gaussian but the Gaussian likelihood is often used as the basis for an approximate estimation scheme). For a Gaussian mean zero stationary process with ACVF $\gamma_X(h)$, we have that X follows a n-variate normal distribution with zero mean, and covariance Σ (remember that the (j,k) element of Σ is equal to $\gamma_X(j-k)$, which is now a function of ω). Thus the likelihood function is

$$L_n(\omega) = (2\pi)^{-n/2}(\det \Sigma)^{-1/2}\exp\left(-\frac{1}{2}x^T\Sigma^{-1}x\right). \tag{3.7}$$

The *maximum likelihood (ML)* estimate of ω, when it exists, is equal to

$$\widehat{\omega}_{ML} \quad = \quad \arg\max_{\omega} L_n(\omega),$$

or equivalently $\widehat{\omega}_{ML} = \arg\max_{\omega} l_n(\omega)$, where $l_n(\omega) = \log L_n(\omega)$ is the *log-likelihood* function.

Calculating the likelihood and log-likelihood functions for a general multivariate normal RV is an $O(n^3)$ calculation because of the matrix inversion. What becomes more useful for stationary processes is when we rewrite the log-likelihood function using one-step-ahead predictions, calculated using a recursive prediction algorithm. This turns the likelihood into a $O(n^2)$ calculation for the most general stationary process – we can speed the likelihood calculation up further for simpler time series models such as AR(p) or MA(q). More precisely, letting $U_t = X_t - P_{t-1}X_t$ denote the *innovations* (the one-step-ahead prediction errors), the log-likelihood is

$$l_n(\omega) \quad = \quad -\frac{n}{2}\log(2\pi) - \frac{1}{2}\sum_{t=1}^{n}\log(\nu_{t-1,1}^2) - \frac{1}{2}\sum_{t=1}^{n}\left(\frac{u_t^2}{\nu_{t-1,1}^2}\right)$$

where u_t is the observed innovation and $\nu_{t-1,1}^2$ is the MSPE for $P_{t-1}X_t$ defined by (3.6) at time $t = 1, \ldots, n$.

Typically the ML estimates for time series models are not available in a closed form; we use numerical optimization algorithms to find the ML estimates (e.g., [11, 37, 64]). This is the approach taken by the `arima` function in R. While there are large sample estimates of the covariance of the ML estimator, it is more common to estimate the covariance using the Hessian matrix. Let $H(\omega)$ denote the matrix of second derivatives of the log-likelihood with respect to the parameters ω. Then statistical inference for the model parameters is based on assuming that $\widehat{\omega}_{ML}$ is approximately a m-variate normal distribution with mean ω and covariance matrix $[-H(\omega)]^{-1}$ (e.g., [59]). In practice the true value of the parameter ω is unknown and we evaluate the Hessian at the ML estimates. This typically results in a loss of statistical efficiency. We should be wary of inferring time series parameters marginally, as usually there exists collinearity among the parameter estimates.

We can also use Bayesian methods of inference for time series models; see e.g., [15, 29, 31, 49].

3.4.6 Model assessment and comparison

Overfitting in time series models leads to highly uncertain and correlated parameter estimates which degrades inference and negatively impacts predictive performance. Thus, it is important to assess the fit of our model, being able to identify when to simplify the model, or to make the dependence structure more complicated when the situation warrants it. In Section 3.4.4 we suggested that we should try not to identify plausible time series models that are too complicated for the dependence that we observe in the data. But, once a model is fit there are a number of ways to assess its fit.

First we can test for significance of the model parameters, cognizant of the warnings above about marginal testing of the model parameters. Based on our inferences, we can drop insignificant parameters from the model, next investigating the fit of the simplified model.

Secondly we can check the fit of our models by examining the *time series residuals*, $\{R_t : t = 1, \ldots, n\}$, defined for each t by

$$R_t = \frac{U_t}{\sqrt{\nu_{t-1,t}^2}};$$

again U_t is the innovation (one-step-ahead predictor error) and $\nu_{t-1,1}^2$ is the MSPE for the prediction – these quantities are calculated from the data and the estimated model parameters. Depending on what we assume about the time series model, the properties of $\{R_t\}$ can vary. With a Gaussian process assumption, the time series residuals are approximately IID normal with zero mean and constant variance; we often verify assumptions using time series plots, the sample ACF and PACF, and normal Q-Q plots of the time series residuals. We can back up our graphical assessments with IID noise tests, for example.

Thirdly, we can compare measures of model fit. Mimicking what happens in regression models, we should not choose the time series model that minimizes the innovation variance as this often leads to overfitting. Commonly we penalize the fit of the statistical model using a term that is related to number of parameters, m, in the model.

Many measures of model fit are an approximation to the Kullbach-Leibler information, a likelihood-based measure that compares our estimated model to the true, but unknown, model. This includes the *Akaike information criterion (AIC)* [1],

$$\text{AIC} = -2l_n(\widehat{\omega}) + 2m,$$

the bias corrected version,

$$\text{AICC} = -2l_n(\widehat{\omega}) + 2m \left[\frac{n}{n-1-m} \right]$$

(AIC tends to overestimate p for AR(p) models; [33, 56]), and the *Bayesian information criterion (BIC)*, also known as the *Schwarz criterion* [55],

$$\text{BIC} = -2l_n(\widehat{\omega}) + m\log n.$$

(Throughout $\widehat{\omega}$ is our estimator of ω.)

These criteria are often used for *order selection*, where we compare a number of time series models fit to the data, choosing those models that minimize the chosen criteria. A commonly used rule-of-thumb with AIC, AICC, and BIC is to consider models are equivalent (with respect to the model fit) if the criterion lies within 2 units.

The use of model selection criteria works best in combination with the other assessment methods described above, to fully understand how a given statistical model fits. Another

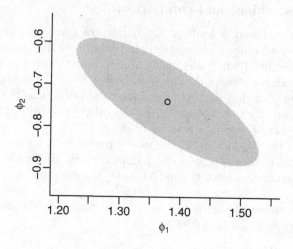

FIGURE 3.3

A 95% confidence region for the AR(2) model parameters, based on the ML estimates. The circle denotes the ML estimates.

useful strategy is to compare simulated realizations from the estimated model with the observed data to see if we are capturing all the interesting features of the time series sufficiently.

3.4.7 Statistical inference for the Canadian lynx series

We now return to our analysis of the Canadian lynx counts around the MacKenzie river from 1821–1934. We previously claimed that an AR(2) model was reasonable for the $\log_{10}$-transformed data $\{x_t : t = 1, \ldots, 114\}$. Fitting this model using the Y-W method with the **ar** function in R, we get $\widehat{\mu}_X = 2.90$, $\widehat{\phi}_1 = 1.35$, $\widehat{\phi}_2 = -0.72$ with $\widehat{\sigma}^2 = 0.058$. This is similar to the ML estimates calculated using the **arima** function: $\widehat{\mu}_X = 2.90 \ (0.06)$, $\widehat{\phi}_1 = 1.35 \ (0.06)$, $\widehat{\phi}_2 = -0.74 \ (0.06)$ and $\widehat{\sigma}^2 = 0.051$ (the numbers in parentheses are the standard errors estimated for the ML estimates). Each parameter in the AR(2) model is significantly different from zero, indicating that that we should not consider a simplified model. (Figure 3.3 confirms that a 95% confidence region for the AR(2) model parameters does not contain zero for either AR coefficient; see, e.g., [12], p.142, for how to calculate joint confidence regions for model parameters).

We next assess the ML fit of the AR(2) model by examining plots of the time series residuals – see Figure 3.4. We see that the series residuals are centered around zero with fairly constant spread. The normal Q-Q plot indicates that assuming Gaussianity is reasonable. The only issue is that we may not fully capture the time series dependence (the lag 10 ACF is outside the IID noise 95% confidence bands; the lags 10, 12 and 13 are outside the Quenouille–based 95% confidence bands for the PACF). This is confirmed by examining the Box-Pierce test: using 20 lags of the ACF, our test statistic is $Q = 35.0$. With a P-value of 0.056 we fail to reject the IID noise assumption with significance level $\alpha = 0.05$, but we could imagine that there may be dependence still left to model.

We calculate the AICC criterion for various model orders, p, of AR(p) model to see if it is worth making the statistical model more complicated (see Figure 3.5; we omit presenting the AICC of 84.2 for the AR(1) model since this model fits so poorly). The AR(10) and AR(11) models fit best according to the AICC, and time series residual plots are indistinguishable

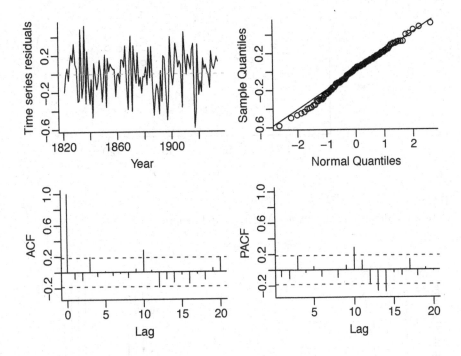

FIGURE 3.4
From left to right, top to bottom, a time series plot, a normal Q-Q plot, and the sample ACF and PACF of the time series residuals, for the AR(2) model fit.

between these two models. Examining, in this case, pointwise CIs for the AR(11) model parameters, we see that $\phi_3, \ldots, \phi_9$ are not significantly different from zero (again, we should be wary of making multiple comparisons). We then fit a subset model with most of the model parameters set to zero to obtain $\widehat{\mu}_X = 2.90$ (0.07), $\widehat{\phi}_1 = 1.19$ (0.07), $\widehat{\phi}_2 = -0.47$ (0.07), $\widehat{\phi}_{10} = 0.39$ (0.07), $\widehat{\phi}_{11} = -0.39$ (0.07), $\widehat{\sigma}^2 = 0.04$, with the lowest AICC found so far of -25.2.

In a similar fashion we could examine ARMA(p, q) model for the log-lynx numbers. But instead we investigate two interesting questions on the basis of our best fitting statistical model found so far. First, we infer upon the center of the distribution for the lynx series. Using the subset model, a 95% CI for the process mean for the log series is

$$2.90 \pm 1.96 \times 0.07;$$

that is between 1.53 and 3.03. With the assumption of a symmetric distribution for the log lynx counts, the mean is equal to the median of the log series, and transforming back to the original scale, a 95% CI for the median number of lynx captured on the MacKenzie river from 1821–1934 is between $10^{1.53} = 33.8$ and $10^{3.03} = 1071.5$.

Using the subset model, Figure 3.6 shows pointwise 95% prediction intervals for the period 1935–1949. (We transform back the endpoints of the prediction intervals on the $\log_{10}$ scale to obtain the prediction limits on the original scale.) These predictions indicate that the subset AR time series model we have fit is able to capture the quasi-periodicity of around 10 years observed in the lynx numbers, and clearly demonstrate higher uncertainty around the peaks, as one would expect.

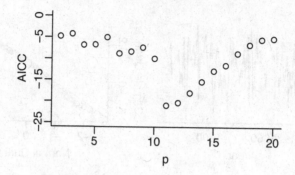

FIGURE 3.5
AICC criterion calculated for various AR(p) model orders.

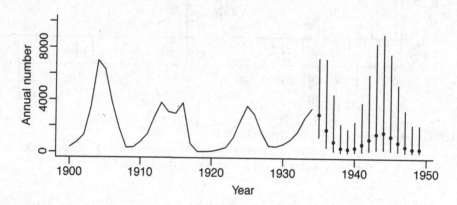

FIGURE 3.6
Time series plots of the lynx counts from 1900–1934. The vertical bars denote pointwise 95% prediction intervals from 1935–1944 on the original scale. The circles denote the point predictions.

3.5   Nonstationary time series

3.5.1   A classical decomposition for nonstationary processes

In practice it may be unlikely that a given time series is stationary. There are often components that we observe that are time-varying. One simple representation that separates nonstationary and stationary components observed in a time series $\{X_t : t \in T\}$ is the *classical decomposition* given by the additive model:

$$X_t = \mu_t + s_t + \eta_t, \quad t \in T. \tag{3.8}$$

In the additive model $\{\mu_t\}$ is a *trend component* that captures (usually) smooth changes in the mean, $\{s_t\}$ is as *seasonal or periodic component* that captures changes in the mean (over and above the trend) that repeats regularly with some period d say, and $\{\eta_t\}$ captures the *(random or irregular) noise or error*. For given data, we may not need a trend and/or seasonal component.

The definition of trend, seasonality, and noise is not unique for any given dataset, which leads to interesting modeling challenges. Regression models and filtering methods are commonly used to estimate the trend and seasonal components in the additive model (e.g., [12]). We often use a three-step strategy. For example: (i) We estimate the trend and seasonality using ordinary least squares regression (this does not account for possible time series dependence in the errors); (ii) Using the regression residuals we find plausible time series models for the error process $\{\eta_t\}$; (iii) We refit the regression model while accounting for time series dependence in the errors. (See the climate trends chapter for more information and a further discussion of the lack on uniqueness of the different model components.)

In other scenarios, statistical methods are used to manipulate the series to isolate specific components of the process while trying to eliminate others. For example in estimating monthly fish stocks, inference may be on isolating the deseasonalized count series; that is the original counts minus the estimated seasonal component.

Some methods of removing components such as the trend may alter the statistical properties of the remaining components. For example, *differencing* is a commonly used to remove a trend: differencing the series $\{X_t : t \in T\}$, yields the series $\{Y_t\}$ defined by

$$Y_t = X_t - X_{t-1}, \quad t \in T.$$

Differencing removes linear and locally linear trends, but can introduce more dependence in the differenced errors. *Lag-differencing of order d*, defined by $Y_t = X_t - X_{t-d}$, can be used to remove a seasonality of period d. Both versions of differencing presented here are a form of linear filtering – this fact can be used to derive the statistical properties of differenced time series.

3.5.2 Stochastic representations of nonstationarity

Typically in the additive model (3.8) we assume that the trend and seasonality are deterministic functions of time or other covariates. Alternatively we can model the trend (and possibly the seasonality) stochastically. The most basic nonstationary process is the *random walk* process $\{X_t : t = 1, 2, \ldots\}$, defined by

$$X_t = \sum_{s=1}^{t} Z_s, \quad t = 1, 2, \ldots, \tag{3.9}$$

where $\{Z_t\}$ is a WN$(0, \sigma^2)$ process. This process is the discrete approximation to Brownian motion. Sample paths for this series exhibit local trends. This is a coarse model for a system which exhibits increasing variance as time progresses. We note that the differenced process, $X_t - X_{t-1}$ (which equals Z_t in this case) is stationary.

Replacing $\{Z_t\}$ in (3.9) with a correlated stationary time series process leads to more interesting nonstationary time series models. A classical example is the *autoregressive integrated moving average (ARIMA)* model [8]. We say that $\{X_t\}$ is an ARIMA(p, d, q) process if it is an ARMA(p, q) process after the series is differenced d times. Extending the ARIMA model to allow for random seasonal components we get the *seasonal ARIMA* model.

Methods of estimation and prediction for ARIMA and seasonal ARIMA are widely available (see e.g., [12] and [64] for a discussion of the methods; software for fitting these models are available in R and SAS).

3.6   Long memory processes

Long memory (LM) or *long-range dependent* processes are models in which the ACVF decays slowly with increasing lags. Stationary LM processes exhibit significant local deviations. Looking over short periods of time, there is evidence of trends and seasonality in such series, but they disappear from view over longer time periods. [4] and [46] are good general references for LM processes.

A reason for choosing LM models for environmental applications is that they can arise from the slow mixing of many short-range time series (such as the AR(1) process) [25]; there are famous examples in hydrology [32, 44], turbulence [41], thermodynamics [14], and in climate science [84]. For statistical modeling, LM processes may fit better to time series with long range correlations, as compared to using high order AR, MA and ARMA models. Additionally, we can develop LM models that build on the ARIMA model framework given above to simultaneously capture long and short range dependence.

The *autoregressive fractionally integrated moving average, ARFIMA(p, d, q)*, process [26, 30] is defined from a simple idea: we replace the nonnegative integer differencing parameter d in the ARIMA process with a real-valued parameter d called the *fractionally differencing or LM parameter*. An ARFIMA(p, d, q) process $\{X_t : t \in T\}$ is an ARMA(p, q) process $\{Y_t\}$ after fractional differencing of order d, as defined by the following filtering operation:

$$Y_t = \sum_{j=0}^{\infty} \frac{\Gamma(j+d)}{\Gamma(j+1)\Gamma(d)} X_{t-j}, \quad t \in T;$$

(The definition of the filter follows from an application of the binomial theorem.) Any ARFIMA$(0, d, 0)$ process is also called a *fractionally differenced, FD(d)*, process [30]. An ARFIMA process is stationary as long as the ARMA process is stationary and $d < 1/2$.

Notwithstanding that the theory of estimation and prediction is more challenging for LM processes (e.g., [4]), there are also practical problems. Gaussian ML can be used for parameter estimation in ARFIMA processes, but ML is computer intensive, can be numerically instable, and the estimation can be sensitive to the choice of the ARMA model components. [18] provides the theoretical properties of ML estimates (and the Whittle estimator, a large sample approximation to Gaussian ML based on a Fourier transform of the time series). There are many approximate, computationally efficient, methods for estimating the parameters of LM processes; e.g., approximate AR schemes, Fourier methods, and wavelet methods. These methods have less numerical computing issues, and can be robust to assumptions made about the model. For further details about estimation and prediction with LM processes, read [4] and [46]; see [23] for a comparison of Fourier and wavelet methods.

3.7   Changepoint methods

In many environmental and ecological problems interest lies in determining whether or not the statistical properties of a time series process changes at a given time point or time points. The point at which the statistical properties change is called a *changepoint*.

When a changepoint occurs at time τ, the statistical properties of the time series process before τ, $\{X_t : t < \tau\}$, is different from the statistical properties of the process starting at τ, $\{X_t : t \geq \tau\}$. Extending the definition to multiple changepoints follows naturally. There is an extensive literature in building test statistics for detecting changes in the distribution of a time series process. For normally distributed observations, [28] derived a likelihood ratio test for detecting changes in the mean and [36] detected changes in the variance. There is also a body of work on building, e.g., segmented regression models for which the properties of the model change at a given set of time points (see e.g., [45]).

For an early review of the theory and methods for time series, see [17]. [35] provides a review of changepoint methodology in environmental science, [3, 54] for climate science, and [60] for identifying so-called thresholds in ecology. [39] provides a review of changepoint methods in the context of introducing the `changepoint` R package (they also review other R packages for detecting changepoints).

3.8   Discussion and conclusions

Time series analysis is being used to analyze datasets in many areas, such as environmental science and ecology. One chapter cannot do the entire discipline justice.

This chapter focused heavily on measures of dependence that are defined over time, typically using lags of the ACVF or ACF of a stationary process. Even for the nonstationary processes of Section 3.5, the nonstationarity is driven by some usually hidden ("latent") stationary process. In the search for more interesting time series processes, the use of latent time series models have become popular. Hierarchical statistical time series models, such as state space models (e.g., [17, 38, 49]) or dynamic linear models (e.g., [45]), combine hierarchies of time series models to create more complicated structures. The different hierarchies can have practical interpretation (e.g., the first level of the hierarchy could indicate how we observe some latent process with measurement error, and the second level could be a model for how this latent process evolves over time). Another reason for their popularity is the increasing availability of accessible and computationally efficient methods to fit hierarchical models. More complicated time series models, such as nonlinear processes (e.g., [61, 97]) and non-Gaussian processes (e.g., [17, 20]), are often are created from latent variable models. (See [62] for an analysis of the Canadian lynx series using a class of nonlinear processes called *threshold models*; also read the dynamic models chapter of this book for further details on latent variable time series models.)

We close by noting that dependence need not be defined in terms of the unit of time. There is a rich history of taking Fourier transforms of time series, re-interpreting dependence in terms of the contributions of random sinusoids of different frequencies. The leads to the *spectral analysis* of time series; see e.g., [9, 50, 51, 77]). More recently other transformations have been developed to model dependence. For example, *wavelet analysis* [64, 78] examines dependence both over time and scale (approximately frequency). Modeling processes that vary over time drive us to investigate how we can efficiently model (both statistically and computationally) time-varying time series dependence. An interesting compromise assumes the process is *locally stationary*, with a dependence structure that slowly evolves over time; see e.g., [19].

Acknowledgments

Craigmile is supported in part by the US National Science Foundation under grants DMS-1407604 and SES-1424481. My thanks to Peter Guttorp, Donald Percival, Alexandra Schmidt, and Marian Scott for providing comments that greatly improved this chapter.

Bibliography

[1] H. Akaike. A new look at the statistical model identification. *IEEE Transactions on Automatic Control*, 19:716–723, 1974.

[2] M. Bartlett. On the theoretical specification and sampling properties of autocorrelated time-series. *Supplement to the Journal of the Royal Statistical Society*, 8:27–41, 1946.

[3] C. Beaulieu, J. Chen, and J. L. Sarmiento. Change-point analysis as a tool to detect abrupt climate variations. *Philosophical Transactions of the Royal Society of London A: Mathematical, Physical and Engineering Sciences*, 370:1228–1249, 2012.

[4] J. Beran. *Statistics for Long Memory Processes*. Chapman and Hall, New York, 1994.

[5] S. Bochner. Monotone funktionen, Stieltjessche integrale und harmonische analyse. *Mathematische Annalen*, 108:378–410, 1933.

[6] S. Bochner. *Harmonic analysis and probability theory*. University of California Press, Berkeley, CA, 1955.

[7] G. Box and D. Pierce. Distribution of residual autocorrelations in autoregressive-integrated moving average time series models. *Journal of the American Statistical Association*, 65:1509–1526, 1970.

[8] G. E. P. Box, G. M. Jenkins, G. C. Reinsel, and G. M. Ljung. *Time Series Analysis: Forecasting and Control, 5th edition*. Wiley, New York, NY, 2015.

[9] D. R. Brillinger. *Time Series: Data Analysis and Theory*. Holt, New York, NY, 1981.

[10] P. J. Brockwell. Continuous-time ARMA processes. In D. N. Shanbhag and C. R. Rao, editors, *Handbook of Statistics*, volume 19, pages 249–276. Elsevier, 2001.

[11] P. J. Brockwell and R. A. Davis. *Time Series. Theory and Methods (Second Edition)*. Springer, New York, 1991.

[12] P. J. Brockwell and R.A. Davis. *Introduction to Time Series and Forecasting (Second Edition)*. Springer, New York, 2002.

[13] M. J. Campbell and A. M. Walker. A survey of statistical work on the Mackenzie River series of annual Canadian lynx trappings for the years 1821-1934 and a new analysis. *Journal of the Royal Statistical Society. Series A (General)*, 140:411–431, 1977.

[14] M. Cassandro and G. Jona-Lasinio. Critical point behaviour and probability theory. *Advances in Physics*, 27:913–941, 1978.

[15] S. Chib and E. Greenberg. Bayes inference in regression models with ARMA(p,q) errors. *Journal of Econometrics*, 64:183–206, 1994.

[16] W. L. B. Constantine, D. B. Percival, and P. G. Reinhall. Inertial range determination for aerothermal turbulence using fractionally differenced processes and wavelets. *Physical Review E*, 64, 2001.

[17] M. Csörgö and L. Horváth. *Limit theorems in change-point analysis*, volume 18. John Wiley & Sons Inc, 1997.

[18] R. Dahlhaus. Efficient parameter estimation for self similar processes. *The Annals of Statistics*, 17:1749–1766, 1989.

[19] R. Dahlhaus. Locally stationary processes. In D. N. Shanbhag and C. R. Rao, editors, *Handbook of Statistics*, volume 30, pages 351–412. 2012.

[20] R. A. Davis, S. H. Holan, R. Lund, and N. Ravishanker. *Handbook of Discrete-Valued Time Series*. CRC Press, Boca Raton, FL, 2016.

[21] J. Durbin and S. J. Koopman. *Time series analysis by state space methods*. Oxford University Press, Oxford, United Kingdom, 2012.

[22] C. Elton and M. Nicholson. The ten year cycle in numbers of Canadian lynx. *Journal of Animal Ecology*, 11:215–244, 1942.

[23] G. Faÿ, E. Moulines, F. Roueff, and M. S. Taqqu. Estimators of long-memory: Fourier versus wavelets. *Journal of Econometrics*, 151:159–177, 2009.

[24] T. Gneiting. Power-law correlations, related models for long-range dependence, and their fast simulation. *Journal of Applied Probability*, 37:1104–1109, 2000.

[25] C. W. J. Granger. Long memory relationships and the aggregation of dynamic models. *Journal of Econometrics*, 14:227–238, 1980.

[26] C. W. J. Granger and R. Joyeux. An introduction to long-memory time series models and fractional differencing. *Journal of Time Series Analysis*, 1:15–29, 1980.

[27] J. Harrison and M. West. *Bayesian Forecasting and Dynamic Models*. Springer, New York, NY, 1999.

[28] D. V. Hinkley. Inference about the change-point in a sequence of random variables. *Biometrika*, 57:1–17, 1970.

[29] S. Holan, T. McElroy, and S. Chakraborty. A Bayesian approach to estimating the long memory parameter. *Bayesian Analysis*, 4:159–190, 2009.

[30] J. R. M. Hosking. Fractional differencing. *Biometrika*, 68:165–176, 1981.

[31] N-J Hsu and F. J. Breidt. Bayesian analysis of fractionally integrated ARMA with additive noise. *Journal of Forecasting*, 22:491–514, 2003.

[32] H. E. Hurst. The problem of long-term storage in reservoirs. *Transactions of the American Society of Civil Engineers*, 116:776–808, 1956.

[33] C. M. Hurvich and C-L Tsai. Regression and time series model selection in small samples. *Biometrika*, 76:297–307, 1989.

[34] R. Hyndman, A. B. Koehler, J. K. Ord, and R. D. Snyder. *Forecasting with exponential smoothing: the state space approach*. Springer-Verlag, Berlin, Germany, 2008.

[35] D. Jarušková. Some problems with application of change-point detection methods to environmental data. *Environmetrics*, 8:469–484, 1997.

[36] T. Jen and A. K. Gupta. On testing homogeneity of variances for Gaussian models. *Journal of Statistical Computation and Simulation*, 27:155–173, 1987.

[37] R. H. Jones. Maximum likelihood fitting of ARMA models to time series with missing observations. *Technometrics*, 22:389–395, 1980.

[38] R. E. Kalman. A new approach to linear filtering and prediction problems. *Journal of Basic Engineering*, 82:35–45, 1960.

[39] R. Killick and I. Eckley. changepoint: An R package for changepoint analysis. *Journal of Statistical Software*, 58:1–19, 2014.

[40] A. N. Kolmogorov. Wienersche spiralen und einigeandere interessante kurven in Hilbertschen raum. *Comptes Rensu (Doklady) Acad Sci. URSS (N.S.)*, 26:115–118, 1940.

[41] A. N. Kolmogorov. The local structure of turbulence in incompressible viscous fluid for very large reynolds numbers. In *Dokl. Akad. Nauk SSSR*, volume 30, pages 301–305, 1941.

[42] G. Ljung and G. Box. On a measure of lack of fit in time series models. *Biometrika*, 65:297–303, 1978.

[43] H. Lütkepohl. *New introduction to multiple time series analysis*. Springer-Verlag, Berlin, Germany, 2005.

[44] B. B. Mandelbrot. *Fractals: form, change and dimension*. WH Freemann and Company, San Francisco, CA, 1977.

[45] V. Muggeo. Estimating regression models with unknown break-points. *Statistics in medicine*, 22:3055–3071, 2003.

[46] W. Palma. *Long-Memory Time Series: Theory and Methods*. Wiley-Interscience, Hoboken, NJ, 2007.

[47] D. B. Percival and A. Walden. *Spectral Analysis for Physical Applications*. Cambridge University Press, Cambridge, 1993.

[48] D. B. Percival and A Walden. *Wavelet Methods for Time Series Analysis*. Cambridge University Press, Cambridge, 2000.

[49] R. Prado and M. West. *Time Series: Modeling, Computation, and Inference*. CRC Press, Boca Raton, FL, 2010.

[50] M. B. Priestley. *Spectral Analysis and Time Series. (Vol. 1): Univariate Series*. Academic Press, London, United Kingdom, 1981.

[51] M. B. Priestley. *Spectral Analysis and Time Series. (Vol. 2): Multivariate Series, Prediction and Control*. Academic Press, London, United Kingdom, 1981.

[52] M. Quenouille. A large-sample test for the goodness of fit of autoregressive schemes. *Journal of the Royal Statistical Society*, 110:123–129, 1947.

[53] R Core Team. *R: A Language and Environment for Statistical Computing*. R Foundation for Statistical Computing, Vienna, Austria, 2016.

[54] J. Reeves, J. Chen, X. L. Wang, R. Lund, and Q. Q. Lu. A review and comparison of changepoint detection techniques for climate data. *Journal of Applied Meteorology and Climatology*, 46:900–915, 2007.

[55] G. Schwarz. Estimating the dimension of a model. *The Annals of Statistics*, 6:461–464, 1978.

[56] R Shibata. Selection of order of an autoregressive model by Akaikes information criterion. *Biometrika*, 63:117–126, 1976.

[57] R. H. Shumway and D. S. Stoffer. *Time series analysis and its applications: with R examples*. Springer, New York, NY, 2010.

[58] R. Smith. Long-range dependence and global warming. In V. Barnett and F. Turkman, editors, *Statistics for the Environment*, pages 141–161. John Wiley, Hoboken, NJ, 1993.

[59] M. Taniguchi and Y. Kakizawa. *Asymptotic theory of statistical inference for time series*. Springer, New York, NY, 2012.

[60] J. D. Toms and M. L. Lesperance. Piecewise regression: a tool for identifying ecological thresholds. *Ecology*, 84:2034–2041, 2003.

[61] H. Tong. *Non-linear Time Series: a Dynamical System Approach*. Oxford University Press, Oxford, United Kingdom, 1990.

[62] H. Tong and K. S. Lim. Threshold autoregression, limit cycles and cyclical data. *Journal of the Royal Statistical Society. Series B (Methodological)*, 42:245–292, 1980.

[63] R. S. Tsay. *Analysis of Financial Time Series (Third Edition)*. Wiley, New York, NY, 2010.

[64] B. Vidakovic. *Statistical Modeling by Wavelets*. Wiley, New York, NY, 1998.

[65] G. U. Yule. On the time-correlation problem, with special reference to the variate-difference correlation method. *Journal of Royal Statistical Society*, 84:497–537, 1921.

[66] G. U. Yule. On a method of investigating periodicities in disturbed series, with special reference to Wolfer's sunspot numbers. *Philosophical Transactions of the Royal Society of London. Series A, Containing Papers of a Mathematical or Physical Character*, 226:267–298, 1927.

4

Dynamic models

Alexandra M. Schmidt

McGill University, Canada

Hedibert F. Lopes

INSPER, Brazil

CONTENTS

4.1 Introduction

Dynamic models are, in a broad sense, probabilistic models that describe a set of observable measurements conditionally on a set of latent or hidden state-space variables whose time and/or space dynamics are driven by a set of time-invariant parameters. This inherently hierarchical description renders dynamic models to the status of one of the most popular statistical structures in many areas of applied science, including neuroscience, marketing, oceanography, financial markets, target-tracking, signal process, climatology and text analysis, to name just a few. Borrowing the notation from [3], also used by [13], the general structure of a dynamic model can be written as

Measurements model: [data | state-space, parameters]

State-space dynamics: [state-space | parameters]
Parameters prior: [parameters]

In time series analysis, dynamic models are simultaneously an alternative to and an extension of Box & Jenkins' autoregressive integrated moving average (ARIMA) models (See, for instance, [47] and [11]). Commonly, the observed process is decomposed as the sum of different time series components such as deterministic or stochastic trends, seasonality and covariate effects. Moreover, different from ARIMA models, in dynamic models the parameters might change with time in order to accommodate *local* structures of the temporal process under study. More importantly, dynamic models deal rather naturally with non-stationary time series and structural changes.

In this chapter we review various aspects of dynamic models. We start by describing univariate normal dynamic linear models (DLM) and their main properties. Dynamic models for non-normal processes are also discussed. Section 4.3 extends the univariate DLM to the multivariate and matrix-variate normal cases. Therein, we note that multivariate DLM naturally accommodate time series of observations made at different spatial locations. Dimension reduction through dynamic factor models is also discussed. Finally, Section 4.4 discusses the use of the state-space framework to approximate convolution models, stochastic partial differential and integro-difference equations. For further reading on DLMs the readers are referred to [60], [46], [47] and [17], and their references.

4.2   Univariate Normal Dynamic Linear Models (NDLM)

Assume that $\{y_t, t \in T\}$ is a stochastic process observed at discrete time, such that $T = \{0, \pm 1, \pm 2, \pm 3, \ldots\}$. Let D_t be the set of information available at time t. Thus D_t includes the observation y_t, and covariates observed at time t (if any), and all previous information D_{t-1}. A normal dynamic linear model (NDLM) decomposes an univariate time series y_t as the sum of two components, an overall time-varying trend, $\mathbf{F}_t'\boldsymbol{\theta}_t$, and an error component, ϵ_t, that follows a zero mean normal distribution with variance V_t. The latent k-dimensional vector $\boldsymbol{\theta}_t$, is known as the *state vector* at time t, and it evolves smoothly with time. More specifically, a NDLM is specified by the following equations:

$$
\begin{aligned}
y_t &= \mathbf{F}_t'\boldsymbol{\theta}_t + \epsilon_t, & \epsilon_t &\sim N(0, V_t) & (4.1) \\
\boldsymbol{\theta}_t &= \mathbf{G}_t\boldsymbol{\theta}_{t-1} + \boldsymbol{\omega}_t, & \boldsymbol{\omega}_t &\sim N(0, \mathbf{W}_t), & (4.2) \\
\boldsymbol{\theta}_0 &\sim N(\mathbf{m}_0, \mathbf{C}_0), & & & (4.3)
\end{aligned}
$$

where $\boldsymbol{\theta}_0$ is the initial information, $\mathbf{m}_0$ and $\mathbf{C}_0$ are known k-dimensional mean vector and $k \times k$ covariance matrix, respectively. $\mathbf{F}_t$ is a k-dimensional column vector of covariates, $\mathbf{G}_t$ is a $k \times k$ matrix, known as the *evolution* matrix, and $\mathbf{W}_t$ is a $k \times k$ covariance matrix describing the covariance structure among the components of $\boldsymbol{\theta}_t$. Equations (4.1) and (4.2) are known as the *observation* and *system*, or *state*, equations, respectively.

Usually, it is further assumed that ϵ_t and $\boldsymbol{\omega}_t$ are independent and mutually independent, and independent of $\boldsymbol{\theta}_0$. Note that, given $\boldsymbol{\theta}_t$, y_t is conditionally independent from past observations.·

A NDLM is completely specified through the quadruple $\{\mathbf{F}_t, \mathbf{G}_t, V_t, \mathbf{W}_t\}$. If it is further assumed that $V_t = V$ and $\mathbf{W}_t = \mathbf{W}$, $\forall t \in T$ this is known as the constant model. Classical examples are the local level and local trend models, the trigonometric seasonal model and the time-varying parameter AR model, all illustrated below.

Example: *(First order polynomial model or time-varying level model)* The simplest NDLM is defined by the quadruple $\{1, 1, V, W\}$, such that $y_t = \mu_t + \epsilon_t$, and $\mu_t = \mu_{t-1} + w_t$, with $\epsilon_t \sim N(0, V)$, and $w_t \sim N(0, W)$. Here, $\theta_t = \mu_t$ and $k = 1$. The ratio W/V, known as the signal-to-noise ratio, plays an important role in determining the behavior of the time varying level μ_t. As $W/V \to 0$, the first order model tends to a constant mean model, and as $W/V \to \infty$ the model reduces to a pure random walk. This model is suitable for short term forecast as it can be shown that the forecast function is constant.

Example: *(Linear trend)* A linear trend model can be defined by assuming that $y_t = \mu_t + \epsilon_t$, $\mu_t = \mu_{t-1} + \beta_{t-1} + w_{1t}$ and $\beta_t = \beta_{t-1} + w_{2t}$, where β_t is the slope of the local level μ_t. In this case, $\theta_t = (\mu_t, \beta_t)'$ and the NDLM is defined by the quadruple $\{\mathbf{F}, \mathbf{G}, V, \mathbf{W}\}$, where $\mathbf{F}' = (1\ 0)$,

$$\mathbf{G} = \begin{pmatrix} 1 & 1 \\ 0 & 1 \end{pmatrix} \quad \text{and} \quad \mathbf{W} = \begin{pmatrix} W_{11} & W_{12} \\ W_{12} & W_{22} \end{pmatrix},$$

for some known values of W_{11}, W_{12}, W_{22} and V. If $W_{22} = 0$ the model results in a constant trend model.

Example: *(Fourier representation of seasonality)* A seasonal structure can be accommodated by a Fourier representation of seasonality. Assuming the period of seasonality is m, then $\mathbf{F}' = (1\ 0)$,

$$\mathbf{G} = \begin{pmatrix} \cos(2\pi/m) & \sin(2\pi/m) \\ -\sin(2\pi/m) & \cos(2\pi/m) \end{pmatrix},$$

and $\mathbf{W} = \mathrm{diag}(W_1, W_2)$.

Example: *(Time-varying auto-regressive process of order L, TV-AR(L))* Assume that $y_t = x_t + \epsilon_t$, where $x_t = \sum_{l=1}^{L} \phi_{tl} x_{t-l} + w_t$. The model is a NDLM where $\theta_t = (\phi_{t1}, \ldots, \phi_{tL})$, $\mathbf{F}' = (1, 0, \ldots, 0)$,

$$\mathbf{G}_t = \begin{pmatrix} \phi_{t1} & \phi_{t2} & \phi_{t3} & \cdots & \phi_{t,L-1} & \phi_{tL} \\ 1 & 0 & 0 & \cdots & 0 & 0 \\ 0 & 1 & 0 & \cdots & 0 & 0 \\ \vdots & \vdots & \vdots & \ddots & \vdots & \vdots \\ 0 & 0 & 0 & \cdots & 1 & 0 \end{pmatrix}$$

and $\omega_t = (w_t, 0, \ldots, 0)'$. If the components of θ_t are fixed across time and $V = 0$, the time-series x_t follows a standard AR(L) process. Loosely speaking, the whole class of well-known ARIMA models can be thought of as special cases of DLMs. See [47] for further details.

4.2.1 Forward learning: the Kalman filter

In a Bayesian NDLM, from Equations (4.1)–(4.3), the posterior distribution of any parameter at time t is based on all available information up to time t, D_t. One aspect of the sequential Bayesian learning in dynamic models is that the posterior distribution for the parameters at time $t - 1$, $p(\theta_{t-1}|D_{t-1})$, when propagated through the system equation (4.2) becomes the prior distribution of the parameters at time t, $p(\theta_t|D_{t-1})$. Should θ_t be time-invariant, this step would vanish and $p(\theta_t|D_{t-1}) = p(\theta_{t-1}|D_{t-1})$. As is standard in Bayesian sequential learning, the prior at time t is then combined with the likelihood $p(y_t|\theta_t)$ from the observation equation (4.1) to produce both predictive density $p(y_t|D_{t-1})$

and posterior density $p(\boldsymbol{\theta}_t|D_t)$. Schematically, the sequential Bayesian inference in dynamic models cycles through three main steps: Evolution, Prediction and Updating, as follows:

$$p(\boldsymbol{\theta}_{t-1}|D_{t-1}) \overset{\text{Evolution}}{\Longrightarrow} p(\boldsymbol{\theta}_t|D_{t-1}) \overset{\text{Updating}}{\Longrightarrow} p(\boldsymbol{\theta}_t|D_t)$$

$$\Big\Downarrow \text{Prediction}$$

$$p(y_t|D_{t-1})$$

Below, following Theorem 4.3 in [60], we specify each of the distributions for each of the steps described above. The resultant distributions for each step depend on the distribution of the initial information $\boldsymbol{\theta}_0|D_0 \sim N(\boldsymbol{m}_0, \boldsymbol{C}_0)$ and the knowledge or not of the observational variance V_t and the variance of the evolution noise, $\boldsymbol{W}_t$. We modify slightly the definition of the NDLM in equations (4.1) and (4.2) to accommodate the more general case when V is unknown but follows an inverse gamma prior distribution, and the distribution of the noise component follows a normal distribution conditional on the value of V, that is, $\boldsymbol{\omega}_t \mid V \sim N(0, V\boldsymbol{W}_t^*)$.

Let $\phi = 1/V$ be the observational precision. Define, at time $t = 0$, a normal-gamma prior distribution for $p(\boldsymbol{\theta}_0, \phi|D_0)$ by $\boldsymbol{\theta}_0|D_0, V \sim N(\boldsymbol{m}_0, V\boldsymbol{C}_0^*)$, and assume $\phi|D_0 \sim Ga(n_0/2, n_0S_0/2)$. The derivation is done by induction, by assuming that $\boldsymbol{\theta}_{t-1}|V, D_{t-1} \sim N(\boldsymbol{m}_{t-1}, V\boldsymbol{C}_{t-1}^*)$ and $\phi|D_{t-1} \sim Ga(n_{t-1}/2, n_{t-1}S_{t-1}/2)$. It can be shown that the steps above are given by [60, pages 109-110]:

- *Evolving the state:* The prior distribution of $\boldsymbol{\theta}_t$ is given by $\boldsymbol{\theta}_t|V, D_{t-1} \sim N(\boldsymbol{a}_t, V\boldsymbol{R}_t^*)$ where $\boldsymbol{a}_t = \boldsymbol{G}_t\boldsymbol{m}_{t-1}$ and $\boldsymbol{R}_t^* = \boldsymbol{G}_t\boldsymbol{C}_{t-1}^*\boldsymbol{G}_t' + \boldsymbol{W}_t^*$. Therefore, $\boldsymbol{\theta}_t|D_{t-1} \sim t_{n_{t-1}}(\boldsymbol{a}_t, S_{t-1}\boldsymbol{R}_t^*)$. Here $t_\nu(\mu, \sigma^2)$ denotes a Student's t distribution with ν degress of freedom, location μ and scale σ^2.

- *Predicting the next observation:* The one-step ahead forecast is given by $y_t|V, D_{t-1} \sim N(f_t, VQ_t^*)$ where $f_t = \mathbf{F}_t'\boldsymbol{a}_t$ and $Q_t^* = 1 + \mathbf{F}_t'\boldsymbol{R}_t^*\mathbf{F}_t$. Therefore, $y_t|D_{t-1} \sim t_{n_{t-1}}(f_t, S_{t-1}Q_t^*)$.

- *Updating the variance:* $\phi|D_t \sim Ga(n_t/2, n_tS_t/2)$, where $n_t = n_{t-1} + 1$, and $n_tS_t = n_{t-1}S_{t-1} + (y_t - f_t)^2/Q_t^*$.

- *Updating the state:* The posterior distribution of $\boldsymbol{\theta}_t$ given the information at time t, is given by $\boldsymbol{\theta}_t|V, D_t \sim N(\boldsymbol{m}_t, V\boldsymbol{C}_t^*)$, with $\boldsymbol{m}_t = \boldsymbol{a}_t + \boldsymbol{A}_t(y_t - f_t)$ and $\boldsymbol{C}_t^* = \boldsymbol{R}_t^* - \boldsymbol{A}_t\boldsymbol{A}_t'Q_t^*$, where $\boldsymbol{A}_t = \boldsymbol{R}_t\mathbf{F}_tQ_t^{-1}$. Therefore, $\boldsymbol{\theta}_t|D_t \sim t_{n_t}(\boldsymbol{m}_t, S_t\boldsymbol{C}_t^*)$.

These recursions are commonly known as *Kalman recursions* or simply *the Kalman filter algorithm* (see, for instance, [60]). If interest lies on forecasting the process h steps ahead, based on the information up to time D_t, the forecast distribution, $y_{t+h}|D_t$, can be obtained through the Kalman filter. It can be shown that $y_{t+h}|D_t \sim t_{n_t}(f_t(h), Q_t(h))$, with $f_t(h) = \mathbf{F}_{t+h}'\boldsymbol{a}_t(h)$, $Q_t(h) = \mathbf{F}_{t+h}'\boldsymbol{R}_t(h)\mathbf{F}_{t+h} + S_{t+h}$, $\boldsymbol{a}_t(0) = \boldsymbol{m}_t$, and $\boldsymbol{R}_t(0) = \boldsymbol{C}_t$.

4.2.2   Backward learning: the Kalman smoother

The Kalman filter is one of the most popular algorithms for the sequential update of hidden/latent states in dynamic systems. However, in the above form, it only provides posterior distribution of a given state $\boldsymbol{\theta}_t$ conditionally on the past observations, $y_1, \ldots, y_t, p(\boldsymbol{\theta}_t|D_t, V)$. For simplicity, we are omitting the dependence on the state variances $\boldsymbol{W}_1^*, \ldots, \boldsymbol{W}_n^*$. In many instances, however, one may want to obtain the posterior distribution of the states given the whole set of observations, $y_1, \ldots, y_n$, i.e. $p(\boldsymbol{\theta}_1, \ldots, \boldsymbol{\theta}_n | V, D_n)$; for instance, to understand the dynamics driving observations as opposed to simply forecasting its future values.

Full joint distribution of states

By the Markov property of Equations (4.1)–(4.3), this joint posterior can be rewritten as

$$p(\boldsymbol{\theta}_1, \ldots, \boldsymbol{\theta}_n | V, D_n) = p(\boldsymbol{\theta}_n | V, D_n) \prod_{t=1}^{n-1} p(\boldsymbol{\theta}_t | \boldsymbol{\theta}_{t+1}, V, D_t). \tag{4.4}$$

First, the forward learning scheme of the previous Section (Kalman filter) is run to obtain $p(\boldsymbol{\theta}_t | V, D_t)$, for $t = 1, \ldots, n$. Then, an analogous backward learning scheme (Kalman smoother) is run to obtain $p(\boldsymbol{\theta}_t | \boldsymbol{\theta}_{t+1}, V, D_t)$, for $t = n - 1, \ldots, 1$. More precisely, another (backwards) application of Bayes' theorem leads to

$$
\begin{aligned}
p(\boldsymbol{\theta}_t | \boldsymbol{\theta}_{t+1}, V, D_t) &\propto f_N(\boldsymbol{\theta}_{t+1}; G_t \boldsymbol{\theta}_t, V W_t^*) f_N(\boldsymbol{\theta}_t; m_t, V C_t^*) \\
&\propto f_N(\boldsymbol{\theta}_{t+1}; \widetilde{m}_t, V \widetilde{C}_t^*),
\end{aligned}
\tag{4.5}
$$

where $\widetilde{m}_t = \widetilde{C}_t^* (G_t' W_t^{-1} \boldsymbol{\theta}_{t+1} + C_t^{*-1} m_t)$ and $\widetilde{C}_t^* = (G_t' W_t^{*-1} G_t + C_t^{*-1})^{-1}$, for $t = n - 1, n - 2, \ldots, 1$. In practice, for some models it may be better to use the Sherman-Morrison-Woodbury [25] formulae for evaluating C_t^*. From the Kalman filter, $V | D_n \sim IG(n_n/2, n_n S_n/2)$, so it follows that $\boldsymbol{\theta}_n | D_n \sim t_{n_n}(\widetilde{m}_n, S_n \widetilde{C}_n^*)$ and

$$\boldsymbol{\theta}_t | \boldsymbol{\theta}_{t+1}, D_t \sim t_{n_n}(\widetilde{m}_t, S_n \widetilde{C}_t^*), \tag{4.6}$$

for $t = n - 1, n - 2, \ldots, 1$.

Marginal distributions of the states

Similarly, it can be shown that the marginal distribution of $\boldsymbol{\theta}_t$ is

$$\boldsymbol{\theta}_t | D_n \sim t_{n_n}(\overline{m}_t, S_n \overline{C}_t^*), \tag{4.7}$$

where, for $t = n - 1, n - 2, \ldots, 1$, $\overline{m}_t = m_t + C_t^* G_{t+1}' R_{t+1}^{*-1}(\overline{m}_{t+1} - a_{t+1})$ and $\overline{C}_t^* = C_t^* - C_t^* G_{t+1}' R_{t+1}^{*-1}(R_{t+1}^* - \overline{C}_{t+1}^*) R_{t+1}^{*-1} G_{t+1} C_t^*$.

Full conditional distributions of the states

Let $\boldsymbol{\theta}_{-t} = \{\boldsymbol{\theta}_1, \ldots, \boldsymbol{\theta}_{t-1}, \boldsymbol{\theta}_{t+1}, \ldots, \boldsymbol{\theta}_n\}$ and $t = 2, \ldots, n - 1$, it follows that the full conditional distribution of $\boldsymbol{\theta}_t$ is

$$
\begin{aligned}
p(\boldsymbol{\theta}_t | \boldsymbol{\theta}_{-t}, V, D_n) &\propto f_N(y_t; \mathbf{F}_t' \boldsymbol{\theta}_t, V) f_N(\boldsymbol{\theta}_{t+1}; G_{t+1} \boldsymbol{\theta}_t, V W_{t+1}^*) \\
&\times f_N(\boldsymbol{\theta}_t; G_t \boldsymbol{\theta}_{t-1}, V W_t^*) = f_N(\boldsymbol{\theta}_t; b_t, V B_t^*),
\end{aligned}
\tag{4.8}
$$

where $b_t = B_t^*(\mathbf{F}_t y_t + G_{t+1}' W_{t+1}^{*-1} \boldsymbol{\theta}_{t+1} + W_t^{*-1} G_t \boldsymbol{\theta}_{t-1})$ and $B_t^* = (\mathbf{F}_t \mathbf{F}_t' + G_{t+1}' W_{t+1}^{*-1} G_{t+1} + W_t^{*-1})^{-1}$. The endpoint parameters $\boldsymbol{\theta}_1$ and $\boldsymbol{\theta}_n$ also have full conditional distributions $N(b_1, V B_1^*)$ and $N(b_n, V B_n^*)$, respectively, where $b_1 = B_1^*(\mathbf{F}_1 y_1 + G_2' W_2^{*-1} \boldsymbol{\theta}_2 + R^{*-1} a_1)$, $B_1^* = (\mathbf{F}_1 \mathbf{F}_1' + G_2' W_2^{*-1} G_2 + R^{-1})^{*-1}$, $b_n = B_n^*(\mathbf{F}_n y_n + W_n^{*-1} G_n \boldsymbol{\theta}_{n-1})$ and $B_n^* = (\mathbf{F}_n \mathbf{F}_n' + W_n^{*-1})^{-1}$. Again,

$$\boldsymbol{\theta}_t | \boldsymbol{\theta}_{-t}, D_n \sim t_{n_n}(b_t, S_n B_t^*) \qquad \text{for all } t. \tag{4.9}$$

Full posterior inference

When the evolution variances $\boldsymbol{W}_1^*, \ldots, \boldsymbol{W}_n^*$ are unknown, closed-form analytical full posterior inference is infeasible and numerical or Monte Carlo approximations are needed. Numerical integration, in fact, is only realistically feasible for very low dimensional settings. Markov Chain Monte Carlo methods have become the norm over the last few decades for state space modelers. In particular, the full joint of Equation (4.4) can be combined with full conditional distributions for V and $\boldsymbol{W}_1^*, \ldots, \boldsymbol{W}_n^*$. This is the well known *forward filtering, backward sampling (FFBS)* algorithm of [9] and [19]. The FFBS algorithm is commonly used for posterior inference in Gaussian and conditionally Gaussian DLMs. The main steps needed to fit DLM's using the software R [50] are in the package dlm, detailed in [45].

4.2.3   Integrated likelihood

Another very important result of the sequential Bayesian depicted above is the derivation of the marginal likelihood of $\boldsymbol{W}_1^*, \ldots, \boldsymbol{W}_t^*$ given $y_1, \ldots, y_n$. Without loss of generality, assume that $\boldsymbol{W}_t = \boldsymbol{W}$ for all $t = 1, \ldots, n$, where n is the sample size, and that $p(\boldsymbol{W})$ denotes the prior distribution of $\boldsymbol{W}$. In this case,

$$p(y_1, \ldots, y_n | \boldsymbol{W}) = \prod_{t=1}^{n} p(y_t | D_{t-1}, \boldsymbol{W}), \tag{4.10}$$

where $y_t | D_{t-1}, \boldsymbol{W} \sim t_{n_{t-1}}(f_t, S_{t-1} Q_t^*)$. Therefore, the posterior distribution of $\boldsymbol{W}$, $p(\boldsymbol{W} | D_n) \propto p(D_n | \boldsymbol{W}) p(\boldsymbol{W})$, can be combined with $p(V | \boldsymbol{W}, D_n)$ to produce the joint posterior of $(V, \boldsymbol{W})$:

$$
\begin{aligned}
p(V, \boldsymbol{W} | D_n) &\propto p(D_n | \boldsymbol{W}) p(V | \boldsymbol{W}, D_n) p(\boldsymbol{W}) \\
&= \left[\prod_{t=1}^{n} p(y_t; f_t, S_{t-1} Q_t^*, n_{t-1}) \right] p(V; n_n/2, n_n S_n/2) p(\boldsymbol{W}).
\end{aligned}
$$

In words, Gaussianity and linearity lead to a posterior distribution for V and $\boldsymbol{W}$ by integrating out all state space vectors $\boldsymbol{\theta}_1, \boldsymbol{\theta}_2, \ldots, \boldsymbol{\theta}_n$. If M independent Monte Carlo draws $\{(V, \boldsymbol{W})^{(1)}, \ldots, (V, \boldsymbol{W})^{(M)}\}$ are obtained from $p(V, \boldsymbol{W} | D_n)$, then M independent Monte Carlo draws from $p(\boldsymbol{\theta}_1, \ldots, \boldsymbol{\theta}_n | D_n)$ are easily obtained by repeating the FFBS of Equation (4.4) (or Equation (4.5)) M times. This leads to M independent Monte Carlo draws from $p(\boldsymbol{\theta}_1, \ldots, \boldsymbol{\theta}_n, V, \boldsymbol{W} | D_n)$, hence there is no need for iterative Markov chain Monte Carlo (MCMC) schemes.

4.2.4   Some properties of NDLMs

Superposition of models

More general time series structures can be modeled through a composition of DLM components. This is possible because of a theorem proved in [60]. Consider m time series y_{it} generated by NDLMs identified by the quadruples $M_i = \{\mathbf{F}_{it}, \mathbf{G}_{it}, V_{it}, \boldsymbol{W}_{it}\}$, $i = 1, 2, \ldots, m$. Under model M_i, the state vector $\boldsymbol{\theta}_{it}$ is of dimension k_i, and the observation and evolution error series are respectively ϵ_{it} and $\boldsymbol{\omega}_{it}$. The state vectors are distinct and, for all distinct $i \neq j$, the series ϵ_{it} and $\boldsymbol{\omega}_{it}$ are mutually independent of the series ϵ_{jt} and $\boldsymbol{\omega}_{jt}$. Then the series $y_t = \sum_{i=1}^{m} y_{it}$ follows a k-dimensional DLM $\{\mathbf{F}_t, \boldsymbol{G}_t, V_t, \boldsymbol{W}_t\}$, where $k = k_1 + \cdots + k_m$, the state vector $\boldsymbol{\theta}_t$ is given by $\boldsymbol{\theta}_t = (\boldsymbol{\theta}_{1t}', \ldots, \boldsymbol{\theta}_{mt}')'$, $\mathbf{F}_t = (\mathbf{F}_{1t}', \ldots, \mathbf{F}_{mt}')'$, $\boldsymbol{G}_t = \text{block diag}(\boldsymbol{G}_{1t}, \ldots, \boldsymbol{G}_{mt})$, $\boldsymbol{W}_t = \text{block diag}(\boldsymbol{W}_{1t}, \ldots, \boldsymbol{W}_{mt})$, and $V_t = \sum_{i=1}^{m} V_{it}$.

Example: *(Dynamic regression)* If at each time t a pair of observations (y_t, x_t) is available, and there is a local linear relationship between y_t and x_t, a simple dynamic regression is defined by making $\mathbf{F}'_t = (1 \; x_t)$, $\mathbf{G} = \mathbf{I}_2$, the 2-dimensional identity matrix, and $\mathbf{W} = \mathrm{diag}(W_1, W_2)$.

Example: *(Time varying level plus an annual seasonal component)* $\mathbf{F}'_t = (1 \; 1 \; 0)$, $\mathbf{W} = \mathrm{diag}(W_1, W_2, W_3)$, and

$$G = \begin{pmatrix} 1 & 0 & 0 \\ 0 & \cos(2\pi/12) & \sin(2\pi/12) \\ 0 & -\sin(2\pi/12) & \cos(2\pi/12) \end{pmatrix}.$$

Discounting factors

The sequential updating steps described in Section 4.2.1 follow normal or Student-t distributions because we assume that $\mathbf{W}_t$ are known for all time t. In practice, this is rarely true. One way to avoid estimating the elements of the covariance matrix $\mathbf{W}_t$ is through the use of discounting factors. As detailed in [60], let $\mathbf{P}_t = \mathbf{G}_t \mathbf{C}_{t-1} \mathbf{G}'_t = Var[\mathbf{G}_t \boldsymbol{\theta}_t | D_{t-1}]$, which can be viewed as the prior variance in a NDLM with $\mathbf{W}_t = \mathbf{0}$, that is, with no evolution error term. The usual NDLM leads to $Var[\mathbf{G}_t \boldsymbol{\theta}_t | D_{t-1}] = \mathbf{P}_t + \mathbf{W}_t$. The idea of discounting factor is introduced by making $\mathbf{W}_t = \mathbf{P}_t(1 - \delta)/\delta$, for some $\delta \in (0, 1]$. Then, $\mathbf{R}_t = \mathbf{P}_t/\delta$, resulting in an increase in the prior variance of $\boldsymbol{\theta}_t$, at time $t - 1$. Note that, for given values of δ and $\mathbf{C}_0$ the series $\mathbf{W}_1, \ldots, \mathbf{W}_n$ is identified. If a NDLM has different components following the superposition theorem, a possible strategy is to specify different values of the discount factor for the different components. Note that the higher the discount factor, the higher the degree of smoothing. For this reason, the value of δ typically used is in the range $[0.8, 1]$ for polynomial, regression, and seasonal components [60].

Intervention and monitoring

Because of the sequential learning structure of NDLMs (Section 4.2.1), they naturally accommodate changes in the observed time series. For example, if an observation is missing at time t, then $D_t = D_{t-1}$, and $\boldsymbol{\theta}_t | D_t \sim N(\mathbf{m}_t, \mathbf{C}_t)$, with $\mathbf{m}_t = \mathbf{a}_t$ and $\mathbf{C}_t = \mathbf{R}_t$. On the other hand, if a change has occurred and it is difficult to attribute it to a particular component, the prior variance matrix $\mathbf{R}_t$ can be changed to reflect the increased uncertainty about all parameters, without a change in the prior mean $\mathbf{a}_t$ that would anticipate the direction of the change.

The adequacy of NDLMs can be monitored online through the comparison between the observed value and the one-step ahead forecast distribution. In particular, Bayes' factors [15] can be used, and are usually based on the predictive densities of the forecast errors, $e_t = y_t - f_t$. See [60, Chapter 11] for more details both on intervention and monitoring in NDLMs.

4.2.5   Dynamic generalized linear models (DGLM)

The DLM was extended by [61] to the case wherein observations belong to the exponential family. Assume that observations y_t are generated from a dynamic exponential family (EF), defined as

$$p(y_t | \eta_t, \phi) = \exp\{\phi[y_t \eta_t - a(\eta_t)]\} b(y_t, \phi), \tag{4.11}$$

where $a(.)$ and $b(.)$ are known functions, η_t is the canonical parameter, and ϕ is a scale parameter, usually time invariant. We denote the distribution of y_t as $EF(\eta_t, \phi)$. Let $\mu_t =$

$E(y_t|\eta_t, \phi)$ and $g(.)$ be a known link function assumed at least twice differentiable, which relates the linear predictor with the canonical parameter, that is

$$g(\mu_t) = \mathbf{F}(\psi_1)'\boldsymbol{\theta}_t, \tag{4.12}$$

where $\mathbf{F}$ and $\boldsymbol{\theta}_t$ are vectors of dimension k, and ψ_1 denotes unknown quantities involved in the definition of $\mathbf{F}(\psi_1)$. Following the parameterization of the exponential family in equation (4.11) it follows that the mean and variance of y_t are, respectively, given by $E(y_t|\eta_t, \phi) = \mu_t = \frac{da(\eta_t)}{d\eta_t} = \dot{a}(\eta_t)$ and $V(y_t|\eta_t, \phi) = V_t = \phi^{-1}\ddot{a}(\eta_t)$.

Usually, in practice, assuming $g(.)$ to be the natural link function provides good results [61]. Some examples, where the link function is suggested by the definition of the canonical function, include the log-linear Poisson and logistic-linear Bernoulli models. The state parameters, $\boldsymbol{\theta}_t$, evolve through time via a Markovian structure, that is

$$\boldsymbol{\theta}_t = \mathbf{G}(\psi_2)\boldsymbol{\theta}_{t-1} + \boldsymbol{\omega}_t, \quad \boldsymbol{\omega}_t \sim N(\mathbf{0}, \mathbf{W}). \tag{4.13}$$

The hyperparameter vector ψ_2 represents possible unknown quantities in $\mathbf{G}(\cdot)$, the $k \times k$ evolution matrix. Lastly, $\boldsymbol{\omega}_t$ is the disturbance associated with the system evolution with covariance structure $\mathbf{W}$, commonly a diagonal matrix which can vary over time. The initial information of the model is denoted by $\boldsymbol{\theta}_0$, and its prior distribution is defined through a k-variate normal distribution, that is, $\boldsymbol{\theta}_0|D_0 \sim N(\mathbf{m}_0, \mathbf{C}_0)$, where D_0 denotes the initial information set. Typically, we assume that the components of $\boldsymbol{\theta}_0$ and $\boldsymbol{\omega}_t$ are independent for all time periods.

Posterior inference

[61] specify only the first and second moments of $\boldsymbol{\omega}_t$. They perform inference taking advantage of conjugate prior and posterior distributions in the exponential family, and used linear Bayes' estimation to obtain estimates of the moments of $\boldsymbol{\theta}_t|D_t$. This approach is appealing as no assumption is made about the shape of the distribution of the disturbance component $\boldsymbol{\omega}_t$. However, we are limited to learn only about the first two moments of the posterior distribution of $\boldsymbol{\theta}_t|D_t$. The assumption of normality of $\boldsymbol{\omega}_t$ allows us to write down a likelihood function for $\boldsymbol{\theta}_t$ and all the other parameters in the model. Inference procedure can be performed using Markov chain Monte Carlo algorithms. However, care must be taken when proposing an algorithm to obtain samples from the posterior distribution of the state vectors $\boldsymbol{\theta}_t$. [61], [33] and [18], among others, proposed different methods to obtain approximations of the posterior distribution of parameters belonging to DGLMs. Estimation methods based on MCMC can be seen at [56], [20], [24], and [40].

The Poisson case

Time series of counts are often encountered in ecological and environmental problems. It is well known that the Poisson distribution assumes that the mean and the variance are equal, which is hardly true in practice. Usually, the variance is much greater than the mean. When this happens it is said that the data are *overdispersed*. Overdispersion is commonly tackled via a mixture of Poisson and gamma distributions, which marginally results in a negative binomial distribution. Another mixture which is quite used is between Poisson and log-normal distributions [6, 36].

Assume that y_t represents the count process under study, for discrete time t. Let $y_t|\lambda_t, \delta_t \sim Poi(\lambda_t\delta_t)$ with $\log\lambda_t = \mathbf{F}_t'\boldsymbol{\theta}_t$ and $\boldsymbol{\theta}_t$ following the evolution in time as in Equation (4.13). Note that, when $\delta_t = 1 \ \forall t$, a dynamic Poisson model arises. From the evolution equation for $\boldsymbol{\theta}_t$, we have $\log\lambda_t = \mathbf{F}_t'\mathbf{G}_t\boldsymbol{\theta}_{t-1} + \mathbf{F}_t'\boldsymbol{\omega}_t$, or $\lambda_t = \exp\{\mathbf{F}_t'\mathbf{G}_t\boldsymbol{\theta}_{t-1} + \mathbf{F}_t'\boldsymbol{\omega}_t\}$. Since $\boldsymbol{\omega}_t$ is normally distributed and independent of $\boldsymbol{\theta}_{t-1}$, we have that $\delta_t^* = \exp\{\mathbf{F}_t'\boldsymbol{\omega}_t\}$ follows a log-normal distribution, whose associated normal has zero mean and variance $\mathbf{F}_t' \mathbf{W} \mathbf{F}_t$. Therefore, this generalized dynamic Poisson model not only accounts for the temporal structure in the time series but is also able to capture overdispersion, as for every time t the marginal variance of y_t (with respect to ω_t) is greater than its mean.

If it is believed that there is some extra variability in the data, one alternative is to assume a continuous mixture such that δ_t follows a continuous distribution assuming positive values, and δ_t is independent of λ_t, $\forall t$. If δ_t follows a Gamma distribution with mean α/β and variance α/β^2, denoted by $\delta_t|\alpha, \beta \sim Ga(\alpha, \beta)$, then the distribution of $y_t|\lambda_t, \alpha, \beta$.

$$p(y_t|\lambda_t, \alpha, \beta) = \int p(y_t|\lambda_t, \delta_t) p(\delta_t|\alpha, \beta) \, d\delta_t, \qquad (4.14)$$

follows a negative binomial distribution with mean and variance given, respectively, by $E(y_t|\lambda_t, \alpha, \beta) = \lambda_t \alpha/\beta = \varphi_t$ and $V(y_t|\lambda_t, \alpha, \beta) = \varphi_t + \lambda_t^2 \alpha/\beta^2$.

Another possibility is to consider a log-normal mixture; that is, to assume $\delta_t|a_\delta, V \sim LN(a_\delta, V)$, where $LN(a_\delta, V)$ stands for the log-normal distribution whose associated normal on the transformed scale has mean a_δ and variance V. Using the properties of conditional expectation, it can be shown that $E(y_t|\lambda_t, a_\delta, V) = \lambda_t \exp\{a_\delta + V/2\} = \phi_t$ and $V(y_t|\lambda_t, a_\delta, V) = \phi_t + \lambda_t^2 \exp\{2a_\delta + V\}(e^V - 1)$.

Notice that both mixtures capture overdispersion as the variances in both cases are the means plus positive quantities. It is worth mentioning that writing y_t as following a Poisson distribution with mean $\lambda_t^* = \lambda_t \delta_t$ is equivalent to assuming $\log \lambda_t^* = \log \lambda_t + \log \delta_t$ which is equal to $\mathbf{F}_t' \boldsymbol{\mu}_t + \log \delta_t$. Therefore, $\log \delta_t$ can be viewed as a random effect present to capture any extra variation in the count process. [60], page 556, discuss this model as an alternative to capture possible extra variations.

Zero-inflated dynamic Poisson model

Another common issue with the modeling of time-varying counts might be the presence of a great amount of zeros relative to what is observed with a Poisson model, for example. In this case, one might use a zero inflated version of the Poisson dynamic model. Following [1], let x_t be a random variable representing the presence ($x_t = 1$) or absence ($x_t = 0$) of the process being observed. Assume that $x_t|\zeta$ follows a Bernoulli distribution with probability of success ζ. Let $\pi(y_t|\lambda_t, \delta_t)$ be a model for the process being observed *given* it is *present* ($x_t = 1$). In particular, here $\pi(\cdot|\lambda_t, \delta_t)$ is the probability distribution function of the Poisson distribution with mean $\lambda_t \delta_t$. By definition, $P(y_t|\lambda_t, \delta_t, x_t = 1) = \pi(y_t|\lambda_t, \delta_t)$ and $P(x_t = 0|\lambda_t, \delta_t, x_t = 0) = 1$. The joint density function of x_t and y_t is given by

$$p(y_t, x_t|\lambda_t, \delta_t, \zeta) = \{\zeta \, \pi(y_t|\lambda_t, \delta_t)\}^{x_t} (1 - \zeta)^{1-x_t} . \qquad (4.15)$$

Therefore, the marginal distribution of y_t, is given by $p(y_t|\lambda_t, \delta_t, \zeta) = \zeta \, \pi(y_t|\lambda_t, \delta_t) + (1 - \zeta)\delta_0(y_t)$, where $\delta_0(y_t)$ is an indicator function, such that $\delta_0(y_t) = 1$ if $y_t = 0$, and zero otherwise. Clearly, $p(y_t = 0|\lambda_t, \delta_t, \zeta) > \pi(y_t = 0|\lambda_t, \delta_t)$.

Notice that x_t is a latent variable, its inclusion in the model results that $P(y_t = 0|\lambda_t, \delta_t, X_t = 1) = \pi(y_t = 0|\lambda_t, \delta_t)$, $P(y_t > 0|\lambda_t, \delta_t, x_t = 0) = 0$, $P(x_t = 1|y_t > 0, \zeta, \lambda_t, \delta_t) = 1$ and,

$$p_t = P(x_t = 1|y_t = 0, \zeta, \lambda_t, \delta_t) = \frac{\zeta \, \pi(y_t = 0|\lambda_t, \delta_t)}{\zeta \, \pi(y_t = 0|\lambda_t, \delta_t) + (1 - \zeta)}, \qquad (4.16)$$

and marginally, as previously defined, x_t follows a Bernoulli distribution with probability of success ζ, denoted by $x_t|\zeta \sim Ber(\zeta)$. Notice that p_t provides an estimate of the probability of presence of the process, at time t, given that no cases were observed. [55] makes comparison of these class of models for artificial datasets and weekly number of cases of dengue fever between 2001 and 2002, in the city of Rio de Janeiro, Brazil. A natural extension of this model is to allow ζ, the probability of presence, to evolve with time.

4.3 Multivariate Dynamic Linear Models

The multivariate extension of the dynamic linear models introduced in Section 4.2 leads to the broad classes of multivariate and matrix variate Gaussian DLMs, both of which we briefly revisit here. We then focus on two quite popular and general classes of multi-dimensional DLMs: spatio-temporal models and dynamic factor models. Here, we closely follow the notation of [60, Chapter 16] and [47, Chapter 10].

4.3.1 Multivariate NDLMs

For a p-dimensional vector $\boldsymbol{y}_t = (y_{t1}, \ldots, y_{tp})'$, a multivariate normal DLM can be written as

$$\boldsymbol{y}_t = \mathbf{F}'_t\boldsymbol{\theta}_t + \boldsymbol{\epsilon}_t, \qquad \boldsymbol{\epsilon}_t \sim N(0, \boldsymbol{V}_t), \tag{4.17}$$

$$\boldsymbol{\theta}_t = \boldsymbol{G}_t\boldsymbol{\theta}_{t-1} + \boldsymbol{\omega}_t, \qquad \boldsymbol{\omega}_t \sim N(0, \boldsymbol{W}_t), \tag{4.18}$$

where $\boldsymbol{\theta}_0 \sim N(\boldsymbol{m}_0, \boldsymbol{C}_0)$. In what follows, for simplicity, it is assumed that both sequences of observation and evolution covariances matrices $\boldsymbol{V}_t$ and $\boldsymbol{W}_t$ are known and of dimensions ($p \times p$) and ($k \times k$), respectively. This assumption is relaxed later on when dealing with the special cases. Therefore, the model is completely specified by the quadruple $\{\mathbf{F}_t, \boldsymbol{G}_t, \boldsymbol{V}_t, \boldsymbol{W}_t\}$, where $\mathbf{F}_t$s are ($k \times p$) matrices and $\boldsymbol{G}_t$s are ($k \times k$) matrices. As an illustration, if for all i, y_{ti} followed the same linear trend introduced in Section 4.2, then $k = 2$. On the other hand, if the linear trends were different across the time series, then $\mathbf{F}' = \boldsymbol{I}_2 \otimes (1, 0)$, a $p \times k$ matrix with $k = 2p$.

It follows directly from Section 4.2.1 that, for $t = 1, 2, \ldots$,

$$(\boldsymbol{\theta}_t | D_{t-1}) \sim N(\boldsymbol{a}_t, \boldsymbol{R}_t),$$

$$(\boldsymbol{y}_t | D_{t-1}) \sim N(\boldsymbol{f}_t, \boldsymbol{Q}_t),$$

$$(\boldsymbol{\theta}_t | D_t) \sim N(\boldsymbol{m}_t, \boldsymbol{C}_t),$$

where $\boldsymbol{a}_t = \boldsymbol{G}_t\boldsymbol{m}_{t-1}$, $\boldsymbol{R}_t = \boldsymbol{G}_t\boldsymbol{C}_{t-1}\boldsymbol{G}'_t + \boldsymbol{W}_t$, $\boldsymbol{f}_t = \mathbf{F}'_t\boldsymbol{a}_t$, $\boldsymbol{Q}_t = \mathbf{F}'_t\boldsymbol{R}_t\mathbf{F}_t + \boldsymbol{V}_t$, with $\boldsymbol{m}_t = \boldsymbol{a}_t + \boldsymbol{A}_t(\boldsymbol{y}_t - \boldsymbol{f}_t)$ and $\boldsymbol{C}_t = \boldsymbol{R}_t - \boldsymbol{A}_t\boldsymbol{Q}_t\boldsymbol{A}'_t$, with $\boldsymbol{A}_t = \boldsymbol{R}_t\mathbf{F}_t\boldsymbol{Q}_t^{-1}$.

On the one hand, in this parametrization the i-th time series $\{y_{ti}\}$ has its own set of regressors, represented the i-th column of the matrices $\mathbf{F}_t$, but shares the dynamic regression coefficients, $\boldsymbol{\theta}_t$, with all other $p-1$ time series. On the other hand, [49] proposed the *common components* multivariate DLM, where each time series $\{y_{ti}\}$ has its own dynamic regression coefficients, $\boldsymbol{\theta}_{ti}$, but shares the set of regressors in $\mathbf{F}_t$.

4.3.2 Multivariate common-component NDLMs

In this case, the model is written as p univariate DLMs sharing the sample quadruple $\{\mathbf{F}_t, \boldsymbol{G}_t, \boldsymbol{V}_t, \boldsymbol{W}_t\}$:

$$y_{ti} = \mathbf{F}'_t\boldsymbol{\theta}_{ti} + \epsilon_{ti}, \qquad \epsilon_{ti} \sim N(0, v_{ii}V_t), \tag{4.19}$$

$$\boldsymbol{\theta}_{ti} = \boldsymbol{G}_t\boldsymbol{\theta}_{t-1,i} + \boldsymbol{\omega}_{ti}, \qquad \boldsymbol{\omega}_{ti} \sim N(0, v_{ii}\boldsymbol{W}_t), \tag{4.20}$$

or, more compactly, y_{ti} follows a DLM with quadruple $\{\mathbf{F}_t, \mathbf{G}_t, v_{ii}V_t, v_{ii}\mathbf{W}_t\}$. Besides a common set of regressors $\mathbf{F}_t$, the univariate models are also linked through the covariance matrix $\mathbf{V}$, where $\mathrm{Cov}(\epsilon_{ti}, \epsilon_{tj}) = v_{ij}V_t$ and $\mathrm{Cov}(\boldsymbol{\omega}_{ti}, \boldsymbol{\omega}_{tj}) = v_{ij}\mathbf{W}_t$, for $i, j = 1, \ldots, p$ and $i \neq j$.

Time-varying parameter vector autoregressions (TVP-VAR)

A quite popular special case are time-varying parameter vector autoregressive models where $\mathbf{F}_t$ contains several lags of all time time series in $\mathbf{y}_t$ (See, for instance, [48], [41] and [34]). More precisely, a standard time-varying parameter VAR(L) can be written as

$$\mathbf{y}_t = \sum_{l=1}^{L} \mathbf{B}_{lt}\mathbf{y}_{t-l} + \boldsymbol{\epsilon}_t, \tag{4.21}$$

where $\boldsymbol{\epsilon}_t \sim N(\mathbf{0}, \mathbf{V}_t)$, $\mathbf{y}_t$ is a p-dimensional vector of ecological or epidemiological measurements, for instance, and $\mathbf{B}_{lt}$ is the time t and lag l autoregressive coefficient matrix. Let $\mathbf{B}_{lt} = (\mathbf{b}_{lt,1}, \ldots, \mathbf{b}_{lt,p})'$, so $\mathbf{b}'_{lt,i}$ is the p-dimensional vector corresponding to the ith row of $\mathbf{B}_{lt}$. Now let $\boldsymbol{\beta}_{it} = (\mathbf{b}'_{1t,i}, \mathbf{b}'_{2t,i}, \ldots, \mathbf{b}'_{kt,i})'$ be the pL-dimensional vector of coefficients corresponding to the ith equation. Therefore, the above TVP-VAR(L) model can be rewritten as

$$\mathbf{y}_t = \mathbf{F}'_t\boldsymbol{\theta}_t + \boldsymbol{\epsilon}_t, \tag{4.22}$$

where $\mathbf{F}'_t = (\mathbf{I}_p \otimes \mathbf{x}'_t)$, $\mathbf{x}_t = (\mathbf{y}'_{t-1}, \mathbf{y}'_{t-2}, \ldots, \mathbf{y}'_{t-L})'$ is a pL-dimensional vector of lagged observations and $\boldsymbol{\theta}_t = (\boldsymbol{\beta}'_{1t}, \ldots, \boldsymbol{\beta}'_{pt})'$ is a k-dimensional vector of autoregressive coefficients, where $k = p^2 L$. In other words, Equation (4.21) has been rewritten as an observation equation for a multivariate NDLM as in Equation (4.17).

4.3.3 Matrix-variate NDLMs

The model is an extension of the previous common-component model to allow for $q \times p$ matrices of observations $\mathbf{y}_t$:

$$\begin{aligned} \mathbf{y}_t &= (\mathbf{I}_q \otimes \mathbf{F}'_t)\boldsymbol{\theta}_t + \boldsymbol{\epsilon}_t, & \boldsymbol{\epsilon}_t &\sim N(\mathbf{0}, \mathbf{U}_t, \mathbf{V}_t), & (4.23) \\ \boldsymbol{\theta}_t &= (\mathbf{I}_q \otimes \mathbf{G}_t)\boldsymbol{\theta}_{t-1} + \boldsymbol{\omega}_t, & \boldsymbol{\omega}_t &\sim N(\mathbf{0}, \mathbf{U}_t \otimes \mathbf{W}_t, \mathbf{V}_t), & (4.24) \end{aligned}$$

where the matrix-variate normal $\boldsymbol{\epsilon} \sim N(\mathbf{0}, \mathbf{U}, \mathbf{V})$ basically means that the ith row of $\boldsymbol{\epsilon}$ is $N(\mathbf{0}, u_{ii}\mathbf{V})$, while its jth column is $N(\mathbf{0}, v_{jj}\mathbf{U})$. For matrix-variate normal distributions see [14]. See [35] for the matrix-variate extension of the dynamic hierarchical models of [21]. Dynamic matrix-variate graphical models are introduced by [10]. Sequential updating, forecasting and retrospective smoothing follow directly from the previous derivations. Similarly, when the prior for $\mathbf{V}$ is inverse Wishart, sequential updating is an extension of the derivations presented in Section 4.2.1. See [47, Chapter 10] for additional details.

4.3.4 Hierarchical dynamic linear models (HDLM)

In certain applications, a natural hierarchical order arises in the dynamics of time series and their interdependences. For example, monthly levels of a set of pollutants might be contemporaneous related and their conditional expectations functions of regional characteristics, such as climate, industrial developments and other demographic aspects. A hierarchy, in this context, might arise to connect regressions across regions by pooling their coefficients, either via a common mean vector or through yet another set of explanatory variables.

Here we briefly review the seminal paper by [21] on dynamic hierarchical linear models

(DHLM) for time series of cross-sectional data, which generalizes the hierarchical model introduced by [37]. See, for instance, [1] and [13] for additional examples of hierarchical models in spatial and spatio-temporal models.

The basic HDLM of [21] is composed of three equations: observational, structural and system, with the structural equations accounting for the levels of hierarchy. For example, a normal DHLM (NDHLM) with three hierarchical levels can be written as

$$
\begin{array}{llll}
\boldsymbol{y}_t & = & \mathbf{F}'_{1t}\boldsymbol{\theta}_{1t} + \boldsymbol{\epsilon}_{1t} & \boldsymbol{\epsilon}_{1t} \sim N(\mathbf{0}, \boldsymbol{V}_{1t}) & (4.25) \\
\boldsymbol{\theta}_{1t} & = & \mathbf{F}'_{2t}\boldsymbol{\theta}_{2t} + \boldsymbol{\epsilon}_{2t} & \boldsymbol{\epsilon}_{2t} \sim N(\mathbf{0}, \boldsymbol{V}_{2t}) & (4.26) \\
\boldsymbol{\theta}_{2t} & = & \mathbf{F}'_{3t}\boldsymbol{\theta}_{3t} + \boldsymbol{\epsilon}_{3t} & \boldsymbol{\epsilon}_{3t} \sim N(\mathbf{0}, \boldsymbol{V}_{3t}) & (4.27) \\
\boldsymbol{\theta}_{3t} & = & \boldsymbol{G}_t\boldsymbol{\theta}_{3,t-1} + \boldsymbol{\omega}_t & \boldsymbol{\omega}_t \sim N(\mathbf{0}, \boldsymbol{W}_t). & (4.28)
\end{array}
$$

The model can be easily reduced to a two-level NDHLM by making $\mathbf{F}_{3t} = \boldsymbol{I}$ and $\boldsymbol{V}_{3t} = \mathbf{0}$. If, in addition, $\mathbf{F}_{2t} = \boldsymbol{I}$ and $\boldsymbol{V}_{2t} = \mathbf{0}$, the above NDHLM reduces to a standard NDLM. A standard normal hierarchical linear model (NHLM) is obtained when $\boldsymbol{G}_t = \boldsymbol{I}$ and $\boldsymbol{W}_t = \mathbf{0}$. As pointed out by [21], a distinct feature of NDHLM is that $\dim(\boldsymbol{\theta}_{1t}) > \dim(\boldsymbol{\theta}_{2t}) > \dim(\boldsymbol{\theta}_{3t})$.

Example: *(Cross-section of random samples of linear growing exchangeable means)* For $\boldsymbol{y}_t = (y_{t1}, \ldots, y_{tp})'$, $\boldsymbol{\theta}_{1t} = (\theta_{1t,1}, \ldots, \theta_{1t,p})'$, $\boldsymbol{\theta}_{2t} = (\mu_t, \beta_t)'$ and $\mathbf{F}' = (\mathbf{1}_p \ \mathbf{0}_p)$, the model is written as a two-stage NDHLM:

$$
\begin{array}{lll}
\boldsymbol{y}_t | \boldsymbol{\theta}_{1t} & \sim & N(\boldsymbol{\theta}_{1t}, \sigma^2 \boldsymbol{I}_n), \\
\boldsymbol{\theta}_{1t} | \boldsymbol{\theta}_{2t} & \sim & N(\mathbf{F}'\boldsymbol{\theta}_{2t}, \tau^2 \boldsymbol{I}_n), \\
\boldsymbol{\theta}_{2t} | \boldsymbol{\theta}_{2,t-1} & \sim & N(\boldsymbol{G}\boldsymbol{\theta}_{2,t-1}, \boldsymbol{W}_t),
\end{array}
$$

where $\boldsymbol{G}$ follows the same structure of the linear trend introduced at the beginning of the chapter.

4.3.5   Spatio-temporal models

Environmental and ecological processes usually vary across space and time simultaneously. Spatio-temporal observations are naturally accommodated in the DLM framework. We start the discussion by considering that locations vary continuously across the region of interest (geostatistical modeling). Then we discuss a spatial dynamic factor model. Next we discuss the problem where the spatial region under study is divided into a finite number of regular or irregular subregions (areal level modeling), resulting in observations for each subregion. Spatial interpolation is the main interest in the geostatistical case, while in the areal modeling case the main interest is on spatial smoothing.

Geostatistical setup: observed locations varying continuously in space

Let $\{y_t(\boldsymbol{s}), \boldsymbol{s} \in D \subset \mathbb{R}^d, t \in T\}$ be a spatial random field at discrete time t, and usually, $d = 1, 2$ or 3. If a partial realization of the spatial random field is available at each time t, the model specification in equation (4.17) is useful to describe the behaviour of the spatial process for each time t.

Following [13], let the observed data $y_t(\boldsymbol{s})$, be a noisy version of the process of interest, $z_t(\boldsymbol{s})$, that is, assume:

$$
\begin{array}{lll}
y_t(\boldsymbol{s}) & = & z_t(\boldsymbol{s}) + v_t(\boldsymbol{s}) & (4.29) \\
z_t(\boldsymbol{s}) & = & \mathbf{F}'_t\boldsymbol{\theta}_t + \epsilon^*_t(\boldsymbol{s}) & (4.30) \\
\boldsymbol{\theta}_t & = & \boldsymbol{G}_t\boldsymbol{\theta}_{t-1} + \boldsymbol{\omega}_t, & (4.31)
\end{array}
$$

where $\epsilon_t^*(\cdot)$ is assumed to follow a zero mean Gaussian process with some covariance matrix Σ_t, while $v_t(s)$ represents measurement error assumed to follow an independent zero mean normal distribution with variance τ^2. It is assumed further that $\epsilon_t^*(s)$ are independent across time, and independent of $v_t(s)$ and ω_t, for all t and $s \in D$.

For simplicity and without lack of generality, we will assume that y_t is observed for time periods $t = 1, \ldots, n$ and locations $s_1, \cdots, s_p$. If we substitute equation (4.30) into (4.29), we obtain $y_t(s) = F_t'\theta_t + \epsilon_t^*(s) + v_t(s)$. If we stack the observations of the process at time t onto a vector $y_t = (y_t(s_1), y_t(s_2), \ldots, y_t(s_p))'$, and assume a similar structure for ϵ_t^* and v_t, respectively, we can write $y_t | F_t, \theta_t, v_t, \tau^2 \sim N(F_t'\theta_t + v_t, \tau^2 I_p)$. In other words, *conditional on* v_t, the elements of y_t are independent across space, for each time t. We can marginalize the distribution of y_t with respect to v_t. As v_t follows a zero mean multivariate normal distribution, then the distribution of $p(y_t | \theta_t, F_t, \tau^2, \Sigma_t) = \int_{v_t} p(y_t | F_t, \theta_t, v_t, \tau^2) \, p(v_t | \Sigma_t) dv_t$ follows a multivariate normal distribution with mean $F_t'\theta_t$ and covariance matrix $V_t = \Sigma_t + \tau^2 I_p$.

Note that, if spatio-temporal covariates are present in the columns of F_t the mean of the process is also varying across space for each time t. On the other hand, if spatio-temporal covariates are not available, then the temporal trend is fixed across space, and ϵ_t captures, at each location, deviations from this overall temporal structure. The main issue is the specification of the covariance matrix Σ_t. The simplest alternative to account for spatial correlation at each time t is to model each element of the covariance matrix $\Sigma_t = \Sigma \ \forall t$, as a function of a common variance times a correlation function that depends on the Euclidean distance among the monitoring locations, that is, $\Sigma_{ij} = \sigma^2 \rho(d_{ij}, \phi)$, where $d_{ij} = \|s_i - s_j\|$ is the Euclidean distance. The parameters in ϕ are typically scalars or low dimensional vectors. For example, ϕ is univariate when the correlation function is exponential $\rho(d, \phi) = \exp\{-d/\phi\}$ or spherical $\rho(d, \phi) = (1 - 1.5(d/\phi) + 0.5(d/\phi)^3) 1_{\{d \leq \phi\}}$, and bivariate when the correlation function is power exponential $\rho(d, \phi) = \exp\{-(d/\phi_1)^{\phi_2}\}$ or Matérn $\rho(d, \phi) = 2^{1-\phi_2} \Gamma(\phi_2)^{-1} (d/\phi_1)^{\phi_2} \mathcal{K}_{\phi_2}(d/\phi_1)$, where $\mathcal{K}_{\phi_2}(\cdot)$ is the modified Bessel function of the second kind of order ϕ_2. In both bivariate cases, $\phi_1 > 0$ is a range parameter controlling the speed of correlation decay between locations, while ϕ_2 is a smoothness parameter that controls the differentiability of the underlying process (see [1] for details). If it is believed that the variance of v_t changes with time, one alternative is to allow σ^2 to vary smoothly with time, for example, by making $\log\sigma_t^2 = \log\sigma_{t-1}^2 + \epsilon_{1t}$, with ϵ_{1t} following a zero mean normal distribution, with possibly unknown but constant variance W_σ, and initial information $\log\sigma_0^2 \sim N(0, b)$, for some known constant b. This will lead to a covariance matrix V_t that changes with time. When inference is performed using MCMC methods, the algorithm must be adapted to include steps to sample from the posterior full conditional distributions of the elements of ϕ, and the variances $\sigma_0^2, \sigma_1^2, \ldots, \sigma_n^2$. This can be achieved, for example, through individual Metropolis-Hastings steps for each of these parameters.

An alternative in the modelling of the covariance matrix is to assign an inverse Wishart prior distribution for Σ_t, which results on a nonstationary (free form) covariance matrix. The discount ideas discussed in Section 4.2.4 were extended by [38] for variance matrices discounting. [53] on the other hand, propose to define Σ_t as a function of covariates. Care must be taken when using covariate information in the covariance structure as it is necessary to guarantee that the resultant covariance structure is positive definite.

Spatial interpolation and temporal prediction

In spatio-temporal settings the aim usually is to perform spatial interpolation of the process for unmonitoring locations or to perform temporal prediction. Initially, assume we want to predict the process for a vector of dimension h, of unobserved locations at observed time $t \in T$, say $\tilde{y}_t = (y_t(\tilde{s}_1), \ldots, y_t(\tilde{s}_h))'$. Assuming that $y = (y_1(s_1), \ldots, y_1(s_p), \ldots, y_n(s_1), \ldots, y_n(s_p))'$ represents the vector of the time series observed at p monitoring locations, we need to obtain the predictive posterior distribution, $p(\tilde{\boldsymbol{y}}_t|\boldsymbol{y})$, which is given by

$$p(\tilde{\boldsymbol{y}}_t|\boldsymbol{y}) = \int_{\boldsymbol{\theta}} p(\tilde{\boldsymbol{y}}_t|\boldsymbol{y}, \boldsymbol{\theta})p(\boldsymbol{\theta}|\boldsymbol{y})d\boldsymbol{\theta}, \tag{4.32}$$

where $\boldsymbol{\theta}$ is a parameter vector comprising all the unknowns in the model.

Spatial interpolations are obtained by considering the distribution of $(\boldsymbol{y}_t, \tilde{\boldsymbol{y}}_t)$ conditional on the parameters, and the initial information $\boldsymbol{\theta}_0$. This distribution is given by

$$\left(\begin{array}{c} \boldsymbol{y}_t \\ \tilde{\boldsymbol{y}}_t \end{array} \middle| \boldsymbol{\theta}, V, W \right) \sim N \left(\left(\begin{array}{c} \mathbf{F}_t'\boldsymbol{\theta}_t \\ \tilde{\mathbf{F}}_t'\boldsymbol{\theta}_t \end{array} \right); \left(\begin{array}{cc} \mathbf{V}_y & \mathbf{V}_{y\tilde{y}} \\ \mathbf{V}_{\tilde{y}y} & \mathbf{V}_{\tilde{y}\tilde{y}} \end{array} \right) \right),$$

where $\tilde{\mathbf{F}}_t$ corresponds to the regression matrix for unobserved locations. Similar notation is used to split the covariance matrix into four blocks. We then have that

$$\tilde{\boldsymbol{y}}_t|\boldsymbol{y}_t, \boldsymbol{\theta} \sim N \left(\tilde{\boldsymbol{\mu}}_t + \mathbf{V}_{\tilde{y}y}(\mathbf{V}_y)^{-1}(\boldsymbol{y}_t - \boldsymbol{\mu}_t), (\mathbf{V}_{\tilde{y}\tilde{y}} - \mathbf{V}_{\tilde{y}y}(\mathbf{V}_y)^{-1}\mathbf{V}_{\tilde{y}y}) \right),$$

where $\boldsymbol{\mu}_t = \mathbf{F}_t'\boldsymbol{\theta}_t$ and $\tilde{\boldsymbol{\mu}}_t = \tilde{\mathbf{F}}_t'\boldsymbol{\theta}_t$ for all t. Once samples from the posterior distribution of the parameter vector are available, samples from the posterior predictive distribution of $\tilde{\boldsymbol{y}}_t$ are easily obtained by composition sampling using the conditional distribution above.

Temporal h-steps ahead prediction at monitoring locations is given by the following posterior predictive distribution

$$p(\boldsymbol{y}_{n+h}|D_n) = \int \prod_{h=1}^{k} p(\boldsymbol{y}_{n+h}|\mathbf{F}_{n+h}, \boldsymbol{\theta}_{n+h}, \boldsymbol{V})p(\boldsymbol{\theta}_{n+h}|\boldsymbol{\theta}_{n+h-1}, \boldsymbol{W})$$

$$\times p(\boldsymbol{\theta}|D_n)p(\boldsymbol{W})p(\boldsymbol{V}) \, d\tilde{\boldsymbol{\theta}} \, d\boldsymbol{\theta} \, d\boldsymbol{W} \, d\boldsymbol{V},$$

where $D_n = \{\boldsymbol{y}_1, \ldots, \boldsymbol{y}_n\}$, $\boldsymbol{\theta} = (\boldsymbol{\theta}_1, \ldots, \boldsymbol{\theta}_n)$ and $\tilde{\boldsymbol{\theta}} = (\boldsymbol{\theta}_{n+1}, \ldots, \boldsymbol{\theta}_{n+h})$. The integral above can be approximated by

$$p(\boldsymbol{y}_{n+h}|D_n) \approx \frac{1}{M} \sum_{m=1}^{M} p(\boldsymbol{y}_{n+h}|\mathbf{F}_{n+h}, \boldsymbol{\theta}_{n+h}^{(m)}, \boldsymbol{V}^{(m)}).$$

Here the superscript (m) denotes samples from the posterior of $\boldsymbol{\theta}_{n+1}, \ldots, \boldsymbol{\theta}_{n+h}$, and $\boldsymbol{V}$. Samples from the distribution above can be obtained by propagating $\boldsymbol{\theta}_{n+h}$ following the system equation in (4.18), and using the samples from the posterior distribution of the parameter vector.

The use of multivariate DLM to model spatio-temporal processes has been widely used in the literature. For example, [31] propose a spatio-temporal model for hourly ozone levels in Mexico City. The model is formulated within the state-space framework, and allows for uncertainty on any missing values of ozone concentrations and covariates. [22] extend the multivariate DLM to allow the coefficients $\boldsymbol{\theta}_t$ to vary smoothly with space. They extend this proposal to model multivariate spatio-temporal processes. [42] extend the class of DLM to allow the error term ϵ_t to follow autoregressive processes with spatially varying coefficients.

Spatio-temporal models for non-normal data

For many environmental processes that vary across space and time, the assumption of normality of the observations is unreasonable. The univariate generalized dynamic model described in Section 4.2.5 is naturally extended to accommodate spatio-temporal observations. Commonly, the spatio-temporal process is built upon the assumption of conditional independence given the parameter vector defining the family of distribution under interest. More specifically, let $y_t(s)$ denote a spatio-temporal process under study, and we assume that

$$
\begin{aligned}
y_t(s)|\eta_t(s), \phi &\sim EF(\eta_t(s), \phi) \\
g(\mu_t(s)) &= \mathbf{F}'_t(s)\boldsymbol{\theta}_t + \epsilon_t(s), \quad \epsilon_t(s) \sim N(\mathbf{0}, \boldsymbol{\Sigma}_t) \\
\boldsymbol{\theta}_t &= \boldsymbol{G}_t\boldsymbol{\theta}_{t-1} + \boldsymbol{\omega}_t, \quad \boldsymbol{\omega}_t \sim N(\mathbf{0}, \boldsymbol{W}) \\
\boldsymbol{\theta}_0|D_0 &\sim N(\mathbf{m}_0, \boldsymbol{C}_0),
\end{aligned}
\tag{4.33}
$$

where, again, $\eta_t(s)$ is the canonical parameter of the exponential family, ϕ is a scale parameter, usually time invariant, $\mu_t(s) = E(y_t(s)|\eta_t(s), \phi)$, $g(.)$ is a known link function, at least twice differentiable, which relates the linear predictor with the canonical parameter, and $\mathbf{F}'_t(s)$ is k-dimensional vector with the covariates observed at location s.

Example: This example shows the resultant properties of the model in equations (4.33) for the Poisson case, where $y_t(s) \mid \lambda_t(s) \sim Poisson(\lambda_t(s))$. This can be viewed as a Poisson-lognormal mixture model. Using the results based on conditional expectation, it can be shown that marginal expectation, variance and covariance structures for $y_t(s)$ are given by:

$$
\begin{aligned}
E(y_t(s)) &= \beta_t(s) = \exp\{\mathbf{F}_t(s)'\boldsymbol{\theta}_t + 0.5\boldsymbol{\Sigma}(s, s)\} \\
V(y_t(s)) &= \beta_t(s) + \beta_t^2(s)[\exp\{\boldsymbol{\Sigma}(s, s)\} - 1] \\
C(y_t(s), y_t(s')) &= \beta_t(s)\beta_t(s')[\exp\{\boldsymbol{\Sigma}(s, s')\} - 1],
\end{aligned}
\tag{4.34}
$$

which is able to capture overdispersion, as $V(y_t(s)) > E(y_t(s))$. If $\mathbf{F}_t(s)$ has covariates that vary with s, regardless of the covariance structure $\boldsymbol{\Sigma}$, $y_t(s)$ is a nonstationary process in space.

Note that in the case of spatio-temporal generalized dynamic models, the MCMC algorithm for the state vector $\boldsymbol{\theta}_t$ can take advantage of the FFBS algorithm to sample from the posterior full conditional of the state vectors, because of the presence of the latent component $\epsilon_t(s)$ in equation (4.33).

A practical example using a generalized spatio-temporal dynamic linear model is proposed by [51], who propose a joint model for rainfall-runoff, two of the most important processes in hydrology. The challenge is to account for the different observational spatial scales of rainfall and runoff. This is done by using generalized dynamic linear models, and the effect of rainfall on runoff is modelled through a transfer function model [61]. [54] explore a Poisson-lognormal mixture spatio-temporal model for outbreaks of malaria in the state of Amazonas, Brazil. They compare models that assume different structures for the covariance matrix $\boldsymbol{\Sigma}$, including an inverse-Wishart prior for $\boldsymbol{\Sigma}$.

Spatial dynamic factor models

A new class of nonseparable and nonstationary space-time models that resembles a standard dynamic factor model (see [44]) is proposed by [32]. In one of its simplest form, the model can be written using the previous notation as

$$
\begin{aligned}
\boldsymbol{y}_t &= \mathbf{F}'\boldsymbol{\theta}_t + \epsilon_t, \quad \epsilon_t \sim N(0, \boldsymbol{V}) \tag{4.35} \\
\boldsymbol{\theta}_t &= \boldsymbol{G}\boldsymbol{\theta}_{t-1} + \boldsymbol{\omega}_t, \quad \boldsymbol{\omega}_t \sim N(0, \boldsymbol{W}) \tag{4.36}
\end{aligned}
$$

where $\boldsymbol{y}_t = (y_{t1}, \ldots, y_{tp})'$ is the p-dimensional vector of observations (locations $s_1, \ldots, s_p$ and times $t = 1, \ldots, n$), $\boldsymbol{\theta}_t$ is an k-dimensional vector of common factors, for $k < p$ (k is potentially several orders of magnitude smaller than p) and $\mathbf{F}' = (\mathbf{F}_{(1)}, \ldots, \mathbf{F}_{(k)})$ is the $p \times k$

matrix of factor loadings. The matrix G characterizes, as before, the evolution dynamics of the common factors,

Equations (4.35) and (4.36) define the first level of the proposed dynamic factor model. Similar to standard factor analysis, it is assumed that the k conditionally independent common factors θ_t capture all time-varying covariance structure in y_t. The conditional spatial dependencies are modeled by the columns of the factor loadings matrix β. More specifically, the jth column of $\mathbf{F}'$, denoted by $\mathbf{F}_{(j)} = (f_{(j)}(s_1), \ldots, f_{(j)}(s_p))'$, for $j = 1, \ldots, k$, is modeled as a conditionally independent, distance-based Gaussian process or a Gaussian random field (GRF), i.e.

$$\mathbf{F}_{(j)} \sim GP(\mathbf{F}_0, \tau_j^2 \rho_{\phi_j}(\cdot)) \equiv N(\mu_j, \tau_j^2 \mathbf{R}_{\phi_j}), \tag{4.37}$$

where $\mathbf{F}_0$ is a p-dimensional mean vector. The (l, l')-element of $\mathbf{R}_{\phi_j}$ is given by $r_{ll'} = \rho(\|s_l - s_{l'}\|, \phi_j)$, $l, l' = 1, \ldots, p$, for suitably defined correlation functions $\rho(\cdot, \phi_j)$, $j = 1, \ldots, k$. The proposed model could, in principle, accommodate nonparametric formulations for the spatial dependence, such as the ones introduced by [23], for instance. The spatial dynamic factor model is defined by equations (4.35)–(4.37).

Areal level spatio-temporal processes

In this scenario, for each time t, spatial observations are made at a (regular or irregular) partition of the region of interest D. Now, both temporal and spatial indices, vary discretely. This process is denoted as y_{ti} for $t = 1, 2, \ldots, n$ and $i = 1, 2, \ldots, p$.

Usually, for each time t, observations are available for all subregions and, commonly, interest lies on spatial smoothing and temporal prediction. In this set up, it is common practice to capture spatial effects through Markov random fields. The idea of the Markov random field is to specify the joint distribution of the spatial effects through local specifications. [5] introduced the conditionally autoregressive (CAR) models. Let ϵ_i be a latent spatial effect at location i, and define the conditional distribution

$$\epsilon_i | \epsilon_j, j \neq i \sim N \left(\sum_j w_{ij} \epsilon_j / w_{i+}, \tau^2 / w_{i+} \right) \quad \text{for } i = 1, 2, \ldots, p.$$

Through Brook's Lemma (see e.g. [1]), it can be shown that the joint distribution of the spatial effects is proportional to $p(\epsilon_1, \ldots, \epsilon_p) \propto \exp\left\{ -\frac{1}{2\tau^2} \epsilon'(\mathbf{D}_w - \mathbf{W})\epsilon \right\}$, where $\mathbf{D}_w$ is diagonal with elements $(\mathbf{D}_w)_{ii} = w_{i+}$. The matrix $\mathbf{W}$ is a proximity matrix defining the neighbourhood structure. In practice, it is common to assume $\mathbf{W}$ as following a first order neighbouring structure, that is, $W_{ij} = 1$ if locations i and j are neighbors, and 0 otherwise. The joint distribution of $\epsilon = (\epsilon_1, \ldots, \epsilon_p)'$ defined in this way is improper because $(\mathbf{D}_w - \mathbf{W})\mathbf{1}_n = 0$. For this reason, this distribution cannot be used as a model for data. One way to make the distribution of ϵ proper is to introduce a parameter ρ in $\mathbf{D}_w - \rho \mathbf{W}$ with $\rho \in (1/\lambda_{(1)}, 1/\lambda_{(p)})$ where $\lambda_{(1)}$ and $\lambda_{(p)}$ are, respectively, the smallest and biggest eigenvalues of $\mathbf{D}_w^{-1/2} \mathbf{W} \mathbf{D}_w^{1/2}$. See [1, pages 80-84] for details.

These models are commonly used in epidemiological studies [16]. For example, [43] propose a Poisson spatio-temporal model for the number of cases of malaria observed at municipalities of the state of Pará, Brazil as a function of rainfall. They propose a dynamic regression model similar to equation (4.33), and compare models that assume ϵ_{it} to be independent across space for each time t, vis-a-vis a CAR prior distribution for ϵ_t. They also explore models that capture a temporal correlation among the ϵ_t by allowing the logarithm of the variance of the conditional autoregressive process to vary smoothly with time.

[59] extend the multivariate DLM to accommodate spatial structures for areal data processes. This is done by assuming proper Gaussian Markov random fields as the distribution of the innovations of both observational and evolution equations, that is, ϵ_t and ω_t in equation (4.33). They discuss inference procedure under the Bayesian paradigm and they propose the forward information filter backward sampler algorithm, a modification of the FFBS sampler algorithm.

4.4   Further aspects of spatio-temporal modeling

The spatio-temporal DLM framework previously discussed can be extended to accommodate more complex processes that result on more flexible space-time covariance structures and/or allow to account for physical information. In the following subsections we briefly discuss spatio-temporal models based on process convolution approaches, stochastic partial differential (SPDE) and integro-difference (IDE) equations. The discussion about SPDEs and IDEs follow closely Chapters 6 and 7 of [13].

4.4.1   Process convolution based approaches

A Gaussian process can be obtained through a constructive approach by convolving a continuous white noise process $\theta(s)$, $s \in D \subset \mathbb{R}^2$ with a smoothing kernel $k(s)$. [29] allow the smoothing kernel to vary with location resulting on a nonstationary Gaussian process. On the other hand, [26] extend the spatial convolution approach to model spatio-temporal processes, and [28] extend this spatio-temporal convolution approach to allow the spatial kernel to evolve smoothly with time. This proposal is as follows. Let $Z_t(s)$ be a stochastic process defined in continuous space D and discrete time t, such that

$$Z_t(s) = \int_D \mathcal{K}(\mathbf{u} - s)\theta(\mathbf{u}, t)d\mathbf{u}, \tag{4.38}$$

where $\mathcal{K}(s)$ is a smoothing kernel. To reduce the dimension of the problem, consider $\theta(\cdot, t)$ to be nonzero at k spatial locations, $\mathbf{l}_1, \dots, \mathbf{l}_k$, and now

$$Z_t(\mathbf{s}) = \sum_{j=1}^{k} \mathcal{K}(\mathbf{l}_j - \mathbf{s})\theta_{jt}, \tag{4.39}$$

where each sequence $\{\theta\}_{jt}$ follows a Gaussian random walk, such that $\theta_{jt} = \theta_{j\,t-1} + \omega_{jt}$ with $\omega_{jt} \sim N(0, W)$. Note that $\mathcal{L} = \{\mathbf{l}_1, \dots, \mathbf{l}_l\}$ plays the role of the spatial support that approximates the continuous process in D, and reduces the spatial dimension of the problem. If we add an independent measurement error component in equation (4.39) together with the random walk assumption for the θ_{jt} we have a DLM:

$$\begin{aligned} \mathbf{y}_t &= \mathbf{F}'\boldsymbol{\theta}_t + \mathbf{v}_t \\ \boldsymbol{\theta}_t &= \mathbf{G}_t\boldsymbol{\theta}_{t-1} + \boldsymbol{\omega}_t \end{aligned}$$

where $\mathbf{y}_t = (Z_t(\mathbf{s}_1), \dots, Z_t(\mathbf{s}_p))'$, $\mathbf{F}_{ji} = \mathcal{K}(\mathbf{l}_j - \mathbf{s}_i)$ and $\boldsymbol{\theta}_t = (\theta_{1t}, \dots, \theta_{kt})'$.

Higdon (2002) [28] uses this approach to model daily concentrations of ozone across 30 days. [57] use a similar approach by describing the mean of the process as a locally-weighted mixture of linear regressions, with the coefficients evolving smoothly with time. [52] consider a class of models for spatio-temporal processes based on convolving independent processes with a discrete kernel that is represented by a lower triangular matrix. This was inspired by coregionalization models for multivariate spatial data. [8] extend the dynamic process convolution model to describe multivariate spatio-temporal processes. This convolution approach can also be used to model non-normal data. For example, [30] use the dynamic process convolution to model the location parameter of a generalized extreme value distribution to describe extreme levels of ozone at some monitoring stations in Mexico City.

4.4.2 Models based on stochastic partial differential equations

Commonly, when modeling spatio-temporal processes, scientific theory is available, e.g. through the specification of deterministic models, to describe a wide class of spatio-temporal processes in ecology, environment and epidemiology. This theory might serve as prior information to describe the process under study. [77] mention that one should make use of the information in the developed science but should also recognize that there are uncertainties associated with that knowledge. For example, [26] uses stochastic partial differential equations to develop general classes of covariance functions for spatial processes. [7] proposes a non-separable space-time covariance matrix based on a physical dispersion model, that is suitable to describe processes that spread or disperse over time, such as air pollution. [4] discusses the combination of physical reasoning and observational data through the use of Bayesian hierarchical models.

The DLM framework previously discussed can be used to accommodate the use of partial differential equations (PDEs) to account for scientific knowledge of the process. Moreover, it accounts for the uncertainty about the specification of the dynamics of the process, as the PDE is substituted by a stochastic PDE. Broadly speaking, this is done by allowing the state evolution to be defined through a partial differential equation model (e.g. reaction-diffusion processes), and the spatial process is approximated through a spatial grid and partial differential equations are solved using finite-difference methods [65]. [64] mentions that focus has been given on reaction-diffusion processes modeled via PDEs, integrodifference equations and discrete-time contact models. The choice among these models depend on whether one is considering discrete time and/or space [64].

Consider the stochastic process $\{Z(s,t), s \in D_s \subset \mathbb{R}, t \in D_t \subset \mathbb{R}\}$. [26] considers models for two-dimensional processes based on the general second-order linear SPDE which leads to three types of models: parabolic, elliptic and hyperbolic forms. Following [13], we focus on the parabolic SPDE [26, 32] which is governed by

$$\frac{\partial Z(s,t)}{\partial t} - \beta \frac{\partial^2 Z(s,t)}{\partial s^2} + \alpha Z(s,t) = \delta(s,t), \tag{4.40}$$

where $\alpha > 0$, $\beta > 0$, and $\delta(s,t)$ is a zero mean random error process. Although these models are difficult to deal with analytically, [26] shows that $Z(s,t)$ has a stationary covariance function and derives an analytical expression for the corresponding spatio-temporal correlation function. Equation (4.40) can be approximated through finite differences,

$$Z(s;t+\Delta_t) \approx \delta_1 Z(s,t) + \delta_2 Z(s+\Delta_s;t) + \delta_2 Z(s-\Delta_s;t), \tag{4.41}$$

by assuming that the first-order partial derivative in time is approximated with a forward difference, the partial second-order in one-dimensional space is approximated with a centered difference, with $\delta_1 = (1 - \alpha\Delta_t - 2\beta\Delta_t/\Delta_s^2)$, $\delta_2 = \beta\Delta_t/\Delta_s^2$, and $\alpha\Delta_t < 1$ and $2\beta\Delta_t/\Delta_s^2 < 1$.

According to equation (4.41), the future value of Z at location s is related to current values at s and neighboring spatial locations. If we discretize space by defining $D_s = \{s_0, \ldots, s_{p+1}\}$, where $s_j = s_0 + j\Delta_s$ and $j = 0, 1, \ldots, p+1$, for each $k = 1, \ldots, p$, we can write

$$Z(s_k; t+\Delta_t) = \delta_1 Z(s_k; t) + \delta_2 Z(s_{k+1}; t) + \delta_2 Z(s_{k-1}; t).$$

Defining $\mathbf{Z}_t = (Z(s_1;t),\ldots,Z(s_p;t))'$, and $\mathbf{Z}_t^{(b)} = (Z(s_0;t),Z(s_{p+1};t))'$, we have

$$\mathbf{Z}_{t+\Delta_t} = \mathbf{G}(\delta)\mathbf{Z}_t + \mathbf{G}^{(b)}(\delta_2)\mathbf{Z}_t^{(b)}, t = 0,1,\cdots \tag{4.42}$$

where $\mathbf{G}(\delta)$ is a $p \times p$, tridiagonal propagator matrix, and $\mathbf{G}^{(b)}(\delta_2)$ is a $p \times 2$ boundary propagator matrix. Let $D_t = \{0,1,2,\cdots\}$, then from a given initial condition $\mathbf{Z}_0$ and boundary conditions $\{\mathbf{Z}_t^{(b)} : t = 0,1,\ldots,\}$ we can obtain a numerical solution to the difference equation in (4.41), which in turn is an approximation to the SPDE in (4.40). If a mean-zero, random error component ϵ_{t+1}, is added to the right-hand side of equation (4.42) we obtain a stochastic spatio-temporal difference equation corresponding to the parabolic SPDE in equation (4.40). We assume that $\{\epsilon_t\}$ is i.i.d. [62] calls this a *diffusion-injection difference equation*. As the boundary term is given, this stochastic difference equation can be seen as a multivariate first-order auto-regressive model.

The idea above can be considered to model counts that vary across space and time. [64] proposes a Poisson-lognormal mixture model to estimate the abundance of House Finches over the eastern USA. The proposed model is similar to the one shown in equations (4.33) with an additional component in the $\log(\lambda_t(s))$ that follows a reaction-diffusion model, which is approximated through first-order forward differences in time and centered differences in space.

4.4.3   Models based on integro-difference equations

As described in [13, Section 7.2] integro-difference equations (IDEs) might be useful in describing spatio-temporal processes that have long-range dependence, requiring a more general form of the propagator matrix $\mathbf{G}$. When time is discrete and space continuous, in the case of linear dynamics, the IDE is given by

$$Z_t(s) = \int_{D_s} g(s,\mathbf{x};\delta_s)Z_{t-1}(\mathbf{x})d\mathbf{x}, \quad s \in D_s \tag{4.43}$$

where D_s is the spatial domain of interest. The function $g(s,\mathbf{x};\delta)$ is called a *redistribution kernel* which can have a different shape for each spatial location s, controlled by parameters δ_s. If we add a random noise term, $\omega_t(s)$, which is independent in time but may be correlated in space and is independent of $Z_{t-1}(s)$, such that

$$Z_t(s) = \int_{D_s} g(s,\mathbf{x};\delta_s)Z_{t-1}(\mathbf{x})d\mathbf{x} + \omega_t(s),$$

a stochastic IDE results.

As proposed by [87], typically one has to discretize the integral equation to reformulate the model. Considering a discretized grid, the kernel defined for locations s_i, $g(s_i,\mathbf{x},\delta_s)$ corresponds to the $i-th$ row of the propagator matrix $\mathbf{G}(\delta)$, such that

$$\mathbf{Z}_t = \mathbf{G}(\delta)\mathbf{Z}_{t-1} + \omega_t$$

where $\omega_t \sim N(\mathbf{0},\mathbf{R})$.

This discretized version of the model is suitable to achieve state-space dimension reduction and can be put onto a state-space framework [12, 87]. [13] mentions that IDE dynamical models can accommodate complicated spatio-temporal dynamics with relatively few parameters. This IDE approach is naturally extented to accommodate non-normal processes. [83] assumes a conditional spatio-temporal Poisson intensity process for cloud intensity data, and uses the kernel-based integro-difference approach as a latent component in the mean of the logarithm of the Poisson intensity process.

The processes described in the previous subsections can be incorporated as components in hierarchical models to accommodate more complex structures. Typically, inference procedure is performed under the Bayesian paradigm. See [13] and, references therein, for more details.

Acknowledgments

A. M. Schmidt was partially supported by CNPq and *Fundação de Amparo à Pesquisa do Estado do Rio de Janeiro* (FAPERJ), Brazil, and the Natural Sciences and Engineering Research Council (NSERC) of Canada. The research of H. F. Lopes is partially financed by Insper, CNPq and FAPESP, Brazil.

Bibliography

[1] D. K. Agarwal, A. E. Gelfand, and S. Citron-Pousty. Zero inflated models with application to spatial count data. *Environmental and Ecological Statistics*, 9:341–355, 2002.

[2] S. Banerjee, B. P. Carlin, and A. E. Gelfand. *Hierarchical modeling and analysis for spatial data*. CRC Press, 2014.

[3] L. M. Berliner. Hierarchical Bayesian time series models. In K. Hanson and R. Silver, editors, *Maximum Entropy and Bayesian Methods*, pages 15–22. Kluwer Acad., Norwell, Mass, 1996.

[4] L. M. Berliner. Physical-statistical modeling in geophysics. *Journal of Geophysical Research: Atmospheres*, 108(D24), 2003.

[5] J. Besag. Spatial interaction and the statistical analysis of lattice systems (with discussion). *Journal of the Royal Statistical Society. Series B (Methodological)*, pages 192–236, 1974.

[6] N. E. Breslow. Extra-Poisson variation in log-linear models. *Journal of the Royal Statistical Society, Series C (Applied Statistics)*, 33:38–44, 1984.

[7] P. E. Brown, G. O. Roberts, K. F. Kåresen, and S. Tonellato. Blur-generated non-separable space–time models. *Journal of the Royal Statistical Society: Series B (Statistical Methodology)*, 62(4):847–860, 2000.

[8] C. A. Calder. Dynamic factor process convolution models for multivariate space–time data with application to air quality assessment. *Environmental and Ecological Statistics*, 14(3):229–247, 2007.

[9] C. K. Carter and R. Kohn. On Gibbs sampling for state space models. *Biometrika*, 81:541–553, 1994.

[10] C. M. Carvalho and M. West. Dynamic matrix-variate graphical models. *Bayesian Analysis*, 2:69–98, 2007.

[11] P. Craigmile. Time series methodology. In A. E Gelfand, M. Fuentes, J. Hoeting, and R. Smith, editors, *Handbook of Environmental and Ecological Statistics*, Chapman & Hall, 2018.

[12] N. Cressie and C. K. Wikle. Space-time Kalman filter. *Encyclopedia of Environmetrics*, 2002.

[13] N. Cressie and C. K. Wikle. *Statistics for spatio-temporal data*. John Wiley & Sons, 2015.

[14] A.P. Dawid. Some matrix-variate distribution theory: Notational considerations and a Bayesian application. *Biometrika*, 68:265–274, 1981.

[15] T.J. DiCiccio, R.A. Kass, A.E. Raftery, and L. Wasserman. Computing Bayes' factors by combining simulation and asymptotic approximations. *Journal of the American Statistical Association*, 92:903–915, 1997.

[16] F. Dominici and A. Wilson. Dynamics of environmental epidemiology. In A. E. Gelfand, M. Fuentes, J. Hoeting, and R. Smith, editors, *Handbook of Environmental and Ecological Statistics*, Chapman & Hall, 2018.

[17] J. Durbin and S.J. Koopman. *Time Series Analysis by State Space Methods*. Oxford Statistical Science Series. Clarendon Press, 2001.

[18] L. Fahrmeir. Posterior mode estimation by extended Kalman filtering for multivariate dynamic linear models. *Journal of the American Statistical Association*, 87:501–509, 1992.

[19] S. Frühwirth-Schnater. Data augmentation and dynamic linear models. *Journal of Time Series Analysis*, 15(2):183–202, 1994.

[20] D. Gamerman. Markov chain Monte Carlo for dynamic generalised linear models. *Biometrika*, 85(1):215–227, 1998.

[21] D. Gamerman and H. S. Migon. Dynamic hierarchical models. *Journal of the Royal Statistical Society, Series B*, 55:629–642, 1993.

[22] A. E. Gelfand, S. Banerjee, and D. Gamerman. Spatial process modelling for univariate and multivariate dynamic spatial data. *Environmetrics*, 16(5):465–479, 2005.

[23] A. E. Gelfand, A. Kottas, and S. N. MacEachern. Bayesian nonparametric spatial modeling with Dirichlet process mixing. *Journal of the American Statistical Association*, 100:1021–35, 2005.

[24] J. Geweke and H. Tanizaki. Bayesian estimation of state space models using Metropolis-Hastings algorithm within Gibbs sampling. *Computational Statistics and Data Ananlysis*, 37:151–170, 2001.

[25] G. H. Golub and C. F. Van Loan. *Matrix Computations*. The Johns Hopkins University Press, Third edition, 1996.

[26] V Heine. Models for two-dimensional stationary stochastic processes. *Biometrika*, 42(1-2):170–178, 1955.

[27] D. Higdon. A process-convolution approach to modelling temperatures in the North Atlantic Ocean. *Environmental and Ecological Statistics*, 5(2):173–190, 1998.

[28] D. Higdon. Space and space-time modeling using process convolutions. In Anderson C., Barnett V., Chatwin P.C., and El-Shaarawi A.H., editors, *Quantitative methods for current environmental issues*, pages 37–56. Springer Verlag, London, 2002.

[29] D. Higdon, J. Swall, and J. Kern. Non-stationary spatial modeling. In J. M. Bernardo, J. O. Berger, A. P. Dawid, and A. F. M. Smith, editors, *Bayesian statistics 6: Proceedings of the Sixth Valencia International Meeting*, pages 761–768. Oxford University Press, 1999.

[30] G. Huerta and B. Sansó. Time-varying models for extreme values. *Environmental and Ecological Statistics*, 14(3):285–299, 2007.

[31] G. Huerta, B. Sansó, and J. R. Stroud. A spatiotemporal model for mexico city ozone levels. *Journal of the Royal Statistical Society: Series C (Applied Statistics)*, 53(2):231–248, 2004.

[32] R. H. Jones and Y. Zhang. Models for continuous stationary space-time processes. In *Modelling longitudinal and spatially correlated data*, pages 289–298. Springer, 1997.

[33] G. Kitagawa. Non-Gaussian state-space modeling of non-stationary time series. *Journal of the American Statistical Association*, 82:1032–1041, 1987.

[34] G. Koop and D. Korobilis. Large time-varying parameter VARs. *Journal of Econometrics*, 177:184–198, 2013.

[35] F. Landim and D. Gamerman. Dynamic hierarchical models: an extension to matrix-variate observations. *Computational Statistics & Data Analysis*, 35:11–42, 2000.

[36] J.F. Lawless. Negative binomial and mixed Poisson regression. *Canadian Journal of Statistics*, 15:209–225, 1987.

[37] D.V. Lindley and A.F.M. Smith. Bayes estimates for the linear model. *Journal of the Royal Statistical Society. Series B*, pages 1–41, 1972.

[38] F. Liu. *Bayesian Time Series: Analysis Methods Using Simulation Based Computation*. PhD thesis, Institute of Statistics and Decision Sciences, Duke University, Durham, North Carolina, USA., 2000.

[39] H. F. Lopes, E. Salazar, and D. Gamerman. Spatial dynamic factor analysis. *Bayesian Analysis*, 3(4):759–792, 2008.

[40] H. S. Migon, A. M. Schmidt, R. E. R. Ravines, and J. B. M. Pereira. An efficient sampling scheme for dynamic generalized models. *Computational Statistics*, 28(5):2267–2293, 2013.

[41] J. Nakajima, M. Kasuya, and T. Watanabe. Bayesian analysis of time-varying parameter vector autoregressive model for the Japanese economy and monetary policy. *Journal of The Japanese and International Economies*, 25:225–245, 2011.

[42] A. A. Nobre, B. Sansó, and A. M. Schmidt. Spatially varying autoregressive processes. *Technometrics*, 53(3):310–321, 2011.

[43] A. A. Nobre, A. M. Schmidt, and H. F. Lopes. Spatio-temporal models for mapping the incidence of malaria in Pará. *Environmetrics*, 16(3):291–304, 2005.

[44] D. Peña and P. Poncela. Forecasting with nonstationary dynamic factor models. *Journal of Econometrics*, 119:291–321, 2004.

[45] G. Petris, S. Petrone, and P. Campagnoli. *Dynamic linear models*. Springer, 2009.

[46] A. Pole, M. West, and J. Harrison. *Applied Bayesian forecasting and time series analysis*. CRC Press, 1994.

[47] R. Prado and M. West. *Time series: modeling, computation, and inference*. CRC Press, 2010.

[48] G. E. Primiceri. Time varying structural vector autoregressions and monetary policy. *Review of Economic Studies*, 72:821–852, 2005.

[49] J.M. Quintana and M. West. Multivariate time series analysis: new techniques applied to international exchange rate data. *The Statistician*, 36:275–281, 1987.

[50] R Development Core Team. *R: A Language and Environment for Statistical Computing*. R Foundation for Statistical Computing, Vienna, Austria, 2008. ISBN 3-900051-07-0.

[51] R. R. Ravines, A. M. Schmidt, H. S. Migon, and C. D. Rennó. A joint model for rainfall-runoff: The case of Rio Grande basin. *Journal of Hydrology*, 353(1):189 – 200, 2008.

[52] B. Sansó, A. M. Schmidt, and A. A. Nobre. Bayesian spatio-temporal models based on discrete convolutions. *Canadian Journal of Statistics*, 36(2):239–258, 2008.

[53] A. M. Schmidt, P. Guttorp, and A. O'Hagan. Considering covariates in the covariance structure of spatial processes. *Environmetrics*, 22(4):487–500, 2011.

[54] A. M. Schmidt, J. A. Hoeting, J. B. M. Pereira, and P. P. Vieira. Mapping malaria in the Amazon rainforest: a spatio-temporal mixture model. In A. O'Hagan and M. West, editors, *The Oxford Handbook of Applied Bayesian Analysis*, pages 90–117. Oxford University Press, 2010.

[55] A. M. Schmidt and J. B. M. Pereira. Modelling time series of counts in epidemiology. *International Statistical Review*, 79(1):48–69, 2011.

[56] N. Shephard and M. Pitt. Likelihood analysis of non-Gaussian measurement time series. *Biometrika*, 84:653–667, 1997.

[57] J.R. Stroud, P. Müller, and B. Sansó. Dynamic models for spatiotemporal data. *Journal of the Royal Statistical Society: Series B (Statistical Methodology)*, 63(4):673–689, 2001.

[58] J.R. Stroud, M.L. Stein, B.M. Lesht, D.J. Schwab, and D. Beletsky. An ensemble Kalman filter and smoother for satellite data assimilation. *Journal of the American Statistical Association*, 105(491):978–990, 2010.

[59] J. C. Vivar and M. A. R. Ferreira. Spatiotemporal models for Gaussian areal data. *Journal of Computational and Graphical Statistics*, pages 658–674, 2012.

[60] M. West and J. Harrison. *Bayesian Forecasting and Dynamic Models*. Springer Series in Statistics, second edition, 1997.

[61] M. West, J. Harrison, and H. S. Migon. Dynamic generalized linear models and Bayesian forecasting. *Journal of the American Statistical Association*, 80(389):73–83, 1985.

[62] P. Whittle. *Systems in stochastic equilibrium*. John Wiley & Sons, Inc., 1986.

[63] C. K. Wikle. A kernel-based spectral model for non-Gaussian spatio-temporal processes. *Statistical Modelling*, 2(4):299–314, 2002.

[64] C. K. Wikle. Hierarchical Bayesian models for predicting the spread of ecological processes. *Ecology*, 84(6):1382–1394, 2003.

[65] C. K. Wikle and N. Cressie. A dimension-reduced approach to space-time Kalman filtering. *Biometrika*, 86(4):815–829, 1999.

[66] C. K Wikle and M. B. Hooten. A general science-based framework for dynamical spatio-temporal models. *Test*, 19(3):417–451, 2010.

5

Geostatistical Modeling for Environmental Processes

Sudipto Banerjee

UCLA, University of California, Los Angeles

CONTENTS

5.1 Introduction

Geographical referencing of environmental data is often achieved using a coordinate system such as longitude-latitude or some planar projections. Each variable of interest is measured at a point whose coordinates are assumed to be fixed. Such data are referred to as *point-referenced* data. This is to be contrasted with *areally-referenced* spatial data which arise as aggregates or summaries over regions or areas such as states, counties, zip-codes, census tracts and so on. Point-referenced data are often referred to as *geostatistical* data due to their long history in geological sciences and are perhaps most conspicuous in the environmental sciences today. This chapter presents some essential exploratory and modeling tools for point-referenced data. More detailed and comprehensive treatments for spatial modeling and data analysis can be found in numerous texts such as [8], [7], [16], and [3].

The fundamental concept underlying statistical inference for environmental data is a *stochastic process* (see, e.g., [18]). A stochastic process is a (possibly uncountable) collection of random variables indexed over space and/or time. We usually denote this by $\{W(\mathbf{s}) : \mathbf{s} \in \mathcal{D}\}$, where $\mathcal{D}$ is usually a fixed subset of $\Re^d$, the d-dimensional Euclidean space. For example, stochastic processes in the time series literature usually are defined over the positive real line. In the spatial context, usually we encounter d to be 2 (say, northings and eastings) or 3 (e.g., northings, eastings, and altitude above sea level). For situations where $d > 1$, the process is often referred to as a *spatial process*. For example, $W(\mathbf{s})$ may represent the level of temperature at a location $\mathbf{s}$. Conceptually, it is sensible to assume the existence of temperature measurements over the entire domain over an uncountable set of locations. In practice, temperature data will be available to us as measurements only from a finite set of locations. We envision temperature as a spatial process and the data to be a *partial realization* of that process.

The statistician seeks inference about the entire spatial process $W(\mathbf{s})$ including interpolation and prediction at arbitrary locations based upon observing only a partial realization of the process. This is usually more challenging than customary statistical inference for a random sample from one population. A key element of a well-defined spatial process is that any partial realization, i.e., any finite subset of random variables, can be described by a valid finite-dimensional joint distribution. This remarkable feature allows us to introduce spatial association through *structured* dependence and carry out full inference about the surface at an uncountable number of locations, in spite of having seen the process only at a finite set of locations.

This chapter focuses upon some key elements in constructing spatial processes. We discuss features such as covariance functions, stationarity, isotropy, and classical geostatistical tools such as the variogram to explore point-referenced data. There is an enormous, and still growing, literature on these topics that will be impossible to cover in one chapter. Further details can be found in the aforementioned books and references therein.

5.2 Elements of point-referenced modeling

5.2.1 Spatial processes, covariance functions, stationarity and isotropy

For our discussion we assume that our spatial process has a mean, say $\mu(\mathbf{s}) = E[W(\mathbf{s})]$, associated with it and that the variance of $W(\mathbf{s})$ exists for all $\mathbf{s} \in \mathcal{D}$. As mentioned above, a spatial process is described by the joint distribution of its finite-dimensional realizations. For example, the process $W(\mathbf{s})$ is said to be *Gaussian* if, for any $n \geq 1$ and any set of sites $\{\mathbf{s}_1, \ldots, \mathbf{s}_n\}$, $\mathbf{W} = (W(\mathbf{s}_1), \ldots, W(\mathbf{s}_n))^T$ has a multivariate normal distribution. A multivariate normal distribution is completely specified by its mean vector and covariance matrix. Since a Gaussian process is endowed with a multivariate normal distribution for any finite realization, the process itself will be specified by a mean function and covariance function. We write $W(\mathbf{s}) \sim GP(\mu(\mathbf{s}), C(\cdot, \cdot))$, where $C(\mathbf{s}, \mathbf{t}) = \text{Cov}(W(\mathbf{s}), W(\mathbf{t}))$ for any $\mathbf{s}, \mathbf{t} \in \mathcal{D}$. This implies $\mathbf{W} \sim N(\boldsymbol{\mu}, C_{\mathbf{W}})$, where $\boldsymbol{\mu}$ is the $n \times 1$ vector with i-th element $\mu(\mathbf{s}_i)$ and $C_{\mathbf{W}}$ is the $n \times n$ covariance matrix for $\mathbf{W}$ with (i, j)-th entry $C(\mathbf{s}_i, \mathbf{s}_j)$.

Any valid covariance function $C(\mathbf{s}, \mathbf{t})$ must ensure that the resulting covariance matrix $C_{\mathbf{W}}$ is positive definite. This is a consequence of the simple fact that the variance of any (non-degenerate) random variable must be positive. Therefore,

$$0 < \text{Var}\left(\sum_{i=1}^{n} a_i W(\mathbf{s}_i)\right) = \sum_{i,j=1}^{n} a_i C(\mathbf{s}_i, \mathbf{s}_j) a_j = \mathbf{a}^T C_{\mathbf{W}} \mathbf{a} \, \forall \mathbf{a} \in \Re^d \setminus \{0\}, \qquad (5.1)$$

where a_i's are scalars and $\mathbf{a}$ is the $n \times 1$ vector of a_i's. We require the above inequality to be strict, i.e., $C_{\mathbf{W}}$ be positive definite whenever the $\mathbf{s}_i$'s are distinct. For the process to be valid, the covariance function needs to satisfy (5.1) for *any* finite, but otherwise arbitrary, set of locations. To find such functions we first consider *stationarity* of the process.

A spatial stochastic process, Gaussian or otherwise, is said to be *strictly stationary* (or *strong* stationarity) if, for any given $n \geq 1$, any set of n sites $\{\mathbf{s}_1, \ldots, \mathbf{s}_n\}$ and any $\mathbf{h} \in \Re^d$, the distribution of $(W(\mathbf{s}_1), \ldots, W(\mathbf{s}_n))$ is the same as that of $(W(\mathbf{s}_1 + \mathbf{h}), \ldots, W(\mathbf{s}_n + \mathbf{h}))$. A spatial process is called *weakly stationary* or *second order stationary* if (i) $\mu(\mathbf{s}) \equiv \mu$ (i.e., it has a constant mean) and (ii) the covariance function $C(\mathbf{s}, \mathbf{s} + \mathbf{h}) = C(\mathbf{h})$ depends only upon the separation vector $\mathbf{h}$ for all $\mathbf{h} \in \Re^d$ such that $\mathbf{s}$ and $\mathbf{s} + \mathbf{h}$ both lie within $\mathcal{D}$. In practice, the spatial process is often used as a 0 mean residual process after adjusting for first-order trends in the means (e.g., using predictors or explanatory variables). Therefore, the first condition is often implicit and stationarity of the process is ascertained from the nature of the covariance function. If we assume that all variances exist, then strong stationarity implies weak stationarity. The converse is not true in general, but does hold for Gaussian processes.

If $\{W(\mathbf{s}) : \mathbf{s} \in \mathcal{D}\}$ is a weakly stationary spatial process, then the covariance matrix for the joint distribution of $\mathbf{W} = (W(\mathbf{s}_1), W(\mathbf{s}_2), \ldots, W(\mathbf{s}_n))^T$ is given by the matrix $C_{\mathbf{W}} = \{C(\mathbf{s}_i - \mathbf{s}_j)\}$ with (i, j)-th entry $C(\mathbf{s}_i - \mathbf{s}_j)$. A remarkable result in classical harmonic analysis attributed to Bochner (1933, 1955) says that a real-valued continuous function is a valid covariance function if and only if it is a characteristic function (Fourier transform) of a random variable symmetric about zero. That any characteristic function satisfies (5.1) is particularly easy to verify for all complex (which subsumes real) scalars a_i. Let $\mathbf{X}$ be a d-variate random variable and let $C(\mathbf{h}) = \mathrm{E}[e^{\iota \mathbf{h}^T \mathbf{X}}]$ be its characteristic function, where $\iota = \sqrt{-1}$ and argument $\mathbf{h} \in \mathfrak{R}^d$. Then,

$$\sum_{i,j=1}^{n} a_i C(\mathbf{s}_i - \mathbf{s}_j) \bar{a}_j = \sum_{i,j=1}^{n} a_i \mathrm{E}[e^{\iota(\mathbf{s}_i - \mathbf{s}_j)^T \mathbf{X}}] \bar{a}_j = \mathrm{E}\left[\sum_{i,j=1}^{n} a_i e^{\iota \mathbf{s}_i^T \mathbf{X}} e^{-\iota \mathbf{s}_j^T \mathbf{X}} \bar{a}_j\right]$$

$$= \mathrm{E}[Z\bar{Z}] = \mathrm{E}[|Z|^2] \geq 0 ,$$

where $Z = \sum_{i=1}^{n} a_i e^{\iota \mathbf{s}_i^T \mathbf{X}}$ and $\bar{Z}$ denotes its complex conjugate. Thus, $C(\mathbf{h})$ is a positive definite function. When the distribution of $\mathbf{X}$ is symmetric about zero, then $C(\mathbf{h})$ is real-valued. Therefore, one way to construct a stationary covariance function is to start with any bounded, positive measure $F(\mathbf{w})$ on $\mathfrak{R}^d$ that is symmetric about $\mathbf{0}$ and let $C(\mathbf{h})$ be its Fourier transform. This yields

$$C(\mathbf{h}) = \int_{\mathfrak{R}^d} e^{\iota \mathbf{h}^T \mathbf{w}} F(d\mathbf{w}) = \int_{\mathfrak{R}}^{d} \cos(\mathbf{h}^T \mathbf{w}) F(d\mathbf{w}) . \tag{5.2}$$

Since $e^{\iota \mathbf{w}^T \mathbf{h}} = \cos(\mathbf{h}^T \mathbf{w}) + \iota \sin(\mathbf{h}^T \mathbf{w})$, the imaginary term disappears due to the symmetry of F around 0. From (5.2), it is evident that $C(\mathbf{0}) = \int_{\mathfrak{R}}^{d} F(d\mathbf{w})$ is the normalizing constant and $F(d\mathbf{w})/C(\mathbf{0})$ is called the *spectral distribution* that induces $C(\mathbf{h})$. If $F(d\mathbf{w})$ admits a density $f(\mathbf{w})$ with respect to Lebesque measure, i.e., $f(\mathbf{w})d\mathbf{w} = F(\mathbf{w})$, then $f(\mathbf{w})$ is called the *spectral density* and (5.2) can be expressed as

$$C(\mathbf{h}) = \int_{\mathfrak{R}}^{d} \cos(\mathbf{h}^T \mathbf{w}) f(\mathbf{w}) d\mathbf{w} . \tag{5.3}$$

If $C(\mathbf{h})$ is integrable over $\mathfrak{R}^d$, then the spectral density can be obtained from $C(\mathbf{h})$ using the *inversion formula* $f(\mathbf{w}) = (1/(2\pi)^d) \int_{\mathfrak{R}}^{d} \cos(\mathbf{h}^T \mathbf{w}) C(\mathbf{h}) d\mathbf{h}$.

A further simplification over weakly stationary processes is obtained when the covariance function $C(\mathbf{h})$ depends upon the separation vector $\mathbf{h}$ only through its Euclidean length $h = \|\mathbf{h}\|$. The covariance function is then called *isotropic* and, sometimes, the process itself is also described by the same name. These are the most frequently adopted choice within stationary covariance functions. [30] investigated the representation in (5.2) for isotropic covariance functions and showed that every isotropic *correlation function* $\rho(h) = C(h)/C(0)$ (note that $\rho : [0, \infty) \to \mathfrak{R}^1$ and $\rho(0) = 1$) over $\mathfrak{R}^1$ can be expressed as

$$\rho(h, \phi) = \int_{0}^{\infty} \Omega_r(zh) G_\phi(dz) , \tag{5.4}$$

where G_ϕ is nondecreasing integrable, $\Omega_r(x) = \left(\frac{2}{x}\right)^{\frac{r-2}{2}} \Gamma\left(\frac{r}{2}\right) J_{\left(\frac{r-2}{2}\right)}(x)$, and $J_\upsilon(\cdot)$ is the Bessel function of the first kind of order υ (see, e.g., [2]). For $r = 1, \Omega_1(x) = \cos(x)$;

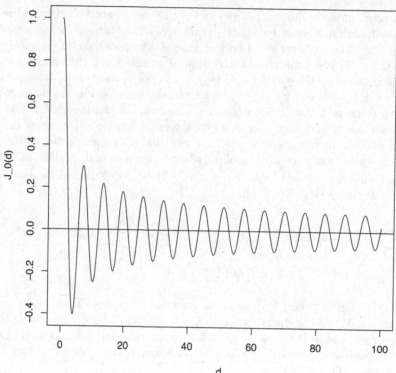

FIGURE 5.1

A plot of $J_0(d)$ out to $d = 100$.

for $r = 2, \Omega_2(x) = J_0(x)$; for $r = 3, \Omega_3(x) = \sin(x)/x$; for $r = 4, \Omega_4(x) = \frac{2}{x}J_1(x)$; and for $r = \infty, \Omega_\infty(x) = \exp(-x^2)$. Specifically, $J_0(x) = \sum_{k=0}^{\infty} \frac{(-1)^k}{k!^2} \left(\frac{x}{2}\right)^{2k}$ and $\rho(d, \phi) = \int_0^\infty J_0(zd)dG_\phi(z)$ provides the class of all permissible correlation functions in $\Re^2$. Note that $J_0(x) = \sum_{k=0}^{\infty} \frac{(-1)^k}{(k!)^2} \left(\frac{x}{2}\right)^{k/2}$. J_0 decreases from 1 at $x = 0$ and will oscillate above and below 0 with amplitudes and frequencies that are diminishing as x increases. Figure 5.1 provides a plot of $J_0(x)$ versus x, revealing that it is not monotonic. This is necessary for $\rho(h, \phi)$ above to capture all correlation functions in $\Re^2$. Typically, correlation functions that are monotonic and decreasing to 0 are chosen but, apparently, valid correlation functions can permit negative associations in distance space.

It is not difficult to see that isotropic correlation functions valid over $\Re^d$ will also be valid over dimensions less than d. However, the converse is not necessarily true: covariance functions that are valid on $\Re^d$ need not be valid on $\Re^{d+1}$ or higher dimensions. A classic example is the isotropic triangular (or tent-shaped) covariance function that is valid in one-dimension, but not in higher dimensions (see, e.g., p71 of [3]). In practical settings, it usually suffices to restrict attention to a fixed dimension d and choose valid correlation functions in $\Re^d$. Nevertheless, it is instructive to note that covariance functions may exist over all dimensions. In fact, [30] further showed that a function is a valid isotropic correlation function over all dimensions if and only it is of the form $\rho(h) = \int_0^\infty \exp(-u^2 h^2)F(du)$, where F is a probability measure over $[0, \infty)$.

The characterizations in (5.2) and (5.3) offer some simple strategies to construct valid covariance functions. Three approaches are mixing, products, and convolution. Mixing notes simply that if $C_1, \ldots, C_m$ are valid correlation functions in $\Re^d$ and if $\sum_{i=1}^{m} p_i = 1$, $p_i > 0$, then $C(\mathbf{h}) = \sum_{i=1}^{m} p_i C_i(\mathbf{h})$ is also a valid correlation function in $\Re^d$. This follows since $C(\mathbf{h})$

is the characteristic function associated with $\sum p_i f_i(\mathbf{x})$, where $f_i(\mathbf{x})$ is the symmetric (about 0) density in d-dimensional space associated with $C_i(\mathbf{h})$. In fact, the sum $\sum_{i=1}^{\infty} a_i C_i(\mathbf{h})$ yields a valid covariance function as well, provided the a_i are all greater than 0 and $\sum_{i=1}^{\infty} a_i < \infty$. Using product forms simply notes that again if $C_1, \ldots, C_m$ are valid in $\Re^r$, then $\prod_{i=1}^{m} C_i$ is a valid correlation function in $\Re^d$. This follows since $\prod_{i=1}^{m} C_i(\mathbf{h})$ is the characteristic function associated with $Z = \sum_{i=1}^{m} X_i$ where the X_i's are independent of each other and have respective characteristic function $C_i(\mathbf{h})$.

The above theoretical results reveal several strategies to construct valid covariance functions, but how do we choose among them in practice? Usual model selection criteria will typically struggle to distinguish among different correlation functions. For example, [12] provide some graphical illustration showing that, through suitable alignment of parameters, the correlation curves from one-parameter isotropic scale choices such as the exponential, Gaussian, or Cauchy will be very close to each other. An alternative perspective is to make the selection based upon theoretical considerations. This possibility arises from the powerful fact that the choice of correlation function determines the smoothness of realizations from the spatial process. More precisely, a process realization is viewed as a random surface over the region. The choice of the covariance function can ensure that these realizations will be almost surely continuous, or mean square continuous, or mean square differentiable, and so on. Elegant theory, developed in [21], [18] and extended in [4], clarifies the relationship between the choice of correlation function and such smoothness.

A particularly popular choice for building spatial models is the Matérn class of covariance functions ([1]; [18]) which arises as the characteristic function from a Cauchy spectral density and takes the form

$$C(h) = \begin{cases} \frac{\sigma^2}{2^{\nu-1}\Gamma(\nu)} (h\phi)^{\nu} K_{\nu}(h\phi) & \text{if } h > 0 \\ \tau^2 + \sigma^2 & h = 0 \end{cases}, \tag{5.5}$$

where $\Gamma(\cdot)$ is the Eulerian Gamma function and K_{ν} is the modified Bessel function of order ν. Implementations of this function are available in several C/C++ libraries and also in R packages such as spBayes and geoR. Note that special cases of the above are the exponential ($\nu = 1/2$) and the Gaussian ($\nu \to \infty$). The parameters σ^2 and τ^2 are variance parameters. The latter, often referred to as the *nugget* captures variation at very small scales, perhaps arising from measurement error, while the former can be regarded as the spatial variance component and is called the *partial sill*; the sum $\sigma^2 + \tau^2$ is called the *sill*. The parameter ϕ controls the rate at which the spatial correlation decays with increase distance. These parameters are revisited in the context of *variograms* in the next section. The parameter ν is, in fact, a smoothness parameter. In two-dimensional space, the greatest integer in ν indicates the number of times the process realizations will be mean square differentiable. In fact, the use of the Matérn covariance function as a model enables the data to inform about ν; we can legitimately infer about process smoothness despite observing the process at only a finite number of locations (see [6]).

Table 5.1 presents a selection of some of the more commonly used isotropic covariance functions. The spherical covariance function is valid in $d = 1, 2$, or 3 dimensions, but for $\Re^4$ or higher it fails to correspond to a spatial variance matrix that is positive definite (as required to specify a valid joint probability distribution). Special cases of (5.5) are the exponential ($\nu = 1/2$) and the Gaussian ($\nu \to \infty$). Since $\nu = \infty$ corresponds to the Gaussian correlation function, the implication is that the use of the Gaussian correlation function results in process realizations that are mean square analytic, which may be too smooth to be appropriate in practice. At $\nu = 3/2$, a closed form expression is obtained, which is used in spatial gradient analysis or "wombling" ([6] and [5]) since the process realizations are mean-square differentiable once but no more. The wave form in Table 5.1 is an example of a valid covariance function that is not monotonic.

TABLE 5.1
Summary of common isotropic parametric covariance functions.

Model	Covariance function, $C(h)$
Spherical	$C(h) = \begin{cases} 0 & \text{if } h \geq 1/\phi \\ \sigma^2\left[1 - \frac{3}{2}\phi h + \frac{1}{2}(\phi h)^3\right] & \text{if } 0 < h \leq 1/\phi \\ \tau^2 + \sigma^2 & h = 0 \end{cases}$
Exponential	$C(h) = \begin{cases} \sigma^2 \exp(-\phi h) & \text{if } h > 0 \\ \tau^2 + \sigma^2 & h = 0 \end{cases}$
Powered exponential	$C(h) = \begin{cases} \sigma^2 \exp(-|\phi h|^p) & \text{if } h > 0 \\ \tau^2 + \sigma^2 & h = 0 \end{cases}$
Gaussian	$C(h) = \begin{cases} \sigma^2 \exp(-\phi^2 h^2) & \text{if } h > 0 \\ \tau^2 + \sigma^2 & h = 0 \end{cases}$
Rational quadratic	$C(h) = \begin{cases} \sigma^2\left(1 - \frac{h^2}{(\phi + h^2)}\right) & \text{if } h > 0 \\ \tau^2 + \sigma^2 & h = 0 \end{cases}$
Wave	$C(h) = \begin{cases} \sigma^2 \frac{\sin(\phi h)}{\phi h} & \text{if } h > 0 \\ \tau^2 + \sigma^2 & h = 0 \end{cases}$
Matérn	$C(h) = \begin{cases} \frac{\sigma^2}{2^{\nu-1}\Gamma(\nu)}(\phi h)^\nu K_\nu(\phi h) & \text{if } h > 0 \\ \tau^2 + \sigma^2 & h = 0 \end{cases}$
Matérn at $\nu = 3/2$	$C(h) = \begin{cases} \sigma^2(1 + \phi h)\exp(-\phi h) & \text{if } h > 0 \\ \tau^2 + \sigma^2 & h = 0 \end{cases}$

5.2.2 Anisotropy and nonstationarity

One can extend stationary correlation functions from isotropy, where association depends only upon distance, to *anisotropy* where association depends upon direction also. One particularly simple example is the class where we *separate* the coordinates of the separation vector and assign a possibly different correlation function to each coordinate. Therefore, if the separation vector is $\mathbf{h} = (h_1, h_2)$, then a valid stationary (but not isotropic) covariance function is $C(\mathbf{h}) = \sigma^2 \rho_1(h_1)\rho_2(h_2)$, where $\rho_1(\cdot)$ and $\rho_2(\cdot)$ are valid correlation functions on $\Re^1$. If ρ_1 and ρ_2 are not identical, then switching h_1 and h_2 will produce different values for the correlation although $\|\mathbf{h}\|$ will not change. One particular choice is to take $\rho_1(h) = e^{-\phi_1 h}$ and $\rho_2(h) = e^{-\phi_2 h}$. This form has been used to construct Gaussian process approximations for computer model output (see, e.g., [29] and [26] and references therein) and also to achieve computational benefits in the analysis of large spatial datasets in agricultural trials (e.g., [23] and [13]). These component-wise separable structures may be appropriate for modeling response surfaces over the often high-dimensional space of inputs in computer models because the inputs reside on their own subspaces with their own scales. In spatial settings, however, the coordinates have the same scale so component-wise dependence structures may be less appropriate.

Another popular class of anisotropic models arises by replacing the usual Euclidean distance in an isotropic correlation function with an ellipsoidal distance. For example, if $C(h) = \sigma^2 \rho(h; \phi)$ is a valid isotropic covariance function in $\Re^r$ (say, from Table 5.1), then we can construct an anisotropic model as $C(\mathbf{h}) = \sigma^2 \rho(\mathbf{h}^T B\mathbf{h}; \phi)$, where B is any $r \times r$ positive definite matrix. When $r = 2$ we obtain a specification with three parameters rather than one. Contours of constant association arising from this $C(\mathbf{h})$ are ellipses and one obtains effective ranges in each spatial direction. These functions are said to capture *geometric anisotropy*, which assumes than the spatial field is obtained by a geometric linear transformation of the original coordinates. [12] provide the details for Bayesian modeling and inference using geometric anisotropic covariance models and discuss some further extensions; also see Chapter 2 of [3] for further details on geometric anisotropy.

Moving one step further, one may wish to shed the assumption of stationarity altogether and construct covariance functions $C(\mathbf{s}, \mathbf{s}')$ that are symmetric and that will still ensure positive-definite covariance matrices for any finite collection of spatial locations. While it is possible to derive theoretical characterizations for valid nonstationary covariance functions (see, e.g., [27]), simpler constructive approaches are preferred for practical modeling, computational tractability, inference and interpretation. There is a massive literature ranging from a variety of existence theorems to practical formulations and a full review is beyond the scope of the current chapter. Popular approaches for building nonstationary models include the deformation approach outlined in the landmark paper by [28], different kernel mixing approaches by [19], [14] and [15], and kernel mixing of processes as developed by [11] and [17]. We refer the reader to Chapter 3 of [3] for a more comprehensive account of nonstationarity.

We conclude this section with a brief discussion of one particular class of nonstationary covariance functions introduced by [25] that has been shown to be a rich, flexible and computationally tractable option for analyzing complex data sets. Extending earlier work by [19], [25] convolve spatially-varying kernels $K_{\mathbf{s}}(\mathbf{u})$ over $\Re^r$ and provide an elegant and direct proof that $C(\mathbf{s}, \mathbf{s}') = \int_{\Re^r} K_{\mathbf{s}}(\mathbf{u}) K_{\mathbf{s}'}(\mathbf{u}) d\mathbf{u}$ is a valid covariance function for $\mathbf{s}, \mathbf{s}' \in \Re^r$ for any positive integer r. With an appropriate choice of kernel functions, they establish that

$$C(\mathbf{s}, \mathbf{s}') = \sigma^2 |\Sigma(\mathbf{s})|^{1/4} |\Sigma(\mathbf{s}')|^{1/4} \left| \frac{\Sigma(\mathbf{s}) + \Sigma(\mathbf{s}')}{2} \right|^{-1/2} \rho(\sqrt{h}) , \qquad (5.6)$$

is a valid nonstationary covariance function, where $\Sigma(\mathbf{s})$ is an $r \times r$ positive definite matrix that varies across the domain, $\rho(\cdot)$ is any valid correlation function and $h = (\mathbf{s} - \mathbf{s}')^T (\Sigma(\mathbf{s}) + \Sigma(\mathbf{s}'))(\mathbf{s} - \mathbf{s}')/2$ is the Mahalonobis distance between $\mathbf{s}$ and $\mathbf{s}'$ over $\Re^r$. An especially versatile special case is the *nonstationary Matérn* covariance function on $\Re^r$,

$$C(\mathbf{s}, \mathbf{s}') = \frac{\sigma^2}{\Gamma(\nu)2^{\nu-1}} |\Sigma(\mathbf{s})|^{1/4} |\Sigma(\mathbf{s}')|^{1/4} \left| \frac{\Sigma(\mathbf{s}) + \Sigma(\mathbf{s}')}{2} \right|^{-1/2} (2\sqrt{\nu h})^\nu \kappa_\nu(2\sqrt{\nu h}) , \qquad (5.7)$$

where $\Gamma(\nu)$ is the Gamma function, κ_ν is a second kind Bessel function and ν is the smoothness parameter as in the isotropic Matérn covariance function. When $\Sigma(\mathbf{s}) = \Sigma$ is constant over space, we obtain a stationary but anisotropic covariance function. Finally, the familiar isotropic Matérn covariance function results when $\Sigma = I$.

5.2.3   Variograms

Apart from strong and weak stationarity, which are equivalent for Gaussian processes, there is a third type of stationarity called *intrinsic* stationarity. Here we assume $E[W(\mathbf{s} + \mathbf{h}) - W(\mathbf{s})] = 0$ and define

$$E[W(\mathbf{s} + \mathbf{h}) - W(\mathbf{s})]^2 = \text{Var}\,(W(\mathbf{s} + \mathbf{h}) - W(\mathbf{s})) = 2\gamma(\mathbf{h}) . \qquad (5.8)$$

Equation (5.8) implies that the expected squared difference between two measurements depends *solely* on their separation vector $\mathbf{h}$ (so that the right-hand side can be written at all), and not the particular choice of $\mathbf{s}$. If this is the case, we say the process is *intrinsically stationary*. The function $2\gamma(\mathbf{h})$ is then called the *variogram*, and $\gamma(\mathbf{h})$ is called the *semivariogram*. Note that intrinsic stationarity defines only the first and second moments of the differences $W(\mathbf{s}+\mathbf{h})-W(\mathbf{s})$. It says nothing about the joint distribution of a finite collection of variables $W(\mathbf{s}_1), \ldots, W(\mathbf{s}_n)$ and, therefore, provides no likelihood. It only describes the behavior of differences rather than the behavior of the data that we observe. This is less appealing from the perspective of data analysis.

The variogram appears in classical kriging ([8]; [7]) where one seeks the best linear unbiased predictor (BLUP). The variogram and its variants have been the subject of much research in classical geostatistics. While more modern model-based approaches for environmental processes have focused on understanding the nature of spatial processes and how covariance functions affect their behavior, variograms are still deployed widely as an exploratory tool to evince what the data has to say about the process. It is easy to compute, is available in a variety of geostatistical software products including in several packages within the R statistical environment such as geoR and gstat, and has some appealing behavioral features. For example, at short distances (i.e., small $\|\mathbf{h}\|$), one would expect $W(\mathbf{s}+\mathbf{h})$ and $W(\mathbf{h})$ to be very similar, that is, $(W(\mathbf{s}+\mathbf{h}) - W(\mathbf{s}))^2$ to be small. As $\|\mathbf{h}\|$ grows larger, we expect less similarity between $W(\mathbf{s}+\mathbf{h})$ and $W(\mathbf{h})$, i.e., we expect $(W(\mathbf{s}+\mathbf{h}) - W(\mathbf{s}))^2$ to be larger. So, a plot of $\gamma(\mathbf{h})$ would be expected to increase with $\|\mathbf{h}\|$, providing some insight into spatial behavior.

It is not difficult to relate the variogram and the covariance function. It is easy to show that

$$\gamma(\mathbf{h}) = C(\mathbf{0}) - C(\mathbf{h}) , \tag{5.9}$$

where $\gamma(\mathbf{h})$ is the semi-variogram defined in (5.8) and $C(\mathbf{h}) = \mathrm{Cov}(W(\mathbf{s}+\mathbf{h}), W(\mathbf{s}))$. From (5.9), it is clear that given C, we are able to recover γ easily. The converse, however, requires a bit more: if the spatial process is *ergodic*, then $C(\mathbf{h}) \to 0$ as $\|\mathbf{h}\| \to \infty$. This means means that the covariance between the values at two points disappears as the points become further separated in space. This assumption appeals to intuition and allows us to arrive at the following relationship

$$C(\mathbf{h}) = \lim_{\|\mathbf{u}\| \to \infty} \gamma(\mathbf{u}) - \gamma(\mathbf{h}) . \tag{5.10}$$

Therefore, for ergodic spatial processes we are able to recover the covariance function from the variogram. More generally, without the assumption of ergodicity the limit on the right hand side of (5.10) need not exist, but if it exists, then the process is weakly (second-order) stationary with $C(\mathbf{h})$ obtained as in (5.10). From (5.9) and (5.10) it is clear that weak stationarity implies intrinsic stationarity, but the converse is not true; see the example of the linear variogram later.

A valid variogram necessarily satisfies a negative definiteness condition. For any set of locations $\mathbf{s}_1, \ldots, \mathbf{s}_n$ and any set of constants $a_1, \ldots, a_n$ such that $\sum_i a_i = 0$, if $\gamma(\mathbf{h})$ is valid, then

$$\sum_i \sum_j a_i a_j \gamma(\mathbf{s}_i - \mathbf{s}_j) \leq 0 . \tag{5.11}$$

[8] discusses further necessary conditions for a valid variogram. While (5.11) resembles the positive-definiteness condition for covariance functions, the motivation is quite different. The positive-definiteness condition for covariances ensures well-defined variance-covariance matrices and, hence, valid joint distributions. The condition in (5.11) arises naturally in the derivation of the kriging equations.

Analogous to covariance functions, we say that the variogram is isotropic if $\gamma(\mathbf{h})$ depends upon the separation vector only through its length $\|\mathbf{h}\|$. Now the semivariogram function is a real-valued function of a scalar argument $h = \|\mathbf{h}\|$ and we write $\gamma(\mathbf{h}) = \gamma(h)$. Otherwise, we say that the variogram is anisotropic. Isotropic variograms are popular because of their

TABLE 5.2
Summary of common parametric isotropic variograms.

model	Semivariogram, $\gamma(h)$
Linear	$\gamma(h) = \begin{cases} \tau^2 + \sigma^2 h & \text{if } h > 0 \\ 0 & h = 0 \end{cases}$
Spherical	$\gamma(h) = \begin{cases} \tau^2 + \sigma^2 & \text{if } h \geq 1/\phi \\ \tau^2 + \sigma^2 \left[\frac{3}{2}\phi h - \frac{1}{2}(\phi h)^3 \right] & \text{if } 0 < h \leq 1/\phi \\ 0 & h = 0 \end{cases}$
Exponential	$\gamma(h) = \begin{cases} \tau^2 + \sigma^2(1 - \exp(-\phi h)) & \text{if } h > 0 \\ 0 & h = 0 \end{cases}$
Powered exponential	$\gamma(h) = \begin{cases} \tau^2 + \sigma^2(1 - \exp(-|\phi h|^p)) & \text{if } h > 0 \\ 0 & h = 0 \end{cases}$
Gaussian	$\gamma(h) = \begin{cases} \tau^2 + \sigma^2(1 - \exp(-\phi^2 h^2)) & \text{if } h > 0 \\ 0 & h = 0 \end{cases}$
Rational quadratic	$\gamma(h) = \begin{cases} \tau^2 + \frac{\sigma^2 h^2}{(\phi + h^2)} & \text{if } h > 0 \\ 0 & h = 0 \end{cases}$
Wave	$\gamma(h) = \begin{cases} \tau^2 + \sigma^2(1 - \frac{\sin(\phi h)}{\phi h}) & \text{if } h > 0 \\ 0 & h = 0 \end{cases}$
Power law	$\gamma(h) = \begin{cases} \tau^2 + \sigma^2 h^\lambda & \text{if } h > 0 \\ 0 & h = 0 \end{cases}$
Matérn	$\gamma(h) = \begin{cases} \tau^2 + \sigma^2 \left[1 - \frac{(\phi h)^\nu}{2^{\nu-1}\Gamma(\nu)} K_\nu(\phi h) \right] & \text{if } h > 0 \\ 0 & h = 0 \end{cases}$
Matérn at $\nu = 3/2$	$\gamma(h) = \begin{cases} \tau^2 + \sigma^2 \left[1 - (1 + \phi h)\exp(-\phi h) \right] & \text{if } h > 0 \\ 0 & h = 0 \end{cases}$

simplicity, interpretability, and, in particular, because a number of relatively simple parametric forms are available as candidates for the semivariogram. Table 5.2 lists a selection of more commonly used parametric isotropic variogram models. The linear and the powered law variograms do not correspond to weakly stationary processes. For the others, the underlying process is ergodic and the associated covariance function can be derived using (5.10). The explicit forms for these covariance functions correspond to those provided in Table 5.1.

Some of the forms in Table 5.2 merit further remarks. Figure 5.2 plots three popular variogram functions. The spherical variogram clearly illustrates the *nugget, sill,* and *range,* three characteristics traditionally associated with variograms. Figure 5.2(b), which plots the spherical semivariogram using the parameter values $\tau^2 = 0.2$, $\sigma^2 = 1$, and $\phi = 1$. While $\gamma(0) = 0$ by definition, the nugget captures the discontinuity as $\lim_{h \to 0^+} \gamma(h) = \tau^2$ and the sill is the asymptotic value of the semivariogram $\lim_{h \to \infty} \gamma(h) = \tau^2 + \sigma^2$. The sill minus the nugget, which is simply σ^2 in this case, is called the *partial sill.*) Finally, the value $h = 1/\phi$ at which $\gamma(h)$ first reaches its ultimate level (the sill) is called the *range.* Variograms are often parametrized through $R \equiv 1/\phi$. Confusingly, both R and ϕ are sometimes referred to as the *range* parameter, although ϕ is more accurately described as the *decay* parameter.

Note that for the linear semivariogram, the nugget is τ^2 but the sill and range are both infinite. For other variograms, such as the exponential, the sill is finite, but only reached asymptotically. In cases like this, the notion of an *effective range* is often used, i.e., the distance at which there is essentially no lingering spatial correlation. This is best achieved for processes that are not only intrinsically but also weakly stationary. We can, then, obtain the covariance function $C(h)$ from $\gamma(h)$ and find the the smallest distance h_0 beyond which

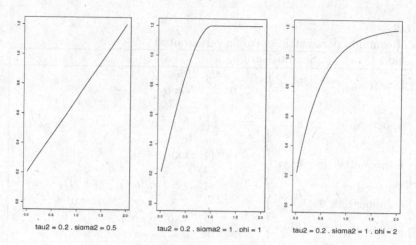

tau2 = 0.2 . sigma2 = 0.5   tau2 = 0.2 . sigma2 = 1 . phi = 1   tau2 = 0.2 . sigma2 = 1 . phi = 2

FIGURE 5.2
Theoretical semivariograms for three models: (a) linear, (b) spherical, and (c) exponential.

the spatial correlation is *negligible*, customarily taken to be below 0.05. For the exponential, we solve $\exp(-h_0\phi) = 0.05$, which yields $h_0 \approx 3/\phi$. The exponential has an advantage over the spherical in that it is simpler in functional form while still being a valid variogram in all dimensions (and without the spherical's finite range requirement). Finally, we point to the wave function as an example of a variogram that is not monotonically increasing. The associated covariance function is $C(h) = \sigma^2 \sin(\phi h)/(\phi h)$. Bessel functions of the first kind include the wave covariance function.

Fitting theoretical variogram models to observed data is also of interest. In practice a variogram model is chosen by plotting a simple nonparametric estimate of the semivariogram, which is often referred to as the *empirical semivariogram* (Matheron, 1963). The customary empirical semivariogram is

$$\widehat{\gamma}(h) = \frac{1}{2N(h)} \sum_{(\mathbf{s}_i,\mathbf{s}_j) \in N(h)} [Y(\mathbf{s}_i) - Y(\mathbf{s}_j)]^2 \,, \tag{5.12}$$

where $N(h)$ is the set of pairs of points such that $\|\mathbf{s}_i - \mathbf{s}_j\| = h$, and $|N(h)|$ is the number of pairs in this set. Unless the observations fall on a regular grid, the distances between the pairs will all be different, so this will not be a useful estimate as it stands. A more informative version of (5.12) is obtained by first "gridding up" the h-space into intervals $I_1 = (0, h_1), I_2 = (h_1, h_2)$, and so forth, up to $I_K = (h_{K-1}, h_K)$ for some (typically regular) grid $0 < h_1 < \cdots < h_K$. Representing the h values in each interval by the midpoint of the interval, we then alter our definition of $N(h)$ to

$$N(h_k) = \{(\mathbf{s}_i, \mathbf{s}_j) : \|\mathbf{s}_i - \mathbf{s}_j\| \in I_k\} \,, \ k = 1, \ldots, K \,.$$

Selection of an appropriate number of intervals K and location of the upper endpoint h_K is reminiscent of similar issues in histogram construction. The empirical semivariogram has been extensively studied. For example, [20] explored sensitivities to the number and width of bins and recommend bins wide enough to capture at least 30 pairs per bin. The sample average of the squared differences may exhibit poor behavior because a Gaussian assumption for the $Y(\mathbf{s}_i)$ would imply that the squared differences in (5.12) will have a distribution that is a scale multiple of the heavily skewed χ_1^2 distribution. In this regard, [9] proposed a robust variant of (5.12) that uses sample averages of $\|Y(\mathbf{s}_i) - Y(\mathbf{s}_j)\|^{1/2}$; this estimate is available in several software packages and is usually less sensitive to outliers.

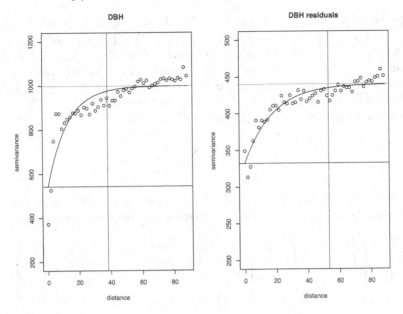

FIGURE 5.3
Isotropic semivariograms for DBH and residuals of a linear regression of DBH onto tree species.

Variograms can be very useful in exploratory data analysis. Figure 5.3 shows some preliminary exploration of the so called "WEF" forest inventory data from a long-term ecological research site in western Oregon. These data are available as a part of the spBayes package within the R statistical software environment and consist of a census of all trees in a 10 ha stand. Diameter at breast height (DBH) and tree height (HT) have been measured for all trees in the stand. The figure on the left shows an empirical semivariogram (small circles) computed using DBH as $Y(\mathbf{s})$ and an exponential variogram model (solid curve) fitted to the empirical semivariogram using a nonlinear weighted least squares approach to estimate the sill, the nugget and the range. The upper and lower horizontal lines are the *sill* and *nugget*, respectively, and the vertical line is the effective range (i.e., that distance at which the correlation drops to 0.05). The figure on the right produces a variogram for the residuals obtained by regressing (using ordinary least squares) DBH on tree species (a categorical variable with five classes). These plots are often useful in assessing the extent of spatial variability in the data and how much of it can be explained using available predictors. Further details on exploratory data analysis with variograms including detection of anisotropy, can be found in Chapter 2 of [3].

5.3   Spatial interpolation and kriging

Spatial interpolation and prediction refers to the problem of estimating the value of the outcome at arbitrary locations (possibly new locations with no available measurements) using available measurements. This exercise is called "kriging", so named by [24] after the South African mining engineer D.G. Krige [22], whose seminal work on empirical geostatistical methods built the foundations of modern geostatistics.

Consider measurements $\mathbf{Y} = (Y(\mathbf{s}_1), Y(\mathbf{s}_2), \ldots, Y(\mathbf{s}_n))^T$ from a set of n fixed locations $\mathcal{S} = \{\mathbf{s}_1, \mathbf{s}_2, \ldots, \mathbf{s}_n\}$. Suppose we are interested in a function $f(\mathbf{s})$, defined for any arbitrary $\mathbf{s} \in \mathcal{D}$, such that $f(\mathbf{s}_i) = Y(\mathbf{s}_i)$ for each of the observed locations. Any such function $f(\mathbf{s})$ is said to *interpolate* the observed values, i.e., it represents a surface that passes exactly through the observed measurements, and the the function is said to be a *spatial interpolator*. For now, consider deterministic interpolation ignoring the distribution of $\mathbf{Y}$. Suppose we restrict attention to generic linear interpolators of the form

$$f(\mathbf{s}) = \mu(\mathbf{s}) + \sum_{j=1}^{n} b_j(\mathbf{s})\beta_j \,, \tag{5.13}$$

where $\mu(\mathbf{s})$ is any known function capturing trend, $b_i(\mathbf{s})$'s are a set of *basis functions* and β_i's are the associated coefficients. The condition for interpolation implies

$$\begin{pmatrix} Y(\mathbf{s}_1) \\ Y(\mathbf{s}_2) \\ \vdots \\ Y(\mathbf{s}_n) \end{pmatrix} = \begin{pmatrix} f(\mathbf{s}_1) \\ f(\mathbf{s}_2) \\ \vdots \\ f(\mathbf{s}_n) \end{pmatrix} = \begin{pmatrix} \mu(\mathbf{s}_1) \\ \mu(\mathbf{s}_2) \\ \vdots \\ \mu(\mathbf{s}_n) \end{pmatrix} + \begin{pmatrix} b_1(\mathbf{s}_1) & b_2(\mathbf{s}_1) & \cdots & b_n(\mathbf{s}_n) \\ b_1(\mathbf{s}_2) & b_2(\mathbf{s}_2) & \cdots & b_n(\mathbf{s}_2) \\ \vdots & \vdots & \ddots & \vdots \\ b_1(\mathbf{s}_n) & b_2(\mathbf{s}_n) & \cdots & b_n(\mathbf{s}_n) \end{pmatrix} \begin{pmatrix} \beta_1 \\ \beta_2 \\ \vdots \\ \beta_n \end{pmatrix}. \tag{5.14}$$

We can succinctly express (5.14) as $\boldsymbol{\mu} + B\boldsymbol{\beta} = \mathbf{Y}$, where $\boldsymbol{\mu}$ is the $n \times 1$ vector with i-th element $\mu(\mathbf{s}_i)$, B is the $n \times n$ matrix with (i,j)th element $b_j(\mathbf{s}_i)$ and $\boldsymbol{\beta}$ is the $n \times 1$ vector with β_j as the j-th element.

It is easy to see that if $\hat{\boldsymbol{\beta}}$ is any solution for $B\boldsymbol{\beta} = \mathbf{Y} - \boldsymbol{\mu}$, then $f(\mathbf{s}) = \mu(\mathbf{s}) + \sum_{j=1}^{n} b_j(\mathbf{s})\hat{\beta}_j$ is a spatial interpolator. Matters are especially simple if the matrix B is nonsingular. Then, $\hat{\boldsymbol{\beta}} = B^{-1}(\mathbf{Y} - \boldsymbol{\mu})$ is the unique solution for (5.14) and we write (5.13) as

$$f(\mathbf{s}) = \mu(\mathbf{s}) + \sum_{j=1}^{n} b_j(\mathbf{s})\hat{\beta}_j = \mu(\mathbf{s}) + \mathbf{b}^T(\mathbf{s})\hat{\boldsymbol{\beta}} = \mu(\mathbf{s}) + \mathbf{b}^T(\mathbf{s})B^{-1}(\mathbf{Y} - \boldsymbol{\mu}) \,, \tag{5.15}$$

where $\mathbf{b}(\mathbf{s})$ is the $n \times 1$ vector with j-th element $b_j(\mathbf{s})$. Clearly the $f(\mathbf{s})$ in (5.15) is an interpolator because $\hat{\boldsymbol{\beta}}$ is obtained from the very conditions for interpolation in (5.14). Nevertheless, a direct verification is also instructive. Since B is nonsingular it satisfies $BB^{-1} = I$. Equating the i-th rows of both matrices yields $\mathbf{b}^T(\mathbf{s}_i)B^{-1} = \mathbf{e}_i^T$ because $\mathbf{b}^T(\mathbf{s}_i)$ is the i-th row of B, where $\mathbf{e}_i^T$ denotes the i-th row of the identity matrix. Plugging in $\mathbf{s} = \mathbf{s}_i$ in the above, we obtain

$$f(\mathbf{s}_i) = \mu(\mathbf{s}_i) + \mathbf{b}^T(\mathbf{s}_i)B^{-1}(\mathbf{Y} - \boldsymbol{\mu}) = \mu(\mathbf{s}_i) + \mathbf{e}_i^T(\mathbf{Y} - \boldsymbol{\mu})$$
$$= \mu(\mathbf{s}_i) + Y(\mathbf{s}_i) - \mu(\mathbf{s}_i) = Y(\mathbf{s}_i) \quad \text{for } i = 1, 2, \ldots, n \,.$$

The above framework can yield a rich class of interpolators specified with the basis functions. Given a finite collection of spatial locations where the outcome has been observed, any choice of basis functions that ensures a nonsingular B will yield a spatial interpolator of the form (5.15). Also, the interpolation property holds for *any* choice of $\mu(\mathbf{s})$.

Kriging can be looked upon as a special case of the generic (5.13) by constructing the basis functions from a valid covariance function. Indeed, if we choose $b_j(\mathbf{s})$ to be a valid covariance function $C(\mathbf{s}, \mathbf{s}_j)$, then B is a covariance matrix with elements $C(\mathbf{s}_i, \mathbf{s}_j)$. Therefore, B is positive definite and, hence, nonsingular so the resulting $f(\mathbf{s})$ is an interpolator for any $\mathbf{s}_i$ in the observed set of locations $\mathcal{S}$. In this case, we write (5.15) as

$$f(\mathbf{s}) = \mu(\mathbf{s}) + \mathbf{c}^T(\mathbf{s}; \boldsymbol{\theta})C^{-1}(\boldsymbol{\theta})(\mathbf{Y} - \boldsymbol{\mu}) \,, \tag{5.16}$$

where $\mathbf{c}(\mathbf{s}; \boldsymbol{\theta})$ is the $n \times 1$ vector with $C(\mathbf{s}, \mathbf{s}_j)$ as its elements and $C(\boldsymbol{\theta})$ is the spatial covariance matrix with elements $C(\boldsymbol{\theta})$. The $f(\mathbf{s})$ constructed in (5.16) is a spatial interpolator over $\mathcal{S}$ for *all* values of $\boldsymbol{\theta}$ supporting the covariance function and for any choice of $\mu(\mathbf{s})$.

Equation 5.16 also emerges from the joint probability law implied by a Gaussian spatial process model. If $Y(\mathbf{s})$ is a Gaussian spatial process with mean $\mu(\mathbf{s})$ and a covariance function $C(\cdot, \cdot; \boldsymbol{\theta})$, then $\mathbf{Y} = (Y(\mathbf{s}_1), Y(\mathbf{s}_2), \ldots, Y(\mathbf{s}_n))^T$ is the observed outcomes with a $N(\boldsymbol{\mu}, C(\boldsymbol{\theta}))$ probability law implied by the Gaussian process. The joint distribution of $\mathbf{Y}$ and an unobserved $Y(\mathbf{s}_0)$ at a new location $\mathbf{s}_0$ is given by

$$\begin{pmatrix} \mathbf{Y} \\ Y(\mathbf{s}_0) \end{pmatrix} \sim N\left(\begin{bmatrix} \boldsymbol{\mu} \\ \mu(\mathbf{s}_0) \end{bmatrix}, \begin{bmatrix} C(\boldsymbol{\theta}) & \mathbf{c}(\mathbf{s}_0) \\ \mathbf{c}^T(\mathbf{s}_0) & C(\mathbf{s}_0, \mathbf{s}_0) \end{bmatrix} \right) .$$

Standard properties of the multivariate normal distribution immediately implies the following:

$$E[Y(\mathbf{s}_0) \,|\, \mathbf{Y}] = \mu(\mathbf{s}_0) + \mathbf{c}^T(\mathbf{s}_0; \boldsymbol{\theta}) C^{-1}(\boldsymbol{\theta})(\mathbf{Y} - \boldsymbol{\mu})$$
$$\text{Var}[Y(\mathbf{s}_0) \,|\, \mathbf{Y}] = C(\mathbf{s}_0, \mathbf{s}_0) - \mathbf{c}^T(\mathbf{s}_0; \boldsymbol{\theta}) C^{-1}(\boldsymbol{\theta}) \mathbf{c}(\mathbf{s}_0; \boldsymbol{\theta}) .$$

Note that the above conditional expectation is exactly $f(\mathbf{s}_0)$ in (5.16), but assuming a probability law enables us to quantify the uncertainty using the expression for the variance. These equations are called the "kriging estimate" and "kriging variance" for $Y(\mathbf{s}_0)$ at a new location $\mathbf{s}_0$. If $\mathbf{s}_0$ is the same as an observed location $\mathbf{s}_i$, then $E[Y(\mathbf{s}_0) \,|\, \mathbf{Y}] = f(\mathbf{s}_i) = Y(\mathbf{s}_i)$ so the kriging estimator interpolates observed values. Furthermore, it is easy to verify that $\text{Var}[Y(\mathbf{s}_0) \,|\, \mathbf{Y}] = 0$ whenever $\mathbf{s}_0$ is one of the observed locations. Therefore, the kriging estimator is a *deterministic* interpolator, i.e., it reproduces the value of the outcome at an observed location and there is no uncertainty there.

In practice, the process parameters $\boldsymbol{\theta}$ will not be known and have to be estimated. They can be fixed using eyeball estimates obtained from variograms, or can be formally estimated using likelihood-based or Bayesian methods. We do not discuss such inference here and refer the reader to the books by [8], [10] or [3]. In any event, the kriging estimator retains its interpolating property irrespective of the value of $\boldsymbol{\theta}$. Figure 5.4 presents the spatial surface constructed by interpolating log metric tons of biomass over a region in Bartlett, New Hampshire, using data from the Bartlett Experimental Forest (BEF) of the U.S. Department of Agriculture and Forest Service. The full dataset is available in the spBayes R package. This spatial surface was obtained by first estimating $\boldsymbol{\theta}$ using Bayesian methods and then carrying out spatial interpolation and prediction using the spBayes software within the R statistical computing environment. Further details on the precise method and the BEF data set can be found in [3].

In the above formulation, it is clear that the interpolation property of kriging holds irrespective of the presence or absence of a nugget. For instance, consider the covariance function $C(\mathbf{s}, \mathbf{t}) = \sigma^2 \rho(\|\mathbf{s} - \mathbf{t}\|; \phi) + \tau^2 \delta_{\{\mathbf{s}=\mathbf{t}\}}$, where $\delta_{\{\mathbf{s}=\mathbf{t}\}} = 1$ if $\mathbf{s} = \mathbf{t}$ and 0 otherwise. Then $\boldsymbol{\theta} = \{\sigma^2, \phi, \tau^2\}$ and $C(\boldsymbol{\theta}) = \sigma^2 H(\phi) + \tau^2 I_n$, where $H(\phi)$ is a spatial correlation matrix with (i, j)-th entry $\rho(\|\mathbf{s}_i - \mathbf{s}_j\|; \phi)$. Irrespective of whether $\tau^2 = 0$ or not, the vector $\mathbf{c}(\mathbf{s}_i, \boldsymbol{\theta}) = (C(\mathbf{s}_i, \mathbf{s}_1; \boldsymbol{\theta}), C(\mathbf{s}_i, \mathbf{s}_2; \boldsymbol{\theta}), \ldots, C(\mathbf{s}_i, \mathbf{s}_n; \boldsymbol{\theta}))^T$ and will equal the i-th row of $C(\boldsymbol{\theta})$ whenever $\mathbf{s}_i$ is an observed location and the kriging estimate will interpolate the value at $\mathbf{s}_i$.

We point out that some authors, (e.g., [18] and [3]) distinguish spatial interpolation in the presence of a nugget from that without a nugget. The latter is sometimes referred to as *noiseless kriging*. The underlying idea here is to note that even if there is measurement error in the process, for an unobserved location there are no measurements, hence no measurement error. Therefore, for a new location $\mathbf{s}_0$ the kriging estimator is

$$E[Y(\mathbf{s}_0) \,|\, \mathbf{Y}] = \mu(\mathbf{s}_0) + \boldsymbol{\gamma}^T(\mathbf{s}_0; \boldsymbol{\theta}) C^{-1}(\boldsymbol{\theta})(\mathbf{Y} - \boldsymbol{\mu}) , \qquad (5.17)$$

Mean predicted log metric tons of biomass

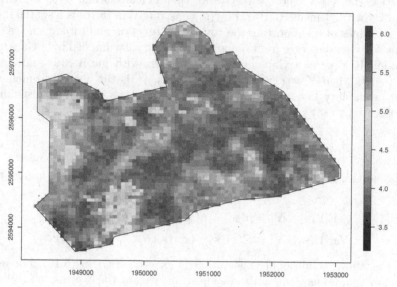

FIGURE 5.4

Interpolated surface of observed log metric tons of biomass from the Bartlett Experimental Forest data

where $\boldsymbol{\gamma}(\mathbf{s}_0; \boldsymbol{\theta})$ is the $n \times 1$ vector with i-th element $\sigma^2 \rho(\|\mathbf{s}_0 - \mathbf{s}_i\|; \phi)$. Simply plugging in $\mathbf{s}_0 = \mathbf{s}_i$, where $\mathbf{s}_i$ is an observed location, will not yield $Y(\mathbf{s}_i)$. This is because $\boldsymbol{\gamma}^T(\mathbf{s}_i; \boldsymbol{\theta})$ does not equal the i-th row of $C(\boldsymbol{\theta})$—the i-th entry of the i-th row of $C(\boldsymbol{\theta})$ is $\sigma^2 + \tau^2$, while the i-th entry in $\boldsymbol{\gamma}^T(\mathbf{s}_i; \boldsymbol{\theta})$ is simply σ^2. Thus, (5.17) is an interpolator only if $\tau^2 = 0$, which corresponds to noiseless kriging. We refer the reader to Chapter 2 in [3] for further details and some intuitive explanation for noiseless kriging.

Finally, we point out that kriging can be developed purely using the classical minimum mean-squared error approach to prediction. In fact, this is how it was originally formulated by Matheron (1963) using intrinsic stationarity. A linear predictor for $Y(\mathbf{s}_0)$ based on $\mathbf{Y}$ would take the form $\sum \ell_i Y(\mathbf{s}_i) + \delta_0$. The best linear prediction under squared error loss would minimize $\mathrm{E}[Y(\mathbf{s}_0) - (\sum \ell_i Y(\mathbf{s}_i) + \delta_0)]^2$ over δ_0 and the ℓ_i. Under the intrinsic stationarity, we would take $\sum \ell_i = 1$ in order that $\mathrm{E}[Y(\mathbf{s}_0) - \sum \ell_i Y(\mathbf{s}_i)] = 0$. This boils down to a fairly routine optimization for a quadratic form under a linear constraint and can be easily solved using Lagrange multipliers and the optimal linear predictor can be seen to be a function of the variogram. Under weak stationarity, we can derive the covariance function from the variogram and the optimal linear predictor coincides with the the right hand side of (5.17). Further details on deriving the kriging equations under intrinsic stationarity can be found, for example, in the texts by [8], [7] and [3] and references therein.

5.4   Summary

This chapter has attempted to provide an overview of the main ingredients for constructing spatial processes and carrying out spatial interpolation. The chapter has *not* dealt with statistical inference (e.g., statistical estimation of model parameters, estimation of uncertainty, prediction and so on). Inferential methods and models for spatial data analysis can be found in the texts by [8], [10] and [3], where the last text focuses primarily on Bayesian methods. Computational developments have contributed enormously to the popularity of spatial data analysis and today there are a number of software packages freely available to users via the Comprehensive R Archive Network (CRAN) (`http://cran.r-project.org`) that can complete tasks ranging from quick exploratory data analysis to more sophisticated Bayesian inference using Markov chain Monte Carlo (MCMC) methods. Packages such as `geoR` and `gstat` offer several accessible functions for computing and plotting variograms. More formal modeling and inference, including Bayesian methods, are offered by `geoR`, `geoRglm`, `spTimer`, `spBayes`, `spate`, and `ramps`. Other packages that offer tools for further exploratory data analysis, processing of spatial data such as easily computing geodesic distances, map projections, interfaces with Geographical Information Systems, and spatial interpolation include `akima`, `fields`, `spatstat`, `RGoogleMaps` and `mapproject`,

Bibliography

[1] *Spatial Variation, 2nd ed.* Springer Verlag, Berlin, Germany, 1960.

[2] M. Abramowitz and I.A. Stegun. *Handbook of Mathematical Functions.* Dover, New York, 1965.

[3] S. Banerjee, B. P. Carlin, and A. E. Gelfand. *Hierarchical Modeling and Analysis for Spatial Data.* Chapman & Hall/CRC, Boca Raton, FL, second edition, 2014.

[4] S. Banerjee and A.E. Gelfand. On smoothness properties of spatial processes. *Journal of Multivariate Analysis*, 84:85–100, 2003.

[5] S. Banerjee and A.E. Gelfand. Bayesian wombling. *Journal of the American Statistical Association*, 101:1487–1501, 2006.

[6] S. Banerjee, A.E. Gelfand, and C.F. Sirmans. Directional rates of change under spatial process models. *Journal of the American Statistical Association*, 98:946–954, 2003.

[7] J.P. Chiles and P. Delfiner. *Geostatistics.* Wiley, New York, 1999.

[8] N.A.C. Cressie. *Statistics for Spatial Data.* Wiley-Interscience, revised edition, 1993.

[9] N.A.C. Cressie and D. Hawkins. Robust estimation of the variogram: I. *Mathematical Geology*, 12:115–125, 1980.

[10] Noel A. C. Cressie and Christopher K. Wikle. *Statistics for spatio-temporal data.* Wiley Series in Probability and Statistics. Wiley, Hoboken, NJ, 2011.

[11] J. Duan and A.E. Gelfand. Finite mixture model of nonstationary spatial data. *Technical report, Institute for Statistics and Decision Sciences, Duke University*, 2003.

[12] M. Ecker and A.E. Gelfand. Bayesian variogram modeling for an isotropic spatial process. *Journal of Agricultural, Biological and Environmental Statistics*, 2:347–369, 1997.

[13] A.O. Finley, S. Banerjee, P. Waldmann, and T. Ericsonn. Hierarchical spatial modeling of additive and dominance genetic variance for large spatial trial datasets. *Biometrics*, 65:441–451, 2009.

[14] M. Fuentes. A high frequency kriging approach for non-stationary environmental processes. *Environmetrics*, 12:469–483, 2001.

[15] M. Fuentes. Spectral methods for nonstationary spatial processes. *Biometrika*, 89:197–210, 2002.

[16] A.E. Gelfand, P.J. Diggle, M. Fuentes, and P Guttorp. *Handbook of Spatial Statistics*. Boca Raton, FL: CRC Press, 2010.

[17] A.E. Gelfand, A. Kottas, and S.N. Mac Eachern. Bayesian nonparametric spatial modeling with dirichlet process mixing. *Journal of the American Statistical Association*, 100:1021–1035, 2005.

[18] T. Gneiting and P. Guttorp. Continuous-parameter spatio-temporal processes. In A. E. Gelfand, P. J. Diggle, M. Fuentes, and P Guttorp, editors, *Handbook of Spatial Statistics*, pages 427–436. Chapman & Hall/CRC, Boca Raton, FL, 2010.

[19] D. Higdon, J. Swall, and J. Kern. Non-stationary spatial modeling. In Berger J.O. Dawid A.P. Bernardo, J.M. and A.F.M. Smith, editors, *Bayesian Statistics 6*, pages 761–768. Oxford University Press, Oxford, UK, 1999.

[20] A.G. Journel and C.J. Huijbregts. *Mining Geostatistics*. Academic Press, New York, 1978.

[21] J.T. Kent. Continuity properties for random fields. *Annals of Probability*, 17:1432–1440, 1989.

[22] D.G. Krige. A statistical approach to some basic mine valuation problems on thewitwatersrand. *Journal of Chemical, Metallurgical and Mining Society of South Africa*, 52:119–139, 1951.

[23] R.J. Martin. The use of time-series models and methods in the analysis of agricultural field trials. *Communications in Statistics: Theory and Methods*, 19:55–81, 1990.

[24] G. Matheron. Principles of geostatistics. *Economic Geology*, 58:1246–1266, 1963.

[25] C.J. Paciorek and M.J. Schervish. Spatial modelling using a new class of nonstationary covariance functions. *Environmetrics*, 17:483–506, 2006.

[26] C.E. Rasmussen and C.K.I. Williams. *Gaussian Processes for Machine Learning*. MIT Press, Cambridge, MA, 2006.

[27] S.U. Rehman and A. Shapiro. An integral transform approach to cross-variograms modeling. *Computational Statistics and Data Analysis*, 22:213–233, 1996.

[28] P.D. Sampson and Guttorp P. Nonparametric estimation of nonstationary spatial covariance structure. *Journal of the American Statistical Association*, 87:108–119, 1992.

[29] T.J. Santner, B.J. Williams, and W.I. Notz. *The Design and Analysis of Computer Experiments*. Springer, New York, 2003.

[30] I.J. Schoenberg. Metric spaces and completely monotone functions. *Annals of Mathematics*, 39:811–841, 1938.

[31] M. L. Stein. *Interpolation of Spatial Data: Some Theory for Kriging*. Springer, New York, NY, first edition, 1999.

6

Spatial and spatio-temporal point processes in ecological applications

Janine B Illian

University of St Andrews, Scotland, UK

CONTENTS

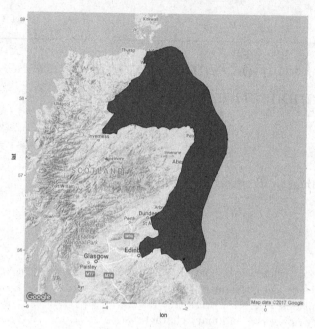

FIGURE 6.1
Locations of harbour porpoise sightings off the East Coast of Scotland.

6.1 Introduction – relevance of spatial point processes to ecology

Generally speaking, ecology focuses on understanding interactions of organisms with their biotic and abiotic environments, i.e., interactions of organisms with individuals from the same or from different species as well as with the physical characteristics of the environment. These interactions take place in space and time, and are reflected in the local spatial structure of the physical and biotic environment as well as in that formed by the locations of the individuals of a species of interest. Therefore, this spatial structure may represent underlying processes, and the extent to which these processes may be uncovered by analysis has been a subject of debate in ecology for many years [57]. The obvious practical implication should be that analyzing the spatial pattern of an ecological community may help the understanding of the underlying ecological processes [20]. As an example, Figure 6.1 shows a spatial point pattern, i.e. the locations of sightings of harbor porpoise off the East Coast of Scotland. Similarly, Figure 6.2 shows the spatial point pattern formed by reported sightings of the Loch Ness monster (often referred to as "Nessie") in Loch Ness, Scotland in the years 1930 to 2016. In both cases there is an interest gaining an understanding of why the individuals have been observed in some parts of the observation area relative to local environmental conditions as well as the characteristics of sighted individuals, and their distribution in space.

In the past, data on the exact locations of individuals in space and time were rarely available, which meant that statistical methods as well as ecological research dealt with data that were either not spatially explicit or that were aggregated into spatial units such as grid cells. These days, technical advances have made it easier to collect data on the exact spatial locations of individuals, and spatially explicit data-sets have become commonly available. This enables us to consider the locations of individuals directly, rather than aggregated

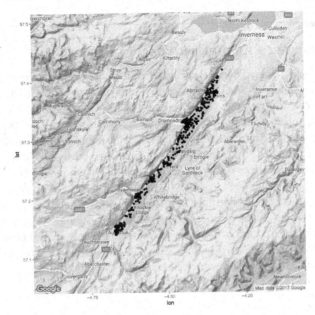

FIGURE 6.2
Locations of reported sightings of the Loch Ness Monster, Loch Ness, Scotland.

counts across space, and hence to analyse the spatial structure and information contained in it [4]. In other words, the data that reflect spatial structure in ecosystems are patterns formed by the locations of individuals in space, i.e. spatial point patterns. The statistical representation of these is a *spatial point process*, which models the spatial location of objects– here individuals or groups of individuals–in space. However, despite often being motivated as models of spatial patterns observed in nature, specifically of trees, publications on spatial point process modeling mainly resided within the statistical literature until rather recently, with only limited applications outside the statistical literature that directly concern or answer ecological (or other scientific) questions. This has changed over the last decade or so, partly due to the improved accessibility of suitable software such as the R-library `spatstat` [7] as well as to an improved dialogue between statisticians and ecologists interested in the respective fields [19, 20, 47, 93, 95].

However, point processes models are still not considered part of the standard toolbox of statistical methods used by ecologists and have only made their way into the ecological literature and statistical ecology community very recently [12, 33, 50, 78, 91]. For this reason, the aims of this chapter are two-fold. As a first aim we present the fundamentals of spatial point process methodology and its relevance in the context of ecology (Part I), providing a short overview of basic statistical theory, exploratory analysis and modeling methodology as discussed in classical point process literature. As a second aim we discuss potential reasons for the lack of awareness of the methodology in the ecological community arguing that there is a discrepancy between the data structures assumed in standard point process methodology and the specific data structures common in ecological studies (Part II). In particular, we discuss that due to data collection restrictions, spatial point patterns are often are not observed as fully mapped patterns so that classical point process methodology is not applicable. Method and model development has to account for sampling processes that often lead to a thinned point process.

6.2   Point processes as mathematical objects

Spatial point patterns are a data structure that is fundamentally different from geo-referenced data. The aim of an analysis of a point pattern is to model the locations of discrete objects in continuous space, while geo-referenced data are samples from a spatially continuous process, observed at discrete locations. These locations have typically been deliberately chosen and are fixed. The locations in a spatial point process analysis, however, are being analyzed and modeled, and the spatial pattern formed by these is the very structure of interest. Spatial point processes are random variables that model these spatial patterns formed by points representing the locations of objects or events. In an ecological context, the spatial structure formed by individuals or groups of individuals is relevant; spatial point patterns are analyzed to improve our understanding of ecological processes, for instance competition for resources or habitat preferences, which result in the observed spatial patterns. In this section, we present a general introduction to some classical point processes methodology, highlighting their relevance in ecological applications.

6.3   Basic definitions

A point pattern as a data structure consists of the spatial locations of objects represented as points $y = \{s_i, i = 1, \ldots, n\}$ in a bounded domain Ω. In many applications $\Omega \subset \mathbb{R}^2$, but general patterns in d dimensions, such as in $\mathbb{R}^d$, or on other manifolds can also be considered. A manifold that might be relevant in practice is for example the sphere $\mathbb{S}^2$ if Ω represents a large subarea of the earth's surface, or even the whole earth [66, 74, 84, 97]. To represent point patterns by a mathematical object and to do inference on this object we need to define a random variable, samples of which have the data structure we are interested in. Since a point pattern can be represented by d vectors of length n, one might naively think that a random variable that generates vectors of length n should do the job. However, recall that we are not deliberately choosing these locations – implying that we are also not choosing their number. The number of points in a point pattern is unknown *a priori* – and hence n is not fixed. It turns out, that a convenient way of defining a random variable that generates patterns of points in space where the number of points in the pattern is not fixed can be described by considering the number of points in every subset of the $\mathbb{R}^d$. Technically, this implies that a point process is a **random measure** and is defined as follows.

Definition 1. A **point process** is denoted by N. It is a random variable that can be seen as a function N operating on sets or, a **random counting measure** and for a subset B of $\mathbb{R}^d$, $N(B)$ is the random number of points in B, i.e. the set B is assigned the number $N(B)$, where B is a Borel set. We assume that $N(B) < \infty$ for all bounded sets B, i.e. that N is "locally finite".

As a function of B, $N(B)$ has the property of additivity, i.e. $N(B_1 \cup B_2) = N(B_1) + N(B_2)$ for disjoint B_1 and B_2, and similarly for countably-many sets. A common assumption is that point processes are invariant under translations and rotations, i.e. that a process N is *stationary* (or *homogeneous*) and isotropic, respectively. N is stationary, if N and the translated point process N_x have the same distribution for all translations x. Similarly, a process N is isotropic if $N = \{x_1, x_2, \ldots\}$ and $R_\alpha N = \{R_\alpha x_1, R_\alpha x_2, \ldots\}$ have the same distribution for all rotations α.

In practice, and in particular in an ecological context, a more complex random variable is often relevant, a **marked point process**. Point patterns might not merely consist of the locations of objects in space as often additional qualitative or quantitative information on the individuals represented by the points is available and relevant. Referred to as *marks*, these data on, e.g., the size or the type of the object may provide an improved understanding of spatial structures if included in an analysis. In other words, each point x_i is assigned a further quantity $m(x_i)$, which provides additional information on the object represented by the point. Typically, the $m(x_i)$ are usually integers or real numbers, but much more general marks may also be considered. A marked point process is denoted by M. We define a random variable, a marked point process, M as a function operating on sets or a random counting measure. For a subset B of $\mathbb{R}^d$ and a subset C of $\mathbb{R}$, $M(B \times C)$ denotes the random number of marked points $[x; m(x)]$ with $x \in B$ and $m(x) \in C$. Note that in this chapter, the term marks is used exclusively for properties of the objects represented by the points themselves; examples of marks include the size of trees in a forest or the gender of a particular animal. Variables, that refer to any location in space such as sea surface temperature or altitude are referred to as (spatial) *covariates*.

We now have defined two suitable random variables that generate samples of the type that we are interested in, i.e., either point patterns or point patterns with marks. The next step is to characterise these random variables by considering suitable summary characteristics which may then be estimated to provide a characterisation of an empirically sampled point pattern (Section 6.4). The information gained from these may then be used to identify a suitable model or class of models. In Section 6.5 we discuss a number of different model classes, realizations of which have different properties with regard to the spatial structure, for example clustering of points or repulsion among the points.

6.4   Exploratory analysis – summary characteristics

In this section we discuss characterisations of point processes which yield tools for the exploratory analysis of point pattern data. Similar to summary characteristics in a non-spatial context, these summarize properties of the data–here the pattern and may be used to suggest a suitable approach to analyzing the data further through a suitable spatial point process model.

6.4.1   The Poisson process–a null model

A point process may be characterized by–amongst other things–an intensity function, i.e. the expected number of points per unit area or formally

$$\lambda(s) = \lim_{r \to 0} \frac{E(N[B_r(s)])}{|B_r(s)|},$$

where B_r is a circle or ball of radius r. The intensity function describes how the point density–or intensity–varies through space and can take on many different forms. It plays an important role in the context of point process models as different types of intensity functions result in different model classes, as discussed in Section 6.5. For the time being let us consider the simple example of a point processes model whose intensity function is constant in space, the spatial **homogeneous Poisson process**. The Poisson process is the classical null model as it may be used to model patterns that are often termed **random patterns** and that are more correctly referred to as patterns exhibiting **complete spatial randomness (CSR)**. Generally speaking, a homogeneous Poisson process X has the following two properties:

(1) The intensity of points is constant in the area under investigation, i.e. $\lambda(x) = \lambda$.

(2) The location of any point in the pattern is independent of the location of any of the other points in the pattern, i.e. there is no interaction among the points.

As a null model it also serves as a reference model when the spatial characteristics of a specific empirical pattern are analyzed. For this purpose, summary characteristics for the empirical pattern may be compared to those known analytically for the Poisson process. In addition, many other, more complicated models may be viewed as a generalization of the homogeneous Poisson process. These are likely to be more realistic models in practical applications. For instance, the pattern in Figure 6.1 is unlikely to be a realization from a homogeneous Poisson process as the intensity of points seems to vary strongly in space and the point might also be clustered; we investigate this formally further below.

6.4.2   Descriptive methods

Descriptive methods are used to analyze the structures of spatial point patterns and are the most common methods from spatial point process theory that have been successfully used both within the statistical and the ecological literature. Descriptive methods form part of any statistical analysis and play a very important role in point process statistics. We briefly review these to provide a perspective and to illustrate the relation to relevant models discussed here. We otherwise refer the reader to the literature, where many of these methods have been discussed in detail and reviewed in the ecological literature by [57] and [94]. While descriptive summarize various aspects of the spatial structure of a point pattern either as an index or as a function (of distance), we focus on functional summary characteristics here [45].

In general, a simple spatial point pattern can exhibit a number of characteristics. The characteristics that are most relevant in practice fall broadly into two classes, first and second order characteristics, with higher orders being very rarely discussed.

"first order" characteristics

First order summary characteristics describe the expected density of individuals (the intensity function $\lambda(x)$ of the point process) in space, and reflect the probability of an individual being found at a given location in the sample space. The intensity may either be constant in space, resulting in a stationary (or homogeneous) pattern, or vary in space, for instance by having a spatial trend, resulting in an inhomogeneous pattern. The intensity function of individuals near location x, $\lambda(x)$, is a function of location, since the density of individuals may be affected by environmental conditions at location x. We see below that considering a non-deterministic but random intensity function, specifically a (Gaussian) random field $\Lambda(x)$, yields flexible and relevant models, for which descriptive characteristics can often be analytically determined.

"second order" characteristics

Second order characteristics relate to the relative position or interaction among points. Points may be not interacting, as in a Poisson process, show regularity or a clustered (or aggregated) structure. Regularity results from negative interaction (inhibition) among the points, while aggregation might be due to either positive interaction (attraction) among points or spatial covariates that attract many points in some areas and fewer in others.

Second order summary statistics hence relate to the relative positions of pairs of individuals and are constructed based on the second-order product density $\rho(x,y)$ obtained by considering $\rho(x,y)dxdy$, that is, the probability that there is a point of point process X in each of two disjoint discs b_1 and b_2 with centers x and y and infinitesimal areas dx and dy.

For stationary patterns, the most prominent summary characteristics are the **Ripley's K-function** and the **pair correlation function,** $g(.)$. The earlier considers the expected number of pairs of points found within a distance r and the expected number of pairs of points found at a distance r relative to the intensity of the point pattern, with $g(r) = \frac{K'(r)}{2\pi r}$. [45] discuss the advantages and disadvantages of the two functions in practical applications. This concerns in particular the issue that the pair correlation function is better suited than the K-function for assessing the distance at which clustering or regularity operates due to the cumulative nature of the later. Both functions are known analytically for the Poisson process, which is exploited in the data analysis where the shape of a function estimated for an empirical point patterns is compared to that of analytical function for the Poisson process.

For instance, the pair correlation function is a constant function and takes on the value $g(r) = \frac{2\pi r}{2\pi r} = 1$ for all distances for a Poisson process. Estimated pair correlation functions for empirical patterns can then be compared to the function with constant value 1. An estimated function above the Poisson reference line for some distances shows that there are more points in the pattern at these distances than expected from the Poisson process. This indicates clustering, while an estimated pair correlation function below the reference line indicates regularity, i.e., we observe fewer points than expected at a distance r. The literature discusses a number of approaches to estimating these functions as well as to accounting for edge effects, see for example classical papers work by [73] and [79] and the discussion in [45]; many of these are available through the versatile R library spatstat [9, 11]; see Section 6.5.4.

For illustration, Figure 6.3 (a) shows an example of the estimated pair correlation function for the pattern formed by sightings of harbor porpoise off the Northeast of Scotland, as shown in Figure 6.1. The estimation uses Ripley's isotropic edge correction [80]. The estimated pair correlation function is above the reference line for the Poisson process for all distances, clearly reflecting the clustering in the pattern. However, the fact that the function does not converge to that for the Poisson process at larger distances may indicate that the assumption of stationarity is not justified–with the implication that using the pair correlation for stationary patterns might not be justified here. It might be more useful to consider the inhomogeneous pair correlation function, which takes the local intensity of the pattern into account [8]. Figure 6.3 (b) shows a plot of the estimated inhomogeneous pair correlation function again for the pattern in Figure 6.1. Here the intensity was estimated by fitting an inhomogeneous Poisson process to the pattern, where the intensity function is assumed to be a second order polynomial (see Section 4). The inhomogeneous pair correlation function converges and indicates clustering up to a distance of $r = 30$ kilometers. Neither the homogeneous nor the inhomogeneous Poisson process seem to be appropriate models here. There is more clustering in the pattern than what would be expected from a pattern generated from either of these processes.

A standard approach to providing a means of deciding whether the estimated function for an empirical pattern deviates from what would be expected for a Poisson pattern, simulation envelopes may be considered that are based on the estimated functions for a set of patterns simulated from a Poisson process or another suitable null-model [93]. Note that these envelopes do not provide a formal significance test [71].

The summary characteristics can of course be generalized to be applicable to marked point patterns in an analysis of relationships between a pattern and the associated marks. Again, various suitable summary characteristics have been considered in the literature for

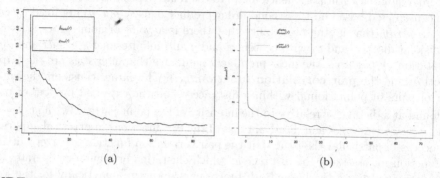

(a)   (b)

FIGURE 6.3
(a) Estimated pair correlation function $\hat{g}(r)$ for the spatial pattern in Figure 6.1 and (b) estimated inhomogeneous pair correlation function g_{inhom}-function for the same pattern. The distance r is measured in kilometres.

this purpose. For example, one might consider a pattern with qualitative marks i and j i.e., points of two different types (e.g. a species) and assess if there is attraction or repulsion between the two types. To this end, a pair density $\rho_{ij}(x, y)$ may be considered that is similar to the one above, where $\rho_{ij}(x, y)dxdy$ is the probability that there is a point of type i in a disc b_1 with center x and a point of type j in a disc b_2 with center y, the discs again having infinitesimal areas dx and dy, respectively. If individuals have quantitative marks, the mark pair density $\rho_{mm}(x, y)$ is an appropriate summary and considers $\rho_{mm}(x, y)dxdy$, the expected value of the product of the marks in discs b_1 and b_2 with centers x and y and infinitesimal areas dx and dy. The resulting summary characteristic is referred to as (Stoyan's) mark correlation function $k_{mm}(r)$. The function takes on values below (above) 1 if the objects represented by the points have smaller (larger) than average marks at a given distance r. Functions other than the product of the marks may be considered here, depending on the questions of interest; [87] and [45] refer to these as *test functions*. For illustration, a mark correlation function with the standard product test function has been estimated for the harbor porpoise data using Ripley's isotropic edge correction, Figure 6.4. Here the marks are the length of the observed porpoises in meters (with values ranging from roughly 0.80 to 1.70 meters). The marks fluctuate around the value of 1 indicating that the marks may be independent of the spatial pattern, i.e. the product of the lengths of two porpoises does not vary with the distance at which these porpoises have been observed.

6.4.3   Usage in ecology

Summary characteristics have been used in many contexts in ecology, mainly with applications in plants and tree studies, and most notably in the context of addressing the question of theories of the coexistence of several species, see for example [54, 76]. These theories are explicitly or implicitly spatial, implying that coexistence of species is directly linked to the nature of the interaction of individuals with their local environment, both in terms of competition with other individuals and their relationship with abiotic properties of the environment [42, 56]. Hence considering the local spatial structure rather than averaging across space better reflects the processes that support and maintain species coexistence. However, while ecologists have realized this when simulating communities using individual-based models, when describing communities and comparing different simulation approaches they have traditionally used summary statistics that relate to first order properties but do not characterize the spatial behavior of individuals. Attempts to clarify which mechanisms

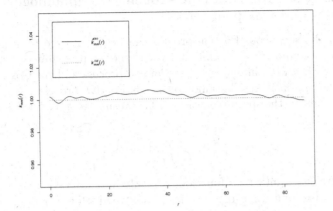

FIGURE 6.4
Mark correlation function for harbor porpoise data (length of the porpoises in meters) observed off the East Coast of Scotland.

operate using first order summary characteristics, however, failed to do so [26, 64, 100]. Recent work has demonstrated that the second order spatial structure of communities provides greater insights for the mechanisms maintaining species richness [15, 18, 19, 20].

While summary characteristics have been used in ecology, point process models have rarely been fitted to data [18, 38, 45, 57, 93, 94, 95]. This may be partly due to the fact that the models themselves are mathematically complex objects and the fact fitting them can be very difficult. We discuss point process models in the next section and discuss the slow pick-up in Part II of this chapter.

6.5   Point process models

So far we have considered the simple example of a Poisson point process, and have seen in the example in Figure 3 that this simple model might often not be relevant in practice–other than in its role as a null model. In this section, we hence present some of the main classes of more complex point process models to provide an overview of some of the relevant structures. We first return to the homogeneous Poisson point process and gradually modify it to provide more complex and hence more flexible models. The Poisson process may be generalised by varying or relaxing properties (1) and (2) in Section 6.4.1. The *inhomogeneous Poisson process* generalizes property (1) by assuming a non-constant (but fixed) intensity function that varies in space. In practice, an inhomogeneous Poisson process may be used to assess if intensity varies with spatial covariates, i.e., if point intensity depends on local environmental conditions as described by the covariates. *Cox processes* relax this further by assuming a *random* intensity function. This allows to account for spatial aggregation or overdispersion not explained by covariates, see Section 6.5.1. The class of *Gibbs processes* relaxes property (2) – the points may attract or repulse each other and hence form regular or clustered patterns.

6.5.1 Modelling environmental heterogeneity – inhomogeneous Poisson processes and Cox processes

As mentioned, the homogeneous Poisson process can be generalized to a process that still assumes independence amongst the points but where the intensity function now depends on the spatial location. Both the inhomogeneous Poisson process and the Cox process account for spatially varying intensities. The inhomogeneous Poisson process has a fixed intensity function that depends on the spatial location s. This results in a likelihood of the following form:

$$\pi(y|\lambda) = \exp\left(|\Omega| - \int_\Omega \lambda(s)ds\right) \prod_{i=1}^{N(\Omega)} \lambda(s_i).$$

The intensity function may be a function of locally measured covariates–properties of a species' habitat for example–that explain the spatial trend. However, in practice the available covariates often do not sufficiently explain the spatial structure as other mechanisms operate that the covariates cannot capture or because relevant covariates have not been measured for a specific point pattern. In a non-spatial context, when modeling count data in a generalized linear model it is common to assume a Poisson distribution. Often the observed variance in a dataset is bigger than the mean, $E[x] \leq \text{Var}[x]$, and hence the assumption of a Poisson distribution might not be justified. One way of accounting for this overdispersion is to assume that the rate parameter λ is random. With a similar idea in mind, we can make the intensity function in a Poisson process random, i.e., assume a random field that models the *spatial* overdispersion, or clustering. This results in a doubly-stochastic point process, which is a point process with two levels randomness, often referred to as a *Cox process* and which follows a Poisson process conditional on a realization of the random field. Due to the stochastic nature of the field these processes are particularly flexible.

In a Cox process, the intensity function $\Lambda(s)$ is a random process and we can consider the likelihood of the point pattern conditional on this random process, which again is a Poisson process. This yields the following likelihood:

$$\pi(y|\Lambda) = \exp\left(|\Omega| - \int_\Omega \Lambda(s)ds\right) \prod_{i=1}^{N(\Omega)} \Lambda(s_i). \tag{6.1}$$

In particular, the case where $\log(\lambda(s))$ is modeled as a Gaussian random field is referred to as a log-Gaussian Cox process [69] and the likelihood is analytically intractable. In practice, the log of the latent field is typically modeled as a linear model of covariates x and a Gaussian random field ξ, i.e.,

$$\log(\Lambda(s)) = x(s)^T\beta + \xi(s). \tag{6.2}$$

Classically, the role of the random field in this construction is to account for spatial structure in the pattern that is unexplained by the covariates [86], however, in more complex models the random field might represent more complex processes and have a different role in a model's interpretation [43].

As an illustration, we fitted a log-Gaussian Cox process model as in Equation (6.2) with two spatially continuous covariates, depth and bottom temperature, to the harbor porpoise data in Figure 6.1(a). The results indicate that both covariates do not significantly explain the spatial structure in the pattern (results not shown). Figure 6.5 shows the estimated spatial effect for this model, indicating that having taken into account these covariates strong clustering is still evident. Note that details on model fitting are discussed in Section 6.5.4.

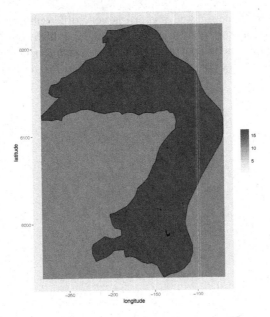

FIGURE 6.5
Estimated Gaussian random field for log-Gaussian Cox model fitted to the harbor porpoise data. The field reflects the expected number of individuals per unit area that could not be accounted for by the covariates in the model.

6.5.2　Modelling clustering – Neyman Scott processes

Neyman Scott processes are a class of models that is often also referred to as Poisson cluster processes. They are constructed by considering a set of so-called "parent points" around which "daughter" points are generated. Specifically, the parent points are generated from a homogeneous Poisson process with intensity λ_p. For each parent point a random number is generated from a Poisson distribution, reflecting the number of daughters in each of the clusters. In a final step, the daughters are distributed around the parents with the number of daughters around each parent point is Poisson distributed with rate parameter μ. The locations of the points follow a probability density that depends on the distance to the parent. Different distributions may be chosen for that purpose, generating different types of cluster processes, including the Matérn process (uniform distribution), the Thomas process (normal distribution) and the Cauchy cluster process (with a circular Cauchy distribution, [35]. For many Neyman Scott processes, the functional summary characteristics (e.g. the K-function or the pair-correlation function) are known analytically. This is used to apply minimum contrast estimation methods for model fitting, which choose the parameters of the process by minimizing the distance between the resulting analytical function and the empirical function as estimated from the data.

In particular, for the Thomas process the points are independently and isotropically scattered following a normal distribution with standard deviation s. The result of a fit of a Thomas process to the pattern in Figure 6.1(a), based on a minimum contrast estimation method using the pair correlation function as a summary characteristic (see Section 6.5.4) is shown in Figure 6.6 (a). Here the green line indicates the pair-correlation function for the fitted model and the black line indicates that estimated for the empirical data. It is clear that there are substantial discrepancies, in particular at larger distances. This is most likely due to the inhomogeneous nature of the pattern. A simulation from the fitted model (Figure

6.6 (b)) also shows that the simulated pattern is very different from the original pattern. Informed by this as well as the analysis in Section 6.4.2, which indicated that the pattern is a clustered and inhomogeneous pattern, we fitted an inhomogeneous Thomas process to the pattern. Figure 6.7 shows the result of this fit indicating a clear improvement to the previous result.

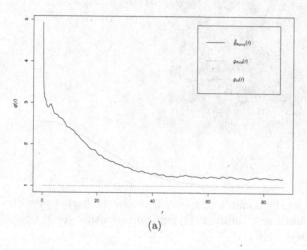

(a)   (b)

FIGURE 6.6
Estimated pair correlation function of the pattern in Figure 6.1(a) and analytical function with parameters for a Thomas process, derived by minimum contrast estimation (a), and simulated pattern with parameters from the same fit.

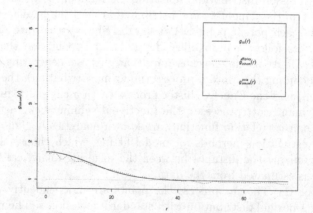

FIGURE 6.7
Estimated inhomogeneous pair correlation function and analytical pair correlation function with parameters for an inhomogeneous Thomas process, derived by minimum contrast estimation.

There have been a few applications of Neyman Scott processes within ecology, probably motivated by the obvious interpretation of the model as reflecting a parent-child relationship, and hence mimicking seed dispersal patterns [95]. Neyman Scott processes have recently also been used in the context of capture-recapture data [33].

6.5.3　Modelling inter-individual interaction – Gibbs processes

So far we have discussed point process models that are defined through their intensity, i.e. the first order behavior of the points. This has been possible because these specific processes are uniquely defined through their intensity as the points are independent of each other in both the homogeneous and the inhomogeneous Poisson process, and conditionally independent in the Cox process. Clearly, only considering the intensity as above is not sufficient if we want to construct a model class that allows for interaction among the points, e.g. second order behavior.

The class of Gibbs processes is therefore often presented based on the *conditional intensity*, which can be considered for a point $y \in \boldsymbol{y}$ or a point $u \notin \boldsymbol{y}$ and may be interpreted as the expected number of points per (infinitesimally small) unit area in a location, given the point pattern. For the homogeneous Poisson process the conditional intensity is equal to the intensity, which is intuitively clear since whether a point is observed in a location does not depend on whether points have been observed nearby or not.[1]

We will discuss Gibbs processes briefly here considering their conditional intensity, which is defined through the probability density of a point pattern $f(\boldsymbol{y})$. This probability density might also be considered directly, where the density of a Gibbs process is defined relative to that of a Poisson process of unit intensity [68]. However, discussing this density here is beyond the scope of this chapter as we aim to focus on the practical relevance of the methods rather than technicalities. Details may be found in many of the standard textbooks on point processes, with a particularly clear introduction given in [9].

The conditional intensity of a point pattern $\boldsymbol{y}$ with probability density $f(\boldsymbol{y})$ is defined as

$$\lambda(u|\boldsymbol{y}) = \frac{f(\boldsymbol{y} \cup \{u\}}{f(\boldsymbol{y})}$$

for a location $u \notin \boldsymbol{y}$ and as

$$\lambda(y_i|\boldsymbol{y}) = \frac{f(\boldsymbol{y})}{f(\boldsymbol{y} \setminus y_i)}$$

for $y_i \in \boldsymbol{y}$.

Now, Gibbs processes are a family of models, i.e. different types of models may be defined that are all constructed in a related way with different types of models differing by the form of their conditional intensity (and hence probability density). In particular, they differ in as to how interaction among points is expressed through a term in the conditional intensity. As special cases of Gibbs processes, in the homogeneous and the inhomogeneous Poisson process the points do not interact and the interaction term (or parameter) is equal to 1, yielding $\lambda(u|\boldsymbol{y}) = \lambda_o$ and $\lambda(u|\boldsymbol{y}) = \lambda_o(u)$, respectively. For the sake of brevity we consider two simple examples of Gibbs processes where interaction actually happens. The first one

---

[1]Note that the treatment of Gibbs processes presented here is referring to Gibbs processes with a finite number of points only [9, 45]. Gibbs processes with an infinite number of points may also be considered, but are omitted from the discussion here as they are, while being mathematically interesting, less relevant in practice in the context of ecological research.

is the *hard core process* with parameter h, with

$$\lambda(u|\mathbf{y}) = \begin{cases} \lambda_o & \text{if there are no points in the pattern that are as close to } u \text{ or} \\ & \text{less than } h \\ 0, & \text{otherwise.} \end{cases}$$

In realizations from a hard core process no two points are closer than the distance h, i.e. one can visualize this as a point pattern with no other points within a circle of radius R around each point. Another simple model is the Strauss process with parameter R, where the conditional intensity is

$$\lambda(u|\mathbf{y}) = \lambda_o \gamma^{t(u,R,\mathbf{y})}.$$

Here $t(u, R, \mathbf{y})$ denotes the number of points within distance R. Strength of interaction varies with variation in the value of γ, where $\gamma = 0$ yields the hard core process and $\gamma = 1$ the Poisson process. Many other Gibbs process models can be constructed by varying the way in which interaction is expressed in the conditional intensity. This yields further pair-wise interaction processes with, for instance, piece-wise linear interaction functions, smoothly decaying functions; these models may typically be used in the context of patterns where negative interaction, or repulsion is to be expected [49]. The area interaction process, which is considers more general interaction than pair-wise ones can also accommodate both negative and positive interactions [10, 49]. In an ecological context, positive and negative interactions might reflect inter-individual facilitation or competition, respectively [17, 32, 61]. Hence a spatial point process analysis might inform about the dynamics within an existing as expressed through the communities spatial footprint. However, despite this obvious interpretation and the potential relevance of Gibbs models, there have been few direct applications to ecological data, but see [27] for an application of Gibbs processes in the context of bird ecology and [37] for an application to human respiratory syncytial virus cells in an *in vitro* experiment.

6.5.4 Model fitting – approaches and software

Most point process models are analytically intractable, typically due to integrals that cannot be analytically solved. Hence standard statistical parameter estimation methods, such as maximum likelihood estimation, are often not available. As a result specific model fitting strategies have been developed that provide model fitting strategies that are often only suitable for specific model classes.

6.5.4.1 Approaches

Broadly speaking two main types of model fitting strategies can be distinguished, those based on summary characteristics and those based on (approximations of) the likelihood. The earlier approach makes use of the fact that functional summary characteristics are known analytically for a number of point process models, such as some Neyman-Scott processes and some Cox processes. In these cases, a minimum contrast approach can be evoked that seeks to minimize the difference between the analytical function and the estimated function for a specific dataset by choosing the model parameters accordingly [67, 68, 89, 92].

The latter approaches typically approximate the likelihood of the point process. The seminal paper by [6] showed that the log-pseudolikelihood of some Gibbs processes, most notably pair-wise interaction processes (with the homogeneous and inhomogeneous Poisson processes as special cases), is formally equivalent to that of a generalized linear model with Poisson error structure and hence model fitting approaches–as well as standard software– developed for these can be used in this context. For Cox processes and some Gibbs processes, modeling in a Bayesian context has been discussed, often using MCMC approaches for model fitting [69]. For instance, the Metropolis adjusted Langevin algorithm has often been applied, to improve acceptance rates and computationally efficient methods such as Variational Bayes, MCMC based approximate inference scheme using pseudo-marginal MCMC or integrated nested Laplace approximation (INLA) [47, 81, 83, 84, 98] are commonly used. Many of the approaches discussed in this section have been implemented in publicly available software so are accessible to the non-specialist user; the following section provides some guidance on packages that might be relevant in the context of ecological modeling.

6.5.4.2 Relevant software packages

A number of R packages have been developed that are relevant for the analysis of point pattern data. The most prominent package, `spatstat`, is a large R-library that has been specifically developed for the use with point pattern data [7]. It originally provided a wrapper around the approach to fitting Gibbs processes using generalized linear modeling software, as introduced in [6]. By now it has grown into a very large library that can handle a number of data structures in addition to point patterns and provides a wide range of analysis methods [9]. These comprise a large selection of summary characteristics, model fitting methods in a frequentist context for Cox and Neyman-Scott approaches using minimum contrast methods, for Gibbs processes as discussed and some MCMC approaches. The summary characteristics in Section 6.4 were estimated with and the models in Sections 6.5.2 and 6.5.3 were fitted with function provided in `spatstat`.

Point process models, in particular log-Gaussian Cox processes, can also be fitted with the model fitting software `R-INLA`, which makes the INLA algorithm available to users [59]. It can be used to fit a large range of general (and hence not necessarily spatial) statistical models in a Bayesian context, but much of the development has been done with spatial and spatio-temporal modeling in mind, due to the computational complexity of these models. The spatially continuous Gaussian random field in the Cox process has to be approximated. In `R-INLA` this can be done using Gauss Markov Random Fields, by either discretizing space using a regular grid or by using a flexible spatially continuous approximation based on the so-called stochastic partial differential equation (SPDE) approach of [48]. The software has facilities for the fitting of complex marked Cox processes as well as spatio-temporal processes and processes on the sphere. A recently emerging sister software package is the library `inlabru`, which is based on `R-INLA`, but aims to provide functionality that facilitates model fitting in practice and is particularly relevant for ecological surveys [5]. The model in Section 6.5.1 as well as those discussed in Section 6 below were fitted using the `inlabru` software. Similarly, the R-library `lgcp` provides software for the fitting of spatial and spatio-temporal log Gaussian Cox processes using MCMC methodology. A number of other smaller packages for specific purposes such as `spatgraph` and `spptest` are also around [70, 75].

6.6 Point processes in ecological applications

As mentioned in the introduction, ecological research focuses on gaining an understanding of the interaction of individuals with their environment in space and time, which may be reflected in the pattern formed by individuals in space and time. Despite point processes being the obvious mathematical representation of this pattern, they are rarely used in the ecological literature. In this section we argue that the potential reasons for this are two-fold, partly due to a lack of suitably complex methodology and partly since ecological data rarely come as the fully-mapped point patterns that point process methodology assumes as the standard data structure.

More specifically, point process methodology is only gradually being picked up by the ecological user community. While summary characteristics have been used in ecological applications and have found their way into the ecological literature to some degree [88], point process models have been very rarely considered and very few publications in the ecological literature exist that use them. [43] speculate as to why this is the case, but it is likely that the main reasons for the slow uptake are that the data structures, the practicalities of data collection, as well as the associated questions in a typical ecological study are often more complicated than those considered in the classical point process literature.

Research that aims at answering an actual ecological question–rather than merely using a dataset as an illustration of a mathematical structure or statistical method–collects data with the main objective of answering that very question without mathematical structures or assumptions such as stationarity or independence in mind. This implies, for instance (a) that, in addition to their locations, data (marks) are collected on the properties of the study objects that seem relevant to a specific research question, (b) that the data are collected in an area that is relevant to the specific research question, (c) and that the quality of the data is limited by practicalities of data collection.

In other words, (a) implies that studies do not tend to focus on merely considering the locations of individuals but also their properties, i.e., the associated marks. Hence, while modeling the point pattern itself might provide some insight, modeling the marks along with the pattern while not assuming independence of the marks and the pattern will provide further insight in underlying ecological processes [48]. With regard to (b) and (c), the resulting spatial scale at which the data are collected reflects the ecological processes of interest. This comes with the practical implication that the size of the spatial area of interest is dictated by the study system. This, in turn, implies that this area is can be very big, often so big that it cannot be sampled in its entirety and, what is worse, the probability of observing an individual may vary in space. As a result, sampling approaches are determined by practicalities of data collection and the available resources–the observation process has to be taken into account in the modeling approach. This is where classical point process methodology usually struggles as limited work has been done on models where only a (very) small subareas of the area of interest has been surveyed–and where detection probabilities vary as is, for instance, common in surveys based on transects, where detection probabilities depend on the distance from the observer [21].

Here we present three data structures, which we consider particularly relevant and common in ecological modeling, and discuss approaches that address a), b) and c) above. The three cases cover:

- complex marked point process models (Section 6.7)

- models for partially observed point patterns (Section 6.8.1)

- models which allow for varying detection probabilities (Section 6.8.2)

6.7   Marked point processes – complex data structures

The statistical literature has often assumed what is referred to as *geostatistical marking* [45], i.e., marked point patterns where the marks are independent of the spatial pattern and hence the values of marks do not depend on local spatial structures. This is akin to the situation in geostatistics where it is common to assume that sampling locations have been chosen independently of the sampled values. This is done to avoid bias resulting from preferential sampling (see Section 7.3.2). In the context of ecology, one can easily think of examples where marks do vary with spatial structure, e.g. trees that grow in close proximity to other trees are smaller than those which have very few neighbors or of larger animals being more threatening in their territorial behavior and hence showing a more repulsive behavior than other individuals. However, while it seems natural that in ecological data sets the assumption of independence might not actually hold, very few modeling approaches have so far been discussed where there is an interest in modeling the marks alongside the pattern, or taking the marks into account as they impact on the pattern. If marks have been included in an analysis, qualitative marks have typically been considered, with different marks representing different species, while very little work has been published that involves quantitative marks [41, 46, 72].

Abstracting the footprint of an ecological community to merely the locations of the individuals is mathematically neat but the pattern alone will almost certainly not reflect all the information that may be contained in an ecological data set. Ecological studies are rarely only interested in analyzing the locations of the objects represented by the points in a point pattern but the objects' properties are also of as much interest as the spatial pattern formed by them. This additional information is often available and may be used to inform on underlying ecological processes. This implies that the objects–i.e. individuals–represented by points are no longer merely analyzed with reference to their location in space but also as individuals with properties, i.e., as points carrying a "mark". These marks may not only help explaining the spatial structure formed by the points, but may themselves exhibit a spatial structure of interest that varies with the spatial structure of the locations. Clearly, methods for the analysis of marked spatial point patterns and the dependence between marks and points are relevant [48].

6.7.1 Different roles of marks in point patterns

Before considering a few examples of models of marked point patterns it is important to initially reflect on the role marks can have in spatial point pattern model as this role may differ. They can provide a better understanding of the pattern, where, for instance, the intensity varies with the value of a mark, or they might be a relevant response variable that we aim to model alongside the pattern–or both. In the context of the interpretation of summary characteristics the two scenarios of *superposition* and *labeling* have been considered [45] and a similar distinction has to be made when modeling point pattern data. In the literature this distinction has been discussed very little and it is important to clarify prior to an analysis which role the marks have in a specific model, in relation to the associated scientific question.

Specifically, we can distinguish two fundamentally different situations: (a) models of the *points* in the data set taking the marks into account and (b) models of the *marks* in a point pattern data set that take into account spatial structure (and potentially other marks). In the context of qualitative marks the distinction between (a) and (b) above implies that one can either (i) consider models of several (sub-) patterns formed by different types of points and the interaction amongst these types, or (ii) consider a single pattern with different qualitative characteristics. In the former situation ("superposition"), the patterns formed by the different types have been generated by separate but not necessarily independent mechanisms. Examples of this situation are patterns formed by different plant species. In the latter situation ("labeling") the points are assumed to originate from a single pattern and some underlying mechanism has lead to different qualitative properties of the points.

As an example, consider trees in a forest that have been either infected or not infected by a specific disease. Note that the distinction made here is analogous to a distinction made for summary characteristics for quantitatively marked point patterns, where the appropriate null models are that of "random superposition" and that of "random labeling" [45]. For quantitative marks the (a) and (b) above distinguish between (i) models where marks are a relevant response variable that is modeled along with the the pattern, or (ii) where the marks are believed to influence the pattern and are an explanatory variable in the model.

6.7.2   Complex models – dependence between marks and patterns

In practice, it is unlikely that marks and patterns are independent as is assumed in geostatistical marking. This is especially the case if the data have been collected for the specific purpose of understanding the relationship between a particular mark and the spatial structure [40, 46], or when the dependence between the mark of interest and the point pattern is (trivially) clear. In this situation, the dependence needs to be accounted for, so that more interesting relationships may be revealed [58].

Here we discuss a modeling approach that treats the marks and the point pattern in a joint (log-Gaussian) Cox process model including two (or more) likelihoods, which are linked through common elements, in particular by sharing the same random field [47]. One (or more) additional linear predictor(s) are considered in addition to the linear predictor for the spatial pattern and hence the intensity of the point pattern in Equation (6.2), which had the following form: $\log(\Lambda(s)) = x(s)^T \beta + \xi(s)$. The additional linear predictor models the marks η as

$$\eta(s) = \tilde{x}(s)^T \tilde{\beta} + \kappa \xi(s), \tag{6.3}$$

where $\tilde{x}$ is a set of covariates that might include some or all of the covariates x used in the model for the spatial pattern and $\xi(s)$ is the same–and hence joint–spatial field as in the model in Equation (6.2) and κ is a parameter. With the appropriate link function, the mark (i.e., the response variable in the mark model) can take on any relevant distribution.

This construction may be generalized to more than one mark [47] and to replicated patterns [46]. [47] use it in a model of a spatial point pattern with two marks, the dependence between the chemical properties of leaves of eucalyptus trees in a koala reserve and the frequency of koalas visiting these trees. The joint spatial field accounts for the spatial structure formed by the trees, reflecting both spatial autocorrelation of the chemical properties of trees that are in close proximity to each other and areas with a higher intensity of trees being potentially attractive to koalas, independent of the leave chemistry. The results indicate that koalas prefer to visit trees whose leaves have a high degree of palatability.

[48] use a similar approach to distinguish between long-term versus short-term suitability of the local environment for plant species in a study with replicated patterns in a sand dune. The study considered two response variables, a qualitative mark, the health status of the plants– reflecting short term survival – and the spatial pattern formed by the plants–reflecting long term survival. There was a difference in the spatial structuring and the dependencies on covariates for the two response variables, indicating different survival strategies at different time scales. A similar approach is used in [58] to model above ground biomass in a tropical rainforest. Trivially so, the amount of above ground biomass in a forest depends on the density of trees (i.e., the local intensity of the point pattern formed by them). By jointly modeling the tree density and above ground biomass this trivial relationship is represented by the joint spatial field while the relationship of interest, i.e., that between the covariates $\tilde{x}(s)$ and the response (here assumed to be normally distributed) can be analyzed and focused on.

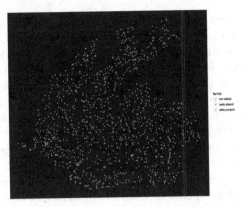

FIGURE 6.8
Location of farms in Slovakia indicating those that have been visited and where an owl was present (blue) or absent (green), and those that have not been visited (red).

6.7.3   Marked point pattern models reflecting the sampling process

Another use of marked point process models that has not been discussed much in the literature is to let the point pattern reflect a complex sampling process. Here the marks are the object of interest, but the sampling process cannot take place uniformly in space (or in any other appropriate spatial sampling design) for practical reasons. Modeling the sampling process alongside the quantities of interest can help account for sampling bias. An example of this concerns the data collection for the owl species *Athena nocturna* (or "little owl") in a study undertaken in Slovakia [30]. The owls are very shy and rare animals whose population is in decline. If one was to try to simply look for them in randomly chosen places or along transects it would be very unlikely that an individual would ever be observed. However, since the little owls tend to nest in farms, the researchers visited a subset of the farm buildings in the area of interest and used territorial call recordings to detect owl presence if those calls were answered, see Figure 6.8.

A marked point process point process model may be considered consisting of a log-Gaussian Cox process modeling the locations of all farms in the area of interest (which are known) with a linear predictor as in Equation (6.2). This then is modeled jointly with the owl absence/presence with a second linear predictor as in Equation (6.3) and again with a shared spatial effect. Since there is neither an interest in understanding the spatial distribution of farms nor their association with spatial covariates in this specific application, the linear predictor for the point process model may only consist of a spatial random field reflecting farm density in space. However, it reflects the modeling process as well as accounts for the local intensity of farms which might impact on owl presence directly. One may assume a second spatial field $\xi_2(s)$ in addition to that shared between the pattern and the observed owl presence $(\xi_1(s))$ that accounts for spatial structures in the owl distribution itself, such that the linear predictor results in

$$\eta(s) = \tilde{x}(s)^T \tilde{\beta} + \kappa \xi_1(s) + \xi_2(s), \tag{6.4}$$

with notation as in Equation (6.3). If the purpose of the study is mainly predicting owl presence at unobserved farms the additional spatial field $\xi_2(s)$ will have little effect on the quality of the prediction. However, the construction in general allows us to include covariates in the linear predictor for the observation locations and hence to assess the dependence of the observation process on these covariates. Here care has to be taken in choosing the priors for the spatial fields to avoid issues with identifiability and overfitting [86]. [51] considers these data, fitting a different approach that models the spatial distribution of little owl given the locations of all farms jointly with a misaligned covariate.

Note that a related modeling approach may be used to account for preferential sampling [22]. Here the points are the sampling locations, i.e. the point pattern reflect the sampling process rather than the biological process of interest as it is opportunistic and not planned and hence deterministic. Geostatistical modeling (see Chapter 5) usually assumes that the locations where the quantity of interest has been sampled is independent of the values observed in the locations. However, in practice this is sometimes not possible, especially if researchers rely on opportunistic sampling methods, such as sampling from commercial fishing boats. With this in mind, one might see geostatistical data are a special case of marked point pattern data, where the sampling process is deterministic and where there is an assumption of independence between sampling process and marks (values of interest).

In all the cases discussed in this and the previous section the role of the mark is what we refer to under (b) in Section 6.7.1, i.e. we aim to model the marks in dependence on the point pattern. In the case of the qualitative marks in the model fitted in [48] this constitutes superposition. All models referenced in Sections 6.7.2 and 7.3.2 were fitted using the software package R-INLA.

6.8   Modelling partially observed point patterns

In practice, in ecological data that have been collected to answer a relevant ecological question for example for conservation purposes [36] the "window of interest" often covers a very large area, which cannot be sampled in its entirety. This implies that it is typically practically infeasible to observe the entire pattern but information on the pattern in some small subareas is available. Studies, however, aim to make inference about the whole area of interest by inferring the spatial behavior in the unobserved area from the observed pattern, and in what follows we will consider such a situation.

Clearly, this needs careful consideration of the appropriate methodology and relevant spatial scales that may be used to "fill in" in the information. More importantly though, this is fundamentally different from the data structures envisaged as the typical application of classical spatial point process methodology. More specifically, point process methodology usually deals with one of two situations that [9] refer to as "Window sampling" and "Small world Model", respectively. In the earlier it is acknowledged that only a small portion of a relevant, infinite process (such as in a plot of trees in an "infinite" forest) has been observed, but it is assumed that any inferred dependences and structures are the same outside the observed area and continue beyond the observation window. The latter assumes that the process is finite and takes place within a fixed and bounded region. Neither of these situations apply here—and this has important consequences for method development and modeling assumptions. Both in window sampling and in the small world model, and hence in most methodology that has been developed so far, a strong assumption is made that (a) the whole area of interest has been surveyed and (b) that none of the objects represented by the points have been missed [9], or if they have, the probability that an object has been missed is known, fixed and the same everywhere in space. However, in many relevant ecological data sets this is not the case and neither of the two situations apply. In this section we consider approaches that do not rely on these implicit assumptions, but explicitly deal with data structures where incomplete detection is to be expected.

To put this differently, the pattern of locations of harbor porpoises (Figure 6.1) is not a real data set, it is a simulated pattern. A data set like this cannot exist, as it is practically impossible to take a snapshot of the locations of all individuals from an animal species in such a large area of the sea. Harbor porpoises exist in these waters, but human beings are currently unable to observe the pattern formed by these in its entirety. In other words, when looking at Figure 6.1, a statistician, unaware of any sampling issues, would probably just say "Well, ok, here's a bunch of points let's model it". An ecologist however, used to thinking about practicalities of data collection, would probably have wondered straight away how anyone would have been able to collect these data. This is a good reflection of what has happened in the point process literature so far–statisticians have developed methodology that assumes a sampling process that is unrealistic in an ecological context and hence methodology is irrelevant for anything but a few smaller data sets or plant data [24].

In contrast to the harbor porpoise data, the pattern of sightings of Nessie (Figure 6.2), a most likely fictitious creature, is a real data set. While the creature is probably fictitious, the point pattern is not; it is based on a real data set where each point in the pattern constitutes a reported sighting of said monster. The harbor porpoise data have been motivated by existing studies on harbor porpoises in the area [96], so while Nessie probably does not exist in Scotland, other than in hundreds of reincarnations as a stuffed animal, harbor porpoises do exist in its coastal waters. However, in order to be able to analyze the spatial behavior of the species, method development has to realize that assuming a "fictitious" sampling process is futile.

Due to the size of the relevant area, data structures in ecology typically violate assumption (a) above, in particular in studies of animals observed in very large spatial areas which, for practical or financial reasons cannot be fully sampled. Commonly, researchers position a number of plots or transects in an area and observe individuals within or along these, for example data collected using digital areal surveys along transects [22, 53]. In many cases the transects cover only a very small fraction of the total area [39]. In addition, it is also very common that the animals are not detected uniformally within these transects as detection becomes less likely with distance from the observer–inadvertently violating assumption (b) [21]. We discuss both these data structures and some modeling approaches in the following, considering recent approaches to dealing with these data structures, based on appropriately modified Cox process models. In particular, the approaches develop thinned log-Gaussian Cox process models that accommodate these data structures by modifying the likelihood in Equation (6.1) and the linear predictor in Equation (6.2) accordingly.

6.8.1   Point patterns observed in small subareas

In large study areas, in particular in the marine environment the spatial scale of interest is large and it is difficult for the observer to move around in the environment. As a result, it is common that only a relatively small subarea of the entire area of interest can be sampled and that this sampling takes place over at least several days, often weeks or months and hence time has to be accounted for. Classically, when collecting data in the marine environment, observers are typically located on a boat, moving along predefined transects looking for individuals. While in this setting there are issues with varying detection probabilities (see Section 6.8.2), other studies use areal surveys, for example, where there are no issues with spatially detection probabilities. Figure 6.9 shows the locations of transects used in a (fictitious) survey of harbor porpoises using an areal survey.

Clearly, this sampling method still yields a point pattern, but when modeling these data with a point process approach the fact that there has been no sampling effort outside the transects has to be taken into account–and fact that some of these plots or transects overlap in space (but not in time) has to be taken into account. Specifically, when constructing the likelihood it is no longer possible to integrate over the whole domain, as in Equation (6.1),

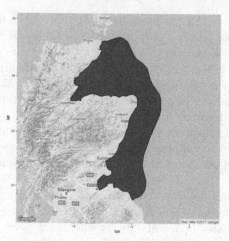

FIGURE 6.9
Example of a simulated data set where the locations of individuals have been recorded on non-overlapping transects.

but we now need to integrate over each of the transects. Assume we have K rectangular transect strips $\{C_1, \ldots, C_K\}$ and the kth transect strip surveyed at time t is denoted by $C_{k(t)}$, then the new likelihood has the following form:

$$\pi(y|\Lambda) = \exp\left(\sum_{k=1}^{K} |C_k| - \sum_{k=1}^{K} \int_{C_k} \Lambda(s; t_{C_k}) ds\right) \prod_{i=1}^{N_y(\Omega)} \Lambda(s_i) \tag{6.5}$$

The log-linear predictor in Equation (6.2) now changes slightly since both the random field and the covariates may depend on time t as not all transects can be visited at the same time

$$\log(\Lambda(s; t)) = \boldsymbol{x}(s; t)^T \beta + \xi(s; t). \tag{6.6}$$

Figure 6.9 shows the locations of transects used to subsample the (simulated) data in Figure 6.1. The model in Equations (6.5) and (6.6) was applied to these data using two covariates, depth and slope. Note that in this illustration and since we are working with simulated data we have not considered the time at which data were collected along each of the transects, pretending they have all been collected at the same time and not considering any temporally changing covariates, which implies that the index t in Equations (6.5) and (6.6) is the same for each observation and can be dropped. In practice, this is unrealistic and such an assumption would not be made, see [97]. Figure 6.10 shows the estimated intensity of harbor porpoises, based on this model.

This approach still assumes that detection along the transects is perfect. With digital areal surveys or other plot sampling techniques this is the case, however, in visual surveys, where human observers are used, detection probability decreases with distance from the observer. This results in a further thinning by a detection function modeling the probability of detecting an individual relative to the object's distance from the observer. The shape and hence the parameters of the detection function are unknown and have to be estimated in addition to the point process parameters.

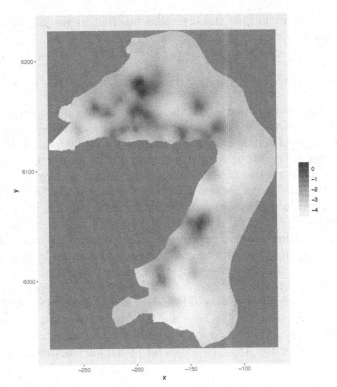

FIGURE 6.10
Estimated intensity for the (simulated) harbor porpoise data based on the model in Equations (6.5) and (6.6) .

6.8.2 Distance sampling

This data structure is very common in ecological studies and the statistical ecology literature discusses it under the term of *distance sampling* ([21] and Chapter 9 of this handbook). Most statistical analyses of distance sampling data so far have mainly focused on abundance estimation across space, without explicitly modelling spatial distributions or considering spatial aggregation in the data. The few spatial analyses have typically not modelled these data through a point process approach but are based on generalised additive models as in [65], but see [90]. Recent work has taken a point process perspective by constructing a log-Gaussian Cox process that incorporates detection [97]. As in classical distance sampling, the assumption is made that detection decreases with distance from the observer, as expressed through a detection function. The detection function is now included in the linear predictor, reflecting the observation process as a thinning of the underlying point process that generated the pattern. Parameters of the detection function have to be jointly estimated along with the parameters of the model that describes the spatial distribution of individuals in space.

The form of the likelihood as given in Equation (6.5) remains the same, however, we need to distinguish between $\lambda(s;t)$ – the intensity of the point process of potentially observable objects that we do not observe in practice and the intensity of the observational point process resulting from a thinning due to incomplete detection, $\Lambda(s;t)$. Combining this with the a transect-dependent detection functions $g_{k(t)}(s)$, yields

$$\Lambda(s;t) = \lambda(s;t)g_{k(t)}(s).$$

Here we now need to include the detection function in the log-linear predictor, i.e. Equation (6.2) now becomes

$$\log(\Lambda(s;t)) = \log(\lambda(s;t)) + \log(g_{k(t)}(s)) = x(s;t)^T \beta + \xi(s;t) + \log(g_{k(t)}(s)). \qquad (6.7)$$

Clearly, detection probabilities might also be dependent on properties of the observed objects, i.e. on marks. An example that is relevant in animal studies is the case where the points represent groups of animals and where the size of a group impacts on detectability, as it is likely that larger groups are easier to detect than smaller groups. Ongoing work by Bachl et al. incorporates properties such as group size in the model. Here, the role of the marks is what we refer to under (a) in Section 6.7.1; i.e. the marks influence the spatial pattern. This may happen in two different ways – the marks may influence the (true, unobserved) underlying spatial distribution of individuals or groups of individuals and they may also impact on the observation process. Specifically, the modeling approach allows the mark, i.e., group size to vary in space, with larger groups sizes observed in some areas and small one in other areas, i.e., the intensity of the spatial pattern varies relative to the mark. In addition, the detection function may have a different shape depending on group size, where the probability for larger groups to be detected is higher at larger distances than for smaller groups. All models in Sections 6.8.1 and 6.8.2 were fitted using the software package `inlabru`, which has been designed to make specifying the observation process convenient for the user; the software also has capabilities for fitting models that incorporate marks in the detection function.

6.9 Discussion

Spatial and spatio-temporal point processes are stochastic processes that describe patterns formed by the locations of individuals in space and time and may be used as models that link these patterns to spatial and temporal conditions. In ecology, where one aims to understand the interaction of individual organisms with their environment, these stochastic processes should be highly relevant in practical applications. However, they are much less frequently applied in the ecological literature than one would expect. While point process methods, and in particular summary characteristics for point processes, have contributed to specific discussions in theoretical ecology in the context of theories of species coexistence, especially in plant communities [19, 20, 43, 62, 99] point process models have rarely been applied and are not part of the standard statistical toolbox available to ecologists. This chapter has argued that this is largely due to the statistical literature lagging behind in the development of methodology that caters for the needs of ecologists, allowing them to answer scientific questions of interest (Section 6.7), and acknowledging that data structures available to ecologists result from complex observation processes (Section 6.8).

6.9.1 Spatial point processes and geo-referenced data

In particular, we have illustrated how recent developments have tried to improve the relevance of the methodology. To provide point process models that can help answer topical scientific questions, we have discussed realistically complex models, such as marked point processes that allow for dependence between marks and points. The examples we have discussed benefited from the flexibility of Cox processes resulting from the random intensity function, coupled with access to fast model fitting software that has made it feasible to fit models within reasonable time. In addition to providing flexible modeling approaches for marked point pattern data, marked point process models may also be used to account for the sampling process of a spatially continuous variable when preferential sampling is an issue.

In both cases, a Cox process bridges the fields of geostatistics and point process modeling, combining a random field–a stochastic process developed for geo-referenced data–and a Poisson processes–a stochastic process reflecting point process data. Our discussion has shown that this has enriched both fields; marked point process models have become more flexible through a model that jointly models non-point pattern data along with a point pattern, and marked point process models aid the analysis of geo-referenced data, relaxing the assumption of random sampling.

This development has been supported by the availability of model fitting algorithms that have not been developed purely for one type of spatial data structure, but provide methods for fitting general spatial models, a strong argument in favor of thinking outside the "box". The INLA approach has unified separate strands of spatial statistics as it concentrates on similarities across these. In other words, rather than focusing on one very specific data structure only, or rather than providing algorithms for any data structure–as MCMC approaches do– the INLA approach has benefited from focusing optimizing a model fitting algorithm in order to address issues common to all spatial modeling approaches, i.e. high computational costs due to spatial dependence. As a result, the flexibility of spatial modeling has been increased.

6.9.2   Spatial point process modeling and statistical ecology

We have argued that in addition to providing flexible and relevant modeling approaches it is also important to acknowledge the specific data structures– and hence observation processes– that ecologists work with, especially in the context of animal studies. This includes situations where only some small subareas of the window of interest have been sampled as a result of the area of interest being too large to be sampled in its entirety. To address this, we have discussed a Cox process model spanning the entire area of interest, based on the information contained in the small subareas, taking into account the increased uncertainty in areas where no data have been collected through a Bayesian modeling approach. In addition, an additional level of complexity–and hence stochasticity–has to be taken into account when detection probabilities vary in space and is unknown, and hence has to be inferred. This is the case, for example, in distance sampling, where detection probabilities decrease with distance from the observer, and hence have to be inferred.

Distance sampling is a classical example of a data structure that has been discussed primarily within the field of *statistical ecology* [23], where method development has been typically motivated by specific data structures common in–and often unique to–ecological research. This is done with the aim to address the relevant needs in the ecological community, traditionally focusing on providing methodology for the analysis of data resulting from a specific sampling regime. For instance, data that the spatial statistical community would refer to as point pattern data have often been discussed in the ecological literature as presence-only data, implying that data on whether individuals from a certain species where absent in a location is not available [25]. Recently, there has been some work unifying different areas of statistical ecological research that had previously been treated separately in the literature by considering common underlying structures. In particular, work has been done that unifies methods for analyzing data resulting from distance sampling and capture re-capture sampling in a class of models based on in inhomogeneous Poisson process [16]. In a similar unifying development to that in spatial statistics, statistical ecology has started moving away from separate methodological approach for each data structure linked to a specific sampling processes. Again, this has been done through considering similarities across the different branches of the field. Here the similarity may be found in the spatial representation of the actual ecological processes of interest, with the sampling regime representing different ways of thinning that actual process.

In summary, both within spatial statistics and in statistical ecology the observation process is currently being reconsidered. On the one hand, developers of spatial point process methodology, who used to not put much emphasis on the observation process–or rather assumed a perfect observation process–now increasingly include it in method development. Statistical ecologists, on the other hand, realize the role of the observation process as a link between the data structure they are dealing with and the underlying ecological processes that they would see if they were able to see everything. In summary, while spatial statistical literature has been moving away from assuming a completely mapped pattern and taking the sampling process into account, statistical ecology literature has increasingly acknowledged that the underlying processes are relevant in model development.

6.9.3   Other data structures

While we have focused our discussion on data structures common in ecological studies, we certainly have not covered all data structures relevant in ecology that might be captured by a spatial point process model, and we certainly cannot be exhaustive here. However, in what follows we briefly consider data structures that are currently getting increasing attention and where point process methodology might be relevant. This concerns the increasingly common data structure where movement patterns of animals has been tracked using GPS tracking devices, often referred to as telemetry data. Here, the points in the point pattern reflect locations where the tracker has emitted a signal.

6.9.3.1   Telemetry data

As mentioned, if a study question is linked to a specific and large area of interest and if the animals are difficult to see from a distance, due to their size or behavior (e.g. frequent diving), the observation process needs to be considered as part of the model. The approaches to dealing with this issues we have discussed so far subsample small areas in which we can assume that all individuals are seen or where detection probability varies with dependence on distance from an observer. In both cases the area of interest is often (much) larger than the subsampled areas and approaches then extrapolate into the entire region. Recently a technology has become increasingly commonplace that does not involve an observer at all and is typically cheaper while being able to generate more information as data are collected by a tagging device [2, 63]. Using different types of technology, typically individual animals are tagged and their movement history is recorded on the basis of a signal that is emitted at regular or irregular intervals specifying the locations of the animal at a given time, see Figure 6.11 for an illustration. Other information along with the location (such as water temperature or salinity) is sometimes recorded as well. The use of this technology has increased over the last decade and information on a large range of different animal species, from butterflies via seals to wolves has been collected. It is beyond the scope of this chapter to discuss the many different scientific questions that motivate data collection on tagged animals. However, it is certainly fair to say that there are at least two main themes within the literature–those studies that are primarily interested in the movement process itself, i.e. the temporal process of what can be quantified as distances traveled and turning angles ("movement ecology") [55] and those that focus on modeling habitat preferences to gain an understanding of which habitats the animals prefer [1].

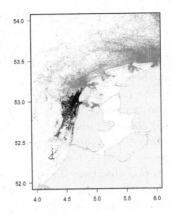

FIGURE 6.11
Distribution of harbour seals tracked in the Netherlands between 2007 and 2017 (transparent grey). In black, locations of all seals departing from a single haul-out site (Noorderhaaks). In red, a single foraging trip. Figure provided by S. Brasseur and G. Aarts (Wageningen Marine Research).

It is the latter case where spatial point processes have made an appearance [34]. The locations of the individuals as recorded by the tagging device form a point pattern, but current standard methodology cannot be readily applied to these patterns. Clearly, recorded patterns are likely to exhibit a strong temporal correlation among successive points and the intensity of observed points depends on the frequency with which the individual device records the location. In addition, the observation window has to be chosen particularly carefully. The observation window plays an important role in any point pattern analysis since the size of the window and hence the size of the pattern need to be big enough to reflect the phenomenon of interest and the shape of the observation window should not be arbitrarily chosen to avoid large areas around the pattern devoid of any points, with the study merely revealing that the data are clustered [45]. In the specific case of telemetry data we have the additional complication that the window cannot be chosen *a priori* as part of the study but is essentially chosen by the animals that are being analyzed. As a result, while an analysis aims to identify the areas in space that the animals prefer, one cannot distinguish whether they are not going to a specific location because they are actively avoiding it due to its unsuitability or whether it is just beyond the range that they can travel–and there might be the absolutely perfect habitat out there, but the animals can never reach it as it is too far away. This is particularly true for central foragers, such as seals [82] which need to haul out at specific locations. Clearly, an individual might also not move to otherwise preferable habitat due to competition with other individuals. Since only a small subset of the population can be tagged in many studies it is difficult to distinguish between areas that are not visited by the individuals under study as they are unsuitable, and those that are visited by other, unobserved, individuals.

6.9.3.2 Spatio-temporal patterns

This brings us to the next and final topic of this paper. The discussion so far has focused almost entirely on spatial point process modeling. Clearly, in practice, the ecology of individuals also concerns their behavior in time. In the point process literature, there has been relatively little discussion on spatio-temporal point processes in general. Developments so far have focused mainly on very specific data structures, typically in epidemiology [29], but there have been individual cases of point processes being used in other contexts such as housing construction [31] or armed conflict [98]. This might be partly due to a lack of data and hence little explicit motivation for the construction of models for the analysis spatio-temporal point pattern data. What is more, considering spatio-temporal models implies an increasing model complexity, and hence computational cost, which is likely to have hindered development. Certainly, however, another complication has been that it is difficult to say what a standard simple spatio-temporal data structure looks like, making it particularly difficult to consider a general class of standard spatio-temporal point process models. Specifically, this concerns, for instance, the issue of object persistence and the relative roles of spatial and temporal structures.

The spatio-temporal point process discussed in [97] is a (log Gaussian) Cox process based on a spatio-temporal random field, making the usual assumption of independence among points given the field. However, this assumption might not always be justifiable. Objects represented by the points in a point pattern vary in terms of their persistence relative to the time-scale of observations. For example individual trees in a forest live for many years and hence many of the trees are observed repeatedly in repeated censuses, while instantaneous events such as earthquakes have little or no duration. Very different issues of dependence among the points need to be considered in these cases. In fact the spatial dependence of the events might themselves be the subject of interest, for example, whether mortality is spatially structured, as in [13]. Taking time into account is less difficult when spatial pattern observed at different points in time may be seen as independent replicates of the same process [44, 52].

A related type of dependence that is likely to exist is that between the spatial and temporal structures in the pattern. While one question here concerns the validity of separability assumptions, the roles of the spatial and temporal structures vary among different studies and so does the relevant type of data aggregation. If there is an interest in changes in the overall spatial structure over time, aggregating the data into patterns in discrete time might be appropriate [3]. In contrast, if the global spatial pattern across, say, a number of years is of interest rather than its change through time temporal dependence is merely a nuisance, e.g. the local clustering in space at a small temporal resolution such as in earthquakes. When developing relevant models it is again important to capture the different dependence structures appropriately relating the model structure to the relevant scientific question. In terms of practicality, the role of the now spatio-temporal field clearly reflects these different dependence structures.

6.9.4   Conclusion

This chapter has highlighted that there clearly is a lot of work to do if one aims to make spatial point process modeling relevant and hence used in ecological research with the approaches discussed being just the beginning process. As part of this process, appropriate communication between method developers and users is important; this includes communication of the methods and of what they imply, in particular in terms of statistical inference. Specifically, communicating methodology across to users might be facilitated by demonstrating and pointing out the equivalence of data structures or methods familiar to ecologists or [25, 77] or by explicitly considering the relevance of the methodology [43].

Communication the implication of the specific choice of model and its assumptions, such as second order stationarity of the spatial field [14] or the issue of prior choice in a Bayesian context is also crucial. For instance, prior choice has been entirely absent from the above discussion, even though priors for Gaussian field in particular have to be carefully chosen to avoid one of two extremes–overfitting on the one hand or the spatial field covering up significant covariate effects due collinearity with the covariates on the other hand. Despite its obvious impact on inference and hence conclusions drawn by the user, in ecological applications an beyond, this discussion is almost entirely absent from the point process literature. As a result of interaction with potential users in the context of training workshops a discussion of the issue has begun, see for example [85, 86], providing an example of how interaction with users can motivate new statistical research that is not only relevant to a specific group of users, but also to those in other fields.

6.10   Acknowledgments

I would like to thank the following people for their comments and support: Geert Aarts, Fabian Bachl, Sophie Brasseur, David Burslem, Martin Dobrý, Esther Jones, Charlotte Jones-Todd, Alicia Ledo, Charles Paxton, Theoni Photopoulou, André Python, Daniel Simpson and Sophie Smout.

Bibliography

[1] G. Aarts, M. MacKenzie, B. McConnell, M. Fedak, and J. Matthiopoulos. Estimating space-use and habitat preference from wildlife telemetry data. *Ecography*, 31(1):140–160, 2008.

[2] G. Aarts, J. Fieberg, S. Brasseur, and J. Matthiopoulos. Quantifying the effect of habitat availability on species distributions. *Journal of Animal Ecology*, 82(6):1135–1145, 2013.

[3] L. Altieri, E. M. Scott EM, D. Cocchi, and J. B. Illian. A changepoint analysis of spatio-temporal point processes. *Spatial Statistics*, 2015. doi: 10.1016/j.spasta.2015.05.005.

[4] K. J. Anderson-Teixeira, S. J. Davies, A. C. Bennett, E. B. Gonzalez-Akre, H. C. Muller-Landau, S. J. Wright, K. Abu Salimand A. M. Almeyda Zambrano, A. Alonso, and J. L. Baltzer. CTFS-ForestGEO: a worldwide network monitoring forests in an era of global change. *Global Change Biology*, 21(2):528–549, 2015.

[5] F. E. Bachl. *inlabru web page*. https://www.inlabru.org, 2016. URL https://www.inlabru.org.

[6] A. Baddeley and R. Turner. Practical maximum pseudolikelihood for spatial point patterns. *Australian & New Zealand Journal of Statistics*, 42(3):283–322, 2000.

[7] A. Baddeley and R. Turner. Spatstat: an R package for analyzing spatial point patterns. *Journal of statistical software*, 12(6):1–42, 2005.

[8] A. Baddeley, J. Møller, and R. Waagepetersen. Non- and semiparametric estimation of interaction in inhomogenous point patterns. *Statistica Neerlandica*, 54:329–350, 2000.

[9] A. Baddeley, E. Rubak, and R. Turner. *Spatial point patterns: methodology and applications with R.* CRC Press, 2015.

[10] A. J. Baddeley and M. N. M. Van Lieshout. Area-interaction point processes. *Annals of the Institute of Statistical Mathematics*, 47(4):601–619, 1995.

[11] A. J. Baddeley and R. Turner. Spatstat: an R package for analyzing spatial point patterns. *Journal of Statistical Software*, 12:1–42, 2005.

[12] R. Bagchi and J. B. Illian. A method for analysing replicated point patterns in ecology. *Methods in Ecology and Evolution*, 6(4):482–490, 2015.

[13] R. Bagchi, P. A. Henrys, P. E. Brown, D. F. R. P. Burslem, P. J. Diggle, C. V. S. Gunatilleke, I. A. U. N. Gunatilleke, A. R. Kassim, R. Law, and S. Noor. Spatial patterns reveal negative density dependence and habitat associations in tropical trees. *Ecology*, 92(9):1723–1729, 2011.

[14] H. Bakka. *Modeling Spatial Dependencies using Barriers and Different Terrains*. PhD thesis, Norwegian University of Science and Technology, Trondheim, 2017.

[15] C. A. Baldeck, K. E. Harms, J. B. Yavitt, R. John, B. L. Turner, R. Valencia, H. Navarrete, S. J. Davies, G. B. Chuyong, D. Kenfack, et al. Soil resources and topography shape local tree community structure in tropical forests. *Proceedings of the Royal Society of London B: Biological Sciences*, 280(1753):20122532, 2013.

[16] D. L. Borchers, B. C. Stevenson, D. Kidney, L. Thomas, and T. A. Marques. A unifying model for capture–recapture and distance sampling surveys of wildlife populations. *Journal of the American Statistical Association*, 110(509):195–204, 2015.

[17] R. W. Brooker, F. T. Maestre, R. M. Callaway, C. L. Lortie, L. A. Cavieres, G. Kunstler, P. Liancourt, K. Tielborger, J. M. J. Travis, F. Anthelme, C. Armas, L. Coll, E. Corcket, S. Delzon, E. Forey, Z. Kikvidze, J. Olofsson, F. Pugnaire, Q. L. Quiroz, P. Saccone, K. Schiffers, M. Seifan, B. Touzard, and R. Michalet. Facilitation in plant communities: the past, the present, and the future. *Journal of Ecology*, 96:18–34, 2008.

[18] C. Brown, R. Law, J. B. Illian, and D. F. R. P. Burslem. Linking ecological processes with spatial and non-spatial patterns in plant communities. *Journal of Ecology*, 99 (6):1402–1414, 2011.

[19] C. Brown, D. F. R. P. Burslem, J. B. Illian, L. Bao, W. Brockelman, M. Cao, L. W. Chang, H. S. Dattaraja S. Davies, and C. V. S. Gunatilleke. Multispecies coexistence of trees in tropical forests: spatial signals of topographic niche differentiation increase with environmental heterogeneity. *Proceedings of the Royal Society of London B: Biological Sciences*, 280(1764):20130502, 2013.

[20] C. Brown, J. B. Illian, and D. F. R. P. Burslem. Success of spatial statistics in determining underlying process in simulated plant communities. *Journal of Ecology*, 104(1):160–172, 2016.

[21] S. T. Buckland, D. R. Anderson, K. P. Burnham, and J. L. Laake. *Distance sampling*. Wiley Online Library, 2005.

[22] S. T. Buckland, M. L. Burt, E. A. Rexstad, M. Mellor, A. E. Williams, and R. Woodward. Aerial surveys of seabirds: the advent of digital methods. *Journal of Applied Ecology*, 49(4):960–967, 2012.

[23] S. T. Buckland, E. A. Rexstad, C. S. Marques, and C. S. Oedekoven. *Distance sampling: methods and applications*. Springer, 2015.

[24] D. F. R. P. Burslem, N. C. Garwood, and S. C. Thomas. Tropical forest diversity – the plot thickens. *Science*, 291:606–607, 2001.

[25] A. Chakraborty, A. E. Gelfand, A. M. Wilson, A. M. Latimer, and J. A. Silander. Point pattern modelling for degraded presence-only data over large regions. *Journal of the Royal Statistical Society: Series C (Applied Statistics)*, 60(5):757–776, 2011.

[26] J. Chave. Neutral theory and community ecology. *Ecology Letters*, 7(3):241–253, 2004.

[27] T. Cornulier and V. Bretagnolle. Assessing the influence of environmental hetero-geneity on bird spacing patterns: a case study with two raptors. *Ecography*, 29(2): 240–250, 2006.

[22] P. J. Diggle, R. Menezes, and T. Su. Geostatistical inference under preferential sam-pling. *Journal of the Royal Statistical Society: Series C (Applied Statistics)*, 59(2): 191–232, 2010.

[29] P. J. Diggle, P. Moraga, B. Rowlingson, and B. M. Taylor. Spatial and spatio-temporal log-Gaussian Cox processes: Extending the geostatistical paradigm. *Statistical Science*, 28(4):542–563, 2013.

[30] M. Dobrý. The abundance of the little owl (*Athene noctua*) in Podunajská Rovina lowland in 2009 and 2010. *Slovak Raptor Journal*, 5:121–126, 2011.

[31] J. A. Duan, A. E. Gelfand, and C. F. Sirmans. Modeling space-time data using stochastic differential equations. *Bayesian Analysis*, 4(4):733–758, 2009.

[32] R. Durrett and S. Levin. Spatial aspects of interspecific competition. *Theoretical Population Biology*, 53:30–43, 1998.

[33] R. M. Fewster, B. C. Stevenson, and D. L. Borchers. Trace-contrast models for capture–recapture without capture histories. *Statistical Science*, 31(2):245–258, 2016.

[34] J. R. Fieberg, J. D. Forester, G. M. Street, D. H. Johnson, A. A. ArchMiller, and J. Matthiopoulos. Used-habitat calibration plots: A new procedure for validating species distribution, resource selection, and step-selection models. *Ecography*, 2017.

[35] M. Ghorbani. Cauchy cluster process. *Metrika*, 76(5):697–706, 2013.

[36] A. Gilles, S. Viquerat, E. A. Becker, K. A. Forney, S. C. V. Geelhoed, J. Haelters, J. Nabe-Nielsen, M. Scheidat, U. Siebert, and S. Sveegaard. Seasonal habitat-based density models for a marine top predator, the harbor porpoise, in a dynamic environ-ment. *Ecosphere*, 7(6), 2016.

[37] J. Goldstein, M. Haran, I. Simeonov, J. Fricks, and F. Chiaromonte. An attraction–repulsion point process model for respiratory syncytial virus infections. *Biometrics*, 71(2):376–385, 2015.

[38] P. Haase. Spatial pattern analysis in ecology based on Ripley's K-function. *Journal of Vegetation Science*, 6:575–582, 1995.

[39] P. S. Hammond, K. Macleod, P. Berggren, D. L. Borchers, L. Burt, A. Cañadas, G. Desportes, G. P. Donovan, A. Gilles, and D. Gillespie. Cetacean abundance and distribution in european atlantic shelf waters to inform conservation and management. *Biological Conservation*, 164:107–122, 2013.

[40] L. P. Ho and D. Stoyan. Modelling marked point patterns by intensity-marked Cox processes. *Statistical Probability Letters*, 78:1194ñ1199, 2008.

[41] H. Högmander and A. Särkkä. Multitype spatial point patterns with hierarchical interactions. *Biometrics*, 55:1051–1058, 1999.

[42] J. B. Illian and D. F. R. P. Burslem. Contributions of spatial point process modelling to biodiversity theory. *Journal de la Société Française de Statistique*, 148:9–29, 2007.

[43] J. B. Illian and D. F. R. P. Burslem. Improving the usability of spatial point processes methodology – an interdisciplinary dialogue between statistics and ecology. *Advances in Statistical Analysis*, 2017.

[44] J. B. Illian and D. K. Hendrichsen. Gibbs point processes with mixed effects. *Environmentrics*, 21:341–353, 2010.

[45] J. B. Illian, A. Penttinen, H. Stoyan, and D. Stoyan. *Statistical Analysis and Modelling of Spatial Point Patterns*. Wiley, Chichester, 2008.

[46] J. B. Illian, S. H. Soerbye, H. Rue, and D. Hendrichsen. Using INLA to fit a complex point process model with temporally varying effects–a case study. *Journal of Environmental Statistics*, 2012.

[47] J. B. Illian, S. H. Sørbye, and H. Rue. A toolbox for fitting complex spatial point process models using integrated nested Laplace approximation (INLA). *The Annals of Applied Statistics*, 6(4):1499–1530, 2012.

[48] J. B. Illian, S. Martino, S. H. Sørbye, J. B. Gallego-Fernández, M. Zunzunegui, M. P. Esquivias, and J. M. J. Travis. Fitting complex ecological point process models with integrated nested laplace approximation. *Methods in Ecology and Evolution*, 4(4): 305–315, 2013.

[49] J. B. Illian, J. Møller, and R. P. Waagepetersen. Hierarchical spatial point process analysis for a plant community with high biodiversity. *Journal of Environmental and Ecological Statistics, first online, DOI: 10.1007/s10651-007-0070-8*, 2007.

[50] D. S. Johnson, M. B. Hooten, and C. E. Kuhn. Estimating animal resource selection from telemetry data using point process models. *Journal of Animal Ecology*, 82(6): 1155–1164, 2013.

[51] C. M. Jones-Todd. *Modelling Complex Dependencies Inherent in Spatial and Spatio-Temporal Point Pattern Data*. PhD thesis, University of St Andrews, 2017.

[52] R. King, J. B. Illian, S. E. King, G. F. Nightingale, and D. K. Hendrichsen. A bayesian approach to fitting gibbs processes with temporal random effects. *Journal of agricultural, biological, and environmental statistics*, pages 1–22, 2012.

[53] W. R. Koski, T. A. Thomas, D. W. Funk, and A. M. Macrander. Marine mammal sightings by analysts of digital imagery versus aerial surveyors: a preliminary comparison. *Journal of Unmanned Vehicle Systems*, 1(01):25–40, 2013.

[54] G. Lan, S. Getzin, T. Wiegand, Y. Hu, G. Xie, H. Zhu, and M. Cao. Spatial distribution and interspecific associations of tree species in a tropical seasonal rain forest of china. *PloS one*, 7(9):e46074, 2012.

[55] R. Langrock, R. King, J. Matthiopoulos, L. Thomas, D. Fortin, and J. M. Morales. Flexible and practical modeling of animal telemetry data: hidden Markov models and extensions. *Ecology*, 93(11):2336–2342, 2012.

[56] R. Law, U. Dieckmann, and J.A.J. Metz. Introduction. In U. Dieckmann, R. Law, and J.A.J. Metz, editors, *The geometry of ecological interactions: Simplifying spatial complexity*, pages 1–6. Cambridge University Press, Cambridge, 2000.

[57] R. Law, J. B. Illian, D. F. R. P. Burslem, G. Gratzer, C. V. S. Gunatilleke, and I. A. U. N. Gunatilleke. Ecological information from spatial patterns of plants: insights from point process theory. *Journal of Ecology*, 97:616–628, 2009.

[58] A. Ledo, J. B. Illian, S. A. Schnitzer, S. J. Wright, J. W. Dalling, and D. F. R. P. Burslem. Lianas and soil nutrients predict fine-scale distribution of above-ground biomass in a tropical moist forest. *Journal of Ecology*, 104(6):1819–1828, 2016.

[59] F. Lindgren and H. Rue. Bayesian spatial modelling with R-INLA. *Journal of Statistical Software*, 63(19), 2015.

[48] F. Lindgren, H. Rue, and J. Lindström. An explicit link between Gaussian fields and Gaussian Markov random fields: The stochastic partial differential equation approach (with discussion). *Journal of the Royal Statistical Society. Series B. Statistical Methodology*, 73(4):423–498, September 2011.

[61] E. Lingua, P. Cherubini, R. Motta, and P. Nola. Spatial structure along an altitudinal gradient in the italian central alps suggests competition and facilitation among coniferous species. *Journal of Vegetation Science*, 19(3):425–436, 2008.

[62] M. De Luis, J. Raventós, T. Wiegand, and Jose J. C. González-Hidalgo. Temporal and spatial differentiation in seedling emergence may promote species coexistence in mediterranean fire-prone ecosystems. *Ecography*, 31(5):620–629, 2008.

[63] J. Matthiopoulos and G. Aarts. In I. L. Boyd, D. Bowen, and S. Iverson, editors, *Marine Mammal Ecology and Conservation: A handbook of techniques*.

[64] B. J. McGill, R. S. Etienne, J. S. Gray, D. Alonso, M. J. Anderson, H. K. Benecha, M. Dornelas, B. J. Enquist, J. L. Green, and F. He. Species abundance distributions: moving beyond single prediction theories to integration within an ecological framework. *Ecology Letters*, 10(10):995–1015, 2007.

[65] D. L. Miller, M. L. Burt, E. A. Rexstad, and L. Thomas. Spatial models for distance sampling data: recent developments and future directions. *Methods in Ecology and Evolution*, 4(11):1001–1010, 2013.

[66] J. Møller and E. Rubak. Functional summary statistics for point processes on the sphere with an application to determinantal point processes. *Spatial Statistics*, 18: 4–23, 2016.

[67] J. Møller and R. Waagepetersen. *Statistical Inference and Simulation for Spatial Point Processes*. Chapman & Hall, London, 2004.

[68] J. Møller and R. P. Waagepetersen. Modern statistics for spatial point processes. *Scandinavian Journal of Statistics*, 34(4):643–684, 2007.

[69] J. Møller, A. R. Syversveen, and R. P. Waagepetersen. Log Gaussian Cox processes. *Scandinavian Journal of Statistics*, 25:451–482, 1998.

[70] M. Myllymäki. The spptest software, 2017.

[71] M. Myllymäki, P. Grabarnik, H. Seijo, and D. Stoyan. Deviation test construction and power comparison for marked spatial point patterns. *Spatial Statistics*, 11:19–34, 2015.

[72] G. F. Nightingale, J. B. Illian, and R. King. Pairwise interaction point processes for modelling bivariate spatial point patterns in the presence of interaction uncertainty. *Journal of Environmental Statistics*, 2015.

[73] J. Ohser and D. Stoyan. On the second-order and orientation analysis of planar stationary point processes. *Biometrical Journal*, 23(6):523–533, 1981.

[74] A. Python, J. B. Illian, C. M. Jones-Todd, and M. Blangiardo. A Bayesian approach to modelling fine-scale spatial dynamics of non-state terrorism: World study, 2002-2013. *arXiv preprint arXiv:1610.01215*, 2016.

[75] T. Rajala. The spatgraph software, 2017.

[76] A. P. Rayburn and T. Wiegand. Individual species–area relationships and spatial patterns of species diversity in a great basin, semi-arid shrubland. *Ecography*, 35(4): 341–347, 2012.

[77] I. W. Renner and D. I. Warton. Equivalence of MAXENT and Poisson point process models for species distribution modeling in ecology. *Biometrics*, 69(1):274–281, 2013.

[78] I. W. Renner, J. Elith, A. Baddeley, W. Fithian, T. Hastie, S. J. Phillips, G. Popovic, and D. I. Warton. Point process models for presence-only analysis. *Methods in Ecology and Evolution*, 6(4):366–379, 2015.

[79] B. D. Ripley. Modelling spatial patterns. *Journal of the Royal Statistical Society. Series B (Methodological)*, pages 172–212, 1977.

[80] B. D. Ripley. *Statistical inference for spatial processes*. Cambridge University Press, 1991.

[81] H. Rue, S. Martino, and N. Chopin. Approximate Bayesian inference for latent Gaussian models using integrated nested Laplace approximations (with discussion). *Journal of the Royal Statistical Society, Series B*, 71(2):319–392, 2009.

[82] R. J. Sharples, M. L. MacKenzie, and P. S. Hammond. Estimating seasonal abundance of a central place forager using counts and telemetry data. *Marine Ecology Progress Series*, 2009.

[83] S. Shirota and A. E. Gelfand. Inference for log Gaussian Cox processes using an approximate marginal posterior. *arXiv preprint arXiv:1611.10359*, 2016.

[84] D. Simpson, J. B. Illian, F. Lindgren, S. H. Sørbye, and H. Rue. Going off grid: computationally efficient inference for log-Gaussian Cox processes. *Biometrika*, 103 (1):49–70, 2016.

[85] S. H. Sørbye and H. Rue. Scaling intrinsic Gaussian Markov random field priors in spatial modelling. *Spatial Statistics*, 8:39–51, 2014.

[86] S. H. Sørbye, J. B. Illian, D. P. Simpson, and D. F. R. P. Burslem. Careful prior specification avoids incautious inference for log-Gaussian Cox point processes. *arXiv preprint arXiv:1709.06781*, 2017.

[87] D. Stoyan and H. Stoyan. *Fractals, random shapes and point fields: Methods of geometrical statistics.* 1994.

[88] E. Velázquez, I. Martínez, S. Getzin, K. A. Moloney, and T. Wiegand. An evaluation of the state of spatial point pattern analysis in ecology. *Ecography*, 2016.

[89] R. Waagepetersen and Y. Guan. Two-step estimation for inhomogeneous spatial point processes. *Journal of the Royal Statistical Society, Series B*, 71:685–702, 2009.

[90] R. Waagepetersen and T. Schweder. Likelihood-based inference for clustered line transect data. *Journal of Agricultural, Biological, and Environmental Statistics*, 11(3): 264–279, 2006.

[91] R. Waagepetersen, Y. Guan, A. Jalilian, and J. Mateu. Analysis of multispecies point patterns by using multivariate log-gaussian cox processes. *Journal of the Royal Statistical Society: Series C (Applied Statistics)*, 65(1):77–96, 2016.

[92] R. P. Waagepetersen. An estimating function approach to inference for inhomogeneous Neyman–Scott processes. *Biometrics*, 63(1):252–258, 2007.

[93] T. Wiegand and K. A. Moloney. Rings, circles, and null-models for point pattern analysis in ecology. *Oikos*, 104(2):209–229, 2004.

[94] T. Wiegand and K. A. Moloney. *Handbook of spatial point-pattern analysis in ecology.* CRC Press, 2013.

[95] T. Wiegand, S. Gunatilleke, N. Gunatilleke, and T. Okuda. Analyzing the spatial structure of a sri lankan tree species with multiple scales of clustering. *Ecology*, 88 (12):3088–3102, 2007.

[96] L. D. Williamson, K. L. Brookes, B. E. Scott, I. M. Graham, G. Bradbury, P. S. Hammond, and P. M. Thompson. Echolocation detections and digital video surveys provide reliable estimates of the relative density of harbour porpoises. *Methods in Ecology and Evolution*, 2016.

[97] Y. Yuan, F. E. Bachl, D. L. Borchers, F. Lindgren, J. B. Illian, S. T. Buckland, H. Rue, and T. Gerrodette. Point process models for spatio-temporal distance sampling data. *Annals of Applied Statistics*, 11:2270–2297, 2017.

[98] A. Zammit-Mangion, M. Dewar, V. Kadirkamanathan, and G. Sanguinetti. Point process modelling of the Afghan War Diary. *Proceedings of the National Academy of Sciences*, 109(31):12414–12419, 2012.

[99] S. Zhang, Y. Huang, and R. Zang. The assembly and interactions of tree species in tropical forests based on spatial analysis. *Ecosphere*, 8(7):e01903, 2017.

[100] T. Zillio and R. Condit. The impact of neutrality, niche differentiation and species input on diversity and abundance distributions. *Oikos*, 116(6):931–940, 2007.

7

Data assimilation

Veronica J. Berrocal

Department of Biostatistics, University of Michigan, School of Public Health, Ann Arbor, MI

CONTENTS

7.1 Introduction

The evolution in time of environmental systems, in particular, geophysical ones, can be described via systems of partial differential equations. In most cases, these equations, that can be either linear or non-linear, embody the physical laws that are assumed to govern the temporal dynamics of the system, and need initial, and in some cases, boundary conditions in order to be integrated forward in time and thus predict the next state of the system. Often, observations of the initial states are not completely available. This was recognized in the seventeenth and eighteenth centuries by mathematicians and astronomers, among which Euler, Lagrange and Laplace, who were trying to determine the orbits of comets and planets using Newton's laws of gravitation [43]. In particular, in his book "Theory of Motion of Heavenly Bodies", Gauss elaborated how available observations of the physical system - in his case, the location of the comet Ceres with respect to the Sun - were not easily translatable into initial conditions for the variables of the governing dynamic equations. In particular, he noted that in order to obtain the best approximation to the (dynamical) behavior of environmental systems, more observations than the minimum number of observations required to solve the (partial) differential equations were needed. More specifically, Gauss wrote "[..] since all our observations and measurements are nothing more than approximations to the truth, the same must be true of all calculations resting on them, and the highest of all computations made concerning concrete phenomena must be to approximate, as nearly as practicable, to the truth. But this can be accomplished in no other way than by *suitable combination of more observations* than the number absolutely requisite for the determination of unknown quantities" [43], thus, in some ways, setting the basis for data assimilation, or combination of various sources of information.

In this chapter, we provide an overview of methods used for data assimilation, starting with a review of some of the algorithms more commonly used in numerical weather forecasting, where data assimilation is operationally used, and concluding with a review of some of the more standard statistical approaches for data assimilation or data fusion proposed in the environmental statistical literature. We highlight that a vast literature on data assimilation is available in the atmospheric sciences: our review does not claim to be exhaustive nor complete in this regard. It is simply an attempt to introduce the interested statistical readership to the topic; for more details on methods for data assimilation from a numerical weather forecasting or, more generally, from a dynamic atmospheric science perspective we turn readers to the excellent books by Evensen [21], Kalnay [39], Lahoz et al. [42], and Lewis et al. [43] among others.

7.2   Algorithms for data assimilation

Data assimilation, that is, the combination of different sources of information, has been very successful and widely used in atmospheric sciences [10, 39, 53], particularly in numerical weather forecasting where it is routinely and operationally implemented to derive weather forecasts. Traditionally, in the geophysical sciences, data assimilation has been carried out through numerical algorithms that differ depending on whether the various data sources are integrated sequentially or not. In most cases it is possible to establish a parallel between some of the classical data assimilation algorithms and statistical estimators obtained under particular frameworks.

We now review the most classical algorithms for data assimilation, grouping them in different categories, depending on whether the environmental system is static vs. dynamic, the dynamical model is deterministic vs. stochastic and the observations are assimilated sequentially vs. all at once.

Let $\mathcal{D}$ be a spatial domain, subset of $\mathbf{R}^d$, and $\mathbf{x}_t \in \mathbf{R}^n$ denote the state of an environmental system operating in $\mathcal{D}$ at discrete times t, $t = 0, \ldots, T$ and at a given location in $\mathcal{D}$. Through discretization of space, it is customary to consider the state of the system at time t only at a finite set $\mathcal{R}$ of locations within $\mathcal{D}$, usually on a grid. For ease of notation, in the remainder we will not formally express the dependence of the state of the system on location, and we will simply adopt the notation $\mathbf{x}_t$ to denote the state of the system at a gridded location $\mathbf{r}$ in $\mathcal{R}$ at time t. Using physical principles and laws, the state of the environmental system at any time t can be mathematically described using n state-space variables governing its dynamics, so that its state at time $t + 1$, $\mathbf{x}_{t+1}$, is linked to the state at time t, $\mathbf{x}_t$, via the following relationship:

$$\mathbf{x}_{t+1} = \mathcal{M}_t(\mathbf{x}_t) \tag{7.1}$$

where $\mathcal{M}_t$ indicates the set of partial or ordinary differential equations involving the n state-space variables according to which the system evolves from time t to time $t + 1$ (e.g. the conservation law of mass, Newton's laws of motion, laws of radiation, convection, etc.). Thus, $\mathcal{M}_t$ denotes the mathematical and numerical model of the system dynamics and (7.1) is referred to as the *state-space model*. If $\mathcal{M}_t$ does not vary with time, we have a static system, otherwise the system is deemed dynamic. The model $\mathcal{M}_t$ might not be linear. In what follows, we will present data assimilation algorithms under the assumption that $\mathcal{M}_t$

is linear, which implies that (7.1) can be expressed as

$$\mathbf{x}_{t+1} = \mathbf{M}_t \mathbf{x}_t \tag{7.2}$$

with $\mathbf{M}_t$ $n \times n$ matrix. In practical implementations, if $\mathcal{M}_t$ is not linear, it is often replaced and approximated by its linearization [42].

The model in (7.2) is a deterministic one since knowing the initial state of the system $\mathbf{x}_0$ uniquely determines its evolution in time. If we assume that the state-space model is not perfect, we have a stochastic model where

$$\mathbf{x}_{t+1} = \mathbf{M}_t \mathbf{x}_t + \boldsymbol{\eta}_{t+1} \tag{7.3}$$

with $\boldsymbol{\eta}_{t+1}$ serially uncorrelated white noise with $n \times n$ covariance matrix $\mathbf{Q}_{t+1}$, $t = 0, 1, \ldots, T$. In the atmospheric science literature, the error terms $\boldsymbol{\eta}_t$ are often called external forcings.

In addition to the state-space model in (7.2) or (7.3), at time t we also have access to observations of the system at a set $\mathcal{S}$ of locations that might not coincide with the gridded locations in $\mathcal{R}$, i.e. we have access to a vector $\mathbf{y}_t \in \mathbf{R}^m$ at locations in $\mathcal{S} \subset \mathcal{D}$. Again, as for the vector $\mathbf{x}_t$, we do not explicitly represent the dependence of $\mathbf{y}_t$ on space.

It is important to notice that the observations might not be observations of the state space variables themselves. For example, in a weather forecasting model, the state-space variables might include temperature and precipitation, while the observations available are observations of earth and atmosphere radiations measured by a satellite, the reflected energy measured by a radar, etc. As the locations where the observations are collected might not coincide with the locations where the model in (7.2) or (7.3) is evaluated, a mapping at time t, $\mathcal{H}_t : \mathbf{R}^n \longrightarrow \mathbf{R}^m$, that relates the state-space ($\mathbf{R}^n$) to the observation space ($\mathbf{R}^m$) is introduced, and it is assumed that:

$$\mathbf{y}_t = \mathcal{H}_t \left(\mathbf{x}_t \right) + \boldsymbol{\epsilon}_t^{(O)}$$

The mapping $\mathcal{H}_t$ is often derived using empirical or physical laws and it usually includes an interpolation operator to account for the fact that $\mathcal{R} \neq \mathcal{S}$. As for the dynamical model $\mathcal{M}_t$, the operator $\mathcal{H}_t$ might not be linear, but it can be approximated with its linearization. In what follows, we will assume that $\mathcal{H}_t$ is linear, thus yielding the following *observation equation* or relationship between observations $\mathbf{y}_t$ and state of the system $\mathbf{x}_t$:

$$\mathbf{y}_t = \mathbf{H}_t \mathbf{x}_t + \boldsymbol{\epsilon}_t^{(O)} \tag{7.4}$$

with $\mathbf{H}_t$ $m \times n$ matrix and $\boldsymbol{\epsilon}_t^{(O)}$ errors accounting for additive instrumentation or representativeness errors in the observations at time t [42] . The $\boldsymbol{\epsilon}_t^{(O)}$ are often modeled to have mean $\mathbf{0}$ and $m \times m$ covariance matrix $\mathbf{R}_t$, $t = 0, 1, \ldots, T$.

In addition to observations, a prior estimate $\mathbf{x}_0^{(B)}$, also called a *background estimate* of the initial state is often available through a previous forecast. The main assumption in data assimilation is that, due to various sources of uncertainties (mainly in the initial state of the system), even under a perfect model (e.g. under equation (7.2)), the estimated states of the system $\mathbf{x}_t^{(B)}, t = 0, 1, \ldots, T$, obtained from $\mathbf{x}_0^{(B)}$ through repeated applications of (7.2) are not the true states $\mathbf{x}_t$, $t = 0, \ldots, T$; rather they contain errors and are related to the true states via

$$\mathbf{x}_t^{(B)} = \mathbf{x}_t + \boldsymbol{\epsilon}_t^{(B)} \qquad t = 0, 1, \ldots, T \tag{7.5}$$

with $\boldsymbol{\epsilon}_t^{(B)}$ serially uncorrelated errors with covariance matrix $\mathbf{P}_t^{(B)}$. Further, the observation errors $\boldsymbol{\epsilon}_t^{(O)}$ and the errors $\boldsymbol{\epsilon}_t^{(B)}$ are assumed to be uncorrelated for each t, e.g. $E\left(\boldsymbol{\epsilon}_t^{(O)} \cdot \boldsymbol{\epsilon}_t^{(B)} \right) = \mathbf{0}$.

Given these settings, the problem that data assimilation aims to solve is to find the estimate of the true states x_t, $t = 0, 1, \ldots, T$ that best fit the observations subject to the constraints imposed by the dynamics of the system.

7.2.1 Optimal interpolation

We start by describing the algorithm of optimal interpolation (OI), which can be used for non-sequential assimilation of observations under the assumption of a perfect state-space model (e.g. assuming that (7.2) holds).

The root of OI is in the concept of least squares, as it is clearly exemplified by the following example. Consider the problem of estimating an unobserved univariate variable x when two independent, noisy, measurements of x, y_1 and y_2, are available, and $y_i = x + \epsilon_i$ with $E(\epsilon_i) = 0$ and $E(\epsilon_i^2) = \sigma_i^2$ for $i = 1, 2$. If the unbiased estimate $x^{(a)}$ of x is constrained to be of the form $x^{(a)} = a_1 y_1 + a_2 y_2$ with a_1, a_2 such that the mean squared error $E\left(x^{(a)} - x\right)^2$ is minimized, it results that $x^{(a)}$ must be equal to

$$x^{(a)} = \frac{\sigma_1^2}{\sigma_1^2 + \sigma_2^2} y_1 + \frac{\sigma_2^2}{\sigma_1^2 + \sigma_2^2} y_2 \tag{7.6}$$

which can also be re-expressed as

$$x^{(a)} = y_1 + \frac{\sigma_1^2}{\sigma_1^2 + \sigma_2^2}(y_2 - y_1)$$

In other words, $x^{(a)}$ can be obtained by adding to the measurement y_1 the difference between the two measurements, y_2 and y_1, appropriately weighted.

Extending this framework to a multivariate setting: assume that a background estimate $x_0^{(B)}$ of the true state of the system x_0 is available along with observations y_0 relative to the same time period (or over a short enough time period to be considered contemporary). Then, the optimal interpolation algorithm seeks an estimate $x_0^{(a)}$ of the true state x_0 of the form

$$x_0^{(a)} = x_0^{(B)} + K\left(y_0 - H_0 x_0^{(B)}\right)$$

with K $m \times n$ matrix, called the *gain matrix*, chosen so that the mean squared error $E\left(x_0 - x_0^{(a)}\right)^2$ is minimized.

Under the assumption that the observation errors $\epsilon_0^{(O)}$ have mean 0 and covariance matrix R_0, the background errors $\epsilon_0^{(B)}$ have mean 0 and covariance matrix $P_0^{(B)}$, and they are uncorrelated, the *optimal linear* estimate $x_0^{(a)}$ of x_0, or the *optimal interpolation* estimate $x_0^{(a)}$ of x_0 is given by:

$$x_0^{(a)} = x_0^{(B)} + Kd \tag{7.7}$$

where $d = y_0 - H_0 x_0^{(B)}$ is called the *innovation vector*. Minimizaton of the mean squared error over K, leads to the following expression for the gain matrix K:

$$K = P_0^{(B)} H_0' \cdot \left(H_0 P_0^{(B)} H_0' + R_0\right)^{-1} \tag{7.8}$$

Under such choice for $x_0^{(a)}$ and K as in (7.7) and (7.8), the uncertainty in the OI estimate $x_0^{(a)}$, also called *analysis*, is quantified by the covariance of the analysis errors and is equal to:

$$A = (I_n - KH_0) P_0^{(B)}$$

where $\mathbf{I}_n$ is the $n \times n$ identity matrix. We note that the if the dynamical model $\mathcal{M}_0$ is linear and so is the mapping $\mathcal{H}_0$ (e.g. equations (7.2) and (7.4) hold) at time $t = 0$, then the expression in (7.7) provides the Best Linear Unbiased Estimator (BLUE) of the true state $\mathbf{x}_0$ given the two independent sources of information, $\mathbf{x}_0^B$ and $\mathbf{y}_0$, and it has minimum variance. We also note the parallel between OI and simple kriging [3, 16], which also yields a Best Linear Unbiased Estimator.

For a practical implementation of the Optimal Interpolation algorithm, it is necessary to know the background error covariance matrix $\mathbf{P}_0^{(B)}$ and the observation error covariance matrix $\mathbf{R}_0$. Usually the latter is estimated from estimates of the instrument errors, whereas for the former, which has a crucial impact on the results, different approaches are used. Some choices include: to use an isotropic covariance function (e.g. Gaussian covariance function as in Gandin [26]), or, define the matrix $\mathbf{P}_0^{(B)}$ so that only points within a radius of influence of a grid point $\mathbf{r} \in \mathcal{R}$ are contributing to the assimilated state $\mathbf{x}^{(a)}$ (e.g. see successive corrections in the Bratseth [9] scheme).

7.2.2 Variational approaches

In variational algorithms for data assimilation, the analysis $\mathbf{x}_0^{(a)}$ at time $t = 0$ is obtained by minimizing an objective function. Two different types of variational algorithms are commonly used, the 3D-Var and the 4D-Var: the former is employed for non-sequential assimilation of observations and background estimate at a given time point, while the latter is used for assimilation of a time series of observations. Both algorithms can be employed under the assumption of a perfect dynamical model leading to what is called a variational algorithm with *strong constraints*. In addition, the 4D-Var algorithm can be applied also to stochastic state-space models leading to what is called a 4D-Var algorithm with *weak constraints*.

In the 3D-Var algorithm, given a background estimate $\mathbf{x}_0^{(B)}$, observations $\mathbf{y}_0$, and a perfect dynamic model, the analysis $\mathbf{x}_0^{(a)}$ is obtained by minimizing, with respect to $\mathbf{x}_0$, the following objective function

$$J(\mathbf{x}_0) = \frac{1}{2} \left(\mathbf{x}_0 - \mathbf{x}_0^{(B)} \right)' \mathbf{P}_0^{(B)-1} \left(\mathbf{x}_0 - \mathbf{x}_0^{(B)} \right) + \frac{1}{2} (\mathbf{y}_0 - \mathbf{H}_0 \mathbf{x}_0)' \mathbf{R}_0^{-1} (\mathbf{y}_0 - \mathbf{H}_0 \mathbf{x}_0) \quad (7.9)$$

which measures the deviations of $\mathbf{x}_0$ from the background estimate $\mathbf{x}_0^{(B)}$ in the state space, and from the observations $\mathbf{y}_0$ in the observation space. Under the assumption that the background errors $\boldsymbol{\epsilon}_0^{(B)}$ and the observation errors $\boldsymbol{\epsilon}_0^{(O)}$ are uncorrelated, have both mean $\mathbf{0}$ and covariance matrices $\mathbf{P}_0^{(B)}$ and $\mathbf{R}_0$, respectively, $\mathbf{x}_0^{(a)}$ is given by:

$$\mathbf{x}_0^{(a)} = \mathbf{x}_0^{(B)} + \mathbf{K} \left(\mathbf{y}_0 - \mathbf{H}_0 \mathbf{x}_0^{(B)} \right) \quad (7.10)$$

where the gain matrix $\mathbf{K}$ is given by

$$\mathbf{K} = \left(\mathbf{P}_0^{(B)-1} + \mathbf{H}_0' \mathbf{R}_0^{-1} \mathbf{H}_0 \right)^{-1} \mathbf{H}_0' \mathbf{R}_0^{-1}. \quad (7.11)$$

This latter expression is equivalent to the formulation of $\mathbf{K}$ under the OI algorithm as per the Sherman-Morrison-Woodbury formula [62, 72], thus leading to the same closed form expression for the analysis $\mathbf{x}_0^{(a)}$ under OI and 3D-Var.

Despite this apparent equivalence, it is important to notice that while in the OI algorithm the minimization of the mean squared error is done with respect to the weight matrix (and thus with respect to $\mathbf{K}$), in the 3D-Var algorithm the minimization of (7.9) is performed with respect to $\mathbf{x}_0$. Moreover, since in most geophysical systems, the dimensions of $\mathbf{x}_0^{(B)}$ and

y_0 are, respectively, of the order of 10^7 and 10^6, computing the gain matrix $\mathbf{K}$ in closed form using either (7.8) or (7.11) is computationally prohibitive. Given the large dimension of $\mathbf{x}_0$, minimization of the objective function in (7.9) is performed by solving the corresponding gradient equations using the so-called *adjoint method* [52]. For more details on the adjoint method and other numerical strategies to solve the gradient equations, the reader is referred to Lewis et al. [43].

As the dimension of the observation space is smaller than that of the state space, Silva et al. [63] introduced the Physical Space Analysis System (PSAS) which minimizes the objective function in (7.9) in the observation space rather than in the state space.

We highlight how, as for the OI algorithm, if both the dynamical model $\mathcal{M}_0$ and the observation mapping $\mathcal{H}$ are linear, the 3D-Var analysis $\mathbf{x}_0^{(a)}$ in (7.10) is the BLUE with minimum variance. In addition, it is also the maximum a posterior Bayesian estimate (or posterior mode) of the true state $\mathbf{x}_0$ if the data assimilation problem is formulated in the following way: let the background estimate $\mathbf{x}_0^{(B)}$ be the prior guess of the true state $\mathbf{x}_0$; in other words, let $\mathbf{x}_0^{(B)}$ be the mean of a Gaussian prior distribution on $\mathbf{x}_0$ with covariance matrix $\mathbf{P}_0^{(B)}$, and let the observations y_0 contribute to a Gaussian likelihood with mean $\mathbf{H}_0\mathbf{x}_0$ and covariance matrix $\mathbf{R}_0$. Then, the posterior distribution of $\mathbf{x}_0$ is Gaussian with a posterior mean equal to $\left(\mathbf{H}_0'\mathbf{R}_0\mathbf{H}_0 + \mathbf{P}_0^{(B)-1}\right)^{-1} \cdot \left(\mathbf{H}_0'\mathbf{R}_0^{-1}\mathbf{y} + \mathbf{P}_0^{(B)-1}\mathbf{x}_0^{(B)}\right)$, which, with algebraic manipulations, can be re-expressed as in (7.10)-(7.11).

If observations $y_0, y_1, \ldots, y_T$ are available at times $0, 1, \ldots, T$, the 3D-Var algorithm can be extended to the 4D-Var algorithm which assimilates the entire flow of observations and yields an estimate $\mathbf{x}_0^{(a)}$ of the initial state of the system whose trajectory in time, according to the dynamical model, best fits the data. More specifically, if the dynamical model is assumed perfect (e.g. (7.2) holds), the 4D-Var algorithm with *strong constraints* [61] seeks an analysis $\mathbf{x}_0^{(a)}$ by minimizing the objective function

$$J(\mathbf{x}_0) = \frac{1}{2}\left(\mathbf{x}_0 - \mathbf{x}_0^{(B)}\right)'\mathbf{P}_0^{(B)-1}\left(\mathbf{x}_0 - \mathbf{x}_0^{(B)}\right) + \frac{1}{2}\sum_{t=0}^{T}(\mathbf{y}_t - \mathbf{H}_t\mathbf{x}_t)'\mathbf{R}_t^{-1}(\mathbf{y}_t - \mathbf{H}_t\mathbf{x}_t)$$

with respect to $\mathbf{x}_0$ where $\mathbf{x}_{t+1} = \mathbf{M}_t\mathbf{x}_t$, $t = 0, \ldots, T$.

On the other hand, if the state-space model is stochastic and (7.3) holds with $E(\boldsymbol{\eta}_t) = 0$ and $E(\boldsymbol{\eta}_t^2) = \mathbf{Q}_t$ for $t = 1, \ldots, T$, the 4D-Var algorithm with weak constraints [60, 74] estimates the initial state $\mathbf{x}_0$ by minimizing the following objective function

$$\begin{aligned}
J(\mathbf{x}_0) &= \frac{1}{2}\left(\mathbf{x}_0 - \mathbf{x}_0^{(B)}\right)'\mathbf{P}_0^{(B)-1}\left(\mathbf{x}_0 - \mathbf{x}_0^{(B)}\right) \\
&+ \frac{1}{2}\sum_{t=0}^{T}(\mathbf{y}_t - \mathbf{H}_t\mathbf{x}_t)'\mathbf{R}_t^{-1}(\mathbf{y}_t - \mathbf{H}_t\mathbf{x}_t) \\
&+ \frac{1}{2}\sum_{t=0}^{T-1}(\mathbf{x}_{t+1} - \mathbf{M}_t\mathbf{x}_t)'\mathbf{Q}_{t+1}^{-1}(\mathbf{x}_{t+1} - \mathbf{M}_t\mathbf{x}_t)
\end{aligned}$$

Variational algorithms have been very successful in producing physically realistic results for meteorological and oceanographic data [13, 20, 66, 67], and since 1997 they have been operationally used at several meteorological and weather service centers, including the European Centre for Medium-Range Weather Forecasts [40], the Japan Meteorological Agency, the Meteorological Office in the UK, and the Meteorological Service of Canada.

7.2.3 Sequential approaches: the Kalman filter

Even though the 4D-Var algorithm allows to assimilate observations over time, it does not allow to update the variance in the assimilated state over time. This is instead possible in the Kalman filter, originally proposed by Kalman [37] and Kalman and Bucy [38] in an engineering context. In particular, in the Kalman filter the covariance matrix of the analysis is advanced in time according to the dynamical model. Suppose that the state of the system evolves in time according to the stochastic dynamic model equation (7.3) with $\boldsymbol{\eta}_t$ white noise with $E(\boldsymbol{\eta}_t) = 0$ and $\text{Var}(\boldsymbol{\eta}_t) = \mathbf{Q}_t$ for $t = 0, 1, \ldots, T$. Additionally, suppose that observations $\mathbf{y}_0, \mathbf{y}_1, \ldots, \mathbf{y}_T$ are available, respectively, at time $t = 0, 1, \ldots, T$ with equation (7.4) linking the observations to the states $\mathbf{x}_t$, $t = 0, 1, \ldots, T$. Furthermore, assume that a background estimate $\mathbf{x}_0^{(B)}$ is available and (7.5) holds. The Kalman filter algorithm typically used for data assimilation consists of two main steps, repeated sequentially: (i) a *forecast step* that advances the forecast and the forecast covariance matrix in time; and (ii) an *update* or *assimilation step* which integrates the available forecast with the new set of observations available.

Describing the first cycle of steps in the Kalman filter algorithm, given the background estimate $\mathbf{x}_0^{(B)}$ yields: in the foreast step at time $t = 1$ a new state of the system is forecast using the system model dynamics, that is

$$\mathbf{x}_1^{(f)} = \mathbf{M}_0 \mathbf{x}_0^{(B)} \tag{7.12}$$

As $\mathbf{x}_0^{(B)}$ is assumed to be an unbiased estimate of $\mathbf{x}_0$, $\mathbf{x}_1^{(f)}$ is also an unbiased estimate of the true state $\mathbf{x}_1$ at time $t = 1$ with covariance matrix given by

$$\mathbf{P}_1^{(f)} = \mathbf{M}_0 \mathbf{P}_0^{(B)} \mathbf{M}_0' + \mathbf{Q}_0 \tag{7.13}$$

since (7.3) holds and $E(\boldsymbol{\eta}_t \cdot \boldsymbol{\epsilon}_0^B) = 0$ for $t = 0, 1, \ldots, T$.

At time $t = 1$, two sources of information on the true state are available: the forecast state, $\mathbf{x}_1^{(f)}$, and the observation, $\mathbf{y}_1$. In the assimilation step, they are combined to yield an unbiased linear estimate $\mathbf{x}_1^{(a)}$ of $\mathbf{x}_1$

$$\mathbf{x}_1^{(a)} = \mathbf{x}_1^{(f)} + \mathbf{K}_1 \left(\mathbf{y}_1 - \mathbf{H}_1 \mathbf{x}_1^{(f)}\right) = \mathbf{x}_1^{(f)} + \mathbf{K}_1 \mathbf{d}_1 \tag{7.14}$$

where $\mathbf{d}_1 = \mathbf{y}_1 - \mathbf{H}_1 \mathbf{x}_1^{(f)}$ is the innovation vector, and $\mathbf{K}_1$, the Kalman gain matrix, is determined so that the following quantity

$$\frac{1}{2}\left(\mathbf{x}_1 - \mathbf{x}_1^{(f)}\right)' \mathbf{P}_1^{(f)}\left(\mathbf{x}_1 - \mathbf{x}_1^{(f)}\right) + \frac{1}{2}(\mathbf{y}_1 - \mathbf{H}_1\mathbf{x}_1)' \mathbf{R}_1^{-1}(\mathbf{y}_1 - \mathbf{H}_1\mathbf{x}_1) \tag{7.15}$$

is minimized. As in OI, the Kalman gain matrix $\mathbf{K}_1$ is obtained by minimizing (7.15) with respect to $\mathbf{K}_1$. However, differently from OI, the forecast error covariance matrix is allowed to change in time.

Minimization of (7.15) yields

$$\mathbf{K}_1 = \mathbf{P}_1^{(f)} \mathbf{H}_1' \left(\mathbf{H}_1 \mathbf{P}_1^{(f)} \mathbf{H}_1' + \mathbf{R}_1\right)^{-1} \tag{7.16}$$

while the covariance matrix $\mathbf{P}_1^{(a)}$ of the analysis $\mathbf{x}_1^{(a)}$ si given by:

$$\mathbf{P}_1^{(a)} = (\mathbf{I}_N - \mathbf{K}_1 \mathbf{H}_1) \mathbf{P}_1^{(f)} \tag{7.17}$$

Given $\mathbf{x}_1^{(a)}$, the Kalman filter algorithm derives $\mathbf{x}_2^{(f)}$ and $\mathbf{P}_2^{(f)}$ in the forecast step as

$$
\begin{aligned}
\mathbf{x}_2^{(f)} &= \mathbf{M}_1 \mathbf{x}_1^{(a)} \\
\mathbf{P}_2^{(f)} &= \mathbf{M}_1 \mathbf{P}_1^{(a)} \mathbf{M}_1' + \mathbf{Q}_1
\end{aligned}
$$

followed by the update step that yields $\mathbf{x}_2^{(a)}$ and $\mathbf{P}_2^{(a)}$ via appropriate modification of (7.14)-(7.16) and (7.17), respectively, and so forth for $t = 2, \dots, T$.

The Kalman filter has been shown to provide the best linear estimate of the true state of the atmosphere even if the initial background guess $\mathbf{x}_0^{(B)}$ is poor. In particular, Miller et al. [47] has shown that unless the system is unstable and the observations are not frequent, in which case the Kalman filter drifts aways from the true state, the Kalman filter can be considered the gold standard of data assimilation [39].

In the case of a linear dynamical model $\mathcal{M}_t$ (e.g. $\mathcal{M}_t = \mathbf{M}_t$) and a linear observation mapping $\mathcal{H}_t$ (e.g. $\mathcal{H}_t = \mathbf{H}_t$), the Kalman filter can be interpreted within a Bayesian framework. Specifically, if we assume a multivariate Gaussian prior on the unobserved true state $\mathbf{x}_t$ at time t with mean $\mathbf{x}_t^{(f)}$ and covariance matrix $\mathbf{P}_t^{(f)}$, and we postulate that the observations $\mathbf{y}_t$ at time t, $t = 0, \dots, T$, are distributed according to a multivariate Gaussian distribution with mean $\mathbf{H}_t \mathbf{x}_t$ and covariance matrix $\mathbf{R}_t$, the update step can be intepreted as deriving the posterior distribution $p(\mathbf{x}_t | \mathbf{y}_t)$. The analysis state $\mathbf{x}_t^{(a)}$ at time t, $t = 0, \dots, T$, is the posterior mean $E(\mathbf{x}_t | \mathbf{y}_t)$ of $\mathbf{x}_t$, while the covariance matrix of the analysis, $\mathbf{P}_t^{(a)}$, is the posterior variance $\mathrm{Var}(\mathbf{x}_t | \mathbf{y}_t)$. Analogously, Nychka and Anderson [48] point out that the forecast step at time $t + 1$ can be interpreted as deriving the following distribution

$$
p(\mathbf{x}_{t+1} | \mathbf{y}_t) = \int p(\mathbf{x}_{t+1} | \mathbf{x}_t) \cdot p(\mathbf{x}_t | \mathbf{y}_t) \, d\mathbf{x}_t \tag{7.18}
$$

where $p(\mathbf{x}_t | \mathbf{y}_t)$ is derived in the update step while $p(\mathbf{x}_{t+1} | \mathbf{x}_t)$ is provided by the dynamical model.

We conclude by noting that, as in 3D-Var and 4D-Var, the matrix operations needed to implement the Kalman filter algorithm are computationally cumbersome since the magnitude of the vector $\mathbf{x}_t$ is often of the order 10^7. Hence, in the late 90's a series of efforts have been proposed in the literature to approximate the posterior distribution $p(\mathbf{x}_t | \mathbf{y}_t)$ in the update step and the forecast distribution $p(\mathbf{x}_{t+1} | \mathbf{x}_t)$ using ensembles [1, 31, 32, 33, 34, 35]. In Ensemble Kalman Filter (EnKF), the forecast error covariance matrix, which is used to derive the Kalman gain matrix and the analysis covariance matrix at each time step, is approximated by the sample covariance of an ensemble of forecasts obtained by carrying out J assimilation cycles simultaneously. More precisely, to generate a set of J analyses in the J data assimilation cycles, the forecasts for time t are combined with J sets of *perturbed observations* obtained by adding to the observations J sets of random mean-zero perturbations with an appropriate covariance matrix. This leads to a set of J analyses at time t, $\mathbf{x}_{t,1}^{(a)}, \dots, \mathbf{x}_{t,J}^{(a)}$. These analyses are then propagated forward in time during the forecast step, yielding J forecast states, $\mathbf{x}_{t+1,1}^{(f)}, \dots, \mathbf{x}_{t+1,J}^{(f)}$, that can be used to estimate the forecast error covariance matrix $\mathbf{P}_{t+1}^{(f)}$ via the sample covariance matrix. A different way to generate an ensemble of forecasts is pursued in the *Square Root Ensemble Kalman Filter* [68, 69], where ensemble members are obtained by applying a deterministic linear transformation to the analysis so that the originated ensemble of forecasts has approximately the correct forecast error covariance matrix.

An evaluation of the Ensemble Kalman Filter has revealed that, especially in small ensembles, the forecast error covariance matrix has the tendency to underestimate the true forecast error covariance matrix leading in turn to Kalman gain matrices that do not weigh the observations properly. To address this issue, various strategies have been proposed, including a hybrid approach between 3D-Var and Ensemble Kalman Filter [31] or a localization of the Kalman filter, yielding the *Local Ensemble Kalman Filter* [2, 24, 49, 65] obtained by localizing the sample forecast covariance using an approach that is reminiscent of covariance tapering [25] in spatial statistics. An exhaustive list and discussion of the new developments in Ensemble Kalman Filter can be found in Evensen [21] and Lahoz et al. [42].

7.3 Statistical approaches to data assimilation

As noted in Nychka and Anderson [48], Wikle and Berliner [70] and hinted in the previous sections, data assimilation is inherently a statistical problem. A growing body of literature has been published in the last 15-20 years on statistical methods for data assimilation or data fusion, as it is often referred to in the environmental statistical literature. In the next section we provide an overview of these modeling efforts, limiting ourselves only to papers where one of the data sources considered is the output of a geophysical numerical model. Papers presenting data fusion statistical methods for data arising from different monitoring networks (e.g.Cowles and Zimmerman [14], Cowles et al. [15], Rappold et al. [54], Zimmerman and Holland [73]), various data sources that do not include a geophysical model output (e.g. Chang et al. [11], Kloog et al. [41], Paciorek et al. [51], Smith and Cowles [64]), or papers whose main focus is to compare the spatio-temporal distribution of an environmental process as estimated from observational data and from a numerical model output (e.g. Davis and Swall [18], Jun and Stein [36], Mannshardt-Shamseldin et al. [45], Sansó and Guenni [59]) will not be reviewed here.

We observe that while data assimilation algorithms in the geophysical sciences were developed to combine different sources of information on environmental systems over time, with the systems being governed by a multitude of environmental processes interacting among each other, in the statistical data fusion literature, the focus has been mostly on developing methods for fusing observations with the output of a numerical model relative to one, at maximum a few environmental processes. In addition, differently from the data assimilation algorithms reviewed in Section 7.2, in the statistical literature for data fusion, the main point of contention is to address the *change of support* problem [3, 29], that is, the difference in spatial resolution between the observational data and the output of the geophysical model. The first is often interpreted as the value of the environmental proces at given points, while the latter, due to the discretization of space, is often interpreted as the average value of the environmental process over a grid cell, or, in some cases, as the value of the process at the centroid of the grid cell. To highlight the intrinsic spatial aspect of the statistical data fusion literature, we will modify slightly the notation introduced previously. Thus, we will indicate with $y_t(\mathbf{s})$ an observation of the environmental process at location $\mathbf{s}$ at time t, $t = 0, 1, \ldots, T$, and we will use the notation $X_t(B)$ or $X_t(\mathbf{r})$ to denote the geophysical model output relative to the environmental process in consideration over grid cell B or at grid cell centroid $\mathbf{r}$, respectively, at time t.

Two major approaches have emerged in statistical data fusion: joint modeling approaches and regression-based approaches.

7.3.1 Joint modeling approaches

In the joint modeling approach, the observed environmental process $Y_t(\mathbf{s}), \mathbf{s} \in \mathcal{S}$ at time t, $t = 0, 1, \ldots, T$, and the output, or the estimate, of the environmental process at time

t, $X_t(B_g), g = 1, \ldots, G$, provided by a numerical model at (centroids $\mathbf{r}_g$ of) grid cells B_g are postulated to be noisy representations of the "true" environmental process, which we indicate with $Z_t(\mathbf{s}), \mathbf{s} \in \mathcal{S}$. The latter is in turn modeled as a spatio-temporal stochastic process, inducing a stochastic specification on $Y_t(\mathbf{s}), \mathbf{s} \in \mathcal{S}$, and $X_t(B_g), g = 1, \ldots, G$. The goal of these type of models is then to estimate $Z_t(\mathbf{s})$ given $X_t(B_g), g = 1, \ldots, G$, and observations $y_t(\mathbf{s}_1), y_t(\mathbf{s}_2), \ldots, y_t(\mathbf{s}_N)$.

Perhaps the most known modeling approach in this category is the Bayesian hierarchical model called Bayesian Melding proposed by Fuentes and Raftery [23] in a purely spatial setting. Fixed time t (which we will not use in the notation as a consequence), the Bayesian Melding model of Fuentes and Raftery [23] assumes that the observed process $Y(\mathbf{s})$ is related to the true latent point-referenced process $Z(\mathbf{s})$ via a measurement error model, e.g.

$$Y(\mathbf{s}) = Z(\mathbf{s}) + \epsilon_Y(\mathbf{s}) \qquad \epsilon_Y(\mathbf{s}) \overset{iid}{\sim} N(0, \tau_Y^2), \tag{7.19}$$

whereas the gridded output of the numerical model (in their application, the output of the air quality model Models-3), is linked to $Z(\mathbf{s})$ via the following relationship

$$X(B) = \frac{1}{|B|} \int_B a(\mathbf{s}) d\mathbf{s} + \frac{1}{|B|} \int_B b(\mathbf{s}) Z(\mathbf{s}) d\mathbf{s} + \frac{1}{|B|} \int_B \epsilon_X(\mathbf{s}) d\mathbf{s} \quad \epsilon_X(\mathbf{s}) \overset{iid}{\sim} N(0, \tau_X^2) \tag{7.20}$$

with $a(\mathbf{s})$ and $b(\mathbf{s})$ representing, respectively, spatially-varying calibration parameters for the numerical model ouput. We note that in a practical implementation of the model, Fuentes and Raftery [23] keep the multiplicative calibration term $b(\mathbf{s})$ constant in space, while the additive term $a(\mathbf{s})$ is left to vary in space. At the second stage of Bayesian Melding, Fuentes and Raftery [23] model the true latent process $Z(\mathbf{s})$ as a Gaussian process with a spatially varying mean $\mu(\mathbf{s})$, expressed as a polynomial function of $\mathbf{s}$, and a potentially non-stationary covariance function $C_Z(\cdot, \cdot; \boldsymbol{\theta})$, with covariance parameter $\boldsymbol{\theta}$. As $\mu(\mathbf{s})$, the additive bias term $a(\mathbf{s})$ in (7.22) is modeled as a polynomial in $\mathbf{s}$. Finally, the third stage of the model provides priors for all model parameters.

Establishing a parallel with data assimilation algorithms typically used in the atmospheric sciences, in Bayesian Melding the "true" environmental process $Z(\mathbf{s})$ is estimated by deriving the posterior distribution $p(Z(\mathbf{s})|\mathbf{y}, \mathbf{x})$ where $\mathbf{y} = \{y(\mathbf{s}_i); i = 1, \ldots, N\}$ and $\mathbf{x} = \{X(B_g), g = 1, \ldots, G\}$ is the entire model output over the G grid cells. We note that the posterior distribution $p(Z(\mathbf{s})|\mathbf{y}, \mathbf{x})$ is given by

$$p(Z(\mathbf{s})|\mathbf{y}, \mathbf{x}) \propto p(\mathbf{y}|Z(\mathbf{s})) \cdot p(\mathbf{x}|Z(\mathbf{s})) \cdot p(Z(\mathbf{s}))$$

and is different from the expression of the posterior distribution $p(\mathbf{x}_t|\mathbf{y}_t)$ in the updating step of the Kalman filter algorithm.

Liu et al. [44] performed an empirical evaluation of the Bayesian Melding model of Fuentes and Raftery [23] and reported that despite the overall good predictive and inferential performance of the model in simulation experiments, the model is computationally burdensome mostly due to the introduction of the stochastic integral in (7.22), thus inhibiting the extension of the model to a space-time setting in its current form.

Different strategies were undertaken to alleviate the computational burden associated with fitting a joint modeling approach that explicitly address the change of support problem in a space-time setting. Motivated by an environmental epidemiology application, Choi et al. [12] extend Bayesian Melding to a space-time setting in an effort to fuse monitoring data on $PM_{2.5}$ concentration with an air quality model output (the Community Multiscale Air Quality – CMAQ) and obtain estimates of $PM_{2.5}$ concentrations in counties in North Carolina in year 2001 to subsequently relate to mortality counts. Specifically, dealing with two monitoring networks, Choi et al. [12] express the space-time monitoring data as

$$Y_t^{(k)}(\mathbf{s}) = Z_t(\mathbf{s}) + \epsilon_{Y,t}^{(k)}(\mathbf{s}) \qquad \epsilon_{Y,t}^{(k)}(\mathbf{s}) \overset{iid}{\sim} N(0, \tau_Y^{2(k)}) \tag{7.21}$$

with superscript $k = 1, 2$ to denote the two different monitoring networks, and the model output data as:

$$X_t(B) = \frac{1}{|B|} \int_B a(\mathbf{s}) d\mathbf{s} + \frac{1}{|B|} \int_B Z_t(\mathbf{s}) d\mathbf{s} + \frac{1}{|B|} \int_B \epsilon_{X,t}(\mathbf{s}) d\mathbf{s} \quad \epsilon_{X,t}(\mathbf{s}) \overset{iid}{\sim} N(0, \tau_X^2) \quad (7.22)$$

To facilitate computation, Choi et al. [12] model the true, latent point-referenced spatio-temporal process $Z_t(\mathbf{s})$ as

$$Z_t(\mathbf{s}) = \mathbf{V}_t(\mathbf{s})\boldsymbol{\beta} + \epsilon_{Z,t}(\mathbf{s}) \quad \epsilon_{Z,t}(\mathbf{s}) \overset{iid}{\sim} N(0, \tau_Z^2)$$

with $\mathbf{V}_t(\mathbf{s})$ spatially and temporally-varying covariates. Thus, Choi et al. [12] account for any spatio-temporal structure in the mean of $Z_t(\mathbf{s})$, leading to more manageable computation as no covariance computation is needed in the model fitting.

A different approach is undertaken by McMillan et al. [46] who define the latent spatio-temporal process not at the point level but at the numerical model grid cell level, hence replacing $Z_t(\mathbf{s})$ with $Z_t(B)$, and modeling the latter as a Gaussian Markov random field with an AR(1) dependence structure in time. The observations, that in the application considered by McMillan et al. [46] refer to concentrations of $PM_{2.5}$, and are modeled on the log-scale, are linked to the latent process in the following way:

$$Y_t(\mathbf{s})|Z_t(B_\mathbf{s}) \sim N(Z_t(B_\mathbf{s}), \tau_Y^2) \qquad (7.23)$$

with $B_\mathbf{s}$ numerical model grid cell containing site $\mathbf{s}$ (in McMillan et al. [46], the B_g's are grid cells of the CMAQ air quality model). Conversely, the numerical model grid cell output at time t is linked to the latent spatio-temporal process $Z_t(B)$ via the following equation

$$X_t(B) = Z_t(B) + \mathbf{V}_t(B)\boldsymbol{\beta} + \epsilon_{X,t}(B) \qquad \epsilon_{X,t}(B) \overset{iid}{\sim} N(0, \tau_X^2)$$

with $\mathbf{V}_k(B)$ set of spatially-varying bias covariates and $\boldsymbol{\beta}$ vector of regression coefficients.

We conclude this review of the joint modeling approaches by discussing the modeling efforts of Foley and Fuentes [22] and Wikle et al. [71] who present versions of the joint modeling approach for data fusion of various sources of wind fields data. Specifically, Foley and Fuentes [22] develop a method to fuse wind fields data from buoys and gridded analysis from the Hurricane Research Division (HRD) of the National Oceanic and Atmospheric Administration (NOAA), whereas Wikle et al. [71] melds gridded wind fields from the National Center for Environmental Predictions (NCEP) with NASA scatterometer data. Both authors employ physics-based deterministic formulas to specify the mean structure of the latent process $Z_t(\mathbf{s})$: namely, Foley and Fuentes [22] employ the Holland model for wind speed while Wikle et al. [71] invoke thin-fluid approximation of large scale tropical dynamics. In both cases, the gridded wind surface data is not expressed via integrals as in (7.22), but simply through: (i) a linear regression on the latent point-referenced process $Z_t(\mathbf{s})$ that included spatially-varying calibration terms to address the change of support problem [22]; and (ii) a linear regression involving projection matrices $\mathbf{H}_t$ to address the mismatch between the grids where data is available and the prediction grids where the latent process is inferred to [71]. In a similar spirit to Wikle, in an analysis of $PM_{2.5}$ concentration data collected at monitors and estimated by the air quality model CMAQ, Paciorek [50] handles the different spatial resolution between observations and CMAQ output by introducing a latent Markov random field, defined over a fine grid. Additionally, as Wikle et al. [71], he deals with the spatial misalignment between the fine grid and the CMAQ grid and the fine grid and the monitor locations through projection matrices.

7.3.2 Regression-based approaches

In regression-based approaches, the geophysical model output is considered as data and is used as a covariate in a linear regression model where the observational data is, in most cases, the outcome. Differences in spatial scale between the two sources of data are handled by allowing the regression coefficients to be spatially-varying. Thus, in this type of approach only the observed environmental process has a stochastic specification.

A simple example of the regression-based approach is provided by the so-called *downscaler model* of Berrocal et al. [6] who, extending the work of Guillas et al. [30], relates observations of the environmental process $Y_t(\mathbf{s})$ at time t, $t = 1, \ldots, T$, to the output, $\{X_t(B_g), g = 1, \ldots, G\}$, of a geophysical numerical model via:

$$Y_t(\mathbf{s}) = a_t(\mathbf{s}) + b_t(\mathbf{s})X_t(B_{\mathbf{s}}) + \epsilon_t(\mathbf{s}) \qquad \epsilon_t(\mathbf{s}) \overset{iid}{\sim} N(0, \tau_Y^2) \qquad (7.24)$$

where $B_{\mathbf{s}}$ is the numerical model grid cell that contains $\mathbf{s}$. The terms $a_t(\mathbf{s})$ and $b_t(\mathbf{s})$ in (7.24) are interpreted as calibration terms for the numerical model output and are modeled as correlated Gaussian processes through the linear method of coregionalization [3, 28]. In practical and subsequent implementations of the downscaler model, the multiplicative bias term $b_t(\mathbf{s})$ is taken to be constant in space but varying in time, while the spatio-temporal term $a_t(\mathbf{s})$ is still assumed to be a spatio-temporal process and it's what allows to *downscale* the gridded output $\{X_t(B_g), g = 1, \ldots, G\}$ at time t to point level, thus handling the change of support problem. Predictions of the environmental process at time t and at any location $\mathbf{s}_0$ in the spatial domain are obtaned by sampling from the posterior predictive distribution $p(Y_t(\mathbf{s}_0)|\mathbf{x}_t)$ where $\mathbf{x}_t = \{X_t(B_g), g = 1, \ldots, G\}$.

To account for any temporal correlation in the observational data, Berrocal et al. [6] consider two scenarios: (i) provide the two calibration terms $a_t(\mathbf{s})$ and $b_t(\mathbf{s})$ with an AR(1) temporal dependence structure and (ii) assume that the two terms are independent in time. In a real data application and cross-validation studies, Berrocal et al. [6] observe a better out-of-sample predictive performance for the latter specification. We note that with $b_t(\mathbf{s})$ constant in space and with an independence structure in the $a_t(\mathbf{s})$, the downscaler model of Berrocal et al. [6] can be considered as an example of a Bayesian universal kriging model.

Similar in spirit to the model of Berrocal et al. [6] is the model of Liu et al. [44] who extends the downscaler model in (7.24) to: (i) allow for the inclusion of additional covariates besides the numerical model output, and (ii) include an AR(1) structure in the error terms $\epsilon_t(\mathbf{s})$ by assuming that $\epsilon_t(\mathbf{s}) = \rho_t \epsilon_{t-1}(\mathbf{s}) + \nu_t(\mathbf{s})$ with $\nu_t(\mathbf{s}) \overset{iid}{\sim} N(0, \sigma_\nu^2)$. Analogously, with the goal of forecasting ozone concentration for the next day, Sahu et al. [58] modifies the equation in (7.24) to account for temporal dependence in the observations by adding to (7.24) a term of the form $\rho Y_{t-1}(\mathbf{s})$.

The downscaler model of Berrocal et al. [6] only deals with one environmental process. Working with observations and CMAQ model output for daily ozone and PM$_{2.5}$ concentrations over a subregion of the SouthEast US during a summer season, Berrocal et al. [5] extend the univariate downscaler model to a multivariate setting. Specifically, given p observed environmental processes $Y_t^{(k)}(\mathbf{s})$, $k = 1, \ldots, p$, at time t, $t = 1, \ldots, T$, and given p corresponding numerical model outputs $\left\{X_t^{(k)}(B_g), g = 1, \ldots, G\right\}$, the Bayesian hierarhical multivariate downscaler model of Berrocal et al. [5] regresses each individual observed environmental process $Y_t^{(k)}(\mathbf{s})$ on all the available numerical model output, e.g. for $\mathbf{s} \in B_{\mathbf{s}}$

$$Y_t^{(k)}(\mathbf{s}) = a_t^{(k)}(\mathbf{s}) + \sum_{l=1}^{p} b_t^{(l)} X_t^{(l)}(B_{\mathbf{s}}) + \epsilon_t^{(k)}(\mathbf{s}) \qquad \epsilon_t^{(k)}(\mathbf{s}) \overset{iid}{\sim} N(0, \tau_Y^{2^{(k)}}) \quad k = 1, \ldots, p \quad (7.25)$$

with the $a_t^{(k)}(\mathbf{s})$ spatio-temporal Gaussian processes, modeled to be independent over time, but correlated across the p processes through the method of coregionalization [27]. In the case study examined by Berrocal et al. [5] on daily ozone and $PM_{2.5}$ concentrations, exploiting the correlation between pollutants as well as leveraging the information in all the numerical model output yielded prediction gains when compared to applying the univariate downscaler model on each individual pollutant.

While Berrocal et al. [5] explicitly model the correlation among the observed concentrations of pollutants by introducing co-dependence in the spatially-varying intercept terms $a_t^{(k)}(\mathbf{s})$, Rundel et al. [56] and Crooks and Özkaynak [17] propose data fusion statistical modeling approaches for speciated $PM_{2.5}$ that induce a correlation among the species of $PM_{2.5}$ via constraints imposed on the sum of the concentrations. Specifically, using observations $Y_t^{(I,1)}(\mathbf{s}), Y_t^{(C,1)}(\mathbf{s}), \ldots, Y_t^{(I,5)}(\mathbf{s}), Y_t^{(C,5)}(\mathbf{s})$ of concentrations of the 5 major components of $PM_{2.5}$ obtained at two monitoring networks (denoted with I and C), observations $Y_t^{(I,PM)}(\mathbf{s}), Y_t^{(C,PM)}(\mathbf{s})$ of total $PM_{2.5}$ mass, and corresponding CMAQ model outputs $\left\{X_t^{(1)}(B_g), g = 1, \ldots, G\right\}, \ldots, \left\{X_t^{(5)}(B_g), g = 1, \ldots, G\right\}, \left\{X_k^{(PM)}(B_g), g = 1, \ldots, G\right\}$, Rundel et al. [56] introduce 6 latent point-referenced spatio-temporal processes $V_t^{(1)}(\mathbf{s}), \ldots, V_t^{(5)}(\mathbf{s}), V_t^{(PM)}(\mathbf{s})$ representing, respectively, the true unobserved concentrations for the five major species of $PM_{2.5}$ and $PM_{2.5}$ total mass. These latent processes are constrained to be such that $\sum_{l=1}^5 V_k^{(l)}(\mathbf{s}) \leq V_k^{(PM)}(\mathbf{s})$ at every point $\mathbf{s}$ and every time t, $t = 1, \ldots, T$. The observed concentrations are linked to the corresponding latent processes $V_t^{(1)}(\mathbf{s}), \ldots, V_t^{(5)}(\mathbf{s}), V_t^{(PM)}(\mathbf{s})$ through measurement error models on the non-negative latent processes defined as $\max(0, V_t^{(1)}(\mathbf{s})), \ldots, \max(0, V_t^{(5)}(\mathbf{s})), \max(0, V_t^{(PM)}(\mathbf{s}))$. Finally, the latent point-referenced spatio-temporal processes $V_t^{(1)}(\mathbf{s}), \ldots, V_t^{(5)}(\mathbf{s}), V_t^{(PM)}(\mathbf{s})$ are linked to the corresponding CMAQ model output through models of the form

$$V_t^{(k)}(\mathbf{s}) = a_t^{(k)}(\mathbf{s}) + b_t^{(k)} X_t^{(k)}(B_\mathbf{s}) \qquad k = 1, 2, \ldots, 5, PM \qquad (7.26)$$

with $B_\mathbf{s}$ grid cell containing $\mathbf{s}$. The additive calibration terms $a_t^{(k)}(\mathbf{s})$ are postulated to be independent across pollutants and over time, and equation (7.26) contains no i.i.d. error terms $\epsilon_t^{(k)}(\mathbf{s})$ as in (7.24) since the latent processes $V_t^{(k)}(\mathbf{s})$ are envisioned to be smooth spatio-temporal processes. Correlation among the latent spatio-temporal processes $V_t^{(k)}(\mathbf{s})$ is induced by the correlation among the pollutants built in the CMAQ model and the constraints on the sum of the $V_t^{(k)}(\mathbf{s})$ for $k = 1, \ldots, 5$. To note that even though this model introduces latent processes as in the joint modeling approaches reviewed in Section 7.3.1, only the observed processes have a stochastic specification and the numerical model outputs are again used simply as covariates.

Crooks and Özkaynak [17] offer a slightly different approach to handle daily speciated concentrations. Specifically, considering daily concentrations of five $PM_{2.5}$ components along with total $PM_{2.5}$ mass over a spatial domain slightly larger than New Jersey over a 5-year period, Crooks and Özkaynak [17] propose a Bayesian hierarchical model that at the first stage models the observed concentrations as conditionally independent given some latent point-referenced spatio-temporal processes. Specifically, they use Gamma distributions for the observed concentrations with parameters given by the latent spatio-temporal processes. The latter are in turn expressed as linear function of the air quality model output with formulation similar in spirit to (7.26).

All the regression-based approaches reviewed thus far establish a link between the observed value of the environmental process at a location with the geophysical model output at the grid cell that contains it. To account for displacement errors in the numerical model output and to account for uncertainty in the correspondance between the observational data at a given location and the numerical model output at the grid cell that contains the location, Berrocal et al. [7] propose a new Bayesian hierarhical downscaler model where the observation $Y_t(\mathbf{s})$ of the environmental process at a location $\mathbf{s}$ at time t is regressed at the first stage onto a new regressor $\tilde{X}_t(\mathbf{s})$. For each $\mathbf{s}$, $\tilde{X}_t(\mathbf{s})$ is constructed by taking a weighted average of the entire model output $\{X_t(B_g), g = 1, \ldots, G\}$ with weights $w_{t,g}(\mathbf{s})$ that are random, and spatially and temporally varying. Thus, the new downscaler model of Berrocal et al. [7] has the following formulation at the first stage

$$Y_t(\mathbf{s}) = a_t(\mathbf{s}) + b_t \tilde{X}_t(\mathbf{s}) + \epsilon_t(\mathbf{s}) \qquad \epsilon_t(\mathbf{s}) \stackrel{iid}{\sim} N(0, \tau_Y^2) \qquad (7.27)$$

with $\tilde{X}_t(\mathbf{s}) = \sum_{g=1}^{G} w_{t,g}(\mathbf{s}) X_t(B_g)$ and $w_{t,g}(\mathbf{s})$ random weights obtained by convolving replicates of a latent mean-zero unit-variance Gaussian process $Q_t(\mathbf{s})$ with a kernel function. Berrocal et al. [7] compares the predictive performance of the proposed downscaler with that of the original downscaler model given in (7.24) in a case study of daily ozone concentrations over the Eastern US and they report a significant improvement in the predictions of ozone concentrations particularly at those locations that are farther from other monitoring sites.

Working with a different set of data, Berrocal et al. [4] applies the downscaler model with random weights to observations and output from a regional climate model, while Berrocal et al. [7] extend it to an extreme value setting. Specifically, with the goal of examining spatio-temporal variations in the exceedance of the National Ambient Air Quality standard for ozone in the US, which is formulated in terms of the fourth highest yearly ozone concentration, Berrocal et al. [8] replace the first stage formulation of the downscaler model of Berrocal et al. [7] with an extreme value distribution. Finally, working in a purely spatial setting, Reich et al. [55] extend the model of Berrocal et al. [7] by proposing a *spectral downscaler*, which admits as special case the downscaler model of Berrocal et al. [6]. More specifically, building upon results from spectral theory in spatial statistics, the Bayesian hierarchical model of Reich et al. [55] regresses the observed environmental process $Y(\mathbf{s})$ onto multiple spectral covariates $\tilde{X}^{(k)}(\mathbf{s}), k = 1, \ldots, p$, each defined as

$$\tilde{X}^{(k)}(\mathbf{s}) = \int A_k(\mathbf{w}) \exp(-i\mathbf{w}'\mathbf{s}) Z(\mathbf{w}) d\mathbf{w} \qquad (7.28)$$

with $Z(\mathbf{w})$ Fourier transform of the CMAQ model output $X(B_g), g = 1, \ldots, G$, taken as a point-referenced process defined at the CMAQ grid cell centroids. The terms $A_k(\mathbf{w})$, $k = 1, \ldots, p$ in (7.28) are basis functions constructed so that $\sum_{k=1}^{p} A_k(\mathbf{w}) = 1$, while $\mathbf{w}$ denotes a set of frequencies determined using the grid cell centroids coordinates. In simulation experiments, the spectral downscaler of Reich et al. [55], that admits both the downscaler model of Berrocal et al. [6] and a particular formulation of the downscaler model with random weigths of Berrocal et al. [7] as special cases, provides a slight predictive improvement over both models.

We conclude this review by citing the work of Denby et al. [19] who compare two data assimilation methods, the ensemble Kalman filter and a universal kriging model, for the estimation of levels of PM_{10} over Europe, and the work of Sahu et al. [57] who present a sophisticated model for downscaling wet deposition of nitrates and sulfates.

Bibliography

[1] Anderson, J. L. (2001). An ensemble adjustment Kalman filter for data assimilation. *Monthly Weather Review 129*, 2894–2903.

[2] Anderson, J. L. (2007). Exploring the need for localization in ensemble data assimilation using a hierarchical ensemble filter. *Physica D 230*, 99–111.

[3] Banerjee, S., B. P. Carlin, and A. E. Gelfand (2014). *Hierarchical Modeling and Analysis for Spatial Data*. Chapman & Hall/CRC. Second edition. Boca Raton, Fla.

[4] Berrocal, V. J., P. Craigmile, and P. Guttorp (2012). Regional climate model assessment using statistical upscaling and downscaling techniques. *Environmetrics 23*, 482–492.

[5] Berrocal, V. J., A. E. Gelfand, and D. M. Holland (2010a). A bivariate spatio-temporal downscaler under space and time misalignment. *Annals of Applied Statistics 4*, 1942–1975.

[6] Berrocal, V. J., A. E. Gelfand, and D. M. Holland (2010b). A spatio-temporal downscaler for outputs from numerical models. *Journal of Agricultural, Biological and Environmental Statistics 15*, 176–197.

[7] Berrocal, V. J., A. E. Gelfand, and D. M. Holland (2012). Space-time data fusion under error in computer model output: an application to modeling air quality. *Biometrics 68*, 837–848.

[8] Berrocal, V. J., A. E. Gelfand, and D. M. Holland (2014). Assessing exceedance of ozone standards: a space-time downscaler of fourth highest ozone concentrations. *Environmetrics 25*, 279–291.

[9] Bratseth, A. M. (1986). Statistical interpolation by means of successive corrections. *Tellus 38A*, 439–447.

[10] Canizeres, R., A. W. Heemink, and H. J. Vested (1998). Application of advanced data assimilation methods for the initialization of storm surge models. *Journal of Hydraulic Research 36*, 655–674.

[11] Chang, H. H., X. Hu, and Y. Liu (2014). Calibrating MODIS aerosol optical depth for predicting daily $PM_{2.5}$ concentrations via statistical downscaling. *Journal of Exposure Science and Environmental Epidemiology 24*, 398–404.

[12] Choi, J., M. Fuentes, and B. J. Reich (2009). Spatial-temporal association between fine particulate matter and daily mortality. *Computational Statistics and Data Analysis 53*, 2989–3000.

[13] Courtier, P. and O. Talagrand (1987). Variational assimilation of meteorological observations with the adjoint vorticity equation. II: Numerical results. *Quarterly Journal of the Royal Meteorological Society 113*, 1329–1347.

[14] Cowles, M. K. and D. L. Zimmerman (2003). A Bayesian space-time analysis of acid deposition data combined from two monitoring networks. *Journal of Geophysical Research 108*, NO. D24, 9006, doi:10.1029/2003JD004001.

[15] Cowles, M. K., D. L. Zimmerman, A. Christ, and D. L. McGinnis (2002). Combining snow water equivalent data from multiple sources to estimate spatio-temporal trends and compare measurement systems. *Journal of Agricultural, Biological and Environmental Statistics 7*, 536–557.

[16] Cressie, N. A. C. (1993). *Statistics for Spatial Data*. Wiley. New York.

[17] Crooks, J. and H. Özkaynak (2014). Simultanous statistical bias correction of multiple $PM_{2.5}$ species froma regional photochemical grid model. *Atmospheric Environment 95*, 126–141.

[18] Davis, J. M. and J. L. Swall (2006). An examination of the CMAQ simulations of the wet deposition of ammonium from a Bayesian perspective. *Atmospheric Environment 40*, 4562–4573.

[19] Denby, B., M. Schapp, A. Segers, P. Builtjes, and J. Horàlek (2008). Comparison of two data assimilation method for assessing PM_{10} exceedances on the European scale. *Atmospheric Environment 42*, 7122–7134.

[20] Dimet, F.-X. L. and O. Talagrand (1968). Variational algorithms for analysis and assimilation of meteorological observations: theoretical aspects. *Tellus A 38*, 97–110.

[21] Evensen, G. (2009). *Data assimilation: the ensemble Kalman filter.* Springer. Second Edition.

[22] Foley, K. M. and M. Fuentes (2008). A statistical framework to combine multivariate spatial data and physical models for hurricane wind prediction. *Journal of Agricultural, Biological and Environmental Statistics 13*, 37–59.

[23] Fuentes, M. and A. E. Raftery (2005). Model evaluation and spatial interpolation by Bayesian combination of observations with outputs from numerical models. *Biometrics 61*, 36–45.

[24] Furrer, R. and T. Bengtsson (2007). Estimation of high-dimensional prior and posterior covariance matrices in Kalman filter variants. *Journal of Multivariate Analysis 98*, 227–255.

[25] Furrer, R., M. G. Genton, and D. Nychka (2006). Covariance tapering for interpolation of large spatial datasets. *Journal of Computational and Graphical Statistics 15*, 502–523.

[26] Gandin, L. S. (1963). Objective analysis of meteorological fields. Gidrommeteoro-logicheskoe Izdatelstvo (GIMIZ).

[27] Gelfand, A. E., H.-J. Kim, C. F. Sirmans, and S. Banerjee (2003). Spatial modeling with spatially varying coefficient processes. *Journal of the American Statistical Association 98*, 387–396.

[28] Gelfand, A. E., A. M. Schmidt, S. Banerjee, and C. F. Sirmans (2004). Nonstationary multivariate process modeling through spatially varying coregionalization. *TEST 13*, 263–312.

[29] Gotway, C. A. and L. J. Young (2002). Combining incompatible spatial data. *Journal of the American Statistical Association 97*, 632–648.

[30] Guillas, S., J. Bao, Y. Choi, and Y. Wang (2008). Statistical correction and downscaling of chemical transport model ozone forecasts over Atlanta. *Atmospheric Environment 42*, 1338–1348.

[31] Hamill, T. and C. Snyder (2000). A hybrid ensemble Kalman filter-3D variational analysis scheme. *Monthly Weather Review 128*, 2905–2919.

[32] Hamill, T., J. Whitaker, and C. Snyder (2001). Distance-dependent filtering of background error covariance estimates in an ensemble Kalman filter. *Monthly Weather Review 129*, 2776–2790.

[33] Houtekamer, P., L. Lefaivre, J. Derome, H. Ritchie, and H. Michell (1996). A system simulation approach to ensemble prediction. *Monthly Weather Review 124*, 1225–1242.

[34] Houtekamer, P. and H. Michell (1998). Data assimilation using an ensemble Kalman filter technique. *Monthly Weather Review 126*, 796–811.

[35] Houtekamer, P. and H. Michell (2001). A sequential ensemble Kalman filter for atmospheric data assimilation. *Monthly Weather Review 129*, 123–137.

[36] Jun, M. and M. L. Stein (2004). Statistical comparison of observed and CMAQ modeled daily sulfate levels. *Atmospheric Environment 38*, 4427–4436.

[37] Kalman, R. (1960). A new approach to linear filtering and prediction problems. *Journal of Basic Engineering 82*, 35–45.

[38] Kalman, R. and R. Bucy (1961). New results in linear filtering and prediction theory. *Transactions of the ASME - Journal of Basic Engineering 83*, 95–107.

[39] Kalnay, E. (2002). *Atmospheric Modeling, Data Assimilation and Predictability*. Cambridge University Press.

[40] Klinker, E., F. Rabier, G. Kelly, and J.-F. Mahfouf (2000). The ECMWF operational implementation of four-dimensional variational assimilation. III: Experimental results and diagnostics with operational configuration. *Quarterly Journal of the Royal Meteorological Society 126*, 1191–1215.

[41] Kloog, I., P. Koutrakis, B. Coull, H. Lee, and J. Schwartz (2011). Assessing temporally and spatially resolved $PM_{2.5}$ exposures for epidemiological studies using satellite aerosol optical depth measurements. *Atmospheric Environment 45*, 6267–6275.

[42] Lahoz, W., B. Khattatov, and R. Ménard (2010). *Data assimilation: making sense of observations*. Springer.

[43] Lewis, J. M., S. Lakshmivarahan, and S. Dhall (2006). *Dynamic data assimilation: a least squares approach*. Cambridge University Press.

[44] Liu, Z., N. Le, and J. V. Zidek (2012). Combining data and simulated data for space-time fields: application to ozone. *Environmental and Ecological Statistics 19*, 37–56.

[45] Mannshardt-Shamseldin, E. C., R. L. Smith, S. R. Sain, L. O. Mearns, and D. Cooley (2010). Downscaling extremes: a comparison of extreme value distributions in point-source and gridded precipitation data. *The Annals of Applied Statistics 4*, 484–502.

[46] McMillan, N., D. M. Holland, M. Morara, and J. Feng (2010). Combining numerical model output and particulate data using Bayesian space-time modeling. *Environmetrics 21*, 48–65.

[47] Miller, R., M. Ghil, and F. Gauthiez (1994). Advanced data assimilation in strongly nonlinear dynamical systems. *Journal of Atmospheric Sciences 51*, 1037–1056.

[48] Nychka, D. and J. Anderson (2010). Data assimilation. In A.E. Gelfand and P.J. Diggle and M. Fuentes and P. Guttorp (Ed.), *Handbook of Spatial Statistics*, pp. 477–492. CRC Press.

[49] Ott, E., B. R. Hunt, I. Szunyogh, A. V. Zimin, E. J. Kostelich, M. Corazza, E. Kalnay, D. J. Patile, and J. A. Yorke (2004). A local ensemble Kalman filter for atmospheric data assimilation. *Tellus A 56*, 415–428.

[50] Paciorek, C. (2012). Combining spatial information sources while accounting for systematic errors in proxies. *Journal of the Royal Statistical Society, Series C. Applied Statistics. 61*, 429–451.

[51] Paciorek, C., Y. Liu, H. Moreno-Macias, and S. Kondragunta (2008). Spatiotemporal association between GOES aerosol optical depth retrievals and ground level $PM_{2.5}$. *Environmental Science Technology 42*, 5800–5806.

[52] Penenko, V. V. and N. N. Obraztsov (1976). A variational initialization method for the fields of the meteorological elements (english translation). *Soviet Meteorol. Hydrol. 11*, 1–11.

[53] Peng, S. and L. Xie (2006). Effect of determining initial conditions by four-dimensional variational data assimilation on storm surge forecasting. *Ocean Modeling 14,* 1–18.

[54] Rappold, A., A. E. Gelfand, and D. M. Holland (2008). Modeling mercury deposition through latent space-time processes. *Journal of the Royal Statistical Society, Series C 57*, 187–205.

[55] Reich, B. J., H. H. Chang, and K. M. Foley (2014). A spectral method for spatial downscaling. *Biometrics 70,* 932–942.

[56] Rundel, C. W., E. M. Schliep, A. E. Gelfand, and D. M. Holland (2015). A data fusion approach for spatial analysis of speciated $PM_{2.5}$ across time. *Environmetrics 26*, 515–525.

[57] Sahu, S. K., A. E. Gelfand, and D. M. Holland (2010). Fusing point and areal level space-time data with application to wet deposition. *Journal of the Royal Statistical Society, Series C 59*, 77–103.

[58] Sahu, S. K., S. Yip, and D. M. Holland (2009). Improved space-time forecasting of next day ozone concentrations in the eastern US. *Atmospheric Environment 43*, 494–501.

[59] Sansó, B. and L. Guenni (2004). A Bayesian approach to compare observed rainfall data to deterministic simulations. *Environmetrics 15*, 597–612.

[60] Sasaki, Y. (1970a). Numerical variational analysis formulated with weak constraints and application to surface analysis of severe storm gust. *Monthly Weather Review 98*, 899–910.

[61] Sasaki, Y. (1970b). Some basic formalism in numerical variational analysis. *Monthly Weather Review 98*, 875–883.

[62] Sherman, J. and W. Morrison (1950). Adjustment of an inverse matrix correspond to a change in one element of a given matrix. *Annals of Mathematical Statistics 21*, 124–127.

[63] Silva, A. D., J. Pfaendtner, J. Guo, M. Sienkiewciz, and S. Cohn (1995). Assessing the effects of data selection with DAO's physical-space-statistical analysis system. In *Proceedings of the second international symposium on the assimilation of observations in meteorology and oceanography.* World Meteorological Organization and Japan Meteorological Agency, Tokyo, Japan.

[64] Smith, B. J. and M. K. Cowles (2007). Correlating point-referenced radon and areal uranium data arising from a common spatial process. *Applied Statistics 56*, 313–326.

[65] Szunyogh, I., E. J. Kostelich, G. Gyarmati, E. Kalnay, B. R. Hunt, E. Ott, E. Satterfield, and J. A. Yorke (2080). A local ensemble transform Kalman filter data assimilation system for the NCEP global model. *Tellus A 60*, 113–130.

[66] Talagrand, O. and P. Courtier (1987). Variational assimilation of meteorological observations with the adjoint vorticity equations. I: Theory. *Quarterly Journal of the Royal Meteorological Society 113*, 1311–1328.

[67] Thacker, W. C. and R. B. Long (1988). Fitting dynamics to data. *Journal of Geophysical Research 93*, 1227–1240.

[68] Tippett, M. K., J. L. Anderon, C. H. Bishop, T. M. Hamill, and J. S. Whitaker (2003). Ensemble square root filters. *Monthly Weather Review 131*, 1485–1490.

[69] Whitaker, J. S. and T. M. Hamill (2002). Ensemble data assimilation without perturbed observations. *Monthly Weather Review 130*, 1913–1924.

[70] Wikle, C. K. and L. M. Berliner (2006). A Bayesian tutorial for data assimilation. *Physica D: nonlinear phenomena 230*, 1–16.

[71] Wikle, C. K., R. F. Miliff, D. Nychka, and L. M. Berliner (2001). Spatiotemporal hierarchical Bayesian modeling: tropical ocean surface winds. *Journal of the American Statistical Association 96*, 382–397.

[72] Woodbury, M. (1950). Inverting modified matrices. Memorandum Report 42, Statistical Research Group, Princeton University.

[73] Zimmerman, D. L. and D. M. Holland (2005). Complementary co-kriging: spatial prediction using data from several environmental monitoring networks. *Environmentrics 16*, 219–234.

[74] Zupanski, D. (1996). A general weak constraint applicable to operational 4D-VAR data assimilation systems. *Monthly Weather Review 125*, 2274–2283.

8

Univariate and Multivariate Extremes for the Environmental Sciences

Daniel Cooley

Colorado State University

Brett D. Hunter

George Mason University

Richard L. Smith

University of North Carolina, Chapel Hill

CONTENTS

8.1 Extremes and Environmental Studies

The underlying philosophy for the statistical analysis of extreme values is to "let the tail speak for itself." In practice, this is often done by analyzing only data that are considered to be extreme. Additionally, the goal of an extreme value (EV) analysis is often extrapolation; one may have data with a record length of 50 years, but need to estimate the magnitude of an event which has an annual exceedance probability (AEP) of .01; that is the so-called

'100-year event'. Because extreme events are rare, EV methods generally result in large uncertainties being associated with estimated quantities.

If the philosophy is to let the tail speak, guidance for practice is to use methods specifically designed for extremes. Fortunately, there are well-developed probability results that provide a solid foundation for extremes-specific statistical methods. Most importantly, these results are quite general and imply that one does not need to know the distribution from which the data arise to be able to characterize the distribution's (joint) tail.

Extremes are of great interest in environmental sciences, as extreme environmental events often have tremendous societal and economic consequences. Much of extreme value theory can trace its roots to applications in hydrology, where researchers sought to use stream and river flow records to estimate the magnitudes of '100-year floods' and other rare events. Extremes studies are now relatively common in atmospheric science, where they have been used to study extreme precipitation, air pollution, heat waves, and other events with societal implications. Our presentation here will present some current techniques for analyzing extreme events. We provide enough background for the unfamiliar reader to follow our presentation, but we do not aim to give a complete overview of theory and methods as there exist very good sources for such material such as [8], [3], and [17]. In Section 8.2, we will introduce the ideas that underlie EV analysis of univariate data, concluding with an application accounting for possible nonstationarity. In Section 8.3, we will discuss multivariate extremes, including how dependence in the tail can be described and modeled.

8.2   Univariate Extremes

8.2.1   Theoretical underpinnings

The foundation of EV theory are results which characterize the limiting distribution of sample maxima. Let Y_t be an iid sequence of random variables, and let $M_n = \bigvee_{t=1}^{n} Y_t$, where $\bigvee$ denotes maximum. If there exist sequences $\{a_n\} > 0$ and $\{b_n\}$ and non-degenerate distribution G such that

$$\frac{M_n - b_n}{a_n} \xrightarrow{d} G \tag{8.1}$$

as $n \to \infty$, then G is said to be an extreme value distribution (EVD). The three-types theorem [25, 27] states EVDs must be a location-scale family of the Frechét with cdf $G(y) = \exp(-y^{-1/\xi})$ for $y \geq 0, \xi > 0$, Gumbel with cdf $G(y) = \exp\{-\exp(-y)\}$ for $y \in \mathbb{R}$, or (reverse) Weibull with cdf $G(y) = \exp\{(-y)^{-1/\xi}\}$ for $y < 0, \xi < 0$. These three types can be characterized by the single representation

$$G(y) = \exp\left\{-(1 + \xi y)^{-1/\xi}\right\}, \tag{8.2}$$

for $1 + \xi y > 0$ and where the right-hand side is interpreted as the limit in the case when $\xi = 0$. The extreme value distributions are equivalent to the max-stable distributions, i.e., the distributions for which there exist sequences such that $G^n(a_n y + b_n) = G(y)$ for all positive integers n and all $y \in \mathbb{R}$.

If Y_t has distribution F and (8.1) holds, then F is said to be in the max-domain of attraction of G. The distributions in the Frechét domain of attraction are the regularly varying distributions, roughly speaking distributions whose tails behave like power functions such as the t, Pareto, and Cauchy distributions. Distributions in the Gumbel domain of attraction have exponentially-decaying tails and include the normal, gamma, and lognormal distributions. The Weibull domain of attraction includes distributions with bounded tails like the beta distribution. In practice, estimates of the parameter ξ indicate to which domain of attraction the underlying distribution belongs. The $\xi > 0$, $\xi = 0$, and $\xi < 0$ cases are respectively termed the heavy-, light-, and bounded-tailed cases. In the environmental sciences, precipitation and stream flows are often found to have heavy-tailed distributions, whereas temperatures tend to have bounded-tailed distributions.

In the environmental sciences, rarely is an iid assumption justified. When Y_t is not iid but is stationary, the class of distributions which satisfy (8.1) are still the EVDs, so long as a certain mixing condition is met [36]. While the mixing condition is difficult to explain, it is not overly restrictive, and environmental processes are often assumed to meet this condition. However, dependence in the sequence $\{Y_t\}$ often implies that the effective block size is less than n, and thus the rate of convergence of (8.1) to an EVD is likely slower than if Y_t were iid.

8.2.2 Modeling Block Maxima

The convergence of renormalized block maxima to a known class of distributions suggests a strategy for modeling block maxima which does not require knowledge or estimation of the underlying distribution of Y_t. From (8.1) and (8.2) if (fixed) n is large, then

$$P\left(\frac{M_n - b_n}{a_n} \leq y\right) \approx \exp\left\{-(1 + \xi y)^{-1/\xi}\right\}.$$

With n fixed, a_n and b_n can be moved to the right-hand side of the approximate equality and treated as parameters (respectively scale parameter τ and location parameter η), thereby obtaining the generalized EV distribution (GEV)

$$P(M_n \leq y) \approx \exp\left[-\left\{1 + \xi\left(\frac{y - \eta}{\tau}\right)\right\}^{-1/\xi}\right], \tag{8.3}$$

for y such that $1 + \xi\left(\frac{y-\eta}{\tau}\right) > 0$. In addition to being theoretically justified, the three-parameter GEV is quite flexible, and the parameter ξ provides information about the nature of the tail of Y_t. More familiar distributions such as the normal or lognormal lack both theoretical justification and the GEV's flexibility for modeling tail behavior.

For a selected block size n, assume there are B blocks of stationary observations $y_{b,t}, b = 1, \ldots, B; t = 1, \ldots, n$. Let $m_b = \bigvee_{t=1}^{n} y_{b,t}$; that is, m_b is the maximum of the observations in the bth block (we have suppressed the dependence of m_b on the block size n). A sensible method for estimating the distribution of M_n is to use $m_1, \ldots, m_B$ to estimate the three parameters of the GEV distribution. Later, we will use numerical maximum-likelihood estimation, but other moment-based methods [31] are also widely used.

The fact that the modeling procedure extracts block maxima and estimates the distribution of these maxima agrees with EV philosophy. However, the GEV parameter estimates themselves are rarely of primary interest (although $\hat{\xi}$ does convey useful information about the nature of the tail). Often the quantity of interest is a high quantile such as a particular AEP level, which can be calculated from the GEV parameters. The AEP(p) level is the quantile which the annual maximum exceeds with probability p. A more familiar term is the r-year return level; that is, the level for which the expected time between exceedances of this level is r years, assuming stationarity. Thus, under stationarity an AEP(.01) level corresponds to a 100-year return level. We use AEP level as it more clearly conveys the probabilistic nature of an exceedance and remains interpretable under non-stationarity due to changing conditions such as climate change or urban development. The asymptotic arguments leading to the GEV provide arguments for its use when estimating these high quantiles requires extrapolation.

Modeling block maxima is also appealing in environmental science because often environmental data exhibit seasonality. In the presence of seasonality, one can no longer assume the data are identically distributed. In practice, when the annual maximum occurs in a particular season (e.g., temperature in summer), one can assume that the annual maximum arises from a season during which the process is approximately stationary, but one should recognize that the effective block length is not $n = 365$. The usual argument against modeling block maxima is that by retaining only the maximum of each block, other data which are useful for characterizing extreme behavior is not utilized. Confidence intervals of GEV parameters are often found to be quite wide, resulting in large uncertainties for quantities such as the AEP(p) level. [24] examine efficiency of the block maximum method in the iid case.

8.2.3　Threshold exceedances

An alternative to modeling block maxima which still follows the EV philosophy is to model exceedances over a high threshold. As with block maxima, asymptotic results suggest a model for exceedances over a suitably high threshold. [2] and [41] showed that if Y_t is in the domain of attraction of an EVD, then the distribution of exceedances over an increasing threshold converge to a generalized Pareto distribution (GPD). A somewhat heuristic justification of modeling the distribution of a random variable conditioned on exceeding a high threshold follows. Assuming $Y_1, Y_2, \ldots$ are iid and starting from (8.3) which assumes n is fixed and large, then for any t

$$P^n(Y_t \leq y) \approx \exp\left[-\left\{1 + \xi\left(\frac{y-\eta}{\tau}\right)\right\}^{-1/\xi}\right]$$

$$\Rightarrow n\log[1 - P(Y_t > y)] \approx -\left\{1 + \xi\left(\frac{y-\eta}{\tau}\right)\right\}^{-1/\xi}$$

$$\Rightarrow nP(Y_t > y) \approx \left\{1 + \xi\left(\frac{y-\eta}{\tau}\right)\right\}^{-1/\xi},$$

where the last line assumes y is large enough such that $\log[1 - P(Y_t > y)] \approx -P(Y_t > y)$. For u sufficiently large and $y > u$,

$$P(Y_t > y \mid Y_t > u) = \frac{nP(Y > y)}{nP(Y > u)}$$

$$\approx \frac{\left\{1 + \xi\left(\frac{y-\eta}{\tau}\right)\right\}^{-1/\xi}}{\left\{1 + \xi\left(\frac{u-\eta}{\tau}\right)\right\}^{-1/\xi}}$$

$$= \left\{1 + \xi\left(\frac{y-u}{\psi_u}\right)\right\}^{-1/\xi},$$

where $\psi_u := \tau + \xi(u - \eta)$ and the final expression is the GPD survival function. The parameter ξ continues to indicate the type of tail and the tail weight, and ψ_u is a scale parameter which depends on the threshold u. To estimate high quantiles of Y_t's distribution, an estimate of $\zeta_u = P(Y_t > u)$ is also required. Importantly, the only assumption on the distribution of Y_t is that it is in the domain of attraction of a EVD.

The above result suggests a strategy for modeling threshold exceedances by first selecting a threshold u above which the distribution's tail is well approximated by a GPD, and then

Threshold	$u = 0$	$u = 0.5$	$u = 2$	$u = 3$	$u = 5$	Truth
$\hat{\xi}$	-.002(.010)	.062(.016)	.192(.048)	.207(.084)	.349 (.224)	.25
$\hat{q}_{.9999}$	8.41(.30)	9.46(.52)	12.42(1.59)	12.75(2.05)	13.60 (3.32)	13.03
$P(T > u)$	.500	.322	.058	.020	.004	–
# Exc	4987	3221	571	193	38	–

TABLE 8.1
Results from fitting a GPD to threshold exceedances of 10000 simulated random variables with a t-distribution with 4 df. $\hat{q}_{.9999}$ is the estimated .9999 quantile and standard errors are in parentheses. When low thresholds such as 0 or 0.5 are selected, biased estimates of ξ and the .9999 quantile result. When high thresholds such as 5 are selected, large uncertainties are associated with parameter estimates.

using the observations which exceed u to estimate the parameters ψ_u and ξ. A challenge with this approach is finding a suitable threshold. Threshold selection involves a bias versus variance tradeoff often found in statistics. If the chosen threshold is too low the GPD approximation will be poor, resulting in bad estimates for ψ_u and ξ, and consequently bad estimates for the high quantiles which are of interest. If the threshold is too high, there will be few exceedances, resulting in large uncertainties associated with the parameters and high quantiles. These qualities are illustrated in Table 8.1, which gives results from fitting a GPD to the exceedances above various thresholds to 10000 realizations of a t-distribution with 4 df. Table 8.1 shows that for this distribution, thresholds of 0 or 0.5 are too low resulting in estimates for ξ and $q_{.9999}$ which are biased low, and the threshold of 5 is too high resulting in large uncertainty associated with the parameter estimates. Thresholds are often selected by resorting to diagnostic plots such as the mean residual life (alternatively mean excess) plot, or plots of $\hat{\xi}$ [8, Sections 4.3.1 and 4.3.4]. However, these plots often do not yield clear-cut values for adequate thresholds [44, Section 4.4.2]. After the model is fit, QQ plots or return level plots [8, Section 4.3.5] can be used to assess model adequacy. Due to the difficulty in selecting a threshold above which a GPD is suitable, it is tempting to instead estimate such a threshold. Several modeling approaches have attempted to estimate a threshold ([48] provide a review), and most propose some sort of model for the data in the 'bulk' of the distribution (i.e., those which lie below the threshold). Recently, [39] proposed an extended-GPD model and used it to model the entire distribution of non-zero rainfall amounts.

As with the GEV, estimation of the GPD parameters can be done via numerical maximum likelihood or by moment-based approaches. A likelihood specified as the product of densities evaluated at each data point assumes independence of the data used for estimation, and moment-based estimation methods similarly assume independence. For many environmental processes, threshold exceedances occur in clusters, contradicting the independence assumption. If estimating the AEP or some other quantity of interest which can be calculated from the marginal distribution, this dependence does not need to be explicitly modeled, but inference should account for it. Peaks-over-threshold approaches [c.f., 7, and references therein] identify clusters, and use only the maximum observation of each cluster for inference, thereby alleviating data dependence. Alternatively, [23] advocate using a simple likelihood incorporating all exceedances and appropriately adjusting uncertainty estimates arising from the misspecified likelihood [14, Section 4.6]. Another way to obtain appropriate parameter estimate uncertainty is to use a block-bootstrap approach to account for temporal dependence. However, if one's interest involves quantities not given by the marginal, such as the sum of multiple days' measurements, then the dependence must be modeled. An approach taken by Smith et al. [52] was to assume the data arise from a Markov chain, thereby allowing the dependence to be specified by a bivariate relationship between Y_t and Y_{t+1}. Importantly, Smith et al. [52] employ a bivariate dependence structure appropriate for modeling tail dependence, as we will discuss in Section 8.3.

Modeling block maxima with the GEV or modeling threshold exceedances with the GPD are the most prevalent EV modeling methods, but they are not the only models suggested by EV theory. Sometimes interest is in neither the block maximum nor exceedances of a some threshold level. For example, the current ozone air quality standard set by the US Environmental Protection Agency is based on the fourth-highest observation in a year.[1] The theory which leads to the GEV being the limiting distribution for the block maximum can be extended to give the limiting distribution of the kth-largest order statistic [8, Section 3.5]. [4] used this approach to model the fourth-largest annual ozone measurement in the Eastern US. A similar argument can be used to model the vector of k-largest order statistics [8, Section 3.5]. Extreme value theory can be connected to point-processes ([51], [8, Chapter 7]), and the GEV can be seen the Poisson-derived probability of observing no points above the level of the distribution's argument. Via the point-process formulation, an alternative parametrization can be given for threshold exceedances [8, p. 132]. This parametrization can be described by the parameters η, τ, and ξ of the GEV distribution, and has the advantage that its parameters are independent of the threshold u, unlike the GPD parametrization's scale and exceedance rate parameters ψ_u and ζ_u.

8.2.4   Regression models for extremes

A major theme throughout statistics is constructing regression-like models to relate a response Y to covariates x. The underlying argument is that the value of x induces a distributional change on Y. In extremes studies, and environmental extremes studies in particular, there have been many examples of building regression extremes models where the parameters of the GEV or GPD are modeled as functions of covariate values, or alternatively functions of time in order to model trends. Importantly, the standard arguments that lead to modeling block maxima with a GEV or threshold exceedances with a GPD require that the block maximum or threshold exceedances are taken from a sequence of identically distributed data. Here we briefly review arguments leading to regression extremes models.

Let $y_{b,t}, t = 1, \ldots, n; b = 1, \ldots, B$ be observations, where n denotes the block size and b is used to index the block. Assume these observations arise from corresponding random variables $Y_{b,t}$ and further that given covariate value x_b, $Y_{b,t} \mid x_b$ are identically distributed for $t = 1, \ldots, n$. Then, from results in Sections 8.2.1 and 8.2.2, if n is large enough, the distribution of $M_{b,n} = \bigvee_{t=1}^{n} Y_{b,t} \mid x_b$ should be well approximated by a GEV. Given that x_b influences the distribution of $Y_{b,t}$, a natural modeling approach is to assume that the GEV parameters are functions of the covariate; that is

$$[M_{b,n} \mid x_b] \stackrel{.}{\sim} GEV(\eta = f_\eta(x_b), \tau = f_\tau(x_b), \xi = f_\xi(x_b)).$$

One can also formulate regression approaches for threshold exceedances by allowing the GPD parameters ψ_u and ξ to vary with the covariate. Since the probability of exceeding a threshold also likely depends on the covariate, ζ_u and/or the threshold u itself could be modeled in terms of the covariate. As both ψ_u and ζ_u are themselves functions of the threshold, a perhaps cleaner formulation for threshold exceedances is via the point process representation and GEV parameters as was done in Smith [51]. Often the functions relating

---

[1]https://www.epa.gov/ozone-pollution/2015-national-ambient-air-quality-standards-naaqs-ozone#rule-summary

EV distribution parameters to covariates are simple linear or perhaps polynomial functions, although [6] employed spline-based functions.

The above argument suggests that any regression extremes model should only employ a covariate that varies more slowly than the response variable, so that the block maximum or threshold exceedances can be assumed to arise from a larger set of observations that can be assumed identically distributed given the covariate. Thus, building a regression model where extreme summer temperatures are regressed on an annual value for the El Niño Southern Oscillation index seems reasonable. Likewise, modeling trends due to climate change by regressing extreme values on year seems reasonable, given that nonstationarity induced by climate change is negligible within a year. There is a desire to relate extreme behavior to variables which vary on the same timescale as the response variable. One approach is to preprocess the data before extracting extreme observations as is done by [21]. [46] uses the bivariate extremes framework presented in Section 8.3.2 to relate daily extreme ozone pollution measurements to daily meteorological data.

8.2.5   Application: Fitting a time-varying GEV model to climate model output

GCMs (formally 'general circulation models', but colloquially 'global climate models') are tools for understanding Earth's climate via simulation. As it is impossible to perform experiments on the Earth itself, GCMs are the best tools available for learning how different forcings, such as increased greenhouse gas concentrations, affect climate. GCMs are deterministic models which employ discretized solutions of the differential equations governing atmospheric circulation and other Earth processes. Coupled models, like the one whose output we will study, combine models for the atmosphere, ocean, land-surface, and sea ice to allow these different Earth processes to interact and influence one another. Running the most state-of-the-art GCMs requires massive computing capability and is done by a number of large institutions around the globe. GCMs can be run with historical forcings in order to compare the GCM-produced climate to historical observations, and also can be run with hypothesized future forcings to produce projections for future climate.

GCMs produce weather-like output for hundreds of meteorological variables on a discretized grid at regular time intervals. GCM output can be treated like data and analyzed in order to understand how well the GCM mimics Earth's processes or to understand how changes in forcings could alter Earth's climate. Although GCM output is indexed by date, GCMs should not be thought of weather prediction models. Rather, it is convenient to think of the produced output as a possible draw from the distribution of weather given the climate of the indexed time. That is, the output from a perfect GCM would not correspond to observed weather, but the distributions of output and observations would match. (Reanalysis output, unlike that from a GCM, should exhibit temporal correspondence [18].) Interpretation of GCM output and comparison to weather-station observations can be tricky, as the spatial resolution of GCM output corresponds to the grid cell resolution and not the point-referenced information recorded at weather stations. Differences between GCM output and weather station observations seem to be exacerbated when considering extreme behavior; for instance Figure 8.1 of [26] shows a clear difference in the distribution of annual maximum precipitation for weather station observations and GCM output. Despite differences between GCM output and observations, the GCM output can provide useful information about how climate is likely to change in response to changes in forcings, and downscaling methods can be used to relate GCM output to observations.

The output we study is from the Community Earth System Model 1 (CESM1) produced by the National Center for Atmospheric Research. CESM1 is a Coupled Model Intercomparison Project 5 [55] model and has a spatial resolution of approximately 1°. The output we study is somewhat unique in that it is from an initial condition ensemble experiment [34]. Because GCMs are deterministic, multiple simulations can be produced by perturbing the conditions used to initialize the model run. Initial condition ensembles are used to study 'internal climate variability'. From a statistics perspective, internal climate variability is analogous to sample variability: any individual simulation produces an incomplete picture of the climate and multiple simulations (ensemble members) provide an idea of the extent of the differences between members. We have 30 ensemble members, each of which can be viewed as an independent realization of the output produced by this GCM under the given forcings. We will restrict our attention to output produced using historical forcings for the 86-year period 1920-2005, although the ensemble experiment was extended for the period 2005-2080 for two climate projections: RCP4.5 (a projection where emissions peak mid-century) and RCP8.5 (emissions continue to grow through the century) [38, 53]. We aim to illustrate a univariate regression model by investigating whether extreme precipitation shows a significant time trend at a particular location. Further, we use the ensemble to investigate internal variability and discuss borrowing strength across locations.

8.2.5.1   Analysis of individual ensembles and all data

We study annual maximum daily precipitation produced by the GCM. The independent replicates of the ensemble provide us with an artificially large data set where for each year we have 30 annual maximum observations. For the time being, we will model each ensemble member's time series of annual maximum precipitation individually to get an idea of how the GCM's interval variability affects statistical models of extreme precipitation. We initially restrict attention to a single grid cell, centered at 105 W longitude and 41 N latitude, which is the grid cell whose center is closest to Fort Collins, Colorado.

As we analyze block maxima, our model is a time-varying GEV. Let $M_b^{(j)}(s) = \bigvee_{t=1}^{n} Y_{b,t}^{(j)}(s)$, where $Y_{b,t}^{(j)}(s)$ denotes the precipitation for day t in year $b = 1920, \ldots, 2005$ at location s for ensemble member j. We assume

$$M_b^{(j)}(s) \sim GEV(\eta_b(s) = \beta_0(s) + \beta_1(s)(b - 1919), \tau(s), \xi(s)). \qquad (8.4)$$

We choose to capture nonstationary behavior in only the location parameter, as increasing the model complexity by adding trends to the scale or shape parameters did not seem warranted.

Parameters were estimated by numerical maximum likelihood and standard errors were obtained from the numerically computed Hessian of the likelihood surface at the ML estimates. The left panel of Figure 8.1 shows the 30 estimates of β_1 along with normal-based 95% confidence intervals, and the right panel shows the same for ξ. The main message from these figures is that there are very large uncertainties associated with these parameter estimates. Most ensembles do not find $\hat{\beta}_1$ to be significantly different from zero. Most of the point estimates for $\hat{\xi}$ are positive, which is generally what is found for weather-station precipitation data, so this gives some confidence that the GCM can produce heavy-tailed behavior. However, more than half of the CIs include zero and individual ensemble point estimates for $\hat{\xi}$ range from -.11 to .33 Distributions with these values of ξ have dramatically different tail behaviors with -.11 corresponding to a bounded tail and .33 a quite heavy tail ($\xi \geq 1/3$ implies an infinite third moment).

Given that ensemble member from a GCM are effectively iid realizations, we have the luxury of utilizing data from all ensemble members to estimate the parameters of model (8.4). ML estimates (standard errors) are $\hat{\beta}_0 - 1919 = 24.3(0.24), \hat{\beta}_1 = -.0067(.0045), \log(\tau) = 1.82(.018)$, and $\hat{\xi} = 0.15(.017)$. Not surprisingly, as shown in Figure 8.1, the uncertainty is dramatically reduced when block maxima from the entire ensemble are used. The 95% CI for β_1 continues to include zero.

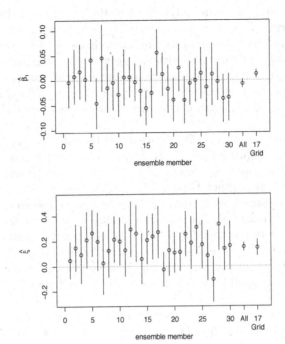

FIGURE 8.1
Estimates for β_1 and ξ for the Fort Collins grid cell from individual ensemble members. "All" on horizontal axes indicates when all ensemble members were used. "17 Grid" indicates only data from ensemble member 17 was used and the method to borrow strength across location in Section 8.2.5.2 was implemented.

8.2.5.2   Borrowing strength across locations

When analyzing observational data, and even in most cases when analyzing climate model output, the practitioner does not have the luxury of utilizing multiple ensembles to reduce uncertainty in the parameter estimates. In this section we will focus on analyzing only ensemble member 17, which we select specifically because it has the highest point estimate $\hat{\beta}_1$ and whose 95% CI for this parameter does not include zero. Of course, if data were generated from model (8.4) with no trend, it would not be surprising that one of 30 ensembles would lead one to erroneously conclude that there was in fact a positive trend. Still, if one only had the output from ensemble 17, one would likely reach different conclusions about the behavior of extreme precipitation for this grid cell. The top left panel of Figure 8.2 shows ensemble 17's annual maximum data and the estimate .95 quantile of the annual maximum (i.e., .05 AEP level) along with a 95% confidence interval generated by the delta method. The top right panel of Figure 8.2 shows the annual maximum data from all ensembles and the estimated .05 AEP level using the data from all ensembles.

Given only output from ensemble 17, we can still consider ways in which parameter uncertainty can be reduced and estimation can be improved. One way to likely decrease parameter uncertainty would be to utilize more data by employing a threshold exceedance approach rather than a block maximum approach. However, threshold exceedance approaches include the additional challenges of determining a threshold and dealing with short-term temporal dependence.

It is somewhat silly to think of this grid cell in isolation, as there is additional information contained in nearby grid cells which could be employed to aid estimation. Extremes studies in environmental sciences have a long history of borrowing strength across locations, dating at least to [13]. A widely-used method with a long history is regional frequency analysis, which defines regions over which data are pooled to estimate parameters [32]. Another approach is to construct hierarchical models which employ statistical spatial models on EV parameters [e.g., 12, 20, 47]. Bayesian inference is often employed for these hierarchical models.

Looking at the behavior of $\hat{\beta}_1(s)$ for adjacent locations to the Fort Collins grid cell is likely to lead one to be suspicious of the strength of the trend found at the Fort Collins grid cell when analyzing only ensemble member 17. Figure 8.3 shows the point estimate for $\hat{\beta}_1(s)$ and the lower and upper bounds of the 95% CIs. The point estimates for the eastern cells east are negative, while those for the central and western cells are positive. It seems unlikely that extreme precipitation would have opposite trends over such a short spatial distance, even given the fact that Fort Collins likes at the eastern boundary of the Rocky Mountains. Furthermore, values for the lower CI bound show that the Fort Collins cell and cell immediately north would have significant positive trends, and the upper CI bound values show that the cell immediately east of Fort Collins would have a significantly negative trend.

While doing RFA or constructing a hierarchical model lies outside the scope of this review, we wish to illustrate how borrowing strength across locations can both reduce uncertainty and aid comprehension. We consider only the Fort Collins grid cell and the eight neighboring cells, but a more thorough analysis would employ many more over a study region of interest. Furthermore, a more thorough analysis would likely employ a spatial model on the parameters rather than the admittedly simple approach we use below.

Let $s_1, \ldots, s_9$ denote the grid cell locations as shown in the left panel of Figure 8.3. We will assume that local differences in the distribution of annual maxima can be captured in the intercept term of the location parameter $\beta_0(s_i)$ and the scale parameter $\tau(s_i)$. We will assume that $\beta_1(s_i) = \beta_1$ and that $\xi(s_i) = \xi$. That is, we will assume that the temporal trend is shared by these nine grid cells, as we believe that it seems reasonable that even if there are local differences in the distributions of annual maxima, the way these distributions change due to changes in forcings will likely be regional. We also assume a common shape parameter for the region, as we believe the fundamental nature of the tail will be nearly constant over such a small region. Furthermore, ξ is difficult to estimate and so borrowing strength to estimate this parameter is advantageous. Let $m_b(s_i)$ be the annual maximum precipitation measurement for block (year) b at location s_i. Our likelihood is

$$\ell(\boldsymbol{\beta}_0, \beta_1, \boldsymbol{\tau}, \xi) = \prod_{i=1}^{9} \prod_{b=1920}^{2005} g(m_b(s_i); \eta_b(s_i) = \beta_0(s_i) + \beta_1(b - 1919), \tau(s_i), \xi),$$

where g denotes the GEV density function, $\boldsymbol{\beta}_0 = (\beta_0(s_1), \ldots, \beta_0(s_9))^T$, and $\boldsymbol{\tau} = (\tau(s_1), \ldots, \tau(s_9))^T$.

The estimated AEP(.05) level is shown in the bottom panel of Figure 8.2. The uncertainty in this quantile estimate has been reduced and the trend is less pronounced than when only output for the Fort Collins grid cell was used. This reduction in trend is confirmed by the point estimate, which appears in the left panel of Figure 8.1 at the location marked "17 Grid". Although the magnitude of the estimated trend is reduced, normal-based confidence intervals would still lead to the conclusion of a signifcant positive trend. The right panel of Figure 8.1 shows a noteworthy reduction in the width of the CI for $\hat{\xi}$.

The idea of borrowing strength across locations involves a bias/variance tradeoff different from the one associated with threshold selection mentioned earlier. When estimating the

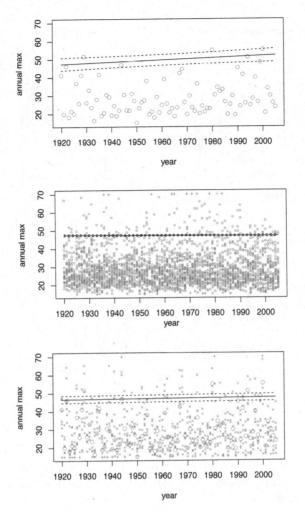

FIGURE 8.2
Top: Circles denote annual maximum daily precipitation (mm) for Fort Collins grid cell from ensemble member 17. Middle: Same but for all ensembles. Bottom shows ensemble 17's annual maximum precipitation for Fort Collins grid cell with circles and annual maxima from eight surrounding grid cells with X's. Plotted in each figure is the point estimate for the .95 quantile (equivalently AEP(.05) level) along with 95% confidence intervals. Point estimates for the top left, top right, and bottom center respectively use only ensemble 17's data for the Fort Collins grid cell, all ensembles' data for the Fort Collins grid cell, and the borrowing strength estimate described in Section 8.2.5.2. In the middle and bottom plots, values greater than 70 mm or less than 15 mm fall outside the plotting window and have been denoted by crosses.

parameters associated with a particular location, using data from other locations leads to bias, and one needs to think carefully about how one's model is pooling information from other locations. In our simple example, we relied on intuition about parameter behavior to decide for which parameters we would borrow strength. Both RFA and hierarchical models deal with data-pooling issues, but in different ways. RFA uses diagnostic tools to pre-define regions over which to pool data, while hierarchical modeling typically allows the

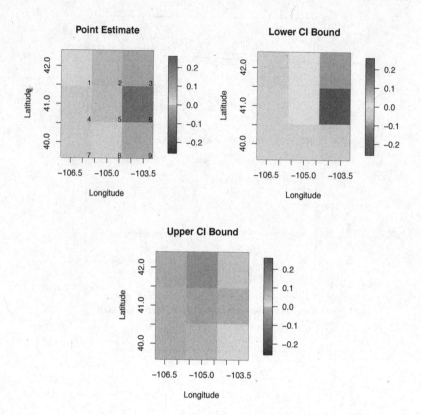

FIGURE 8.3
Point estimates (top), lower 95% CI bound (top right), and upper 95% CI bound (bottom center) for $\beta_1(s)$ (mm/year). Yellow-to-red colors indicate positive values while turqoise-to-blue indicate negative values.

spatial model to dictate how pooling will be accomplished. We note that borrowing strength across locations for the purpose of better estimating marginal parameters is fundamentally different both in approach and aim from spatially modeling dependence in the data, and these differences are examined in [16].

When we used the entire ensemble to estimate the trend for the Fort Collins grid cell, we found no significant trend and the point estimate was negative, contrary to the expected behavior of extreme precipitation in a warming climate. [26] studied this same GCM output more extensively. There are two important differences between the analysis in [26] and that done here. First, [26] more sensibly regressed the GEV location and log-scale parameters on global mean temperature rather than year, and this allows the trend in extreme precipitation to be non-linear in time. Second, [26] modeled the output for the period 1920-2080, using both the historical forcings we studied for the period 1920-2005, and additionally output forced by future projections (RCP4.5 and RCP8.5). [26] found a significant positive relationship between the GEV parameters and increasing global mean temperature, and much of the change in model-produced extreme precipitation behavior occurred in the latter portion of the studied period due to accelerated warming.

8.3   Multivariate Extremes

We now turn our attention to modeling multivariate extreme values. We will restrict our attention to the bivariate case as this is most easily discussed and visualized. The ideas presented herein can be extended to dimensions greater than two, but modeling extremes of even moderate dimension (e.g., 4 or 5) remains a challenging problem.

Our underlying guiding principle continues to be "let the tail speak for itself," but now we must accurately account for the *dependence* which is found in the joint tail of the distribution. A full accounting of dependence requires one to know the entire joint distribution. In the absence of such knowledge, correlation is widely used as a summary metric of bivariate dependence. However, correlation alone does not typically yield complete information about dependence unless the distribution is assumed to be Gaussian. Correlation will not be useful for our purpose in characterizing dependence in the tail, as its very definition implies that it measures dependence at the center of a distribution.

We will present a framework which will asymptotically characterize the dependence in the tail of a bivariate distribution. This framework arises from asymptotic results, but is also motivated by practical modeling goals which could require extrapolation. Before we develop this framework it is useful to present a summary metric of bivariate tail dependence. Suppose $\boldsymbol{Y} = (Y_1, Y_2)$ is a bivariate vector with marginal cdfs $F_1(y)$ and $F_2(y)$. [9] define

$$\chi = \lim_{u \to 1} P(F_1(Y_1) > u \mid F_2(Y_2) > u).$$

Similar to correlation, χ is a useful summary metric, but does not give complete information about dependence in the tail. There are several other related bivariate metrics of tail dependence such as the extremal coefficient [15, 49] and the madogram [11]. An advantage of χ is that its definition is readily interpretable and clearly focuses on dependence in the joint tail. We note that similar to correlation, the definition of χ requires one to account for differences in marginal behavior (here by applying F_i) before quantifying dependence.

When $\chi = 0$, the components Y_1 and Y_2 are termed asymptotically independent. Asymptotic independence is a degenerate case as many bivariate distributions with different dependence structures have this property. In this review, we focus on the asymptotic dependence case ($\chi > 0$) for two reasons. First, positive values of χ indicate that largest values of Y_1 and Y_2 can occur concurrently, and we are particularly interested in accurately modeling phenomena where risk is exacerbated by joint occurrences of large values. Second, modeling extreme behavior under asymptotic independence is much more difficult, though it continues to be an area of interest [e.g., 37, 43, 56].

8.3.1   Multivariate EVDs and componentwise block maxima

The classical approach for multivariate extremes is to describe the class of limiting distributions of componentwise maxima. Let $\boldsymbol{Y}_t = (Y_{t,1}, Y_{t,2})$ be an iid sequence of random vectors, and let $\boldsymbol{M}_n = (\bigvee_{t=1}^n Y_{t,1}, \bigvee_{t=1}^n Y_{t,2})$. If there exist sequences $\boldsymbol{a}_n > 0, \boldsymbol{b}_n \in \mathbb{R}^2$ such that

$$\frac{\boldsymbol{M}_n - \boldsymbol{b}_n}{\boldsymbol{a}_n} \xrightarrow{d} G,$$

then G is said to be a multivariate extreme value distribution (MVEVD) (arithmetic operations on vectors are applied componentwise throughout). The MVEVDs can be shown to be max-stable, that is $G^n(\boldsymbol{a}_n \boldsymbol{y} + \boldsymbol{b}_n) = G(\boldsymbol{y})$ for all $n \in \mathbb{Z}_+$. As in the univariate case, if $\boldsymbol{Y}_t$ is not iid but is stationary and mixes sufficiently, the limit remains a MVEVD [3, Section 10.5].

Because convergence of multivariate distributions requires convergence of the univariate marginals, the univariate marginals of a MVEVD G must be EVDs. However, marginal convergence to EVDs does not guarantee that a multivariate distribution belongs to the class of MVEVDs. For example, the distribution $F(\boldsymbol{y}) = \exp\{-(y_1^{-1} + y_1^{-1}y_2^{-1} + y_2^{-1})\}$ has unit Frechét marginals, and hence $\boldsymbol{a}_n = (n,n)^T$ and $\boldsymbol{b}_n = \boldsymbol{0}$. However, $F^n(\boldsymbol{a}_n\boldsymbol{y} + \boldsymbol{b}_n) = \exp\{-(y_1^{-1} + n^{-1}y_1^{-1}y_2^{-1} + y_2^{-1})\} \neq F(\boldsymbol{y})$. F is in the domain of attraction of the MVEVD $G(\boldsymbol{y}) = \exp\{-(y_1^{-1} + y_2^{-1})\}$.

Characterizing the class of MVEVDs is more difficult than the univariate case because it cannot be reduced to a simple parametric form. Generally, the family is characterized by assuming some canonical form for the marginal distributions. We will assume that the marginals of G are unit Frechét, which is sometimes referred to as the 'simple' case [17, p. 217]. One can show that the class of simple MVEVDs has the form

$$G(\boldsymbol{y}) = \exp\{-V(\boldsymbol{y})\}, \tag{8.5}$$

where the exponent measure function V has the property $V(s\boldsymbol{y}) = s^{-1}V(\boldsymbol{y})$ for any $s > 0$ ([3] gives a full development). With this property, G is easily shown to be max-stable. Furthermore, the tail dependence metric $\chi = 2 - V(1,1)$.

Given that n is large enough, the distribution of $\boldsymbol{M}_n$ should be well-approximated by a MVEVD. Let n denote a selected block length and let $\boldsymbol{y}_{b,t} = (y_{b,t,1}, y_{b,t,2}), b = 1,\ldots,B; t = 1,\ldots,n$ be a sequence of observations from a stationary process. Let $\boldsymbol{m}_b = (\bigvee_{t=1}^n y_{b,t,1}, \bigvee_{t=1}^n y_{b,t,2})$ be the B componentwise block maximum observations. One method for estimating the distribution of $\boldsymbol{M}_n$ is to fit a parametric model for a MVEVD to the block maxima data $\boldsymbol{m}_1,\ldots\boldsymbol{m}_B$. Several bivariate parametric models for $V(\boldsymbol{x})$ have been suggested [35, Section 3.4].

The cdf in (8.5) assumes the marginals are unit Frechét, which is generally not true in practice. It may be useful to envision a two-step process, where one first fits GEVs to the univariate marginals, transforms the data to have unit Frechét margins, then fits a parametric form for $V(\boldsymbol{x})$. A marginal transformation to Frechét can be defended by Proposition 5.10 of Resnick [42], which states that the domain of attraction is unchanged by monotone transformations of the marginals. This result can be interpreted as an extremes result of Sklar's theorem about copulas [40]. In practice, inference for the marginals and dependence structure can be done all-at-once, allowing one to better capture the full uncertainty than a two-step method. [8] provides an illustrative example where the annual sea levels at two Australian locations are modeled by a bivariate EVD with a logistic [28] dependence structure. The R package evd [54] provides a tool to perform bivariate maximum likelihood estimation.

Although the aforementioned statistical approach is nicely justified by the argument that the MVEVDs are the limiting distribution of a vector of block maxima, one might ask why one would want to model such data. Constructing a vector of componentwise maxima may seem unnatural, as the constructed block-maximum $\boldsymbol{m}_b$ likely does not correspond to any observation $\boldsymbol{y}_{b,t}, t \in \{1,\ldots,n\}$. When confronted with such an approach, our experience is that cooperating scientists are at first dubious. There are questions of interest which are naturally answered by an analysis of block maxima. The sea level example of [8] could be motivated by an insurance application where the company's reserves are calculated on an annual basis, and one needs to assess the probability that the maximum sea level in any given year could exceed some damage threshold resulting in claims from both of the two cities. More importantly, a vector of componentwise maxima which occur at different times still retains information about tail dependence. Heuristically, this can be understood as follows. Suppose the first component of block b's maximum, $m_{b,1}$ occurs at time t^* and is unusually large for a block maximum. If the components of $\boldsymbol{Y}_t$ exhibit asymptotic dependence, then we would expect $y_{b,t^*,2}$ to be quite large. Since $m_{b,2} > y_{b,t^*,2}$, even if $m_{b,2}$ does not occur at time t^* we would still expect it to be larger than usual. Another reason which could motivate the analysis of componentwise block maxima is that extreme observations due to the same event might not be recorded contemporaneously [33]; for example, a single storm might affect two locations on different days.

8.3.2   Multivariate threshold exceedances

In the univariate case, the GPD used to model threshold exceedances was intricately tied to results for block maxima. In the multivariate setting, we would like any modeling approach to likewise be related to the MVEVDs. From a modeling point-of-view, we desire an approach which utilizes only observations deemed extreme, and which can exhibit asymptotic dependence. This second requirement precludes most familiar multivariate models, including any model which exhibits Gaussian dependence (e.g., a Gaussian copula) [50].

To model multivariate threshold exceedances, one must first define what is meant by a multivariate exceedance. One approach is to set threshold values for each univariate marginal distribution which naturally leads one to describe events in terms of Cartesian coordinates. The multivariate GPD of [45] is defined in terms of Cartesian coordinates, and can be tied to the exponent measure function $V(\boldsymbol{x})$ in the $\xi = 1$ case. In the bivariate case with thresholds for each marginal, an observation can have one component exceed the threshold while other does not. [52] proposed using a censored likelihood approach for such settings. Below, we will define a threshold exceedance in terms of the norm of the vector, and we will model via the framework of regular variation, which again can be related to the representation for MVEVDs in (8.5). [30] give an alternative, conditional approach to modeling multivariate extreme values. This approach was related to regular variation in [29], and while we do not discuss it here, the conditional approach has been shown to be particularly useful for modeling the joint tail in the case of asymptotic independence.

Regular variation implies that the tail of the distribution behaves like a power function. A comprehensive treatment of regular variation is given by [44]. The MV regularly varying distributions classify the family of distributions which are in the domain of attraction of a MVEVD with heavy-tailed marginals with a common tail index $\xi > 0$. Fundamental to our approach is that regular variation is easily described in terms of polar coordinates.

Before we formally define regular variation, let us investigate behavior via a specific example. $\boldsymbol{Z} = (Z_1, Z_2)$ has a bivariate logistic distribution [28] if $P(\boldsymbol{Z} \leq \boldsymbol{z}) = \exp\{-(z_1^{-1/\beta} + z_2^{-1/\beta})^\beta\}$, for $z_1, z_2 > 0$ and some $\beta \in (0, 1]$. This distribution has unit Fréchet marginals and also happens to be max-stable, although only the fact that it is regularly varying is important for our purposes here. We construct iid replicates $\boldsymbol{z}_1, \dots, \boldsymbol{z}_n$ of $\boldsymbol{Z}$ where $\beta = .5$, and in the left and center panels of Figure 8.4 we plot $\boldsymbol{z}_i/n$ for $n = 500$ and $n = 5000$. The plots appear similar because the realizations have been normalized by n, and although the right panel has ten times as many points as the left, the normalization causes most of these points to pile up near the origin.

The definition of regular variation involves a convergence statement and is not straightforward, but our example above helps explain. A nonnegative random vector $\boldsymbol{Z} = (Z_1, Z_2)$ is said to be regularly varying if there exists a sequence a_n such that

$$nP\left(\frac{\boldsymbol{Z}}{a_n} \in A\right) \xrightarrow{v} \nu(A), \qquad (8.6)$$

where ν is a Radon measure on $\mathcal{C} = [0, \infty]^2 \setminus \{\boldsymbol{0}\}$, A is any ν-continuity set in $\mathcal{C}$, and v denotes vague convergence (a type of weak convergence) for measures [44, p.49]. For the logistic simulation the normalizing sequence $a_n = n$. The left-hand side of (8.6) reminds one of convergence of a binomial to a Poisson, and the measure ν on the right-hand side

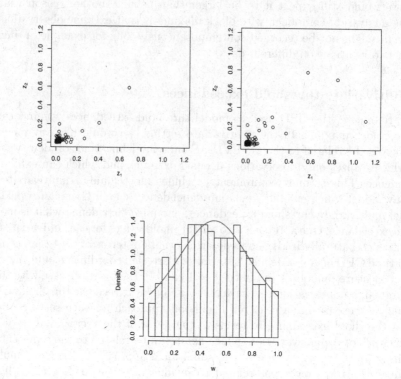

FIGURE 8.4

Scatterplots of z_i/n, where z_i are realizations from a bivariate logistic random vector and $n = 500$ (top left) and $n = 5000$ (top right). The bottom panel shows the angular density h along with a histogram of the angular components of the largest 5% of the realizations.

yields the expected number of points falling in a set A. Although we do not develop it here, there is an equivalent point-process representation of regular variation, and [15] use a similar argument to develop spatial extremes models in Chapter 9. The origin is excluded from $\mathcal{C}$ because the limit measure ν would be infinite for any set which included it, as more and more points pile up at the origin as $n \to \infty$.

It can be shown that for any scalar c, the measure ν has the property

$$\nu(cA) = c^{-1/\xi}\nu(A), \tag{8.7}$$

where $1/\xi > 0$ is the index of regular variation and from which we see the aforementioned power law tail behavior. For our logistic example, $\xi = 1$.

The radial scaling property (8.7) gives rise to an equivalent definition to (8.6), but in terms of polar coordinates. For any norm, let $R = \|\boldsymbol{Z}\|$ and $\boldsymbol{W} = \boldsymbol{Z}/R$,

$$nP\left(R/a_n > r, \boldsymbol{W} \in B\right) \xrightarrow{v} r^{-1/\xi}H(B), \tag{8.8}$$

where H is some Radon measure on $\mathcal{S} = \{\boldsymbol{x} \in \mathcal{C} : \|\boldsymbol{x}\| = 1\}$, and B is any H-continuity set in $\mathcal{S}$. Critically for modeling, (8.8) implies that the radial component R becomes independent of the angular component $\boldsymbol{W}$ as the random vector becomes large (in terms of the radial component). The measure H describes the dependence found in the joint tail of $\boldsymbol{Z}$ and is termed the spectral or angular measure; we prefer the lesser-used 'angular' terminology so as not to lead to confusion with Fourier-based methods. More mass of H toward the center of S implies stronger dependence and that Z_1 and Z_2 are more likely to be large at the same time. The special case of asymptotic independence corresponds to H consisting of point masses at the points $(0,1)$ and $(1,0)$. If H is continuously differentiable, let $h(\boldsymbol{w})$ denote this derivative. Depending on the choice of $\{a_n\}$, H may or may not be a probability measure; for our logistic example with $a_n = n$, $H(S) = 2$.

Expression (8.8) implies that the large values of $\boldsymbol{Z}$ can be well modeled by a univariate radial component R whose (large) values follow a power law described by ξ and an independent angular component $\boldsymbol{W}$ whose behavior is determined by H. In the $\xi = 1$ case, it is convenient to use the L_1 norm: $\|\boldsymbol{Z}\|_1 = Z_1 + Z_2$ to define these polar coordinates. Furthermore in the bivariate case, S is isomorphic to $[0, 1]$. For our logistic example, the form of h is known. Letting $r_i = \|\boldsymbol{z}_i\|_1$ and $w_i = z_{i,1}/r_i$, the right panel of Figure 8.4 shows a histogram of the angular components w_i that correspond to the largest 5% of the radial components along with the density given by (normalized) h.

To tie regular variation to the MVEVDs, assume $\boldsymbol{Y}_t$ (introduced in Section 8.3.1) is in the domain of attraction of a simple MVEVD. Then $\boldsymbol{Y}_t$ is regularly varying with $\xi = 1$ and limiting measure ν (alternatively angular measure H) such that the exponent measure function

$$V(\boldsymbol{y}) = \nu([\boldsymbol{0}, \boldsymbol{y}]^c) = \int_{(\boldsymbol{w},r)\in[\boldsymbol{0},\boldsymbol{y}]^c} r^{-2}dH(w) = \int_S \max\left(\frac{w}{y_1}, \frac{1-w}{y_2}\right) dH(w).$$

If the regularly varying random vector has common marginals, then there exists a balance condition on the angular measure:

$$\int_S wdH(w) = \int_S (1-w)dH(w). \tag{8.9}$$

If h exists and is symmetric about 0.5, then this balance condition will be met, but symmetry is not required.

Returning to the modeling objectives listed at the beginning of this section, MV regular variation is a suitable modeling framework for multivariate threshold exceedances as its definition only describes the tail of the distribution, it is a framework that can accommodate asymptotic dependence, and it has natural connections to the MVEVDs. Similar to modeling with parametric MVEVD models, one can envision a two step process where the marginals are transformed to a convenient heavy-tailed distribution, and then dependence is described by modeling H. Proposition 5.10 of Resnick [42] continues to imply that this marginal transformation does not affect the fundamental nature of dependence. This two step procedure is similar in spirit to copula methods, but with important differences: (1) only a small subset of large observations (defined in terms of the radial component after transformation) are used in the analysis, and (2) a model specifically suited for describing tail dependence is used. [10] introduced a likelihood-based inference method for exceedances based on a point-process representation of regular variation.

8.3.3 Application: Santa Ana winds and dryness

Risk for wildfires is related to many conditions, among which are air temperature, wind-speed, humidity, and supply of fuel. An atmospheric regime commonly referred to as the Santa Ana winds is known to be related to wildfire occurrence in southern California. The Santa Ana winds are actually a multivariate meteorological regime. Because the regime originates in the desert east of southern California, they are associated with warm temperatures, low humidity, and high winds. This regime mostly occurs in autumn and winter, but we restrict our attention to autumn (September, October, and November) as these months have historically had more fires than the winter months. Two of the most destructive fires on record in California, the Cedar Fire in 2003 and the Witch Fire in 2007, were associated with Santa Ana conditions..

We obtain data from the March AFB station[2] from the HadISD dataset [19]. The station was chosen because of its location (Riverside County CA, longitude -117.3, latitude 33.9, elevation 468 m), its length and relative completeness of record, and because this station's variables appeared to show a signal with known Santa Ana events. The data encompass the years 1973-2015, and our analysis will assume the data are temporally stationary.

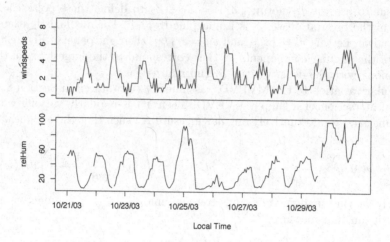

FIGURE 8.5

Time series of hourly windspeed (m/s) and relative humidity (%) for 10/21/2003 through 10/31/2003. The Cedar fire occurred on 10/25/2003.

We wish to have a bivariate time series that conveys a daily risk of fire, and we focus on meterological variables which summarize windspeed and the dryness of the air. As is often the case in environmental studies, the data we analyze result from some preprocessing. The HadISD data are recorded hourly, and provide a windspeed (m/s) and dewpoint measurement. We choose to work with relative humidity (%) rather than dewpoint, as dewpoint is directly affected by air temperature and displays a decreasing trend over the course of the autumn season. Relative humidity is calculated as a function of temperature and dewpoint via a commonly used formula employing constants suggested by [1]. Figure 8.5 shows the time series of hourly windspeeds and relative humidity for 10 days in October 2003, including the day of occurrence of the Cedar Fire. Both time series show clear diurnal cycles: windspeeds tend to be largest during the day and relative humidity increases at night. On October 25, we see the pattern is disrupted by the Santa Ana event which corresponded with the Cedar Fire. Windspeeds are stronger than on previous days, and the diurnal cycle of relative humidity is disrupted with very little increase in humidity that night. Based on exploratory analysis of this and several other known Santa Ana events, we construct our daily summaries as follows. Our daily windspeed measurement is taken to be the mean of the four maximum hourly windspeeds during the 24 hour period beginning at midnight the day of recording. Taking the mean of the four maxima not only helps to provide a daily

---

[2]Station number 722860-23119.

windspeed summary but also helps to create daily data that are more continuous than the hourly data. Our daily dryness measurement is taken to be the mean of the six relative humidity measurements beginning at 11 pm the day of recording and ending at 4 am the following day, which is then negated so that the greatest values correspond to the driest conditions. Although our daily dryness measure includes several relative humidity measurements taken from the first few hours of the following day, our hope is that this signal is associated with the recorded day's dryness, and we believe this signal is clearer than one which would employ daytime relative humidity measurements which are often quite dry at this location regardless of whether in the Santa Ana regime or not.

The left panel of Figure 8.6 shows our daily summary values for windspeed and dryness. The scatterplot shows that high windspeed does not have to occur when the air is particularly dry, nor do dry conditions only occur when the windspeed is high. However, the plot also shows that very high windspeed and very dry conditions can occur simultaneously, and putting these two conditions together leads to the explosive fire conditions about which we are concerned. The red dot and orange dot respectively indicate October 25, 2003 and October 21, 2007, the initial days of the Cedar and Witch Fires. Both are days which experienced high winds and dry conditions (the Witch Fire in fact corresponds to the day with the driest conditions in our record), but a number of days show similar dryness and higher windspeed. Of course, high windspeeds and dry air do not themselves cause large fires to happen; a wildfire requires an ignition source, the size of a fire depends on the immediacy of response and accessibility of location, and fire risk is not fully captured by these two variables.

8.3.3.1 Assessing tail dependence

We wish to assess the tail dependence between our daily and dryness variables, and ultimately to assess how often these high-risk conditions occur simultaneously. The center panel of Figure 8.6 shows $\hat{\chi}$ for u increasing from 0.70 to almost 1. If we attribute the drop at the end of the plot to very low sample size (which seems justified by the uncertainty shown by the approximate 95% confidence intervals), the plot shows that estimates for χ seem to level off at about 0.3 for high values of u. This can loosely be interpreted by saying that when windspeeds are at their highest levels, our dryness measure is also at its highest levels about 30% of the time. Based on the plot of $\hat{\chi}$ and the scatterplot, we conclude that our data exhibit asymptotic dependence.

Let $\boldsymbol{Y}_t = (Y_{t,1}, Y_{t,2})$ be the random vector whose variates denote our windspeed and dryness measures. We will estimate the probability of an event occurring in two different risk regions. Based somewhat on the values associated with the Cedar Fire, we define $\mathcal{R}_1 = \{y \in \mathbb{R}^2 : y_1 > 7.8, y_2 > -19.5\}$. The risk region $\mathcal{R}_1$ is shown in the left panel of Figure 8.6. As the Cedar Fire is in $\mathcal{R}_1$, we know that conditions in this region are conducive to large fires. We define $\mathcal{R}_2 = \{y \in \mathbb{R}^2 : y_1 > 12.75, y_2 > -10.5\}$. This region, also shown in the left panel of Figure 8.6, is very extreme and contains no observations in our data record. We aim to obtain estimates of $p_1 = P(\boldsymbol{Y}_t \in \mathcal{R}_1)$ and $p_2 = P(\boldsymbol{Y}_t \in \mathcal{R}_2)$. It is clear that it will be necessary to accurately account for the dependence in the joint tail to assess p_1 and p_2.

Let $\boldsymbol{y}_t = (y_{t1}, y_{t2}), t = 1, \ldots, n$ denote our bivariate windspeed and dryness observations. Based on mean residual life plots [8, Section 4.3], GPD's were fit to the upper 5% of the data of each marginal. Estimates of ξ were -0.15 (se $= 0.07$) for windspeed and -0.41 (0.05) for dryness. Thus, the univariate marginals appear to have bounded tails, and these fitted GPD's yield point estimates of the upper endpoints for the respective distributions of 17.8 and -5.3. Bounded tails agree with basic intuition and knowledge about these variates; for instance, it would be impossible for our dryness variate to return a positive value.

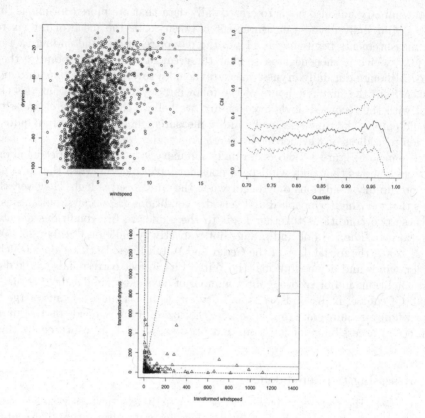

FIGURE 8.6

Top left: scatterplot of daily summary values for windspeed and dryness. Top right: plot of $\hat{\chi}$. Bottom: scatterplot of daily summaries after marginal transformation to unit Frechét. The red solid lines region define risk regions $\mathcal{R}_1$ and $\mathcal{R}_2$ in the original scale and $\mathcal{R}_1^*$ and $\mathcal{R}_2^*$ in the Frechét scale. The red dot corresponds to the day of the Cedar Fire, and the orange dot corresponds to the day of the Witch Fire. Dashed rays in the Frechét scale plot point to events which lie outside the plot window.

Using the regular variation framework to model tail dependence requires a transformation of the marginal distributions. We choose to use a combination of parametric and nonparametric methods to transform the marginal distributions to be approximately unit Frechét. Let $\hat{u}_i, \hat{\psi}_{u_i}$, and $\hat{\xi}_i$ denote the empirical .95 quantile, and GPD scale and shape estimates for the ith marginal, $i = 1, 2$. We let

$$\hat{F}_i(y) = \begin{cases} (n+1)^{-1} \sum_{t=1}^n \mathbb{I}(y_{t,i} \leq y) & \text{for } y \leq \hat{u}_i, \\ 1 - .05\big(1 + \hat{\xi}_i(y - \hat{u}_i)/\hat{\psi}_{u_i}\big)^{-1/\hat{\xi}_i} & \text{for } y > \hat{u}_i. \end{cases} \tag{8.10}$$

Then, letting $t(\boldsymbol{y}) = (t_1(y_1), t_2(y_2))$ where $t_i(y) = G^{-1}(\hat{F}_i(y))$ and G is the cdf of the unit Frechét, we construct transformed observations $\boldsymbol{z}_t = t(\boldsymbol{y}_t)$. We also produce the polar transformed observations $r_t = z_{t,1} + z_{t,2}$ and $w_t = z_{t,1}/r_t$. Additionally, we can transform the risk regions to the new space: $\mathcal{R}_i^* = t(\mathcal{R}_i), i = 1, 2$.

The right panel of Figure 8.6 shows the data after transformation, as well as the transformed risk regions. Shown with a red dot is the day corresponding to the Cedar Fire. Plots

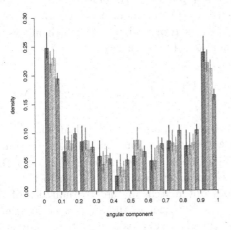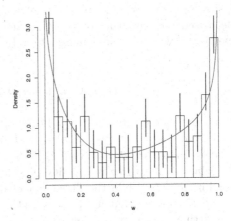

FIGURE 8.7

Left: histograms of the angular components of the events whose radial components exceed the .97 (black), .95 (red), .90 (green), and .75 (blue) quantiles. Right fitted parametric angular measure model of a mixture of two beta distributions.

on the Frechét scale can be hard to interpret because the transformed data are so heavy-tailed. In the figure, events that have radial components R_t larger than the .99 empirical quantile have been plotted with triangles, events with R_t between the .98 and .99 empirical quantiles are shown with X's, and all other events are plotted with circles. We see that on this Frechét scale, one's eye is naturally drawn to only the very largest events, as it is difficult to distinguish any points other than the largest 1%. Because the largest transformed points are so large, we have chosen the plot window to show only the region $[0, 1400]^2$, and five points lie outside this region. One of these points lies near the x-axis at $(15381, 10.9)$, and another lies distant from both axes at $(309, 1894)$. The other three points are near the y-axis at $(14.8, 6977)$, $(2.5, 2257)$, and $(52, 9265)$, and this last point corresponds to the Witch Fire. Rays are shown in the figure extending to these five points; we have indicated with an orange dashed line the ray to the Witch Fire, which lies inside $\mathcal{R}_1^*$.

We next investigate and model the angular measure. Histograms of angular components for events whose radial components exceed the .97, .95, .90, and .75 quantiles are shown in the left panel of Figure 8.7. Accounting for the uncertainty associated with these histograms (shown by .95 confidence intervals based on the multinomial distribution), it appears that the distributions of angular components whose radial components exceed the .90, .95, and .97 quantiles are very similar. Not surprisingly, one sees a noteworthy change in the distribution of angular components whose corresponding radial components only exceed the .75 quantile. We decide to model the angular measure by retaining angular components whose radial component exceeds the empirical .95 quantile, which coincides with the quantiles used for parametric modeling of the marginal tail, though there is no theoretical reason why these quantiles should match. Furthermore, as shown in the right panel of Figure 8.6, many points with large radial components will have either $z_{t,1}$ or $z_{t,2}$ below the marginal .95 quantile. The right panel of Figure 8.7 shows a histogram of the values of $w_{t,1}$ for the events whose radial component r_t exceeds the .95 empirical quantile, along with .95 confidence intervals based on the multinomial distribution.

Using the likelihood-based approach of [10], we fit parametric models to the angular components of the large points. We fit a few angular measure models, including the angular measure corresponding to the logistic model and one given by a beta density, both of which have a single parameter and yield symmetric densities. Of the models we tested, we select a model based on a mixture of two beta densities as proposed by [5], which allows for asymmetry and which has four parameters. Letting $\boldsymbol{\theta} = (\pi_1, \alpha_1, s_1, s_2)$,

$$h(w; \boldsymbol{\theta}) = \sum_{j=1}^{2} \pi_j \frac{\Gamma(s_j)}{\Gamma(s_j \alpha_j) \Gamma(s_j(1 - \alpha_j))} w^{s_j \alpha_j - 1} (1 - w)^{s_j(1 - \alpha_j) - 1}, \ w \in [0, 1]$$

where the mixture weights $\pi_1 + \pi_2 = 1$ and $\pi_j > 0$, $s_j > 0$, and $0 \leq \alpha_j \leq 1$ all for $j = 1, 2$. The additional constraint $\pi_1 \alpha_1 + \pi_2 \alpha_2 = 1/2$ assures condition (8.9) is met. Maximum likelihood estimates are $\hat{\boldsymbol{\theta}} = (0.39, 0.12, 7.19, 2.78)$. Choice of this model is also supported by AIC values of the three model fits. The fitted angular measure is shown in the right panel of Figure 8.7. While the fitted model does not match all the irregularities of the histogram, we contend that it captures much of the behavior of the angular components, and that fitting even more complex parametric models would risk over-fitting. For more flexibility, it is possible to model angular measures for bivariate vectors non-parametrically and maintain the balance condition (8.9) [22], but we do not pursue that here. To make sense of the asymmetry, note that if w is close to zero then the event has high/extreme dryness and non-high windspeed.

8.3.3.2   Risk region occurrence probability estimation

As p_1 and p_2 are equivalent to $P(\boldsymbol{Z}_t \in \mathcal{R}_1^*)$ and $P(\boldsymbol{Z}_t \in \mathcal{R}_2^*)$, we work in the Frechét transformed space and perform estimation of these probabilities in two different ways. First, using our parametric model fit above, for $i = 1, 2$,

$$\nu(\mathcal{R}_i^*) = \int_{(\boldsymbol{w}, r) \in \mathcal{R}_i^*} r^{-2} h(\boldsymbol{w}, \boldsymbol{\theta}) d\boldsymbol{w} = \int_{w=0}^{1} \min\left(\frac{w}{z_{i,1}^*}, \frac{1 - w}{z_{i,2}^*}\right) h(w, \boldsymbol{\theta}) dw, \qquad (8.11)$$

where $(z_{1,1}^*, z_{1,2}^*) = (29.2, 72.5)$ and $(z_{2,1}^*, z_{2,2}^*) = (1266.4, 820.0)$ are the lower-left corners of $\mathcal{R}_1^*$ and $\mathcal{R}_2^*$ respectively. Plugging in $\hat{\boldsymbol{\theta}}$, we obtain $\hat{\nu}(\mathcal{R}_1^*) = 3.5 \times 10^{-3}$ and $\hat{\nu}(\mathcal{R}_2^*) = 1.7 \times 10^{-4}$. To obtain our probability estimate, we begin from (8.6), setting $a_n = 2n$ as this is the correct normalizing sequence for $\boldsymbol{Z}$ with Frechét marginals and h a probability density. Assuming n is fixed and large enough such that the convergence implies approximate equality, and letting $A = \mathcal{R}_i^*/2n$, we obtain

$$nP\left(\frac{\boldsymbol{Z}_t}{2n} \in \frac{\mathcal{R}_i^*}{2n}\right) \approx \nu\left(\frac{\mathcal{R}_i^*}{2n}\right) \qquad (8.12)$$
$$\Rightarrow P(\boldsymbol{Z}_t \in \mathcal{R}_i^*) \approx 2\nu(\mathcal{R}_i^*).$$

Plugging in $\hat{\nu}(\mathcal{R}_1^*)$ and $\hat{\nu}(\mathcal{R}_2^*)$ above yields estimates $\hat{p}_1 = 7.0 \times 10^{-3}$ and $\hat{p}_2 = 3.4 \times 10^{-4}$; uncertainty for these estimates is discussed below. Given that our studied season consists of 91 autumn days, the first probability estimate implies that we expect to see about 6 days in 10 seasons which fall in $\mathcal{R}_1$ (with conditions roughly as or more extreme than those associated with the Witch Fire). The second probability estimate implies that we would expect to see 3 days in 100 seasons which fall in $\mathcal{R}_2$.

Alternative to employing a parametric angular measure model, a nonparametric estimate can be obtained via the scaling property (8.7). We will scale by a factor of 10, and we define $\mathcal{R}_i^\dagger = \{z \in [0, \infty]^2 \setminus \{0\} : 10z \in \mathcal{R}_i\}$, for $i = 1, 2$. Via (8.7) and assuming the approximation in (8.12), one can obtain

$$P(\boldsymbol{Z}_t \in \mathcal{R}_i^*) \approx \frac{1}{10} P(\boldsymbol{Z}_t \in \mathcal{R}_i^\dagger).$$

Employing empirical estimates for $P(\boldsymbol{Z}_t \in \mathcal{R}_1^\dagger)$ and $P(\boldsymbol{Z}_t \in \mathcal{R}_2^\dagger)$ yields estimates $\tilde{p}_1 = 6.8 \times 10^{-3}$ and $\tilde{p}_2 = 1.5 \times 10^{-4}$.

Fully accounting for uncertainty in multivariate extremes methods requires accounting for uncertainty in the marginal estimation, and in the modeling of the tail dependence structure. We proceed with a straightforward paired bootstrap method which mimics our estimation methods above. Although we are not attempting to model temporal dependence, we use a block bootstrap approach with block length of five days in order to account for the increased parameter uncertainty due to temporal dependence. Let $\boldsymbol{y}_t^{(b)}, t = 1, \ldots, n$ represent a block-resample of $\boldsymbol{y}_t, t = 1, \ldots, n$. Marginally, we fit a GPD above the .95 quantile of this resample, and transform the marginals via $\boldsymbol{z}_t^{(b)} = t^{(b)}(\boldsymbol{y}_t^{(b)})$, where $t^{(b)}$ is as before, but with the data and GPD parameter estimates from the resample. Additionally, we create $\mathcal{R}_i^{*(b)} = t^{(b)}(\mathcal{R}_i)$. For the parametric method, we retain the points whose radial components are in the largest 5%, obtain parameter estimates $\hat{\boldsymbol{\theta}}^{(b)}$ for the mixture-of-betas angular measure model, and integrate (8.11). For the non-parametric method, we calculate the number of bootstrap resampled points falling in $\mathcal{R}_i^{*(b)}/10$ and scale as above. To obtain 95% confidence intervals, we repeat these procedure for $b = 1, \ldots, B = 1000$, and take the .025 and .975 quantiles of the estimates $\hat{p}_i^{(b)}$. Our parametric bootstrap-estimated 95% confidence interval for p_1 is $(5.2 \times 10^{-3}, 8.9 \times 10^{-3})$ and for p_2 is $(8.1 \times 10^{-5}, 5.5 \times 10^{-4})$; our nonparametric intervals are respectively $(5.2 \times 10^{-3}, 8.8 \times 10^{-3})$ and $(2.5 \times 10^{-5}, 4.1 \times 10^{-4})$. Standard empirical estimates are $\check{p}_1 = 4.9 \times 10^{-3}$ and $\check{p}_2 < 2.5 \times 10^{-4}$. While the empirical estimate for p_1 doesn't quite fall in the 95% confidence interval, it does fall in a 98% interval.

For comparison, we obtain an estimate for p_1 using a simple Gaussian copula approach. Via rank transform, we obtain uniform marginals, and then estimate the correlation parameter of the Gaussian copula to be $\hat{\rho} = .23$. Then by transforming $\mathcal{R}_1$ to the corresponding region in $[0, 1]^2$, we obtain an estimate of 1.4×10^{-3} for p_1, much lower than either of the estimates via the regular variation framework, and also less than the empirical estimate for p_1. The Gaussian copula model is different from the regular variation-based approach in that it is asymptotically independent. Likely more important in terms of affecting the estimate of p_1, the Gaussian copula's dependence parameter estimate was based on the entire data set rather than focusing on extreme observations.

Our example illustrates a method for modeling dependence in the tail and using this model to estimate the probabilities associated with pre-defined risk regions. Our method relies on the framework of regular variation, which is inherently linked to multivariate EV theory, and which utilizes only information found in the extreme observations. Employing this framework on this windspeed and dryness data requires marginal transformation to a heavy-tailed distribution, and the transformation we employed is rather dramatic, as seen in the right panel of Figure 8.6. This could be considered a drawback, as interpreting dependence in this transformed heavy-tailed data set requires familiarity with the framework. Importantly, the regular variation framework allows one to model tail dependence in the asymptotically dependent case. If data exhibit asymptotic dependence, it is particularly important to employ a model which can capture asymptotic dependence when extrapolating beyond the range of the data, as we do to estimate p_2.

Our example assumed the data were stationary, ignoring changes occurring due to climate change. It is of great interest to better understand how fire risk will change due to an altered climate. This data set is likely too short to accurately model how these variates have changed over the study period, and even if such a model were obtained, projecting a nonstationary model into the future should be done cautiously. Nevertheless, one could imagine applying a method such as this to climate model output, run under both current climate and a projected future climate. By comparing the bivariate extremes models, one could assess how extreme behavior differs between the two periods, both marginally and in its tail dependence structure.

8.4   Conclusions

In this chapter, we aimed to provide an introduction to extreme value modeling both in the univariate and bivariate cases. Although our coverage of these topics is necessarily incomplete, hopefully the philosophy of letting the tail speak for itself was made clear in both our review of methods and our illustrative examples.

In the introduction we mentioned that extremes studies are relatvely prevalent in hydrology and atmospheric science. These disciplines benefit from data records from recording stations which take measurements at regular intervals and whose records can span a large number of years. Such data records make extreme analyses feasible, since a subset of extreme data can be extracted. Other environmental areas such as ecology have seen little extremes work done, and one reason for this is likely a lack of long, sustained data records.

In our applications, we studied extreme precipitation and windspeed/humidity, the extremes of which can be viewed as exhibiting short-lived, 'spike'-like behavior. Although there has been much work describing temporal dependence in extremes, existing EV methods seem best suited for describing short-lived, spike-like phenomena. Prolonged extreme environmental events such as droughts or heat waves remain challenging to model. One can still imagine that with such phenomena, it would continue to be beneficial to focus on and model only the tail, and not allow inference about such phenomena to be contaminated by non-extreme data.

Acknowledgments

D. Cooley is partially supported by NSF grant DMS-1243102. R.L. Smith is partially supported by NSF grants DMS-1127914, DMS-1638521, and DMS-1242957.

Bibliography

[1] Alduchov, O. A. and Eskridge, R. E. (1996). Improved Magnus form approximation of saturation vapor pressure. *Journal of Applied Meteorology*, 35(4):601–609.

[2] Balkema, A. and de Haan, L. (1974). Residual life time at great age. *The Annals of Probability*, 2(5):792–804.

[3] Beirlant, J., Goegebeur, Y., Segers, J., Teugels, J., Waal, D. D., and Ferro, C. (2004). *Statistics of Extremes: Theory and Applications*. Wiley, New York.

[4] Berrocal, V., Gelfand, A., and Holland, D. (2014). Assessing exceedance of ozone standards: a space-time downscaler for fourth highest ozone concentrations. *Environmetrics*, 25(4):279–291.

[5] Boldi, M.-O. and Davison, A. C. (2007). A mixture model for multivariate extremes. *Journal of the Royal Statistical Society, Series B*, 69:217–229.

[6] Chavez-Demoulin, V. and Davison, A. C. (2005). Generalized additive models for sample extremes. *Journal of the Royal Statistical Society, Series C (Applied Statistics)*, 54(1):207–222.

[7] Chavez-Demoulin, V. and Davison, A. C. (2012). Modelling time series extremes. *REVSTAT-Statistical Journal*, 10:109–133.

[8] Coles, S. G. (2001). *An Introduction to Statistical Modeling of Extreme Values*. Springer Series in Statistics. Springer-Verlag London Ltd., London.

[9] Coles, S. G., Heffernan, J., and Tawn, J. A. (1999). Dependence measures for extreme value analysis. *Extremes*, 2:339–365.

[10] Coles, S. G. and Tawn, J. A. (1991). Modeling multivariate extreme events. *Journal of the Royal Statistical Society, Series B*, 53:377–92.

[11] Cooley, D., Naveau, P., and Poncet, P. (2006). Variograms for spatial max-stable random fields. In Bertail, P., Doukhan, P., and Soulier, P., editors, *Dependence in Probability and Statistics*, Springer Lecture Notes in Statistics. Springer, New York.

[12] Cooley, D., Nychka, D., and Naveau, P. (2007). Bayesian spatial modeling of extreme precipitation return levels. *Journal of the American Statistical Association*, 102:824–840.

[13] Dalrymple, T. (1960). Flood frequency analyses. Water supply paper 1543-a, U.S. Geological Survey, Reston, VA.

[14] Davison, A. C. (2003). *Statistical Models*, volume 11. Cambridge University Press.

[15] Davison, A. C., Huser, R., and Thibaud, E. (2018). Spatial extremes chapter, this volume. In Gelfand, A., editor, *The Handbook of Environmental Statistics*, pages XX–XX. Chapman and Hall.

[16] Davison, A. C., Padoan, S., and Ribatet, M. (2012). Statistical modeling of spatial extremes. *Statistical Science*, 27(2):161–186.

[17] de Haan, L. and Ferreira, A. (2006). *Extreme Value Theory*. Springer Series in Operations Research and Financial Engineering. Springer, New York.

[18] Dee, D., Fasullo, J., Shea, D., Walsh, J., and National Center for Atmospheric Research Staff (2016). The climate data guide: Atmospheric reanalysis: Overview and comparison tables. Retrieved from https://climatedataguide.ucar.edu/climate-data/atmospheric-reanalysis-overview-comparison-tables.

[19] Dunn, R. J., Willett, K. M., Thorne, P. W., Woolley, E. V., Durre, I., Dai, A., Parker, D. E., and Vose, R. E. (2012). HadISD: a quality-controlled global synoptic report database for selected variables at long-term stations from 1973–2011. *Climate of the Past*, 8:1649–1679.

[20] Dyrrdal, A. V., Lenkoski, A., Thorarinsdottir, T. L., and Stordal, F. (2015). Bayesian hierarchical modeling of extreme hourly precipitation in norway. *Environmetrics*, 26(2):89–106.

[21] Eastoe, E. F. and Tawn, J. A. (2009). Modelling non-stationary extremes with application to surface level ozone. *Journal of the Royal Statistical Society: Series C (Applied
Statistics)*, 58(1):25–45.

[22] Einmahl, J. H. and Segers, J. (2009). Maximum empirical likelihood estimation of
the spectral measure of an extreme-value distribution. *The Annals of Statistics*, pages
2953–2989.

[23] Fawcett, L. and Walshaw, D. (2007). Improved estimation for temporally clustered
extremes. *Environmetrics*, 18(2):173–188.

[24] Ferreira, A. and de Haan, L. (2015). On the block maxima method in extreme value
theory:PWM estimators. *The Annals of Statistics*, 43(1):276–298.

[25] Fisher, R. A. and Tippett, L. H. C. (1928). Limiting forms of the frequency distribution of the largest or smallest members of a sample. *Proceedings of the Cambridge
Philosophical Society*, 24:180–190.

[26] Fix, M., Cooley, D., Sain, S., and Tebaldi, C. (2016). A comparison of US precipitation
extremes under RCP8.5 and RCP4.5 with application to pattern scaling. *Climatic
Change*. DOI: 10.1007/s10584-016-1656-7.

[27] Gnedenko, B. (1943). Sur la distribution limite du terme maximum d'une série
aléatoire. *The Annals of Mathematics*, 44(3):423–453.

[28] Gumbel, E. (1960). Distributions des valeurs extrêmes en plusieurs dimensions. *Publ.
Inst. Statis. Univ. Paris*, 9:171–173.

[29] Heffernan, J. E. and Resnick, S. I. (2007). Limit laws for random vectors with an
extreme component. *The Annals of Applied Probability*, pages 537–571.

[30] Heffernan, J. E. and Tawn, J. A. (2004). A conditional approach for multivariate
extreme values (with discussion). *Journal of the Royal Statistical Society, Series B*,
66:497–546.

[31] Hosking, J. R. M. (1990). L-moments: Analysis and estimation of distributions using
linear combinations of order statistics. *Journal of the royal statistical society, Series
B*, 52:105–124.

[32] Hosking, J. R. M. and Wallis, J. R. (1997). *Regional Frequency Analysis: An Approach
based on L-Moments*. Cambridge, University Press, Cambridge., U.K.

[33] Johndrow, J. and Wolpert, R. (2016). Tail waiting times and the extremes of stochastic
processes. *arXiv:1512.07848*.

[34] Kay, J., Deser, C., Phillips, A., Mai, A., Hannay, C., Strand, G., Arblaster, J., Bates,
S., Danabasoglu, G., Edwards, J., et al. (2015). The community earth system model
(CESM) large ensemble project: A community resource for studying climate change
in the presence of internal climate variability. *Bulletin of the American Meteorological
Society*, 96(8):1333–1349.

[35] Kotz, S. and Nadarajah, S. (2000). *Extreme Value Distributions*. Imperial College
Press, London.

[36] Leadbetter, M., Lindgren, G., and Rootzén, H. (1983). *Extremes and Related Properties
of Random Sequences and Processes*. Springer-Verlag, New York.

[37] Ledford, A. W. and Tawn, J. A. (1996). Statistics for near independence in multivariate extreme values. *Biometrika*, 83:169–187.

[38] Moss, R. H., Edmonds, J. A., Hibbard, K. A., Manning, M. R., Rose, S. K., Van Vuuren, D. P., Carter, T. R., Emori, S., Kainuma, M., Kram, T., et al. (2010). The next generation of scenarios for climate change research and assessment. *Nature*, 463(7282):747–756.

[39] Naveau, P., Huser, R., Ribereau, P., and Hannart, A. (2016). Modeling jointly low, moderate, and heavy rainfall intensities without a threshold selection. *Water Resources Research*, 52:27532769.

[40] Nelsen, R. (2006). *An Introduction to Copulas, 2nd Edition*. Lecture Notes in Statistics No. 139. Springer, New York.

[41] Pickands, J. (1975). Statistical inference using extreme order statistics. *Annals of Statistics*, 3:119–131.

[42] Resnick, S. (1987). *Extreme Values, Regular Variation, and Point Processes*. Springer-Verlag, New York.

[43] Resnick, S. (2002). Hidden regular variation, second order regular variation and asymptotic independence. *Extremes*, 5(4):303–336.

[44] Resnick, S. (2007). *Heavy-Tail Phenomena: Probabilistic and Statistical Modeling*. Springer Series in Operations Research and Financial Engineering. Springer, New York.

[45] Rootzen, H. and Tajvidi, N. (2006). Multivariate generalized Pareto distributions. *Bernoulli*, 12:917–930.

[46] Russell, B. T., Cooley, D., Porter, W. C., Reich, B. J., and Heald, C. L. (2016). Data mining to investigate the meteorological drivers for extreme ground level ozone events. arXiv:1504.08080v4.

[47] Sang, H. and Gelfand, A. E. (2010). Continuous spatial process models for spatial extreme values. *Journal of Agricultural, Biological, and Environmental Statistics*, 15:49–65.

[48] Scarrott, C. and MacDonald, A. (2012). A review of extreme value threshold estimation and uncertainty quantification. *REVSTAT–Statistical Journal*, 10(1):33–60.

[49] Schlather, M. and Tawn, J. (2003). A dependence measure for multivariate and spatial extreme values: Properties and inference. *Biometrika*, 90:139–156.

[50] Sibuya, M. (1960). Bivariate extreme statistics, i. *Ann. Inst. Statist. Math.*, 11:195–210.

[51] Smith, R. L. (1989). Extreme value analysis of environmental time series: An application to trend detection in ground-level ozone. *Statistical Science*, 4:367–393.

[52] Smith, R. L., Tawn, J. A., and Coles, S. G. (1997). Markov chain models for threshold exceedances. *Biometrika*, 84:249–268.

[53] Stephenson (2018). Climate projections reference, this volume. In Gelfand, A., editor, *The Handbook of Environmental Statistics*, pages XX–XX. Chapman and Hall.

[54] Stephenson, A. G. (2002). evd: Extreme value distributions. *R News*, 2(2):0.

[55] Taylor, K. E., Stouffer, R. J., and Meehl, G. A. (2012). An overview of CMIP5 and the experiment design. *Bulletin of the American Meteorological Society*, 93(4):485–498.

[56] Weller, G. and Cooley, D. (2013). A sum characterization of hidden regular variation with likelihood inference via expectation-maximization. *Biometrika*, 101:17–36.

9

Environmental Sampling Design

Dale L. Zimmerman

University of Iowa

Stephen T. Buckland

University of St Andrews, Scotland, UK

CONTENTS

9.1 Introduction

Environmental and ecological data occur in space and time, and precisely where and when they occur may affect what can be inferred from the data. For example, if precipitation is measured at gauges widely separated in space, average precipitation amounts over large regions may be estimated quite well but little may be inferred about the local or small-scale spatial variability of those amounts. Consequently, considerable effort has been put into the development of *sampling designs*, i.e., selections of spatial locations and times, that maximize or otherwise enhance the quality of inferences that can be made from sampling the environment. This chapter considers such designs. In particular, it reviews designs used to monitor an environmental variable (such as precipitation, or the ambient level of an atmospheric pollutant) over space and time, and designs used to estimate the abundance, or population size, of an organism in the environment.

Among the designs we will describe are those that are probability-based, i.e., that involve the selection of samples according to a random mechanism that can be described by a probability distribution. Other designs to be considered are not probability-based; nevertheless, they are chosen to meet certain well-defined statistical or geometric objectives. This sets them apart from other types of non-probability-based designs, such as those that arise from *haphazard sampling*, in which samples are chosen purely out of convenience, or *judgment sampling*, in which samples are chosen with a goal of being "representative" in some sense.

9.2 Sampling Design for Environmental Monitoring

9.2.1 Design framework

We describe sampling design for environmental monitoring within the following framework for spatio-temporal data. Assume that the environmental variable of interest, Z, is continuous (or discrete with so many levels that it can be approximated as such) and varies over a region $\mathcal{S}$ and a time interval $\mathcal{T}$ of interest. Let $Z(\mathbf{s}, t)$ denote the value of Z at point site $\mathbf{s} \in \mathcal{S}$ and time point $t \in \mathcal{T}$. Let $\mathcal{D}$ denote the *design space*, i.e., the set of site-time pairs where it is possible to take observations on Z, and suppose that the sample size n, i.e., the number of site-time pairs at which Z will be observed, is predetermined. The sampling design problem is then to select n site-time pairs from $\mathcal{D}$ in such a way as to best meet some objective or combination of objectives. A wide variety of objectives are possible, including such things as estimating a spatial average or temporal trend, detecting noncompliance with regulatory limits, determining the environmental impact of an event, measuring the effects of environmental mitigation, and detecting extreme levels (e.g., floods or smog) so that public alerts can be issued.

Although $\mathcal{D}$ could in principle be a continuum coinciding with $\mathcal{S} \times \mathcal{T}$, administrative, practical, and economic considerations may restrict $\mathcal{D}$ to a finite subset of points in $\mathcal{S} \times \mathcal{T}$. Even in cases where there are no reasons such as these for restricting $\mathcal{D}$, for some approaches to sampling design it may be necessary to discretize $\mathcal{D}$ in order to simplify the search for the best, or a least a good, design.

It is possible for the costs of sampling to vary across $\mathcal{D}$; for example, some sites may be more expensive to travel to than others. However, variable costs are often difficult to incorporate formally into the design criteria to be reviewed here, and we will not attempt to do so. In practice, a design of one of the types described here, which are all obtained via statistical considerations, may need to be modified to meet budgetary or other non-statistical constraints.

In some situations where a good sampling design is sought, data have already been collected at some sites and times, perhaps using haphazard or judgment sampling, and the job of the designer is to select sites where additional observations should be taken (augmentation) or sites that, perhaps due to reduced budgets, must be deleted from the network (contraction). In other situations, however, the designer has the opportunity to create the design *de novo*. Designs of the latter type may be constructed all at once or built in stages, using parameter estimates or other information from earlier stages to guide the selection of sites and times for subsequent stages.

Some of the sampling design approaches to be described require the use of a design criterion function and a computational algorithm to optimize that function. A variety of algorithms have been proposed. For relatively small augmentation and contraction problems, enumeration of all possible designs may be feasible. In problems too large for complete enumeration, greedy algorithms, exchange algorithms, branch and bound algorithms, simulated annealing, and gradient-based methods may be used.

In what follows, we break the design problem for environmental monitoring into two parts. The first part, spatial sampling design, deals only with *where* to sample; the temporal aspect is completely ignored. Sections 9.2.2, 9.2.3, and 9.2.4 in turn review model-based, probability-based, and space-filling approaches to spatial sampling design, and Section 9.2.5 describes extensions of the approaches to multivariate and stream network data. Because attention in those sections is limited to the spatial aspect of sampling design, we suppress the time index in $Z(\mathbf{s}, t)$ therein. The second part of the problem, space-time sampling design, considers not only where but *when* to sample and is the topic of Section 9.2.6. The time index will reappear in that section.

9.2.2 Model-based design

Model-based sampling design, as its name suggests, applies to situations in which a statistical model is assumed for $Z(\mathbf{s})$. The model specifies that the observations on Z are a spatially incomplete sample of one realization of a random field, or stochastic process, $\{Z(\mathbf{s}) : \mathbf{s} \in \mathcal{S}\}$. Some assumptions typically are made about the first- and second-order moment structure of the random field. The mean function of the random field is assumed to exist and be linear in its parameters, i.e., $E[Z(\mathbf{s})] = [\mathbf{x}(\mathbf{s})]'\boldsymbol{\beta}$ where $\mathbf{x}(\mathbf{s})$ is a vector of known functions of the spatial coordinates or other covariates observable at $\mathbf{s}$ and $\boldsymbol{\beta}$ is a vector of unknown parameters. Also, the covariance function, $\text{cov}[Z(\mathbf{s}), Z(\mathbf{u})] = C(\mathbf{s}, \mathbf{u}; \boldsymbol{\theta})$ is assumed to exist, where $C(\cdot)$ is a known positive definite function of pairs of location vectors and $\boldsymbol{\theta}$ is a vector of unknown parameters; it may be assumed further that $C(\mathbf{s}, \mathbf{u}; \boldsymbol{\theta}) = C(\|\mathbf{s} - \mathbf{u}\|; \boldsymbol{\theta})$, i.e., that $\{Z(\mathbf{s}) - E[Z(\mathbf{s})] : \mathbf{s} \in \mathcal{S}\}$ is second-order stationary. The problem is to choose a design to optimize a criterion that measures how suitable the design is for for making precise inferences about either the parameters of the model or predictions of unobserved values of $Z(\cdot)$. The design criterion may thus depend on the object(s) of primary inferential interest, how the chosen object is estimated/predicted, and how to measure the quality of the chosen estimator/predictor.

We describe model-based design with respect to criteria focused respectively on covariance parameter estimation, spatial prediction, mean parameter estimation, and spatial entropy. Further details may be found in the book of Müller (2007). Several quantities play an important role in our description. For a given design $D = \{\mathbf{s}_1, \ldots, \mathbf{s}_n\}$, let $\mathbf{X}$ denote the matrix whose ith row is $[\mathbf{x}(\mathbf{s}_i)]'$, let $\boldsymbol{\Sigma}$ denote the matrix whose ijth element is $C(\mathbf{s}_i, \mathbf{s}_j; \boldsymbol{\theta})$, and let $\mathbf{z}$ denote the vector with ith element $Z(\mathbf{s}_i)$. Each of $\mathbf{X}$, $\boldsymbol{\Sigma}$, and $\mathbf{z}$ depend on the design D, but this is not explicitly indicated by the notation. It is assumed that both $\boldsymbol{\Sigma}$ and $\mathbf{X}'\boldsymbol{\Sigma}^{-1}\mathbf{X}$ are invertible, which may slightly restrict the set of candidate designs.

9.2.2.1 Covariance estimation-based criteria

Sometimes understanding the second-order spatial dependence, or covariance, structure of a model is the primary inferential objective. Even when it is not, and estimation of mean parameters or spatial prediction of unobserved values of Z is the primary inferential goal, the success of these other goals may depend, in part, on how well the covariance parameters are estimated. So in either case it would seem appropriate to pay some attention to the quality of the design for estimating covariance parameters. Relevant early work on this topic considered design criteria for good estimation of the semivariogram, a close cousin of the covariance function. Russo (1984) considered a design to be optimal if it minimized the dispersion of intersite distances within bins used for the classical method-of-moments estimator of the semivariogram, while Warrick and Myers (1987) proposed a criterion that measured how well the intersite distribution corresponding to a design conforms to a user-specified distribution. Neither of these criteria are directly related to estimation quality of semivariogram or covariance parameters, however.

Criteria that are more relevant to estimation quality were introduced by Müller and Zimmerman (1999), Zhu and Stein (2005), and Zimmerman (2006). Müller and Zimmerman's approach was to maximize the determinant of the information matrix of the weighted or generalized least squares estimators of semivariogram parameters. Zhu and Stein's and Zimmerman's approaches were similar, except that it was the determinant of the information matrix of maximum likelihood (ML) or residual maximum likelihood (REML) estimators of covariance parameters that was maximized, under an assumption of a Gaussian random field model. For example, for REML estimation, the criterion to be maximized is

$$\phi_{CPE}(D; \boldsymbol{\theta}) = |\mathbf{I}(\boldsymbol{\theta}, D)|$$

where subscripts "CPE" denote covariance parameter estimation, the ijth element of the information matrix $\mathbf{I}(\boldsymbol{\theta}, D)$ is given by $(1/2)\text{tr}[\mathbf{P}(\partial\boldsymbol{\Sigma}/\partial\theta_i)\mathbf{P}(\partial\boldsymbol{\Sigma}/\partial\theta_j)]$, and $\mathbf{P} = \boldsymbol{\Sigma}^{-1} - \boldsymbol{\Sigma}^{-1}\mathbf{X}(\mathbf{X}'\boldsymbol{\Sigma}^{-1}\mathbf{X})^{-1}\mathbf{X}'\boldsymbol{\Sigma}^{-1}$. Designs that are optimal with respect to this criterion and others like it generally consist of several small clusters or linear strands of sites. Many of the clusters lie along the periphery of S, with the remainder spaced more or less evenly across D. The leftmost panel in Figure 9.1 displays a design of size 5 that maximizes $\phi_{CPE}(\cdot; \boldsymbol{\theta})$ for a scenario in which $\mathcal{D}$ is a 5×5 square grid and the underlying random field has mean 0 and isotropic exponential covariance function $C(\mathbf{s}, \mathbf{u}) = 0.5^{\|\mathbf{s}-\mathbf{u}\|}$. Such a design produces relatively many small lags and large lags, and fewer intermediate lags, than most other designs, which facilitates more precise estimation of covariance parameters.

One practical difficulty in choosing a single best design for covariance parameter estimation is that the optimal design (corresponding to a given estimator and a given measure of the utility of the design for optimizing the quality of that estimator) may depend on the unknown parameter vector $\boldsymbol{\theta}$. In other words, a globally optimal design may not exist, and we may have to settle for a *locally* optimal design (a design that is optimal for a specific value of $\boldsymbol{\theta}$ and presumably for values in a neighborhood of it). There are various ways to circumvent this difficulty. One is to use a two-stage adaptive sampling approach wherein some sites are sampled without regard to their optimality in order to provide an initial estimate of $\boldsymbol{\theta}$ and then subsequent sites are chosen to yield the locally optimal design corresponding to this estimate. Alternatively, a maximin approach chooses the design that maximizes the minimum value of $\phi_{CPE}(D, \boldsymbol{\theta})$ across all values of $\boldsymbol{\theta}$. Still another approach is Bayesian, in which the uncertainty about $\boldsymbol{\theta}$ in the design criterion is incorporated into the model via a prior distribution, about which we will say more in the next subsection.

9.2.2.2 Prediction-based criteria

Let $Z(\mathbf{s}_0)$ denote the realized but unobserved value of Z at an arbitrary point $\mathbf{s}_0 \in \mathcal{S}$. If the covariance parameter vector $\boldsymbol{\theta}$ is known, the best (in the sense of minimizing the prediction error variance) linear unbiased predictor (BLUP) of $Z(\mathbf{s}_0)$ is given by

$$\hat{Z}(\mathbf{s}_0; \boldsymbol{\theta}, D) = [\mathbf{c} + \mathbf{X}(\mathbf{X}'\boldsymbol{\Sigma}^{-1}\mathbf{X})^{-1}(\mathbf{x}_0 - \mathbf{X}'\boldsymbol{\Sigma}^{-1}\mathbf{c})]'\boldsymbol{\Sigma}^{-1}\mathbf{z}.$$

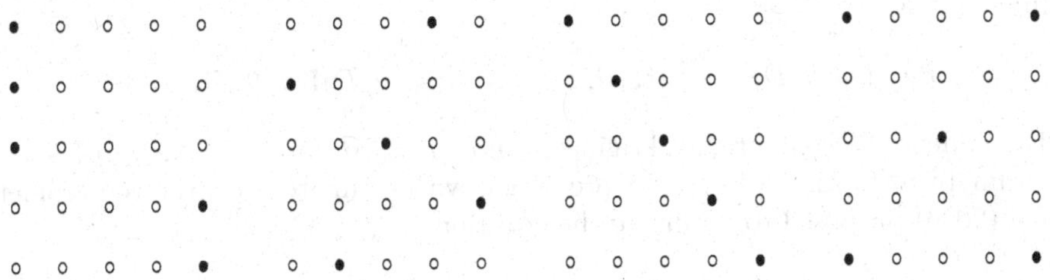

FIGURE 9.1
Optimal model-based designs of size 5 for the examples described in Section 9.2.2. First (leftmost) panel, optimal design for covariance estimation; second panel, optimal design for prediction with known covariance parameters; third panel, optimal design for empirical prediction; fourth panel, optimal design for mean estimation. Candidate design points within each grid are represented by open circles, and points actually belonging to the design are represented by closed circles.

Here, $\mathbf{c}$ is the vector whose ith element is $C(\mathbf{s}_i, \mathbf{s}_0; \boldsymbol{\theta})$ and $\mathbf{x}_0 = \mathbf{x}(\mathbf{s}_0)$. The prediction error variance, or *kriging variance*, associated with $\hat{Z}(\mathbf{s}_0; \boldsymbol{\theta}, D)$ is given by

$$
\begin{aligned}
\sigma_K^2(\mathbf{s}_0; \boldsymbol{\theta}, D) &= \operatorname{var}_{\boldsymbol{\theta}}[\hat{Z}(\mathbf{s}_0; \boldsymbol{\theta}, D) - Z(\mathbf{s}_0)] \\
&= C(\mathbf{s}_0, \mathbf{s}_0; \boldsymbol{\theta}) - \mathbf{c}'\boldsymbol{\Sigma}^{-1}\mathbf{c} \\
&\quad + (\mathbf{x}_0 - \mathbf{X}'\boldsymbol{\Sigma}^{-1}\mathbf{c})'(\mathbf{X}'\boldsymbol{\Sigma}^{-1}\mathbf{X})^{-1}(\mathbf{x}_0 - \mathbf{X}'\boldsymbol{\Sigma}^{-1}\mathbf{c}).
\end{aligned}
$$

Observe that the kriging variance depends on neither the mean parameters $\boldsymbol{\beta}$ nor the observed data $\mathbf{z}$; it does, however, depend on the prediction site $\mathbf{s}_0$, the covariance parameter $\boldsymbol{\theta}$, and the design D. A common measure of the global performance of a design for prediction under a model with known covariance parameters is the maximum prediction error variance over a set $\mathcal{P}$ where prediction is desired, i.e.,

$$
\phi_K(D; \boldsymbol{\theta}) = \max_{\mathbf{s} \in \mathcal{P}} \sigma_K^2(\mathbf{s}; \boldsymbol{\theta}, D).
$$

Here, subscript "K" stands for "kriging" and $\mathcal{P}$ is typically either $\mathcal{D}$ or $\mathcal{S}$. If the mean function is constant, designs that minimize $\phi_K(\cdot; \boldsymbol{\theta})$ tend to be "regular," i.e., their sites are rather uniformly dispersed over the study region. If, instead, the mean function is planar, the sites again have good spatial coverage but tend to be denser near the periphery of the study region. In either case, the design is qualitatively much different than an optimal design for estimating covariance parameters. A design that minimizes $\phi_K(\cdot; \boldsymbol{\theta})$ for the same scenario described in the previous subsection is shown in the second panel of Figure 9.1.

Although designs that minimize $\phi_K(\cdot)$ are of interest in their own right, they do not necessarily perform well for the prediction problem of greatest practical importance, which is to predict unobserved values of Z when the spatial covariance parameters are unknown. The standard predictor in this situation, known as the empirical BLUP (E-BLUP), is given by an expression identical to the BLUP when $\boldsymbol{\theta}$ is known but with the covariance function evaluated at an estimate $\hat{\boldsymbol{\theta}}$, rather than at the hitherto-assumed-known $\boldsymbol{\theta}$. That is, the E-BLUP at $\mathbf{s}_0$ is given by $\hat{Z}(\mathbf{s}_0; \hat{\boldsymbol{\theta}}, D)$. The E-BLUP is an unbiased predictor under fairly general conditions, but an exact expression for its prediction error variance, $\sigma_{EK}^2(\mathbf{s}_0; D)$, is unknown (except in very special cases) and simulation-based approaches to approximate it at more than a few sites are computationally prohibitive. Harville and Jeske (1992) proposed to approximate it by

$$\sigma_{EK}^2(\mathbf{s}_0; \boldsymbol{\theta}, D) \doteq \sigma_K^2(\mathbf{s}_0; \boldsymbol{\theta}, D) + \text{tr}[\mathbf{A}(\mathbf{s}_0, \boldsymbol{\theta}, D)\mathbf{I}^{-1}(\boldsymbol{\theta}, D)]$$

where "EK" denotes "empirical kriging" and $\mathbf{A}(\mathbf{s}_0; \boldsymbol{\theta}, D) = \text{var}_{\boldsymbol{\theta}}[\partial \hat{Z}(\mathbf{s}_0; \boldsymbol{\theta}, D)$ $/\partial \boldsymbol{\theta}]$. Accordingly, for optimal design for prediction with estimated covariance parameters, Zimmerman (2006) proposed to minimize the criterion

$$\phi_{EK}(D; \boldsymbol{\theta}) = \max_{\mathbf{s} \in \mathcal{P}} \sigma_{EK}^2(\mathbf{s}; \boldsymbol{\theta}, D),$$

which combines a measure of design quality for prediction under a model having known covariance parameters (the first term in Harville and Jeske's approximation) with a measure of quality for covariance parameter estimation (the second term). Not surprisingly, then, designs that are locally optimal for empirical kriging represent a compromise between designs that are optimal for each of its component objectives. That is, they often have relatively good spatial coverage while also having a few clusters. A design that minimizes $\phi_{EK}(\cdot; \boldsymbol{\theta})$ for the aforementioned scenario is displayed in the third panel of Figure 9.1. How good the spatial coverage is in this particular design is debatable.

The prediction-based design approach based on $\phi_{EK}(\cdot)$ is focused on good empirical point prediction. Zhu and Stein (2006) carried this approach further by constructing a criterion that gives some attention to the utility of the design for interval prediction in addition to point prediction. The spatial configuration of sites in designs that are good with respect to this extended criterion do not appear to be markedly different from designs that are good for empirical point prediction, however.

A more natural way to account for covariance parameter uncertainty when constructing efficient designs for spatial prediction is a Bayesian approach, as expounded by Diggle and Lophaven (2006). They put a prior distribution on the unknown parameters and then minimized an estimate of the spatially averaged prediction variance obtained by Monte Carlo sampling from the posterior distribution for $\boldsymbol{\theta}$ and hence from the predictive distribution of the (assumed) Gaussian random field. The designs obtained in this fashion for the examples they considered are qualitatively very similar to those obtained by the empirical prediction-based approaches of Zimmerman (2006) and Zhu and Stein (2006).

9.2.2.3 Mean estimation-based criteria

Suppose that greatest interest lies not in covariance parameter estimation or prediction but in estimating the model's mean parameters, $\boldsymbol{\beta}$, as well as possible. Although a well-developed classical theory exists for exact optimization of an experimental design for the estimation of mean (regression function) parameters (e.g., Federov, 1972; Silvey, 1980), among the assumptions on which it is based are independence of the observations and the possibility of sampling repeatedly at any combination of the regressors. These assumptions are not met in the spatial sampling design problem, so attempts to apply the theory to this problem have faced significant challenges and are still evolving. We will describe instead an approximate approach to optimal design for mean parameter estimation that was studied by Müller (2005) for use when the covariance parameters are known, and a natural extension of it to the case of unknown covariance parameters.

If $\boldsymbol{\theta}$ is known, the best (in the sense of minimizing the estimator's variance) linear unbiased estimator (BLUE) of $\boldsymbol{\beta}$ is $\hat{\boldsymbol{\beta}} = (\mathbf{X}'\boldsymbol{\Sigma}^{-1}\mathbf{X})^{-1}\mathbf{X}'\boldsymbol{\Sigma}^{-1}\mathbf{z}$. A standard measure of a design's quality for estimating $\boldsymbol{\beta}$, known as the D-optimality criterion, is the determinant of the information matrix or equivalently (assuming a Gaussian process) the determinant of the inverse of the covariance matrix of $\hat{\boldsymbol{\beta}}$, i.e., $\phi_{ME}(D; \boldsymbol{\theta}) = |\mathbf{X}'\boldsymbol{\Sigma}^{-1}\mathbf{X}|$, where "ME" denotes "mean estimation." Provided that the model is second-order stationary and the mean function is constant, designs that maximize this criterion tend to consist of regularly spaced sites that cover the study region, which makes sense since such a configuration maximally reduces the dependence among pairs of observations. On the other hand, if the mean function is a planar trend, i.e., $E[Z(\mathbf{s})] = \mathbf{s}'\boldsymbol{\beta}$, then the optimal design tends to have sites on or near the boundary of the study region. These differences are likely due to the higher leverage that boundary sites have when the mean function is planar rather than constant, with the result that estimation is improved by sampling more points near the boundary in spite of the strong correlation among observations taken there. The rightmost panel of Figure 9.1 displays the design that maximizes $\phi_{ME}(\cdot; \boldsymbol{\theta})$ for the aforementioned scenario.

If, more realistically, $\boldsymbol{\theta}$ is not known and is estimated by $\hat{\boldsymbol{\theta}}$, then a natural estimator of $\boldsymbol{\beta}$ is the *empirical* best linear unbiased estimator $\tilde{\boldsymbol{\beta}} = (\mathbf{X}'[\boldsymbol{\Sigma}(\hat{\boldsymbol{\theta}})]^{-1}\mathbf{X})^{-1}\mathbf{X}'[\boldsymbol{\Sigma}(\hat{\boldsymbol{\theta}})]^{-1}\mathbf{z}$. An approximation for the covariance matrix of $\tilde{\boldsymbol{\beta}}$ akin to that given for $\sigma^2_{EK}(\mathbf{s}_0; \boldsymbol{\theta}, D)$ may be given, or one may again use a Bayesian approach to incorporate uncertainty about $\boldsymbol{\theta}$. In the former approach, the determinant of the inverse of the approximate covariance matrix of $\tilde{\boldsymbol{\beta}}$ may serve as a design criterion. Designs that maximize this criterion usually do not differ greatly from those that maximize the D-optimality criterion for the case of known $\boldsymbol{\theta}$ and planar mean function.

9.2.2.4 Multi-objective and entropy-based criteria

Rather than optimizing a criterion measuring the quality of a design with respect to a single, narrowly focused objective, we may optimize a criterion that combines several objectives. In fact, as noted previously, the EK criterion of Zimmerman (2006) combines criteria for covariance parameter estimation and prediction, and the criterion of Zhu and Stein (2006) combines criteria for point and interval prediction. Two general strategies for combining criteria yield what are called compound designs (Müller and Stehlík, 2010) and Pareto-optimal designs (Lu, Anderson-Cook, and Robinson, 2011; Müller et al., 2015). A compound design criterion is a weighted linear combination of two or more component design criteria, while Pareto-optimal designs (also called the Pareto frontier) are those designs for which one of the component design criteria cannot be improved without worsening the design with respect to another component criterion. Naturally, both approaches tend to yield compromise optimal designs having some features of the designs optimal with respect to each of the component criteria.

In some situations, it may be difficult to elicit precise design objectives from the users of the environmental monitoring data, and it is also possible that the objectives may change over time. In fact, future uses of the data, and therefore future design objectives, may not necessarily be foreseen at the outset of a long-term monitoring program. For these situations, perhaps the best thing to do is try to minimize the uncertainty of the actual responses, rather than of model parameters or predictions based upon them. In a series of papers published over the last 30+ years, J. Zidek and various co-authors proposed to measure this uncertainty by the entropy of the joint predictive distribution of the responses at "ungauged" sites (i.e., unobserved $Z(\mathbf{s})$) given those at "gauged" sites. For further details on this approach, we refer the reader to summaries of this literature in Chapter 11 of Le and Zidek (2006) or Chapter 13 of this volume, plus Fuentes et al. (2007). The approach tends to add sites to an existing design that correspond to responses that are highly unpredictable, either because of their lack of correlation with responses at gauged sites or because of their greater intrinsic variability.

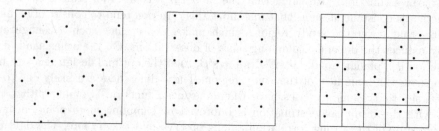

FIGURE 9.2
Probability-based spatial sampling designs, for 25 sites in a square study region. Left panel, simple random sample; middle panel, systematic sample; right panel, stratified random sample using square strata resulting from a 5×5 grid partition of the study region.

9.2.3   Probability-based spatial design

In probability-based sampling design, no model is assumed for the spatial attribute process $Z(\cdot)$; in fact, for fixed $\mathbf{s}$, $Z(\mathbf{s})$, though it is unknown (unless $\mathbf{s}$ is included in the sample), is regarded as nonrandom. Instead, randomness is introduced into the system by the investigator through a random process for selecting the sites in the design from S. The locations of sampled sites are determined entirely by this random process; no objective function is optimized, hence no numerical optimization algorithm is needed. The random process is such that the probability of a site being selected for the design, called the *inclusion probability* if $\mathcal{D}$ is finite and the *inclusion density function* if $\mathcal{D} = S$, is known for every site. Inclusion densities need not be constant across sites, and different joint inclusion densities correspond to different designs. The inclusion densities provide a mathematical foundation for an inferential paradigm known as design-based inference, which is much different than model-based inference. The objects of interest in design-based inference are usually the mean or cumulative distribution function of Z. Predicting unobserved values of $Z(\mathbf{s})$ or estimating the parameters of a model for $Z(\cdot)$ are not purposes for which probability-based sampling was developed (though one can, of course, still perform those activities for such samples).

9.2.3.1   Simple random sampling

In simple random sampling, sites are selected independently and without replacement from a uniform inclusion distribution on $\mathcal{D}$. One such design is displayed in the left panel of Figure 9.2. The advantages of this type of sampling include that it is seen as "fair" or unbiased; it requires no knowledge about Z or how it is distributed in space; and it yields a relatively simple theoretical basis for inference. The main disadvantages are that it could, by chance, result in a design in which some sample sites are quite close to one another (clusters), yielding redundant information if Z is positively spatially autocorrelated, and that some subregions of appreciable size may not be sampled at all. That the locations of these clusters and empty regions would likely change from one simple random sample to another implies that estimators of population quantities of interest, though they may be unbiased, could vary a lot between samples. As a result, simple random sampling is almost never used in environmental monitoring studies.

9.2.3.2 Systematic random sampling

In systematic random sampling, a single initial site is chosen according to simple random sampling and the remaining $n - 1$ sites are located nonrandomly, according to some regular pattern (to the extent possible within $\mathcal{D}$). The pattern could be a square, equilateral triangular, or hexagonal grid, for example. The middle panel of Figure 9.2 shows a systematic square grid design for which the grid is oriented with the axes of the square study region. It would also be possible to choose the angle of orientation randomly, according to a uniform distribution on $(0°, 90°)$.

The characteristics of a systematic design differ in important ways from those of the simple random design described previously and the stratified random design to be described next. First and foremost, neither large unsampled subregions nor clusters of sites can occur, so the design has a property known as *spatial balance*. Spatial balance may be desirable for several reasons. For example, in air or groundwater pollution studies it may be politically unacceptable to not sample at all from large areas where significant numbers of citizens reside. Also, from a model-based perspective, spatial balance ensures that no two sites are so close together that they provide essentially redundant information. On the other hand, as noted previously, some clusters in the design may be desirable for the purpose of estimating the covariance parameters of a model. Furthermore, from a pure probability-based perspective, a disadvantage of a systematic design is that it carries with it a risk of producing poor estimates if there is a periodicity in $Z(\cdot)$ that coincides with the orientation and spacing of the grid.

Among the various regular grids, which is best? Although site determination is generally easiest for the square grid, the triangular and hexagonal grids facilitate estimation of the covariance function in three directions, which is useful for checking for isotropy. Furthermore, the triangular grid appears to be slightly more efficient than the others for estimating the overall mean of Z (Matérn, 1960) as well as for purposes of covariance parameter estimation and spatial prediction (Olea, 1984; Yfantis et al., 1987).

9.2.3.3 Stratified random sampling

A sampling scheme that overcomes the weaknesses of simple random sampling without incurring problems due to periodicity is stratified random sampling. In stratified random sampling, $\mathcal{S}$ is partitioned into strata (subregions) and simple random samples, typically of size one or two, are taken independently within each stratum; see the right panel of Figure 9.2 for an example. Although this can still result in some small clusters, if the strata are chosen appropriately it will make it impossible for large subregions to be unsampled. Consequently, a stratified random design can be more efficient (i.e., quantities of interest are estimated with smaller variance) than a simple random design.

How should the strata be chosen? Efficiency gains are largest when the variability within strata is as small as possible, relative to the variability between strata. Thus, for example, for sampling a pollutant around a point source for which there is no favored diffusion direction, annular regions centered at the point source would be a sensible choice. For spatial sampling design more generally, it makes sense to choose strata that are geographically compact since $Z(\cdot)$ is usually positively spatially correlated, hence more homogeneous, within such strata. In the absence of any additional information, approximately square, triangular, or hexagonal partitionings are reasonable possibilities. A particular stratified random design that has received considerable attention is the randomized-tessellation stratified design (Overton and Stehman, 1993). The strata for this design are hexagons formed by the tessellation of a triangular grid. The regularity of the hexagonal strata confers good spatial balance upon the design, while the randomization within strata greatly diminishes any phase-correspondence with a periodic surface over $\mathcal{S}$. Consequently, the randomized-tessellation stratified design is more efficient than a systematic design of the same size for estimating the mean of a population; moreover, it yields a better estimate of the variance of the mean estimator (Overton and Stehman, 1993). If S is so irregularly shaped that it is not possible to form strata that are approximate hexagons, a k-means clustering algorithm may be applied to a fine discretization of the study region to construct geographically compact strata (Walvoort et al., 2010).

In some sampling situations, information may be available on an auxiliary variable (e.g., elevation) that is correlated with Z. In those situations it would often be desirable to choose strata to minimize within-strata variation of the auxiliary variable, subject to some geographical compactness constraints. A variance quadtree algorithm that accomplishes this for square strata of varying sizes was proposed by Minasny et al. (2007).

9.2.3.4   Variable probability sampling

More complex sampling designs are possible, including some with unequal marginal inclusion probabilities resulting in what is called variable probability sampling. Such a design might be appropriate for sampling lakes for levels of mercury in fish, for example, if one wants the probability that a lake is selected to be proportional to its size because larger lakes tend to hold more fish. The key to making unbiased inferences from a variable probability sampling design is the Horvitz-Thompson Theorem (Horvitz and Thompson, 1952) and its analogue for sampling in a continuum (Cordy, 1993). By this theorem, an unbiased estimator — known as the Horvitz-Thompson estimator — of the population total is given by the weighted sum of all the sampled values, where the weights are the reciprocals of the inclusion probabilities. Furthermore, the variance of the Horvitz-Thompson estimator is given by an expression involving the marginal and pairwise inclusion probabilities and can be estimated unbiasedly by a function of the data and those same probabilities; for details see Stevens (1997).

Stevens (1997) also extended the randomized-tessellation stratified design to allow the inclusion probability density function to vary across strata while retaining good spatial balance. The design, which he called the multiple-density, nested, random-tessellation stratified (MD-NRTS) design, achieves variable site inclusion densities by using a series of nested grids. The choices of sampling intensity in this design are limited, however. The generalized random-tessellation stratified (GRTS) design (Stevens and Olsen, 2004) overcomes this limitation by using a hierarchical randomization method to map points in two-dimensional space into one-dimensional space, while preserving proximity relationships between points to a considerable degree, and then samples systematically along the single dimension. MD-NRTS and GRTS designs have been used with considerable success in numerous monitoring studies carried out within the United States Environmental Protection Agency's Environmental Monitoring and Assessment Program (EMAP).

9.2.4   Space-filling designs

Space-filling designs are, as their name suggests, designs for which no convex subregion of the study region having appreciable area (relative to the sampling intensity) is devoid of sampling points. The concept is essentially the same as that of spatial balance described previously for probability-based designs. In contrast to a probability-based design, however, the sample locations of a space-filling design are determined not by randomization but by optimizing a mathematical criterion. Furthermore, optimization of a space-filling design criterion does not require any prior knowledge of or reference to a model governing the observations, and is usually less computationally intensive than optimization of a model-based criterion. Space-filling designs have been developed primarily for use in computer

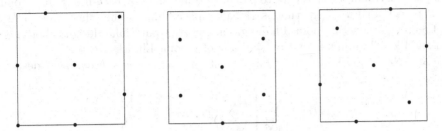

FIGURE 9.3

Space-filling designs, for 7 sites in a square study region. Left panel, maximin design; middle panel, minimax design; right panel, maximin (and minimax) Latin hypercube design.

experiments, but they are also relevant to environmental monitoring studies. They may be particularly useful when the investigator cannot define a suitable model-based criterion or when there are multiple competing design objectives. Five types of space-filling designs have received the most attention: maximin distance designs, minimax distance designs, Latin hypercube designs, spatial coverage designs, and regular grids.

An n-point *maximin distance design* (Johnson et al., 1990) is an n-point design for which the shortest Euclidean distance between sampled sites is maximized, i.e., a design for which

$$\phi_{Mm}(D) = \min_{\mathbf{s}_i, \mathbf{s}_j \in D, i \neq j} \|\mathbf{s}_i - \mathbf{s}_j\|$$

is maximized over all possible n-point designs D. Not surprisingly, such a design necessarily includes points on the boundary of the study region. Maximin distance designs are solutions to the problem of packing circles (in two dimensions) or spheres (in three dimensions) inside a region, about which much is known when the region is square, rectangular, or circular; see, e.g., http://www.packomania.com/. An example of a 7-point maximin distance design in the unit square is displayed in the left panel of Figure 9.3.

An n-point *minimax distance design* (Johnson et al., 1990), in contrast to a maximin distance design, is a design for which the maximum distance from any point in the study region to the closest point in the design is minimized, i.e., a design for which

$$\phi_{mM}(D) = \max_{\mathbf{s} \in \mathcal{S}} \min_{\mathbf{s}_i \in D} \|\mathbf{s} - \mathbf{s}_i\|$$

is minimized over all possible n-point designs D. It may be more appropriate to call such a design a "minimaximin" distance design in light of the form of $\phi_{mM}(\cdot)$, but the shorter name has prevailed. For fixed n, minimax distance designs often have fewer points on the boundary than maximin distance designs. Minimax designs are related to so-called covering problems in geometry. An example of a 7-point minimax distance design in the unit square is displayed in the middle panel of Figure 9.3.

An arguably undesirable feature of maximin and minimax distance designs is that they are not necessarily space-filling when projected to their lower-dimensional subspaces. This can be visualized in the maximin and minimax designs of Figure 9.3, where the projections to either the bottom or the left side of the square would result in both coincident points and relatively large gaps between points. An n-point *Latin hypercube (Lh) design* (McKay et al., 1979; Pistone and Vicario, 2010) addresses this shortcoming by having the property that all of its one-dimensional projections are n-point maximin designs; this latter property is equivalent to each one-dimensional projection comprising the sequence $\{0, 1/(n-1), 2/(n-1), \ldots, 1\}$ (when $\mathcal{S}$ is a unit square or cube). For any n there are many Lh designs, some of which may be very poor with respect to global (d-dimensional) space-filling. It is common, therefore, to seek an Lh design that is globally space-filling within that class. The right panel of Figure 9.3 shows a 7-point Lh design; in fact, this particular design is both globally maximin and globally minimax within the class of 7-point Lh designs.

Royle and Nychka (1998) defined an n-point *spatial coverage design* as a design that minimizes

$$\phi_{C,r,q}(D) = \left[\sum_{s \in \mathcal{S}} \left\{ \left(\sum_{s_i \in D} \|s - s_i\|^r \right)^{q/r} \right\} \right]^{1/q}$$

over all possible n-point designs D, where r and q are negative and positive integers, respectively. Note that $\phi_{C,r,q}(D)$ is a function of the same quantities involved in the minimax criterion $\phi_{mM}(D)$; more specifically, $\phi_{C,r,q}(D)$ is an L_q-average of generalized distances between candidate points and the design. Since $r < 0$, the generalized distance $\left(\sum_{s_i \in D} \|s_i - s\|^t \right)^{1/r}$ converges to 0 as s converges to any site in D. The design that minimizes $\phi_{C,r,q}(\cdot)$ depends on r and q, but appears to be insensitive to these values if r is very negative. In the limit as $r \to -\infty$ and $q \to \infty$, $\phi_{C,r,q}(\cdot)$ converges to the minimax distance criterion. Royle and Nychka (1998) featured spatial coverage designs with $r = -5$ and $q = 1$.

The last category of space-filling designs, *regular grids*, has already been discussed in the context of probability-based design. The only difference between regular grids of the probability-based variety and those of the space-filling variety is that the locations of the gridpoints of the latter are not determined randomly but are selected either to maximize the minimum distance between them (a maximin grid design) or to minimize the maximum distance between a point in $\mathcal{S}$ and the nearest gridpoint (a minimax grid design). If $\mathcal{S}$ is rectangular, some points of a maximin grid design will necessarily lie on the boundary of $\mathcal{S}$, whereas points of a minimax grid design will lie entirely within the $\mathcal{S}$.

In practice, irregularity of $\mathcal{S}$, constraints on the candidate sampling sites and/or the sheer size of the design problem will preclude an investigator from finding an optimal maximin, minimax, or Lh design within the catalog of known ones, in which case the design must be found using a numerical optimization algorithm. Many of the same algorithms used to optimize model-based design criteria may be used successfully to optimize space-filling criteria (Pronzato and Müller, 2012). Exchange algorithms seem to be especially popular in this regard. Royle and Nychka (1998) presented an exchange algorithm for the construction of their spatial coverage designs and demonstrated empirically that it could obtain near-optimal designs efficiently for reasonably large problems.

How do space-filling designs compare to model-based designs? A comparison of Figure 9.3 with Figure 9.1 makes it plain that the site configuration of a globally space-filling design is similar to that of a design that optimizes a model-based prediction-error variance criterion like $\phi_K(\cdot)$, but very different from that of a design that optimizes some other model-based criteria. Consequently, space-filling designs perform very well for spatial prediction with known covariance parameters; in fact, Zimmerman and Li (2013) and Li and Zimmerman (2015) showed, for cases with $n = 4$ or 5 and $\mathcal{D}$ a 4×4 or 5×5 grid (as it is in Figure, that the ϕ_K-optimal design is a maximin or minimax Lh design, and Johnson et al. (1990) established a theoretical connection between maximin and minimax distance designs and designs that optimize certain other prediction-oriented criteria. However, for objectives other than spatial prediction, especially for estimation of covariance parameters, space-filling designs can perform very poorly.

For a more detailed review of space-filling designs, including some not described here, see Pronzato and Müller (2012). For applications of space-filling approaches to the design of real environmental monitoring networks, see Nychka and Saltzman (1998) and Holland et al. (1999).

9.2.5 Design for multivariate data and stream networks

Our presentation to this point has assumed implicitly that the environmental monitoring program is focused on a single variable (Z). Often, however, it is of interest to monitor two or more variables. The variables are usually correlated with each other, both non-spatially (at the same location) and spatially (at different, especially nearby sites), and the best linear unbiased multivariate prediction procedure known as co-kriging exploits both types of correlation to improve prediction quality over that which is possible by kriging each variable separately. Contributions to the literature on probability-based and space-filling sampling design generally have not been concerned with how many variables are of interest, but have assumed, either implicitly or explicitly, that if more than one variable is of interest then every such variable is observed at each site in the design. This multivariate spatial sampling design feature, known as *collocation*, is often practically and economically advantageous, relative to observing some of the variables at some sites and other variables at other sites. Consequently, most treatments of multivariate sampling design from a model-based perspective (e.g., Le and Zidek, 1994; Bueso et al., 1999; Yeh et al., 2006; Vašát et al., 2010) have also restricted attention to collocated designs. Within the model-based design paradigm, however, collocation is not necessarily statistically optimal.

Li and Zimmerman (2015) present an extensive investigation of multivariate model-based optimal sampling design, allowing for collocation but not requiring it. They found that, generally, the within-variable characteristics of optimal designs with respect to multivariate criteria are similar to those of designs that are optimal with respect to the analogous univariate criteria. For example, designs optimal for covariance and cross-covariance parameter estimation consist of a few clusters or strands, while designs optimal for co-kriging have excellent spatial coverage. Figures 9.4 and 9.5, reprinted from Li and Zimmerman (2015), display examples of such designs, as determined by a simulated annealing algorithm, for a scenario in which there are two variables of interest, $\mathcal{D}$ is a 25×25 square grid with unit spacing, and $n = 15$ for each variable. The observations are assumed to arise from a second-order stationary, isotropic, separable (across variables) bivariate Gaussian process with unknown constant means and a bivariate Matérn covariance function with equal correlation decay parameters and both smoothness parameters equal to 0.5. Nine cases, corresponding to all combinations of three correlation decay parameters and three cross-correlation parameters, are displayed in the figure for each design criterion. The design criteria for which the displayed designs are optimal are, in Figure 9.4, the determinant of the information matrix associated with the REML estimator of covariance and cross-covariance parameters (which is maximized), and in Figure 9.5, the maximum determinant of the 2×2 matrix of prediction errors over a fine discretization of the unit square (which is minimized). What is at least as interesting as the spatial configuration of sites in these optimal designs is the extent of collocation. The optimal design for covariance and cross-covariance estimation tends to be completely collocated, while the optimal design for co-kriging (with either known or unknown covariance and cross-covariance parameters) can have much less collocation. The extent of collocation in the optimal co-kriging design tends to decrease as either the spatial correlation gets stronger or the cross-correlation gets weaker. On the other hand, in most cases in which collocated designs are not optimal for co-kriging, it turns out that they are reasonably efficient. These results, plus the aforementioned practical and economic considerations, provide support for using a completely collocated design in multivariate spatial sampling.

Another application area for the principles and methods of environmental sampling design described herein is stream networks. The same broad categories of designs (model-based, probability-based, and space-filling) apply, but there are some important differences as a result of the unique topology of stream networks (Dobbie et al., 2008) and the distinct

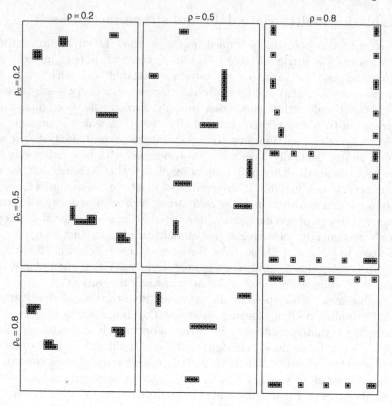

FIGURE 9.4
Optimal designs for covariance and cross-covariance estimation for the example described in Section 9.2.5. Closed circles correspond to one variable, and open squares to the other.

characteristics of models for spatial dependence on streams, namely, the tail-up and tail-down models of Ver Hoef et al. (2006) and Ver Hoef and Peterson (2010); see also Chapter 18 of this volume. Stream networks have a hierarchical branching structure and a flow direction and volume, so for variables such as point-source pollutants it may be sensible to regard observations taken at flow-connected sites to be correlated in such a way that the correlation decays with hydrologic distance (i.e., distance within the stream network) rather than with Euclidean distance, while regarding observations taken at flow-unconnected sites to be uncorrelated; this is what a tail-up model does. Alternatively, for variables like counts of fish (which can move both upstream and downstream) it can be argued that one should allow for the possibility of correlation between observations at all sites, be they flow-connected or flow-unconnected; this corresponds to a tail-down model. Furthermore, it stands to reason that an observation at a site downstream of the confluence of two tributaries of unequal flow volume would be more highly correlated with observations on the tributary having greater flow volume.

These distinctives of stream network variables may affect the relevance and utility of various sampling designs. For example, for stratified random sampling, Liebetrau (1979) proposed taking stream segments (sections of streams between confluences) as the sampling units and stratifying not on geography but on stream order (number of upstream tributaries), drainage area, or mean annual flow. A space-filling approach was proposed by Dixon et al. (1999), with the wrinkle that the maximum subcatchment area (defined

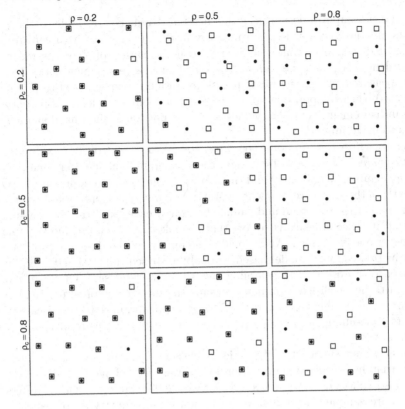

FIGURE 9.5
Optimal designs for co-kriging with known covariance and cross-covariance parameters for the example described in Section 9.2.5. Closed circles correspond to one variable, and open squares to the other.

as the drainage area of a sampling site minus the drainage area of any other sampling sites further upstream), rather than maximum intersite distance, was minimized. Som et al. (2014) and Falk, McGree, and Pettitt (2014) investigated model-based sampling design for stream networks from frequentist and (pseudo-)Bayesian perspectives, respectively. Som et al. (2014) found in particular that: for estimating the covariance parameters of tail-up models, the best designs have clusters on flow-unconnected segments; for estimating the covariance parameters of tail-down models, the best designs tend to have triadic clusters at stream confluences (one site just downstream of the confluence and one on each tributary just above the confluence); and for all other inference objectives, the best designs usually have a site on the outlet segment and many headwater segments, with the remaining sites evenly dispersed throughout the remainder of the network. Cost considerations may preclude using a design with too many headwater segments, however, as those segments are usually the least accessible. The findings of Falk, McGree, and Pettitt (2014) were similar.

9.2.6 Space-time designs

Environmental monitoring studies typically occur in space *and* time, and often have as one of their aims the detection of change or trend in Z over time. Consequently, their sampling design may have a temporal component in addition to a spatial component. That said, for some studies temporal design need not be considered. For example, if the monitoring instruments measure more or less continuously in time (e.g., water temperature loggers in streams) or aggregate over time (air filters), or if the nature of Z or the use to which it will be put demands that it be sampled at specific times (e.g. annual April 1 snow water equivalent measurements in the western U.S. mountain snowpack used to predict summer water supplies), then temporal design is not relevant. Temporal design is potentially useful only when measurements represent values of Z on a portion of the time domain $\mathcal{T}$ to which one desires to make inferences.

Designs for which the temporal component is given consideration may be classified as either *static* or *dynamic*. In a static design, the sites do not change over time (equivalently, the times are common across sites) and the temporal component of design may be considered independently of the spatial component. A static design may be appropriate when measuring stations are expensive to set up and cannot be moved easily and the cost of sampling and measurement at those stations is so low that it makes little sense to not sample at all of them whenever one is sampled. Analogous to the purely spatial design approaches reviewed previously in this chapter, model-based, probability-based, and "time-filling" approaches may be taken to choose the n_T sampling times of a static design. Of these, model-based designs seem to be much less common. Perhaps the most common static temporal design consists simply of regular time points $1, 2, \ldots, n_T$ (in appropriate time scale units), due partly to its time-filling properties and partly to the relative convenience of a constant sampling interval.

If a design is not static, it is dynamic. A class of dynamic designs for environmental monitoring that has received much attention is the class of *panel* designs (Urqhhart et al., 1993), also called *revisit designs* (McDonald, 2003). Here, a "panel" refers to a set of sites that are sampled contemporaneously. Two important types of panel designs are *rotating panel designs* and *serially alternating designs*. For concreteness suppose that the basic sampling interval is a year. A rotating panel design prescribes that each panel will be sampled in each of several consecutive years and then removed permanently from future consideration; as each panel is removed, another is started. The top portion of Table 9.1a shows a rotating panel design in which three panels are sampled in each of four years. In contrast, a serially alternating design prescribes that each panel is sampled in every rth year, where r is an integer. The top portion of Table 9.1b displays a serially alternating design in which there are four panels, each of which is sampled every fourth year, for a total sampling period of 8 years.

Both types of panel designs just described may be augmented with a panel that is sampled every year. The designs in Table 9.1, with the bottom portions included, are an augmented rotating panel design and an augmented serially alternating design, respectively. Augmentation improves the designs for certain purposes; for example, without augmentation, the estimation of a change in Z from one year to the next is nonestimable in the serially alternating design of Table 9.1b. Urquhart et al. (1993) recommended allocating 10-20% of the sites to the augmented panel for best performance of either panel design. Urquhart and Kincaid (1999) showed that an augmented serially alternating design has greater power to detect trend than an augmented rotating panel design with the same total number of sites, but that the latter estimates current "status" (the mean of Z over space at a particular time) more precisely than the former.

Rotating and serially alternating panel designs have obvious advantages over static designs, but their structure requires that every site be sampled in multiple years. A fully dynamic design allows each panel to be sampled only once. Wikle and Royle (1999) considered fully dynamic model-based design for monitoring spatiotemporal processes. For a relatively simple Gaussian separable model with first-order Markovian temporal dependence, they found that the degree to which the average prediction error variance at time

(a)

Panel	Year			
	1	2	3	4
1	*			
2	*	*		
3	*	*	*	
4		*	*	*
5			* .	*
6				*
Augmented	*	*	*	*

(b)

Panel	Year							
	1	2	3	4	5	6	7	8
1	*				*			
2		*				*		
3			*				*	
4				*				*
Augmented	*	*	*	*	*	*	*	*

TABLE 9.1
(a) An augmented rotating panel design, and (b) an augmented serially alternating panel design. An asterisk indicates that the panel is sampled in that year.

t is reduced by a dynamic design, from what it is for a static design, is largest when the temporal dependence is strong. Furthermore, in that circumstance the spatial locations of the optimal dynamic design for spatiotemporal prediction at time t are typically quite far removed from those at time $t - 1$. This is intuitively reasonable, since $Z(\mathbf{s}, t - 1)$ is strongly informative of $Z(\mathbf{s}^*, t)$ when $\mathbf{s}^*$ is at or near $\mathbf{s}$ and the temporal dependence is strong. Not surprisingly, when the spatiotemporal process has more complicated dynamics or is nonseparable or non-Gaussian, the dynamic prediction-optimal design can be more complicated, even nonintuitive, but is still much more efficient than the optimal static design (Wikle and Royle, 1999, 2005).

A more highly specialized space-time design is the Before-After-Control-Impact (BACI) design (Green, 1979; Stewart-Oaten, 1986). The objective of a BACI design is to assess the impact on the environment of some human activity or disturbance. For example, it may be desirable to determine the effect that effluent from a newly built municipal sewage treatment plant, discharged into a river, has on aquatic organisms downstream. In a standard BACI design, samples are taken at two classes of sites, one designated as impact sites where the effects of the activity or disturbance, if any, would occur, and the other as control sites. At each site, observations are taken at two sets of times, one before the activity begins and the other after. Whether the activity has impacted the environment is assessed by acting as though the BACI design is a classical randomized experiment and carrying out an F-test for Time by Treatment interaction in a linear model that has fixed effects for two times (before and after) and two treatments (control and impact) and their interaction. However, since the treatments are not actually randomly assigned to sites, inferences must be model-based rather than design-based. This and several other issues have led to the development of many variations on the basic BACI design. For an overview of these variations, see Smith (2002).

9.2.7   Discussion

Thus far we have described three main approaches to design for environmental monitoring — model-based, probability-based, and space-filling — and we have indicated which approaches are applicable and likely to be effective for various objectives. A fundamental requirement of the model-based approach is, of course, that a model be assumed for the spatial process. If there are several competing models and uncertainty about which is best, one possibility would be to select, at the first stage of sampling at least, a design that optimally discriminates among them. A large literature exists on the topic of optimal design for model discrimination in a classical regression context, but virtually no attention has been given to this problem in a spatial context. Accordingly, if the model is uncertain or unknown, presently we recommend using a probability-based, space-filling, or "hybrid" design. Hybrid designs are probability-based designs that have good spatial coverage over the region of interest, yet have some sites close together. An example of a class of hybrid designs is the "lattice plus close pairs" design of Diggle and Lophaven (2006), which has recently been studied further by Chipeta et al. (2016).

Although spatial sampling design has matured considerably within the past 20 years, research continues along several fronts. For example, as geostatistics for functional data has developed, so has spatial sampling design for such data (Bohorquez, Giraldo, and Mateu, 2016). For the future, there is surely a need for the development of designs that can discriminate among competing spatial models, and for more work on adaptive (multi-stage), multivariate, and space-time design.

9.3   Sampling for Estimation of Abundance

When estimating the abundance of individuals in a biological population, the issue of detectability of objects of interest (usually animals or animal groups) typically arises. That is, we cannot usually assume that all objects present on a sampling unit are detected. If we can, then simple plot sampling methods can be adopted, in which some form of probability-based design (Section 9.2.3) is adopted, and design-based methods yield the required abundance estimates. The abundance estimator for any given design can be formulated as a special case of the unbiased Horvitz-Thompson estimator (Horvitz and Thompson, 1952).

When detection of an object on a plot is uncertain, we must estimate the probability of detection, so that the inclusion probability of the Horvitz-Thompson estimator must be estimated. Provided we adopt a consistent estimator of the detection probability, the resulting "Horvitz-Thompson-like" estimator (Borchers and Burnham, 2004) is asymptotically unbiased.

We first consider distance sampling, which may be considered an extension of plot sampling in which not all objects on a sampled plot are counted. In the case of point transect sampling, the plots are circles of radius w centered on the point, and for line transect sampling, the plots are strips of width $2w$, extending out a distance w either side of the transect. An observer travels to each point, or along each line, recording distance from the point or line of each object detected. If bearing (points) or distance along the line (lines) is also recorded for each detected object, then the data may be regarded as a thinned spatial point process. Conventional distance sampling methods are a hybrid of model-based methods to estimate probability of detection and design-based methods to estimate abundance in the wider survey region, given estimates of detection probabilities, but fully model-based methods can also be adopted. We only address design issues here; see Buckland et al. (2015) for analysis methods.

Survey design has received less attention for mark-recapture methods, but this is starting to change, with the rise of spatial capture-recapture methods (Efford, 2004; Borchers and Efford, 2008; Royle et al., 2014). In spatial capture-recapture, sensors are located according to some design through the study region. With the need to capture a number of animals at more than one location, the spacing of sensors becomes crucial, especially for large study areas, when sensors may have to be distributed in clusters, with large gaps between clusters. Note the use of the term 'sensor' here. These may be the more traditional traps of conventional capture-recapture, but they may represent some other means of detecting (and re-detecting) animals, for example camera traps, acoustic sensors (for which the 'objects' sampled might be individual animals, but might be individual calls from animals), or observers.

9.3.1 Distance sampling

Distance sampling comprises a suite of methods based on a design of lines or points. Line transect sampling is the most widely-used technique, while point transect sampling is sometimes preferred, especially for songbird surveys. There are several variants on these two schemes. The sampling units are the plots centered on the lines (strips of half-width w) or points (circles of radius w). A complication is that w is often not known in advance, but is estimated from the data. Further, in multi-species surveys, w may vary by species. Thus a strategy of pre-defining all possible plots in the study region, and of selecting a sample of these plots, is often not an option.

9.3.1.1 Standard probability-based designs

Point designs

Any of the designs of Figure 9.2 may be used. Systematic designs are generally preferred, as they give abundance estimates with higher precision when object density through the survey region exhibits spatial trends (Fewster et al., 2009). Sometimes, strata are defined, and a systematic sample of points selected in each stratum.

With either simple random sampling or stratified random sampling, because the sampling units are circles of radius w, it is possibly for units to overlap when the design is based on points. (Systematic sampling avoids this minor complication.) If w is known, the study region can be divided into squares of side $2w$, and a simple random or stratified random sample of squares selected. If w is estimated from data, the squares can be made larger, so that the length of one side is certain to exceed $2w$. The sample points are then located at the centers of the sampled squares.

If travel between points is costly or time-consuming, a cluster of points might be positioned at each sampling location. The analyst can then either pool data across each cluster, or adopt cluster sampling methods to estimate abundance.

In some surveys, a line design is used, and points are placed systematically along the lines. See Strindberg et al. (2004) for details of this and other variants.

Line designs

Line designs are more problematic than point designs, as the plots are typically much longer than they are wide. If it is practical to have long lines traversing the width of the study area, a simple design is a systematic sample of parallel lines spanning the full length of the study area (Figure 9.6, top left panel). For surveys in which lines can be traversed quickly but surveyed slowly, such as in dung surveys (Marques et al., 2001), short, systematically-spaced segments are often covered along each line (Figure 9.6, top right). If the distance between segments is the same as the distance between successive parallel lines, this gives a systematic sample of line segments through the study region.

Typically, the number of lines in a line transect survey, or within a single stratum of a stratified design, is modest, so that a given realization of a simple random design can give uneven cover. A systematic random design addresses this issue (Fewster et al., 2009). Another option is shown in the bottom left panel of Figure 9.6; here, the survey region is partitioned by a grid of rectangles, each of width $2w$, where w is the distance from the line beyond which detections are not recorded. A simple random sample of rectangles can then be selected, and a line placed down the center of each sampled rectangle. A similar method is to have fewer, wider rectangles, and a line is randomly located within each rectangle. As noted in Section 9.2.3, this is a form of stratified sampling with one unit sampled per stratum.

Especially for shipboard surveys of large regions, but also for aerial surveys when successive lines would ordinarily be widely-separated, zig-zag or saw-tooth designs are commonly used. These ensure that search effort is more-or-less continuous, without long gaps while the vessel travels to the next line. Thus for a given level of resources, additional surveying can be done. Such designs are similar to a systematic random design, but coverage probability varies if the width of the study region varies. The bottom right panel of Figure 9.6 illustrates how such designs can be implemented so that the variability in coverage probability is smaller than for a design in which the sampler angle is constant (Strindberg et al., 2004). A major axis is defined along the length of the survey region, and a systematic grid of parallel lines is placed at right-angles. Waypoints are defined where these lines intersect the region boundary, and the transects are formed by straight lines linking waypoints, alternating from one boundary to the other.

Automated survey design

Automated survey design can be very useful when used in conjunction with a Geographic Information System. Survey regions are often complex shapes, and unconstrained random designs can be inefficient, for example when a ship must travel long distances to travel from one line to another. An automated design algorithm can be used to quantify coverage probability through the survey region for constrained designs, zig-zag designs, or other designs that do not given uniform coverage. This also allows different designs to be compared for efficiency, allowing the user to evaluate options before committing resources.

Edge effects

For small survey regions, edge effects can be problematic. A simple solution is to adopt "plus-sampling": the samplers (circles or strips) are located through an enlarged region that includes a "bufferzone" extending a distance w beyond the boundary of the survey region. Observers then cover points or line segments located in the bufferzone, recording any detected objects that are within the survey region boundaries. If it is not possible or cost-effective to do this, then 'minus-sampling' is used, resulting in reduced coverage along the boundary. When using an automated survey design, simulation can be used to quantify coverage probability through the region, so that bias is not introduced to abundance estimates.

Model-based versus design-based methods

When design-based analysis methods are used to extrapolate from the sampled plots to

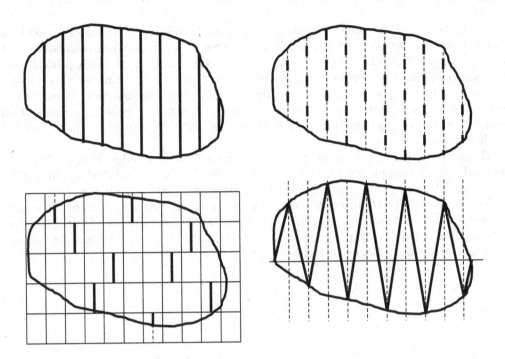

FIGURE 9.6
Four options for line transect designs. Top left: systematic sample of parallel lines. Top right: systematic grid of line segments (thick lines), placed along lines traversing the study region (dashed lines). Bottom left: a simple random sample of rectangular sampling units; the line is placed down the center of each selected unit. Bottom right: a zig-zag or saw-tooth sampler, determined by joining systematically-spaced waypoints alternating between top and bottom boundary of the study region.

the survey region, careful consideration needs to be given to the survey design, to ensure that abundance and precision are reliably estimated. When model-based methods are used to model object density through the survey region (Buckland et al., 2016), the survey design is less influential, giving greater flexibility for designing the survey. However, the design should seek to ensure that samplers are spread throughout the survey region, and that they span the range of values for any covariate used to model animal density.

9.3.1.2 Adaptive distance sampling designs

Distance sampling is especially effective for surveying objects that are sparsely distributed through a large region. However, if those objects tend to occur in a relatively small number of patches, standard distance sampling may yield both a low number of detections (because much effort is expended away from these patches) and high variance.

Adaptive cluster sampling (Thompson, 2002) was designed for surveying objects whose spatial distribution is patchy, providing a mechanism for conducting additional survey effort in localities where objects have been found, while giving design-unbiased estimation. The method was adapted for distance sampling by Pollard (2002) and Pollard and Buckland (2004).

An initial sample of units is selected at random. If the number of observations in a sampled unit satisfies some trigger condition, then units in a pre-defined neighborhood of the triggering unit are also sampled. If any of the adaptive units in the neighborhood meet the condition, then the neighborhood of each of these units is also sampled. The process repeats until no newly-sampled units meet the condition. The combination of an initial unit and its associated adaptive units is termed a cluster. Within the cluster, any units which do not meet the trigger condition are termed edge units, whilst any units which meet the condition form a network.

Consider first point transect sampling. In Figure 9.7, we show a definition of the neighborhood around a sample point based on square units. Designs can be based on other units, for example triangles or hexagons (Pollard and Buckland, 2004). Figure 9.8 shows a realization of a design based on square units, in which an initial systematic sample of squares is selected, and squares in the neighborhood are surveyed if at least one object is detected in a sampled square. Further detections trigger additional effort, so that initially separate units can merge into a single cluster.

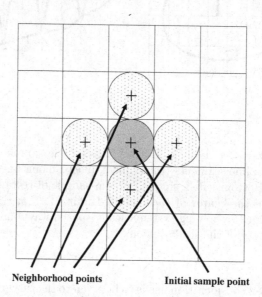

FIGURE 9.7

An adaptive design based on square units. The shaded circle is one of the original sample of plots, and the neighborhood is defined here as the four plots immediately adjacent to the shaded plot. The point in the shaded plot is surveyed, and if extra effort is triggered, the point at the center of each of the four plots in the neighborhood is surveyed.

For line transect surveys, we can define units as in the bottom left panel of Figure 9.6. Thus the units are rectangles, and for sampled units, the line passes through the center of the rectangle. A realization of a design of this type is shown in Figure 9.9. Here, the initial line transect is taken as the centerline of a strip of units placed end-to-end, and the neighborhood of a unit is defined as the two units to either side of that unit, above and below the strip. In this example, a single detection triggers effort in the neighborhood, and if at least one further detection is made in a neighborhood unit, the next unit away from the initial strip is also surveyed.

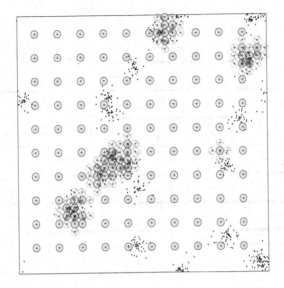

FIGURE 9.8
Simulation of an adaptive point transect survey on a highly clustered population. There were 100 initial sample points in a systematic grid and 82 adaptive points. The crosses identify the points and the circles the area covered at each point. The solid (yellow) circles identify points which belong to a network and the diagonally hatched (blue) circles are edge units. Objects are black dots and detected objects are surrounded by a red square. Pollard, 2002.

A major limitation of design-unbiased adaptive distance sampling methods is that the amount of survey effort is not known in advance. For marine surveys carried out by ship or aircraft, the platform is typically available for a pre-determined length of time. If the survey finishes early, resources are wasted, while if it is not possible to complete the survey in the time available, some of the survey region will remain unsurveyed, compromising abundance estimation. For this reason, Pollard et al. (2002) developed a form of adaptive distance sampling in which the additional effort triggered is a function of how much total survey effort is still available, thus ensuring that the survey finishes on schedule. While no longer design-unbiased, bias was found to be small.

Pollard et al. (2002) tested this fixed-effort adaptive distance sampling method on harbour porpoise in the Gulf of Maine and Bay of Fundy region. Adaptive and conventional methods gave very similar estimates of porpoise density, with slightly improved precision for the adaptive method. Perhaps the most significant gain from the adaptive method however was a substantial increase in the number of detections (551, compared with 313 for the conventional survey).

9.3.1.3 Designed distance sampling experiments

In distance sampling, the objective is usually to estimate abundance of the population of interest. Sometimes, we might instead wish to assess the effects (if any) of a management regime or an impact on population densities. For example, does spraying a forest block with pesticide to eradicate an insect pest have a detrimental effect on songbird densities?

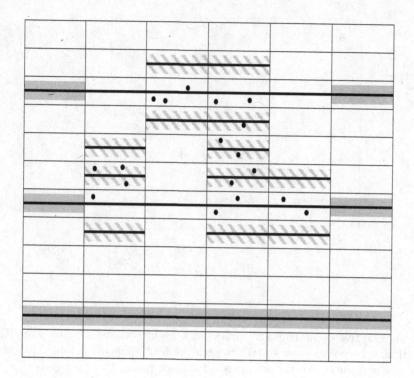

FIGURE 9.9

Part of a design comprising an initial systematic sample of lines. Each line is divided into segments, and the rectangles formed by extending each segment out to a distance w either side of the line form the units. A detection within a unit triggers effort in the neighborhood, which is formed by the units immediately above and below it.

Or does a proposed conservation measure in farmland have the desired effect on species of conservation concern?

Typically in a designed experiment using distance sampling, there is only one treatment, and we wish to compare that treatment with a control. There might be no time element (we simply compare densities on treatment plots with those on control plots at the time of the surveys), or we may have repeat visits to plots, and are interested in change over time. We cannot assume that repeat counts at the same plot are independent.

Design considerations are similar to those for the design of agricultural experiments, while analysis methods need to account for imperfect detection of objects on plots, and for the analysis of counts (Buckland et al., 2015). In the case of environmental impact assessments, a BACI design (before/after control/impact; McDonald et al., 2000; Buckland et al., 2009) can be useful when it is possible to have adequate replication. Thus plots are established on both control and treatment sites, and monitored before and after the treatment is applied.

Matched-pairs designs are useful for comparing a single treatment with a control. Where the treatment can readily be applied to a single plot, the design can utilize paired plots, as in a large-scale study to assess the effect of field conservation buffers on bird densities (Evans et al., 2013; Oedekoven et al., 2014). Where the treatment is applied at the level of site, with replicate plots at each site, then sites may be paired, as in a study assessing the effect on bird densities of controlled burns in ponderosa pine forest (Buckland et al., 2009).

9.3.2 Capture-recapture

9.3.2.1 Standard capture-recapture

There has been relatively little work done on survey design in non-spatial capture-recapture. When the goal of the study is to estimate the size of a population, the requirements of the approach are quite demanding. For example, no animal in the population should have zero probability of capture; in practice, for reliable estimation, we seek a design that ensures no probabilities are close to zero. Thus traps need to be spread throughout the range of the population of interest. Realistically, the approach is usually only useful for populations that are contained within a small study area.

When the population extends beyond the study area in which traps are set, animal movement onto the study area results in over-estimation of abundance, unless the effective area trapped can be estimated reliably. The bias reduces as the size of the study area increases; study area size should be appreciably larger than average home range size. For a camera trap survey of ocelots, Maffei and Noss (2008) concluded that the study area should be at least four times the size of the average home range.

More commonly, capture-recapture methods are used to estimate survival rates. In this context, survey design is seldom considered in detail. It is assumed that marked animals are representative of animals in the population, or at least of those that have the same values of any modeled covariates. There is no longer any requirement that all animals in the population have non-zero probability of capture, and study area size is only important if death of a marked animal is indistinguishable from emigration from the study area.

9.3.2.2 Spatial capture-recapture

To estimate abundance using spatial capture-recapture methods, it is no longer necessary for all animals in the population to have non-zero probability of capture, nor is it necessary for the study area to be large relative to the size of average home range. Of course, the abundance being estimated is defined by the boundaries of the study area, rather than by the biological population of interest.

As noted by Royle et al. (2014), solutions to the problem of survey design in spatial capture-recapture are largely ad hoc. They recommend simulation to explore different trap configurations. Successful design relies on knowledge of the spatial behavior of the species, to ensure that simulations identify an efficient design.

Royle et al. (2014) note that model-based design (Section 9.2.2) is a standard approach in classical sampling, but has not seen widespread use in ecology. They believe that the approach has great potential for spatial capture-recapture, and give a preliminary formulation. They note that normal linear models, as described in Section 9.2.2, are unsuitable for spatial capture-recapture, for which a Poisson or binomial model is more appropriate. They thus develop methods adopting the formulation of generalized linear models, allowing a link function $g(\cdot)$. Thus they take

$$g(E(\mathbf{y})) = \alpha_0 + \alpha_1 d(\mathbf{x}, \mathbf{s})^2 = \mathbf{M}'\boldsymbol{\alpha}$$

where $\mathbf{y}$ is the vector of length J of encounter frequencies in the J traps, $d(\mathbf{x}, \mathbf{s})$ is the distance between trap location $\mathbf{x}$ and individual activity center $\mathbf{s}$, $\boldsymbol{\alpha}$ is a vector of parameters of length two, and $\mathbf{M}$ is the $J \times 2$ design matrix. They then proceed to develop an optimal design criterion for spatial capture-recapture.

9.3.3 Discussion

The rapid rise in the use of spatial capture-recapture methods is likely to result in renewed interest in the design of capture-recapture studies. Most such studies are carried out within small study areas, and the issue of optimal design when the methods are applied to large study areas is a difficult one that requires effective solutions, if such studies are to become commonplace. To ensure adequate recaptures at distinct locations in a large study area, sensors must be placed according to a clustered design, and the optimal trade-off between number of clusters and number of sensors within each cluster is unclear. In distance sampling, large-scale designed experiments are starting to be used to assess the impact of a treatment, development or management action, and again, optimal design strategies are needed to ensure efficient use of resources.

Bibliography

[1] Bohorquez, M., Giraldo, R., and Mateu, J. (2016). Optimal sampling for spatial prediction of functional data. *Statistical Methods and Applications*, 25:39–54.

[2] Borchers, D.L. and Burnham, K.P. 2004. General formulation for distance sampling. In: S.T. Buckland, D.R. Anderson, K.P. Burnham, J.L. Laake, D.L. Borchers, and L. Thomas (eds.), *Advanced Distance Sampling*, Oxford University Press, Oxford, 6–30.

[3] Borchers, D.L. and Efford, M. 2008. Spatially explicit maximum likelihood methods for capture-recapture studies. *Biometrics*, 64:377–385.

[4] Buckland, S.T., Oedekoven, C.S., and Borchers, D.L. 2016. Model-based distance sampling. *Journal of Agricultural, Biological, and Environmental Statistics*, 21:58–75.

[5] Buckland, S.T., Rexstad, E.A., Marques, T.A., and Oedekoven, C.S. 2015. *Distance Sampling: Methods and Applications*. Springer, New York.

[6] Buckland, S.T., Russell, R.E., Dickson, B.G., Saab, V.A., Gorman, D.N., and Block, W.M. 2009. Analysing designed experiments in distance sampling. *Journal of Agricultural, Biological, and Environmental Statistics*, 14:432–442.

[7] Bueso, M.C., Angula, J.M., Cruz-Sanjulián, J., and García-Aróstegui, J. 1999. Optimal spatial sampling design in a multivariate framework. *Mathematical Geology*, 5:507–525.

[8] Chipeta, M.G., Terlouw, D.J., Phiri, K.S., and Diggle, P.J. (2016). Inhibitory geostatistical designs for spatial prediction taking account of uncertain covariance structure. arXiv:1605.00104v1.

[9] Cordy, C. 1993. An extension of the Horvitz-Thompson theorem to point sampling from a continuous universe. *Statistics and Probability Letters*, 18:353–362.

[10] Diggle, P. and Lophaven, S. 2006. Bayesian geostatistical design. *Scandinavian Journal of Statistics*, 33:53–64.

[11] Dixon, W., Smyth, G.K., and Chiswell, B. 1999. Optimized selection of river sampling sites. *Water Resources*, 33:971–978.

[12] Dobbie, M.J., Henderson, B.L., and Stevens, D.L. 2008. Sparse sampling: Spatial design for monitoring stream networks. *Statistics Surveys*, 2:113–153.

[13] Efford, M. 2004. Density estimation in live-trapping studies. *Oikos*, 106:598–610.

[14] Evans, K.O., Burger, L.W., Oedekoven, C.S., Smith, M.D., Riffell, S.K., Martin, J.A., and Buckland, S.T. 2013. Multi-region response to conservation buffers targeted for northern bobwhite. *Journal of Wildlife Management*, 77:716–725.

[15] Falk, M.G., McGree, J.M., and Pettitt, A.N. 2014. Sampling designs on stream networks using the pseudo-Bayesian approach. *Environmental and Ecological Statistics*, 21:751–773.

[16] Federov, V.V. 1972. *Theory of Optimal Experiments*. Academic Press, New York.

[17] Fewster, R.M., Buckland, S.T., Burnham, K.P., Borchers, D.L., Jupp, P.E., Laake, J.L., and Thomas, L. 2009. Estimating the encounter rate variance in distance sampling. *Biometrics*, 65:225–236.

[18] Fuentes, M., Chaudhuri, A., and Holland, D.M. 2007. Bayesian entropy for spatial sampling design of environmental data. *Environmental and Ecological Statistics*, 14:323–340.

[19] Green, R.H. 1979. *Sampling Design and Statistical Methods for Environmental Biologists*. Wiley, Chichester.

[20] Harville, D.A. and Jeske, D.R. 1992. Mean squared error of estimation or prediction under a general linear model. *Journal of the American Statistical Association*, 87:724–731.

[21] Holland, D.M., Saltzman, N., Cox, L., and Nychka, D. 1999. Spatial prediction of sulfur dioxide in the eastern United States. In: J. Gómez-Hernández, A.O. Soares, and R. Froidevaux (eds.), *geoENV II: Geostatistics for Environmental Applications*, Kluwer, Dordrecht, 65–76.

[22] Horvitz, D.G. and Thompson, D.J. 1952. A generalization of sampling without replacement from a finite universe. *Journal of the American Statistical Association*, 47:663–685.

[23] Johnson, M.E., Moore, L.M., and Ylvisaker, D. 1990. Minimax and maximin distance designs. *Journal of Statistical Planning and Inference*, 26:131–148.

[24] Le, N.D. and Zidek, J.V. 1994. Network designs for monitoring multivariate random spatial fields. In: J.P. Vilaplana and M.L. Puri (eds.), *Recent Advances in Statistics and Probability*, VSP, Zeist, 191–206.

[25] Le, N.D. and Zidek, J.V. 2006. *Statistical Analysis of Environmental Space-Time Processes*. Springer, New York.

[26] Li, J. and Zimmerman, D.L. 2015. Model-based sampling design for multivariate geostatistics. *Technometrics*, 25:75–86.

[27] Liebetrau, A.M. 1979. Water quality sampling: Some statistical considerations. *Water Resources Research*, 15:1717–1725.

[28] Lu, L., Anderson-Cook, C.M., and Robinson, T.J. 2011. Optimization of designed experiments based on multiple criteria utilizing a Pareto frontier. *Technometrics*, 53:353–365.

[29] Maffei, L. and Noss, A.J. 2008. How small is too small? Camera trap survey areas and density estimates for ocelots in the Bolivian Chaco. *Biotropica*, 40:71–75.

[30] Marques, F.F.C., Buckland, S.T., Goffin, D., Dixon, C.E., Borchers, D.L., Mayle, B.A., and Peace, A.J. 2001. Estimating deer abundance from line transect surveys of dung: sika deer in southern Scotland. *Journal of Applied Ecology*, 38:349–363.

[31] Matérn, B. 1960. *Spatial Variation*. Springer-Verlag, New York.

[32] McDonald, T.L. 2003. Review of environmental monitoring methods: survey designs. *Environmental Monitoring and Assessment*, 85:277–292.

[33] McDonald, T.L., Erickson, W.P., and McDonald, L.L. 2000. Analysis of count data from before-after control-impact studies. *Journal of Agricultural, Biological, and Environmental Statistics*, 5:262-279.

[34] McKay, M., Beckman, R., and Conover, W. 1979. A comparison of three methods for selecting values of input variables in the anlysis of output from a computer code. *Technometrics*, 21:239–245.

[35] Minasny, B., McBratney, A.B., and Walvoort, D.J.J. 2007. The variance quadtree algorithm: Use for spatial sampling design. *Computers & Geosciences*, 33:383–392.

[36] Müller, W.G. 2005. A comparison of spatial design methods for correlated observations. *Environmetrics*, 16:495–505.

[37] Müller, W.G. 2007. *Collecting Spatial Data*, 3rd ed. Springer, Berlin.

[38] Müller, W.G., Pronzato, L., Rendas, J., and Waldl, H. 2015. Efficient prediction designs for random fields (with Discussion). *Applied Stochastic Models in Business and Industry*, 31:178–203.

[39] Müller, W.G. and Stehlík, M. 2011. Compound optimal spatial designs. *Environmetrics*, 21:354–364.

[40] Müller, W.G. and Zimmerman, D.L. 1999. Optimal designs for variogram estimation. *Environmetrics*, 10:23–37.

[41] Nychka, D. and Saltzman, N. 1998. Design of air quality networks. In: *Case Studies in Environmental Statistics*, D. Nychka, W. Piegorsch, and L. Cox (eds.), Springer-Verlag, New York, 51–76.

[42] Oedekoven, C.S., Buckland, S.T., Mackenzie, M.L., King, R., Evans, K.O., and Burger, L.W. 2014. Bayesian methods for hierarchical distance sampling models. *Journal of Agricultural, Biological, and Environmental Statistics*, 19:219–239.

[43] Olea, R.A. 1984. Sampling design optimization for spatial functions. *Mathematical Geology*, 16:369–392.

[44] Overton, W.S. and Stehman, S.V. 1993. Properties of designs for sampling continuous spatial resources from a triangular grid. *Communications in Statistics - Theory and Methods*, 22:2641–2660.

[45] Pistone, G. and Vicario, G. 2010. Comparing and generating Latin Hypercube designs in Kriging models. *Advances in Statistical Analysis*, 94:353–366.

[46] Pollard, J.H. 2002. *Adaptive Distance Sampling*. Ph.D. thesis, University of St Andrews.

[47] Pollard, J.H. and Buckland, S.T. 2004. Adaptive distance sampling surveys. In: *Advanced Distance Sampling*, S.T. Buckland, D.R. Anderson, K.P. Burnham, J.L. Laake, D.L. Borchers, and L. Thomas (eds), Oxford University Press, Oxford, 229–259.

[48] Pollard, J.H., Palka, D., and Buckland, S.T. 2002. Adaptive line transect sampling. *Biometrics*, 58: 862–870.

[49] Pronzato, L. and Müller, W. 2012. Design of computer experiments: space filling and beyond. *Statistics and Computing*, 22:681–701.

[50] Royle, J.A., Chandler, R.B., Sollmann, R., and Gardner, B. 2014. *Spatial Capture-Recapture*. Academic Press, Amsterdam.

[51] Royle, J.A. and Nychka, D. 1998. An algorithm for the construction of spatial coverage designs with implementation in S-Plus. *Computers & Geosciences*, 24:479–488.

[52] Russo, D. 1984. Design of an optimal sampling network for estimating the variogram. *Soil Science Society of America Journal*, 48:708–716.

[53] Silvey, S.D. 1980. *Optimal Design*. Chapman and Hall, London.

[54] Smith, E.P. 2002. BACI design. In: *Encyclopedia of Environmetrics*, A.H. El-Shaarawi and W.W. Piegorsch (eds), Chichester, Wiley, 141–148.

[55] Som, N., Monestiez, P., Ver Hoef, J.M., Zimmerman, D.L., and Peterson, E.E. 2014. Spatial sampling on streams: General principles for inference on aquatic networks. *Environmetrics*, 25:306–323.

[56] Stevens, D.L. 1997. Variable density grid-based sampling designs for continuous spatial populations. *Environmetrics*, 8:167–195.

[57] Stevens, D.L. and Olsen, A.R. 2004. Spatially balanced sampling of natural resources. *Journal of the American Statistical Association*, 99:262–278.

[58] Stewart-Oaten, A., Murdoch, W.W., and Parker, K.R. 1986. Environmental impact assessment: pseudoreplication in time? *Ecology*, 67:929–940.

[59] Strindberg, S., Buckland, S.T., and Thomas, L. 2004. Design of distance sampling surveys and Geographic Information Systems. In: *Advanced Distance Sampling*. S.T. Buckland, D.R. Anderson, K.P. Burnham, J.L. Laake, D.L. Borchers, and L. Thomas (eds), Oxford University Press, Oxford, 190–208.

[60] Thompson, S.K. 2002. *Sampling*. 2nd ed. Wiley, New York.

[61] Urquhart, N.S. and Kincaid, T.M. 1999. Designs for detecting trend from repeated surveys of ecological resources. *Journal of Agricultural, Biological, and Environmental Statistics*, 4:404–414.

[62] Urquhart, N.S., Overton, W.S., and Birkes, D.S. 1993. Comparing sampling designs for monitoring ecological status and trends: impace of temporal patterns. In: *Statistics for the Environment*, V. Barnett and K.F. Turkman, eds., Chichester, Wiley, 71–85.

[63] Vašát, R., Heuvelink, G.B.M., and Boruvka, L. 2010. Sampling design optimization for multivariate soil mapping. *Geoderma*, 155:147–153.

[64] Ver Hoef, J.M. and Peterson, E.E. 2010. A moving average approach for spatial statistical models of stream networks. *Journal of the American Statistical Association*, 105:6–18.

[65] Ver Hoef, J.M, Peterson, E.E., and Theobald, D. 2006. Spatial statistical models that use flow and stream distance. *Environmental and Ecological Statistics*, 13:449-464.

[66] Walvoort, D.J.J., Brus, D.J., and de Gruijter, J.J. 2010. An R package for spatial coverage sampling and random sampling from compact geographical strata by k-means. *Computers & Geosciences*, 36:1261–1267.

[67] Warrick, A.W. and Myers, D.E. 1987. Optimization of sampling locations for variogram calculations. *Water Resources Research*, 23:496–500.

[68] Wikle, C.W. and Royle, J.A. 1999. Space-time dynamic design of environmental monitoring networks. *Journal of Agricultural, Biological, and Environmental Statistics*, 4:489–507.

[69] Wikle, C.W. and Royle, J.A. 2005. Dynamic design of ecological monitoring networks for non-Gaussian spatio-temporal data. *Environmetrics*, 16:507–522.

[70] Yeh, M.S., Lin, Y.P., and Chang, L.C. 2006. Designing an optimal multivariate geostatistical groundwater quality monitoring network using factorial kriging and genetic algorithms. *Environmental Geology*, 50: 101–121.

[71] Yfantis, E.A., Flatman, G.T., and Behar, J.V. 1987. Efficiency of kriging estimation for square, triangular, and hexagonal grids. *Mathematical Geology*, 19:183–205.

[72] Zhu, Z. and Stein, M.L. 2005. Spatial sampling design for parameter estimation of the covariance function. *Journal of Statistical Planning and Inference*, 134:583–603.

[73] Zhu, Z. and Stein, M.L. 2006. Spatial sampling design for prediction with estimated parameters. *Journal of Agricultural, Biological, and Environmental Statistics*, 11:24–44.

[74] Zimmerman, D.L. 2006. Optimal network design for spatial prediction, covariance parameter estimation, and empirical prediction. *Environmetrics*, 17:635–652.

[75] Zimmerman, D.L. and Li, J. 2013. Model-based frequentist design for univariate and multivariate geostatistics. In: *Spatio-Temporal Design*, Mateu, J. and Müller, W.G. (eds), Chichester, Wiley, 37–53.

10

Accommodating so many zeros: univariate and multivariate data

James S. Clark

Nicholas School of the Environment, Duke University, Durham, NC

Alan E. Gelfand

Department of Statistical Science, Duke University, Durham, NC

CONTENTS

Zeros play a unique role in ecological and environmental data and often need special modeling consideration to accommodate them. The purpose of this chapter is to elaborate this challenge. Two illustrative examples are the 0 resulting from no precipitation at a location (an environmental variable) and the 0 resulting from species absence at a location (an ecological variable).

Modeling choices we discuss include Tobit models, zero-inflated models along with hurdle models, and ordinal and nominal categorical data models. We consider these choices in both the univariate and multivariate settings. We formalize these models and discuss their properties, fitting, and associated inference. Contexts can be spatial, temporal, and spatio-temporal.

Keywords: hurdle model; logit; nominal data; ordinal data; probit; Tobit model; zero-inflation

10.1   Introduction

How do zeros arise in environmental data? What happens when there are too many of them? Zeros play a unique role in ecological and environmental data and often require special modeling specification to accommodate them. The relevant modeling literature has grown substantially in the past twenty years. The purpose of this chapter is to elaborate this development. As a concrete example, temperature data are recorded as a continuous variable, measured to a level of precision that depends on the study. The degree of precision is given by the number of significant digits. A temperature value of "0" is interpreted like any other value in the support/domain of temperatures (unless it's on the Kelvin scale). A temperature observation takes a value in R^1, a continuous variable with no point masses. However, consider the following common environmental situations where, in each case, a zero has special meaning:

1. Like temperature, precipitation is recorded as a continuous variable. However, a zero means something different; it is the event of "no precipitation" and, unlike any other value, a zero has positive probability of occurring anywhere, any time. This is an example of *continuous data with point mass at zero*.

2. Presence of a species is recorded in sample plots based on the degree of effort spent on finding it. For sessile organisms, effort could be plot area. For mobile animals, it could be search time, search area, or both. In species distribution modeling (SDM) such data are termed *presence-absence data* with 0 denoting absence. Although such data may be recorded as zeros and ones, in fact they have no scale at all. They are just binary outcomes; we could just as well call them 'failure' and 'success'.

3. Abundance, in the form of counts, is tallied in sample plots as $(0, 1, 2, ...)$ and again depends on effort. Here too, zeros can have special meaning. If we observe a positive count, we could expect that, with greater effort, the count will increase. We do not necessarily make the same assumption if we observe a 0 (for example, think of fishing). These are *discrete count data*.

4. Abundance of a species might be recorded as either 'absent', 'rare', or 'abundant', scored as 0, 1, or 2, respectively. Here again, there is no absolute scale; the difference between 0 and 1 is not the same as the difference between 1 and 2. We refer to such data as *ordinal count data*; the categories have an order but the numerical labeling is arbitrary. However, with the above labeling, again 0 has a special meaning: absent.

5. Discrete attributes of an organism are recorded. Several states are possible, only one of which can be assigned to a given observation. Examples include leaf type, color, disease symptoms, phenological state, and so forth. In the absence of order,

they are referred to as *nominal* categorical data. The *levels* refer to possible states. The observed level is recorded as a 1. All others are recorded as 0. These are referred to as *multinomial trials*. The observation is a vector having one entry as a "1", the others as "0"s. The multiple possible levels are related by the fact that only one level can be 1; rather than "nothing", zero means "not the level observed".

6. Multiple species are counted in a sample where the sum of the counts is not related to actual abundance. Examples include numbers of microfossils in sediment cores and numbers of OTUs in microbiome data[1]. We refer to such count data as *count composition data*. When the data have been normalized to a fraction of the total for each species, we refer to the data as *fractional composition data*. In both cases zeros have special meaning, for the same reasons as given in situation 3 above.

These disparate examples demonstrate the frequent occurrence of zeros in environmental and ecological data and also have in common a distinctive meaning for 0, a different interpretation from other values in the data, and often requiring a specialized model.

If we consider discrete count data, there are many modeling options which include a point mass at 0, e.g., Poisson, negative binomial, binomial. However, the incidence of 0's can be overwhelming (Fig. 10.1b, c), more than these models can capture while at the same time explaining the non-zero counts well. For example, the Poisson distribution provides a theoretical basis for catch per unit effort, but only if there are *poissons*[2] in the lake. A sample from multiple lakes needs to accommodate the fact that, perhaps only some lakes have *poissons*. Furthermore, no two fishermen expect the same success due to different gear, different bait - technically, we refer to this as different *detection probability*. Here, we are exemplifying the notion in the literature of a "structural zero" vs. a "random zero." Empty or zero cells in a table of counts can be classified as structural zeros or random (sampling) zeros. A structural zero in a cell means that no observation could fall into that cell (Berger and Zhang 2005). The notion of a structural zero could apply to lakes without poissons. The classical illustration for categorical data is a 2×2 table with row variable sex (Male or Female) and column variable pregnancy state (Yes or No); one of the cells is always empty.

When both are incorporated into a model as a discrete mixture, the models are termed *zero-inflated models* (Lambert 1992, Martin et al. 2007, Wenger and Friedman 2008). They are formulated to allow that one component *inflates* the probability of 0 in the other. That is, the probability of falling in the 0 class under the structural component inflates the chance of obtaining a 0 under the random/chance component. In particular, there can be a presence/absence model and a second model for abundance given presence. In the capture-recapture setting (see Chapter 13 and references therein), failure to recapture can be viewed as a 0 and can arise for two reasons. The animal was dead (structural). Alternatively, it is alive, but not detected (random). A hurdle model (Mullahy 1986, Faddy 1997) is used to model only non-zero observations.

Zero-inflation is not the only way to model zeros. The point mass at zero needed for a continuous response like precipitation cannot be handled by zero-inflation, because a continuous model has no point mass to inflate. Likewise, fractional composition data (e.g., the proportions of individuals from different species rather than counts of individuals from different species) are continuous on $[0, 1]$, again, with a point mass needed needed not only at 0, but also at 1, the latter for observations containing only one species. Another example would consider land use classifications where the proportions associated with a given areal unit indicate the proportions of the area of that unit assigned to different land

---

[1] An operational taxonomic unit (OTU) is an operational definition of a species or group of species often used when only DNA sequence data is available. It is the most commonly used microbial diversity unit.

[2] Apologies!

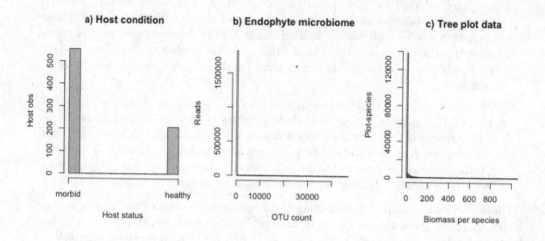

FIGURE 10.1
Zeros in three data types. a) Seedling hosts can be in a 'healthy' or 'morbid' state, scored as 0 and 1. b) Composition count data for their endophytic microbiome ($S = 175$ OTUs occurred in at least 50 observations) are 96% zeros. c) Continuous abundance with point mass at zero–the biomass per species on each plot in FIA data is 82% zeros. From Hersh et al (in prep) and Clark et al. (2014).

use classifications, e.g., urban, agriculture, forest, water, or fractional vegetation cover types (Alig et al. 2005, Leininger et al. 2013, Schwantes et al. 2016).

Composition data introduce the further complication of multiple response variables. Species abundances are typically recorded for multiple species. There is dependence among species. Likewise, trait studies rarely focus on single traits, and traits co-vary (Westoby et al. 2002, Wright et al. 2007, Swenson and Weiser 2010, Clark 2016). There is a small but growing literature that looks at joint species across environmental gradients (Ovaskainen et al. 2010, de Valpine and Harmon-Threatt 2013, Clark et al. 2014, Pollock et al. 2014, Thorson et al. 2015), but still with only limited attention to problems posed in the case of massive over-representation of zeros. At a given site, there is often a long list of species with potential to occur, only a few of which are actually observed.

Variables like precipitation can involve space and space-time observations, suggesting models that incorporate dependence in space, in time, or both. If it is not raining at location **s** then it is likely not to be raining at a location near **s**; if it is not raining at time t, then it is not likely to be raining at a time near t. A similar argument can be made for space or space-time dependence with regard to species abundance (therefore presence-absence), hence with occurrence of 0's.

The objective of this chapter is to provide an overview of the modeling literature for the above issues. Again, we focus on models for environmental and ecological processes where modeling of 0's receives special attention. We present their stochastic specifications along with associated implications. While attempting to point out the range of approaches that have been applied, we attempt some synthesis by showing how a general framework can accommodate many of these problems. We note that model assessment - adequacy and comparison - are not considered here. Such assessment is discussed in Chapter 2 and in the context of specific applications in the ensuing chapters.

The format of the chapter is as follows. In Section 2 we present basic univariate modeling for zeros. Section 3 extends this to the setting of multinomial trials where the response is a vector having a single '1' and the rest '0's. Section 4 briefly points to extending these formulations to data collected over space and time. Section 5 focuses on multivariate modeling, i.e., on joint modeling of responses, with an example. Section 6 focuses on two examples of joint attribute modeling. Finally, Section 7 offers a brief summary.

10.2 Basic univariate modeling ideas

Models for environmental data that contain zeros typically have two requirements: (i) a point mass at 0 and (ii) the remainder of support on the *positive* part of the real line. There is a substantial literature on atomic distributions or mixed distributions, i.e., distributions that have a portion of their mass on a continuous domain and the remainder of their support allocated to points also called atoms (Koopmans 1969). A first approach provides a point mass at 0, with the remainder captured by a continuous distribution on $(0, \infty)$. These models are often referred to as Tobit models (Cameron and Trivedi 2005) and are taken up in Section 2.3.

The second approach applies to count variables, i.e., models with entirely discrete support $\{0, 1, 2, ..., \}$. Furthermore, the support can be rescaled, e.g., if there is a fixed total count, N (which might be interpreted as 'effort'), we can rescale to say $\{0, 1/N, 2/N, ...1\}$. Regardless, the objective is to fit all of the observed counts well, including an excess of 0 observations. These models are customarily referred to a zero-inflated models (Welsh et al. 1996, Cameron and Trivedi 1998) and are taken up in Section 2.2. However, first we review the binary data setting.

10.2.1 Zeros and ones

The Bernoulli distribution for a binary response, zeros and ones, is a point of departure, i.e., $y_i \sim Bernoulli(\theta_i)$. For example, species distribution models (SDMs) are often built from generalized linear models (GLMs), where the Bernoulli likelihood is coupled with a regression to explain how the probability that a species is present is influenced by predictors. It applies when data are assigned to two states, as can occur with pollination (Moran and Clark 2010), phenology (Polansky and Robbins 2013, Gleim et al. 2014), reproduction (Patrick, and Simons 2014), predation (Bell and Clark 2015), site occupancy (Dorazio 2007), survival and/or morbidity (Augspurger 1984, Ibanez et al. 2007, Metcalf et al. 2009, Cox et al. 2010, Bowlby and Gibson 2015), and infection status (Clark and Hersh 2009, Hooten et al. 2010, Valle et al. 2011, Haas et al. 2016). In mark-recapture studies it describes recapture (Chapter 13).

The probability model requires a scale transformation, typically accomplished with a link function, which takes the regression on $(-\infty, \infty)$ and transforms it to the probability scale $(0, 1)$. The most commonly used links are the logit,

$$\theta_i = \frac{exp(\mathbf{x}_i^T \beta)}{1 + exp(\mathbf{x}_i^T \beta)} \tag{10.1}$$

and probit,

$$\theta_i = \Phi(\mathbf{x}_i^T \beta) \tag{10.2}$$

where $\Phi(\cdot)$ is the standard normal cumulative distribution function, $\mathbf{x}_i$ is a vector of covariates, and β is a vector of corresponding coefficients. Figure 10.2a shows predicted status of seedling hosts from a study of infection by fungal endophytes using the probit link and data from Figure 10.1a. Note that the model tends to predict the mean probability of morbidity; this is an expected finding for a model that explains little variation in the data.

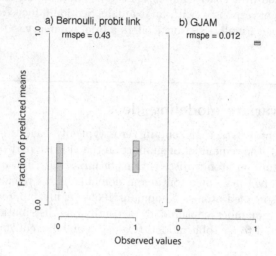

FIGURE 10.2
Host condition from Figure 10.1a predicted using a probit link in the R function `glm` (a) and as a joint model with microbiome data of Figure 10.1b in the package `gjam` (Section 10.6.1). (b). In both cases, predictor variables are winter temperature, moisture deficit, host species, canopy gap, and polyculture, the last three variables being factors. Boxes bound 95% of predictions for observed zeros and ones, with a horizontal line at the median prediction. Root mean square prediction errors (rmspe) are shown for both models (from Hersh, Benetiz, Vilgalys, and Clark, in preparation).

The non-linear link function means that the sensitivity to a predictor x_{iq} depends on all of the predictors in the model. That is, there is an implicit assumption that all predictors interact, the form of those interactions imposed by the form of the link function. Said in a different way, coefficients must be interpreted with caution. The logit link is commonly used in species distribution models where interest lies in predicting where a species will occur on the basis of environmental variables (Elith and Leathwick, 2009). Link functions and zero inflation are described in the more general setting of count data, in the next subsection.

10.2.2   Zero-inflated count data

Count data with an excess of zeros arise in many contexts (Ver Hoef and Boveng 2007, Cunningham and Lindenmayer 2008, Sileshi et al. 2009, Linden and Mntyniemi 2011, Smith et al. 2012, Lynch et al. 2014). A basic example considers abundance data for species of trees (adults or seedlings) where an excess of zeros arises because, for many plots, the environment is not suitable for the species. As a more elaborate illustration, soil microbiome data (Fig. 10.1b) are *count composition* data – the total number of reads for each operational taxonomic unit (OTU) is not related to overall abundance. Only relative values are meaningful. In fact, the total count can be viewed as a metric of observation *effort*. In this case, the higher the total count, the greater the effort, and the more representative the counts. A sample may range from 10^2 to 10^6 counts, distributed over thousands of OTUs. The number of zeros can be overwhelming (Fig. 10.1b).

In an influential paper, Lambert (1992) describes an industrial application where a manufacturing process moves between a reliable state in which defects are extremely rare and an unreliable state in which the number of defects follow a Poisson distribution. A natural way to model such data is with an additional point mass p at 0. That is, with probability p, we sample a degenerate distribution at 0 and with probability $(1-p)$ we sample a non-degenerate distribution $G(\cdot|\Theta)$. When $G(\cdot|\Theta)$ is Poisson, $Po(\lambda)$, we obtain the well-known zero-inflated Poisson (ZIP) models, which we denote as $ZIP(p, \lambda)$. This model has been studied extensively. Cohen (1963) and Johnson and Kotz (1969) discuss ZIP model without covariates, Lambert (1992) employs the ZIP model in a linear regression setup, using canonical links for both p and λ. She obtains the maximum likelihood estimates of the parameters using the E-M algorithm. Ghosh et al. (1998) discuss fully Bayesian methods for fitting ZIP models.

We start with some notation. Let Y be a count random variable. Let $G(y|\Theta)$, $y = 0, 1, 2, \ldots$ denote a probability mass function associated with these counts. The zero-inflated distribution associated with $G(y|\Theta)$ is then defined as

$$
\begin{aligned}
P(Y = 0|p, \Theta) &= p + (1-p)G(0|\Theta) \\
P(Y = y|p, \Theta) &= (1-p)G(y|\Theta), \ y > 0.
\end{aligned}
\tag{10.3}
$$

Equivalently, we can write

$$
\pi(y|p, \Theta) = p\delta_0 + (1-p)G(y|\Theta).
\tag{10.4}
$$

Usual choices for $G(y|\Theta)$ include the Poisson and negative binomial distributions. McMurdle and Holmes (2014) apply the negative binomial distribution to individual taxon from OTU data like those in Figure 10.1b.

10.2.2.1   The k-ZIG

Here, we address two limitations of the ZIP. First, in all of this literature, either a logit or probit model is used for p as a function of covariate information to explain the occurrence of the excess zeros. Both of these links are limited in their ability to explain an extreme number of zeros. That is, both of these links will struggle to find the covariates providing significant explanation. Their symmetry (i.e., $\log\frac{(1-p)}{p} = -\log\frac{p}{(1-p)}$ and $\Phi(1-p) = -\Phi(p)$) is a further restriction. Wang and Dey (2010) and references therein provide a summary of flexible parametric link functions developed to model categorical response data. However, these links are not purpose-built for handling an excessive amount of zeros in the context of modeling zero-contaminated count data.

Secondly, the choice of G to be Poisson implies limited shape and tail behavior. Alternative possibilities include finite Poisson mixtures and continuous Poisson mixtures, e.g., the Poisson Gamma (equivalently, negative binomial). The negative binomial model is more flexible than its Poisson counterpart in accommodating overdispersion (Lawless, 1987). Cameron and Trivedi (1986) study a variety of stochastic models, including the zero-inflated negative binomial model, for count data. Gurmu et al. (1999) develop a continuous mixture model for count data based on series expansion for the unknown density of the unobserved heterogeneity. They also discuss how to modify the mixture model to accommodate excess zeros. Cui and Wang (2009) develop a discrete mixture model for zero-inflated count data.

A convenient umbrella under which we can consider a rich range of zero-inflated count data models is called the k-ZIG (Ghosh et al. 2012). With the above notation, suppose we assume $G(y|\Theta)$ to be a zero-inflated distribution itself with a point mass q at zero and $G_0(y|\Theta_0)$ is the probability mass function corresponding to the non-degenerate part. Denote this zero-inflated distribution as $ZIG(q, \Theta_0)$. Inserting this G into (10.3) implies that a further proportion p is taken from the $\text{ZIG}(q, \Theta_0)$ and assigned to the event $\{y = 0\}$ thereby further enhancing the probability of this event. Writing out (10.3) explicitly with $G(y|\Theta)$ replaced by $\text{ZIG}(q, \Theta_0)$, we get

$$
\begin{aligned}
\pi(y|p, q, \Theta_0) &= p\delta_0 + (1 - p)(q\delta_0 + (1 - q)G_0(y|\Theta_0)) \\
&= (p + (1 - p)q)\delta_0 + (1 - p)(1 - q)G_0(y|\Theta_0)
\end{aligned}
\tag{10.5}
$$

It can be easily shown that p and q are not identifiable in the model described in (10.5). So we reparametrize (10.5) by assuming $(1 - p) = (1 - \theta)^{k-1}$ and $(1 - q) = (1 - \theta)$, yielding

$$
\pi(y|\theta, \Theta_0, k) = (1 - (1 - \theta)^k)\delta_0 + (1 - \theta)^k G_0(y|\Theta_0)
\tag{10.6}
$$

We refer to the distribution in (4) as a k-$\text{ZIG}(\theta, \Theta_0)$. Now, the expected number of zeros is controlled by the parameters k and θ. When $k = 1$, and $G_0(.)$ is $\text{Po}(\lambda)$, (10.6) yields the familiar ZIP model. With the goal of modeling excess zeros, we would imagine $k > 1$. However, in principle, k need not be integer valued and can be < 1; we can use the data to learn about k. In subsequent sections and through simulations we show that, under a suitable prior, the full conditional distribution for k has a closed form and the data provides consequential information about k. From (10.6) one can immediately notice that defining $\tilde{\theta} = 1 - (1 - \theta)^k$, we obtain the standard expression for a zero-inflated model as in (2):

$$
\pi(y|\tilde{\theta}, \Theta_0) = \tilde{\theta}\delta_0 + (1 - \tilde{\theta})G_0(y|\Theta_0)
\tag{10.7}
$$

Thus the k-$\text{ZIG}(\theta, \Theta_0)$ can be interpreted as a $\text{ZIG}(\tilde{\theta}, \Theta_0)$. However, this does not mean that fitting (5) is equivalent to fitting (2), as we clarify below.

10.2.2.2 Properties of the k-ZIG model

From (10.6) it is clear that the overall probability of zeros under the k-ZIG model is given by

$$
\begin{aligned}
P(Y = 0|\theta, \Theta_0, k) &= (1 - (1 - \theta)^k) + (1 - \theta)^k G_0(0|\Theta_0) \\
&= 1 - (1 - \theta)^k (1 - G_0(0|\Theta_0))
\end{aligned}
\tag{10.8}
$$

Since $0 \leq (1 - G_0(0|\Theta_0)) \leq 1$, it is clear that $P(Y = 0|\theta, \Theta_0, k) \to 1$ as $k \to \infty$, regardless of $\theta < 1$. From $\text{ZIG}(\tilde{\theta}, \Theta_0)$ in (5), we can see that unconditional moments are given by

$$
\begin{aligned}
E(Y|\tilde{\theta}, \Theta_0) &= (1 - \tilde{\theta})E_{G_0}(Y|\Theta_0) \\
var(Y|\tilde{\theta}, \Theta_0) &= \tilde{\theta}(1 - \tilde{\theta})[E_{G_0}(Y|\Theta_0)]^2 + (1 - \tilde{\theta})var_{G_0}(Y|\Theta_0)
\end{aligned}
$$

where $E_{G_0}(Y|\Theta_0)$ and $var_{G_0}(Y|\Theta_0)$ denote the expectation and variance of the random variable Y with probability mass function given by $G_0(.|\Theta_0)$. If we assume that $G_0(y|\Theta_0)$ is $\text{Po}(\lambda)$, then the moments turn out to be

$$
\begin{aligned}
E(Y|\theta, \lambda, k) &= (1 - \theta)^k \lambda \\
var(Y|\theta, \lambda, k) &= (1 - \theta)^k \lambda(1 + (1 - (1 - \theta)^k)\lambda)
\end{aligned}
$$

Note that the expectation under the k-ZIP model is a decreasing function of k. Hence for $k \geq 1$, the expectation under the k-ZIP model is less than that under the standard ZIP model. The variance, however, does not show this monotonicity. Controlling the value of θ, we can induce greater (smaller) variability in the k-ZIP as compared to the standard ZIP. This allows increased flexibility in modeling the non-zero counts.

10.2.2.3 Incorporating the covariates

As a result of the connection between the k-ZIG(θ, Θ_0) model and the standard ZIG$(\tilde{\theta}, \Theta_0)$ model, do we want to specify a regression model on θ or on $\tilde{\theta}$? Suppose we adopt a logit link on θ. Denoting the set of covariates that influence the occurrences by $\mathbf{x}_1$ and the corresponding regression parameters by $\boldsymbol{\beta}_1$, we get $\log \frac{\theta}{1-\theta} = \mathbf{x}_1 \boldsymbol{\beta}_1$ so

$$\tilde{\theta} = 1 - \left(\frac{1}{1 + \exp(\mathbf{x}_1^T \boldsymbol{\beta}_1)} \right)^k. \tag{10.9}$$

So, with a logit link on θ, we can fit the k-ZIG model described in (10.6) as a ZIG$(\tilde{\theta}, \Theta_0)$. In either case, the likelihood is a function of k and it is possible to draw inference about it.

If, however, we specify a logit link on $\tilde{\theta}$, we get $\log \frac{\tilde{\theta}}{1-\tilde{\theta}} = \mathbf{x}_1^T \boldsymbol{\beta}_1$ so

$$\theta = 1 - \left(\frac{1}{1 + \exp(\mathbf{x}_1^T \boldsymbol{\beta}_1)} \right)^{1/k}. \tag{10.10}$$

Now, regardless of whether we fit the ZIG model with a logit link on $\tilde{\theta}$ or k-ZIG model with link (10.10) on θ, the likelihood is independent of k. This parametrization makes it impossible to learn about k from the data. The above discussion clarifies the remark below (5); evidently, it is better to work with (10.9)).

10.2.2.4 Model fitting and inference

Suppose a sample such that the Y_i are independently distributed as k-ZIG$(\theta_i, \Theta_{0,i})$. Then for $(Y_i, \mathbf{x}_{1i}, \mathbf{x}_{2i}), i = 1, 2, ..., n$, the conditional likelihood is given by $L(\theta, \Theta_0, k; \mathbf{Y}) \equiv$

$$
\begin{aligned}
[\mathbf{Y}|\theta, \Theta_0, k] &= \Pi_{i=1}^n \left\{ (1 - (1 - \theta_i)^k)\delta_0 + (1 - \theta_i)^k G_0(y_i|\Theta_{0,i}) \right\} \\
\text{with } \theta_i &= \frac{\exp(\mathbf{x}_{1i}^T \boldsymbol{\beta}_1)}{1 + \exp(\mathbf{x}_{1i}^T \boldsymbol{\beta}_1)}.
\end{aligned}
\tag{10.11}
$$

where $\mathbf{x}_1$ denotes the covariates that are used to explain the Bernoulli zeros. With the ZIP model in (2), we assume $G(y|\Theta)$ to be Po(λ) and define a parametric form for λ,

$$\lambda = \exp(\mathbf{x}_2^T \boldsymbol{\beta}_2) \tag{10.12}$$

where $\mathbf{x}_2$ denotes the covariates that influence abundance and the $\boldsymbol{\beta}_2$ are corresponding regression parameters. A more flexible model is obtained when, instead of defining a parametric form for λ, we assume that the λs are driven by a mixture model. In particular, suppose we assume $\lambda \sim$ Gamma(a, b) and incorporate the covariates by setting

$$b = \exp(\mathbf{x}_2^T \boldsymbol{\beta}_2). \tag{10.13}$$

The marginal model obtained after integrating over λ is, therefore, the zero-inflated negative binomial. Two reasons for specifying G as a Poisson-Gamma mixture rather than using the negative binomial parametrization are: (i) modeling the integer valued parameter in the negative binomial using covariates is awkward and (ii) modeling the mean of the negative binomial through covariates is difficult due to the combinatoric expressions in its form.

So, returning to (10.11), we consider $G_0(y_i|\Theta_{0,i})$ as either a Po(λ_i) with $\lambda_i = \exp(\mathbf{x}_{2i}\boldsymbol{\beta}_2)$ or a Poisson-Gamma(a, b_i) with $b_i = \exp(\mathbf{x}_{2i}\boldsymbol{\beta}_2)$. Likelihood inference using the specification in (10.11) - (10.13) could be attempted but would prove very challenging. Since the model is hierarchical, it would more conveniently be fitted within a Bayesian framework. This results in a Bayesian model

$$\Pi_i[Y_i|\boldsymbol{\beta}_1, \boldsymbol{\beta}_2, k, X_{1i}, X_{2i}][\boldsymbol{\beta}_1][\boldsymbol{\beta}_2][k]. \tag{10.14}$$

which yields the hierarchical form

$$\Pi_i[Y_i|\boldsymbol{\beta}_1, \lambda, k, X_{1i}]\Pi_i[\lambda_i|a, \boldsymbol{\beta}_2, X_{2i}][\boldsymbol{\beta}_1][\boldsymbol{\beta}_2][a][k]. \tag{10.15}$$

10.2.2.5 Hurdle models

An important distinction in the literature is between the ZIP model and the Poisson hurdle model (Cameron and Trivedi, 1998; Welsh et al., 1996, Falk et al. 2015). The difference is simple to describe but its implications are primarily at the process level. That is, practically, it will be difficult to distinguish them in model fitting, because the latter can be viewed as a special case of the former.

The ZIP is a mixture model which allows zeros to come from two sources, as noted above. One source is the implicit Bernoulli trial, the other is through the Poisson law. A natural setting is modeling abundance for a rare species. We would expect zeros because some sampled habitats are unsuitable for the species. However, even within a suitable habitat, the species might not be present due to its rarity. The Poisson hurdle model allows only one source for zeros, a Bernoulli model, and then models the non-zeros as a Poisson truncated greater than 0. Once a "hurdle" is passed, presences will be observed. The natural setting here is a habitat suitability model which, for a given species, distinguishes the processes of colonization (presence/absence) from growth (abundance). These models are easier to fit because they enable separation of the likelihood for these two processes. In particular, we might adopt a logit or probit link for the Bernoulli component and a suitable choice of $G(y|\theta)$, truncated to $y = 1, 2, \ldots$ for the non-zero counts.

10.2.3 Zeros with continuous density $G(y)$

Tobit models were originally developed to model non-negative economic data (Tobin, 1958, Cameron et al. 2005). They are employed to model data that have been truncated in some way (see Amemiya, 1984, for a detailed review in the economic setting). Most frequently, they are used to model a truncation at 0 and in this regard are finding increased application to environmental settings, e.g., for modeling precipitation (Sahu et al. 2010), species distributions (Clark et al., 2014), and ecological traits (Clark, 2016a). Customarily, the truncated distribution has been a normal distribution function, with a regression specification incorporated.

Returning to the probit model of Section 2.1, we equate the event $Y = 1$ with the event that a normally-distributed latent variable W with unit variance is positive, $W > 0$. Hence, the event $Y = 0$ is associated with the event $W \leq 0$. In other words, Y is a function of W. With a probabilistic regression model, we have

$$W_i \sim N(\mathbf{x}_i^T \boldsymbol{\beta}, 1)$$
$$P(Y_i = 1) = P(W_i > 0) = \Phi(\mathbf{x}_i^T \boldsymbol{\beta})$$

where $\Phi(\cdot)$ is the standard normal cumulative distribution function, $\mathbf{x}_i$ is a vector of covariates, and β are corresponding coefficients. In terms of the vector of latent variables, $\mathbf{W}$, associated with the vector of observations, $\mathbf{Y}$,

$$[\mathbf{W}|\beta, \mathbf{Y}] = \prod_i^n N(W_i; \mathbf{x}_i^T \boldsymbol{\beta}) \times \mathcal{I}_i$$
$$\mathcal{I}_i = I(W_i > 0)^{Y_i} I(W_i \leq 0)^{1-Y_i}.$$

This model is fitted in the likelihood setting most easily using the E-M algorithm, treating the W_i's as missing data (Dempster et al. 1977). In the context of Bayesian model fitting using Gibbs sampling, the W_i are introduced as latent variables. They can be sampled from the truncated normal,

$$[W_i|\beta, Y_i = 0] = N(\mathbf{x}_i^T\beta, 1)I(W_i < 0)$$
$$[W_i|\beta, Y_i = 1] = N(\mathbf{x}_i^T\beta, 1)I(W_i \geqslant 0).$$

The probit is attractive in this context, because, with a normal prior distribution on β, the conjugacy with the normal distribution for $\mathbf{W}$ allows direct sampling of $\beta|\mathbf{W}$. Unit variance is assigned to the W_i in order to identify the β vector; otherwise, only β/σ is identified. This non-identifiability arises because with binary data there is no inherent scale.

The Tobit model extends the probit to continuous treatment of non-zero values. It entails a latent variable, W, through $Y = max(0, W)$, such that W is a random variable on R^1. The Tobit model differs from the probit in two regards. The indicator for the Tobit is $\mathcal{I}_i = I(w_i = y_i)^{I(y_i>0)}I(w_i \leq 0)^{I(y_i=0)}$. So, latent variables are drawn only for observed zeros, i.e., $w_i|\beta, y_i = 0 \sim N(\mathbf{x}_i^T\beta, \sigma^2)1(w_i \leq 0)$ while, given $y_i > 0$, $w_i = y_i$. The second departure from probit models is the need to include σ^2, hence to sample $\sigma^2|\beta, \mathbf{W}$. This arises because now the non-zero data have a *scale*. Again, within a Bayesian framework, Gibbs sampling is direct for both $\beta|\sigma, \mathbf{W}$ and $\sigma^2|\beta, \mathbf{W}$.

While the normal distribution for W is customary, other models, e.g., skew normals or t's can be considered in order to provide asymmetry and heavier tails. For instance, precipitation, when it occurs, might exhibit skewness to the right with regard to amount. Model fitting for skew normal or t models can again be easily handled within Gibbs sampling. For the t distribution, we merely scale mix normals. That is, if we define $W = \sigma U/\sqrt{V}$ where $U \sim N(0,1)$ and, independently, V comes from an inverse Gamma distribution, i.e., $V \sim IG(r/2, r/2)$, then $W \sim \sigma t_r$. For the skew normal, we take advantage of the representation of a skew normal through convolution of a unit normal and an absolute normal (Azzalini 2005). That is, if $W = \delta|V_0| + (1 - \delta)V_1$ where $V_0 \sim N(0,1)$ and, independently, $V_1 \sim N(0, \sigma^2)$, then W is said be distributed as skew normal, $SN(0, \sigma^2, \delta)$. $\delta \in (-1, 1)$ is the skewness parameter; $\delta = 0$ provides a $N(0, \sigma^2)$ distribution. Hence, we are always able to implement convenient normal-normal conjugacy.

A more general latent version specifies $Y = Y_{obs} \quad if \quad W \geq 0, Y = 0 \quad if \quad W < 0$. Now, Y is not a deterministic function of W; rather, Y follows a conditional distribution given W. This specification allows the outcome process, the data Y, to now be driven by levels of say a latent environmental process, W. An illustration to model wet deposition of sulfates and nitrates through precipitation is presented in Sahu et al. (2010). The point mass at 0 arises because there is no wet deposition without precipitation. Precipitation is driven by a latent environmental process.

10.3 Multinomial trials

Multinomial trials produce a response that is a vector whose entries consist of a single '1' with the rest being '0's. So, again, we will have the problem of zeros. In environmental and ecological settings the categories in the multinomial trials arise from one of two settings. In the first, categories are created in an ordinal fashion. In some versions there will be a scale upon which interval censoring has been imposed. In other versions, there will be an ordering of the categories which results from a latent, unobservable scale. This is the topic of Section 3.1. In the second setting the categories are nominal. Only one category can arise but there is no order to the categories. In the absence of a scale, models must be specified differently. This is the topic of Section 3.2.

A noteworthy point here is that the '0's we consider here are entries in the multinomial trial vectors. This is apart from the "labels" of the categories and whether one of these labels is '0'. Also, here we consider a multinomial trial as an observation on a single species or individual at say a given site. When we turn to multivariate modeling in Section 5, we are thinking in terms of observations for multiple species or individuals at a site, anticipating dependence between these site-level observations.

10.3.1   Ordinal categorical data

Recalling the introduction to Section 2, following either of the two paths there, for non-negative data we can partition the positive values, i.e., $(0, \infty)$ into say, K ordered sets, e.g., $(0, a_1], (a_1, a_2], \ldots, (a_{K-1}, \infty)$. This setting can arise when positive observations are aggregated based on a partition, but their exact values are lost (hence, they are interval-censored). Then, we must build a model to explain incidence on the partition. In order to work with standard distributions, we usually introduce them using latent variables with a latent partition. Again, we still seek to capture the point mass at 0 well in addition to explaining the observed counts over the partition. As a different version, consider species abundance data, which is recorded as 'absent', 'rare', 'intermediate', 'dominant'. Podani (2005) discusses data of this type. Here, most ecologists would agree that the first class fits the concept of zero, or 'none'.

However, the concept of zero is not always relevant for ordinal data. Consider animal body condition recorded as, say, 'poor', 'fair' and 'good' (Schick et al. 2013). We can think of these classes as ordered but there is no notion of 'zero.' The five stages of budbreak identified by Norby et al. (2003) are: (no activity, buds swelling, buds just opening, leaves unfolding, leaves curled). Here the concept of zero is unclear. We might view the first stage as 'zero development', i.e., 'none'. Alternatively, we might see it as simply a different stage of development, 'dormant', that does not entail the concept of zero. The ordinal classes for shade-tolerance, flood-tolerance, and drought-tolerance (Baker 1949, Niinemets and Valladares 2006) could arguably assign to the first class the notion of 'zero-tolerance', but typically it would be interpreted as simply the 'lowest' class of tolerance and not involving the notion of zero. Finally, recovery from treatment, recorded as 'much worse', 'worse', 'no change', 'better', or 'much better', implies order. Here, 'no change' might be viewed as a zero. In any event, in each of these cases, data can be modeled as ordinal data, but the concept of zero differs; with ordered data, zero may or may not conform to the notion of 'nothing'.

Ordinal data can be modeled by extending the probit. We can think of the ordinal scores as associated with intervals that create a partition that segments the real line. With K ordinal categories, we specify the partition as $(u_0 = -\infty, u_1, u_2, \ldots, u_{K-1}, u_K = \infty)$. Because there may be no inherent location to the observations, the scale can be anchored at zero, with the first interval having a partition that bounds negative values, i.e., $u_1 = 0$ and $(u_0, u_1] = (-\infty, 0]$. (In fact, this is needed for identifiability. Otherwise, the latent partition could only be identified up to translation.) Whether or not the first level is called 'zero', this partition can be used. The interval for the second level has partition $(u_1, u_2] = (0, u_2]$. Continuing in this fashion, the last interval has partition (u_{K-1}, ∞) This is McCullagh's (1980) model for ordinal values, where all but the first two and the last elements must be inferred (Lawrence et al. 2008). As with the probit modeling, we envision a latent variable that lies within an interval. Together, the contiguous intervals provide support over the real line, which, recalling the notation of w's for these variables, could be written this way:

$$w_i|y_i \sim N(\mathbf{x}_i\boldsymbol{\beta}, 1)\mathcal{I}_i$$

$$\mathcal{I}_i = \prod_{i=1}^{K} I_{ik}^{I(y_i=k)}(1 - I_{ik})^{I(y_i \neq k)}.$$

where $I_{ik} = I(u_{k-1} < w_i \leqslant u_k)$. Like with the probit, the unit variance scale is imposed.

As with the probit and Tobit models, under Bayesian model fitting, the latent w_i can be generated by Gibbs sampling from truncated normals. In other words, the data are an interval censored version of the latent w_i. However, the partition is unknown and thus is sampled conditional on the latent variables, i.e.,

$$u_k|W, Y \sim unif\left(max_i(w_i|y_i = k), min_i(w_i|y_i = k+1)\right) \tag{10.16}$$

In cases where 'zero' (or any other) class dominates, this latent-variable approach can describe data well. The discrete observations can be asymmetric and skewed, accommodated by the partition (Lawrence et al. 2008).

Ordinal data are part of the analysis described in Section 10.6.2. We preface that analysis here to show examples of an estimated partition for three ordinal tolerance classes: flood, drought, and shade (Fig. 10.3). The figure shows the posterior distribution of the cut points that create the partition. Clear separation for the first three levels obtains, because these levels are abundant in the data. The upper classes are rare and, thus, poorly resolved. We note that this figure does not show joint realizations of the u's; rather, it shows the marginal distributions of the realizations, obtained by marginalizing the joint distribution.

10.3.2 Nominal categorical data

A categorical response with only two outcomes/levels can be modeled analogously to presence-absence data (Section 2.1). When there are multiple unordered outcomes/levels we are observing multinomial trials. With K outcomes/levels, each observation is a $K \times 1$ vector. One of the levels is redundant, because only one level can have a 1, the other $K-1$ levels being 0. Therefore, with n observations, designating one level as a *reference*, we can organize the observed responses as an $n \times K - 1$ matrix $\mathbf{Y}$, each row $\mathbf{y}_i$ having at most one 1. In terms of a latent variables, for each i, we now have one for each k. That is, we have $w_{i,k}$ such that

$$y_{i,k} = \begin{cases} 1, & w_{i,k} > 0 \wedge w_{i,k} \geq max_{k'}(w_{i,k'}) \\ 0, & w_{i,k} < max_{k'}(0, w_{i,k'}) \end{cases} \tag{10.17}$$

where $k' = 1, \ldots, K-1$. Rows having all zeros indicate that the reference level was observed. Collecting the $w_{i,k}$ into a vector $\mathbf{w}_i$, we now adopt a multivariate normal specification,

$$\mathbf{w}_i|\mathbf{y}_i, \mathbf{x}_i, \sim MVN(\mathbf{Bx}_i, \boldsymbol{\Sigma})\mathcal{I}_i \tag{10.18}$$

where $\mathbf{B}$ is a $K - 1 \times Q$ matrix of coefficients, $\boldsymbol{\Sigma}$ is a $(K-1) \times (K-1)$ correlation matrix, and $\mathcal{I}_i$ is defined below. This is a multinomial probit specification (Greene 2000), adopting the assumption that the observed level has the largest value of w_{ik}. The indicator can be written as

$$\mathcal{I}_i = \prod_{k=1}^{K-1} [I(w_{ik} > 0)I(w_{ik} \geq max_{k'}(w_{ik'})]^{y_{ik}} \times I\left(w_{ik} < max_{k'}(0, w_{ik'})\right)^{1-y_{ik}} \tag{10.19}$$

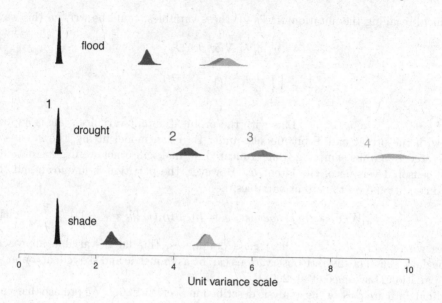

FIGURE 10.3
The estimated partition for ordinal traits, tolerance of flood, drought, and shade. Each has five classes from 'no tolerance' to 'high tolerance'. Posterior estimates of (u_1, u_2, u_3, u_4) are labeled as $1, 2, 3, 4$ and separate the five classes. Poor estimates of upper values result from the fact that they rarely occur in the data set. Despite overlap for upper values, they are ordered – in any given realization the u's are ordered so u_3 is always less than u_4 (Clark 2016a).

The multinomial probit for categorical responses is most easily fit within the Bayesian framework treating the $w_{i,k}$ as latent variables (McCulloch et al. 2000, Zhang et al. 2008). However, the model is not identified. The scale is centered by the fact that the observed response is bounded by zero, and all other levels must be less than that observed. Still, the parameters are not identified because the likelihood is unchanged upon multiplication by a constant. One approach entails fixing one element of the covariance matrix (McCulloch et al. 2000). Alternatively, the latent variables can be drawn on the correlation scale. Let $\tilde{\mathbf{w}}_i = \mathbf{V}^{-1/2}\mathbf{w}_i$ with $\mathbf{V} = diag(\mathbf{\Sigma})$. Then,

$$\tilde{\mathbf{w}}_i | \mathbf{y}_i, \sim MVN(\mathbf{V}^{-1/2}\mathbf{B}\mathbf{x}_i, \mathbf{R})\mathcal{I}_i \qquad (10.20)$$

where $\mathbf{R} = \mathbf{V}^{-1/2}\mathbf{\Sigma}\mathbf{V}^{-1/2}$. Parameters are sampled as $\mathbf{B}|(\mathbf{W}, \mathbf{\Sigma})$ and $\mathbf{\Sigma}|(\mathbf{W}, \mathbf{B})$. As a result, the covariance matrix can be sampled as an inverse Wishart, then transformed to a correlation matrix $\mathbf{R}$. This *parameter expansion* is discussed in the Section 5.

10.4   Spatial and spatio-temporal versions

In many environmental and ecological applications, data is collected over space and/or time. Here too the challenge of over-abudant zeros is an issue (Lyashevska et al. 2016). Now, extending Section 2, our responses will be denoted by $Y(\mathbf{s})$ or $Y(\mathbf{s}, t)$. All of the modeling issues of that section are still in play. However, additionally, we will now need to incorporate dependence in space and in time between the responses. Chapter 5 develops general modeling for the spatial setting, so-called geostatistical data. The theme that emerges is the use of Gaussian processes to introduce spatial random effects which, in turn, create spatial dependence among the $Y(\mathbf{s})$. The Gaussian process provides a valid covariance function which, in the simplest case, captures decay in dependence as a function of distance. In the context of Section 2, since the latent W's are assumed normally distributed, this leads to the introduction of a latent Gaussian process, $W(\mathbf{s})$, over a suitable spatial region. If we need a latent vector at a location as in Section 3, then we will need a multivariate Gaussian process requiring a valid cross-covariance function (again, see Chapter 5). For example, we might observe species abundance at a collection of plots within a region. We would anticipate that presence/absence or abundance at one site might be stochastically dependent on presence/absence or abundance at a nearby site.

If we bring in time, we find two paths. One extends the geostatistical model to space × time. Then, time is viewed as continuous over some window. Now, we introduce a space-time Gaussian process with a valid space-time covariance function. This function enables inclusion of space-time dependence (See Chapter 5 and also Banerjee et al. 2014). Again, we provide this process for the latent variables, now $W(\mathbf{s}, t)$. For example, with monitoring stations, we may record whether or not it is raining (a *0-1* outcome) at a given location and time. We would expect that, if it is not raining at time t at location $\mathbf{s}$, then it is probably not raining at time $t + \Delta$ for Δ small (temporal dependence). We would expect that, if it is not raining at time t at location $\mathbf{s}$, then it is probably not raining at location $\mathbf{s}'$ at time t if $\mathbf{s}'$ is near $\mathbf{s}$.

The alternative path arises when time is discretized, e.g., daily precipitation. Then, we find ourselves in the setting of dynamic or state space models (See Chapter 4 as well as West and Harrison 1997). We now denote the latent variables by $W_t(\mathbf{s})$. Again, Gaussian processes will be employed to capture spatial dependence while temporal dependence will typically be introduced autoregressive specifications (see Chapter 3). As an example, we might consider a demographic setting. Suppose we look at the size distribution of individuals at a given site in a given year. Then, in the next year, we would expect the size distribution to depend upon the current size distribution. Additionally, we might expect similar size distributions at sites that are near to each other in space.

Each of the foregoing settings, in the context of the problem of zeros, will lead to model specifications that are hierarchical (see Chapter 2). We will have a first stage data model with a second stage Gaussian process specification and third stage (hyper)parameters. Model fitting and inference will most naturally be implemented using Markov chain Monte Carlo in a Bayesian framework.

10.5   Multivariate models with zeros

Most environmental data sets are multivariate, in the sense that more than one response variable is collected in an observation at a location. The responses could be observations associated with multiple species at a sample location. They could be multiple traits, where trait values are taken to be weighted mean values, the weights coming from the relative abundances of species that possess those traits (Doledec et al. 1996, Ackerly and Cornwell 2007, Albert et al. 2010, Kleyer et al. 2012). Alternatively, the responses could be multiple attributes of an individual, such as its health status (Fig. 10.1*a*) or its microbiome composition (Fig. 10.1*b*). In environmental settings, the responses could include pollutant concentrations of chemical species or particle size classes, such as $PM_{2.5}$. In these last settings, perhaps upon transformation, we often can use multivariate normal models, as we briefly discuss in Section 5.1. In Sections 5.2 and 5.3, we focus on multivariate versions of the models in those sections.

The emergence of heterogeneous datasets fuels a growing need for models that can accommodate many types of data jointly, including an excess of zeros (Fig. 10.1b, c). The joint distribution is complicated by the fact that the different response variables are often measured on different scales, e.g., a combination of presence-absence, discrete counts, continuous abundance, composition, ordinal scores, and categorical attributes. In Figure 10.1a host status is binary, while its microbiome is compositional (Fig. 10.1b). Zero-accommodating models that extend to multivariate data, include the logit (Ovaskainen et al. 2010), probit (Chib and Greenberg 1998, Pollock et al. 2014), Tobit (Sahu et al. 2010, Clark et al. 2014), categorical (Zhang et al. 2008), and mixed response types (Clark et al. 2014, 2016a). To motivate this synthetic view, we begin with the multivariate normal distribution for continuous responses, clarifying its limitations as a data model in the context of the problem of zeros. Then, we briefly consider joint species distribution models. Finally, we move to an overarching framework that accommodates these various data types in a general way.

10.5.1   Multivariate Gaussian models

In this subsection, we focus on the situation of a vector of continuous observations taken at every site. Again, an example of an application would be the collection measurements of environmental contaminants at at given site. In particular, the components of particulate matter, say $PM_{2.5}$ - called species - include sulfates, nitrates, total carbonaceous matter, ammonium, and fine soil, along with "other" might be recorded (Rundel et al. 2016). Upon suitable transformation of the components of the vector, we might adopt a multivariate normal distribution for the measurements at each site. Similarly, for an individual, we might measure a set of continuous characteristics or traits, e.g., for a tree, we might measure diameter, height, leaf characteristics, phenological status, and so forth. Also, with community-weighted trait data for a set of continuous traits, we would obtain multivariate continuous measurements. Again, upon transformation, we might adopt a multivariate normal model. Introducing site-level predictors, we could then specify

$$\mathbf{y}_i \sim MVN(\mathbf{B}\mathbf{x}_i, \Sigma) \quad i = 1, 2, ..., n \tag{10.21}$$

This model is identical to (10.18) except it is written for observed rather than latent data. This model is standard in the literature providing easily interpreted coefficents and familiar pairwise dependence structure.

The key point here is that we have no point masses. This model is not directly applicable to the 'problem of zeros.' To be more precise, suppose we are interested in species distributions and we wish to model them jointly. (Certainly, with regard to presence/absence or abundance there would be interspecific dependence at a given site.) Then, even if we observed continuous measurements associated with each species, e.g., biomass, basal area, or percent coverage, the model in (10.5.1) could never be used. If species s was not observed at site i, we would assign a '0' biomass, basal area, or percent cover to species s. If the measurement was viewed as some continuous trait associated with the species, it would have to be recorded as 'NA', not available. Hence, for the problem of zeros, we can only imagine (10.18) as a latent model.

10.5.2 Joint species distribution models

Customarily, species distribution models (SDMs) are fitted independently across a collection of species (Elith et al. 2006, Ferrier et al. 2007, Chakraborty et al. 2010, Calabrese et al. 2014). For multiple species, to make predictions at the community level, individual models are aggregated or *stacked* Calabrese et al. 2014). However, collectively, the independent models tend to imply too many species per location (Baselga and Araujo 2010). It is easy to find further situations in which this approach may yield misleading results (Clark et al. 2014). The problem is evident. Individual SDMs ignore the fundamental notion that the distribution and abundance of species is a joint process which involves species simultaneously (e.g., through competition, mutualism, etc.) rather than an independent one for each. Modeling species individually does not allow underlying joint relationships to be exploited (Latimer et al. 2009, Ovaskainen et al. 2011).

Joint species distribution models (JSDMs) that incorporate species dependence have only recently been considered, including applications to presence-absence (Ovaskainen et al. 2010, Ovaskainen and Soininen 2011, Pollock et al. 2014), continuous or discrete abundance (Latimer et al. 2009, Thorson et al. 2015), abundance with large numbers of zeros (Clark et al. 2014), multinomial models for count composition data (Haslett et al. 2006, Paciorek and McLachlan 2009, de Valpine and Harmon-Threatt 2013). Very recently, methods extending to discrete, ordinal, and compositional data have been proposed (Clark 2016a). JSDMs directly address the problem of zeros by jointly characterizing the presence and/or abundance of multiple species at a set of locations. They partition the drivers into two components, one associated with environmental suitability, the other accounting for species dependence through the *residuals*, i.e., adjusting for the environment. These recent JSDMs are specified through hierarchical models, introducing latent multivariate normal structure, capturing dependence through the associated covariance matrices.

There are two strategies for introducing latent multivariate normal structure. One is to introduce dependence between species at the *first* stage, viewing the data vector at site i, $\mathbf{Y}_i$ as a componentwise function, $Y_{ij} = g(w_{ij})$, of $\mathbf{w}_i$ where $\mathbf{w}_i$ is latent multivariate normal. In practice, the data will have at least one atom - at 0. For example, with biomass, $Y_{ij} = w_{ij}$ if $w_{ij} > 0$ and $w_{ij} = 0$ if $w_{ij} \leq 0$. This is the foregoing Tobit specification. With binary response (presence/absence), $Y_{ij} = 1$ if and only if $w_{ij} > 0$, i.e., $g(w_{ij}) = 1(w_{ij} > 0)$. The second, say with presence/absence, is to model dependence between the (random) probabilities of presence, e.g., between $P(y_{ij} = 1)$ and $P(y_{ij'} = 1)$. This would move the dependence to the *second* modeling stage, i.e., in the *mean* specification, introducing a probit link.

10.5.3 A general framework for zero-dominated multivariate data

The problem of zeros has been addressed as part of a general framework for heterogeneous data with generalized joint attribute modeling (GJAM) (Clark et al. 2016a, Taylor-Rodriguez et al. 2016). As noted above, in ecology, an observation vector $\mathbf{y}_i$ can consist of both continuous and discrete observations on, say $s = 1, ..., S$ species or attributes. It only makes sense to link them through a covariance matrix if they can be represented on a similar scale. Moreover, if we can essentially avoid non-linear link functions, there is an added advantage for interpretation. We summarize a framework that accommodates differences in effort, both between response variables s in the same observation i, and between observations i. The framework not only allows zeros as a point mass, but can also inflate the probability of a zero.

10.5.3.1   Model elements

An observation vector $\mathbf{y}_i$ can include both continuous and discrete observations. At the first stage there is an $S \times 1$ vector $\mathbf{w}_i \in \Re^S$ that is multivariate normal. Each element of $\mathbf{w}_i$ corresponds to a matching element in vector $\mathbf{y}_i$. That is, $\mathbf{y}_i$ is a *function* of $\mathbf{w}_i$. For continuous observations in $\mathbf{y}_i$ the matching element w_{is} is known and equal to y_{is}. For discrete y_{is} the matching element w_{is} is known only to lie within an interval. The set of all intervals is a partition of continuous.space, $\mathcal{P}_{is}$ at points $u_{is,k}$. Each response y_{is} is assigned a label $k = 0, \dots, K_{is}$ for an interval in the partition. In other words, w_{is} occupies the kth interval in a partition. The number of intervals can differ between observations and between species, because each response can be observed in different ways. For discrete observations k is a censored interval, and w_{is} is a latent variable. Thus, a subset of intervals can be censored, $\mathcal{C} \in \{k = 0, \dots, K\}$, including only those associated with discrete y_{is}. For nominal variables in $\mathbf{y}_i$ the vector $\mathbf{w}_i$ is increased in length to hold each of the non-independent categories, or 'levels', in $\mathbf{y}_i$ (see below).

Latent variables have the model

$$\mathbf{w}_i | \mathbf{x}_i, \mathbf{y}_i, \mathbf{u}_i \quad \sim \quad MVN(\boldsymbol{\mu}_i, \boldsymbol{\Sigma}) \times \prod_{s=1}^{S} \mathcal{I}_{is} \tag{10.22}$$

$$\boldsymbol{\mu}_i = \mathbf{Bx}_i$$

$$\mathcal{I}_{is} = \prod_{k \in \mathcal{C}} I_{is,k}^{I(y_{is}=k)} (1 - I_{is,k})^{I(y_{is} \neq k)}$$

where $I_{is,k} = I(w_{is} \in (u_k, u_{k+1}])$, $\mathbf{B}$ is an $S \times Q$ matrix of coefficients and $\boldsymbol{\Sigma}$ is a $S \times S$ covariance matrix.

Returning to the Tobit example, the first interval $k = 0$ is censored by $(u_0, u_1] = (-\infty, 0]$. It is a censored interval, because $y_{is} = 0$ is a discrete observation. The second interval has partition $(u_1, u_2) = (0, \infty)$. This second interval is not censored, because $y_{is} > 0$ is continuous. In the Tobit example, the set $\mathcal{C}$ is the single interval $k = 0$. Differences between data types are discussed later in this section.

10.5.3.2   Specific data types

When multivariate data are all of one type, previous models arise as special cases of GJAM. For *presence-absence data* GJAM is Chib and Greenberg's (1998) multivariate probit model, with partition $\mathbf{u} = (-\infty, 0, \infty)$. This is a multivariate extension of (10.2). The fitted correlation matrix $\boldsymbol{\Sigma}$ in (10.22) describes the tendency for species to be present or absent together, after accounting for the effects of $\mathbf{X}$, which are captured by coefficients in $\mathbf{B}$. This model has been applied to ecological data by Pollock et al. (2014).

Multivariate *categorical data* fit within the same framework. The univariate case of Section 10.3.2 already has a $K - 1$ correlation matrix, where K is the number of levels in a categorical response (10.19). Multivariate categorical responses arise when each species is assigned a categorical value. GJAM is similar to Zhang et al.'s (2008) model, using a set of columns for each categorical response s. Let $w_{is,k}$ refer to the latent value for the kth level of response s. For multiple responses,

$$y_{is,k} = \begin{cases} 1, & w_{is,k} > 0 \wedge w_{is,k} = max_{k'}(w_{is,k'}) \\ 0, & w_{is,k} < max_{k'}(0, w_{is,k'}) \end{cases} \tag{10.23}$$

where $k' = 1, \dots, K_s - 1$, and K_s is the number of levels for response s. The first line of eqn 10.23 specifies that the observed class must not only be positive, but also the largest $w_{is,k}$ value in the $K_s - 1$ columns. If all values are negative, then the reference class is the observed class. The matrix $\mathbf{W}$ has columns for each level of each response, $\mathbf{w}_i = (w_{i1,1}, \ldots, w_{i1,K_1-1}, \ldots, w_{iS,1} \ldots, w_{iS,K_S-1})$. The $\sum_s K_s - S$ rows and columns of correlation matrix $\boldsymbol{\Sigma}$ hold correlations for all combinations of s and k. The indicator corresponding to 10.19 for the multivariate model (10.22) is

$$\mathcal{I}_{is} = \prod_{k=1}^{K-1} [I(w_{is,k} > 0)I(w_{is,k} = max_{k'}(w_{is,k'}))]^{y_{is,k}} \tag{10.24}$$

$$\times I\left(w_{is,k} < max_{k'}(0, w_{is,k'})\right)^{1-y_{is,k}} \tag{10.25}$$

For multivariate *ordinal data* GJAM is Lawrence et al.'s (2008) model, having partition $\mathbf{u}_{is} = (-\infty, 0, u_{is,2}, u_{is,3}, \ldots, u_{is,K}, \infty)$. Section 10.3.1 describes a partition for a single species. The multivariate partition has elements $u_{is,k}$ that differ for each species, both in number and location. Because there is no scale, all but the first two and the last elements must be inferred. In fact, the ordinal partition estimates in Figure 10.3 come from GJAM (Section 10.6).

The overwhelming number of zeros in count data (Fig. 10.1) can likewise be modeled under this framework. For *discrete abundance data* the partition for observation i can be defined to account for sample effort. Where effort $E = 1$, the partition for counts $0, 1, 2, \ldots, max(y_{is})$ is set at midpoints between these values, $\mathbf{u} = (-\infty, 1/2, 3/2, \ldots, max(y_{is}) - 1/2, \infty)$. When effort varies between observations the partition shifts to the 'effort scale',

$$\mathbf{u}_i = \left(-\infty, \frac{1}{2E_i}, \frac{3}{2E_i}, \ldots, \frac{max_s(y_{is}) - 1/2}{E_i}, \infty\right) \tag{10.26}$$

where effort E_i is plot area, search time, and so forth. This partition places W on the scale of Y/E. For example, if plot area is measured in hectares, then W has units of ha^{-1}. Note that the second value in (10.26), $\frac{1}{2E_i}$, allows zero inflation. Because this value is greater than zero, the latent value W can assume positive values when observed $Y = 0$. As effort E_i increases, this value approaches zero and the probability of a false-negative declines. To exclude this type of zero-inflation, the second value can simply be assigned the value 0.

For *composition count data* effort is the total count for the observation, $E_i = \sum_s y_{is}$. Only the last finite value from (10.26) is simplified to reflect the fact that effort is the total count for observation i,

$$\mathbf{u}_i = \left(-\infty, \frac{1}{2E_i}, \frac{3}{2E_i}, \ldots, 1 - \frac{1}{2E_i}, \infty\right) \tag{10.27}$$

This partition places W on the composition scale. Although the observations range from 0 to E_i, only relative abundance is important. Thus, the composition scale for W is interpretable in terms of relative abundance.

For *fractional composition data*, where only the fraction of the total is known, data are continuous, with point mass at 0 (any response that is absent) and 1 (when only one response is present), with partition

$$\mathbf{u}_i = (-\infty, 0, 1, \infty) \tag{10.28}$$

In this case, only the first and last intervals are censored. Taking $E = 1$ for all observations we have $Y = W$ for $W \in (0, 1 - \epsilon)$ for a small ϵ and introduce a nonlinear transformation $Y = g(W)$ from $(1 - \epsilon, \infty)$ to $(1 - \epsilon, 1]$.

In summary, the GJAM framework can combine data types, accommodate zeros, and model on the observation scale. Each data type involves a coefficient matrix $\mathbf{B}$ and a covariance matrix $\boldsymbol{\Sigma}$. Depending on a partition $\mathcal{P}$, which can incorporate effort E, parameters generate continuous W. In the case of ordinal data the partition is also estimated. Evidently, GJAM specifies a hierarchical model which can be readily fitted with Markov chain Monte Carlo. Due to linearity at the first stage, the coefficients, covariance, and latent W can be simulated directly with Gibbs sampling. For presence-absence, ordinal, and categorical data, Σ is a correlation matrix (diagonal elements are 1's). This 'unit-variance scale' is imposed because these data types have no absolute scale; they do have a relative scale and, thus, a meaningful correlation matrix. Direct samples are still available through parameter expansion (Section 10.3.2). We see that the partition and selective use of parameter expansion allows modeling with (10.22), where each column of $\mathbf{Y}$ can be a different data type. In the application that follows we show how it applies to combined data.

10.6   Joint Attribute Modeling Application

Here, we present two examples involving many zeros and multiple data types. First, we offer a simple application, with only two data types - presence/absence and effort based fractional composition. Then, a second application considers tree abundance and 12 community-weighted traits comprised of three different data types.

10.6.1   Host state and its microbiome composition

A study of environmental effects on seedling host status and endophytic microbiome illustrates a multivariate response with massive zeros. To determine the relationship between host plants and the endophytes that infect them, seedlings were grown under a range of moisture, temperature, light, and competitive settings (Hersh, Benetiz, Vilgalys, and Clark, in prep). We mention this example only briefly to show the benefits of this framework for prediction.

Responses include host condition and the endophyte microbiome (Table 10.1). The condition of each of $n = 762$ seedlings was recorded as 'healthy' or 'morbid' (Fig. 10.2a). The endophyte microbiome was quantified as count composition data by molecular sequencing of each seedling. Zeros are massively overrepresented in these data (Fig. 10.2b). The response matrix $\mathbf{Y}$ includes one column for the binary host state (presence-absence), followed by 175 columns of composition data, one for each operational taxonomic unit (OTU). The reference class for OTU composition data is taken to be the sum of counts for all OTUs that occur in less than 50 samples. In GJAM the full matrix is modeled with eqn 10.22. For the first column the indicator $I_{is,k}$ depends on the partition for presence-absence ($\mathbf{u} = (-\infty, 0, \infty), \mathcal{C} = \{0, 1\}$). For remaining columns it depends on the partition for composition count data ($\mathbf{u}_i$ from eqn 10.27, $\mathcal{C}_i = \{0, 1, \ldots, E_i\}$).

Focusing on prediction for this first example, Figure 10.2, shows the improvement for host status when it is modeled jointly with the microbiome. The two panels in Figure 10.2 are both based on the probit, (a) being a standard univariate GLM and (b) as discussed in section 10.5.3.2. They have the same design matrix.

10.6.2   Forest traits

The geographic distribution of tree species and of 12 trait values for trees on $n = 1617$ plots in eastern North America combines variables that include large numbers of zeros. Two analyses address multivariate data with large numbers of zeros. First, tree data were modeled as composition abundance data, with biomass per species being a continuous variable, with

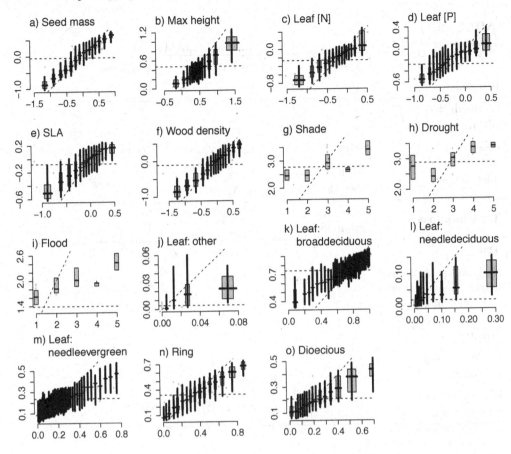

FIGURE 10.4
Predicted versus observed CWMM trait values for data types described in Table 28.1 for the joint model using `gjam`. Poor predictive capacity for ordinal data in this case (g, h, i) results from the small variation in modal values for each plot. The four fractional composition data types for leaf habit are predicted well for common (k, m), but not rare (j, l) data types.

point mass at zero (Table 10.1). Tree abundance data come from the USFS Forest Inventory and Analysis (FIA) plots in 31 eastern states, aggregated in covariate space to 1 ha aggregate plots. All trees with a diameter at breast height (DBH) of at least 12.7 cm were measured on four 7.2 m radius subplots. Collectively, the species abundances are mostly zeros (Figure 10.1c). Environmental predictors include a moisture index from the FIA, the hydrothermal deficit HTD (degree hours with a negative water balance, $P < PET$), and winter temperature (Clark et al. 2016b).

A second analysis of trait data was completed on community weighted mean/modal (CWMM) trait values (Table 10.1), obtained by weighting trait values for a species by the abundance of each species in each sample plot, i.e.. The first six *continuous* traits in Table 10.1 (seed mass, maximum height, ...) are simply centered and standardized by the standard deviation, as is common practice for such variables. Three *ordinal* tolerance classes follow. Because these scores have no scale, there is no community weighted mean value. However, there is a community-weighted modal value, the modal score for each plot. Four leaf types, two dioecy categories (one sex or two), and two xylem categories (ring-porous or

TABLE 10.1

Response variables and data types for applications of GJAM

Response	Data type	Partition $\mathcal{P}$	Censored $\mathcal{C}$
Endophyte microbiome			
host status	binary	$(-\infty, 0, \infty)$	$\{0, 1\}$
microbiome	count composition	$(-\infty, \frac{1}{2E_i}, \ldots,$ $1 - \frac{1}{2E_i}, \infty)$	$\{0, 1, \ldots, E_i\}$
Tree biomass	continuous abundance	$(-\infty, 0, \infty)$	$\{0\}$
Traits			
seed mass, maximum height, leaf nitrogen, leaf phosphorus, specific leaf area, wood density[2]	continuous	$(-\infty, \infty)$	$\varnothing^1$
shade, drought, flood tolerance	ordinal	$(-\infty, 0, u_{s,2}, u_{s,3},$ $u_{s,4}, \infty)^3$	$\{0, \ldots, 4\}$
leaf type, dioecy, ring-porous xylem[4]	fractional composition	$(-\infty, 0, 1, \infty)\}$	$\{0, 2\}$

[1] No censored intervals
[2] Centered and standardized variables
[3] Includes points that are estimated
[4] Include 4 leaf types, 2 levels of dioecy, and 2 levels of xylem anatomy

not) are *categorical* variables–each individual is assigned to one of the potential classes. As community-weighted mean values for plots, categorical variables become *fractional composition*. The four composition classes for leaf type occupy three columns in $\mathbf{Y}$, the 'other' class serving as the reference. The two binary variables, dioecy and xylem anatomy, each occupy one column in $\mathbf{Y}$. As in the foregoing example with two data types (Section 10.6.1), data of all types in Table 10.1 are combined in response matrix $\mathbf{Y}$, with the partition differing by data type as discussed in Section 10.5.3.2. The variables 'leaf type', 'dioecy', and 'ring-porous' include large numbers of zeros. Parameter expansion is used in Gibbs sampling of ordinal tolerance classes (Section 10.3.2). The joint model accommodates the zeros, while still providing excellent predictions of traits.

Prediction capacity from the model is substantial (Fig. 10.4). All continuous variables are accurately predicted (Fig. 10.4$a - f$). Ordinal tolerance classes are not well predicted (Fig. 10.4$g - i$), largely because community weighted modal values vary little across the data set. This limited variation explains the inability to resolve the partition for the upper tolerance classes in Figure 10.3. Fractional composition data are well predicted for the abundant classes (Fig. 10.4k, m, n, o), but not for rare classes (Fig. 10.4j, l).

To compare how coefficients for the tree biomass model relate to those for the trait model, fitted coefficients are shown for temperature in Figure 10.5. Recall, tree biomass in the tree model represent the weights used to obtain CWMM values in the trait model. The agreement between them is apparent. High wood density and ring-porous anatomy are

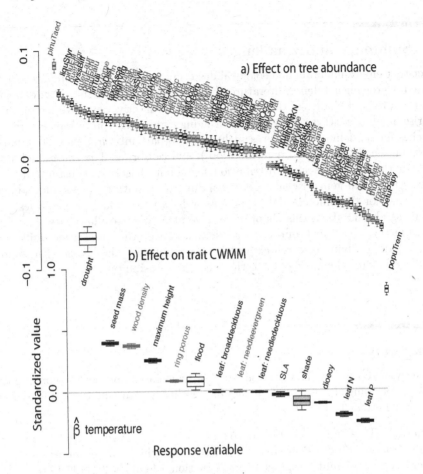

FIGURE 10.5
a) Regression coefficients for tree biomass responses to temperature, highlighting species that have ring-porous anatomy (brown), evergreen leaves (green) and high shade tolerance (grey). b) The coefficients for CWMM trait values as response variables (Table 28.1). Traits follow shading in part (a). Part (a) only includes species for which zero is outside the 95% credible interval.

most common at high temperatures (positive values in Figure 10.5b), as are the species that possess those traits (brown shading in Figure 10.5a). Needle-leaf evergreens show no effect of temperature (Fig. 10.5b), despite the fact that species that possess those traits have strong temperature affiliations, albeit offsetting in direction (green in Figure 10.5a). Shade-tolerant species (grey) do not strongly align with temperature, with a slight tendency toward cool climates. As a trait value, shade tolerance has a modest negative response (Fig. 10.5b).

By modeling close to the observation scale GJAM coefficients have direct interpretation. For data that lack an absolute scale (presence-absence, ordinal, categorical) unit variance is imposed. For other data types, the 'link function' is effectively linear (Clark et al. 2016). Regression coefficients **B** have units of Y/X. The coefficients in Figure 10.5 are standardized by the standard deviation in Y, thus showing relative responses The covariance matrix likewise has elements with different scales, each element $\Sigma_{s',s}$ being a product of scales $Y_{s'}Y_s$. Comparisons can be done on the correlation scale.

10.7   Summary and Challenges

The fact that environmental and ecological data sets are frequently dominated by zeros has motivated a growing modeling literature. In addition, this literature is now beginning to confront the related challenge that most datasets are multivariate. We have clarified that zeros arise in many contexts, each context bringing its own interpretation. The two principal approaches for modeling excess zeros include a two-part mixture, where one part inflates the zeros present in the other ('zero-inflation'), and censoring. The traditional alternatives of inflating the zero class in a distribution for counts has been substantially broadened to address the more general challenge of handling joint data types for individuals (where '0's can occur for the different data types) as well as joint modeling of species (where '0's can occur as absences). In this chapter we have attempted to illuminate a portion of the literature in this area. In the previous section, we have attempted some unification. Still, this is a rapidly evolving research area and the presentation here along with the references we have offered only open the door to the modeling possibilities.

Bibliography

[1] Ackerly, D. D. and W. K. Cornwell. 2007. A trait-based approach to community assembly: partitioning of species trait values into within- and among-community components. Ecology Letters 10:135-145.

[2] Albert, C.H., Thuiller, N. G. Yoccoz, R. Douzet, S. Aubert, and S. Lavorel. 2010. A multi trait approach reveals the structure and the relative importance of intra vs. interspecific variability in plant traits. Functional Ecology 24:1192-1201.

[3] Albert, J. H. and Chib, S. 1993. Bayesian analysis of binary and polychotomous response data. Journal of the American Statistical Association, 88:669-679.

[4] Alig, R., Lewis, D.J., and Swenson, J.J. 2005. Is forest fragmentation driven by the spatial configuration of land quality? The case of western Oregon. Forest Ecology and Management 217: 266-274.

[5] Augspurger, C. K. 1984. Seedling survival of tropical tree species - interactions of dispersal distance, light-gaps, and pathogens. Ecology 65:1705-1712.

[6] Azzalini, A. 2005. The skew-normal distribution and related multivariate families, Scandinavian Journal of Statistics, 32, 159-188.

[7] Baker, F.S. 1949. A revised tolerance table. Journal of Forestry 47:179-181.

[8] Banerjee, S., B P Carlin, and A E Gelfand. 2014. Hierarchical Modeling and Analysis for Spatial Data, 2nd ed., Chapman & Hall/CRC, Boca Raton.

[9] Baselga, A. and Araujo, M.B. 2010. Do community models fail to project community variation effectively? Journal of Biogeography. 37: 1842-1850.

[10] Bell, D.M. and J.S. Clark 2015. Seed predation and climate impacts on reproductive variation in temperate forests of the southeastern USA. Oecologia, in press.

[11] Berger, V.W. and J. Zhang. 2005. Structural Zeros, Encyclopedia of Statistics in Behavioral Science, J. Wiley and Sons, New York.

[12] Bowlby, H. D., and Gibson, A. J. F. 2015. Environmental effects on survival rates: robust regression, recovery planning and endangered Atlantic salmon. Ecology and Evolution, 5(16), 3450-3461. http://doi.org/10.1002/ece3.1614

[13] Brynjarsdottir, J. and A.E. Gelfand. 2014. Collective sensitivity analysis for ecological regression models with multivariate response. Journal of Biological, Environmental, and Agricultural Statistics, 19, 481-502.

[14] Calabrese, J. M., Certain, G., Kraan, C. and Dormann, C. F. (2014). Stacking species distribution models and adjusting bias by linking them to macroecological models. Global Ecol.Biogeogr., 23, 99-112.

[15] Cameron, A. C. and P. K. Trivedi. 1998. Regression Analysis of Count Data. New York, Cambridge University Press.

[16] Cameron, A. C. and P. K. Trivedi. 2005. Microeconometrics: Methods and Applications, Cambridge University Press, New York.

[17] Chakraborty, A., Gelfand, A.E., J.A. Silander, Jr., A.M. Latimer, and A.M. Wilson. (2010) Modeling large scale species abundance through latent spatial processes. Annals of Applied Statistics, 4, 1403-1429.

[18] Chib, S. 1998. Analysis of multivariate probit models. Biometrika, 85(2):347-361.

[19] Chib, S. and E. Greenberg. 1998. Analysis of multivariate probit models. Biometrika 85, 347-361.

[20] Clark, J.S. 2016a. Why species tell us more about traits than traits tell us about species: Predictive models. Ecology, in press.

[21] Clark, J.S. 2016b. gjam: Generalized Joint Attribute Modeling . Comprehensive R Archive Network (CRAN). https://cran.r-project.org/web/packages/gjam/index.html

[22] Clark, J.S., D. M Bell, M. Kwit, A. Powell, And K. Zhu. 2013. Dynamic inverse prediction and sensitivity analysis with high-dimensional responses: application to climate-change vulnerability of biodiversity. Journal of Biological, Environmental, and Agricultural Statistics, 18:376-404.

[23] Clark, J. S., Gelfand, A. E., Woodall, C. W., and Zhu, K. (2014). More than the sum of the parts: forest climate response from joint species distribution models. Ecological Applications, 24(5):990-999.

[24] Clark, J.S. and M. H. Hersh. 2009. Inference when multiple pathogens affect multiple hosts: Bayesian model selection. Bayesian Analysis 4:337-366.

[25] Clark, J.S., D. Nemergut, B. Seyednasrollah, P. Turner, and S. Zhang. 2016. Generalized joint attribute modeling for biodiversity analysis: Median-zero, multivariate, multifarious data, in revision.

[26] Cox R. M., Parker E. U., Cheney D. M., Liebl A. L., Martin L. B., Calsbeek R. 2010 Experimental evidence for physiological costs underlying the trade-off between reproduction and survival. Funct. Ecol. 24, 1262-1269. doi:10.1111/j.1365-2435.2010.01756.x (doi:10.1111/j.1365-2435.2010.01756.x)

[27] Cunningham, R.B., and D. B. Lindenmayer. (2008) Modeling count data of rare species: some statistical issues. Ecology 86, 1135-1142.

[28] Dempster, AP, Laird, NM and Rubin, DB 1977. Maximum likelihood from incomplete data via the EM algorithm. J. Royal Stat. Soc. 39, 1-38.

[29] de Valpine, P. and A. N. Harmon-Threatt 2013. General models for resource use or other compositional count data using the Dirichlet-multinomial distribution. Ecology 94:2678-2687.

[30] Doledec, S., Chessel, D., ter Braak, C.J.F. and Champely, S. 1996. Matching species traits to environmental variables: a new three-table ordination method. Environmental and Ecological Statistics 3:143-166.

[31] Dorazio, R. M., 2007, On the choice of statistical models for estimating occurrence and extinction from animal surveys: Ecology, 88, 2773- 2782.

[32] Elith, J. and Leathwick, J. R. 2009. Species distribution models: ecological explanation and prediction across space and time. Annual Review of Ecology, Evolution, and Systematics, 40:677.

[33] Elith J, et al. 2006. Novel methods improve prediction of species' distributions from occurrence data. Ecography 29: 129-151.

[34] Faddy, M. 1997. Extended Poisson process modelling and analysis of count data. Biometrical Journal, 39 , 431-440.

[35] Falk, M.G., R. O'Leary, M. Nayak, P. Collins, snf S. Low Choy. 2015. A Bayesian hurdle model for analysis of an insect resistance monitoring database, Environmental and Ecological Statistics, 22, 207.

[36] Ferrier, S., Manion, G., Elith, J., and Richardson, K. (2007). Using generalized dissimilarity modeling to analyse and predict patterns of beta diversity in regional biodiversity assessment. Diversity and Distributions, 13:252-264.

[37] Ghosh, S., A.E. Gelfand, K. Zhu, and J.S. Clark, 2012 The k-ZIG: flexible modeling for zero-inflated counts, Biometrics, 68, 3, 878-885.

[38] Gleim ER, Conner LM, Berghaus RD, Levin ML, Zemtsova GE, and Yabsley MJ. 2014. The Phenology of Ticks and the Effects of Long-Term Prescribed Burning on Tick Population Dynamics in Southwestern Georgia and Northwestern Florida. PLoS ONE 9: e112174. doi:10.1371/journal.pone.0112174

[39] Greene, W. H. 2000. Econometric Analysis, Fourth edition ed. Upper Saddle River, NJ: Prentice Hall.

[40] Haas, S. E., Hall Cushman, J., Dillon, W. W., Rank, N. E., Rizzo, D. M. and Meentemeyer, Ross K. (2016), Effects of individual, community, and landscape drivers on the dynamics of a wildland forest epidemic. Ecology, 97: 649-660.

[41] Haslett J, Whiley M, Bhattacharya S. 2006. Bayesian palaeoclimate reconstruction. Journal of the Royal Statistical Society, Series A 169:395-438.

[42] Hooten, M. B., Anderson, J., and Waller, L. A. (2010). Assessing North American Influenza Dynamics with a Statistical SIRS Model. Spatial and Spatio-Temporal Epidemiology, 1, 177-185. http://doi.org/10.1016/j.sste.2010.03.003

[43] Ibanez, I., J.S. Clark, S. LaDeau, and J. Hille Ris Lambers 2007. Exploiting temporal variability to understand tree recruitment response to climate change, Ecological Monographs, 77:163-177.

[44] Iverson, L. R., Prasad, A. M., Matthews, S. N., and Peters, M. (2008). Estimating potential habitat for 134 eastern US tree species under six climate scenarios. Forest Ecology and Management, 254:390-406.

[45] Kleyer M., Dray S., Bello F., Lep J., Pakeman R.J., Strauss B., Thuiller W., and Lavorel S. 2012. Assessing species and community functional responses to environmental gradients: which multivariate methods? Journal of Vegetation Science 23:805-821.

[46] Koopmans, L.H. 1969. Some Simple Singular and Mixed Probability Distributions, The American Mathematical Monthly, 76, 297-299

[47] Latimer, A., Banerjee, S., Sang Jr, H., Mosher, E., and Silander Jr, J. 2009. Hierarchical models facilitate spatial analysis of large data sets: a case study on invasive plant species in the northeastern united states. Ecology Letters, 12:144-154

[48] Lawrence, E., Bingham, D., Liu, C., and Nair, V. N. 2008. Bayesian inference for multivariate ordinal data using parameter expansion. Technometrics, 50:182-191.

[49] Leininger TJ, Gelfand AE, Allen JM, and Silander JA 2013. Spatial regression modeling for compositional data with many zeros, Journal of Agricultural, Biological, and Environmental Statistics, 18, 314-334.

[50] Lindén, A. and Mntyniemi, S. (2011), Using the negative binomial distribution to model overdispersion in ecological count data. Ecology, 92: 1414-1421.

[51] Lyashevska, O., Brus, D. J. and van der Meer, J. (2016), Mapping species abundance by a spatial zero-inflated Poisson model: a case study in the Wadden Sea, the Netherlands. Ecol Evol, 6: 532-543. doi:10.1002/ece3.1880

[52] Lynch, H. J., Thorson, J. T. and Shelton, A. O. (2014), Dealing with under- and over-dispersed count data in life history, spatial, and community ecology. Ecology, 95: 3173-3180.

[53] Martin, T. G., Wintle, B. A., Rhodes, J. R., Kuhnert, P. M., Field, S. A., Low-Choy, S. J., Tyre, A. J. and Possingham, H. P. (2005), Zero tolerance ecology: improving ecological inference by modelling the source of zero observations. Ecology Letters, 8: 1235-1246.

[54] McCullagh, P. 1980, Regression models for ordinal data, Journal of the Royal Statistical Society, Ser. B, 42, 109-142.

[55] McCulloch R., Polson N., Rossi P. 2000. Bayesian analysis of the multinomial probit model with fully identified parameters. J. Econometrics 99:173-193

[56] McMurdle, P.J. and S. Holmes. 2014. Waste not, want not: why rarifying microbiome data is inadmissible. Plos Computational Biology DOI: 10.1371/journal.pcbi.1003531.

[57] Metcalf, C.J.E., J. S. Clark, and S. M. McMahon. 2009. Overcoming data sparseness and parametric constraints in modeling of tree mortality: a new non-parametric Bayesian model. Canadian Journal of Forest Research, 39:1677-1687.

[58] Moran, E.V. and J.S. Clark. 2010. Estimating seed and pollen movement in a monoecious plant: a hierarchical Bayesian approach integrating genetic and ecological data. Molecular Ecology, 20, 1248-1262.

[59] Mullahy, J. 1986. Specification and testing of some modified count data models. Journal of Econometrics 33: 341-365.

[60] Niinemets,U., and Valladares, F., 2006. Tolerance to shade, drought, and waterlogging of temperate northern hemisphere trees and shrubs. Ecol. Monogr. 76, 521-547.

[61] Norby RJ, Hartz-Rubin JS, Verbrugge MJ (2003) Phenological responses in maple to experimental atmospheric warming and CO2 enrichment. Global Change Biology, 9, 1792-1801.

[62] Ovaskainen, O., J. Hottola, and J. Siitonen (2010) Modeling species co-occurrence by multivariate logistic regression generates new hypotheses on fungal interactions. Ecology, 91, 2514-2521.

[63] Ovaskainen, O., and J. Soininen. (2011) Making more out of sparse data: hierarchical modeling of species communities. Ecology 92, 289-295.

[64] Paciorek CJ, McLachlan JS. 2009. Mapping ancient forests: Bayesian inference for spatio-temporal trends in forest composition using the fossil pollen proxy record. Journal of the American Statistical Association. 104:608-622.

[65] Patrick, P.W., and A.M. Simons. 2014. Secondary reproduction in the herbaceous monocarp Lobelia inflata: time-constrained primary reproduction does not result in increased deferral of reproductive effort. BMC Ecology 14 14:1, DOI: 10.1186/1472-6785-14-15.

[66] Podani, J. (2005), Multivariate exploratory analysis of ordinal data in ecology: Pitfalls, problems and solutions. Journal of Vegetation Science, 16: 497?510. doi: 10.1111/j.1654-1103.2005.tb02390.x

[67] Polansky L, and Robbins MM. 2013. Generalized additive mixed models for disentangling long-term trends, local anomalies, and seasonality in fruit tree phenology. Ecology and Evolution. 3:3141-3151. doi:10.1002/ece3.707.

[68] Pollock, L. J., Tingley, R., Morris, W. K., Golding, N., O'Hara, R. B., Parris, K. M., Vesk, P. A., and McCarthy, M. A. (2014). Understanding co-occurrence by modelling species simultaneously with a Joint Species Distribution Model (JSDM). Methods in Ecology and Evolution, 5:397-406.

[69] Royle, J.A. 2004. N-mixture models for estimating population size from spatially replicated counts. Biometrics 60, 108-115.

[70] Rundel, C., E. M. Schliep, A.E. Gelfand, and D.M. Holland 2016 A data fusion approach for spatial analysis of speciated PM2.5 across time, Environmetrics (forthcoming).

[71] Sahu, S.K., A.E. Gelfand, and D.M. Holland 2010. Fusing Point and Areal Level Space-time Data with Application to Wet Deposition, Journal of the Royal Statistical Society - C , 59,1, 77-103.

[72] Schick, R.S., S. D. Kraus, R. M. Rolland, A. R. Knowlton, P. K. Hamilton, H. M. Pettis, R. D. Kenney, and J. S. Clark. 2013. Using hierarchical Bayes to understand movement, health, and survival in critically endangered marine mammals. PLOS One, 8: e64166. doi:10.1371/journal.pone.0064166.

[73] Schwantes, A.M., Swenson, J.J. and Jackson, R.B. 2016. Quantifying Drought-Induced Tree Mortality in the Open Canopy Woodlands of Central Texas. Remote Sensing of Environment. 181: 54-64.

[74] Sileshi, G., G. Hailu, and G. I. Nyadzi. 2009. Traditional occupancy?abundance models are inadequate for zero-inflated ecological count data. Ecological Modelling 220: 1764-1775.

[75] Smith, A. N. H., Anderson, M. J. and Millar, R. B. (2012), Incorporating the intraspecific occupancy?abundance relationship into zero-inflated models. Ecology, 93: 2526-2532.

[76] Swenson N.G. and M. D. Weiser 2010. Plant geography upon the basis of functional traits: an example from eastern North American trees. Ecology 91:2234-2241.

[77] Taylor-Rodriguez, D., K. Kaufeld, E. Schliep, J. S. Clark, and A. Gelfand, 2016. Joint Species distribution modeling: dimension reduction using Dirichlet processes, in revision.

[78] Thorson, J.T., M. D. Scheuerell, A. O. Shelton, K. E. See, H. J. Skaug and K. Kristensen. (2105) Spatial factor analysis: a new tool for estimating joint species distributions and correlations in species range. Methods in Ecology and Evolution, on-line view.

[79] Valle D., J.S. Clark, and K. Zhao. 2011. Enhanced understanding of infectious diseases by fusing multiple datasets: a case study on malaria in the western Brazilian Amazon Region. PLoS ONE 6(11): e27462. doi:10.1371/journal.pone.0027462

[80] Ver Hoef, J. M. and Boveng, P. L. 2007. Quasi-poisson vs. negative binomial regression: how should we model overdispersed count data? Ecology, 88: 2766-2772.

[81] Wenger, S. J. and Freeman, M. C. 2008. Estimating species occurrence, abundance, and detection probability using zero-inflated distributions. Ecology, 89: 2953-2959.

[82] Welsh, A. H., Cunningham, R.B., Donnelly, C. F. and Lindenmayer, D. B. 1996. Modelling the abundance of rare species: statistical models for counts with extra zeros. Ecological Modelling, 88, 297 - 308.

[83] West, M and J Harrison. 1997. Bayesian Forecasting and Dynamic Models, 2nd ed., Springer-Verlag, New York

[84] Westoby M., Falster D.S., Moles, A.T. 2002. Plant ecological strategies: some leading dimensions of variation between species. Annual Review of Ecology System 33:125-159.

[85] Wright, I. J., D. D. Ackerly, F. Bongers, K. E. Harms, G. Ibarra-Manriquez, M. Martinez-Ramos, S. J. Mazer, H. C. Muller-Landau, H. Paz, N. C. A. Pitman, L. Poorter, M. R. Silman, C. F. Vriesendorp, C. O. Webb, M. Westoby, and S. J. Wright. 2007. Relationships among ecologically important dimensions of plant trait variation in seven Neotropical forests. Annals of Botany 99:1003-1015.

[86] Zhang, X., W.J. Boscardin, and T.R. Belin. 2008. Bayesian analysis of multivariate nominal measures using multivariate multinomial probit models. Computational Statistics and Data Analysis 52, 3697-3708.

11

Gradient Analysis of Ecological Communities (Ordination)

Michael W. Palmer

Oklahoma State University

CONTENTS

11.1   Introduction

Quantitative community ecology is one of the most challenging branches of modern environmetrics. Community ecologists typically need to analyze the effects of multiple environmental factors on dozens (if not hundreds) of species simultaneously, and statistical errors (both measurement and structural) tend to be large and ill-behaved. It is not surprising, therefore, that community ecologists have employed a variety of multivariate approaches. These approaches have been both endogenous and borrowed from other disciplines. The majority of techniques fall into two main groups: classification and ordination. Classification is the placement of species or sample units into groups, and ordination is the arrangement or 'ordering' of species and/or sample units along gradients. In this chapter, I will describe the use and properties of the most widely used ordination methods.

11.2   History of ordination methods

Although community ecology is a fairly young science, the application of quantitative methods began fairly early (McIntosh 1985). In 1930, Ramensky began to use informal ordination techniques for vegetation. Such informal and largely subjective methods became widespread in the early 1950's (Whittaker 1967). In 1951, Curtis and McIntosh 1951 developed the 'continuum index', which later lead to conceptual links between species responses to gradients and multivariate methods. Shortly thereafter, Goodall (1954) introduced the term 'ordination' in an ecological context for Principal Components Analysis. Bray and Curtis (1957) developed polar ordination, which became the first widely-used ordination technique in ecology. Austin (1968) used canonical correlation to assess plant-environment relationships in what may have been the first example of multivariate direct gradient analysis in ecology. In 1973, Hill introduced correspondence analysis, a technique originating in the 1930's, to ecologists. Correspondence analysis gradually supplanted polar ordination, which today has few practitioners. Fasham (1977) and Prentice (1977) independently discovered and demonstrated the utility of Kruskal's (1964) nonmetric multidimensional scaling, originally intended as a psychometric technique, for community ecology. Hill (1979) corrected some of the flaws of Correspondence Analysis and thereby created Detrended Correspondence Analysis, which is the most widely used indirect gradient analysis technique today. The software to implement Detrended Correspondence Analysis, DECORANA, became the backbone of many later software packages. Gauch's (1982) book "Multivariate Analysis in Community Ecology" described ordination in non-technical terms to the average practitioner, and allowed ordination techniques to enter the mainstream. Ter Braak (1986) ushered in a revolution in ordination methods with Canonical Correspondence Analysis. This technique coupled Correspondence Analysis with regression methodologies, and provides for hypothesis testing. Ter Braak and Prentice (1988) developed a theoretical unification of ordination techniques, hence placing gradient analysis on a firm theoretical foundation.

The more recent packages PC-ORD (McCune and Mefford 1999), Vegan (Oksanen et al. 2013), and CANOCO (ter Braak and Šmilauer 2012) have increased the accessibility of gradient analysis to ecologists, and the textbook *Multivariate Analysis of Ecological Data using Canoco 5* by Šmilauer and Lepš (2014) presents basic through advanced methods to practicing community ecologists.

11.3 Theory and background

11.3.1 Properties of community data

Ordination methods are essentially operations on a community data matrix (or species by sample matrix). A community data matrix has taxa (usually species) as rows and samples as columns (Table 11.1) or vice versa. In community ecology, the term "sample" has diverged from its usage in statistics, and refers to the basic unit of observation. In most studies of vegetation, the sample is a quadrat, relevé, or transect – though it may consist of a compilation of subsamples (as is the case with Table 11.1). Samples in animal ecology may consist of traps, seine sweeps, or survey routes. Biogeographic studies may rely on the cells of large grids or political units as samples.

The elements in community data matrices are abundances of the species. 'Abundance' is a general term that can refer to density, biomass, cover, or even incidence (presence/absence) of species. The choice of an abundance measure will depend on the taxa and the questions under consideration. Any of the matrix's constituent column vectors is considered the *species composition* for the corresponding sample. Species composition is sometimes expressed in terms of relative abundance; i.e. constrained to a constant total such as 1 or 100%. The purpose of ordination methods is to interpret patterns in species composition.

Regardless of the scale or taxa involved, most community data matrices share some general properties:

- They tend to be sparse: a large portion (often the majority) of entries consists of zeros.

- Most species are infrequent. That is, the majority of species is typically present in a minority of locations, and contributes little to the overall abundance.

- The number of factors influencing species composition is potentially very large. For example, forest tree density can be influenced by time since fire, elevation, nutrients, soil depth, soil texture, water availability and many other factors.

- The number of important factors is typically few. That is, a few factors can explain the majority of the explainable variation. Another way of saying this is that the *intrinsic dimensionality* is low.

- There is much noise. Even under ideal circumstances, replicate samples will vary substantially from each other. This is largely due to stochastic events and contingency (*sensu* Parker and Pickett 1998), though observer error may also be appreciable.

- There is much redundant information: species often share similar distributions. For example, the abundance of *Haplohymenium triste* gives some insights into the abundance of *Anomodon rostratus*, and the abundance of *Dicranum montanum* helps predict the abundance of *Leucobryum albidum* (Table 11.1). It is this property of redundancy that allows us to make sense of compositional data.

For any ordination method to be generally useful, it must be able to cope with the above properties of community data matrices.

11.3.2 Coenospace

As mentioned in the introduction, *ordination* is the arrangement of species and/or samples along gradients. Indeed, ordination can be considered a synonym for multivariate gradient

TABLE 11.1

a) Importance values of bryophyte species (mosses and liverworts) growing on the trunks of trees in three sites within the Duke Forest, North Carolina (Palmer, 1986). Importance values are a measure of relative abundance. Each sample represents the average of 10 trees in a given site. A) Data sorted alphabetically. B) Data with rows and columns sorted in order of Correspondence Analysis species scores and sample scores, respectively. Sample codes indicate the species of tree and the site; BN = *Betula nigra*, birch; LT = *Liriodendron tulipifera*, tulip tree; PE = *Pinus echinata*, shortleaf pine; PO = *Platanus occidentalis*, sycamore; PT = *Pinus taeda*, loblolly pine; QA = *Quercus alba*, white oak. Numbers 1 through 3 equal sites 1 through 3, respectively. a)

Bryophyte species	BN2	LT1	LT2	PE3	PO2	PT1	PT3	QA1	QR1
Amblystegium serpens	1	0	5	0	0	0	0	3.2	2.3
Anomodon attenuatus	0.9	17.6	26.4	0	41.2	0	0	27.3	22.4
Anomodon minor	2.1	1	5.4	0	9.4	0	0	2.4	0.6
Anomodon rostratus	0	0	1.4	0	4.7	0	0	14	13.6
Brachythecium acuminatum	0	3.1	0	0	0	0	0	5.3	2.9
Brachythecium oxycladon	0.9	1.8	0.8	0	1.6	0	0	1.7	0.5
Bryoandersonia illecebra	1.8	3.7	2	0	0.7	0	0	5.9	1.6
Campylium hispidulum	0.9	1.2	0	0	0	0	0	2.2	2.6
Clasmatodon parvulus	0	1	8.1	0	8.3	0	0	1.1	0.4
Dicranum montanum	1.8	0	0	6.8	0	5.8	9	0	0
Dicranum scoparium	0	0	0	2	0	7.1	0	0	0
Entodon seductrix	0	2.8	5.4	0	0.8	0	0	7.7	7.5
Frullania eboracensis	3.7	13.1	10.1	0	7	0	0	3.8	1.3
Haplohymenium triste	0	0.6	0.6	0	2.2	0	0	0.6	2.4
Isopterygium tenerum	16.1	4.6	2.2	30.9	1.4	26.8	18.2	0.6	2.4
Leucobryum albidum	3.1	0	0	44.6	0	35.9	59	0	0
Leucodon julaceus	1.8	5.5	6.7	0	6.8	0	0	7.2	9.7
Lophocolea heterophylla	20.2	0.5	1.4	7.4	0	8.4	0	0	0.9
Platygyrium repens	29.4	5.5	9.5	0	2.1	3.2	0	1.8	2.3
Porella platyphylla	0	1.1	0.6	0	0.7	0	0	2.1	3.2
Radula complanata	0	4	0	0	0.9	0	0	2.3	6.2
Radula obconica	0	6.1	2.5	0	3.5	0	0	2.2	6.1
Sematophyllum adnatum	11	7.9	6.2	6.4	3.9	7.4	6.1	4.6	5.9
Thelia asprella	0	3.3	0	0	0	0	0	0.5	0.9
Thuidium delicatulum	0.9	15.6	1.6	0	0.7	0	0	1.5	0.4

b)

Bryophyte species	QA1	PO2	QR1	LT2	LT1	BN2	PT1	PE3	PT3	CA Axis 1 species scores
Anomodon rostratus	14	4.7	13.6	1.4	0	0	0	0	0	-0.90
Haplohymenium triste	3.8	2.2	1.3	0.6	0.6	0	0	0	0	-0.90
Brachythecium acuminatum	5.3	0	2.9	0	3.1	0	0	0	0	-0.86
Porella platyphylla	2.1	0.7	3.2	0.6	1.1	0	0	0	0	-0.86
Anomodon attenuatus	27.3	41.2	22.4	26.4	17.6	0.9	0	0	0	-0.85
Clasmatodon parvulus	1.1	8.3	0.4	8.1	1	0	0	0	0	-0.84
Radula complanata	2.3	0.9	6.2	0	4	0	0	0	0	-0.84
Radula obconica	2.2	3.5	6.1	2.5	6.1	0	0	0	0	-0.83
Leucodon julaceus	7.2	6.8	9.7	6.7	5.5	1.8	0	0	0	-0.80
Thelia asprella	0.5	0	0.9	0	3.3	0	0	0	0	-0.78
Entodon seductrix	0	0.8	0	5.4	2.8	0	0	0	0	-0.77
Anomodon minor	2.4	9.4	0.6	5.4	1	2.1	0	0	0	-0.75
Frulania eboracensis	7.7	7	7.5	10.1	13.1	3.7	0	0	0	-0.74
Amblystegium serpens	3.2	0	2.3	5	0	1	0	0	0	-0.74
Bryoandersonia illecebra	5.9	0.7	1.6	2	3.7	1.8	0	0	0	-0.72
Thuidium delicatulum	1.5	0.7	0.4	1.6	15.6	0.9	0	0	0	-0.71
Campylium hispidulum	2.2	0	2.6	0	1.2	0.9	0	0	0	-0.71
Brachythecium oxycladon	1.7	1.6	0.5	0.8	1.8	0.9	0	0	0	-0.71
Platygyrium repens	1.8	2.1	2.3	9.5	5.5	29.4	3.2	0	0	-0.08
Sematophyllum adnatum	4.6	3.9	5.9	6.2	7.9	11	7.4	6.4	6.1	0.10
Lophocolea heterophylla	0	0	0.9	1.4	0.5	20.2	8.4	7.4	0	0.61
Isopterygium tenerum	0.6	1.4	2.4	2.2	4.6	16.1	26.8	30.9	18.2	0.93
Dicranum scoparium	0	0	0	0	0	0	7.1	2	0	1.26
Dicranum montanum	0	0	0	0	0	1.8	5.8	6.8	9	1.27
Leucobryum albidum	0	0	0	0	0	3.1	35.9	44.6	59	1.33
CA Axis 1 sample scores	-0.94	-0.93	-0.86	-0.76	-0.73	0.29	1.24	1.32	1.45	

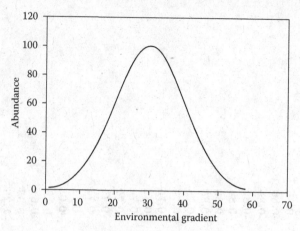

FIGURE 11.1

Species response curve. For this curve, the optimum is 30, and the tolerance is 10. Real curves will have much noise, and will not necessarily be symmetrical.

analysis. Therefore, before discussing ordination, it is necessary to describe the fundamental underlying model of species responses to gradients. Although ecologists had a basic understanding environmental control of species composition since the beginning of ecology (McIntosh 1985), Whittaker (1967, 1969) provided a formalization of terms and concepts for the *unimodal model*. Simply put, the unimodal modal states that species response functions (i.e. the relationship between the abundance of species as a function of position along a gradient) are unimodal, or one-peaked (Figure 11.1). In other words, there is a unique set of optimal conditions for a species, at which the species has maximal abundance. As conditions differ from this optimum, to the extent of the difference, abundance will decrease. Although Figure 11.1 displays the response of a species to a single gradient, the unimodal modal is readily extended to multiple gradients.

A *coenocline* is a pictorial representation of all species response functions combined along a single gradient (Figure 11.2). Given the large number of species and the high noise in most studies, coenoclines are usually only displayed in highly simplified form. Nevertheless, they are useful heuristic concepts. *Coenoplanes* (2 environmental gradients) and *coenospaces* (>2 gradients) are even more difficult to display. However, an ordination *biplot* (discussed later) is an abstracted depiction of coenospace.

It must be stressed that the coenospace model does not assume that species have unimodal responses through *space*. Indeed, as the environment may exhibit discontinuous, irregular, or even fractal spatial patterns (Palmer 2007) then species necessarily are distributed in complex ways throughout the landscape. This implies no inherent contradiction with the study of gradient analysis. Also it should be noted that the coenospace model implies a conceptual continuity of ecological communities, and thus one finds discontinuities only if the environment is discontinuous or (more commonly) incompletely sampled (Gleason 1917). Thus, the process of numerical classification or clustering, unlike gradient analysis, is arbitrary and intrinsically subject to sampling irregularities.

Although there are occasionally exceptions to the unimodal model (e.g. bimodal distributions, or qualitative noise due to vicariance events), the model is reasonable for most ecological systems. However, refining the model by assigning functional forms to species response functions has proven difficult (Austin 1987). Gaussian functions (or Gaussian logit functions; ter Braak and Looman 1987) are attractive because they are controlled by relatively few parameters. In addition, a Gaussian assumption leads to elegant proofs and

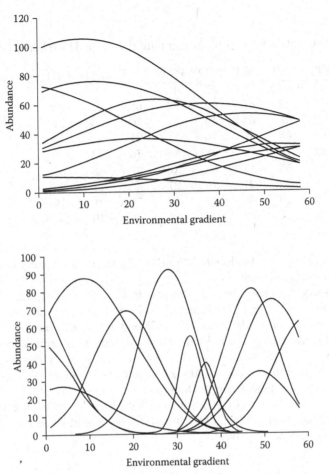

FIGURE 11.2
Two coenoclines. Note these are is hypothetical examples; real examples would have much noise. The top example has lower beta diversity than the second example.

simplifications (ter Braak and Looman 1986). However, other functions (e.g. the Beta function, Minchin 1987) are needed to allow skewed, platykurtic, and leptokurtic forms. The price paid for adopting such models is the larger number of parameters needed. In any case, ecological theory is mute regarding the form of species response functions, and choices have typically been made on empirical grounds. It is important to note that unimodality alone is a sufficient assumption for the unimodal methods to work, so the details of the shape of species response curves are seldom important.

11.3.3 Alpha, beta, gamma diversity

Whittaker (1967, 1969) tied the unimodal model to levels of diversity, three of which have become central to community ecology: *alpha diversity* is the diversity (either measured in terms of species richness or a synthetic index such as H') of a community; *beta diversity* (also known as 'species turnover' or 'differentiation diversity') is the rate of change in species composition from one community to another along gradients; *gamma diversity* is

TABLE 11.2

Euclidean distance matrix for bryophyte communities of the Duke Forest. For abbreviations, see Table 11.1.

	BN2	LT1	LT2	PE3	PO2	PT1	PT3	QA1	QR1
BN2	0	42	43	55	57	46	67	50	46
LT1	42	0	22	61	32	53	68	25	23
LT2	43	22	0	63	19	55	70	20	21
PE3	55	61	63	0	70	11	21	65	62
PO2	57	32	19	70	0	64	76	22	26
PT1	46	53	55	11	64	0	27	58	55
PT3	67	68	70	21	76	27	0	71	69
QA1	50	25	20	65	22	58	71	0	10
QR1	46	23	21	62	26	55	69	10	0

the diversity of a region or a landscape (Figure 11.2). Gamma diversity can be measured in the same units as alpha diversity.

Ecological theory does not offer guidelines as to the proper spatial scale for distinguishing "alpha diversity" from "gamma diversity". Indeed, these scales are arbitrary and depend upon the objectives of the study (Palmer and White 1994). In practice, we consider alpha diversity to be the diversity of the individual sample unit or observation, and gamma diversity to be the diversity of all sample units combined. Beta diversity then becomes a measure of how distinct the sampling units are along gradients. A gradient with high beta diversity is considered a 'long' gradient because there is much change in species composition. Ecologists have proposed a number of beta diversity indices (e.g. Whittaker 1969, Wilson and Mohler 1983, Oksanen and Tonteri 1995); I will discuss one of these later in the context of Detrended Correspondence Analysis.

11.3.4   Ecological similarity and distance

We consider two samples with similar species composition to be ecologically similar, and two samples which share few species to be ecologically distant. The concept of ecological distance is akin to beta diversity, but it deviates from it in important respects: samples can be ecologically distant due to noise rather than environmental differences, and ecological distance is not measured along gradients. However, some ordination techniques such as NMDS require measures of ecological distance. Numerous measures of ecological distance (or its complement, ecological similarity) are in use (see Legendre and Legendre 1998). Table 11.2 is an example of a distance matrix, calculated from the data matrix in Table 11.1. Note several things:

- The distance matrix is square and symmetric, i.e. its rows are the same as its columns.

- The diagonals are zero, meaning that there is no difference between a sample and itself. Because of this and the previous observation, distance matrices are frequently represented as a triangular matrix, ignoring the values above and including the diagonal.

- Some ecological insights can be derived from the matrix. For example, within a genus (e.g. comparing *Quercus alba* and *Quercus rubra*) or species (e.g. comparing *Liriodendron tulipifera* in two sites) of host trees, epiphytic bryophyte communities are similar (low values), but between genera (e.g. between *Quercus* and *Pinus*), communities are dissimilar.

- All information about particular bryophyte species is lost – so any analyses relying on the distance matrix alone will have limits to its interpretability.

11.4   Why ordination?

According to Gauch (1982): "Ordination primarily endeavors to represent sample and species relationships as faithfully as possible in a low-dimensional space". But why is this objective desirable? There are a number of answers, but most are derived from the 'properties of community' data as described above:

- It is impossible to visualize multiple dimensions simultaneously. While physicists grumble if space exceeds four dimensions, ecologists typically grapple with dozens or even hundreds of dimensions (species and/or samples).

- A single multivariate analysis saves time, in contrast to a separate univariate analysis for each species.

- Ideally and typically, dimensions of this 'low dimensional space' will represent important and interpretable environmental gradients.

- If statistical tests are desired, problems of multiple comparisons are diminished when species composition is studied in its entirety.

- Statistical power is enhanced when species are considered in aggregate, because of redundancy.

- By focusing on 'important dimensions', we avoid interpreting (and misinterpreting) noise. Thus, ordination is a 'noise reduction technique' (Gauch 1982).

- We can determine the relative importance of different gradients; this is virtually impossible with univariate techniques.

- Community patterns may differ from population patterns.

- Some techniques provide a measure of beta diversity.

- The graphical results from most techniques often lead to ready and intuitive interpretations of species-environment relationships.

11.5   Exploratory analysis and hypothesis testing

Reduction of dimensionality is not the only reason to use ordination. At one point, the primary goal of ordination was considered "exploratory" (Gauch, 1982). It was the job of the ecologist to use his or her knowledge and intuition to collect and interpret data (e.g. Peet, 1980); pure objectivity could potentially interfere with the ability to distinguish important gradients. Ordination was often considered as much an art as a science, and early critiques derided it as 'just telling you what you knew anyway'. With the introduction of

TABLE 11.3

Common ordination techniques, by category (largely derived from ter Braak and Prentice, 1988). The names of the techniques and their acronyms are given in bold. For further explanation, see text.

Informal techniques

Indirect gradient analysis

Distance-based approaches
- **Polar ordination, PO (Bray-Curtis ordination)**
- **Principal Coordinates Analysis, PCoA (Metric multidimensional scaling)**
- **Nonmetric Multidimensional Scaling, NMDS**

Eigenanalysis-based approaches

Linear model

Principal Components Analysis, PCA

Unimodal model

Correspondence Analysis, CA (Reciprocal Averaging)

Detrended Correspondence Analysis, DCA

Direct gradient analysis
- Linear model
  Redundancy Analysis, RDA
- Unimodal model
  Canonical Correspondence Analysis, CCA
  Detrended Canonical Correspondence Analysis, DCCA

CCA, testing statistical hypotheses became routine, and it was possible to go beyond mere "exploratory" analysis. (ter Braak, 1985). However, rigorous hypothesis testing requires complete objectivity, which results in repeatability and falsifiability. Thus the two basic motivations for ordination, hypothesis testing and exploratory analysis, can potentially conflict with each other. The two approaches can be reconciled with a cross-validation approach, as discussed later.

11.6 Ordination vs. Factor Analysis

Factor Analysis is a body of techniques that makes inferences about factors, or latent variables, underlying a multivariate data set. Thus in the broadest sense, ordination is a form of factor analysis. However, most ecologists do not frame ordination in the context of factor analysis. One philosophical explanation for this is that ecologists tend to recognize (Stemming from Hutchinson's (1957) concept of the hyperdimensional niche) that there are an indefinitely large number of factors influencing species, and thus they are uncomfortable with models that imply intrinsic dimensionality is low (even if they do recognize that few gradients are *important*). Most leading textbooks on ordination do not mention factor analysis. Legendre and Legendre (1998) briefly describe and promote factor analysis but admit *"with the exception of the varimax rotation ... factor analysis is not currently used in ecology ... Confirmatory factor analysis ... is not currently used in ecology."*

11.7 A classification *of* ordination

Numerous ordination methods have been put forward, but the most common ones are organized in Table 11.3. The dichotomy between indirect and direct gradient analysis (Gauch, 1982; ter Braak and Prentice, 1988), while sometimes blurred in practice, is crucial. Indirect gradient analysis utilizes only the species by sample matrix (e.g. Table 11.1). If there is any information about the environment, it used after indirect gradient analysis, as an interpretative tool. When we perform an indirect analysis, we are essentially asking the species what the most important gradients are. It is entirely possible that the most important gradients are ones for which we have no external data (e.g. intensity of past disturbance), yet indirect analysis will take advantage of redundancy in the data set and display such gradients.

Direct gradient analysis, in contrast, utilizes external environmental data in addition to the species data. In its simplest form, direct gradient analysis is a regression technique. Direct analysis tells us if species composition is related to our measured variables. Ideally, it will be able to do this even if we did not measure the most important gradients (Palmer, 1993). Direct analysis allows us to test the null hypothesis that species composition is unrelated to measured variables. A special case of direct gradient analysis is when our 'measured variables' are experimentally imposed treatments.

Table 11.3 also distinguishes between distance-based techniques (derived from distance matrices such as Table 11.2) and eigenanalysis-based techniques. This distinction is somewhat arbitrary because Principal Coordinates Analysis can be solved through eigenanalysis, and eigenanalysis-based techniques can usually be described in a 'distance framework' (for example, correspondence analysis can be described in terms of chi-squared distances). Eigenanalysis-based methods are further subdivided into linear models and unimodal models (ter Braak and Prentice, 1988), although unimodal models appear to perform well even with linear data (ter Braak and Šmilauer, 2012).

The techniques in Table 11.3 are described below.

11.8 Informal techniques

Subjective ordering of communities along one or more axes can be heuristically useful. For example, Whittaker (1967) arranged communities on axes of exposure and elevation. Fuhlendorf and Smeins (1997; Figure 11.5) placed communities in the context of fire frequency and grazing intensity. Likewise, species can be placed along axes of their ecological characteristics (e.g. Grime, 1979). Such informal techniques need not be quantitative, but to be effective they do need to communicate relevant concepts.

11.9 Distance-based techniques

Polar ordination, Principal Coordinates Analysis, and Nonmetric Multidimensional Scaling differ considerably in their algorithms and properties, yet all rely on a distance matrix as input. Thus, they are all highly sensitive to the choice of the distance metric, and they all

TABLE 11.4

Polar Ordination Axis scores for the bryophyte data.

	Axis 1	First candidate 2 for Axis 2	Second candidate for Axis 2
Endpoint 1	PT3	QA1	QR1
Endpoint 2	PO2	PE3	LT2
BN2	46	28	18
LT1	62	9	12
LT2	68	4	21
PE3	9	65	7
PO2	76	−2	17
PT1	16	57	8
PT3	0	68	7
QA1	68	0	3
QR1	65	3	0
Correlation with Axis 1	1.000	−0.9961	0.292841

'hide information'. That is, when ordinating samples, the information about species in such methods is collapsed.

11.9.1　Polar ordination

Polar ordination (PO; Bray and Curtis, 1957) arranges samples between endpoints or 'poles' according to the distance matrix. In the earliest versions of PO, these endpoints were the two samples with the highest ecological distance between them, *or* two samples which are suspected of being at opposite ends of an important gradient (thus introducing a degree of subjectivity).

Using the first of these criteria, and the example in Table 11.2, we define PT3 and PO2 as endpoints of the first axis. We assign PT3 (endpoint 1) a score of zero, and PO2 (endpoint 2) a score of 76 (its distance of separation from endpoint 1). We arrange the remaining samples along the first axis according to their dissimilarity to PT3 and their similarity to PO2, using

$$\text{Axis 1 score} = (D^2 + D1^2 - D2^2)/2D \tag{1}$$

Where D is the distance between the endpoints, D1 is the distance between a sample and the first endpoint, and D2 is the distance between a sample and the second endpoint. Table 11.4 shows the resulting PO axis scores.

The selection of endpoints for higher axes is a bit more involved. The simplest method is to choose the pair of samples, not including the previous endpoints, with the maximum distance of separation. However, this criterion selects QA1 and PE3, which results in an axis that has a strong negative correlation with axis 1 (Table 11.4; Figure 11.13a). This is undesirable, because the second axis contains little information that is not already contained in axis 1. Instead of these endpoints, we choose two samples (QR1 and LT2), which by tedious calculation results in a low correlation with axis 1 (Figure 11.3b). The ordination diagram is readily interpretable, with the first axis distinguishing bryophyte communities on pine trees from communities on other trees, and the second axis distinguishing oaks from the other hardwoods. These patterns are consistent with others in the literature (cited and reanalyzed in Palmer, 1986).

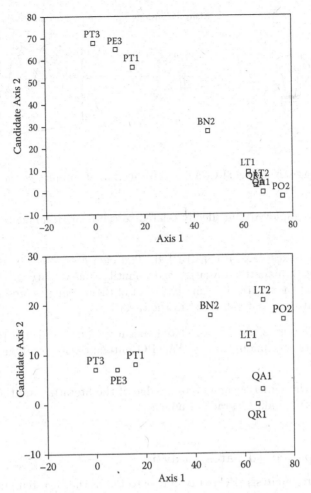

FIGURE 11.3

Polar ordination of the moss example. The second example has a more reasonable second axis than the first.

Beals (1984) extended Bray-Curtis ordination and discussed its variants, and is thus a useful reference.

11.9.1.1 Interpretation of ordination scatter plots

At this point, it is worth making several observations concerning the interpretation of ordination diagrams (not just PO diagrams):

- The direction of the axes (e.g. left vs. right; up vs. down) is arbitrary and should not affect the interpretation.

- The numeric scale on the axis is not very useful for the interpretation (an exception for this is DCA, in which the scales are in units of beta diversity).

- In PO and most other techniques (but not NMDS), the order of the axes is important. Thus, axis 1 is more important than axis 2, etc. The meaning of 'importance' depends

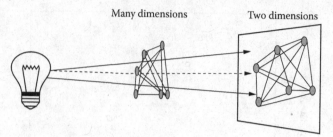

Many dimensions   Two dimensions

FIGURE 11.4
Principal Coordinates Analysis (PCoA) as a projection of samples connected by distances.

on the technique employed, but ideally related to the relative influence of environmental gradients.

- Third and higher axes can be constructed. The choice of 'when to stop' interpreting new axes is largely a matter of taste, the quantity and quality of the data, and the ability to interpret the results. Fortunately, most of the techniques presented later provide supplemental statistics that can assist in the task.

- It is desirable that axes not be correlated, because you would like them to represent different gradients. Techniques other than PO automatically result in uncorrelated (or *orthogonal*) axes.

- A biologist's insight, experience, and knowledge of the literature are the most important tools for interpreting indirect gradient analysis.

11.9.2   Principal coordinates analysis

Principal coordinates analysis (PCoA) is similar to PO in that it attempts to represent the distances between samples. In particular, it maximizes the linear correlation between the distances in the distance matrix, and the distances in a space of low dimension (typically, 2 or 3 axes are selected). PCoA is perhaps best understood geometrically. The distance between two items can be faithfully represented by one dimension (a line). The distances between three items are faithfully represented by 2 dimensions (a plane): that is, the items will form the vertices of a triangle, a planar object. Distances between four objects define a tetrahedron (a 3-dimensional object). To generalize, distances between N objects can be faithfully represented in N-1 dimensions. Unfortunately, it is difficult for the human mind to grasp more than 3 dimensions simultaneously, so we need to project such multidimensional objects onto lower dimensional space (Figure 11.4). The PCoA algorithm is analogous to rotating the multidimensional object such that the distances (lines) in the 'shadow' are maximally correlated with the distances (connections) in the object.

Although PCoA is based on a distance matrix, the solution can be found by eigenanalysis. When the distance metric is Euclidean, PCoA is equivalent to Principal Components Analysis (thus the bryophyte analysis for PCoA will not be presented here, but rather in the PCA section).

11.9.3   Nonmetric Multidimensional Scaling

PCoA suffers from a number of flaws, in particular the arch effect (discussed later in the context of PCA and CA). These flaws stem, in part, from the fact that PCoA maximizes a *linear* correlation. Nonmetric Multidimensional Scaling (NMDS) rectifies this by maximizing the *rank order* correlation. The algorithm in brief outline, proceeds as follows:

- The user selects the number of dimensions (N) for the solution, and chooses an appropriate distance metric.

- The distance matrix is calculated.

- An initial configuration of samples in N dimensions is selected. This configuration can be random, though the chances of reaching the correct solution are enhanced if the configuration is derived from another ordination method.

- A measure of 'stress' (mismatch between the rank order of distances in the data, and the rank order of distances in the ordination) is calculated

- The samples are moved slightly in a direction that decreases the stress.

- 4 and 5 are repeated until 'stress' appears to reach a minimum. The final configuration of points may be rotated if desired.

The final configuration of points represents your ordination solution. The configuration is dependent on the number of dimensions selected; e.g. the first two axes of a 3-dimensional solution does not necessarily resemble a 2-dimensional solution. The stress will typically decrease as a function of the number of dimensions chosen; this function can aid in the selection of the results. For the bryophyte data of Table 11.2, the stress is 7.565, 0.881, and 0.001 for the 1, 2, and 3-dimensional solutions, respectively. Thus, the huge drop from the first to the second solution implies that a second axis is useful in explaining species composition. The third dimension is not quite as necessary (i.e. the drop in stress is not as dramatic), but since the stress of a 3D solution is negligible, we will adopt it (Figure 11.5).

Note that the same gradient in bryophyte species composition appears, as in the case of PO (Figure 11.3), although in NMDS (Figure 11.5), the gradient from *Pinus* species to other host species is reflected in Axis 2. However, recall that in NMDS the order of the axes is arbitrary: the first axis is not necessarily more important than the second axis, etc. This is why it is sometimes useful to rotate the solution (such as by the Varimax method) – although there is no theory that states that the final solution will represent a 'gradient.' Other problems and advantages of NMDS will be discussed later, when comparing it to Detrended Correspondence Analysis.

11.10   Eigenanalysis-based indirect gradient analysis

An introduction to eigenanalysis is beyond the scope of this chapter. However, in the context of ordination there are several points worth making. For eigenanalysis-based methods:

- An eigenanalysis is performed on a square, symmetric matrix derived from the data matrix (e.g. Table 11.1).

- There is a unique solution to the eigenanalysis, no matter the order of data.

- Each ordination axis is an eigenvector, and is associated with an eigenvalue. The coordinates for the ith sample along a given axis is the ith element of the axis' eigenvector.

- Axes are ranked by their eigenvalues. Thus, the first axis has the highest eigenvalue, the second axis has the second highest eigenvalue, etc.

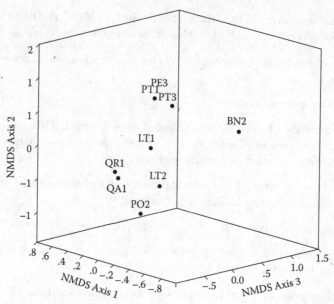

FIGURE 11.5

NMDS of the bryophyte data. Except when there are few samples, such as with this case, two 2-d plots (e.g. axis 2 vs. axis 1 and axis 3 vs. axis 1) may be better than one 3-d plot.

- Eigenvalues have mathematical meaning that can aid in interpretation. In principal components analysis, eigenvalues are 'variance extracted'. In methods related to correspondence analysis, eigenvalues are 'inertia extracted', or equivalently, correlation coefficients.

- Axes are orthogonal to each other.

- There are a potentially large number of axes (usually, the number of samples minus one, or the number of species minus one, whichever is less) so there is no need to specify the dimensionality in advance. However, the number of dimensions worth interpreting is usually very low.

- Species and samples are ordinated simultaneously, and can hence both be represented on the same ordination diagram (if this is done, it is termed a *biplot*).

11.10.1 Principal Components Analysis

The simplest and oldest eigenanalysis-based method is Principal Components Analysis (PCA). It is used for many purposes, but I will only discuss its applicability as an ordination method here. Geometrically, PCA is a rigid rotation of the original data matrix, and can be defined as a projection of samples onto a new set of axes, such that the maximum variance is projected or "extracted" along the first axis, the maximum variation uncorrelated with axis 1 is projected on the second axis, the maximum variation uncorrelated with the first and second axis is projected on the third axis, etc. Figure 11.6 illustrates the similarities between PCA and PCoA (Figure 11.4).

One of the biggest differences between PCA and PCoA is that the variables (i.e. species) representing the original axes are projected as biplot arrows. In the bryophyte communities (Figure 11.7), these biplot arrows greatly aid in interpretation. The first axis represents a gradient from communities on *Pinus* (on the right) to hardwood trees (on the left), with

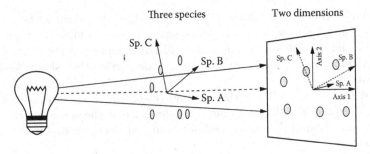

FIGURE 11.6
PCA as a projection of data points. Typical data sets will have many more than 3 species.

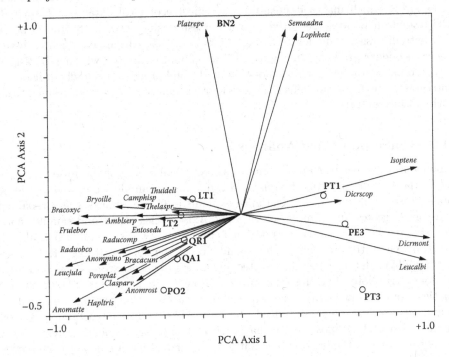

FIGURE 11.7
A PCA biplot of species and samples. The species are represented by arrows and the first four letters of the genus and species name (Table 11.1)

Betula (in the middle) being intermediate. The bryophyte species that point to the lower left are those that dominate on *Populus* and *Quercus*, those that dominate on *Betula* point up, and those that dominate on *Pinus* point to the right (see Table 11.1).

The eigenvalues represent the variance extracted by each axis, and are often conveniently expressed as a percentage of the sum of all eigenvalues (i.e. total variance). In the bryophyte example, The first four axes explain approximately 73%, 17%, 4%, and 3% of the variance, respectively. Since the first two axes explain (cumulatively) about 90% of the variance, we deem the 2-dimensional solution of Figure 11.7 adequate.

In most applications of PCA (e.g. as a factor analysis technique), variables are often measured in different units. For example, PCA of taxonomic data may include measures of size, shape, color, age, numbers, and chemical concentrations. For such data, the data must be standardized to zero mean and unit variance (the typical default for most computer programs). For ordination of ecological communities, however, all species are measured in the same units, and data should not be standardized. In matrix algebra terms, most PCAs are eigenanalyses of the correlation matrix, but for ordination they should be PCAs of the covariance matrix.

In contrast to Correspondence Analysis and related methods (see below), species are represented by arrows. This implies that the abundance of the species is continuously increasing in the direction of the arrow, and decreasing in the opposite direction. Thus PCA is a 'linear method'.

Although the discussion above implies that PCA is distinctly different from PCoA, the two techniques end up being identical, if the distance metric is Euclidean.

Unfortunately, this linear assumption causes PCA to suffer from a serious problem, the *horseshoe effect*, which makes it unsuitable for most ecological data sets (Gauch, 1982). The PCA solution is often distorted into a horseshoe shape (with the toe either up or down) if beta diversity is moderate to high. The horseshoe can appear even if there is an important secondary gradient. In Figure 11.7 we cannot easily tell whether BN2 is at one end of a secondary gradient, or if its position at the end of axis 2 is merely a distortion. In extreme cases of the horseshoe effect, the gradient extremes are incurved, resulting in great difficulties of interpretation.

11.10.2   Correspondence Analysis

Correspondence Analysis (CA) is also known as reciprocal averaging, because one algorithm for finding the solution involves the repeated averaging of sample scores and species scores (citations). Instead of maximizing 'variance explained', CA maximizes the correspondence between species scores and sample scores. First Axis species scores and sample scores are assigned such that the weighted correlation between the two is maximized (Table 11.1b, Figure 11.8), where the 'weight' is the abundance of the species. Table 11.1b shows the original bryophyte data matrix, but sorted in order of species scores (rows) and sample scores (columns). Note that the structure of Table 11.1b is more apparent than in an alphabetical sort (Table 11.1a). The largest abundances fall on the diagonal, with small values and zeros off the diagonal. Indeed, the resorted table becomes a tabular version of a coenocline (Figure 11.2): most species have, with some noise, a unimodal response to CA axis 1. As with some of the previous ordinations, the first axis is a gradient from hardwoods to pines, with birch being intermediate.

The eigenvalue of the CA axis is equivalent to the correlation coefficient between species scores and sample scores (Gauch, 1982; Pielou, 1984). For the bryophyte data, the first eigenvalue is 0.805, which is fairly strong (indeed, the strong correlation can be visualized in Figure 11.8). It is not possible to arrange rows and/or columns in such a way that makes the correlation higher. The second and higher axes also maximize the correlation between species scores and sample scores, but they are constrained to be uncorrelated with (orthogonal to) the previous axes. The 2nd through 4th axes' eigenvalues are 0.284, 0.162, and 0.141, implying that the first axis is by far the most important.

If species scores are standardized to zero mean and unit variance, the eigenvalues also represent the variance in the sample scores (but not, as is often misunderstood, the variance in species abundance). In the context of CA, we term this variance the *inertia* of an axis. The sum of all eigenvalues is the total inertia (1.511 for the bryophyte data). Thus the percentage of inertia 'extracted' by the first two axes is 100 * (0.805 + 0.284)/1.511 = 72.1%.

Since CA is a unimodal model, species are represented by a point rather than an arrow (Figure 11.9). This is (under some choices of scaling; see ter Braak and Šmilauer, 2012)

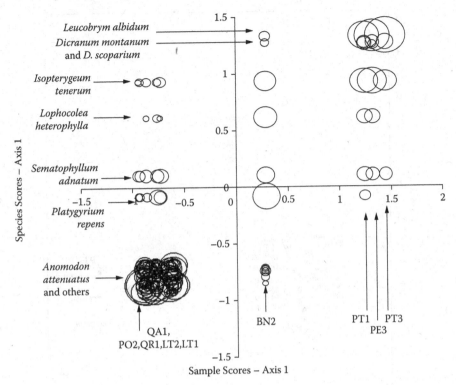

FIGURE 11.8

Correspondence Analysis species scores as a function of sample scores for the bryophyte data, illustrating the correlation that is maximized between species and samples. The sample weight (abundance) is indicated by the size of the circle.

the weighted average of the samples in which that species occurs. With some simplifying assumptions (ter Braak and Looman, 1987), the species score can be considered an estimate of the location of the peak of the species response curve (Figure 11.1).

The 2nd and higher axes of the CA solution, like those of PCA, can be distorted for data sets of moderate to high beta diversity (Figure 11.9). The CA distortion is called the *arch effect*, which is not as serious as the *horseshoe effect* of PCA because the ends of the gradients are not incurved. Nevertheless, the distortion is prominent enough to seriously impair ecological interpretation.

In addition to the arch, the axis extremes of CA can be compressed. In other words, the spacing of samples along an axis may not affect true differences in species composition. We suspect this is the case for our data (Figures 11.8, 11.9) because the hardwood trees and the pine trees form tight clusters at the opposite end of the first axis – much tighter than would be expected on the basis of dissimilarity (Table 11.2). Gradient compression can be quite blatant in simulated data sets (Figure 11.11). The problems of gradient compression and the arch effect led to the development of Detrended Correspondence Analysis.

11.10.3 Detrended Correspondence Analysis

Detrended Correspondence Analysis (DCA) eliminates the arch effect by *detrending* (Hill and Gauch, 1982). There are two basic approaches to detrending: by polynomials and by segments (ter Braak and Šmilauer 2012). Detrending by polynomials is the more elegant of

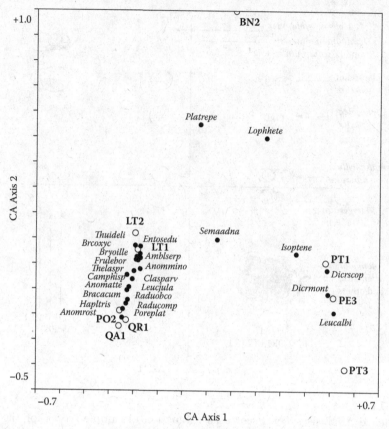

FIGURE 11.9
Correspondence Analysis of the bryophyte data.

the two: a regression is performed in which the second axis is a polynomial function of the first axis, after which the second axis is replaced by the residuals from this regression. Similar procedures are followed for the third and higher axes. Unfortunately, results of detrending by polynomials can be unsatisfactory and hence detrending by segments is preferred. To detrend the second axis by segments, the first axis is divided up into segments, and the samples within each segment are centered to have a zero mean for the second axis (see illustrations in Gauch 1982). The procedure is repeated for different 'starting points' of the segments. Although results in some cases are sensitive to the number of segments (Jackson and Somers, 1991), the default of 26 segments is usually satisfactory. Detrending of higher axes proceeds by a similar process.

The compression of the ends of the gradients is corrected by *nonlinear rescaling*. Rescaling shifts sample scores along each axis such that the average width (or 'tolerance'; Figure 11.1) is equal to 1. Figure 11.11 shows not only how the compression of CA disappears, but also how the species tolerances are equalized (without changing sample order). Rescaling has a beneficial consequence: the axes are scaled in units of beta diversity (SD units, or units of species standard deviations). Thus if the underlying gradient is important well known, it is possible to plot the DCA scores as a function of the gradient, and thereby determine whether the species 'perceive' the gradient differently than we measure it (Figure 11.11). Steeper slopes indicate zones of high beta diversity in such graphs.

Note that the shape of the species response curves may change if axes are rescaled (Figure

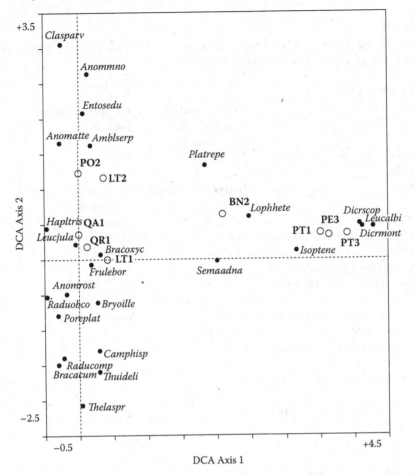

FIGURE 11.10
DCA of the bryophyte data.

11.11). Thus, skewness and kurtosis are largely artifacts of the units of measurement for which we choose to measure the environment. Since such measures are arbitrary with respect to nature, we are usually not too concerned if the Gaussian model (Figure 11.1) does not work too well.

For the bryophyte example, DCA no longer shows an arch effect (Figure 11.10). Because of the rescaling, the minimum sample score is zero for each axis. The maximum sample score is 3.9 along the axis, indicating that approximately 4 standard deviations of species response curves fits along the dominant gradient. With a beta-diversity this high, the samples at the left extreme of the gradient share few species with those at the right (confirmed in Table 11.1b). The first axis species scores correspond with what we know about the biology of the species: for example, *Anomodon attenuatus* and *Anomodon rostratus* are restricted to hardwood trees, and *Dicranum scoparium* and *Leucobryum albidum* are restricted to pines and birch. The second DCA axis has a beta diversity of 1.2 standard deviation units, reflecting low beta diversity. Thus, the opposite ends of the second axis are rather similar. A tentative interpretation is that the second axis represents a site effect, with forest #1 having lower scores than forest #2. Such an interpretation would not have been possible with the arch effect in CA (Figure 11.9). Now that the axes are scaled in units of beta

TABLE 11.5

Some of the major differences between NMDS and DCA. Bold face indicates what can be considered (in most cases) a better characteristic.

	NMDS	DCA
Computation time	High	*Low*
Distance metric	Highly sensitive to choice of distance metric	*Do not need to specify*
Simultaneous ordering of species and samples	No	**Yes**
Arch effect	**Rarely occurs**	Artificially and inelegantly removed
Related to direct gradient analysis methods	No	**Yes**
Need to pre-specify numbers of dimensions prior to interpretation	Yes	**No**
Need to specify parameters for number of segments, etc.	**No**	Yes
Solution changes depending upon number of axes viewed	Yes	**No**
Handles samples with high noise levels	No(?)	**Yes**
Guaranteed to reach the global solution	No	**Yes**
Results in measures of beta diversity	No	**Yes**
Used in other disciplines (e.g. psychometry)	**Widely**	No(?)
Axes interpretable as gradients	No	Yes
Derived from a model of species response to gradients	No	Yes

diversity, we can interpret the distances separating samples more easily. For example, the three pine samples remain close together in DCA, indicating their similarity is not merely a result of the gradient compression of CA.

11.10.4 Contrast between DCA and NMDS

DCA and NMDS are the two most popular methods for ordination without explanatory variables. The reason they have remained side-by-side for so long is because, in part, they have different strengths and weaknesses. While the choice between the two is not always straightforward, it is worthwhile outlining a few of the key differences (Table 11.5). Some of the issues are relatively minor: for example, computation time is rarely an important consideration, except for the hugest data sets. Some issues are not entirely resolved: the degree to which noise affects NMDS, and the degree to which NMDS finds local rather than global options still need to be determined (in the case of the bryophyte data, it took several iterations, with different optimization criteria, before the solution in Figure 11.5 was reached – so blind acceptance of the first solution is not recommended). Since NMDS is a distance-based method, all information about species identities is hidden once the distance matrix is created. For many, this is the biggest disadvantage of NMDS.

Note that the last two entries in Table 11.5 do not indicate which method has the advantage. This is perhaps the biggest difference between the two methods: DCA is based on an underlying model of species distributions, the unimodal model, while NMDS is not. Thus, DCA is closer to a theory of community ecology. However, NMDS may be a method of choice if species composition is determined by factors other than position along a gradient: For example, the species present on islands may have more to do with vicariance biogeography and chance extinction events than with environmental preferences – and for such a system, NMDS would be a better *a priori* choice. As De'ath (1999) points out, there are two classes of ordination methods - 'species composition representation' (e.g. NMDS) and 'gradient analysis' (e.g. DCA, CCA). The choice between the methods should ultimately be governed by this philosophical distinction.

11.11   Direct gradient analysis

In direct gradient analysis (DGA), species are directly related to measured environmental factors. Although DGA can be as simple as a scatterplot of species abundance as a function of position along a measured gradient, community data typically have many species and multiple gradients. Thus, DGA is best coupled with a dimension-reduction technique, i.e. ordination. Since multivariate DGA results in axes that are *constrained* to be a function of measured factors, *constrained ordination* is a synonym of DGA. In the methods described here, sample scores are constrained to be linear combinations of explanatory variables. As in regression, explanatory (environmental) variables can be continuous or nominal. Unlike ordinary least squares regression, significance is assessed with a Monte Carlo Permutation Procedure (Manly, 1992), and hence does not rely on distributional assumptions of a test statistic.

The two most commonly used constrained ordination techniques are Redundancy Analysis (RDA) and Canonical Correspondence Analysis (CCA). RDA is the constrained form of PCA, and is inappropriate under the unimodal model. CCA is the constrained form of CA, and therefore is preferred for most ecological data sets (since unimodality is common). CCA also is appropriate under a linear model, as long as one is interested in species composition rather than absolute abundances (ter Braak and Šmilauer, 2012). Since most of the discussion concerning CCA also relates to RDA, I will discuss the unique features of RDA briefly after the discussion of CCA.

11.11.1   Canonical Correspondence Analysis

Simply put, Canonical Correspondence Analysis is the marriage between CA and multiple regression. Like CCA, CA maximizes the correlation between species scores and sample scores (Figure 11.8). However, in CCA the sample scores are constrained to be linear combinations of explanatory variables. Because of the 'constraint', eigenvalues in CCA will be lower than in CA.

We can also describe the maximization in CCA as finding the best dispersion of species scores (Figures 11.12, 11.13). This view of CCA makes its link to unimodal models clear. If a combination of environmental variables is strongly related to species composition, CCA will create an axis from these variables that makes the species response curves (e.g. Figure 11.1) most distinct. The second and higher axes will also maximize the dispersion (or *inertia*) of species, subject to the constraints that these higher axes are linear combinations of the explanatory variables, *and* that they are orthogonal to all previous axis.

There are as many constrained axes as there are explanatory variables. The total 'explained inertia' is the sum of the eigenvalues of the constrained axes. The remaining axes are unconstrained, and can be considered 'residual'. The total inertia in the species data is the sum of eigenvalues of the constrained and the unconstrained axes, and is equivalent to the sum of eigenvalues, or total inertia, of CA. Thus, explained inertia, compared to total inertia, can be used as a measure of how well species composition is explained by the variables. Unfortunately, a strict measure of 'goodness of fit' for CCA is elusive, because the arch effect itself has some inertia associated with it – and it is not always clear whether this inertia belongs in the 'explained' or 'unexplained' portion.

CCA benefits from the advantages of multiple regression, including:

- It is possible that patterns result from the combination of several explanatory variables; these patterns would not be observable if explanatory variables are considered separately.

- Many extensions of multiple regression (e.g. stepwise analysis and partial analysis) also apply to CCA.

- It is possible to test hypotheses (though in CCA, hypothesis testing is based on randomization procedures rather than distributional assumptions).

- Explanatory variables can be of many types (e.g. continuous, ratio scale, nominal) and do not need to meet distributional assumptions.

- Of course, as with multiple regression, one needs to be aware of some caveats:

- In observational studies one cannot necessarily infer direct causation.

- The independent effects of highly correlated variables are difficult to disentangle. However, CCA (and regression) can test the null hypothesis that such variables are completely redundant.

- It is possible to 'overfit' the data as the number of variables approaches the number of samples (instead of $r^2=1$, the explained inertia will equal the total inertia and the CCA solution equals the CA solution). The solution is no longer 'constrained' by the variables.

- Noise in explanatory variables will have an effect on the predicted values (McCune, 1997). This is not usually a serious problem, because we are typically more interested in environmental variables and species than we are with these predicted values (i.e. sample scores).

- The interpretability of the results is directly dependent on the choice and quality of the explanatory variables.

- Although both multiple regression and CCA find the best linear combination of explanatory variables, they are not guaranteed to find the true underlying gradient (which may be related to unmeasured or unmeasurable factors), nor are they guaranteed to explain a large portion of variation in the data. Some ecologists have rejected CCA and other direct gradient analysis techniques because of this, but finding relationships between *measured* variables and species composition is actually a desirable attribute.

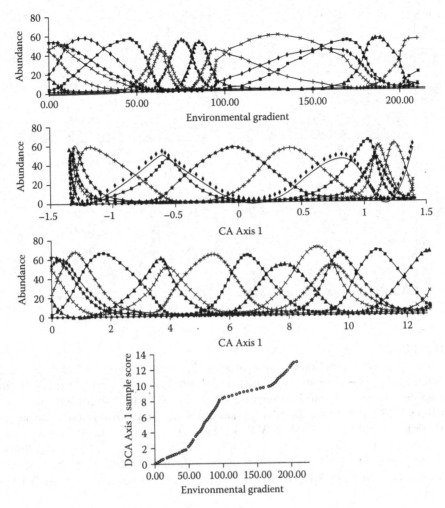

FIGURE 11.11
A hypothetical coenocline, illustrating the compression of CA and the rescaling of DCA.

One of the biggest advantages of CCA lies in the intuitive nature of its ordination diagram, or *triplot*. It is called a triplot because it simultaneously displays three pieces of information: samples as points, species as points, and environmental variables as arrows (or points). Figure 11.14 is a triplot of a CCA for the forested vegetation of the Tallgrass Prairie Preserve in Osage County, Oklahoma. Certain species (such as *Asimina triloba, Quercus muehlenbergii, Fraxinus muehlenbergii,* towards the right of the diagram, are found in conditions of high pH and calcium. Three tree species typical of crosstimbers forests (*Quercus stellata, Quercus marilandica,* and *Carya texana*) are found on the left. Crosstimbers forests are generally found on relatively acid (i.e. low pH), sandstone-derived soils (Francaviglia, 2000). The arrow representing % cover of water points upwards. Not surprisingly, wetland tree species (*Cephalanthus occidentalis* and *Salix nigra*) are located towards the top.

In many, if not most, data sets, CCA triplots can get very crowded. Solutions for this include:

- Separate the parts of the triplot into biplots or scatterplots (e.g. plotting the arrows in a different panel of the same figure).

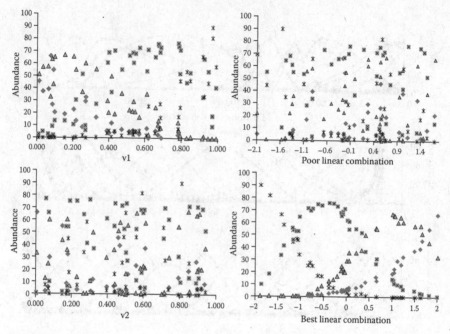

FIGURE 11.12

Species abundance as a function of explanatory variables in a hypothetical 4-species coen-ocline. Each species is represented by a different symbol. V1 is a 'better' variable than V2, because species are more clearly segregated along the V1 axis. Species have no apparent relationship to V2. Species also have no apparent relationships to the 'poor' linear combination $(0.1V1 - 0.2V2 + 0.3V3 + 0.7V4)$ but have very strong unimodal relationships to the 'best' linear combination $(-1.0V1 - 0.3V2 + 0.1V3 - 0.1V4)$ – which, by definition, is the CCA first axis. Note that the the 'best linear combination' appears to be a cleaned-up version of a mirror image of V1. The mirror image is because the coefficient for V1 is large and negative.

- Rescaling the arrows so that the species and sample scores are more spread out.

- Only plotting the most abundant species (but by all means, keep the rare species in the analysis).

- Omitting sample scores. After all, they are merely linear combinations of the environment. The samples are, in a sense, 'tools' for determining species-environment relationships – so their value as scores is limited. However, it is important to view sample scores to ascertain whether there may be outliers, or gaps in the data.

- Some combination of the above. Whatever is chosen, it is best to keep the particular objective of your study in mind.

- Noise in the species abundance data set is not much of a problem for CCA (Palmer, 1993). However, it has been argued that noise in the environmental data can be a problem (McCune, 1997). It is not at all surprising that noise in the predictor variables will cause noise in the sample scores, since the latter are linear combinations of the former. Species scores are much less sensitive to noise in the environment than are sample scores.

- It is important to point out that there is a common myth that 'symmetric, evenly spaced

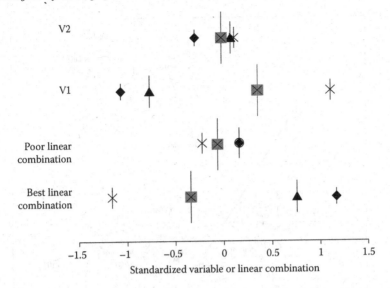

FIGURE 11.13
The weighted average location (i.e. species scores) of the 4 species in Figure 11.12, as a function of variables 1, 2, or linear combinations of four explanatory variables. Symbols for species are the same as in Figure 11.12, and vertical bars represent the summed abundance or 'weight' of the species. Note that the spread of scores is greater for V1 than for V2, and better for the 'best linear combination' than for the 'poor linear combination'. CCA chooses the coefficients for the best linear combination such that the dispersion (or *inertia*) of the species scores is maximized.

Gaussian curves' are critical assumptions for CCA (or DCA). The critical assumption is that species are arranged along gradients in unimodal functions.

11.11.2   Environmental variables in CCA

A concern is often expressed about the use of highly correlated variables. Such redundant variables are very common in ecology. For example, soil pH, and calcium are typically highly correlated with each other. As with multiple regression, it is difficult to disentangle the independent effects of such variables (as in Figure 11.14). However, they represent no major obstacle for graphical display. They are unlikely to affect the position of species and samples much, and the fact that they all end up pointing the same direction immediately makes their intercorrelations obvious. In general, small angles imply high positive correlations between variables, and arrows pointing in opposite directions will be negatively correlated. It is probably obvious that the choice of variables in CCA is crucial for the output. Meaningless variables will produce meaningless results. However, a meaningful variable that is not necessarily related to the most important gradient may still yield meaningful results (Palme,r 1993).

There are only as many 'constrained' or 'canonical' axes as there are independent environmental variables. Thus, if there are only two variables, the CCA solution is 2-dimensional. However, software packages such as CANOCO will present higher axes. These axes represent the 'residual' variation. It is possible for the first residual axis to have a higher eigenvalue than the first constrained axis. Residual axes are very useful in exploratory analyses: they can provide you with hints of what important variables might be missing.

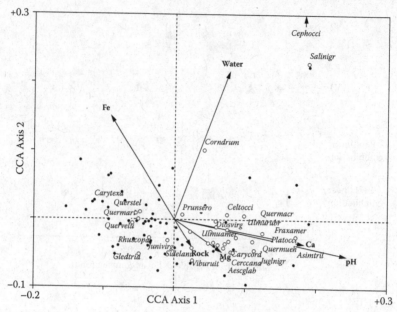

FIGURE 11.14

Triplot from a CCA of the forests in the Tallgrass Prairie Preserve, Oklahoma. Environmental variables are represented by blue arrows, samples (quadrats) by small open circles, and species by closed blue circles. The species are listed by the first four letters of the genus and the specific epithet, they include: *Aesculus glabra, Asimina triloba, Carya cordiformis, Carya texana, Celtis occidentalis, Cephalanthus occidentalis, Cercis canadensis, Cornus drummondii, Diospyros virginiana, Fraxinus americana, Gleditsia triacanthos, Juglans nigra, Juniperus virginiana, Platanus occidentalis, Prunus serotina, Quercus macrocarpa, Quercus marilandica, Quercus muehlenbergii, Quercus stellata, Quercus velutina, Rhus copallina, Salix nigra, Sideroxylon lanuginosa, Ulmus americana, Ulmus rubra, Viburnum rufidulum.*

If many variables are included in an analysis, much of the inertia becomes 'explained'. This is an analogous situation to multiple regression: the multiple r^2 or 'variance explained' increases as a function of the number of variables included. As the number of variables approaches the number of samples, then the 'explained inertia' approaches the total inertia, and the CCA solution approaches the CA solution. In other words, the ordination is no longer 'constrained' by the variables. It is very likely that the arch effect, which rarely occurs with low numbers of variables, will appear with higher numbers of variables.

In multiple regression, it is typical to include quadratic terms for explanatory variables. For example, if you expect a response variable to reach a maximum at an intermediate value of an explanatory variable, including this explanatory variable AND the square of the explanatory variable may allow a concave-down parabola to provide a reasonable fit. However, quadratic terms are not to be encouraged in CCA. This would be asking for trouble, as it may force an arch effect to appear.

Explanatory variables need not be continuous in CCA. Indeed, factors are very useful. In older programs (and internally within newer programs) they are dummy-encoded. A dummy variable takes the value 1 if the sample belongs to that category, and 0 otherwise. Factors are used if you have discrete experimental treatments, year effects, different bedrock types, or in the case of the bryophyte example (Table 11.1), host tree species.

Note that the use of dummy or factor variables *does not contradict* the assumption of gradient analysis. There may still be underlying gradients, along which the different levels of a factor can be arranged. Even an extreme case of a two-level factor (*e.g.* fertilization vs. not) represents two points along a gradient (e.g. a nutrient gradient).

As with regression, the outcome of CCA is highly dependent on the scaling of the explanatory variables. Unfortunately, we cannot know *a priori* what the best transformation of the data will be, and it would be arrogant to assume that our measurement scale is the same scale used by plants and animals. Nevertheless, we must make intelligent guesses. For example, it is likely that plants do not respond to soil chemical concentrations in a linear way. A 10 ppm difference is much more meaningful at low concentrations than it is at high concentrations. A logarithmic transformation (which emphasizes orders of magnitude of difference rather than absolute difference) is therefore likely to be much closer to the 'truth' than a linear scale.

Aspect (compass direction of a slope) clearly must be converted before it can be used. 359 degrees is almost the same direction as 2 degrees. Conversion to a factor indicating direction (N, S, E, W) or more detailed (NE, N, NW, W, SW, S, SE, E) or even a 16-point scale might be useful. Alternatively, a trigonometric conversion to an exposure index can be valuable (Roberts, 1986).

Any linear transformation of variables (e.g. kilograms to grams, meters to inches, Fahrenheit to Centigrade) will not affect the outcome of direct gradient analysis whatsoever.

There are many limitations to CCA (some of these were pointed out by McCune, 1997). However, most of these limitations are identical to the limitations of multiple regression. Foremost among these limitations is that correlation does not imply causation, and a variable that appears to be strong may merely be related to an unmeasured but 'true' gradient. As with any technique, results should be interpreted in light of these limitations.

11.11.3   Hypothesis testing

Hypothesis testing is straightforward with CCA by means of a randomization test (Manly, 1992). The observed first eigenvalue, or the sum of all eigenvalues is calculated for the data. Then this value is compared to the corresponding statistic calculated from each of many random permutations of the data. These permutations keep the actual data intact, but randomly associate the environmental data with the species data. If the true statistic is greater than or equal to 95% of the statistics from the permuted data, we can reject the null hypothesis that species are not related to the environment. The first eigenvalue test determines whether the first CCA axis is stronger than random expectation, and the trace statistic (sum of all canonical axes) tests whether there is an overall relationship between species and environment. Both tests usually yield similar results, but exceptions do occur. It is important to note that permutations tests in CCA and RDA are special cases of the popular technique MRPP (Palmer et al., 2008), and thus MRPP as a standalone operation is rarely necessary.

11.11.4   Redundancy Analysis

As mentioned previously, most of the discussion of CCA pertains to Redundancy Analysis (RDA). However, note that RDA is a linear method. Some of the special properties of RDA include:

- Since it is a linear method, species as well as environmental variables are represented by arrows. In most cases, it is best to represent the two sets of arrows in two figures for ease of display.

- CCA focuses more on species composition, i.e. *relative* abundance. Thus, if you have a gradient along which *all* species are positively correlated, RDA will detect such a gradient while CCA will not.

- With RDA, it is possible to use 'species' that are measured in different units. If so, the data must be centered and standardized. But in general, as an ordination technique, the species should not be standardized.

- RDA can useful when gradients are short. In particular, RDA may be the method of choice in a short-term experimental study. In such cases, the treatments are the coded as factors or dummy variables. The sample ID or block might be a covariable in a partial RDA, if one wishes to factor out local effects.

- 'variance explained' is actually a variance explained, and not merely inertia. Thus, variation partitioning, and interpretation of eigenvalues, are more straightforward than for CCA.

Note that CCA has a *linear face* and is thus perfectly acceptable for short gradients. In such cases, RDA will uncover overall patterns in abundance. For example, fertilizing a grassland might result in an increase in abundance of all plant species, and RDA will reveal that. However, CCA would pick out the more subtle distinctions in *relative* abundance, i.e. which species perform differentially better or worse after fertilizer application.

11.12   Extensions of direct ordination

CCA and RDA are useful techniques in their own right, but it is their extensions that make them particularly useful for community ecologists. Here they are only described briefly, but the reader is referred to Šmilauer and Lepš (2014) for fuller explanations with examples and case studies.

- *Partial ordination:* Partial ordination is useful when one wishes to 'factor out' nuisance effects such as block effects, known biases, or uninteresting gradients, and examine the 'residual' variation in species composition. More commonly, it is useful in sequentially 'factoring out' groups of variables in factorial designs, so one can examine independently the effects of different factors. A simple case would be when multiple sites are sampled at multiple times. Partial analysis allows you to independently examine variation due to sites and variation due to time. Partial ordination is the backbone for other more advanced methods such as stepwise ordination, variation partitioning, etc.

- *Stepwise Ordination*: Stepwise ordination is useful when an ecologist has a large number of explanatory variables but wishes to choose a smaller subset of them for explaining compositional variation. It chooses variables in the order that explains the most residual variation. It is a useful exploratory technique but its value for statistical inference is questionable due to multiple comparisons. It also greatly declutters ordination triplots.

- *Cross validation*: Many if not most applications of ordination are used in an exploratory mode, in which finding patterns is of greater importance than testing *a priori* hypotheses objectively. Thus, a number of *ad hoc* decisions and 'data diving' are employed during analysis. However, dividing the data set into two (typically halves) allows part to be used for an exploratory portion to discover patterns, and a second part to be used as a confirmatory analysis to test these patterns against null expectations. Cross validation as applied to ordination is discussed in Hallgren et al. (1999)

- *Variation partitioning*: Very often our explanatory variables fall into different categories that are difficult to disentangle. For example, both land use and topography may influence species composition, yet land use may be highly correlated with topography. Variation partitioning allows you to ascertain that amount of variation in composition uniquely explained by topography, that amount uniquely explained by land use, and that variation which is impossible (using the current data) to disentangle (Peres-Neto et al., 2006)

- *Principal Coordinates of Neighbor Matrices (PCNM)*: Spatial autocorrelation is often considered a nuisance factor for statistical inferences. Variation partitioning using spatial variables (e.g. x and y coordinates and simple polynomial functions thereof) as one set of variables is one way to assess this problem. However, spatial autocorrelation does not necessarily behave as a smooth function. PCNM is a more sophisticated way that incorporates intersample distances, to more cleanly separate spatial and nonspatial components of the data (Borcard et al., 2004).

- *Principal Response Curves*: Long-term ecological experiments typically involve following how treatments affect composition through time, in contrast to a control that is also sampled through time. Principal Response Curves provide an elegant way to analyze AND display treatment effects on species composition through time (ter Braak and Šmilauer, 2014).

- *Utilizing additional matrices*: While community ecology largely studies the effects of treatment on species composition, there is increasing interest in the analysis of species traits, or species evolutionary history (e.g. Pillar and Duarte, 2010). Relatively simple extensions to eigenvector methods allow the analysis AND display of such traits or history. There also exists a diversity of techniques (e.g. Proctrustes rotations) to compare the results of one ordination to another (see Šmilauer and Lepš, 2014).

11.13 Conclusions

Ecologists have typically taken one of two approaches to multivariate analysis: to assume almost nothing of variation in community structure, or to assume a basic model that species respond to gradients. The first approach has often led to a reliance on classification and NMDS, while the second has led to eigenanalysis-based gradient analysis methods. Yet other ecologists promote pluralism and advocate trying all approaches simultaneously (Økland, 2007). Nevertheless, the debate remains heated (e.g. Austin, 2013). As ter Braak and Šmilauer (2015) put it, *"Some users may be happy with an ordination ... [that] ... correctly reflects the calculated dissimilarities, but we are happier with an ordination ... [that] ... represents the true underlying ecological gradients."* It is important to realize that the disputes are not merely quibbles over methodology, but represent a healthy scientific discussion of the fundamental nature of ecological communities.

Bibliography

[1] Austin, M. P. 1968. An ordination study of a chalk grassland community. *Journal of Ecology*, 56:739-757.

[2] Austin, M. P. 1987. Models for the analysis of species' response to environmental gradients. *Vegetatio*, 69:35-45.

[3] Austin, M. P. 2013. Inconsistencies between theory and methodology: a recurrent problem in ordination studies. *Journal of Vegetation Science*, 24:251-268.

[4] Beals, E. W. 1984. Bray-Curtis ordination: an effective strategy for analysis of multivariate ecological data. *Advances in Ecological Research*, 14:1-55.

[5] Borcard, D., Legendre, P., Avois-Jacquet, C, and Tuomisto, H. 2004. Dissecting the spatial structure of ecological data at multiple scales. *Ecology*, 85:1826-1832.

[6] Bray, J. R., and Curtis, J. T. 1957. An ordination of the upland forest communities of southern Wisconsin. *Ecological Monographs*, 27:325-349.

[7] Curtis, J. T., and McIntosh, R. T. 1951. An upland forest continuum in the prairie-forest border region of Wisconsin. *Ecology*, 32:476-496.

[8] De'ath, G. 1999. Principal curves: a new technique for indirect and direct gradient analysis. *Ecology*, 80:2237-2253.

[9] Fasham, M. J. R. 1977. A comparison of nonmetric multidimensional scaling, principal components and reciprocal averaging for the ordination of simulated coenoclines and coenoplanes. *Ecology*, 58:551-561.

[10] Francaviglia, R. V. 2000. The Cast Iron Forest. University of Texas Press, Austin.

[11] Fuhlendorf, S. D., and Smeins, F. E. 1997. Long-term vegetation dynamics mediated by herbivores, weather and fire in a Juniperus-Quercus savanna. *Journal of Vegetation Science*, 8:819-828.

[12] Gauch, H. G., Jr. 1982. Multivariate Analysis and Community Structure. Cambridge University Press, Cambridge.

[13] Gauch, H. G., Jr. 1982. Noise reduction by eigenvalue ordinations. *Ecology*, 63:1643-1649.

[14] Gleason, H. A. 1917. The structure and the development of the plant association. *Bulletin of the Torrey Botanical Club*, 43:463-481.

[15] Goodall, D. W. 1954. Objective methods for the classification of vegetation. III. An essay in the use of factor analysis. *Australian Journal of Botany*, 1:39-63.

[16] Grime, J. P. 1979. Plant strategies and vegetation processes. Wiley & Sons, Chichester.

[17] Hallgren, E., Palmer, M. W. and Milberg, P. 1999. Data diving with cross validation: an investigation of broad-scale gradients in Swedish weed communities. *Journal of Ecology*, 87:1037-1051.

[18] Hill, M. O. 1973. Reciprocal averaging: an eigenvector method of ordination. *Journal of Ecology*, 61:237-249.

[19] Hill, M. O. 1979. DECORANA - A FORTRAN program for detrended correspondence analysis and reciprocal averaging. Cornell University, Ithaca, New York.

[20] Hutchinson, G. E. 1957. Concluding remarks. *Cold Spring Harbor Symposium on Quantitative Biology,* 22:415-427.

[21] Jackson, D. A., and Somers, K. M. 1991. Putting things in order: the ups and downs of detrended correspondence analysis. *American Naturalis,t* 137:704-712.

[22] Kruskal, J. B. 1964. Nonmetric multidimensional scaling: a numerical method. *Psychometrika,* 29:115-129.

[23] Legendre, P., and Legendre, L. 1998. Numerical Ecology 2nd English edition. Elsevier, Amsterdam.

[24] Manly, B. F. J. 1992. Randomization and Monte Carlo methods in biology. Chapman and Hall, New York.

[25] McCune, B. 1997. Influence of noisy environmental data on canonical correspondence analysis. *Ecology,* 78:2617-2623.

[26] McCune, B., and Mefford, M. J. 1999. PC-ORD. Multivariate Analysis of Ecological Data, Version 4. MjM Software Design, Glendenen Beach.

[27] McIntosh, R. P. 1985. The Background of Ecology. Cambridge University Press, Cambridge, Great Britain.

[28] Minchin, P. R. 1987. Simulation of multidimensional community patterns: towards a comprehensive model. *Vegetatio,* 71:145-156.

[29] Økland, R. H. 2007. Wise use of statistical tools in ecological field studies. *Folia Geobotanica,* 42:123-140.

[30] Oksanen, J., Blanchet, F. G., Kindt, R. Legendre, P., Minchin, P. R., O'Hara, R. B., Simpson, G. L., Solymos, P., Stevens, M. H. H., and Wagner, H. 2013. vegan: Community Ecology, Package. R package version 2.0-8.

[31] Oksanen, J., and Tonteri, T. 1995. Rate of compositional turnover along gradients and total gradient length. *Journal of Vegetation Science,* 6:815-824.

[32] Palmer, M. W. 1986. Pattern in corticolous bryophyte communities of the North Carolina piedmont: Do mosses see the forest or the trees? *Bryologist,* 89:59-65.

[33] Palmer, M. W. 1993. Putting things in even better order: the advantages of canonical correspondence analysis. *Ecology,* 74:2215-2230.

[34] Palmer, M. W. 2007. Species-area curves and the geometry of nature. Pages 15-31 *in* D. Storch, P. A. Marquet, and J. H. Brown, editors. Scaling Biodiversity. Cambridge University Press, Cambridge.

[35] Palmer, M. W., and White, P. S. 1994. On the existence of communities. *Journal of Vegetation Science,* 5:279-282.

[36] Palmer, M. W., McGlinn, D. J., Westerberg, L., and Milberg, P. 2008. Indices for detecting changes in species composition: some simplifications of RDA and CCA. *Ecology,* 89:1769-1771.

[37] Parker, V. T., and Pickett, S. T. A. 1998. Historical contingency and multiple scales of dynamics within plant communities. Pages 171-191 *in* D. L. Peterson and V. T. Paker, editors. Ecological Scale. Columbia University Press, New York.

[38] Peet, R. K. 1980. Ordination as a tool for analyzing complex data sets. *Vegetatio*, 42:171-174.

[39] Peres-Neto, P. R., Legendre, P., Dray, S., and Borcard, D. 2006. Variation partitioning of species data matrices: Estimation and comparison of fractions. *Ecology*, 87:2614-2625.

[40] Pielou, E. C. 1984. The Interpretation of Ecological Data: A Primer on Classification and Ordination. Wiley, New York.

[41] Pillar, V. D., and Duarte, L. D. S. 2010. A framework for metacommunity analysis of phylogenetic structure. *Ecology Letters* 13:587-596.

[42] Prentice, I. C. 1977. Non-metric ordination methods in ecology. *Journal of Ecology*, 65:85-94.

[43] Roberts, D. W. 1986. Ordination on the basis of fuzzy set theory. *Vegetatio*, 66:123-131.

[44] Šmilauer, P., and Lepš, J. 2014. Multivariate Analysis of Ecological Data using CANOCO 5. Cambridge University Press, Cambridge.

[45] ter Braak, C. J. F. 1985. CANOCO - A FORTRAN program for canonical correspondence analysis and detrended correspondence analysis. IWIS-TNO, Wageningen, The Netherlands.

[46] ter Braak, C. J. F. 1986. Canonical correspondence analysis: a new eigenvector technique for multivariate direct gradient analysis. *Ecology*, 67:1167-1179.

[47] ter Braak, C. J. F., and Looman, C. W. N. 1986. Weighted averaging, logistic regression and the Gaussian response model. *Vegetatio*, 65:3-11.

[48] ter Braak, C. J. F., and Looman, C. W. N. 1987. Regression. Pages 29-77 *in* R. H. G. Jongman, C. J. F. ter Braak, and O. F. R. van Tongeren, editors. Data Analysis in Community and Landscape Ecology. Pudoc, Wageningen, The Netherlands.

[49] ter Braak, C. J. F., and Prentice, I. C. 1988. A theory of gradient analysis. *Advances in Ecological Research*, 18:271-313.

[50] ter Braak, C. J. F., and Šmilauer, P.. 2012. Canoco Reference Manual and User's Guide. Software for Ordination (version 5.0) Biometris, Wageningen.

[51] Whittaker, R. H. 1967. Gradient analysis of vegetation. *Biological Reviews* 42:207-264.

[52] Whittaker, R. H. 1969. Evolution of diversity in plant communities. *Brookhaven Symposia on Biology*, 22:178-195.

Part II

Topics in Ecological Processes

12

Species distribution models

Otso Ovaskainen

Helsinki: University, Helsinki, Finland

CONTENTS

12.1 Aims of species distribution modelling

During the past decades there has been a great increase in the development and use of species distribution models [2, 4, 18], also referred to as bioclimate envelope models [3, 40], habitat suitability models [5, 27] or ecological niche models [27]. These models generally aim at explaining and predicting variation in the occurrences or abundances of species as a function of environmental and spatial variables.

Common purposes of applying species distribution models include those related to inference, prediction, and projection [18, 27]. Questions related to inference (or explanation) ask which environmental variables (including e.g. climatic, geographic, anthropogenic and spatial variables) control the distributions of species. The distinction between prediction and projection is that prediction relates to interpolation (the training data are from similar conditions as the situation to which the predictions are to be made), whereas projection relates to extrapolation (the training data are outside the conditions to which the projections are to be made), making the latter a more challenging task.

When using species distribution models for inference, ideally one would like to establish causal links between environmental variables and species occurrences, i.e. to learn about ecological species niches as well as ecological interactions among species. However, most species distribution models, as well as data used to parameterize them, allow only for correlative inference. Due to this mismatch, species distribution models have been criticized to be poorly linked to ecological theory [4, 27], and to be prone to misuse due to confusion between what the models actually deliver and what users wish that they would express [3]. It has been argued that many of these caveats could be overcome by integrating ecological processes, such as dispersal and species interactions, more explicitly into species distribution modelling [21, 22, 41, 48]. However, as developing mechanistic yet general modelling approaches is not easy, correlative approaches continue to be most widely used.

Predictions and projections are commonly needed in the situation where species occurrences have been sampled in parts of the study region, and one would like to use these data to predict species occurrences in un-sampled areas. The motivation for making such predictions include the need to produce species distribution maps, or to pinpoint areas where the climatic and habitat conditions are suitable, thus identifying regions where the species of interest is highly likely to occur [3, 5]. Other examples where predictions and projections are needed include the evaluation of the consequences of climate or land use change, e.g. the evaluation of management options aimed to increase the viability of a species of economical or conservation interest, or to decrease the abundance of an invasive species or a pest species [6, 24, 25, 40].

Often the interest is not at the species level but at the level of a species community. In the 'predict first, assemble later' strategy [19], species communities are simply viewed as sums of their component species. This is the logic behind so called stacked species distribution models [9], in which community-level characteristics, such as species richness or community composition, are obtained by summing up the predictions of species-level models. However, it has become increasingly acknowledged that ecological interactions among the species influence their distributions and consequently that the distributions of different species cannot be considered as independent of each other [51]. Consequently, there has been an increasing interest in the development and application of models that follow the 'assemble and predict together' strategy [19] by accounting for dependencies among the species during the modelling process, e.g. by applying so called joint species distribution models [12, 15, 28, 36, 38, 39, 45, 47, 49].

Community-level species distribution modelling is used to explain which environmental variables control variation in community-level characteristics [28, 39], which species interact with each other and in which ways [12, 36, 37, 38, 45, 47], and how species traits [1, 7, 44] and phylogenetic relationships [1, 29] influence species interactions and the responses of the species to the environment. Predictions of community-level characteristics to areas where the species have not been sampled can used e.g. to identify areas of high species richness [12, 38], or to decompose the study area into regions that have similar species communities and that thus form e.g. natural management units [20]. If the species show strong co-occurrence patterns, one may use joint species distribution models to generate improved predictions for the occurrences of a focal species [36, 47] by utilizing information about the occurrences of the other species, the other species of which are called indicator species in this context [10].

There is a large array of statistical methods applicable for species distribution modelling, including many parametric and non-parametric alternatives for the functional relationships between environmental and species variables, such as generalized linear models, generalized additive models, regression tree models, and maximum entropy models [16, 17, 18, 24, 25, 35, 42, 43]. The diversity of methods has generated a rich literature on methods comparisons, where performances of different methods have been compared by applying them to real or artificial datasets [16, 26, 35]. Consensus methods that combine the predictions of different types of models have also been proposed [33].

In this chapter, we will exemplify both species-level and community-level approaches to species distribution modeling. To keep the treatment concise, we have necessarily needed to make choices among the many kinds of data types and modelling techniques that we could have illustrated. Concerning the data types, we will focus on presence-absence data, which data consist of sampling units (e.g. plots) that are scored for the presences or absences of the species. Another commonly applied data type is so called presence-only data, which consist of the locations of known observations, without systematic information indicating areas where the species are absent. While presence-absence data typically arise from scientific surveys, presence-only data are often based on museum records or citizen science projects. As observations can be missing from an area either because the species does not occur there, or because there is lack of observation effort, statistically rigorous analysis of presence-only data is challenging [17, 18, 50]. For a reader interested in point pattern analyses with presence-only data, we recommend the review of reference [46].

As is the case with almost any statistical exercise, the researcher aiming to apply species distribution models faces the challenge of choosing among the many kinds of possible statistical modelling frameworks. In this chapter we will restrict the treatment to one particular family of models, namely generalized linear mixed models. The reason for this choice is that they form a flexible, general and widely applied modeling framework that allows one to incorporate many features, such as hierarchical and multivariate models, that we wish to illustrate in this chapter. We start by introducing example data that will be used thorough this chapter.

12.2   Example data used in this chapter

To illustrate different techniques and uses of species distribution modeling, we use plant data from the state of Victoria in Australia. The data are a subset from the Victorian Biodiversity Atlas, which is part of the Atlas of Living Australia (http://www.ala.org.au/). The data used here include the presence-absences of the 100 most common herb species in 3000 sampling locations. These locations are a randomly selected subset of the total of ca. 30,000 locations included in the Victorian Biodiversity Atlas data. Fig. 12.1 illustrates the distribution of one of the most common and one of the rarest species among these 100 herb species. Measuring commonness by prevalence p (fraction of occupied sites), we have $p = 0.28$ for the common species (Ivy leaf Violet; *Viola hederacea*; panel A) and $p = 0.02$ for the rare species (Small Purslane; *Calandrinia eremaea*; panel B). Panel C of Fig. 12.1 shows the variation in species richness, i.e. the total number of species observed in any site. In these data, species richness varies between 0 and 30 species. To keep the analyses simple, we will relate the occurrences of the species to two environmental covariates only, which measure salinity (Fig. 12.1D) and precipitation (Fig. 12.1E). In all analyses of this chapter, we normalized the values of these environmental covariates to zero mean and unit variance. To address questions related both to inference and prediction, we have divided the 3000 sampling sites to those to be used to parameterize the model (to be called training sites) and those that are not used for model parameterization but for which model predictions are to be compared to the data (to be called validation sites). To make the task of prediction challenging, we randomly selected 10% of the sites (300 sites) as training sites, and left the remaining 90% of the sites (2700 sites) as validation sites (Fig. 12.1F).

12.3   Single species distribution models

Within the generalized linear modelling framework [13, 23], a response variable (y) is assumed to be generated by a particular distribution, such as normal, Bernoullli or Poisson. In the context of species distribution modelling, y is usually the occurrence (i.e., presence or absence) or abundance of the species. We assume that the species occurrence or abundance

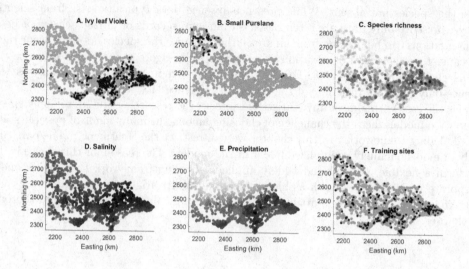

FIGURE 12.1

Example data used in this chapter to illustrate species distribution modelling. In all panels, the dots show 3000 sampling locations from which environmental and species data have been acquired. In panels A and B the black dots show the presences and grey dots the absences of two species that we consider in the context of single species modelling: the Ivy leaf Violet (A) and the Small Purslane (B). Panel C shows variation in species richness, the lightest color corresponding to the presence of none and the darkest to the presence of 30 species out of the community of 100 herbs considered here. Panels D and E show variation in two environmental covariates, namely salinity (D) and precipitation (E), that we use as predictors for species distribution modelling. The lightest and the darkest colors correspond to the smallest and highest values in the data. Panel F shows 300 randomly selected training sites (black dots) used to parameterize the models. The remaining 2700 validation sites (grey dots) were used to test the model predictions.

has been measured in a set of sites $i = 1, ..., n$, and denote the occurrence or abundance of the species in site i by y_i. We assume that in these sites, also a set of v explanatory variables (x_{ik}) have been measured, with $k = 1, ...v$. The expectation of y_i is related to the explanatory variables as

$$E(y_i) = g^{-1}(L_i),\tag{12.1}$$

where

$$L_i = \Sigma_{k=1}^{v} x_{ik}\beta_k \tag{12.2}$$

is the linear predictor for the site i, and g is a link function. The regression parameter β_k measures how the occurrences or abundances of the species depend on the covariate k. The $n \times v$ matrix $\mathbf{X}$ with elements x_{ik} is called the design matrix. To include an intercept, which models the mean occurrence probability of the species, we set the values in the first column of the design matrix to $x_{i1} = 1$ for all i.

The choice of the statistical distribution and the related link function and error distribution depend on the nature of the data, i.e. if species occurrence is measured as presence-absence or as abundance, and if abundance is measured as the counted number of individuals, or at a continuous scale such as biomass. The Victorian herb data are of presence-absence nature. We thus denote occupied sites by $y_i = 1$ and empty sites by $y_i = 0$, and model the probability of a site being occupied using the Bernoulli distribution. To convert the linear predictor into the probability scale, we use the probit link function $g(p) = \Phi^{-1}(p)$, where Φ is the cumulative distribution function of the standard normal distribution $N(0,1)$, and thus write

$$\Pr(y_i = 1) = \Phi(L_i). \tag{12.3}$$

For binary data, the choice of the probit link function instead of e.g. a logit or some other link function is somewhat arbitrary from the biological point of view. One mathematical reason why the probit regression is convenient is that Eq. 12.3 can be written equivalently as $y_i = 1_{z_i > 0}$, where $z_i = L_i + \epsilon_i$, and $\epsilon_i \sim N(0,1)$. This formulation makes it possible to parameterize the model with the help of the many techniques available for normally distributed data.

We will denote model parameters generically by $\boldsymbol{\theta}$. In the present model, the parameters consist just of the regression coefficients $\boldsymbol{\beta}$, $\boldsymbol{\theta} = (\boldsymbol{\beta})$. The likelihood of observing the data $\boldsymbol{y}$, given the model parameters, is given by

$$p(\boldsymbol{y}|\boldsymbol{\theta}) = \Pi_{i=1}^n \left[\Phi(L_i) y_i + (1 - \Phi(L_i))(1 - y_i) \right], \tag{12.4}$$

where the linear predictor L_i depends on the model parameters $\boldsymbol{\theta} = (\boldsymbol{\beta})$ as specified by Eq. 12.2. We note that Eq. 12.4 assumes that the species occurrences are independent in space after accounting for the effects of the covariates. With spatially explicit data, such as considered here, this assumption is not likely hold. However, let us start by assuming spatial independence, and then later relax this assumption.

In this chapter we will perform model fitting with Bayesian inference. For doing so, a prior distribution for the model parameters needs to be defined. Denoting by $p(\boldsymbol{\theta})$ the density of the prior distribution, the density of the posterior distribution $p(\boldsymbol{\theta}|\boldsymbol{y})$ can be obtained through the Bayes rule as $p(\boldsymbol{\theta}|\boldsymbol{y}) \propto p(\boldsymbol{\theta})p(\boldsymbol{y}|\boldsymbol{\theta})$, where for the present model $p(\boldsymbol{y}|\boldsymbol{\theta})$ is given by Eq. 12.4. As we wish to compare inference obtained from single- and joint-species distribution models, we will define the prior so that the single species distribution model is a special case of the joint species distribution model. For this reason, we specify the prior disribution for the present model, as well as for all other models of this chapter, in the section *Prior distributions*, found in the end of this chapter.

For the case study with the Victorian herb data, we use as predictors salinity and precipitation. We include as predictors also their squared values to allow the species occurrence probabilities to peak at an intermediate value of the environmental covariates. Thus the design matrix $\mathbf{X}$ has $v = 5$ columns, which correspond to the intercept, and the main and the squared effects of each of the two environmental variables.

We fitted the above described model separately for the two focal species of Ivy leaf Violet and Small Purslane. Their responses to the environmental covariates are illustrated in Fig. 12.2. The occurrence probability of Ivy leaf Violet peaks with an intermediate level of both salinity and precipitation (Fig. 12.2AC). In contrast, the occurrence probability of Small Purslane decreases with both environmental variables (Fig. 12.2EG). However, these responses have much more uncertainty than those of the Ivy leaf Violet, simply due to the fact that the data contain only few occurrences of Small Purslane and thus there is limited statistical power to infer the influence of the covariates.

A central question with any statistical modelling exercise relates to the explanatory power of the model: how much of the variation in the response variable does the model explain? One simple measure for quantifying the explanatory power of a model fitted to binary data is Tjur's R^2, which is defined as the mean model prediction for those sites where the species occurs, minus the mean model prediction for those sites where the species does not occur [48]:

$$R^2 = \frac{\sum_i y_i \Phi(L_i)}{\sum_i y_i} - \frac{\sum_i (1 - y_i) \Phi(L_i)}{\sum_i 1 - y_i} \tag{12.5}$$

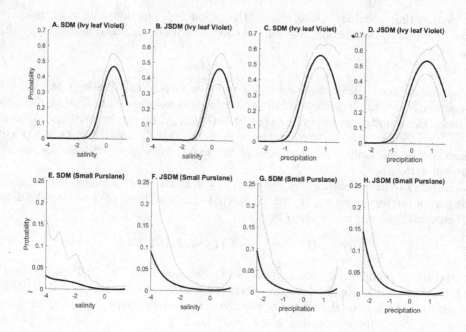

FIGURE 12.2

Responses of the two focal species to environmental covariates. Shown are model-predicted occurrence probabilities as a function of salinity (with precipitation set to its mean value) and precipitation (with salinity set to its mean value). The black lines show the posterior mean of the response curve and the grey lines the 90% posterior quantiles. The upper panels (A-D) correspond to Ivy leaf Violet and the lower panels (E-H) to Small Purslane. The first and third columns (panels ACEG) are based on single species distribution models, whereas the second and fourth columns (panels BDFH) are based on a joint species distribution model.

The values of Tjur's R^2 fall between -1 and 1, with the value of $R^2 = 1$ corresponding to the ideal case where the model predicts both presences and absences with certainty. The value of $R^2 = 0$ corresponds to the case where the model does not discriminate presences and absences any better than at random, and $R^2 < 0$ to the unlikely case in which the model discriminates presences and absences worse than at random. If we evaluate the model predictions over the same training sites to which the model has been fitted to, we obtain Tjur's $R^2 = 0.28$ for Ivy leaf Violet and Tjur's $R^2 = 0.12$ for Small Purslane (Table 12.1). However, as the same data are used both to fit the model and to evaluate its performance, such computed R^2 values may be inflated due to overfitting. If computing Tjur's R^2 values over the validation sites, we obtain the somewhat smaller but more relevant values of $R^2 = 0.24$ for Ivy leaf Violet and $R^2 = 0.10$ for Small Purslane.

Species distribution models are always likely to include only a subset of those variables that in reality influence the distribution of a species [5, 14, 27, 30], the influence of the missing variables thus remaining unexplained. It can be difficult to improve the model by adding more explanatory variables, because including too many potential explanatory variables makes it difficult to identify their effects, because the most relevant covariates may not be known, or because their measurements may not be available. One way of accounting for un-modelled processes is to use a random effect, which may have a spatial or other such structure.

TABLE 12.1

Variation in species occurrence explained by different versions of species distribution models. We report the Tjur's R^2 [48] values for two focal species: Ivy leaf Violet (a common species shown in Fig. 12.1A) and Small Purslane (a rare species shown in Fig. 12.1B), computed for the training (validation) data. SDM refers to a single species distribution model that contains only the focal species, JSDM to a joint species distribution model that models the occurrences and co-occurrences of all the 100 species simultaneously. The models include either covariates (salinity and precipitation) or space (spatially structured residual for SDMs and spatially structured latent factors for JSDMs), or both of these. For JSDMs, the predictions have been done either without accounting for the occurrences of the other species, or assuming that the occurrences of all other 99 species than the focal species are known in all sites and predicting the occurrence probabilities of the focal species conditional on those.

Model	Ivy leaf Violet	Small Purslane
SDM (covariates)	0.28 (0.24)	0.12 (0.10)
SDM (space)	0.36 (0.27)	0.15 (0.09)
SDM (covariates+space)	0.38 (0.29)	0.11 (0.09)
JSDM (covariates)	0.26 (0.23)	0.11 (0.09)
JSDM (space)	0.51 (0.29)	0.21 (0.10)
JSDM (space+other species)	0.64 (0.34)	0.63 (0.16)
JSDM (covariates+space)	0.46 (0.30)	0.14 (0.10)
JSDM (covariates+space+other species)	0.64 (0.33)	0.68 (0.19)

There are many methods for accounting for spatial autocorrelation, such as autocovariate regression, spatial eigenvector mapping, generalized least squares methods, autoregressive models and generalized estimating equations [11, 14, 31]. We will assume a spatially structured residual, and thus extend the model as

$$L_i = \Sigma_{k=1}^{v} x_{ik}\beta_k + \eta_i, \qquad (12.6)$$

where η_i is the random effect for site i. To impose a spatial structure, we assume that the vector of random effects follows the multivariate normal distribution $\eta \sim N(\mathbf{0}, \mathbf{V})$. We assume an exponentially decaying spatial covariance function, so that $V_{ij} = \lambda^2 \exp(-d_{ij}/\alpha)$, where λ^2 is the variance, d_{ij} is the distance between sites i and j, and the parameter α measures the spatial scale of autocorrelation. For this model, the model parameters include $\theta = (\beta, \lambda, \alpha, \eta)$, and the prior density decomposes as $p(\theta) = p(\beta, \lambda, \alpha)p(\eta|\lambda, \alpha)$, illustrating the hierarchical nature of the model. We define the prior $p(\beta, \lambda, \alpha)$ that we assumed for the plant case study in the section *Prior distributions*.

Including a spatial random effect improves the explanatory power of the models, as seen by comparing the performance of the model variant SDM (covariates + space) to that of SDM (covariates) for Ivy leaf Violet (Table 12.1). However, the improvement is much greater if the model's predictive power is evaluated against the training data than if it is evaluated against the validation data, indicating partial overfitting. To avoid overfitting, the prior for λ has been chosen so that it shrinks the values of η_i towards zero (see *Prior distributions*). Here we have deliberately chosen somewhat arbitrary priors rather than optimizing them to maximize the model's performance. In more refined analysis, the prior for the spatial random effect could be tuned to improve the model's predictive performance for the validation data. In addition to improving the model fit, the inclusion of a spatial random effect can bring information about the spatial scale α at which the data involves unexplained variation. For Ivy leaf Violet, the posterior median (95% credibility interval) of the spatial scale α is 300 (200...800) km, whereas for the Small Purslane it is 400 (50...800) km. As illustrated by the wide credibility intervals, with limited data the accurate estimation of the spatial scale parameter is difficult.

We may also consider a model which includes only the spatial random effect η but no environmental predictors, and thus the design matrix consists of only the intercept column. With our example data, such a model has essentially equally good predictive performance as the model with covariates (Table 12.1). With spatially structured residual only, the model produces a smoothing of the data, and thus its predictive ability is based solely on interpolation. However, a model with space only does not explain why species occurrences vary in space, and thus unlike a model with predictors, it is not useful for questions related to inference.

12.4   Joint species distribution models

In the previous section, we applied single species distribution models to examine the occurrences of two species of plants. Often one is however not interested in questions related to a single species, but in questions related to a community consisting of many species. In this section we extend the single-species model of the previous section to a joint species distribution model. The response variable to be modelled is a $n \times m$ matrix $\mathbf{Y}$, with the element y_{ij} describing the presence ($y_{ij} = 1$) or absence ($y_{ij} = 0$) of species $j = 1, ...m$ in site $i = 1, ...n$.

With the probit link function, the model for species j is given by

$$\Pr(y_{ij} = 1) = \Phi(L_{ij}), \tag{12.7}$$

where the linear predictor is given by $L_{ij} = \Sigma_k x_{ik}\beta_{kj}$. While the environmental covariates x_{ik} are the same for all species, the regression parameters are species-specific, β_{kj} measuring the influence of the covariate k on the occurrence of the species j.

The likelihood for observing the data y, given the model parameters $\theta = (\beta)$, is obtained by multiplying the single-species likelihoods (Eq. 12.4) over the species,

$$p(\mathbf{Y}|\theta) = \Pi_{j=1}^m \Pi_{i=1}^n \left[\Phi(L_{ij})y_{ij} + (1 - \Phi(L_{ij}))(1 - y_{ij})\right]. \tag{12.8}$$

Eq. 12.8 assumes that the species occur independently in space and independently of each other after accounting for the covariates, both of which assumptions we will relax later. But before extending the model, let us ask how the basic joint species distribution model of Eq. 12.8 differs from a collection of single-species models? As such the models are identical; we have just added the index j to have separate parameters for each species. The difference between single and joint species distribution models is not in the likelihood $p(\mathbf{Y}|\theta)$, but in the prior $p(\theta)$ that will be assumed for the model parameters. One option would be simply to use a stacked species distribution model, i.e. to parameterize the models separately for each species [9]. This is the case if not only the likelihood but also the prior can be decomposed as a product over the species, $p(\theta) = \Pi_{j=1}^m p(\theta_j)$, where θ_j denotes the prior parameter vector for species j. In this case, the posterior density decomposes into a product over the species, implying that the parameters can be estimated separately for each species. However, to bring more insight at the community level, we construct a joint species distribution model. Joint species distribution models do not parameterize models separately for each species, but they do it in a way that shares information among the species-specific models [12, 15, 28, 36, 38, 39, 45, 47, 49]. In this case, the prior $p(\theta)$ cannot be decomposed into a product over the species.

There are three reasons why it can be beneficial to model all species jointly. The first reason is that the use of a joint species distribution model can improve the parameterization of the species level models. While estimating the parameters for a given species, the joint model can borrow information from the other species. The second reason is that a joint species distribution model provides a compact and parameter sparse description of the entire community. The third reason for using joint species distribution models is that they make it possible to analyze co-occurrence patterns among the species. The first and second advantages of joint species distribution models relate to shared responses among the species to environmental covariates, whereas the third reason relates to statistical co-occurrence. We next discuss these two themes in more detail in separate subsections.

12.4.1 Shared responses to environmental covariates

Data acquired for species rich communities are typically characterized by many rare species, making it difficult to obtain accurate parameter estimates. To allow the parameterizations of species-specific models to borrow information from the other species, and to provide a compact summary of community-level features, we may connect the singles-species models of Eq. 12.7 together by the multivariate normal model [39]

$$\beta_{\cdot j} \sim N(\mu, \Sigma). \tag{12.9}$$

Here $\beta_{\cdot j}$ is the vector of all regression coefficients for species j, and thus it describes how that species responds to environmental variation. By Eq. 12.9 we assume that $\beta_{\cdot j}$ is a sample from a community-level distribution, which has a mean vector μ and a variance-covariance matrix Σ. The dimension of the mean vector μ is $v \times 1$. In our example model it has thus $v = 5$ parameters, which describe how the occurrence probability of a typical species depends on the environmental variables. The diagonal elements of the $v \times v$ variance-covariance matrix Σ measure how much variation there is among the species in their responses to the environmental covariates, whereas the off-diagonal elements measure covariation between responses to different covariates. In this model, the model parameters are given by $\theta = (\beta, \mu, \Sigma)$, and the prior distribution decomposes as $p(\theta) = p(\mu, \Sigma)p(\beta|\mu, \Sigma)$. The prior $p(\mu, \Sigma)$ that we assumed for the plant case study is described in the section *Prior distributions*.

We note that the choice of the multivariate normal distribution in Eq. 12.9 is only one of the many kind of choices one could make to let species-specific models share information among each other. One alternative is the classification of species into distinct groups that are assumed to respond similarly to the environment, an approach called species archetype modelling [15, 28].

While the estimated responses of Ivy leaf Violet to environmental variation are essentially identical whether they are estimated by the single species distribution model or the joint species distribution model, the estimated responses of the Small Purslane are somewhat different between these two modelling approaches (Fig. 12.2). This is because the training data contain few occurrences of this species, and thus the parameterization of the model is substantially influenced also by the community-level model of Eq. 12.9.

Fig. 12.3 illustrates the response of the typical species to the two environmental covariates, as well as the responses of all the 100 species making up the community. The figure suggests that most species respond positively to salinity, whereas that the responses to precipitation are more mixed, some species showing a positive and some a negative response, while others peaking at an intermediate value.

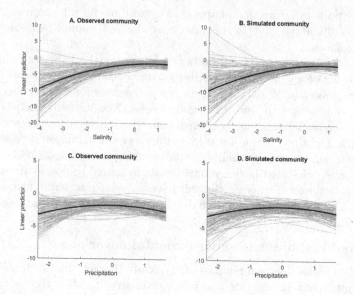

FIGURE 12.3

Community-level responses of 100 herb species to environmental covariates. Shown are posterior mean estimates of linear predictors (probit-transformed occurrence probabilities) as a function of the two environmental covariates of salinity (AB) and precipitation (CD). The value of the covariate not varied in each panel is set to its mean value. In the left-hand panels (AC), the grey lines show the linear predictors for each of the 100 species (computed based on the species-specific β parameters), whereas the black lines show the mean linear predictor for the community (computed based on the μ parameter). In the right-hand panels (BD) the grey lines represent hypothetical species, the parameters of which have been randomized from the community-level distribution of Eq. 12.9.

While the left-hand panels of Fig. 12.3 are based on the estimated species-specific responses, the right-hand panels show the responses for 100 hypothetical species generated from the community-level distribution of Eq. 12.9. The left-hand and right-hand panels appear to be largely similar, suggesting that the community-level parameters μ and Σ are sufficient for describing how species within this community generally respond to the environment. Thus, if one is not interested in the responses of the individual species but on the community-level properties, one may simplify the analyses by basing them on the 20 parameters included in the community-level parameters μ and Σ, instead on the 500 parameters included in the species-specific parameters β.

Two community-level properties of major interest are variation in species richness (i.e. the number of species) and variation in species composition (i.e., which species are present). Both these can be inferred from the estimates of μ and Σ. For an environment described by the $v \times 1$ vector x of environmental covariates, the expected species richness S can be computed as

$$S = n\Phi(x^T\mu/\sqrt{x^T\Sigma x + 1}). \tag{12.10}$$

To see this, note that with probit regression the probability of occurrence for a randomly selected species is $\Pr(z > 0)$, where $z = x^T\beta + \epsilon$, with $\beta \sim N(\mu, \Sigma)$ and $\epsilon \sim N(0, 1)$. Basic properties of the multivariate normal distribution imply that $z \sim N(\mu_z, \sigma_z^2)$, where $\mu_z = x^T\mu$ and $\sigma_z^2 = x^T\Sigma x + 1$. Thus $\Pr(z > 0) = 1 - \Phi((0 - \mu_z)/\sigma_z) = \Phi(\mu_z/\sigma_z)$, yielding Eq. 12.10.

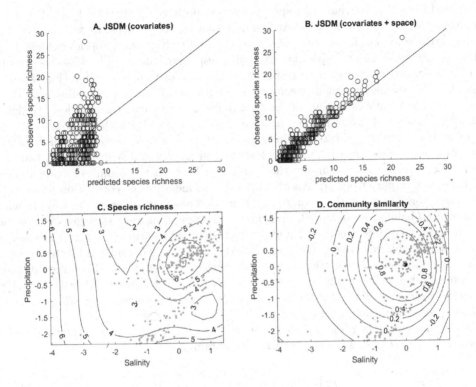

FIGURE 12.4
Patterns of species richness and community similarity predicted by a joint species distribution model. The upper panels compare the observed and predicted species richness among the 300 training sites, each shown by a dot. The results are shown for joint species distribution models including covariates only (A) or covariates and spatial latent factors (B). In these panels, the lines show the identities $y = x$ corresponding to the ideal cases where predictions would match the observations. The lower panels show how species richness (panel C, based on Eq. 12.10) and community similarity (panel D, based on Eq. 12.11) are predicted to vary with the two environmental covariates. In panel D the black dot shows the environmental conditions of the reference community to which the similarity of the focal community is compared to. The reference environmental conditions are set to the mean values over the data, hence both salinity and precipitation equal zero. The grey dots in panels CD show the environmental conditions present in the training data.

Fig. 12.4A compares the species richness predicted by Eq. 12.10 to the observed species richness over the training sites. While there is a positive correlation between these two, there is a mismatch in the sense that the observed species richness shows more variation than the predicted one. This is partly so because we have assumed that the occurrences of the species are statistically independent of each other. We will return to this issue when considering statistical co-occurrence in the next section.

Equipped with Eq. 12.10, we may examine how the expected species richness varies with environmental conditions. In Fig. 12.4C we have predicted the expected species richness for sites varying in their salinity and precipitation. Interestingly, the response to salinity is non-monotonic, so that sites with an intermediate amount of salinity are predicted to have the lowest number of species. The predicted species richness is highest for high values of salinity, as expected from the fact that most species respond positively to it (Fig. 12.3). The reason why sites with low salinity are predicted to have more species than sites with intermediate salinity is that some species are specialized to low salinity (Fig. 12.3). Species richness varies much less with respect to precipitation than with respect to salinity (Fig. 12.4), as expected from the generally less pronounced responses of the species to precipitation (Fig. 12.3).

Let us then consider community composition, with the aim of understanding what drives variation in the identities of the species that make up the local communities. We denote by $L(x)$ the vector of linear predictors for all species for a site with environmental conditions x, and by $p(x) = \Phi(L(x))$ the corresponding vector of occurrence probabilities. One option for measuring the expected level of community similarity between two environments x_A and x_B is to compute the correlation between the occurrence probabilities, $\rho_p(x_A, x_B) = \mathrm{Cor}(p(x_A), p(x_B))$. Another option is to measure the correlation between the linear predictors, $\rho_L(x_A, x_B) = \mathrm{Cor}(L(x_A), L(x_B))$. These two differ in the scale at which occurrence probability is measured, and thus in the weighting that they give for common and rare species. As it does not make a big difference for the measure $\rho_p(x_A, x_B)$ whether the occurrence probabilities are e.g. 0.001 or 0.01, this measure is dominated by the responses of the common species. In contrast, at the scale of the linear predictor also variation among small probabilities makes a difference, as e.g. $\Phi^{-1}(0.01) = -2.3$ and $\Phi^{-1}(0.001) = -3.1$ are considerably different. Thus, the measure $\rho_L(x_A, x_B)$ gives more weight to the responses of the rare species than the measure $\rho_p(x_A, x_B)$.

As the measure $\rho_L(x_A, x_B)$ is defined at the linear scale, it is easy to analyze mathematically. As $\mathrm{Cov}(L(x_A), L(x_B)) = x_A^T \Sigma x_B$, we obtain [39]

$$\rho_L(x_A, x_B) = \frac{x_A^T \Sigma x_B}{\sqrt{x_A^T \Sigma x_A \, x_B^T \Sigma x_B}}. \qquad (12.11)$$

Community similarity $\rho_L(x_A, x_B)$ examines if the same species are especially likely to occur in two sites characterized by covariates x_A and x_B, independently of their absolute occurrence probabilities. The measures of species richness (Eq. 12.10) and community similarity (Eq. 12.11) provide complementary insight into community structure. Two sites may differ in their species richness, in their community similarity, or in both of these.

Eq. 12.11 provides a convenient mapping for examining how environmental similarity translates to community similarity. Fig. 12.4D illustrates this mapping by showing how community composition depends on variation in salinity and precipitation in the case study of Victorian herbs. As expected, the further the environmental conditions are from those of the reference site (shown by the black dot in Fig. 12.4D), the lower is the predicted community similarity. While species richness (Fig. 12.4C) is rather insensitive to variation in precipitation, community composition (Fig. 12.4D) is not. Thus, while moving from sites with much precipitation to sites with little precipitation, the number of species does not change much, but there is considerable turnover in the identities of the species. As we measure community similarity by correlation, the values of $\rho_L(x_A, x_B)$ are constrained to the range between minus one and one. Some of the values in Fig. 12.4D are smaller than zero, indicating very substantial turnover in the species community: species that are more common than average under some environmental conditions are less common than average under some other environmental conditions.

As another application of community similarity, Fig. 12.5A illustrates the partitioning of the study into regions that are predicted to have similar communities [20]. To construct such a decomposition, we applied a k-means clustering based on the predicted probabilities of species occurrence. As we predict species occurrence probabilities based on their responses to variation to salinity and precipitation, the regions with similar community compositions consist of sites that are similar in these two environmental variables. However, the end

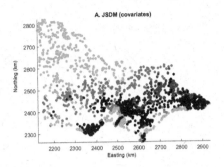
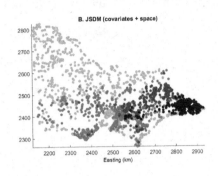

FIGURE 12.5
Classification of the study area into regions with similar community compositions. The four shades of grey classify the sites to areas which differ from each other in terms of their predicted species communities. The classification is based on k-means clustering applied to probabilities of species occurrence predicted by the joint species distribution model with, covariates only (A) or covariates and spatial latent factors (B). The number of clusters was set to four for illustrative purposes.

result is different from what could be expected by clustering the environmental variables directly, as now the clustering is weighted on how relevant these variables are for the species community.

The joint species distribution modelling framework allows for many kinds of extensions, such as considerations of how species traits [1, 7, 44] or phylogenetic relationships [1, 29] influence the responses of the species to the environment. Let us briefly illustrate how species traits can be incorporated by adding another hierarchical layer to the model. Instead of the common mean μ in Eq. 12.9, we may assume that the expected response $\mu_{.j}$ of species j to the environment depends on its traits. We thus write [1]

$$\beta_{.j} \sim N(\mu_{.j}, \Sigma), \tag{12.12}$$

and model the species-specific expectation $\mu_{.j}$ as a regression to the species traits. Denoting by T_{jl} the trait $l = 1, ..., t$ measured for species j, we thus write $\mu_{kj} = \Sigma_l T_{jl}\gamma_{lk}$, where γ_{lk} measures how trait l influences the response of the species to the environmental covariate k. In this model, the model parameters consist of $\theta = (\beta, \gamma, \Sigma)$, where we have omitted μ as it is a deterministic function of γ. The prior density decomposes as $p(\theta) = p(\gamma, \Sigma)p(\beta|\gamma, \Sigma)$. Our choice for $p(\gamma, \Sigma)$ used to fit the model to the plant data is given in the section *Prior distributions*.

We consider a classification of the Victorian herbs to three groups: (i) large herbs, (ii) medium herbs, and (iii) small or prostate herbs. In this case the $m \times t$ trait matrix **T** has the dimension 100×3, with the values consisting of zeros or ones indicating to which group each species belongs to. Fig. 12.6 illustrates the group-specific responses to the environmental parameters based on the estimated γ parameters. While all the three species groups respond positively to salinity, the occurrence probabilities of large herbs are much higher in high salinity sites than the occurrence probabilities of the other kind of herbs. The responses of all groups of herbs peak similarly at an intermediate value of precipitation, and thus variation in response to precipitation takes place mainly at the species level rather than at the group level. While in this example the traits were categorical, we note that the regression setting allows equally well for the inclusion of continuously valued traits.

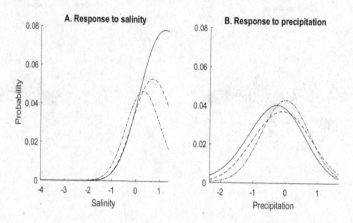

FIGURE 12.6
Group-level responses of the 100 Victorian herb species to environmental covariates. Shown are posterior mean occurrence probabilities as a function of the two environmental covariates of salinity (A) and precipitation (B). The environmental covariate not varied in each panel was set to its mean value. The tree types of lines show the group-specific responses for large herbs (continuous lines), medium herbs (dashed lines), and small or prostate herbs (dash-dotted lines), computed based on the γ parameters of the trait-based joint species distribution model (Eq. 12.12).

With the inclusion of the traits, the covariance between linear predictors for two communities occupying sites characterized by environmental conditions $\boldsymbol{x}_A$ and $\boldsymbol{x}_B$ becomes

$$\mathrm{Cov}(\boldsymbol{L}(\boldsymbol{x}_A), \boldsymbol{L}(\boldsymbol{x}_B)) = \boldsymbol{x}_A^T(\gamma^T \mathrm{Cov}(\mathbf{T})\gamma + \boldsymbol{\Sigma})\boldsymbol{x}_B, \qquad (12.13)$$

where the $t \times t$ variance-covariance matrix $\mathrm{Cov}(\mathbf{T})$ measures how different the species are in terms of their trait composition. In Eq. 12.13, the terms $\gamma^T \mathrm{Cov}(\mathbf{T})\gamma$ and $\boldsymbol{\Sigma}$ partition the variation among species in their responses to the environmental conditions to components that can and cannot be attributed to the measured traits. Thus, with the inclusion of traits, community similarity is not to be measured by Eq. 12.11, but by scaling Eq. 12.13 to the correlation scale.

12.4.2   Statistical co-occurrence

The models discussed in the previous subsection combine species-specific models by assuming that the species show shared responses to environmental covariates (Eq. 12.9), which responses may depend on their traits (Eq. 12.12). However, the models of the previous subsection still assume that, conditional on the environmental conditions, the species occur independently of each other. Thus, if in a given site the occurrence probability of species 1 is p_1 and the occurrence probability of species 2 is p_2, the probability that both of these species are found simultaneously from the site is given by the product $p_1 p_2$. There are two principal reasons why this may not be a valid assumption. The first is that the model may lack some relevant environmental covariates which influence the occurrences of both species, and thus cause an association among them. The second is that the two species may interact ecologically with each other, making them co-occur either more or less often than expected at random. From "snapshot data" of species occurrence, such as the Victorian herb data considered here, it is unfortunately not possible to disentangle among these two reasons behind co-occurrence [31, 37, 49]. But even if the identification of the underlying causal reason remains at the level of interpretation, it is still of major interest to quantify patterns of co-occurrence and use such patterns for prediction. We will next define a joint species distribution model that relaxes the assumption of independence, and that can thus be used to ask if the two species are found more or less often together than would be expected by their responses to the environmental covariates [12, 36, 38, 39, 45, 47].

We will quantify statistical co-occurrence with the help of a correlation matrix $\mathbf{R}$, the element $R_{j_1 j_2}$ being the correlation in occurrence (at the level of linear predictor) between species j_1 and j_2. Estimating such a correlation matrix for a large species community is challenging simply due to the dimensionality of the problem. For example, for the 100 Victorian herbs, $\mathbf{R}$ will be a 100×100 matrix. As the matrix is symmetric, and as the diagonal elements of a correlation matrix are fixed to one, the number of species-to-species correlations to be estimated is 4950, which is a very large number compared to the amount of data available to parameterize the model. To make the estimation of high dimensional correlation matrices possible, we utilize latent variable modeling [7, 47, 49], and thus extend the linear predictor to

$$L_{ij} = \Sigma_{k=1}^{v} x_{ik}\beta_{kj} + \Sigma_{k=1}^{f} \eta_{ik}\lambda_{kj}. \tag{12.14}$$

Here the η parameters are the latent factors, which can be considered to represent unobserved environmental variables. Unlike the measured environmental covariates $\boldsymbol{x}$, the unobserved environmental variables $\boldsymbol{\eta}$, including their number f, are not known a priori, and hence they need to be estimated. The λ parameters are called factor loadings, and they measure how the species respond to the unobserved environmental variables. The main motivation of using the factor model is that the number of parameters to be estimated is proportional to $m \times f$, which is much smaller than the dimension of the correlation matrix $\mathbf{R}$ if the number of factors is much smaller than the number of species, i.e. if $f << m$.

In the model of Eq. 12.14, the parameters consist of $\boldsymbol{\theta} = (\boldsymbol{\beta}, \boldsymbol{\gamma}, \boldsymbol{\Sigma}, \boldsymbol{\eta}, \boldsymbol{\lambda})$, and the prior density can be decomposed as $p(\boldsymbol{\theta}) = p(\boldsymbol{\gamma}, \boldsymbol{\Sigma}, \boldsymbol{\eta}, \boldsymbol{\lambda}) p(\boldsymbol{\beta} | \boldsymbol{\gamma}, \boldsymbol{\Sigma})$. Out of the parameters, the latent factors ($\boldsymbol{\eta}$) and their loadings ($\boldsymbol{\lambda}$) appear as a product. As it is not possible to identify the absolute values of two terms of a product, we may normalize one of these. In factor analysis, it is common to make the assumption $\eta_{ik} \sim N(0, 1)$, which assumption we will follow here. For the factor loadings $\boldsymbol{\lambda}$, we assumed the multiplicative gamma shrinkage prior [7, 36], which shrinks the factor loadings towards zero, the amount of shrinkage increasing with the number k of the factor (see section *Prior distributions*).

Consider two species j_1 and j_2. At the level of the linear predictor, the covariance in their occurrence among the sites is

$$\text{Cov}(\boldsymbol{L}_{\cdot j_1}, \boldsymbol{L}_{\cdot j_2}) = \Sigma_{k,l=1}^{v} \text{Cov}(\boldsymbol{x}_{\cdot k}, \boldsymbol{x}_{\cdot l})\beta_{lj_1}\beta_{kj_2} + \Sigma_{k=1}^{f} \lambda_{kj_1}\lambda_{kj_2}. \tag{12.15}$$

The first part in the right hand side of this equation describes the co-occurrence induced by the shared responses of the species to the measured environmental conditions x_{ik}, whereas the second part describes the residual co-occurrence, which is of our main interest here. We define the species-to-species covariance matrix $\boldsymbol{\Omega}$ as $\boldsymbol{\Omega} = \boldsymbol{\lambda}^T \boldsymbol{\lambda}$, so that $\Omega_{ij} = \Sigma_{k=1}^{f} \lambda_{kj_1}\lambda_{kj_2}$. The correlation matrix $\mathbf{R}$ is obtained by scaling $\boldsymbol{\Omega}$ by the standard deviations, so that $R_{ij} = \Omega_{ij}/\sqrt{\Omega_{ii}\Omega_{jj}}$. A negative correlation $(-1 < R_{j_1 j_2} < 0)$ indicates that the two species co-occur less often than expected from their responses to the environmental covariates, a positive correlation $(0 < R_{j_1 j_2} < 1)$ indicates that the two species co-occur more often than expected at random, whereas a zero correlation $R_{j_1 j_2} = 0$ indicates that the occurrences of the two species are statistically independent.

In the case of spatial data, the measured environmental covariates typically show a spatial pattern (e.g. Fig. 12.1DE), and thus it can be expected that also the unmeasured environmental covariates do so. To allow for such a possibility, we relax the assumption that the latent variables are independent among the sites. Instead, we assume that they show a spatial autocorrelation of the form $\mathrm{Cov}(\eta_{i_1 k}, \eta_{i_2 k}) = f_k(d_{i_1 i_2})$, where f_k is a spatial covariance function and $d_{i_1 i_2}$ is the distance between the sites i_1 and i_2. This assumption induces residual covariance between species not only within sites, but also among sites. If the two sites are at distance d from each other, the covariance between species j_1 and j_2 becomes $\Sigma_{k=1}^{f} \lambda_{k j_1} \lambda_{k j_2} f_k(d)$ [38]. Like we did in the case of the single-species distribution models, we assume an exponentially decaying correlation structure, and thus that $f_k(d) = \exp(-d_{ij}/\alpha_k)$, where α_k is the spatial scale at which the factor k varies. The parameters of this model consist of $\boldsymbol{\theta} = (\boldsymbol{\beta}, \boldsymbol{\gamma}, \boldsymbol{\Sigma}, \boldsymbol{\eta}, \boldsymbol{\lambda}, \boldsymbol{\alpha})$, and the prior density can be decomposed as $p(\boldsymbol{\theta}) = p(\boldsymbol{\gamma}, \boldsymbol{\Sigma}, \boldsymbol{\lambda}, \boldsymbol{\alpha}) p(\boldsymbol{\beta}|\boldsymbol{\gamma}, \boldsymbol{\Sigma}) p(\boldsymbol{\eta}|\boldsymbol{\alpha})$.

Figure 12.7 shows the association network as measured by the correlation matrix $\mathbf{R}$ for the Victorian herb data. In a model that includes the environmental covariates of salinity and precipitation, most of the estimated residual correlations are positive (Fig. 12.7A), meaning that many species pairs tend to co-occur more often than expected at random. This implies that there is more variation in species richness than one would expect by independent occurrences, as the co-occurring species are simultaneously either present or absent from a site. As a result, a model that accounts for co-occurrence predicts the amount of variation in species richness much better (Fig. 12.4B) than a model that assumes statistical independence (Fig. 12.4A). A model that includes only the spatial latent factors but no environmental covariates identifies also species pairs with negative co-occurrence (Fig. 12.7B). This is because e.g. species which occur predominantly at sites with high salinity are not found at the same sites as species which occur predominantly at sites with low salinity. The difference between panels A and B in Fig. 12.7 illustrates how the inclusion of environmental covariates may change the inferred species association networks. Finding out whether the positive associations in Fig. 12.7A are due to ecological interactions or due to missing covariates relating either to the environmental conditions or the observation process is not possible without some additional data.

Accounting for co-occurrences in the statistical model makes it possible to utilize information about other species when constructing model predictions. To exemplify, let us consider e.g. Small Purslane as our focal species. Let us assume that in addition to the environmental covariates, we would also know the presences and absences of all other species in the validation sites. Knowledge about other species in a validation site i (and in a spatial model, also in nearby sites) yields information about the value of the unobserved environmental variables η_{ik}, which in turn influences the model's prediction on the occurrence probability of the focal species (see [36] for technical details). Accounting for the occurrences of the other species greatly increases the model's ability to predict the occurrences of Small Purslane in the validation sites, with Tjur's R^2 increasing from 0.10 to 0.19 (Table 12.1). Similarly, knowledge about the occurrences of the other species increases the model's ability to predict the occurrences of Ivy leaf Violet on the validation sites.

The above example illustrates how knowledge about the occurrences of (a subset of) other species can be used to improve predictions about a focal species. Species of conservation concern or of other special interest may be difficult to survey, e.g. due to their rarity or difficulties related to detection or species identification. Assuming that the species of interest show co-occurrence with other species that are easier to survey, the occurrences of those other species (called indicator species in the conservation context [10]) can be used to generate improved predictions about sites in which the species of interest are likely to occur. The multivariate approach illustrated here allows one to incorporate dependencies among all species within the community. A technically simpler way for accounting for species interactions in the context of single species modelling is the inclusion of the occurrences of e.g. dominant species as predictors [32].

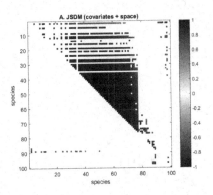 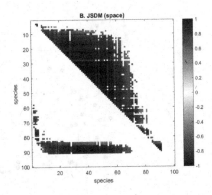

FIGURE 12.7

Species-to-species association matrices estimated by joint species distribution models. The dots above the diagonal indicate species pairs that show a positive correlation with at least 95% posterior probability, whereas the dots below the diagonal indicate species pairs that show a negative correlation with at least 95% posterior probability. The correlation matrices were estimated with the help of latent factors (Eqs. 12.14 and 12.15). The two panels show the estimated association matrices based on the joint species distribution model with covariates and spatial latent factors (A) and with spatial latent factors only (B). The species have been ordered to visualize the structure of the association network.

As illustrated above, the inclusion of latent factors enables one to quantify the level of statistical co-occurrence among the species. In the example constructed above, we assumed that the latent factors are of spatial nature. The inclusion of spatially structured latent factors facilitates also many kinds of other analyses. As one example, in Fig. 12.8 we have used the fitted model to examine how much of the distance decay of community similarity in the Victorian herb data can be explained by spatial variation in salinity and precipitation. The figure shows a steep decline in the realized community similarity (the black dots and line) as a function of distance between the sites. A part of this pattern is generated by the spatial latent factors (the grey line), but most of it is attributable to spatial variation in salinity and precipitation (the difference between the black and the grey lines).

Let us conclude by noting that in this chapter we have assumed that there is no direct information about species interactions, and thus we have inferred e.g. the association networks illustrated in Fig. 12.7 from data on species occurrences. Sometimes e.g. a food web interaction network among the species can be inferred directly. How such direct information on food web structures can be best integrated into species distribution modelling is an active area of research [41].

12.5 Prior distributions

We describe here the prior distributions assumed in the example analyses. To ensure comparability among the results, we fitted all models using the full multivariate model with traits (Eq. 12.12). In this model, the parameters consist of $\theta = (\beta, \gamma, \Sigma, \eta, \lambda, \alpha)$, and the prior density decomposes as $p(\theta) = p(\gamma, \Sigma, \lambda, \alpha)p(\beta|\gamma, \Sigma)p(\eta|\alpha)$.

We assume that the parameters γ, Σ, λ and α are independent in the prior distribution, and thus that $p(\gamma, \Sigma, \lambda, \alpha) = p(\gamma)p(\Sigma)p(\lambda)p(\alpha)$.

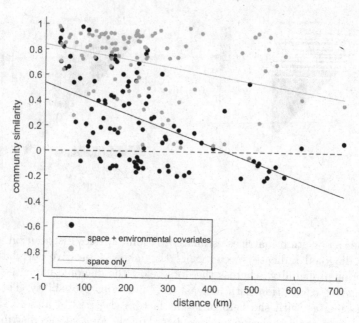

FIGURE 12.8

Distance decay in community similarity decomposed into the influences of environmental covariates and spatial effects. Shown is community similarity as a function of distance between sites, the dots corresponding to 100 randomly selected pairs of sites. Community similarity was measured by $\rho_p(\boldsymbol{x}_A, \boldsymbol{x}_B)$, defined as the correlation between predicted occurrence probabilities. The black dots correspond to actual community similarity that depends both on the similarity in environmental conditions (captured by the regression part of a joint species distribution model) as well as the spatial distance between the sites (captured by the spatial latent factors). The grey dots are based on setting the environmental conditions to their mean values over the sites, and thus they show the distance decay in community similarity generated by spatial latent factors only. The continuous lines show linear regression models fit to the data, and the dashed line shows zero correlation.

For the $\boldsymbol{\gamma}$ parameters we assumed the normally distribution prior $\gamma_{lk} \sim N(0,1)$ independently among the (l,k). For the variance covariance matrix $\boldsymbol{\Sigma}$ we assumed the inverse Wishart prior $\boldsymbol{\Sigma} \sim \mathrm{W}^{-1}(\boldsymbol{\Sigma}_0, f_0)$ with $f_0 = v + 1 = 6$ degrees of freedom and the variance covariance matrix set to the identity matrix, $\boldsymbol{\Sigma}_0 = \mathbf{I}$.

For the factor loadings $\boldsymbol{\lambda}$, we assumed the multiplicative gamma shrinkage prior [7, 36], which shrinks the factor loadings towards zero, the amount of shrinkage increasing with the factor identity k. The density of this prior can be written with the help of auxiliary parameters $\boldsymbol{\phi}$ and $\boldsymbol{\tau}$ as $p(\boldsymbol{\lambda}) = p(\boldsymbol{\lambda}|\boldsymbol{\phi},\boldsymbol{\tau})p(\boldsymbol{\phi})p(\boldsymbol{\tau})$, with the distributional assumptions

$$\lambda_{kj} \sim N(0, \phi_{kj}^{-1}\tau_k^{-1}), \quad \phi_{kj} \sim \mathrm{Gamma}(v/2, v/2), \quad \tau_k = \Pi_{l=1}^{k}\delta_l, \qquad (12.16)$$

where

$$\delta_1 \sim \mathrm{Gamma}(a_1, b_1), \quad \delta_l \sim \mathrm{Gamma}(a_2, b_2) \text{ for } l > 1. \qquad (12.17)$$

The parameters were set to $v = 3$, $b_1 = 1$ and $b_2 = 1$. For single species analyses, we set $a_1 = a_2 = 5$ and included one latent factor only. For analyses with all $m = 100$ species, we set $a_1 = a_2 = 50$ and estimated the number of latent factors. The reason for setting less shrinkage for the single species case is that with a single species the data involve less evidence about a spatial pattern in the residual, and thus a prior with very high amount of shrinkage would be likely to suppress not only noise but also the signal. While technically the model has an infinite number of factors, in practice their number is restricted by including only those first f factors whose influence is not negligible [7].

For the spatial scale parameters, we set half of the probability mass to $\alpha_k = 0$ corresponding to the assumption of non-spatial latent factors, and distributed the remaining probability uniformly in the range from zero to the maximal distance between the sampling locations.

All models considered in this chapter can be obtained as special cases of the full model. In models without species traits, we included in the trait matrix $\mathbf{T}$ the intercept only. In single species models, we set the number of species to $m = 1$.

The posterior distributions were sampled with Matlab using the Markov chain Monte Carlo sampling scheme described in the references [36, 38].

12.6 Acknowledgments

We thank Graeme Newell and Matt White for providing a cleaned version of the Victorian Biodiversity Atlas data, and Jennifer Hoeting for providing helpful comments on an earlier versions of this chapter. This study was funded by the Academy of Finland (CoE grant no. 250444 and grant no. 309581) and the Research Council of Norway (CoE grant no. 223257).

Bibliography

[1] N. Abrego, A. Norberg, and O. Ovaskainen. Measuring and predicting the influence of traits on the assembly processes of wood-inhabiting fungi. *Journal of Ecology*, 105:1070–1081, 2017

[2] M. B. Araujo and A. Guisan. Five (or so) challenges for species distribution modelling. *Journal of Biogeography*, 33(10):1677–1688, 2006.

[3] M. B. Araujo and A. Townsend Peterson. Uses and misuses of bioclimatic envelope modeling. *Ecology*, 93(7):1527–1539, 2012.

[4] M. Austin. Species distribution models and ecological theory: A critical assessment and some possible new approaches. *Ecological Modelling*, 200(1-2):1–19, 2007.

[5] S. Barry and J. Elith. Error and uncertainty in habitat models. *Journal of Applied Ecology*, 43(3):413–423, 2006.

[6] L. J. Beaumont, A. J. Pitman, M. Poulsen, and L. Hughes. Where will species go? incorporating new advances in climate modelling into projections of species distributions. *Global Change Biology*, 13(7):1368–1385, 2007.

[7] A. Bhattacharya and D. B. Dunson. Sparse bayesian infinite factor models. *Biometrika*, 98(2):291–306, 2011.

[8] A. M. Brown, D. I. Warton, N. R. Andrew, M. Binns, G. Cassis, and H. Gibb. The fourth-corner solution - using predictive models to understand how species traits interact with the environment. *Methods in Ecology and Evolution*, 5(4):344–352, 2014.

[9] J. M. Calabrese, G. Certain, C. Kraan, and C. F. Dormann. Stacking species distribution models and adjusting bias by linking them to macroecological models. *Global Ecology and Biogeography*, 23(1):99–112, 2014.

[10] T. Caro. *Conservation by proxy: indicator, umbrella, keystone, flagship, and other surrogate species*. Island Press, Washington, 2010.

[11] A. Chakraborty, A. E. Gelfand, A. M. Wilson, A. M. Latimer, and Jr. Silander, J. A. Modeling large scale species abundance with latent spatial processes. *Annals of Applied Statistics*, 4(3):1403–1429, 2010.

[12] J. S. Clark, A. E. Gelfand, C. W. Woodall, and K. Zhu. More than the sum of the parts: forest climate response from joint species distribution models. *Ecological Applications*, 24(5):990–999, 2014.

[13] A. J. Dobson and A. Barnett. *An Introduction to Generalized Linear Models*. Chapman and Hall/CRC, third edition, 2008.

[14] C. F. Dormann, J. M. McPherson, M. B. Araujo, R. Bivand, J. Bolliger, G. Carl, R. G. Davies, A. Hirzel, W. Jetz, W. D. Kissling, I. Kuehn, R. Ohlemueller, P. R. Peres-Neto, B. Reineking, B. Schroeder, F. M. Schurr, and R. Wilson. Methods to account for spatial autocorrelation in the analysis of species distributional data: a review. *Ecography*, 30(5):609–628, 2007.

[15] P. K. Dunstan, S. D. Foster, and R. Darnell. Model based grouping of species across environmental gradients. *Ecological Modelling*, 222(4):955–963, 2011.

[16] J. Elith and C. H. Graham. Do they? how do they? why do they differ? on finding reasons for differing performances of species distribution models. *Ecography*, 32(1):66–77, 2009.

[17] J. Elith, C. H. Graham, R. P. Anderson, M. Dudik, S. Ferrier, A. Guisan, R. J. Hijmans, F. Huettmann, J. R. Leathwick, A. Lehmann, J. Li, L. G. Lohmann, B. A. Loiselle, G. Manion, C. Moritz, M. Nakamura, Y. Nakazawa, J. M. Overton, A. T. Peterson, S. J. Phillips, K. Richardson, R. Scachetti-Pereira, R. E. Schapire, J. Soberon, S. Williams, M. S. Wisz, and N. E. Zimmermann. Novel methods improve prediction of species' distributions from occurrence data. *Ecography*, 29(2):129–151, 2006.

[18] J. Elith and J. R. Leathwick. *Species Distribution Models: Ecological Explanation and Prediction Across Space and Time*, volume 40 of *Annual Review of Ecology Evolution and Systematics*, pages 677–697. 2009.

[19] S. Ferrier and A. Guisan. Spatial modelling of biodiversity at the community level. *Journal of Applied Ecology*, 43(3):393–404, 2006.

[20] S. D. Foster, G. H. Givens, G. J. Dornan, P. K. Dunstan, and R. Darnell. Modelling biological regions from multi-species and environmental data. *Environmetrics*, 24(7):489–499, 2013.

[21] J. Franklin. Moving beyond static species distribution models in support of conservation biogeography. *Diversity and Distributions*, 16(3):321–330, 2010.

[22] L. Gallien, T. Muenkemueller, C. H. Albert, I. Boulangeat, and W. Thuiller. Predicting potential distributions of invasive species: where to go from here? *Diversity and Distributions*, 16(3):331–342, 2010.

[23] A. Gelman, J. B. Carlin, H. S. Stern, D. B. Dunson, A. Vehtari, and D. B. Rubin. *Bayesian Data Analysis.* Chapman and Hall/CRC, London, third edition, 2013.

[24] A. Guisan and W. Thuiller. Predicting species distribution: offering more than simple habitat models. *Ecology Letters*, 8(9):993–1009, 2005.

[25] A. Guisan and N. E. Zimmermann. Predictive habitat distribution models in ecology. *Ecological Modelling*, 135(2-3):147–186, 2000.

[26] P. A. Hernandez, C. H. Graham, L. L. Master, and D. L. Albert. The effect of sample size and species characteristics on performance of different species distribution modeling methods. *Ecography*, 29(5):773–785, 2006.

[27] A. H. Hirzel and G. Le Lay. Habitat suitability modelling and niche theory. *Journal of Applied Ecology*, 45(5):1372–1381, 2008.

[28] F. K. C. Hui, D. I. Warton, S. D. Foster, and P. K. Dunstan. To mix or not to mix: comparing the predictive performance of mixture models vs. separate species distribution models. *Ecology*, 94(9):1913–1919, 2013.

[29] A. R. Ives and M. R. Helmus. Generalized linear mixed models for phylogenetic analyses of community structure. *Ecological Monographs*, 81(3):511–525, 2011.

[30] C. S. Jarnevich, T. J. Stohlgren, S. Kumar, J. T. Morisette, and T. R. Holcombe. Caveats for correlative species distribution modeling. *Ecological Informatics*, 29:6–15, 2015.

[31] A. M. Latimer, S. Banerjee, Jr. Sang, H., E. S. Mosher, and Jr. Silander, J. A. Hierarchical models facilitate spatial analysis of large data sets: a case study on invasive plant species in the northeastern united states. *Ecology Letters*, 12(2):144–154, 2009.

[32] P. C. le Roux, L. Pellissier, M. S. Wisz, and M. Luoto. Incorporating dominant species as proxies for biotic interactions strengthens plant community models. *Journal of Ecology*, 102(3):767–775, 2014.

[33] M. Marmion, M. Parviainen, M. Luoto, R. K. Heikkinen, and W. Thuiller. Evaluation of consensus methods in predictive species distribution modelling. *Diversity and Distributions*, 15(1):59–69, 2009.

[34] C. Merow, N. LaFleur, Jr. Silander, J. A., A. M. Wilson, and M. Rubega. Developing dynamic mechanistic species distribution models: Predicting bird-mediated spread of invasive plants across northeastern north america. *American Naturalist*, 178(1):30–43, 2011.

[35] C. N. Meynard and J. F. Quinn. Predicting species distributions: a critical comparison of the most common statistical models using artificial species. *Journal of Biogeography*, 34(8):1455–1469, 2007.

[36] O. Ovaskainen, N. Abrego, P. Halme, and D. Dunson. Using latent variable models to identify large networks of species-to-species associations at different spatial scales. *Methods in Ecology and Evolution*, 7:549–555, 2016

[37] O. Ovaskainen, J. Hottola, and J. Siitonen. Modeling species co-occurrence by multivariate logistic regression generates new hypotheses on fungal interactions. *Ecology*, 91(9):2514–2521, 2010.

[38] O. Ovaskainen, D. B. Roy, R. Fox, and B. J. Anderson. Uncovering hidden spatial structure in species communities with spatially explicit joint species distribution models. *Methods in Ecology and Evolution*, 7(4):428–436, 2016.

[39] O. Ovaskainen and J. Soininen. Making more out of sparse data: hierarchical modeling of species communities. *Ecology*, 92(2):289–295, 2011.

[40] R. G. Pearson and T. P. Dawson. Predicting the impacts of climate change on the distribution of species: are bioclimate envelope models useful? *Global Ecology and Biogeography*, 12(5):361–371, 2003.

[41] L. Pellissier, R. P. Rohr, C. Ndiribe, J.-N. Pradervand, N. Salamin, A. Guisan, and M. Wisz. Combining food web and species distribution models for improved community projections. *Ecology and Evolution*, 3(13):4572–4583, 2013.

[42] S. J. Phillips, R. P. Anderson, and R. E. Schapire. Maximum entropy modeling of species geographic distributions. *Ecological Modelling*, 190(3-4):231–259, 2006.

[43] S. J. Phillips and M. Dudik. Modeling of species distributions with maxent: new extensions and a comprehensive evaluation. *Ecography*, 31(2):161–175, 2008.

[44] L. J. Pollock, W. K. Morris, and P. A. Vesk. The role of functional traits in species distributions revealed through a hierarchical model. *Ecography*, 35(8):716–725, 2012.

[45] L. J. Pollock, R. Tingley, W. K. Morris, N. Golding, R. B. O'Hara, K. M. Parris, P. A. Vesk, and M. A. McCarthy. Understanding co-occurrence by modelling species simultaneously with a joint species distribution model (JSDM). *Methods in Ecology and Evolution*, 5(5):397–406, 2014.

[46] I. W. Renner, J. Elith, A. Baddeley, W. Fithian, T. Hastie, S. J. Phillips, G. Popovic, and D. I. Warton. Point process models for presence-only analysis. *Methods in Ecology and Evolution*, 6(4):366–379, 2015.

[47] J. T. Thorson, M. D. Scheuerell, A. O. Shelton, K. E. See, H. J. Skaug, and K. Kristensen. Spatial factor analysis: a new tool for estimating joint species distributions and correlations in species range. *Methods in Ecology and Evolution*, 6(6):627–637, 2015.

[48] T. Tjur. Coefficients of determination in logistic regression models-a new proposal: The coefficient of discrimination. *American Statistician*, 63(4):366–372, 2009.

[49] D. I. Warton, F. G. Blanchet, R. B. O'Hara, O. Ovaskainen, S. Taskinen, S. C. Walker, and F. K. C. Hui. So many variables: Joint modeling in community ecology. *Trends in Ecology & Evolution*, 30(12):766–779, 2015.

[50] D. I. Warton, I. W. Renner, and D. Ramp. Model-based control of observer bias for the analysis of presence-only data in ecology. *Plos One*, 8(11), 2013.

[51] M. S. Wisz, J. Pottier, W. D. Kissling, L. Pellissier, J. Lenoir, C. F. Damgaard, C. F. Dormann, M. C. Forchhammer, J.-A. Grytnes, A. Guisan, R. K. Heikkinen, T. T. Hoye, I. Kuehn, M. Luoto, L. Maiorano, M.-C. Nilsson, S. Normand, E. Ockinger, N. M. Schmidt, M. Termansen, A. Timmermann, D. A. Wardle, P. Aastrup, and J.-C. Svenning. The role of biotic interactions in shaping distributions and realised assemblages of species: implications for species distribution modelling. *Biological Reviews*, 88(1):15–30, 2013.

13

Capture-Recapture and distance sampling to estimate population sizes

Richard J. Barker

University of Otago, Otago, New Zealand

CONTENTS

13.1 Basic ideas

In this chapter we describe models that underpin the study of animal distributions in space and time. By 'model' we mean a statistical model for data, an idealized description of how the data might have been generated built using probability distributions.

Ideally ecologists and environmental scientists would be able to model directly how animal populations are distributed in space and how this evolves over time. For such models we need need data that we can think of as population summaries such as time- and location-specific abundances, or densities if we scale these by the area occupied, and quantities such as birth and death rates. Armed with such data we could then describe population changes in terms of dynamic models. These models will describe how the population evolves through additions and losses and may also describe how locations of individuals change.

In this chapter we focus on a different aspect of modeling populations. We treat the population summaries such as abundance, or birth and death rates as fixed attributes of our study population that must be inferred. That is, we treat these attributes as parameters. A key feature of the models we consider is that not all animals can be seen or caught, and are instead sampled. The sampling process usually involves limitations imposed by nature, and these aspects must be accounted for in describing any uncertainties we have about inferred parameters.

In almost all the models we consider, auxiliary data are used to help model nuisance aspects of the sampling process. Such data might include captures and recaptures of marked animals, or visual or audio records from plots or transects. These records might be collected by observers or automatically via cameras or recording devices. They might include samples of DNA obtained from shed material such as hair, skin or feces. We also consider data collected from sampled plots, and data that represent simple counts of individuals. In all cases we require a sampling model to account for the fact that such data represent some sort of sample.

The models for the sampling process conditional on distribution in space or time (i.e., treating the distribution as fixed at the time of sampling) can be quite complex but analyses based on these models should not be seen as an endpoint. Rather, they are a means to an end, that end being inference about the underlying population processes. Many of the techniques we discuss include some elements of modeling the space/time process but for most this is in the form of fairly basic summaries. At the end of this chapter we will provide a brief summary of how state-space modeling [8] can be used to allow more formal embedding of space/time processes into the analysis.

Notation

Generally, we will use $f(\cdot|\cdot)$ to denote conditional densities and $f(\cdot)$ marginal densities; we will not discriminate between continuous or discrete random variables. Thus, $f(y|x)$ is shorthand for the more standard notation $f_Y(y|X = x)$ or $\Pr(Y = y|X = x)$, and represents our model for y given other quantities x. We use this shorthand for brevity. Lower case Roman letters in italics such as y will denote realisations of random variables. Parameters will usually be denoted by Greek letters but we follow historical departures from this convention. For example, p will usually represent a capture probability. Quantities such as y and θ might be scalar, vectors or matrices as determined by context; we will not use bold fonts for this sort of discrimination.

13.2 Inference for closed populations

In closed population modeling we focus on descriptions of the population at a particular point in time. The closed population models we consider here mostly focus on abundance. Formally we regard this as a summary of a population associated with some region $\mathcal{S}$. Usually this is a latent feature of the analysis rather than a specific focus, although the spatial capture recapture models allow some inference about features of $\mathcal{S}$. Sampling is treated as though it occurs instantaneously, although this will be rarely possible. More realistically, we operate under the assumption that any population summaries are an adequate description of the population over the period in which sampling occurred.

13.2.1 Censuses and finite population sampling

It is rare for a census to be possible for a population in S at a time t. Instead it may be feasible to subdivide S into units that can be sampled with a census carried out within a unit. Units may be grouped according to some measure of similarity facilitating stratified sampling, or if these groups are numerous, cluster sampling might be used instead. A feature of these methods is that exclusion of some units is by design and provided this is done using probability sampling, the exclusion of units can be modeled appropriately. We follow [29] (Chapter 8) and argue that all aspects of the data collection process should be modeled and that models should be formulated for the *complete data* an approach pioneered by [60]. Aspects that are left out the model should be justified according to the concept of *ignorability* [60].

Full model-based analysis of sample survey data under simple, stratified and cluster sampling is illustrated by [41], Section 8.5, including comparison of finite and infinite population inference. Inference about quantities such as N are an example of finite population inference. For frequentist inference this involves finite population correction. Bayesian inference involves modeling the complete data in which posterior prediction is used for inferring values on unsampled units [also, see 29, Chapter 8]. Infinite population inference, typically based on an estimate of μ the mean value across sampling units, represents the expected value across a theoretical infinite population of sampling units. Parameters such as μ represent some function of parameters governing the underlying spatial point process, and can be thought of as a simple summary of the abundance process.

13.2.2 The problem of imperfect detection

With censuses on subplots discussed in the previous section we assume that all animals in a survey unit can be sampled. More usually, it is virtually impossible to capture or see all animals we so must model the detection process. Options for dealing with imperfect detection include:

(i) Using auxiliary data, such as from marked animals via capture-recapture or distances to animal sightings, to help model an explicit detection process in the model

(ii) Adding structure to models for incomplete counts as in the N-mixture approach of [57]

(iii) Modeling incomplete counts using regression approaches that allow use of covariates to control for detectabilty while restricting inference to relative abundance as in [43].

Each of these approaches, which we discuss in detail below, has advantages and disadvantages. The sampling models are more complex than for sample surveys and require assumptions about the capture process some of which may be difficult to satisfy. One such example is homogeneity of capture probabilities among identifiable sub-units of the population. Failure of this assumption, termed 'heterogeneity' remains problematic.

Capture-recapture data leads to a very rich array of models, covering both open and closed populations, and for open populations allowing inference about birth rates, survival probabilities, movement probabilities, and abundance or population growth rate $\lambda_j = N_{j+1}/N_j$ [52].

A drawback of capture-recapture studies is that they can be expensive owing to difficulties of catching and marking animals, although recent advances allowing for non-invasive approaches based on natural marks such as DNA from shed material such as skin, feathers or feces have proven helpful [44, 74]. Distance sampling methods also can involve intense and expensive fieldwork.

The N-mixture approach replaces auxiliary data with replication of the counts and introduces an assumption that detection probability p is constant, after adjusting for covariates (hereafter 'the constant p assumption). Inference depends on this constant p assumption and a hierarchical model for abundance. As we do not need to mark animals the N-mixture approach can be carried out much more cheaply than capture-recapture.

Like N-mixture modeling, the regression approach also depends on count replication and the constant p assumption. With both N-mixtures and Poisson regression we have little or no ability to test the constant p assumption [5]. Therefore while cheaper we do not expect inference to be as robust as methods that make use of auxiliary data to help infer detection probabilities.

13.2.3   Capture-recapture on closed populations

Capture-recapture is a method for inferring population summaries or parameters using releases and recaptures of marked animals. Inference for closed populations usually focuses on abundance N. For closed population capture-recapture modeling our starting point is model A of [25] carried out over k sampling occasions. The complete data we represent by an $(N \times k)$ matrix x of indicators x_{ij} denoting capture by individual i in sample j. Without loss of generality we order x so that the first n rows correspond to the n distinct animals that were caught during the study. This upper portion of x we refer to as x_{obs}. The lower portion of dimension $N - n \times k$ is a matrix of zeros; the only unknown is its dimension.

Let

- Ω = the set of all possible capture histories with elements of the form $\omega_1 \omega_2 \ldots \omega_k$ where ω_j is an indicator for capture in sample j. For example, in a $k = 2$ study $\Omega = \{11, 10, 01, 00\}$. In this example element $\omega_1 \omega_2 = 11$ corresponds to capture on both occasions, $\omega_1 \omega_2 = 10$ corresponds to capture on the first occasion but not the second, etc.

- p_j = the probability of capture in sample j ($j = 1, \ldots, k$).

- π_ω = the probability of capture history ω expressed as a simple function of the collection of capture probabilities $\{p_j\}$. For example, $\pi_{11} = p_1 p_2$.

- y_ω = the number of animals with capture history ω at the end of the experiment

- m_j = the number of marked animals caught in sample j.

- u_j = the number of unmarked animals caught in sample j.

- $n'_j = u_j + m_j$ = the number of animals caught in sample j.

Then

$$f(x|N, p) = \frac{N!}{\prod_{\omega \in \Omega} y_\omega!} \prod_{\omega \in \Omega} \pi_\omega^{y_\omega}. \tag{13.1}$$

We can rearrange (13.1) as

$$f(x|N, p) = \underbrace{\prod_{j=2}^{k} \frac{\binom{M_j}{m_j}\binom{U_j}{u_j}}{\binom{N}{n'_j}}}_{f(m|n', N)} \times \underbrace{\prod_{j=1}^{k} \binom{N}{n'_j} p_j^{n'_j} (1 - p_j)^{N - n'_j}}_{f(n'|N, \{p_j\})}. \tag{13.2}$$

The first term, the product of hypergeometric distributions may be used as a partial likelihood for N conditional on the numbers caught in each sample n. [70] shows that the n'_j are

Bayes-ancillary for N for the prior $f(p_1, \ldots, p_k) \propto \prod_{j=1}^{k} p_j^{-1}$. Under this prior the likelihood contribution to the posterior for N is just the term $f(m|n', N)$. Apart from knowing that $N > \max(\{n'_j\})$, the second term contributes only a small amount of information about N that is entangled with information about the capture-probabilities, up to the choice of prior.

An alternative way to rearrange (13.2) is as

$$f(x|N, p) = \underbrace{\binom{N}{n} (\pi^+)^n (1 - \pi^+)^{N-n}}_{f(n|N, \pi^+)} \times \underbrace{\frac{n!}{\prod_{\omega \in \Omega^+} x_\omega!} \prod_{\omega \in \Omega^+} \xi_\omega^{y_\omega}}_{f(\{x_\omega\}_{\omega \in \Omega^+}|n, p)},$$

where Ω^+ represents the set of capture histories excluding the null history $00\ldots0$, $\pi^+ = 1 - \prod_{j=1}^{k}(1 - p_j)$ and $\xi_\omega = \pi_\omega/\pi^+$. Given the collection $\{\xi_\omega; \omega \in \Omega^+\}$ we can solve for the p_j. Since $1 - \pi^+ = \prod_j(1 - p_j)$, we have complete confounding of the p_j in π^+ and we can see that it is the full capture histories of those individuals caught at least once that provide the information for inference about the capture probabilities. That is, it is recapture histories that contain the augmenting information that can be used to infer parameters of the detection process.

The model (13.2) with the assumption that detection probabilities vary only by sampling occasion is commonly referred to as model M_t, the 't' denoting time. Starting from the most general case where detection probabilities p_{ij} are distinct among individuals indexed by $i = 1, \ldots, N$ and occasions indexed by j, [48] describe a sequence of eight families of models M_0, M_t, M_b, M_h, M_{th}, M_{tb}, M_{bh}, and M_{tbh} that make different assumptions about catchability.

In model M_0 we make the constraint $p_1 = p_2 = \ldots p_k$ and all animals have the same capture probability which is the same in all samples. In M_b ('b' for behaviour) there are two distinct capture probabilities, one for first capture and another for recapture. These are otherwise constant through time and among individuals. Model M_h ('h' for heterogeneity) allows capture probabilities to vary among individuals. That is, $p_{i1} = p_{i2} = \ldots p_{ik}$ for each $i \in \{1, \ldots, N\}$.

The models in the 'h' family, such as M_h is of particular interest because it is widely believed that capture probabilities can vary markedly between individuals. Indexing individuals by i, we can model the individual detection probabilities as completely determined by covariates via a generalized linear model. However, we then have to consider what to do about the missing covariates which are unobserved for the animals that are not caught.

One approach is to write the model in terms of a complete data likelihood and integrate missing components from the model. If we denote covariates for individual i by z_i, modeled as drawn from a distribution $f(z_i|\gamma)$ with support on $\mathcal{Z}$, and specify the model for p_{ij}, the capture probability for individual i in sample j, as $p_{ij} = g_{\theta_j}(z_i)$, then we can write the complete data model as

$$f(x|z, \beta, \gamma) = \binom{N}{n} \prod_{i=1}^{N} \prod_{j=1}^{k} f(x_{ij}|z_i, \theta_j) f(z_i|\gamma).$$

The covariates z are only observed for individuals 1 through n. Integrating out the unobserved covariates, and noting that the integrals are identical among individual that are not caught, we find that the observed data likelihood is given by

$$\mathcal{L}(\theta, \gamma | n, z_1, \ldots, z_n) = \binom{N}{n} \prod_{i=1}^{n} \prod_{j=1}^{k} p_{ij}^{x_{ij}} (1 - p_{ij})^{1-x_{ij}} f(z_i|\gamma)$$

$$\times \left(\int_{z \in \mathcal{Z}} \prod_{j=1}^{k} (1 - g_{\theta_j}(z)) f(z|\gamma) dz \right)^{N-n}$$

$$= \binom{N}{n} \prod_{i=1}^{n} \prod_{j=1}^{k} p_{ij}^{x_{ij}} (1 - p_{ij})^{1-x_{ij}} f(z_i|\gamma) \times (1 - \pi)^{N-n},$$

$$= \binom{N}{n} \pi^n (1 - \pi)^{N-n} \prod_{i=1}^{n} f(z_i|\gamma) \frac{\prod_{j=1}^{k} p_{ij}^{x_{ij}} (1 - p_{ij})^{1-x_{ij}}}{\pi}, \qquad (13.3)$$

where $\pi = 1 - \mathbb{E}_z \left[\prod_{j=1}^{k} (1 - g_{\theta_j}(z)) \right]$. This likelihood is closely related to the model for distance sampling discussed in Section 13.2.4 below.

An alternative approach used by [33] uses a likelihood obtained by conditioning on n, which means that covariates are not required for $N - n$ uncaught animals. Estimation of N under this model is based on the Horwitz-Thompson estimator [32], substituting estimated sampling probabilities for true values:

$$\hat{N} = \sum_{i=1}^{n} \hat{\eta}_i^{-1}$$

where $\hat{\eta}_i = 1 - \prod_{j=1}^{k} (1 - g_{\hat{\theta}_j}(z_i))$, and $\hat{\theta}_j$ is the maximum likelihood estimator for θ_j which is estimated using the conditional likelihood [see 33, for details].

The assumption that the capture probabilities are fully determined by the covariates is a strong one. More generally, we might model the detection probabilities in terms of a deterministic component and a random component. In the simplest such case we have no covariates and the p_i are described by a probability distribution $f(p_i|\theta)$. For this version of M_h the sufficient statistics are f_j $(j = 1, \ldots, k)$, the numbers of animals caught exactly j times. Unknown is $f_0 = N - n$. [39] has shown that different models with different expectations $\mathbb{E}(f_0)$ can have near identical values for $f_1, \ldots, f_k$ and thus the model is non-identifiable in the sense that the data are insufficient for the analyst to distinguish among different choices of family for $f(p_i)$.

A major source of heterogeneity is associated with delineation of the study area. Rarely is it possible to prescribe $\mathcal{S}$ in such a way that capture probabilities are equal among individuals. More typically, individuals with activity centers near areas of intense capture effort within $\mathcal{S}$ will have a higher probability of capture than individuals in more remote or less intensively sampled areas of $\mathcal{S}$. This makes heterogeneity of capture problematic as in effect we have an unobserved random variable, location of an activity center, that governs catchability.

13.2.4　Distance sampling methods on closed populations

The idea that detection probability depends on distance from the detector is exploited in distance sampling techniques. Data used in distance sampling typically comprises distances from a transect centerline (line transects) or from a single location (point transects). For line transects, the basic design is that an observer moves along a (usually) straight line recording the perpendicular distance from the center of the line to the detection.

If we denote the distance to the transect line by x, and the indicator for detection of individual i by d_i ($d_i = 1$ if individual i is detected and zero otherwise), then detections are modeled by the detection function $g_\theta(x) = \Pr(d_i = 1|x)$, with $g_\theta(0) = 1.0$. That is, we assume perfect detection at the center.

To derive the likelihood we consider a line transect of length L and width $2w$ in which animals can be seen if within distance w of the centerline and not otherwise. We also assume random placement of the transect line in our study region S of area A so that the probability an individual is included withing the quadrat defined by our line transect is $p_c = 2wL/A$. At the end of the study we can describe each animal by the latent variables I the indicator for inclusion in the quadrat, d for detections and x for the distance of the animal from the transect centerline. All of this assumes instantaneous sampling. Animals are seen if $I = d = 1$, and therefore if there are N animals in S, m within the transect, and n seen we can write the complete data model as

$$f(I, d, x|p_c, \theta, \gamma) = \prod_{i=1} f(I_i, d_i, x_i|p_c, \theta, \gamma)$$

$$= \binom{N}{n} \prod_{i=1}^{n} p_c g_\theta(x_i) f(x_i|\gamma) \prod_{i=n+1}^{N} (1 - p_c g_\theta(x_i)) f(x_i|\gamma).$$

Integrating out the unknown distances for animals not detected, we obtain

$$\mathcal{L}(N, \theta, \gamma|n, x_1, \ldots, x_n) = \binom{N}{n} p_c^n (1 - p_c \mathbb{E}_x[g_\theta(x_i)])^{N-n} \prod_{i=1}^{n} g_\theta(x_i) f(x_i|\gamma), \quad (13.4)$$

the likelihood given by [8].

In fitting the model, the usual assumption is that the x_i can be modeled as Uniform$(0, w)$, this model justified through the random placement of the transects. In effect we are assuming a Poisson process for locations in $\mathbb{R}^2$; this requires the further assumption that animals are distributed independently in space. A common violation of this assumption is when animals occur in groups. If it is reasonable that groups are distributed in space independently, then modeling based on detections of groups is a way to relax the uniform assumption for x [for further details see 8, 12, 14, for more comprehensive treatment of distance sampling methods].

The detection function $g_\theta(x)$ is critical and must be specified by the user for the analysis. There is a lot of flexibility in this choice. [47] recommend that detection functions:

1. be flexible, so that they can take a wide variety of shapes;

2. require a relatively small number of parameters;

3. be flat at zero distance (i.e., have a slope of zero at $x = 0$).

As it is unusual for objects to become more detectable with increasing x they also recommend that detection functions be monotonic non-increasing with increasing x.

Guidance on choice of selection functions is given by [15] who describe models of the form $g_\theta(x) = \kappa_{\theta_1} \psi_{\theta_2}(x)$, where κ_{θ_1} is "key function' chosen as a starting point for $g_\theta(x)$ with flexibility added through a series expansion model $\psi_{\theta_2}(x)$. For example, a uniform key function together with a cosine series expansion give the Fourier series model advocated by [18]. Another choice for κ is the half-normal model

$$g_\theta(x) = e^{\frac{-x^2}{2\sigma^2}} x^2.$$

Both the half-normal and uniform model have the advantage that they meet the shape criterion, that the derivative $g'_\theta(0) = 0$. This is a requirement that the detection function have a "shoulder" at $x = 0$ related to the fact that estimation depends on being able to estimate $f(0)$, where $f(\cdot)$ is the probability density function of the detection distances conditional on detection.

Two important assumptions that must be considered carefully are (i) $g_\theta(0) = 1.0$ and (ii) that animals are detected at their initial locations, prior to any movement in response to observers [14]. How to relax the first assumption we consider next. The second assumption is more a question of careful survey design when monitoring mobile animals [see 14, Section 7.7 for further discussion].

Combined capture-recapture and distance sampling models

Standard distance sampling models require the assumption that $g_\theta(0) = 1$; perfect detection on the transect line or at the center of the plot. For relatively immobile animals, or in habitat that is easily searched, this assumption may be reasonable. In many applications, this assumption may be problematic. Marine mammal applications of line transect methodology has led to considerable thought on the problem of imperfect detection on the transect lines as animals below the surface may be missed even though they are on the transect line. One approach is to use multiple independent observers, usually two [10]. Observations by the two observers are matched at the end of the survey so that we know which animals were seen by both, which by only observer 1 and which only by observer 2. With just two observers we require complete independence of the observers. However, heterogeneity of capture or availability will result in failure of the independence assumption [3], although [9] showed that this assumption could be relaxed to the case where there is just one distance at which this correlation is zero. More than two observers are required to fully resolve this.

An alternative viewpoint is to consider distance simply as a covariate that governs detection. This approach has been exploited by [1], [46] and [54] who model the capture-recapture data in terms of detection functions that may depend on the observer. However, these approaches depend on the assumption that heterogeneity of detection is fully explained by the distance x. If this is not the case then we have a heterogeneity of capture with consequences for inference as discussed by [39].

Spatial capture recapture models

Theory for combining capture-recapture and distance sampling ideas has become fully developed in the spatial capture recapture (SCR) methodology. The focus of these models is on the underlying spatial process governing animal distributions in space, usually expressed in terms of an 'activity center' [see 7, for a recent review].

In spatial capture recapture locations of detectors are used to model detection probabilities relative to animal activity centers. Unlike line transects where observers move along a transect to detect animals, spatial capture-recapture analyzes data from a static array of detectors at known locations that can be used to identify repeated encounters of individual animals. These detector might be traps that physically capture animals or they could be passive detectors such as cameras that identify animals from recorded images.

[7] develop a comprehensive model to unify capture-recapture and distance sampling that includes non-spatial capture-recapture and capture-recapture distance sampling as special cases, with sufficient flexibility to accommodate spatial capture recapture. In this approach capture histories x represent 'detections', and a model $f(x|\theta, z)$ is used to describe the detections of animals conditional on the locations z of animals. If animals are able to move during the survey then z represents an activity center. Another model component describes distribution of the the activity centers. Provided the detectors are at known locations then it is possible to infer aspects of the animal distribution model.

In the model of [7] detection histories $x = \{x_{ijk}\}$ are indexed by individual ($i = 1, \ldots, N$), occasion ($j = 1, \ldots, J$) and detector ($k = 1, \ldots, K$), and modeled as exchangeable Bernoulli observations. If detection probability $p_{ijk} = g_{\theta_{jk}}(z_i)$ then the complete model for both detections and locations can be written as

$$f(z, x|\theta, \phi) = f(z|\phi)f(x|z, \theta)$$

$$= \binom{N}{n} \prod_{i=1}^{N} f(z_i|\phi) \prod_{j=1}^{J} \prod_{k=1}^{K} f(x_{ijk}|\theta, z_i)$$

$$= \binom{N}{n} \prod_{i=1}^{N} f(z_i|\phi) \prod_{j=1}^{J} \prod_{k=1}^{K} p_{ijk}^{x_{ijk}} (1 - p_{ijk})^{1-x_{ijk}}.$$

The locations are unobserved for the animals we don't detect. Using the notation x_{obs} and z_{obs} to denote the observed capture histores and locations, we can write the observed data likelihood as

$$\mathcal{L}(\phi, \theta|x_{\text{obs}}, z_{\text{obs}}) = \binom{N}{n} \left\{ \prod_{i=1}^{n} f(z_i|\phi) \prod_{j=1}^{J} \prod_{k=1}^{K} p_{ijk}^{x_{ijk}} (1 - p_{ijk})^{1-x_{ijk}} \right.$$

$$\left. \times \left[\int_{z \in \mathcal{Z}} \prod_{j=1}^{J} \prod_{k=1}^{K} (1 - g_{\theta_{jk}}(z)) f(z|\phi) dz \right]^{N-n} \right\}$$

$$= \binom{N}{n} \prod_{i=1}^{n} f(z_i|\phi) \prod_{j=1}^{J} \prod_{k=1}^{K} p_{ijk}^{x_{ijk}} (1 - p_{ijk})^{1-x_{ijk}} \times (1 - \pi)^{N-n} \quad (13.5)$$

where π is the probability that an animal is detected at least once during the experiment. We can rewrite the model (13.5) as

$$\mathcal{L}(\phi, \theta|x_{\text{obs}}, z_{\text{obs}}) = \underbrace{\binom{N}{n} \pi^n (1 - \pi)^{N-n}}_{f(n|\phi, \theta)}$$

$$\times \underbrace{\prod_{i=1}^{n} f(z_i|\phi) \frac{\prod_{j=1}^{J} \prod_{k=1}^{K} p_{ijk}^{x_{ijk}} (1 - p_{ijk})^{1-x_{ijk}}}{\pi}}_{f(x_{\text{obs}}, z_{\text{obs}}|n, \phi, \theta)}. \quad (13.6)$$

This is corresponds to the model given by [7] (equation 3). Comparison with model (13.3) above shows that this model is a simple extension of model M_{th} and provides a complete description of capture-recapture distance sampling based on exact distances to the detected individuals, with choice of $g_{\theta_{jk}}(z_i)$ to be specified.

For SCR, the data do not include distances to animal locations but rather spatial locations of the detectors. Distances from these to the animal locations is a latent feature of the model. To derive the SCR model from the model (13.6) and add a set of detector-specific covariates u_k. These are the spatial locations of the detectors. The detection probabilities in (13.6) are now given by $p_{ijk} = g_{\theta_{jk}}(z_i, u_k)$. Again, $g(\cdot)$ must be specified and since now none of the z's are observed these are integrated from the model leading to the likelihood of [7].

13.2.5   N-mixture models for closed populations

N-mixture models were developed by [57] as an alternative to estimating abundance using tools such as capture-recapture or distance sampling that can be difficult and expensive, or impractical.

Capture-recapture and distance sampling rely on auxiliary data to help estimate detection probabilities. N-mixture modeling instead relies on modeling variability among replicated counts to infer abundance, along with an assumption that p is constant among sites and replicates (or is a deterministic function of known covariates). N-mixture modeling exploits the fact that if we assume that we have iid replicates of Binomial(N, p) observations then we can infer both N and p.

For N-mixture modeling we rely on counts that are replicated in space and time. We assume that the replicated counts (indexed by $j = 1, \ldots, J$) at a particular site (indexed by $i = 1, \ldots, I$) represent a single closed population comprising N_i distinct animals. The collection $\{N_i\}_{i=1}^I$ are modeled as a random sample from a Poisson distribution.

If we let y_{ij} denote the count of animal obtained at site i and visit or replicate j, the model introduced by [57] is

$$y_{ij}|N_i \overset{iid}{\sim} \text{Binomial}(N_i, p) \; i = 1, \ldots, I; \; j = 1, \ldots, J;$$

and

$$N_i \sim \text{Poisson}(\lambda).$$

[26] show that marginally, the y_{ij} are distributed as multivariate Poisson observations with marginal means and variances identical, and with non-zero correlation among counts within sites. Correlations are zero among sites.

The model can be expanded to allow for covariates associated with detection or abundance [58], or to allow for more hierarchical structure than implied by the Poisson model for N [57] including excess zeros [35].

Because of the ability to forgo collecting potentially expensive auxiliary data such as recaptures of marked animals or distances in line transects, while still making inference about abundance and detection, the N-mixture model has proven popular. However, the model can be difficult to fit when detection probabilities and/or the number of sampling occasions are small [26]. Validity of the analysis also depends critically on the assumption that p is constant.

An problem with N-mixture models is that abundance N_i may be ill-defined. By prescribing that such a population exists we may introduce individual heterogeneity in p. For example, as formally expressed in distance sampling models, we expect detection probability to decline with increasing distance from the detector. Such variation in detectability if unmodeled will lead to the nonidentifiability problems discussed by [39]. Another potential issue with N-mixture models is that the binomial counting model presupposes that animals are not counted more than once. This is difficult to avoid without individually identifiable marks on the animals.

[5] have recently criticized N-mixture modeling because under plausible alternative models that fit may the data equally well, N_i can be non-identifiable or not even feature in the model. Moreover, they argue that the only reliable information that can be extracted from replicate count data is relative abundance and the data used in N-mixture model can easily be analyzed by standard Poisson regression which we now describe.

13.2.6   Count regression

Under the assumption that detection rates are either constant or deterministic functions of fully-observed covariates, an alternative to N-mixture modeling is simple Poisson regression. In this approach we abandon the idea that we can reliably estimate N and p and instead model the relative abundances $\gamma_i = N_i/N_0$ where N_0 represents abundance on a reference site. Instead of the conditionally independent Binomial model used for N-mixtures, [42] model the counts as

$$y_{ij}|\mu_i \overset{iid}{\sim} \text{Poisson}(\mu_{ij})$$

where the μ_{ij} can themselves be modeled in terms of fixed and random effects expressed in terms of hierarchical structure [e.g. 41, section 5.3].

Like the N-mixture model, these models have sufficient flexibility to allow for covariates that control detectability, and for covariates that influence relative abundance, both in space and time. As these models are based on Poisson regression they can be implemented using standard glm regression software such as is found in R. The key constraint is that we restrict our inference to $\gamma_i = N_i/N_0$ which is expressed through parameters in the model for μ_{ij} [5]. However, as noted in Section 13.2.5, relative abundance is all that we can expect to make robust inference about based on counts in the presence of imperfect detection.

13.3   Inference for open populations

We now switch focus to open population models, ones that are subject to additions through birth and immigration, and deletions through death and emigration. In the closed population case we assume that our population parameters represent a meaningful summary of the study population at the time of sampling. Now we want to describe changes to the population, changes that will reflect movements of animals in space as well as evolution of the spatial processed over time. Thinking in terms of our hypothetical study region S movement of animals during the study period will see some animals displaced from within S to outside of S and vice versa. Also, we will have animals that that are born into our study region and we will have animals initially present within our study region that die. In combination movements of animals and processes governing birth and death will generate dynamics that we wish to summarize.

One of the simplest descriptions of population dynamics is trend modeling. The Poisson regression models discussed in the last section allow inference for open population through trend model. For example, [42], used log-linear modeling to infer trends in population size over time. All of the methods discussed above can in fact be applied in a similar way. If we have a sequence of capture-recapture studies or distance sampling studies, with the population subject to change between study occasions then we can extend these models to allow abundance to change between studies possibly described in terms of a hierarchical model. For example, [63] describe 13 consecutive studies of a grizzly bear population in Montana from 1986-1995, where in each year a closed population capture-recapture study was carried out but without individuals being identified across years. They added a hierarchical model for the abundances $N_1, \ldots N_{13}$ describing these as Poisson realizations following a trend in time with error. To illustrate flexibility of this approach, they also demonstrated trend-modeling of spina bifida prevalence in New York city between 1969 and 1974 using a hierarchical extension of a log-linear model for closed population capture recapture described by [31]

The focus for the remainder of this chapter will be on open population capture-recapture studies in which individually identifiable animals can be followed through time.

13.3.1   Crosbie-Manly-Schwarz-Arnason model

In open population capture recapture the intervals between sampling occasions are sufficiently long for it to be likely that significant changes occur in the population between occasions. We build our description around the model or [24] and [66] that fully extends Model A of [25] to open populations (model 13.1 above).

Extending the notation of section 13.2.3, let

- ν = the number of distinct animals ever available for capture during the study.

- β_j = the probability that an individual ever available for capture, first becomes available between occasions k, and $j+1$ $(j = 0, 1, \ldots, k-1)$; β_0 is the probability that an individual was available at the start of the study.

- ϕ_j = the probability that an animal in the population at j is in the population at $j+1$ $(j = 1, \ldots, k-1)$.

- η_j = the probability that an animal caught at j is returned to the population $(j = 1, \ldots, k)$.

- ψ_j = the probability that an animal ever available for capture has not yet been caught by occasion j; $\psi_1 = \beta_0$ and $\psi_{j+1} = \psi_j(1-p_j)\phi_j + \beta_j$ $(j = 2, \ldots, k-1)$.

- λ_j = the probability than at animal released at j is ever recaptured; $\lambda_{k-1} = \phi_{k-1}p_k$, $\lambda_j = \phi_j p_{j+1} + \phi_j(1-p_{j+1})\lambda_{j+1}$ $(j = 1, \ldots, k-2)$.

- τ_j = the probability of capture at j for an animal marked before j and caught at or after j $(j = 2, \ldots, k)$.

- U_j = the number of unmarked animals in the population at the time of sample j.

- u_j = the number of unmarked animals caught in sample j $(j = 1, \ldots, k)$; $n = \sum_{j=1}^{k} u_i$.

- R_j = the number of animals released from the sample caught at j $(j = 1, \ldots, k)$.

- r_j = the number of the R_j that are recaptured.

- m_j = the number of marked animals caught in sample j $(j = 2, \ldots, k)$.

- n'_j = the number of animals caught in sample j; $n'_1 = u_1$; $n'_j = u_j + m_j$ $(j = 2, \ldots, k)$.

- T_j = the number of animals marked and released prior to j and recaptured at some occasion $h, j \leq h \leq k$; $T_2 = r_1$, and $T_{j+1} = T_j m_j + r_j$, $(j = 2, \ldots, k2)$.

Starting with the description (13.1) and deriving expressions for π_ω we can show [e.g. see 40, 66] that the full open population model can be factored into terms depending on sufficient statistics. If we let $B(y; n, p)$ denote a binomial probability mass function evaluated at y with parameters n and p, and $M(y; n, \pi)$ denote a multivariate pmf evaluated for vector y

with parameters n and π, then

$$f(x|\nu,\beta,\phi,p,\eta) = h(x) \times \underbrace{B(n;\nu,\alpha)}_{f(n|\nu,\beta,\phi,p)} \times \underbrace{M(u;n,\xi)}_{f(u|n,\beta,\phi,p)} \times \underbrace{\prod_{j=1}^{k} B(R_j;n'_j,\eta_j)}_{f(R|n',\eta)}$$

$$\times \underbrace{\prod_{j=1}^{k-1} B(r_j;R_j,\lambda_j)}_{f(r|R,\phi,p)} \times \underbrace{\prod_{j=2}^{k} B(m_j;T_j,\tau_j)}_{f(m|T,\phi,p)} \qquad (13.7)$$

where $\alpha = \sum_{j=1}^{k} \psi_j p_j$ and $\xi_j = \psi_j p_j/\alpha$ $(j=1,\ldots,k)$. The constant $h(x)$ is the distribution of the capture histories conditional on sufficient statistics and contains no parameters. It does play a role in goodness of fit assessment [51].

Not all parameters in the model (13.7) are identifiable or able to be reliably estimated [see 40, 64, for a fuller discussion of identifiability]. The CMSA model is closely related to a number of historically important models including the Jolly-Seber model [34, 68] and the Cormack-Jolly-Seber (CJS) [23, 34, 68].

13.3.2 Cormack-Jolly-Seber model and tag-recovery models

In relation to model (13.7) the CJS model is given by

$$f(m,r|R,\phi,p) = f(r|R,\phi,p) \times f(m|T,\phi,p)$$

noting that the elements of T are functions of elements of R and r. Thus, the CJS model is a partial likelihood embedded within the full CMSA model, ignoring information from captures of unmarked animals. If the capture and marking process are very different from the recapture process, such that capture probabilities are modeled as distinct from recapture probabilities, then the CJS model can be regarded as the full likelihood for parameters ϕ and recapture probabilities p_j for $j=2,\ldots,k$ as in the case of the fulmar petrel study considered by [23].

With the CJS model focus is on the survival process as we have insufficient information to model entry of animals into the population. A parallel set of model developed independently by [69] and [55] is the tag-recovery model, in which cohorts of tagged animals of size R_j are released at times indexed by $j=1,\ldots,k$, with the numbers recovered dead r_j for the intervals $(j,j+1)$ for $j=1,\ldots,k$. The model is parameterized in terms of survival and tag-recovery probabilities, where

- S_j = the probability an animal alive at the time of sampling occasion j is still alive at sampling occasion $j+1$ $(j=1,\ldots,k-1)$.

- f_j = the probability that an animal alive at the time of sampling occasion j is recovered dead in $(j,j+1)$ and its tag reported.

If r_{jh} denotes the numbers released at j that are recovered in $(h,h+1)$, and $r_j = \sum_{h=j}^{k} r_{jh}$, then for release cohort j, the vector $(r_{jj},r_{jj+1},\ldots,r_{jk},R_j-r_j)$ is modeled as multinomial with index R_j and parameter vector

$$\pi_j = c(f_j, S_j f_{j+1}, \ldots, S_j \cdots S_{k-1} f_k, S_j S_{j+1} \cdots S_k).$$

When this model was published it was not realized at the time that the CJS model and the tag-recovery model are in fact the same model but for different types of data. If for the CJS model we define m_{jh} as the numbers of animals released at j that are next captured at h, and noting that $r_j = \sum_{h=j+1}^{k} m_{jh}$ then the factorization

$$f(m, r|R, \phi, p) = f(r|R, \phi, p) \times f(m|T, \phi, p)$$

in (13.7) can be shown to be identical to

$$f(m, r|R, \phi, p) = \prod_{j=1}^{k-1} \mathrm{M}((m_{j,j+1}, \ldots, m_{j,k}, R_j = r_j); R_j, \pi_j)$$

where

$$\pi_j = c(\alpha_j, \beta_j \alpha_{j+1}, \ldots, \beta_j \cdots \beta_{k-1} \alpha_k, \beta_j \beta_{j+1} \cdots \beta_k),$$

$$\alpha_j = \phi_{j-1} p_j,$$

and

$$\beta_j = \phi_j (1 - p_{j+1}).$$

Thus, the models differ in their parameterization and the type of data: (i) live-recapture for the CJS model and (ii) dead-recovery for the tag-recovery model. They also differ in their interpretation of the survival parameter which explains the different notation. The tag-recovery models assume that animals are exposed to tag-recovery throughout their range, in which case animals exit the population solely through death. In the CJS model, animals are assumed to be exposed to capture only in a restricted area of their range, usually ill-defined as the area in which the animal is at risk of capture. Thus, animals can leave the population either through death or through emigration. A complication, is that animals may leave permanently or they may leave and later return. Following [2] we assume that movement may be Markovian, depending on two parameters:

- F_j = the probability an animal alive at occasions j and $j+1$, and at risk of capture at j, remains at risk of capture at $j+1$.

- F_i' = the probability an animal alive at occasions j and $j+1$, and not at risk of capture at j, becomes at risk of capture at $j+1$.

Under permanent emigration, $F_i' = 0$ and the CJS survival parameter is confounded as $\phi_j = F_j S_j$, where S_j is the true survival probability. However, if we allow temporary emigration but with $F_i = F_i'$, we can show that the CJS capture probability is now confounded with the movement probability so that the capture probability now equals $F_{i-1} p_i$, the joint probability of remaining at risk of capture and being caught. Algebraically we cannot distinguish between the two forms as they lead to identical likelihood but with different parameterizations [2, 36].

It is possible that in a single study we have both types of data (live-recapture and dead-recovery). With live-recaptures by definition deemed to occur at the time of sample j, it is also possible to have a third type of data, that of live-resightings that occur in the interval $(j, j+1)$ during which the population is exposed to losses; these might be natural deaths and, say, deaths due to hunting which generate the tag-recovery information.

Expansion of the CJS model for live-recaptures to incorporate dead-recoveries was described by [17]. [2] completed the generalization to allow all three data types. With these joint models we are now able to discriminate between the various types of temporary emigration.

13.3.3 Pollock's robust design

The robustness of the survival estimators to individual heterogeneity in capture was noted by [19]. This observation led [50] to suggest a design in which capture-recapture studies are carried out at two time scales. The first comprises periods of short-term sampling, called secondary periods, during which the population is assumed closed. The second called, primary periods, comprises intervals during which the population is open that occur between bouts of secondary sampling [73]. The idea is that in the analysis, abundance estimation could be largely based on closed population models accommodating some forms of heterogeneity in p, while extracting information about survival from open population, for example, fitted to data collapsed across the secondary occasions.

The CMSA model (13.7) is sufficiently general to accommodate the robust design. Between secondary occasions we simply set entry probabilities to zero and survival probabilities to 1, and leave them unconstrained between primary occasions. [36] described a restricted version of this likelihood in which modeling is conditional on u. This leads to an slight loss in efficiency owing to the fact that u is not an ancillary statistic for the parameters of the restricted model [64]. However, the model of [36] is generalized to allow for Markovian temporary emigration. Adding temporary emigration to the CMSA model would complete the description in (13.7) to incorporate animal movement.

13.3.4 Capture recapture models for population growth rate

[52] introduced a restricted version of the CMSA model that reparameterizes the model in terms of population growth rate. This allows the user to focus on trends in population growth, similar to the model discussed in the Section 13.2.6 on count regression.

An unusual feature of Pradel's model is that he conditions on n (i.e., treats this random feature of the data as though it were specified in advance.). Although conditioning on aspects of the data has the potential to discard information Pradel's intuition was that the effect of this would be slight. [40] showed that α can be rewritten as $\beta_0 \Omega$ where Ω is a function of model parameters that is functionally independent of β_0. They concluded that in consequence, n contains no information about N other than that $\nu \geq n$. [64] identified more precisely what is meant by 'no information; they showed that n is Bayes-ancillary for the parameters p, ϕ and β in the presence of a prior distribution $f(\nu) \propto \nu^{-1}$. This means that conditioning on n in place of using the full CMSA model can be justified with respect to use of a marginal likelihood for p, ϕ and β with marginalization across ν using this scale scale prior. Inference based on this marginal likelihood and likelihood conditioned on n will be identical.

With his conditional likelihood, [52] considered three different reparameterizations, replacing β with (i) λ_j, an index to N_{j+1}/N_j ($j = 2, \ldots, k$) or (ii) γ_j = the probability that an animal in the population at j was in the population at $j - 1$ ($j \leqq 2, \ldots, k$), or a 'fecundity rate' parameter $f_j = 1/\gamma_j - 1$. [40] reformulated Pradel's model and redefined $f_j = \beta_j/d_j$ where $d_{j+1} = d_j \phi_j + \beta_j$ for $j = 1, \ldots, k - 1$, with $d_1 = \beta_0$. Since $d_j = \mathbb{E}[N_j|\nu]/\nu$ and $\beta_j = \mathbb{E}[B_j|\nu]/\nu$, it follows that under the parameterization of [40] $f_j = \mathbb{E}[\beta_j|\nu]/\mathbb{E}[N_j|\nu] \approx \mathbb{E}[B_j/N_j|\nu]$ and is an index to per-capita recruitment.

As [40] show this formulation has advantages, including that it leads to explicit maximum likelihood estimation. It also means we can fully express the factorization of the CMSA model (13.7) in terms of the parameters ϕ and f which facilitates hierarchical extensions of the model. For example, [40] extended the model to allow a stochastic relationship between $\log(f)$ and $\text{logit}(\phi)$ to illustrate how hierarchical modeling allows us to describe demographic dependencies among parameters of the model.

13.3.5 Capture recapture models in terms of latent variable

[61], [62], and [65] describe hierarchical mark-recapture models for open populations expressed in terms of latent variables. This approach has a number of advantages including allowing expression of of complicated models in terms of simple components, (ii) providing a convenient way to modeling missing data, and (iii) as a way of introducing parsimony by partial pooling in models with large numbers of parameters.

The framework described by [61] is sufficiently general to accommodate wide-ranging choice of capture-recapture design including all of the designs in the previous sections and some extensions we have not discussed. One important class of models is the multistate model, an extension of the CJS model in which capture and survival probabilities are modeled in terms of a time-varying categorical covariate 'state' which is unobserved for animals that are not caught [11, 67]. Two important extensions are to continuous time-varying covariates [6] and to uncertain categorical covariates [the 'multievent' model of 53].

A key feature of the hierarchical representation in terms of latent variables is the separation of the model into a process describing births and deaths and an observation process. This so-called state-space representation while thought of as a modern development was in fact a key feature of the approach followed by [34] in deriving his solution to the JS model [64]. Jolly described his model in terms of latent variables M_{jh}, the number of marked animals last seen at j that are still alive at h, and the parameters U and ϕ. Treating the latent variables as parameters he was able to find maximum likelihood solutions. Here we follow [61] and describe the model in terms of latent times of birth and death. Let

- b_i = the index of the occasion when the animal entered the population: $b_i = 0$ if the animal was in the population at the start of the study, $b_i = j$ if the animal entered the population between sample j and $j + 1$ $(j = 1, \ldots, k - 1)$.

- d_i = the index of the occasion when the animal left the population: $d_i = j$ if the animal left the population between samples j and $j + 1$ $(j = 1, \ldots, k - 1)$; $d_i = k$ if the animal was still in the population at occasion k.

- l_{ij} an indicator that takes the value 1 if animal i is removed from the population in sample j and zero otherwise, with $l_j = \sum_{i=1}^{N} l_{ij}$. l_j = the number of animals lost on capture in sample j; $l_j = n'_j - R_j$ $(j = 1, \ldots, k)$.

We build the model for the capture-histories x_{ij} by modeling the latent variable b_i as categorical outcome with support $0, 1, \ldots, k - 1$ and probability $\beta = (\beta_0, \ldots, \beta_{k-1})$. Similarly, we model the latent variable d_i as categorical with support $b_i, \ldots, k$ and parameter vector $\zeta = (\zeta_{b_i}, \ldots, \zeta_k)$ where

$$\zeta_j = \begin{cases} 1 - \phi_j & j = b_i \\ (1 - \phi_j) \prod_{h=b_i}^{j-1} \phi_h & b_i < j < k \\ \prod_{h=b_i}^{k} \phi_h & j = k \end{cases}$$

In terms of these latent variables, the model is

$$f(b, d | N, \beta, \phi) = f(b | \beta, N) f(d | b, \phi, N);$$

from the latent variables we can construct population summaries of interest such as N_j, the number of animals alive on occasion j, and the numbers B_j that entered the population between j and $j + 1$, or the numbers D_j that left between j and $j + 1$.

The description above is incomplete as we need to add an observation process as well as account for the partial determination of b and d. Reading from left to right, the index of the first non-zero value of $x_i = x_{i1}, \ldots, x_{ik}$ gives us a left censored value for b_i except that if $x_{i1} = 1$ we know that $b_i = 0$. Similarly we obtain right censored values for d_i according to the last non-zero value for x_{ij} except that if $x_{ik} = 1$ we know that $d_i = k$.

Given b and d the model for the capture histories is straight forward. If a_{ij} is the indicator for availability of capture (i.e., takes the value 1 if $b_i \le j \le d_i$ except that if the animal was removed from the population in sampling at j, a_{ij} is 0 in all subsequent occasions). Provided the parameters for losses on capture are distinct from remaining parameters, then we can condition on any losses on capture l through the description of a. Alternatively, we can break the model for a_{ij} up into a deterministic component that is a function of b_i and d_i, and a stochastic component for l_{ij}. The model for $x|a$ is simply

$$f(x|b,d,l) = \binom{N}{n} \prod_{i=1}^{N} \prod_{j=1}^{k} (a_{ij}p_j)^{x_{ij}} (1 - a_{ij}p_j)^{1-x_{ij}}.$$

If, for simplicity, we condition on l, a complete data likelihood (CDL) is given by

$$f(x|N,\beta,\phi,p) = f(b,d|N,\beta,\phi) = f(b|\beta,N)f(d|b,\phi,N)f(x|b,d)$$

subject to restrictions on what is known about a. The CDL is useful for Bayesian inference and also hierarchical extensions of the CMSA model. Using software such as JAGS it is straightforward to fit the model using MCMC to generate posterior distributions on the unknown elements of b and d (or equivalently a), as well as the parameters N, β, ϕ, p and ν.

13.4 Combining observation and process models

The state-process representation of the previous section brings us back to the focus of animal population studies: inference about the distribution of animals in space and their dynamics through time. Already we have considered cases where the underlying process model has been an integral part of inference. These include:

- Where we have a sequence of closed population capture-recapture models. At the start of Section 13.3 we described an example where closed population studies were replicated in time for trend modeling. Replication could instead be spatial or it could be both both spatial and temporal.

- In the SCR models of Section 13.2.4 where the underlying process model for animal locations is required to model heterogeneity in capture

- In the N-mixture models of Section 13.2.5 and the count regression models of Section 13.2.6 where a process model for animal abundance (or expected counts) in space and/or time is used to provide replication to improve inference about factors that influence abundance

A general treatment of state-process modeling in ecology is given by [8], primarily for closed populations, and [13] for open populations. The state-space model is also discussed in Chapter X (Ken's chapter). Interest is in describing spatial aspects of the population and on temporal dynamics. For the latter we might be interested in predicting future population trajectories. Applications discussed by [13] include count surveys of seal pups, line transects observations of fecal pellets for sika deer, or fisheries data from vessel surveys. [4] used a Bayesian hierarchical model to joint model demographic processes and sampling error in an integrated analysis of various data sources for the New Zealand saddleback. The primary aim of this analysis was to predict future population trajectories of a translocated population.

13.5   Software and model fitting

Fitting many of the models of this chapter require specialized software. For frequentist inference, Program MARK [72], with an R interface R-MARK [37] contain all of the major capture recapture models for closed an open population including the robust design, multi-state models and for models including auxiliary data. Alternatives include eSURGE [22] which allows fitting of multistate models as well as the multievent model of [53]. The flexibility of the multistate model to accommodate a wide array of capture recapture data types is discussed by [38]. Program DISTANCE [71] is the primary software tool for distance sampling and DENSITY [28] for the SCR models of [27].

For Bayesian inference, the software BUGS [45] and JAGS [49] employ Markov chain Monte Carlo methods for fitting models described by a similar syntax. Both can be implemented using an R interface of which there are several. Code for model fitting using BUGS/JAGS is described in [56], [41], [61], and [65]. More recently the development of STAN [20] allows fitting by Hamiltonian MCMC which offers advantages for some models.

Bibliography

[1] Alpízar-Jara, R. and Pollock, K. H. (1996). A combination line transect and capture-recapture sampling model for multiple observers in aerial surveys. *Environmental and Ecological Statistics* **3**, 311–327.

[2] Barker, R. J. (1997). Joint modeling of live-recapture, tag-resight, and tag-recovery data. *Biometrics* **53**, 666 – 677.

[3] Barker, R. J. (2008). Theory and application of mark-recapture and related techniques to aerial surveys of wildlife. *Wildlife Research* **35**, 268–274.

[4] Barker, R. J., Schofield, M. R., Armstrong, D. P., and Davidson, R. S. (2009). Bayesian hierarchical models for inference about population growth. In *Modeling Demographic Processes in Marked Populations*, pages 3–17. Springer.

[5] Barker, R. J., Schofield, M. R., Link, W. A., and Sauer, J. (2017). On the reliability of n-mixture models for count data. *Biometrics* page In press.

[6] Bonner, S. J. and Schwarz, C. J. (2006). An extension of the Cormack-Jolly-Seber model for continuous covariates with application to *Microtus pennsylvanicus*. *Biometrics* **62**, 142 – 149.

[7] Borchers, D., Stevenson, B., Kidney, D., Thomas, L., and Marques, T. (2015). A unifying model for capture–recapture and distance sampling surveys of wildlife populations. *Journal of the American Statistical Association* **110**, 195–204.

[8] Borchers, D. L., Buckland, S. T., and Zucchini, W. (2002). *Estimating animal abundance: closed populations*. Springer, London.

[9] Borchers, D. L., Laake, J. L., Southwell, C., and Paxton, C. G. M. (2006). Accommodating unmodeled heterogeneity in double-observer distance sampling surveys. *Biometrics* **62**, 372–378.

[10] Borchers, D. L., Zucchini, W., and Fewster, R. M. (1998). Mark-recapture models for line transect surveys. *Biometrics* **54**, 1207–1220.

[11] Brownie, C., Hines, J., Nichols, J., Pollock, K., and Hestbeck, J. (1993). Capture-recapture studies for multiple strata including non-Markovian transitions. *Biometrics* **49**, 1173–1187.

[12] Buckland, S., Anderson, D., Burnham, K., and Laake, J. (2008). *Advanced distance sampling*. Oxford University Press, Oxford, U. K.

[13] Buckland, S., Newman, K., Thomas, L., and Koesters, N. (2004). State-space models for the dynamics of wild animal populations. *Ecological modelling* **171**, 157–175.

[14] Buckland, S. T., Anderson, D. R., Burnham, K. P., and Laake, J. L. (1993). *Distance Sampling: Estimating Abundance of Biological Populations*. Chapman and Hall, London.

[15] Buckland, S. T., Anderson, D. R., Burnham, K. P., Laake, J, L., Borchers, D. L., and Thomas, L. (2001). *Introduction to distance sampling: Estimating abundance of biological Populations*. Oxford University Press, Ocford.

[16] Buckland, S. T., Newman, K. B., Fernández, C., Thomas, L., and Harwood, J. (2007). Embedding population dynamics models in inference. *Statistical Science* pages 44–58.

[17] Burnham, K. P. (1993). A theory for combined analysis of ring recovery and recapture data. In Lebreton, J.-D. and North, P., editors, *Marked individuals in bird population studies.*, pages 199–213. Birkhauser Verlag, Basel.

[18] Burnham, K. P., Anderson, D. R., and Laake, J. L. (1980). Estimation of density from line transect sampling of biological populations. *Wildlife Monographs* **72**, 1–202.

[19] Carothers, A. D. (1973). The effects of unequal catchability on Jolly-Seber estimates. *Biometrics* **29**, 79–100.

[20] Carpenter, B., Gelman, A., Hoffman, M., Lee, D., Goodrich, B., Betancroft, M., Brubaker, M. A., Guo, J., Li, P., and Riddell, A. (2015). Stan: A probabilistic programming language. In press,. *Journal of Statistical Software* .

[21] Caswell, H. (2001). *Matrix population models, second edition.* Sunderland, Massachusetts, Sinauer Associates.

[22] Choquet, R., Rouan, L., and Pradel, R. (2009). Program e-surge: a software application for fitting multievent models. In *Modeling demographic processes in marked populations*, pages 845–865. Springer.

[23] Cormack, R. M. (1964). Estimates of survival from the sighting of marked animals. *Biometrika* **51**, 429–438.

[24] Crosbie, S. F. and Manly, B. F. J. (1985). Parsimonious modelling of capture-mark-recapture studies. *Biometrics* **41**, 385–398.

[25] Darroch, J. (1958). The multiple-recapture census. I estimation of a closed population. *Biometrika* **45**, 343–59.

[26] Dennis, E. B., Morgan, B. J. T., and Ridout, M. S. (2015). Computational aspects of N-mixture models. *Biometrics* **71**, 237–246.

[27] Efford, M. (2004). Density estimation in live-trapping studies. *Oikos* **106**, 598–610.

[28] Efford, M. G. (2012). Density 5.0: software for spatially explicit capture–recapture. *Department of Mathematics and Statistics, University of Otago, Dunedin, New Zealand http://www. otago. ac. nz/density* .

[29] Gelman, A., Carlin, J. B., Stern, H. S., Dunson, D. B., Vehtari, A., and Rubin, D. B. (2014). *Bayesian Data Analysis (Third edition)*. Chapman and Hall/CRC, Boca Raton, Florida, USA.

[30] Gimenez, O., Rossi, V., Choquet, R., Dehais, C., Doris, B., Varella, H., Vila, J.-P., and Pradel, R. (2007). State-space modelling of data on marked individuals. *Ecological Modelling* **206**, 431–438.

[31] Hook, E., Albright, S., and Cross, P. (1980). Use of Bernoulii census and log-linear methods for estimating the prevalence of spina bifida in livebirths and the completeness of vital record reports in New York state. *American Journal of Epidemiology* **112**, 750–8.

[32] Horvitz, D. G. and Thompson, D. J. (1952). A generalization of sampling without replacement from a finite universe. *Journal of the American Statistical Association* **47**, 663 – 685.

[33] Huggins, R. M. (1989). On the statistical analysis of capture experiments. *Biometrika* **76**, 133–140.

[34] Jolly, G. (1965). Explicit estimates from capture-recapture data with both death and immigration-stochastic model. *Biometrika* **52**, 225–247.

[35] Joseph, L. N., Elkin, C., Martin, T. G., and Possingham, H. P. (2009). Modeling abundance using N-mixture models: the importance of considering ecological mechanisms. *Ecological Applications* **19**, 631–642.

[36] Kendall, W., Nichols, J., and Hines, J. (1997). Estimating temporary emigration and breeding proportions using capture-recapture data with Pollock's robust design. *Ecology* **78**, 563–578.

[37] Laake, J. L. (2013). RMark: An r interface for analysis of capture-recapture data with MARK. AFSC Processed Rep. 2013-01, Alaska Fish. Sci. Cent., NOAA, Natl. Mar. Fish. Serv., Seattle, WA.

[38] Lebreton, J.-D., Nichols, J. D., Barker, R. J., Pradel, R., and Spendelow., J. A. (2009). Modeling individual animal histories with multistate capture-recapture models. *Advances in Ecological Research* **41**, 87–173.

[39] Link, W. A. (2003). Nonidentifiability of population size from capture-recapture data with heterogeneous detection probabilities. *Biometrics* **59**, 1123–1130.

[40] Link, W. A. and Barker, R. J. (2005). Modeling association among demographic parameters in analysis of open population capture-recapture data. *Biometrics* **61**, 46–54.

[41] Link, W. A. and Barker, R. J. (2010). *Bayesian inference with ecological applications*. Academic Press, London, UK.

[42] Link, W. A. and Sauer, J. R. (1997). Estimation of population trajectories from count data. *Biometrics* **53**, 488–497.

[43] Link, W. A. and Sauer, J. R. (2002). A hierarchical analysis of population change with application to cerulean warblers. *Ecology* **83**, 2832–2840.

[44] Lukacs, P. M. and Burnham, K. P. (2005). Review of capture-recapture methods applicable to noninvasive genetic sampling. *Molecular Ecology* **14**, 3909–3919.

[45] Lunn, D., Thomas, A., Best, N., and Spiegelhalter, D. (2000). WinBUGS – a Bayesian modelling framework: concepts, structure, and extensibility. *Statistics and Computing* **10**, 325–337.

[46] Manly, B. F. J., McDonald, L. L., and Garner, G. W. (1996). Maximum likelihood estimation for the double-count method with independent observers. *Journal of Agricultural, Biological, and Environmental Statistics.* **1**, 170–189.

[47] Miller, D. L. and Thomas, L. (2015). Mixture models for distance sampling detection functions. *PLoS ONE* **10**, e0118726.

[48] Otis, D. L., Burnham, K. P., and White, G. C. anderson, D. R. (1978). Statistical inference from capture data on closed animal populations. *Wildlife Monographs* **62**, 1–135.

[49] Plummer, M. (2003). JAGS: A program for analysis of Bayesian graphical models using Gibbs sampling. In *Proceedings of the 3rd international workshop on distributed statistical computing (DSC 2003), March 20-22, Vienna, Austria*, pages 20–22.

[50] Pollock, K. H. (1982). A capture-recapture design robust to unequal probability of capture. *Journal of Wildlife Management* **46**, 752 – 757.

[51] Pollock, K. H., Hines, J. E., and Nichols, J. D. (1985). Goodness-of-fit tests for open capture-recapture models. *Biometrics* **41**, 399–410.

[52] Pradel, R. (1996). Utilization of capture-mark-recapture for the study of recruitment and population growth rate. *Biometrics* **52**, 703–709.

[53] Pradel, R. (2005). Multievent: an extension of multistate capture-recapture models to uncertain states. *Biometrics* **61**, 442–447.

[54] Quang, P. X. and Becker, E. F. (1997). Combining line transect and double count sampling techniques for aerial surveys. *Journal of Agricultural, Biological , and Environmental Statistics* **2**, 230–242.

[55] Robson, D. and Youngs, W. (1971). Statistical analysis of reported tag-recaptures in the harvest from an exploited population. *Unpublished report, Cornell University Biometry Unit* **369**,.

[56] Royle, J. and Dorazio, R. (2008). *Hierarchical modeling and inference in ecology: the analysis of data from populations, metapopulations and communities.* Academic Press.

[57] Royle, J. A. (2004). N-mixture models for estimating population size from spatially replicated counts. *Biometrics* **60**, 108–115.

[58] Royle, J. A., Nichols, J. D., Kéry, M., and Ranta, E. (2005). Modelling occurrence and abundance of species when detection is imperfect. *Oikos* **110**, 353–359.

[59] Royle, J. A. and Young, K. V. (2008). A hierarchical model for spatial capture-recapture data. *Ecology* **89**, 2281–2289.

[60] Rubin, D. B. (1976). Inference and missing data. *Biometrika* **63,** 581–592.

[61] Schofield, M. R. and Barker, R. J. (2008). A unified capture-recapture framework. *Journal of Agricultural, Biological and Environmental Statistics* **13,** 458–477.

[62] Schofield, M. R. and Barker, R. J. (2009). A further step toward the mother-of-all-models: Flexibility and functionality in the modeling of capture-recapture data. In Thompson, D. L., Cooch, E. G., and Conroy, editors, *Modeling demographic processes in marked populations*, pages 677–689. Springer.

[63] Schofield, M. R. and Barker, R. J. (2014). Hierarchical modeling of abundance in closed population capture-recapture models under heterogeneity. *Environmental and Ecological Statistics* **21,** 435 – 451.

[64] Schofield, M. R., Barker, R. J., Gelman, A., Cook, E. R., and Briffa, K. R. (2016). A model-based approach to climate reconstruction using tree-ring data. *Journal of the American Statistical Association* **111,** 93–106.

[65] Schofield, M. R., Barker, R. J., and MacKenzie, D. I. (2009). Flexible hierarchical mark-recapture model for open populations using WinBUGS. *Envirnmental and Ecological Statistics* **16,** 369–387.

[66] Schwarz, C. J. and Arnason, A. N. (1996). A general methodology for the analysis of capture-recapture experiments in open populations. *Biometrics* **52,** 860–873.

[67] Schwarz, C. J., Schweigert, J., and Arnason, A. (1993). Estimating migration rates using tag-recovery data. *Biometrics* **49,** 177–193.

[68] Seber, G. A. F. (1965). A note on the multiple-recapture census. *Biometrika* **52,** 249–259.

[69] Seber, G. A. F. (1970). Estimating time-specific survival and reporting rates for adult birds from band returns. *Biometrika* **57,** 313–318.

[70] Severini, T. A. (1995). Information and conditional inference. *Journal of the American Statistical Association* **90,** 1341–1346.

[71] Thomas, L., Buckland, S. T., Rexstad, E. A., Laake, J. L., Strindberg, S., Hedley, S. L., Bishop, J. R., Marques, T. A., and Burnham, K. P. (2010). Distance software: design and analysis of distance sampling surveys for estimating population size. *Journal of Applied Ecology* **47,** 5–14.

[72] White, G. C. and Burnham, K. P. (1999). Program mark: survival estimation from populations of marked animals. *Bird Study* **46,** S120–S139.

[73] Williams, B. K., Nichols, J. D., and Conroy, M. J. (2002). *Analysis and management of animal populations*. Academic Press, San Diego, California, USA.

[74] Wright, J. A., Barker, R. J., Schofield, M. R., Frantz, A. C., Byrom, A. E., and Gleeson, D. M. (2009). Incorporating genotype uncertainty into mark-recapture-type model for estimating animal abundance. *Biometrics* **65,** 833–840.

14

Animal Movement Models

Mevin Hooten

U.S. Geological Survey and Colorado State University

Devin Johnson

National Oceanic and Atmospheric Administration

CONTENTS

14.1 Introduction

Advancements in biotelemetry technology have led to a rich set of statistical tools for analyzing data arising as observations of spatio-temporal trajectories of individual animals [32]. In this chapter, we focus primarily on biotelemetry ("telemetry" hereafter) data that were recorded as positions in geographical space, denoted as $\mathbf{s}_i$, for a set of observation times (termed "fixes") $t_1, \ldots, t_i, \ldots, t_n$. Methodologically, we review statistical models for satellite derived telemetry data that were collected with fixes close enough together in time that a movement signal is present in the data. The movement signal is a representation of the true path, denoted as $\boldsymbol{\mu}(t)$. The characteristics of the true path and its relationship with the environment and other individual paths serves as the primary target for statistical inference in movement ecology. The inherent structure in the individual's path is often unknown perfectly and presents the primary challenge for statistical modeling. The structure of the path may depend on endogenous correlative properties based on the physics of movement as well as exogenous drivers such as terrain, habitat, and the social structure of populations and communities of animals [6].

The past two decades have given rise to thousands of publications and new journals specifically devoted to animal movement research. The techniques that have been developed for studying animal movement include a huge variety of approaches, from theoretical, to simulation-based, to statistical. Of the existing statistical approaches for studying animal movement, we review parametric models that explicitly incorporate temporal dependence in telemetry data.

Most animal movement models focus on the physical process that gives rise animal trajectories, however the data model is also an important consideration because of the unique sources of measurement error associated with telemetry data. Thus, a hierarchical modeling framework [8] provides an excellent way to incorporate structure and uncertainty associated with the data and latent movement process. In what follows, we highlight several data and process models that can be combined in a hierarchical modeling framework using conditional probability distributions, such as

$$\mathbf{s}_i \sim [\mathbf{s}_i | \boldsymbol{\mu}(t_i), \boldsymbol{\theta}_s] \tag{14.1}$$

$$\boldsymbol{\mu}(t_i) \sim [\boldsymbol{\mu}(t_i) | \{\boldsymbol{\mu}(t), \forall t \in \mathcal{T}_{-t_i}\}, \boldsymbol{\theta}_\mu] \tag{14.2}$$

where probability distributions are denoted as '[·]' and $\{\boldsymbol{\mu}(t), \forall t \in \mathcal{T}_{-t_i}\}$ is the set of all positions (i.e., the path) in the underlying movement process except at time t_i and $\boldsymbol{\theta}_s$ and $\boldsymbol{\theta}_\mu$ represent the data- and process-level parameters, respectively. State-space model specifications like (14.1) and (14.2) have become popular in animal movement ecology (e.g., [39] [51]). However, there are many other ways to write a conditional process model (14.2) and we return to those in Section 14.3–14.5. In fact, process models for animal movement are familiar to most statisticians and often fall into one of three main categories: Point process models, discrete-time models, and continuous-time models. We describe each type of process model in what follows, but first, we turn our attention to data models (14.1) for various sources of telemetry data.

14.2 Data Models

Data uncertainty is an important component of animal movement models because all sources of telemetry data are measured with error. Thus, the true underlying position process $\boldsymbol{\mu}(t)$ is unknown and usually considered in a two-dimensional geographical support $\mathcal{M}$. The observed telemetry data $\mathbf{s}(t_i)$ for $i = 1, \ldots, n$ are measurements of $\boldsymbol{\mu}(t_i)$ and may have different support $\mathcal{S}$ due to constraints on either $\boldsymbol{\mu}(t)$ or $\mathbf{s}(t_i)$. A simple, but commonly assumed, model for telemetry data is a bivariate Gaussian distribution such that $\mathbf{s}(t_i) \sim \mathrm{N}(\boldsymbol{\mu}(t_i), \boldsymbol{\Sigma}_i)$. For example, global positioning system (GPS) telemetry data are often assumed to have Gaussian error distributions. We allow for temporal variation in error distributions by indexing the error covariance $\boldsymbol{\Sigma}_i$ with i because individual position relative to satellite positions and environmental and behavioral conditions vary over time. For example, the arrangement of overhead satellites allows GPS telemetry device manufactures to report a "dilution of precision" (DOP) metric that provides some indication of the quality of each observation. Thus, a simple time-varying model for error covariance is $\boldsymbol{\Sigma}_i \equiv \sigma_i^2 \mathbf{I}$ where $\mathbf{I}$ is a 2×2 identity matrix and $\log(\sigma_i) = \alpha_0 + \alpha_1 \mathrm{DOP}_i$, for example. This error covariance model can be extended by parameterizing it as

$$\boldsymbol{\Sigma}_i \equiv \sigma_i^2 \begin{bmatrix} 1 & \rho_i \sqrt{a_i} \\ \rho_i \sqrt{a_i} & a_i \end{bmatrix}. \tag{14.3}$$

In situations where it can be reasonably assumed that the measurement error is temporally homogeneous, we let $\boldsymbol{\Sigma}_i \equiv \boldsymbol{\Sigma}, \forall i$.

Another common form of telemetry data are collected using the Service Argos system (hereafter "Argos"). Argos telemetry data are measured by exploiting properties of the Doppler effect based on polar orbiting satellites. The individual animal is fitted with a similar type of telemetry device, but rather than triangulating the position using a network of overhead satellites as in GPS telemetry data, Argos tracks the individual as it passes overhead with a single satellite, using its own trajectory as a frame of reference. The polar orbiting nature of Argos satellites results in a unique error structure that resembles an X-pattern [17][20] (Figure 14.1). A flexible model that can account for the X-shaped pattern

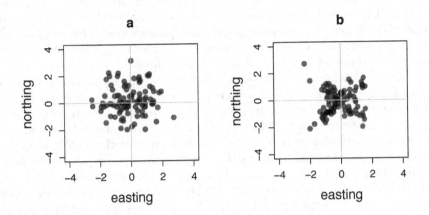

FIGURE 14.1
Two examples of telemetry position errors (i.e., $\mathbf{s}_i - \boldsymbol{\mu}(t_i)$, $i = 1, \ldots, n$). (a) independent Gaussian with single variance parameter $\sigma^2 = 1$ and (b) mixture Gaussian with non-diagonal covariance (14.3) with mixture probability $p = 0.5$, variance $\sigma^2 = 1$, and covariance parameters $\rho = 0.9$ and $a = 1$.

in Argos telemetry data involves a mixture of Gaussian distributions (or t distributions for heavier tails). For earlier Argos data (before year 2007), Brost et al. [13] and Buderman et al. [14] proposed the model

$$\mathbf{s}_i \sim p \cdot \mathrm{N}(\boldsymbol{\mu}(t_i), \boldsymbol{\Sigma}_i) + (1 - p) \cdot \mathrm{N}(\boldsymbol{\mu}(t_i), \boldsymbol{\Psi}\boldsymbol{\Sigma}_i\boldsymbol{\Psi}') , \tag{14.4}$$

where the covariance matrix $\boldsymbol{\Sigma}_i$ is defined as in (14.3) and the rotation matrix

$$\boldsymbol{\Psi} \equiv \begin{bmatrix} 1 & 0 \\ 0 & -1 \end{bmatrix} \tag{14.5}$$

rotates the second distribution to form the X-pattern (Figure 14.1). Brost et al. [13] and Buderman et al. [14] treated the covariance parameters as unknown and estimated them simultaneously while fitting hierarchical animal movement models. For more recent Argos data, McClintock et al. [43] use a similar model specification as (14.4) but parameterized the covariance matrix using additional known information about the shape of error ellipses provided by Service Argos.

There are many other forms of potentially useful measurement error models for telemetry data and they may depend directly on other environmental or behavioral information associated with the individual animal position. Appropriately accommodating measurement error in telemetry data is a rapidly expanding area of research and was often ignored in early statistical modeling efforts due to theoretical and computational challenges. However, hierarchical models for animal movement have been essential for facilitating inference while incorporating inherent uncertainty associated with telemetry data. Now that we have a means for accommodating most forms of measurement error, we turn our attention to statistical models for the underlying position process associated with animal movement paths.

14.3   Point Process Models

Most animal movement models rely heavily on concepts from spatial statistics and time series. However, because many animal movement models originated in other disciplines, they often are referred to with different names. For example, one of the most commonly used statistical models for animal movement is the point process model, but animal ecologists are more familiar with the phrase "resource selection function" (RSF). We describe why point process models are referred to as RSFs in what follows.

Point process models for animal movement treat the individual positions $\boldsymbol{\mu}(t_i)$ as random quantities arising from a multivariate distribution in geographic space. The goal for inference is often to learn about the probability distribution that gave rise to a finite set of positions $\boldsymbol{\mu}_i$ for $i = 1, \ldots, n$, associated with the times for which data were collected. Animal ecologists often refer to the distribution of $\boldsymbol{\mu}(t_i)$ as the utilization distribution (UD). A simple point process model for the individual positions is one with a well-known parametric form, such as the bivariate Gaussian distribution, $\boldsymbol{\mu}_i \sim \mathrm{N}(\boldsymbol{\mu}^*, \boldsymbol{\Sigma}_\mu)$, where $\boldsymbol{\mu}^*$ represents a central place and $\boldsymbol{\Sigma}_\mu$ controls the shape and scale of the UD. The multivariate Gaussian distribution as a point process model can only represent simple UDs and does not provide an obvious way to incorporate environmental information, such as landscape features and habitat. Thus, we can construct a more realistic UD by defining a flexible surface that depends on spatial features.

Rather than step through the various types of named point processes (see [34]; [5]), we begin with the basic heterogeneous point process represented as a weighted distribution

$$[\boldsymbol{\mu}_i|\boldsymbol{\beta},\boldsymbol{\theta}] \equiv \frac{g(\mathbf{x}(\boldsymbol{\mu}_i),\boldsymbol{\beta})f(\boldsymbol{\mu}_i,\boldsymbol{\theta})}{\int g(\mathbf{x}(\boldsymbol{\mu}),\boldsymbol{\beta})f(\boldsymbol{\mu},\boldsymbol{\theta})d\boldsymbol{\mu}} , \tag{14.6}$$

where $f(\boldsymbol{\mu},\boldsymbol{\theta})$ is referred to as the "availability" distribution and $g(\mathbf{x}(\boldsymbol{\mu}),\boldsymbol{\beta})$ is referred to as the "resource selection function" [42]. Letting both f and g be non-negative functions over $\mathcal{M}$, their product will be non-negative and can be normalized such that $[\boldsymbol{\mu}_i|\boldsymbol{\beta},\boldsymbol{\theta}]$ integrates to one, resulting in a proper probability density function for each $\boldsymbol{\mu}_i$. The weighted distribution provides a convenient way to construct a custom probability distribution for point processes as long as we condition on the number of points n [49] [50].

The availability function $f(\boldsymbol{\mu},\boldsymbol{\theta})$ in (14.6) specifies what "resources" are available to the individual and the resource selection function $g(\mathbf{x}(\boldsymbol{\mu}_i),\boldsymbol{\beta})$ relates the environmental covariates $\mathbf{x}$ to the probability density of the individual occupying location $\boldsymbol{\mu}_i$. The product of the availability and resource selection functions define the general shape of the probability density function for $\boldsymbol{\mu}_i$ over the support $\mathcal{M}$. Larger values for $[\boldsymbol{\mu}_i|\boldsymbol{\beta},\boldsymbol{\theta}]$ increase the likelihood of the individual choosing position $\boldsymbol{\mu}_i$ over other positions that were available.

Historically, telemetry data were assumed to be temporally independent and a bivariate uniform distribution on $\mathcal{M}$ was specified as the availability distribution, where $f(\boldsymbol{\mu}_i,\boldsymbol{\theta}) \equiv \mathrm{Unif}(\mathcal{M})$ and $\boldsymbol{\theta}$ is empty. The most common form of RSF is $g(\mathbf{x}(\boldsymbol{\mu}_i),\boldsymbol{\beta}) = \exp(\mathbf{x}(\boldsymbol{\mu}_i)'\boldsymbol{\beta})$, resembling a log link in Poisson regression. The model can then be fit using either likelihood or Bayesian methods (assuming a suitable prior for $\boldsymbol{\beta}$, such as a multivariate Gaussian). In fact, standard generalized linear models (i.e., Poisson and logistic regression) can be used to fit the model in (14.6) [60] [1].

Our focus is on temporally dependent data and processes, thus, following Johnson et al. [38] [36], we generalize the point process model such that

$$\boldsymbol{\mu}_i \sim \frac{\exp(\mathbf{x}'(\boldsymbol{\mu}_i)\boldsymbol{\beta})f(\boldsymbol{\mu}_i|\boldsymbol{\mu}_{i-1},\boldsymbol{\theta})}{\int \exp(\mathbf{x}'(\boldsymbol{\mu})\boldsymbol{\beta})f(\boldsymbol{\mu}|\boldsymbol{\mu}_{i-1},\boldsymbol{\theta})d\boldsymbol{\mu}} , \tag{14.7}$$

where the availability function contains an explicit temporal dependence. The temporal dependence in $f(\boldsymbol{\mu}_i|\boldsymbol{\mu}_{i-1}, \boldsymbol{\theta})$ implies that the availability changes as the animal moves and as the time between telemetry fixes $\Delta_i = t_i - t_{i-1}$ increases [4] [53] [16] [22]. For example, Brost et al. [13] used the availability function

$$f(\boldsymbol{\mu}_i|\boldsymbol{\mu}_{i-1}, \boldsymbol{\theta}) \propto \exp\left(-\frac{||\boldsymbol{\mu}_i - \boldsymbol{\mu}_{i-1}||}{\Delta_i \phi}\right), \tag{14.8}$$

where $||\boldsymbol{\mu}_i - \boldsymbol{\mu}_{i-1}||$ represents the distance between $\boldsymbol{\mu}_i$ and $\boldsymbol{\mu}_{i-1}$, and ϕ controls the range of temporal structure in the movement process (e.g., large ϕ corresponds to smoother trajectories).

Embedding the spatio-temporal point process model (14.7) into the hierarchical framework (14.1)–(14.2) and using a mixture data model (14.4), Brost et al. [13] obtained resource selection inference for a harbor seal (*Phoca vitulina*) in the Gulf of Alaska (Figure 14.2). Brost et al. [13] based their inference on two spatial covariates thought to be important for

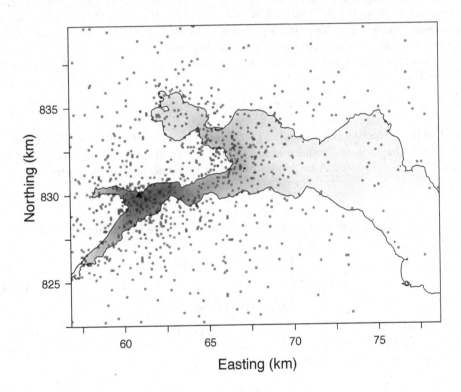

FIGURE 14.2
Argos telemetry data (i.e., $\mathbf{s}_i$, for $i = 1, \ldots, n$; shown as points) for an individual harbor seal and the estimated posterior mean UD (i.e., utilization distribution) (i.e., $E(\{\boldsymbol{\mu}_i, \forall i\}|\{\mathbf{s}_i, \forall i\}))$ based on true underlying positions $\boldsymbol{\mu}_i$ and known covariates $\mathbf{X}$. The shaded region represents an inlet in Alaska while the white region represents land. Recall that the true positions $\boldsymbol{\mu}_i$ are constrained to occur in the water regions only.

harbor seal space use, bathymetry (ocean depth) and distance from haul out (site used by the individual to rest on shore). The natural constraints of the marine environment allowed Brost et al. [13] to estimate measurement error covariance parameters without any further prior information about Argos telemetry error.

Similar modeling ideas to the spatio-temporal point process have arisen in the animal ecology literature. In particular, the idea of a "step-selection" function provides an animal movement modeling framework containing a temporally varying availability component [10] [23] [52]. While point process models have been used in statistics for decades [18], there have been recent efforts to reconcile them with step-selection function models.

14.4 Discrete-Time Models

This section focuses on animal movement models that are expressed as time series models, rather than spatial point processes. Thus, for convenience, we use a standard time series notation and index our multivariate temporal process $\boldsymbol{\mu}_t$ by t (for $t = 1, \ldots, T$).

A simple discrete-time model for the position process is a vector autoregressive (VAR) model $\boldsymbol{\mu}_t \sim N(\mathbf{M}\boldsymbol{\mu}_{t-1}, \boldsymbol{\Sigma})$ where the 2×2 propagator matrix $\mathbf{M}$ can be parameterized in a variety of ways. For example, setting $\mathbf{M} \equiv \mathbf{I}$ results in an "intrinsic" conditional autoregressive model. The VAR model form is a natural spatio-temporal model [19], but with a very low dimensional space (i.e., two-dimensional).

Several useful generalizations of the VAR model are easily obtained. For example, an attracting VAR process can be written as

$$\boldsymbol{\mu}_t \sim N(\mathbf{M}\boldsymbol{\mu}_{t-1} + (\mathbf{I} - \mathbf{M})\boldsymbol{\mu}^*, \boldsymbol{\Sigma}), \tag{14.9}$$

where $\boldsymbol{\mu}^*$ is the attracting point and, if $\mathbf{M} \equiv \rho\mathbf{I}$ with $0 < \rho < 1$, the process will be stationary (i.e., the individual will not wander aimlessly, instead staying within the proximity of $\boldsymbol{\mu}^*$). The VAR model in (14.9) can be generalized to include two or more attractors as

$$\boldsymbol{\mu}_t = \mathbf{M}\boldsymbol{\mu}_{t-1} + (\mathbf{I} - \mathbf{M})\boldsymbol{\mu}_t^* + \boldsymbol{\varepsilon}_t, \tag{14.10}$$

where the attracting point $\boldsymbol{\mu}_t^*$ varies over time as a mixture of a finite set of J attracting points

$$\boldsymbol{\mu}_t^* = \begin{cases} \boldsymbol{\mu}_1^* & \text{with probability } p_1 \\ \quad \vdots \\ \boldsymbol{\mu}_J^* & \text{with probability } p_J \end{cases}. \tag{14.11}$$

Alternatively, the VAR model can be reparameterized to model turning angles and step lengths explicitly. Jonsen et al. [39] modeled the velocity process as a VAR using the relationship $\mathbf{v}_t = \boldsymbol{\mu}_t - \boldsymbol{\mu}_{t-1}$, where $\mathbf{v}_t$ is the velocity vector and

$$\mathbf{v}_t \sim N(\mathbf{M}\mathbf{v}_{t-1}, \boldsymbol{\Sigma}). \tag{14.12}$$

In (14.12), the propagator matrix is parameterized as

$$\mathbf{M} \equiv \gamma \begin{pmatrix} \cos(\theta) & -\sin(\theta) \\ \sin(\theta) & \cos(\theta) \end{pmatrix}, \tag{14.13}$$

which has a geometric interpretation because θ represents the turning angle and γ ($0 < \gamma < 1$) dampens the contribution of the dynamics in the velocity. In all of these VAR models, the covariance matrix can be parameterized as $\boldsymbol{\Sigma} \equiv \sigma_\mu^2\mathbf{I}$ or allowed to be fully unknown. In case of the latter, Jonsen et al. [39] used an inverse Wishart prior for $\boldsymbol{\Sigma}$ in a Bayesian setting to fit the model.

For the simulated data in Figure 14.3, we fit the model in (14.13) assuming that $\boldsymbol{\Sigma} \equiv \sigma_\mu^2\mathbf{I}$

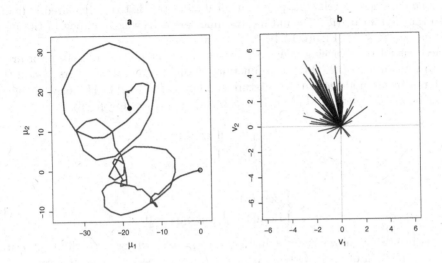

FIGURE 14.3
Simulated position process for an individual (i.e., $\boldsymbol{\mu}_t = \sum_{\tau \leq t} \mathbf{v}_\tau$) using (14.12) for $T = 100$ equally spaced time steps and $\theta = \pi/8$, $\gamma = 0.9$, and $\sigma^2 = 1$. (a) shows the trajectory (i.e., $\boldsymbol{\mu}_t$ or position process) with open and closed circles denoting the starting and ending positions, respectively. (b) shows the velocity vectors ($\mathbf{v}_t$).

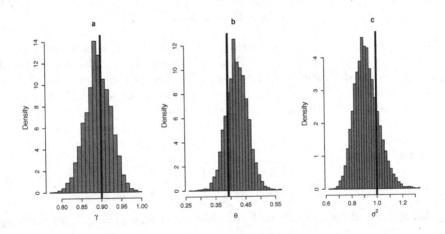

FIGURE 14.4
Marginal posterior distributions for model parameters (a) γ, (b) θ, and (c) σ^2) based on the simulated position processes in Figure 14.3. True parameter values used to simulated data are shown as vertical lines.

and a uniform prior on $(-1, 1)$ for θ, an inverse gamma prior for σ_μ^2, and uniform prior on $(0, 1)$ for γ. Resulting inference (Figure 14.4) showed that the velocity model in (14.12) recovers the parameters well. The posterior for θ indicates that the simulated individual is prone to take left turns ($\theta > 0$) and its dynamics are somewhat damped in the resulting velocity process ($\gamma < 1$) (Figure 14.4).

To account for time varying behavior, Jonsen et al. [39] allow the propagator matrix in (14.12) to vary in time (i.e., $\mathbf{M}_t$) as a function of time varying parameters γ_t and θ_t. The parameters can be modeled with a mixture similar to that in (14.11), or a change-point model, or based on a hidden Markov model. In the latter, we can specify

$$\theta_t = \begin{cases} \theta_1 & \text{if } z_t = 1 \\ \theta_2 & \text{if } z_t = 0 \end{cases}, \tag{14.14}$$

where

$$z_t \sim \begin{cases} \text{Bern}(p) & \text{if } z_{t-1} = 1 \\ \text{Bern}(1-p) & \text{if } z_{t-1} = 0 \end{cases}. \tag{14.15}$$

The model in (14.14) allows the turning angle to switch between two different states, θ_1 and θ_2, as a function of an underlying binary Markov process z_t. The smoothness in z_t is governed by the parameter p. If p is large or small (i.e., close to 0 or 1), the turning angles will be fairly consistent over time, whereas if p is closer to 0.5, the turning angles will switch back and forth between θ_1 and θ_2 more rapidly. Inference for z_t can suggest when changes in behavior occur, providing insight about energy budgets and general movement patterns for animals.

Finally, many researchers prefer to model animal movement in polar coordinates [58], specifying statistical models for turning angles (or bearings) and movement rates (or step lengths) directly [47]. For example, Morales et al. [47] modeled movement rates (r_t) and turning angles (θ_t) with

$$r_t \sim \text{Weib}(a_t, b_t), \tag{14.16}$$
$$\theta_t \sim \text{WrapCauchy}(m_t, \rho_t), \tag{14.17}$$

where 'Weib' stands for Weibull distribution and 'WrapCauchy' stands for the wrapped Cauchy distribution. The Weibull distribution is a generalization of the exponential distribution and the wrapped Cauchy is a circular distribution with support $-\pi \leq \theta_t \leq \pi$ that converges to a uniform on a circle as $\rho_t \to 0$. To provide structure on the process model parameters, Morales et al. [47] suggested several potential model specifications similar to (14.14) and (14.15), but for a_t, b_t, m_t, and ρ_t. Many of their proposed formulations use hidden Markov models [61], and they referred to one such specification as the "double-switch with covariates" where

$$z_t \sim \begin{cases} \text{Bern}(p_{1,t}) & \text{if } z_{t-1} = 1 \\ \text{Bern}(p_{2,t}) & \text{if } z_{t-1} = 0 \end{cases}, \tag{14.18}$$

and $p_{1,t}$ and $p_{2,t}$ are allowed to vary with covariates as $\text{logit}(p_{1,t}) = \mathbf{x}_t' \boldsymbol{\beta}_1$ and $\text{logit}(p_{2,t}) = \mathbf{x}_t' \boldsymbol{\beta}_2$.

McClintock et al. [45] reparameterized and generalized the models of [47] to incorporate more complicated hidden Markov processes that govern switches between J states as

$$\mathbf{z}_t \sim \text{MN}(1, \mathbf{P}\mathbf{z}_{t-1}), \tag{14.19}$$

where the parameter of interest, say a_t, is modeled as

$$a_t = \begin{cases} a_1 & \text{if } z_{1,t} \\ \vdots \\ a_J & \text{if } z_{J,t} \end{cases} , \tag{14.20}$$

and $\mathbf{P}$ is a matrix of transition probabilities.

Parsimony in the model can be increased by letting all parameters depend on the same $\mathbf{z}_t$. Similar types of latent state-transition models are becoming common in population modeling studies (e.g., [27]). McClintock et al. [45] also used a central attracting place in their modeling framework which resulted in a similar effect as that we described in (14.9).

While Morales et al. [47] and McClintock et al. [45] focused on ways to cluster movement behavior into categories for improved inference about activity budgets, Tracey et al. [57] developed similar models aiming to infer an individual's response to spatial features. For example, consider the model for the true position

$$\boldsymbol{\mu}_t = \boldsymbol{\mu}_{t-1} + d_t \mathbf{a}_t , \tag{14.21}$$

where d_t is the distance and $\mathbf{a}_t$ is the angle between successive positions (i.e., $\mathbf{a}_t \equiv (\cos(\theta_t), \sin(\theta_t))'$). Tracey et al. [57] used a similar model specification as [47] and [45] such that

$$d_t \sim \text{Gamma}(a_t, b_t) , \tag{14.22}$$
$$\theta_t - m_t^* \sim \text{vonMises}(m, \rho_t) , \tag{14.23}$$

where the von Mises provides an alternative circular distribution to the wrapped Cauchy and the gamma is an alternative to the Weibull. In (14.23), m_t^* is the response angle to a given feature of interest. The concentration parameter ρ_t can then be linked to the distance (d_t^*) from the individual to the feature by $\log(\rho_t) = \beta_0 + \beta_1 d_t^*$. Similarly, Tracey et al. [57] linked the parameters in the model for displacement distance to the distance in the same fashion.

Overall, discrete-time approaches to modeling animal movement have been, and will continue to be, popular due to their intuitive nature, flexibility, and similarity to existing time series methods. Discrete-time models are readily generalized and can be coupled with models for auxiliary data such as accelerometer measurements from the telemetry device or additional information about the individual's position in a third dimension (i.e., altitude or depth; [46] [44] [43] [54] [54] [35]).

14.5 Continuous-Time Models

An intuitive way to transition into continuous-time models for animal movement trajectories is to use a limiting form of a discrete time model. Thus, consider a simple discrete time VAR model

$$\mathbf{b}(t_i) = \mathbf{b}(t_{i-1}) + \boldsymbol{\varepsilon}(t_i) , \tag{14.24}$$

where the displacement vector is $\boldsymbol{\varepsilon}(t_i) \sim N(\mathbf{0}, \sigma^2 \Delta t \mathbf{I})$, $\mathbf{b}(t_i)$ represents a dynamic process influencing position, and $\mathbf{b}(t_0) = \mathbf{0}$. Letting $\Delta t = t_i - t_{i-1}$ for all i and writing the dynamic process as a sum of successive steps, we have

$$\mathbf{b}(t_i) = \sum_{j=1}^{i} \mathbf{b}(t_j) - \mathbf{b}(t_{j-1}) \tag{14.25}$$

$$= \sum_{j=1}^{i} \boldsymbol{\varepsilon}(t_j) . \tag{14.26}$$

That is, the current position of an individual $\mathbf{b}(t_i)$ is the accumulation of all former steps. As the time interval between positions (Δt) approaches zero in (14.26), we arrive at

$$\mathbf{b}(t_i) = \lim_{\Delta t \to 0} \sum_{j=1}^{i} \boldsymbol{\varepsilon}(t_j) , \tag{14.27}$$

which, using Ito notation, can be written as

$$\mathbf{b}(t) = \int_0^t \frac{d\mathbf{b}(\tau)}{d\tau} d\tau , \tag{14.28}$$

the scaled Weiner process (or Brownian motion, hence the 'b' notation). In (14.28), $d\mathbf{b}(t) = \boldsymbol{\varepsilon}(t)$, thus, $d\mathbf{b}(t) = \mathbf{b}(t) - \mathbf{b}(t - \Delta t)$ as $\Delta t \to 0$. To convert this back to the actual position process $\boldsymbol{\mu}(t)$, consider the discrete-time formulation for $\boldsymbol{\mu}(t_i)$ where

$$\boldsymbol{\mu}(t_i) = \boldsymbol{\mu}(t_{i-1}) + \boldsymbol{\varepsilon}(t_i) , \tag{14.29}$$

then assume an initial position $\boldsymbol{\mu}(0)$, consider the limit as $\Delta t \to 0$, and substitute in the Weiner process to yield

$$\boldsymbol{\mu}(t) = \boldsymbol{\mu}(0) + \mathbf{b}(t) . \tag{14.30}$$

A natural extension of the basic Weiner process is to allow for attraction and drift in the process. Thus, a similar continuous-time stochastic process arising as a limiting process of (14.9) can be obtained using the same techniques so that

$$\boldsymbol{\mu}(t) = \boldsymbol{\mu}(0) + \int_0^t (\mathbf{M} - \mathbf{I})(\boldsymbol{\mu}(\tau) - \boldsymbol{\mu}^*)d\tau + \mathbf{b}(t) , \tag{14.31}$$

where the attracting point is $\boldsymbol{\mu}^*$ and the integral in (14.31) induces a form of stationarity in the process under certain conditions for $\mathbf{M}$. Differentiating both sides of (14.31) leads to the stochastic differential equation (SDE)

$$\frac{d\boldsymbol{\mu}(t)}{dt} = (\mathbf{M} - \mathbf{I})(\boldsymbol{\mu}(t) - \boldsymbol{\mu}^*) + \frac{\boldsymbol{\varepsilon}(t)}{dt} , \tag{14.32}$$

which is known as the Ornstein-Uhlenbeck process (OU; e.g., [21] [9] [37]). Some of the earliest animal movement models in the statistical literature assumed a similar form as (14.30) (e.g., [21] [9]). In fact, Dunn and Gipson [21] expressed the OU process using exponentials in a conditional formulation as

$$\mu(t)|\mu(\tau) \sim \mathrm{N}\left(\mu(\tau)e^{-\theta(t-\tau)}, \frac{\sigma^2}{2\theta}(1 - e^{-2\theta(t-\tau)})\right) , \tag{14.33}$$

for $t < \tau$ and where we have reduced the process to be one-dimensional for simplicity. In (14.33), $\mu(\tau)$ is the position at time τ and θ is an autocorrelation parameter. As $|t - \tau|$ decreases, $\mu(t)$ will be similar to $\mu(\tau)$.

SDEs are valuable for modeling animal trajectories because animal movement is naturally a continuous-time process, but they may not be as intuitive as discrete-time models. However, when implementing continuous-time models for animal movement, they must be discretized for computational purposes regardless. A primary interest in point process models and discrete-time models involves the incorporation of covariate information about the individual's environment. To make use of covariate information in the continuous-time setting Brillinger [11] recommends the use of potential functions.

Potential functions can be incorporated into SDE models by replacing the attraction component in the OU (14.32) with a gradient function $\mathbf{g}(\boldsymbol{\mu}(t))$ such that

$$\frac{d\boldsymbol{\mu}(t)}{dt} = \mathbf{g}(\boldsymbol{\mu}(t)) + \frac{\varepsilon(t)}{dt} . \qquad (14.34)$$

where $\mathbf{g}(\boldsymbol{\mu}(t))$ is the gradient of a potential function in geographical space (i.e., $\mathbf{g}(\boldsymbol{\mu}(t)) = -\nabla p(\boldsymbol{\mu}(t))$; [12]). The potential function serves as a map, upon which the individual animal is more likely to move "downhill" like a marble on a wavy surface. For example, Figure 14.5 shows a simulated individual trajectory based on a circular-shaped potential surface with two divots. This potential function can be expressed as

$$p(\boldsymbol{\mu}(t)) = \begin{cases} -\theta_1(\frac{1}{2}[\boldsymbol{\mu}(t)|\boldsymbol{\mu}_1^*, \sigma_1^2] + \frac{1}{2}[\boldsymbol{\mu}(t)|\boldsymbol{\mu}_2^*, \sigma_2^2]) & \text{if } \boldsymbol{\mu}(t) \in \mathcal{R}^c \\ \theta_2\sqrt{(\boldsymbol{\mu}(t) - \boldsymbol{\mu}_3^*)'(\boldsymbol{\mu}(t) - \boldsymbol{\mu}_3^*)} & \text{if } \boldsymbol{\mu}(t) \in \mathcal{R} \end{cases} , \qquad (14.35)$$

where, $[\boldsymbol{\mu}(t)|\boldsymbol{\mu}_1^*, \sigma_1^2]$ and $[\boldsymbol{\mu}(t)|\boldsymbol{\mu}_2^*, \sigma_2^2]$ are multivariate Gaussian density functions, $\boldsymbol{\mu}_3^*$ is the overall space use center, and the multipliers θ_1 and θ_2 control the strength of boundary and attraction. Figure 14.5 shows a simulated trajectory based on the potential function in (14.35). The simulated individual trajectory is attracted to $\boldsymbol{\mu}_1^*$ and $\boldsymbol{\mu}_2^*$ and, if it wanders outside of $\mathcal{R}^c$, it slides back in due to the steepness of potential at the boundary.

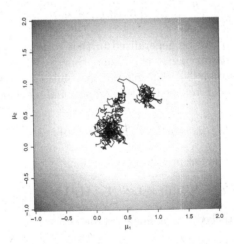

FIGURE 14.5
Simulated individual trajectory based on the potential function in (14.35) which is comprised of two attracting points and a steeply rising boundary condition delineating a circular region of space use. The potential function appears as the background image with joint trajectory simulation (black line).

To implement statistical SDE models with potential function components, Brillinger [11] recommends a statistical model similar to

$$\boldsymbol{\mu}(t_i) - \boldsymbol{\mu}(t_{i-1}) \sim \mathrm{N}((t_i - t_{i-1})\mathbf{g}(\boldsymbol{\mu}(t_{i-1})), (t_i - t_{i-1})\sigma^2 \mathbf{I}) . \tag{14.36}$$

For example, we fit a Bayesian SDE model based on (14.36) with the potential function

$$p(\boldsymbol{\mu}(t), \boldsymbol{\beta}) = \beta_1 \mu_1(t) + \beta_2 \mu_2(t) + \beta_3 \mu_1^2(t) + \beta_4 \mu_2^2(t) + \beta_5 \mu_1(t)\mu_2(t) , \tag{14.37}$$

to the GPS telemetry data analyzed by Hooten et al. [29], arising from the trajectory of an individual mountain lion (*Puma concolor*) in Colorado, USA (Figure 14.6). We used a Gaussian prior for the coefficients in the potential function ($\boldsymbol{\beta} \sim \mathrm{N}(\mathbf{0}, 100 \cdot \mathbf{I})$) and an inverse gamma prior for the variance component ($\sigma^2 \sim \mathrm{IG}(0.001, 0.001)$). Figure 14.6a shows the posterior mean potential function and Figure 14.6b shows the posterior standard deviation of the potential function. The latter indicates our level of uncertainty in the shape of the potential function. Estimated potential functions, like that shown in Figure 14.6a can be useful for describing the shape of the UD as well as allowing us to learn about the parameters that may affect the UD shape (especially in the case where spatial covariates are used to characterize the shape).

Like discrete-time movement models, continuous-time formulations can also be used to model velocity rather than position direction. Johnson et al. [37] developed a hierarchical model with an OU process on velocity (referred to as a continuous-time correlated random walk model, CTCRW). Hooten and Johnson [30] showed that the CTCRW model can be expressed as a convolution

$$\boldsymbol{\eta}(t) = \int_0^T h(t, \tau)\mathbf{b}(\tau)d\tau , \tag{14.38}$$

where $\boldsymbol{\mu}(t) = \boldsymbol{\mu}(0) + \boldsymbol{\eta}(t)$ and the basis functions $h(t, \tau)$ are

$$h(t, \tau) = \begin{cases} 1 & \text{if } 0 < \tau \le t \\ 0 & \text{if } t < \tau \le T \end{cases} . \tag{14.39}$$

The convolution in (14.38) can be generalized further to incorporate many types of basis functions $h(t, \tau)$, leading to varying levels of smoothness and ecological interpretations of animal memory and perception [30]. The general convolution specification for a continuous-time animal movement model, referred to as a "functional movement model" [14] [30], can be written as

$$\boldsymbol{\mu}(t) = \boldsymbol{\mu}(0) + \int_0^T \mathbf{H}(t, \tau)\mathbf{b}(\tau)d\tau , \tag{14.40}$$

which leads to the joint Gaussian process specification

$$\boldsymbol{\mu} \sim \mathrm{N}(\boldsymbol{\mu}(0) \otimes \mathbf{1}, \sigma^2(\mathbf{I} \otimes \tilde{\mathbf{H}})(\mathbf{I} \otimes \tilde{\mathbf{H}})') , \tag{14.41}$$

where, the positions are stacked into a single vector as

$$\boldsymbol{\mu} \equiv (\mu_1(t_1), \dots, \mu_1(t_n), \mu_2(t_1), \dots, \mu_2(t_n))'$$

and the elements of $\tilde{\mathbf{H}}$ are

$$\tilde{h}(t, \tau) = \int_\tau^T h(t, u)du , \tag{14.42}$$

for each dimension (i.e., latitude and longitude). The advantage of (14.41) is that, by shifting the dependence in the movement process from the first-order (i.e., mean) to the second-order (i.e., covariance), (14.41) leads to efficient computational algorithms. Gaussian process

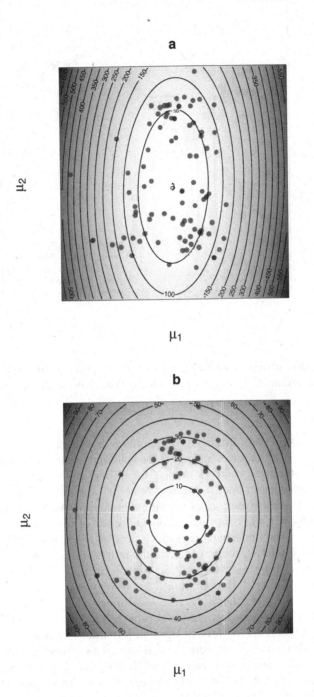

FIGURE 14.6
Posterior (a) mean and (b) standard deviation of the potential function based on fitting the Bayesian SDE model (using the potential function specification in (14.37)) to the mountain lion telemetry data (dark points). Isopleth contours are shown as dark lines.

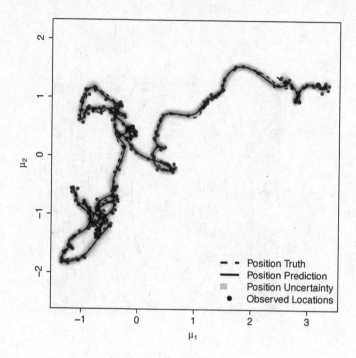

FIGURE 14.7

A simulated stochastic process (dashed line) and data (points) from the FMM in (14.41) with Gaussian measurement error with posterior realizations of the position process (gray lines).

specifications using kernel convolutions have become very popular in spatial statistics (e.g., [7] [26] [41] [48] [15]) and the same benefits can be achieved in models for animal movement.

Continuous-time stochastic process models for animal movement based on integrated Weiner processes, as first described by Johnson et al. [37], provide computationally efficient and realistic statistical reconstructions of individual animal paths. For example, using Gaussian basis functions for $h(t, \tau)$, a Bayesian implementation of the model in (14.41) produces excellent predictions of the underlying trajectory process (Figure 14.7).

As a result of the predictive ability of functional movement models, numerous secondary modeling procedures have been developed that rely on posterior predictive distributions of individual paths. For example, Hooten et al. [31] developed a discrete-space model that linked transition probabilities among discrete areal units to landscape covariates in those units. Hanks et al. [25] developed a similar approach but linked velocities in continuous-space to gradients in landscape covariates through potential functions.

One primary challenge with these complicated animal movement models is that they can be slow to implement for realistic data sets. Thus, Hanks et al. [24] derived a reduced dimensional version of the secondary model in [31] using sufficient statistics of the discrete-space representation of the data. They showed that the multinomial model with Gaussian latent variables of [31] can be fit using a generalized linear model with a Poisson likelihood and offset based on residence time. As with the secondary models developed by [31] and [25], the discrete-space Poisson specification of [24] is paired with a multiple imputation procedure to account for the path uncertainty resulting from a first-stage functional movement model fit.

14.6   Conclusion

Since the early animal movement models appeared in prominent statistical journals (e.g., [21] [3] [2]), there has been an explosion of new approaches for modeling telemetry data. We highlighted the dominant types of models (i.e, point processes, discrete-time, and continuous-time models) in this chapter, but we have only scratched the surface of statistical models for animal movement. Hooten et al. [32] provide further detail on these main model types as well as several specific examples of each.

Movement is an essential element of individual, population, and community health for animals and their interactions among each other and humans [40]. Despite the critically important role that animal movement plays in ecosystems and the increasing rate of methodological developments to study animal movement, technological advances are still outpacing statistical modeling efforts. It is clear that ongoing efforts to track and monitor animals are fruitful, and that hierarchical modeling is a fundamental statistical tool to obtain accurate and honest inference [56] [28]. However, further advances in statistical methodology are needed to provide rigorous inference for massive telemetry data sets involving many interacting individuals (e.g., [55]), at sub-minute temporal resolutions, and spanning months or years, as well as real-time monitoring protocols (e.g., [59]). The past decade of environmental science has seen similar developments in spatial and spatio-temporal statistics (e.g., [19]) and the statistical emulation of computer models (e.g., [33]) and, as models for animal trajectories increase in their importance, we will see the same types of computationally efficient statistical models arise in movement ecology.

Acknowledgments

The authors thank Brett McClintock and Juan Morales for numerous discussions and comments on earlier versions of this chapter. Hooten acknowledges support from NSF DMS 1614392. Any use of trade, firm, or product names is for descriptive purposes only and does not imply endorsement by the U.S. Government.

Bibliography

[1] G. Aarts, J. Fieberg, and J. Matthiopoulos. Comparative interpretation of count, presence-absence, and point methods for species distribution models. *Methods in Ecology and Evolution*, 3:177–187, 2012.

[2] R. Anderson-Sprecher. Robust estimates of wildlife location using telemetry data. *Biometrics*, pages 406–416, 1994.

[3] R. Anderson-Sprecher and J. Ledolter. State-space analysis of wildlife telemetry data. *Journal of the American Statistical Association*, 86:596–602, 1991.

[4] S.M. Arthur, B.F.J. Manly, L.L. McDonald, and G.W. Garner. Assessing habitat selection when availability changes. *Ecology*, 77:215–227, 1996.

[5] A. Baddeley, E. Rubak, and R. Turner. *Spatial Point Patterns: Methodology and Applications with R.* CRC Press, 2016.

[6] F. Barraquand and S. Benhamou. Animal movements in heterogeneous landscapes: identifying profitable places and homogeneous movement bouts. *Ecology*, 89:3336–3348, 2008.

[7] R. Barry and J.M. Ver Hoef. Blackbox kriging: spatial prediction without specifying variogram models. *Journal of Agricultural, Biological and Environmental Statistics*, 1:297–322, 1996.

[8] L.M. Berliner. Hierarchical Bayesian time series models. In K. Hanson and R. Silver, editors, *Maximum Entropy and Bayesian Methods*, pages 15–22. Kluwer Academic Publishers, 1996.

[9] P.G. Blackwell. Random diffusion models for animal movement. *Ecological Modelling*, 100:87–102, 1997.

[10] M.S. Boyce, J.S. Mao, E.H. Merrill, D. Fortin, M.G. Turner, J.M. Fryxell, and P. Turchin. Scale and heterogeneity in habitat selection by elk in Yellowstone National Park. *Ecoscience*, 10:321–332, 2003.

[11] D.R. Brillinger. Modeling spatial trajectories. In A.E. Gelfand, P.J. Diggle, M. Fuentes, and P. Guttorp, editors, *Handbook of Spatial Statistics*. Chapman & Hall/CRC, Boca Raton, Florida, USA, 2010.

[12] D.R. Brillinger, H.K. Preisler, A.A. Ager, and J.G. Kie. The use of potential functions in modeling animal movement. In E. Saleh, editor, *Data Analysis from Statistical Foundations*. Nova Science Publishers, Huntington, New York, USA, 2001.

[13] B.M. Brost, M.B. Hooten, E.M. Hanks, and R.J. Small. Animal movement constraints improve resource selection inference in the presence of telemetry error. *Ecology*, 96:2590–2597, 2015.

[14] F.E. Buderman, M.B. Hooten, J. Ivan, and T. Shenk. A functional model for characterizing long distance movement behavior. *Methods in Ecology and Evolution*, 7:264–273, 2016.

[15] C.A. Calder. Dynamic factor process convolution models for multivariate spacetime data with application to air quality assessment. *Environmental and Ecological Statistics*, 14:229–247, 2007.

[16] A. Christ, J.M. Ver Hoef, and D.L. Zimmerman. An animal movement model incorporating home range and habitat selection. *Environmental and Ecological Statistics*, 15:27–38, 2008.

[17] D.P. Costa, P.W. Robinson, J.P.Y. Arnould, A.-L. Harrison, S. E. Simmons, J. L. Hassrick, A. J. Hoskins, S. P. Kirkman, H. Oosthuizen, S. Villegas-Amtmann, and D. E. Crocker. Accuracy of Argos locations of pinnipeds at-sea estimated using Fastloc GPS. *PLoS One*, 5:e8677, 2010.

[18] N.A.C. Cressie. *Statistics for Spatial Data: Revised Edition.* John Wiley and Sons, New York, New York, USA, 1993.

[19] N.A.C. Cressie and C.K. Wikle. *Statistics for Spatio-Temporal Data.* John Wiley and Sons, New York, New York, USA, 2011.

[20] D.C. Douglas, R. Weinzierl, S.C. Davidson, R. Kays, M. Wikelski, and G. Bohrer. Moderating Argos location errors in animal tracking data. *Methods in Ecology and Evolution*, 3:999–1007, 2012.

[21] J.E. Dunn and P.S. Gipson. Analysis of radio-telemetry data in studies of home range. *Biometrics*, 33:85–101, 1977.

[22] J.D. Forester, H.K. Im, and P.J. Rathouz. Accounting for animal movement in estimation of resource selection functions: sampling and data analysis. *Ecology*, 90:3554–3565, 2009.

[23] D. Fortin, H.L. Beyer, M.S. Boyce, D.W. Smith, T. Duchesne, and J.S. Mao. Wolves influence elk movements: Behavior shapes a trophic cascade in Yellowstone National Park. *Ecology*, 86:1320–1330, 2005.

[24] E.M. Hanks, M.B. Hooten, and M. Alldredge. Continuous-time discrete-space models for animal movement. *Annals of Applied Statistics*, 9:145–165, 2015a.

[25] E.M. Hanks, M.B. Hooten, D.S. Johnson, and J. Sterling. Velocity-based movement modeling for individual and population level inference. *PLoS One*, 6:e22795, 2011.

[26] D. Higdon. A process-convolution approach to modeling temperatures in the North Atlantic Ocean. *Environmental and Ecological Statistics*, 5:173–190, 1998.

[27] N.T. Hobbs, C. Geremia, J. Treanor, R. Wallen, P.J. White, M.B. Hooten, and J.C. Rhyan. State-space modeling to support adaptive management of brucellosis in the Yellowstone bison population. *Ecological Monographs*, 85:525–556, 2015.

[28] M.B. Hooten, F.E. Buderman, B.M. Brost, E.M. Hanks, and J.S. Ivan. Hierarchical animal movement models for population-level inference. *Environmetrics*, 27:322–333, 2016.

[29] M.B. Hooten, E.M. Hanks, D.S. Johnson, and M.W. Aldredge. Reconciling resource utilization and resource selection functions. *Journal of Animal Ecology*, 82:1146–1154, 2013b.

[30] M.B. Hooten and D.S. Johnson. Basis function models for animal movement. *Journal of the American Statistical Association*, In Press.

[31] M.B. Hooten, D.S. Johnson, E.M. Hanks, and J.H. Lowry. Agent-based inference for animal movement and selection. *Journal of Agricultural, Biological and Environmental Statistics*, 15:523–538, 2010a.

[32] M.B. Hooten, D.S. Johnson, B.T. McClintock, and J.M. Morales. *Animal Movement: Statistical Models for Telemetry Data*. Chapman and Hall/CRC, 2017.

[33] M.B. Hooten, W.B. Leeds, J. Fiechter, and C.K. Wikle. Assessing first-order emulator inference for physical parameters in nonlinear mechanistic models. *Journal of Agricultural, Biological and Environmental Statistics*, 16:475–494, 2011.

[34] J. Illian, A. Penttinen, H. Stoyan, and D. Stoyan. *Statistical Ananlysis of Spatial Point Patterns*. Wiley-Interscience, 2008.

[35] S. Isojunno and P.J.O. Miller. Sperm whale response to tag boat presence: biologically informed hidden state models quantify lost feeding opportunities. *Ecosphere*, 6(1):1–46, 2015.

[36] D.S. Johnson, M.B. Hooten, and C.E. Kuhn. Estimating animal resource selection from telemetry data using point process models. *Journal of Animal Ecology*, 82:1155–1164, 2013.

[37] D.S. Johnson, J.M. London, M.A. Lea, and J.W. Durban. Continuous-time correlated random walk model for animal telemetry data. *Ecology*, 89:1208–1215, 2008b.

[38] D.S. Johnson, D.L. Thomas, J.M. Ver Hoef, and A. Christ. A general framework for the analysis of animal resource selection from telemetry data. *Biometrics*, 64:968–976, 2008a.

[39] I.D. Jonsen, J. Flemming, and R. Myers. Robust state-space modeling of animal movement data. *Ecology*, 45:589–598, 2005.

[40] R. Kays, M.C. Crofoot, W. Jetz, and M. Wikelski. Terrestrial animal tracking as an eye on life and planet. *Science*, 348(6240):aaa2478, 2015.

[41] H.K.H. Lee, D.M. Higdon, C.A. Calder, and C.H. Holloman. Efficient models for correlated data via convolutions of intrinsic processes. *Statistical Modelling*, 5:53–74, 2005.

[42] BFL Manly, Lyman McDonald, Dana Thomas, Trent L McDonald, and Wallace P Erickson. *Resource selection by animals: statistical design and analysis for field studies.* Springer Science & Business Media, 2007.

[43] B.M. McClintock, J.M. London, M.F. Cameron, and P.L. Boveng. Modelling animal movement using the argos satellite telemetry location error ellipse. *Methods in Ecology and Evolution*, 6:266–277, 2015.

[44] B.T. McClintock, D.S. Johnson, M.B. Hooten, J.M. Ver Hoef, and J.M. Morales. When to be discrete: the importance of time formulation in understanding animal movement. *Movement Ecology*, 2:21, 2014.

[45] B.T. McClintock, R. King, L. Thomas, J. Matthiopoulos, B.J. McConnell, and J.M. Morales. A general discrete-time modeling framework for animal movement using multistate random walks. *Ecological Monographs*, 82:335–349, 2012.

[46] B.T. McClintock, D.J.F. Russell, J. Matthiopoulos, and R. King. Combining individual animal movement and ancillary biotelemetry data to investigate population-level activity budgets. *Ecology*, 94(4):838–849, 2013.

[47] J.M. Morales, D.T. Haydon, J. Friar, K.E. Holsinger, and J.M. Fryxell. Extracting more out of relocation data: building movement models as mixtures of random walks. *Ecology*, 85:2436–2445, 2004.

[48] C.J. Paciorek and M.J. Schervish. Spatial modelling using a new class of nonstationary covariance functions. *Environmetrics*, 17:483–506, 2006.

[49] G.P. Patil and C.R. Rao. On size-biased sampling and related form-invariant weighted distributions. *Indian Journal of Statistics*, 38:48–61, 1976.

[50] G.P. Patil and C.R. Rao. The weighted distributions: a survey of their applications. In P.R. Krishnaiah, editor, *Applications of Statistics*. North Holland Publishing Compan, 1977.

[51] T.A. Patterson, L. Thomas, C. Wilcox, O. Ovaskainen, and J. Matthiopoulos. State-space models of individual animal movement. *Trends in Ecology and Evolution*, 23:87–94, 2008.

[52] J.R. Potts, G. Bastille-Rousseau, D.L. Murray, J.A. Schaefer, and M.A. Lewis. Predicting local and non-local effects of resources on animal space use using a mechanistic step selection model. *Methods in Ecology and Evolution*, 5:253–262, 2014.

[53] J.R. Rhodes, C.A. McAlpine, D. Lunney, and H.P. Possingham. A spatially explicit habitat selection model incorporating home range behavior. *Ecology*, 86:1199–1205, 2005.

[54] D.J. Russell, S.M. Brasseur, D. Thompson, G.D. Hastie, V.M. Janik, G. Aarts, B.T. McClintock, J. Matthiopoulos, S.E. Moss, and B. McConnell. Marine mammals trace anthropogenic structures at sea. *Current Biology*, 24:R638–R639, 2014.

[55] H. Scharf, M.B. Hooten, B.K. Fosdick, D.S. Johnson, J.M. London, and J.W. Durban. Dynamic social networks based on movement. *Annals of Applied Statistics*, In Press.

[56] R.S. Schick, S.R. Loarie, F. Colchero, B.D. Best, A. Boustany, D.A. Conde, P.N. Halpin, L.N. Joppa, C.M. McClellan, and J.S. Clark. Understanding movement data and movement processes: current and emerging directions. *Ecology letters*, 11:1338–1350, 2008.

[57] J.A. Tracey, J. Zhu, and K. Crooks. A set of nonlinear regression models for animal movement in response to a single landscape feature. *Journal of Agricultural, Biological and Environmental Statistics*, 10:1–18, 2005.

[58] P. Turchin. *Quantitative Analysis of Animal Movement*. Sinauer Associates, Inc. Publishers, Sunderland, Massachusettes, USA, 1998.

[59] J. Wall, G. Wittemyer, B. Klinkenberg, and I. Douglas-Hamilton. Novel opportunities for wildlife conservation and research with real-time monitoring. *Ecological Applications*, 24:593–601, 2014.

[60] D.I. Warton and L.C. Shepherd. Poisson point process models solve the "pseudo-absence problem" for presence-only data in ecology. *Annals of Applied Statistics*, 4:1383–1402, 2010.

[61] W. Zucchini, I.L. MacDonald, and R. Langrock. *Hidden Markov Models for Time Series: An Introduction Using R, Second Edition*. CRC press, 2016.

15

Population Demography for Ecology

Ken Newman

US Fish and Wildlife Service, Lodi, CA

CONTENTS

15.1   Introduction

The word "Demography" is a combination of the ancient Greek words demos, meaning "the people", and "graphy", which refers to the "the writing or recording or study of". One definition of demography is "the science of vital and social statistics, as of births, deaths, diseases, marriages, etc, of populations" [10]. Our focus here is on ecology and ecological populations, and demography will be defined as the scientific study and characterization of biological populations' structure and dynamics. The simplest structure is total abundance at arbitrary points in time, while more complex structure includes abundances for multiple partitions of a population, e.g., numbers by sex, age, and spatial location. Dynamics refers to changes in structure and abundances over time as well as processes, sometimes called vital rates, which include reproduction, growth, maturity, movement, and mortality, that cause these changes.

People are interested in demography for a variety of reasons. One is inherent curiosity about abundances and dynamics. Why do the numbers of wolves (*Canis lupus*) on Isle Royale (in Lake Superior) fluctuate the way that they do? What effect will decreased snowpack levels have on the geographic range of American pika (*Ochotona princeps*) in Yosemite National Park? Answers to such questions require not only estimates of abundances of the species but also understanding of the factors that affect the abundances and dynamics.

For species harvested commercially, for sport, or for subsistence, e.g., salmon (*Oncorhynchus spp.*), red deer (*Cervus elaphus*), morel mushrooms (*Morchella spp.*), and black duck (*Anas rubripes*), people want to know how harvest affects population abundances and dynamics. Comparison of alternative harvest regulations is facilitated by predictions of the magnitude and sustainability of harvest levels. Predicting the effects of setting harvest regulations, e.g., a bag limit of 10 black ducks for a one month hunting season, requires some understanding of how this mortality might interact with other sources of mortality and other processes, like reproduction or movement. Estimates of the degree to which harvest mortality will be compensatory (removes individuals that would have died anyway from other factors) and additive (the number of animals that will be removed over and above those that would have died from other factors) are useful.

For species declared threatened or endangered by a government agency there are legal mandates for actions to be taken, or avoided, by managers of land or water resources inhabited by the species. Those actions can pertain directly to the population, such as to not take actions that could kill, harm, or harass the species, or indirectly to the species's habitat. To recover the population, interest is in identifying actions to increase the species abundance, e.g., by restoring habitat, and predicting the effects of actions. For example, the United States Fish and Wildlife Service (USFWS) has a mandate to develop "Conservation Management Plans" for species listed as threatened or endangered under the US Endangered Species Act. Such plans must include (a) specification of management actions to conserve the species, (b) measurable criteria which would lead to a determination that the species can be "delisted", no longer declared threatened, and (c) estimates of the time and cost to carry out such actions. Demographic models are central to identifying such actions, to predicting the effects of actions, and to prioritizing multiple actions.

Questions about demographics split into questions about *abundances* and about *processes*. *How many* individuals, or what volume or mass, are there, and were there previously, in the entire population and in subpopulations distinguished by sex, location, age, or genotype? Answering this question is often quite challenging depending on the magnitude of the abundances, geographic location and range, physical size, mobility, degree of elusiveness, and ability to detect individuals. A variety of statistical sampling methods, e.g., mark-recapture, and technological tools and devices, e.g., radio tracking, have been developed to help provide answers to the *how many* question. A variety of methods of estimating population abundances are discussed in Williams et al. [70], Borchers et al. [1], Buckland et al. [4] and Elzinga et al. [21], with the latter focused on plant populations.

Even if population abundances were known with certainty, questions about population processes remain. Why were the numbers what they were last year and why are they what they are now? What are the relative effects of each process on abundances at specific points in time? For example, how do adult female fecundity rates of salmon, egg hatching success rates, and larval to juvenile survival combine to affect the abundance of juveniles? How do environmental conditions, both natural and anthropogenic, affect these processes?

The focus of this chapter is on mathematical and statistical approaches to answering such *process* questions. Answering these questions involves a population dynamics model (PDM), a quantification of the relationship between past abundance and current abundances. PDMs can characterize how changes in environmental and anthropogenic factors influence population processes, and how changes in these processes translate into changes in population abundances. Measures of the degree of uncertainty as to the consequences are critical as well. For endangered species, PDMs are central to population viability analysis [PVA, 46]. PVAs use PDMs to make predictions about population trajectories, typically via computer simulation. PDMs are used to estimate extinction probabilities as a function of environmental conditions and anthropogenic factors, including accidents, like oil spills, and deliberate actions, like habitat restoration.

Answers to these initial what, why, and how questions often lead to further what, why, and how questions. Answers at the end of sequence of questions can lead to ideas about management actions to take and implementation of a particular action may then be justified by reversing the direction to yield a so-called results chain [42]. For example, a proposed management action is to plant riparian vegetation along a stream where juvenile salmon rear. The results chain is the vegetation grows and provides increasing shade along the stream, the shade reduces water temperatures, lowered temperatures increases juvenile survival, and population abundance increases. This conceptual understanding guides data collection and long term biological monitoring programs [54], and further model development. To assess the effects of planting riparian vegetation, a monitoring program collects a time series of measurements of vegetation biomass, hours of shade, stream temperatures, juvenile abundances before and after the month of May (to estimate survival) at both treatment sites and control sites where no planting is done [Before-After-Control-Impact BACI designs, 60].

The organization of the remainder of this chapter is the following. Section 15.2 is an overview of components of demography, including subpopulations and processes, while Section 15.3 is a progression of mathematical models more or less corresponding to these components. The next four sections discuss different approaches to modeling population dynamics. Section 15.4 discusses matrix population models (MPMs) which project the abundances of a finite and discrete set of sub-populations forward at discrete points in times. Section 15.5 is on integral projection models (IPMs), which can be viewed as extensions of MPMs where a continuous valued covariate, e.g., length, can be used to characterize sub-populations without arbitrary discretization of the covariate into disjoint intervals. Individual based models

(IBMs), discussed in Section 15.6, are the ultimate partitioning of a population into multiple sub-populations where the life history of each individual member of the population is modeled separately. Section 15.7 is on state-space models (SSMs) which are statistical time series models that separate stochastic variation in processes from statistical sampling error in estimates of population components, and can, in principle, contain MPMs, IPMs, and IBMs. Section 15.8 concludes the chapter with pointers to further literature on MPMs, IPMs, IBMs, and SSMs, comments on topics of demography that were minimally or not at all discussed, and thoughts about the future of biological demography.

For convenience some of the more frequently used acronyms are shown in Table 15.1.

TABLE 15.1
Listing of frequently used acronyms and their meaning.

Acronym	Meaning
PDM	Population Dynamics Model
MPM	Matrix Projection Model
IPM	Integral Projection Model
IBM	Individual-Based Model
SSM	State-Space Model
PVA	Population Viability Analysis

15.2 Components of demography

The basic components of demography are abundances and processes. Total abundances at evenly spaced points in time are denoted n_t, $t=1,2,\ldots,T$. The simplest process is the change in abundance from one time point to the next. Such changes can be expressed either in an absolute sense, $n_t - n_{t-1}$, or a relative sense, n_t/n_{t-1}, and in both cases we refer to the change as population growth.

If population abundance can be enumerated, then a succinct and completely accurate characterization of the population and its dynamics is trivial. For example, the numbers of fish in an aquarium on July 1, 2011, July 1, 2012, and July 1, 2013 were $n_{2011} = 70$, $n_{2012} = 61$, and $n_{2013} = 82$, respectively. The additive abundance changes were -9 and 21, and the relative changes were 0.87 and 1.34.

Exact enumeration is relatively rare and uninteresting in isolation. Complexity in demographic modeling arises in several ways: (1) multiple subpopulations, (2) multiple processes, (3) environmental and demographic stochasticity, (4) density dependence, (5) competition and predation, (6) human manipulation of process dynamics, (7) uncertainty in abundances.

15.2.1 Multiple subpopulations

Multiple subpopulations are subsets of a populations that are distinguished by attributes, including sex, age, sexual maturity level, spatial location, genotype, and phenotype. Such partitioned populations are sometimes called structured population, e.g., age-structured or stage-structured populations, and, in the case of spatially distinct populations, metapopulations [38].

Partitioning can be subjective and arbitrary, and depends on the available data. Arbitrariness occurs when the distinguishing attributes are continuous variables, such as measures of individual size like weight, height, length. For example, if the variable is weight, the number of partitions can vary as can the labeling of the partitions; e.g., small $= < 10$ kg, $10 \leq$ medium < 20kg, and large ≥ 20kg. The partitioning of continuous attributes is an important distinction between MPMs (Section 15.4) and IPMs (Section 15.5).

The finest partitioning of a population is at the individual entity level as the values of each individual's characteristics throughout its entire existence are the most complete description possible. This may be conceptually possible, but usually not practically possible. As a mathematical exercise, however, the modeling of individuals in a population can be useful for elucidating population level dynamics and will be discussed in Section 15.6 on IBMs.

15.2.2   Multiple processes

The process of population growth can be partitioned into multiple processes that include at least survival and reproduction, but can also include movement, individual growth, and maturation. Partitioning a population into multiple subpopulations can lead to additional process partitioning, e.g., age class specific survival probabilities. Partitioning by sex and size affects handling of reproduction, while spatial partitioning requires a movement process and location-specific movement probabilities.

Conversely, the temporal nature of processes, sequential, overlapping, or simultaneous, can lead to population partitioning. For example, a sequence of life cycle processes for salmon is egg fertilization in freshwater, egg hatching and larval emergence, survival to fry stage, smoltification, migration to the ocean, survival in the ocean, migration back to the freshwater, spawning, and death. Subpopulations of a cohort are then distinguished by life stage.

If size is a distinguishing characteristic, defined ordinally (e.g., small, medium, and large) or continuously (e.g., length in cm), then individual growth is a process affecting dynamics. Individual growth dynamics are quantified in terms of the probability of moving from one size class to another (as in MPMs, section 15.4) or by a conditional probability density function for size z'_{t+1} given previous size z_t (as in IPMs, section 15.5).

15.2.3   Stochasticity

Population dynamics are complicated by environmental and demographic stochasticity. Environmental stochasticity is between year (or any time period) variation in underlying vital rates, such as survival or reproduction, that is typically due to variation in environmental conditions such as air temperature or precipitation. Demographic stochasticity is between-individual variability conditional on a specific vital rate; e.g., if the survival probability for 100 fish is 0.7, the number surviving will not be exactly 70 and variation in that number is due to demographic stochasticity. Unless population numbers are relatively low, as for a severely endangered species, demographic stochasticity has little effect on population dynamics compared to environmental stochasticity. A rule of thumb when doing PVA, [46] is that demographic variation can be ignored in the case of a single population with at least 100 individuals, and in the case of multiple subpopulations, or life stages, there are at least 20 individuals in the most important subpopulations.

15.2.4   Density dependence

As any population increases in abundance, resource limits necessarily reduce population growth; e.g., values of $n_t/n_{t-1} > 1$ cannot be sustained. Resource limits directly affect survival and reproduction, and influence growth and movement processes, as well, in other words, these vital rates are abundance or density dependent. While decreasing abundance or density typically leads to increases in survival and reproduction, there are situations where decreases in abundance beyond a threshold lower vital rates; e.g., individuals have difficulty finding mates and cannot reproduce, what is known as an Allee effect, a problem for critically endangered species.

15.2.5   Competitors, predators, and prey

Vital rate processes for a given species, say species A, can be affected by the abundance of other species in several ways. If another species, species B, uses the same resources, e.g., consumes the same prey items, the the species are in competition, and the increased abundance of the competitor B lowers the survival and reproduction of species A. If a third species, species C, preys upon species A, then the abundance of the predator C obviously affects survival of A. If a fourth species, species D, is a prey item, then its abundance can also affect the vital rates of A. An important consideration in mathematical modeling is whether abundances of competitors, predators, or prey are treated as covariates, i.e., input variables for vital rates of a given species, or the abundances of these other species are modeled simultaneously in a multi-species PDM.

15.2.6   Human manipulation of dynamics

Human activities affecting population dynamics include harvest and species protection. Survival probabilities in PDMs need to modified by harvest, and reproduction and movement can also be affected. PDMs can be used to evaluate alternative harvest regulations including cases of selective harvest of subpopulations; e.g., only mature male red deer can be harvested during a summer time period. For endangered populations, dynamics are manipulated by regulating human activities and carrying out actions to increase and improve habitat. Projections of the effects of such regulations and actions on population dynamics are central to PVA.

15.2.7   Uncertainty in abundances

Uncertainties about abundance, or vital rates, introduce uncertainty in PDMs over and above the environmental and demographic stochasticity, what [50] label "partial observability". The time at which samples are taken can also affect the ability to estimate various process parameters, and can affect mathematical model formulation. Rees et al. [53] give an example of a sequence of processes: reproduction, followed by mortality, and then growth. If abundance estimates are made just before reproduction, abundance change includes a term for the probability of the previous year's reproduction (recruits) living an entire year. If abundance estimates are made just after reproduction, the annual abundance change does not reflect the survival of this year's reproduction as the estimates were made before subsequent mortality, and the survival of the previous year's reproduction is entangled with the survival of the previous year's abundance of old entities (non-recruits). Inserting additional sampling or estimation points in the year is one means of disentangling the effects of multiple processes.

15.3   General mathematical features of PDMs

Here we present various mathematical and probabilistic formulations of demographic models paralleling some of the features of Section 15.2. The simplest demographic model is for a single population with a single deterministic and density independent process. Such a model can be expressed in terms of absolute or relative changes in abundance. Absolute changes, $n_t - n_{t-1}$, translate into additive models,

$$n_t \;=\; n_{t-1} + \Delta_t, \tag{15.1}$$

with $\Delta_t < 0$ and $\Delta_t > 0$ indicating decline and growth, respectively, while relative changes, n_t/n_{t-1}, translate into multiplicative models,

$$n_t \;=\; \lambda_t n_{t-1}, \tag{15.2}$$

with $0 \leq \lambda_t < 1$ or $\lambda_t > 1$ for decline or growth.

15.3.1   Multiple subpopulations

Partitioning a single population into two or more populations extends the scalar n_t to a vector $\mathbf{n}_t$. For example, if a population of deer is distinguished by three life stages, young, immature, and mature, then the abundance vector at time t is

$$\mathbf{n}_t \;=\; \begin{bmatrix} n_{y,t} \\ n_{i,t} \\ n_{m,t} \end{bmatrix}$$

The length of the abundance vector over time need not remain fixed. The effects of a sequence of processes may cause the vector to expand, e.g., following reproduction, or to shrink, following an aggregation of age classes [5].

15.3.2   Multiple processes

Survival and reproduction.

With the additive single population model (15.1), Δ_t can be partitioned into survival and reproduction components,

$$n_t \;=\; n_{t-1} + R_t - M_t \tag{15.3}$$

where R_t is the number of (surviving) young produced between $t-1$ and t and M_t is the number of mortalities from the n_{t-1}. R_t and M_t may or may not be functions of n_{t-1}. In contrast, multiplicative models make explicit the dependence of change on previous abundance. Assume that in the interval $(t-1, t)$ mortality occurs first (the fraction surviving begin ϕ_t), followed by reproduction (with rate ρ_t), and there is no additional mortality before time t. Then the growth rate, λ_t (15.2), is simply the product of survival and reproduction:

$$n_t \;=\; (1 + \rho_t)\phi_t n_{t-1}. \tag{15.4}$$

The order of processes, mortality and reproduction, relative to the time of measurement (t) does not affect λ_t in this case but the following cases demonstrate when order does matter.

A more complex model with subpopulations of young and mature individuals has different survival fractions for just born young and the mature individuals, $\phi_{y,t}$ and $\phi_{m,t}$, and the time t at which abundances are counted relative to the reproductive process affects model formulation. For one scenario, t occurs immediately after reproduction, the young subpopulation are those just born (denoted $n_{0,t}$ in Scenario 1 below). Under a second scenario, t occurs just before reproduction, and, assuming the time interval is one year, the young will be nearly age 1 at the time of counting (denoted $n_{1,t}$ in Scenario 2).

Scenario 1: t just after reproduction
$$\left[\begin{array}{l} n_{m,t} = \phi_{m,t} n_{t-1} + \phi_{y,t} n_{0,t-1} \\ n_{0,t} = \rho_t(\phi_{m,t} n_{t-1} + \phi_{y,t} n_{0,t-1}) \end{array} \right]$$

Scenario 2: t just before reproduction
$$\left[\begin{array}{l} n_{m,t} = \phi_{m,t} n_{t-1} \\ n_{1,t} = \phi_{y,t} \rho_t n_{t-1} \end{array} \right]$$

Immigration and emigration.

The scalar additive model with reproduction and survival (15.3) can be extended to include immigration and emigration,

$$n_t = n_{t-1} + R_t - M_t + I_t - E_t$$

where I_t is the number immigrating into the population and E_t is the number emigrating from the population. The scalar multiplicative model (15.4) can be extended but does not necessarily remain multiplicative. The ordering of processes is again important. Assuming that immigrants arrive, and emigrants leave after mortality occurs, but prior to reproduction, the model is

$$n_t = (1 + \rho_t)(\zeta_t \phi_t n_{t-1} + I_t)$$

where ζ_t is the fraction of the survivors from n_{t-1} that stay and I_t is again the number of immigrants. If the order of processes change, the model changes. For example, suppose that immigrants arrive and emigrants leave after mortality and reproduction, then

$$n_t = \zeta_t \rho_t \phi_t n_{t-1} + I_t.$$

Movement.

For spatially-defined subpopulations, the process of movement is relevant. Immigration and emigration is of course a movement process but where the individuals are coming from or going to are not distinguished. A multiplicative formulation is more natural than an additive model, and a movement transition matrix can be inserted into the dynamics equation, say $\mathbf{n}_t = M_t \mathbf{n}_{t-1}$ where survival and reproduction are ignored. For example with three regions labeled A, B, and C, a time invariant transition matrix has the following structure.

$$M = \left[\begin{array}{ccc} \mu_{A \to A} & \mu_{A \to B} & \mu_{A \to C} \\ \mu_{B \to A} & \mu_{B \to B} & \mu_{B \to C} \\ \mu_{C \to A} & \mu_{C \to B} & \mu_{C \to C} \end{array} \right]$$

where $\mu_{i \to j}$ is the probability of moving from area i to area j in one time step, and the rows sum to 1.

Individual animal growth.

For subpopulations distinguished by size classes, transition between classes can be modeled as the fractions moving from one class to another. The process is analogous to that for movement between spatial regions. For populations partitioned to the individual entity level, growth from the size, e.g., length or weight, at time t, z_t, to another size at time $t+1$, z_{t+1}, can be modeled by the addition of an individual growth increment, x_{t+1},

$$z_{t+1}|z_t \;=\; z_t + x_{t+1}$$

x_{t+1} could be a function of the size at time t, z_t. Such fine scale handling of growth is central to IPMs (Section 15.5) and can be a part of IBMs (Section 15.6).

15.3.3 Stochasticity

The mathematical distinction between demographic and environmental stochasticity is demonstrated using the scalar multiplicative model (15.2). Demographic stochasticity arises when, for a given year t, there is constant underlying annual population growth rate, denoted λ, but there is between-individual variation in the growth rate contribution. Environmental stochasticity reflects between year variation in that underlying rate λ_t. Environmental and demographic variation typically coincide, and a hierarchical model makes clear the relationships:

$$\text{Environmental stochasticity} \qquad \lambda_t \sim \text{Gamma}(\alpha, \beta)$$
$$\text{Demographic stochasticity} \qquad n_t | n_{t-1}, \lambda_t \sim \text{Poisson}(n_{t-1}\lambda_t)$$

Asymptotic results for environmentally stochastic growth rates.

The long term, or asymptotic, behavior of a single population trajectory with environmentally stochastic annual growth rates is tractable and has similarities with deterministic exponential growth models. Consider the following single population model with environmental stochasticity only (ignoring the issue of abundances necessarily being discrete values):

$$n_t \;=\; \lambda_t n_{t-1}, \text{ where } \lambda_t \overset{iid}{\sim} \text{Distribution}(\mu, \sigma^2) \tag{15.5}$$

where $E[\lambda_t]=\mu$ and $V[\lambda_t]=\sigma^2$. Given an initial abundance $n_0 > 0$, n_t can be rewritten as

$$n_t \;=\; n_0 \prod_{i=1}^{t} \lambda_i$$

Taking the natural logarithm of both sides of the equation,

$$\ln(n_t) \;=\; \ln(n_0) + \sum_{i=1}^{t} \ln(\lambda_i),$$

which can be re-expressed as

$$\frac{\ln(n_t) - \ln(n_0)}{t} \;=\; \frac{1}{t}\sum_{i=1}^{t} \ln(\lambda_i) \tag{15.6}$$

The righthand side of (15.6) is the mean of a sequence of independent random variables, $\ln(\lambda_i), i = 1, \ldots, t$. Adding the assumption that the $E(\ln(\lambda_t)^2) < \infty$, the strong law of large numbers says that the average converges to $E[\ln(\lambda)]$. Further, by the Central Limit Theorem, the asymptotic distribution the mean of the log of the "annual" growth rates is normal. Denoting the sample average log growth rate by $\overline{\ln(\lambda)})$

$$\overline{\ln(\lambda)} = \frac{1}{t}\sum_{i=1}^{t} \ln(\lambda_i) \;\sim\; \text{Asymptotic Normal}\left(E[\ln(\lambda)], V(\ln(\lambda))\right)$$

Another way to express this result, using the lefthand side of (15.6),

$$\ln(n_t) \sim \text{Asymptotic Normal} \left(\ln(n_0) + tE[\ln(\lambda)], tV(\ln(\lambda)) \right)$$

or

$$n_t \sim \text{Asymptotic Lognormal} \left(n_0 \exp(tE[\ln(\lambda)]), tV(\ln(\lambda)) \right) \tag{15.7}$$

Thus, the median population abundance at t is identical to a deterministic exponential growth model.

Stochasticity in individual processes.

Survival, reproduction, movement, individual animal growth, and other processes can be made stochastic. An example is a survival process for a scalar population with a logit-normal model for environmental stochasticity and a binomial distribution for demographic stochasticity. Letting $\phi_{c,t}$ be the survival probability for subpopulation c at time t,

$$logit(\phi_{c,t}) \sim \text{Normal} \left(\beta_{0,\phi,c}, \sigma_{\phi,c}^2 \right)$$
$$n_{c,t} \sim \text{Binomial} \left(n_{c,t-1}, \phi_c \right)$$

where $logit(x) = \ln(x/(1-x))$.

15.3.4   Density dependence

In the ecological literature, there are several well-known single population, deterministic and discrete time-indexed models with density dependent population growth rates including the Gompertz [15], Ricker, Beverton-Holt, and logistic models [27]. Here we just present a deterministic Ricker model formulation [taken from 27]. The Ricker model originated with fish populations, but is now applied many other kinds of populations.

$$\text{Ricker model} \quad : \quad n_t = (\phi_a + \phi_y b \exp(-cn_{t-1})) n_{t-1}, \quad b > 0, c > 0, \tag{15.8}$$

where n_t can be viewed as the sum of surviving adults from the previous year ($\phi_a n_{t-1}$) and surviving progeny, with ϕ_y the survival fraction for offspring produced at rate $b\exp(-cn_{t-1})$. The parameter b is the maximum number of offspring per adult, theoretically possible in the absence of any resource limitations, while $\exp(-cn_{t-1})$ is a density dependent dampening of that maximum.

In the case of multiple subpopulations, if the vital rates and abundances for one subpopulation do not affect another subpopulation, then the above univariate density dependent models can be applied on a per subpopulation basis. If subpopulations occupy the same geographic area and compete for resources, then density dependent formulations will include the abundances of other subpopulations. Density dependent dynamics also arise for populations of different species that are either in predator-prey relationships or competing for an in-common resource.

In the case of multiple processes, e.g., survival, reproduction, movement, and individual animal growth, process-specific density dependence relationships can arise. In the Ricker model, for example, reproduction is density dependent while survival is density independent. Density dependence in movement processes for spatially distinct subpopulations (metapopulations) is likely as the probability of movement from one region to another could be a function of the relative densities of individuals in each region, e.g., the probability of moving from a high density region to a low density region increases as the difference in densities increases. Of course, stochasticity can be incorporated into density dependent formulations for different processes.

Density dependence both within a single population and for populations of two or more different species, e.g., predator and prey populations, can lead to relatively complex population dynamics. The Lotka-Volterra predator-prey model [27] can with certain parameter combinations lead to periodic oscillations in the abundances of each population. Within a single population, discrete time single population models like the Ricker and discrete logistic model can lead to damped or expanding oscillations, different periodicities, or chaos (no periodicity and apparently random fluctuations; [see, for example, Figure 2.6 in 27].

15.3.5   Inclusion of covariates

Mathematical formulations of population processes often include covariates, one of the earliest examples being the modeling of survival as a function of weather data [51]. The effects of deliberate human manipulations or incidental anthropogenic consequences, e.g., the erection of a wind turbine and subsequent bird mortality, can be translated into covariates for process models. Abundances of predators, competitors, or prey can also be used as covariates in models for survival and reproduction of a single species population dynamics model in contrast to jointly modeling the population dynamics of several species. The legitimacy of such handling of these other populations may depend upon the degree to which other populations are affected by the abundance of the population of interest.

15.3.6   Remarks: Estimability and Data Collection.

It is easy to formulate a population dynamics model where the parameters cannot be estimated given the available data. For example, annual surveys alone do not allow separate estimation of the survival probability, ϕ_t, and reproductive rate, γ_t, in the simple univariate model (15.4). Intuitively given estimates of n_t and n_{t-1} one can just estimate the combination $(1 + \gamma_t)\phi_t$.

One way to disentangle such combinations of parameters, in the case of sequential processes, is to have abundance estimates at time points immediately after the end each process. For example, in the model (15.4) abundance should be measured twice a year, once immediately following the survival process, and once after reproduction. The reality of the processes is typically more complicated, with such sharp demarcations unlikely, but formulating such models can provide guidance for data collection.

15.4   Matrix Projection Models, MPMs

One of the oldest and most popular types of population dynamics models are matrix projection models (MPMs). Lewis [39] and Leslie [37] independently proposed MPMs as a means of modeling the population dynamics of age-structured populations (age-specific subpopulations). Let $n_{0,t}$ denote the number of young at time t and $n_{a,t}$ be the abundance for ages 1 to $A - 1$, and $n_{A+,t}$ be the abundance of age A and older individuals. A deterministic formulation for the dynamics can be written as

$$
\begin{bmatrix}
n_{0,t} \\
n_{1,t} \\
n_{2,t} \\
\vdots \\
n_{A,t} \\
n_{A+,t}
\end{bmatrix}
=
\begin{bmatrix}
\gamma_0 & \gamma_1 & \gamma_2 & \cdots & \gamma_{A-1} & \gamma_A \\
\phi_1 & 0 & 0 & \cdots & 0 & 0 \\
0 & \phi_2 & 0 & \cdots & 0 & 0 \\
\vdots & 0 & 0 & \cdots & \phi_A & \phi_{A+}
\end{bmatrix}
\begin{bmatrix}
n_{0,t-1} \\
n_{1,t-1} \\
n_{2,t-1} \\
\vdots \\
n_{A,t-1} \\
n_{A+,t-1}
\end{bmatrix}
\tag{15.9}
$$

or more compactly as $n_t = Ln_{t-1}$, where L is referred to as a Leslie matrix, and is analogous to the scalar multiplicative model (15.2). Lefkovitch [36] proposed MPMs where subpopulations are distinguished by life stage, e.g., young, immature, and mature, thus a stage-structured model in contrast to an age-structured model. Of course, partitioning by gender, genotype, and many other subpopulation identifiers is possible. This simple structure, $n_t = Ln_{t-1}$, has been extended in many ways including time varying L, the use of covariates to model the components of L, adding stochasticity and density dependence.

15.4.1 Analysis of MPMs

Apparently simple MPMs, such as (15.9), can yield complex dynamics depending upon the components of L, and the many extensions of MPMs have added to this complexity. To gain deeper understanding of the dynamics of MPMS, Caswell [6, p. 18] developed four sets of questions, which have been paraphrased below.

1. *What is the asymptotic behavior of the MPM?* As time increases, does the total population grow or decline exponentially? Do the relative proportions of each subpopulation become constant? Does the population approach an upper bound (carrying capacity)? Do the total population and individual subpopulation abundances oscillate (in a damped or undamped manner)? Do the abundances display periodicity? Do the abundances become chaotic?

2. *Is the MPM ergodic?* In other words, are the asymptotic dynamics independent of the initial conditions, e.g., independent of the actual values of n_0?

3. *What are the transient dynamics?* What are the dynamics like in the short term as opposed to the asymptotic or limiting results?

4. *How sensitive are the results to the values of the elements of L?* The survival probabilities and fecundity rates, for example, are estimates, and will have some degree of estimation error. How much would the population dynamics, including asymptotic and transient dynamics, change if some elements of the matrix were changed "slightly"?

We will not address all these questions further here and refer the interested reader to Caswell [6]. However we will briefly discuss one type of asymptotic behavior, for both deterministic and stochastic MPMs, which is analogous to single population exponential growth models.

15.4.2 Limiting behavior of density independent, time invariate MPMs

Results from matrix algebra can be used to describe the asymptotic behavior of a time invariant projection matrix [see 6, chap 4.5]. If the matrix is (a) nonnegative (all elements are ≥ 0), (b) irreducible (e.g., every age class can contribute to every other age class at some point in time), (c) primitive (there is some positive integer k such that every element in the matrix raised to the power k, L^k, is a positive number), then in the limit the population dynamics are either exponential growth or decay, i.e., $A^T n_t = \lambda n_t$, where λ is a scalar value that is multiplied against each component of the vector n_t. Further, the relative proportions of each component of n_t will remain constant.

For example, consider an MPM with three age classes (Young, Adult, Adult) and an initial abundance $n'_0 = (100,50,10)$ and the following Leslie matrix

$$\mathbf{L} = \begin{bmatrix} \gamma_{Young} & \gamma_{Adult} & \gamma_{Old} \\ \phi_{Young} & 0 & 0 \\ 0 & \phi_{Adult} & \phi_{Old} \end{bmatrix} = \begin{bmatrix} 0.0 & 1.2 & 1.4 \\ 0.3 & 0.0 & 0.0 \\ 0.0 & 0.5 & 0.9 \end{bmatrix} \tag{15.10}$$

The population abundances over 9 iterations are:

Stage	1	2	3	4	5	6	7	8	9	10
Young	100	74	84	90	103	116	131	148	167	189
Adult	50	30	22	25	27	31	35	39	44	50
Old	10	34	46	52	59	67	76	86	97	109

The population growth rates, per stage, over time:

Stage	2	3	4	5	6	7	8	9	10
Young	0.74	1.13	1.08	1.14	1.12	1.13	1.13	1.13	1.13
Adult	0.60	0.74	1.13	1.08	1.14	1.12	1.13	1.13	1.13
Old	3.40	1.34	1.14	1.14	1.13	1.13	1.13	1.13	1.13

Thus after six generations the annual growth rate reaches 13% and stays there. The fraction of the population in each stage class stabilizes as well:

Stage	1	2	3	4	5	6	7	8	9	10
Young	0.62	0.54	0.55	0.54	0.54	0.54	0.54	0.54	0.54	0.54
Adult	0.31	0.22	0.15	0.15	0.14	0.14	0.14	0.14	0.14	0.14
Old	0.06	0.25	0.30	0.31	0.31	0.31	0.31	0.31	0.31	0.31

Thus, after six generations the fractions in the Young, Adult, and Old stages remain 0.54, 0.14, and 0.31.

The limiting population growth rate and proportions of each category can be determined analytically using matrix algebra, in particular, by carrying out an eigen analysis of $\mathbf{L}$. For a p by p matrix $\mathbf{L}$, the eigen analysis yields p eigenvalues, $\lambda_1, \ldots, \lambda_p$, and p corresponding right eigenvectors, $v_1, \ldots, v_p$. An eigenvalue and its corresponding eigenvector have the relationship, $\mathbf{L}v_i = \lambda_i v_i$. Denote the largest eigenvalue λ_1 and its corresponding eigenvector v_1. Then λ_1 is equal to limiting population growth rate, in the example 1.13 (more precisely, 1.12938), and dividing each element of v_1 by its total yields the limiting fractions, here (0.54, 0.14, 0.31).

15.4.3   Stochasticity

One way to add stochasticity to MPMs is to randomly draw elements of the matrix from probability distributions, e.g., randomly draw survival probabilities for age a individuals, thereby introducing environmental stochasticity. Under some conditions, in the absence of density dependence for example, the introduction of environmental, or demographic, stochasticity will not appreciably alter the asymptotic dynamics from that of a deterministic

MPM. In other words, the above eigen analysis results more or less hold: in the limit there is an average growth rate and stable population structure. Caswell [6, Chap. 14] provides details of these results [with some of earliest work from 9, 65]. Below we closely follow [6, p. 393] and somewhat mimic the derivation of the asymptotic distribution of the stochastic univariate model shown in (15.5 - 15.7). We start with a stochastic process of matrices, $\mathbf{L}_1, \mathbf{L}_2, \ldots$, which satisfy certain regularity conditions, including being stationary (the joint distribution for $(\mathbf{L}_{t_1}, \mathbf{L}_{t_2}, \ldots, \mathbf{L}_{t_n})$ is the same as that for $(\mathbf{L}_{t_1+h}, \mathbf{L}_{t_2+h}, \ldots, \mathbf{L}_{t_n+h})$ for any finite $n > 0$, $t_1, t_2, \ldots, t_n$, and $h > 0$), and ergodic (roughly put, the initial value of $\mathbf{L}_1$ does not affect the eventual behavior of the sequence). Further assume an upper bound on the magnitude of the initial matrix, $E(ln^+\|\mathbf{L}\|_1) < \infty$, where $\|\mathbf{L}\| = \sup_{\mathbf{n}\neq 0} \frac{\|\mathbf{Ln}\|}{\|\mathbf{n}\|}$ and $ln^+(x) = \max(0, \ln(x))$. The total population size at time t, denoted $N(t)$, is the vector norm of $\mathbf{n}_t$ ($\|\mathbf{n}_t\| = \sum_i |n_{t,i}|$). Given an initial vector $\mathbf{n}_0$:

$$N(t) \quad = \quad \|\mathbf{n}_t\| = \left\|\prod_{i=1}^{t} \mathbf{L}_i \mathbf{n}_0\right\| \tag{15.11}$$

$$\Rightarrow$$

$$\frac{1}{t}\ln(N(t)) \quad = \quad \frac{1}{t}\ln\left\|\prod_{i=1}^{t}\mathbf{L}_i\mathbf{n}_0\right\|. \tag{15.12}$$

Furstenberg and Kesten [23] proved that, with probability 1, the limit of (15.12) exists:

$$\lim_{t\to\infty}\frac{1}{t}\ln(N(t)) \quad = \quad \lim_{t\to\infty}\frac{1}{t}\ln\left\|\prod_{i=1}^{t}\mathbf{L}_i\mathbf{n}_0\right\| = \ln(\lambda_s), \tag{15.13}$$

where λ_s is called the stochastic growth rate. Lower and upper bounds on λ_s can be calculated from the average minimum row sums and average maximum row sums of the matrices, namely,

$$\sum_i \pi_i R_{\min}^{(i)} \quad \leq \ln(\lambda_s) \leq \quad \sum_i \pi_i R_{\max}^{(i)} \tag{15.14}$$

where π_i is the asymptotic probability of environment i occurring (corresponding to matrix $\mathbf{L}_i$) and $R_{\min}^{(i)}$ and $R_{\max}^{(i)}$ are the minimum and maximum row sums of $\mathbf{L}_i$ [6, p. 395].

With further conditions on the matrices, $\mathbf{L}_i$, including nonnegativity, the asymptotic distribution of the population total is lognormal:

$$N(t) \quad \sim \quad \text{Asymptotic Lognormal}\left(\exp(t\ln(\lambda_s)), t\sigma^2\right) \tag{15.15}$$

where σ^2 is some constant. Thus, similar to (15.7), the asymptotic median of the population total is the same as for a univariate exponential population growth model, and λ_s is analogous to the largest eigenvalue, λ_1, of a deterministic MPM.

15.4.4 Building block approach to matrix construction

Deterministic skeletons for relatively complex MPMs can be constructed using a building block approach (Buckland et al. [5]; Newman et al. [47]). A crucial assumption is the approach is that there is a particular sequence to processes which operate on a vector of population abundances, such as survival, then movement, then reproduction. An example from Newman et al. [47, eq. 2.11, p. 18] has two size-class subpopulations, small and large, and a sequence of three processes: survival, followed by growth (from small to large), and then reproduction. The survival probabilities are size specific (ϕ_S and ϕ_L), the probability that a small individual becomes large is π, and only large individuals can reproduce and they do so with rate ρ.

$$
\begin{bmatrix} n_{S,t} \\ n_{L,t} \end{bmatrix} = \begin{bmatrix} 1 & \rho \\ 0 & 1 \end{bmatrix} \begin{bmatrix} 1-\pi & 0 \\ \pi & 1 \end{bmatrix} \begin{bmatrix} \phi_S & 0 \\ 0 & \phi_L \end{bmatrix} \begin{bmatrix} n_{S,t-1} \\ n_{L,t-1} \end{bmatrix}
$$

$$
= \begin{bmatrix} (1-\pi+\rho\pi)\phi_S & \rho\phi_L \\ \pi\phi_L & \phi_L \end{bmatrix} \begin{bmatrix} n_{S,t-1} \\ n_{L,t-1} \end{bmatrix} = \mathbf{L}\mathbf{n}_{t-1} \tag{15.16}
$$

The matrix in (15.16) is an example of a Lefkovitch matrix which is arguably more simply constructed by using such a building block approach than by trying to construct the final matrix in a single operation.

15.4.5 Determining the elements of projection matrices

The most common way to use MPMs has been to plug in estimates of matrix components from various, and often independent, studies, and then make population projections using those point estimates. Caswell [6, p. 22], for example, states that, to fill the elements of the matrix, life tables are used. Life tables contain mortality probabilities, the probability that an individual of age a will die before reaching age $a+1$, and maternity functions, the expected number of offspring that an age a individual will produce in the next year, from which survival probabilities ϕ and reproductive rates γ (15.9) can be calculated. However, how mortality probabilities and maternity functions are constructed in the first place may be no trivial task. With wildlife populations, mark-recapture studies where animals are aged at time of marking can provide estimates of age-specific survival, and, in some situations, estimates of reproductive success. Of course, the addition of more subpopulations and processes increases the "data requirements and mathematical complexities [which] can quickly overwhelm an investigation of these parameter-rich models" [70, p161].

An alternative to the above approach of estimating matrix elements separately from inference about population abundances is to combine stochastic population dynamics with statistical sampling error, or estimation uncertainty, in matrix elements and population abundances. The SSM framework provides a structure for doing this and is discussed in Section 15.7.

15.4.6 Density dependent MPMs

Density dependence can be introduced into MPMs by simply making some of the elements of the projection matrix density dependent. For example, referring to (15.10), the fecundity of the old group could be expressed as a function of the total abundance of adult and old individuals, $\gamma_{Old,t} = (n_{Adult,t} + n_{Old,t}) \exp(-c(n_{Adult,t} + n_{Old,t}))$, a variant of the Ricker model. The linearity aspect of the MPM is subsequently altered and the analyses carried out for density independent MPMs do not directly apply, e.g., the eigen analysis is no longer directly applicable. See Caswell [6, Chap. 16] for detailed discussion of a variety of density dependent models, subsequent dynamics, and analytical approaches.

15.5 Integral Projection Models, IPMs

The partitioning of a population into discrete subpopulations, namely formulating a structured population, may be arbitrary when natural divisions are lacking. For example, suppose individual weight (in kg) is the feature used to subdivide the population for an MPM. The specified weight classes, small, medium, and large, necessarily have arbitrary boundaries, say, (0,5), [5,10), [10+]. An animal weighing 4.99 kg is labeled small and one weighing 5.0 kg is medium. Those two individuals will be treated differently in terms of population processes, e.g., the survival probability is 0.5 for small individuals and 0.8 for medium individuals, while the actual survival probabilities for both individuals may be much more similar. Integral Projection Models [IPMS; 18], sometimes called integrodifference equation models [see 6, for historical references], are a modeling approach that maintains the continuous nature of a factor that distinguishes population members, while (generally) maintaining the discrete time step characteristic of MPMs.

15.5.1 Kernel structure of IPMs.

The core of an IPM is the *kernel*, denoted $K(z\prime_{t+1}|z_t)$, which is analogous to an element in the transition matrix of an MPM. The kernel can be viewed as a conditional probability density function for the "probability" that an animal of size z at time t, denoted z_t, is size $z\prime$ at time $t+1$, denoted $z\prime_{t+1}$. The word probability is put in quotation marks as this is a density not a probability. More accurately $K(z\prime_{t+1}|z_t)\Delta$ is an approximate probability for such a movement from size z_t to a size in an interval of width Δ containing $z\prime_{t+1}$, e.g., $z\prime_{t+1} \pm 0.5\Delta$. The number of individuals in a given size class at time $t+1$ is then the sum of all individuals of *any* size class at time t, $n(z_t^*)$, that survive, grow, and/or contribute to individuals of size class z_{t+1} at time $t+1$ where $z_t^* \in \Omega$ and Ω is a suitably large range of sizes, so

$$n(z\prime_{t+1}) = \int_{z_t \in \Omega} K(z\prime_{t+1}|z_t)n(z_t)dz_t \qquad (15.17)$$

A simpler version of the kernel is time invariant, $F(z\prime|z)$, where the conditional density for the contribution to size class $z\prime$ at time $t+1$ from size class z_t is the same for all times t.

The population growth process is the result of multiple processes, including survival and reproduction. So the kernel K can be decomposed into survival of the current population and reproduction entering the population. Here, however, individual size is also a factor and survival and reproduction is into a specific size class, z_{t+1}. Thus growth from size class z_t to z_{t+1} is a third process to account for. The resulting partitioning of the kernel is

$$K(z\prime_{t+1}|z_t) = P(z\prime_{t+1}|z_t) + F(z\prime_{t+1}|z_t) \qquad (15.18)$$

where P is the survival/growth kernel, the combined conditional density for surviving to time $t+1$ and changing to size class z_{t+1}, and F is the fecundity kernel, the conditional density for recruits at time $t+1$ of size z_{t+1} [43].

There are a wide variety of formulations for the survival/growth kernel. One formulation is to treat the two processes as independent, the result being the product of the conditional probability of surviving, $\phi(z_t)$, and the conditional density of moving to size class $z\prime$, $g(z\prime_{t+1}|z_t)$:

$$P(z\prime_{t+1}|z_t) = \phi(z_t)g(z\prime_{t+1}|z_t) \qquad (15.19)$$

In principle, a joint density for survival and growth could be used; e.g., movement to a much larger size class is linked with lowered survival probability. The survival probability could be a more complicated function of competing or sequential mortality factors; e.g., there are two mortality processes occurring in sequence, $\phi(z_t)= \phi_{1,z_t}\phi_{2,z_t}$.

The fecundity kernel can be made complex as well. For example, it could be a function of four processes: a size dependent probability distribution for the number of eggs produced, $f(E|z_t)$, a probability that the eggs are fertilized, p_E, a probability that the fertilized eggs will hatch, p_h, and a density function for the size of hatched larvae, $h(z\prime)$. Then

$$F(z\prime_{t+1}|z_t) = f(E|z_t)p_E p_h h(z\prime) \tag{15.20}$$

Merow et al. [43] note that a common feature of the survival/growth and fecundity kernel formulations is an *individual component*, e.g., $\phi(z_t)$ in (15.19) and $(f(E|z_t)p_E p_h)$ in (15.20), and a *size redistribution component*, e.g., $g(z\prime_{t+1}|z_t)$ in (15.19) and $h(z\prime)$ in (15.20).

15.5.2 Implementation of an IPM

Equation (15.17) is analogous to the generation of a single component in the state vector of an MPM. With an MPM, the entire state vector at time $t+1$ is $\mathbf{n}_{t+1} = L\mathbf{n}_t$, where the ith entry in $\mathbf{n}_{t+1}$, denoted $n_{i,t+1}$, is the following sum:

$$n_{i,t+1} = \sum_{j=1}^{p} L_{i,j} n_{j,t} \tag{15.21}$$

where L has p columns. Each $L_{i,j}$ in the summation is akin to a kernel function as it is the per individual contribution from "size" class i at time t to "size" class j from time $t+1$. If the vector $\mathbf{n}_t$ is further partitioned into a relative large number of size classes, the summation operation in (15.21) approaches an integration operation.

Implementation of an IPM is in practice the reverse operation. Referring to the integral in (15.17), the interval Ω, which contains the range of size classes that can contribute to size class $z\prime$, is partitioned into m size classes. A finite sum approximation to integration, e.g., the midpoint rule, the trapezoid rule, or Simpson's rule, is used calculate the number of individuals in size class $z\prime$. An example of the midpoint rule: suppose Ω is an interval $[L, U]$ which is partitioned into m intervals of equal length $(U - L)/h$, and let z_i be the midpoint of the ith size class, also known as mesh points [53], where

$$z_i = L + (i - 0.5) * j, \quad i = 1, 2, \ldots, m$$

The integral (15.17) can be approximated by

$$n(z\prime_{t+1}) \approx \sum_{i=1}^{m} K(z\prime|z_i) h n(z_{i,t}) \tag{15.22}$$

15.5.3 Estimation of kernel components

The problem of specifying kernel components parallels the problem of determining components of the transition matrices in MPMs. Assuming that relevant data on size, survival, reproduction success, etc, are available, there are many standard statistical model fitting procedures, linear regression, nonlinear regression, generalized linear models including logistic regression, and generalized additive models, that can be used to construct the components of $K(z\prime_{t+1}|z_t)$. Likewise, many of the associated model fit diagnostic procedures could, and should be, used to assess the quality of the estimated components of the kernel [53].

A number of probability and density functions are needed to calculate the transition densities of the survival/growth kernel (e.g., (15.19)), and the fecundity kernel (e.g., (15.20)). For individual components that are probabilities, e.g., the conditional probability of survival, sample data on size conditional outcomes can be used to calculate estimates. For example, a mark-recovery study of banded ducks could provide size-specific annual survival probabilities based on a smooth fitted survival function, e.g., $\log\left(\phi/(1-\phi)\right)|z = \beta_0 + \beta_1 z$. Whether or not time-specific functions could be fit may depend upon the number of years of data available. Survival probabilities can be a function of size and environmental covariates, e.g., winter temperatures. For size redistribution components, such as the conditional density for moving from size class z to z^* in (15.19), size measurements made over time on multiple individuals are required.

Inference methods for IPMs are continually developing. For example, Ghosh et al. [24] use Bayesian hierarchical models where the size distribution is a point pattern on some interval and carry out an integrated analysis that combines the parameter estimation/model fitting stage and the projection stage.

15.5.4   Application, use and analysis of IPMs

Plant species were the most common organisms in early applications of IPMs, e.g., Northern Monkhood [18], with growth transitions between different plant sizes, e.g., stem diameter, and processes like flowering strategies. The scope of applications has since expanded to include birds [Great tits, 7], arachnids [soil mites, 3], mammals (Soay sheep), diseases [hosts and parasites, 44].

The questions asked of MPMs in Section 15.4.1 can be asked of IPMs. Is there a limiting population growth rate, a dominant eigenvalue λ_1 and corresponding stable "size" class distribution? Ellner and Rees [20] gives examples of sensitivity analyses of IPMs. Software for IPMs includes the R package IPMpack.

In addition to analysis of population dynamics, ecological inference using IPMs includes analysis of evolutionary strategies [20]. Brooks et al. [3] separated out the effects of individual body size on developmental rates from the effects of environmental conditions on reproductive rates. Metcalf et al. [44] examined the feedback between host and parasite in an epidemiological analysis.

15.6   Individual Based Models, IBMs

Individual based models in ecology [IBMs; 14] are computer simulation procedures that can track the entire life history of multiple individuals simultaneously. Variables tracked include emergence into the population (date of birth, germination, hatch date), size at birth, sex, size over time, time and duration of sexual maturity and reproduction, spatial location and movement, senescence, and death. A central feature is the modeling of interactions of individuals with each other, including individuals of the same species, e.g., reflecting competition for resources and density dependence, and individuals of different species, e.g., reflecting predator-prey dynamics or, more broadly, ecological community interactions. Another key feature is the simulation of interactions of individuals with their abiotic environment, e.g., air temperature and precipitation, and their biotic environment excluding like individuals, e.g., vegetative browse and zooplankton.

The opportunity to insert complexity into dynamic processes underlying demographics is relatively unlimited, constrained primarily by computer storage and processing speed. Population level properties can be examined at any time in the simulation process by aggregating the states of individuals in arbitrary ways. For example, a simulation starts at time t_0 with a vector of 1000 individuals where each individual has an associated vector of initial conditions such as age, weight, sex, spatial location, and maturity. Survival, growth, movement, and reproduction processes are then applied to each individual and, at time t_1, numbers of individuals in different spatial regions further distinguished by sex and age class, say, are tallied to yield abundances of multiple subpopulations. Repeating the simulation and aggregation K times yields a multivariate time series of subpopulation abundances, $n_{t_1}, n_{t_2}, \ldots, n_{t_K}$. Analysis of population level dynamics can then be conducted, studying such things as the effects of region-specific harvest regulations on different sub-populations of deer, for example. If the effects of environmental and anthropogenic factors on the population dynamics cannot be readily examined analytically, IBM output can provide some experiential, albeit simulated, insight.

15.6.1 Statistical designs for and analysis of IBMs

The simulation nature of IBMs with multiple attributes and multiple levels to attributes lends itself to using methods from the statistical design of experiments to construct a time series of any length with an arbitrary number of sub-populations. For example, if the attributes of interest are sex, spatial location, and age class with corresponding levels of (female, male), (I, II, III, IV) regions, and ages (0,1,2,3+), then a factorial design with $2 \times 4 \times 4 = 32$ "treatment" combinations can be conducted with r replications of each combination. Statistical methods such as analysis of variance or response surface modeling can then be used to examine the effects of the factors and treatment combinations. Aggregated data can be used to construct simple MPMs, like year-specific Leslie matrices, and methods for assessing MPMs, such as calculating annual finite population growth rates for multiple years can be employed [for such an example, see 55].

The computational burden of IBMs can grow in a number of ways. First, as the number of attributes of interest and the levels of each attribute increases, the number of treatment combinations can grow rapidly. Second, as the level of environmental stochasticity (or demographic) increases, the number of replicates required to provide a desired level of precision for estimates of average population level responses increases as well. Third, questions about the effects of the distribution of initial attribute values at time t_0 as well as questions about the nature of the processes, e.g., density dependent or density independent recruitment success or the chosen value, or distribution of values, for juvenile survival probabilities, can lead to extensive sensitivity analyses.

15.6.2 Comparison with population models

In contrast to population models, such as the Ricker model (15.8), for which long term population dynamics such as exponential growth, an asymptotic upper bound, or periodicity can sometimes be determined analytically or by elementary computer simulation, population-level behavior for IBMs is an *emergent* property. The dynamics are the result of potentially complex interactions of individuals with each other and with their abiotic environment [14], and can demonstrate "the importance of local interactions between individuals in ecological systems" [30].

A succinct way to contrast population-level models and IBMs is top-down versus bottom-up. Population-level models are *top-down* in that they predict what happens to individuals as function of population level characteristics, e.g., fecundity of the individual decreases as the total population abundance increases (density dependence exists). Conversely, IBMs are *bottom-up* in that modeling begins with the characteristics of multiple individuals and manifests characteristics of the population as a whole. An interesting example of the latter is with *Anolis* lizards in the Caribbean [discussed in 56] where an IBM simulated energy gained per unit time after a lizard consumed a prey item as a function of distance from the prey and the optimal foraging distance could then be determined. From that model for the "energy capture" the daily growth rate of the lizard was predicted, with distinction made between growth prior to reproductive stage and during the reproductive stage. Using these results an optimal growth rate, as a function of age, was calculated, which was then used with information on survival probabilities and maternity rates to determine that optimal female body size was 45mm. As Roughgarden [56] said "[t]his example illustrates a complete and successful modeling protocol that begins with the properties of an individual and culminates in the an evolutionary prediction of the adult body size for lizards on an island in the absence of congeneric competitors".

15.6.3   Applications of IBMs

The earliest applications of IBMs in ecology were mostly in forestry, and such applications remain common. In the IBM JABOWA [2], individual trees were the fundamental entities and the central measure on each tree was its stem diameter (at some height on the tree). Other tree measures such as volume and crown biomass can be functions of diameter. Emergence, growth, and death of a tree are functions of interactions with neighboring trees, their size and proximity and the degree to which they compete for resources like light and water, for example, and functions of interactions with the abiotic environment, e.g., soil type and chemistry, precipitation, temperature, and light. Forestry IBMs have been used for management purposes, e.g., to predict growth and yield of commercially harvested species, as well as purely scientific reasons, i.e., to "explore ecological mechanisms and patterns of structure and functional dynamics in natural forest ecosystem" [40].

Applications to fish populations are common as well, where IBMs "track the attributes of individual fish through time and aggregate them to generate insights into population function" [66]. IBMs simulate how fish of different phenotypes interact with their biotic and abiotic environment. Differences in phenotype can refer to differences in length, weight, sex, and age, the biotic environment can include prey items, such as zooplankton or vegetation, and the abiotic environment can include water temperature, salinity, water clarity. An IBM for a small estuarine fish, delta smelt [*Hypomesus transpacificus*, 55] also included bioenergetics considerations, namely the transformation of consumed prey into fish growth.

IBMs in ecology can be broadly divided into applications for (individual) populations, communities and ecosystems. Single population-level IBMs have been mentioned above, e.g., *Anolis* lizards and Delta Smelt, but IBMs have used to model predator-prey dynamics [11]. A community-level application by Weiss et al. [69] used an IBM to simulate how the dynamics of a community assembly of 90 hypothetical plant types were affected by soil attributes and grazing intensities. The results were then compared to field-based observations of species richness and diversity. Least common are ecosystems level applications; a hypothetical food web system used an IBM to model interactions between three trophic levels, plant, herbivore and carnivore [58].

15.6.4   Data needs and structure

IBMs have at least three levels of data needs. One is an initial individual attribute vector [66], and initial values for components of the biotic and abiotic environment. When proximity to other individuals is a factor in the dynamics, an initial spatial distribution is needed and locations might be randomly placed as in a Poisson process, systematically placed, clustered, or placed with probabilities proportional to particular habitat conditions. Other individual attributes, e.g., size, sex, age, need to be assigned. To achieve greater realism, the actual multivariate distribution of such attributes should be mimicked. Initial biotic attributes can include type, abundance, and spatial location of competitors, predators, and food items. Initial abiotic features may be relatively static, e.g., soil types, water sources, or dynamic, e.g., air temperature and precipitation.

A second data need is for information about how the individuals interact with each other and with their environment. For example, how is the probability of survival affected by the availability and proximity to food items? How is movement affected by population density, biotic and abiotic features?

A third data need is field-based observations to verify that IBM output, and apparent emergent population level properties, e.g., collective survival, reproduction, and movement rates, are reasonable.

Given these data needs, IBMs, particularly those designed for specific applied problems have been criticized as being too "data hungry" [26]. Available data may thus constrain and guide IBM formulation, affecting things like the time step resolution, spatial scope, number of attributes followed, and number of interactive processes simulated.

15.6.5 Relationship with IPMs

Longitudinal data on individuals are central to both IPMs and IBMs. IPMs *use* such data to model population, or sub-population, level probabilities of transitions from one attribute value to another. In contrast, IBMs, starting at time $t = 0$ with a vector of n_0 individuals each with an associated attribute vector, *generate* longitudinal data per individual. Such data generated by IBMs can be used to evaluate fitting procedures for IPMs and the subsequent performance of IPMs can be evaluated by comparing IPM predictions to the "true" values generated by simulated IBM output [53].

15.7 State-Space Models, SSMs

State-space models (SSMs) are models for two parallel time series, a state process and an observation time series. The state process time series describes the temporal evolution of the true, but generally unknown, state of nature; it is here denoted n_t, t=0, 1, 2, ..., T, where n_t can be a vector of varying length. The state n_0 is referred to as the initial state. The observation time series, denoted y_t with t=1,2,..., T, is a sequence of imperfect or inexact measurements of the state process time series. The integer valued subscripting of both time series is used here, $t_1, t_2, \ldots, t_T$, but arbitrary time points are possible. The time series indexing for both time series do not necessarily coincide, e.g., there could be half the observations if the state is only observed every other time point, although statistical estimation limitations might occur. Also, the dimensions of n_t and y_t need not be the same, although situations where the dimensions differ can affect estimability. For an ecological example: n_t is a vector of true abundances of subpopulations at time t and the components of y_t are estimates of one or more components of n_t.

The probabilistic structure of a SSM is a paired sequence of probability distributions (probability mass functions for integer valued components or probability density functions for continuous valued components) that characterize the evolution of n_t and the relationship between y_t and n_t. The the probability distribution for n_t is typically first order Markov, i.e., n_t given n_{t-1} is conditionally independent of all other states, and y_t given n_t is independent of all other state vectors and all other observation vectors.

15.7.1   Normal dynamic linear models

A classic SSM, originating from Kalman [32], is the normal dynamic linear model (NDLM);
for example,

$$
\begin{aligned}
\mathbf{n}_0 &\sim D(\theta) \\
\mathbf{n}_t | \mathbf{n}_{t-1} &\sim \mathrm{MVN}\left(\mathbf{L}\mathbf{n}_{t-1}, \Sigma\right), \quad t = 1, 2, \ldots, T \\
\mathbf{y}_t | \mathbf{n}_t &\sim \mathrm{MVN}\left(\mathbf{B}\mathbf{n}_t, \Omega\right), \quad t = 1, 2, \ldots, T
\end{aligned}
$$

where $D(\theta)$ denotes an arbitrary probability distribution with parameter θ which may be
degenerate, i.e., $\mathbf{n}_0$ is a fixed value, MVN is multivariate normal, L and B are matrices,
and Σ and Ω are variance-covariance matrices. As denoted here all the matrices are time
invariant, but that is not necessary. Given $\mathbf{y}_t$, $t=1,2,\ldots,T$, and the values of $\mathbf{n}_0$, $\mathbf{L}$, $\mathbf{B}$, Σ,
and Ω, the conditional distribution of $\mathbf{n}_t$, which is multivariate normal, can be determined
using an algorithm known as the Kalman filter. The Kalman filter also yields the calculated
value of the likelihood (the joint marginal distribution of $\mathbf{y}_t$, $t=1,2,\ldots,T$), which can then, in
principle, be used to estimate unknown parameters of the transition and variance-covariance
matrices. However, in practice there are considerable restrictions on the estimability of
the parameters, and potentially high correlations between estimates of Σ and Ω [15]. The
notation $\mathbf{L}$ for the state transition matrix was selected to suggest the notion of a Leslie
matrix (15.9) as SSM extensions of MPMs are not uncommon (Newman [48], Sullivan [61],
and see the gray whale example in section 6.4.2.2 of Newman et al. [47]).

15.7.2   Non-normal, nonlinear SSMs

The NDLM structure is often too constricting and unrealistic for population dynamics
modeling. More realistic state-space models can on occasion be "shoe-horned" into the
NDLM framework by a mathematical transformation of states or observations, e.g., a log
transformation, and thus allow usage of the Kalman filter. For example, Dennis et al. [15]
used a stochastic Gompertz model for the state process distribution.

$$
n_t | n_{t-1} = \lambda n_{t-1}^{1+\alpha} \exp \epsilon_t
$$

where $\alpha \leq 0$ and $\epsilon_t \sim \mathrm{Normal}(0, \sigma_\epsilon^2)$. A natural log transform yields a linear normal state
distribution.

$$
\ln(n_t) | \ln(n_{t-1}) \sim \mathrm{Normal}\left(\ln(\lambda) + (1+\alpha)\ln(n_{t-1}), \sigma_\epsilon^2\right)
$$

Another way to modify an otherwise non-normal, and perhaps nonlinear SSM, into a NDLM
approximation is to work with just the first two moments of the state process distribution
and then use the mean and covariance structure as the normal mean vector and covariance
matrix. Newman [48] and Newman et al. [47] give examples of such substitutions. A sim-
plistic univariate example is to suppose that a scalar valued state n_t is Binomial(n_{t-1}, ϕ_t),
where ϕ_t is the survival probability, perhaps a function of covariates. The conditional ex-
pected value of n_t is of course $\phi_t n_{t-1} \equiv L_t n_{t-1}$, and the conditional variance is $n_{t-1}\phi_t(1-\phi_t)$
$\equiv Q_t$. Other, perhaps somewhat slight, departures from the NDLM formulation can be ac-
commodated by Taylor series transformations of the process, using the Extended Kalman
Filter [EKF; 19]. A more recent alternative to the EKF, which has been shown to have at
least equal and often far superior performance [17, p. 236] is the Unscented Kalman Filter
[31].

Computer intensive Monte Carlo methods such as Markov chain Monte Carlo [MCMC,
25] and Sequential Monte Carlo [SMC 16] offer the ultimate flexibility for fitting nonlinear,

non-normal SSMs. With the MC procedures applied to such SSMs, Bayesian inference has been the dominant approach, but not always [see 12, 28, for exceptions]. One of the first ecological applications using MC methods was by Meyer and Millar [45], who used the program BUGS (Bayesian inference Using Gibbs Sampling) to fit an SSM with scalar states and observations. The state was scaled biomass ($p_t = B_t/K$), rather than abundance, where biomass (B_t) was divided by carrying capacity, K, thus $0 < p_t \leq 1$), and the observation was a biased measure of scaled biomass, an index (y_t):

$$p_t|p_{t-1} \sim \text{Lognormal}\left(\ln\left(p_{t-1} + rp_{t-1}(1-p_t) - \frac{C_{t-1}}{K}\right), \sigma_p^2\right)$$

$$y_t|p_t \sim \text{Lognormal}\left(\ln\left(qKp_t\right), \sigma_o^2\right)$$

Thus the SSM was intrinsically nonlinear (no transformation of the state would linearize the mean structure) and non-normal.

15.7.3 Hierarchical and continuous time SSMs

An extension of SSMs is a hierarchical state-space model (HSSM). A general formulation for an HSSM in a Bayesian framework is the following

$$\text{Prior distribution} \quad : \quad \pi(\eta, \omega) \tag{15.23}$$

$$\text{Stochastic variation in parameter} \quad : \quad h(\Theta_t, \eta) \tag{15.24}$$

$$\text{State process model} \quad : \quad g_t(\mathbf{n}_t|\mathbf{n}_{t-1}, \Theta_t) \tag{15.25}$$

$$\text{Observation model} \quad : \quad f_t(\mathbf{y}_t|\mathbf{n}_t, \Omega) \tag{15.26}$$

where π, h, g_t, and f_t denote probability distribution functions. Newman and Lindley [49] used Sequential Monte Carlo to fit a Bayesian HSSM to salmon data which included both environmental and demographic stochasticity. The environmental stochasticity was modeled as above with separate distributions for year-specific survival and maturation probabilities. Demographic stochasticity was incorporated in the state process equations using multinomial distributions to reflect between individual variation in survival and maturation (although given the population size, the influence of demographic stochasticity on the results was likely minimal).

Durbin and Koopman [17] discuss continuous time SSMs for a couple cases including what is called a continuous time local level SSM. Here $n(t) = n(0) + \sigma_\epsilon \omega(t)$, where ω_t arises from a Brownian motion process, which means $\omega(0)=0$, $\omega(t) \sim \text{Normal}(0,t)$ for $0 < t < \infty$, and "jumps" or increments without common endpoints are independent, e.g., $\omega(2) - \omega(1)$ is independent of $\omega(4) - \omega(3)$. For an ecological application of continuous time SSMs see Johnson et al. [29] who model the location of marine mammals using telemetry data.

15.8 Concluding Remarks

15.8.1 Omissions and sparse coverage

Continuous time demographic models have been largely ignored here, excepting the Lotka-Volterra predator-prey model. Williams et al. [70] provides an introduction to continuous Markov processes, including birth and death processes, and Brownian motion in the context of models for animal populations. Differences in the ecological dynamics of discrete time and continuous time models are examined by Gurney and Nisbet [27]. Durbin and Koopman [17] and Johnson et al. [29] are references for continuous time SSMs.

Some aspects of ecological theory which have demographic implications that were omitted include fitness, adaptation, and mutation. Effective population size, N_e, of an existing population, here defined as the minimum number of individuals necessary in a hypothetical population that would represent existing populations ability to retain the genetic diversity present, is an important concept for endangered species, and methods for calculating N_e were not addressed. Coverage of the demography of multiple populations, communities, and ecosystems was scanty, and measures of community structure such as species richness and models for changes in such measures were not mentioned at all. Demographic modeling of ecosystems has been popular in fisheries [8, 67, 68] with Ecopath with Ecosim and Ecospace the leading software.

15.8.2   Recommended literature

For MPMs, Caswell [6] remains an outstanding reference with near encyclopedic coverage of material to 2001. For stochastic MPMs, the Tuljapurkar [64] book is a classic.

For IPMs, there are two "How To" papers, Rees et al. [53] and Merow et al. [43] which provide the basic components of IPMs, ways of estimating the kernel components, and ways of making the projections (using numerical integration methods). The original paper [18] includes detailed discussion of the advantages of IPMs over MPMs, while Ellner and Rees [20] include detailed examples of stable population analyses often done with MPMs. More sophisticated and integrated IPM fitting and projection approaches are described by Ghosh et al. [24].

For IBMs, Grimm and Railsback [26] provide a book length treatment, with DeAngelis and Grimm [14] a more recent overview paper. Roughgarden [56] gives an alternative perspective on the definition of and uses of IBMs, viewing agent-based models as a special case, for example.

For SSMs, Durbin and Koopman [17] is an extremely thorough book length treatment of SSMs. Two thirds of the book covers linear SSMs, including classical treatment with the Kalman algorithms and extensions. The remainder discusses nonlinear, non-normal SSMs including special cases and quite general formulations that are typically fit by Monte Carlo procedures. Specific focus on the use of SSMs for population dynamics modeling is given by Newman et al. [47].

15.8.3   Speculations on future developments

Data.

The volume and complexity of data on individual organisms continues to grow as the life spans of biological monitoring programs extend, as new monitoring programs are established, and as data collection technology advances. Electronic monitoring devices, e.g., radio tag collars, acoustic tags, tags that record the diving depths of marine animals, provide increasingly fine temporal and spatial resolution information on individual animal movement. Chemical analyses of organisms yield more information about individual life histories, e.g., chemical analyses of bony structures in fish, such as otoliths, can pinpoint birth place and migration paths [59]. Environmental DNA (eDNA) is an emerging tool for indirectly detecting species presence [62]. Remote sensing is adding increasingly detailed data abiotic environments. In short, "attribute vectors" [66] for individuals, populations, and abiotic and biotic environments are getting longer and longer.

Model formulations.

Appreciation of the need to separately model process variation (environmental and demographic stochasticity) and observation noise (e.g., sampling errors) will increase. Consequently, formulation of SSMs, and, more generally, hierarchical models [33, 35], for demographic data will increase. Extensions of MPMs, IPMs, and IBMs that explicitly distinguish both types of variation will likely become more common, too.

Hierarchical extensions of MPMs within the normal dynamic linear model framework of SSMs date back to the 1990s, e.g., Sullivan [61] and Newman [48]. More recently, Newman et al. [47], in an application to the Eastern North Pacific gray whales (*Eschrichtius robustus*) population, contrasted an MPM with observation error only with a NDLM extension. Differences in some of the parameter estimates were considerable, e.g., juvenile survival probability was estimated to be 0.9999 (upper bound) for the observation error only model and 0.8281 for the SSM. Advances in model fitting procedures lessen the need to restrict process models to linear formulations, implicit to MPMs, with additive normal (or multiplicative lognormal) distributions. More biologically realistic nonlinear, and non-Gaussian formulations may make applications in the MPM framework less common. However, the MPM structure will remain valuable for formulating approximate deterministic skeletons underlying more realistic models [5].

For IPMs and IBMs, process and observation uncertainty can be readily partitioned and accounted for by computer simulation. With IPMs, bootstrapping the kernel density components yields measures of parameter estimate uncertainty as well as between animal variation. For example, uncertainty about parameters of the growth density model, $g(z\prime_{t+1}|z_t)$, in the survival/growth kernel (15.19), can be assessed by resampling the longitudinal data on sizes to generate a bootstrapped distribution of growth densities. For a given fitted growth density model, simulated variation of individual sizes around the expected size at time $t+1$ reflects demographic variation. For IBMs, computer simulation of between individual variation and parameter uncertainty can be carried out within a designed experiment structure to (a) determine the relative import of specific factors on the model predictions and (b) quantify the degree of uncertainty in model predictions.

Model fitting.

Extended attribute vectors for increasingly large numbers of individuals, along with increasingly complex demographic model formulations, necessitate increasingly complex model fitting procedures. The pace of development for fitting such models is rapid and the variety of model fitting options available is increasing. Here we focus on options for fitting dynamic hierarchical demographic models.

Mathematical integration and numerical optimization are at the heart of hierarchical model fitting procedures, with the integration being over the unobserved state process. In special cases, algorithms exist for analytic evaluation of the integrals, e.g., NDLMs and the Kalman filter. As discussed in section 15.7.2, numerical approximations to nonlinear, but Gaussian, population dynamics models yield models amenable to such analytic solutions. For general hierarchical dynamic models approximate analytic solutions to the integration problems include the Laplace approximation [63] and the Integrated Nested Laplace Approximation [INLA, 70]. The software packages, AD Model Builder [ADMB, 22] and Template Model Builder (https://github.com/kaskr/adcomp/), use Laplace approximations to integrate over the state process to yield the likelihood and then automatic differentiation for calculating maximum likelihood estimates of the parameters. Widely used and well established software for carrying out the integration using Monte Carlo procedures such as MCMC and sequential Monte Carlo (section 15.7.2 includes WinBUGS [41] and JAGS [52]. Two recent software additions are NIMBLE [13] and the R package pomp, both of which allow users to choose from a variety of computer intensive model fitting procedures. NIMBLE extends the BUGS software and allows estimation within Bayesian or likelihood frameworks. The R package `pomp`, for "partially observed Markov processes" , contains a variety of procedures for fitting state-space models, with including "sequential Monte Carlo, iterated filtering, particle Markov chain Monte Carlo, approximate Bayesian computation, maximum synthetic likelihood estimation, nonlinear forecasting, and trajectory matching" [34].

Bibliography

[1] Borchers, D. L., Buckland, S. T., and Zucchini, W. (2002). *Estimating animal abundance: closed populations*, volume 13. Springer Science & Business Media.

[2] Botkin, D. B., Janak, J. F., and Wallis, J. R. (1972). Some ecological consequences of a computer model of forest growth. *The Journal of Ecology*, pages 849–872.

[3] Brooks, M. E., Mugabo, M., Rodgers, G. M., Benton, T. G., and Ozgul, A. (2015). How well can body size represent effects of the environment on demographic rates? disentangling correlated explanatory variables. *Journal of Animal Ecology*.

[4] Buckland, S. T., Anderson, D. R., Burnham, K. P., Laake, J. L., Borchers, D., and Thomas, L. (2001). *Introduction to distance sampling estimating abundance of biological populations*. Oxford University Press.

[5] Buckland, S. T., Newman, K. B., Fernández, C., Thomas, L., and Harwood, J. (2007). Embedding population dynamics models in inference. *Statistical Science*, pages 44–58.

[6] Caswell, H. (2001). *Matrix population models*. Wiley Online Library.

[7] Childs, D. Z., Sheldon, B. C., and Rees, M. (2016). The evolution of labile traits in sex-and age-structured populations. *Journal of Animal Ecology*, 85(2):329–342.

[8] Christensen, V. and Pauly, D. (1992). Ecopath iia software for balancing steady-state ecosystem models and calculating network characteristics. *Ecological modelling*, 61(3):169–185.

[9] Cohen, J. E. (1976). Ergodicity of age structure in populations with markovian vital rates, i: countable states. *Journal of the American Statistical Association*, 71(354):335–339.

[10] Companies, M.-H. (2005). *Random House Webster's college dictionary*. Random House Incorporated.

[11] Cuddington, K. and Yodzis, P. (2002). Predator-prey dynamics and movement in fractal environments. *The American Naturalist*, 160(1):119–134.

[12] De Valpine, P. (2003). Better inferences from population-dynamics experiments using monte carlo state-space likelihood methods. *Ecology*, 84(11):3064–3077.

[13] de Valpine, P., Turek, D., Paciorek, C. J., Anderson-Bergman, C., Lang, D. T., and Bodik, R. (2015). Programming with models: writing statistical algorithms for general model structures with nimble. *arXiv preprint arXiv:1505.05093*.

[14] DeAngelis, D. L. and Grimm, V. (2013). Individual-based models in ecology after four decades. *F1000prime reports*, 6:39–39.

[15] Dennis, B., Ponciano, J. M., Lele, S. R., Taper, M. L., and Staples, D. F. (2006). Estimating density dependence, process noise, and observation error. *Ecological Monographs*, 76(3):323–341.

[16] Doucet, A. d. F. and Gordon, N. (2001). *Sequential Monte Carlo methods in practice.* Springer-Verlag.

[17] Durbin, J. and Koopman, S. J. (2012). *Time series analysis by state space methods.* Number 38 in Oxford Statistical Science Series. Oxford University Press.

[18] Easterling, M. R., Ellner, S. P., and Dixon, P. M. (2000). Size-specific sensitivity: applying a new structured population model. *Ecology*, 81(3):694–708.

[19] Einicke, G. A. and White, L. B. (1999). Robust extended kalman filtering. *IEEE Transactions on Signal Processing*, 47(9):2596–2599.

[20] Ellner, S. P. and Rees, M. (2006). Integral projection models for species with complex demography. *The American Naturalist*, 167(3):410–428.

[21] Elzinga, C. L., Salzer, D. W., Willoughby, J. W., and Gibbs, J. P. (2009). *Monitoring plant and animal populations: a handbook for field biologists.* John Wiley & Sons.

[22] Fournier, D. A., Skaug, H. J., Ancheta, J., Ianelli, J., Magnusson, A., Maunder, M. N., Nielsen, A., and Sibert, J. (2012). Ad model builder: using automatic differentiation for statistical inference of highly parameterized complex nonlinear models. *Optimization Methods and Software*, 27(2):233–249.

[23] Furstenberg, H. and Kesten, H. (1960). Products of random matrices. *The Annals of Mathematical Statistics*, 31(2):457–469.

[24] Ghosh, S., Gelfand, A. E., and Clark, J. S. (2012). Inference for size demography from point pattern data using integral projection models. *Journal of agricultural, biological, and environmental statistics*, 17(4):641–677.

[25] Gilks, W., Richardson, S., and Spiegelhalter, D. (1996). *Markov chain Monte Carlo in practice.* Chapman & Hall Nueva York.

[26] Grimm, V. and Railsback, S. F. (2013). *Individual-based modeling and ecology.* Princeton university press.

[27] Gurney, W. and Nisbet, R. M. (1998). *Ecological dynamics.* Oxford University Press.

[28] Ionides, E., Bretó, C., and King, A. (2006). Inference for nonlinear dynamical systems. *Proceedings of the National Academy of Sciences*, 103(49):18438–18443.

[29] Johnson, D. S., London, J. M., Lea, M.-A., and Durban, J. W. (2008). Continuous-time correlated random walk model for animal telemetry data. *Ecology*, 89(5):1208–1215.

[30] Judson, O. P. (1994). The rise of the individual-based model in ecology. *Trends in Ecology & Evolution*, 9(1):9–14.

[31] Julier, S. J. and Uhlmann, J. K. (2004). Unscented filtering and nonlinear estimation. *Proceedings of the IEEE*, 92(3):401–422.

[32] Kalman, R. E. (1960). A new approach to linear filtering and prediction problems. *Journal of basic Engineering*, 82(1):35–45.

[33] Kery, M. and Royle, J. (2016). *Applied Hierarchical Modeling in Ecology: Analysis of distribution, abundance and species richness in R and BUGS, Vol. 1.* Academic Press.

[34] King, A., Nguyen, D., and Ionides, E. (2016). Statistical inference for partially observed markov processes via the r package pomp. *Journal of Statistical Software.*

[35] King, R., Morgan, B., Gimenez, O., and Brooks, S. (2009). *Bayesian analysis for population ecology.* CRC Press.

[36] Lefkovitch, L. (1965). The study of population growth in organisms grouped by stages. *Biometrics*, pages 1–18.

[37] Leslie, P. H. (1945). On the use of matrices in certain population mathematics. *Biometrika*, 33(3):183–212.

[38] Levins, R. (1969). Some demographic and genetic consequences of environmental heterogeneity for biological control. *Bulletin of the Entomological society of America*, 15(3):237–240.

[39] Lewis, E. (1942). On the generation and growth of a population. *Sankyha*, pages 93–96.

[40] Liu, J. and Ashton, P. S. (1995). Individual-based simulation models for forest succession and management. *Forest Ecology and Management*, 73(1):157–175.

[41] Lunn, D. J., Thomas, A., Best, N., and Spiegelhalter, D. (2000). Winbugs-a bayesian modelling framework: concepts, structure, and extensibility. *Statistics and computing*, 10(4):325–337.

[42] Margoluis, R., Stem, C., Swaminathan, V., Brown, M., Johnson, A., Placci, G., Salafsky, N., and Tilders, I. (2013). Results chains: a tool for conservation action design, management, and evaluation. *Ecology and Society*, 18(3):22.

[43] Merow, C., Dahlgren, J. P., Metcalf, C. J. E., Childs, D. Z., Evans, M. E., Jongejans, E., Record, S., Rees, M., Salguero-Gómez, R., and McMahon, S. M. (2014). Advancing population ecology with integral projection models: a practical guide. *Methods in Ecology and Evolution*, 5(2):99–110.

[44] Metcalf, C., Graham, A., Martinex-Bakker, M., and Childs, D. (2016). Opportunities and challenges of integral projection models for modelling hostparasite dynamics. *Journal of Animal Ecology*, 85:343–355.

[45] Meyer, R. and Millar, R. B. (1999). Bugs in bayesian stock assessments. *Canadian Journal of Fisheries and Aquatic Sciences*, 56(6):1078–1087.

[46] Morris, W. F., Doak, D. F., et al. (2002). Quantitative conservation biology. *Sinauer, Sunderland, Massachusetts, USA.*

[47] Newman, K., Buckland, S., Morgan, B., King, R., Borchers, D., Cole, D. J., Besbeas, P., Gimenez, O., and Thomas, L. (2014). *Modelling population dynamics: Model formulation, fitting and assessment using state-space methods.* Springer.

[48] Newman, K. B. (1998). State-space modeling of animal movement and mortality with application to salmon. *Biometrics*, pages 1290–1314.

[49] Newman, K. B. and Lindley, S. T. (2006). Accounting for demographic and environmental stochasticity, observation error, and parameter uncertainty in fish population dynamics models. *North American Journal of Fisheries Management*, 26(3):685–701.

[50] Nichols, J. D., Johnson, F. A., and Williams, B. K. (1995). Managing north american waterfowl in the face of uncertainty. *Annual review of ecology and systematics*, pages 177–199.

[51] North, P. M. and Morgan, B. J. (1979). Modelling heron survival using weather data. *Biometrics*, pages 667–681.

[52] Plummer, M. et al. (2003). Jags: A program for analysis of bayesian graphical models using gibbs sampling. In *Proceedings of the 3rd international workshop on distributed statistical computing*, volume 124, page 125. Technische Universit at Wien Wien, Austria.

[53] Rees, M., Childs, D. Z., and Ellner, S. P. (2014). Building integral projection models: a user's guide. *Journal of Animal Ecology*, 83(3):528–545.

[54] Reynolds, J. H., Knutson, M. G., Newman, K. B., Silverman, E. D., and Thompson, W. L. (2016). A road map for designing and implementing a biological monitoring program. *Environmental Monitoring and Assessment*, 188(7):1–25.

[55] Rose, K. A., Kimmerer, W. J., Edwards, K. P., and Bennett, W. A. (2013). Individual-based modeling of delta smelt population dynamics in the upper san francisco estuary: I. model description and baseline results. *Transactions of the American Fisheries Society*, 142(5):1238–1259.

[56] Roughgarden, J. (2012). Individual based models in ecology: An evaluation, or how not to ruin a good thing. In *Philosophy of Science Assoc. 23rd Biennial Mtg (San Diego, CA); PSA 2012 Symposia*.

[70] Rue, H., Martino, S., and Chopin, N. (2009). Approximate bayesian inference for latent gaussian models by using integrated nested laplace approximations. *Journal of the royal statistical society: Series b (statistical methodology)*, 71(2):319–392.

[58] Schmitz, O. J. and Booth, G. (1997). Modelling food web complexity: the consequences of individual-based, spatially explicit behavioural ecology on trophic interactions. *Evolutionary Ecology*, 11(4):379–398.

[59] Secor, D. H., Henderson-Arzapalo, A., and Piccoli, P. (1995). Can otolith microchemistry chart patterns of migration and habitat utilization in anadromous fishes? *Journal of experimental marine Biology and Ecology*, 192(1):15–33.

[60] Smith, E. P. (2002). Baci design. *Encyclopedia of environmetrics*.

[61] Sullivan, P. J. (1992). A kalman filter approach to catch-at-length analysis. *Biometrics*, pages 237–257.

[62] Thomsen, P. F. and Willerslev, E. (2015). Environmental dna–an emerging tool in conservation for monitoring past and present biodiversity. *Biological Conservation*, 183:4–18.

[63] Tierney, L. and Kadane, J. B. (1986). Accurate approximations for posterior moments and marginal densities. *Journal of the american statistical association*, 81(393):82–86.

[64] Tuljapurkar, S. (1990). *Population dynamics in variable environments.* Springer-Verlag.

[65] Tuljapurkar, S. and Orzack, S. H. (1980). Population dynamics in variable environments i. long-run growth rates and extinction. *Theoretical Population Biology*, 18(3):314–342.

[66] Van Winkle, W., Rose, K. A., and Chambers, R. C. (1993). Individual-based approach to fish population dynamics: an overview. *Transactions of the American Fisheries Society*, 122(3):397–403.

[67] Walters, C., Christensen, V., and Pauly, D. (1997). Structuring dynamic models of exploited ecosystems from trophic mass-balance assessments. *Reviews in fish biology and fisheries*, 7(2):139–172.

[68] Walters, C., Pauly, D., and Christensen, V. (1999). Ecospace: prediction of mesoscale spatial patterns in trophic relationships of exploited ecosystems, with emphasis on the impacts of marine protected areas. *Ecosystems*, 2(6):539–554.

[69] Weiss, L., Pfestorf, H., May, F., Körner, K., Boch, S., Fischer, M., Müller, J., Prati, D., Socher, S. A., and Jeltsch, F. (2014). Grazing response patterns indicate isolation of semi-natural european grasslands. *Oikos*, 123(5):599–612.

[70] Williams, B. K., Nichols, J. D., and Conroy, M. J. (2002). *Analysis and management of animal populations.* Academic Press.

16

Statistical Methods for Modeling Traits

Matthew E. Aiello-Lammens

Pace University, New York

John A. Silander, Jr.

University of Connecticut

CONTENTS

16.1 Introduction

For more than 200 years ecologists have appreciated that an organism's phenotypic traits dictate its interactions with the physical environment and other organisms [28, 62]. Community ecologists in particular have long appreciated that an examination of variation in species traits expressed in different environments can offer insights into patterns of occurrences, assuming environmental factors dictate which traits are best suited for resource acquisition (*sensu* [32]) and hence species survival and reproduction. A frequently used analogy is that of a 'filter', whereby environmental conditions are said to 'filter' out a set of species from a regional species pool, resulting in the community observed at a particular location (e.g., [76]). More recently, it's been noted that a focus on species traits, rather than species identity, can help alleviate challenges associated with reliance on any taxonomic naming systems [47]. After all, selection acts on a species phenotype, not its identity. This has the advantage of allowing for application of trait-based analyses on collected data sets of community composition with observations that have incomplete or inconsistent names. This is particularly useful in the age of large, but often incomplete data sets (e.g., [31]). Applications of trait-based modeling, or trait-based ecology, have been proposed as a promising mechanism to advance the field of community ecology, leading to general principles that can be applied to make predictions regarding community assembly and/or ecosystem function [38, 47, 76]. Moving in this direction will no doubt increase the utility of community ecology in solving pressing ecological problems associated with changing environmental conditions globally.

Another reason that trait-based ecology has seen utility throughout the history of ecology is that it provides a way to simplify the complexities entailed in observing multiple species. As noted by [21] "Species traits are useful in searching for functional types (combination of attributes) in order to reduce the diversity of species to a diversity of function". In part, this idea is the conceptual underpinning of dynamical global vegetation models. At present, there remains huge challenges in predicting with confidence the distributions of all plant species on earth, and thus precise community compositions (though strides are being made in this direction - e.g. the Botanical Information and Ecology Network and Map of Life projects). However, predicting species' trait distributions given environmental conditions may be much more tenable and yield better large-scale (e.g. global) ecological community maps [71].

Though great advances have been made in trait-based ecology, there are limitations to the utility of current methods. As with any analysis, care must be taken when using a trait-based approach to make sure that the analytical method chosen matches well with the question being explored and the data available. In many situations it may be difficult to separate the effects of physico-chemical environmental filtering from the influences of biotic interactions, leading to misleading or invalid interpretations of the results of some trait-based analyses [9, 35]. Further, concerns about the theoretical underpinnings of trait-based ecology have been noted [64], as well as the fact that several of the fundamental assumptions of the field continue to lack adequate empirical support. These assumptions include the claim that functional traits are directly associated with individual fitness, that intra-specific variability in trait values can be ignored when considering functional traits in a community context, and that functional trait values can be predicted based on measured environmental conditions. Despite these concerns, there is a general consensus that trait-based ecology offers a promising path toward increasing the predictive abilities of community ecology [47] and ecosystem function [76]. Additionally, advances in trait-based analytical methods, including the use of statistical, model-based approaches [38, 74, 75], are beginning to address some of the major challenges outlined above.

In this chapter, we provide an overview of major statistical methods for trait-based ecological modeling. We begin with a description of the data commonly used in trait-based analyses and highlight some issues to consider when collecting these data. We then discuss common 'classic' trait-based analytical methods, which have been predominantly algorithmic. These methods are demonstrated using two data sets, one simulated and one from field collections, with analyses carried out using the R statistics language [60]; all code can be acquired at https://github.com/mlammens/Trait-Modeling. We conclude by briefly describing some of the advances being made in the move toward statistical, model-based approaches. The methods we present in this chapter are described with plant communities in mind and the examples are botanical, reflecting our own research foci. We must note that historically trait-based approaches to understand patterns of occurrences have been developed in both the plant and animal ecology literature, however there are some differences in terminology and research objectives. In some respects, this has lead to parallel development of two bodies of literature. Increasing connections between trait-based analyses in plant and animal research will surely benefit both fields.

16.1.1 What Are We Modeling?

We must begin by considering exactly *what* we intend to model. Are we interested in traits and/or the statistical distribution of traits, or rather traits as they are useful for either understanding or predicting some ecological pattern or process? Our goal may be to understand trait-by-environment relationships [52], trait-trait relationships [49, 79], the extent and causes of trait variability [26], or to impute trait values [67]. On the other hand, we may be more interested in using traits as a source of information that helps us to predict species occurrences (e.g. [59]), understand processes of community assembly (e.g., [2]), or identify trait-by-environment relationships, as mediated by species occurrences (e.g., [7, 18]). In simple modeling terms, we should first acknowledge if we are treating trait measurements as response variables or predictor variables. The methods we focus on here are primarily associated with determining trait-by-environment relationships, but can be useful in addressing several questions in trait ecology.

16.1.2 Traits Versus "Functional Traits"

Throughout most of this chapter, we use the terms 'trait' and 'functional trait' synonymously. However, it is worth distinguishing these two briefly. Violle and colleagues [73] define a trait as "any morphological, physiological or phenological feature measurable at the individual level, from the cell to the whole-organism level, without reference to the environment or any other level of organization." A trait **attribute** is a measurement collected for an individual at a given place and time. The average (or another appropriate statistical measure) of a group of attributes can be considered the trait value for a population. From a biological perspective, We can define a population as a group of individuals of the same species, a definition useful for investigating trait by species relationships. Note that species do not have specific trait values, but rather aggregations or distributions of trait attributes. While this is straightforward for quantitative traits, this is also the case for categorical trait values. For example, individual plants may show variation in their degree of serotiny, but a species might be classified as "serotinous" (i.e., releasing seeds in response to fire) more generally.

A 'functional trait' is a trait that indirectly effects an organism's fitness [73]. The degree to which a functional trait is related to fitness can vary, from a trait such as leaf photosynthetic rate, which is strongly related to resource acquisition, to a trait such as specific leaf area, which is positively associated with some measures of photosynthetic rate [56]. A distinction that is often used to describe the extremes of this spectrum is "hard" versus "soft" functional traits [27, 77]. Most studies reporting either novel ecological findings resulting from trait-based analyses, or methodological advances in this field typically use "soft" trait data. Often, though not exclusively, soft traits can also be "integrative", or "composite", traits. That is, they are traits that integrate a large number of physiological processes, which in turn effect fitness. Given the lack of firm evidence on the connections between functional traits (soft traits in particular) and organismal fitness [64], there is a need for further research that applies these methods to traits with verified fitness consequences.

16.2 Overview of Data for Trait Modeling

Morphological measurements of trait values taken on individual organisms are the most basic kind of data for trait-based analyses. However, replicate sample data sufficient to estimate statistical distributions are difficult to come by, because they require intensive field sampling that is beyond the scope of most research projects. This is particularly true as the spatial scale of any research question increases, for it is often impractical to intensively sample individuals beyond a regional scale. Database projects combining data from multiple small-scale research projects, such as the Global Plant Network (GLOPNET) database of plant traits [79], the TRY database of plant traits [31], and the Botanical Information and Ecology Network (BIEN) database of New World plant occurrences and trait values [44], are beginning to address this challenge (e.g., [16]). Nevertheless, many established and novel trait modeling methods address this by ignoring intraspecific variation, assigning an aggregate trait value, usually the mean, to specific species. This aggregation is particularly prevalent in published analytical approaches connecting community assembly with trait-by-environment relationships. This is justified simply based on the assumption that interspecific trait variation is much greater than intraspecific variation, but there is evidence that this is not always the case (e.g., [12, 51]). Such assumptions may be appropriate in the case of analyses focusing on assembly of groups of species of the same functional type, for example, temperate forest tree species [66], but this is an area of potential further investigation. Another justification for ignoring intraspecific variation is that in systems with very high species turnover, interspecific variation should be greater, but again, further investigations into this claim are needed.

Information on environmental conditions at the geographic locations where trait measurements were made and the abundance of individuals at these locations are also frequently used in trait modeling. In many cases, particularly for studies focused on community assembly, information on the abundance (or simply presence/absence) of species in a defined plot is available, as are associated environmental conditions at these sample locations. However, environmental conditions at the locations where trait measurements were taken may also be missing. These data can be separated into three tables, or matrices. Following the terminology of [18], these tables are:

- **L** - an n x p table of abundance (or presence/absence) values for p species at n sites.
- **R** - an n x m table of values for m environmental variables at n sites.
- **Q** - an p x s table of values for s traits for p species.

Note that in this case, each species is assigned a single trait value and intra-specific trait variation is ignored, as noted above. Most established trait-based community analyses are based on different ways of associating these three data tables, as demonstrated below (and see [34] for a review).

The ideal data set for investigating community level processes via traits is one where traits are measured on all individuals in multiple plots spanning environmental gradients. Such data could be readily analyzed using generalized linear mixed-effect models (e.g., an expansion on methods used in [29]). However, complete sampling like this is rare (see [4] as an example). A more realistic scenario is that there is some amount of replication at the species level and that the locations where trait sampling was done could be associated with environmental conditions. In such cases trait values are typically aggregated at the species level (e.g., [14]). However, in some cases are used to examine the partitioning of trait variation within species may be modeled (e.g., [1]). Trait modeling methods that incorporate information on intraspecific variability are needed (see [10, 38] as examples).

16.2.1   Example data sets

We use two data sets to illustrate various ecological trait modeling methods. The first is a simulated trait data set and the second a well-studied data set on plant species abundance in South Africa.

For the simulated data, data were generated to follow the **L**, **R**, **Q** form described above for $n = 100$ sample plots, $p = 20$ species, $m = 2$ environmental gradients, and $s = 3$ traits. To generate abundance values for each plot and each species, we first created two environmental gradients, x_1 and x_2 which were evenly spaced between -2.6 and 2.6 and -1.9 and 1.9, respectively. Then abundance z for each sample plot was generated in a manner similar to that presented in [34]. We assumed a Gaussian-shaped response curve to create an abundance associated with each environmental gradient. For example, the abundance associated with the first environmental gradient is given by

$$z_1 = c \exp^{-(x_1-u)^2/2t^2} \tag{16.1}$$

where c is the maximum abundance (set to 5 here), x_1 is the first environmental gradient, u is the optimum environmental gradient value for a given species, and t is the niche breadth for each species, i.e. the width of the range of environmental conditions the species thrives in, here set to 1. The optimum environmental gradient values, u, for each species were generated by drawing 20 values along each environmental gradient axis, assuming a uniform random probability distribution where $u \in [\min(x), \max(x)]$, respectively. The resulting z values then represent species abundance values at a given environmental condition, and were then used to draw random values from a Poisson distribution with mean $\lambda = z$. Abundance values were generated independently for each environmental gradient and the total abundance for each species in each plot was determined by the sum of these values.

Trait data, y, for three traits were generated for each species assuming that trait-by-environment relationships were the same for all species. We simulated data for three commonly used traits, leaf mass area (LMA), leaf carbon to nitrogen ratio (C:N ratio), and leaf succulence. Descriptions of the ecological relevance of these traits can be found in [2], and more generally in [56]. The simulated relationships between each trait and the environmental conditions were given by

$$
\begin{aligned}
y_{LMA} &= -3.9 - 0.8x_1 + 0.5x_2 + \epsilon_1 \\
y_{C:N} &= 3.9 - 0.8x_1 - 0.5x_2 + \epsilon_2 \\
y_{Succ} &= -3.4 + 0.5x_1 + 0.5x_2 + \epsilon_3
\end{aligned} \tag{16.2}
$$

The vector of random errors, ϵ, were drawn from a multi-variate normal distribution, with variance:

$$
\begin{bmatrix}
0.2 & 0.2 & 0.0 \\
0.2 & 0.4 & 0.0 \\
0.0 & 0.0 & 0.4
\end{bmatrix}
$$

thus inducing a correlation between LMA and C:N ratio, but not with succulence. While the three traits are simulated, the general relationships among traits, and between traits and environmental factors, were based on those observed in our second, real-world data set. Further, these are traits that appear in the literature frequently and are ecological relevant: LMA is related to leaf growth rates and leaf toughness, C:N ratio is related to photosynthetic rates, and succulence is related to water storage.

By using identical trait-by-environment relationships for all species, we are assuming an extreme environmental filter on traits. This ignores several ecological and evolutionary processes that allow species co-existence and assumes that there are not different trait combinations that can result in establishment and survival in a given environment. This last assumption is not well tested, but there is evidence that it is not universally valid [39, 72]. Despite these shortcomings, our simulated data set allows for demonstrations in the situation where trait-by-environment relationships are strong and trait data is complete for all individuals sampled in plots along environmental gradients.

The second data set is from vegetation surveys carried out in the Baviaanskloof sub-region of the Cape Floristic Region (CFR) of South Africa. This region and these data are more fully described in [2], but in brief, the CFR is an area near the southwestern tip Africa that is extraordinarily rich in plant taxonomic and functional diversity, contains extensive environmental gradients over relatively short distances, and is well studied [3, 45]. In 2011 and 2012, vegetation surveys of 118, 5 x 10 m plots, were carried out where esti-mates of percent cover (a proxy for abundance) for each species in each plot were collected. Environmental conditions at the plots were either recorded in the field or extracted using the geolocations of each plot and GIS layers from [78]. These environmental conditions included minimum temperature in July (MTmin.07), maximum temperature in January (MTmax.01), solar radiation, and elevation. Trait measurements were collected from 1934 individual plants, representing 314 species, following protocols in [13]. These traits included plant height, canopy area, leaf-mass area (LMA), leaf length-to-width ratio (LWR), leaf lamina thickness, maximum leaf lamina width, leaf freshwater content (FWC), and stem freshwater content (twFWC). These measurements were taken on individuals near to, but not in, the survey plots. For each species, multiple individuals of that species were measured, and in many cases, at multiple locations. While extensive, this data set does not represent a complete sampling of all individuals in all plots. Rather, if trait values are aggregated (e.g., mean values), then this data set is comprised of an **R**, **L**, and **Q** table. Note that because the locations where trait measurements were collected was noted, there is important spa-tial information that can be used to inform trait-by-environment relationships, but using traditional analyses, this information is ignored when addressing community level questions.

16.3 Exploratory Data Analysis

Exploratory data analysis is an important step in any modeling endeavor, but particularly useful in trait modeling, where data sets are often large, relatively messy, and multivariate in nature. For the CFR data set, a simple plot of trait-by-trait relationships and trait densities demonstrated the need to log transform all the trait values. Subsequent plotting reveals that there is extensive variation in trait values and some strong correlations between traits (Figure 16.1). Similar exploratory analyses of trait-by-environment relationships are also useful, and presented later in the chapter.

16.3.1 Dimension Reduction of Trait Data

Functional traits are often correlated, particularly if we consider commonly measured traits in plant functional ecology (e.g., canopy area and height, fresh water content and leaf lamina thickness) [50]. Thus, it may be useful to apply dimension reduction techniques as part of an exploratory data analysis. This can be done with principal component analysis (PCA) or other forms of dimension reduction (principal coordinate analysis, correspondence analysis, etc.), depending on the data sets being considered Chapter 11. Later in the chapter we discuss extensions of ordination techniques that can also be used to investigate trait-by-environment relationships. Here we simply highlight its utility in identifying relationships between variables. These approaches may also help in identifying variables, either response

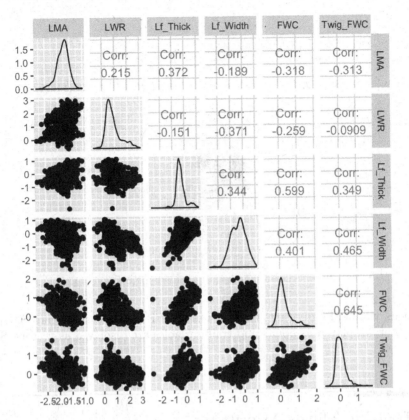

FIGURE 16.1
Trait-trait scatter plots for CFR data set. Plots on diagonal show the trait distributions and upper panels show trait-trait Pearson correlation values. All trait measurements were log transformed.

or explanatory, to include in subsequent analyses. When measurement for many traits are available, such as is the case when working with some large databases, it may be tempting to include as many traits as possible in any analysis. However, there is evidence that the dimensionality of trait-space is limited. Laughlin [37] analyzed three large trait data sets, each containing multiple traits, and using dimension reduction found at most **six dimensions** provided unique information. It behooves the analyst to use exploratory data analysis to check whether there is redundant information in the data set being considered, and if so, carefully choose variables to include based on the biology associated with the question at hand.

We applied PCA to the CFR trait measurements. The first two PCs explain 66% of the variance, while the first three PCs explain 82% of the variance. An examination of the first two principal components shows that PC1 is comprised mostly of traits associated with water storage - FWC, twig FWC, and maximum leaf width (Figure 16.2). PC2 is dominated by leaf mass area and leaf thickness. PC3 is comprised mostly of leaf length-width ratio and begins to show distinctions among FWC, twig FWC, and maximum leaf width. These results suggest that the major axes of variation are associated with a plant's ability to store water and leaf shape, which in this case is likely related to water-loss via transpiration.

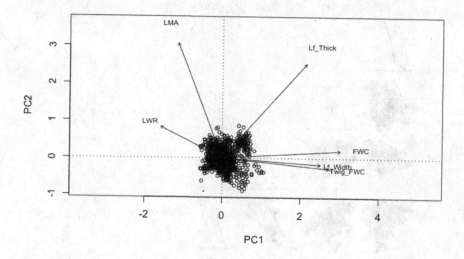

FIGURE 16.2

Multi-dimensional bi-plot of results from a PCA of CFR trait measurements. Samples (i.e., individuals) are displayed as points and trait variables are displayed as vectors. Scaling = 2.

16.4   Trait Modeling - Algorithmic

Investigations of relationships between multiple traits and multiple environmental conditions, considering multiple species are obviously multivariate in nature. While statistical theory regarding multivariate analyses is rich, in practice most multivariate methods used in ecological studies are algorithmic. That is, there is no proper statistical model specified for the observations (i.e., trait values or species abundances), but rather most analyses entail examination of measures of dissimilarity [74] and use of randomization techniques to assess whether observed associations are likely explained by chance alone [24]. In this Section we describe five commonly used algorithmic approaches to trait-based modeling: individual-level redundancy analysis (RDA), community-level redundancy analysis (CWM-RDA), community-level randomization/ permutation, fourth-corner analysis, and RLQ analysis. Later, in Section 16.5 we highlight some of the recent advances that bring model-based statistical methods to trait-based ecology. There are a plethora of possible dissimilarity measures to calculate and algorithms to apply (see [43] for an extensive review), and choices of each depend on the data being examined and the questions being asked. A more specific review of several algorithmic-based methods used to relate species-level and community-level traits to environmental conditions is presented by [34]. In their review, Kleyer et al. briefly describe multiple methods to associate the **R**, **L**, and **Q** data tables, and compare the interpretations of the results of these different methods as applied to both a simulated and real data set. Most of these methods are two-step processes, where first species presence/absence (or abundance) is related to environmental conditions, then species traits are used to explain the presence-by-environment relationships. It is important to note that in the approaches reviewed by [34], fixed trait values are assigned to each species. That is, all of the methods considered ignore intraspecific trait variability. Again, this is common with most trait-by-environment analyses.

16.4.1 Redundancy Analysis on Individual-Level Trait Data

Redundancy analysis (RDA) is a canonical ordination method [5, 43], initially described by Rao [61]. The goal of RDA is to examine the relationships between a set of predictor variables, $\mathbf{X}$, and response variables $\mathbf{Y}$, with particular emphasis on understanding how much of the variation in $\mathbf{Y}$ is described by $\mathbf{X}$. Thus, conceptually it is a multivariate analog to multi-predictor ordinary least squares regression [23]. However, mathematically it bridges canonical correlation analysis and multivariate regression [53].

As in other ordination analyses, the goal of RDA is to find a linear combination of a set of observed variables that describes a new space, with orthogonal axes where variation is maximized along each axis. The major distinction in RDA is that the ordination axes describe a space that is maximumly related to a linear combination of a set of explanatory variables, $\mathbf{X}$ [43]. That is, assume we have n observations of p variables, that comprise an $n \times p$ matrix $\mathbf{Y}$. In PCA, we set out to find a new matrix, $\mathbf{Z}$, that is a linear combination of $\mathbf{Y}$, and could write the first principal component as

$$Z_{i1} = c_{11}Y_{i1} + c_{12}Y_{i2} + ... + c_{1p}Y_{ip}$$

where c are the coefficients for each original column of $\mathbf{Y}$. In RDA, we also have a set of m predictor variables, comprising an $n \times m$ matrix $\mathbf{X}$, and wish to find

$$Z_{i1} = a_{11}X_{i1} + a_{12}X_{i2} + ... + a_{1m}X_{im}$$

where a are the coefficients for each original column of $\mathbf{X}$.

Carrying out an RDA is a multi-step process. The first step is to perform multi-predictor regression for each of the response variables individually. Following the notation of [43], this can be written as

$$\hat{\mathbf{Y}} = \mathbf{X}[\mathbf{X}'\mathbf{X}]^{-1}\mathbf{X}'\mathbf{Y} \tag{16.3}$$

where $\mathbf{X}'$ refers to the transpose of $\mathbf{X}$. Then, a matrix is constructed comprising the fitted response variables (i.e. $\hat{\mathbf{Y}}$) as determined from the separate regression analyses. PCA is then applied to the fitted values, using the covariance matrix of $\hat{\mathbf{Y}}$

$$\mathbf{S}_{\hat{\mathbf{Y}}'\hat{\mathbf{Y}}} = [1/(n-1)]\hat{\mathbf{Y}}'\hat{\mathbf{Y}} \tag{16.4}$$

Replacing $\hat{\mathbf{Y}}$ in Equation (16.4) with Equation (16.3), and reducing yields

$$\mathbf{S}_{\hat{\mathbf{Y}}'\hat{\mathbf{Y}}} = \mathbf{S}_{\mathbf{YX}}\mathbf{S}_{\mathbf{XX}}^{-1}\mathbf{S}_{\mathbf{YX}}' \tag{16.5}$$

where $\mathbf{S}_{\mathbf{YX}}$ is a $p \times m$ covariance matrix among $\mathbf{Y}$ and $\mathbf{X}$ and $\mathbf{S}_{\mathbf{XX}}$ is the $m \times m$ covariance matrix of $\mathbf{X}$. The PCA is completed by eigen decomposition of Equation (16.5), yielding a matrix of eigenvectors $\mathbf{U}$, which can then be used to find the ordination of objects either in the space of the response variables or the predictor variables, using equations

$$\mathbf{F} = \mathbf{Y_c}\mathbf{U}$$
$$\mathbf{Z} = \hat{\mathbf{Y}}\mathbf{U} \tag{16.6}$$

where $\mathbf{Y_c}$ is a centered $\mathbf{Y}$ matrix. An additional PCA of the residuals matrix, $\mathbf{Y_{res}} = \mathbf{Y} - \hat{\mathbf{Y}}$, yielding a matrix of eigenvectors $\mathbf{U_{res}}$, can also be completed. Examining any patterns or structure in the biplot of PCA on the residuals could lead to an understanding of the variance in $\mathbf{Y}$ that is not explained by $\mathbf{X}$.

There are three major results of an RDA: the redundancy statistic R^2, a measure of statistical significance of the relationship between Y and X, and a visualization of the results via a tri-plot. The redundancy statistic is analogous to the R^2 in ordinary least squares regression, and thus represents the proportion of variation in Y explained by X. It is calculated as

$$R^2_{\mathbf{Y}|\mathbf{X}} = \frac{SS(\hat{\mathbf{Y}})}{SS(\mathbf{Y})} \tag{16.7}$$

where $SS(\hat{\mathbf{Y}})$ and $SS(\mathbf{Y})$ are the sums of squares of $\hat{\mathbf{Y}}$ and $\mathbf{Y}$, respectively. An adjusted R^2 is calculated as

$$R_{adj} = 1 - (1 - R^2_{\mathbf{Y}|\mathbf{X}})\frac{(n-1)}{(n-m-1)} \tag{16.8}$$

where as above, m is the number of explanatory variables (i.e., columns of $\mathbf{X}$).

In some cases, a F statistic can be calculated and used to provide a measure of statistical significance of the relationship between $\mathbf{Y}$ and $\mathbf{X}$, but it is often the case that the data in $\mathbf{Y}$ and/or $\mathbf{X}$ do not meet the necessary assumptions for this to be valid. Thus, permutation tests are used to assess the significance of the relationship. In fact, this is the default behavior when implementing RDA in the R statistical programming language.

The results of RDA can be visualized as a tri-plot, where either $\mathbf{F}$ or $\mathbf{Z}$ from (Equation (16.6)), the response variables ($\mathbf{Y}$), and the explanatory variables ($\mathbf{X}$) are plotted in the ordination space, usually along the first two-axes. As in PCA, there are two forms of scaling of $\mathbf{U}$ that can be used to make the plot, type 1 where the eigenvectors are scaled to have unit length 1 or type 2 where each eigenvector is scaled to the length of its corresponding eigenvalue. There are a number of differences between these plots, which are explained nicely in [43] and [5], but one significant difference is that with scaling type 1, the angles between the response and explanatory variables reflect their correlations, but not the angles among response variables. With scaling type 2, on the other hand, both sets of angles, as well as those among response variables, reflect correlation.

Given an appropriate data set, we can ignore any information that might be derived from the $\mathbf{L}$ matrix (i.e., the community context), and investigate trait-by-environment relationships directly. A data set conducive to such an analysis is one that includes trait measurements taken on individual plants and measurements of environmental conditions at the location of these plants. Our CFR trait data set is a an example of this kind of data. Provided that multiple traits and multiple environmental conditions were measured, redundancy analysis (RDA) can be used to examine trait-by-environment relationships.

We applied RDA to both of our example data sets, taking advantage of the fact that we know the environmental conditions at the locations were individuals were measured. The results of RDA applied to the simulated data set confirm that there are strong trait-by-environment relationships, as indicated by an adjusted R^2 value of 0.80. Examining the RDA tri-plot (Figure 16.3), we see that the known relationships are returned. LMA and C:N ratio are negatively associated with x_1, while succulence is positively associated. LMA and succulence are positively associated with x_2, while C:N ratio is negatively associated.

Applying RDA to the CFR trait data set, we find that relatively little of the variation in traits is explained by the measured environmental conditions - the adjusted R^2 value of the RDA is 0.04. Examining the RDA tri-plot (Figure 16.4), we see that there are positive associations between FWC, twig-FWC, and maximum leaf width and MTmax.01 and MTmin.07. We also see positive associations between LMA and solar radiation and elevation.

There are two potential problems with the RDA approach as applied here. First, ignoring all information from the $\mathbf{L}$ data matrix, that is the community context, essentially ignores the effects of biotic interactions on species (co-)occurrence patterns. It could be argued that if a large number of individuals are sampled in a broad spatial area, the biotic interactions in aggregate results in noise, but no specific trends, as we would expect from physio-chemical environmental effects. This is sometimes termed the Eltonian Noise hypothesis [58], but more research is needed to establish its generality. Second, by carrying out these analyses

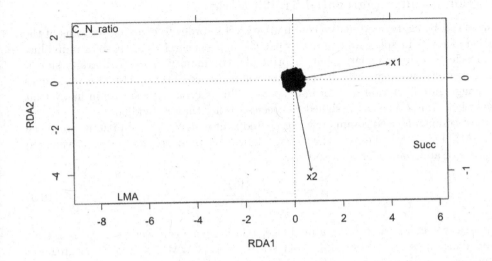

FIGURE 16.3
Tri-plot showing results of RDA of the simulated data set. Samples (i.e., individuals) are displayed as points, environmental variables (i.e., predictor variables) are displayed as vectors, and trait variables (i.e., response variables) are displayed as text labels. Scaling = 2.

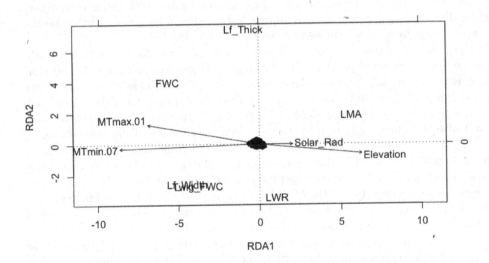

FIGURE 16.4
Tri-plot showing results of RDA of CFR data set. Samples (i.e., individuals) are displayed as points, environmental variables (i.e., predictor variables) are displayed as vectors, and trait variables (i.e., response variables) are displayed as text labels. Trait variables Lf_Width and Twig_FWC are overlapping near -5, -2. Scaling = 1.

on several species simultaneously, we may be confounding trait-by-environment with trait-by-evolution associations (i.e., phylogentic constraints).

16.4.2   Community Aggregated Trait Metrics

Trait values can be aggregated at the community level in order to examine patterns of the distribution of traits in space and/or time that are more general than those of individual species. Species can be enigmatic, but in principle, community aggregated traits should represent general trends, primarily driven by environmental conditions. Also, it is sometimes helpful to aggregate trait values if data are sparse. This is typically the case in many trait databases when trait values are assigned to species, rather than individuals.

The most commonly used community aggregated trait metric is the community weighted mean (CWM). This is calculated as the average trait value of all species in a plot, weighted by a measure of abundance for each species:

$$CWM = \frac{\sum_{i=1}^{n} W_i T_i}{\sum_{i=1}^{n} W_i} \tag{16.9}$$

where W is a weighting value (e.g., abundance, relative percent cover, etc.), T is a trait value, and i indicates the ith species. Analyses that use CWMs, or another aggregated trait value, are sometimes considered mean-field approaches. Aggregated trait values can certainly be calculated in the case where we have measurements from all individuals in each community plot, but these data are rare and such an aggregation ignores potentially interesting information on intra-specific variation, as well as variation among plots. In the case that there is complete sampling, or partial sampling, it is possible that calculating CWM using plot specific values will yield different CWMs than if these values are first aggregated at the species level [11]. Thus, the standard way of calculating CWMs using the **L** and **Q** matrices can result in relatively inaccurate or biased values. While the analyst may be constrained to the less information rich approach, they should be aware of the potential inaccuracies inherent in species-level aggregations.

In principle, the CWM should represent some optimal trait value for a plot, however this interpretation is contested and particularly problematic when a community weighted trait distribution is not unimodal and approximately normal [54]. In the case where visual data analysis confirms that these distributions are in fact unimodal and normally distributed, or readily transformed to ensure normality, then the CWM values could be interpreted as an indicator of trait values likely to 'pass through' any environmental filters constraining species occurrence [36]. That is, we should expect that the trait values most favored by environmental conditions would be most prominent in a plot, and less favorable values will be less represented, proportional to their distance from the optimum.

Few aggregation metrics beyond the CWM appear in the literature. Community mean (i.e., non-weighted) values are sometimes calculated, as is appropriate when **L** comprises of presence/absence data only. In principle, other metrics should yield interesting ecological information. For example, community weighted standard deviations may be indicative of the variation in trait values present in communities, though other community metrics address trait variation, generally referred to as functional diversity (e.g., [8, 46, 57]). Similarly, community quantile trait values may be indicative of limits to some ecological strategies. At present, an exploration of the utility of other aggregation metrics is needed.

16.4.2.1   CWM-RDA

Regression techniques can be applied to determine CWM trait-by-environment relationships, including one-at-a-time CWM trait-by-environment regression (i.e., a single predictor and response variable), multiple linear regression relating a single CWM trait value to multiple environmental variables, or multivariate multiple linear regression (e.g., RDA - as used in Section 16.4.1). Applying RDA to CWM trait values is one of the approaches reviewed

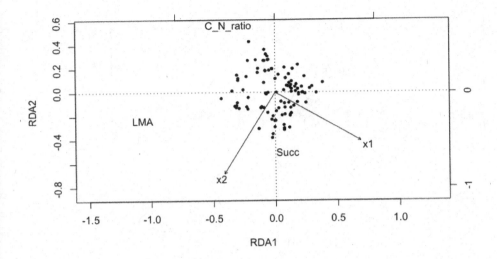

FIGURE 16.5
Tri-plot showing results of RDA of the CWM values of the simulated data set. Arrows represent environmental conditions (predictor variables), labels without arrows represent trait measurements (response variables), and points are observations of CWMs at 100 plots. Scaling = 2.

by [34], who noted it's utility in providing "a good overview of trait-by-environment relationships". As is standard with applying regression techniques, these approaches result in estimates of the trend or shape of relationships between environmental conditions and traits, and provide measures of the amount of observed trait variation explained by environmental gradients.

We applied CWM-RDA to both of our example data sets. The results from both largely concurred with those of the Individual-level RDAs. CWM-RDA applied to the simulated dataset resulted in the known associations (compare Figures 16.3 and 16.5.

The differences between individual-level RDA and CWM-RDA are more apparent for the CFR data set. Most notable is the change in the position of the environmental condition solar radiation with respect to the first two RDA axes (compare Figures 16.4 and 16.6). The results of CWM-RDA position solar radiation and elevation perpendicular to each other, in contrast to their position of being nearly in parallel in the individual-level RDA. Interestingly, the results of CWM-RDA show that LWR is more strongly positively associated with solar radiation than with elevation, a distinction that is hard to see in the individual-level RDA results. As with the individual-level RDA results, the predictor variables explain a large amount of the variation in the response variables for the simulated data set ($R^2 = 0.80$), and very little for the CFR data set ($R^2 = 0.05$). Given the latter finding, it is possible that the distinctions between the individual-level and CWM RDAs noted above can be attributed to random variation, and any biological interpretations should be tentative, pending further analysis.

16.4.2.2 CWM Randomization Approaches

Randomization or permutation methods can be used to examine whether CWM values differ from null expectations, thus allowing us to examine support for environmental filtering of species occurrences based on traits. This approach stems from a more general use of

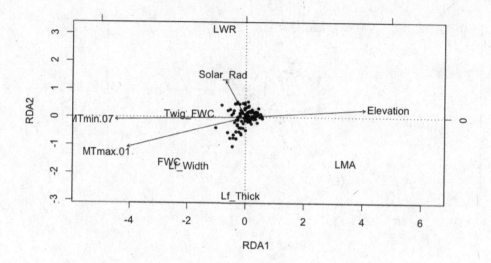

FIGURE 16.6
Tri-plot showing results of RDA of the CWM values of the CFR data set. Arrows represent environmental conditions (predictor variables), labels without arrows represent trait measurements (response variables), and points are observations of CWMs at 120 vegetation plots. Scaling = 1.

randomization methods applied in community ecology [22, 24]. The general steps involved are to randomize (permute) the original data a large number of times, calculate the desired statistic or measurement each time (here, CWM trait values for each trait in each plot), and to calculate a standardized effect size [25] as:

$$SES = \frac{X_{obs} - \bar{X}_{null}}{SD(X_{null})} \tag{16.10}$$

where SD is standard deviation.

There are several examples of using randomization approaches with functional traits to detect signals of community assembly mediated by environmental conditions (e.g., [2, 15, 66, 69]). One challenge in applying these methods is that the randomization (permutation) methods applied to the data is not necessarily consistent, and difference can lead to inconsistent results. Recent work has started to standardize this analysis via development of an R package specifically on this topic [68].

We applied a randomization technique to our example data sets where we permuted the species richness values within community plots, as implemented in the R package **picante**, [33]). The simulated data set shows strong deviations from the null distribution, which could be interpreted as evidence for environmental filtering. Of the 100 plots, for each trait nearly half of the plots had CWM values were SES was ≥ 1.96 or ≤ -1.96 (LMA = 48, C:N ratio = 44, and succulence = 41). There was no strong pattern of SES values in aggregate deviating from a distribution with mean = 0 (Figure 16.7).

Overall, our CFR example data set shows weak signals of environmental filtering (Figure 16.8), which fits with our results of individual-level RDA and CWM-RDA. Very few plots had CWM values where SES was ≥ 1.96 or ≤ -1.96 (max leaf width = 2, leaf thickness = 1, and FWC = 2). However, looking at SES values in aggregate for each trait, we see that CWM FWC and twig-FWC values are overall significantly negative (One-sample t-tests,

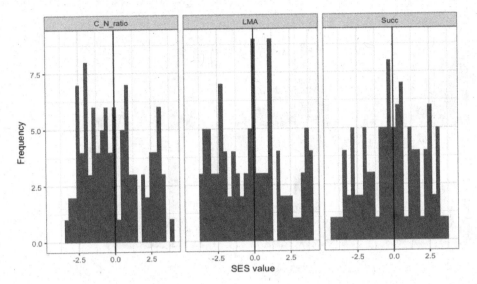

FIGURE 16.7
Histograms of SES values for each CWM trait considering 100 simulated plots and 999 randomizations of within-site species richness values (i.e., permutation within rows of the **L** matrix).

$p = 0.008$ and 0.049, respectively), suggesting that there may be some locations where plot trait values may reflect environmental filtering, thus influencing community assembly.

16.4.2.3 Concerns with CWM approaches

CWM approaches are noted as being among the most commonly applied methods in analyses of trait by environment relationships, specifically when considering a community-level context [55]. Unfortunately, they have also been shown to result in inaccurate results, either because of variation in the methods used to measure and estimate species abundances at the community level [40] or because of the influence of data standardization and weighting used in different analysis techniques [55]. With regards to the latter, Peres-Neto and colleagues [55] recently showed that a strong correlation between CWM trait values and environmental conditions can be found in a simulated dataset where species distributions were strongly influenced by environmental conditions, but trait values were randomly generated. These inconsistencies are partly due to the way trait values are weighted in the CWM approach. This is related to another potential issue with CWM analyses, which is the influence of ecological correlations that are not fully considered. For example, it is possible that trait by environment relationships differ within communities versus between communities, but CWMs aggregate the within community relationships, concealing their effects. Recent investigations of the consistency, or lack there of, of the global leaf-economic spectrum relationships [50] provide an indication that local, within-community relationships are likely to differ from those at larger spatial scales. Failing to consider these differences may result in committing ecological fallacies, yielding inaccurate interpretations of trait by environment relationships. A full investigation into the effects of ecological correlations, and the overall robustness of the CWM analysis approaches, is sorely missing at present.

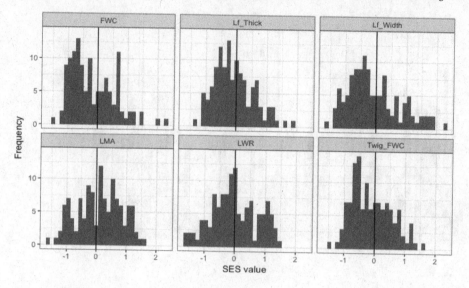

FIGURE 16.8
Histograms of SES values for each CWM trait at 120 plots and 999 randomizations of within-site species richness values (i.e., permutation within rows of the **L** matrix).

16.4.3   Fourth-corner Methods

When presented with the **R**, **L**, and **Q** matrices, it is common to view the problem of determining trait-by-environment relationships as that of 'finding' the missing corner, or **Ω**, in the matrix in Figure 16.9. Fourth-corner methods set out to 'fill' in the values of **Ω** (i.e., the Fourth-corner problem Section 16.4.3.1) or provide other means to examine trait-by-environment relationships, mediated by species abundances (i.e., RLQ analysis in Section 16.4.3.2).

16.4.3.1   Fourth-corner Problem

Conceptually, and in the case where **L** is a presence/absence matrix, the fourth-corner matrix can be calculated as:

$$\Omega = \mathbf{Q}'\mathbf{L}'\mathbf{R} \tag{16.11}$$

or

$$\Omega' = \mathbf{R}'\mathbf{L}\mathbf{Q} \tag{16.12}$$

These tables are combined to form an 'inflated data table' [42, 43]. If the values in **R** and **Q** are categorical, then the inflated table can be used to form **Ω** as a contingency table. A χ^2 or G statistic can then be calculated to test for an association between traits in **R** and environmental conditions in **Q**. However, because there is no known reference distribution for the resulting fourth-corner statistics, the resulting χ^2 (or G) cannot be interpreted by itself. Rather, matrix **L** is permuted and the statistic is recalculated with the new matrix many times, to created a null distribution. Legendre et al. [42] suggested four different ways to permute **L** and [21] and [6] presented two additional methods. These methods are outlined in Appendix B of [19] and described in greater detail, with ecological interpretations in [43]. In brief:

- Model 1: Permute species presence/absence (abundance) values independently (i.e., permute values in the columns of **L** independently).

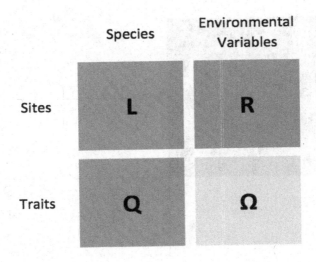

FIGURE 16.9
A visual representation of the relationships between the **R**, **L**, **Q**, and missing fourth-corner, Ω, matrices.

- Model 2: Permute the sites (i.e., shuffle rows of **L**).

- Model 3: Permute species presence/absence (abundance) values at each site independently (i.e., permute values in the rows of **L** independently).

- Model 4: Permute the site-level species assemblages (i.e., shuffle the columns of **L**).

- Model 5: Permute both the rows and columns of **L**.

- Model 6: Perform both Model 2 and Model 4 permutations and retain the higher p-value from the two tests.

The fourth-corner approach can also be used to examine relationships among quantitative trait and environment variables, as well as mixed quantitative and categorical variables. In the first case, the resulting Ω matrix will be comprised of Pearson correlation coefficients, provided values are first standardized and the resulting correlation values are divided by the total number of rows in the inflated matrix minus 1 [43]. In the second case, categorical variables must first be coded into dummy variables. In the situation where one quantitative variable and one qualitative variable are being considered, then Ω is equivalent to computing a F statistic or pseudo-F statistic [21]. Dray and Legendre [21] present a test statistic for the multivariate case where there is a combination of quantitative and categorical variables. They also provided an important expansion of the fourth-corner approach by presenting a framework to apply this method to abundance data, rather than being limited to presence/absence data. Lastly, they noted the mathematical overlaps between the fourth-corner statistics and RLQ analysis. More recently Peres-Neto and colleagues [55] proposed a new fourth-corner statistic named the Chessel fourth-corner correlation which they contend will be a replacement for past forth-corner statistics, as well as CWM approaches.

The resulting Ω matrix provides the relationships between a single trait and a single environmental condition, for all combinations of the traits and environmental conditions

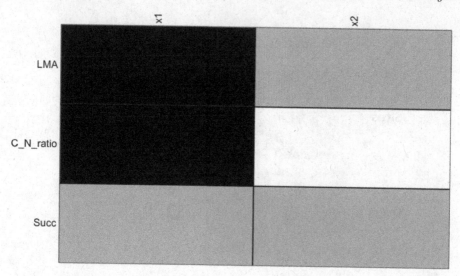

FIGURE 16.10
Visualization of the results of a Fourth-corner analysis of the simulated data. Grey boxes indicated a significant positive relationship, black boxes indicated a significant negative relationship, and white boxes indicated no significant relationship. (Compare these results to Figure 16.3)

(Pearson correlation coefficient, χ^2 statistic, or correlation ratio, depending on the variable types). The significance of these values are then determined via permutation tests, as noted above. The results can be visualized as a matrix of colored boxes coded to show positive and negative significant associations.

We applied fourth-corner analysis on both the simulated and CFR data sets. The fourth-corner analysis returned accurately the positive and negative trait-by-environment relationships of the simulated data set, showing significant positive associations between Succulence and x_1 and LMA and Succulence and x_2, and negative associations between LMA and C:N ratio and x_1 (Figure 16.10). The modeled negative relationship between C:N ratio and x_2 was not found to be significantly different from 0, but this was also the noisiest of the simulated associations. Interestingly, while the general relationships were returned, the strength of the associations ranges from -0.26 to 0.21, which is much weaker than the simulated associations, which ranged from -0.8 to 0.5 (see Equation (16.2)).

The fourth-corner analysis of the CFR data set showed that there were some significant trait-by-environment associations (Figure 16.11, though the values of these associations were relatively weak, ranging from -0.18 to 0.17. It is particularly interesting to note that FWC showed significant associations with environment, in light of the fact that this trait showed some signs of being subject to environmental filtering based on the CWM-randomization method applied above.

16.4.3.2   RLQ Analysis

One shortcoming of the fourth-corner approach described in Section 16.4.3.1 is that the results largely ignore the potential interactions among multiple traits and/or environmental conditions. Biologically speaking, this is a problem, as we may expect some degree of interactions among trait values due to fundamental trade-offs [79]. Interactions among environmental conditions are also common. (Note that based on these statements the importance and

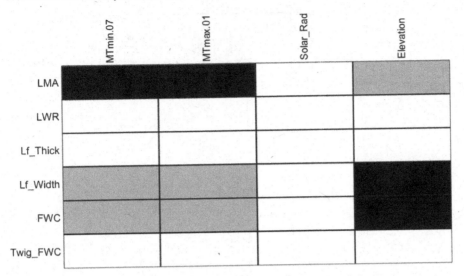

FIGURE 16.11
Visualization of the results of a Fourth-corner analysis of the CFR data set. Grey boxes indicated a significant positive relationship, black boxes indicated a significant negative relationship, and white boxes indicated no significant relationship.(Compare these results to Figure 16.4)

appropriateness of using multivariate methods when examining trait-by-environment relationships is generally assumed, however there is at least some evidence suggesting a benefit to the study of one trait at a time [70].) RLQ analysis is an ordination approach introduced by Doledec et al. [18], and since expanded upon [19, 21], that addresses this shortcoming in the fourth-corner problem analysis. As with other ordination approaches (e.g., PCA, CA, RDA), RLQ analysis allows for a visualization of these multivariate relationships.

Methodologically, RLQ analysis is an extension of co-inertia analysis [17]. Co-inertia analysis provides an alternative approach to ordination methods based on analysis of what's termed the duality diagram [20]. As described in [43] and [5], co-inertia analysis is a two-step process where a matrix of co-variances between elements of two data matrices is calculated, then decomposed to determine the eignenvalues and eigenvectors of this co-variance matrix. The points and variables of the initial two data matrices are then projected on the co-intertia axes. RLQ analysis extends this from a two-table to a three-table analysis, specifically dealing with the $\mathbf{R}$, $\mathbf{L}$, and $\mathbf{Q}$ matrices, resulting in a linear combination of the variables of $\mathbf{R}$ and $\mathbf{Q}$ that yield a maximal covariance, weighted by the species abundance information in $\mathbf{L}$ [18, 20]. Mathematically this corresponds to the generalized singular value decomposition of the Ω matrix [19]. A brief description of the associated matrix algebra is presented in Appendix B of [19], and more fully described in [18]. As noted by Kleyer et al. [34], a similar method is double Canonical Correspondence Analysis (CCA), first presented in [41]. We do not go into detail on this method here, but mention it to point out other methods the reader may investigate.

The result of an RLQ analysis is a series of plots indicating the direction and strength of relationships between species, sampling sites (i.e., plots), traits, and environmental conditions. For our purposes, we focus on the bi-plot of the relationships between trait variables and environmental conditions. We applied RLQ analysis to both of our example data set. As with the fourth-corner approach, RLQ returned the simulated trait-by-environment re-

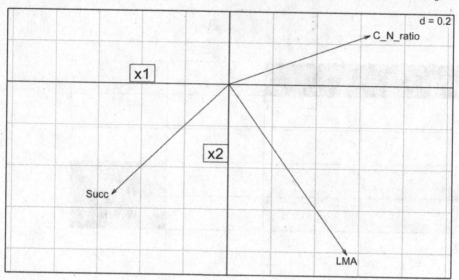

FIGURE 16.12
Bi-plot showing results of RLQ analysis of simulated data set. Environmental variables depicted as arrows and traits as boxes along the first two axes of RLQ analysis.

lationships (Figure 16.12). Succulence and LMA are positively associated with x_2. LMA and C:N ratio are negatively associated with x_1, while succulence is positively associated with x_1. The approximately equal angles between succulence and x_1 and x_2 suggests that these associations are approximately equal, which is in fact what was simulated (Equation (16.2). In this plot, we also see that succulence and C:N ratio are negatively correlated, though this was not included in the co-variance structure of these traits during simulation.

The results of RLQ analysis of the CFR data set also largely concurred with those of the fourth-corner and the CWM-RDA approaches (Figure 16.13. FWC is positively associated with MTmax.01 and MTmin.07 and negatively associated with elevation. LMA shows opposite relationships to this, and is generally negatively associated with LMA. From the fourth-corner results, we see that maximum leaf-width has the same relationships with the environmental variables as FWC, however the RLQ results suggest that these relationships may not be as strong or as direct. Note that the RLQ bi-plot is largely similar to the CWM-RDA tri-plot, though the axes are rotated and the vector lengths differ for the environmental conditions in the two plots.

A major advantage of the RLQ method over other modeling approaches, particularly the fourth-corner approach , is that the resulting tri-plot and trait and environment scores are calculated considering the multi-variate relationships between multiple traits and multiple environmental conditions. However, as with most multi-variate methods, RLQ is considered largely exploratory, and thus it does not provide a robust framework to test the significance, or importance, of associations between individual traits and environmental conditions. Further, the resulting ordination plots can be quite challenging to interpret. Dray et al. [19] recently proposed a method of combining RLQ and fourth-corner approaches. This method seems promising and is already relatively well cited, but remains an algorithmic approach and lacks some of the advantages of the novel statistical model-based approaches now emerging [63, 74].

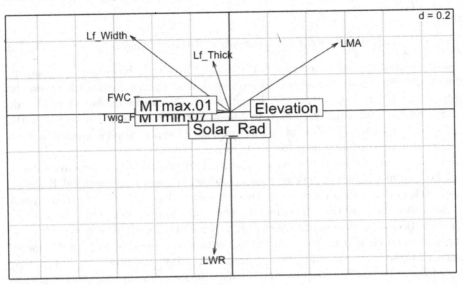

FIGURE 16.13
Bi-plot showing results of RLQ analysis of CFR data set. Environmental variables depicted as arrows and traits as boxes along the first two axes of RLQ analysis.

16.4.3.3 Maximum Entropy

As noted in Section 16.1, trait-based modeling is a promising approach to better understanding and predicting community assembly. To this end, Shipley and colleagues [65] proposed a method of predicting species abundances in a community based on their functional trait values. Their method relied on principles of statistical mechanics, specifically entropy maximization, which can be argued is an algorithmic approach. We do not go into the details of this method here, but Merow and colleagues [48] present a robust examination of the utility and limitations of this method, as applied in analyzing diverse floras such as our CFR data set. Recently, Warton et al. [75] leverage the connections between maximum entropy and Poisson regression and Poisson point-process regression to incorporate regression model based tools and approaches into a maximum entropy approach. This statistical model-based view on an algorithm method provides a bridge between these analytical methods.

16.5 Trait Modeling - Statistical Model-Based

Current advances in trait-based ecology emphasize the need to move from algorithmic approaches to statistical model based approaches. This is particularly true for areas of community ecology that involve functional trait analyses (*sensu* [30, 74, 75]). In these approaches, traits are often treated as predictor variables for species presence (or abundance). Thus, these methods can help us understand trait-by-environment relationships as they relate to the presence of a species, but trait values are not modeled directly. Brown et al. [7] present what they term as 'the fourth-corner solution', which is directly related to the approach discussed in Section 16.4.3.1. Essentially, they propose that rather than calculate the Ω matrix (see Equation (16.11)) via algorithmic matrix-methods (i.e., data matrix inflation or generalized singular value decomposition), one could build a multi-species logistic regression model of the form:

$$\text{logit}(p_{ij} = \beta_0 + \beta_1 \text{env}_i + \beta_2 \text{trait}_j + \beta_3(\text{env} * \text{trait})_{ij} \tag{16.13}$$

where p_{ij} is the probability of occurrence for species j at site i, based on environmental variables at the site i and trait values for species j, and the interactions between the environment and trait variables. Of course, Equation (16.13) can be expanded to include multiple traits and environmental variables. In essence, Ω is the interactions between the two different sets of predictor variables of species presence/absence. Similar methodological approaches have also been proposed by [30] and [59].

Laughlin et al. [38] also present a novel approach to investigating trait-by-environment relationships by way of predicting species presence/absence using a hierarchical Bayesian modeling approach. A major factor separating this work from that of [7, 30, 59] is the explicit inclusion of intra-specific variation. In principle, the other approaches could incorporate intra-specific variation via inclusion of a species level factor in a hierarchical model, however this is an area of research still to be explored and advanced. One challenge that appears to be common to these methods, and also requires further investigation, is that fact that the accuracy of these approaches becomes increasingly questionable as the number of taxa and the relative endemism of taxa increases (i.e., the number of 0s in the L matrix for a particular species increases). Both of these limitations are presented by the CFR data, as well as other biological rich systems.

Some new approaches do treat trait values as the response variable(s). For example, Mitchell and colleagues [52] used an approach that allows for the investigation of the relationships between multiple traits and multiple environmental conditions simultaneously, while also considering the influences of phylogentic relationships among species. However, they did not account for species abundance, and thus lack a full community perspective.

A newly developed model-driven approach has been recently developed by Schliep and colleagues [63] which uses flexible, joint trait distribution models across environments and species while incorporating intra-taxon variability as well as inter-site variability. This has been implemented in a Bayesian framework, and the joint trait distribution models allow for mixed continuous, binary, and ordinal trait variables while also incorporating dependence among traits. This enables both joint and conditional trait prediction at unobserved sites. These models can thus provide a alternative model-based solution to the fourth-corner problem and RLQ approaches. This approach also implements mixture models wherein the actual distribution of a trait at a site is a mixture of the trait distributions for each species at the site. This provides a more realistic summary of community-level trait distributions than simple CWMs and avoids the assumption of unimodality and normality. For more details, see [63].

16.6 Conclusion

Trait-based ecological analysis has the potential to uncover important insights into the processes governing species occurrences and community assembly. There is a rich and diverse set of well established approaches, most of which stem from the algorithmic tradition followed in multivariate statistical analysis. The results from these methods provide an excellent base from where we can establish hypotheses regarding ecological processes. Novel methods that explicitly incorporate statistical model-based thinking offer great promise in taking the next steps in trait-based ecology to establishing this subfield as one based on prediction. However, model-based approaches will not eliminate the need to continue to address more basic biological and ecological aspects of trait-based modeling. These include robust investigations of the linkage between soft and hard functional traits and individual- and population-level fitness, better understanding and modeling of interactions among trait variables, methods to incorporate microenvironment conditions and identify when these are needed, and movement beyond the most commonly measured functional traits, to traits that are more closely connected to the question at hand. Trait-based ecology has a long and rich history and should have an equally long and rich future.

16.7 Acknowledgments

The authors thanks Jennifer Hoeting for valuable feedback on earlier versions of this chapter. Portions of this work were supported by a grant from the National Science Foundation Dimensions of Biodiversity program (DEB-1046328). The authors thank Jasper Slingsby, the South African Environmental Observation Network Fynbos Node, and personal at Eastern Cape Parks and Tourism, SANParks, and Cape Nature for assistance in data access and collection.

Bibliography

[1] David D. Ackerly and William K. Cornwell. A trait-based approach to community assembly: partitioning of species trait values into within- and among-community components. *Ecology Letters*, 10(2):135–45, feb 2007.

[2] Matthew E. Aiello-Lammens, Jasper A. Slingsby, Cory Merow, Hayley Kilroy Mollmann, Douglas Euston-Brown, Cynthia S. Jones, and John A. Silander. Processes of community assembly in an environmentally heterogeneous, high biodiversity region. *Ecography*, 40(631):10.1111/ecog.01945, 2016.

[3] Nicky Allsopp, Jonathan F. Colville, and G. Anthony Verboom, editors. *Fynbos: Ecology, Evolution, and Conservation of a Megadiverse Region*. Oxford University Press, Oxford, UK, 2014.

[4] Christopher Baraloto, C. E. Timothy Paine, Sandra Patiño, Damien Bonal, Bruno Hérault, and Jerome Chave. Functional trait variation and sampling strategies in species-rich plant communities. *Functional Ecology*, 24(1):208–216, 2010.

[5] Daniel Borcard, François Gillet, and Pierre Legendre. *Numerical Ecology with R*. Springer, New York, 2011.

[6] Cajo J. F. Ter Braak, Anouk Cormont, and Stéphane Dray. Improved testing of species traits-environment relationships in the fourth-corner problem. *Ecology*, 93(7):1525–1526, 2012.

[7] Alexandra M. Brown, David I. Warton, Nigel R. Andrew, Matthew Binns, Gerasimos Cassis, and Heloise Gibb. The fourth-corner solution - using predictive models to understand how species traits interact with the environment. *Methods in Ecology and Evolution*, 5(4):344–352, apr 2014.

[8] Marc W. Cadotte. The new diversity: Management gains through insights into the functional diversity of communities. *Journal of Applied Ecology*, 48(5):1067–1069, 2011.

[9] Marc W. Cadotte and Caroline M. Tucker. Should Environmental Filtering be Abandoned? *Trends in Ecology & Evolution*, In Press, 2017.

[10] Carlos P Carmona, Francesco De Bello, Norman W H Mason, and Jan Lepš. Traits Without Borders: Integrating Functional Diversity Across Scales. *Trends in Ecology & Evolution*, 31(5):382–394, 2016.

[11] Carlos P. Carmona, Cristina Rota, Francisco M. Azcárate, and Begoña Peco. More for less: sampling strategies of plant functional traits across local environmental gradients. *Functional Ecology*, 29(4):579–588, oct 2015.

[12] Martin L. Cody. Niche theory and plant growth form. *Vegetatio*, 97(1):39–55, 1991.

[13] J. H. C. Cornelissen, S. Lavorel, E. Garnier, Sandra Díaz, N. Buchmann, D. E. Gurvich, P. B. Reich, H. Ter Steege, H. D. Morgan, M. G. A. Van Der Heijden, J. G. Pausas, and H. Poorter. A handbook of protocols for standardised and easy measurement of plant functional traits worldwide. *Australian Journal of Botany*, 51(4):335, 2003.

[14] WK Cornwell and David D. Ackerly. Community assembly and shifts in plant trait distributions across an environmental gradient in coastal California. *Ecological Monographs*, 79(1):109–126, 2009.

[15] Francesco de Bello, Sandra Lavorel, Sébastien Lavergne, Cécile H. Albert, Isabelle Boulangeat, Florent Mazel, and Wilfried Thuiller. Hierarchical effects of environmental filters on the functional structure of plant communities: a case study in the French Alps. *Ecography*, 36(3):393–402, mar 2013.

[16] Sandra Díaz, Jens Kattge, Johannes H C Cornelissen, Ian J Wright, Sandra Lavorel, Stéphane Dray, Björn Reu, Michael Kleyer, Christian Wirth, I Colin Prentice, Eric Garnier, Gerhard Bönisch, Mark Westoby, Hendrik Poorter, Peter B Reich, Angela T Moles, John Dickie, Andrew N Gillison, Amy E Zanne, Jerome Chave, S. J. Wright, Serge N Sheremet'ev, Hervé Jactel, Baraloto Christopher, Bruno Cerabolini, Simon Pierce, Bill Shipley, Donald Kirkup, Fernando Casanoves, Julia S Joswig, Angela Günther, Valeria Falczuk, Nadja Rüger, Miguel D. Mahecha, and Lucas D. Gorné. The global spectrum of plant form and function. *Nature*, 529(7585):167–171, 2015.

[17] Sylvain Dolédec and Daniel Chessel. Co-inertia analysis: an alternative method for studying species-environment relationships. Freshwater Biology 31:277-294. *Freshwater Biology*, 31:277–294, 1994.

[18] Sylvain Dolédec, Daniel Chessel, Cajo J. F. Ter Braak, and S. Champely. Matching species traits to environmental variables: a new three- table ordination method. *Environmental and Ecological Statistics*, 3:143–166, 1996.

[19] Stéphane Dray, Philippe Choler, Sylvain Dolédec, Pedro R. Peres-Neto, Wilfried Thuiller, Sandrine Pavoine, and Cajo J F ter Braak. Combining the fourth-corner and the RLQ methods for assessing trait responses to environmental variation. *Ecology*, 95(1):14–21, jan 2014.

[20] Stéphane Dray and Anne-Béatrice Dufour. The ade4 Package: Implementing the Duality Diagram for Ecologists. *Journal of statistical software*, 22(4), 2007.

[21] Stéphane Dray and Pierre Legendre. Testing the species traits-environment relationships: the fourth-corner problem revisited. *Ecology*, 89(12):3400–3412, dec 2008.

[22] Nicholas J. Gotelli. Null Model Analysis of Species Co-Occurrence Patterns. *Ecology*, 81(9):2606, sep 2000.

[23] Nicholas J. Gotelli and Aaron M. Ellison. *A Primer of Ecological Statistics*. Sinauer Associates, Sunderland, MA, 2004.

[24] Nicholas J. Gotelli and Gary R. Graves. *Null Models in Ecology*. Smithsonian Institution Press, Washington, DC, 1996.

[25] Nicholas J. Gotelli and Declan J McCabe. Species Co-Occurrence : A Meta-Analysis of J. M. Diamond's Assembly Rules Model. *Ecology*, 83(8):2091–2096, 2002.

[26] Catherine H. Graham, Juan L. Parra, Boris A. Tinoco, F. Gary Stiles, and Jim A. McGuire. Untangling the influence of ecological and evolutionary factors on trait variation across hummingbird assemblages. *Ecology*, 93(8):S99–S111, 2012.

[27] J. G. Hodgson, P. J. Wilson, R. Hunt, J. P. Grime, and K. Thompson. Allocating C-S-R plant functional types: a soft approach to a hard problem. *Oikos*, 85:282–294, 1999.

[28] Alexander Von Humboldt and Aimé Bonpland. *Essay on the Geography of Plants*. University of Chicago Press, Chicago, reprint ed edition, 2013.

[29] Stephen T. Jackson. Natural, potential and actual vegetation in North America. *Journal of Vegetation Science*, 24(4):772–776, oct 2012.

[30] Tahira Jamil, Wim A. Ozinga, Michael Kleyer, and Cajo J. F. ter Braak. Selecting traits that explain species-environment relationships: a Generalized Linear Mixed Model approach. *Journal of Vegetation Science*, 24(6):988–1000, 2013.

[31] J. Kattge, Sandra Díaz, S. Lavorel, I. C. Prentice, P. Leadley, G. Bönisch, E. Garnier, M. Westoby, P. B. Reich, I. J. Wright, J. H C Cornelissen, C. Violle, S. P. Harrison, P. M. Van Bodegom, M. Reichstein, B. J. Enquist, N. A. Soudzilovskaia, David D. Ackerly, M. Anand, O. Atkin, M. Bahn, T. R. Baker, D. Baldocchi, R. Bekker, C. C. Blanco, B. Blonder, W. J. Bond, R. Bradstock, D. E. Bunker, F. Casanoves, J. Cavender-Bares, J. Q. Chambers, F. S. Chapin, J. Chave, D. Coomes, W. K. Cornwell, J. M. Craine, B. H. Dobrin, L. Duarte, W. Durka, J. Elser, G. Esser, M. Estiarte, W. F. Fagan, J. Fang, F. Fernández-Méndez, A. Fidelis, B. Finegan, O. Flores, H. Ford, D. Frank, G. T. Freschet, N. M. Fyllas, R. V. Gallagher, W. A. Green, A. G. Gutierrez, T. Hickler, S. I. Higgins, J. G. Hodgson, A. Jalili, S. Jansen, C. A. Joly, A. J. Kerkhoff, D. Kirkup, K. Kitajima, M. Kleyer, S. Klotz, J. M H Knops, K. Kramer, I. Kühn, H. Kurokawa, Daniel C. Laughlin, T. D. Lee, M. Leishman, F. Lens, T. Lenz, S. L. Lewis, J. Lloyd, J. Llusià, F. Louault, S. Ma, M. D. Mahecha, P. Manning, T. Massad, B. E. Medlyn, J. Messier, A. T. Moles, S. C. Müller, K. Nadrowski, S. Naeem, Ü Niinemets, S. Nöllert, A. Nüske, R. Ogaya, J. Oleksyn, V. G. Onipchenko, Y. Onoda, J. Ordoñez, G. Overbeck, W. A. Ozinga, S. Patiño, S. Paula, J. G. Pausas, J. Peñuelas, O. L. Phillips, V. Pillar, H. Poorter, L. Poorter, P. Poschlod, A. Prinzing, R. Proulx, A. Rammig, S. Reinsch, B. Reu, L. Sack, B. Salgado-Negret, J. Sardans, S. Shiodera, B. Shipley, A. Siefert, E. Sosinski, J. F. Soussana, E. Swaine, N. Swenson, K. Thompson, P. Thornton, M. Waldram, E. Weiher, M. White, S. White, S. J. Wright, B. Yguel, S. Zaehle, A. E. Zanne, and C. Wirth. TRY - a global database of plant traits. *Global Change Biology*, 17(9):2905–2935, 2011.

[32] Paul A. Keddy. Assembly and response rules: two goals for predictive community ecology. *Journal of Vegetation Science*, 3(2):157–164, 1992.

[33] Steven W. Kembel, Peter D. Cowan, Matthew R. Helmus, William K. Cornwell, Helene Morlon, David D. Ackerly, Simon P. Blomberg, and Campbell O. Webb. Picante: R tools for integrating phylogenies and ecology. *Bioinformatics*, 26(11):1463–1464, 2010.

[34] Michael Kleyer, Stéphane Dray, Francescode Bello, Jan Lepš, Robin J. Pakeman, Barbara Strauss, Wilfried Thuiller, and Sandra Lavorel. Assessing species and community functional responses to environmental gradients: which multivariate methods? *Journal of Vegetation Science*, 23(5):805–821, oct 2012.

[35] Nathan J. B. Kraft, Peter B. Adler, Oscar Godoy, Emily James, Steve Fuller, and Jonathan M. Levine. Community assembly, coexistence, and the environmental filtering metaphor. *Functional Ecology*, 29(5):592–599, sep 2015.

[36] Geneviève Lajoie and Mark Vellend. Understanding context dependence in the contribution of intraspecific variation to community trait environment matching. *Ecology*, 96(11):2912–2922, 2015.

[37] Daniel C. Laughlin. The intrinsic dimensionality of plant traits and its relevance to community assembly. *Journal of Ecology*, 102:186–193, nov 2014.

[38] Daniel C. Laughlin, Chaitanya Joshi, Peter M van Bodegom, Zachary a Bastow, and Peter Z Fulé. A predictive model of community assembly that incorporates intraspecific trait variation. *Ecology letters*, 15(11):1291–9, nov 2012.

[39] Daniel C. Laughlin and Julie Messier. Fitness of multidimensional phenotypes in dynamic adaptive landscapes. *Trends in Ecology & Evolution*, 30(8):487–496, aug 2015.

[40] Sandra Lavorel, Karl Grigulis, Sue McIntyre, Nick S. G. Williams, Denys Garden, Josh Dorrough, Sandra Berman, Fabien Quétier, Aurélie Thébault, and Anne Bonis. Assessing functional diversity in the field methodology matters! *Functional Ecology*, 22:134–147, nov 2008.

[41] Sandra Lavorel, C. Rochette, and Jean-Dominique Lebreton. Functional groups for response to disturbance in Mediterranean old fields. *Oikos*, 84(3):480–498, 1999.

[42] Pierre Legendre, R Galzin, and ML Harmelin-Vivien. Relating behavior to habitat: solutions to the fourth-corner problem. *Ecology*, 78(2):547–562, 1997.

[43] Pierre Legendre and Louis Legendre. *Numerical Ecology*. Elsevier, 3 edition, 2012.

[44] Brian S. Maitner, Brad Boyle, Nathan Casler, Rick Condit, John Donoghue, Sandra M. Durán, Daniel Guaderrama, Cody E. Hinchliff, Peter M. Jørgensen, Nathan J.B. Kraft, Brian McGill, Cory Merow, Naia Morueta-Holme, Robert K. Peet, Brody Sandel, Mark Schildhauer, Stephen A. Smith, Jens-Christian Svenning, Barbara Thiers, Cyrille Violle, Susan Wiser, and Brian J. Enquist. The <scp>BIEN R</scp> package: A tool to access the Botanical Information and Ecology Network (BIEN) database. *Methods in Ecology and Evolution*, 2017(March):1–7, 2017.

[45] John C. Manning and Peter Goldblatt. *Plants of the Greater Cape Floristic Region, Volume 1: The Core Cape Flora*. South African National Biodiversity Institute, 2013.

[46] Norman W. H. Mason and Francesco De Bello. Functional diversity: A tool for answering challenging ecological questions. *Journal of Vegetation Science*, 24:777–780, 2013.

[47] Brian J McGill, Brian J Enquist, Evan Weiher, and Mark Westoby. Rebuilding community ecology from functional traits. *Trends in ecology & evolution*, 21(4):178–85, apr 2006.

[48] Cory Merow, Andrew M Latimer, and John A Silander Jr. Can entropy maximization use functional traits to explain species abundances? A comprehensive evaluation. *Ecology*, 92(7):1523–37, jul 2011.

[49] Julie Messier, Martin J. Lechowicz, Brian J. McGill, Cyrille Violle, and Brian J. Enquist. Interspecific integration of trait dimensions at local scales: the plant phenotype as an integrated network. *Journal of Ecology*, 105(6):1775–1790, 2017.

[50] Julie Messier, Brian J. McGill, Brian J. Enquist, and Martin J. Lechowicz. Trait variation and integration across scales: Is the leaf economic spectrum present at local scales? *Ecography*, 2016.

[51] Julie Messier, Brian J McGill, and Martin J Lechowicz. How do traits vary across ecological scales? A case for trait-based ecology. *Ecology Letters*, 13(7):838–48, jul 2010.

[52] Nora Mitchell, Timothy E. Moore, Hayley Kilroy Mollmann, Jane E. Carlson, Kerri Mocko, Hugo Martinez-Cabrera, Christopher Adams, John a. Silander, Cynthia S. Jones, Carl D. Schlichting, and Kent E. Holsinger. Functional Traits in Parallel Evolutionary Radiations and Trait-Environment Associations in the Cape Floristic Region of South Africa. *The American Naturalist*, pages 000–000, 2015.

[53] Keith E. Muller. Relationships between redundancy analysis, canonical correlation, and multivariate regression. *Psychometrika*, 46(2):139–142, 1981.

[54] Robert Muscarella and María Uriarte. Do community-weighted mean functional traits reflect optimal strategies? *Proceedings of the Royal Society B: Biological Sciences*, 283:20152434, 2016.

[55] Pedro R. Peres-Neto, Stéphane Dray, and Cajo J.F.ter Braak. Linking trait variation to the environment: critical issues with community-weighted mean correlation resolved by the fourth-corner approach. *Ecography*, 40(7):806–816, 2017.

[56] N. Pérez-Harguindeguy, Sandra Díaz, E. Garnier, S. Lavorel, H. Poorter, P. Jaureguiberry, M. S. Bret-Harte, W. K. Cornwell, J. M. Craine, D. E. Gurvich, C. Urcelay, E. J. Veneklaas, P. B. Reich, L. Poorter, I. J. Wright, P. Ray, L. Enrico, J. G. Pausas, a. C. de Vos, N. Buchmann, G. Funes, F. Quétier, J. G. Hodgson, K. Thompson, H. D. Morgan, H. ter Steege, L. Sack, B. Blonder, P. Poschlod, M. V. Vaieretti, G. Conti, a. C. Staver, S. Aquino, and J. H. C. Cornelissen. New handbook for standardised measurement of plant functional traits worldwide. *Australian Journal of Botany*, 61(3):167–234, 2013.

[57] Owen L. Petchey and Kevin J. Gaston. Functional diversity (FD), species richness and community composition. *Ecology Letters*, 5(3):402–411, may 2002.

[58] A Townsend Peterson, Jorge Soberón, Richard G Pearson, Robert P. Anderson, Enrique Martínez-Meyer, Miguel Nakamura, and Miguel B. Araújo. *Ecological niches and geographic distributions (MPB-49)*. Princeton University Press, Princeton, 2011.

[59] Laura J. Pollock, William K. Morris, and Peter A. Vesk. The role of functional traits in species distributions revealed through a hierarchical model. *Ecography*, 35(8):716–725, 2012.

[60] R Core Team. R: A language and environment for statistical computing, 2012.

[61] C. Radhakrishna Rao. The use and interpretation of principal component analysis in applied research. *Sankhy: The Indian Journal of Statistics, Series A*, 26:329–358, 1964.

[62] A. F. W. Schimper. *Plant-Geography Upon a Physiological Basis*. Clarendon Press, Oxford, UK, 1903.

[63] Erin M Schliep, Alan E. Gelfand, Rachel M. Mitchell, Matthew E. Aiello-Lammens, and John A. Silander. Assessing the joint behaviour of species traits as filtered by environment. *Methods in Ecology and Evolution*, 00:1–12, oct 2017.

[64] Bill Shipley, Francesco de Bello, Johannes H. C. Cornelissen, Étienne Laliberté, Daniel C. Laughlin, and Peter B. Reich. Reinforcing loose foundation stones in trait-based plant ecology. *Oecologia*, 180(4):922–931, 2016.

[65] Bill Shipley, Denis Vile, and Eric Garnier. From plant traits to plant communities: a statistical mechanistic approach to biodiversity. *Science*, 314(5800):812–4, nov 2006.

[66] Andrew Siefert, Catherine Ravenscroft, Michael D. Weiser, and Nathan G. Swenson. Functional beta-diversity patterns reveal deterministic community assembly processes in eastern North American trees. *Global Ecology and Biogeography*, 22(6):682–691, jun 2013.

[67] Nathan G. Swenson. Phylogenetic imputation of plant functional trait databases. *Ecography*, 37:105–110, sep 2014.

[68] Adrien Taudiere and Cyrille Violle. cati: An R package using functional traits to detect and quantify multi-level community assembly processes. *Ecography*, (May), 2015.

[69] Wilfried Thuiller, Jasper A Slingsby, Sean D J Privett, and Richard M. Cowling. Stochastic species turnover and stable coexistence in a species-rich, fire-prone plant community. *PLoS ONE*, 2(9):e938, jan 2007.

[70] Christopher H. Trisos, Owen L. Petchey, and Joseph a. Tobias. Unraveling the interplay of community assembly processes acting on multiple niche axes across spatial scales. *The American Naturalist*, 184(5):593–608, sep 2014.

[71] P. M. Van Bodegom, J. C. Douma, J. P. M. Witte, J. C. Ordoñez, R. P. Bartholomeus, and R. Aerts. Going beyond limitations of plant functional types when predicting global ecosystem-atmosphere fluxes: exploring the merits of traits-based approaches. *Global Ecology and Biogeography*, 21(6):625–636, jun 2012.

[72] W. C. E. P. Verberk, C. G. E. van Noordwijk, and a. G. Hildrew. Delivering on a promise: integrating species traits to transform descriptive community ecology into a predictive science. *Freshwater Science*, 32(2):531–547, 2013.

[73] Cyrille Violle, Marie-Laure Navas, Denis Vile, Elena Kazakou, Claire Fortunel, Irène Hummel, and Eric Garnier. Let the concept of trait be functional! *Oikos*, 116(5):882–892, may 2007.

[74] David I. Warton, Scott D. Foster, Glenn De'ath, Jakub Stoklosa, and Piers K. Dunstan. Model-based thinking for community ecology. *Plant Ecology*, 216:669–682, 2015.

[75] David I. Warton, Bill Shipley, and Trevor Hastie. CATS regression a model-based approach to studying trait-based community assembly. *Methods in Ecology and Evolution*, 6:389–398, 2015.

[76] Colleen T. Webb, Jennifer A. Hoeting, Gregory M. Ames, Matthew I. Pyne, and N. LeRoy Poff. A structured and dynamic framework to advance traits-based theory and prediction in ecology. *Ecology Letters*, 13(3):267–283, 2010.

[77] Evan Weiher, Adrie Van Der Werf, Ken Thompson, Michael Roderick, Eric Garnier, and Ove Eriksson. Challenging Theophrastus : A common core list of plant traits for functional ecology. *Journal of Vegetation Science*, 10(5):609–620, 1999.

[78] Adam M. Wilson and John A. Silander Jr. Estimating uncertainty in daily weather interpolations: a Bayesian framework for developing climate surfaces. *International Journal of Climatology*, 34(8):2573–2584, jun 2014.

[79] Ian J Wright, Peter B Reich, Mark Westoby, David D. Ackerly, Zdravko Baruch, Frans Bongers, Jeannine Cavender-Bares, Terry Chapin, Johannes H C Cornelissen, Matthias Diemer, Jaume Flexas, Eric Garnier, Philip K Groom, Javier Gulias, Kouki Hikosaka, Byron B Lamont, Tali Lee, William Lee, Christopher Lusk, Jeremy J Midgley, Marie-Laure Navas, Ulo Niinemets, Jacek Oleksyn, Noriyuki Osada, Hendrik Poorter, Pieter Poot, Lynda Prior, Vladimir I Pyankov, Catherine Roumet, Sean C Thomas, Mark G Tjoelker, Erik J Veneklaas, and Rafael Villar. The worldwide leaf economics spectrum. *Nature*, 428(6985):821–827, 2004.

17

Statistical models of vegetation fires: Spatial and temporal patterns

J.M.C. Pereira

Forest Research Centre, School of Agriculture, University of Lisbon (JMC Pereira)

K.F. Turkman

Center of Statistics and Applications of the University of Lisbon, Faculty of Sciences, University of Lisbon (KF Turkman)

CONTENTS

17.1 Introduction

17.1.1 The global relevance of vegetation fires

Vegetation burning is a global scale process and became evident in the geological record soon after the emergence of terrestrial plants. It affects the global distribution and structure of vegetation, the major biogeochemical cycles, and the climate system (Bowman et al. 2009). In a recent analysis of global area burned during the years 2006-2008, and using 300m spatial resolution satellite imagery, Alonso-Canas and Chuvieco [2] mapped 3.6×10^6 - 3.8×10^6 km^2 burned annually, an area larger than India. Vegetation burning releases substantial amounts of greenhouse gases to the atmosphere. Schultz et al. [80] analyzed atmospheric emissions due to biomass burning for the period 1960-2000 and estimated global carbon emissions ranging from 1410 to 3140 teragrams (Tg) of carbon per year, with a mean annual value of 2078 Tg. Greenhouse gas emissions from vegetation burning may account for about $12-13\%$ of total (anthropogenic plus natural) annual emissions of nitrogen oxides (NOx) to $32-34\%$ for carbon monoxide (CO). Besides their global impacts on the atmosphere and climate, emissions from biomass burning also affect human health. Johnston et al. [39] estimated the annual global human mortality that can be attributed to smoke from vegetation fires at 339,000 deaths, with an interquartile range of $260,000 - 600,000$. The most affected regions were sub-Saharan Africa (157,000) and Southeast Asia (110,000). Annual mortality was 262,000 during La Nina years, compared with 532,000 during El Nino years.

A growing body of research suggests the possibility of an increase in fire occurrence and area burned in response to global climate change. However, spatial variability is very large with some areas remaining stable or even showing decreasing fire incidence [30, 42]. Krawchuk et al. [44] developed multivariate statistical models of the relationship between various environmental drivers and the current distribution of fire activity. They applied those models to project changes in future global pyrogeography, identifying regional hotspots of change in fire probability under future climate conditions prescribed by a global climate model. They found a complex spatial pattern of fire probability change, with areas of increase counter-balanced by areas of decrease, and stressed that changes in fire activity may occur rapidly and further complicate environmental conditions for species trying to cope with changing climate. Jolly et al. [40] used various global climate data sets and meteorological fire danger indices to develop an annual measure of fire weather season length and mapped geographical and temporal trends from 1979 to 2013. They found that fire weather seasons have increased in length across $29.6 \times 10^6 km^2$, or 25.3% of the vegetated land surface of the Earth, yielding an increase of 18.7% in global mean fire weather season length. Such fire weather changes, if coupled with availability of ignition sources and plant fuels, may have substantial impacts on global ecosystems and economies. Knorr et al. [43] compared the impact of climate change and increasing atmospheric CO_2 concentration on burned area with the impact of demographic changes, using ensembles of climate simulations combined with historical and projected population changes and various socio-economic development pathways for 1901 - 2001. In the past, humans have extensively suppressed vegetation fires, and for future scenarios global burned area is expected to decline under a moderate emissions scenario but start to increase again from around mid-century under high greenhouse gas emissions development pathways. Knorr et al. [43] found that human exposure to vegetation fires is likely to increase in the future mainly due to population growth in areas with high fire incidence, rather than by an increase in burned area.

17.1.2 Fire likelihood, intensity, and effects

Fire plays a complex role in its relation with vegetation. It is a natural ecological factor, important to maintain ecosystem dynamics, productivity, and biodiversity, and is also a land management tool widely used in croplands, rangelands, and forests all over the world. On the other hand, vegetation fires annually affect millions of hectares of forests, woodlands and other vegetation, endangering human populations, originating substantial economic losses in terms of resources destroyed and costs resulting from prevention and suppression activities [26]. Fire, in its two roles in terrestrial ecosystems, is a pervasive disturbance, especially important in regions with abundant ignition sources, enough biomass to sustain combustion, and a long enough dry season to allow burning. Most land management activities involve disturbing ecosystems, such as harvesting crops, logging forests, and prescribed burning, or suppressing disturbance, as in pest control, disease control, and fire-fighting. The boundary between fire as a land management tool and wildfire is permeable, since managed burning of croplands and rangelands may escape control and damage forests, while lightning-caused fires started in wildlands may spread into urban areas and croplands, with undesirable consequences.

Managing fire, so that society and the environment reap its benefits while avoiding potential harm, is subject to many sources of uncertainty. Fire behavior is hard to predict;

required data often are inaccurate or missing; value metrics to prioritize location/allocation of personnel and equipment across fires and resources at risk are scarce and controversial; there is limited understanding of the effectiveness of preventive fuel treatments, and of ecosystem response to fire [88]. Risk assessment is a central element of decision-making under uncertainty, and Miller and Ager [52] recognize three components of fire risk: likelihood of the event; expected behavior, including fire intensity; and effects, which depend on fire behavior. Likelihood of the event is influenced by precedent conditions, such as fire weather, fuel moisture and abundance, and availability of ignition sources. Fire likelihood can be measured either as ignition probability, the probability that a fire will start at a given location, or burn probability (the probability that a given area will be affected by fire), regardless of starting point. Ignition probability typically is statistically modelled using fire occurrence data, while burn probability can be estimated via simulation, or using multi-year fire perimeter data. Estimates of ignition probability are particularly useful to plan rapid response, or initial attack operations, while burn probabilities are more often applied in to plan fuels management interventions [52]. Fire intensity measures the rate of aboveground fuel consumption. It describes the physical process of energy production from combustion of organic matter as the rate of heat release per linear meter of the flame front (kW.$^{m-1}$) [41]. Its major environmental drivers are fuel load, fuel moisture content, and wind speed. Fire effects on soil, water, and biological resources that control ecosystem sustainability are described by fire severity, a qualitative, ecosystem-specific, measure. It is a function of environmental factors that affect the combustion of vegetation, such as load, type, and moisture of live and dead fuel, air temperature and humidity, wind speed, and topography [16].

17.1.3 Acquisition of fire data from Earth observation satellites

The research reviewed in section 17.2 is mostly based on satellite remote sensing, a data acquisition technology with very specific characteristics worthy of a brief overview. Vegetation fires generate various types of signals that can be observed from space: thermal radiation, smoke emissions, charcoal and ash residues, and altered vegetation structure. In this chapter, we focus on active fire data obtained from analysis of the thermal signal, and burned area maps derived from analyzing charcoal deposition. Figure 17.1 shows the fire thermal signal as lines of red/orange pixels forming active flame fronts along the edges of recently burned areas (dark purple surfaces). The thermal signal produced by active flame fronts generates point/linear data that are snapshots of each fire's history, acquired at the time of satellite overpass. Alternatively, the surface darkening generated by charcoal deposition is more persistent and provides a fuller overview of the areas affected by fire.

Earth observation (EO) satellite image data are characterized by their resolution along four dimensions: spatial, temporal, spectral, and radiometric. Spatial resolution is described by the image pixel size, representing the extent of an area being observed on the ground; temporal resolution is the amount of time between consecutive passages of the satellite over the same location at the surface; spectral resolution is defined by the number and width of observation windows (bands, or channels) along the electromagnetic spectrum; radiometric resolution refers to the quantization level of the electromagnetic signal and corresponds to the number of distinct grey levels present in the image.

Satellite imagery are recorded in a raster data structure, a rectangular tessellation of the plane into cells, organized into rows and columns where each cell, or pixel, contains a value representing the amount of energy reflected or emitted from the corresponding location on the ground. Due to the multispectral nature of most EO data, each image will contain one raster layer per spectral channel. The frequency of acquisition of such images ranges from a 26-day revisit cycle for the SPOT High Resolution Geometric radiometers, to 96 images per

FIGURE 17.1
The fire thermal signal as lines of red/orange pixels forming active flame fronts along the edges of recently burned areas (dark purple surfaces). Shortwave infrared (red color plane), near-infrared (green color plane), red (blue color plane) composite Landsat satellite image of a tropical savanna in Angola during the fire season.

day with the Meteosat Second Generation (MSG) Spinning Enhanced Visible and Infrared Imager (SEVIRI). Thus, detection of land surface changes, such as fire monitoring and burned area mapping, often relies on the analysis of extensive time series of multispectral images.

Detection and classification of land surface changes using multitemporal approaches is complicated by the angular dependence of the surface reflected radiation as function of both solar illumination and satellite viewing directions. Scattering and absorption of solar radiation by atmospheric gases and aerosols introduce additional spatial and temporal variability in the data, and cloud cover often prevents observation of the Earth surface. Algorithms to detect and map active fires typically rely on a set of land surface temperature thresholds set at different wavelengths, and on spatial contrast tests to identify and map the positive thermal anomalies generated by vegetation combustion [31, 33]. Active fires are represented as point data, with information on location, time of detection, and fire intensity. A broader variety of techniques has been used to detect and map burned areas, including multiple thresholding [8], supervised classification [63], and time series change detection [71]. Remote sensing imagery are employed to map fires at scales ranging from the high spatial resolution analysis of a single fire event to the development of coarser resolution global, multi-year, active fire and burned area atlases. Burned area data typically are made available as dichotomously classified (burned/unburned) raster grids, sometimes with indication of the degree of confidence associated with the classification label. Those data can be converted to polygons in a vector data structure, and the centroids of the polygons may be labeled with the polygon area to generate a spatial marked point process.

17.1.4   Research overview

Vegetation fires are inherently random in the timing of their location and occurrence, in the detailed behavior of each individual event, and in the particularities of their effects on soils, water, flora, fauna, and air. Therefore, substantial efforts have been directed towards statistical modeling of several fire-related processes. Such efforts may be structured according to the risk assessment framework of Miller and Ager [52]:

- Fire likelihood, which deals with pre-fire events, and aims at predicting the probability of fire occurrence, or the extent of area burned over a given spatial extent and temporal period, or planning fire suppression resource deployment and initial attack dispatch.

- Fire behavior itself, i.e. predicting the fire rate of spread, intensity, and size and shape of the area burned.

- Fire effects, addressing post-fire events, such as modeling fire-induced tree mortality, increased soil erosion, altered watershed hydrology, and changes in biological communities, including vegetation recovery.

Statistical methods employed for fire occurrence modeling and prediction include logistic regression and Poisson regression to predict human-caused fires in Ontario [50, 95], beta regression to estimate the proportion of burned area in Portugal [82] and Bayesian logistic regression to estimate the probability of occurrence of large fires in the Sidney region using fire weather data [11]. Fire ignition modeling needs to consider biophysical, societal, and management variables. Analyses need to be disaggregated by type of cause, especially for the main distinction between natural (lightning) and anthropogenic causes, which have different drivers and display distinct spatial and temporal patterns [68]. Fire ignition analysis raises specific problems, such as modeling specific atmospheric conditions conducive to dry thunderstorms, to predict lightning-caused fires [15]; space-time clustering of fires with specific causes, e.g. arson [67]; or analyse the effect of wildfire prevention efforts on the number of wildfire occurrences [14].

Modeling and prediction of area burned has relied on a wide variety of techniques, including Multivariate Adaptive Regression Splines to assess the response of area burned to climate change in western North America [7], and generalized additive models to characterize the current and future distribution of global fire activity [44]. Amaral-Turkman et al. [4] modeled fire occurrence and percent area burned in Australia using hierarchical Bayesian space-time models, and Chen et al. [17] assessed seasonal predictability of global burned area using sea surface temperature data and multiple regression analysis. Sá et al. [74] modeled fire incidence in Portugal using vegetation, precipitation and anthropogenic covariates and generalized additive models for location, scale and shape (GAMLSS).

Sullivan [86] reviewed two-dimensional fire spread models and identified two main classes: simulation models, and mathematical analogue models. The former are further subdivided into raster-based and vector-based, according to the type of geographical information data structure supporting their implementation. Raster-based fire spread simulation models use methods similar to cellular automata [1] or rely on minimum travel time [29] to spread the fire via contagion to adjacent grid cells, while vector-based simulation relies on the Huygens wavelet principle to expand a linear fire perimeter [5, 28]. Both approaches implement existing empirical or quasi-empirical models for frontal fire rate of spread, converting one-dimensional models [70] to two dimensions and then propagating a fire perimeter across a modelled landscape [86]. Mathematical analogue models are not based on a physical representation of fire behavior but on some kind of mathematical concept appropriate to simulate fire spread, such as the already mentioned cellular automata, cell-based discrete event system specification [92], percolation theory [23], and diffusion-limited aggregation [19].

Fire effects models typically address smaller areas than fire prediction models, often the area burned in a single event, focus on short-term consequences, and deal with a single ecosystem component or resource affected. Some of the statistical modeling techniques used for fire prediction are also employed to model fire consequences, with fire behavior characteristics such as intensity and flame residence time as independent variables and a broad variety of effects on soils, water, flora, and fauna as dependent variables. Recent examples of statistical analyses linking fire behavior with its environmental effects are provided by Wooley et al. [94], who reviewed applications of logistic regression to predict post-fire tree mortality of western North American conifer species, and by the Brando et al. [12] analysis of the role of bark characteristics, tree size, wood density, and fire behavior on fire-induced mortality in a Brazilian tropical forest with generalized linear mixed models. Hyde et al. [38] used generalized linear models to assess the effects of vegetation disturbance by fire on thresholds for channel initiation in Montana and Idaho watersheds, relying on remotely sensed data to map fire severity. Impacts of fire extent and severity on bird communities of SE Australia were modeled by Lindenmayer et al. [47] with hierarchical Bayesian analysis. Analysis of post-fire vegetation recovery has become more popular with the increased availability of time series of satellite imagery that allow for frequent observation of fire-affected areas at various spatial resolutions. Gouveia et al. [35] modeled the time dynamics of normalized difference vegetation index (NDVI) over areas burned in Portugal, and Gouveia et al. [36] used the same model to analyze the combined effect of drought and fire on vegetation recovery in the Iberian Peninsula. A recent advance in this area is provided by Paci et al. [58], who proposed a hierarchical Bayesian space-time model for post-fire vegetation recovery in Spain. A review of statistical methods for vegetation fire prediction based on different data sources is given in [87].

Also worthy of a brief mention is research on the interface between statistics and operations research. Lee et al. [45] combined an optimization model and stochastic simulation to manage deployment of initial attack resources for fire suppression in California, while Rytwinski and Crowe [73] used a combinatorial simulation-optimization approach for selecting the location of fuel-breaks to minimize expected losses from vegetation fires. Ferreira et al. [27] developed a stochastic dynamic programming approach to optimize stand management scheduling under fire risk for pine forests in Portugal. Other interesting problems are terrestrial and airborne fire detection patrol route planning and initial attack resource deployment [49].

17.2   Statistical methods and models for vegetation fire studies

The topics addressed in this section fit the risk framework previously introduced in section 17.1.2. The research reviewed relied on fire maps derived from remote sensing data with diverse time span and geographical extent to analyze spatial, temporal, and spatio-temporal patterns of fire occurrence. We focus on issues related to pre-fire circumstances, which determine fire likelihood and intensity and on post-fire aspects, reflecting the interplay between fire behavior and environmental conditions. Analyzing the environmental conditions under which past fires occurred and the spatio-temporal patterning of fires over several years and broad geographical areas is key to develop predictive ability for future fire events. Thus, modeling post-fire vegetation recovery contributes to assess the regrowth of fire risk, while characterizing fire return intervals using historical data provides a tool to estimate the likelihood of future events. The existence of weekly cycles in vegetation burning is a very recent finding and reveals a clear anthropogenic fingerprint on fire activity. Such cycles may also contribute to generate observed broad geographical scale weekly cycles in atmospheric pollution and meteorological variables.

17.2.1 Spatio-temporal point patterns of vegetation fires

Data coming from satellite images and ground sources can be point referenced and therefore are well suited to understand the spatial point patterns of fire events, as well as fire sizes. In particular, data coming from satellite imagery take the form

$$\{x(s_j, t), j = 1, ..., n_t, t = 1, ..., T\},$$

where $s \in D \subset \mathcal{R}^2$ represents the spatial location (centroid) of the fire scar observed at the end the annual fire season t, D the area under study and $x(s, t)$ is the size of the fire scar with centroid at s, observed in year t.

Ideally, the data on point patterns, as described above, should be treated as a realization of a spatio-temporal marked point process, discrete in time and continuous in space. Typically such point processes are modeled by Poisson point processes with intensity function $\lambda(s, t, x)$; see for example, Turner [91], Serra et al. [81], Xu and Schoenberg [96], Moller and Diaz-Avalos [53], Schoenberg [77], Brillinger et al. [13] and Pereira et al. [64]. However, at present, inference on such a model is not computationally feasible unless marks, namely fire sizes, are independent of the local point density, in which case the marks and the point patterns can be modeled separately and the intensity function takes the simpler form $\lambda(s, t) f(x)$, where $\lambda(s, t)$ is the intensity function of the points and $f(x)$ is the density of fire sizes.

Assuming this simplification, point patterns of fire incidences can be modeled by a Cox process [54]. Cox processes, in particular log Gaussian Cox processes and shot noise Cox processes have found applications in vegetation fire studies [53, 64].

A spatio temporal log-Gaussian Cox process $N(s, t)$ can be represented hierarchically in the following form

$$N(s, t) | \lambda(s, t) \sim \text{inhomogeneous Poisson process with intensity} \lambda(s, t), \qquad (17.1)$$

where,

$$\log \lambda(s, t) = \mu(s, t) + W(s, t),$$
$$\mu(s, t) = z(s, t)' \beta,$$

and $W(s, t)$ is a zero mean Gaussian process with covariance matrix $\Sigma(s, t)$.

Here, $z(s, t)$ represents the covariate information to explain the sources of inhomogeneity. $W(s, t)$ is a latent spatio-temporal process quantifying the variation not captured by the set of covariates available for the study. In Pereira et al. [64], a simple structure for $W(s, t)$ is assumed:

$$W(s, t) = \phi W(s, t - 1) + \eta(s, t),$$

where $\eta(s, t)$ are isotropic Gaussian field with Matern covariance, which are independent and identically distributed over time and $|\phi| < 1$ is the temporal correlation parameter, so that $W(s, t)$ is a stationary, isotropic Gaussian process in space and stationary first order autoregressive process in time. The likelihood, conditional on a realization of point patterns, the latent Gaussian random field W and the covariates is given by

$$\mathcal{L}(\boldsymbol{\theta} | w(s_i, t), z(s_i, t), (s_i, t), i = 1, ..., n_t, t = 1, ...T)$$

$$\propto \exp(-\sum_{t=1}^{T} \int_d \lambda(s, t) ds) \prod_{t=1}^{T} \prod_{i=1}^{n_t} \lambda(s_i, t),$$

where θ is the vector of model parameters. Inference based on such a likelihood brings many computational challenges due to the integral of the intensity function that appears in the likelihood, whose computation depends on costly matrix algebra operations on the dense covariance matrix of the Gaussian field W. The stochastic partial differential equation method (SPDE) proposed by Lindgren et al. [48], which approximates the Gaussian field W by a Gaussian Markov Random Field with local neighbourhood and sparse precision matrix, permits efficient Bayesian inferential methods for such models. The model (17.1), using the SPDE method, is implemented in Pereira et al. [64] within the powerful INLA framework [72]. Moller and Diaz-Avalos [53] suggest a shot-noise Cox point process as an alternative to log Gaussian Cox process in modeling daily fire patterns, together with specific inferential methods based on partial likelihoods. Schoenberg et al. [79] and Xu and Schoenberg [96] suggest non-parametric and semi-parametric methods of inference for estimating the intensity function of vegetation fire point patterns. See also Stoyan and Penttinen [84] for a review of point process methods in forestry.

17.2.2   Models for fire sizes

A variety of models such as log normal, half normal and exponential distributions have been suggested as models for fire sizes [85]. However, it is generally an established fact that a small number of fires cause most of the damage, often expressed by the phrase " *1% of the fires do 99% percent of the damage*" [85], therefore distributions with heavy upper tail should in principle give better fit as models to fire sizes. Indeed, different forms of the Pareto (or power law) distribution [3, 22, 69] are also suggested, particularly for regions where fire regimes produce very large fires. However, these models often lead naturally to the question whether they fail to reflect reality due to restrictions imposed by geography and consequently finite fire sizes. Truncated distributions, such as the truncated Pareto distribution, are suggested to bridge these two conflicting modeling strategies. For example, Cumming [21] reports that the truncated Pareto gives the best fit for lighting caused wildfires in the boreal mixed wood forests of Alberta, reporting that the end point of this fitted distribution is 650.000 ha, and restricting the size of the possible observable maximum fire size. Reed and McKelvey [69] studied fire sizes as part of stochastic fire growth models and relate fire sizes to the extinguishment-growth rate ratio. They also showed that under certain conditions on this ratio, the Pareto distribution appears as the natural model for fire sizes. The Pareto distribution

$$F(x) = 1 - (\beta/x)^\eta \quad , \quad \beta \leq x < \infty$$

belongs to the max-domain of attraction of the extreme value distribution Frecht [10] with slow decay of upper tail. For values $\eta < 2$, this distribution has infinite variance, whereas for $\eta < 1$, it has infinite mean. Heavy tails consistent with Pareto models with $\eta < 2$ have been widely observed [22] and give good empirical fit to the data. Truncation on this distribution is introduced to bring this model in line with reality so as to produce models with finite end point. However, Schoenberg et al. [78] argue that restricting the frequency of large fires by truncation is unrealistic and is inconsistent with extreme value theory. Alternatively, they suggest a more flexible tapered Pareto distribution

$$F(x) = 1 - (\beta/x)^\eta \exp(\frac{\beta - x}{\theta}) \quad , \quad \beta \leq x < \infty.$$

The tapered Pareto distribution has moderately heavy tails and belong to the max-domain of attraction of the Gumbel extreme value distribution. Schoenberg et al. [78] also give a comparison of many different models commonly used for vegetation fire distributions based on data coming from Los Angeles county.

Vegetation fire sizes depend on many factors, exhibiting a wide variety of regimes varying over space and time. The combination of the abundance of fire prone vegetation together with adverse meteorological conditions seems to impact on the upper tail behaviour of the vegetation fire size distribution. Lighter tailed models, such as the lognormal or exponential are recommended by Li et al. [46], whereas models with moderately heavy tails such as the tapered Pareto as reported by Schoenberg et al. [78]. Models consistent with heavy upper tails are also reported [22]. See Cui and Perera [20] for a review of models for fire sizes, factors that influence forest fire sizes and how this knowledge may be useful in forest and fire management.

Extreme value theory is the natural inferential tool to quantify the risk of large fires typically beyond the range of the observed data [10]. Within this theory, the most commonly used tool is the Peaks over threshold (POT) method, and the Generalized Pareto distribution (GPD) appears as the natural model for the excess fire sizes above a suitably chosen high threshold. In this method, the tail of the conditional distribution of fire size, which we denote by X is modeled by the GPD given by

$$P(X > x + u | X > u) = \left(1 + k\frac{x}{\sigma}\right)^{-1/k},$$

where $\sigma > 0$, $k \in (-\infty, \infty)$ and $1 + k\frac{x}{\sigma} > 0$. Here σ is the scale parameter and depends on the chosen threshold u. The shape parameter k is invariant with the threshold u. This model represents the tails of light, moderate as well as heavy tailed distributions through the proper choice of the shape parameter k, and there is a close relationship between the GPD and the generalized extreme value distribution(GEV) [10]. Mendes et al. [51] and Turkman et al. [89] suggest the GPD model for the tail of the vegetation fire size distributions; spatio-temporal variations are introduced into the model through the model parameters using Bayesian hierarchical modeling techniques.

17.2.3 Models for fire incidence and fire frequency data

Treating satellite data as point patterns in space and time and modelling them by using an appropriate marked point process as explained in section 17.2.1 has some drawbacks. There is strong empirical evidence [64] that the density of the marks is not independent from the local point density, so that simple structures of the form $\lambda(s, t, x) = \lambda(s, t)f(x)$ may not be adequate for vegetation fire point patterns. Further, upon applying Schoenberg's test of separability [77], complex space-time dependence structures are observed in the data, indicating that latent Gaussian process $W(s, t)$ given in (17.1) with separable space time covariance structures may not be adequate to capture these structures. Although latent process models with more complex dependence structures are known [34], such modeling strategies require extra computational and modeling skills and may not bring the degree of desired benefit in these modeling strategies. Hence, it may be more feasible to aggregate data over suitably chosen areal units and work with mathematically more tractable models at the cost of some loss of information. For example, modeling counting processes over areal units is relatively easier than modeling point patterns over continuous space. In these studies typically, generalized linear models with areal unit specific information, together with Gaussian Markov random fields based on neighbourhood structures as latent random factor to account for space time dependence, are employed. See for example, [64]. See also Serra et al. [81] for modelling counts of vegetation fires with specific fire sizes at the county level.

Further simplifications, at the cost of additional loss of information, can be achieved by transforming the point pattern data into fire incidence data through the transformation

$$Y(i,t) = \begin{cases} 1, & \text{if at least one fire is observed in areal unit } i \text{ at time unit } t; \\ 0, & \text{otherwise.} \end{cases}$$

The marks, namely sizes of the individual fires, can be aggregated into burned area fraction of each areal unit during the temporal unit t. Study of the fire incidence data is centered on two related random variables defined over each areal unit: time since the last fire, that is, the survival function, and the time between two consecutive fires, that is, mortality. Often the hazard function, which is the rate of mortality conditional on survival until time t, is the target quantity for modeling [32, 55, 66, 93]. Modeling time since last fire is more flexible than modeling inter-arrival times between fires due to the possibility of introducing into the model dynamic covariables observed at each unit of time. Right and left censoring can equally be incorporated into the model. Assuming that $T_j(i)$ represents the time of the last (jth fire) in areal unit i, Wilson et al. [93] modeled the conditional probability $P(T_{j+1}(i) \geq t | T_j(i) > t - 1)$ and show how right and left censoring can be taken into account in the model. Their model does not take into account the dependence of these random variables among the areal units; Gelfand and Monteiro [32] introduced a latent Markov Random field into their model to take into account the spatial dependence among areal units, but report that, for their data, introduction of such spatial structure did not improve upon the results of Wilson et al. [93]. Turkman et al. [90] assumed that within each areal unit, annual fire incidence can be modeled as a two state nonhomogeneous Markov chain, with states representing the situation of having at least one fire event and not having a fire event. Thus, for each areal unit i, the fire incidences $Y(i,t), t = 1, 2, ..., T$ are assumed to be a nonhomogeneous Markov chain with transition probabilities

$$M = \begin{pmatrix} p_{00}(i,t) & p_{01}(i,t) \\ p_{10}(i,t) & p_{11}(i,t). \end{pmatrix} \tag{17.2}$$

Here, the logit of the transition probabilities is assumed to be a linear function of dynamic and static covariates observed on an annual basis. Further, spatial dependence among areal units is taken into consideration by introducing a Markov random field in these link functions. This Markov model is somewhat similar to the model introduced in Gelfand and Monteiro [32].

An important component characterizing a fire regime is the fire interval distribution, that is, the distribution of the time until a given location reburns [57]. Inter-arrival times between fire incidences in each areal unit can be studied by transforming fire incidence data into fire frequency data, in which inter-arrival times between observed fires are stored as vectors of different sizes for each areal unit. Thus, for a specific grid cell i where n_i fires were observed during the study period, let

$$\mathbf{Z}(i) = \left(Z_1(i), ..., Z_{g(n_i)}(i) \right),$$

where $Z_j(i)$ is the number of years between the jth and the $j+1$th fires, for $j = 1, 2, .., g(n_i)$. Here, $Z_1(i)$ is the number of years before a fire is observed in pixel i since the beginning of the observation period, whereas $Z_{g(n_i)}(i)$ is the time period since the last fire is observed in pixel i, by the end of the study period. Note that

$$g(n_i) = \begin{cases} n_i - 1, & \text{if there is neither left nor right censoring;} \\ n_i, & \text{if there is only one censoring, either left or right;} \\ n_i + 1, & \text{if there is both left and right censoring.} \end{cases} \tag{17.3}$$

reflects the effect of the type of censoring on the dimension of the vector $\mathbf{Y}(i)$.

Typically, for each areal unit i the vector of inter-arrival times $\mathbf{Z}(i)$ are assumed to be i.i.d. realizations of lifetimes and the continuous Weibull distribution is used as the model to characterize time since fire and fire interval distributions [18, 37, 65]. The assumption of continuous models for inter-arrival times where fire interval data are acquired, at best, on an annual time frame, was questioned by Falk and Swetnam [25] who argue that, instead, discrete distribution such as the negative binomial to model the years before the first success should be used. However, recognizing that the assumption of independence required for Bernoulli trials is not realistic, they suggest the use of the lognormal distribution for fire intervals, based on the consideration that the probability (total) that a fire event will occur, is the contingent product of the probabilities of n constituent factors and hence the logarithm of that total probability is the sum of n components. Oliveira et al. [56] go further by suggesting the use of the discrete lognormal distribution to model fire interval data acquired on annual time frame. In their work they compare three models for fire frequency data, namely the continuous Weibull, discrete Weibull and discrete lognormal. They conclude that discrete the lognormal is more appropriate than the more popular Weibull model for fire frequency analysis in fire-prone tropical savannas and potentially in other regions experiencing more frequent fires.

Turkman et al. [90] suggested treating inter-arrival times as conditionally independent but not identical realizations of discrete lifetimes, having discrete Weibull distribution and represent the inter areal unit dependency through a latent spatial Markov random field introduced in the model thorough the model parameters within an Bayesian hierarchical modeling strategy.

17.2.4 Measures of risk for vegetation fires and fire risk maps

Risk is defined as expected loss, therefore a good measure of fire risk should infer to the probability that a fire will happen and its consequences [6]. However, there seems to be no agreement on the definition of fire risk [75]. Most often, fire risk is defined as the probability of fire initiation or ignition. Other definitions tend to include the potential damages fire can cause. As a proxy to fire damage, often fire size or fire intensity are used as measures of damage, together with the probability of ignition. Therefore, a typical fire danger measure is given in terms of the probability of ignition at any given place and time and the probability that the fire size, or intensity will exceed over a certain threshold conditional on ignition [51]. If point patterns of vegetation fires are modeled by a point processes of the sort given in (17.1), then fire risk for year $t + 1$ based on all the information up to year t can then be defined as the posterior predictive mean of the (random) intensity function $\lambda(s, t + 1)$ [64]. Alternatively, the predictive distribution of the random variable $N(A)$, the number of fire occurrences above a certain size in year $t + 1$ in region A, can be defined as the fire risk for region A [64, 90]. If one bases inferences on the fire incidence or fire frequency data over areal units, then fire risk can be defined in alternative forms:

- The probability of a fire in any areal unit in year $t + 1$ given the time since the last fire is observed in that unit, in the spirit of survival analysis [32, 56, 93]

- The probability of fire in year $t + 1$ in any areal unit given the observed fire incidences up to the year $t - 1$. In relation to the Markov model (17.2), this probability can be defined in terms of the transition probabilities $p_{j,1}(t)$, $j = 0, 1$ given in (17.2) [90].

A plot of such probabilities as a function of spatial coordinates or areal units results in (annual) probability risk maps and are fundamental decision support tools for global fire management [64, 90]

17.2.5   Post-fire vegetation recovery

The rate of post-fire vegetation recovery is a very good indicator of the accumulation of fire-prone vegetation in any areal unit and, therefore, can be used as a good measure of fire risk. Most plant growth models, specific to plant species, look at the change in biomass M as a function of time. Absolute growth rate, defined as the derivative of M with respect to time, and relative growth rate, defined as the exponential increase in size relative to the plant at the start of a given time period, are central in these studies. Most traditional studies of growth are based on linear and nonlinear regression analysis and on the study of growth curves [59]. See also Schliep et al. [76] for a species-level modeling strategy to quantify biomass change within a Bayesian framework. Most often, post fire vegetation recovery has to be monitored at areal units and therefore measures of growth have to be established at the scale of areal units rather on individual species.

The normalized difference vegetation index (NDVI) is a measure of greenness obtained from satellite imagery and provides a good indication of vegetation accumulation [35, 58]. Landsat derived NDVI values compare well with field data. Typically, post fire vegetation recovery analysis is based on comparing post fire NDVI values to pre-fire conditions. Gouveia et al. [35] and Bastos et al. [9] modeled post fire NDVI dynamics at fixed areal units by first defining

$$y(t) = NDVI(t) - NDVI^*(t),$$

then modeling $y(t)$ by monotone curves of the form

$$y(t) = a \exp(-bt)$$

where $NDVI(t)$, $NDVI^*(t)$ represent respectively the vegetation greenness and the ideal healthy state of vegetation at time t at a specific areal unit. In this case, $y(t)$ can be defined as the greenness deficit in any areal unit at time time t. The parameter a represents the greenness deficit at the time of the occurrence of the fire event, whereas b characterizes the vegetation recovery rate. $NDVI^*(t)$ is a function with annual cycle to be estimated based on the maximum NDVI observed under ideal fire-free conditions in each month at a given areal unit. Paci et al. [58] suggest an alternative spatio-temporal Bayesian hierarchical method. This method first models the entire space-time NDVI surface based on observations on unburned areal units and this surface is then interpolated to areal units where fire activity was observed. Finally observed NDVI values on these areal units are compared with the predicted NDVI values, providing recovery rates both at areal unit level and at individual fire scar level over the study region. Specifically, assume that $NDVI(s,t)$ and $NDVI_F(s,t)$ represent, respectively, the NDVI values at areal units s in the absence of fire and in the presence of fire at time unit t. First, these NDVI values are transformed using the probit function

$$Z(s,t) = \Phi^{-1}(NDVI(s,t)), \tag{17.4}$$

where $\Phi(x)$ is the standard normal cumulative distribution function. $Z(s,t)$ is then modeled by

$$Z(s,t) = \mu(s,t) + w(s,t) + \epsilon(s,t),$$

where $\epsilon(s,t) \sim N(0,\sigma^2)$ are i.i.d. pure errors, $\mu(s,t)$ represents the dynamic covariate information at each areal units and $w(s,t)$ is a spatio-temporal random effect to account for the space time variations by other covariates not included in the model. Once the model is fitted, and $Z(t,s)$ values are interpolated to areal units with fire activity, $NDVI(s,t)$ can be recovered from the inverse of the transform in (17.4) and the recovery rate $r(s,t)$ at areal unit s and time t is calculated by

$$r(s,t) = \frac{NDVI_F(s,t)}{E(NDVI(t,s)|data)},$$

where $E(NDVI(t,s)|data)$ represents the posterior predictive mean of the space time NDVI process at the areal units fire activity.

17.2.6 Weekly cycles of vegetation burning as anthropogenic fingerprint

Vegetation fires display strong spatial and temporal patterns at different scales due to ecoclimatic drivers and land management practices. Since natural processes do not display weekly cycles, a natural approach to detect and quantify anthropogenic influences on vegetation fire patterns is to look for weekly cycles in vegetation fire activity [24, 60]. Several studies are available on the detection of weekly cycles in various meteorological processes and there is an ongoing debate about whether the reported weekly cycles are statistically significant or not. Since biomass burning emissions have important atmospheric effects, the study of weekly cycles in vegetation fires may contribute towards understanding and explanation of weekly cycles in meteorological variables.

Pereira et al. [61] looked at global daily number of fire activities and reported a statistically significant peak in the spectrum at the angular frequency corresponding to a 7-day cycle, apart from the expected peaks at angular frequencies corresponding to one year and its first three harmonics. They also report strong correlations at lag 7 in the deseasonalized and detrended time series. They fitted a model $Y_t = M_t + X_t$, for the logarithm of the counts, where M_t is a deterministic mean function with a linear trend and seasonal component and X_t an AR(7) process. Using an integrated nested Laplace approximation introduced by Rue et al. [72], they observed that the estimated mean function corresponding to Sunday fire counts is smaller than the mean functions corresponding to the other days of the week, almost uniformly for every week. Besides, for several periods of the year, mean functions corresponding to week days are outside the 95% simultaneous credible bound for Sunday. Pereira et al. [60] also tested the hypotheses that the week day with the fewest fires in sub-Saharan Africa depends on regionally predominant religious affiliation. They modeled fire density (fire counts per km square) observed per week day in each of 372 administrative regions using a negative binomial regression model with fire counts as response variable, region area as offset and a structured random effect to account for spatial dependence. Anthropogenic biomes ("anthromes"), predominant religious affiliation in each region, the weekday and their 2-way, 3-way interactions were used as independent variables. A Bayesian methodology was used to fit this spatial generalized linear model, and by using posterior contour probabilities [83] it was found that fire activity in African croplands is significantly lower on Sunday in predominantly Christian regions, and on Friday in predominantly Muslim regions. A similar study was carried out for the observed fire activity in the USA at the county level [62], where the significance of Sunday minima was found to vary as a function of the predominant religious affiliation.

Acknowledgments

Research was partially funded by FCT (Fundação para a Ciência e a Tecnologia), Portugal, through the projects UID/MAT/00006/2013 and UID/AGR/00239/2013.

Bibliography

[1] Alexandridis, A., Russo, L., Vakalis, D., Bafas, G. V., and Siettos, C. I. (2011). Wildland fire spread modelling using cellular automata: evolution in large-scale spatially heterogeneous environments under fire suppression tactics. *International Journal of Wildland Fire*, 20:633–647.

[2] Alonso-Canas, I. and Chuvieco, E. (2015). Global burned area mapping from envisat-meris and modis active fire data. *Remote Sensing of Environment*, 163:140–152.

[3] Alvarado, E., Sandberg, D. V., and Pickford, S. W. (1998). Modeling large forest fires as extreme events. *Northwest Science*, 72:66–74.

[4] Amaral-Turkman, M. A., Turkman, K. F., Le Page, Y., and Pereira, J. M. C. (2011). Hierarchical space-time models for fire ignition and percentage of land burned by wildfires. *Environmental and Ecological Statistics*, 18:601–617.

[5] Anderson, D., Catchpole, E., de Mestre, N., , and Parkes, T. (1982). Modelling the spread of grass fires. *Journal of Australian Mathematics Society, Series B*, 23:451–466.

[6] Bachmann, A. and Allgower, B. (1998). Framework for the assessment of wildfire risk. In *Proceedings of the 3rd International Conference on Forest Fire Research, 14th Conference on Fire and Forest Meteorology*.

[7] Balshi, M. S., McGuire, A. D., Duffy, P., Flannigan, M., Walsh, J., and Melillo, J. (2009). Assessing the response of area burned to changing climate in western boreal north america using a multivariate adaptive regression splines (mars) approach. *Global Change Biology*, 15:578–600.

[8] Barbosa, P., Grgoire, J. M., and Pereira, J. M. C. (1999). An algorithm for extracting burned areas from time series of avhrr gac data applied at a continental scale. *Remote Sensing of Environment*, 69:253–263.

[9] Bastos, A., Gouveia, C., DaCamara, C. C., and Trigo, R. M. (2011). Modelling post-fire vegetation recovery in portugal. *Biogeoscience*, 8:3593–3607.

[10] Beirland, J., Goegebeur, Y., Segers, J., and Teugels, J. (2004). *Statistics of Extremes: Theory and applications*. J Wiley, Chichester.

[11] Bradstock, R. A., Cohn, J. S., Gill, A. M., Bedward, M., and Lucas, C. (2009). Prediction of the probability of large fires in the sydney region of south-eastern australia using fire weather. *Journal of Wildland Fire*, 18:932–943.

[12] Brando, P. M., Nepstad, D. C., Balch, J. K., Bolker, B., Christman, M. C., Coe, C., and Putz, F. E. (2012). Fire-induced tree mortality in a neotropical forest: the roles of bark traits, tree size, wood density and fire behavior. *Global Change Biology*, 18:630–641.

[13] Brillinger, D. R., Haiganoush, K. P., and Benoit, J. W. (2008). Risk assessment: a forest fire example. In Goldstein, D. R., editor, *Statistics and science: a Festchrift for Terry Speed*.

[14] Butry, D. T., Prestemon, J. P., and Abt, K. L. (2010). Optimal timing of wildfire prevention education. *WIT Transactions on Ecology and the Environment*, 137:197–206.

[15] Calef, M. P., Mcguire, A. D., and Chapin, F. S. (2008). Human influences on wildfire in alaska from 1988 through 2005: An analysis of the spatial patterns of human impacts. *Earth Intereactions*, 12:1–17.

[16] Certini, G. (2005). Effects of fire on properties of forest soils: a review. *Oecologia*, 143:1–10.

[17] Chen, Y., Morton, D. C., Andela, N., Giglio, L., and Randerson, J. T. (2016). How much global burned area can be forecast on seasonal time scales using sea surface temperatures? *Environmental Research Letters*, 11:045001.

[18] Clark, J. S. (1989). Ecological disturbance as a renewal process: theory and application to fire history. *Oikos*, 56:17–30.

[19] Conti, M., Marconi, U. M. B., Perona, G., and Brebbia, C. A. (2010). Diffusion limited propagation of burning fronts. In *In Second International Conference on Modelling, Monitoring and Management of Forest Fires*, Kos, Greece. WIT Press.

[20] Cui, W. and Perera, A. H. (2008). What do we know about forest fire size distribution, and why is this knowledge useful for forest management? *Int J of Wildland Fire*, 17:234–244.

[21] Cumming, S. G. (2001). A parametric model of the fire-size distribution. *Can J For Res*, 31:1297–1303.

[22] de Zea Bermudez, P., Mendes, J., Pereira, J. M. C., Turkman, K. F., , and Vasconcelos, M. J. P. (2009). Spatial and temporal extremes of wildfire sizes in portugal (19842004). *Int J Wildland Fire*, 18:983–991.

[23] Drossel, B. and Schwabl, F. (1992). A review of logistic regression models used to predict post-fire tree mortality of western north american conifers. *Photogrammetric Engineering and Remote Sensing*, 69:1629–1632.

[24] Earl, N., Simmonds, I., and Tapper, N. (2015). Weekly cycles of global fires - associations with religion, wealth and culture, and insights into anthropogenic influences on global climate. *Geophysical Research Letters*, 42.

[25] Falk, D. A. and Swetnam, T. W. (2003). Scaling rules and probability models for surface fire regimes in ponderosa pine forests. Technical report, Rocky Mountain Research Station, General Technical Report RMRS-P-29:301317 Fort Collins, CO).

[26] FAO (2007). Fire management global assessment 2006. fao forestry paper 151,. Technical report, Food and Agriculture Organization of the United Nations, Rome.

[27] Ferreira, L., Constantino, M., and Borges, J. G. (2014). A stochastic approach to optimize maritime pine (pinus pinaster) stand management scheduling under fire risk: An application in portugal. *Annals of Operations Research*, 219:359–377.

[28] Finney, M. A. (1998). Farsite: Fire area simulatormodel development and evaluation. Technical report, Technical Report Research Paper RMRS-RP-4, USDA Forest Service.

[29] Finney, M. A. (2002). Fire growth using minimum travel time methods. *Canadian Journal of Forest Research*, 32:1420–1424.

[30] Flannigan, M. D., Krawchuk, M. A., de Groot, W. J., Wotton, M., and Gowman, L. M. (2009). Implications of changing climate for global wildland fire. *International Journal of Wildland Fire*, 18:483–507.

[31] Flasse, S. and Ceccato, P. (1996). A contextual algorithm for avhrr fire detection. *International Journal of Remote Sensing*, 17:419–424.

[32] Gelfand, A. E. and Monteiro, V. D. (2013). Explaining return times for wildfires. *Journal of Statistical Theory and Practice*, 8:534–545.

[33] Giglio, L., Descloitres, J., Justice, C. O., and Kaufman, Y. J. (2003). An enhanced contextual fire detection algorithm for modis. *Remote Sensing of Environment*, 87:273–282.

[34] Gneiting, T. (2002). Nonseparable, stationary covariance functions for space-time data. *J Am Stat Assoc*, 97:590–600.

[35] Gouveia, C., Da Camara, C. C., and Trigo, R. M. (2010). Post-fire vegetation recovery in portugal based on spot/vegetation data. *Nat Hazards Earth Syst Sci*, 10:673–684.

[36] Gouveia, C. M., Bastos, A., Trigo, R. M., and Da Camara, C. C. (2012). Drought impacts on vegetation in the pre- and post-fire events over iberian peninsula. *Natural Hazards and Earth System Science*, 12:3123–3137.

[37] Grissino-Mayer, H. D. (1999). Modeling fire interval data from the american southwest with the weibull distribution. *International Journal of Wildland Fire*, 9:37–50.

[38] Hyde, K. D., Wilcox, A. C., Jencso, K., and Woods, S. (2014). Effects of vegetation disturbance by fire on channel initiation thresholds. *Geomorphology*, 214:84–96.

[39] Johnston, F. H., Henderson, S. B., Chen, Y., Randerson, J. T., Marlier, M., DeFries, R. S., Kinney, P., Bowman, D. M. J. S., and Brauer, M. (2012). Estimated global mortality attributable to smoke from landscape fires. *Environmental Health Perspectives*, 120:695–7019.

[40] Jolly, W. M., Cochrane, M. A., Freeborn, P. H., Holden, Z. A., Brown, T. J., Williamson, G. J., and Bowman, D. M. J. S. (2015). Climate-induced variations in global wildfire danger from 1979 to 2013. *Nature Communications*, 6:7537.

[41] Keeley, J. E. (2009). Fire intensity, fire severity and burn severity: a brief review and suggested usage. *International Journal of Wildland Fire*, 18:116–126.

[42] Keeley, J. E. and Syphard, A. (2016). Climate change and future fire regimes. *Geosciences*, 6.

[43] Knorr, W., Arneth, A., and Jiang, L. (2016). Demographic controls of future global fire risk. *Nature Climate Change*, doi:10.1038/nclimate2999.

[44] Krawchuk, M. A., Moritz, M. A., Parisien, M. A., Van Dorn, J., and Hayhoe, K. (2009). Global pyrogeography: the current and future distribution of wildfire. *PLoS ONE*, 4:e5102.

[45] Lee, Y., Fried, J. S., Albers, H. J., and Haight, R. G. (2013). Deploying initial attack resources for wildfire suppression: spatial coordination, budget constraints, and capacity constraints. *Canadian Journal of Forest Research*, 43:56–65.

[46] Li, C., Corns, I. G. W., and Yang, R. C. (1999). Fire frequency and size distribution under natural conditions; a new hypotheses. *landscape Ecology*, 14:533–542.

[47] Lindenmayer, D. B., Blanchard, W., McBurney, L., Blair, D., Banks, S. C., Driscoll, D. A., Smith, A. L., and Gill, A. M. (2014). Complex responses of birds to landscape-level fire extent, fire severity and environmental drivers. *Diversity and Distributions*, 20:467–477.

[48] Lindgren, F., Rue, H., and Lindström, J. (2011). An explicit link between Gaussian fields and Gaussian Markov random fields: the stochastic partial differential equation approach (with discussion). *Journal of the Royal Statistical Society, Series B*, 73(4):423–498.

[49] Martell, D. L. (2007). Forest fire management: Current practices and new challenges for operational researchers. In Weintraub, A., Romero, C., Bjrndal, T., and Epstein, R., editors, *Handbook of Operations Research in Natural Resources*. Springer Science.

[50] Martell, D. L., Otukol, S., and Stocks, B. J. (1987). A logistic model for predicting daily people-caused forest fire occurrence in ontario. *Canadian Journal of Forest Research*, 17:394–401.

[51] Mendes, J. M., de Zea Bermudez, P., Pereira, J. M. C., Turkman, K. F., and Vasconcelos, M. J. P. (2013). Spatial extremes of wildfire sizes: Bayesian hierarchical models for extremes. *Environ Ecol Stat*, 17:1–28.

[52] Miller, C. and Ager, A. A. (2013). A review of recent advances in risk analysis for wildfire management. *International Journal of Wildland Fire*, 22:1–14.

[53] Moller, J. and Diaz-Avalos, C. (2010). Structured spatio-temporal shot-noise cox point process models, with a view to modelling forest fires. *Scandinavian Journal of Statistics*, 37:2–25.

[54] Moller, J. and Waagepetersen, R. P. (2003). *Statistical Inference and Simulation for Spatial Point Processes*. Chapman and Hall/CRC.

[55] Moritz, M. A. (2000). Spatiotemporal analysis of controls on shrubland fire regimes: age dependency and fire hazard. *Ecology*, 84:351–361.

[56] Oliveira, S. L. J., Amaral Turkman, M. A., and Pereira, J. M. C. (2013). An analysis of fire frequency in tropical savannas of northern australia, using a satellite-based fire atlas. *International Journal of Wildland Fire*, 22:479–492.

[57] Oliveira, S. L. J., Pereira, J. M. C., and Carreiras, J. M. B. (2012). Fire frequency analysis in portugal (19752005), using landsat-based burnt area maps. *Int J Wildland Fire*, 21:48–60.

[58] Paci, L., Gelfand, A. E., Beamonte, M. A., Rodrigues, M., and Prez-Cabello, F. (2015). Space-time modeling for post-fire vegetation recovery. *Stochastic environmental Research and Risk Assessment*, DOI:10 1007/s00477-015-1182-6:1–13.

[59] Paine, C. E. T., Marthews, T. R., Vogt, D. R., Purves, D., Rees, M., Hector, H., and Turnbull, L. A. (2012). How to fit nonlinear plant growth models and calculate growth rates: an update for ecologists. *Methods in Ecology and Evolution*, 3:245–256.

[60] Pereira, J. M., Oom, D., Pereira, P., Amaral Turkman, A., and Turkman, K. F. (2015). Religious affiliation modulates weekly cycles of cropland burning in sub-saharan africa. *Plos One*, Doi:10 1371/journal pone 0139189:2.

[61] Pereira, J. M. C., Oom, D., Amaral Turkman, M. A., and Turkman, K. F. (2016a). Weekly cycles in daily global fire count time series. Technical report, CEAUL-University of lisbon, Report No 3/2016.

[62] Pereira, J. M. C., Oom, D., Pereira, P., Amaral Turkman, M. A., and Turkman, K. F. (2016b). The effect of religious affiliation on vegetation burning in usa counties. Technical report, CEAUL-University of lisbon, Report No 4/2016.

[63] Pereira, J. M. C., Vasconcelos, M. J. P., and Sousa, A. M. (2000). A rule-based system for burnt area mapping in temperate and tropical regions using noaa/avhrr imagery. In Innes, J. L., Verstraete, M. M., and Beniston, M., editors, *Biomass Burning and its Inter-Relationships with the Climate System*. Kluwer Academic Publishers, Dordrecht.

[64] Pereira, P., Turkman, K. F., Amaral Turkman, M. A., Sa, A., and Pereira, J. M. (2013). Quantification of annual wildfire risk; a spatio-temporal point process approach. *Statistica*, 1:55–68.

[65] Polakow, D. A. and Dunne, T. T. (1999). Modelling fire-return interval t: stochasticity and censoring in the two-parameter weibull model. *Ecological Modelling*, 121:79–102.

[66] Preisler, H. K., Brillinger, D. R., Burgan, R. E., and Benoit, J. W. (2004). Probability based models for estimation of wildfire risk. *International Journal of Wildland Fire*, 13:133–142.

[67] Prestemon, J. P. and Butry, D. T. (2010). Wildland arson: a research assessment. In Pye, J. M., Rauscher, H. M., and Sands, Y., editors, *Advances in Threat Assessment and their Application to Forest and Rangeland management*. US Department of Agriculture Forest Service, Pacific Northwest Reseach Station.

[68] Prestemon, J. P., hawbaker, T. J., Bowden, Carpenter, M. J., Brooks, M. t., Abt, K. L., Sutphen, R., and Screnton, S. (2013). Wildfire ignitions: A review of the science and recommendations for empirical modeling. srs-171. Technical report, US Department of Agriculture Forest Service, Southern research Station, Asherville, NC.

[69] Reed, W. J. and McKelvey, K. S. (2002). Power-law behaviour and parametric models for the size-distribution of forest fires. *Ecological Modelling*, 150:239–254.

[70] Rothermel, R. C. (1972). A mathematical model for predicting fire spread in wildland fuels. Technical report, USDA Forest Service Research Paper INT-115. Intermountain Forest and Range Experiment Station, Ogden, UT.

[71] Roy, D. P., Lewis, P. E., and Justice, C. O. (2002). Burned area mapping using multi-temporal moderate spatial resolution dataa bi-directional reflectance model-based expectation approach. *Remote Sensing of Environment*, 83:263–286.

[72] Rue, H., Martino, S., and Chopin, N. (2009). Approximate bayesian inference for latent gaussian models using inte- grated nested laplace approximations (with discussion). *Journal of the Royal Statistical Society, Series B*, 71:319–392.

[73] Rytwinski, A. and Crowe, K. A. (2010). A simulation-optimization model for selecting the location of fuel-breaks to minimize expected losses from forest fires. *Forest Ecology and Management*, 260:1–11.

[74] Sá, A. C., Turkman, M. A., & Pereira, J. M. (2018). Exploring fire incidence in Portugal using generalized additive models for location, scale and shape (GAMLSS). *Modeling Earth Systems and Environment*, 4(1):199–220.

[75] San-Miguel-Ayanz, J., Carlson, J. D., Alexander, M. E., Tolhurst, K., Morgan, G., Sneeuwjagt, R., and Dudfield, M. (2003). Current methods to assess fire danger potential. In Chuvieco, E., editor, *Wildland Fire Danger Estimation and Mapping - The Role of Remote Sensing Data*. World Sci Publish Co Pte Ltd.

[76] Schliep, E. M., Gelfand, A. E., Clark, J. M., and Zhu, K. (2016). Modeling change in biomass across the eastern us. *Environ Ecol Stat*, 23.

[77] Schoenberg, F. P. (2004). Testing separability in spatial-temporal marked point processes. *Biometrics*, 60:471–481.

[78] Schoenberg, F. P., Peng, R., and Woods, J. (2003). On the distribution of wildfire sizes. *Environmetrics*, 14:583–592.

[79] Schoenberg, F. P., Pompa, J., and Chang, C. H. (2009). A note on non-parametric and semiparametric modeling of wildfire hazard in los angeles county, california. Technical report, doi:10 1 1 125 3561.

[80] Schultz, M. G., Heil, A., Hoelzemann, J. J., Spessa, A., Thonicke, K., Goldammer, J. G., Held, A. C., Pereira, J. M. C., and van het Bolscher, M. (2008). Global wildland fire emissions from 1960 to 2000. *Global Biogeochemical Cycles*, 22:doi:10 1029/2007GB003031.

[81] Serra, L., Saez, M., Juan, P., Varga, D., and Mateu, J. (2014). A spatio-temporal poisson hurdle point process to model wildfires. *Stoch Environ Res Risk Assess*, 24:1671–1684.

[82] Silva, G. L., Soares, P., Marques, S., Dias, M. I., Oliveira, M. M., and Borges, J. G. (2015). A bayesian modelling of wildfires in portugal. In Bourguignon, J. P., Jeltsch, R., Pinto, A., and Viana, M., editors, *Dynamics, Games and Science*. Springer.

[83] Sørbye, S. H. and Rue, H. (2011). Simultaneous credible bands for latent gaussian models. *Scandinavian Journal of Statistics*, 38:712–725.

[84] Stoyan, D. and Penttinen, A. (2000). Recent applications of point process methods in forestry statistics. *Statistical Sciences*, 15:61–78.

[85] Strauss, D., Bednar, L., and Mees, R. (1989). Do one percent of the forest fires cause ninety-nine percent of the damage? *Forest science*, 35:319–328.

[86] Sullivan, A. L. (2009). Wildland surface fire spread modelling. *International Journal of Wildland Fire*, 18:387–403.

[87] Taylor, S. W., Woolford, D. G., Dean, C. B., and L, D. (2013). Wildfire prediction to inform fire management: statistical science challanges. *Statistical Science*, 28:586–615.

[88] Thompson, M. P. and Calkin, D. E. (2011). Uncertainty and risk in wildland fire management: A review. *Journal of Environmental Management*, 92:1895–1909.

[89] Turkman, K. F., Amaral Turkman, M. A., and Pereira, J. M. (2010). Asymptotic models and inference for extremes of spatio-temporal data. *Extremes*, 13:375–397.

[90] Turkman, K. F., Amaral Turkman, M. A., Pereira, P., Sa, A., and Pereira, J. M. C. (2014). Generating annual fire risk maps using bayesian hierarchical models. *Journal of Statistical Theory and Practice*, 8:509–533.

[91] Turner, R. (2009). Point patterns of forest fire locations. *Environ Ecol Stat*, 16:197–223.

[92] Vasconcelos, M. J. P., Gonalves, A., Catry, F. X., Paul, J. U., and Barros, F. (2002). A working prototype of a dynamic geographical information system. *International Journal of Geographical Information Science*, 16:69–91.

[93] Wilson, A. M., Latimer, A. M., Silander, J. A., Gelfand, A. E., and de Klerk, H. (2010). A hierarchical bayesian model of wildfire in a mediterranean biodiversity hotspot: Implications of weather variability and global circulation. *Ecological Modelling*, 221:106–112.

[94] Wooley, T., Shaw, D. C., Ganio, L. M., , and Fitzgerald, S. (2001). A review of logistic regression models used to predict post-fire tree mortality of western north american conifers. *Photogrammetric Engineering and Remote Sensing*, 21:1–35.

[95] Wotton, B. M., Martell, D. l., and Logan, K. A. (2003). Climate change and people-caused forest fire occurrence in ontario. *Climatic Change*, 60:275–295.

[96] Xu, H. and Schoenberg, F. P. (2011). Point process modeling of wildfire hazard in los angeles county, california. *The Annals of Applied Statistics*, 5:684–704.

18

Spatial Statistical Models for Stream Networks

Jay M. Ver Hoef
National Oceanic and Atmospheric Administration

Erin E. Peterson
Queensland University of Technology

Daniel J. Isaak
U.S. Forest Service, Boise, Idaho

CONTENTS

18.1 Introduction

Spatial statistical models are broadly applied within the fields of geology, geography, ecology, real estate, and medicine, just to name a few. Spatial statistical models often have two broad goals: 1) to describe how covariates affect a response variable, and 2) to make predictions at unsampled locations. For example, data might be used to fit a model establishing a relationship between elevation and air temperature, and also to make a map of air temperature based on a dense set of predictions. When fitting the model and making predictions, a spatial statistical model uses a covariance matrix composed of autocovariance values that rely on Euclidean distances (from straight lines in two dimensions) among all sites. Distances for stream networks, however, are poorly represented by Euclidean distances because fish and water quality attributes do not move over land but follow the sinuous pattern of the network. To apply traditional spatial statistical modeling techniques, those distances must be accurately represented in a covariance matrix structure for stream networks. This chapter provides an overview of recent research on this topic and a new class of spatial-stream network (SSN) model that is being used for an increasing number of applications.

Ver Hoef et al. [49] and Cressie et al. [9] initially described new models for stream networks based on moving average constructions. The models used stream distance, which was defined as the shortest distance between two locations computed only along the stream network. Peterson et al. [34] provide in-depth discussion on the statistical and ecological consequences of substituting stream distance measures for Euclidean distance. The original work defined a class where the moving average function was termed "tail-up," and further models with a "tail-down" moving average function were described by Ver Hoef and Peterson [47], and we give further details below. Both types of models were necessary to cover a range of autocorrelation possibilities, so a variance component approach, combining tail-up, tail-down, and classical geostatistical models based on Euclidean distance, was promoted in Garreta et al. [21], Peterson and Ver Hoef [35] and Ver Hoef and Peterson [47]. Although SSN models are most often used for streams, they can be generalized to other dendritic networks [37].

Because each stream topology is unique, prior to fitting the SSN models, data are preprocessed to develop distance and weighting matrices in a geographical information system (GIS) using the STARS (Spatial Tools for the Analysis of River Systems) custom toolset [36] for ArcGIS [16]. STARS produces the spatial, topological, and attribute information needed to fit a SSN model using the SSN package [48] with the R statistical software [38].

Many applications of the SSN models have appeared in recent years for objectives such as temperature and climate change [e.g., 22, 39, 44], water quality attributes [e.g., 7, 20, 31, 32], aquatic macroinvertebrates [19, 29], fish distribution and abundance [23, 24], variography [30, 53] and sampling designs [17, 42]. In this chapter, we review the class of models developed in Ver Hoef and Peterson [47] and bring it up to date.

18.1.1 Motivating Example

Organisms that live in streams and rivers are ectothermic and so temperature profoundly affects their ecology. Concerns about climate change and habitat alteration that degrade thermal environments have led to extensive stream temperature monitoring in previous decades by dozens of natural resources agencies throughout North American and Europe. In the American West, the NorWeST (Northwest Stream Temperature) project has aggregated and organized, from >100 natural resource agencies, most of the digital temperature records collected using miniature sensors into a massive online database (http://www.fs.fed.us/rm/boise/AWAE/projects/NorWeST.html)
The database hosts >200 million hourly temperature recordings measured at >20,000 unique stream sites. SSN models have been used with a subset of the data to create prediction maps and develop a consistent set of high-resolution climate scenarios. The project is on-going, but prediction map scenarios at a 1-km resolution have been developed for >500,000 kilometers of streams and rivers (Fig. 18.1). Detailed maps of higher resolution are available for download, and an interactive map that allows zooming and panning are available at the NorWeST website.

Based on leave-one-out crossvalidation there was an r^2 of 0.90 between true and predicted temperature values at observed sites, with a root-mean-squared prediction error (RMSPE) of 1.0degC across all streams. The accuracy of the SSN temperature model used to develop the scenarios, and the models ability to extract reliable information from large, non-random datasets collected by the management community, has led to the rapid adoption

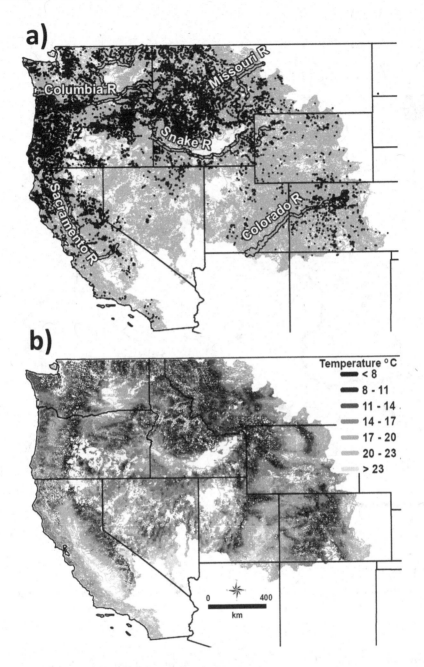

FIGURE 18.1
a) Locations of temperature records measured with sensors at >20,000 unique stream sites in the NorWeST database, where the blue lines are streams, and the solid black circles are locations. b) Mean August temperature predictions throughout most of the western United States using methods described in this chapter.

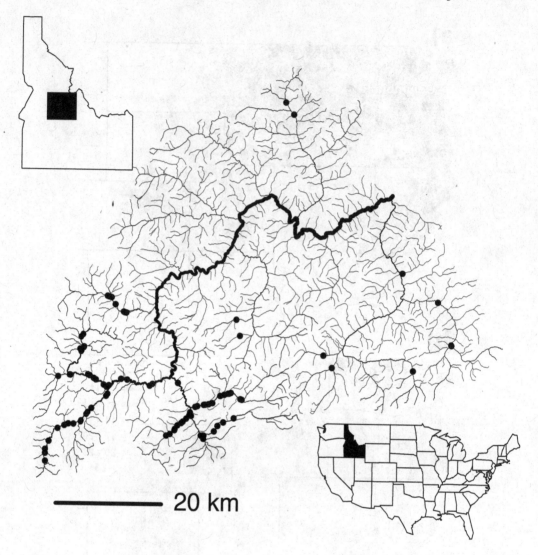

FIGURE 18.2
The study area in central Idaho, where average August temperature data were collected from the Middle Fork of the Salmon River in 2004. The sample locations are shown as solid black circles.

and broad use of NorWeST scenarios for conservation planning. The scenarios have been used to precisely identify climate refugia for cold-water fishes such as trout and char [26, 27], characterize thermal niches of fish and amphibian species [2, 25], describe how salmon migrations are affected by temperature [50], and estimate the distribution and abundance of fish populations [14, 24].

To illustrate statistical methods for stream networks in this Chapter, we use a very small subset of the NorWeST data. The stream network and sampling locations for the example are shown Fig. 18.2.

18.1.2 Why Stream Network Models?

Many researchers have called for using stream distance, rather than Euclidean distance, when developing spatial statistical models for stream networks [15, 43, 52]. There is a problem, however, if we simply use stream distance, rather than Euclidean distance, in a standard geostatistical autocovariance model [e.g., a large number geostatistical models based on Euclidean distance can be found in 8, p. 80–97]. To illustrate, we computed stream distances between all pairs of stream locations shown in Fig. 18.2, and then used stream distances (rather than Euclidean distances) in five commonly-used autocovariance models from geostatistics. For all 5 models, the "nugget" effect was set to zero and the "partial sill" was set to 1. The parameter that controls the amount of autocorrelation, often called the "range," was allowed to vary. For each value of the range for each model, a spatial covariance matrix (a description of a spatial covariance matrix can be found in Chapter ...) was determined based on stream distance among the 90 locations in Fig. 18.2. The minimum eigenvalue as a function of the range is shown in Fig. 18.3. Notice that negative eigenvalues mean that the covariance matrix is not positive definite, and hence not valid. Fig. 18.3 demonstrates that the linear-with-sill, rational quadratic, and Gaussian models are not valid when using stream distance for this data set for certain range values. This illustrates that simple substitution of stream network distance for Euclidean distance does not guarantee valid models, in the sense that the covariance matrix is not guaranteed to be positive definite. Even when a model appears to be valid for all range values for a given data set, such as the exponential or spherical models in Fig. 18.3, there is no guarantee that additional points, e.g. for prediction purposes, will keep covariance matrices positive definite. Similarly, a change in network configuration could cause these models to fail; indeed, Ver Hoef et al. [49] show a network where the spherical model has negative eigenvalues. Despite this, some researchers continue to erroneously substitute network distance into models designed for geostatistics. Thus, in order to use stream distance rather than Euclidean distance, another approach is indicated.

Even though substituting stream distance for Euclidean distance does not guarantee valid models, it is fair to ask why use stream distance at all? Fitting SSN models requires more effort than fitting standard geostatistical methods. Calculating stream distances, and weighting stream segments (as we will describe below) requires considerable GIS computation, whereas Euclidean distance can be readily calculated in R (or any software) without a GIS. What is gained by the stream network models? Consider a simple example using the data in Fig. 18.2. We withheld one measurement, the open circle in Fig. 18.4, which also shows its two closest locations (solid circles), and their values, within a small section of Fig. 18.2. We predicted the withheld observation using ordinary kriging [12] based on Euclidean distance, and also predicted it based on a stream-network model [47] that we will discuss in greater detail below. Ordinary kriging, based on Euclidean distance, predicted a value for the withheld location of 13.1degC, with a 90% prediction interval of ± 2.74degC. A tail-up stream-network model predicted a lower value of 9.4degC, with a prediction interval of ± 1.75degC. Both intervals capture the true value, but the stream network prediction is more accurate. However, there is more of interest. Recall that ordinary kriging generally weights nearby locations most heavily, and the prediction, 13.1, is somewhere between observed values 12.3 and 15.0. From this perspective, the Euclidean model produces a sensible prediction (Fig. 18.4). Yet, this is not logically consistent for a flowing stream. Notice that the temperature decreased from 15.0degC to 12.3degC downstream between the two observed locations on the main stream segment (thicker line), even though temperature generally increases while going downstream. This suggests that the unobserved tributary added cold water, causing the drop in temperature. Logically, the downstream temperature of 12.3degC should be some weighted average of temperatures from the two upstream seg-

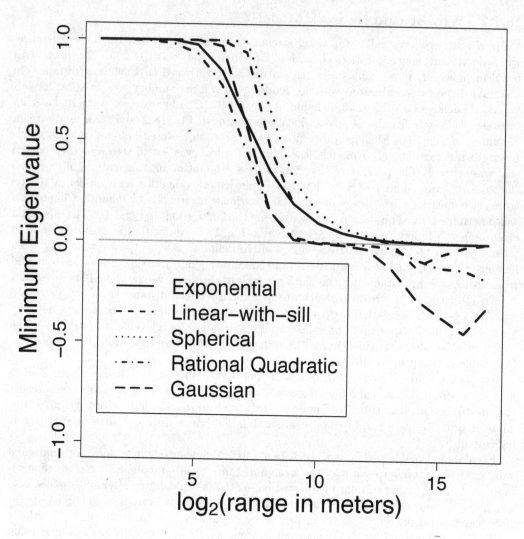

FIGURE 18.3
Minimum eigenvalues of spatial covariance matrices for classic geostatistical models based
on Euclidean distances, for various range parameters (in log base 2 meters), but when using
using stream distance, rather than Euclidean distance, among all locations shown in the
Study Area (Fig. 18.2).

ments; one with a temperature of 15.0degC, and while the other temperature is unknown,
it is surely less than 12.3degC. Thus, the ordinary kriging estimate of 13.1degC is not logically consistent, whereas the estimate from the stream-network model, 9.4degC, is logically
consistent. Although this is but a single prediction, we have observed this type of difference
between ordinary kriging and stream network model predictions on many occasions, where
the stream network model tends to reflect the topological constraints and the effect of flow
on correlation structure. A similar example is given by Peterson et al. [37].

While it is desirable to build models using stream distance, generating measures of
distance is more complex in a SSN model than a model based on Euclidean distance. Stream

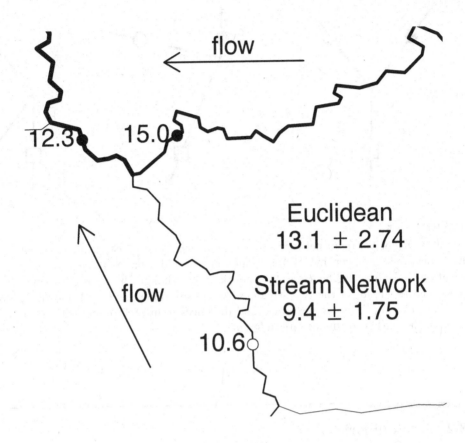

FIGURE 18.4
A comparison of ordinary kriging predictions, using a covariance matrix based on Euclidean distance, and predictions from a stream network model, using a covariance matrix based on stream distance and flow information. The values at the solid circles are used as observed data to predict the value at the open circle. The real value at the open circle is given, along with predictions for the two models with 90% prediction invervals.

network data have a dual spatial coordinate system because streams are embedded within the terrestrial landscape they flow through. As a result, points within a stream are located in 2-D space (geographical location), as well as within the network (topological location) (Fig. 18.5). Although models based on either Euclidean distance *or* network distance may be simpler to implement, critical information about environmental or ecological processes and correlation structures may be lost when models do not adequately account for the dual coordinate system [5, 13, 18, 37, 41, 45, 46]. Therefore, we will promote combining classical geostatistical models with those developed for stream networks. The details are given below. First, we give some preliminaries on notation and computing distance within networks, as these are less familiar than typical 2-D coordinate systems and computing Euclidean distance.

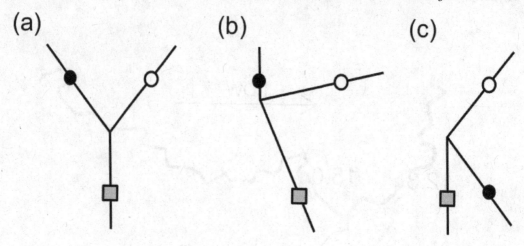

FIGURE 18.5
Locations within a stream network are characterized by a dual spatial coordinate system.
Three locations are given by solid and open circles and a grey square. In (a) and (b) the
3 locations have exactly the same 2-D coordinates, although their positions (and distances
from each other) within the two stream networks are different. In (a) and (c) the 3 locations have exactly the same positions (and distances from each other) within the network,
although their 2-D coordinates are different.

18.2   Preliminaries

The stream network topology requires some preliminaries in notation and distance concepts
and computation, which we give here, prior to actually building models.

18.2.1   Notation

We define stream segments as lines between junctions in a stream network [47, 49]. Our
convention is that a location downstream is assigned a lower real number for position than
a location upstream. A dendritic stream network has a single most-downstream point (the
outlet), which we set to 0, and it is the point from which all distances are computed. We
define "distance upstream" to be the length of the line from any location on a stream
network that can be connected by a continuous line to the network outlet (lowest point in
that network). For model development it will be convenient to extend all terminal upstream
segments to ∞ and a line downstream of the outlet to $-\infty$.

There are a finite number of stream segments in an SSN, and we index them arbitrarily
with $i = 1, 2, \ldots$. Many locations will have the same distance from the outlet (our 0 point)
in an SSN, so to uniquely define locations using distance upstream, we denote locations as
x_i, where x is distance upstream and i indicates that it is on the ith stream segment. In
Fig. 18.6, r_1 is distance r upstream and it is located on the segment labeled 1. Note the
arbitrary labeling of segments r_1, s_2, and t_3 as seen in Fig. 18.6. We use l_i for the smallest,
and u_i for the largest, upstream distance on the ith segment (Fig. 18.6).

Let A be the set of all stream segment indices. We denote U_i, a subset of A, as the set of stream segments upstream of x_i, including the ith segment, and we denote U_i^* as the set excluding the ith segment. We denote D_i, also a subset of A, as the set of stream segments downstream of x_i, including the ith segment, and we denote D_i^* as the set excluding the ith segment. This notation make precise the definition of "flow-connected" (FC) for two locations, r_i and s_j, if $U_i \cap U_j \neq \emptyset$, and they are "flow-unconnected" (FU) if $U_i \cap U_j = \emptyset$. We can also define the set of stream segments "between" two FC locations, $r_i \leq s_j$, exclusive of the ith and jth segments, as $D_j^* \setminus D_i$. For point locations, we use $\vee_s$ to denote the network only upstream of point s, including all branchings, and $\wedge_s$ to denote the network downstream of s that follows flow only (i.e., it does not go downstream and then back upstream).

Modeling covariance requires total stream distance between pairs of points, so for two FU locations we will use a to indicate the shorter distance from one location to the nearest junction downstream that shares flow with the other location (Fig. 18.6), and we use b to indicate the longer distance to the same junction. We use h for the distance between two FC locations (Fig. 18.6).

FIGURE 18.6
Three locations on a stream network, r_1, s_2, t_3. The location of the farthest upstream distance on segment 1 is u_1. The locations of the farthest downstream distances on segments 2 and 3 are l_2 and l_3, respectively. Effectively, $u_1 = l_2 = l_3$, but it is convenient to use the distinct notation. For FU sites (s_2 and t_3), a is used for the shorter distance to their common junction, and b is used for the longer distance. When sites are FC (s_2 and r_1), h is used to denote the distance between them.

18.2.2   Computing Distance in a Branching Network

In general, a straightforward approach for keeping track of network structure in a directed graph, such as a stream network, is to label segments (we adopt the graph terminology "edge"). A ForwardStar data structure [1] is commonly used to keep track of which edge leads to another edge. This data structure is based on a set of "to-from" tables, which store information about the coincidence of features and can be searched to generate new spatial information. However, traversing to-from tables generated from a large stream network is computationally intensive, especially in an interpreted programming language like R. As an alternative, we developed a system based on a binary identifier (ID), which is used to rapidly assess flow-connectivity and stream distance between any two locations on the dendritic stream network.

The process of assigning binary IDs is conceptually simple. First, the outlet edge (i.e., the most downstream edge in the network) is identified and assigned a binary ID equal to 1 (Fig. 18.7). The information stored in the to-from table is then used to identify edges that are directly upstream from the outlet edge. Binary IDs are assigned to the upstream edge(s) by arbitrarily appending a 0 or 1 to the downstream binary ID. For example, binary IDs 10 and 11 are directly upstream from binary ID 1 in Fig. 18.7. This process of moving upstream and assigning binary IDs continues until every edge in the stream network has been assigned a binary ID. Note that the binary ID is only unique within a stream network. Also, the binary ID is not the binary number that is equivalent to the segment number. Stream segments are numbered arbitrarily and sequentially, but binary IDs actually represent the branching structure.

The binary IDs are useful because they provide a way to rapidly determine whether locations have an FC or FU relationship. Two locations are considered FC when water flows from an upstream location to a downstream location. In contrast, two locations are FU when they reside on the same network, but water does not flow between them. In Fig. 18.7, site r_1 resides on the most downstream segment in the network (segment 1). Thus, the binary ID for r_1 is completely nested within the binary ID for the edge containing site s_3 ("1" is nested within "110"), which indicates that the two locations are FC. If the binary IDs for two edges are not nested, as is the case for s_3 and t_6 ("110" is not nested within "1111"), the two locations are FU. In addition, the binary IDs for FU locations contain information about the closest common downstream location. As an example, the most common downstream junction between sites s_3 and t_6 is "11" (Fig. 18.7) and so "11" is the binary ID of the stream segment where sites s_3 and t_6 diverge.

One of the advantages of SSN models is that they can be used to represent both FC and FU relationships within a stream network. In order to do this, information about directional relationships must be preserved within an asymmetric distance matrix, which contains the downstream-only distance between each pair of locations. Four pieces of spatial information are generated using the STARS toolset and subsequently used to calculate the downstream-only stream distance matrix: 1) the network ID, 2) the binary IDs, 3) the upstream distance from the network outlet to the most upstream location on each edge, u_i and 4) the upstream distance from the outlet to each location, x_i. As mentioned previously, the binary IDs are only unique within a network and so the first step is to determine whether two locations reside on the same network using the network IDs. When this is true, the binary IDs are used to determine whether two locations are FC or FU. If they are FU, the downstream-only distance between t_3 and s_2 (Fig. 18.6), a, is $t_3 - u_1$, while the downstream-only distance between s_2 and t_3, b, is $s_2 - u_1$. For FC sites, again determined by the binary IDs, the downstream-only distance from s_2 to r_1 is $s_2 - r_1 > 0$, while $r_1 - s_2 = 0$. In fact, for any set of n locations, the directional distances can be stored in an asymmetric $n \times n$ matrix, which stores the distance to the common junction. In addition, the total stream distance between locations is generated by adding the transpose of the $n \times n$ matrix, to itself. Thus, everything required to construct the models described below is contained in this asymmetric matrix.

FIGURE 18.7
A binary identifier (ID) is assigned to each segment in a stream network. Use of binary IDs allows rapid assessment of flow-connectedness, distance to junctions, and stream distance among points on the network.

18.3   Moving Average Construction for Spatial Error Process

Yaglom [51] shows that a large class of autocovariances can be developed by creating random variables as the integration of a moving-average function over a white-noise random process,

$$Z(s|\boldsymbol{\theta}) = \int_{-\infty}^{\infty} g(x - s|\boldsymbol{\theta})dW(x), \tag{18.1}$$

where x and s are locations on the real line, $g(x|\boldsymbol{\theta})$ is called the moving-average function defined on $\mathcal{R}^1$ and is square-integrable (so $E[Z(s|\boldsymbol{\theta})^2] = \int_{-\infty}^{\infty} g(x|\boldsymbol{\theta})^2 dx$), and we assume that $W(x)$ is Brownian motion. Streams are dichotomous and have flow, and this creates the need for a unique set of models not seen in either time-series or spatial statistics.

The models are created using moving average constructions. In the "tail-up" models, the moving average function starts at some location and is non-zero only upstream of that location(Fig. 18.8(a,c)). Because the tail points in the upstream direction, spatial dependency only occurs between FC locations. In addition, the moving average function must split when it reaches an upstream junction and some weighting must occur to produce stationary models. In contrast, "tail-down" models are constructed when the moving average function starts at some location and is non-zero only downstream of that location (Fig. 18.8(b,d)). In this model, spatial dependency may occur between both FC and FU locations. However, the amount of autocorrelation will generally be different between FC and FU pairs of locations even when they are separated by the same stream distance. A full development of these models is given in Ver Hoef and Peterson [47].

18.3.1   Tail-up models

For the following development, let r_i and s_j denote two locations on a stream network, and let h be the stream distance between them. For FC locations, from Eq. 18.1, the unweighted covariance between two such locations is

$$C_t(h|\boldsymbol{\theta}) = \int_h^\infty g(x|\boldsymbol{\theta})g(x - h|\boldsymbol{\theta})dx, \qquad (18.2)$$

where h is the stream distance between locations r_i and s_j. As mentioned earlier, a unique feature of tail-up stream network models is the splitting of $g(x|\boldsymbol{\theta})$ as it goes upstream (Fig. 18.8(a)), which is achieved by assigning a weighting attribute to each stream segment. To account for the splitting [9, 49], (18.1) is modified to construct a spatial process on a stream network as

$$Z(s_i|\boldsymbol{\theta}) = \int_{V_{s_i}} g(x - s_i|\boldsymbol{\theta})\sqrt{\frac{\Omega(x)}{\Omega(s_i)}}dW(x),$$

where $\Omega(x)$ is an additive function that ensures stationarity in variance; that is, $\Omega(x)$ is constant within a stream segment, but then $\Omega(x)$ is the sum of each segment's value when two segments join at a junction. (Fig. 18.9). This definition leads to (18.3) where $\pi_{i,j} = \sqrt{\Omega(s_j)/\Omega(r_i)}$. If two sites are FU, then their covariance is zero, by construction (Fig. 18.8(c)). Then the following tail-down covariance models have been developed using the moving average construction [49]:

$$C_u(r_i, s_j|\boldsymbol{\theta}_u) = \begin{cases} \pi_{i,j}C_t(h|\boldsymbol{\theta}_u) & \text{if } r_i \text{ and } s_j \text{ FC,} \\ 0 & \text{if } r_i \text{ and } s_j \text{ FU,} \end{cases} \qquad (18.3)$$

where $\pi_{i,j}$ are weights due to branching characteristics of the stream, and the function $C_t(h|\boldsymbol{\theta}_u)$ can take the following forms:

- Tail-up Linear-with-Sill Model,

$$C_t(h|\boldsymbol{\theta}_u) = \sigma_u^2 \left(1 - \frac{h}{\alpha_u}\right) I\left(\frac{h}{\alpha_u} \leq 1\right),$$

- Tail-up Spherical Model,

$$C_t(h|\boldsymbol{\theta}_u) = \sigma_u^2 \left(1 - \frac{3}{2}\frac{h}{\alpha_u} + \frac{1}{2}\frac{h^3}{\alpha_u^3}\right) I\left(\frac{h}{\alpha_u} \leq 1\right),$$

- Tail-Up Exponential Model,

$$C_t(h|\boldsymbol{\theta}_u) = \sigma_u^2 \exp(-3h/\alpha_u),$$

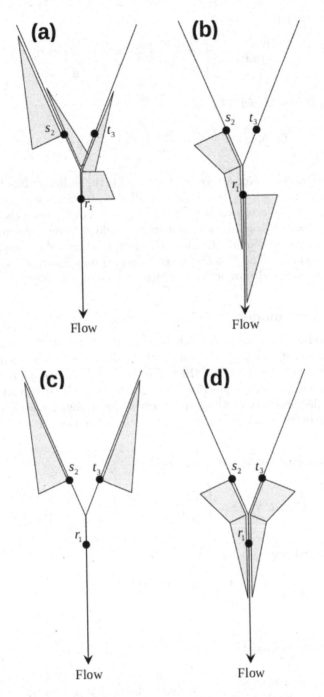

FIGURE 18.8
Three locations on a stream network, r_1, s_2, and t_3. Moving average functions are shown as grey triangles, where the function value is imagined as the width of the triangle. The construction of the random variable is obtained piece-wise by integrating the moving average function against a white noise process on each line segment. Overlapping moving average functions create autocorrelation in different ways: for (a) tail-up FC locations (b) tail-down FC locations, (c) tail-up FU locations, and (d) tail-down FU locations.

- Tail-up Mariah Model,

$$C_t(h|\boldsymbol{\theta}_u) = \begin{cases} \sigma_u^2 \left(\frac{\log(90h/\alpha_u+1)}{90h/\alpha_u} \right) & \text{if } h > 0, \\ \sigma_u^2 & \text{if } h = 0, \end{cases}$$

- Tail-up Epanechnikov Model [21],

$$C_t(h|\boldsymbol{\theta}_u) = \frac{\sigma_u^2(h-\alpha_u)^2 f_{eu}(h;\alpha_u)}{16\alpha_u^5} I\left(\frac{h}{\alpha_u} \le 1 \right),$$

where $f_{eu}(h;\alpha_u) = 16\alpha_u^2 + 17\alpha_u^2 h - 2\alpha_u h^2 - h^3$, $I(\cdot)$ is the indicator function (equal to one when the argument is true), $\sigma_u^2 > 0$ is an overall variance parameter (also known as the partial sill), $\alpha_u > 0$ is the range parameter, and $\boldsymbol{\theta}_u = (\sigma_u^2, \alpha_u)^\top$. Note the factors 3, and 90 for the exponential and Mariah models, respectively, which cause the autocorrelation to be approximately 0.05 when h equals the range parameter, which helps compare range parameters (α_u) across models. (The distance at which autocorrelation reaches 0.05 is sometimes called the effective range when models approach zero asymptotically.)

18.3.2   Tail-down models

For tail-down models, we also distinguish between the FC and FU situation. When two sites are FU, recall that b denotes the longer of the distances to the common downstream junction, and a denotes the shorter of the two distances. Then the model is the integral of the overlapping moving average functions seen in Fig. 18.8(d). If two sites are FC, again use h to denote their total separation distance via the stream network, and the model is the integral of the overlapping moving average functions seen in Fig. 18.8(b). The following are tail-down models:

- Tail-Down Linear-with-Sill Model, $b \ge a \ge 0$,

$$C_d(a,b,h|\boldsymbol{\theta}_d) = \begin{cases} \sigma_d^2 \left(1 - \frac{h}{\alpha_d} \right) I\left(\frac{h}{\alpha_d} \le 1 \right) & \text{if FC}, \\ \sigma_d^2 \left(1 - \frac{b}{\alpha_d} \right) I\left(\frac{b}{\alpha_d} \le 1 \right) & \text{if FU}, \end{cases}$$

- Tail-Down Spherical Model, $b \ge a \ge 0$,

$$C_d(a,b,h|\boldsymbol{\theta}_d) = \begin{cases} \sigma_d^2(1 - \frac{3}{2}\frac{h}{\alpha_d} + \frac{1}{2}\frac{h^3}{\alpha_d^3}) I\left(\frac{h}{\alpha_d} \le 1 \right) & \text{if FC} \\ \sigma_d^2 \left(1 - \frac{3}{2}\frac{a}{\alpha_d} + \frac{1}{2}\frac{b}{\alpha_d} \right) \left(1 - \frac{b}{\alpha_d} \right)^2 I\left(\frac{b}{\alpha_d} \le 1 \right) & \text{if FU} \end{cases}$$

- Tail-down Exponential Model,

$$C_d(a,b,h|\boldsymbol{\theta}_d) = \begin{cases} \sigma_d^2 \exp(-3h/\alpha_d) & \text{if FC}, \\ \sigma_d^2 \exp(-3(a+b)/\alpha_d) & \text{if FU}, \end{cases}$$

- Tail-down Mariah Model,

$$C_d(a,b,h|\boldsymbol{\theta}_d) = \begin{cases} \sigma_d^2 \left(\frac{\log(90h/\alpha_d+1)}{90h/\alpha_d} \right) & \text{if FC}, h > 0, \\ \sigma_d^2 & \text{if FC}, h = 0, \\ \sigma_d^2 \left(\frac{\log(90a/\alpha_d+1)-\log(90b/\alpha_d+1)}{90(a-b)/\alpha_d} \right) & \text{if FU}, a \ne b, \\ \sigma_d^2 \left(\frac{1}{90a/\alpha_d+1} \right) & \text{if FU}, a = b, \end{cases}$$

• Tail-down Epanechnikov Model, $b \geq a \geq 0$,

$$C_d(a, b, h|\boldsymbol{\theta}_d) = \begin{cases} \frac{\sigma_d^2(h-\alpha_d)^2 f_{eu}(h;\alpha_d)}{16\alpha_d^5} I\left(\frac{h}{\alpha_u} \leq 1\right) & \text{if FC,} \\ \frac{\sigma_d^2(b-\alpha_d)^2 f_{ed}(a,b;\alpha_d)}{16\alpha_d^5} I\left(\frac{b}{\alpha_u} \leq 1\right) & \text{if FU,} \end{cases}$$

where f_{eu} was defined for tail-up models, $f_{ed}(a, b; \alpha_d) = 16\alpha_d^3 + 17\alpha_d^2 b - 15\alpha_d^2 a - 20\alpha_d a^2 - 2\alpha_d b^2 + 10\alpha_d ab + 5ab^2 - b^3 - 10ba^2$, $\sigma_d^2 > 0$ and $\alpha_d > 0$, and $\boldsymbol{\theta}_d = (\sigma_d^2, \alpha_d)^{\top}$. Although not necessary to maintain stationarity, the weights used in the tail-up models can be applied to the tail-down models as well. Note that h is unconstrained, because for model-building we imagine that the headwater and outlet segments continue to infinity, as first described by Ver Hoef et al. [49]. Also note that a does not appear in the tail-down linear-with-sill model, but is used indirectly because the model depends on the point that is farthest from the junction; i.e., b, and so a is the shorter of the two distances.

FIGURE 18.9
The additive function, $\Omega(x)$, is constant within a segment, and takes on value $\sum_i w_i$ for that segment. As seen in the figure, the w's accumulate while going downstream.

18.3.3 Spatial Linear Mixed Models

The most general linear model that we consider is

$$\mathbf{Y} = \mathbf{X}\boldsymbol{\beta} + \mathbf{z}_u + \mathbf{z}_d + \mathbf{z}_e + \mathbf{W}_1\boldsymbol{\gamma}_1 + \ldots + \mathbf{W}_p\boldsymbol{\gamma}_p + \boldsymbol{\epsilon}, \tag{18.4}$$

where $\mathbf{X}$ is a design matrix of fixed effects and $\boldsymbol{\beta}$ is a vector of parameters. The vector $\mathbf{z}_u$ contains spatially-autocorrelated random variables with a tail-up autocovariance, with

$\text{var}(\mathbf{z}_u) = \sigma_u^2 \mathbf{R}(\alpha_u)$, where $\mathbf{R}(\alpha_u)$ is a correlation matrix that depends on the range parameter α_u as described in Section 18.3.1. The vector $\mathbf{z}_d$ contains spatially-autocorrelated random variables with a tail-down autocovariance, $\text{var}(\mathbf{z}_d) = \sigma_d^2 \mathbf{R}(\alpha_d)$ as described in Section 18.3.2. The vector $\mathbf{z}_e$ contains spatially-autocorrelated random variables with a Euclidean distance autocovariance, $\text{var}(\mathbf{z}_e) = \sigma_e^2 \mathbf{R}(\alpha_e)$ as described in traditional textbooks on geostatistics. We can include non-spatial random effects, where $\mathbf{W}_k$ is a design matrix for random effects $\boldsymbol{\gamma}_k; k = 1, \ldots, p$ with $\text{var}(\boldsymbol{\gamma}_k) = \sigma_k^2 \mathbf{I}$, and $\boldsymbol{\epsilon}$ contains independent random variables with $\text{var}(\boldsymbol{\epsilon}) = \sigma_0^2 \mathbf{I}$. When used for spatial prediction, this model is referred to as "universal" kriging [28, p. 107] , with "ordinary" kriging being the special case where the design matrix $\mathbf{X}$ is a single column of ones [12, p. 119]. The most general covariance matrix we form is,

$$\text{cov}(\mathbf{Y}) = \boldsymbol{\Sigma} = \sigma_u^2 \mathbf{R}(\alpha_u) + \sigma_d^2 \mathbf{R}(\alpha_d) + \sigma_e^2 \mathbf{R}(\alpha_e) + \sigma_1^2 \mathbf{W}_1 \mathbf{W}_1^\top + \ldots + \sigma_p^2 \mathbf{W}_p \mathbf{W}_p^\top + \sigma_0^2 \mathbf{I}. \tag{18.5}$$

18.4   Example

Now that spatial covariance matrices for stream networks are available, statistical inference occurs in exactly the same way as for 2-D geostatistics based on Euclidean distance. In other words, (18.4) is a spatial linear mixed model, and a data analysis can proceed in much the same way as for geostatistical analysis. We provide an example, showing some subtle distinctions that arise for the stream network models, using the data shown in Fig. 18.2 and described in Section 18.1.1. Model covariates were generated in a GIS and imported into R for analysis. Two covariates, elevation and slope, are included in the example.

18.4.1   Estimation

We will focus on restricted maximum likelihood (REML), but maximum likelihood, Bayesian methods, etc., could also be used. REML creates a likelihood of just the covariance parameters by taking appropriate contrasts to eliminate covariate effects (also, this can be viewed as integrating over the fixed effects parameters). That is, we minimized,

$$\mathcal{L}(\boldsymbol{\theta}|\mathbf{y}) = (\mathbf{y} - \mathbf{X}\boldsymbol{\beta}_g)' \boldsymbol{\Sigma}_{\boldsymbol{\theta}}^{-1} (\mathbf{y} - \mathbf{X}\boldsymbol{\beta}_g) + \log(|\boldsymbol{\Sigma}_{\boldsymbol{\theta}}|) + \log(|\mathbf{X}' \boldsymbol{\Sigma}_{\boldsymbol{\theta}}^{-1} \mathbf{X}|) + (n - p)\log(2\pi) \tag{18.6}$$

for $\boldsymbol{\theta}$, where

$$\boldsymbol{\beta}_g = (\mathbf{X}' \boldsymbol{\Sigma}_{\boldsymbol{\theta}}^{-1} \mathbf{X})^{-1} \mathbf{X}' \boldsymbol{\Sigma}_{\boldsymbol{\theta}}^{-1} \mathbf{y}. \tag{18.7}$$

is the generalized least squares estimator of $\boldsymbol{\beta}$. Note that we show the dependence of $\boldsymbol{\Sigma}$ on $\boldsymbol{\theta}$ with a subscript. For illustration, we fit a model with all four covariance components: a linear-with-sill tail-up model, a spherical tail-down model, an exponential Euclidean distance model, and a nugget effect. The REML estimates for the covariance parameters in the temperature model (with elevation and slope as fixed effects) were $\sigma_u^2 = 2.011$ with a range of $\alpha_u = 777619$ m for the tail-up linear-with-sill model, $\sigma_d^2 = 2.356$ with a range of $\alpha_d = 83782$ m for the tail-down spherical model, $\sigma_e^2 = 0.006$ with a range of $\alpha_e = 225244$ m for the Euclidean distance exponential model, and a nugget effect of $\sigma_0^2 = 0.009$ (Table 18.1). Note that the variance components are approximately equal for the tail-up model and tail-down model, but we could consider dropping the Euclidean distance model because of the small variance component ($\sigma_e^2 = 0.006$).

The covariance matrix, $\hat{\boldsymbol{\Sigma}}$ in Eq. 18.5, was constructed using the estimated covariance parameters obtained from minimizing Eq.18.6 (Table 18.1), and generalized least squares was used to estimate the fixed effects β in Eq. 18.4;

$$\hat{\beta} = (\mathbf{X}'\hat{\mathbf{\Sigma}}^{-1}\mathbf{X})^{-1}\mathbf{X}'\hat{\mathbf{\Sigma}}^{-1}\mathbf{y}$$

with a covariance matrix of $(\mathbf{X}'\mathbf{\Sigma}_\theta^{-1}\mathbf{X})^{-1}$. The estimated fixed effects were $\hat{\beta}_1 = -17.947$ with a standard error of 5.3344 for elevation, and $\hat{\beta}_2 = -7.07$ with a standard error of 11.671 for slope (Table 18.2).

With a spatial covariance matrix constructed (Eq. 18.5) for a linear model (Eq. 18.4), many of the same tools are available for data analysis, including many methods for model selection and comparison, covariate selection, model diagnostics, etc. Extensions to generalized linear models, hierarchical Bayesian models, etc., proceed in a way similar to geostatistics based on Euclidean distance. We describe an empirical semivariogram next, that is unique to stream networks, and is part of model exploration and diagnostics.

TABLE 18.1
Estimated covariance parameters using REML. The label "Var. Component" is the variance component from Eq. 18.5, "Parm. Type" is the type of parameter, either a partial sill (parsill, σ_i^2 for $i = u$, d, e, or 0), or a range parameter (α_i for $i = u$, d, e), and the "Estimate" is the value from the maximized likelihood.

Var. Component	Parm. Type	Estimate
LinearSill.tailup	parsill	2.0110
LinearSill.tailup	range	777619.3832
Spherical.taildown	parsill	2.3559
Spherical.taildown	range	83781.8091
Exponential.Euclid	parsill	0.0063
Exponential.Euclid	range	225243.7755
Nugget	parsill	0.0091

TABLE 18.2
Estimated fixed effects.

Parameter	Estimate	Std.Err.	t-value	$\Pr(t|H_0)$
(Intercept)	145.8829	40.4160	3.610	0.0005
Elevation	-17.9474	5.3344	-3.364	0.0011
Slope	-7.0734	11.6707	-0.606	0.5461

18.4.2   Torgegram

Many users of spatial statistics are familiar with empirical semivariogram plots, which are used to understand how covariance changes with Euclidean distance. A new but similar graphic, called a Torgegram, was used for stream network data. The Torgegram computes the average squared-differences like an empirical semivariogram, except that it is based on stream distance with the semivariance plotted separately for FC and FU pairs (see Fig. 18.10, where the size of the circles are proportional to the number of pairs for each binned distance class). Such figures were given in [47, 49]. The rationale behind a Torgegram

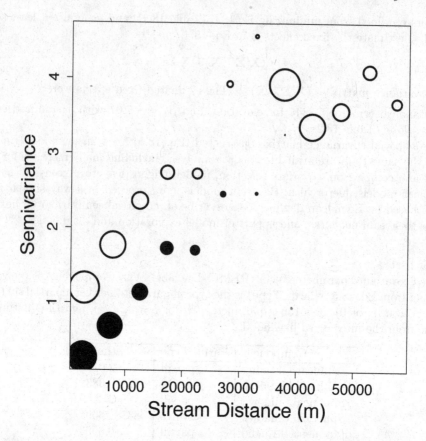

FIGURE 18.10

Toregram for the example data, computed on the residuals from the fitted model. The solid circles form the empirical semivariogram for FC sites, and the open circles form the empirical semivariogram for FU sites. The sizes of the circles are proportional to the number of pair that were averaged for each distance class.

is that autocorrelation may be evident only between FC sites (e.g., for passive movement of chemical particles), while in other cases autocorrelation may be evident between FU sites (e.g., for fish abundance because fish can move both downstream and upstream). Moreover, Section 18.3 showed models that are more natural for autocorrelation among FC sites only (Subsection 18.3.1), those that also allow autocorrelation among FU sites (Subsection 18.3.2), and variance component approaches that combine them, along with models based on Euclidean distance (Subsection 18.3.3). The Torgegram is a visual tool for evaluating autocorrelation separately for FC and FU sites, and can help inform the selection of candidate models for fitting. Further examples and discussion of the Torgegram are given in Peterson et al. [37], with a careful consideration of weighting and asymmetry in Zimmerman and Ver Hoef [53].

18.4.3 Prediction

Predictions at unsampled locations are affected by the mean (i.e., covariate) model and autocorrelation with nearby sites. The distinguishing features of stream networks are branching

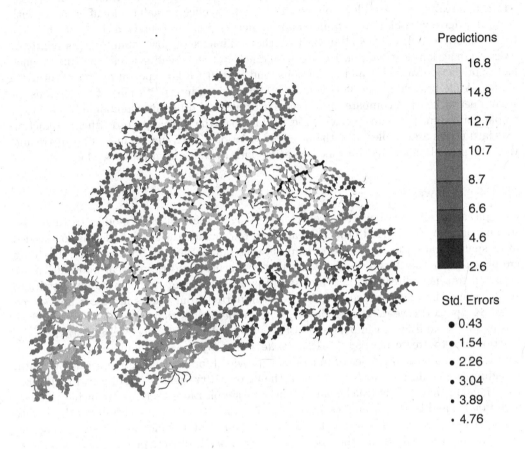

FIGURE 18.11
Temperature predictions for the example data set. The observed data are plotted as open circles, and colored by the same break points as the predictions. The predictions are solid circles. Note that the sizes of the solid circles are inversely proportional to the standard error, so more confidence is assigned to larger circles.

and flow volume (segment weights), which alter autocorrelation relationships so that they are not simple functions of distance, and this in turn affects predictions. Fig. 18.11 shows predictions of temperature throughout the study area at approximately 1 km intervals along the network, but predictions are possible for any location in the network. Notice how the standard errors increase when going up branches. Without any other data nearby that is upstream, an observed value is obviously influenced by flow from the two upper branches, which increases uncertainty. For example, temperature will be largely determined by the mixing (i.e. averaging) of waters from the two upstream branches. This average could be produced from two similar temperature values, or two widely differing values that yield the same average. Thus, branching increases prediction uncertainty, unless there are other data on the upstream branches. Flow volume also has a large effect. Small terminal branches with small flow have very high uncertainty, and predictions tend to center on the expectation of the covariate model. Fig. 18.11 shows that the dominance of flow volume allows relatively precise prediction upstream on the main channel, but the small side channels have much higher prediction variances and predictions tend to drift to the expectation of the covariate model. While branching and flow volume require unique interpretations of predictions for stream networks when compared to classical geostatistics, other ideas remain intact. For example, for the main stream segment, Fig. 18.11 shows the typical pattern where prediction standard errors are smaller near the observed locations (both upstream and downstream), with the prediction standard errors increasing with distance from these two locations.

18.4.4 Software

Two pieces of software have been developed to help users fit SSN models to their own streams datasets. The STARS tools [36] are used to calculate the spatial information needed to fit spatial statistical models to stream network data using the SSN package. The data are pre-processed in ArcGIS and the attributes, features, and topological information are exported to a new directory (i.e. the .ssn object) in a format that can be imported into R [38] using the *SSN* package [48]. Data are imported from the .ssn object and stored as an S4 `SpatialStreamNetwork` object, a new object class that builds on the spatial *sp* classes [6]. In addition to importing existing spatial data, the SSN package provides the functionality to create new `SpatialStreamNetwork` objects and simulate new data on a `SpatialStreamNetwork`. A set of functions is provided for modeling stream network data, including spatial linear models (SLMs) for the `SpatialStreamNetwork` object. In addition, traditional models that use Euclidean distance and simple random effects are included, along with Poisson and binomial families for a generalized linear mixed model framework. A suite of helper functions are provided in the *SSN* package including functions for subsetting and accessing data stored within the `SpatialStreamNetwork` object, plotting and diagnostic functions, and conversion to other *sp* objects are provided. Prediction (kriging) can be performed for missing data or for a separate set of unobserved locations, or block prediction (block kriging) can be used over sets of stream segments. Both of these toolsets, along with example datasets and tutorials can be downloaded from,
http://www.fs.fed.us/rm/boise/AWAE/projects/SpatialStreamNetworks.shtml.

18.5 Summary

Since inception in 2006 [9, 49], these models have matured rapidly as a subfield of spatial statistics and share many of its features. It also shares features with one dimensional time series. However, the branching structure, along with flow, adds interesting complexity and features not found in either 2-dimensional spatial statistics nor time series. Other authors have elaborated on the basic models. O'Donnell et al. [33] describe a functional approach to capturing spatial and temporal structure on stream networks using the same basic construction, drawing on ideas from the semiparametric modelling literature by using penalized splines. A comparison to the geostatistical models presented here can be found in Rushworth et al. [40]. Models are being developed for extreme values [3], and several authors discuss sample design considerations specific to these models [17, 42]. Yet, there are many opportunities for further research; spatio-temporal models [e.g., 11] and methods for massive data sets [e.g., 4, 10] are obvious candidates. Almost any topic of research in geostatistics will have a counterpart for stream networks.

Bibliography

[1] Ahuja, R., Magnanti, T., and Orlin, J. (1993), *Network Flows: Theory, Algorithms, and Applications*, New Jersey: Prentice-Hall.

[2] Al-Chokhachy, R., Schmetterling, D. A., Clancy, C., Saffel, P., Kovach, R. P., Nyce, L. G., Liermann, B., Fredenberg, W. A., and Pierce, R. (2016), "Are brown trout replacing or displacing bull trout populations in a changing climate?" *Canadian Journal of Fisheries and Aquatic Sciences*, 73, 1395–1404.

[3] Asadi, P., Davison, A. C., and Engelke, S. (2015), "Extremes on river networks," *arXiv preprint arXiv:1501.02663*.

[4] Banerjee, S., Gelfand, A. E., Finley, A. O., and Sang, H. (2008), "Gaussian predictive process models for large spatial data sets," *Journal of the Royal Statistical Society: Series B (Statistical Methodology)*, 70, 825–848.

[5] Benda, L., Poff, N. L., Miller, D., Dunne, T., Reeves, G., Pess, G., and Pollock, M. (2004), "The network dynamics hypothesis: how channel networks structure riverine habitats," *BioScience*, 54, 413–427.

[6] Bivand, R. S., Pebesma, E. J., and Gomez-Rubio, V. (2008), *Applied Spatial Data Analysis with R*, Springer, NY.

[7] Brennan, S. R., Torgersen, C. E., Hollenbeck, J. P., Fernandez, D. P., Jensen, C. K., and Schindler, D. E. (2016), "Dendritic network models: Improving isoscapes and quantifying influence of landscape and in-stream processes on strontium isotopes in rivers," *Geophysical Research Letters*, 43, 5043–5051.

[8] Chiles, J.-P. and Delfiner, P. (1999), *Geostatistics: Modeling Spatial Uncertainty*, New York: John Wiley & Sons.

[9] Cressie, N., Frey, J., Harch, B., and Smith, M. (2006), "Spatial Prediction on a River Network," *Journal of Agricultural, Biological, and Environmental Statistics*, 11, 127–150.

[10] Cressie, N. and Johannesson, G. (2008), "Fixed rank kriging for very large spatial data sets," *Journal of the Royal Statistical Society, Series B*, 70, 209–226.

[11] Cressie, N. and Wikle, C. K. (2011), *Statistics for Spatio-temporal Data*, Hoboken, New Jersey: John Wiley & Sons.

[12] Cressie, N. A. C. (1993), *Statistics for Spatial Data, Revised Edition*, New York: John Wiley & Sons.

[13] Dale, M. and Fortin, M.-J. (2010), "From graphs to spatial graphs," *Annual Review of Ecology, Evolution, and Systematics*, 41, 21.

[14] Dauwalter, D. C., Fesenmyer, K. A., and Bjork, R. (2015), "Using Aerial Imagery to Characterize Redband Trout Habitat in a Remote Desert Landscape," *Transactions of the American Fisheries Society*, 144, 1322–1339.

[15] Dent, C. I. and Grimm, N. B. (1999), "Spatial Heterogeneity of Stream Water Nutrient Concentrations Over Successional Time," *Ecology*, 80, 2283–2298.

[16] ESRI (2014), "ArcGIS Desktop: Release 10.3." Tech. rep., Environmental Systems Research Institute, Redlands, CA.

[17] Falk, M. G., McGree, J. M., and Pettitt, A. N. (2014), "Sampling designs on stream networks using the pseudo-Bayesian approach," *Environmental and ecological statistics*, 21, 751–773.

[18] Fausch, K. D., Torgersen, C. E., Baxter, C. V., and Li, H. W. (2002), "Landscapes to riverscapes: bridging the gap between research and conservation of stream fishes a continuous view of the river is needed to understand how processes interacting among scales set the context for stream fishes and their habitat," *BioScience*, 52, 483–498.

[19] Frieden, J. C., Peterson, E. E., Webb, J. A., and Negus, P. M. (2014), "Improving the predictive power of spatial statistical models of stream macroinvertebrates using weighted autocovariance functions," *Environmental Modelling & Software*, 60, 320–330.

[20] Gardner, K. K. and McGlynn, B. L. (2009), "Seasonality in Spatial Variability and Influence of Land Use/Land Cover and Watershed Characteristics on Stream Water Nitrate Concentrations in a Developing Watershed in the Rocky Mountain West," *Water Resources Research*, 45, DOI:10.1029/2008WR007029.

[21] Garreta, V., Monestiez, P., and Ver Hoef, J. M. (2009), "Spatial Modelling and Prediction on River Networks: Up model, down model, or hybrid," *Environmetrics*, 21, 439–456, DOI: 10.1002/env.995.

[22] Isaak, D., Luce, C., Rieman, B., Nagel, D., Peterson, E., Horan, D., Parkes, S., and Chandler, G. (2010), "Effects of Climate Change and Recent Wildfires on Stream Temperature and Thermal Habitat for Two Salmonids in a Mountain River Network," *Ecological Applications*, 20, 1350–1371.

[23] Isaak, D. J., Peterson, E. E., Ver Hoef, J. M., Wenger, S. J., Falke, J. A., Torgersen, C. E., Sowder, C., Steel, E. A., Fortin, M.-J., Jordan, C. E., et al. (2014), "Applications of spatial statistical network models to stream data," *Wiley Interdisciplinary Reviews: Water*, 1, 277–294.

[24] Isaak, D. J., Ver Hoef, J. M., Peterson, E. E., Horan, D., and Nagel, D. (2017), "Scalable population estimates using spatial-stream-network (SSN) models, fish density surveys, and national geospatial database frameworks for streams," *Canadian Journal of Fisheries and Aquatic Sciences*, 74, 147–156.

[25] Isaak, D. J., Wenger, S., and Young, M. K. (2017), "Big biology meets microclimatology: Defining thermal niches of aquatic ectotherms at landscape scales for conservation planning," *Ecological Applications*, 27, 977–990.

[26] Isaak, D. J., Young, M. K., Luce, C. H., Hostetler, S. W., Wenger, S. J., Peterson, E. E., Ver Hoef, J. M., Groce, M. C., Horan, D. L., and Nagel, D. E. (2016), "Slow climate velocities of mountain streams portend their role as refugia for cold-water biodiversity," *Proceedings of the National Academy of Sciences*, 113, 4374–4379.

[27] Isaak, D. J., Young, M. K., Nagel, D. E., Horan, D. L., and Groce, M. C. (2015), "The cold-water climate shield: delineating refugia for preserving salmonid fishes through the 21st century," *Global change biology*, 21, 2540–2553.

[28] Le, N. D. and Zidek, J. V. (2006), *Statistical Analysis of Environmental Space-Time Processes*, New York: Springer.

[29] Lois, S., Cowley, D. E., Outeiro, A., San Miguel, E., Amaro, R., and Ondina, P. (2015), "Spatial extent of biotic interactions affects species distribution and abundance in river networks: the freshwater pearl mussel and its hosts," *Journal of Biogeography*, 42, 229–240.

[30] McGuire, K. J., Torgersen, C. E., Likens, G. E., Buso, D. C., Lowe, W. H., and Bailey, S. W. (2014), "Network analysis reveals multiscale controls on streamwater chemistry," *Proceedings of the National Academy of Sciences*, 111, 7030–7035.

[31] Money, E., Carter, G., and Serre, M. L. (2009), "Modern Space/Time Geostatistics using River Distances: Data Integration of Turbidity and E.coli Measurements to Assess Fecal Contamination Along the Raritan River in New Jersey," *Environmental Science and Technology*, 43, 3736–3742.

[32] — (2009), "Using River Distances in the Space/Time Estimation of Dissolved Oxygen Along Two Impaired River Networks in New Jersey," *Water Research*, 43, 1948–1958.

[33] O'Donnell, D., Rushworth, A., Bowman, A. W., Scott, E. M., and Hallard, M. (2014), "Flexible regression models over river networks," *Journal of the Royal Statistical Society: Series C (Applied Statistics)*, 63, 47–63.

[34] Peterson, E. E., Theobald, D. M., and Ver Hoef, J. M. (2007), "Geostatistical Modeling on Stream Networks: Developing Valid Covariance Matrices Based on Hydrologic Distance and Stream Flow," *Freshwater Biology*, 52, 267–279.

[35] Peterson, E. E. and Ver Hoef, J. M. (2010), "A mixed-model moving-average approach to geostatistical modeling in stream networks," *Ecology*, 93, 644–651.

[36] — (2014), "STARS: An ArcGIS toolset used to calculate the spatial information needed to fit spatial statistical models to stream network data," *Journal of Statisticla Software*, 56, 1–17.

[37] Peterson, E. E., Ver Hoef, J. M., Isaak, D. J., Falke, J. A., Fortin, M.-J., Jordan, C., McNyset, K., Monestiez, P., Ruesch, A. S., Sengupta, A., Som, N., Steel, A., Theobald, D. M., Torgersen, C. E., and Wenger, S. J. (2013), "Stream networks in space: concepts, models, and synthesis," *Ecology Letters*, 16, 707–719.

[38] R Core Team (2015), *R: A Language and Environment for Statistical Computing*, R Foundation for Statistical Computing, Vienna, Austria.

[39] Ruesch, A. S., Torgersen, C. E., Lawler, J. J., Olden, J. D., Peterson, E. E., Volk, C. J., and Lawrence, D. J. (2012), "Projected climate-induced habitat loss for salmonids in the John Day River network, Oregon, USA," *Conservation Biology*, 26, 873–882.

[40] Rushworth, A., Peterson, E., Ver Hoef, J., and Bowman, A. (2015), "Validation and comparison of geostatistical and spline models for spatial stream networks," *Environmetrics*, 26, 327–338.

[41] Schlosser, I. J. and Angermeier, P. (1995), "Spatial variation in demographic processes of lotic fishes: conceptual models, empirical evidence, and implications for conservation," *American Fisheries Society Symposium*, 17, 360–370.

[42] Som, N. A., Monestiez, P., Ver Hoef, J. M., Zimmerman, D. L., and Peterson, E. E. (2014), "Spatial sampling on streams: principles for inference on aquatic networks," *Environmetrics*, 25, 306–323.

[43] Torgersen, C. E., Gresswell, R. E., and Bateman, D. S. (2004), "Pattern Detection in Stream Networks: Quantifying Spatial Variability in Fish Distribution," in *Proceedings of the Second Annual International Symposium on GISSpatial Analyses in Fishery and Aquatic Sciences*, eds. Nishida, T., Kailola, P. J., and Hollingworth, C. E., Saitama, Japan: Fishery GIS Research Group, pp. 405–420.

[44] Turschwell, M. P., Peterson, E. E., Balcombe, S. R., and Sheldon, F. (2016), "To aggregate or not? Capturing the spatio-temporal complexity of the thermal regime," *Ecological Indicators*, 67, 39–48.

[45] Urban, D. and Keitt, T. (2001), "Landscape connectivity: a graph-theoretic perspective," *Ecology*, 82, 1205–1218.

[46] Urban, D. L., Minor, E. S., Treml, E. A., and Schick, R. S. (2009), "Graph models of habitat mosaics," *Ecology Letters*, 12, 260–273.

[47] Ver Hoef, J. M. and Peterson, E. (2010), "A Moving Average Approach for Spatial Statistical Models of Stream Networks (with discussion)," *Journal of the American Statistical Association*, 105, 6–18.

[48] Ver Hoef, J. M., Peterson, E., Clifford, D., and Shah, R. (2014), "SSN: An R package for spatial statistical modeling on stream networks," *Journal of Statistical Software*, 56, 1–45.

[49] Ver Hoef, J. M., Peterson, E. E., and Theobald, D. (2006), "Spatial Statistical Models That Use Flow and Stream Distance," *Environmental and Ecological Statistics*, 13, 449–464.

[50] Westley, P. A., Dittman, A. H., Ward, E. J., and Quinn, T. P. (2015), "Signals of climate, conspecific density, and watershed features in patterns of homing and dispersal by Pacific salmon," *Ecology*, 96, 2823–2833.

[51] Yaglom, A. M. (1987), *Correlation Theory of Stationary and Related Random Functions. Volume I*, New York: Springer-Verlag.

[52] Yuan, L. L. (2004), "Using Spatial Interpolation to Estimate Stressor Levels in Unsampled Streams," *Environmental Monitoring and Assessment*, 94, 23–38.

[53] Zimmerman, D. L. and Ver Hoef, J. M. (2017), "The Torgegram for Fluvial Variography: Characterizing Spatial Dependence on Stream Networks," *Journal of Computational and Graphical Statistics*, 26, 253–264.

Part III

Topics in Environmental Exposure

19

Statistical methods for exposure assessment

Montse Fuentes

Virginia Commonwealth University

Brian J Reich

North Carolina State University, Raleigh, North Carolina

Yen-Ning Huang

Indiana University, Bloomington, Indiana

CONTENTS

19.1 Defining Exposure

Assessing the exposure of an individual or population is a challenging statistical task [21]. Borrowing from the concept of the genome in genetics, recently the concept of an *exposome* [13] has emerged to help explain the complexities of exposure assessment and guide research in this area. According to the Centers for Disease Control and Prevention [1], "The exposome can be defined as the measure of all the exposures of an individual in a lifetime and how those exposures relate to health." A statistical analysis of exposure must consider complexities such as simultaneous exposure from multiple contaminants, spatiotemporal aggregation, and multiple sources of uncertainty. A data collection plan and subsequent statistical analysis relies on clearly defining the exposure of interest, and the precise definition of exposure and exposure metrics is crucial for interpreting the results and limitations of a study.

The receptor-based definition of exposure begins by defining the *concentration* of the contaminant at location $\mathbf{s}$ and time t, $C(\mathbf{s}, t)$. The exposure of an individual between times

t_0 and t_1 is then defined as the cumulative amount of the contaminant that reaches the individual,

$$\int_{t_0}^{t_1} C[\mathbf{u}(t), t]dt,\tag{19.1}$$

where $\mathbf{u}(t)$ is the spatial location of the individual at time t. Taking this concept a step further, the *dose* is the amount of the contaminant that is actually ingested by the individual. Depending on the nature of the contaminant and individual, these three summaries can be quite different, and thus the concentration-response, exposure-response, and dose-response curves used to quantify the health effect of the contaminant can have substantially different meanings.

For example, in Section 19.2 we consider the example of exposure to fine particulate matter ($PM_{2.5}$). In this example, we use $PM_{2.5}$ concentrations measured at several locations in the US to estimate the concentration surface for all $\mathbf{s}$. We note however that these estimated concentrations may only be crude estimates of exposure and dose. Even two individuals living in the same residence may have different exposures depending on their commuting pattern and amount of time spent outdoors. For a thorough discussion of individual exposure see Chapter X1. Similarly, two individuals with identical exposures may have different doses depending in their inhalation rate or metabolism.

As indicated by (19.1), the level of aggregation plays a key role in exposure assessment. In some cases, concentration data are only available in aggregates, and in other settings aggregation is performed to reduce uncertainty in the concentration estimates or to better represent exposure. The most common form of aggregation is temporal, which can be used to distinguish between *acute exposure* over a narrow time window, usually close to the health event of interest, and *cumulative exposure* over a longer period. For example, we may compute the average concentration over the previous few days to measure acute exposure to explain asthma events, or the average concentration over the first trimester to measure a mother's cumulative exposure during pregnancy. Spatial aggregation has similar benefits. A common motivation for spatial aggregation is to match the scale of a health endpoint defined at a regional level. For example, the county-average concentration may be deemed the most relevant measure of exposure to analyze the number of events in the county. Even in cases where the objective is to measure exposure for an individual, if the exact location of the individual over time (i.e., $\mathbf{u}(t)$ in (19.1)) is unknown, a spatial average may be a reasonable proxy for exposure.

In addition to spatiotemporal aggregation, other forms of aggregation can be used depending on the nature of the contaminant. For example, $PM_{2.5}$ is defined as the total concentration of particles with diameter below 2.5 micrometers, and is thus an aggregate of a complex mixture of contaminants. The total $PM_{2.5}$ concentration can be decomposed in several interesting and useful ways. Recent results [12] suggest that ultrafine particles with diameter less than 100 nanometers may in fact have stronger health effect than larger particles. From a policy perspective, it may be more relevant to decompose $PM_{2.5}$ by source rather than size, e.g., concentration of particles attributed to motor vehicle traffic versus biomass burning [19]. Finally, $PM_{2.5}$ is a complex mixture of different types of particles, such as nitrate, sulfate, and organic carbon. There is no reason to assume that all types of particles have the same health effects, and so recent work has begun to study the effects of individual particle types [11].

The type of exposure that can be assessed is determined by the nature of the data collected. Concentration is arguably the easiest quantity to estimate. In response to the U.S. Clean Air Act, the U.S. EPA has established an extensive network to monitor the ambient concentration of several air pollutants. These networks provide hourly or daily measurements at fixed spatial locations, and combined with spatiotemporal interpolation methods give estimates of the concentration surface $C(\mathbf{s}, t)$ or aggregated summaries. However, with these *indirect* measurements it is difficult to assess the exposure of an individual. *Direct measurements* such as wearable devices are less abundant and more expensive, but provide valuable information about individual exposure, and the relationship between concentration and exposure. Measurements of the dose ingested by the individual typically require even more invasive sampling, such as hair or blood samples, but provide the most relaible data to establish dose-response curves.

19.2   Spatiotemporal Mapping of Monitoring Data

A key task in an exposure assessment is to estimate the concentration function $C(\mathbf{s}, t)$ (denoted in this section as $Y(\mathbf{s}, t)$ as is common in the spatiotemporal literature) for relevant locations $\mathbf{s}$ and times t. One approach is to develop a method to estimate the concentration for all $\mathbf{s}$ and t. The estimated concentration function can then be used to monitor changes over time, identify spatial hotspots, extract concentrations for specific locations, and compute aggregate measures such as (19.1). Collecting data for all spatiotemporal locations is often infeasible (with some notable exceptions such as remote sensing, satellite, or computer simulation data; see Section 19.4), and we focus on the case where data are collected at a finite number of locations and statistical algorithms are used to interpolate the measurements to create a concentration function applicable for all $\mathbf{s}$ and t. In this section we discuss three popular spatial interpolation methods: K-nearest neighbors, inverse distance weighting, and Kriging. Spatial methods for a single time point are discussed in general, with emphasis on quantifying uncertainty, and then demonstrated with an application to $PM_{2.5}$ data. Spatiotemporal extensions are also provided.

19.2.1 K-Nearest Neighbor (KNN) interpolation

The simplest algorithm to estimate concentration at an unmonitored locations is to use the measured value for the nearest monitoring location. The estimated nearest-neighbor concentration surface is then piecewice constant over Voronoi polygons [14] around the monitors. This is a special case of K-nearest neighbor (KNN) interpolation (with $K = 1$). Let $Y(\mathbf{s})$ be the observed data at location $\mathbf{s} \subset \mathbb{R}^2$ and $\mathbf{s}_1, \ldots, \mathbf{s}_N$ be the N monitor locations. The KNN estimate at an unobserved location $\mathbf{s}_0$ is the average of K nearest observations. If $d_i(\mathbf{s}_0) = \|\mathbf{s}_0 - \mathbf{s}_i\|$ denotes the distance between $\mathbf{s}_0$ and $\mathbf{s}_i$, then the KNN estimator can be written

$$\hat{Y}(\mathbf{s}_0) = \sum_{i=1}^{N} w_i Y(\mathbf{s}_i) \tag{19.2}$$

where $w_i = \frac{1}{K}$ if $\mathbf{s}_i$ belongs to the nearest K neighbors of $\mathbf{s}_0$ and 0 otherwise. The variance of $\hat{Y}(\mathbf{s}_0)$ is

$$\mathrm{Var}[\hat{Y}(\mathbf{s}_0)] = \sum_{i=1}^{N} \sum_{j=1}^{N} w_i w_j \mathrm{Cov}(\mathbf{s}_i, \mathbf{s}_j) \tag{19.3}$$

where $\mathrm{Cov}(\mathbf{s}_i, \mathbf{s}_j)$ is the spatial covariance function. Applying this variance requires an estimate of the spatial covariance function. As described further in Section 19.2.3, if the covariance function is known then Kriging is the optimal method for spatial prediction. Therefore, in KNN is typically conducted without assuming that the covariance function is known, making uncertainty difficult to quantify. In addition, it has been shown that using the nearest monitoring site to estimate air pollutant exposures performs poorly in complex urban environments [10].

19.2.2   Inverse Distance Weighting (IDW)

Inverse distance weighting (also known as kernel smoothing) is an extension to KNN that weights locations according to their distance from the prediction location. The IDW estimate of $Y(\mathbf{s}_0)$ has the form of (19.2) but with a different choice of weights. To encode spatial structure, the weights decrease with the distance from $\mathbf{s}_0$. The weights are often written as a function of symmetric kernel function $k(\mathbf{s}_i, \mathbf{s}_0) = k(\mathbf{s}_0, \mathbf{s}_i) \geq 0$ and forced to sum to one, i.e., $\sum_{i=1}^{N} w_i(\mathbf{s}_0) = 1$, using the transformation

$$w_i(\mathbf{s}_0) = \frac{k(\mathbf{s}_i, \mathbf{s}_0)}{\sum_{j=1}^{N} k(\mathbf{s}_j, \mathbf{s}_0)}. \tag{19.4}$$

There are many possible kernel functions. Two popular options are the Gaussian (squared exponential) kernel $k(\mathbf{s}_i, \mathbf{s}_o) = \exp\{-[d_i(\mathbf{s}_0)/\rho]^2\}$ and the power kernel $k(\mathbf{s}_i, \mathbf{s}_o) = d_i(\mathbf{s}_0)^{-1/\rho}$, where $\rho > 0$ is the kernel bandwidth and controls the degree of spatial smoothness. For large datasets, these kernels can be slow because they are a global average over all N observations. When N is large, computation can be improved by using a compact kernel that is zero for large $d_i(\mathbf{s})$ so that the IDW estimate becomes a local average. As with the choice of covariance function in Kriging (Section 19.2.3), it is important to carefully choose the parametric form of the kernel function k and bandwidth parameter ρ, and test for sensitivity to these specifications. Also, variance of the IDW estimate is given by (19.3), and is thus difficult to compute without estimates of the spatial covariance function.

19.2.3   Kriging

Kriging is a powerful interpolation method described in detail in Chapter X, and briefly here. Kriging can be viewed as the predictions that arise from modeling the data as a realization of a Gaussian process defined by its mean function $E[Y(\mathbf{s})] = \mu(\mathbf{s})$ and covariance function $\mathrm{Cov}[Y(\mathbf{s}), Y(\mathbf{s}')] = \rho(\mathbf{s}, \mathbf{s}')$; in this section we assume that the covariance function is stationary and isotropic, i.e., $\rho(\mathbf{s}, \mathbf{s}') \equiv \rho(h)$ where h is the distance between $\mathbf{s}$ and $\mathbf{s}'$. This assumption implies that the covariance is the same for all pairs of points separated by the same distance, and is invariant to translations and rotations. A Kriging analysis proceeds by estimating the mean and covariance function in the model-building step, and with estimates in hand the Kriging equations are applied for prediction at new locations.

     We assume parametric forms for both the mean and covariance functions. Let the mean function be $\mu(\mathbf{s}) = \mathbf{X}(\mathbf{s})^T \boldsymbol{\beta}$, where $\mathbf{X}(\mathbf{s})$ is a p-vector of spatial covariates. For example, in Chapter X2 the spatial land-use regression models include local covariates such as distance to a roadway or a point source. Depending on assumptions about the mean function, Kriging can be classified as simple, ordinary, or universal Kriging. In simple kriging we assume the mean function $\mu(\mathbf{s})$ is known at all locations and so Kriging prediction can be applied on the mean-zero residuals. In ordinary Kriging the mean is an unknown constant $\mu(\mathbf{s}) = \beta_0$ that needs to be estimated from the data, and universal Kriging includes covariates $\mathbf{X}(\mathbf{s})$ and requires estimating the coefficient vector $\boldsymbol{\beta}$.

     Many options are available, but we use the Matern spatial covariance function [18]

$$\rho(h) = \begin{cases} \tau^2 + \sigma^2 & \text{for } h = 0 \\ \sigma^2 \frac{2^{1-\nu}}{\Gamma(\nu)} \left(\sqrt{2\nu}\frac{h}{\rho}\right) K_\nu \left(\sqrt{2\nu}\frac{h}{\rho}\right) & \text{for } h > 0 \end{cases} \tag{19.5}$$

where K is the modified Bessel function of the second kind. The Matern has four parameters $\boldsymbol{\theta} = (\tau^2, \sigma^2, \rho, \nu)$: the nugget $\tau^2 > 0$ effects only variance at distance zero, e.g., measurement error; the partial sill $\sigma^2 > 0$ dictates the overall spatial variance; $\rho > 0$ is the spatial range and determines the strength of long-range spatial correlation; and $\nu > 0$ measures the local spatial correlation and smoothness.

Closely related to the covariance function is the variogram, defined as $E\{[Y(\mathbf{s}) - Y(\mathbf{s}')]^2\} = \gamma(\mathbf{s}, \mathbf{s}')$. For the stationary and isotropic model the variogram is a function of only the distance between locations, i.e., $\gamma(\mathbf{s}, \mathbf{s}') \equiv \gamma(h)$. When the covariance exists, it is related to the variogram as $\gamma(h) = \rho(0) - \rho(h)$. For example, the Matern variogram is $\gamma(0) = 0$ and

$$\gamma(h) = \tau^2 + \sigma^2 \left[1 - \frac{2^{1-\nu}}{\Gamma(\nu)} \left(\sqrt{2\nu} \frac{h}{\rho} \right) K_\nu \left(\sqrt{2\nu} \frac{h}{\rho} \right) \right] \tag{19.6}$$

for $h > 0$.

A useful graphical tool to determine whether a particular spatial covariance function such as the Matern fits the data well is the empirical variogram. For example, under the assumption that the process is intrinsically stationary with isotropic variogram $\gamma(h)$, the non-parametric empirical variogram is [5]:

$$\hat{\gamma}(h) = \frac{1}{2|N(h)|} \sum_{N(h)} [Y(\mathbf{s}_i) - Y(\mathbf{s}_j)]^2 \tag{19.7}$$

where $N(h)$ denotes the set of pairs (i, j) such that $\|\mathbf{s}_i - \mathbf{s}_j\| = (h - \epsilon, h + \epsilon)$, $\epsilon > 0$ is the bin width, and $|N(h)|$ is the number of pairs in $N(h)$. The empirical variogram is computed and plotted for a grid of distances $h \in \{h_1, h_2, ...\}$ and compared to the model-based variogram $\gamma(h)$ in (19.6) to determine if the parametric model is appropriate.

With the assumptions of normality, stationarity, and parametric mean and covriance functions, the likelihood of the observations at the N sample locations $\mathbf{Y} = [Y(\mathbf{s}_1), ..., Y(\mathbf{s}_N)]^T$ is simply the multivariate normal with mean vector $\mathbf{X}\boldsymbol{\beta}$ and covariance matrix $\Sigma(\boldsymbol{\theta})$, where $\mathbf{X} = [\mathbf{X}(\mathbf{s}_1)^T, ..., \mathbf{X}(\mathbf{s}_N)^T]^T$ is the $N \times p$ covariate matrix and $\Sigma(\boldsymbol{\theta})$ is the $N \times N$ matrix with (i, j) element equal to $\rho(\|\mathbf{s}_i - \mathbf{s}_j\|)$ evaluated at $\boldsymbol{\theta}$. The likelihood function is then

$$l(\boldsymbol{\beta}, \boldsymbol{\theta}) = -\frac{1}{2}\log(|\Sigma(\boldsymbol{\theta})|) - \frac{1}{2}(\mathbf{Y} - \mathbf{X}\boldsymbol{\beta})^T \Sigma(\boldsymbol{\theta})^{-1}(\mathbf{Y} - \mathbf{X}\boldsymbol{\beta}).$$

The parameters $\boldsymbol{\beta}$ and $\boldsymbol{\theta}$ can therefore be estimated by maximum likelihood analysis, although this can be slow for large N because it requires many operations on the $N \times N$ covariance matrix $\Sigma(\boldsymbol{\theta})$.

Given parameter estimates $\hat{\boldsymbol{\beta}}$ and $\hat{\boldsymbol{\theta}}$, the Kriging prediction for unmonitored location $\mathbf{s}_0$ is the conditional mean of $Y(\mathbf{s}_0)$ given the monitor data $\mathbf{Y}$,

$$\hat{Y}(\mathbf{s}_0) = \mathbf{X}(\mathbf{s}_0)\hat{\boldsymbol{\beta}} + \Sigma_0(\hat{\boldsymbol{\theta}})\Sigma(\hat{\boldsymbol{\theta}})^{-1} \left(\mathbf{Y} - \mathbf{X}\hat{\boldsymbol{\beta}} \right) \tag{19.8}$$

where Σ_0 is the N vector of covariances between $Y(\mathbf{s}_0)$ and $\mathbf{Y}$. Similarly, the prediction variance is simply the variance of the condition distribution of $Y(\mathbf{s})$ given $\mathbf{Y}$,

$$\text{Var}[\hat{Y}(\mathbf{s}_0)] = \rho(0) - \Sigma_0(\hat{\boldsymbol{\theta}})\Sigma(\hat{\boldsymbol{\theta}})^{-1}\Sigma_0(\hat{\boldsymbol{\theta}}). \tag{19.9}$$

Alternatively, the Kriging predictor can be written

$$\hat{Y}(\mathbf{s}_0) = \mathbf{X}(\mathbf{s}_0)^T \hat{\boldsymbol{\beta}} + \sum_{i=1}^{N} w_i(\mathbf{s}_0) \left[Y(\mathbf{s}_i) - \mathbf{X}(\mathbf{s}_i)^T \hat{\boldsymbol{\beta}} \right] \tag{19.10}$$

where the weights $w_i(\mathbf{s}_0)$ are the N elements of $\Sigma_0(\hat{\boldsymbol{\theta}})\Sigma(\hat{\boldsymbol{\theta}})^{-1}$, which can be determined using the variogram function γ evaluated at $\hat{\boldsymbol{\theta}}$. Therefore, as in (19.2) the Kriging prediction (19.8) is a linear combination of the observations. Rather than specifying the weights directly as in KNN or IDW, the Kriging weights are determined indirectly via estimates of the mean and covariance functions. It can be shown that under the assumed mean and covariance functions, these weights are optimal in the sense that they give the best linear unbiased predictions (BLUP).

Kriging naturally leads to spatial-average concentration estimates and uncertainty quantification for these aggregate estimates. Let $Y(B) = \frac{1}{|B|} \int_B Y(\mathbf{s}) ds$ be the spatial-average concentration over region B (with area $|B|$). The integral can be approximated arbitrarily well using a finite average

$$Y(B) \approx \frac{1}{J} \sum_{j=1}^{J} Y(s_{0j})$$

where the J grid points s_{0j} cover B and $\mathbf{Y}_0 = [Y(\mathbf{s}_{01}), ..., Y(\mathbf{s}_{0J})]^T$ are the concentrations at the J grid points. The Kriging estimate for $\mathbf{Y}_0$ is

$$\hat{\mathbf{Y}}_0 = \mathbf{X}_0 \hat{\boldsymbol{\beta}} + \Sigma_0(\hat{\boldsymbol{\theta}}) \Sigma(\hat{\boldsymbol{\theta}})^{-1} \left(\mathbf{Y} - \mathbf{X}\hat{\boldsymbol{\beta}} \right) \tag{19.11}$$

where $\mathbf{X}_0$ is the $J \times p$ covariate matrix for the grid points and $\Sigma_0(\hat{\boldsymbol{\theta}})$ is the $J \times N$ matrix defining the covariance between $\mathbf{Y}_0$ and $\mathbf{Y}$ with (j, i) element defined by ρ with parameters set to $\hat{\boldsymbol{\theta}}$. The predicted value is then

$$\hat{Y}(B) = \frac{1}{J} \sum_{j=1}^{J} \hat{Y}(s_{0j})$$

where $\hat{Y}(s_{0j})$ is the Kriging prediction in (19.8). The variance of this estimate is

$$\text{Var}[\hat{Y}(B)] = \frac{1}{J^2} \sum_{j=1}^{J} \sum_{k=1}^{J} \text{Cov}[\hat{Y}(\mathbf{s}_{0j}), \hat{Y}(\mathbf{s}_{0k})], \tag{19.12}$$

where $\text{Cov}[\hat{Y}(\mathbf{s}_{0j}), \hat{Y}(\mathbf{s}_{0k})]$ is the (j, k) element of $\text{Cov}(\hat{Y}_0) = \Sigma_{00}(\hat{\boldsymbol{\theta}}) - \Sigma_0(\hat{\boldsymbol{\theta}}) \Sigma(\hat{\boldsymbol{\theta}})^{-1} \Sigma_0(\hat{\boldsymbol{\theta}})^T$ for $J \times J$ covariance matrix $\Sigma_{00}(\hat{\boldsymbol{\theta}})$.

19.2.4   Bayesian Interpolation

The covariance matrix, Σ, in the kriging method is treated as known. The estimated covariance, $\hat{\Sigma}$, replaces Σ in the kriging formula resulting in the classical kriging interpolator. In this case the uncertainty in the covariance structure is ignored which leads to overconfidence in the interpolated values. In Bayesian kriging, uncertainty about Σ (as well as all other parameters in the model) is quantified. For instance, [22] apply universal Bayesian kriging to estimate the residential exposure to air pollutants in a high-risk area providing an assessment of model uncertainty.

In classical kriging, θ are estimated using likelihood approaches or empirical methods and then treated as known for interpolation. In Bayesian kriging, we introduce a prior distribution on the parameters and update the distribution using the data. The solution to Bayesian kriging is the posterior predictive distribution of $Y(\mathbf{s}_0, t_0)$ given the observations $\mathbf{Y}$:

$$\pi(Y(\mathbf{s}_0, t_0)|\mathbf{Y}) \propto \int \pi(Y(\mathbf{s}_0, t_0)|\mathbf{Y}, \theta) \pi(\theta|\mathbf{Y}) d\theta \tag{19.13}$$

where $\pi(\theta|\mathbf{Y})$ is the posterior of the model parameters θ. As a helpful reference, [2] provide a tutorial for Bayesian kriging using `WinBUGS`.

Spatial models often assume the outcomes follow normal distributions. However, the Gaussian assumption in the standard kriging approaches may be overly-restrictive for air pollution data, which often display erratic behavior, such as rapid changes in time or space. Bayesian nonparametric methods avoid dependence on distributional parametric assumptions by working with probability models on function spaces. In particular, [16] introduce a nonparametric model extending the stick-breaking prior of [17], which is frequently used in Bayesian modelling to capture uncertainty in the parametric form of an outcome. The stick-breaking prior is extended to the spatial setting by assigning each location a different, unknown distribution, and smoothing the distributions in space with a series of kernel functions. This results in a flexible spatial model, as different kernel functions lead to different relationships between the distributions at nearby locations. This model is also computationally convenient because it avoids inverting large matrices, which often hinders analysis of large spatial data sets.

19.2.5 Comparison of methods

We have described three procedures for spatial prediction: K-nearest neighbors (KNN), inverse distance weighting (IDW), and Kriging. All three methods use a linear combination of the observations as prediction, all three have tuning parameters (number of neighbors, choice of kernel/covariance, etc) that need to be selected carefully. KNN and IDW have the clear advantage of simplicity, and can be computed efficiently for large datasets because (assuming compact kernel for KNN) they use only local measurements for prediction. Limitations of these methods are that it is difficult to incorporate spatial covariates and challenging to quantify uncertainty without declaring a parametric covariance function. If a parametric covariance is assumed, then Kriging is the natural choice because it results in the optimal spatial predictions under the presumed model. Kriging also has the advantage of being able to include spatial covariates and, especially using the Bayesian framework, full characterization of uncertainty. Therefore, despite the mathematical and computational burden, Bayesian Kriging is our preferred method for spatial interpolation.

19.2.6 Case study: PM$_{2.5}$ data in the Eastern US

We use PM$_{2.5}$ data from August 21, 2006 to illustrate the spatial interpolation methods. Figure 19.1 (top) plots the data at the 229 monitoring location in the Eastern US. The data are pooled over the Interagency Monitoring of PROtected Visual Environment (IMPROVE) network and the Speciation Trends Network (STN).

Spatial interpolation based on the nearest neighbor and inverse distance weighting (with Gaussian kernel and bandwidth chosen based on 5-fold cross validation) are plotted in Figure 19.2. The nearest neighbor interpolation is piecewise constant, whereas the IDW interpolation is smooth. Bayesian Kriging estimates are plotted in Figure 19.3 (top row). In this analysis the mean function at location $\mathbf{s} = (s_1, s_2)$ is taken to be $\mu(\mathbf{s}) = \beta_0 + \beta_1 s_1 + \beta_2 s_2$ and the spatial covariance is Matern with nugget σ^2, partial sill τ^2, range ρ and smoothness ν. We select flat prior for the mean coefficients, and $\sigma^2, \tau^2, \rho \sim \text{InvGamma}(0.01, 0.01)$ and $\nu \sim \text{Unif}(0.5, 10)$. The predictions are fairly similar to the IDW predictions. The prediction standard deviation is the largest in Texas, Mississippi, and North Dakota where the monitors are the most sparse.

The interpolated surfaces in Figures 19.2 and 19.3 can be aggregated to estimate the state-average concentrations, $\frac{1}{|B|} \int_B C(\mathbf{s}) d\mathbf{s}$, where B is the region that defines the state,

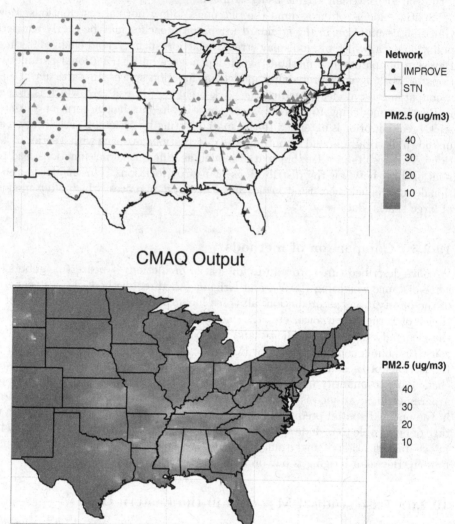

FIGURE 19.1
Fine particulate matter concentrations from the STN and IMPROVE monitoring networks (top) and CMAQ model output (bottom) on August, 21 2006.

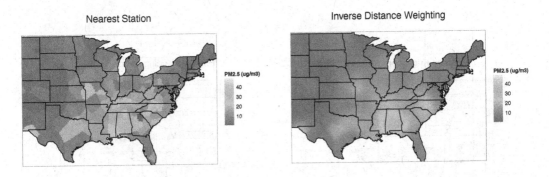

FIGURE 19.2
Fine particulate matter concentrations from the STN and IMPROVE monitoring networks on August, 21 2006 interpolated using the nearest monitoring station and smoothed using inverse distance weighting.

$|B|$ is the state's area, and $C(\mathbf{s})$ is the concentration at location $\mathbf{s}$. The integrals are approximated in Figure 19.4 by computing estimates of $C(\mathbf{s})$ on a 200×200 rectangular grid of points covering the spatial domain and averaging the predictions within each state. The IDW and Kriging estimates both show that the largest state-averages are in the southeast, with the largest average in Alabama. The largest uncertainty is in Texas and Mississippi where the data are sparse. However, standard deviation of the state averages are considerably smaller than the standard deviations at individual locations (Figure 19.3), illustrating a benefit of spatial aggregation. The top left panel of Figure 19.4 shows the simple average of all monitors in the state. The state with the highest simple average is Mississippi, which has only a single monitor (Figure 19.1). The contrast between the simple average and the Kriging estimates illustrates the advantage of spatial smoothing to borrow strength across nearby observations and appropriately weight the monitors.

19.3 Spatiotemporal extensions

Next, we present two possible approaches, suggested by [9], to include temporal interactions in the kriging model, either by considering time as the third dimension, or by fitting a covariance function with both spatial and temporal components.

It is straightforward to generalize the idea of purely spatial kriging to spatiotemporal kriging with time as the third dimension. In the case of purely spatial process, h in (19.7) can be calculated using Euclidean distances. If we add time as the third dimension, we can adapt the calculation of h to accommodate the additional dimension as long as we take into account the different scales between the temporal dimension and the spatial dimension. [9] give more details on methods to optimize the scaling.

The second approach for including temporal interactions is to introduce a spatiotemporal covariance function. The empirical spatiotemporal semivariogram can be extended from (19.7), as follows,

$$\hat{\gamma}(h, u) = \frac{1}{2|N(h, u)|} \sum_{N(h,u)} \{Y(\mathbf{s}_i, t_k) - Y(\mathbf{s}_j, t_l)\}^2 \qquad (19.14)$$

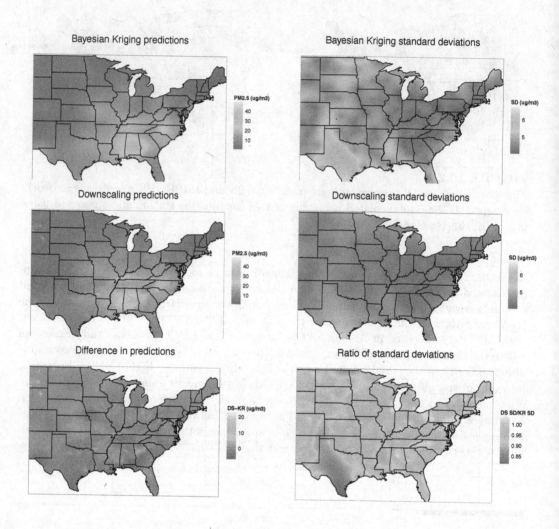

FIGURE 19.3
Estimated (and prediction standard deviation) particulate matter concentration via Bayesian Kriging ("KR"; top) and Bayesian downscaling ("DS"; middle) for August, 21 2006. The bottom row plots the difference in predicted values and ratios of prediction standard deviations

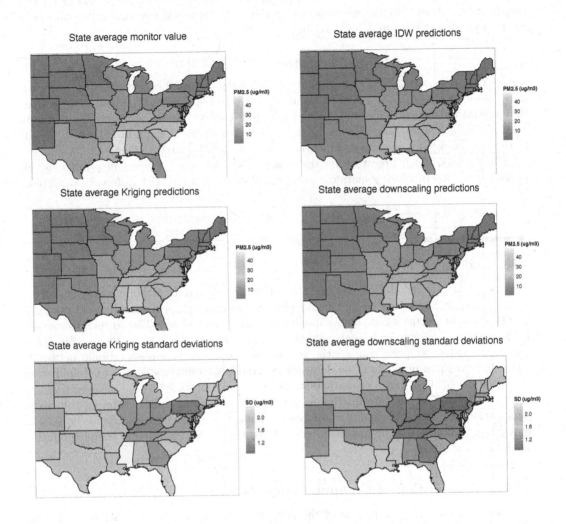

FIGURE 19.4
Estimated state-average fine particulate matter concentration via sample means of the monitor values (top left), and means of interpolated values using inverse distance weighting, Bayesian Kriging and Bayesian downscaling for August, 21 2006. The final row maps standard deviation for the Bayesian methods.

where in this case $N(h)$ denotes the set of pairs of observations separated by distance h and time u. Model fitting of the theoretical parametric spatiotemporal variogram can be performed under different assumptions of the spatiotemporal covariance function. The solution for simple spatiotemporal kriging is:

$$\hat{Y}(\mathbf{s}_0, t_0) = \mu(\mathbf{s}_0, t_0) + \mathbf{k}_0' \Sigma^{-1}(\mathbf{Y} - \boldsymbol{\mu}) \qquad (19.15)$$

where $\mathbf{Y} = (Y(\mathbf{s}_1), \ldots, Y(\mathbf{s}_n))'$ represent all the spatiotemporal observations, $\boldsymbol{\mu} = E(\mathbf{Y})$, $\mu(\mathbf{s}_0, t_0) = E[Y(\mathbf{s}_0, t_0)]$, $\Sigma = \text{Cov}(\mathbf{Y})$ and $\mathbf{k}_0' = \text{Cov}[Y(\mathbf{s}_0, t_0), \mathbf{Y}]$. The solutions for spatiotemporal ordinary kriging and universal kriging can be derived similarly as in purely spatial case. When the observations are measured with error, we can add $\tau^2 I(h = 0, u = 0)$ to the spatiotemporal covariance function to account for the nugget effect.

There exists a rich literature on the study of spatiotemporal covariance funcions, [6] provide examples of various spatiotemporal covariance functions with different properties.

19.4   Data Fusion

The use of monitoring data facilitates the estimation of trends of pollution levels across space and time. However, the air quality physical models can be a valuable and powerful tool to addressing gaps if not enough monitoring data are available. The air quality models, based on the dynamics and mechanics of atmospheric processes, typically provide information at higher temporal and spatial resolution than data from observational networks. Though, errors and biases in these deterministic models are inevitable due to simplified or neglected physical processes or mathematical approximations used in the physical parameterization. Therefore the introduction of statistical models to combine different sources of information, physical models and observations, known as data fusion, is a powerful tool to obtain improved maps of air pollution and reduce the prediction error of the air pollution predictive surfaces.

19.4.1   Calibration of Computer Models

To compensate for the limitations of the field data and understand the behavior of complex physical processes, deterministic model outputs are increasingly used in environmental research. As opposed to statistical models, deterministic models are simulations based on differential equations which attempt to represent the underlying physical processes. Using a large number of grid cells, they generate averaged concentrations with high spatial and temporal resolution and without missing values. Ideally, such model outputs would help fill the space-time gaps in observational data. However, the model output are simulations, and the uncertainty about them should be characterized. The various sources of uncertainty are classified as low quality of emissions data, model inadequacy and residual variability. As result, to use model output, particularly to provide assistance in control strategy selections, it may be necessary first to calibrate the model.

Air quality models, such as EPA's Community Multi-scale Air Quality (CMAQ) model [4], provide one of the most reliable tools for air quality assessment. CMAQ simulates ambient air concentrations of numerous pollutants on a myriad of spatial and temporal scales. These model simulations are designed to support both regulatory assessments by EPA program as well as scientific studies. One particularly important use of numerical models is forecasting, for which model outputs are used as predictor values in statistical models when monitoring measurements are not available in the future. The evaluation of physically based computer models, in particular in the context of air quality applications is crucial to assist in control strategy selection. A rigorous statistical assessment of model performance is needed, since selecting the wrong control strategy has costly economic and social consequences. The objective comparison of mean and variances of modeled air pollution concentrations with the ones obtained from observed field data is the common approach for assessment of model performance.

However, outputs from numerical models of geophysical phenomena are different from observations for a number of reasons, including problems with model inputs, imperfect and/or inadequate descriptions of processes in the model and errors from numerical simulation methods [20]. It is also worth to note that the measurements are made at specific locations, whereas CMAQ concentrations represent a volume-average value (corresponding to the volume of the grid cell). This discrepancy in spatial representativeness is a fundamental source of uncertainty when calibrating models [7]. Thus, quantifying the discrepancies and subsequently adjusting outputs in order to achieve compatibility with the observations are critical to improve and utilize numerical models.

A standard approach to calibration is to simply shift and scale the numerical model output by matching the sample mean and standard deviation. One drawback of this strategy is that it fails to calibrate properly the tails of the modeled air pollution distribution, and improving the ability of these numerical models to characterize high pollution events is of critical interest for air quality management. Therefore, due to the differences in the tail of the distributions and the desire to accurately model extreme events, calibrating sample quantile levels instead of raw data has been applied in the literature [23].

19.4.2 Spatial Downscaler

A spatial downscaler is a statistical framework to combine disparate spatial data from observations and physical models in order to improve the spatial prediction of a process of interest. [3] develop a fully model-based strategy within a Bayesian framework to downscale air quality numerical model output to a point level. In their static spatial model, the observations are regressed on the numerical model output using spatially-varying coefficients that are specified through a correlated spatial Gaussian process.

Let $\tilde{Y}(B, t)$ be the CMAQ values over grid cell B at time t, all the points $\mathbf{s}$ within the same grid cell are assigned the same CMAQ value. The observed data $Y(\mathbf{s}, t)$ and the CMAQ output are related as follows:

$$Y(\mathbf{s}, t) = \beta_0(\mathbf{s}, t) + \beta_1(\mathbf{s}, t)\tilde{Y}(B, t) + \epsilon(\mathbf{s}, t) \tag{19.16}$$

where the bias terms $\beta_0(\mathbf{s}, t)$ and $\beta_1(\mathbf{s}, t)$ are functions of space and time. In order to interpolate to a new location of interest $\mathbf{s}_0$ at time t_0, we obtain the posterior predictive distribution of $Y(\mathbf{s}_0, t_0)$ given the data $Y(\mathbf{s}_i, t_j)$ and $\tilde{Y}(B_k, t_j)$.

19.4.3 Spatial Bayesian Melding

[8] develop a Bayesian approach to address the air quality numerical models evaluation problem, and show how it can be used to remove the bias in the numerical model output. This Bayesian framework provides a natural approach to compare data from very different sources taking into account different uncertainties, and it also provides posterior distributions of quantities of interest that can be used for scientific inference. The monitoring data is not treated as the "ground truth". Instead, it is assumed that there is some smooth underlying (but unobserved) field that measures the "true" concentration/flux of the pollutant at each location. Monitoring data are these "true values" plus some measurement

error. The CMAQ output can also be written in terms of this true underlying (unobservable) process, with some parameters that explain the bias and microscale noise in CMAQ. The *truth* is assumed to be a smooth underlying spatial process with some parameters that explain the large scale and short scale dependency structure of the air pollutants. CMAQ is evaluated by comparing the distribution of the monitoring data at a given location, to the predictive posterior distribution of CMAQ at that given point in space and time. The bias in CMAQ is quantified by obtaining the posterior distribution of the bias parameters given the monitoring data and CMAQ output.

More specifically, we model the true underlying air pollution of interest using both observed data and numerical model output. We do not consider ground measurements to be the "true" values because they are measured with error. Thus, we denote the observed pollution values at location $\mathbf{s} \in D_1$ on day $t \in D_2$ from the monitoring stations by $Y(\mathbf{s}, t)$, where $D_1 = \{\mathbf{s} : \mathbf{s}_1, \dots, \mathbf{s}_N\} \subset \mathbb{R}^2$ and $D_2 = \{t : 1, \dots, T\} \subset \mathbb{R}$, and it is modeled as

$$Y(\mathbf{s}, t) = Z(\mathbf{s}, t) + e_F(\mathbf{s}, t), \tag{19.17}$$

where $Z(\mathbf{s}, t)$ is the unobserved "true" underlying spatial-temporal process at location $\mathbf{s}$ and at time t. The measurement error $e_F(\mathbf{s}, t) \sim N(0, \sigma_F^2)$ is assumed to be independent of the true underlying process.

Since the CMAQ values are averages over grid squares, not point measurements, we model the CMAQ values, $\tilde{Y}(B_i, t)$, where subregions $B_1, \dots, B_b$ cover the spatial domain B, as follows:

$$\tilde{Y}(B_i, t) = a(B_i) + \frac{1}{|B_i|} \int_{B_i} Z(\mathbf{s}, t) d\mathbf{s} + e_N(B_i, t), \tag{19.18}$$

where $a(B_i)$ is the additive bias of the CMAQ output in subregion B_i and is assumed to be a polynomial function of the centroid of the subregion, $\mathbf{s}_i$, with a vector of coefficients, $\mathbf{a}_0$.

The true underlying process Z is modeled as a function of the weather covariates:

$$Z(\mathbf{s}, t) = \mathbf{M}^T(\mathbf{s}, t)\mathbf{z} + e_z(\mathbf{s}, t), \tag{19.19}$$

where $\mathbf{M}(\mathbf{s}, t)$ is a vector of meteorological variables.

In order to predict $Z(\mathbf{s}_0, t_0)$, the true pollution values at space $\mathbf{s}_0$ and time t_0, given the data, we obtain the posterior predictive distribution of $Z(\mathbf{s}_0, t_0)$ given the data.

The hierarchical structure of the Bayesian melding approach, allows a characterization of the different types of error and biases in the sources of information, data and physical models, with regard to an underlying true process. Those different type of error structures and biases due in part to the change of support problem get combined and become unidentifiable in the downscaler approach.

19.4.4   Case Study: Statistical downscale of PM$_{2.5}$

Figure 19.1 (bottom) plots the 12km×12km CMAQ model output for August 21, 2006. Both the CMAQ model output and monitoring data have the same broad-scale trends with the largest concentrations in Kentucky and the southeast. There is a strong linear relationship between a monitor's observation and the CMAQ output from the grid cell that contains the monitor (Figure 19.5, left). However, there are clear differences between these two data products. The CMAQ output displays several local features not seen in the monitor data, such as local hotspots in Montana and Wyoming. Also, the CMAQ output underestimates the extremely large air pollution values (Figure 19.5, right).

To combine the two data sources, we fit the model in (19.16) but with spatially-constant intercept and slope, i.e.,

$$Y(\mathbf{s}) = \beta_0 + \beta_1 \tilde{Y}(B) + \epsilon(\mathbf{s}),$$

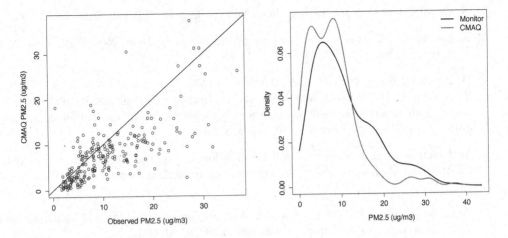

FIGURE 19.5
Fine particulate matter concentrations from the STN and IMPROVE monitoring networks compared with the CMAQ model output from the CMAQ grid cell containing the monitor on August, 21 2006.

where ϵ's Matern covariance function and all priors are the same as the Bayesian Kriging model described in Section 19.2.6. The differences between Bayesian Kriging and Bayesian downscaling methods are shown in Figure 19.3 (bottom row). The advantage of incorporating CMAQ is evident in the 15% reduction in prediction standard deviation in Texas where monitoring data is sparse.

The local CMAQ hotspots in Montana and Wyoming are also present in the downscaling predictions. [15] explore the value of incorporating local CMAQ output by comparing the agreement between monitoring data and model output for different spatial resolutions. They find that CMAQ output and monitor data have similar large-scale spatial patterns, but disparate fine-scale structure, and propose a multi-resolution downscaler to avoid using CMAQ inappropriately at fine scales.

Bibliography

[1] CDC - Exposome and Exposomics - NIOSH Workplace Safety and Health Topic. http://www.cdc.gov/niosh/topics/exposome/. Accessed: 2016-03-25.

[2] S. Banerjee, B. P. Carlin, and A. E. Gelfand. *Hierarchical modeling and analysis for spatial data*. Chapman and Hall, Boca Raton, FL., 2nd edition, 2014.

[3] V.J. Berrocal, A.E. Gelfand, and D.M. Holland. A spatio-temporal downscaler for output from numerical models. *Journal of Agricultural, Biological, and Environmental Statistics*, 15:176–197, 2010.

[4] D. W. Byun and K. L. Schere. Review of the governing equations, computational algorithms and other components of the models-3 community multiscale air quality (cmaq) modeling system. *Applied Mechanics Reviews*, 59:51–77, 2006.

[5] N. Cressie. *Statistics for Spatial Data.* New York: John Wiley & Sons, Inc., 1993.

[6] N. Cressie and C. Wikle. *Statistics for Spatio-Temporal Data.* New York: John Wiley & Sons, Inc., 2011.

[7] B. Eder, D. Kang, S.T. Rao, R. Mathur, S. Yu, T. Otte, K. Schere, R. Wayland, S. Jackson, P. Davidson, J. McQueen, and G. Bridgers. A demonstration of the use of national air quality forecast guidance for developing local air quality index forecasts. *Bull. of the Amer. Meteor. Soc.*, 91:313–326, 2010.

[8] M. Fuentes and A. E. Raftery. Model evaluation and spatial interpolation by bayesian combination of observations with outputs from numerical models. *Biometrics*, 61:36–45, 2007.

[9] B. Gräler, L. Gerharz, and E. Pebesma. Spatio-temporal analysis and interpolation of PM_{10} measurements in Europe. Tech. rep., ETC ACM, 2012.

[10] J. Gulliver, K. de Hoogh, D. Fecht, D. Vienneau, and D. Briggs. Comparative assessment of gis-based methods and metrics for estimating long-term exposures to air pollution. *Atmospheric Environment*, 45:7072–7080, 2011.

[11] Health Effects Institute. National Particle Component Toxicity (NPACT) Initiative Report on Cardiovascular Effects. http://pubs.healtheffects.org/getfile.php?u=946, 2013. Accessed: 2016-03-26.

[12] Health Effects Institute. Understanding the Health Effects of Ambient Ultrafine Particles. http://pubs.healtheffects. org/view.php?id=394, 2013. Accessed: 2016-03-26.

[13] G Miller. *The Exposome: A Primer.* Academic Press, Waltham, MA, 2014.

[14] A. Okabe, B. Boots, K. Sugihara, S. N. Chiu, and D. G. Kendall. *Spatial Tessellations: Concepts and Applications of Voronoi Diagrams.* New York: John Wiley & Sons, Inc., 2nd edition, 2000.

[15] B. J. Reich, H. H. Chang, and K. M. Foley. A spectral method for spatial downscaling. *Biometrics*, 70:932–942, 2014.

[16] B. J. Reich and M. Fuentes. A multivariate semiparametric bayesian spatial modeling framework for hurricane surface wind fields. *Annals of Applied Statistics*, 1:249–264, 2007.

[17] J. Sethuraman. A constructive definition of dirichlet priors. *Statistica Sinica*, 4:639–650, 1994.

[18] M. L. Stein. *Interpolation of Spatial Data: Some Theory for Kriging.* New York: Springer, 1999.

[19] G. D. Thurston, K. Ito, and R. Lall. A source apportionment of U.S. fine particulate matter air pollution. *Atmospheric Environment*, 45:3924–3936, 2011.

[20] B. J. Turpin and H.-J. Lim. Species contributions to pm2.5 mass concentrations: Revisiting common assumptions for estimating organic mass. *Aerosol Sci. Technol.*, 35:602–610, 2001.

[21] U.S. EPA. *Guidelines for Exposure Assessment.* U.S. Environmental Protection Agency, Washington, DC, 1992.

[22] A. M. Vicedo-Cabrera, A. Biggeri, L. Grisotto, F. Barbone, and D. Catelan. A bayesian kriging model for estimating residential exposure to air pollution of children living in a high-risk area in italy. *Atmos. Environ*, 8(1):87–95, 2013.

[23] J. Zhou, M. Fuentes, and J. Davis. Calibration of numerical model output using non-parametric spatial density functions. *Journal of Agricultural, Biological, and Environmental Statistics*, 16:531–553, 2011.

20

Alternative models for estimating air pollution exposures – Land Use Regression and Stochastic Human Exposure and Dose Simulation for particulate matter (SHEDS-PM)

Joshua L. Warren

Yale University, School of Public Health, Department of Biostatistics

Michelle L. Bell

Yale University, School of Forestry and Environmental Studies

CONTENTS

20.1   Introduction

When investigating the associations between human exposure to air pollution and various adverse health outcomes, the majority of past epidemiologic and biostatistical studies have relied on the use of ambient air pollution concentrations measured from a set of stationary air pollution monitors located across a study region to estimate exposure. These studies have successfully uncovered the association between elevated ambient concentrations and a number of adverse health outcomes including risk of mortality and hospital admissions among others (Dockery and Pope 1994; Pope et al. 1995; Schwartz 1994). This evidence from epidemiology and biostatistics, along with that of many other disciplines such as toxicology, have formed the basis for health-based regulatory standards and guidelines to protect human health through improved air quality such as through the United States

(US) National Ambient Air Quality Standards, set the by US Environmental Protection Agency, and the World Health Organization's Air Quality Guidelines. Thus, such research has contributed substantially to improved air quality and public health worldwide.

While these studies have successfully established a link between exposure and health, they do not address all critical questions regarding how air pollution affects health. Exposure estimates based on ambient monitoring have several limitations in regards to application in health studies. In addition to the issue of measurement error, most monitoring networks were established to address compliance with regulatory standards, not exclusively for health-focused scientific study. The monitors are more frequently located in urban areas with far less coverage in rural or semi-urban areas. Some areas do not have monitoring networks at all, especially in less industrialized regions of the world. Even cities with monitors may not have sufficient coverage to provide detailed information on within-city heterogeneity of pollution levels, which can vary by pollutant, source, topography, weather, and land-use. Temporal resolution may also be limited as some pollutants may not be measured year-round (e.g., monitoring of ozone during the warm season in many parts of the US) or every day (e.g., monitoring of fine particles every three to six days in many parts of the US). Further, the use of ambient monitors does not incorporate information on study populations' daily activity patterns, relating to the amount of time spent in various locations indoors and outdoors. Most populations spend a substantial amount of time indoors. Use of ambient monitors also obscures differences in exposures among persons, who also experience occupational exposure.

A variety of alternative exposure methodologies have been developed to address some of these limitations, many of which are discussed in other chapters. Some methods fuse results from ambient air quality modeling, thereby incorporating the chemical transformation and physical transport of pollution, with measurements from ambient monitoring networks (Berrocal et al. 2010 and 2011, Bravo et al. 2012, Chen et al. 2014). Other approaches include Kriging techniques that provide interpolated values at locations without monitors based on estimated spatial heterogeneity patterns (Alexeeff et al. 2015, Sampson et al. 1994, Young et al. 2014) and modeling approaches using satellite imagery data, sometimes in combination with land-use and/or weather data (Di et al. 2016, Hyder et al. 2014, Lee et al. 2014, Mordukhovich et al. 2016).

This chapter describes two such alternative methods for improving local exposure estimates and predictions by incorporating covariates into modeling frameworks and simulation models for area- and individual-level exposure estimation. The Land-Use Regression (LUR) model, primarily addresses issues of within-city spatial heterogeneity of pollution concentrations and offers the ability to improve exposure estimates and predictions at finer spatial scales than traditional methods. The Stochastic Human Exposure and Dose Simulation for particulate matter (SHEDS-PM), primarily addresses issues of differences in exposures among individuals, including their daily activity patterns. Each method provides an innovative approach to estimation of exposure to air pollution for human health studies in situations for which the approach of using ambient monitoring networks is limited.

20.2 Land Use Regression Modeling

20.2.1 Background

Ambient air pollutant levels can vary within cities, which can introduce exposure misclassi-fication when a central monitor or small number of monitors are used to estimate exposure for a health-based study. This exposure misclassification could be differential based on the residential and work locations of various subpopulations in relation to the spatial patterns of pollution levels. The within-city spatial heterogeneity can vary by pollutant and source. Pollutants from regional sources often exhibit less spatial variability than those from local sources. Air pollution from coal-fired power plants, such as particulate matter with aerodynamic diameter no larger than 2.5 μm ($PM_{2.5}$) with a high fraction of ammonium sulfate, is less spatially heterogeneous than more localized sources, such as $PM_{2.5}$ with a high fraction of ammonium nitrate from vehicles. Topography and meteorology as well as building height and the configuration of buildings can affect the distribution of pollutants within a city. The patterns of particulate matter concentrations relate to roadways, wind speed, solar radiation, chemical and physical transformation of particles, and other factors (Moon 2001). The spatial variation of concentration can vary by whether the pollutant is primary, meaning directly emitted, or secondary, formed in the atmospheric through the transformation of precursors. The density of air pollution monitors across a city are often not sufficient to provide a full understanding of the spatial heterogeneity of the surface concentration of pollutants within a community.

A variety of exposure methods have been developed to address this limitation, including the use of multiple monitors within a city although clearly this method is limited by the number and distribution of monitors. Statistical methods such as Kriging interpolate values that allow different exposure levels throughout a study area (Alexeeff et al. 2015, Laurent et al. 2016, Ramos et al. 2015, Sampson et al. 1994, Young et al. 2014). Other exposure study designs to address within-city spatial heterogeneity include the use of regional air quality modeling, or fusion of air quality models with ambient monitors. Some studies have used proximity to roadways as an approximation of exposure to pollutants from the line source of roadways in order to capture the differences of vehicular pollution by location (Puett et al. 2014, Urman et al. 2014). Each of these approaches can estimate concentrations that permit different levels across a city yet have their own limitations, such as the input data required and the expertise necessary to perform air quality modeling.

Another approach to estimate exposure allowing for intra-urban spatial variation is land use regression (LUR) modeling, which has become a prevalent method for epidemiological studies, primarily for urban areas. Variables of land-use are combined with detailed ambient measurements of air pollutants, and other location-based variables such as elevation, in a regression analysis to produce a prediction model that can be used to estimate exposure values throughout the city or at specific locations such as the residences of study participants. The output of LUR is a model that estimates a surface concentration field for a pollutant of interest, which can be used to estimate the pollutant level at a specific location. The amount of smoothness in the final set of predictions often depends on the spatial smoothness of the included explanatory covariates since their associations with the pollutant of interest drive the prediction of concentrations in unobserved spatial locations.

20.2.2   Land Use Regression Methods

No standard framework exists for LUR approaches, and study designs differ although the overall purpose of LUR modeling approaches is generally to create a prediction model to estimate air pollution concentrations at locations for which monitors are not present, or alternatively for time periods for which monitors are not in operation. The approach leverages the associations among air pollution levels and location-specific variables. The name "land use regression" is a bit of a misnomer in that it only partially covers the type of information incorporated; other variables are often included such as traffic data, weather and other factors. LUR is most often applied to traffic-related air pollutants, such as nitrogen dioxide (NO_2) and $PM_{2.5}$, but has also been applied to other air pollutants such as metals, sulfur dioxide, black carbon, and volatile organic compounds (Amini et al. 2014; Clougherty et al. 2013; Oiamo et al. 2015; Smith et al. 2011; Zhang et al. 2015). The Least-squares regression framework is most often utilized along with the relevant modeling assumptions (e.g., linear relationship, normally distributed errors, homoscedasticity).

Other exposure methods can also incorporate such variables, such as air quality modeling that includes data on land use, weather, and traffic. Whereas LUR involves estimating the association among these variables, air quality modeling integrates the actual chemical and physical processes among these relationships, such as the role that meteorological variables play in the atmospheric chemistry of secondary particle formation. Proximity models, such as those that use distance to a major roadways as a proxy for vehicular pollution, incorporate only on aspect of the factors that affect spatial heterogeneity. LUR does not incorporate the underlying physical and chemical processes, but uses regression analysis to include these relationships.

The LUR method uses ambient measures of air pollution as a dependent variable and land use and the other factors as independent variables in a regression model, often through a geographic information system (GIS). The resulting regression model is used to generate an estimated concentration surface field or to predict concentrations at particular receptor points of interest. The predictor variables included in LUR modeling can differ considerably across studies. Common classes and definitions of geographic predictor variables include road types, traffic count, altitude, meteorology, conditions near the monitoring locations, and land cover (Hoek et al. 2008; Ryan and LeMasters 2007). Road type has been defined in different ways such as the number of persons served, traffic count, major versus minor road, etc. Road length is sometimes measured as the length of road of a certain type within a specified radius of the point of interest. Land cover can be categorized in several ways such as household density, population density, industrial use, distance to coast, open space, or household density.

LUR modeling requires data on air pollution measurements as regression model input, and some studies have employed data from available air pollution monitoring networks, although usually such networks do not have sufficient density for this application and are generally designed to gauge compliance with regulatory standards rather than to assess spatial variability of pollution levels. More typically, LUR requires ambient air pollution monitoring campaigns, often multiple campaigns to gauge pollution levels across seasons. These are frequently one to two weeks in length, conducted for multiple seasons, although studies have used a range of other timeframes such as annual averages. For air monitoring campaigns, the authors can purposefully select sites for monitoring that address key areas of concern, such as placing higher monitor density in areas of the city anticipated to have high variance of pollution levels or giving special attention to areas of particular interest such as key industrial sources or areas of interest for exposure such as schools.

No standard approach exists for determining the number of monitoring sites, monitor density (the number of sites per area), or location of sites for the monitoring campaigns for LUR. Some studies have suggested 40 to 80 sites as a reasonable number of monitoring sites, with consideration given to the size of the city (Hoek et al. 2008). The anticipated spatial heterogeneity of the air pollutant of concern is also a factor as surface concentration fields with high variance would require a larger number of monitors. Some studies use pilot campaigns with a smaller number of monitoring locations to perform an initial assessment of spatial heterogeneity of pollution in order to help develop the design of the larger monitoring campaign. Most monitoring sites are typically fixed, with some studies rotating monitoring locations. Early LUR studies often used an ad hoc approach to perform monitoring campaigns. A growing number of studies have presented systematic methods for selecting the number of monitoring sties, optimized to specified criteria, and/or for selecting the locations of the monitoring sites within the city.

Some research has investigated optimization methods as an approach to systematically design monitoring networks for LUR modeling. Such analyses could be structured to minimize cost and achieve measurement of a specified target of spatial heterogeneity of the pollutant of interest. Kanaroglou et al. (2006) developed a method to optimize air pollution monitoring locations using a case study of 100 NO_2 monitors for Toronto, Canada, in which the analysis aims to optimize monitor placement to capture variability in the pollutant concentration and population density. The authors estimate a continuous concentration surface field or "demand surface" for which higher demand represent a need for higher monitor density. The demand surface is calculated such that a higher demand reflects anticipated higher spatial variability of the air pollutant of interest. This requires an approximation of the spatial variability of the pollutant levels, which is the ultimate outcome of interest, and thus not available with detailed information. Data from a pilot monitoring campaign or existing regulatory monitors can be used. The estimated spatial variability of the pollutant, $\widehat{\gamma}(\vec{x}, \vec{h})$, at location $\vec{x}$ is estimated by (Kanaroglou et al. 2006):

$$\widehat{\gamma}\left(\vec{x}, \vec{h}\right) = \frac{\sum_h \left(z\left(\vec{x}\right) - z(\vec{x} + \vec{h})\right)^2}{2}$$

where $\vec{h}$ is the distance between two points. The demand surface is also calculated such that a higher demand reflects areas of high population density, under the assumption that the users are particularly interested in minimizing exposure misclassification in these regions. Rather than population density, the user could focus on other features such as specific subpopulations, including a particular race/ethnicity or age group. A weighting scheme is imposed such that (Kanaroglou et al. 2006):

$$W_R = \frac{P_R/P_T}{\widehat{\gamma}_R/\widehat{\gamma}_T}$$

where P_R is the study population of interest in region R, P_T is the total study population, $\widehat{\gamma}_R$ is the estimated variability for study region R, $\widehat{\gamma}_T$ is estimated variance for the entire study region, and W_R is a weight applied for each location $\vec{x}$ in region R. The numerator portion of the weighting equation (P_R/P_T) gives higher weight to regions with higher population density and is the fraction of the total population that is in region R. The denominator proportion of the weighting equation ($\widehat{\gamma}_R/\widehat{\gamma}_T$) is the proportion of the total variability of the concentration of the pollutant of interest that is attributable to region R. A constrained optimization problem is solved to determine the location of monitors. Using optimization from variations of the classic facility location problem, the authors' method selects where to place monitors based on a specified number of monitors to allocate. The case study uses 100 monitoring locations. Note that this method provides a systematic approach to locating monitors, but not to selecting the number of monitors, and also is dependent on input data related to spatial variation. Other studies have built on the Kanaroglou et al. (2006) method to address spatially autocorrelated data (Kumar 2009).

Evaluation of LUR models include comparison of measured values to predicted estimates from an LUR model that did not include those measured values as input (i.e., cross validation or hold-out validation). Evaluation metrics include the root mean squared prediction error:

$$RMSE = \sqrt{\sum_{i=1}^{n} (\hat{y}_i - y_i)^2/n}$$

where $\hat{y}_i$ is the predicted concentration at location i, y_i is the measured concentration at location i, and n is the number of monitoring locations used to calculate the $RMSE$. The

R^2 of the LUR models has ranged from 0.17 to 0.96, with typical values of 0.60 to 0.70 (Hoek et al. 2008; Ryan and LeMasters 2007). Performance is normally comparable to or better than alternative methods such as Kriging or dispersion models (Hoek et al. 2008).

20.2.3   Examples of Land Use Regression

In Boston, Massachusetts and nearby communities, LUR modeling was used to estimate the levels of several traffic-related pollutants: $PM_{2.5}$, NO_2, and elemental carbon (EC) (Clougherty et al. 2008). A monitoring campaign was conducted at the outside of 44 homes that reflected different communities and traffic conditions. Not all measurements were conducted at the same time. Temporal adjustment was conducted to address differences in background levels and meteorology. The authors performed LUR modeling considering traffic indicators, land use, meteorology, elevation, and population in relation to each air pollution concentrations. Traffic estimates were generated using a kernel function to estimate concentrations under a Gaussian decay function for roads within a certain distance of the homes, and using the resulting values to generate traffic scores divided into tertiles. This and other data were combined to create 25 different traffic indicators including kernel-weighted density within five buffer distances from the homes ranging from 50 to 500 m (units of vehicle-m/day/m^2), as well as unweighted traffic densities at those buffers, total roadway lengths within the five buffer distances from the homes (units of meters), distance to nearest urban road of 8500 or more cars/day, average daily traffic (units of vehicles/day), and characteristics of nearest major road such as trucks/day. Many of these traffic metrics were highly correlated. The LUR method permitted different variables to be associated with different air pollutants, and incorporated traffic metrics using both correlation and clustering approaches. Models were built with stepwise forward regression, with a criteria of keeping variables with a p-value of 0.1 or lower. The general LUR model is:

$$c_{ijt} = \beta_{0j} + \beta_{1j} D_{jt} + \beta_{2j}\, T_i + \beta_{3j}\, T_i M_{it} + \beta_{4j}\, S_{it} + \varepsilon_{ijt}$$

where c_{ijt} is the measured concentration of pollutant j at location i at time t; β_{0j} is the model intercept for pollutant j; D_{jt} is the measured concentration of pollutant j at time t at a central regulatory monitor; T_i is the value of a specified traffic indicator at location i; M_{it} is the value of a modifier variable such as a meteorological or site characteristic at location i at time t; S_{it} is a variable to represent other sources (e.g., smoking or grilling) at measured location i and sampling time period t; ε_{ijt} is an error term for site i, pollutant j, and time t ($\varepsilon_{ijt} \sim N\left(0, \sigma_\varepsilon^2\right)$; independent and identically distributed); and β_{2j}, β_{3j}, and β_{4j} are regression coefficients for the model for pollutant j. $PM_{2.5}$ and EC concentrations were log transformed. The authors also included extensive sensitivity analysis. The patterns of concentrations were associated with local traffic patterns and meteorological factors. Results indicated that spatial variability differed by pollutant with higher variability for EC. For $PM_{2.5}$, the best traffic indicator was total roadway length within a 100 m buffer of the home, with other important variables as smoking or grilling near the monitoring location, and population density. The variables identified in the NO_2 LUR model included population density, roadway length within 50 m of the home, and season.

While no standard approach exists for LUR models, modified versions of the above basic form have been applied to generate exposure estimates for epidemiological studies linking air pollution to human health. Porta et al. (2016) investigated exposure to NO_2 and multiple metrics of particulate matter (PM_{coarse}, $PM_{2.5}$ and $PM_{2.5}$ absorbance) estimated through LUR modeling in relation to cognitive development as measured by an IQ composite score in a cohort of children in Italy. A higher level of NO_2 exposure during pregnancy was associated with lower verbal IQ and verbal comprehension IQ. LUR methods, adjusted to temporally match the period of interest, for NO_2 and benzene were applied to data from a cohort of 2409 pregnant women in Spain (Estarlich et al. 2016). For the total study population, the authors did not observe associations between exposure and risk of preterm birth; associations were observed for both pollutants for the subset of women who spent more time at home. Turner et al. (2016) estimated $PM_{2.5}$ and NO_2 through LUR approaches and identified associations with all-cause and cause-specific mortality in a large prospective study of US adults.

Recent research involves hybrid methods, combining LUR with other approaches such as satellite imagery (Meng et al. 2016; Murdokovich et al. 2015; Rice et al. 2016). LUR and chemical transport modeling were considered together to estimate ozone and $PM_{2.5}$ in Los Angeles, US (Wang et al. 2016). The joint model performed better with respect to RMSE than either the chemical transport model or the LUR alone. Others merged LUR with Kriging and satellite-derived estimates to predict annual NO_2 (Young et al. 2016). Inclusion of either Kriging or satellite estimates produced improved estimates of NO_2 compared to the use of the LUR model alone.

20.2.4 Limitations to Land Use Regression

The types and quality of data used as inputs for LUR differ by study, which can affect results and also hampers comparison of results across studies. For example, LUR modeling incorporates various types of traffic information largely due to variation in the types of data available by study location. Some areas may have data available on traffic count for total vehicles, whereas other locations may have counts for passenger vehicles and trucks separately. Traffic data are often not available for all streets, and road classifications may not fully distinguish between different levels of emissions. Traffic data may have been collected, but not publically available (Hoek et al. 2008). Data may be available at different time scales, such as daily or annual counts. Data availability is a key hindrance of the use of LUR especially for applications in non-developed regions of the world where sufficient input data is lacking. Some of the very areas for which exposure information is particularly needed are the regions for which this method is very difficult to apply. The spatial resolution of land use data can vary, as can the type and number of land use categories. The cost of performing LUR may be a factor for some studies due to the need for air monitoring campaigns. Choices in study design can affect the results from LUR models, such as the number of monitors to employ and their locations, the variables to include in regression analysis, and the buffer size.

Results from LUR studies are rarely applicable to other locations, due to underlying differences such as in the pollutant mixture, topography, or climate. LUR models are local by design; efforts to produce LUR models that can be transferred to other areas may only be possible for areas with similar characteristics as well as similar datasets for the input variables used to generate the model. Cities that have been analyzed for LUR studies include locations in the United Kingdom, Netherlands, Czech Republic, Sweden, Germany, and Canada, which highlights the typical regions of application for LUR modeling as developed nations, primarily in North America and Europe (Hoek et al. 2008; Ryan and LeMasters 2007). Palton et al. (2015) examined transferability by investigating whether location-specific LUR models developed to estimate particle number concentration for one Boston neighborhood could be applied to another Boston neighborhood, and tested generalizability by examining whether a Boston-level LUR model could be applied to each of four Boston neighborhoods. The authors found that the transferability of neighborhood-specific models to other neighborhoods, even within the same general metropolitan area (Boston) was poor, although the general Boston model performed better when applied to localized Boston neighborhoods if calibrated with local-level data. Wang et al. 2014 investigated transferability of LUR models in 23 European cities to other areas by generating models that excluded one area at a time (Wang et al. 2014). With respect to transferability, the models performed better for NO_2 than $PM_{2.5}$.

For health studies aimed at estimating the health effects of acute exposure (i.e., a day or a few days), LUR studies are less applicable as the estimated exposures are typically annual. The models are geared towards estimating spatial variability in pollution levels, with some information for temporal variability. This approach does not perform well to identify particular sources of pollution.

Further, the LUR model may not match the time period of interest for an epidemiological study, especially if the health data to be used are historical. For retrospective studies, results from LUR models must be adjusted to reflect previous time periods.

20.3   Population Exposure Modeling

20.3.1   Background

It is clear that the majority of the population spends most of their typical day indoors (e.g., work, residence, restaurants, retail stores) or in transit. When people are outdoors during a given day, they are not necessarily located near one of the ambient monitors in an area. This is especially true for people located in rural areas where stationary monitors are less likely to be located. Air pollution levels may differ by location within a city, and this spatial heterogeneity can differ by pollutant and source. The difference between indoor and outdoor levels is affected by building structure, ventilation patterns, and outdoor levels, as well as indoor sources (Moon 2001). As a result, personal-level exposures can vary from those defined based on ambient concentrations due to a number of factors including patterns of movement during the day and working environment. Therefore, ambient air pollution concentrations, even when statistically modeled and predicted at new spatial locations and time periods, may not reflect the true nature of exposure experienced by each individual in a study.

Population exposure modeling aims to address this issue by estimating the exposure level for individual subtypes. However, a number of other study designs also have been developed to account for the disconnect between exposures defined using ambient concentrations and personal-level exposure, such as the use of personal monitoring devices. These monitoring devices are typically worn throughout a day and collect measurements over time based on the microenvironment that the individual spends time in and the activities in which he or she participates. These studies have uncovered a number of associations with adverse health outcomes (see Ashmore and Dimitrouloupolou 2009 for a review of these studies in children). While the use of personal monitoring devices can better reflect the exposure experienced by an individual, their use in a study also has shortcomings. First, the use of personal monitoring devices can be expensive, resulting in a smaller sample size and less statistical power to detect associations of interest. Also, the length of time covered by these studies may be shorter due to use of the personal monitoring devices that must be worn by each participant during the timeframe of the study. Participants may be unwilling or unable to wear these devices for long periods of time. Measurements of exposure over long timeframes, such as years, that are needed for some studies are not feasible. Participant dropout is also a potential issue given the active role that the participants play compared to the use of ambient concentration in a study setting. Personal exposure monitoring may also be difficult for some subpopulations of interest such as infants.

Other methods to estimate personal exposures may blend methods such as microenvironment modeling that combines measurements at various locations, such as work and home, with daily activity patterns for each study participant. Exposure estimates are generated with time-weighted averages based on how long each participant spends in each microenvironment. This method shares some of the limitations of personal exposure monitoring, such as the short timeframe and small study population that is most commonly applied. Biomonitoring, to estimate the actual individual-level biologically effective dose, can be extremely useful, but biomarkers are invasive and not available for all pollutants and timeframes of exposure, and further methodologies can be cost-prohibitive for large study populations.

Population exposure modeling offers another potential solution to some of these issues. Unlike exposure estimations that rely on ambient concentrations alone, population exposure modeling are probabilistic methods to provide predictions/estimates of personal-level exposure by relating ambient concentrations to concentrations experienced in different microenvironments. It can also be less expensive and less demanding on participants since they have no active role in the collection of data. This approach also can provide predictions for longer periods of time than direct measurement of personal exposures alone since the population exposure modeling is driven by ambient concentration data. However, this type of modeling relies heavily on pre-collected data sources including, but not limited to, ambient concentrations at different spatial scales and time periods, and daily activity patterns for a collection of individuals from different geographic areas, time periods, and demographic backgrounds.

If the necessary data sources are readily available, then population exposure modeling represents a method for estimating personal-level exposure while simultaneously characterizing the uncertainty in the modeling process and propagating this uncertainty through to the final set of predictions. The models are built on a set of unknown parameters that must either be estimated or defined based on the user's understanding of the values and their associated uncertainty. This is typically handled through the use of probability distributions that reflect the current state of knowledge regarding parameter values and uncertainty. Within a Bayesian framework, these are referred to as prior distributions though in exposure modeling the predictions of personal-level exposures are often based on the selected priors, and it is not possible to update these prior beliefs through calculation of posterior distributions. Population exposure models ultimately provide predictions of average population-level exposures through the simulation of personal-level exposures from a representative yet hypothetical set of individuals. These estimated exposures can then be used to investigate associations of interest as in epidemiologic studies of air pollution exposure and adverse health outcomes.

20.3.2 Stochastic Human Exposure and Dose Simulation for particulate matter (SHEDS-PM)

One of the most popular and utilized population exposure models in past statistical and epidemiologic work is the Stochastic Human Exposure and Dose Simulation for particulate matter (SHEDS-PM) model, introduced by Burke et al. (2001). In the original work, the authors developed a framework to estimate average population-level exposure to $PM_{2.5}$ through the simulation of individual-level exposures from nine different microenvironments for a group of location representative simulated individuals. Through the development of their population exposure model, the authors were able to predict population-level average $PM_{2.5}$ exposure for different geographic regions and time periods, determine the proportion of total $PM_{2.5}$ exposure due to ambient and non-ambient origins, identify the main factors associated with increased/decreased personal exposure, and determine which model inputs lead to high uncertainty in the final set of predicted concentrations. Using a two-dimensional Monte Carlo sampling technique, the authors investigated the inter-individual variability of exposures in the population as well as the variability in population-level exposures. They applied SHEDS-PM to a case study of $PM_{2.5}$ exposures from Philadelphia, Pennsylvania from 1992-1993.

SHEDS-PM estimates population-level exposure distributions by first simulating exposures for individuals within the population of interest. In order to simulate realistic individuals, the method relies on demographic information from the geographic locations in the study. In the original work, individual characteristics were simulated from different US census tracts in Philadelphia based on demographic information available from US census data. The simulated individual-level information included sex, age, employment status, housing type, and smoking status. Smoking status was randomly assigned to each individual based on their age and sex using smoking prevalence data from the US.

Once these individual characteristics are simulated based on the demographics from a specific geographic location, the simulated individuals are matched with human activity pattern data collected from actual (not simulated) individuals who have similar demographics and characteristics of interest. The human activity pattern data consist of detailed daily activity diaries from real individuals that describe where a person spent time during a day and the activities in which they participated. These data were obtained from the Environmental Protection Agency's Consolidated Human Activity Database (CHAD) in the original work. This database combines data from 10 surveys to include over 22,000 diary days of data. When determining personal-level exposures, it is important to account for time spent in different microenvironments and activities performed in these microenvironments as non-ambient exposures can lead to high variability in exposures between people with similar characteristics living in similar spatial regions.

Once a set of individuals are simulated from a geographic region and matched with a specific daily activity pattern diary, personal-level exposures are estimated for each individual from nine different microenvironments. These microenvironments include outdoor, in vehicles, residence, office, school, store, restaurant, bar, and other indoor environments. The SHEDS-PM user must first input ambient concentrations within the geographic regions of interest. In the original work, this represented a set of ambient $PM_{2.5}$ concentrations from different spatial locations within each Philadelphia census tract. These data were originally required to be input at a temporal resolution of 12 or 24 hour averages with a minimum of a single year of seasonal data also required. These ambient concentrations are used to directly define the outdoor microenvironment concentrations and are also used to indirectly inform about exposures from the indoor and in-vehicle microenvironments. Ambient concentrations are often available directly from air pollution monitoring networks or indirectly through deterministic chemistry models. Both sources of information may require advanced statistical modeling to overcome their respective weaknesses. These limitations are discussed further in Section X.3.4.

Estimating personal-level concentrations from the in-vehicle and indoor microenvironments requires knowledge of the association between ambient concentrations and the concentrations in the each of the different microenvironments. The assumed functional form for this relationship is taken to be a linear model such that

$$C_m(h,t) = a_m + b_m C_{amb}(h,t)$$

where $C_m(h,t)$ is the $PM_{2.5}$ concentration from microenvironment m at hour h of day t, $C_{amb}(h,t)$ is the ambient concentration at hour h of day t, a_m is the model intercept for microenvironment m, and b_m is the model slope for microenvironment m. For residential exposure, a single compartment, steady-state mass balance equation is used to define the intercept and slope parameters needed to relate ambient concentrations with residential exposure based on infiltration of ambient exposure inside the home and $PM_{2.5}$ exposure that originates inside the home such that

$$b_{\text{res}} = \frac{P * ach}{ach + k} \quad \text{and}$$

$$a_{\text{res}} = \frac{E_{\text{smk}} N_{\text{cig}} + E_{\text{cook}} t_{\text{cook}} + E_{\text{other}} T}{(ach + k) V T}$$

where P is the penetration factor, k is the deposition rate, ach is the air exchange rate, E_{smk} is the emission rate for smoking, N_{cig} is the number of cigarettes smoked during the model time step, E_{cook} is the emission rate for cooking, t_{cook} is the time spent cooking during the model time step, E_{other} is the emission rate for other sources, T is the model time step, and V is the residential volume.

For the remaining microenvironments, a_m and b_m are unstructured and rely on prior information input by the user. Once the individual-level microenvironment exposures are calculated, the average exposure for individual i during day t is defined as:

$$E_i(t) = \frac{1}{24} \sum_{h=1}^{24} \sum_{m=1}^{9} E_{im}(h, t)$$

where $E_{im}(h, t)$ is the exposure for person i in microenvironment m at hour h of day t and is defined as a function of the microenvironment concentrations and time spent in each microenvironment such that

$$E_{im}(h, t) = C_m(h, t) T_{im}(h, t)$$

where $T_{im}(h, t)$ is the time that person i spends in microenvironment m during hour h of day t.

For each of the linear equations that relate ambient concentrations to the different microenvironment concentrations, there are a number of introduced unknown parameters. For the residential concentration model, this includes the parameters in the balance mass equations and for the remaining microenvironment models this includes a_m and b_m. SHEDS-PM requires the user to input prior information regarding the estimated true values of these parameters along with a measure of the value's uncertainty. To account for this uncertainty, distributions are input for each of the model parameters. This allows the user to control the amount of uncertainty for these parameters based on the available prior knowledge of the relationships of interest. The hyper-parameters are also assigned prior distributions (or hyper-priors) in order to account for additional variability in the modeling process. Therefore, SHEDS-PM accounts for variability due to person-to-person differences as well as variability due to the specification of the variability distributions. Specific details regarding typically used prior distributions that vary by season of the year can be found in Burke et al. (2001) and Reich et al. (2009).

Once the prior distributions are selected and input by the user, SHEDS-PM randomly draws values from these parameter distributions for a simulated individual. These randomly sampled values define the equations and therefore allow for prediction of microenvironment-specific concentrations for the individual. The time spent in the different microenvironments is obtained from the daily activity diaries that were previously matched to each individual. The estimated individual-level exposures on a specific day are then combined to create a distribution of population-level exposures. The final set of predictions represents a cross-sectional distribution of $PM_{2.5}$ exposures across a geographic domain. For each date of the input data, SHEDS-PM simulates a new set of individuals instead of simulating exposure for the same set of individuals over time. The output distribution on a single day within a single geographic region of interest describes the inter-individual variability in personal exposures due to differing daily activities and behaviors.

20.3.3   Examples of SHEDS-PM Use

Predictions from SHEDS-PM have been used directly in a number of epidemiologic studies of air pollution exposure and adverse health outcomes. This includes investigating the association between exposure to $PM_{2.5}$ and emergency department visits due to transmural myocardial infarction in New Jersey, US (Baxter et al. 2013; Hodas et al. 2013); between exposure to $PM_{2.5}$, sulfate, EC, ozone, and daily emergency department visits in Atlanta, Georgia, US (Dionisio et al. 2013; Sarnat et al. 2013); and between exposure to $PM_{2.5}$ and daily respiratory hospitalizations in New York City, US (Jones et al. 2013).

A number of studies have included the SHEDS-PM methodology within a more statistically sophisticated framework in order to better characterize the uncertainty and variability in assigning exposures to individuals and populations when estimating health effects. Berrocal et al. (2011) used SHEDS-PM to define $PM_{2.5}$ exposure for pregnant women in North Carolina, US when studying its impact on birthweight. The authors used the simulated exposure distributions generated from 30 simulated individuals who matched the actual study participant with respect to age and census tract as prior information through the use of a discrete uniform distribution in order to account for within individual exposure variability. While no significant associations were identified for any of the pregnancy time windows of interest, the authors found that the use of SHEDS-PM resulted in improved prediction of birthweight for the resulting births in the study when compared to defining exposure based on ambient concentrations alone.

Chang et al. (2012) used personal-level ambient $PM_{2.5}$ exposures from SHEDS-PM in their modeling framework and found an association between increased exposure and daily mortality in New York City. They also introduced a statistical emulator approach that eased the computational burden of using SHEDS-PM over numerous days. Instead of running the model for each of the study days, it was run for a subset of the days and a statistical model was introduced to relate ambient $PM_{2.5}$ concentrations with the personal-level exposures generated by SHEDS-PM. The statistical emulator approach was used to interpolate the full set of SHEDS-PM concentrations across the entire study period.

Reich et al. (2009) used personal-level exposures from SHEDS-PM within a study of daily mortality and PM of varying sizes (10, 2.5, and 0.01 to 0.40 μm in aerodynamic diameter) in Fresno, California. They introduced a dynamic factor analysis model to study the association of interest and found that particles with diameter between 0.02 and 0.08 μm was associated with daily mortality at 4-day lag periods. SHEDS-PM exposure distributions on each day were approximated with normal distributions where the prior distributions for the mean and variance parameters were defined based on the sample mean and variance of the exposure distribution. This allowed the uncertainty in exposures to propagate through to the estimation of the health effects while also easing the computational burden of working with the exposure distributions directly.

Mannshardt et al. (2013) used a similar framework to that of Reich et al. (2009) to incorporate the SHEDS-PM exposure distributions into the estimation of the health effects. The authors compared the performance of ambient concentrations, estimates from an atmospheric chemistry model, and concentrations from SHEDS-PM when investigating the association between daily emergency hospital admissions in New York and $PM_{2.5}$ exposure. They found that the estimated health effect results were similar among all three metrics but use of the atmospheric chemistry model estimates and SHEDS-PM personal-level concentrations resulted in reduced uncertainty in effect estimation.

Calder et al. (2008) introduced a simplified version of SHEDS-PM, hierarchical exposure simulation, in order to simulate personal-level exposures for their study population. This model was used to explore the association between cardiovascular disease mortality in North

Carolina and PM$_{2.5}$ exposure. Holloman et al. (2004) also used a modified version of the original SHEDS-PM framework in order to examine the same outcome in North Carolina.

These various examples of the use of SHEDS-PM show a range of applications in epidemiologic and statistical studies. SHEDS-PM has been used to incorporate intra-person differences in exposure estimates based on demographics and activity patterns to better estimate PM$_{2.5}$ exposures in a study population and in a variety of statistical purposes to explore the effect of different exposure metrics on resulting effect estimates and on the resulting uncertainty of effect estimates.

20.3.4 Limitations to SHEDS-PM

SHEDS-PM has limitations that a user should be aware of before its application. For instance, the model requires that the user inputs ambient concentrations from multiple locations within a geographic domain of interest on a fine time scale for a long period of time. Air pollution monitoring networks often do not collect these data on a daily basis (e.g., typically every three to six days for PM$_{2.5}$ in the US) or do not have full spatial coverage in an area. A potential solution to fill in these space-time gaps is through statistical modeling and interpolation of the ambient concentrations using Kriging or data fusion techniques such as downscaling (Berrocal et al. 2010). These methods can provide spatially dense predictions of daily ambient concentrations across a domain, given that sufficient monitoring data are available; however, these estimating approaches introduce their own uncertainties. See Chapter 5: Geostatistical Modeling for Environmental Processes, Chapter 7: Data Assimilation, and Chapter 19: Statistical methods for exposure assessment, for detailed information on these topics. It is also possible to use estimated ambient concentrations from chemistry-physical transport models, either directly or through improved estimates that fuse such results with ambient monitoring data, as such methods provide full spatiotemporal coverage for many different pollutants' concentrations. However, these models have their own uncertainties and have been shown to produce biased estimates with respect measured ambient concentrations; such results should be used with caution in this way.

The personal-level exposures generated by SHEDS-PM can also difficult to validate given the lack of availability of measured personal-level exposure data for most populations. When spatiotemporally interpolating ambient concentrations, selected monitors can be omitted from the analysis and predicted values can be compared with these omitted monitoring values in order to provide some sense of validation of the predictions. However, for SHEDS-PM, measured personal-level exposures broken down by microenvironment are rarely available, so validation can be a difficult task.

Finally, a number of modeling assumptions are made in the development of SHEDS-PM that can potentially be difficult to validate. A number of studies have investigated the sensitivity of SHEDS-PM to the required inputs and have focused on individual components of the model. This includes exploring the sensitivity in the variability of personal-level exposure to PM$_{2.5}$ to changes in the penetration factor (P), the deposition rate (k), and the air exchange rate (ach) (Cao and Frey 2011a) and changes in the averaging time of the concentrations (Jiao et al. 2012). Other studies have provided critical assessments of the SHEDS-PM methodology for a number of microenvironments including for indoor residential PM$_{2.5}$ exposures (Deshpande et al. 2009), exposure to environmental tobacco smoke (Cao and Frey 2009; Cao and Frey 2011b), and in-vehicle exposures (Liu et al. 2009).

Overall, SHEDS-PM modeling offers a method of estimating personal-level exposures that incorporates factors not addressed in typical studies based on ambient concentrations, while being less expensive and intrusive than studies that rely on personal monitoring devices. If the requisite data on ambient concentrations and human activity patterns are available, and parameter uncertainty can be described through the use of prior probability distributions, then exposure modeling may represent a useful method for describing personal-level exposures.

20.4 Conclusions

These two alternative models for estimating exposure to air pollution for human health studies, LUR and SHEDS-PM, help counter limitations in the approach that was most commonly applied historically in epidemiological studies, the use of one or a small number of stationary ambient monitors. Specifically, LUR can be used to address issues of within-city spatial heterogeneity of air pollutant concentrations, with some models also providing information on temporal variability such as by season, and SHEDS-PM can be used to address issues of variability among persons that relate to differences by activity patterns.

As noted above, each approach has its own limitations and introduces additional uncertainties related to the quality of the input data. These methods have substantial data requirements that may be prohibitive for some study regions and some research questions. In particular, less industrialized nations often do not have regulatory monitors for air pollution, detailed land-use information, or traffic data to use as inputs into LUR models, or information on hypothetical populations' activity patterns for input into SHEDS-PM models. These areas can have some of the highest levels of air pollution in the world, but may be ineligible for LUR, SHEDS-PM, or other innovative methods to assess air pollution exposure. In other areas, primarily Europe and North American, the growing use of these approaches, and other novel exposure methods, as discussed in other chapters, have been applied to improve exposure estimates for studies of air pollution and human health and have contributed to the body of scientific evidence on air quality and public health. Recent studies have combined the strengths of multiple methods for exposure assessment and may present a path forward to address the limitations of these methods.

Bibliography

[1] Alexeeff, S. E., Schwartz, J., Kloog, I., Chudnovsky, A., Koutrakis, P., and Coull, B. A. 2015. Consequences of kriging and land use regression for $PM_{2.5}$ predictions in epidemiologic analyses: insights into spatial variability using high-resolution satellite data. *Journal of Exposure Science and Environmental Epidemiology*, 25:138–144.

[2] Amini, H., Taghavi-Shahri, S. M., Henderson, S. B., Naddafi, K., Nabizadeh, R. and Yunesian, M. 2014. Land use regression models to estimate the annual and seasonal spatial variability of sulfur dioxide and particulate matter in Tehran, Iran. *Science of the Total Environment*, 488–489:343–53.

[3] Ashmore, M. R., and Dimitroulopoulou, C. 2009. Personal exposure of children to air pollution. *Atmospheric Environment*, 43:128–141.

[4] Baxter, L. K., Burke, J., Lunden, M., Turpin, B. J., Rich, D. Q., Thevenet-Morrison, K., Hodas, N., and Özkaynak, H. 2013. Influence of human activity patterns, particle composition, and residential air exchange rates on modeled distributions of $PM_{2.5}$

exposure compared with central-site monitoring data. *Journal of Exposure Science and Environmental Epidemiology*, 23:241–247.

[5] Berrocal, V. J., Gelfand, A. E., and Holland, D. M. 2010. A spatio-temporal downscaler for output from numerical models. *Journal of Agricultural, Biological, and Environmental Statistics*, 15:176–197.

[6] Berrocal, V. J., Gelfand, A. E., Holland, D. M., Burke, J., and Miranda, M. L. 2011. On the use of a $PM_{2.5}$ exposure simulator to explain birthweight. *Environmetrics*, 22:553–571.

[7] Bravo, M. A., Fuentes, M., Zhang, Y., Burr, M. J., and Bell, M. L. 2012. Comparison of exposure estimation methods for air pollutants: ambient monitoring data and regional air quality simulation. *Environmental Research*, 116:1–10.

[8] Burke, J. M., Zufall, M. J., and Özkaynak, H. 2001. A population exposure model for particulate matter: case study results for $PM_{2.5}$ in Philadelphia, PA. *Journal of Exposure Analysis and Environmental Epidemiology*, 11:470–489.

[9] Calder, C. A., Holloman, C. H., Bortnick, S. M., Strauss, W., and Morara, M. 2008. Relating ambient particulate matter concentration levels to mortality using an exposure simulator. *Journal of the American Statistical Association*, 103:137–148.

[10] Cao, Y., and Frey, H. C. 2011a. Geographic differences in inter-individual variability of human exposure to fine particulate matter. *Atmospheric Environment*, 45:5684–5691.

[11] Cao, Y., and Frey, H. C. 2011b. Assessment of interindividual and geographic variability in human exposure to fine particulate matter in environmental tobacco smoke. *Risk Analysis*, 31:578–591.

[12] Cao, Y., Frey, H. C., Liu, X., and Deshpande, B. K. 2009. Evaluation of the Modeling of Exposure to Environmental Tobacco Smoke (ETS) in the SHEDS-PM Model. *Annual meeting & exhibition proceedings CD-ROM Air & Waste Management Association*, 2009:2009-A-239-AWMA

[13] Chang, H. H., Fuentes, M., and Frey, H. C. 2012. Time series analysis of personal exposure to ambient air pollution and mortality using an exposure simulator. *Journal of Exposure Science and Environmental Epidemiology*, 22:483–488.

[14] Chen, G., Li, J., Ying, Q., Sherman, S., Perkins, N., Rajeshwari, S., and Mendola, P. 2014. Evaluation of observation-fused regional air quality model results for population air pollution exposure estimation. *Science of the Total Environment*, 485–486:563–574.

[15] Clougherty, J. E., Kheirbek, I., Eisl, H. M., Ross, Z., Pezeshki, G., Gorczynski, J. E., Johnson, S., Markowitz, S., Kass, D. and Matt, T. 2013. Intra-urban spatial variability in wintertime street-level concentrations of multiple combustion-related air pollutants: the New York City Community Air Survey (NYCCAS). *Journal of Exposure Science and Environmental Epidemiology*, 23:232–40.

[16] Deshpande, B. K., Frey, H. C., Cao, Y., and Liu, X. 2009. Modeling of the penetration of ambient $PM_{2.5}$ to indoor residential microenvironment. *Proceedings of the 102nd Annual Conference and Exhibition, Air & Waste Management Association*, Detroit, USA.

[17] Di, Q., Kloog, I., Koutrakis, P., Lyapustin, A., Wang, Y., and Schwartz, J. Assessing $PM_{2.5}$ Exposures with high spatiotemporal resolution across the continental United States. *Environmental Science and Technology*, 50:4712–4721.

[18] Dionisio, K. L., Isakov, V., Baxter, L. K., Sarnat, J. A., Sarnat, S. E., Burke, J., Resenbaum, A., Graham, S. E., Cook, R., Mulholland, J., and Özkaynak, H. 2013. Development and evaluation of alternative approaches for exposure assessment of multiple air pollutants in Atlanta, Georgia. *Journal of Exposure Science and Environmental Epidemiology*, 23:581–592

[19] Dockery, D. W., and Pope, C. A. 1994. Acute respiratory effects of particulate air pollution. *Annual Review of Public Health*, 15:107–132.

[20] Estarlich, M., Ballester, F., Davdand, P., Llop, S., Esplugues, A., Fernández-Somoano, A., Lertxundi, A., Guxens, M., Basterrechea, M., Tardón, A., Sunyer, J., and Iñiguez, C. 2016. Exposure to ambient air pollution during pregnancy and preterm birth: A Spanish multicenter birth cohort study. *Environmental Research* 147:50–58.

[21] Hodas, N., Turpin, B. J., Lunden, M. M., Baxter, L. K., Özkaynak, H., Burke, J., Ohman-Strickland, P., Thevenet-Morrison, K., Kostis, J. B., for the MIDAS 21 Study Group, and Rich, D. Q. 2013. Refined ambient $PM_{2.5}$ exposure surrogates and the risk of myocardial infarction. *Journal of Exposure Science and Environmental Epidemiology*, 23:573–580.

[22] Hoek, G. , Beelen, R., de Hoogh, K., Vienneau, D., Gulliver, J., Fischer, P., and Briggs, D. 2008. A review of land-use regression models to assess spatial variation of outdoor air pollution. *Atmospheric Environment*, 42:7561–7578.

[23] Hyder, A., Lee, H. J., Ebisu, K., Koutrakis, P., Belanger, K., and Bell, M. L. 2014. $PM_{2.5}$ exposure and birth outcomes: use of satellite- and monitor-based data. *Epidemiology*, 25:58–67.

[24] Jiao, W., Frey, H. C., and Cao, Y. 2012. Assessment of inter-individual, geographic, and seasonal variability in estimated human exposure to fine particles. *Environmental Science & Technology*, 46:12519–12526.

[25] Jones, R. R., Özkaynak, H., Nayak, S. G., Garcia, V., Hwang, S. A., and Lin, S. 2013. Associations between summertime ambient pollutants and respiratory morbidity in New York City: Comparison of results using ambient concentrations versus predicted exposures. *Journal of Exposure Science and Environmental Epidemiology*, 23:616–626.

[26] Kanaroglou, P. S. , Jerrett, M., Morrison, J., Beckerman, B., Altaf Arain, M., Gilbert, N. L., and Brook, J.R. 2005. Establishing an air pollution monitoring network for intra-urban population exposure assessment: a location-allocation approach. *Atmospheric Environment*, 39:2399–2409.

[27] Kumar, N. 2009. An optimal spatial sampling design for intra-urban population exposure assessment. *Atmospheric Environment* 43:1153. doi: 10.1016/j.atmosenv.2008.10.055

[28] Laurent, O., Hu, J., Li, L., Kleeman, M. J., Bartell, S. M., Cockburn, M., Escobedo, L., and Wu, J. 2016. Low birth weight and air pollution in California: Which sources and components drive the risk? *Environment International*, 92–93:471–477.

[29] Lee, H. J., Kang, C. M., Coull, B. A., Bell, M. L., and Koutrakis, P. 2014. Assessment of primary and secondary ambient particle trends using satellite aerosol optical depth and ground speciation data in the New England region, United States. *Environmental Research*, 133:103–110.

[30] Liu, X., Frey, H. C., Cao, Y., and Deshpande, B. 2009. Modeling of in-vehicle PM$_{2.5}$ exposure using the stochastic human exposure and dose simulation model. *Annual meeting & exhibition proceedings CD-ROM Air & Waste Management Association*, 2:1087–1100.

[31] Mannshardt, E., Sucic, K., Jiao, W., Dominici, F., Frey, H. C., Reich, B., and Fuentes, M. 2013. Comparing exposure metrics for the effects of fine particulate matter on emergency hospital admissions. *Journal of Exposure Science and Environmental Epidemiology*, 23:627–636.

[32] Meng, X., Fu, Q., Ma, Z., Chen. L., Zou, B., Zhang, Y., Xue, W., Wang, J., Wang, D., Kan, H., and Liu, Y. 2016. Estimating ground-level PM(10) in a Chinese city by combining satellite data, meteorological information and a land use regression model. *Environmental Pollution*, 208(Pt A):177–184.

[33] Moon, C. 2015. Exposure assessment of air pollutants: a review on spatial heterogeneity and indoor/outdoor/personal exposure to suspended particulate matter, nitrogen dioxide, and ozone. *Atmospheric Environment*, 35:1–32.

[34] Mordukhovich, I., Coull, B., Kloog, I., Koutrakis, P., Vokonas, P., and Schwartz, J. 2015. Exposure to sub-chronic and long-term particulate air pollution and heart rate variability in an elderly cohort: the Normative Aging Study. *Environmental Health* Nov 6;14:87. doi: 10.1186/s12940-015-0074-z.

[35] Mordukhovich, I., Kloog, I., Coull, B., Koutrakis, P., Vokonas, P., and Schwartz, J. 2016. Association Between Particulate Air Pollution and QT Interval Duration in an Elderly Cohort. *Epidemiology*, 27:284–290.

[36] Oiamo, T. H., Johnson, M., Tang, K., and Luginaah, I. N. 2015. Assessing traffic and industrial contributions to ambient nitrogen dioxide and volatile organic compounds in a low pollution urban environment. *Science of the Total Environment*, 529:149–157.

[37] Patton, A. P., Zamore, W., Naumova, E. N., Levy, J. I., Brugge, D., and Durant, J. L. 2015. Transferability and generalizability of regression models of ultrafine particles in urban neighborhoods in the Boston area. *Environmental Science and Technology*, 49:6051–6060.

[38] Pope 3rd, C. A., Bates, D. V., and Raizenne, M. E. 1995. Health effects of particulate air pollution: time for reassessment? *Environmental Health Perspectives*, 103:472–480.

[39] Porta, D., Narduzzi, S., Badaloni, C., Bucci, S., Cesaroni, G., Colelli, V., Davoli, M., Sunyer, J., Zirro, E., Schwartz, J. and Forastiere, F. 2016. Air pollution and cognitive development at age 7 in a prospective Italian birth cohort. *Epidemiology*, 27:228–236.

[40] Puett, R. C., Hart, J. E., Yanosky, J. D., Spiegelman, D., Wang, M., Fisher, J. A., Hong, B., and Laden, F. 2014. Particulate matter air pollution exposure, distance to road, and incident lung cancer in the nurses' health study cohort. *Environmental Health Perspectives*, 122:926–932.

[41] Ramos, Y., St-Onge, B., Blanchet, J. P., and Smargiassi, A. 2015. Spatio-temporal models to estimate daily concentrations of fine particulate matter in Montreal: Kriging with external drift and inverse distance-weighted approaches. *Journal of Exposure Science and Environmental Epidemiology.* doi: 10.1038/jes.2015.79

[42] Reich, B. J., Fuentes, M., and Burke, J. 2009. Analysis of the effects of ultrafine particulate matter while accounting for human exposure. *Environmetrics*, 20:131–146.

[43] Rice, M. B., Rifas-Shiman, S. L., Litonjua, A. A., Oken, E., Gillman, M. W., Kloog, I., Luttmann-Gibson, H., Zanobetti, A., Coull, B. A., Schwartz, J., Koutrakis, P., Mittleman, M. A., and Gold, D. R. 2016. Lifetime exposure to ambient pollution and lung function in children. *American Journal and Respiratory Critical Medicine*, 193:881–888.

[44] Ryan, P. H., and LeMasters, G., K. 2007. A review of land-use regression models for characterizing intraurban air pollution exposure. *Inhalation Toxicology*, 19(Suppl 1):127–133.

[45] Sampson, P. D., Richards, M., Szpiro, A. A., Bergen, S., Sheppard, L., Larson, T. V., and Kaufman, J. D. 1994. A regionalized national universal kriging model using Partial Least Squares regression for estimating annual $PM_{2.5}$ concentrations in epidemiology. *Atmospheric Environment*, 75:383–392.

[46] Sarnat, S. E., Sarnat, J. A., Mulholland, J., Isakov, V., Özkaynak, H., Chang, H. H., Klein, M., and Tolbert, P. E. 2013. Application of alternative spatiotemporal metrics of ambient air pollution exposure in a time-series epidemiological study in Atlanta. *Journal of Exposure Science and Environmental Epidemiology*, 23:593–605.

[47] Schwartz, J. 1994. Air pollution and daily mortality: a review and meta analysis. *Environmental Research*, 64:36–52.

[48] Smith, L. A., Mukerjee, S., Chung, K. C, and Afghani, J. 2011. Spatial analysis and land use regression of VOCs and NO_2 in Dallas, Texas during two seasons. *Journal of Environmental Monitoring*, 13:999–1007.

[49] Turner, M. C., Jerrett, M., Pope, C.A., 3rd, Krewski, D., Gapstur, S.M., Diver, W. R., Beckerman, B. S, Marshall, J. D., Su, J., Crouse, D. L., and Burnett, R. T. 2016. Long-term ozone exposure and mortality in a large prospective study. 193:1134–1142.

[50] Urman, R., McConnell, R., Islam, T., Avol, E. L., Lurmann, F. W., Vora, H., Linn, W. S., Rappaport, E. B., Gilliland, F. D., and Gauderman, W. J. 2014. Associations of children's lung function with ambient air pollution: joint effects of regional and near-roadway pollutants. *Thorax*, 69:540–547.

[51] Wang, M., Beelen, R., Bellander, T., Birk, M. Cesaroni, G., Cirach, M., Cyrys, J., de Hoogh, K., Declercq, C., Dimakopoulou, K., Eeftens, M., Eriksen, K. T., Forastiere, F., Galassi, C., Grivas, G., Heinrich, J., Hoffmann, B., Ineichen, A., Korek, M., Lanki, T., Lindley, S., Modig, L., Mölter, A., Nafstad, P., Nieuwenhuijsen, M. J., Nystad, W., Olsson, D., Raaschou-Nielsen, O., Ragettli, M., Ranzi, A., Stempfelet, M., Sugiri, D., Tsai, M.Y., Udvardy, O., Varró, M. J., Vienneau, D., Weinmayr, G., Wolf, K., Yli-Tuomi, T., Hoek, G., and Brunekreef, B. 2014. Performance of multi-city land use regression models for nitrogen dioxide and fine particles, *Environmental Health Perspectives*, 122:843—849.

[52] Wang, M., Sampson, P. D., Hu, J., Kleeman, M., Keller, J. P., Olives, C., Szpiro, A. A., Vedal, S., and Kaufman, J. D. 2016. Combining land-use regression and chemical transport modeling in a spatiotemporal geostatistical model for ozone and $PM_{2.5}$. *Environmental Science and Technology*, 50:5111–5118.

[53] Young, M. T., Bechle, M. J., Sampson, P. D., Szpiro, A. A., Marshall, J. D., Sheppard, L., and Kaufman, J. D. 2016. Satellite-based NO_2 and model validation in a national prediction model based on universal kriging and land-use regression. *Environmental Science and Technology*, 50:3686–3694.

[54] Young, M. T., Sandler, D. P., DeRoo, L. A., Vedal, S., Kaufman, J. D., and London, S. J. 2014. Ambient air pollution exposure and incident adult asthma in a nationwide cohort of U.S. women. *American Journal of Respiratory and Critical Care Medicine*, 190:914–921.

[55] Zhang, J. J. Y., Sun, L., Barrett, O., Bertazzon, S., Underwood, F.E., and Johnson, M. 2014. Development of land-use regression models for metals associated with airborne particulate matter in a North American city. *Atmospheric Environment*, 106:165–177.

21

Preferential sampling of exposure levels

Peter J Diggle

CHICAS, Lancaster Medical School, Lancaster University, UK

Emanuele Giorgi

CHICAS, Lancaster Medical School, Lancaster University, UK

CONTENTS

21.1   Introduction

Geostatistical methods have their origins in the South African Mining industry (Krige, 1951) and were subsequently developed extensively by the late Georges Matheron and colleagues at Fontainebleau, France (Matheron, 1963; Chilès and Delfiner, 1999, 2012). To the best of our knowledge, Matheron introduced the term "La Géostatistique."

The canonical geostatistical problem is the following. Given data $Y_i : i = 1, ..., n$ associated with locations $x_i : i = 1, ..., n$ in a spatial region A, such that each Y_i is related to the value at x_i of an unobserved spatially varying phenomenon $\mathcal{S} = \{S(x) : x \in A\}$, what can be inferred about $\mathcal{S}$? In the original setting, $\mathcal{S}$ represented the spatial variation in the amount of valuable mineral that could be extracted from a mining operation over the region A.

Ripley (1981, Section 4.4) re-cast the canonical geostatistical problem as one of stochastic process prediction. Cressie (1991) classified geostatistics as one of three main branches of spatial statistical methods. These methods have now been used in a wide range of application areas including the physical environmental sciences; see for example, Webster and Oliver (2001). Diggle, Moyeed and Tawn (1998) coined the term "model-based geostatistics" to mean the application of general principles of statistical modelling and inference to geostatistical problems. In particular, they proposed a Markov chain Monte Carlo algorithm for Bayesian inference in a class of *generalized linear geostatistical models*. These are simply generalized linear mixed models (Breslow and Clayton, 1993) that include a latent, spatially continuous Gaussian process $S(x)$ in the linear predictor. In this chapter, we shall focus on the *linear Gaussian geostatistical model*, in which the Y_i are conditionally independent and Normally distributed given the latent Gaussian process $S(x)$. Specifically,

$$Y_i|\mathcal{S} \sim \mathrm{N}(d(x_i)'\beta + S(x_i), \tau^2), \tag{21.1}$$

where $d(x)$ denotes a set of spatially indexed covariates that, in principle, could be measured at any $x \in A$.

The term *preferential sampling* was introduced by Diggle, Menezes and Su (2010) to mean that the process that generates the sampling locations x_i is stochastically dependent on the latent process $\{S(x) : x \in A\}$. For reasons that will become clear, we shall call this *weakly preferential* sampling and add a stricter definition, namely *strongly preferential sampling*, to mean that the sampling locations and the latent process remain stochastically dependent conditional on the measurements Y_i.

In Section 2 of this chapter we set out some definitions for different classes of geostatistical design and describe some well-known design constructions. In Section 3 we first review the material in Diggle, Menezes and Su (2010), then consider some other ways of analysing preferentially sampled geostatistical data. Section 4 describes a re-analysis of data on lead pollution levels in Galicia, north-west Spain, previously analysed in Diggle, Menezes and Su (2010).

21.2   Geostatistical sampling designs

Often in practice, a geostatistical data-set consisting of a set of locations x_i and corresponding measurements Y_i is presented to the analyst as a *fait accompli*, with no formal definition of how the locations were selected. Whether or not this is the case, we call the set of sampling locations, $\mathcal{X} = \{x_i \in A : i = 1, ..., n\}$ the geostatistical *design*. We also call $\mathcal{S} = \{S(x) : x \in A\}$ the *signal process* and $\mathcal{Y} = \{Y_i : i = 1, ..., n\}$ the *measurement data*.

21.2.1   Definitions

A design is *deterministic* if it consists of a set of pre-specified points, otherwise it is *stochastic*. Unless the study in question can be replicated, this distinction is somewhat philosophical, but the idea that an actual design can be viewed as a probability-based selection from a suitably defined set of potential designs is fundamental to any understanding of why preferential sampling matters.

Formally, a stochastic design therefore constitutes a realisation of a finite point process on A. A stochastic design is *uniform* on A if the marginal distribution of each sampled location is uniform on A; informally, each point $x \in A$ is equally likely to be included in $\mathcal{S}$. Two examples of a uniform design are a *completely random* design, in which the sampled locations are mutually independent and uniformly distributed on A, and a *regular lattice* design with, strictly, the lattice origin uniformly distributed over one of the lattice cells; see Figure 21.1.

A design is *adaptive* if the elements of $\mathcal{X}$ are chosen sequentially, either singly or in batches, and the choice of the current batch of locations depends on the measurement data obtained from earlier batches. As noted earlier, a design is *strongly preferential* if $\mathcal{X}$ and

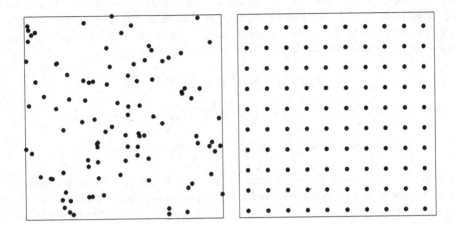

FIGURE 21.1
Two uniform sampling designs. In the left-hand panel the 100 sampling locations are an independent random sample from the uniform distribution on the unit square, in the right-hand panel they form a square lattice with thee bottom-left location sampled randomly from the uniform distribution on the square $(0.0, 0.1) \times (0.0, 0.1)$.

$\mathcal{S}$ are stochastically dependent given $\mathcal{Y}$. Previous authors have not explicitly invoked the conditioning on $\mathcal{Y}$, but we shall show that it can have important practical implications.

Strongly preferential sampling as here defined is the geostatistical version of *informative follow-up* in longitudinal studies whereby, for example in a medical setting, a patient presents for measurement when they are experiencing a particular combination of symptoms; see, for example, Lin, Scharfstein and Rosenheck (2004).

The converse of strongly preferential sampling is *informative missingness*, meaning that the absence of a measurement a particular location x conveys information about the value of $S(x)$. The analogous problem in the context of longitudinal data has been studies extensively; see, for example, Diggle and Kenward (1994), Little (!995), Hogan and Laird (1997), Daniels and Hogan (2008). Missingness can only be identified when the set of intended sampling locations is specified beforehand. In environmental monitoring applications, a common reason for a measurement to be missing is that the value of whatever is being measured falls below a detection limit. In this case, provided that the detection limit is a known constant, the data are not missing, but rather are *right-censored* and should be analysed accordingly, although this may present technical difficulties.

21.3 Preferential sampling methodology

Recall our general notation $\mathcal{S}$, $\mathcal{X}$ and $\mathcal{Y}$ to denote the signal process, design and measurement data, respectively. In what follows, we use $[\cdot]$ to mean "the distribution of," and a vertical bar to denote stochastic conditioning.

A general factorisation of the joint distribution of $\mathcal{S}$, $\mathcal{X}$ and $\mathcal{Y}$ is

$$[\mathcal{S}, \mathcal{X}, \mathcal{Y}] = [\mathcal{S}][\mathcal{X}|\mathcal{S}][\mathcal{Y}|\mathcal{X}, \mathcal{S}], \tag{21.2}$$

and integration with respect to $\mathcal{S}$ gives the likelihood function based on the data $\mathcal{X}$ and $\mathcal{Y}$.

Other factorisations are, of course, available but (21.2) is the most natural from a modelling perspective because the signal process precedes the design and, in the non-adaptive case, the design precedes the measurement data. Integration of (21.2) with respect to S gives the log-likelihood function for the data, $\mathcal{X}$ and $\mathcal{Y}$, as

$$L(\mathcal{X}, \mathcal{Y}) = \log \int [\mathcal{X}|\mathcal{S}][\mathcal{Y}|\mathcal{X}, \mathcal{S}][\mathcal{S}]d\mathcal{S}. \tag{21.3}$$

Under non-preferential sampling, $[\mathcal{X}|\mathcal{S}] = [\mathcal{X}]$, hence (21.3) becomes

$$L(\mathcal{X}, \mathcal{Y}) = \log \int [\mathcal{Y}|\mathcal{X}, \mathcal{S}][\mathcal{S}]d\mathcal{S} + \log[\mathcal{X}], \tag{21.4}$$

from which it follows that the stochastic variation in $\mathcal{X}$ can be ignored for inference about $\mathcal{S}$ or $\mathcal{Y}$. Conventional geostatistical methods do precisely this, by treating the design as a fixed set of locations $x_i : i = 1, ..., n$ and, typically, assuming that the measurements Y_i are conditionally independent given the corresponding $S(x_i)$, hence $[\mathcal{Y}|\mathcal{X}, \mathcal{S}] = \prod_{i=1}^{n}[Y_i|S(x_i)]$.

To explain why the distinction between weakly and strongly preferential sampling matters, we use an alternative factorisation to (21.2), namely

$$[\mathcal{S}, \mathcal{X}, \mathcal{Y}] = [\mathcal{S}][\mathcal{Y}|\mathcal{S}][\mathcal{X}|\mathcal{Y}, \mathcal{S}], \tag{21.5}$$

Under weakly preferential sampling, $[\mathcal{X}|\mathcal{Y}, \mathcal{S}] = [\mathcal{X}|\mathcal{Y}]$ and the log-likelihood becomes

$$L(\mathcal{X}, \mathcal{Y}) = \log \int [\mathcal{Y}|\mathcal{S}][\mathcal{S}]d\mathcal{S} + \log[\mathcal{X}|\mathcal{Y}]. \tag{21.6}$$

Ignoring the term $\log[\mathcal{X}|\mathcal{Y}]$ therefore leads to inferences about $\mathcal{S}$ or $\mathcal{Y}$ that are valid, but may or may not be inefficient depending on the exact specification of $[\mathcal{X}|\mathcal{Y}]$. Under strongly preferential sampling, no factorisation of $[\mathcal{S}, \mathcal{X}, \mathcal{Y}]$ separates $\mathcal{X}$ and $\mathcal{S}$, hence valid inference requires the stochastic nature of $\mathcal{X}$ to be taken into account, i.e. $\mathcal{X}$ is not ignorable.

In the remainder of this chapter, we use the unqualified term *preferential* as a shorthand for *strongly preferential*

21.3.1 Non-uniform designs need not be preferential

It is easy to imagine circumstances in which a non-uniform spatial distribution of locations has practical advantages. These include: stratification of the study-region, with relatively high sampling intensity in sub-regions of particular concern (for example, monitoring air pollution more intensively in areas of high population density); maximising overall efficiency when some sub-regions are relatively inaccessible and therefore cost more to sample; designing to span the range of a potentially important covariate.

All of these circumstances can be captured by adding to our notation $\mathcal{D}$, the set of all design factors. Then, a design is non-preferential if

$$[\mathcal{X}|\mathcal{S}; \mathcal{D}] = [\mathcal{X}; \mathcal{D}]. \tag{21.7}$$

In (21.7), we make a notational distinction between conditional dependence on a stochastic quantity, denoted by the vertical bar, and functional dependence on a non-stochastic quantity, denoted by the semi-colon.

21.3.2 Adaptive designs need not be strongly preferential

The essence of adaptive design is to exploit the information about $\mathcal{S}$ gleaned from an initial set of sampling locations in order to maximise the additional information that can be gained from later sampling locations. It is intuitively clear that this must induce some form of dependence between $\mathcal{X}$ and $\mathcal{S}$ in the complete process. To explain why adaptive designs need not be strongly preferential, let $\mathcal{X}_0$ denote an initial sampling design chosen independently of $\mathcal{S}$, and $\mathcal{Y}_0$ the resulting measurement data. Analysis of the resulting data informs the choice of a further batch of sampling locations $\mathcal{X}_1$, which generate additional data $\mathcal{Y}_1$, and so on. The complete data-set consists of $\mathcal{X} = \{\mathcal{X}_0, \mathcal{X}_1, ... \mathcal{X}_k\}$ and $\mathcal{Y} = \{\mathcal{Y}_0, \mathcal{Y}_1, ..., \mathcal{Y}_k\}$. The associated likelihood for the complete data-set is

$$[\mathcal{X}; \mathcal{Y}] = \int [\mathcal{X}; \mathcal{Y}; \mathcal{S}] d\mathcal{S} \tag{21.8}$$

We consider first the case $k = 1$. The standard factorisation of any multivariate distribution gives

$$[\mathcal{X}, \mathcal{Y}, \mathcal{S}] = [\mathcal{S}, X_0, Y_0, X_1, Y_1] = [\mathcal{S}][\mathcal{X}_0|\mathcal{S}][\mathcal{Y}_0|\mathcal{X}_0, \mathcal{S}][\mathcal{X}_1|\mathcal{Y}_0, \mathcal{X}_0, \mathcal{S}][\mathcal{Y}_1|\mathcal{X}_1, \mathcal{Y}_0, \mathcal{X}_0, \mathcal{S}]. \tag{21.9}$$

On the right-hand side of (21.9), note that by construction, $[\mathcal{X}_0|\mathcal{S}] = [\mathcal{X}_0]$ and $[\mathcal{X}_1|\mathcal{Y}_0, \mathcal{X}_0, \mathcal{S}] = [\mathcal{X}_1|\mathcal{Y}_0, \mathcal{X}_0]$. It then follows from (21.8) and (21.9) that

$$[\mathcal{X}, \mathcal{Y}] = [\mathcal{X}_0][\mathcal{X}_1|\mathcal{Y}_0, \mathcal{X}_0] \int [\mathcal{Y}_0|\mathcal{X}_0, \mathcal{S}][\mathcal{Y}_1|\mathcal{X}_1, \mathcal{Y}_0, \mathcal{X}_0; \mathcal{S}][\mathcal{S}] d\mathcal{S}. \tag{21.10}$$

The term outside the integral in (21.10) is the conditional distribution of X given Y_0. Using the fact, again by construction, that $[\mathcal{Y}_0|\mathcal{X}_1, \mathcal{X}_0; \mathcal{S}] = [\mathcal{Y}_0|\mathcal{X}_0; \mathcal{S}]$ the integral simplifies to

$$\int [\mathcal{Y}|\mathcal{X}, \mathcal{S}][\mathcal{S}] d\mathcal{S} = [\mathcal{Y}|\mathcal{X}].$$

It follows that

$$[\mathcal{X}, \mathcal{Y}] = [\mathcal{X}|\mathcal{Y}_0][\mathcal{Y}|\mathcal{X}]. \tag{21.11}$$

Equation (21.11) shows that the conditional likelihood, $[\mathcal{Y}|\mathcal{X}]$, can legitimately be used for inference although, depending on how $[\mathcal{X}|\mathcal{Y}_0]$ is specified, it may be inefficient. The argument leading to (21.11) extends to $k > 1$ with essentially only notational changes.

21.3.3 The Diggle, Menezes and Su model

Diggle, Menezes and Su (2010, henceforth DMS) used a parametric model to demonstrate some of the practical implications of strongly preferential sampling. Specifically, they assumed that $\mathcal{S}$ is a stationary Gaussian process and that, conditional on $\mathcal{S}$, $\mathcal{X}$ is an inhomogeneous Poisson process with intensity $\Lambda(x) = \exp\{\alpha + \beta S(x)\}$. Unconditionally, $\mathcal{X}$ is a log-Gaussian Cox process (Møller, Syversteen and Waagepeterson, 1998). This corresponds to non-preferential sampling when $\beta = 0$, and to strongly preferential sampling otherwise.

In an application to lead pollution data from Galicia, north-west Spain, likelihood-based analysis within this parametric formulation gave clear evidence of strongly preferential sampling. Also, allowance for this produced predictions of $\mathcal{S}$ that were materially different from predictions based on the standard linear Gaussian model assuming non-preferential sampling. However, the use of a single parameter to capture both the strength of the non-preferentiality and the amount of non-uniformity in $\mathcal{X}$ is somewhat inflexible. In the next section, we describe a more flexible class of models.

21.3.4 The Pati, Reich and Dunson model

Pati, Reich and Dunson (2011, henceforth PRD) propose an extension of DMS by adding a second Gaussian process, as follows.

Firstly, we again assume that $\mathcal{X}$ is a log-Gaussian process with intensity

$$\Lambda(x) = \exp\{\alpha + S_1(x)\}, \tag{21.12}$$

where S_1 is a Gaussian process with mean zero, variance σ_1^2 and correlation function $\rho(u/\phi_1)$. Secondly, we assume that measurements $Y_i : i = 1, ..., n$ at locations x_i follow the model

$$Y_i = \mu + \beta \log\{\Lambda(x_i)\} + S_2(x_i) + Z_i, \tag{21.13}$$

where S_2 is an isotropic Gaussian process, independent of S_1, with mean zero, variance σ_2^2 and correlation function $\rho(u/\phi_2)$, where u is the Euclidean distance between any two locations. Also, the Z_i are mutually independent $N(0, \tau^2)$ variates. The model reduces to a re-parameterised version of DMS when the process S_2 is absent, i.e. $\sigma_2^2 = 0$. Otherwise, for convenience we replace τ^2 by $\nu^2 \sigma_2^2$, so that the parameter ν^2 represents a noise-to-signal ratio.

In the model defined by (21.12) and (21.13), the parameter β controls the degree of preferentiality in the sampling of the Y_i, whilst the process S_2 allows for a component of the spatial variation in the measurement process that is not linked to the sampling process. This allows us to parameterise the model so that the preferentiality parameter β does not feature in the sub-model for the sampling intensity $\Lambda(x)$. As shown in the next section, this in turn allows us to develop an inferential algorithm that uses an approximation to the field S_1 to circumvent the otherwise intractable distribution of $\Lambda(x)$

21.3.4.1 Monte Carlo maximum likelihood using stochastic partial differential equations

We now outline a Monte Carlo maximum likelihood procedure for parameter estimation under the PRD model. By setting $\mathcal{S} = \{S_1, S_2\}$, we rewrite (21.3) as

$$[\mathcal{X}; \mathcal{Y}] = \int \int [\mathcal{X}; \mathcal{Y}; S_1; S_2] \, dS_1 dS_2 = \int \int [S_1][S_2][\mathcal{X}|S_1][\mathcal{Y}|\mathcal{X}; \mathcal{S}] \, dS_1 dS_2.$$

$$\tag{21.14}$$

In the integrand of the expression above, the elements of $\mathcal{X}$ given S_1 are independently and identically distributed with probability density

$$\frac{\Lambda(x)}{\int \Lambda(u) \, du}, x \in A. \tag{21.15}$$

The integral that forms the denominator of (21.15) is intractable. For this reason, we use an approximation of S_1.

Pati *et al.* (2011) approximate S_1 using a low-rank approximation based on a kernel convolution representation (Higdon, 1998). However, this approach has substantial computational advantages only for relative large values of the scale of the spatial correlation parameter ϕ_1. Diggle *et al.* (2013) use an extended regular grid on which they define S_1. By wrapping this grid onto a torus, they then carry out inversion of the spatial covariance matrix using Fourier methods.

In this chapter, we approximate S_1 using a technique based on stochastic partial differential equations. The high computational efficiency and ease of implementation of this approach make it an attractive alternative to either low-rank approximations or Fourier methods. Following Lindgren *et al.* (2011), we use the representation of a Gaussian process with Matérn covariance structure as the solution of the following stochastic partial differential equation (SPDE),

$$(\phi^{-2} - \Delta)^{\alpha/2}\{\omega S_1(x)\} = W(x), x \in A, \tag{21.16}$$

where Δ is the Laplacian and $W(x)$ is Gaussian white noise. The parameter ϕ, in (21.16) is the range parameter of the Matérn covariance function $\gamma(u)$ in its standard parameterisation,

$$\gamma(u) = \sigma^2\{\Gamma(\kappa)2^{\kappa-1}\}^{-1}(u/\phi)^{\kappa}K_{\kappa}(u/\phi) : u \geq 0,$$

where $K_{\kappa}(\cdot)$ is a modified Bessel function of the second kind. The remaining parameters in (21.16) are $\alpha = \kappa + 1$ and

$$\omega^2 = \frac{\phi^{2\kappa}\Gamma(\kappa)}{4\pi\sigma^2\Gamma(\kappa+1)}.$$

We then approximate the field $\mathcal{S}_1$ by $\tilde{\mathcal{S}}_1$, where

$$\tilde{\mathcal{S}}_1(x) = \sum_{k=1}^{n}\psi_k(x)W_k, x \in A. \tag{21.17}$$

In (21.17), the $\psi_k(\cdot)$ are piecewise linear basis functions defined by a triangulation of A, and $W = (W_1, \ldots, W_k)$ is a zero-mean multivariate Gaussian variate with covariance matrix, Q^{-1} say, chosen to give the required approximation. For example, for $\alpha = 2$ and and using a Finite Element method for projection of the SPDE onto the basis representation, the required form of Q is

$$Q = \omega^2(\phi^{-4}C + 2\phi^{-2}G_1 + G_2)$$

where C, G_1 and G_2 are sparse matrices whose explicit expressions can be found in Lindgren *et al.* (2011).

Integration of (21.14) with respect to $\mathcal{S}_2$ is analytically tractable, and leads to the approximation

$$[\mathcal{X}; \mathcal{Y}] = \int [\mathcal{S}_1][\mathcal{X}|\mathcal{S}_1][\mathcal{Y}|\mathcal{X}; \mathcal{S}_1]\, d\mathcal{S}_1 \approx \int [\tilde{\mathcal{S}}_1][\mathcal{X}|\tilde{\mathcal{S}}_1][\mathcal{Y}|\mathcal{X}; \tilde{\mathcal{S}}_1]\, d\tilde{\mathcal{S}}_1. \tag{21.18}$$

Now, let θ denote the vector of parameters of $\tilde{\mathcal{S}}_1$, which we denote by writing $\tilde{\mathcal{S}}_1(\theta)$, and write (21.18) as

$$\int [\tilde{\mathcal{S}}_1(\theta); \mathcal{X}; \mathcal{Y}]\, d\tilde{\mathcal{S}}_1 = \int \frac{[\tilde{\mathcal{S}}_1(\theta); \mathcal{X}; \mathcal{Y}]}{[\tilde{\mathcal{S}}_1(\theta_0); \mathcal{X}; \mathcal{Y}]}[\tilde{\mathcal{S}}_1(\theta_0); \mathcal{X}; \mathcal{Y}]\, d\tilde{\mathcal{S}}_1$$

$$\propto \int \frac{[\tilde{\mathcal{S}}_1(\theta); \mathcal{X}; \mathcal{Y}]}{[\tilde{\mathcal{S}}_1(\theta_0); \mathcal{X}; \mathcal{Y}]}[\tilde{\mathcal{S}}_1(\theta_0)|\mathcal{X}; \mathcal{Y}]\, d\tilde{\mathcal{S}}_1, \tag{21.19}$$

where θ_0 represents a "best guess" for $\hat{\theta}$.

To compute the integral in (21.19), we then use an independence sampler (IS) to simulate N realisations from the conditional distribution $[\tilde{\mathcal{S}}_1(\theta_0)|\mathcal{X}; \mathcal{Y}]$. At each iteration of the IS algorithm, we propose a new value from a multivariate Student's t distribution with 10 degrees of freedom, and location parameter and dispersion matrix given by the mode of $[\tilde{\mathcal{S}}_1(\theta_0); \mathcal{X}; \mathcal{Y}]$ and the inverse of the negative Hessian at the mode, respectively. Let $\tilde{\mathcal{S}}_{1(j)}$ denote the j-th out of the N simulated samples. The final approximation of the likelihood function is then given by

$$\frac{1}{N}\sum_{j=1}^{N}\frac{[\tilde{\mathcal{S}}_{1(j)}(\theta); \mathcal{X}; \mathcal{Y}]}{[\tilde{\mathcal{S}}_{1(j)}(\theta_0); \mathcal{X}; \mathcal{Y}]}. \tag{21.20}$$

To maximize (21.20) with respect to θ, we use a Newton-Raphson algorithm based on analytical expressions of the gradient function and Hessian matrix. After obtaining the Monte Carlo maximum likelihood estimate, say $\hat{\theta}_N$, we reiterate the algorithm setting $\theta_0 = \hat{\theta}_N$ and repeat this procedure until convergence.

21.4   Application: lead pollution monitoring

We now apply the PRD model of Section 21.3.4 to data on lead concentrations in moss samples, previously analysed in DMS. The data derive from two surveys of the same area (Galicia, north-west Spain), conducted in 1997 and 2000 (Figure 21.2). The 1997 survey used a non-uniform, and therefore potentially preferential, sampling design with more intensive sampling in the northern part of Galicia. The 2000 survey used an approximate lattice design, which we assume to be non-preferential, i.e. for the 2000 survey data, $\beta = 0$.

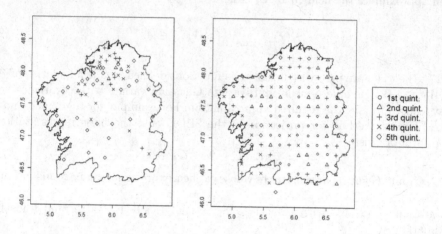

FIGURE 21.2

Sampling locations for the two surveys of lead concentrations in moss samples. The two maps correspond to the 1997 (left panel) and 2000 (right panel) surveys. Unit of distance is 100km. Each point is represented by a symbol corresponding to a quintile class of the observed lead-concentration values as indicated by the legend.

The two panels of Figure 21.2 show the sampling locations of the 1997 and 2000 surveys. Each point in the map is represented by a symbol corresponding to a quintile class of the observed lead-concentration values. We see that the level of lead concentration was higher in 1997 with a North-South trend that is not obviously present in the 2000 data.

A useful exploratory device to look for evidence of preferential sampling is a scatter plot of the response variable against a non-parametric estimate of the sampling density at each measurement location. Figure 21.3 is an example of this plot for the 1997 log-transformed lead pollution data, with sampling density estimated using a simple kernel smoother. The plot shows a weak negative association indicating a higher concentration of monitoring stations in areas with lower lead concentration, suggesting preferential over-sampling of locations with lower-than-average pollution levels. Note, in this context, that Schlather, Ribeiro and Diggle (2004) proposed a formal test that could also be used as an exploratory device.

Following DMS, we specify an exponential correlation function for S_2. We use a Matérn covariance function with $\kappa = 1$ for S_1 in order to have an explicit expression for the SPDE approximation. Figure 21.4 shows the triangulated mesh on which the SPDE representation for S_1 is built. The triangles outside the borders of Galicia are used in order to avoid edge effects. We allow the mean parameter of the measurement process to differ between the two surveys, but assume that other model parameters are common to both; DMS provided some evidence to support this. We then estimate the model parameters using the Monte Carlo

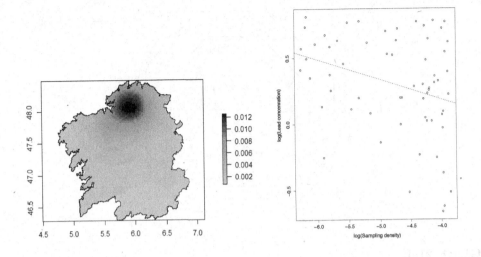

FIGURE 21.3
Kernel density estimate, $\hat{f}(x)$, of the sampling density for the 1997 lead concentration data (left-hand panel), scatter plot of log-transformed lead concentrations, Y_i, against $\log \hat{f}(x_i)$ (right-hand panel, dashed line is the least squares fit.

maximum likelihood algorithm of Section 21.3.4.1, applied to the combined likelihood for the data from both surveys. We simulate 10,000 samples from $[\mathcal{S}_1 | \mathcal{X}; \mathcal{Y}]$ by iterating the IS algorithm 12,000 times, with a burn-in of 2,000. Figure 21.5 shows some diagnostic plots on the convergence of the IS algorithm based on the region-wide average of $\tilde{S}_1$, given by

$$\frac{1}{|A|} \int_A \tilde{S}_1(x) \, dx$$

where A corresponds to the whole of Galicia and $|A|$ is its area. These indicate a satisfactory mixing of the Markov chain.

Table 21.1 shows Monte Carlo maximum likelihood estimates and associated 95% confidence intervals for the fitted model.

To the extent that they are directly comparable, the point estimates in Table 21.1 are qualitatively similar to those reported in DMS. Specifically, mean levels are lower in 2000 than in 1997 ($\mu_{00} < \mu_{97}$), and the estimate of the preferentiality parameter β is negative. However, the confidence interval for β comfortably includes zero, whereas DMS found significant evidence of preferential sampling. An explanation for this can be seen in the values of the estimated spatial variance components, σ_1^2 and σ_2^2, which together with the estimate of β imply that most of the spatial variation in lead concentrations is unrelated to the spatial heterogeneity in the distribution of the 1997 sampling locations.

We estimate the range of the spatial variation to be much larger for $\mathcal{S}_1$ than for $\mathcal{S}_2$. The point estimate of ϕ_2 is comparable to that obtained by DMS. However, the additional flexibility of the PDR model has allowed us to capture the different spatial scales of variation in the relative sampling intensity and the lead concentration surface.

Figure 21.6 shows plug-in predictions for the lead concentration surfaces in 1997 and in 2000, and for the ratio of the two. As expected, the 1997 surface shows considerable spatial variation unrelated to the variation in sampling intensity shown in the left-hand panel of Figure 21.3, whilst the ratio of the 1997 and 2000 surfaces shows the spatial variation about the overall reduction in lead concentrations between the two dates.

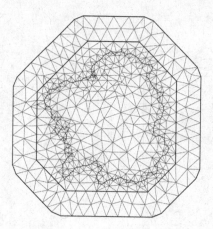

FIGURE 21.4

Triangulated mesh used for the SPDE representation of the field $\mathcal{S}_1$.

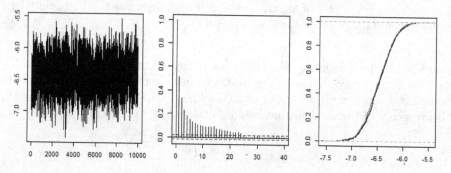

FIGURE 21.5

Trace plot (left panel), correlogram (central panel) and empirical cumulative density functions of the first and second 5000 samples (right panel) of the region-wide average of $\tilde{\mathcal{S}}_1$ obtained from the independence sampler algorithm.

21.5 Discussion

In any model for a preferentially sampled geostatistical data-set, at least one of the parameters is likely to be poorly identified. DMS already experienced this in their analysis of the Galicia lead pollution data, and the extra flexibility of the PRD model can only exacerbate the problem. Nevertheless, we would recommend using the PRD model as a vehicle for investigating preferential sampling effects, rather than the over-rigid DMS model.

In the analysis of the Galicia lead pollution data, our strategy for dealing with poor identifiability has followed that of DMS in arguing that the lattice-like sampling design of the 2000 survey can safely be assumed to be non-preferential, the models for the 1997 and 2000 pollution surfaces can be assumed to have some parameters in common, and parameters can therefore be estimated by pooling the two contributions to the overall likelihood.

In general, several different strategies can be considered. One would be to identify co-

TABLE 21.1

Monte Carlo maximum likelihood parameter estimates and 95% confidence interval for the model fitted to the 1997 and 2000 surveys of lead concentrations in moss samples. The unit of measure for ϕ_1 and ϕ_2 is 100km.

Parameter	Estimate	95% confidence interval
α	-0.356	(-1.006, 0.293)
μ_{97}	0.854	(-0.565, 2.273)
μ_{00}	0.725	(0.527, 0.923)
β	-0.114	(-0.345, 0.117)
σ_1^2	6.610	(1.832, 23.822)
σ_2^2	0.148	(0.093, 0.237)
ϕ_1	1.855	(0.976, 3.525)
ϕ_2	0.235	(0.121, 0.459)
ν^2	0.349	(0.112, 1.089)

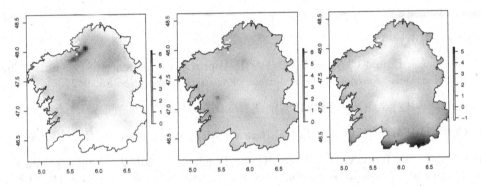

FIGURE 21.6

Plug-in predictions for the spatial variation in lead concentrations in 1997 (left panel) and 2000 (centre panel), and the log-ratio of the two (right panel)

variates that can partially explain the spatial variation in both the sampling intensity and the variable of interest. As in other areas of statistical modelling with random effects, we can think of the unobserved surface S_1 that features in both (21.12) and (21.13) as a proxy for the combined effects of unmeasured covariates. As expressed formally by (21.7), addition of covariates to a model can render what would otherwise be preferential sampling non-preferential. The same applies *a fortiori* to spatially stratified sampling, provided that the stratification forms part of the original study-design.

A third strategy would be to treat the preferential sampling parameter, i.e. β in the PRD model, as a pre-specified constant and conduct a sensitivity analysis. We would recommend this when only a single data-set is available, with no obvious candidate covariates and no knowledge of how the sample locations were chosen.

A fourth would be to use Bayesian inference with an informative prior for the preferential sampling parameter. It is unclear to us how one might elicit this, and an uninformative prior would hide rather than solve the problem. Of course, this is not a reason for avoiding Bayesian methods if that is the analyst's preferred inferential paradigm.

Strongly preferential sampling invalidates conventional geostatistical inferences. Whenever possible it should be avoided using explicit probability-based sampling designs. However, this is a counsel of perfection. When analysing observational geostatistical data, investigating preferential sampling effects is a better strategy than simply ignoring them.

Using a probability-based design should not equate to sampling completely at random with the study-region. In the absence of covariate information, sampling designs that are spatially more regular than random usually lead to more precise predictions. When covariate information is available, spanning the range of important covariates will pay dividends in the same way that it does in simple regression modelling. Stratification of the study-region is also a useful tool, especially in context where precise prediction is more important in some sub-regions, for example in areas of high population density. Adaptive design, which are weakly preferential, can also bring gains in efficiency of prediction by comparison with analogous non-adaptive designs; see, for example, Chipeta *et al* (2016).

The PrevMap R package (Giorgi and Diggle, 2016) provides functionality for parameter estimation and spatial prediction of the PDR model, using the computational methods described in Section 21.3.4.1. The R code for the analysis of the Galicia data can be found at http://www.lancaster.ac.uk/staff/giorgi/.

Bibliography

[1] Breslow, N.E. and Clayton, D.G. (1993). Approximate inference in generalized linear mixed models. *Journal of the American Statistical Association* **88**, 9–25.

[2] Chilès, J-P and Delfiner, P. (1999). *Geostatistics*. New York: Wiley.

[3] Chilès, J-P and Delfiner, P. (2012). *Geostatistics* (second edition). Hoboken: Wiley.

[4] Chipeta, M., Terlouw, D.J., Phiri, K. and Diggle, P.J. (2016). Adaptive geostatistical design and analysis for sequential prevalence surveys. *Spatial Statistics*, **15**, 70–84.

[5] Cressie, N.A.C. (1991). *Statistics for Spatial Data*. New York: Wiley.

[6] Daniels, M.J. and Hogan, J.W. (2008). *Missing Data in Longitudinal Studies: Strategies for Bayesian Modeling and Sensitivity Analysis*. Boca Raton: Chapman and Hall/CRC

[7] Diggle, P.J. and Kenward, M.G. (1994). Informative dropout in longitudinal data analysis (with Discussion). *Appl. Statist.* **43**, 49–93.

[8] Diggle, P.J., Menezes, R. and Su, T.-L. (2010). Geostatistical analysis under preferential sampling (with Discussion).*Applied Statistics*, **59**, 191–232.

[9] Diggle, P.J., Moyeed, R.A. and Tawn, J.A. (1998). Model-based geostatistics (with Discussion). *Applied Statistics*, **47** 299–350.

[10] Diggle, P.J., Moraga, P., Rowlingson, B., Taylor, B.M. (2013). Spatial and spatio-temporal log-Gaussian Cox processes: Extending the geostatistical paradigm. *Statistical Science*, **28**, 542–563.

[11] Giorgi, E., Diggle, P.J. (2016). PrevMap: An R package for prevalence mapping. *Journal of Statistical Software*. To appear.

[12] Hogan, J. and Laird, N. (1997). Model-based approaches to analyzing incomplete longitudinal and failure-time data. *Statistics in Medicine*, **16**, 259–72.

[13] Krige, D.G. (1951). A statistical approach to some basic mine valuation problems on the Witwatersrand. *Journal of the Chemical, Metallurgical and Mining Society of South Africa*, **52**, 119–39.

[14] Higdon, D. (1998). A process-convolution approach to modeling temperatures in the North Atlantic Ocean. *Environmental and Ecological Statistics*, **5**, 173-190.

[15] Lin, H., Scharfstein, D.O. and Rosenheck, R.A. (2004). Analysis of longitudinal data with irregular, outcome-dependent follow-up. *Journal of the Royal Statistical Society*, B **66**, 791-813.

[16] Lindgren, F., Havard, R., Lindstrom, J. (2011). An explicit link between Gaussian fields and Gaussian Markov random fields: the stochastic partial differential equation approach (with discussion). *Journal of the Royal Statistical Society, Series B*, **73**, 423–498.

[17] Little, R.J.A. (1995). Modelling the drop-out mechanism in repeated-measures studies. *Journal of the American Statistical Association*, **90**, 1112–21.

[18] Little, R.J. and Rubin, D.B. (2002). *Statistical Analysis with Missing Data (seond edition)*. New York : Wiley.

[19] Matheron, G. (1963). Principles of geostatistics. *Economic Geology*, **58**, 1246–66.

[20] Møller, J., Syversveen, A. R., and Waagepetersen, R. P. (1998). Log-Gaussian Cox processes. *Scandinavian Journal of Statistics*, **25**, 451-482

[21] Pati, D., Reich, B. J., and Dunson, D. B. (2011). Bayesian geostatistical modelling with informative sampling locations. *Biometrika*, **98**, 35-48.

[22] Ripley, B.D. (1981). *Spatial Statistics*. New York : Wiley.

Schlather, M.S., Ribeiro, P.J. and Diggle, P.J. (2004).

[23] Detecting dependence between marks and locations of marked point processes. *Journal of the Royal Statistical Society*, B **66**, 79–93.

[24] Webster, R. and Oliver, M.A. (2001). *Geostatistics for Environmental Scientists*. Chichester: Wiley.

22

Monitoring network design

James V. Zidek

University of British Columbia, Canada

Dale L. Zimmerman

University of Iowa

CONTENTS

22.1   Introduction

The increasing recognition of the association between adverse human health conditions and many environmental substances as well as processes has led to the need to monitor them. That need has become more acute with the growth in the size of the World's human population that, far from peaking at around nine billion in 2050 as demographer's estimates had suggested, is now projected to keep growing so that by the year 2100, it could grow to as much as fifteen billion (Gerland et al. 2014). Since a lot of the environmental hazards have anthropogenic sources, their levels are thus likely to rise. Add to that are the increasing health risks associated with climate change, notably extreme temperatures (Li et al. 2012), and environmental monitoring networks become a vital necessity for such things as developing mitigation strategies.

A by–product of the negative health impacts of environmental hazards is their burden on society, the subject of the Global Burden of Disease Project study by the World Health Organization being carried out during the period in which this chapter was being written. Early indications are of a substantial Global burden due to air pollution. The project's analysis has had to be rely on satellite data and simulated data from the chemical transport models. However, little "ground truth" from monitoring sites with which to calibrate these alternative sources is available. Thus there is a need for a large number of additional sites over much of the world, as indicated in Figure 22.1.

The vital need for monitoring as well as the challenges involved in siting monitors to measure environmental hazards (see http://nadp.sws.uiuc.edu/ merc-net/MercNetFinalReport.pdf for a modern example), leads to the topic of this chapter, which reviews approaches that have been taken to setting up and modifying monitoring networks. Exploration of the this topic began in the last century, notably by hydrologists, concerning such things as snow melt, flooding as well as water storage, and geostatisticians mapping underground ore depositions. Many approaches have been proposed, some being quite specialized or ad hoc (Bueso et al. 1999a). While whole books have been written on the topic of this chapter (Müller 2007; Ucinski 2004) along with a comprehensive recent survey (Dobbie et al. 2007), the need for brevity limits our scope. In fact, we cover just the basic problems that spatial network designers face and selected approaches to solve them, ones that are supported by well-developed foundations, including principles, theory and implementation strategies. Nevertheless, we hope would-be designers will find useful practical guidance on how to formulate defensible recommendations.

The topic of this chapter lies in the general field of spatial sampling design, a broad subject that includes a rich treasury of specialized topics such as transect and composite sampling. Many epidemiological, biological, ecological, land use and agricultural studies that often rely on survey sampling methods would fall in this domain, one about which entire books have been written. The restriction of this chapter to the design of monitoring networks has led to the restricted set of topics we have included. A more extensive account of this topic in the setting of environmental epidemiology can be found in Shaddick and Zidek (2015), while Le and Zidek (2006) treat this topic abstractly in the context of spatial fields.

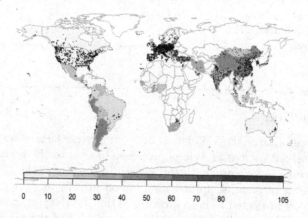

FIGURE 22.1

Locations of ground monitoring sites (black dots) for PM2.5 and PM10 in the World Health Organisations cities database (2014). Colours indicate county level mean values of PM2.5 (using PM2.5 converted from PM10 measurements where appropriate). *Courtesy of Dr Gavin Shaddick.*

We begin by describing some of the many reasons why monitoring networks have been designed.

22.2 Monitoring environmental processes

Important environmental processes have been monitored for a variety of purposes over a very long time. Concerns about climate change have led to the measurement of sea levels and the extent to which polar ice-caps have receded. Concern for human health and welfare and the need to regulate airborne pollutants by the US Environmental Protection Agency has led to the development of urban air shed monitoring networks; cities out of compliance with air quality standards suffer serious financial penalties. The degradation of landscapes, lakes, and monuments led to the establishment of networks for monitoring acidic precipitation as well as to surveys of water quality. Exploratory drilling to find oil reserves on the northern slopes of Alaska generated concern for the health of benthic organisms which feed the fish that feed human populations. The result: the USA National Oceanic and Atmospheric Agency's (NOAA's) decision to monitor the concentrations of trace metals in the seabed before and after the startup of drilling (Schumacher and Zidek 1993). Predicting the height of tsunamis following earthquakes in the Indian Ocean has led NOAA to install monitoring buoys ("tsunameters") that can help assess the type of earthquake that has occurred. Hazardous waste sites must be monitored. Mercury, a cumulative poison, must be monitored and that poses a challenge for designing a monitoring network since it can be transported in a variety of ways. Concerns about flooding along with the need for adequate supplies of water for irrigation has resulted in the monitoring of precipitation as well as snow melt; not surprisingly hydrologists were amongst the earliest to develop rational approaches to design. Insuring water quality leads to programs that collect water samples at specific locations along sea fronts popular with swimmers, with red flags appearing on bad days.

These examples illustrate the importance of monitoring networks, the oldest one being perhaps that constructed in ancient times along the Nile River to measure its height for the purpose of forecasting the extent of the annual flood. It consisted of a series of instruments called "nilometers," each essentially a staircase in a pit next to the river, to measure the height of the river at a specific location. In contrast, modern technology has produced networks of cheap-to-deploy sensors that automatically upload their measurements to central electronic data recorders. Air pollution concentrations can be recorded in this way as well as soil moisture content and snow water equivalent.

Networks have societally important purposes. Moreover, the data they provide are important to modelers such as forecasters of future climate. However, in practice they are seldom designed to maximize the "bang for the buck" from their product. Instead, their construction will often be influenced by administrative, political and other pragmatic considerations. Moreover, they and their purpose may evolve and change over time (Zidek et al. 2000).

However, establishing these sites and maintaining them can be expensive. For tsunameters in NOAA's Pacific Ocean National Tsunami Hazard Mitigation Program, for example, Gonzalez et al. (2005) estimate a cost of USD250k to install a new system and a cost of USD30k per year to maintain it. Designers must therefore find a defensible basis for a design recommendation even if pragmatic considerations ultimately lead to modifications of the "ideal" design for the intended purpose.

In this chapter, we explore a variety of approaches designers have developed, along with their rationales. The choice of an approach depends on such things as context, the objective, "discipline bias" and designer background.

Before we begin, a note on vocabulary: monitored sites are sometimes called "gauged" sites, especially in hydrology. The devices placed at such sites may be called "monitors" or "gauges." Monitoring sites are sometimes called "stations". Optimizing a design requires a "design objective."

22.3   Design objectives

As the examples in Section 22.2 show, networks can be used for a variety of purposes such as:

1. determining the environmental impact of an event, such as a policy-induced intervention or the closure of an emissions source;

2. assessing trends over space or over time;

3. determining the association between an environmental hazard and adverse health outcomes;

4. detecting non-compliance with regulatory limits on emissions;

5. issuing warnings of impending disaster;

6. monitoring a process or medium such as drinking water to ensure quality or safety;

7. monitoring an easy-to-measure surrogate for a process or substance of real concern;

8. monitoring the extremes of a process.

However a network's purpose may change over time, as in an example of Zidek et al. (2000) where a network now monitoring air pollution was formed by combining three networks originally set up to monitor acidic precipitation. Thus the sites were originally placed in rural rather than urban areas where air pollution is of greatest concern. Hence additional sites had to be added.

Moreover, the network's objectives may conflict. For example, non-compliance detection suggests siting the monitors at the places where violations are seen as most likely to occur. But an environmental epidemiologist would want to divide the sites equally between areas of high risk and areas of low risk to maximize the power of their health effects analyses. Even when the objective seems well-defined, such as monitoring to detect extreme values of a process, it may lead to a number of objectives on examination for implementation (Chang et al. 2007).

Often many different variables of varying importance are to be concurrently measured at each monitoring site. The challenges now compound, hence different importance weights may need to be attached. To minimize cost, the designer could elect to measure different variables at different sites. Further savings may accrue from making the measurements less frequently, forcing the designer to consider the inter-measurement times. In combination, these many choices lead to a bewildering set of objective functions to optimize simultaneously. That has led to the idea of designs based on multi-attribute theory, ones that optimize an objective function that embraces many or all of the purposes (Zhu and Stein, 2006; Xia et al., 2006; Sampson, PD, Guttorp, P and DM Holland, http://www.epa.gov/ttn/amtic/files/ambient/pm25/workshop/spatial/sampsn2.pdf; Müller and Stehlik, 2010).

However, that approach will not be satisfactory for long term monitoring programs when the network's future uses cannot be foreseen, as in the example of Zidek et al (2000). Moreover in some situations the "client" may not be even be able to specify the network's purposes precisely (Ainslie et al. 2009). Yet as noted above, the high cost of network construction and maintenance will require the designer to select a defensible justification for the design she or he eventually proposes. This chapter presents a catalogue of approaches that may provide such a justification.

22.4 Design paradigms

In practice the domains in which the monitors are to be sited are "discretized," meaning the possible choices lie in a set $\mathcal{D}$ of finite size N. Practical considerations may make this set quite small. For example, the expensive equipment involved will have to be put in a secure, easily accessible location, one that is away from contaminating sources such as heavy traffic flows. The sites may be located on a geographical grid that follows the contours of a catchment area, for example.

Most approaches to design fall into one of the following categories (Müller, 2005; Le and Zidek 2006; Dobbie et al. 2008):

1. Geometry-based: The approach goes back a long way (Dalenius et al. 1960). It involves heuristic arguments and includes such things as regular lattices, triangular networks, or space-filling designs (Cox et al. 1997; Royle and Nychka 1998; Nychka and Saltzman 1998). The heuristics may reflect prior knowledge about the process. These designs can be especially useful when the design's purpose is exploratory (Müller, 2005). In their survey of spatial sampling, Cox et al. (1997) provide support for their use when certain aggregate criteria are used, for example when the objective function is the average of kriging variances. In fact, Olea (1984) finds in a geological setting when universal kriging is used to produce a spatial predictor, that a regular hexagonal pattern optimally reduces the average standard error amongst a slew of geometrical patterns for laying out a regular spatial design. Cox et al. (1997) conjecture that geometric designs may be good enough for many problems and treat this matter as a "research issue." However, since the approaches in this category tend to be somewhat specialized, our need for brevity precludes a detailed discussion.

2. Probability-based (see Section 22.5): This approach to design has been used for well over half a century by a variety of organizations such as opinion polling companies and government statistical agencies. It has the obvious appeal that sample selection is based on the seemingly "objective" technique of sampling at random from a list of the population elements (the *sampling frame*). Thus in principle (though not in practice) the designers need not have any knowledge of the population or the distribution of its characteristics. Moreover they may see competing methods as biased because they rely on prior knowledge of the population, usually expressed through models. Those models involve assumptions about the nature of the process under investigation, none of which can be exactly correct, thus skewing selection and biasing inference about the process being monitored. Nevertheless, for reasons given below, this approach has not been popular in constructing spatial (network) designs.

3. Model-based (see Section 22.6): The majority of designs for environmental monitoring networks rely on the model-based approach. For although the models do indeed skew the selection process, they do so in accord with prior knowledge and can make the design maximally efficient in extracting relevant information for inferences about the process. In contrast the probability-based approach may be seen as gambling on the outcome of the randomization procedure and hence risking the possibility of getting designs that ignore aspects of the process that are important for inference.

Since each paradigm has merits and deficiencies, the eventual choice will depend on such things as context and the designer's scientific background. However, once selected the paradigms do provide a rational basis for selecting the monitoring sites. In the rest of this chapter, we will see how they can be implemented as well as their strengths and weaknesses.

22.5 Probability-based designs

This paradigm has been widely used for such things as public opinion polling and the collection of official statistics in national surveys. Hence a large literature exists for it, largely outside the domain of network design. Thus we restrict our review to a few of these designs in order to bring out some of the issues that arise, referring the reader interested in details to a more comprehensive recent review (Dobbie et al. 2008).

Simple random sampling (SRS)

In the simplest of these designs, sites are sampled at random from a list of sites called the "sampling frame" with equal probability and without replacement. Responses at each site would then be measured and could even be vector valued, to include a sequence of values collected over time say. They would be (approximately) independent in this paradigm where the responses at each site, both in the sample and out, are regarded as fixed and the probabilities derive entirely from the randomization process. This is an important point. It means that responses from two sites quite close to one another would be (approximately) unrelated, not something that would seem reasonable to a model-based designer.

Randomization in this paradigm yields a basis for an inferential theory including testing, confidence intervals and so on. Moreover, these products of inference are reasonably easy to derive and can be quite similar to those from the other theories of spatial design.

Stratified random sampling

However, in practice SRS designs are almost never used and often stratified random sampling is used instead. For one thing, budgetary constraints often limit the number of sites that can be included in the network and with SRS these could, by chance, end up in the very same geographic neighborhood, an outcome that would generally be considered undesirable. Moreover, practical considerations can rule out SRS designs. For example, maintenance and measurement can entail regular visits to the site and long travel times. Thus, spreading the selected sites out over sub-regions can help divide the sampling workload equitably. Sites monitoring hazardous substances may have to placed in all of a number of administrative jurisdictions such as counties, in response to societal concerns and resulting political pressures. Legislation may also force such a division of sites.

Finally, statistical issues may lead to a subdivision of sites into separate strata. For example, gains in statistical efficiency can be achieved when a region consists of a collection of homogeneous subregions (called strata). Then only a small number of sites needs to be selected from each stratum. Although this is a form of model-based sampling in disguise, the appeal of stratified sampling designs has led to their use in important monitoring programs, for example a survey of US lakes (Eilers et al. 1987) and in EMAP (http://www.epa.gov/emap).

However, even though stratification forces the sites to be spread out geographically, it does not rule out adjacent pairs of sites being close together across stratum boundaries. Moreover, like all designs that rely on a model of some kind, stratified ones may produce no statistical benefits if that model fails to describe the population well. Practical considerations can also rule out their use, leading to other more complex designs.

Variable probability designs

Complex designs with multiple stages can sample sites with varying probabilities. Consider for example, a hypothetical survey of rivers (including streams). Stage 1 begins with the construction of a list (a sampling frame), perhaps with the help of aerial photos, of catchment areas called the "primary sampling units" or PSUs. With the help of that sampling frame, a sample of PSUs is selected using SRS, a process which guarantees equally likely selection of all items on the list, large and small. Then at Stage 2, for each PSU selected during Stage 1 a list (subsampling frame) is constructed of rivers in that PSU. Items on these lists are called "secondary sampling units" or SSUs. From these subsampling frames, a random sample of SSUs is selected with an SRS design.

Naive estimates of population characteristics, using such a design could well be biased. To illustrate suppose an equal number of elements say 10, are sampled from each of the selected PSUs selected in Stage 1. Further imagine that one of those PSUs, call it "A", contained 1,000,000 rivers while another, "B", had just 1,000. Then every single member of A's sample would represent 100,000 of A's streams while each member of B's would represent just 100 streams. The result: responses measured on each of B's elements would (without adjustment) grossly over – contribute to estimates of overall population characteristics compared to A's. In short naive estimates computed from the combined sample would be biased in favor of the characteristics of the over represented small PSUs. How can this flaw be fixed?

The ingenious answer to that question is embraced in an inferential procedure called the Horvitz - Thompson (HT) estimator, named after its inventors. In fact, the procedure contends with the potential estimator bias in a very large class of practical designs.

To describe HT estimator, suppose a fixed number, say n, of sites are to be sampled. Any design can be specified in terms of its sample selection probabilities, $P(S)$, for all $S = \{s_1, \ldots, s_n\} \subseteq \mathcal{D}$. Bias can now be assessed in terms of the chances that any given population element s such as a river in our hypothetical example is included in our random sample S, is $\pi_s = P(s \in S)$. Now denote the response of interest at location s by $Y(s)$, a quantity regarded as non-random in this context, even though it is in fact unknown. Suppose the quantity of inferential interest to be the population mean $\bar{Y} = \sum_{s \in \mathcal{D}} Y(s)/N$. Then a "design-unbiased" estimator (meaning one that is unbiased over all possible samples S that might be selected under the specified design) is given by the HT estimator:

$$\hat{\bar{Y}} = \sum_{i=1}^{n} Y(s_i)/(N\pi_{s_i})$$

$$= \sum_{s \in S} Y(s)/(N\pi_s) \tag{22.1}$$

$$= \sum_{s \in \mathcal{D}} I_S(s)Y(s)/(N\pi_s)$$

where $I_S(s)$ is 1 or 0 according as $s \in S$ or not. It follows that $\pi_s = E\{I_S(s)\}$ and therefore that $\hat{\bar{Y}}$ is unbiased.

One can estimate other quantities such as strata means when spatial trends are of interest in a similar way to compensate for the selection bias. As a more complicated example, when both $Y(s)$ and $X(s)$ are measured at each of the sample sites, one could compensate for selection bias in estimating the population level regression line's slope using:

$$\frac{\sum_{i=1}^{n}[Y(s_i) - \hat{\bar{Y}}][X(s_i) - \hat{\bar{X}}]/(\pi_{s_i})}{\sum_{i=1}^{n}[X(s_i) - \hat{\bar{X}}]^2/(\pi_{s_i})}.$$

In short, the HT estimator addresses the problem of selection bias for a very large class designs.

We now turn to the competing paradigm for which a very rich collection of approaches to design have been developed.

22.6   Model based designs

Broadly speaking, model-based designs optimize some form of inference about the process or its model parameters. We now describe a number of these approaches that get at one or another of these objectives.

22.6.1   Estimation of covariance parameters

Designs may need to provide estimates about the random field's model parameters, like those associated with its spatial covariance structure or variogram, for example, which play a central role in the analysis of geostatistical data. There a valid variogram model is selected and the parameters of that model are estimated before kriging (spatial prediction) is performed. These inference procedures are generally based upon examination of the empirical variogram, which consists of average squared differences of data taken at sites lagged the same distance apart in the same direction. The ability of the analyst to estimate variogram parameters efficiently is affected significantly by the design, particularly by the spatial configuration of sites where measurements are taken.

This leads to design criteria that emphasize the accurate estimation of the variogram. Müller and Zimmerman (1999) consider, for this purpose, modifications of design criteria that are popular in the context of (nonlinear) regression models, such as the determinant of the covariance matrix of the weighted or generalized least squares estimators of variogram parameters. Two important differences in the present context are that the addition of a single site to the design produces as many new lags as there are existing sites and hence also produces that many new squared differences from which the variogram is estimated. Secondly, those squared differences are generally correlated, which precludes the use of many standard design methods that rest upon the assumption of uncorrelated errors. Nevertheless, several approaches to design construction that account for these features can be devised. Müller and Zimmerman (1999) show that the resulting designs are much different than random designs on the one hand and regular designs on the other, as they tend to include several groups of tightly clustered sites. The designs depend on the unknown parameter vector θ, however, so an initial estimate or a sequential design and sampling approach is needed in practice. Müller and Zimmerman also compare the efficiency of their designs to those obtained by simple random sampling and to regular and space-filling designs, among others, and find considerable improvements.

Zhu and Stein (2005) and Zimmerman (2006) consider designs optimal for maximum likelihood estimation of the variogram under an assumption of a Gaussian random field. Since the inverse of the Fisher information matrix approximates the covariance of the variogram's maximum likelihood estimators, they use the determinant of that inverse as their criterion function. This function, like the criterion function of Müller and Zimmerman (1999), depends not only on the set of selected design points S that is to be optimized, but also on the unknown variogram parameter vector θ. Zhu and Stein offer various proposals to address this difficulty:

- (locally optimal design): plug a preliminary estimate of θ into the criterion function;

- (minimax design): a variation of assuming nature makes the worst possible choice of θ and the designer then chooses the best possible S under the circumstances;

- (Bayesian design): put a distribution on θ.

Zhu and Stein propose a simulated annealing algorithm for optimization of their design criterion. They assess these proposals with simulation studies based on use of the Matérn spatial covariance model and make the following conclusions:

1. Although the inverse information approximation to the covariance matrix of maximum likelihood estimators is accurate only if samples are moderately large, the approximation yields a very similar ordering of designs, even for relatively small samples, as when the covariance matrix itself is used; hence the approximation can serve generally as a useful design criterion.

2. The locally optimal approach (which is applicable only when a preliminary estimate of θ is available) yields designs that provide for much more precise estimation of covariance parameters than a random or regular design does.

3. The more widely applicable Bayesian and minimax designs are superior to a regular design, especially when the so-called "range parameter" (which measures the rate at which the spatial correlation between sites declines with distance) is known or a preliminary estimate of it is available.

Zimmerman (2006) makes similar proposals and obtains similar results. In addition, he focusses attention on the finding that for purposes of good estimation of covariance parameters, a design should have a much greater number of small lags than will occur in either a regular or random arrangement of sites. The design should have many large lags as well. Such a distribution of lags is achieved by a design consisting of regularly spaced clusters.

22.6.2   Estimation of mean parameters: The regression model approach

The regression modeling approach to network design focuses on optimizing the estimators of coefficients of a regression model. Development of that approach has taken place outside the context of network design (Smith 1918; Elfving 1952 and Kiefer 1959) and an elegant mathematical theory for this problem has emerged (Silvey 1980; Fedorov and Hackl 1997; Müller 2007) along with numerical optimization algorithms.

The approach as originally formulated concerns continuous sampling domains, $\mathcal{X}$, and optimal designs, ξ, with finite supports $x_1, \ldots, x_m \in \mathcal{X}$ ($\sum_{i=1}^{m} \xi(x_i) = 1$). In all, $n\xi(x_i)$ (suitably rounded) responses would then be measured at x_i for all $i = 1, \ldots, m$ to obtain a total of n responses $y_1, \ldots, y_n$ across all sites (see the example in the following paragraph). Key elements are: a regression model, $y(x) = \eta(x, \beta) + \varepsilon(x)$ relating y to the selected (and fixed) x's; the assumption of independence of the ε's from one sample point x to another. Optimality means maximally efficient estimation of β, that is, designs ξ which optimize $\Phi(M(\xi))$, $M(\xi)$ denoting the Fisher information matrix and Φ, a positive-valued function depending on the criterion adopted.

As an example, in simple linear regression, $y(x) = \alpha + \beta x + \varepsilon(x)$, $x \in [a, b]$ and $M(\xi) = \sigma^2 [\mathbf{X}'\mathbf{X}]^{-1}$. There the optimal design which minimizes the variance of the least squares estimator of β, has the intuitively appealing form, $x_1 = a, x_2 = b$ while $\xi(x_1) = \xi(x_2) = 1/2$.

However the approach does not apply immediately to network design. For one thing, possible site locations are usually quite restricted. Moreover once sited, the monitors must measure the process of interest for an indefinite period. Finally, to measure n responses of a random field at a single time point would require n monitors, so that $\xi \equiv 1/n$ would be completely determined once its support was specified, rendering the theory irrelevant.

Nevertheless, an attempt has been made to adapt the approach for network design (Gibrik, 1976; Fedorov and Müller, 1987, 1989) on the grounds (Fedorov and Müller 1989) that, unlike other approaches which possess only algorithms for finding sub-optimal designs, truly optimal designs could be found with this one. That attempt (Fedorov and Müller 1987) assumes regression models for times $t = 1, \ldots, T$ that capture both temporal and spatial covariance: $y_t(x_i) = \eta(x_i, \beta_t) + \varepsilon_t(x_i)$ where the εs are independent and the β_ts are both random as well as autocorrelated. Moreover, $\eta(x_i, \beta_t) = g^T(x_i)\beta_t$ for a known vector-valued g.

The celebrated Karhunen - Loève (KL) expansion makes their model more general than it may appear at first glance (Federov and Müller 2008). KL tells us that η has an infinite series expansion in terms of the orthogonal eigenvectors of its spatial covariance function. These eigenvectors become the gs in the expansion of η once the series has been truncated (although in practice they may involve covariance parameters whose estimates need to be plugged-in.)

However, using an eigenfunction expansion of the spatial covariance to validate the regression model presents technical difficulties when the proposed network is large (Fedorov 1996). Moreoever, in an unpublished manuscript Spöck and Pilz (2008) point to the difficulty of finding "numerically reliable" algorithms for solving KL's eigenvalue problems as a serious practical limitation to the approach.

Spöck and Pilz (2010) go on to propose a different way of bringing the regression approach into spatial design. Their method is based on the polar spectral representation of isotropic random fields due to Yaglom (1987). That representation equates the spatial covariance (assumed known) as an integral of a Bessel function of the first with respect to a spectral distribution function. That equation in turn yields a representation of the ε process for the regression model postulated above in terms of sines and cosines of arbitrary high precision depending how many terms are kept. That adds a second parametric regression term to the regression model above and like the KL approach puts the spatial design problem into a form susceptible to analysis by methods in the broader domain of Bayesian design (Pilz 1991). This approach also face practical challenges since commonly environmental processes are not isotropic and moreover their covariance matrices are not known. The authors circumvent the latter by finding designs that are minimax against the unknown covariance, where spatial prediction provides the objective function on which the minimax calculation can be based.

Overall the attempt to move the regression modeling theory into spatial design faces a number of challenges in applications:

1. As noted above the feasible design region will usually be a discrete set, not a continuum and use of the so-called "continuous approximation" that has been proposed to address that difficulty, leads to further difficulties. For one thing, the result will not usually be a feasible solution to the original problem. For another, the solution may be hard to interpret (Fedorov and Müller 1988, 1989). Although (Müller 2001) has helped clarify the nature of that approximation, the value of substituting it for the hard-to-solve exact discrete design problem remains somewhat unclear.

2. The design objective inherited from the classical approach to regression - based design, the one based on inference about the regression parameters (the βs) will not always seem appropriate especially when they are mere artifacts of the orthogonal expansion described above.

3. Even when the regression model is genuine (as opposed to one from an eigenfunction expansion) and the objective function is meaningful, the range of spatial covariance kernels will be restricted unless the ε's are allowed to be spatially correlated. That need is met in the extensions of the model (Fedorov 1996; and Müller 2001). However, the resulting design objective function does not have much in common with the original besides notation (Fedorov 1996, p524).

4. The assumed independence of the εs also proves a limitation in this context, although a heuristic way around this difficulty offers some promise (Müller 2005).

5. The complexity of random environmental space-time processes renders their random response fields only crudely related to spatial site co-ordinates. Moreover, its shape can vary dramatically over time and season. In other words, finding a meaningful, known vector-valued function g would generally be difficult or impossible.

To summarize, although the regression modeling approach to network design comes with an evolved theory and a substantial toolbox of algorithms, using that approach in network design will prove challenging in practice.

22.6.3 Spatial prediction

Instead of estimating model parameters as in the two approaches discussed above, the prediction of the random field at unmonitored sites based on measurements at the monitored sites may be taken as the design objective. In some cases the spatial covariance has been taken as known (McBratney 1981; Yfantis et al. 1987; Behenni and Cambinis 1992; Su and Cambanis 1993; and Ritter 1996). In others it was not and in one such case, a Bayesian approach was used to estimate the unknown parameters (Currin 1991).

That is the spirit underlying the approach taken in geostatistical theory that has some natural links with the regression modeling approach described above. That theory has traditionally been concerned with spatial random fields, not space-time fields until very recently

(Myers 2002) and has a large literature devoted to it (see for example, Wackernagel, 2003). So we will not describe this approach in detail here.

Two methods are commonly employed, ordinary kriging and universal kriging. The first concerns the prediction of an unmeasured co-ordinate of the response vector, say $y_1(x_0)$, using an optimal linear predictor based on the observed response vectors at all the sampling sites, assuming that all responses have the same unknown mean. The coefficients of that optimal predictor are found by requiring it to be unbiased and to minimize the mean square prediction error. They will depend on the covariances between responses and the covariances between the prediction and the responses, covariances that are unrealistically assumed to be known and later estimated from the data usually without adequately accounting for the additional uncertainty thereby introduced. In contrast to the first method, the second relies on a regression model precisely of the form given in the previous subsection, i.e. $y(x) = g^T(x)\beta + \varepsilon(x)$ where the ε's are assumed to have a covariance structure of known form. However, unlike the regression-modeling approach above, the goal is prediction of the random response (possibly a vector) at a point where it has not been measured. Moreover, g (which can be a matrix in the multivariate case) can represent an observable covariate process. Optimization again relies on selecting coefficients by minimizing mean squared prediction error subject to the requirement of unbiasedness. Designs are commonly found iteratively one future site at a time, by choosing the site x_0 where the mean squared prediction error of the optimum predictor proves to be greatest. The designs tend to be very regular in nature, strongly resembling space-filling designs.

22.6.4   Prediction and process model inference

Why not combine the goals of predicting the random field at unmeasured sites and the estimation of the field's model parameters in a single design objective criterion? Zhu and Stein (2006) and Zimmerman (2006) make that seemingly natural merger.

They focus on the case of an (isotropic) Gaussian field so if θ, the vector of covariance model parameters, were known the best linear predictor $\hat{Y}(s; S, \theta)$ of the unmeasured response at location s, $Y(s)$, could be explicitly computed as a function of the responses at points in S, $Y = \{Y(s),\ s \in S\}$, and θ. So could its mean-squared prediction error (MSPE) $M(s; S, \theta)$. That quantity could then be maximized over s to get the worst case and a criterion $M(S, \theta)$ to maximize in finding an optimum S. (Alternatively it could be averaged.) This criterion coincides with that described in the previous subsection.

Since θ is unknown, it must be estimated by, say, the REML or ML estimator, $\hat{\theta}$. The optimal predictor could then be replaced by $\hat{Y}(s; S, \hat{\theta})$ in the manner conventional in geostatistics. But then $M(s; S, \theta)$ would not correctly reflect the added uncertainty in the predictor. Moreover, the designer may wish to optimize the performance of the plug-in predictor as well as the performance of $\hat{\theta}$, where each depends on the unknown θ. If only the former were of concern the empirical kriging (EK)-optimality criterion of Zimmerman (2006) could be used. This criterion is given by

$$EK(\theta) = \max_{s \in S}\{M(s; S, \theta) + \operatorname{tr}[A(\theta)B(\theta)]\},$$

where $A(\theta)$ is the covariance matrix of the vector of first-order partial derivatives of $\hat{Y}(s; S, \theta)$ with respect to θ, and $B(\theta)$ is the inverse of the Fisher information matrix associated with $\hat{\theta}$. If one is also interested in accurately estimating the MSPE of $\hat{Y}(s; S, \hat{\theta})$, the EA (for "estimated adjusted") criterion of Zhu and Stein (2006) is more appropriate. The EA criterion is an integrated (over the study region) weighted linear combination of $EK(\theta)$ and the variance of the plug-in kriging variance estimator.

Zimmerman (2006) makes the important observation that the two objectives of optimal covariance parameter estimation and optimal prediction with known covariance parameters are actually antithetical; they lead to very different designs in the cases he considers (although in some special cases, they may agree). It seems that, in general, compromise is necessary. Indeed, Zimmerman's examples and simulations indicate that the EK-optimal design resembles a regular or space-filling design with respect to overall spatial coverage, but that it has a few small clusters and is in this sense "intermediate" to the antithetical extremes. Furthermore, the EK-optimal design most resembles a design optimal for covariance parameter estimation when the spatial dependence is weak, whereas it most resembles a design optimal for prediction with known parameters when the spatial dependence is strong. The upshot of this for the designer is that placing a few sites very close together, while of no benefit for prediction with known covariance parameters, may substantially improve prediction with unknown covariance parameters.

Overall, the EA and EK approaches seem to work quite well and are well worth using when their objectives seem appropriate, albeit with the caveat that their complexity may make them difficult to explain to non-experts and hence to "sell" to them.

Arguably, a more natural way to incorporate uncertainty about the covariance parameters than the approaches of Zhu and Stein (2006) and Zimmerman (2006) is via a Bayesian approach (Diggle and Lophaven 2006). In a Bayesian approach to design for spatial prediction, a natural design criterion is the spatially averaged variance of the Bayesian predictive distribution of the unobserved random field Y given the observed data. This Bayesian predictive distribution is the integral, over θ, of the classical predictive distribution of Y given the observed data and θ, with different values of θ weighted by their posterior probabilities, assuming that a prior distribution has been specified for θ. Diggle and Lophaven's examples, in which the covariance parameters have relatively diffuse priors, suggest that a Bayesian approach to design for prediction with unknown covariance parameters yields designs qualitatively similar to those obtained by the frequentist approaches of Zhu and Stein (2006) and Zimmerman (2006). That is, most of the sites in the optimal design are well-spaced throughout the study region but there are a few close pairs, representing once again a compromise between prediction with known covariance parameters and covariance parameter estimation.

22.6.5 Entropy-based design

Previous subsections covered design approaches that can be viewed as optimally reducing uncertainty about model parameters or about predictions of unmeasured responses. To these we now add another, optimally reducing the uncertainty about responses by measuring them. The question is which are the ones to measure? Surprisingly, these three can be combined in a single framework. Not only that, achieving the third objective, simultaneously achieves a combination of the other two. The use of entropy to represent these uncertainties is what makes this possible and ties these objectives together. Software for finding entropy–based monitoring network designs is in the R–CRAN package EnviroStat along with a vignette showing how it is done.

As noted earlier, specifying exact design objectives can be difficult or impossible while the high cost of monitoring demands a defensible design strategy. Design objectives have one thing in common, the reduction of uncertainty about some aspect of the random process of interest. Bayesian theory equates uncertainty with a probability distribution while entropy says the uncertainty represented by such a distribution may be quantified as entropy. Thus selecting a design to minimize uncertainty translates into the maximization of entropy reduction.

That observation leads to the entropy based theory of network design and provides objective functions for it (Caselton and Husain 1984; Caselton and Zidek 1984; Shewry and Wynn 1987; Sebastiani and Wynn 2000; Bueso et al. 1998, 1999b; Angulo et al. 2000; Angulo and Bueso 2001; Fuentes et al. 2007). The idea of using entropy in experimental design goes back even further (Lindley 1956).

The approach can be used to reduce existing networks in size (Caselton et al. 1990; Wu and Zidek 1992; Bueso et al. 1998) or extend them (Guttorp et al. 1993). The responses can be vector-valued as when each site is equipped with gauges to measure several responses (Brown et al. 1994a). Costs can be included with the possibility that gauges can be added or removed from individual sites before hitherto unmonitored sites are gauged (Zidek et al. 2000). Data can be missing in systematic ways as for example when some monitoring sites are not equipped to measure certain responses (Le et al. 1997) or when monitoring sites commence operation at different times (giving the data a monotone or staircase pattern) or both (Le et al. 2001; Kibria et al. 2002). Software for implementing an entopy-based design approach can be found at http://enviro.stat.ubc.ca, while a tutorial on its use is given by Le and Zidek (2006) who describe one implementation of the theory in detail.

Although the general theory concerns processes with a fixed number $k = 1, 2, \ldots$ of responses at each site, we assume $k = 1$ for simplicity. Moreover we concentrate on the problem of extending the network; the route to reduction will then be clear. Suppose g of the sites are currently gauged (monitored) and u are not. The spatial field thus lies over $u + g$ discrete sites.

Relabel the site locations as $\{s_1, \ldots, s_u, s_{u+1}, \ldots, s_{u+g}\}$ and let:

$$
\begin{aligned}
\mathbf{Y}_t^{(1)} &= (Y(t, s_1), \cdots, Y(t, s_u))' \\
\mathbf{Y}_t^{(2)} &= (Y(t, s_{u+1}), \cdots, Y(t, s_{u+g}))' \\
\mathbf{Y}_t &= (\mathbf{Y}_t^{(1)'}, \mathbf{Y}_t^{(2)'})'
\end{aligned}
$$

Assuming no missing values, the dataset D is comprised of the measured values of $\mathbf{Y}_t^{(2)}$, $t = 1, \ldots, T$.

Although in some applications (e.g. environmental epidemiology) predicted values of the unobserved responses (past exposure) $\{\mathbf{Y}_t^{(1)}, \ t = 1, \ldots, T\}$ may be needed, here we suppose interest focuses on the $u \times 1$ vector, $\mathbf{Y}_{T+1}^{(1)} = (Y(T+1, s_1), \cdots, Y(T+1, s_u))'$, of unmeasured future values at the currently 'ungauged" sites at time $T + 1$.

Our objective of extending the network can be interpreted as that of optimal partitioning of $\mathbf{Y}_{T+1}^{(1)}$ which for simplicity and with little risk of confusion, we now denote by $Y^{(1)}$. After reordering its coordinates, the proposed design would lead to the partition $Y^{(1)} = (Y^{(rem)'}, Y^{(add)'})$, $Y^{(rem)'}$ being a u_1-dimensional vector representing the future ungauged site responses and $Y^{(add)'}$ is a u_2-dimensional vector representing the new future gauged sites to be added to those already being monitored. If that proposed design were adopted, then at time $T + 1$, the set of gauged sites would yield measured values of the coordinates in the vector $(Y^{(add)'}, Y^{(2)'}) = (Y_{T+1}^{(add)'}, Y_{T+1}^{(2)'}) \equiv G$ of dimension $u_2 + g$. But which of these designs is optimal?

Suppose $\mathbf{Y}_t$ has the joint probability density function, f_t, for all t. Then the total uncertainty about $\mathbf{Y}_t$ may be expressed by the entropy of its distribution, i.e. $H_t(\mathbf{Y}_t) = E[-log f_t(\mathbf{Y}_t)/h(\mathbf{Y}_t)]$, where $h(\cdot)$ is a so-called reference density (Jaynes 1963). It need not be integrable but its inclusion makes the entropy invariant under one–to–one transformations of the scale of $\mathbf{Y}_t$. Note that the distributions involved in H_t may be conditional on certain covariate vectors, $\{\mathbf{x}_t\}$, that are regarded as fixed.

Usually in a hierarchical Bayesian model, $\mathbf{Y}_{T+1}$'s probability density function, $f_{(T+1)}(\cdot) = f_{(T+1)}(\cdot \mid \theta)$ will depend on a vector of unspecified model parameters θ in the first stage of modeling. Examples in previous subsections have included parameters in the spatial covariance model. Therefore using that density to compute $H_{T+1}(\mathbf{Y}_{T+1})$

would make it an unknown and unable to play the role in a design objective function. To turn it into a usable objective function we could use instead $\mathbf{Y}_{T+1}$'s marginal distribution obtained by averaging $f_{(T+1)}(\cdot \mid \theta)$ with respect to θ's prior distribution. However as we have seen in previous subsections, inferences about θ may well be a second design objective (Caselton et al. 1992). Thus we turn from $H_{T+1}(\mathbf{Y}_{T+1})$ to $H_{T+1}(\mathbf{Y}_{T+1}, \theta) = H_{T+1}(Y, \theta)$ in our simplified notation for $\mathbf{Y}_{T+1}$.

Conditional on D, the total a priori uncertainty may now be decomposed as

$$H(Y, \theta) = H(Y \mid \theta) + H(\theta).$$

Assuming for simplicity that we take the reference density to be identically 1 (in appropriate units of measurement), we have

$$H(Y \mid \theta) \;=\; E[-\log(f(Y \mid \theta, D)) \mid D]$$

and

$$H(\theta) \;=\; E[-\log(f(\theta \mid D)) \mid D].$$

But for purposes of optimizing design, we need a different decomposition that reflects the partitioning of future observations into ungauged and gauged sites, $Y' = (U, G)$ where $U \equiv Y^{(rem)'}$ and G is defined above. Now represent $H(Y, \theta) = TOT$ as

$$TOT = PRED + MODEL + MEAS$$

where with our unit reference density:

$$
\begin{aligned}
PRED &\;=\; E[-\log(f(U \mid G, \theta, D)) \mid D]; \\
MODEL &\;=\; E[-\log(f(\theta \mid G, D)/) \mid D]; \\
MEAS &\;=\; E[-\log(f(G \mid D)) \mid D].
\end{aligned}
$$

Coming back to an observation above, by measuring G we would know what it is (ignoring measurement error for expository simplicity). Thus the uncertainty about G before the responses were measured ($MEAS$) would become 0 after G is measured. So it is optimal to choose $Y^{(add)'}$ to correspond to the maximum achieveable value of $MEAS$, in order to maximize the drop in our uncertainty from measuring G at time $T + 1$. However, since TOT is fixed that optimum will simultaneously minimize $PRED + MODEL$ representing the combined objective of prediction and inference about model parameters. Incidentally, had we started with $H(Y)$ instead of $H(Y, \theta)$, and made a decomposition analogous to that given above, we would have arrived at the same optimization criterion: the maximization of $MEAS$.

In the above we have presented a very general theory of design. But how can it be implemented in practice? One solution, presented in detail in Le and Zidek (2006) assumes that the responses can, through transformation, be turned into a Gaussian random field approximately, when conditioned on $\theta = (\beta, \Sigma)$, β representing covariate model coefficients and Σ the spatial covariance matrix. In turn, θ has a so-called generalized inverted Wishart distribution (Brown et al. 1994b) which has a hypercovariance matrix Ψ and a vector of degrees of freedom as hyperparameters. These assumptions lead to a multivariate-t posterior predictive distribution for Y amongst other things. However, the hyperparameters such as those in Ψ in the model are estimated from the data, giving this approach an empirical Bayesian character. Finally $MEAS$ can be explicitly evaluated in terms of those estimated hyperparameters and turns out to have a surprisingly simple form.

In fact, for any proposed design the objective function (to be minimized) becomes (Le and Zidek 2006) after multiplication by a minus sign among other things, the seemingly natural:

$$det(\hat{\Psi}_{U|G}), \tag{22.2}$$

the determinant of the estimated residual hypercovariance for U given G (at time $T+1$, since G is not yet known at time T), where det denotes the determinant. The seeming simplicity of this criterion function is deceptive; in fact, it combines the posterior variances of the proposed unmeasured sites and the posterior correlations between sites in a complicated way and the optimum design, G^{opt} often selects surprising sites for inclusion (Ainslie et al. 2007).

However, finding the optimum design is a very challenging computational problem. In fact, the exact optimal design in Equation 22.2 cannot generally be found in reasonable time, it being an NP-hard problem. This makes suboptimal designs an attractive alternative (Ko et al 1995). Among the alternatives are the "exchange algorithms", in particular the (DETMAX) procedure of Mitchell (1974a, b) cited by Ko et al. (1995). Guttorp et al. (1993) propose a "greedy algorithm" which at each step, adds (or subtracts if the network is being reduced in size) the station that maximally improves the design objective criterion. Ko et al. (1995) introduce a greedy plus exchange algorithm. Finally, Wu and Zidek (1992) cluster prospective sites into suitably small subgroups before applying an exact or inexact algorithm so as to obtain suboptimal designs that are good at least within clusters.

Exact algorithms for problems moderate in size are available (Guttorp et al. 1993; Ko et al. 1995). These results have been extended to include measurement error (Bueso et al. 1998) where now the true responses at both gauged and ungauged sites need to be predicted. Linear constraints to limit costs have also been incorporated (Lee 1998), although alternatives are available in that context (Zidek et al. 2000).

While the entropy approach offers an unified approach to network design especially when unique design objectives are hard to specify, like all approaches it has shortcomings. We list some of these below:

1. When a unique objective can be specified, the entropy optimal design will not be optimal. Other approaches like those covered above would be preferred.

2. Except in the case of Gaussian fields with conjugate priors, computing the entropy poses a challenging problem that is the subject of current research but is not yet solved. However, progress has recently been made there as well (Fuentes et al. 2007).

3. Although the field can often be transformed to make it more nearly Gaussian, that transformation may also lead to new model parameters that are difficult to interpret, making the specification of realistic priors difficult. (The entropy itself would however be invariant under such transformations.)

4. Computing the exact entropy optimum design in problems of realistic size is challenging. That could make simpler designs such as those based on geometry and other methods described above more appealing especially when the goals for the network are short term.

A general problem with approaches to design is their failure to take cost into consideration. That will after always be a limiting factor in practice. However, some progress has been made for the entropy theory designs (Zidek et al. 2000). Moreover, Fuentes et al. (2007) provide a very general theory for entropy based spatial design that allows for constrained optimization that is able to incorporate the cost of monitoring in addition to other things.

22.7   From ambient monitors to personal exposures

This chapter would not be complete without some mention of the link between monitoring networks and personal exposures, which in turn form the foundation for the assessment of the human health effects of environmental hazards. The ambient monitors yield the concentration levels that can be used in conjunction with adverse health outcome counts to quantify their relationship for the purpose of determining regulatory standards. The result are the so–called concentration response functions (CRFs). These describe the incremental decrease in the number of counts that would be due to hypothetical decrements in the level of those concentrations. A health risk analysis analysis would calculate those decreases for various scenarios until an acceptable improvement is seen and this analysis might form the basis for new regulatory standards. However, the association described by the CRFs is not necessarily a predictive one leading to a second approach that is. And here we will discuss urban air pollution fields for definiteness, although in principle these ideas can be applied in other contexts as well.

This alternative approach uses exposure response functions (ERFs) that have been estimated through exposure chamber studies, where the physiological reactions of healthy subjects are assessed at safe levels of the pollutant (see Richmond et al. 2001 for example). The ERFs can now be used in a risk analysis of a different sort to predict the decrements lung function that would result in hypothetical reductions in personal exposure.

But how could these hypothetical reductions in personal exposures be translated into the effects of hypothetical changes in regulatory standards that will ineluctably involve the ambient levels? A first step can involve the use of spatial predictors that would enable an urban area's pollution field to be mapped down to a microscale, a fine scale resolution of the local area (Burke et al. 2001). Changes in the ambient levels would then translate into exposures at the local levels. But one step remains.

That step takes the predicted local exposures into the microenvironments in which people are found, in proportions determined by time–activity patterns as well as commuting patterns. These microenvironments would include: car cabins; restaurants; and kitchens. The degree of personal exposures will depend on such things as whether windows are open or closed, air conditioners are on or off and whether gas stoves are being used. Indoor exposures need to be included as well. For example fireplaces can be a source of particulate air pollution in certain seasons. So ultimately simulators have been built that generate random microenvironmental concentrations, given those in the surrounding local area SHEDS-PM (Burke et al. 2001), APEX (Richmond et al. 2001) and pCNEM (Zidek et al. 2005). These simulators generate random cumulative personal exposures over the day that are experienced by random individuals drawn from the relevant population as they pass through different microenvironments. These in turn can be used to generate the personal exposures sustained by random individuals depending on the ambient levels seen in an urban area and through the ERFs, the impact on their physiology. The results can be quite different than what might be predicted from ambient levels alone (Shaddick et al. 2005). However here too the monitoring network plays a fundamental role emphasizing anew the importance of good monitoring designs.

22.8   New directions

As we have seen above, site locations may well be selected to optimize a specific objective function, so in that sense are preferentially sampled rather than being randomly chosen from the space of available sites. da Silva Ferreira and Gamerman (2015) propose a general decision theory within a Bayesian context, for selecting a network of new monitoring sites. It involves an expected utility

$$U(\mathbf{d}) = E_{\theta, \mathbf{y_d}|\mathbf{y}}[u(\mathbf{d}, \theta, \mathbf{y_d})]$$

that represents the value of any proposed set of spatial sites d given the data vector $\mathbf{y}$, where $\mathbf{y}_d$ is the random vector of additional data that would be gained from choosing d. Moreover they provide a solution to the challenging computational problem of maximizing the utility over d. They give as an example the goal of maximally reducing the variance of a spatial predictor of a random field $S(x)$ averaged over at a spatial point x where $\mathbf{y}$ is the vector of values of S observed with error at a number of other points $\mathbf{x}$.

The authors then apply their theory to the topic of another chapter in this volume, the effect of preferential site selection where now we mean the selection depends on S itself. Now the sites $\mathbf{x}$ are assumed to be obtained from a nonhomogeneous Poisson point process over the spatial domain as in Diggle et al. (2010) and the posterior distribution is conditional on both $\mathbf{y}$ and the points at which y was measured. It turns out that optimal design will in this situation change due to the effects of preferential sampling. For example, if you had the information that the original sites had been deliberately placed where air pollution say, was high and you wanted to maxim ally reduce the average prediction variance of the pollution field over the entire space by adding new sites, you would choose these differently than you would if you ignored that information. This work complements nicely other work on preferential sampling where interest lies say on the effects on relative risks in an environmental epidemiology study of the association between an environmental hazard such as air pollution and human health (Shaddick, Zidek and Liu 2015).

An important new direction in network designs stems from the availability of inexpensive wireless monitors that can be deployed in large numbers. These communicate with a parent sensor that can then transmit data to a central station via a satellite link. Design issues different from those discussed earlier now arise, for example the structure of the communication network for the cluster of sensors along with data management issues (Cheng et al. 2007; Katenka 2009). In one case encountered by the first author of this chapter, the sensors were attached to staff in a large hospital for the purpose of discovering interactions of the personnel and ultimately to estimate parameters in an infectious disease transmission model. Limitations of space keep us from going into more detail.

Another area of monitoring network design receiving attention recently recognizes that often more than one environmental contaminant is of interest. For example, respiratory illness may be affected by ambient levels of $PM_{2.5}$ *and* ozone, among possibly other contaminants. This has led to the development of network designs for multivariate observations. Two main tacks have been taken. The first is based on entropy once more, exploiting its status as an all-purpose design criterion (Le and Zidek 1994; Brown et al. 1994a). The second, more recent tack considers criteria focused directly on cross-covariance estimation (the estimation of covariance among different variables) and/or co-kriging (the multivariate extension of kriging that accounts for spatial correlation and spatial cross-correlation among observations) (Li and Zimmerman 2015). Due to the practical and economic advantages of *collocation*, i.e., the observation of all variables at each sampled site, the extent of collocation in designs that are optimal with respect to these criteria is of prime interest. Also of interest is the relative efficiency of collocated designs. Li and Zimmerman (2015) find that designs that are optimal for covariance and cross-covariance estimation are collocated, while designs that are optimal for co-kriging need not be, especially if the spatial correlation is strong and the cross-correlation is weak. On the other hand, the optimal collocated design is usually quite efficient.

22.9 Concluding remarks

This chapter has fairly comprehensively reviewed principled statistical approaches to the design of monitoring networks. However the challenges involved transcend the mere application of these approaches as described here. Some of these challenges are mentioned in our introduction and more detailed reviews consider others (Le and Zidek 2006). For example, although statistical scientists would well recognize the importance of measurement and data quality, they may not accept or even see their vital role in communicating that to those who will build and maintain the network. And while the normative approaches do provide defensible proposals for design or redesign, these cannot be used before thoroughly reviewing things like the objectives and available data. The latter will suggest important considerations not captured in more formal design approaches. For example, the case study presented in Ainslie et al. (2007) benefited greatly from learning this network's evolutionary history and analyzing the network's spatial correlation structure. They helped to understand the results of the entropy-based analysis and strengthened the eventual recommendations based partly on it.

Our review has not covered quite a number of special topics such as mobile monitors (required for example in the event of the failure of a nuclear power generator and the spread downwind of its radioactive cloud) and micro-sensor monitoring networks. The latter, reflecting changing technology, involve a multitude of small monitors that may cost as little as a few cents that can transmit to each other as well as to their base station, which uploads its data to a global monitoring site where users can access the data. Although these networks pose new design challenges, the principles set out in this chapter can still be brought to bear. Furthermore the reader will have recognized that the "future" in our formulation of the entropy - based design is defined as time $T + 1$. Thus the optimum design there will, in principle depend on T, although it should remain stable for a period of time. However, this observation points to the important general point, that at least all long term monitoring networks, however designed, must be revisited periodically and redesigned if necessary, something that is seldom done in practice.

Bibliography

[1] Ainslie, B, Reuten, C, Steyn, DG, Le, ND and Zidek, JV (2009). Application of an entropy-based optimization technique to the redesign of an existing monitoring network for single air pollutants. *Journal of Environmental Management*, 90, 2715–2729.

[2] Angulo, JM, Bueso, MC, Alonso, FJ (2000). A study on sampling design for optimal prediction of space-time stochastic processes. *Stochastic Environmental Research and Risk Assessment*, 14, 412–427.

[3] Angulo, JM, and Bueso, MC (2001). Random perturbation methods applied to multivariate spatial sampling design. *Environmetrics*, 12, 631–646.

[4] Richmond, H. and Palma, T. and Glen, G. and Smith, L.(2001). Overview of APEX (2.0): EPA's pollutant exposure model for criteria and air toxic inhalation exposures. *Annual Meeting of the International Society of Exposure Analysis, Charleston, South Carolina.*

[5] Benhenni, K and Cambanis, S (1992). Sampling designs for estimating integrals of stochastic processes. *Annals of Statistics*, 20, 161–194.

[6] Bueso, MC, Angulo, JM, and Alonso, FJ (1998). A state-space model approach to optimum spatial sampling design based on entropy. *Environmental and Ecological Statistics*, 5, 29–44.

[7] Bueso, MC, Angulo, JM, Qian, G and Alonso, FJ (1999a). Spatial sampling design based on stochastic complexity. *Journal of Multivariate Analysis*, 71, 94–110.

[8] Bueso, MC, Angulo, JM, Curz-Sanjuliàn, J and García-Aróstegui, JL (1999b). Optimal spatial sampling design in a multivariate framework. *Mathematical Geology*, 31, 507–525.

[9] Brown, PJ, Le, ND and Zidek, JV (1994a). Multivariate spatial interpolation and exposure to air pollutants. *Canadian Journal of Statistics* 22, 489–510.

[10] Brown, PJ, Le, ND and Zidek, JV (1994b). Inference for a covariance matrix. In *Aspects of Uncertainty: A Tribute to D.V. Lindley*, Eds AFM Smith and PR Freeman. New York: Wiley.

[11] Burke, J and Zufall, M and Ozkaynak, H (2001). A population exposure model for particulate matter: case study results for $PM_{2.5}$ in philadelphia, PA. *Journal of Exposure Analysis and Environmental Epidemiology*, 11, 470-489.

[12] Caselton, WF and Husain, T (1980). Hydrologic networks: information transmission. *Water Resources Planning and Management Division, ASCE*, 106(WR2), 503–520.

[13] Caselton, WF and Zidek, JV (1984). Optimal monitoring network designs. *Statistics and Probability Letters*, 2, 223–227.

[14] Caselton, WF, Kan, L and Zidek, JV (1992). Quality data networks that minimize entropy. In *Statistics in the Environmental and Earth Sciences*, Eds P Guttorp and A Walden, 10–38. London: Griffin.

[15] Chang, H, Fu, AQ, Le, ND and Zidek, JV (2007). Designing environmental monitoring networks to measure extremes. *Environmetnal and Ecological Statistics*, 14, 301–321.

[16] Cheng, Z, Pan, J, and Chen, S (2007). Optimal design of the wireless sensor network on environmental monitoring. *Applied Mechanics and Materials*, 687–691, 779–782.

[17] Cox, DD, Cox, LH and Ensor, KB (1997). Spatial sampling and the environment: some issues and directions. *Environmental and Ecological Statistics*, 4, 219–233.

[18] Currin, C, Mitchell, T, Morris, M, Ylvisaker, D (1991). Bayesian prediction of deterministic functions, with applications to the design and analysis of computer experiments. *Journal of the American Statistical Association*, 86, 953–963.

[19] da Silva Ferreira, G and Gamerman, D (2015). Optimal design in geostatistics under preferential sampling. *Bayesian Analysis*, 10, 711–735.

[20] Dalenius, T, Hajek, J, Zubrzycki, S (1960). On plane sampling and related geometrical problems. *Proceedings of the Fourth Berkeley Symposium on Mathematical Statistics and Probability, Vol I*, Berkeley: University of California Press, 125–150.

[21] Diggle, P and Lophaven, S (2006). Bayesian geostatistical design. *Scandinavian Journal of Statistics*, 33, 53–64.

[22] Diggle, PJ, Menezes, R, and Su, T (2010). Geostatistical inference under preferential sampling. *Journal of the Royal Statistical Society: Series C (Applied Statistics)*, 59, 191–232.

[23] Dobbie, MJ, Henderson, BL and Stevens Jr, DL (2008). Sparse sampling: spatial design for monitoring stream networks. *Statistics Surveys*, 2, 113-153.

[24] Eilers, JM, Kanciruk, P, McCord, RA, Overton, WS, Hook, L, Blick, DJ, Brakke, DF, Lellar, PE, DeHans, MS, Silverstein, ME and Landers, DH (1987). Characteristics of lakes in the western United States. *Data Compendium for Selected Physical and Chemical Variables, Vol 2,*. Washington, DC: Environmental Protection Agency, EP/600/3-86-054b.

[25] Elfving, G (1952). Optimum allocation in linear regression theory. *Annals of Mathematical Statistics*, 23, 255–262.

[26] Fedorov, VV (1996). Design for spatial experiments: model fitting and prediction. In *Handbook of Statistics*. (Eds) S Ghosh and CR Rao, 13, 515–553.

[27] Fedorov, VV and Hackl, P (1997). *Model-Oriented Design of Experiments, volume 125 of Lecture Notes in Statistics*. New York: Springer-Verlag.

[28] Fedorov, VV and Müller, WG (1988). Two approaches in optimization of observing networks. In *Optimal Design and Analysis of Experiments* , (Eds.) Y Dodge, VV Fedorov, and HP Wynn, 239–256. New York: North Holland.

[29] Fedorov, VV and Müller, W (1989). Comparison of two approaches in the optimal design of an observation network. *Statistics*, 3, 339–351.

[30] Fedorov, VV and Müller, W (2008). Optimum design for correlated fields via covariance kernel expansions. *mODa 8 - Advances in Model-Oriented Design and Analysis Proceedings of the 8th International Workshop in Model-Oriented Design and Analysis held in Almagro, Spain, June 4-8, 2007*. Physica-Verlag HD, 57–66.

[31] Fuentes, M, Chaudhuri, A, and Holland, DM (2007). Bayesian entropy for spatial sampling design of environmental data. *Environmental and Ecological Statistics* 14, 323–340.

[32] Gaudard, M, Karson, M, Linder, E and Sinha, D (1999). Bayesian spatial prediction (with discussion). *Environmental and Ecological Statistics*, 6, 147–171.

[33] Gerland, P, Raftery, AE, Ševčkov, H, Li, N, Gu, D, Spoorenberg, T, Alkema, L, Fosdick, BK, Chunn, J, Lalic, N, Bay, G, Buettner, T, Heilig, GK, and Wilmoth, J (2014). World population stabilization unlikely this century. *Science*, 10, 234–237.

[34] Gonzalez, FI, Bernard, EN, Meinig, C, Eble, MC, Mofjeld, and Stalin, C (2005). The NTHMP Tsunameter Network. *Natural Hazards*, 35, 25–39.

[35] Guttorp, P, Le, ND, Sampson, PD and Zidek, JV (1993). Using entropy in the re-design of an environmental monitoring network. In *Multivariate Environmental Statistics* (Eds) GP Patil, CR Rao and NP Ross. New York: North Holland/Elsevier Science, 175–202.

[36] Jaynes, ET (1963). Information theory and statistical mechanics, *Statistical Physics* , 3, (Ed) KW Ford, New York: Benjamin, 102–218.

[37] Katenka, NV (2009). Statistical problems in wireless sensor networks. PhD Thesis, Department of Statistics, University of Michigan.

[38] Kiefer, J (1959). Optimum experimental design. *Journal of the Royal Statistical Society, Series B*, 21, 272–319.

[39] Kibria, GBM, Sun L, Zidek V, Le, ND (2002). Bayesian spatial prediction of random space-time fields with application to mapping $PM_{2.5}$ exposure. *Journal of the American Statistical Association*, 457, 101–112.

[40] Ko CW, Lee J, Queyranne M (1995). An exact algorithm for maximum entropy sampling. *Oper Res*, 43, 684–691.

[41] Le, ND and Zidek, JV (1992). Interpolation with uncertain spatial covariances: a Bayesian alternative to kriging. *Journal of Multivariate Analysis*, 43, 351–74.

[42] Le, ND and Zidek, JV (1994). Network designs for monitoring multivariate random spatial fields. In *Recent Advances in Statistics and Probability*, (Eds) JP Vilaplana and ML Puri, 191–206.

[43] Le, ND and Zidek, JV (2006). *Statistical Analysis of Environmental Space-Time Processes*. New York:Springer.

[44] Le, ND, Sun, W and Zidek, JV (1997). Bayesian multivariate spatial interpolation with data missing-by-design. *Journal of the Royal Statistical Society, Series B*, 59, 501–510.

[45] Le, ND, Sun, L and Zidek, JV (2001). Spatial prediction and temporal backcasting for environmental fields having monotone data patterns. *Canadian Journal of Statistics*, 29, 516–529.

[46] Lee, J (1998). Constrained maximum entropy sampling. *Operations Research*, 46, 655–664.

[47] Li, B, Sain, S, Mearns, LO, Anderson, HA, Kovats, S, Ebi, KL, Bekkedal, MYV, Kanarek, MS, and Patz, JA (2012). The impact of extreme heat on morbidity in Milwaukee, Wisconsin. *Climatic Change*, 110, 959–976.

[48] Li, J and Zimmerman, DL (2015). Model-based sampling design for multivariate geostatistics. *Technometrics*, 25, 75–86.

[49] Lindley, DV (1956). On the measure of the information provided by an experiment. *Annals of Mathetmatical Statistics*, 27, 968–1005.

[50] Linthurst, RA, Landers, DH, Eilers, JM, Brakke, DF, Overton, WS, Meier, EP and Crowe, RE (1986). *Characteristics of Lakes in the Eastern United States. Volume 1. Population Descriptions and Physico–Chemical Relationships*. EPA/600/4-86/007a, U.S. Environmental Protection Agency, Washington, DC, 1986, 136 pp.

[51] McBratney, AB, Webster, R, and Burgess, TM (1981). The design of optimal sampling schemes for local estimation and mapping of regionalized variables. I - Theory and method. *Computers and Geosciences*, 7, 331–334.

[52] Müller, WG and Zimmerman, DL (1999). Optimal design for variogram estimation. *Environmetrics*, 10, 23–37.

[53] Myers, DE (2002) Space-time correlation models and contaminant plumes. *Environmetrics*, 13, 535–553.

[54] Müller, WG (2005). A comparison of spatial design methods for correlated observations. *Environmetrics*, 16, 495–506.

[55] Müller, WG (2007). *Collecting Spatial Data: Optimum Design of Experiments for Random Fields, Third Edition*. Heidelberg: Physica-Verlag.

[56] Müller, WG and Stehlik, M (2010). Compound optimal spatial designs. *Environmetrics*, 21, 354–364.

[57] Nychka, D and Saltzman, N (1998). Design of air-quality monitoring designs. In *Case Studies in Environmental Statistics*, Eds D Nychka, W Piegorsch and L Cox. New York: Springer, 51–76.

[58] Ozone (2006). Air Quality Criteria for Ozone and Related Photochemical Oxidants (Final). *U.S. Environmental Protection Agency, Washington, DC*, EPA/600/R-05/004aF-cF.

[59] Pilz, J (1991). *Bayesian Estimation and Experimental Design in Linear Regression Models*. New York: Wiley.

[60] Ritter, K (1996). Asymptotic optimality of regular sequence designs. *Annals of Statistics*, 24, 2081-2096.

[61] Royle, JA and Nychka, D (1998). An algorithm for the construction of spatial coverage designs with implementation in Splus. *Computers and Geosciences*, 24, 479–488.

[62] Schumacher, P and Zidek, JV (1993). Using prior information in designing intervention detection experiments. *Annals of Statistics*, 21, 447–463.

[63] Sebastiani, P, and Wynn, HP (2000). Maximum entropy sampling and optimal Bayesian experimental design. *Journal of the Royal Statistical Society, Series B*, 62, 145–157.

[64] Shaddick, G and Zidek, JV and White, R and Meloche, J and Chatfield, C (2005). Results of using a probabilistic model to estimate personal exposure to air pollution on a study of the short term effect of PM10 on mortality. *Epidemiology*, 16(5), S90.

[65] Shaddick, G, Zidek, JV and Liu, Y (2015). Mitigating the effects of preferentially selected monitoring sites for environmental policy and health risk analysis. *Spatial and Spatio-Temporal Epidemiology*, submitted.

[66] Shewry, M, and Wynn, H (1987). Maximum entropy sampling. *Journal of Applied Statistics*, 14, 165–207.

[67] Silvey, SD (1980). *Optimal Design*. London: Chapman and Hall.

[68] Shaddick, G and Zidek, JV (2015). *Spatio-Temporal Methods in Environmental Epidemiology*. London: Chapman and Hall/CRC.

[69] Smith, K (1918). On the standard deviations of adjusted and interpolated values of an observed polynomial function and its constants and guidance they give towards a proper choice of the distribution of observations. *Biometrika*, 12, 1–85.

[70] Spöck, G and Pilz, J (2010). Spatial sampling design and covariance-robust minimax prediction based on convex design ideas. *Stochastic Environmental Researh and Risk Assessment*, 24, 463–482.

[71] Su, Y and Cambanis, S (1993). Sampling designs for estimation of a random process. *Stochastic Processes and their Applications*, 46, 47–89.

[72] Sun, W, Le, ND, Zidek JV, and Burnett, R (1998). Assessment of a Bayesian multivariate interpolation approach for health impact studies. *Environmetrics*, 9, 565–586.

[73] Ucinski, D (2004). *Optimal Measurement Methods For Distributed Parameter System Identification.* Boca Raton, FL: CRC Press.

[74] Wackernagel, H (2003). *Multivariate Geostatistics: An Introduction with Applications, 2nd Edition.* Berlin: Springer.

[75] Wu, S and Zidek, JV (1992). An entropy based review of selected NADP/NTN network sites for 1983-86. *Atmospheric Environment*, 26A, 2089–2103.

[76] Xia, G, Miranda, ML, and Gelfand, AE (2006). Approximately optimal spatial design approaches for environmental health data. *Environmetrics*, 17, 363–385.

[77] Yaglom, AM (1987). *Correlation Theory of Stationary and Related Random Functions.* New York: Springer.

[78] Yfantis, EA, Flatman, GT, and Behar, JV (1987). Efficiency of kriging estimation for square, triangular and hexagonal grids. *Mathematical Geology*, 19, 183–205.

[79] Zhu, Z and Stein, ML (2005). Spatial sampling design for parameter estimation of the covariance function. *Journal of Statistical Planning and Inference*, 134, 583–603.

[80] Zhu, Z and Stein, ML (2006). Spatial sampling design for prediction with estimated parameters. *Journal of Agricultural, Biological, and Environmental Statistics*, 11, 24–44.

[81] Zidek, J and Shaddick, G and White, R and Meloche, J and Chatfield, C (2006). Using a probabilistic model (pCNEM) to estimate personal exposure to air pollution *Environmetrics*, 16, 481-493.

[82] Zimmerman, DL (2006). Optimal network design for spatial prediction, covariance parameter estimation, and empirical prediction. *Environmetrics*, 17, 635–652.

[83] Zidek, JV, Sun, W and Le, ND (2000). Designing and integrating composite networks for monitoring multivariate gaussian pollution fields. *Applied Statistics*, 49, 63–79.

[84] Zidek, JV, Sun, W and Le, ND (2000). Designing Networks for Monitoring Multivariate Environmental Fields Using Data With Monotone Pattern. TR 2003, Statistical and Applied Mathematical Sciences Insitute, North Carolina.

23

Statistical methods for source apportionment

Jenna R. Krall

George Mason University, Fairfax, Virginia

Howard H. Chang

Emory University, Atlanta, Georgia

CONTENTS

23.1 Introduction

Environmental pollutants frequently exist as complex mixtures of chemical and non-chemical constituents. For example, particulate matter (PM) air pollution is a mixture of solid and liquid particles with diverse physical and chemical properties. Complex pollutant mixtures like PM are often formed when multiple sources (e.g. biomass burning or traffic) contribute pollutants to the mixture. Information about which sources contribute to a pollution mixture is important to identifying major polluters and to developing strategies to reduce pollution.

Pollution is frequently observed using receptors, such as ambient monitors and other environmental sensors, that can measure individual constituents within a mixture, but do not have direct information on the sources that generated the constituents. Though pollution from individual sources can sometimes be isolated by measuring pollution from or near a single source, for example PM concentrations near a wildfire, the environmental exposure of interest may be within a community where multiple sources contribute to pollution levels. Identifying the combination of sources that contribute to a community's pollution mixture is challenging because most observed constituents are generated by multiple sources. For example in the study of PM, organic carbon PM is associated with biomass burning, traffic, and other sources. Source apportionment aims to decompose the observed pollution mixture, as represented by measurements of individual constituents, into information about the sources that contribute to the mixture. This chapter describes statistical methods and challenges for source apportionment.

23.1.1   Source apportionment

Most models for source apportionment use a mass balance framework, which assumes constituent mass (i.e. PM measured in $\mu g/m^3$) is conserved from the source to the receptor, specifically that there is no mass lost due to physical removal or environmental chemistry. These models generally represent observed constituents at a receptor as a linear combination of the unknown mixture concentrations from each source and the constituent composition of each source. Let $\mathbf{X}_{T \times P}$ represent mass concentrations of constituents measured over T days for P mixture constituents at one receptor. The usual source apportionment model assumes

$$\mathbf{X} = \mathbf{GF} + \mathbf{e} \qquad (23.1)$$

where $\mathbf{G}_{T \times L}$ represents the mixture concentrations corresponding to L sources over T days and $\mathbf{F}_{L \times P}$ represents the constituent compositions of each source, or the source profiles. Last, $\mathbf{e}$ includes additional unknown sources or measurement error. Frequently, g_{tl} can be interpreted as the total mass on day t emitted from source l and f_{lp} describes the proportion of total mass from source l that is constituent p. Hence the measured constituent mass x_{tp} is the sum of contributions from L individual sources plus error, $x_{tp} = \sum_{l=1}^{L} g_{tl} f_{lp} + e_{tp}$. In practice, the number of measured constituents P is larger than the number of sources L. In some source apportionment methods, the source apportionment model is solved day-by-day for each $\mathbf{x}_t$; however, we write the model in matrix form because it will be more familiar to statisticians.

Commonly, both $\mathbf{G}$ and $\mathbf{F}$ are unknown. In this case, the usual source apportionment model does not have a unique solution without additional constraints. For example, there is rotational ambiguity in the standard source apportionment model. If there exists solutions $\mathbf{G}$ and $\mathbf{F}$, for any $L \times L$ orthogonal matrix $\mathbf{R}$ there are competing solutions $\mathbf{G}^*$ and $\mathbf{F}^*$ because

$$\mathbf{X} = \mathbf{GF} + \mathbf{e} = (\mathbf{GR})(\mathbf{R'F}) + \mathbf{e} = \mathbf{G}^*\mathbf{F}^* + \mathbf{e} \qquad (23.2)$$

There are many proposed source apportionment methods, and each method uses different constraints to estimate $\mathbf{G}$ and $\mathbf{F}$.

Henry [13] provided general conditions for finding acceptable solutions $\mathbf{G}$ and $\mathbf{F}$ to the source apportionment model, namely:

1. The model must explain the observations $\mathbf{X}$.

2. All elements of **G** and **F** must be non-negative.

3. $\sum_{p=1}^{P} f_{lp}$ must be less than or equal to 1.

Condition 1 implies that $\mathbf{X} \approx \mathbf{GF}$. Because negative concentrations in **G** and negative compositions in **F** should not be permitted, condition 2 ensures a plausible solution with respect to mass balance. Last, condition 3 requires that taken together, all constituents cannot contribute more than 100% to **F**.

The total mass corresponding to source l, $\mathbf{g}_l$, describes how pollution emitted from a specific source varies over time. Therefore **G** can help determine those sources that contribute most to total pollution and also may be used to estimate health effects associated with exposure to source-specific pollution. In contrast, **F** is not generally used in subsequent analyses. However, **F** can be used to determine the reasonableness of the source apportionment solution by matching the identified sources in **F** to known sources in a community.

Development of source apportionment models and their application has been primarily focused on sources of PM air pollution, in part because PM constituent data is collected nationally by the US Environmental Protection Agency (US EPA) through the Chemical Speciation Network (CSN). In addition, estimates of source-specific ambient PM are important to understanding the major sources that contribute to total ambient PM and for determining how PM impacts health. Examples of other applications of source apportionment include analysis of polychlorinated biphenyls (PCBs) in aquatic sediment [50] and chemical and non-chemical parameters in water [52].

Generally, source apportionment methods fall into two categories: methods that assume the source profiles, **F**, are known and methods that assume the source profiles are unknown. When **F** is known, least squares techniques can be applied to estimate **G**. However when **F** is unknown, methods that are exploratory in nature or methods that incorporate prior information about the composition of sources can be applied.

23.1.2 Example

To demonstrate the source apportionment problem, consider data from the Queens College US EPA CSN monitor that measures chemical constituents of PM less than 2.5 μm ($PM_{2.5}$) in New York City. From 2003-2005, the dataset contains concentrations for 24 $PM_{2.5}$ chemical constituents and total $PM_{2.5}$ for 267 days. Monitors in the EPA CSN generally measure constituent concentrations every third or sixth day. As shown in Figure 23.1, the $PM_{2.5}$ example data is dominated by ammonium ion, elemental carbon, nitrate, organic carbon, and sulfate. Figure 23.2 shows the time series plots from 2003-2005 for three $PM_{2.5}$ constituents: nickel, vanadium, and lead. The plots illustrate temporal correlation between the constituents, possibly because these constituents are frequently emitted by an oil combustion, or residual oil, source [19]. The correlations between these constituents are 0.62 (nickel/vanadium), 0.55 (nickel/lead), 0.70 (vanadium/lead). One aim of source apportionment is to estimate the time series of $PM_{2.5}$ from residual oil, $\mathbf{g}_{oil}$, and the proportion each constituent contributes to residual oil, $\mathbf{f}_{oil}$.

23.2 Methods when source profiles are known

In source apportionment, sometimes prior information about the constituent composition of **F** is known. When **F** is completely known or a good estimate of **F** is available, then the focus of source apportionment becomes estimating **G**.

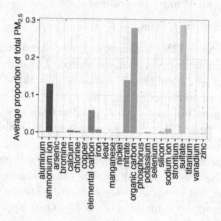

FIGURE 23.1
For 24 PM$_{2.5}$ constituents, the average proportion of total PM$_{2.5}$ mass from one monitor in New York City, 2003-2005

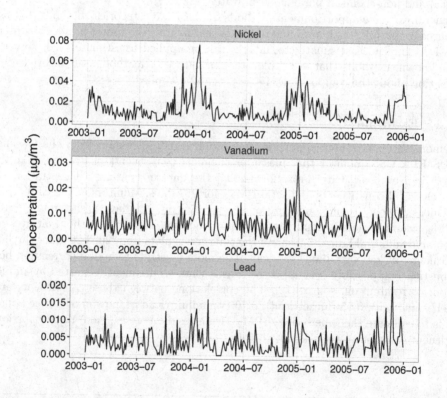

FIGURE 23.2
Time series plots of nickel, vanadium, and lead PM$_{2.5}$ representing columns of **X** from 2003-2005 in New York City. These three PM$_{2.5}$ constituents are associated with residual oil PM$_{2.5}$.

23.2.1 Ordinary least squares approaches

When $\mathbf{F}$ is known without error and the rows of $\mathbf{F}$ are not linearly dependent, $\mathbf{G}$ in equation 23.1 can be estimated by minimizing

$$\sum_{t=1}^{T}\sum_{p=1}^{P}\left(x_{tp} - \sum_{l=1}^{L} g_{tl} f_{lp}\right)^2$$

to find the ordinary least squares (OLS) estimator $\hat{\mathbf{G}} = \mathbf{XF'(FF')}^{-1}$ [51]. The standard OLS approach assumes the error variance is the same for all constituents, however constituent variances often vary substantially in environmental mixtures. To account for known constituent-specific variances σ_p^2, a weighted least squares (WLS) approach can be applied to minimize

$$\sum_{t=1}^{T}\sum_{p=1}^{P}\frac{\left(x_{tp} - \sum_{l=1}^{L} g_{tl} f_{lp}\right)^2}{\sigma_p^2} \tag{23.3}$$

If the rows of $\mathbf{F}$ are not linearly dependent, then the WLS estimator is $\hat{\mathbf{G}} = \mathbf{XV}^{-1}\mathbf{F'(FV}^{-1}\mathbf{F')}^{-1}$ where $\mathbf{V}$ is a $P \times P$ diagonal matrix containing the variances σ_p^2 [51]. If the variances are the same for all constituents, i.e. $\sigma_p^2 = \sigma^2$ for all p, the WLS solution reduces to the OLS solution. OLS and WLS can be fitted using most statistical software packages.

23.2.2 Chemical Mass Balance (CMB)

The OLS and WLS methods assume that the source profiles, $\mathbf{F}$, are known without error. In practice, $\mathbf{F}$ is not usually known and is estimated using prior knowledge or external data. In this case, the corresponding uncertainty associated with estimating $\mathbf{F}$ should be incorporated into the estimate of $\mathbf{G}$. A commonly-applied source apportionment method called Chemical Mass Balance (CMB) estimates $\mathbf{G}$ using an effective variance WLS approach that incorporates uncertainties in $\mathbf{F}$ [57]. Frequently, uncertainties are reported as a percentage for each pollutant in $\mathbf{F}$, for example as in the US EPA SPECIATE database (https://www.epa.gov/air-emissions-modeling/speciate-version-45-through-40). These uncertainties may account for statistical uncertainty resulting from multiple experiments and instrument measurement uncertainty.

An effective variance WLS approach uses an iterative procedure to update estimates of the effective variances $\mathbf{V}_{e,t}$ that are used to subsequently update $\mathbf{G}$. CMB differs from WLS by iteratively re-estimating the weights σ_p in equation 2 to incorporate the uncertainties from $\mathbf{F}$. We show updates for $\mathbf{g}_t$ in place of $\mathbf{G}$ in this section to reflect that CMB is applied day-by-day. In CMB, the matrix $\mathbf{V}_{e,t}$ will be a $P \times P$ diagonal matrix representing the effective variances for each constituent for each day t. Starting with $\mathbf{G}^0 = 0$, at each step k and day t CMB computes

1. $(\mathbf{V}_{e,t}^k)_{pp} = \sigma_p^2 + \sum_{l=1}^{L}(g_{tl}^k)^2 \times \phi_{lp}^2$

2. $\mathbf{g}_t^{k+1} = [\mathbf{x}_t'(\mathbf{V}_{e,t}^k)^{-1}\mathbf{F'}][(\mathbf{F}(\mathbf{V}_{e,t}^k)^{-1}\mathbf{F'})^{-1}]$

where $(\mathbf{V}_{e,t}^k)_{pp}$ is the pth element on the diagonal of $\mathbf{V}_{e,t}^k$ and ϕ_{lp}^2 is the uncertainty associated with f_{lp}. Note that step 1 is derived from $\text{Var}\left(x_{tp} - \sum_{l=1}^{L} g_{tl} f_{lp}\right) = \sigma_p^2 + \sum_{l=1}^{L}\left(g_{tl}^2 \times \text{Var}(f_{lp})\right)$ and step 2 is the usual WLS operator given $(\mathbf{V}_{e,t}^k)_{pp}$. The iterative procedure is generally stopped when no element of $\mathbf{G}$ changes by more than 1%. Once the final estimates $\mathbf{G}^k$ and $\mathbf{V}_{e,t}^k$ have been obtained, then

$$\text{Var}(\mathbf{g}_t^k) = (\mathbf{F}(\mathbf{V}_{e,t}^k)^{-1}\mathbf{F}')^{-1} \tag{23.4}$$

When ϕ_{lp}^2 is zero for all l and p, this reduces to the usual WLS solution. The derivation of the CMB algorithm is outlined in Watson et al. [57].

The standard CMB method does not require estimated concentrations in $\mathbf{g}_t$ to be positive. Negative concentrations in the estimated $\mathbf{G}$ can occur when sources are highly positively correlated in composition (i.e. the rows of $\mathbf{F}$ are correlated). When $\mathbf{G}$ contains negative values, it is recommended to drop one source at a time from the model, starting with the source that has the most negative concentrations in $\mathbf{G}$, until concentrations in $\mathbf{G}$ are mostly positive [7]. For two sources that are highly correlated in composition, this approach aims to remove one of these sources from the model.

Another approach would be to directly impose positivity constraints on the concentrations in $\mathbf{G}$, for example using non-negative least squares [56], though these models are not frequently applied in practice. In source apportionment of PM, CMB has been extended to help resolve secondary sources (PM mixtures that form through chemical reactions in the air) by placing constraints on the estimation of $\mathbf{G}$ based on gas-to-particle ratios [34].

A major limitation of the CMB approach is that it requires estimates of $\mathbf{F}$. In many cases, neither the number of sources nor composition of sources at a receptor are known. Therefore, CMB is limited to applications where prior information is available about sources impacting the receptor. The US EPA has made software available to apply CMB (`https://www3.epa.gov/scram001/receptor_cmb.htm`).

23.3   Methods when source profiles are unknown

When the source profiles $\mathbf{F}$ are unknown, as is common in practice, many source apportionment methods have been proposed. These approaches vary in their assumptions about the source apportionment problem and whether the methods can incorporate prior information.

23.3.1   Principal component analysis (PCA)

Principal component analysis (PCA) is a dimension reduction approach that constructs new variables that explain most of the variation in the constituent data $\mathbf{X}$. PCA-based approaches are useful because they do not require any prior information about $\mathbf{G}$ or $\mathbf{F}$, but they generally do not adhere to all the requirements set forth by Henry [13] and are therefore more exploratory in nature. For example, PCA can be used to determine which constituents are correlated and therefore are likely generated by the same source.

For the observed constituent data $\mathbf{X}$, PCA represents its correlation or covariance matrix Σ as

$$\Sigma = \mathbf{A}\Lambda\mathbf{A}'$$

where $\mathbf{A}$ is a matrix of the eigenvectors of Σ, or principal components (PC), and Λ is a diagonal matrix containing the corresponding eigenvalues [21]. In source apportionment studies, Σ is nearly always the correlation matrix because the constituents frequently have greatly different variances, which will drive the estimated directions of variation in $\mathbf{A}$. The data $\mathbf{X}$ can be approximated using the first L PCs that explain most of the variability in

Σ such that $\Sigma \approx \mathbf{A}^{(L)}\boldsymbol{\Lambda}^{(L)}\mathbf{A}^{(L)'}$, where $\mathbf{A}^{(L)}$ is the matrix of the first L eigenvectors and $\boldsymbol{\Lambda}^{(L)}$ is a diagonal matrix of the first L eigenvalues.

The PC scores based on the correlation matrix Σ are defined as $\mathbf{X}^*\mathbf{A}^{(L)}$, where $\mathbf{X}^*$ is a $T \times P$ matrix with $x_{tp}^* = \frac{x_{tp} - \bar{x}_p}{\sigma_p}$. The PC scores and PC loadings $\mathbf{A}^{(L)}$ are not good representations of $\mathbf{G}$ and $\mathbf{F}$ respectively because the values are not constrained to be positive. In addition, the columns of $\mathbf{A}^{(L)}$ are orthogonal, which is often not reasonable for mixture sources. For example in studies of PM sources, mobile PM and biomass burning PM often contain some of the same chemical constituents and are therefore correlated. Last, since the columns of $\mathbf{A}^{(L)}$ are orthogonal, the estimate of the PC scores is an OLS estimate of equation 23.1 given $\mathbf{A}^{(L)}$. Despite its limitations when applied to source apportionment problems, PCA is easy to apply with implementations in most statistical software packages.

23.3.2 Absolute principal component analysis (APCA)

Absolute principal component analysis (APCA) [54] has been proposed to modify standard PCA to better fit the source apportionment problem. APCA can be summarized in several steps:

1. Apply standard PCA to the correlation matrix Σ and select L sources

2. Rotate the PCs $\mathbf{A}^{(L)}$ using varimax rotation [10]

3. Compute the PC scores $\mathbf{B} = [\mathbf{X}\mathbf{V}^{-1/2}\Sigma^{-1}]\mathbf{A}^{(L)}$, where $\mathbf{V}$ is a diagonal matrix containing the variances σ_p^2 of the columns of $\mathbf{X}$ and Σ is the sample correlation matrix

4. Estimate the source concentrations $\hat{g}_{tl} = \hat{\eta}_l \times b_{tl}$ using linear regression $E(m_t) = \eta_0 + \sum_{l=1}^{L} \eta_l \times b_{tl}$, where m_t is the total mass of the mixture observed on day t.

5. Estimate the source profiles $\hat{f}_{lp}$ using linear regression $E(x_{tp}) = \alpha_0 + \sum_{l=1}^{L} \hat{g}_{tl} \times f_{lp}$

Varimax rotation in step 2 maximizes the sample standard deviation of each PC to obtain more interpretable sources. In step 3, using the scaled but uncentered data to estimate the PC scores instead of $\mathbf{X}^*$ will yield more positive PC scores. APCA regresses total mass observed for each day, m_t, on the estimated PC scores to ensure the sum of daily source concentrations $\sum_{l=1}^{L} \hat{g}_{tl}$ is approximately equal to the observed daily total mass. Last, if $\hat{\alpha}_0$ is substantially different than zero, this may indicate that there are unexplained sources not included in the model.

APCA is a convenient source apportionment method because it can be applied using standard statistical software. However, APCA does not constrain $\mathbf{G}$ or $\mathbf{F}$ to be positive and still suffers from many of the same problems as using PCA for source apportionment.

23.3.3 Unmix

Unmix provides a graphical solution to the source apportionment problem using PCA [14]. PCA is applied to determine the L-dimensional subspace ($L < P$) that is defined by the L sources. By applying PCA, one can find the L-dimensional subspace that explains most of the variability in the constituent data. To further reduce the solution space within the L-dimensional subspace, Unmix finds possible solutions that meet the positivity constraint for the sources. Unmix ultimately estimates $\mathbf{G}$ and $\mathbf{F}$ using "edge points", which represent days when one source contributes minimally to the total pollution mixture. The process of finding edge points and using them to estimate $\mathbf{G}$ and $\mathbf{F}$ is explained in detail in Henry [15]. The graphical description of Unmix in Henry [14] is helpful to conceptualize the Unmix solution to source apportionment. Though software is currently available for US EPA Unmix 6.0 at `http://www.epa.gov/air-research/unmix-60-model-environmental-data-analyses`, it does not run on recent operating systems.

23.3.4   Factor analytic methods

Factor analytic methods, which aim to directly model equation 23.1, have also been applied to estimate $\mathbf{G}$ and $\mathbf{F}$. Exploratory factor analysis (EFA), like PCA, does not require prior information about sources, but often does not yield interpretable results. Confirmatory factor analysis, while more interpretable than EFA, requires some prior information on $\mathbf{F}$ [3]. Target Transformation Factor Analysis (TTFA) [17] modifies estimated sources to make them more interpretable. Specifically, TTFA applies a rotation matrix to the estimated source profiles so they better match a hypothesized source profile matrix.

Iterated confirmatory factor analysis (ICFA) [6] iterates estimates of columns of $\mathbf{F}$ in factor analysis to obtain interpretable results without requiring known information about $\mathbf{F}$. While $\mathbf{F}$ does not need to be known, ICFA requires a starting guess for an $L \times L$ full rank submatrix of $\mathbf{F}$. This submatrix is sufficient to estimate $\mathbf{G}$ and the full factor matrix $\mathbf{F}$ in standard confirmatory factor analysis. To account for possible errors in this submatrix, an iterated procedure is conducted where $P - q$ columns ($q > L$) of $\mathbf{F}$ are updated at each step. This procedure continues until convergence. Once the profiles converge, the source concentrations $\mathbf{G}$ can be estimated using WLS or a similar approach.

Standard factor analytic approaches do not require solutions to meet the positivity constraints or composition constraints of source apportionment (conditions 2-3 from Henry [13]). However, many commonly used standard statistical packages have implemented factor analytic methods. Other algorithms for incorporating positivity constraints into factor analytic models have been proposed [28, 30, 53], though these are not commonly applied in the source apportionment literature.

23.3.5   Positive matrix factorization

Positive matrix factorization (PMF) [41] is a factor analytic method that attempts to meet the non-negativity requirements of source apportionment while accounting for uncertainty. PMF does not require prior information about $\mathbf{F}$, but is flexible in its ability to incorporate available information. PMF minimizes

$$Q = \sum_{t=1}^{T} \sum_{p=1}^{P} \frac{\left(x_{tp} - \sum_{l=1}^{L} g_{tl} f_{lp} \right)^2}{u_{tp}^2} \tag{23.5}$$

subject to $g_{tl} \geq 0, f_{lp} \geq 0$ for all t, p, l. The u_{tp}^2 are known uncertainties corresponding to each x_{tp} and they generally represent method measurement error. The uncertainties allow individual data points to be downweighted in the estimation of sources and are often chosen to be a function of the analytic uncertainty and the detection limit of the measurement method. Several algorithms have been proposed to minimize (23.5) subject to the positivity constraints, including alternating least squares and conjugate gradient algorithms [38].

While the standard application of PMF does not require additional information about $\mathbf{G}$ or $\mathbf{F}$, there may be multiple solutions that lead to similar, small values of Q. Prior information can be incorporated into PMF to reduce rotational ambiguity. The Multilinear Engine (ME-2) [38] fits the PMF model and allows auxiliary equations, sometimes referred to as "pulling equations" in the PMF literature, of the form

$$Q^a = \sum_{j=1}^{J} Q_j^a = \sum_{j=1}^{J} (r_j/s_j)^2 \tag{23.6}$$

where for each auxiliary equation j, r_j represents residuals and s_j controls the degree of penalty and is sometimes refered to as the "softness" of the equation [40]. Then, PMF minimizes $Q + Q^a$. Residuals of the form r_j can be used to constrain the optimization of Q^a in equation 23.6 by penalizing deviations of g_{tl} from a target value $\tilde{g}$ by specifying $r_j = (g_{tl} - \tilde{g})$. The choice of s_j controls the size of the penalty accorded the deviations from $\tilde{g}$. Similar equations can be included for elements of $\mathbf{F}$ or functions of elements of $\mathbf{G}$ or $\mathbf{F}$. Auxiliary equations can also incorporate additional data, including wind direction and speed, into source estimation [23].

Rotational ambiguity can also be explored by a rotational parameter available in PMF that constrains all elements of $\mathbf{G}$ without requiring the user to specify target values $\tilde{g}$ [40]. Graphical methods have also been proposed to determine appropriate rotations for PMF solutions [42]. The technical details that describe how the PMF model is fit in the ME-2 program are provided elsewhere [38, 40].

To date, PMF has been the most commonly-applied source apportionment model to estimate sources of PM. The popularity of PMF has been driven in part by its flexibility, and in part because the US EPA has provided a graphical user interface to PMF ME-2 (`https://www.epa.gov/air-research/positive-matrix-factorization-model-environmental-data-analyses`).

23.3.6 Bayesian approaches

Sources can also be estimated using a Bayesian framework such that,

$$\mathrm{pr}(\mathbf{G}, \mathbf{F} \mid \mathbf{X}) \propto \mathrm{pr}(\mathbf{X} \mid \mathbf{G}, \mathbf{F})\mathrm{pr}(\mathbf{G})\mathrm{pr}(\mathbf{F})$$

where $\mathrm{pr}(\cdot)$ represents a probability distribution. Bayesian source apportionment (BSA) yields the full posterior distribution $\mathrm{pr}(\mathbf{G}, \mathbf{F} \mid \mathbf{X})$ of $\mathbf{G}$ and $\mathbf{F}$ given the available data $\mathbf{X}$. From the posterior distribution, $\mathbf{G}$ and $\mathbf{F}$ could be estimated by their posterior means. The likelihood for the data, $\mathrm{pr}(\mathbf{X} \mid \mathbf{G}, \mathbf{F})$, can be specified to ensure that constituent concentrations will be non-negative. BSA models can incorporate constraints on the source apportionment problem directly by specifying appropriate prior distributions for $\mathbf{G}$ and $\mathbf{F}$, $\mathrm{pr}(\mathbf{G})$ and $\mathrm{pr}(\mathbf{F})$ respectively. These prior distributions can incorporate known or hypothesized information about the mixture sources, such as which constituents are commonly emitted from a specific source. BSA models address some of the limitations in other methods, namely meeting all the conditions set forth by Henry [13] and flexibly incorporating prior information.

The likelihood for each element of $\mathbf{X}$, given $\mathbf{G}$ and $\mathbf{F}$, is often specified as univariate normal or lognormal. Specifying the lognormal likelihood has the advantage of explicitly requiring all elements of $\mathbf{X}$ to be greater than zero, though this approach does not allow concentrations to be exactly zero. The lognormal likelihood for $\mathbf{X}$ also leads to lognormally distributed measurement error $\mathbf{e}$, which is discussed in Section 23.5.3.

Lognormal and truncated normal priors have been proposed for both $\mathbf{G}$ and $\mathbf{F}$, which explicitly require all source concentrations and source compositions to be positive. In general, vague priors are specified or hyperparameters are chosen based on additional information. The estimated $\mathbf{G}$ and $\mathbf{F}$ are frequently rescaled so that $\mathbf{G}$ represents the daily concentration of source-specific total mass and $\mathbf{F}$ represents the percent contribution of each constituent to each source. Another approach is to represent $\mathbf{F}$ using the Generalized Dirichlet distribution [31], which directly models the multivariate nature of each source profile $\mathbf{f}_l$ and provides more flexibility than the standard Dirichlet distribution.

As previously mentioned, the standard source apportionment model (equation 23.1) is not identifiable without additional constraints. Park et al. [45] presented several sets of sufficient, but not necessary, conditions for identifiability that are frequently used in BSA. The most commonly-applied identifiability conditions in source apportionment include:

C1 There are at least $L - 1$ constituents that do not contribute to each source. Specifically for every source l, there are $L - 1$ constituents where $f_{lp} = 0$.

C2 The rank of $\mathbf{F}^{(k)}$ is $L-1$, where $\mathbf{F}^{(k)}$ is created using only the columns with zeros in the kth row (as defined by C1) with those zeros removed.

C3 For each source l, there is one constituent p_l known to be emitted from that source such that $f_{lp_l} = 1$.

Condition C1 specifies constituents not believed to be emitted by a particular source l. Condition C2 implies that the zeroes specified in C1 cannot be the same for all sources; in other words, the sources must be different in composition from each other. Last, C3 is a normalization criteria that does not help to identify sources, but eliminates rotational ambiguity [45]. To meet these identifiability conditions, previous studies have used EFA [35], results from previous analyses [47], or additional external information [9]. Other identifiability conditions, in particular conditions on $\mathbf{G}$, are also outlined in Park et al. [45].

Easy-to-use software for BSA is not widely available, and generally application-specific, customized Markov Chain Monte Carlo (MCMC) methods for BSA are implemented in computational languages such as R. Because BSA requires the estimation of all parameters in $\mathbf{G}$ and $\mathbf{F}$, and therefore requires evaluating each parameter's convergence, BSA can be difficult to implement in practice.

23.3.7 Ensemble approaches

Ensemble approaches aim to combine results from multiple source apportionment approaches into one "ensemble" estimate of sources. A major advantage of these approaches is that they do not require users to select only one source apportionment method. However, these approaches still require that each source apportionment method be fitted appropriately to the data and may require more prior information than might be available.

One ensemble approach computes ensemble-based source profiles (EBSPs) for $\mathbf{F}$ and corresponding uncertainties by combining information across source apportionment methods [27]. To obtain the EBSPs, first each source apportionment method j is applied to estimate source concentrations $\mathbf{G}_j$ for a short time period, such as one month. Then, a weighted average of the source concentrations across methods is obtained, with weights equal to the inverse uncertainties for each g_{tlj}. The weighted averages $\bar{\mathbf{G}}$ are then used with $\mathbf{X}$ in an "inverse CMB" approach to estimate the EBSPs. Last, the EBSPs are used to estimate the final ensembled source concentrations $\mathbf{G}$ for the full time period of interest, generally across several years. This EBSP approach does not require $\mathbf{G}_j$ and $\bar{\mathbf{G}}$ to contain the same number of days as the final source concentrations $\mathbf{G}$, which is helpful when incorporating computationally intensive source apportionment methods in the estimation of the EBSPs. However, the EBSPs can still be used to estimate $\mathbf{G}$ for the entire time period spanned by $\mathbf{X}$.

Balachandran et al. [1] introduced an approach for estimating uncertainties corresponding to the ensemble source concentrations. This approach computes the root mean squared error (RMSE) of source concentrations for each method $\mathbf{G}_j$ from the weighted average $\bar{\mathbf{G}}$.

These RMSEs are used as the updated source apportionment method uncertainties. Last, the weighted average source concentrations are recomputed based on these updated uncertainties to obtain $\bar{\mathbf{G}}^*$, which is then used to estimate the EBSPs in the standard ensemble approach.

A Bayesian-based ensemble averaging approach [2] has also been proposed. This approach selects an inverse-gamma prior for the uncertainties of $\mathbf{G}$ in the standard ensemble method, and then draws from the corresponding posterior using a Monte Carlo approach. These draws are ultimately used to create multiple datasets of source concentrations $\mathbf{G}$ that reflect the variation in uncertainties.

There is not widely available software to implement ensemble source apportionment approaches.

23.4 Comparison of source apportionment methods

Consider the dataset containing $PM_{2.5}$ chemical constituent concentrations from 2003-2005 for one monitor in New York City as described in Section 23.1.2. Positive matrix factorization (PMF) is one of the most commonly-applied source apportionment techniques and PMF can be applied to chemical constituent data without any additional information. We apply PMF and another source apportionment approach, absolute principal component analysis (APCA), to estimate the sources that generated the 24 chemical constituents of $PM_{2.5}$. Separately, PMF and APCA were applied to the entire dataset assuming 6 sources.

Figure 23.3 shows the estimated source profile matrix $\hat{\mathbf{F}}$ obtained from PMF. Specifically, this profile figure illustrates the amount each constituent contributes to each source in concentration units $(\mu g/m^3)$. At this stage, the sources have not been "named" based on their estimated source compositions and so are listed as "Source 1"-"Source 6". For some sources, naming the source based on known sources in New York City is easy; for example, Source 5 contains primarily ammonium ion and nitrate and is likely a secondary nitrate source. For other sources, naming is more challenging. For example, Source 6 contains elemental carbon and zinc, which are frequently emitted from diesel vehicles. However, Source 6 also contains aluminum, silicon, and titanium, which are frequently associated with soil dust. Additional information may be needed to better separate sources in this setting.

The estimated source concentrations $\hat{\mathbf{G}}$ are shown in Figure 23.4 for a subset of the total time frame, 2003-2004. Sources 1 and 3 seem to have a seasonal pattern, with higher concentrations in the summer, while Source 5 shows higher concentrations in the winter months. Each time series $\hat{\mathbf{g}}_l$, $l = 1...5$ could be used in subsequent analyses to estimate health effects associated with source-specific $PM_{2.5}$.

Though PMF is a very popular source apportionment method, one could have chosen another method that does not require prior information, such as APCA. When APCA is applied to the New York City data, a different set of sources is obtained and most sources do not correspond well to sources estimated using PMF. Using the correlation between source concentration time series, two sources estimated using APCA matched estimated PMF sources, namely secondary sulfate $PM_{2.5}$ $(r = 0.84)$ and mobile-related $PM_{2.5}$ $(r = 0.79)$. Because APCA relies on PCA, the sources it estimates may not correspond to known sources in a community.

Figure 23.5 shows estimated source concentrations from PMF and APCA corresponding to secondary sulfate (Source 1 in PMF, Figure 23.4) and mobile-related (Source 6 in PMF, Figure 23.4) $PM_{2.5}$, where mobile-related $PM_{2.5}$ is a mixture of mobile emissions and dust.

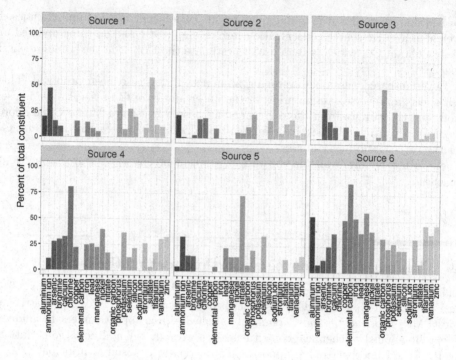

FIGURE 23.3

Estimated source compositions (source profiles $\hat{\mathbf{F}}$) using PMF from one $PM_{2.5}$ constituent monitor in New York City from 2003-2005.

For secondary sulfate $PM_{2.5}$, the methods yield similar source concentration distributions, though more differences are apparent for mobile-related $PM_{2.5}$. Table 23.1 shows the mean and standard deviations for secondary sulfate $PM_{2.5}$ and mobile-related $PM_{2.5}$ using both APCA and PMF.

Source of $PM_{2.5}$	APCA	PMF
Secondary sulfate	5.34 (7.33)	4.41 (5.98)
Mobile-related	4.80 (2.92)	1.95 (1.42)

TABLE 23.1

Table of means (standard deviations) of source-specific $PM_{2.5}$ using PMF and APCA from 2003-2005 in New York City.

23.5 Challenges in source apportionment

There are multiple challenges that arise when estimating sources using source apportionment models [48]. This section outlines some commonly applied approaches to address these challenges.

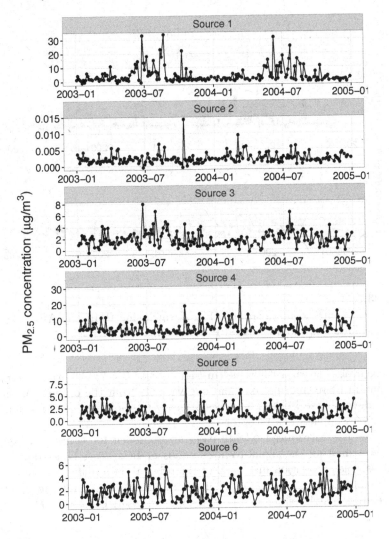

FIGURE 23.4
Estimated PM$_{2.5}$ source concentrations $\hat{\mathbf{G}}$ in $\mu g/m^3$ from Source 1 -Source 6 estimated using PMF from 2003-2004 in New York City.

23.5.1 Number of sources

Most source apportionment methods require the number of sources L to be known, though frequently the number of sources is unknown. In practice, several approaches are used to select L prior to performing source apportionment. Some of these approaches come from the PCA literature and they select L based on the number of PCs that explain most of the variability in the constituent data. A scree plot [21], which plots each eigenvalue against its order, can be used to choose L based on the "elbow", or flattening out, of the points. Figure 23.6 shows the scree plot for the New York City data, which selects $L = 2$. The elbow indicates the point where adding more sources will not explain substantial additional variation in the data, though choosing the elbow can sometimes be subjective. When PCA is applied to the sample correlation matrix, Kaiser's rule or the rule-of-one selects L as the

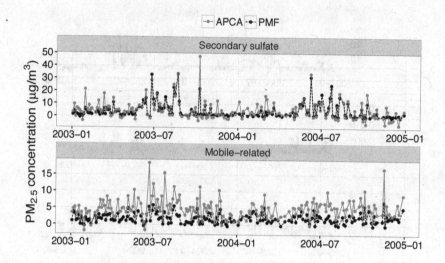

FIGURE 23.5

Estimated $PM_{2.5}$ source concentrations $\hat{G}$ in $\mu g/m^3$ using PMF and APCA corresponding to secondary sulfate and mobile-related $PM_{2.5}$ in New York City.

number of eigenvalues greater than one (in Figure 23.6, $L = 7$). The rule-of-one finds those PCs that contain more information than one variable from the original dataset (e.g. one constituent) alone. These PCA approaches are easy to apply, but may not reflect the true number of sources impacting a receptor.

Henry et al. [16] introduced NUMFACT, an approach for estimating the number of sources L by estimating the signal-to-noise ratios of PCs using bootstrapping. Applying PCA to the sample correlation matrix Σ of the original data, define a_k as the kth sample eigenvector and eigenvalues $\lambda_1, ..., \lambda_P$. Then, let a_{in}^* be the ith sample eigenvector from the sample correlation matrix of the nth bootstrapped dataset. For $i = 1...P - 1$, NUMFACT

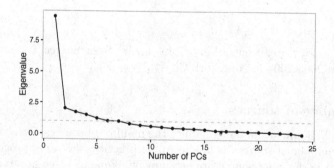

FIGURE 23.6

Scree plot corresponding to PCA applied to the correlation matrix of the New York City $PM_{2.5}$ chemical constituent data.

defines

$$W_i = \frac{\sum_{n=1}^{N} \sum_{k=1}^{i} (\mathbf{a}_{in}^{*\prime} \mathbf{a}_k)^2}{N - \sum_{n=1}^{N} \sum_{k=1}^{i} (\mathbf{a}_{in}^{*\prime} \mathbf{a}_k)^2}$$

$$sn_i = \frac{\frac{\lambda_i \sqrt{W_i}}{1+\sqrt{W_i}}}{\left(\sum_{k=1}^{P-1} \frac{\lambda_k}{1+\sqrt{W_k}}\right)/(P-1)}$$

where $sn_P = 0$. Larger values of the weights W_i indicate less noisy eigenvectors because the W_i's roughly correspond to the ratio of the regression sum of squares to the error sum of squares comparing the bootstrapped eigenvectors to the total sample eigenvectors. Then, sn_i is an estimate of the signal-to-noise ratio for each i. In NUMFACT, L is chosen as the number of sn_i's that exceed some cutoff, often 2. For the New York City dataset, NUMFACT selects $L = 6$ sources. NUMFACT is less intuitive and more computationally intensive compared with other approaches for choosing L, but may select L closer to the true number of sources [16].

Bayesian model comparison, which computes the posterior probabilities of each model given the data, can also be applied to compare the relative performance of likelihood-based models with different numbers of sources L [47].

23.5.2 Incorporating uncertainty

It is important to quantify uncertainties associated with the estimated source contributions $\hat{\mathbf{G}}$ to facilitate propagating uncertainties to subsequent health or air quality analyses. Uncertainties in $\hat{\mathbf{G}}$ may reflect measurement error in the constituent data, uncertainty or error in the source profile matrix, missing sources, or uncertainty in factor rotation. These uncertainties are likely to vary across the identified sources, days, and study areas with differences in meteorology, placement of the receptor, and other attributes.

For methods where $\mathbf{F}$ is assumed known, the estimate of $\mathbf{G}$ is a deterministic solution based on some optimization criteria. Therefore, one can assign uncertainties to $\mathbf{F}$ and propagate the uncertainties to $\hat{\mathbf{G}}$. For example, in CMB, the uncertainty associated with a particular source g_{tl} is given by equation 23.4. One challenge in these approaches is how the uncertainties should be chosen for each f_{lp}. Using CMB, a Monte Carlo approach has been used to assess uncertainties in $\hat{\mathbf{G}}$ under different distributional assumptions for $\mathbf{F}$ [29].

When $\mathbf{F}$ is unknown, uncertainties in $\hat{\mathbf{G}}$ can be obtained by bootstrapping the constituent data and estimating sources for each bootstrapped dataset. Bootstrapped uncertainties are obtained based on the variation in the estimated sources across bootstrapped datasets. Such approaches are useful because they do not require additional assumptions about $\mathbf{G}$ or $\mathbf{F}$ and can be applied to any source apportionment method. To account for temporal correlation in the constituent measurements, a block-bootstrap procedure is typically implemented. Bootstrapped uncertainties have been used with multiple source apportionment methods, such as APCA and PMF [24]. Bootstrapping has also been used to estimate uncertainties in $\hat{\mathbf{F}}$ [43]. One challenge in obtaining bootstrapped uncertainties is that bootstrap samples of the constituent data may not identify the same number of sources or similar source profiles as the original dataset. This limitation requires mapping each source identified from the bootstrap sample to an original source, based on some criteria such as the temporal correlation between the estimated source concentrations.

Bayesian methods naturally provide uncertainty estimates for all the model parameters via posterior distributions. Accounting for uncertainties in model selection, such as the chosen identifiability criteria, is challenging but can be implemented using Bayesian model comparison [47].

23.5.3   Measurement error

Measurement error in source apportionment may be driven by unexplained sources not defined by $\mathbf{F}$ or instrument-related measurement error. Most source apportionment models assume equation 23.1 with additive normal measurement error. Additive measurement error may not be appropriate because for small $\sum_{l=1}^{L} g_{tl} f_{lp}$, adding normal error may make the predicted concentration x_{tp} negative. Another approach is to assume both $\mathbf{X}$ and therefore $\mathbf{e}$ are lognormally distributed,

$$\mathbf{X} = \mathbf{GF} \circ \mathbf{e}$$

$$\log(\mathbf{X}) = \log(\mathbf{GF} \circ \mathbf{e}) = \log(\mathbf{GF}) + \log(\mathbf{e})$$

where $\circ$ is elementwise multiplication [36]. These models can be fit using MCMC techniques.

Accounting for measurement error directly in some source apportionment models can be challenging. Because PMF incorporates uncertainty for each observation u_{tp} into equation 23.5, these uncertainties can be increased for observations x_{tp} known to be uncertain or poor. Paatero and Hopke [39] recommend estimating the signal-to-noise ratio for each constituent and using that to remove or downweight constituents with large amounts of measurement error. In Bayesian source apportionment, similar downweighting to account for measurement error is also possible by increasing uncertainties for individual observations. Christensen and Gunst [4] demonstrate that the CMB method, and other similar approaches, can be viewed within a general measurement error framework.

One common cause of instrument-related measurement error is the method limit of detection (LOD), defined as the point below which it is difficult to determine whether an observed concentration is truly nonzero. Most source apportionment methods cannot directly handle concentrations below the detection limit (BDL) and therefore some approach to handle BDL concentrations must be applied. Approaches proposed to handle BDL values in source apportionment include using the reported BDL concentrations directly, replacing BDL concentrations with a constant proportion of the LOD (e.g. 1/2 LOD), discarding or downweighting constituents with large proportions of BDL concentrations [39], and imputing BDL concentrations [26].

23.5.4   Temporal variation

Most commonly applied source apportionment methods assume that source concentrations g_{tl} are independent over time. If the observations are collected with sufficient temporal gaps, this assumption may be appropriate. However, if temporal dependence exists, driven by meteorology or other time-varying factors, this correlation should be incorporated into the estimation of sources. One way to address temporal dependence is to allow temporally-varying covariates, such as meteorology, to impact estimated source concentrations in a hierarchical model. For example

$$\gamma(E(g_{tl})) = \alpha_0 + \sum_{j=1}^{J} \alpha_j z_{tj} \tag{23.7}$$

where z_{tj} represents time-varying covariates $1..J$ and $\gamma(\cdot)$ is some link function. This model could also be extended to include random effects that account for residual temporal dependence [37]. Christensen and Sain [5] propose accounting for temporal dependence in source apportionment models using a nested block bootstrap approach.

Another approach explicitly introduces temporal dependence in $\mathbf{G}$ from equation 23.1 using an autoregressive process within a dynamic linear model [45] such that

$$\mathbf{g}_t = \boldsymbol{\xi} + (\mathbf{g}_{t-1} - \boldsymbol{\xi})\boldsymbol{\Theta} + \boldsymbol{\omega}_t \tag{23.8}$$

where $\boldsymbol{\xi}$ is the mean source concentrations, $\boldsymbol{\Theta}$ is a diagonal matrix of coefficients $1..L$ that control the autoregressive process, and $\boldsymbol{\omega}_t$ represents measurement error [44]. Then, the source concentrations on day t depend explicitly on the concentrations observed the previous day. This formulation assumes that all temporal correlation comes through the sources, but one could also specify additional temporal correlation due to meteorology or other factors using an autoregressive process on the random error,

$$\mathbf{x}_t = \sum_{l=1}^{L} g_{tl}\mathbf{f}'_l + \boldsymbol{\eta}_t + \boldsymbol{\delta}_t$$

$$\boldsymbol{\eta}_t = \boldsymbol{\eta}_{t-1}\boldsymbol{\Phi} + \boldsymbol{\nu}_t$$

where $\boldsymbol{\delta}_t$, which is independent and normally distributed, represents uncorrelated measurement error and $\mathbf{g}_t$ is represented as in equation 23.8. The component $\boldsymbol{\eta}_t$ then allows for other temporal correlation in the observed constituent data not associated with sources. Models such as these can be fit in a Bayesian framework.

The standard source apportionment model in equation 23.1 assumes the source compositions $\mathbf{F}$ are constant over time. Heaton et al. [12] developed a Dirichlet process model, which uses time-dependent Dirichlet distributions on each element of $\mathbf{F}$. Their approach allows source compositions to vary over time to account for seasonal or long-term trends in source mixture compositions.

23.5.5   Evaluation of source apportionment results

A major challenge in source apportionment modeling is how to validate the source estimates. Direct observations of source concentrations are rarely available and therefore there is no "gold standard" to use when evaluating the source-apportioned estimates of $\mathbf{G}$. Time series of estimated source concentrations can be compared with known temporal trends in the source (e.g. higher secondary sulfate $PM_{2.5}$ in the summer) or with observed temporal trends of pollutants known to be associated with the source (e.g. levoglucosan in biomass burning $PM_{2.5}$). Summary statistics, such as the mean source concentration, can also be used to evaluate the reasonableness of the source concentrations estimated.

For source profiles $\mathbf{F}$, the estimates can sometimes be compared to observed source compositions obtained directly from the source (e.g. $PM_{2.5}$ chemical composition obtained directly from gasoline tailpipe emissions). The US EPA has compiled source profiles for PM and volatile organic compounds (VOCs) in the SPECIATE database (`https://www.epa.gov/air-emissions-modeling/speciate-version-45-through-40`), which could be used as a benchmark for estimated source profiles. However, when source profiles estimated from receptor-oriented data differ from source profiles observed at a source, it is unclear whether these differences are driven by problems with the statistical procedure (e.g. non-convergence in BSA), improper specification of identifiability conditions, or whether the mass balance principle does not hold. In practice, estimates of the source profiles $\mathbf{F}$ are often compared with previously reported source profiles in the literature.

23.5.6   Multiple site data

Generally, source apportionment has been applied to data from individual receptors. When data from multiple receptors are available, source apportionment models may be extended to incorporate spatial information in the estimation of $\mathbf{G}$ and $\mathbf{F}$. However, extending source apportionment methods to multiple receptors is challenging because the number and composition of sources defined by $\mathbf{F}$ may vary between receptors. Therefore, limited research has been conducted in the development of source apportionment methods that use observed data to define how **G** and **F** vary over space.

The simplest approach to extend source apportionment models to multiple receptors is to fix **F** to be identical over space. While this approach facilitates the estimation of **G** at multiple receptors, a major limitation is that it does not allow the composition of sources to vary spatially. Source compositions at a particular receptor will be highly influenced by factors such as distance from the source, landscape, and meteorology. In addition, for point sources such as power plants, the "power plant" mixture composition may vary between different power plants.

When it is reasonable to assume that source compositions **F** do not vary between receptors, several multi-receptor source apportionment methods have been proposed. Thurston et al. [55] extended APCA to multiple receptors over space by concatenating constituent data **X** across locations and applying APCA with a mixed model to account for within-receptor correlation. In a factor analysis framework, Jun and Park [22] also assume **F** does not vary over space, but use a more sophisticated model that assumes **G** follows a multivariate spatial Gaussian process to account for spatial correlations in source concentrations. The most recent version of PMF, PMF 5.0, can estimate sources using data from multiple sites assuming the source compositions do not vary between sites.

Other work has applied source apportionment methods to single pollutants observed over space to estimate spatial locations of major sources [46, 49]. In the simplest approach, measurements at different locations can be treated as the observed variables (i.e. constituents in equation 23.1) and equation 23.1 can be fitted for T days and P spatial locations using existing source apportionment methods. Hence, the **G** matrix represents concentrations of each source (defined by some unknown location) over time, and the **F** matrix reflects which sources of the single pollutant are associated with each unknown spatial location. Under the same single-pollutant framework, Lopes et al. [32] describe a spatio-temporal dynamic factor model to improve predictions, however there are not clear interpretations of the latent factors.

23.6   Estimating source-specific health effects

The estimated source concentrations $\hat{\mathbf{G}}$ can help to determine which sources are the greatest contributors to a pollution mixture, such as those sources of PM that contribute most to total ambient PM. Source concentrations can also be related to health outcomes to estimate source-specific health effects. In PM pollution, determining which sources are most associated with health can help the development of targeted PM regulations to better protect public health.

To estimate short-term health effects associated with source-specific mixture exposure, time series models can be applied relating a temporally-varying health outcome, such as daily mortality or hospitalizations, to temporally-varying concentrations corresponding to each source, $\mathbf{g}_l$. In most cases, the source concentrations $\mathbf{g}_l$ are assumed to be known. Then letting Y_t be a health outcome at time t and $\mu_t = E(Y_t)$,

$$\gamma(\mu_t) = \alpha_0 + \sum_{l=1}^{L} \beta_l g_{tl} + \sum_{j=1}^{J} \alpha_j z_{tj} \tag{23.9}$$

The function $\gamma(\cdot)$ is often chosen as a log link to represent health outcomes that are approximately Poisson distributed, such as daily mortality. Of the remaining terms, β_l represents the source-specific health effect corresponding to source l and z_{tj} represents known time-varying confounders such as temperature. Other health effect regression models, such as spatial models, could also be introduced to estimate long-term health effects of sources.

Generally, source-specific health effects are estimated in a two-stage approach. In the first stage, sources are estimated using any source apportionment method. In the second stage, the estimated source concentrations $\hat{\mathbf{G}}$ are treated as fixed predictors in a health effects regression model similar to equation 23.9. In a two-stage approach, previous work has shown estimated short-term health effects of $PM_{2.5}$ sources do not vary substantially between different source apportionment methods used to estimate $\mathbf{G}$ [20, 33].

One limitation of the two-stage approach is that it does not incorporate uncertainty from estimating $\mathbf{G}$ into the estimated health effects. Alternatively, a Bayesian ensemble source apportionment approach has been proposed, which estimates multiple datasets of source concentrations $\mathbf{G}$ [2]. Each dataset can be used to estimate health effects, and the variation in estimated health effects then represents uncertainty from estimating $\mathbf{G}$ [8].

Another approach to incorporate the uncertainty from estimating $\mathbf{G}$ into the estimated health effects is to use a joint model. Joint models estimate both $\mathbf{G}$ and source-specific health effects by fitting equations 23.1 and 23.9 simultaneously. Fully Bayesian joint models for estimating source-specific health effects have been proposed [35]. A limitation of the joint modeling approach is that it allows source estimates to vary depending on the health outcome considered, since the posterior distribution of $\mathbf{G}$ depends on the health data.

23.7 Conclusions

Other methods besides source apportionment exist for estimating sources, particularly for estimating sources of air pollution. Back trajectory methods [18] estimate backwards in time from the monitor to the source. Instead of working from the monitor to the source, air quality models incorporate source emissions to estimate ambient pollution. Dispersion models can estimate spatially-resolved ambient pollution concentrations using source emissions and meteorology. Chemical transport models such as the Community Multiscale Air Quality (CMAQ) model can also be used along with appropriate constituent indicators, or tracers, to estimate sources [25].

In most applications of source apportionment, the data include constituent concentrations observed at one monitor. Using these data alone in a source apportionment model, it is impossible to determine whether the pollution is generated locally (e.g. within a community) or is transported to a community from another area. Determining the geographic area that generates observed pollution is critical to inform regulation of specific point sources, such as factories. Some source apportionment models can incorporate additional data to help determine locations of sources, such as PMF incorporating wind data [23] or methods using multiple site data for one pollutant [46, 49]. Numerical air quality models including dispersion models may be used to distinguish and estimate the different contributions of local and non-local sources.

Source apportionment methods generally aim to estimate $\mathbf{G}$ and $\mathbf{F}$ such that $\mathbf{X} \approx \mathbf{GF}$. Standard source apportionment methods differ in how $\mathbf{G}$ and $\mathbf{F}$ are estimated and how much information is assumed known about these two matrices. For example, while CMB and similar approaches assume that $\mathbf{F}$ is known, methods such as PMF do not require any prior information about $\mathbf{G}$ or $\mathbf{F}$. Bayesian source apportionment methods are powerful because they can explicitly meet the constraints of source apportionment, provide uncertainties for all model parameters, and incorporate auxiliary information. However because these approaches require substantial tuning and software is not readily available to fit these models, they are not as commonly applied as other methods such as CMB and PMF.

Bibliography

[1] S. Balachandran, J. E. Pachon, Y. Hu, D. Lee, J. A. Mulholland, and A. G. Russell. Ensemble-trained source apportionment of fine particulate matter and method uncertainty analysis. *Atmospheric Environment*, 61:387–394, 2012.

[2] S. Balachandran, H. H. Chang, J. E. Pachon, H. A. Holmes, J. A. Mulholland, and A. G. Russell. Bayesian-based ensemble source apportionment of PM2. 5. *Environmental Science & Technology*, 47(23):13511–13518, 2013.

[3] K. A. Bollen. *Structural Equations with Latent Variables*. John Wiley & Sons, 1989.

[4] W. F. Christensen and R. F. Gunst. Measurement error models in chemical mass balance analysis of air quality data. *Atmospheric Environment*, 38(5):733–744, 2004.

[5] W. F. Christensen and S. R. Sain. Accounting for dependence in a flexible multivariate receptor model. *Technometrics*, 44(4):328–337, 2002.

[6] W. F. Christensen, J. J. Schauer, and J. W. Lingwall. Iterated confirmatory factor analysis for pollution source apportionment. *Environmetrics*, 17(6):663–681, 2006.

[7] C. T. Coulter. EPA-CMB 8.2 users manual. *US Environmental Protection Agency*, 2004.

[8] K. Gass, S. Balachandran, H. H. Chang, A. G. Russell, and M. J. Strickland. Ensemble-based source apportionment of fine particulate matter and emergency department visits for pediatric asthma. *American Jjournal of Epidemiology*, 181(7):504–512, 2015.

[9] A. J. Hackstadt and R. D. Peng. A Bayesian multivariate receptor model for estimating source contributions to particulate matter. pollution using national databases. *Environmetrics*, 25(7):513–527, 2014.

[10] C. W. Harris and H. F. Kaiser. Oblique factor analytic solutions by orthogonal transformations. *Psychometrika*, 29(4):347–362, 1964.

[45] J. Harrison and M. West. *Bayesian Forecasting & Dynamic Models*. Springer, 1999.

[12] M. J. Heaton, C. S. Reese, and W. F. Christensen. Incorporating time-dependent source profiles using the dirichlet distribution in multivariate receptor models. *Technometrics*, 52(1):67–79, 2012.

[13] R. C. Henry. *Multivariate Receptor Models*. Elsevier Science Publishers, Amsterdam, 1991.

[14] R. C. Henry. History and fundamentals of multivariate air quality receptor models. *Chemometrics and Intelligent Laboratory Systems*, 37(1):37–42, 1997.

[15] R. C. Henry. Multivariate receptor modeling by N-dimensional edge detection. *Chemometrics and Intelligent Laboratory Systems*, 65(2):179–189, 2003.

[16] R. C. Henry, E. S. Park, and C. H. Spiegelman. Comparing a new algorithm with the classic methods for estimating the number of factors. *Chemometrics and Intelligent Laboratory Systems*, 48(1):91–97, 1999.

[17] P. K. Hopke. Target transformation factor analysis. *Chemometrics and Intelligent Laboratory Systems*, 6(1):7–19, 1989.

[18] P. K. Hopke. Review of receptor modeling methods for source apportionment. *Journal of the Air & Waste Management Association*, 66(3):237–259, 2016.

[19] K. Ito, N. Xue, and G. Thurston. Spatial variation of $PM_{2.5}$ chemical species and source-apportioned mass concentrations in New York City. *Atmospheric Environment*, 38(31):5269–5282, 2004.

[20] K. Ito, W. F. Christensen, D. J. Eatough, R. C. Henry, E. Kim, F. Laden, R. Lall, T. V. Larson, L. Neas, P. K. Hopke, and G. D. Thurston. PM source apportionment and health effects: 2. An investigation of intermethod variability in associations between source-apportioned fine particle mass and daily mortality in Washington, DC. *Journal of Exposure Science & Environmental Epidemiology*, 16(4):300–310, 2006.

[21] I. T. Jolliffe. *Principal Component Analysis*. New York, NY: Springer New York, 2002.

[22] M. Jun and E. S. Park. Multivariate receptor models for spatially correlated multipollutant data. *Technometrics*, 55(3):309–320, 2013.

[23] E. Kim, P. K. Hopke, P. Paatero, and E. S. Edgerton. Incorporation of parametric factors into multilinear receptor model studies of Atlanta aerosol. *Atmospheric Environment*, 37(36):5009–5021, 2003.

[24] M.-A. Kioumourtzoglou, B. A. Coull, F. Dominici, P. Koutrakis, J. Schwartz, and H. Suh. The impact of source contribution uncertainty on the effects of source-specific PM2.5 on hospital admissions: A case study in Boston, MA. *Journal of Exposure Science & Environmental Epidemiology*, 24(4):365–371, 2014.

[25] B. Koo, G. M. Wilson, R. E. Morris, A. M. Dunker, and G. Yarwood. Comparison of source apportionment and sensitivity analysis in a particulate matter air quality model. *Environmental Science & Technology*, 43(17):6669–6675, 2009.

[26] J. R. Krall, C. H. Simpson, and R. D. Peng. A model-based approach for imputing censored data in source apportionment studies. *Environmental and Ecological Statistics*, 22(4):779–800, 2015.

[27] D. Lee, S. Balachandran, J. Pachon, R. Shankaran, S. Lee, J. A. Mulholland, and A. G. Russell. Ensemble-trained PM2.5 source apportionment approach for health studies. *Environmental Science & Technology*, 43(18):7023–7031, 2009.

[28] D. D. Lee and H. S. Seung. Algorithms for non-negative matrix factorization. In *Advances in Neural Information Processing Systems*, pages 556–562, 2001.

[29] S. Lee and A. G. Russell. Estimating uncertainties and uncertainty contributors of CMB PM 2.5 source apportionment results. *Atmospheric Environment*, 41(40):9616–9624, 2007.

[30] J. Liang and D. Fairley. Validation of an efficient non-negative matrix factorization method and its preliminary application in Centrap California. *Atmospheric Environment*, 40(11):1991–2001, 2006.

[31] J. W. Lingwall, W. F. Christensen, and C. S. Reese. Dirichlet based Bayesian multivariate receptor modeling. *Environmetrics*, 19(6):618–629, 2008.

[32] H. F. Lopes, E. Salazar, D. Gamerman, et al. Spatial dynamic factor analysis. *Bayesian Analysis*, 3(4):759–792, 2008.

[33] T. F. Mar, K. Ito, J. Q. Koenig, T. V. Larson, D. J. Eatough, R. C. Henry, E. Kim, F. Laden, R. Lall, L. Neas, M. Stölzel, P. Paatero, P. K. Hopke, and G. D. Thurston. PM source apportionment and health effects. 3. Investigation of inter-method variations in associations between estimated source contributions of $PM_{2.5}$ and daily mortality in Phoenix, AZ. *Journal of Exposure Science & Environmental Epidemiology*, 16(4): 311–320, 2006.

[34] A. Marmur, A. Unal, J. A. Mulholland, and A. G. Russell. Optimization-based source apportionment of $PM_{2.5}$ incorporating gas-to-particle ratios. *Environmental Science & Technology*, 39(9):3245–3254, 2005.

[35] M. C. Nikolov, B. A. Coull, P. J. Catalano, and J. J. Godleski. An informative Bayesian structural equation model to assess source-specific health effects of air pollution. *Biostatistics*, 8(3):609–624, 2007.

[36] M. C. Nikolov, B. A. Coull, P. J. Catalano, E. Diaz, and J. J. Godleski. Statistical methods to evaluate health effects associated with major sources of air pollution: a case-study of breathing patterns during exposure to concentrated Boston air particles. *Journal of the Royal Statistical Society: Series C (Applied Statistics)*, 57(3):357–378, 2008.

[37] M. C. Nikolov, B. A. Coull, P. J. Catalano, and J. J. Godleski. Multiplicative factor analysis with a latent mixed model structure for air pollution exposure assessment. *Environmetrics*, 22(2):165–178, 2011.

[38] P. Paatero. The multilinear engine: A table-driven, least squares program for solving multilinear problems, including the n-way parallel factor analysis model. *Journal of Computational and Graphical Statistics*, 8(4):854–888, 1999.

[39] P. Paatero and P. K. Hopke. Discarding or downweighting high-noise variables in factor analytic models. *Analytica Chimica Acta*, 490(1–2):277–289, 2003.

[40] P. Paatero and P. K. Hopke. Rotational tools for factor analytic models. *Journal of Chemometrics*, 23(2):91–100, 2009.

[41] P. Paatero and U. Tapper. Positive Matrix Factorization: A non-negative factor model with optimal utilization of error estimates of data values. *Environmetrics*, 5(2):111–126, 1994.

[42] P. Paatero, P. K. Hopke, B. A. Begum, and S. K. Biswas. A graphical diagnostic method for assessing the rotation in factor analytical models of atmospheric pollution. *Atmospheric Environment*, 39(1):193–201, 2005.

[43] P. Paatero, S. Eberly, S. Brown, and G. Norris. Methods for estimating uncertainty in factor analytic solutions. *Atmospheric Measurement Techniques*, 7(3):781–797, 2014.

[44] E. S. Park, P. Guttorp, and R. C. Henry. Multivariate receptor modeling for temporally correlated data by using MCMC. *Journal of the American Statistical Association*, 96 (456):1171–1183, 2001.

[45] E. S. Park, C. H. Spiegelman, and R. C. Henry. Bilinear estimation of pollution source profiles and amounts by using multivariate receptor models. *Environmetrics*, 13(7): 775–798, 2002.

[46] E. S. Park, P. Guttorp, and H. Kim. Locating major PM10 source areas in Ssoul using multivariate receptor modeling. *Environmental and Ecological Statistics*, 11(1):9–19, 2004.

[47] E. S. Park, P. K. Hopke, M.-S. Oh, E. Symanski, D. Han, and C. H. Spiegelman. Assessment of source-specific health effects associated with an unknown number of major sources of multiple air pollutants: a unified Bayesian approach. *Biostatistics*, 15 (3):484–497, 2014.

[48] A. Pollice. Recent statistical issues in multivariate receptor models. *Environmetrics*, 22(1):35–41, 2011.

[49] A. Pollice and G. J. Lasinio. Major PM10 source location by a spatial multivariate receptor model. *Environmental and Ecological Statistics*, 19(1):57–72, 2012.

[50] P. Rachdawong and E. R. Christensen. Determination of PCB sources by a principal component method with nonnegative constraints. *Environmental Science & Technology*, 31(9):2686–2691, 1997.

[51] G. A. Seber and A. J. Lee. *Linear regression analysis*, volume 2. John Wiley & Sons, 2003.

[52] V. Simeonov, J. Stratis, C. Samara, G. Zachariadis, D. Voutsa, A. Anthemidis, M. Sofoniou, and T. Kouimtzis. Assessment of the surface water quality in Northern Greece. *Water research*, 37(17):4119–4124, 2003.

[53] R. Tauler, M. Viana, X. Querol, A. Alastuey, R. Flight, P. Wentzell, and P. Hopke. Comparison of the results obtained by four receptor modelling methods in aerosol source apportionment studies. *Atmospheric Environment*, 43(26):3989–3997, 2009.

[54] G. D. Thurston and J. D. Spengler. A quantitative assessment of source contributions to inhalable particulate matter pollution in metropolitan Boston. *Atmospheric Environment*, 19(1):9–25, 1985.

[55] G. D. Thurston, K. Ito, and R. Lall. A source apportionment of U.S. fine particulate matter air pollution. *Atmospheric Environment*, 45(24):3924–3936, 2011.

[56] D. Wang and P. K. Hopke. The use of constrained least-squares to solve the chemical mass balance problem. *Atmospheric Environment*, 23(10):2143–2150, 1989.

[57] J. G. Watson, J. A. Cooper, and J. J. Huntzicker. The effective variance weighting for least squares calculations applied to the mass balance receptor model. *Atmospheric Environment*, 18(7):1347–1355, 1984.

24

Statistical Methods for Environmental Epidemiology

Francesca Dominici

Harvard University, Cambridge, Massachusetts

Ander Wilson

Colorado State University, Fort Collins, Colorado

CONTENTS

24.1 Introduction

Estimated health risks from short and long term exposure to air pollution from epidemiological studies have provided the necessary evidence base for setting more stringent National Ambient Air Quality Standards (NAAQS) in the United States and around the world [161, 162, 173, 174]. These estimated health risks have carried an enormous weight in setting more stringent NAAQS. The annual cost of implementation and compliance with these NAAQS is now reaching astronomical numbers: billion of dollars in the US alone. Given the need to justify these costs, statistical analyses aimed at addressing questions in environmental epidemiology have been subjected to immense scrutiny both from industry and regulatory agencies. Such statistical analyses need to overcome many data challenges. First, a critical and timely question is whether current levels of air pollution, which in the United States are low and close to background levels, are harmful. This calls for statistical methods that estimate the exposure-response relationship and can reliably estimate health risks at low levels of exposure. Second, health outcomes, air pollution levels and potential confounders vary in time and space, and because of the observational nature of the data, there are many potential measured and unmeasured confounders. Third, exposure is often measured with error. Forth, often the signal-to-noise ratio in the data is low and the target of inference is often the estimation of a health risk that is small and hard to detect. Fifth, often we are interested in estimating health effects associated with simultaneous exposure to multiple pollutants, thus raising the need for a rigorous definition of health risk associated with simultaneous exposure to multiple pollutants, and also the need for a new framework for adjusting for confounding in the context of multiple exposures. Sixth, air pollution data are incomplete and their missing data pattern might not be at random. For example, often ambient levels of ozone are not measured in the winter season. These and many other challenges have led to the development of sophisticated statistical methods with the goal of providing estimates of health effects that are unbiased and precise.

In this chapter, we first introduce the data setting and provide a high level overview of the concept of confounding in air pollution epidemiology. In Section 24.2, we describe the three most common study designs for estimating health effects of short term exposures, long term exposures, and air quality interventions: time series, cohort, and the intervention studies. In Section 24.3, we provide a general overview of methods for estimating the exposure-response relationships in the context of time series and cohort studies. In Section 24.4, we summarize some recent methodological developments for adjustment for confounding that identify key confounders and adjust for model uncertainty using model averaging. Finally, in Section 24.5 we summarize recent developments for estimating the health effects of exposure to multiple pollutants. Please note that, this chapter does not attempt to review the whole body of the literature on statistical contributions in air pollution epidemiology, which is massive. Our goal is to provide a useful summary of the challenges and opportunities in this field.

24.1.1   Data Characteristics

Data in air pollution tend to have some unique features. For example, outdoor levels of air pollution are measured by monitoring stations maintained by government agencies [160]. The levels of these ambient air pollutants are continuous (e.g. particulate matter and ozone) and vary across time and space and for the most part are available at daily levels. Although there are several approaches for measuring exposure (e.g. exposure assessment at individual level using personal monitors), here we will present ideas in the context of exposure to outdoor levels from monitoring stations maintained by the US Environmental Protection Agency (EPA) or other government agencies. In this setting, the most common approach is to assign to each individual or to each geographical area the ambient level from the closest monitoring station or group of stations. More specifically, in the context of time series studies, which will be described more in detail below, the daily average level of air pollution in a geographical area (such as a county or a city) is calculated by taking an average measure of the daily levels from all the monitoring stations that are located in that geographical location [89, 139, 140, 178]. In the context of cohort studies, long term exposure to air pollution at the residential addresses of the members in the cohort need to be calculated. In the last few years there have been numerous contributions in the context of developing spatio-temporal models that use monitoring data, land use regression variables, atmospheric models, and satellite data to estimate these residential exposures at the highest level of accuracy as possible [54, 98, 142, 168, 180, 181]. The development and application of these models for exposure prediction is changing the landscape of air pollution epidemiology. With these approaches, it is now becoming possible to estimate exposure to air pollution at the residential address of each participant in the cohort with a high level of accuracy in terms of spatial resolution (at the address) and temporal resolution (daily). These predictions are cross-validated with ambient levels from monitoring stations and have a high level of prediction accuracy (see Section 24.2.5 for additional details).

Data on health outcomes are often obtained from cohorts [14, 42, 47, 55, 68, 75, 90, 117, 126, 129, 170, 172] and administrative data, such as Medicare claims data [29, 61, 147, 169, 186, 189]. For example, the Harvard Six Cities Study (SCS) and the American Cancer Society (ACS) study are the two landmark epidemiological cohort studies that had an enormous impact on our understanding of the health effects of air pollution [55, 126]. These studies enroll subjects at the baseline, and at baseline collected an extensive list of individual level characteristics (e.g. body mass index, smoking, occupation, income). In these cohort studies, individuals are followed over time to ascertain time to an event (e.g. a heart attack, hospitalization, or death). In the SCS and ACS, average exposure aggregated at the county level or at the larger metropolitan area level was assigned to each individual. This is a fairly coarse description of a subject's exposure to air pollution that lends itself to exposure measurement error. As mentioned above, other more recent cohort studies estimate individual level exposure at the subject's residential address using prediction models that can reduce measurement error [42, 64, 146].

Traditional cohort studies, such as the SCS and the ACS have the advantage of capturing extensive individual level information on the potential confounders but they are limited by the fact that they are "closed" cohort studies in the sense that they do not allow enrollment of new individuals into the cohort. As such, these cohorts cannot be used to estimate the health effects of air pollution using more recent data, nor can be used to track health effects in the future as air pollution levels continue to decline. To overcome this challenge, more recent epidemiological studies have leveraged "open" cohort data, such as Medicare claims, to estimate the effects of long-term exposure to air pollution [72, 94, 153, 189]. Open cohort data permits new enrollees to enter into the cohort every year, thus allowing us to routinely estimate health effects over time as air pollution levels continue to decline. A limitation of the administrative cohorts is that they tend to measure a limited set of potential measured confounders compared to the traditional cohorts. On the other hand, they include large and often nationally representative study populations.

24.1.2 Sources of Confounding Bias

A key concern in environmental epidemiology studies is confounding. The source and degree of confounding bias and the type of statistical methods needed to adjust for confounding depend on the study design. In the context of randomized experiments, because assignment to treatment (that is exposure to high versus low levels of air pollution) is at random, characteristics of the individuals that are "exposed" and "not exposed" to air pollution are automatically balanced between the two groups because of the randomization. In the context of time series studies, the goal is to estimate the association between day-to-day changes in air pollution levels and day-to-day changes in the daily number of deaths (or hospitalizations) within a city and then combine information regarding these estimated associations across cities (or other geographical locations, such as counties or metropolitan areas). Because of the nature of a time series design, city-specific characteristics that do not vary from day to day (e.g. smoking prevalence or average income in that city) are not confounders [122]. In cohort studies, the goal is to estimate the association between long term exposure to air pollution and time to an event for individuals that live in different geographical locations with different levels of air pollution. Therefore individual level and/or area level characteristics that are associated with the exposure and also predictive of the outcome (e.g. smoking, body mass index, occupation, income) can be confounders [111, 115]. In the context of intervention studies we are interested in estimating the effect of a specific intervention on both: 1) air pollution levels and 2) health outcomes. For example, let's consider the intervention designed to reduce traffic congestion during the 1996 Summer Olympic Games in Atlanta, Georgia, and the effect of this intervention on ambient ozone levels [119]. In this context, a potential confounder of the effect of the intervention on ambient ozone levels is time trend. It is possible the ambient air pollution levels could have been declining over time regardless of the intervention. A key questions is, therefore, whether the observed degree of reduction in ambient levels of ozone can be "truly" attributed to the intervention or would this decline have occurred anyway even absent the intervention [10, 58, 119, 192].

In addition, when the goal is estimation of the health effects of a single and pre-specified environmental agent (e.g. ozone), environmental epidemiology studies are particularly prone to confounding bias because of the many co-exposure and covariates that are correlated with the exposure and also influence health. For example, temperature is highly correlated with ozone because ozone formation occurs on warm sunny days. Temperature also affects health and therefore can confound the estimation of health risks associated with short term exposure to ozone [15, 17, 20, 22, 24, 84]. In the context of chronic effect studies, where most of the information comes from variation across geographical locations in long-term exposure to air pollution, an important potential confounder that must be consider is average income. For example, geographical locations with lower average income also have higher levels of air pollution and average income affects health outcomes.

An increasing number of papers in the air pollution epidemiological literature use the term "causal effects" in the abstract and in the summary [74, 147, 169, 193]. In the context where observational data is analyzed to assess the "effect" of an action (such as an intervention, an exposure, or a treatment) on a given outcome, the scientific community is increasingly relying on statistical methods being labeled as "causal," and such methods are frequently contrasted against more traditional methods labeled as "associational" and therefore non-causal. In the specific context of air pollution epidemiology, the promise of identifying causal relationships between, for example, air pollution exposure and human health, has rightly generated ample enthusiasm.

We have argued that causal inference methods can provide a rigorous framework to use data to learn about consequences of specific actions [194]. More specifically, causal inference methods are tools for: 1) formalizing thinking about what data can tell about cause and effect; 2) forcing explicit definitions of familiar notions; 3) clarifying common threats to validity in epidemiological studies, while at the same time providing a remedy. In this chapter, we will not formalize the presentation of the statistical methods using the the formal notation of potential outcomes in causal inference, but we will provide references of recent contribution on this area.

24.2   Epidemiological Designs

In this section, we briefly review three popular designs in environmental epidemiology: multi-site time series studies, cohort studies, and intervention studies. Time series studies have been a popular choice for air pollution epidemiology to estimate acute health effects associated with short term exposure. Cohort studies are a popular choice to study the chronic health effects associated with long term exposure. Intervention studies are very useful to assess the consequences of a specific air quality intervention on both the ambient levels of air pollution and health outcomes. It is important to recognize that several other epidemiological study designs can be used to estimate the health effects of environmental exposures in addition to the those discussed here, other popular choices include case-crossover and panel studies (see for example [21, 51, 87, 96]). We conclude this section by discussing spatial misalignment and exposure prediction modeling as they relate to environmental epidemiology studies.

24.2.1 Multi-Site Time Series Studies

Multi-site time series studies of air pollution and health outcomes (i.e. mortality and morbidity) have provided evidence that daily variation in air pollution levels is associated with daily variation in mortality and morbidity counts [3, 18, 67, 128]. These findings have served as key epidemiological evidence for several reviews of the NAAQS for particulate matter and ozone (for example [20, 61, 138, 144, 149, 187]). Considering the large policy impact of epidemiological studies of short term effects of air pollution, critics have raised concerns about the adequacy of current model formulations. In this section, we briefly review two important methodological contributions for the analyses of multi-site time series studies: 1) model choice to adjust for unmeasured confounding and 2) accounting for the delayed effect of a few days by the development and application of distributed lag models.

Time series studies estimate the association between day-to-day changes in air pollution levels and day-to-day changes in the daily number of events (deaths or hospitalizations) that have occurred within a given geographical location, such as a city or a county [23, 63].

Generalized linear models (GLM) with parametric splines (e.g., natural cubic splines) [113] or generalized additive models (GAM) with nonparametric smoothers (e.g., smoothing splines or loess smoothers) [79], are generally used to estimate effects associated with exposure to air pollution while accounting for smooth fluctuations in mortality that confound estimates of the pollution effect [60].

Time series data on pollution and mortality are generally analyzed by using log-linear, Poisson regression models for over-dispersed counts with the daily number of deaths (or hospitalizations) as outcome, the (possibly lagged) daily level of pollution as a linear predictor and smooth functions of weather variables and calendar time used to adjust for time-varying confounders. Investigators around the world have used different approaches to adjust for confounding, making it difficult to compare results across studies (for example, [89, 141]).

To date, the statistical properties of these different approaches have not been comprehensively compared. Peng et al conducted an extensive simulation study aimed at characterizing model uncertainty and model choice in adjusting for seasonal and long-term trends in time series models of air pollution and mortality (unmeasured confounders) [122]. The authors generated data under several confounding scenarios and systematically compare the performance of the various methods with respect to the mean squared error of the estimated air pollution coefficient. They found that the bias in the estimates generally decreases with more aggressive smoothing and that model selection methods which optimize prediction, which have been the standard approach, may not be suitable for obtaining an estimate with small bias. In another paper, Dominici et al developed a bandwidth selection method to reduce confounding bias in the health risk estimate due to unmeasured time-varying factors, such as season and influenza epidemics [59]. More specifically, in this paper the authors calculate in closed form the asymptotic bias and variance of the air pollution risk estimate as they vary the degree of adjustment for unmeasured confounding factors. The authors then show that confounding bias can be removed by including in the Poisson regression model smooth functions of time and temperature that are sufficiently flexible to predict pollution (instead of tuning to predict the outcome). These asymptotic calculations show that in most situations it is possible to effectively reduce unmeasured confounding bias by estimating the number of degrees of freedom in the smooth functions of time and temperature that best predict pollution levels. Controlling for the potential confounding effects of "measured confounders" (such as weather variables) is a better-identified problem than controlling for "unmeasured confounders." More details on how to adjust for measured confounders and at the same time account for model uncertainty in the confounding adjustment are discussed later in this chapter (see Section 24.4).

24.2.1.1 Distributed Lag Models

Time series studies have provided strong evidence of an association between increased levels of ambient air pollution and increased hospitalizations, typically at a single lag of 0, 1 or 2 days after an air pollution episode. An important scientific objective is to estimate the cumulative risk of hospitalization of an air pollution episode over a few days after the air pollution event. More specifically, if in a given day the level of air pollution increases from x_t to $x_t + \delta$, our goal is to estimate the cumulative risk associated with a potential increase in the adverse health outcome on the same day t but also on the following L days (e.g. $y_t, y_{t+1}, \ldots, y_{t+L}$).

A distributed lag model (DLM) is a regression model that includes lagged exposure variables as covariates; its corresponding distributed lag (DL) function describes the relationship between the lag and the coefficient of the lagged exposure variable. DLMs have recently been used in environmental epidemiology for quantifying the cumulative effects of weather and air pollution on mortality and morbidity (for example [66, 80, 81, 123, 171, 184]).

Standard methods for formulating DLMs include unconstrained, polynomial, and penalized spline DLMs [4, 84, 107, 134, 150, 151, 154, 184, 188]. Standard methods for fitting DL functions may fail to take full advantage of prior information about the shape of the DL function for environmental exposures, or for any other exposure with effects that are believed to smoothly approach zero as lag increases, and are therefore at risk of producing suboptimal estimates. In a paper by Welty et al, the authors propose a Bayesian DLM (BDLagM) that incorporates prior knowledge about the shape of the DL function and also allows the degree of smoothness of the DL function to be estimated from the data [171]. In a simulation study, they compare the proposed Bayesian approach with alternative methods that use unconstrained, polynomial, and penalized spline DLMs. They also illustrate the connection between BDLagMs and penalized spline DLMs.

In a subsequent paper, the same set of authors extended the BDLagMs to multi-site time series studies [123]. Specifically these authors introduce a Bayesian hierarchical distributed lag model (BHDLM) for estimating a national average distributed lag function relating particulate matter (PM) air pollution exposure to hospitalizations for cardiovascular and respiratory diseases. The BHDLM builds on earlier work [171, 184] by smoothing DL function estimates across lags and by providing a method for combining these functions across locations. The specific prior allows for more flexibility in the shape of the DL function at the shorter lags but less flexibility in the shape of the DL function at the longer lags. In addition, the hierarchical model lets us examine the range of shapes in the county-specific distributed lag functions. The authors have established that this proposed methodology is related to penalized spline modeling with a special type of penalty. This connection, along with evidence from simulation studies that were conducted by Welty et al [171], creates a basis for understanding the statistical properties of the approach.

24.2.2 Cohort Studies

Air pollution cohort studies associate long-term exposure with health outcomes. Either a prospective or retrospective design is possible. In a prospective design, participants complete a questionnaire at entry into the study to elicit information about age, sex, weight, education, smoking history, and other subject-specific characteristics. They are followed over time for mortality or other health events. A measure of cumulative air pollution is often used as the exposure variable. A key design consideration for air pollution cohort studies is identifying a cohort with sufficient exposure variation. Individuals from multiple geographic locations must be studied in order to assure sufficient variation in cumulative exposure, particularly when ambient air pollution measurements are used. However, by maximizing the geographical variability of exposure, the relative risk estimates from cohort studies are likely to be confounded by area-specific characteristics.

Survival analysis tools can evaluate the association between air pollution and mortality. Typically the Cox proportional-hazards model [40, 44] is used to estimate mortality rate ratios for airborne pollutants while adjusting for potential confounding variables (for example [6]). Relative risk is estimated as the ratio of hazards for an exposed relative to an unexposed or reference group.

The epidemiological evidence on the long-term effects of air pollution on health has been reviewed by Pope [124]. The Harvard SCS and the ACS [55, 126] are among the largest air pollution prospective cohort studies. In the SCS [55, 105, 106] a random sample of 8111 adults who resided in one of the six US communities at the time of the enrollment was followed for 14 to 16 years. An analysis of all-cause mortality revealed an increased risk of death associated with increases in particulate matter and sulfate air pollution after adjusting or individual-level confounders. Because of the small number of locations, findings of this study cannot be generalized easily.

The ACS study [125, 126, 127] evaluated effects of pollution on mortality using data from a large cohort drawn from 151 metropolitan areas. Ambient air pollution from these areas was linked with individual risk factors for 552,138 adult residents. The ACS study covered a larger number of areas, however, the subjects were not randomly sampled as in the SCS. Both studies reported similar results: the relative risk of all-cause mortality was 1.26 (95% CI 1.08,1.47) for an 18.6 $\mu g/m^3$ change in PM less than 2.5 μm in aerodynamic diameter ($PM_{2.5}$) in the SCS and 1.17 (95% CI 1.09, 1.26) for a 24.5 $\mu g/m^3$ change in $PM_{2.5}$ in the ACS study. A detailed reanalysis of these two studies [102, 103] and a new ACS study including data for a longer period of time [127] replicated and extended these results by incorporating a number of new ecological covariates and applying several models for spatial autocorrelation.

More recently, estimates of the long term effects of air pollution on health have been obtained with statistical methods that resemble the ones of multi-site time series studies [72, 86, 94, 157]. In the papers by Janes et al (2007) and Greven et al (2011), the authors aim at addressing a very specific question: whether there is an association between month-to-month variations in mortality rates and month-to-month variations in the average $PM_{2.5}$ for the previous 12 months (global effect) in the Medicare population [72, 86]. In these studies the authors decompose the global effects into two parts: 1) the association between the national average trend (NAT) in the monthly $PM_{2.5}$ levels averaged over the previous 12 months and the NAT in monthly mortality rates (national effect); and 2) the association between the deviation of the community-specific trend from the NAT of $PM_{2.5}$ and the deviation of the community-specific trend from the NAT of mortality rates (local effect). They decompose the global effect into a national effect plus a local effect because they hypothesize that the national effect is more likely to be affected by unmeasured confounding than the local effect. They also argue that if there are large differences between the local and the national effects then the global effect should not be reported without a more in depth investigation of confounding. They acknowledge in the papers that differences between the local and national effects might be due to measurement error and not necessarily unmeasured confounding.

Kioumourtzoglou et al (2015) introduced a similar modeling approach where the authors specify time-varying Cox proportional hazards models separately in each city [94]. More specifically, by analyzing data within each city, they first assess as whether year-to-year fluctuations in $PM_{2.5}$ concentrations, around their long-term trends, are associated with year-to-year survival variations within cities. City-specific analyses eliminated confounding by factors that do not vary across time but vary across cities. They specify a multi-stage model. First, they fitted separate Cox proportional hazards models in each city, stratifying by age (5-year categories), gender, race (white, black, other) and follow-up time, with follow-up beginning on January 1st after entry in the cohort. They used annual and 2-year total $PM_{2.5}$ mass concentrations, separately, as time-varying exposures. They used the counting process extension of the proportional hazards model by Andersen and Gill [5] and created multiple observations for each subject, with each observation representing a person-year of follow-up. In addition, they adjusted linearly for calendar year, hence controlling for long-term trends and focusing the analysis on variations in exposure around its time trend. In the second stage, they combined the city-specific health effect estimates, using a random effects meta-analysis. With this approach, they eliminated all confounding by covariates that vary across cities, since this is a city-specific analysis, and by covariates whose long-term trends coincide with trends in $PM_{2.5}$ within cities, since those trends were removed.

24.2.3   Intervention Studies

The regulatory and policy environment surrounding air quality management warrants new types of epidemiological evidence above and beyond the one provided by estimating the exposure-response function from time series and cohort studies (often called observational studies). Whereas air pollution epidemiology has typically informed policies with estimates of exposure-response relationships between pollution and health outcomes, new types of evidence can inform current debates about the actual health impacts of air quality regulations. Directly evaluating specific regulatory strategies is distinct from and complements estimating exposure-response relationships and puts increased emphasis on assessing the effectiveness of well-defined regulatory interventions [192, 193]. In this section we want to sharpen the analytic distinctions between studies that *directly* evaluated the effectiveness of specific policies and those that estimated exposure-response relationships between pollution and health from observational studies. A recent report by Zigler et al published by the Health Effects Institute [194] provides a comprehensive overview of statistical methods for causal inferences to assess the health impacts of air quality regulations.

Randomized control trials would be the best way to measure the health benefits of PM reductions, but for obvious reasons, true experiments are generally not feasible. One exception is chamber studies of controlled exposure, but they rely on healthy subjects and focus only on limited subclinical endpoints. An observational study of the health effects of particulates boils down to a comparison of health outcomes across space and/or time among places with differing levels of air pollution (see above the sections on multi-site time series studies and cohort studies). For the cohort studies, one challenge is that the people who live in the more polluted places frequently have differing initial levels of health (e.g., due to differences in smoking rates, diet, or socioeconomic status) from the levels of people who live in the less polluted places. Another challenge is that there may be locational determinants of health (e.g., hospital quality or water pollution) that differ across the places and are correlated with air pollution levels. Further, people may sort to locations based on their (likely unobserved) susceptibility to pollution and other related health problems and/or they may spend greater resources on self-protection in polluted locations in ways that are not measured in available datasets. Statistical methods, mostly based on regression approaches, aim to "adjust" for observed confounders, by including the available measures of behavioral, socioeconomic, and locational differences as covariates in the regression model, but many of the determinants of health are unobserved and can lead to biased estimates of the relationship between health and particulates [58].

Thanks to the rigorous statistical methods that have been developed and applied to the assembled data, and to the enormous effort of government agencies and specific investigators in conducting independent re-analyses (see for example Krewski et al [104]) analyses of observational data have had a large impact on air quality regulations and on the supporting analyses of their accompanying benefits. Nonetheless, legitimate concerns remain. While important progress has been made in adjusting for measured and unmeasured confounding in observational studies, it remains true that there may be unobserved differences across the populations and locations, and the measurable differences may not have been adjusted for sufficiently.

One type of intervention study is the quasi-experimental (QE) studies. In a QE evaluation, the researcher compares outcomes between a treatment group and a control group, just as in a classical experiment; but, treatment status is determined by politics, an accident, or most importantly a regulatory action. The key difference with an observational study in this setting is that the QE approach is devoted to identifying treatment-induced variation in particulates that plausibly mitigates confounding or omitted variables bias in the estimated relationship between human health and particulates, rather than relying on the variation presented by nature and optimizing agents. Despite the "nonrandom" assignment of treatment status, it is possible to draw causal inferences from the differences in outcomes (here by outcomes we refer to both air pollution levels and health risks) between the treatment and control groups in a quasi- or natural experiment, provided certain assumptions are met. This approach has been used extensively in recent years and has permitted more credible inferences about the impacts of a wide range of relationships. In fact, there is an emerging QE literature on the human health effects of air pollution that relies on designs where an "action" has affected the ambient levels and the chemical composition of air pollution. Some of the most well-known examples are: 1) the ban of coal sales in Dublin [39]; 2) the differential reduction in total suspended particulates (TSPs) across the country as a consequence of the 1981-1982 recession [35]; 3) the air pollution reduction interventions before, during, and after the Beijing Olympic games [133]; 4) a steel plant strike [131]; 5) features of the Clean Air Act [53, 193] and, 6) the Chinese policy that provided free coal for heating in cities north of the Huai River [36]. See also [1, 69, 114, 159] for detailed reviews.

The Huai River study illustrates some of the appealing features of QE designs and more specifically of regression discontinuity (RD) [108]. It exploits a Chinese policy that provided free coal for winter heating in areas north of the Huai River and denied coal-based heating to the south of the river. The idea is to compare locations just north and south of the river. In this setting, the RD design relies on the assumption that any confounders (both observed and unobserved) vary smoothly with latitude as one crosses the Huai River, except for the availability of coal-based indoor heating. The authors controlled for these potential confounders through adjustment for a flexible polynomial in distance to the river, measured as degrees latitude that each location is north of the Huai River. The authors find that north of the river the policy led to discrete increases in TSPs and discrete decreases in life expectancy (derived from age-specific mortality rates). Importantly, the effect of TSPs on life expectancy is largely insensitive to whether observable covariates are included in the specification, which would be the case in a randomized control trial.

While QE approaches promise more credible estimates, they are not without limitations. It is important that QE designs are able to demonstrate that observable covariates are balanced by the treatment and credibly explain why unobserved ones are likely to be balanced too. In cases where the covariates are imbalanced and/or the unobserved ones are unlikely to be balanced, quasi-experimental estimates are unlikely to be more credible than associational estimates. Further, QE approaches can often be demanding of the data and lack statistical power. As is the case with associational estimates, applying QE estimates to other settings (e.g., places, periods, and demographic groups) requires careful consideration and in some cases may be inappropriate. This challenge can be greater with QE approaches where the selection of the study population is dictated by the available treatment (e.g. the Chinese policy) and therefore is beyond the researcher's controls.

24.2.4   Spatial Misalignment

Estimating the health risks of environmental exposures often involves examining data at different levels of spatial resolution. This mismatch between data measured at different resolutions results in spatial misalignment [1], which can bias the health risk estimates. Spatial misalignment in environmental health studies is very common because the air pollution data and the data on the health outcomes often come from different and unrelated sources. For example, data on ambient air pollution levels often are based on a network of monitors operated by the US EPA where each monitor measures daily levels of ambient air pollution at a specific point location. On the other hand, data on health outcomes, such as the numbers of hospital admissions for cardiovascular disease and number of deaths, generally are obtained from Centers for Medicare and Medicaid Services (CMS) and other governmental agencies, such as the National Center of Health Statistics (NCHS). Because health and exposure data are often collected independently of each other, they are rarely spatially aligned. Hence, a direct comparison of the exposure and health outcome is not possible without a model (or an assumption) to align the two sources of information both spatially and temporally. In a time series study of air pollution and health, one is interested in estimating associations between daily changes in county-wide hospital admissions or mortality counts and daily changes in county-wide average levels of a specific pollutant. The problem is that we do not directly observe county-wide average pollutant levels. Rather, we have measurements taken at a handful of monitors (sometimes only one) located somewhere inside the county.

For a spatially homogeneous pollutant, the value of the pollutant at a single monitor may be representative of the county-wide average ambient level of that pollutant. Some pollutants, particularly some gases such as ozone, are reasonably spatially homogeneous across the area of a county. The total mass of $PM_{2.5}$, whose health risks have been examined extensively, is fairly spatially homogeneous, and monitor measurements of $PM_{2.5}$ in counties with multiple monitors tend to be highly correlated across both time and space [16, 19, 121].

With a pollutant such as $PM_{2.5}$, the misalignment between the continuous nature of the pollutant process and the aggregated nature of the health data does not typically pose as serious a problem as some other pollutants. In this situation, current approaches for data analysis may provide reasonable estimates of risk. There are many sources of measurement error in the analysis of air pollution and health data, and much previous work has focused on the mismatch between personal and ambient exposures to an airborne pollutant and how to adjust for measurement error in health effect models [56, 95, 152, 190].

This is indeed an important problem, but it is typically not one that can be dealt with using the types of data that are routinely available. Given an aggregated health outcome, the ideal exposure is the average "personal" exposure over the target population [190]. Because it is unrealistic to measure this quantity repeatedly over long periods of time, population studies must estimate this value [31, 34] or, more commonly, resort to suitable proxies such as the average "ambient" concentration. A key assumption made in previous time series analyses of air pollution and health data has been that the pollutant of interest is spatially homogeneous and that the monitor value on a given day (or the average of a few monitors) is approximately equal to the true ambient average concentration over the study population area. Any difference between the monitor value and the true ambient average concentration is what we call "spatial misalignment error."

In past analyses, the assumption that this error was zero may have been reasonable given that most previous analyses focused on pollutants such as the total mass of PM, ozone, and other pollutants that have been shown to be fairly spatially homogeneous over relatively long distances [140]. Recently, data have become available from the US EPA's Chemical Speciation Trends Network (STN) as well as state and local air monitoring stations which provide daily mass concentrations of approximately 60 different chemical elements of $PM_{2.5}$. These data are monitored in over 200 locations around the United States starting from the year 2000. Although the data are promising and are the subject of intense interest and new research, they also raise new statistical challenges. In particular, the usual assumption that the monitor value is approximately equal to the ambient average is less tenable when examining certain components of $PM_{2.5}$ [16]. For example, when estimating the correlations between pairs of monitors in the STN for 7 chemical components of $PM_{2.5}$ mass as a function of the distance between the monitors, these authors found that some of the chemical components of $PM_{2.5}$ are spatially homogeneous with high correlations over long distances (> 50 km), while other components are spatially heterogeneous and exhibit practically no correlation beyond short distances (< 20 km). Much of the spatial heterogeneity in the chemical components measured by the STN can be explained by the nature of the sources of the various components. For example, elemental carbon (EC) and organic carbon matter (OCM) tend to be emitted primarily from vehicle or other mobile sources, and thus their spatial distribution can depend on the localized nature of those sources. Secondary pollutants such as sulfate and nitrate are created in the air by the chemical and physical transformation of other pollutants and tend to be more regional in nature. Hence, for spatially heterogeneous pollutants such as EC or OCM, the daily level of those pollutants at a single monitor may be a poor surrogate for the daily county-wide average ambient level of that pollutant.

In a paper by Peng and Bell the authors describe a general method for estimating health risks associated with PM components from time series models while adjusting for potential spatial misalignment error [120]. They first develop a spatial-temporal model for the exposure of interest and estimate the degree of spatial misalignment error for each component in a location. Then they apply two methods–a regression calibration procedure and a two-stage Bayesian model–to estimate the health risks associated with these components and compare two results to standard approaches. Their findings indicate that the effect of spatial misalignment depends on monitor coverage within a county and the spatial variability of the pollutant of interest.

Although here we are discussing this issue mainly on time series studies, spatial misalignment can also induce error in cross-sectional studies of air pollution and health. Gryparis et al demonstrate in detail how to handle the misalignment errors in these types of studies and compare the performance of a number of different statistical approaches [73]. An alternate modeling approach has been proposed by Fuentes et al for estimating the spatial association between spectated fine particles and mortality [65]. Both approaches introduce a spatial model for the monitored pollutant concentrations and either predict pollutant values at unobserved locations or compute area averages over counties to link with county-level health data.

Ultimately, the best way to address the problem of spatial misalignment might be to move away from the county-based summaries of the outcome of interest when possible and begin using summaries with finer spatial resolution, such as zip codes or census tracts. In general, a decrease in the area covered per monitor is associated with lower spatial misalignment error, and this effect is far more pronounced for pollutants that are inherently heterogeneous such as sodium ion, silicon, and EC. This requires the availability of exposure and health data for the smaller areal units. Exposure data from sparse monitoring networks may not be available for smaller areal units; however, this can be over come by exposure prediction modeling as discussed in Section 24.2.5. Unfortunately, many types of health data are simply not available at finer spatial resolution, and we often must accept what is available. Furthermore, due to activity patterns, a high spatial resolution does not necessarily better capture personal exposure than a larger area when individuals move between areas (e.g. live in one zip code, but work in another). Thus, there is a strong need for methods that address spatial misalignment of air pollutant concentrations used in health studies.

24.2.5   Exposure Prediction Modeling

More recently there have been several contributions in the literature that estimates exposure to $PM_{2.5}$ at small spatial resolution and in geographical areas that are far from monitoring stations. These methods include interpolation or Kriging methods [12] and land-use regression models [64]. Recent studies have focused on improving prediction by combining sophisticated statistical models with multiple data sources. For example, several recent papers have used sophisticated spatiotemporal models and supplemented existing monitoring networks with other data sources including land use characteristics, chemical transport models, and/or supplemental monitoring data [13, 38, 82, 91, 93, 109, 116, 163, 168, 183]. These approaches, in combination of improved validation procedures [167], have proven successful in improving prediction accuracy of $PM_{2.5}$, PM components, and other ambient pollutants. For example, one of these hybrid approaches has focused on use of satellite-retrieved aerosol optical depth (AOD), in addition to monitor and land use data, to predict ground-level $PM_{2.5}$ concentrations [2, 85, 98]. Kloog et al have developed hybrid prediction models that use satellite-based AOD data in conjunction with daily calibration and spatio-temporal statistical models to predict daily $PM_{2.5}$ mass concentrations at $1\text{km} \times 1\text{km}$ grid cells in New England, Mid-Atlantic States, and the Southeastern US. This provides a major advance in the field of particle exposure assessment and epidemiology because it predicts exposure levels in both urban and rural environments. Using these data, they simultaneously estimate chronic and acute effects of $PM_{2.5}$ on cardiovascular hospital admissions, mortality, and other specific outcomes, including examining the effect at low $PM_{2.5}$ concentrations [97, 99, 100]. Even with improved prediction methods, there is still potential for bias due to measurement error and several recent studies have proposed additional adjustment approaches [11, 77, 158].

In a recent paper, Cefalu and Dominici pointed out that in environmental epidemiology, we are often faced with two challenges [32]. First, an exposure prediction model is needed to estimate the exposure to an agent of interest, ideally at the individual level (see above several references). Second, when estimating the health effect associated with the exposure, confounding adjustment is needed in the health-effects regression model (see for example the next section on Bayesian Adjustment for Confounding). The current literature addresses these two challenges separately. That is, methods that account for measurement error in the predicted exposure often fail to acknowledge the possibility of confounding, whereas methods designed to control confounding often fail to acknowledge that the exposure has been predicted.

Recent work by these authors and by Szpiro et al [32, 156] consider exposure prediction and confounding adjustment in a health-effects regression model simultaneously. Using theoretical arguments and simulation studies, they show that the bias of a health-effect estimate is influenced by the exposure prediction model, the type of confounding adjustment used in the health-effects regression model, and the relationship between these two. Moreover, the authors argue that even with a health-effects regression model that properly adjusts for confounding, the use of a predicted exposure can bias the health-effect estimate unless all confounders included in the health-effects regression model are also included in the exposure prediction model. Bergen et al illustrate the importance of accounting for exposure model characteristics when estimating health effects using predicted exposure through a case study using the Multi-Ethnic Study of Atherosclerosis and Air Pollution study [26].

24.3 Estimating the Exposure-Response Relationship

To protect public health and welfare against the dangers of air pollution, the US EPA establishes NAAQS. In 2012, in response to mounting evidence demonstrating the harmful effects of exposure to fine particulate matter, the EPA enacted more stringent NAAQS for $PM_{2.5}$. However, as air pollution levels continue to decrease, regulatory actions are becoming increasingly expensive to maintain. The annual cost of implementation and compliance with the NAAQS has reached 350 million dollars. Given the need to justify these costs, research examining the public health benefits of cleaner air will be subject to immense scrutiny. Yet significant gaps in knowledge remain, particularly with regard to the health effects of long-term exposure to lower levels of air pollution.

Despite a substantial amount of epidemiological literature on the health effects of both short-term and long-term exposure to air pollution, few studies have characterized the health effects of air pollution at levels in accordance with or lower than the most recent NAAQS for $PM_{2.5}$ (now set at 12 $\mu g/m^3$ for annual mean $PM_{2.5}$). As air pollution levels decrease, studies are needed to determine if further reductions will lead to substantial improvements in health. In this setting, estimation of an exposure-response function to estimate the health effects at various exposure levels (say lower than the current standards) is needed.

24.3.1 Generalized Linear Models

The most common approach to estimate the exposure-response relationship is a GLM with linear exposure-response relationship. For observation i, the model is

$$g\{E(Y_i|X_i, Z_i)\} = \beta_0 + X_i\beta + \sum_{j=1}^{p} Z_{ij}\gamma_j, \tag{24.1}$$

where Y_i is the health outcome, X_i is the exposure of interest, $Z_{i1}, \ldots, Z_{ip}$ are a set of subject specific covariates or confounding variables. However, the linear exposure-response function does not answer these important questions about health effects specifically at low (or high) concentrations. In this section we will introduce approaches to address this problem.

When the goal is estimation of the health effects of exposure to $PM_{2.5}$ at low concentrations, a common and intuitive approach is to restrict the analysis to days and/or geographical locations that have low exposure. For example, a recent paper by Shi et al [153] used these types of restrictions to study the effect of long-term exposure to low concentrations of $PM_{2.5}$ on mortality in a New England time-series study. In this setting, (24.1) took the form a Poisson regression with a log link, $g(\mu) = \log(\mu)$, and the exposure was average $PM_{2.5}$ over the previous 365 days. The analysis was conducted on all individuals and restricted to those that were exposed to a average $PM_{2.5}$ less than 10 $\mu g/m^3$. The authors found a 7.52% (95% CI: 1.95,13.40%) increase in morality associated with a 10 $\mu g/m^3$ in $PM_{2.5}$ in the complete data analysis but a 9.28% (95% CI: 0.76, 18.52%) increase in mortality in the restricted analysis. In this case restriction is able to tell us two important things about the exposure-response relationship. First, comparing the restricted analysis to the unrestricted analysis is suggestive (although not statistically significant) of a larger exposure effect at lower concentration levels. Second, the presence of an exposure effect of long-term exposure to $PM_{2.5}$ below 10 $\mu g/m^3$ provides a key piece evidence that additional reduction in $PM_{2.5}$ exposure below 10 $\mu g/m^3$ could reduce the mortality burden of $PM_{2.5}$ exposure.

A key aspect that makes restriction appealing is that it requires no statistical methods beyond GLM, yet by carefully selecting the exposure range for restriction it can estimate a different exposure effects over different concentration ranges. In practice, it can be difficult to choose the cut points. One approach is to estimate the model with several cut points. Then, use an information criterion such as AIC or BIC to find the best fitting model.

24.3.2 Semi-Parametric Approaches

A flexible alternative to a linear exposure-response function or restriction is a nonlinear exposure-response function. Several approaches have been proposed, for example, natural cubic splines [22, 49, 57], penalized splines [148, 153], B-splines [155], or loess [145]. Here, we consider using a spline basis function to estimate a nonlinear exposure-response function f. The regression model then takes the form

$$g\{E(Y_i|X_i, Z_i)\} = \beta_0 + f(X_i; \mathbf{v}) + \sum_{j=1}^{p} Z_{ij}\gamma_j \qquad (24.2)$$

where $f(X; \mathbf{v})$ is modeled as a spline basis with k knots at locations $\mathbf{v} = (v_1, \ldots, v_k)$. The number of knots is generally specified *a priori* based on prior subject knowledge or chosen to maximized model fit as measured by AIC BIC, or GCV. However, this results in inference that is conditional upon the number and location of knots.

To account for uncertainty in knot selection, Dominici et al [57] estimated the number and location of knots from the data using reversible jump Markov chain Monte Carlo (RJM-CMC) [70] in an analysis of the 88 largest US cities. This approach, allows the model to change both the number and location of knots. In addition, RJMCMC accounts for model uncertainty in knot selection. An alternative approach to account for model uncertainty in knot selection is to average the estimates over several candidate models with different number and location of knots [148] using Bayesian model averaging (BMA) [83, 130].

24.3.3 Model Uncertainty in the Shape of the Exposure-Response

Accounting for model uncertainty when estimating the shape of the exposure-response is a critically important aspect of studies of air pollution and health. Bobb et al have proposed a flexible class of time series models to estimate the relative risk of mortality associated with heat waves and conduct BMA to account for the multiplicity of potential models for the exposure-response function [27]. The proposed approach overcomes many of the challenges in estimating the adverse health effects of heat-wave events. More specifically, within each city, the authors specify a semi-parametric model to flexibly capture the nonlinear relation between several weather variables and mortality. The model makes as few assumptions as possible about the shape of the temperature-mortality function and does not require the cities to have the same model or even to include the same temperature predictors (e.g. one city could us maximum daily temperature and another could use mean daily temperature). This allows for heterogeneity of the temperature-mortality association across cities, in accordance with findings in prior studies that the shape of the temperature-mortality curve varies by US region [48]. The authors also incorporate model uncertainty in the specification of the temperature-mortality exposure-response function by conducting BMA. They found that for some cities the estimation of health risks associated with heat waves is sensitive to the specification of the exposure-response function (e.g. specification of the relationship between weather variables and mortality counts).

An important element of this analysis was the comparison of the posterior variance of the log relative risk under BMA to the variance under a single model which emphasized a twofold benefit of model averaging. Under the model averaging approach, each candidate model is weighted by its posterior probability and the uncertainty of estimates from less plausible models (those with low posterior model probability) do not contribute to the model-averaged uncertainty estimate. However, when multiple models are plausible, BMA incorporates the variability from each potential model. Thus conditioning inference on a single model obtained through a model selection procedure likely underestimates statistical uncertainty.

24.4   Confounding Adjustment

Estimating the effect of an exposure on an outcome, while properly adjusting for confounding factors, is a common and challenging goal in the analysis of observational studies. In the context of estimation of health effects of exposure to environmental contaminants, the choice of approaches for adjusting for confounding and approaches for selecting the key confounders among a large set is critical. Indeed health effect estimates can be sensitive to these choices [58].

A common practice is to select a statistical model for the estimation of the effect, and report effect estimates and confidence intervals that are conditional on that model being correct. This does not account for "adjustment uncertainty," that is uncertainty about which variables should be included in the model to properly adjust for confounding. It is possible to effectively convey this uncertainty by sensitivity analysis, showing the variation of the effect estimate and its interval over a range of plausible choices of confounders [59, 122].

BMA has been suggested as a more formal tool to account for model uncertainty [83, 130]. The BMA approach used here is based on augmenting the model with indicator variables of whether each predictor is included in the model as an unknown nuisance parameters. This results in a weighted average of predictions whose weights depend on the support that each selection receives from the data. This principled approach enjoys a number of desirable properties from a frequentist point of view as well, and has performed competitively in out-of-sample prediction comparisons [37, 182]. The conceptual simplicity and solid logic behind treating the unknown confounder subset as an unknown parameter is attractive in adjustment uncertainty as well. BMA estimates the exposure effect as a weighted average of model-specific effect estimates, again using the model's posterior probabilities as weights. Viallefort et al [164] applied this method to estimate an exposure's odds ratio in case-control studies. Other applications include air pollution research [41, 101].

However, while effective in some cases, traditional implementations of BMA, with an uninformative prior on the inclusion indicator of each potential confounder in the regression model can face severe limitations [45, 166]. Most of these can be traced to the fundamental difficulty arising with the fact that the regression coefficient, representing the effect of exposure to air pollution on a health outcome, may have a different interpretation across models that include a different set of covariates to adjust for confounding [43]. Crainiceanu et al [45] noted that model uncertainty methods useful in prediction may not generally perform well in adjustment uncertainty. They introduced a two-step approach (CDP) to estimate an exposure effect accounting for adjustment uncertainty. In the first step, this approach regresses exposure on a large set of potential confounders and selects confounders that are associated with exposure. In the second step, it regresses outcome on exposure, after including the confounders identified in the first step. Compared to this approach, traditional BMA with non-informative priors on whether or not include a covariate into the regression model to adjust for confounding did not perform well. This is because the posterior model probabilities obtained from non-informative priors will assign high weights to models that include covariates that are strong predictors of the outcome only, and will assign low weights to models that include strong predictors of the exposure and weak predictors of the outcome. It is well known that when we omit these variables in the regression model, the health effect estimates will be affected by confounding bias (see [166] for a compelling example that illustrates this point).

Wang et al [165, 166] have developed a novel Bayesian approach which we call Bayesian Adjustment for Confounding (BAC) to adjust for confounding and account for uncertainty in the choice of confounders in the context where we estimate the health effect of air pollution using a GLM and we adjust for confounding by including covariates into the regression model as a linear term. Here, we describe in detail the approach proposed by Wang et al [165, 166]. Other more sophisticated approaches that account for model uncertainty in the confounding adjustment have been developed by Zigler et al in the context of propensity score methods [195] and by Cefalu et al in the context of double robust estimation [33].

24.4.1   Bayesian Adjustment for Confounding

We start by introducing some notation. We denote by X the exposure and Y the outcome. We also assume that we have information on a set of p potential confounders $\mathbf{Z} = \{Z_1, \ldots, Z_p\}$ identified because they are likely to affect Y, though their effects could be weak. *A priori*, there may be uncertainty about whether potential confounders should be adjusted for in effect estimation.

Though many of our ideas are more general, we discuss our approach in the context of two linear regression models: one for exposure and one for outcome. In each equation, potential confounders are either included or excluded, depending on unknown vectors of indicators $\boldsymbol{\alpha}^X \in \{0,1\}^p$ and $\boldsymbol{\alpha}^Y \in \{0,1\}^p$. Here $\alpha_j^X = 1$ (or $\alpha_j^Y = 1$) whenever Z_j is included in the exposure (or outcome) model. For brevity, we refer to the parameters $\boldsymbol{\alpha}^X$ and $\boldsymbol{\alpha}^Y$ as "models". Conditional on unknown parameters, and confounders, the regression equations for exposure X_i and outcome Y_i are

$$E\{X_i\} \;=\; \sum_{j=1}^p \alpha_j^X \delta_j^{\boldsymbol{\alpha}^X} Z_{ij} \tag{24.3}$$

$$E\{Y_i | X_i\} \;=\; \beta^{\boldsymbol{\alpha}^Y} X_i + \sum_{j=1}^p \alpha_j^Y \delta_j^{\boldsymbol{\alpha}^Y} Z_{ij} \tag{24.4}$$

where i indexes the sampling unit. For regression coefficients, β and δ, we use a notation that explicitly keeps track of the fact that those coefficients differ in meaning with $\boldsymbol{\alpha}^X$ and $\boldsymbol{\alpha}^Y$. This is especially important when one attempts to make inferences that involve estimates of the exposure effect obtained using different models. Intercept columns can be included among the Zs. Some α_m^Ys can be fixed with value 1 if confounders are deemed required *a priori*.

In developing a model for effect estimation, when a true confounder is added or removed from the regression model, the interpretation of the exposure coefficient changes; however, when a model includes all true confounders, and one adds an additional variable that is not associated with X or that is neither associated with X nor Y, the interpretation of the exposure coefficient does not change. This is in contrast to prediction, where the predicted quantities typically maintain the same interpretation across models.

Thus, when studying confounding adjustment, it is useful to consider the smallest outcome model that includes all the necessary confounders. We denote it by α_*^Y, and refer to it as the minimal model. The estimand of interest —the true effect of X on Y, is the coefficient of X in this model, or $\beta_* = \beta^{\alpha_*^Y}$. If there are interactions between exposure and confounders, the estimands are model coefficients of both the main effect and the interaction terms. See Wang et al (2015) [165] for details regarding a BMA approach to account for model uncertainty in both the selection of confounders and in the selection of interaction terms between exposure variable and each of the potential confounders.

We will focus on the situation where there are no interaction terms. Our goal is estimation of β_* when α_*^Y is unknown. A key observation is that all models that contain at least the confounders as in the minimal model will provide estimates of the exposure effect that are also interpretable as estimates of β. On the other hand, a model that does not include the minimal model, that is, a model that excludes at least one true confounder, will provide estimates of a parameter that are not the estimated parameter of interest. Hence, the resulting exposure effect estimate is confounded.

The importance of including in the outcome model all the potential confounders that belong to the minimal model suggests that an approach that acknowledges the fact that only a fraction of the models harbor the coefficient of interest with the correct interpretation, could be successful in addressing adjustment uncertainty from a Bayesian standpoint. BAC jointly considers the exposure and outcome models, as in equations (24.3) and (24.4), and includes unknown model selection parameters α^X and α^Y.

The authors specify a prior distribution on $\alpha^Y | \alpha^X$ such that

$$\frac{P(\alpha_j^Y = 1 | \alpha_j^X = 1)}{P(\alpha_j^Y = 0 | \alpha_j^X = 1)} = \omega, \quad \frac{P(\alpha_j^Y = 1 | \alpha_j^X = 0)}{P(\alpha_j^Y = 0 | \alpha_j^X = 0)} = 1, \quad j = 1, \ldots, p, \qquad (24.5)$$

where $\omega \in [1, \infty]$ is a dependence parameter denoting the prior odds of including Z_j into the outcome model when Z_j is included in the exposure model. When $\omega = \infty$, the first equation in (24.5) becomes $P(\alpha_j^Y = 1 | \alpha_j^X = 1) = 1$, and requires that any Z_j for which $\alpha_j^X = 1$ is automatically included in the outcome model. When $1 < \omega < \infty$, our prior on $\alpha^Y | \alpha^X$ provides a chance to rule out the predictors that are only associated with X but not associated with Y.

The conditional prior of α^Y given α^X in (24.5) plays a key role in approximating the marginal posterior distribution of the exposure coefficient under the minimal model, β_*,

$$P(\beta_* | D) = \sum_{\alpha^Y} P(\beta_* | \alpha^Y, D) P(\alpha^Y | D),$$

where $D = (X, Y)$ contains vectors of observed data for the exposure and the outcome. Our analysis is also conditional on observed data for potential confounders $Z_1, \ldots, Z_j$, and they will not be noted in posteriors for simplicity of notation. When ω is large, the conditional prior in (24.5) greatly increases the chance for predictors strongly correlated with X to be included in the outcome model. These predictors are confounders if they are also correlated with Y. Therefore, the prior leads to a posterior distribution of α^Y ($P(\alpha^Y | D)$) that assigns mass mostly to models that are fully adjusted for confounding, that is, models containing the minimal model. For these models, $\beta^{\alpha^Y} = \beta_*$ so that $P(\beta_* | \alpha^Y, D) = P(\beta^{\alpha^Y} | \alpha^Y, D)$.

Therefore, approximately,

$$P(\beta_*|D) \doteq \sum_{\alpha^Y} P(\beta^{\alpha^Y}|\alpha^Y, D)P(\alpha^Y|D), \tag{24.6}$$

where $P(\beta^{\alpha^Y}|\alpha^Y, D)$ can be directly estimated from observed data. This approximation will be further discussed in Section 24.4.2.

Our goal is to calculate the posterior distribution of the parameters of interest $(\alpha^X, \alpha^Y, \beta_*)$ in equations (24.3) and (24.4). Details are in Wang et al 2012 [166].

24.4.2 Relation to BMA

In the context of effect estimation, several authors [83, 130] suggested to calculate the posterior distribution of the effect by taking an average over models, weighted by their posterior probabilities:

$$\sum_{\alpha^Y} P(\beta^{\alpha^Y}|\alpha^Y, D)P(\alpha^Y|D). \tag{24.7}$$

This corresponds to marginalization according to the law of total probabilities, but only if the parameters β^{α^Y} have the same interpretation.

From the perspective of adjustment uncertainty, (24.7) can be decomposed into two parts: the sum over models that include the correct estimand, and the rest. That is

$$\sum_{\alpha^Y \supseteq \alpha^Y_*} P(\beta_*|\alpha^Y, D)P(\alpha^Y|D) + \sum_{\alpha^Y \not\supseteq \alpha^Y_*} P(\beta^{\alpha^Y}|\alpha^Y, D)P(\alpha^Y|D). \tag{24.8}$$

where $\alpha^Y \supseteq \alpha^Y_*$ indicates that model α^Y contains all the variables that are also contained in model α^Y_*. The second term of (24.8) averages across models that do not include α_*^Y, and therefore do not estimate the same effect.

In BMA one needs to be careful about not assigning large weights to the models in the second term of equation (24.8). A common practice in traditional implementations of BMA is to use uniform, or highly dispersed, priors on the α^Ys and often on the effect of interest as well. When the prior is the same for all models, the ratio of the weights given to models α_1 and α_2 is the Bayes Factor $P(Y|\alpha_1)/P(Y|\alpha_2)$ [88] and the posterior model probabilities in BMA are driven by a model's predictive ability, which may differ from its ability to properly adjust for confounding in effect estimation.

24.4.3 Air Pollution Example

In this section, we briefly summarize how the authors applied BAC to daily time series data for Nassau County, NY for the period 1999-2005. Although this data analysis was mainly used as an illustration of BAC, the results clearly illustrate the potential application and impact of BAC in epidemiology studies of observational data. The data include 1,532 daily records of emergency hospital admissions, weather variables, and $PM_{2.5}$ levels. A more extensive description of this data set can be found in [61]. The goal was to estimate the increase in the rate of hospitalizations for cardiovascular disease (CVD) associated with a $10\mu g/m^3$ increase in $PM_{2.5}$, while accounting for age-specific longer-term trends, weather and day of the week. The hospitalization rate was calculated separately for each age group (≥ 75 or not) on each day. In the outcome model, to control for longer-term trends due, for example, to changes in medical practice patterns, seasonality and influenza epidemics, the authors included smooth functions of calendar time. They also included a smooth function to allow seasonal variations to be different in the two age groups. To control for the weather effect, they include smooth functions of temperature and dew point. The outcome model that includes all the necessary confounders can be defined below [59, 62, 122]

$$Y_{at} = \beta \, PM_{2.5t} + DOW + \text{intercept for age group } a$$
$$+ ns(Temp_t, df_{Temp}) + ns(Temp_{t1-3}, df_{Temp}) + ns(Dew, df_{Dew})$$
$$+ ns(Dew_{t1-3}, df_{Dew}) + ns(t, df_t) + ns(t, df_{at}) \times \text{age group} + \epsilon_t,$$

where the outcome $Y_{at} = \sqrt{\text{CVD hospital admissions}/\text{size of population at risk}}$ for each age group a (≥ 75 or not) on day t (=1, ..., 1532). $PM_{2.5t}$ denotes the level of particulate matter having diameter less than 2.5 micrometer on day t. DOW are indicator variables for the day of the week. $Temp_t$ and $Temp_{t1-3}$ are the temperature on day t and the three-day running mean, respectively. Dew_t and Dew_{t1-3} are the dew point on day t and the three-day running mean. The quantity $ns(., df)$ is a natural cubic spline with df degrees of freedom. The authors included $ns(t, df_t)$, $ns(Temp_t, df_{Temp})$, $ns(Temp_{t1-3}, df_{Temp})$, $ns(Dew, df_{Dew})$ and $ns(Dew_{t1-3}, df_{Dew})$ in the outcome model to adjust for the potential nonlinear confounding effects of seasonal variations, temperature and dew point. The quantity $ns(t, df_{at}) \times$ age group is a natural cubic spline of t for the ≥ 75 age group to allow its seasonal variation to be different from the other age group. Similar to [45], df_{Temp} was set to 12, df_{Dew} was set to 12, df_t was set to 16 per year, and df_{at} was set to 4. These degrees of freedom were considered sufficiently large for the full model to include all the potential confounders [45]. The residuals ϵ_t were assumed to be independent and identically distributed with a normal $N(0, \sigma^2)$ distribution. After dropping some potential confounders due to collinearity, the authors considered 164 potential confounders.

Several approaches were used to analyze the data: BAC, CDP [45], FBMA (BMA with the exposure variable always forced into the outcome model), NBMA (BMA with the exposure variable not forced into the outcome model), and stepwise. For BAC, the authors considered priors with $\omega = 2, 4, 10$ or ∞. The estimated $PM_{2.5}$ effect ($\times 10,000$) denoted by $\hat{\beta}$ is listed in Table 24.1: BAC (with $\omega = \infty$) and CDP provide estimates of the short-term effect of $PM_{2.5}$ on CVD hospital admissions with 95% CIs that do not include 0. With $\omega = \infty$, BAC provided similar estimates of the exposure effect as CDP. Moreover, all three methods provided smaller standard errors than the one obtained under the full model. In comparison, FBMA and NBMA provided a very different and not statistically significant estimate of the exposure effect. Some confounders known to be important, such as temperature and dew point, were down weighted in BMA. Both temperature and dew point were positively correlated with $PM_{2.5}$ and negatively correlated with hospitalization rate. Failure to include them in the model diminished the $PM_{2.5}$ effect. This example illustrated that in practical applications BMA and BAC can lead to different conclusions. The key difference lies in the linking strength between the exposure model and the outcome model. As the strength decreases, which corresponds to smaller value of ω, the estimates from BAC become closer to that from BMA.

24.4.4 Concluding Remarks

Estimating an exposure effect, while accounting for the uncertainty in the adjustment for confounding, is of essential importance in observational studies. Building upon earlier work [45, 59], in BAC the authors have developed Bayesian solutions to the estimation of the association between X and Y accounting for the uncertainty in the confounding adjustment. Given a set of potential confounders, the authors simultaneously address model selection for both the outcome and the exposure. While they presented their methods in the setting of linear models, BAC is a general concept and is not constrained to the linear case. These

TABLE 24.1

Comparison of estimates of $PM_{2.5}$ effect on CVD hospitalization rate based on BAC, CDP, FBMA, NBMA, stepwise, and the full model. (Table from Wang et al 2012)

		$\hat{\beta}$	$SE(\hat{\beta})$	95% CI
Full model		0.291	0.092	(0.110, 0.471)
BAC	$\omega = \infty$	0.226	0.081	(0.067, 0.385)
	$\omega = 10$	0.217	0.079	(0.060, 0.371)
	$\omega = 4$	0.186	0.085	(0.019, 0.351)
	$\omega = 2$	0.155	0.079	(0.007, 0.317)
CDP		0.221	0.089	(0.045, 0.396)
FBMA		0.140	0.077	(−0.008, 0.298)
NBMA		0.007	0.033	(0.000, 0.131)
Stepwise		0.106	0.066	(−0.023, 0.234)

approaches have been extended to generalized linear models using relatively well understood computational strategies [165].

Like BMA, BAC take a weighted average over models rather than making inference based on a single model. However, they attempt to provide an estimate of the exposure effect by combining information across regression models that include all the requisite confounders, to ensure that the regression coefficient of interest maintains the same interpretation across models. A nice feature of BMA that is retained by BAC is that the importance of confounders can be evaluated based on posterior inclusion probability. This information may reveal underlying connections between exposure and confounders, which may become of interest for future research. BAC is more computationally intensive than BMA.

Successful application of BAC relies on availability of all confounders. Scientific knowledge is required to ensure that these assumptions are valid. Statistical methods may also help to check whether there is evidence for the existence of unmeasured confounders. If there are no unmeasured confounders, the full model, that is, the model including all variables correlated with X and Y, those correlated with Y only, as well as potentially others that are not associated with either, will provide unbiased estimates of the exposure effect. However, using the full model will generally yield wider confidence intervals compared to BAC. By combining estimations from different smaller models, especially from models that only include requisite confounders but do not include many unnecessary variables, BAC can provide more precise inference than the full model.

In the propensity score literature, it is recommended to include variables that are strongly correlated with Y but only weakly correlated with X into the model for calculating the propensity score, as the bias resulting from their exclusion would dominate any loss of efficiency in modest or large studies [30, 137]. One of the strengths of the BAC method, shared by others such as doubly robust estimation [143], is that BAC can identify these in a data-based way, rather than having to rely on prior knowledge as required in propensity score adjustment.

Zigler et al introduced a new set of approaches to account for adjustment uncertainty when the adjustment for confounding is done by using propensity score (PS) methods [195]. In causal inference, typically, simple or ad hoc methods are employed to arrive at a single propensity score model, without acknowledging the uncertainty associated with the model selection. Zigler et al introduced three Bayesian methods for PS variable selection and model averaging that (a) select relevant variables from a set of candidate variables to include in the PS model and (b) estimate causal treatment effects as weighted averages of estimates under different PS models. The associated weight for each PS model reflects the data-driven support for that model's ability to adjust for the necessary variables [195]. One drawback of application of these approaches is that they assume that the treatment assignment or the exposure is binary, whereas the exposure to air pollution is a continuous variables. Extensions of these approaches in the context of generalized propensity score methods for multi-level treatments and/or continuous treatments [179, 191] are needed.

In this section, while we do not take a causal inference perspective, the methods illustrated here have several points of contacts with causal inference methodologies that are based on joint modeling of exposure and outcome as functions of confounders [135, 136] and with their Bayesian counterparts [112]. This literature strongly emphasizes, as we do, the critical role of model specification and the need for robustness to the choice of confounders [9, 71, 137]. From this perspective, the BAC approach and extensions of these ideas in the context of propensity score matching [195] and double robust estimation [33] achieves a combination of three desirable properties: effect estimation efficiency, via the exposure model; variable selection robustness, achieved by allowing the selection to be a random variable; and bias reduction, achieved by including prior information to favor predictors of exposure in the selection of variables for the outcome model.

24.5 Estimation of Health Effects From Simultaneous Exposure to Multiple Pollutants

The majority of environmental epidemiology has estimated the health effects of exposure to a single pollutant. However, populations are in reality exposed to a diverse mixture of pollutants. Recent research in environmental epidemiology has turned to estimate the health effects of mixtures of multiple pollutants.

In this setting, for each individual i we observe the exposure level for a vector of M pollutants $\mathbf{X}_i = (X_{i1}, \ldots, X_{iM})^T$. As M increases, the exposure-response function $f(\mathbf{X})$, as in (24.2), becomes increasing more complex. In particular, there may be a nonlinear exposure-response relationship for each of the M pollutants as well as interactions between pollutants. In this setting, the curse of dimensionality makes estimating the exposure-response function a challenging statistical problem.

Several methods have been proposed to estimate the health effects of mixtures. In this section we highlight two recently developed methods that take different approaches to the multi-pollutant problem. The first approach was developed in the context of air pollution epidemiology and compares the health effect of ambient exposures on days with different "multi-pollutant profiles" [7, 185]. This can address questions such as are the health effects of $PM_{2.5}$ different when the particulates come from fresh local traffic pollution verses regionally transported pollution, for example. The second approach takes a flexible machine learning approach to estimate a high-dimensional exposure-response surface [28]. This second approach can estimate the health effects at any level of exposure to a high-dimensional vector of pollutants. We conclude with a discussion of confounding adjustment in the context for multi-pollutant models.

24.5.1   Multi-Pollutant Profile Clustering and Effect Estimation

The multi-pollutant profiling approach to estimating the health effect of multi-pollutant mixtures is built on a two step procedure [7, 185]. The first step is to identify clusters of days or locations that have a similar multi-pollutant profile. For example, days with pollution profiles dominated by primary emissions from local traffic or those high in regional power-plant emissions. The second step is to estimate the health effects of exposure to each cluster and compare the relative toxicities of the clusters.

There are ample statistical methods for clustering [25, 52, 78, 92]. Here will will focus on the k-means algorithm as proposed for multi-pollutant clustering [7, 185]. The k-means algorithm was originally developed by Hartigan and Wang [78] and is appealing due to the computational efficiency. The proposed approach clusters days or locations based on the measured concentrations of M pollutants. The general concept is to partition the n observations of M pollutants, $\{\mathbf{X}_i = (X_{i1}, \ldots, X_{iM})^T\}_{i=1}^n$, into k clusters. For each cluster there is a mean points (k means in total) and each observations is nearest in euclidian distance to its clusters mean.

In the context of environmental epidemiology, k-means (or any clustering approach) can identify clusters of similar exposures. For example, Austin et al [7] found five clusters each consisting of days in Boston, MA, with one of five air pollution profiles: 1) "low particles–high ozone" contained lower concentrations of $PM_{2.5}$ and higher than average levels of O_3; 2) "crustal" which had higher levels of Si, CA, Br, and Ti; 3) "winter primary" which contained higher levels of Ni, V, Zn, Br, CO, NO and SO_2; 4) "regional summer" which was heavy in regionally transported pollutants; and 5) "winter–low primary, higher O_3" which is similar to cluster 3 but with lower $PM_{2.5}$ and higher O_3 levels.

The challenge to k-means clustering is determining the optimal number of clusters. Ideally, the researcher would have prior knowledge about the number of clusters. However, this is often unknown and methods have been developed to select k. One such approach is the select the value of k that minimizes the Davies-Bouldin (DB) index [50]. The DB index measures cluster compactness verse the distance between cluster centers. It is expressed as

$$\text{DB} = \frac{1}{n} \sum_{i=1, i\neq j}^{n} \max \left\{ \frac{\sigma_i + \sigma_j}{\|\boldsymbol{\mu}_i - \boldsymbol{\mu}_j\|} \right\}, \tag{24.9}$$

where $\boldsymbol{\mu}_i$ is the center of cluster i, σ_i is the average distance between $\boldsymbol{\mu}_i$ and the observations in cluster i, and n is the number of clusters. The choice of k that minimizes (24.9) will be have compact clusters with well separated cluster means relative to other choices of k.

The second step of the multi-pollutant clustering and effect estimation approach is to estimate the relative toxicities of the clusters. Zanobetti et al [185] provide a nice illustration of this by interacting indicators of membership into five clusters with total $PM_{2.5}$ mass. Here $C_{tj} = 1$ if day t had air pollution consistent with profile j and $C_{tj} = 0$ otherwise. The proposed Poisson time-series model is then

$$g\{E(Y_t | X_t, \mathbf{C}_t, \mathbf{Z}_t)\} = \beta_0 + X_t\beta + \sum_{j=2}^{5} C_{tj}\eta_j + \sum_{j=2}^{5} X_t C_{tj}\delta_j + \sum_{k=1}^{r} Z_{tk}\gamma_j, \tag{24.10}$$

where cluster one is omitted as a reference category. The model can then be interpreted as the effect of total $PM_{2.5}$ mass (X in this model) on mortality (Y in this example) specifically on days with each of the five cluster profiles. Hence, we can estimate and test differences in the toxicity of $PM_{2.5}$ with various compositions.

This approach is appealing on several levels. First both the cluster and analysis stages can be completed with standard methods and existing software. Second, the approach is easily interpretable because the potentially high-dimensional exposure profile is reduced to a categorical variable indicating cluster membership. However, the analysis is limited to the cluster-level and interpretability relies on identifying clusters that are scientifically meaningful.

24.5.2   High-Dimensional Exposure-Response Function Estimation

Another approach to directly estimate the multi-pollutant exposure-response functions. Under this approach, the regression model is

$$g\{E(Y_i|\mathbf{X}_i, \mathbf{Z}_i)\} = f(\mathbf{X}_i) + \sum_{k=1}^{r} Z_{ik}\gamma_j, \tag{24.11}$$

where $f : \mathbb{R}^M \to \mathbb{R}$ is the multi-pollutant exposure-response function. Several approaches have been used to estimate nonlinear and non-additive multivariate exposure-response functions. These approaches include spline or polynomial expansions [36, 175], loess [132], and machine learning approaches [28]. Here, we focus on Bayesian kernel machine regression (BKMR) to estimate the multivariate-exposure-response surface as introduced by Bobb et al [28]. Compared to many alternative approaches, BKMR offers the advantage of easily scaling with the number of pollutants (M) while allowing for nonlinear effects and high order interactions.

Under the BKMR approach, we assume that the exposure-response function f resides in a function space with an associated positive semidefinite reproducing kernel $K : \mathbb{R}^M \times \mathbb{R}^M \to \mathbb{R}$. The exposure-response function f can then be represented in either of two forms. First is the primal form where f can be represented with a basis expansions $f(\mathbf{x}) = \sum_{l=1}^{L} \phi_l(\mathbf{x})\theta_l$, where $\{\phi_l(\mathbf{x})\}_{l=1}^{L}$ is a set of basis functions and $\{\theta_l\}_{l=1}^{L}$ are regression coefficients.

The second form for representing f, which is used here, is the dual form through the kernel $K(\cdot, \cdot)$ as $f(\mathbf{x}) = \sum_{i=1}^{n} K(\mathbf{x_i}, \mathbf{x})\alpha_i$, where $\{\alpha_i\}_{i=1}^{n}$ are unknown regression coeficients. According to Mercer's theorem [46] a kernel $K(\cdot, \cdot)$ in the dual form corresponds to a unique set of orthogonal basis functions in the primal representation. Here, we consider the Gaussian kernel $K(\mathbf{x}, \mathbf{x}') = \exp\{-\sum_{m=1}^{M} \rho_m(X_m - X'_m)^2\}$, where $\{\rho_m\}_{m=1}^{M}$ are tuning parameters. The Gaussian kernel corresponds to a set of radial basis functions.

For the multi-pollutant normal linear regression, the dual representation can be represented as the mixed model [110]

$$Y_i \sim N(f_i + \mathbf{Z}_i^T\boldsymbol{\gamma}, \sigma^2) \tag{24.12}$$

$$\mathbf{f} \sim N(0, \tau\mathbf{K}). \tag{24.13}$$

In (24.13), $\mathbf{f} = (f_1, \ldots, f_n)^T$, $\mathbf{K}$ is a $n \times n$ kernel matrix where the (i,j) element is $K_{i,j} = \exp\{-\sum_{m=1}^{M} \rho_m(X_{im} - X_{jm})^2\}$, and τ is a tuning parameter. The mixed model can then be solved with existing mixed model software [110] or through Bayesian methods [28] to estimate the multi-pollutant exposure-response function.

As illustrated in the original work [28], the dual form naturally scales as the number of exposures increases. At the same time, BKMR allows for a high-dimensional exposure-response function estimation. This exposure-response surface can include nonlinearities and complex interactions between two or more of the M exposures.

24.5.3   Confounding Adjustment in Multiple Pollutant Models

As with this single pollutant models discussed in Section 24.3, confounding is an important issue in multi-pollutant exposure-response modeling. Recent papers have addressed confounding adjustment specifically in the multi-pollutant context [176, 177]. In particular, recent work highlights an important distinction between confounding adjustment when studying the health effects of a single pollutant and when studying the health effects of multiple pollutants [177]. When estimating the effects of multiple pollutants we are often interested in both main effects and interactions between pollutants. In order to properly adjust for confounding it is essential to not only consider confounders that are associated with any one pollutant and the outcome but to identify the set of confounders that are associated with the multivariate exposure (all main effects and interactions of interests) and outcome.

To address confounder adjustment when estimating the health effects of simultaneous exposure to multiple pollutants, a recent paper proposed using a Bayesian model averaging approach for estimating the health effects of simultaneous exposure to multiple pollutants (including main effects and interactions) that, at the same time, account for model uncertainty in the confounding adjustment [177]. More specifically, to adjust for confounding, the authors develop an informative prior on covariate inclusion. They apply the proposed method to the National Health And Nutrition Survey data (NHANES) data to estimate the effects of exposure to 132 nutrients and persistent pesticides grouped into 24 groups on lipid levels. When p is close to n, multivariate exposure effect estimates with the proposed approach are fully adjusted for confounding and has smaller variance than the same estimate obtained under a model that includes all available covariates.

Importantly, their proposed method uses information from the relationship between the potential confounders and the multivariate exposure to construct the prior. This newly proposed method offers several advantages over previously developed confounder adjustment methods. First, this approach scales as the dimension of the multivariate exposure vector increases. Previously proposed confounder adjustment approaches have relied on exposure modeling [165, 166, 176] and would require an additional exposure model for each additional agent. This is impractical when the number of individual agents is large. For example, the one exposure group in above data analysis with 22 agents would require 22 separate exposure models using exposure modeling approaches [165, 166, 176], an infeasible number of exposure models. A second advantage of the proposed method is that it explicitly addresses confounding of the multivariate exposure effect. It is indeed important to point out that the confounders of the single agents (for example exposure to $PM_{2.5}$) are different from the confounders of the multivariate exposure effect (for example health effects for simultaneous exposure to $PM_{2.5}$, O_3, NO_2 and all their pairwise interactions). A third advantage is the specific guidance provided for tuning the strength of the prior on covariate inclusion (in Wang et al this parameter was denoted by ω as shown in (24.5)). In BAC, the authors need to specify *a priori* a value for ω. There is a similar tuning parameter in Wilson et al [177]; however, the authors introduce an approach that allows identification of a suitable value of the tuning parameter that balances model parsimony and confounder adjustment.

With increased availability of high-dimensional exposure data (called exposome) there is growing interest in understanding the effect of the simultaneous exposure to multiple environmental agents on complex diseases. However, estimating a high-dimensional exposure-response function is statistically challenging and there is often uncertainty as to which covariates to include in the model to estimate the multivariate exposure effect. The methods described in this section fill methodological gaps in the area of estimating the health effects of simultaneous exposure to multiple pollutants. These methods are valuables tool that can reliably be used to estimate health effects of mixtures while simultaneously allowing a rigorous adjustment for confounding and guarantee model parsimony.

Bibliography

[1] Accountability Working Group. Assessing Health Impact of Air Quality Regulations: Concepts and Methods for Accountability Research. Executive summary. Technical Report September, Health Effects Institute, Boston, MA, 2003.

[2] Jassim Al-Saadi, James Szykman, R. Bradley Pierce, Chieko Kittaka, Doreen Neil, D. Allen Chu, Lorraine Remer, Liam Gumley, Elaine Prins, Lewis Weinstock, Clinton MacDonald, Richard Wayland, Fred Dimmick, and Jack Fishman. Improving National Air Quality Forecasts with Satellite Aerosol Observations. *Bulletin of the American Meteorological Society*, 86(9):1249–1261, sep 2005.

[3] Brooke A Alhanti, Howard H Chang, Andrea Winquist, James A Mulholland, Lyndsey A Darrow, and Stefanie Ebelt Sarnat. Ambient air pollution and emergency department visits for asthma: a multi-city assessment of effect modification by age. *Journal of Exposure Science & Environmental Epidemiology*, 26(May):1–9, 2015.

[4] Shirley Almon. The Distributed Lag Between Capital Appropriations and Expenditures. *Econometrica*, 33(1):178–196, 1965.

[5] Per Kragh Andersen and Richard D Gill. Cox's Regression Model for Counting Processes: A Large Sample Study. *The Annals of Statistics*, 10(4):1100–1120, 1982.

[6] Richard W Atkinson, Iain M Carey, Andrew J Kent, Tjeerd P van Staa, H Ross Anderson, and Derek G Cook. Long-term exposure to outdoor air pollution and incidence of cardiovascular diseases. *Epidemiology*, 24(1):44–53, 2013.

[7] Elena Austin, Brent Coull, Dylan Thomas, and Petros Koutrakis. A framework for identifying distinct multipollutant profiles in air pollution data. *Environment International*, 45(1):112–121, 2012.

[8] Sudipto Banerjee, Bradley P Carlin, and Alan E Gelfand. *Hierarchical Modeling and Analysis for Spatial Data*. CRC Press, 2014.

[9] Heejung Bang and James M Robins. Doubly Robust Estimation in Missing Data and Causal Inference Models. *Biometrics*, 61(4):962–973, 2005.

[10] Christopher D Barr, David M Diez, Yun Wang, Francesca Dominici, and Jonathan M Samet. Comprehensive smoking bans and acute myocardial infarction among medicare enrollees in 387 US counties: 1999-2008. *American Journal of Epidemiology*, 176(7):642–648, 2012.

[11] Xavier Basagaña, Inmaculada Aguilera, Marcela Rivera, David Agis, Maria Foraster, Jaume Marrugat, Roberto Elosua, and Nino Künzli. Measurement error in epidemiologic studies of air pollution based on land-use regression models. *American Journal of Epidemiology*, 178(8):1342–6, oct 2013.

[12] Hanefi Bayraktar and F Sezer Turalioglu. A Kriging-based approach for locating a sampling sitein the assessment of air quality. *Stochastic Environmental Research and Risk Assessment*, 19(4):301–305, oct 2005.

[13] Rob Beelen, Gerard Hoek, Danielle Vienneau, Marloes Eeftens, Konstantina Dimakopoulou, Xanthi Pedeli, Ming-Yi Tsai, Nino Künzli, Tamara Schikowski, Alessandro Marcon, Kirsten T. Eriksen, Ole Raaschou-Nielsen, Euripides Stephanou, Evridiki

Patelarou, Timo Lanki, Tarja Yli-Tuomi, Christophe Declercq, Grégoire Falq, Morgane Stempfelet, Matthias Birk, Josef Cyrys, Stephanie von Klot, Gizella Nádor, Mihály János Varró, Audrius Ddel, Regina Gražulevičien, Anna Mölter, Sarah Lindley, Christian Madsen, Giulia Cesaroni, Andrea Ranzi, Chiara Badaloni, Barbara Hoffmann, Michael Nonnemacher, Ursula Krämer, Thomas Kuhlbusch, Marta Cirach, Audrey de Nazelle, Mark Nieuwenhuijsen, Tom Bellander, Michal Korek, David Olsson, Magnus Strömgren, Evi Dons, Michael Jerrett, Paul Fischer, Meng Wang, Bert Brunekreef, and Kees de Hoogh. Development of NO2 and NOx land use regression models for estimating air pollution exposure in 36 study areas in Europe The ESCAPE project. *Atmospheric Environment*, 72(2):10–23, jun 2013.

[14] Rob Beelen, Ole Raaschou-Nielsen, Massimo Stafoggia, Zorana Jovanovic Andersen, Gudrun Weinmayr, Barbara Hoffmann, Kathrin Wolf, Evangelia Samoli, Paul Fischer, Mark Nieuwenhuijsen, Paolo Vineis, Wei W. Xun, Klea Katsouyanni, Konstantina Dimakopoulou, Anna Oudin, Bertil Forsberg, Lars Modig, Aki S. Havulinna, Timo Lanki, Anu Turunen, Bente Oftedal, Wenche Nystad, Per Nafstad, Ulf De Faire, Nancy L. Pedersen, Claes Göran Östenson, Laura Fratiglioni, Johanna Penell, Michal Korek, Göran Pershagen, Kirsten Thorup Eriksen, Kim Overvad, Thomas Ellermann, Marloes Eeftens, Petra H. Peeters, Kees Meliefste, Meng Wang, Bas Bueno-De-Mesquita, Dorothea Sugiri, Ursula Krämer, Joachim Heinrich, Kees De Hoogh, Timothy Key, Annette Peters, Regina Hampel, Hans Concin, Gabriele Nagel, Alex Ineichen, Emmanuel Schaffner, Nicole Probst-Hensch, Nino Künzli, Christian Schindler, Tamara Schikowski, Martin Adam, Harish Phuleria, Alice Vilier, Françoise Clavel-Chapelon, Christophe Declercq, Sara Grioni, Vittorio Krogh, Ming Yi Tsai, Fulvio Ricceri, Carlotta Sacerdote, Claudia Galassi, Enrica Migliore, Andrea Ranzi, Giulia Cesaroni, Chiara Badaloni, Francesco Forastiere, Ibon Tamayo, Pilar Amiano, Miren Dorronsoro, Michail Katsoulis, Antonia Trichopoulou, Bert Brunekreef, and Gerard Hoek. Effects of long-term exposure to air pollution on natural-cause mortality: An analysis of 22 European cohorts within the multicentre ESCAPE project. *The Lancet*, 383(9919):785–795, 2014.

[15] Michelle L Bell and Francesca Dominici. Effect modification by community characteristics on the short-term effects of ozone exposure and mortality in 98 US communities. *American Journal of Epidemiology*, 167(8):986–97, apr 2008.

[16] Michelle L Bell, Francesca Dominici, Keita Ebisu, Scott L Zeger, and Jonathan M Samet. Spatial and temporal variation in PM2.5 chemical composition in the United States for health effects studies. *Environmental Health Perspectives*, 115(7):989–995, 2007.

[17] Michelle L Bell, Francesca Dominici, and James M Samet. A meta-analysis of time-series studies of ozone and mortality with comparison to the national morbidity, mortality, and air pollution study. *Epidemiology*, 16(4):436–445, jul 2005.

[18] Michelle L. Bell, Keita Ebisu, Brian P. Leaderer, Janneane F. Gent, Hyung Joo Lee, Petros Koutrakis, Yun Wang, Francesca Dominici, and Roger D. Peng. Associations of PM2.5 Constituents and Sources with Hospital Admissions: Analysis of Four Counties in Connecticut and Massachusetts (USA) for Persons 65 Years of Age. *Environmental Health Perspectives*, 122(2):138–144, nov 2013.

[19] Michelle L Bell, Keita Ebisu, and Roger D Peng. Community-level spatial heterogeneity of chemical constituent levels of fine particulates and implications for epidemiological research. *Journal of Exposure Science & Environmental Epidemiology*, 21(4):372–384, 2011.

[20] Michelle L Bell, Aidan McDermott, Scott L Zeger, Jonathan M Samet, and Francesca Dominici. Ozone and short-term mortality in 95 US urban communities, 1987-2000. *JAMA*, 292(19):2372–8, nov 2004.

[21] Michelle L Bell, Marie S O'Neill, Nalini Ranjit, Victor H Borja-Aburto, Luis a Cifuentes, and Nelson C Gouveia. Vulnerability to heat-related mortality in Latin America: a case-crossover study in Sao Paulo, Brazil, Santiago, Chile and Mexico City, Mexico. *International Journal of Epidemiology*, 37(4):796–804, aug 2008.

[22] Michelle L Bell, Roger D Peng, and Francesca Dominici. The exposureresponse curve for ozone and risk of mortality and the adequacy of current ozone regulations. *Environmental Health Perspectives*, 114(4):532–536, jan 2006.

[23] Michelle L Bell, Jonathan M Samet, and Francesca Dominici. Time-series studies of particulate matter. *Annual Review of Public Health*, 25:247–280, jan 2004.

[24] Michelle L Bell, Antonella Zanobetti, and Francesca Dominici. Who is more affected by ozone pollution? A systematic review and meta-analysis. *American Journal of Epidemiology*, 180(1):15–28, 2014.

[25] Asa Ben-Hur, David Horn, Hava T Siegelmann, and Vladimir Vapnik. Support vector clustering. *Journal of Machine Learning Research*, 2:125–137, 2001.

[26] Silas Bergen, Lianne Sheppard, Paul D Sampson, Sun Young Kim, Mark Richards, Sverre Vedal, Joel D Kaufman, and Adam A Szpiro. A National Prediction Model for PM2.5 Component Exposures and Measurement ErrorCorrected Health Effect Inference. *Environmental Health Perspectives*, 121(9):1017–1025, jun 2013.

[27] Jennifer F Bobb, Francesca Dominici, and Roger D Peng. A Bayesian model averaging approach for estimating the relative risk of mortality associated with heat waves in 105 U.S. cities. *Biometrics*, 67(4):1605–16, dec 2011.

[28] Jennifer F Bobb, Linda Valeri, Birgit Claus Henn, David C Christiani, Robert O Wright, Maitreyi Mazumdar, John J Godleski, and Brent A Coull. Bayesian kernel machine regression for estimating the health effects of multi-pollutant mixtures. *Biostatistics*, 16(3):493–508, jul 2015.

[29] Michael Brauer, Cornel Lencar, Lillian Tamburic, Mieke Koehoorn, Paul Demers, and Catherine Karr. A Cohort Study of Traffic-Related Air Pollution Impacts on Birth Outcomes. *Environmental Health Perspectives*, 116(5):680–686, jan 2008.

[30] M Alan Brookhart, Sebastian Schneeweiss, Kenneth J Rothman, Robert J Glynn, Jerry Avorn, and Til Stürmer. Variable selection for propensity score models. *American Journal of Epidemiology*, 163(12):1149–56, jun 2006.

[31] Janet M Burke, Maria J Zufall, and H Ozkaynak. A population exposure model for particulate matter: case study results for PM2.5 in Philadelphia, PA. *Journal of Exposure Analysis and Environmental Epidemiology*, 11(6):470–489, 2001.

[32] Matthew Cefalu and Francesca Dominici. Does exposure prediction bias health-effect estimation?: The relationship between confounding adjustment and exposure prediction. *Epidemiology*, 25(4):583–90, jul 2014.

[33] Matthew Cefalu, Francesca Dominici, Nils Arvold, and Giovanni Parmigiani. Model averaged double robust estimation. *Biometrics*, 73(2):410–421, jun 2017.

[34] Howard H Chang, Montserrat Fuentes, and H Christopher Frey. Time series analysis of personal exposure to ambient air pollution and mortality using an exposure simulator. *Journal of Exposure Science and Environmental Epidemiology*, 22(5):483–488, sep 2012.

[35] Kenneth Y Chay and Michael Greenstone. The Impact of Air Pollution on Infant Mortality: Evidence from Geographic Variation in Pollution Shocks Induced by a Recession. *The Quarterly Journal of Economics*, 118(3):1121–1167, aug 2003.

[36] Kai Chen, Hai Bing Yang, Zong Wei Ma, Jun Bi, and Lei Huang. Influence of temperature to the short-term effects of various ozone metrics on daily mortality in Suzhou, China. *Atmospheric Environment*, 79:119–128, nov 2013.

[37] Hugh A. Chipman, Edward I. George, and Robert E. McCulloch. Bayesian treed models. *Machine Learning*, 48(1-3):299–320, 2002.

[38] Jungsoon Choi, Brian J. Reich, Montserrat Fuentes, and Jerry M. Davis. Multivariate Spatial-Temporal Modeling and Prediction of Speciated Fine Particles. *Journal of Statistical Theory and Practice*, 3(2):407–418, jun 2009.

[39] Luke Clancy, Pat Goodman, Hamish Sinclair, and Douglas W Dockery. Effect of air-pollution control on death rates in Dublin, Ireland: An intervention study. *Lancet*, 360(9341):1210–1214, 2002.

[40] David Clayton and Michael Hills. *Statistical Models in Epidemiology*. Oxford University Press, 1993.

[41] Merlise A Clyde. Model uncertainty and health effect studies for particulate matter. *Environmetrics*, 11(August 1999):745–763, 2000.

[42] Martin A. Cohen, Sara D. Adar, Ryan W. Allen, Edward Avol, Cynthia L. Curl, Timothy Gould, David Hardie, Anne Ho, Patrick Kinney, Timothy V. Larson, Paul Sampson, Lianne Sheppard, Karen D. Stukovsky, Susan S. Swan, L.-J. Sally Liu, and Joel D. Kaufman. Approach to Estimating Participant Pollutant Exposures in the Multi-Ethnic Study of Atherosclerosis and Air Pollution (MESA Air). *Environmental Science & Technology*, 43(13):4687–4693, jul 2009.

[43] Guido Consonni and Piero Veronese. Compatibility of Prior Specifications Across Linear Models. *Statistical Science*, 23(3):332–353, aug 2008.

[44] David Roxbee Cox and David Oakes. *Analysis of survival data*. CRC Press, 1984.

[45] Ciprian M Crainiceanu, Francesca Dominici, and Giovanni Parmigiani. Adjustment uncertainty in effect estimation. *Biometrika*, 95(3):635–651, sep 2008.

[46] Nello Cristianini and John Shawe-Taylor. *An introduction to support vector machines and other kernel-based learning methods*. Cambridge University Press, 2000.

[47] Dan L. Crouse, Paul A. Peters, Aaron van Donkelaar, Mark S. Goldberg, Paul J. Villeneuve, Orly Brion, Saeeda Khan, Dominic Odwa Atari, Michael Jerrett, C. Arden Pope, Michael Brauer, Jeffrey R. Brook, Randall V. Martin, David Stieb, and Richard T. Burnett. Risk of nonaccidental and cardiovascular mortality in relation to long-term exposure to low concentrations of fine particulate matter: A canadian national-level cohort study. *Environmental Health Perspectives*, 120(5):708–714, 2012.

[48] Frank C Curriero, Karlyn S Heiner, Jonathan M Samet, Scott L Zeger, Lisa Strug, and Jonathan A Patz. Temperature and mortality in 11 cities of the eastern United States. *American Journal of Epidemiology*, 155(1):80–7, jan 2002.

[49] Michael J Daniels, Francesca Dominici, Jonathan M Samet, and Scott L Zeger. Estimating Particulate Matter-Mortality Dose-Response Curves and Threshold Levels: An Analysis of Daily Time-Series for the 20 Largest US Cities. *American Journal of Epidemiology*, 152(5):397–406, sep 2000.

[50] David L Davies and Donald W Bouldin. A Cluster Separation Measure. *IEEE Transactions on Pattern Analysis and Machine Intelligence*, PAMI-1(2):224–227, apr 1979.

[51] Dawn L DeMeo, Antonella Zanobetti, Augusto A Litonjua, Brent A Coull, Joel Schwartz, and Diane R Gold. Ambient air pollution and oxygen saturation. *American Journal of Respiratory and Critical Care Medicine*, 170(4):383–387, 2004.

[52] Arthur P Dempster, Nann Laird, and Dona Rubin. Maximum likelihood from incomplete data via the EM algorithm. *Journal of the Royal Statistical Society, Series B*, 39(1):1–38, 1977.

[53] Olivier Deschenes, Michael Greenstone, and Joseph S Shapiro. Defensive investments and the demand for air quality: Evidence from the nox budget program and ozone reductions (No. w18267). Technical report, National Bureau of Economic Research, 2012.

[54] Qian Di, Petros Koutrakis, and Joel Schwartz. A hybrid prediction model for PM2.5 mass and components using a chemical transport model and land use regression. *Atmospheric Environment*, 131:390–399, 2016.

[55] Douglas W Dockery, C. Arden Pope, Xiping Xu, John D Spengler, James H Ware, Martha E Fay, Benjamin G Ferris, and Frank E Speizer. An Association between Air Pollution and Mortality in Six U.S. Cities. *New England Journal of Medicine*, 329(24):1753–1759, dec 1993.

[56] Francesca Dominici, Scott L Zeger, and James M Samet. A measurement error model for time-series studies of air pollution and mortality. *Biostatistics*, 1(2):157–75, 2000.

[57] Francesca Dominici, Michael Daniels, Scott L Zeger, and Jonathan M Samet. Air Pollution and Mortality: Estimating Regional and National DoseResponse Relationships. *Journal of the American Statistical Association*, 97(457):100–111, mar 2002.

[58] Francesca Dominici, Michael Greenstone, and Cass R Sunstein. Particulate Matter Matters. *Science*, 344(6181):257–259, apr 2014.

[59] Francesca Dominici, Aidan McDermott, and Trevor J Hastie. Improved semiparametric time series models of air pollution and mortality. *Journal of the American Statistical Association*, 99(468):938–948, dec 2004.

[60] Francesca Dominici, Aidan McDermott, Scott L. Zeger, and Jonathan M Samet. On the use of generalized additive models in time-series of air pollution and health. *American Journal of Epidemiology*, 156(3):193–203, 2002.

[61] Francesca Dominici, Roger D Peng, Michelle L Bell, Luu Pham, Aidan McDermott, Scott L Zeger, and Jonathan M Samet. Fine Particulate Air Pollution and Hospital Admission for Cardiovascular and Respiratory Diseases. *JAMA*, 295(10):1127–1134, 2006.

[62] Francesca Dominici, Jonathan M Samet, and Scott L Zeger. Combining evidence on air pollution and daily mortality from the 20 largest US cities: a hierarchical modelling strategy. *Journal of the Royal Statistical Society: Series A*, 163(3):263–302, oct 2000.

[63] Francesca Dominici, Lianne Sheppard, and Merlise Clyde. Health effects of air pollution: A statistical review. *International Statistical Review*, 71:243–276, 2003.

[64] Marloes Eeftens, Rob Beelen, Kees de Hoogh, Tom Bellander, Giulia Cesaroni, Marta Cirach, Christophe Declercq, Audrius Ddel, Evi Dons, Audrey de Nazelle, Konstantina Dimakopoulou, Kirsten Eriksen, Grégoire Falq, Paul Fischer, Claudia Galassi, Regina Gražulevičien, Joachim Heinrich, Barbara Hoffmann, Michael Jerrett, Dirk Keidel, Michal Korek, Timo Lanki, Sarah Lindley, Christian Madsen, Anna Mölter, Gizella Nádor, Mark Nieuwenhuijsen, Michael Nonnemacher, Xanthi Pedeli, Ole Raaschou-Nielsen, Evridiki Patelarou, Ulrich Quass, Andrea Ranzi, Christian Schindler, Morgane Stempfelet, Euripides Stephanou, Dorothea Sugiri, Ming-Yi Tsai, Tarja Yli-Tuomi, Mihály J Varró, Danielle Vienneau, Stephanie von Klot, Kathrin Wolf, Bert Brunekreef, and Gerard Hoek. Development of Land Use Regression Models for PM 2.5 , PM 2.5 Absorbance, PM 10 and PM coarse in 20 European Study Areas; Results of the ESCAPE Project. *Environmental Science & Technology*, 46(20):11195–11205, oct 2012.

[65] Montserrat Fuentes, Hae Ryoung Song, Sujit K. Ghosh, David M. Holland, and Jerry M. Davis. Spatial association between speciated fine particles and mortality. *Biometrics*, 62(3):855–863, 2006.

[66] Antonio Gasparrini. Modeling exposure-lag-response associations with distributed lag non-linear models. *Statistics in Medicine*, 33(5):881–899, 2014.

[67] Antonio Gasparrini, Yue-Liang Leon Yuming Yue Liang Leon Guo, Masahiro Hashizume, Eric Lavigne, Antonella Zanobetti, Joel Schwartz, Aurelio Tobias, Shilu Tong, Joacim Rocklöv, Bertil Forsberg, Michela Leone, Manuela De Sario, Michelle L. Bell, Yue-Liang Leon Yuming Yue Liang Leon Guo, Chang-fu Fu Wu, Haidong Kan, Seung-Muk Muk Yi, Micheline De Sousa Zanotti Stagliorio Coelho, Paulo Hilario Nascimento Saldiva, Yasushi Honda, Ho Kim, and Ben Armstrong. Mortality risk attributable to high and low ambient temperature: a multicountry observational study. *The Lancet*, 386(9991):369–375, jul 2015.

[68] Ulrike Gehring, Olena Gruzieva, Raymond M Agius, Rob Beelen, Adnan Custovic, Josef Cyrys, Marloes Eeftens, Claudia Flexeder, Elaine Fuertes, Joachim Heinrich, Barbara Hoffmann, Johan C. de Jongste, Marjan Kerkhof, Claudia Klümper, Michal Korek, Anna Mölter, Erica S. Schultz, Angela Simpson, Dorothea Sugiri, Magnus Svartengren, Andrea von Berg, Alet H. Wijga, Göran Pershagen, and Bert Brunekreef. Air pollution exposure and lung function in children: The ESCAPE project. *Environmental Health Perspectives*, 121(11-12):1357–1364, 2013.

[69] Thomas A Glass, Steven N Goodman, Miguel A Hernán, and Jonathan M Samet. Causal inference in public health. *Annual Review of Public Health*, 34:61–75, 2013.

[70] Peter J Green. Reversible Jump Markov Chain Monte Carlo Computation and Bayesian Model Determination. *Biometrika*, 82(4):711–732, 1995.

[71] Sander Greenland. Invited commentary: variable selection versus shrinkage in the control of multiple confounders. *American Journal of Epidemiology*, 167(5):523–529, mar 2008.

[72] Sonja Greven, Francesca Dominici, and Scott L Zeger. An Approach to the Estimation of Chronic Air Pollution Effects Using Spatio-Temporal Information. *Journal of the American Statistical Association*, 106(494):396–406, jun 2011.

[73] Alexandros Gryparis, Christopher J. Paciorek, Ariana Zeka, Joel Schwartz, and Brent A. Coull. Measurement error caused by spatial misalignment in environmental epidemiology. *Biostatistics*, 10(2):258–274, 2009.

[74] Amber J. Hackstadt, Elizabeth C. Matsui, D'Ann L. Williams, Gregory B. Diette, Patrick N. Breysse, Arlene M. Butz, and Roger D. Peng. Inference for environmental intervention studies using principal stratification. *Statistics in Medicine*, 33(28):4919–4933, 2014.

[75] Anjum Hajat, Matthew Allison, Ana V Diez-Roux, Nancy Swords Jenny, Neal W Jorgensen, Adam A Szpiro, Sverre Vedal, and Joel D Kaufman. Long-term Exposure to Air Pollution and Markers of Inflammation, Coagulation, and Endothelial Activation A Repeat-measures Analysis in the Multi-Ethnic Study of Atherosclerosis (MESA). *Epidemiology*, 26(3):310–320, 2015.

[76] Chris Hans. Model uncertainty and variable selection in Bayesian lasso regression. *Statistics and Computing*, 20(2):221–229, 2010.

[77] Jamie E Hart, Xiaomei Liao, Biling Hong, Robin C Puett, Jeff D Yanosky, Helen Suh, Marianthi-Anna Kioumourtzoglou, Donna Spiegelman, and Francine Laden. The association of long-term exposure to PM2.5 on all-cause mortality in the Nurses' Health Study and the impact of measurement-error correction. *Environmental Health*, 14:38, 2015.

[78] J A Hartigan and M A Wong. Algorithm AS 136: A K-Means Clustering Algorithm. *Journal of the Royal Statistical Society: Series C*, 28(1):100–108, 1979.

[79] Trevor J Hastie and Robert J Tibshirani. *Generalized Additive Models*. CRC Press, 1990.

[80] Matthew J Heaton and Roger D Peng. Flexible Distributed Lag Models using Random Functions with Application to Estimating Mortality Displacement from Heat-Related Deaths. *Journal of Agricultural, Biological, and Environmental Statistics*, 17(3):313–331, sep 2012.

[81] Matthew J Heaton and Roger D Peng. Extending distributed lag models to higher degrees. *Biostatistics*, 15(2):398–412, apr 2014.

[82] Gerard Hoek, Marloes Eeftens, Rob Beelen, Paul Fischer, Bert Brunekreef, K. Folkert Boersma, and Pepijn Veefkind. Satellite NO2 data improve national land use regression models for ambient NO2 in a small densely populated country. *Atmospheric Environment*, 105(2):173–180, 2015.

[83] Jennifer A Hoeting, David Madigan, Adrian E Raftery, and Chris T Volinsky. Bayesian Model Averaging: A Tutorial. *Statistical Science*, 14(4):382–417, nov 1999.

[84] Yi Huang, Francesca Dominici, and Michelle L Bell. Bayesian hierarchical distributed lag models for summer ozone exposure and cardio-respiratory mortality. *Environmetrics*, 16(5):547–562, aug 2005.

[85] Perry Hystad, Eleanor Setton, Alejandro Cervantes, Karla Poplawski, Steeve Desch-enes, Michael Brauer, Aaron van Donkelaar, Lok Lamsa, Randall Martin, Michael Jerrett, and Paul Demers. Creating national air pollution models for population ex-posure assessment in Canada. *Environmental Health Perspectives*, 119(8):1123–1129, 2011.

[86] Holly Janes, Francesca Dominici, and Scott L. Zeger. Trends in Air Pollution and Mortality: an approach to the assessment of unmeasured confounding. *Epidemiology*, 18(4):416–423, 2007.

[87] Holly Janes, Lianne Sheppard, and Thomas Lumley. Case Crossover Analyses of Air Pollution Exposure Data: referent selection strategies and their implications for bias. *Epidemiology*, 16(6):717–726, 2005.

[88] R Kass and A Raftery. Bayes factors. *Journal of the American Statistical Association*, 90(430):773 – 95, 1995.

[89] Klea Katsouyanni, Jonathan M Samet, H Ross Anderson, Richard Atkinson, Alain Le Tertre, Sylvia Medina, Evangelia Samoli, Giota Touloumi, Richard T Burnett, Daniel Krewski, Timothy Ramsay, Francesca Dominici, Roger D Peng, Joel Schwartz, Antonella Zanobetti, and HEI Health Review Committee. Air pollution and health: a European and North American approach (APHENA). Technical Report 142, Health Effects Institute, Boston, MA, 2009.

[90] Joel D Kaufman, Sara D Adar, R Graham Barr, Matthew Budoff, Gregory L Burke, Cynthia L Curl, Martha L Daviglus, Ana V Diez Roux, Amanda J Gassett, David R Jacobs, Richard Kronmal, Timothy V Larson, Ana Navas-Acien, Casey Olives, Paul D Sampson, Lianne Sheppard, David S Siscovick, James H Stein, Adam A Szpiro, and Karol E Watson. Association between air pollution and coronary artery calcification within six metropolitan areas in the USA (the Multi-Ethnic Study of Atherosclerosis and Air Pollution): a longitudinal cohort study. *The Lancet*, 6736(16):1–9, 2016.

[91] Joshua P. Keller, Casey Olives, Sun-Young Kim, Lianne Sheppard, Paul D. Samp-son, Adam A. Szpiro, Assaf P. Oron, Johan Lindström, Sverre Vedal, and Joel D. Kaufman. A Unified Spatiotemporal Modeling Approach for Predicting Concentra-tions of Multiple Air Pollutants in the Multi-Ethnic Study of Atherosclerosis and Air Pollution. *Environmental Health Perspectives*, 58(12):7250–7, nov 2014.

[92] S. Kim, M. G. Tadesse, and Marina Vannucci. Variable selection in clustering via Dirichlet process mixture models. *Biometrika*, 93(4):877–893, dec 2006.

[93] Sun-Young Kim, Lianne Sheppard, Silas Bergen, Adam A Szpiro, Paul D Sampson, Joel D Kaufman, and Sverre Vedal. Prediction of fine particulate matter chemical com-ponents with a spatio-temporal model for the Multi-Ethnic Study of Atherosclerosis cohort. *Journal of Exposure Science and Environmental Epidemiology*, 26(5):520–528, sep 2016.

[94] Marianthi-Anna Kioumourtzoglou, Joel D Schwartz, Marc G Weisskopf, Steven J Melly, Yan Wang, Francesca Dominici, and Antonella Zanobetti. Long-term PM Exposure and Neurological Hospital Admissions in the Northeastern United States. *Environmental Health Perspectivesealth perspectives*, 124(1):23–29, 2015.

[95] Marianthi-Anna Kioumourtzoglou, Donna Spiegelman, Adam A Szpiro, Lianne Shep-pard, Joel D Kaufman, Jeff D Yanosky, Ronald Williams, Francine Laden, Biling

Hong, and Helen Suh. Exposure measurement error in PM2.5 health effects studies: A pooled analysis of eight personal exposure validation studies. *Environmental Health*, 13(1):2, 2014.

[96] Marianthi-Anna Kioumourtzoglou, Antonella Zanobetti, Joel D Schwartz, Brent a Coull, Francesca Dominici, and Helen H Suh. The effect of primary organic particles on emergency hospital admissions among the elderly in 3 US cities. *Environmental Health*, 12(1):68, 2013.

[97] Itai Kloog, Brent A Coull, Antonella Zanobetti, Petros Koutrakis, and Joel D Schwartz. Acute and chronic effects of particles on hospital admissions in New-England. *PLoS ONE*, 7(4):2–9, 2012.

[98] Itai Kloog, Petros Koutrakis, Brent A Coull, Hyung Joo Lee, and Joel Schwartz. Assessing temporally and spatially resolved PM 2.5 exposures for epidemiological studies using satellite aerosol optical depth measurements. *Atmospheric Environment*, 45(35):6267–6275, 2011.

[99] Itai Kloog, Steven J Melly, William L Ridgway, Brent a Coull, and Joel Schwartz. Using new satellite based exposure methods to study the association between pregnancy PM2.5 exposure, premature birth and birth weight in Massachusetts. *Environmental Health*, 11(1):40, 2012.

[100] Itai Kloog, Francesco Nordio, Antonella Zanobetti, Brent A Coull, Petros Koutrakis, and Joel D Schwartz. Short term effects of particle exposure on hospital admissions in the mid-atlantic states: A population estimate. *PLoS ONE*, 9(2):1–7, 2014.

[101] Gary Koop and Lise Tole. Measuring the health effects of air pollution: to what extent can we really say that people are dying from bad air? *Journal of Environmental Economics and Management*, 47(1):30–54, jan 2004.

[102] Daniel Krewski, Richard T Burnett, Mark S Goldberg, Kristin Hoover, Jack Siemiatycki, Michael Abrahamowicz, and Warren H White. Validation of the Harvard Six Cities Study of Particulate Air Pollution and Mortality. *New England Journal of Medicine*, 350(2):198–199, 2004.

[103] Daniel Krewski, Richard T Burnett, Mark S Goldberg, Kristin Hoover, Jack Siemiatycki, Michael Jerrett, Michael Abrahamowicz, and Warren H White. Reanalysis of the Harvard Six Cities Study and the American Cancer Society Study of particulate air pollution and mortality. Technical report, Health Effects Institute, Cambridge, MA, 2000.

[104] Daniel Krewski, Michael Jerrett, Richard T Burnett, Renjun Ma, Edward Hughes, Yuanli Shi, Michelle C Turner, C Arden Pope, George Thurston, Eugenia E Calle, Michael J Thun, Bernie Beckerman, Pat DeLuca, Norm Finkelstein, Kaz Ito, D K Moore, K Bruce Newbold, Tim Ramsay, Zev Ross, Hwashin Shin, and Barbara Tempalski. Extended follow-up and spatial analysis of the American Cancer Society study linking particulate air pollution and mortality. Technical Report 140, Health Effects Institute, Boston, MA, 2009.

[105] Francine Laden, Lucas M Neas, Douglas W Dockery, and Joel Schwartz. Association of Fine Particulate Matter from Different Sources with Daily Mortality in Six U.S. Cities. *Environmental Health Perspectives*, 108(10):941–947, aug 2000.

[106] Francine Laden, Joel Schwartz, Frank E Speizer, and Douglas W Dockery. Reduction in Fine Particulate Air Pollution and Mortality. *American Journal of Respiratory and Critical Care Medicine*, 173(6):667–672, mar 2006.

[107] Edward E. Leamer. A Class of Informative Priors and Distributed Lag Analysis. *Econometrica*, 40(6):1059, nov 1972.

[108] David S Lee and Thomas Lemieux. Regression Discontinuity Designs in Economics. *Journal of Economic Literature*, 48(2):281–355, jun 2010.

[109] Johan Lindström, Adam A. Szpiro, Paul D. Sampson, Assaf P. Oron, Mark Richards, Tim V. Larson, and Lianne Sheppard. A Flexible Spatio-Temporal Model for Air Pollution with Spatial and Spatio-Temporal Covariates. *Environmental and Ecological Statistics*, 21(3):411–433, sep 2014.

[110] Dawei Liu, Xihong Lin, and Debashis Ghosh. Semiparametric regression of multi-dimensional genetic pathway data: least-squares kernel machines and linear mixed models. *Biometrics*, 63(4):1079–88, 2007.

[111] Muhammad Mamdani, Kathy Sykora, Ping Li, Sharon-Lise T Normand, David L Streiner, Peter C Austin, Paula A Rochon, and Geoffrey M Anderson. Reader's guide to critical appraisal of cohort studies: 2. Assessing potential for confounding. *BMJ (Clinical research ed.)*, 330(7497):960–962, apr 2005.

[112] Lawrence C. McCandless, Paul Gustafson, and Peter C. Austin. Bayesian propensity score analysis for observational data. *Statistics in Medicine*, 28(1):94–112, jan 2009.

[113] Peter McCullagh and John A Nelder. *Generalized Linear Models*. CRC Press, 1989.

[114] Kelly Moore, Romain Neugebauer, Frederick Lurmann, Jane Hall, Victor Brajer, Sianna Alcorn, and Ira Tager. Ambient ozone concentrations and cardiac mortality in southern california 1983-2000: Application of a new marginal structural model approach. *American Journal of Epidemiology*, 171(11):1233–1243, 2010.

[115] Sharon-Lise T Normand, Kathy Sykora, Ping Li, Muhammad Mamdani, Paula A Rochon, and Geoffrey M Anderson. Readers guide to critical appraisal of cohort studies: 3. Analytical strategies to reduce confounding. *BMJ (Clinical research ed.)*, 330(7498):1021–1023, apr 2005.

[116] C Olives, L Sheppard, J Lindstrom, P D Sampson, J D Kaufman, and A A Szpiro. Reduced-Rank Spatio-Temporal Modeling of Air Pollution Concentrations in the Multi-Ethnic Study of Atherosclerosis and Air Pollution. *Annals of Applied Statistics*, 8(4):2509–2537, 2014.

[117] Sung Kyun Park, Marie S. O'Neill, Pantel S. Vokonas, David Sparrow, and Joel Schwartz. Effects of air pollution on heart rate variability: The VA normative aging study. *Environmental Health Perspectives*, 113(3):304–309, 2005.

[118] Trevor Park and George Casella. The Bayesian Lasso. *Journal of the American Statistical Association*, 103(482):681–686, 2008.

[119] Jennifer L Peel, Mitchell Klein, W Dana Flanders, James A Mulholland, Paige E Tolbert, and HEI Health Review Committee. Impact of improved air quality during the 1996 Summer Olympic Games in Atlanta on multiple cardiovascular and respiratory outcomes. Technical Report 148, Health Effects Institute, Boston, MA, 2010.

[120] Roger D. Peng and Michelle L. Bell. Spatial misalignment in time series studies of air pollution and health data. *Biostatistics*, 11(4):720–740, 2010.

[121] Roger D Peng, Howard H Chang, and Michelle L Bell. Coarse particulate matter air pollution and hospital admissions for cardiovascular and respiratory diseases among Medicare patients. *JAMA*, 299(18):2172–9, may 2008.

[122] Roger D Peng, Francesca Dominici, and Thomas A Louis. Model choice in time series studies of air pollution and mortality. *Journal of the Royal Statistical Society: Series A*, 169(2):179–203, mar 2006.

[123] Roger D Peng, Francesca Dominici, and Leah J Welty. A Bayesian hierarchical distributed lag model for estimating the time course of risk of hospitalization associated with particulate matter air pollution. *Journal of the Royal Statistical Society: Series C*, 58(1):3–24, feb 2009.

[124] C Arden Pope. Mortality effects of longer term exposures to fine particulate air pollution: review of recent epidemiological evidence. *Inhalation toxicology*, 19 Suppl 1(October 2006):33–38, 2007.

[125] C Arden Pope, Richard T Burnett, George D Thurston, Michael J Thun, Eugenia E Calle, Daniel Krewski, and John J Godleski. Cardiovascular Mortality and Long-Term Exposure to Particulate Air Pollution: Epidemiological Evidence of General Pathophysiological Pathways of Disease. *Circulation*, 109(1):71–77, 2004.

[126] C Arden Pope, Michael J Thun, Mohan M Namboodiri, Douglas W Dockery, John S Evans, Frank E Speizer, and Clark W. Heath Jr. Particulate air pollution as a predictor of mortality in a prospective study of U.S. adults. *American Journal of Respiratory and Critical Care Medicine*, 151(3 Pt 1):669–674, 1995.

[127] C Arden Pope III, Richard T Burnett, Michael J Thun, Eugenia E Calle, Daniel Krewski, Kazuhiko Ito, and George D Thurston. Lung cancer, cardiopulmonary mortality, and long-term exposure to fine particulate air pollution. *JAMA*, 287(9):1132, mar 2002.

[128] Helen Powell, Jenna R. Krall, Yun Wang, Michelle L. Bell, and Roger D. Peng. Ambient Coarse Particulate Matter and Hospital Admissions in the Medicare Cohort Air Pollution Study, 19992010. *Environmental Health Perspectives*, 123(11):1152–1158, apr 2015.

[129] Ole Raaschou-Nielsen, Zorana J. Andersen, Rob Beelen, Evangelia Samoli, Massimo Stafoggia, Gudrun Weinmayr, Barbara Hoffmann, Paul Fischer, Mark J. Nieuwenhuijsen, Bert Brunekreef, Wei W. Xun, Klea Katsouyanni, Konstantina Dimakopoulou, Johan Sommar, Bertil Forsberg, Lars Modig, Anna Oudin, Bente Oftedal, Per E. Schwarze, Per Nafstad, Ulf De Faire, Nancy L. Pedersen, Claes Göran Östenson, Laura Fratiglioni, Johanna Penell, Michal Korek, Göran Pershagen, Kirsten T. Eriksen, Mette Sørensen, Anne Tjønneland, Thomas Ellermann, Marloes Eeftens, Petra H. Peeters, Kees Meliefste, Meng Wang, Bas Bueno-de Mesquita, Timothy J. Key, Kees de Hoogh, Hans Concin, Gabriele Nagel, Alice Vilier, Sara Grioni, Vittorio Krogh, Ming Yi Tsai, Fulvio Ricceri, Carlotta Sacerdote, Claudia Galassi, Enrica Migliore, Andrea Ranzi, Giulia Cesaroni, Chiara Badaloni, Francesco Forastiere, Ibon Tamayo, Pilar Amiano, Miren Dorronsoro, Antonia Trichopoulou, Christina Bamia, Paolo Vineis, and Gerard Hoek. Air pollution and lung cancer incidence in 17 European cohorts: Prospective analyses from the European Study of Cohorts for Air Pollution Effects (ESCAPE). *The Lancet Oncology*, 14(9):813–822, 2013.

[130] Adrian E Raftery, David Madigan, and Jennifer A Hoeting. Bayesian model averaging for linear regression models. *Journal of the American Statistical Association*, 92(437):179, mar 1997.

[131] Michael R Ransom and C Arden Pope III. External health costs of a steel mill. *Contemporaneous Economic Policy*, 13(2):86–97, 1995.

[132] Cizao Ren, Gail M Williams, Kerrie Mengersen, Lidia Morawska, and Shilu Tong. Does temperature modify short-term effects of ozone on total mortality in 60 large eastern US communities? An assessment using the NMMAPS data. *Environment International*, 34(4):451–8, may 2008.

[133] David Q Rich, Howard M Kipen, Wei Huang, Guangfa Wang, Yuedan Wang, Ping Zhu, Pamela Ohman-Strickland, Min Hu, Claire Philipp, Scott R Diehl, Shou-En Lu, Jian Tong, Jicheng Gong, Duncan Thomas, Tong Zhu, and Junfeng (Jim) Zhang. Association Between Changes in Air Pollution Levels During the Beijing Olympics and Biomarkers of Inflammation and Thrombosis in Healthy Young Adults. *JAMA*, 307(19):2068, may 2012.

[134] Steven Roberts. An investigation of distributed lag models in the context of air pollution and mortality time series analysis. *Journal of the Air & Waste Management Association (1995)*, 55(3):273–282, 2005.

[135] James M Robins, Steven D Mark, and Whitney K Newey. Estimating exposure effects by modelling the expectation of exposure conditional on confounders. *Biometrics*, 48(2):479–95, jun 1992.

[136] Paul R Rosenbaum and Donald B Rubin. The Central Role of the Propensity Score in Observational Studies for Causal Effects. *Biometrika*, 70(1):41, apr 1983.

[137] Donald B Rubin. Estimating causal effects from large data sets using propensity scores. *Annals of Internal Medicine*, 127(8 Pt 2):757–63, oct 1997.

[138] James M Samet and Francesca Dominici. Fine particulate air pollution and mortality in 20 US cities, 19871994. *New England Journal of Medicine*, 343(24):1742–9, dec 2000.

[139] Jonathan M Samet, Francesca Dominici, Scott L Zeger, Joel Schwartz, and Douglas W Dockery. The National Morbidity, Mortality, and Air Pollution Study. Part I: Methods and methodologic issues. Technical Report 94 Pt 1, Health Effects Institute, Cambridge, MA, jun 2000.

[140] Jonathan M Samet, Scott L Zeger, Francesca Dominici, Frank C Curriero, Ivan Coursac, Douglas W Dockery, J Schwartz, and A Zanobetti. The National Morbidity, Mortality, and Air Pollution Study. Part II: Morbidity and mortality from air pollution in the United States. Technical Report Pt 2, Health Effects Institute, Cambridge, MA, jun 2000.

[141] Evangelia Samoli, Roger Peng, Tim Ramsay, Marina Pipikou, Giota Touloumi, Francesca Dominici, Rick Burnett, Aaron Cohen, Daniel Krewski, Jon Samet, and Klea Katsouyanni. Acute effects of ambient particulate matter on mortality in Europe and North America: Results from the APHENA study. *Environmental Health Perspectives*, 116(11):1480–1486, 2008.

[142] Paul D Sampson, Adam A Szpiro, Lianne Sheppard, Johan Lindström, and Joel D Kaufman. Pragmatic estimation of a spatio-temporal air quality model with irregular monitoring data. *Atmospheric Environment*, 45(36):6593–6606, 2011.

[143] Daniel O. Scharfstein, Andrea Rotnitzky, and James M. Robins. Adjusting for Nonignorable Drop-Out Using Semiparametric Nonresponse Models. *Journal of the American Statistical Association*, 94(448):1096–1120, 1999.

[144] Joel Schwartz and Douglas W Dockery. Increased mortality in Philadelphia associated with daily air pollution concentrations. *The American Review of Respiratory Disease*, 145:600–604, 1992.

[145] Joel Schwartz. Assessing Confounding, Effect Modification, and Thresholds in the Association between Ambient Particles and Daily Deaths. *Environmental Health Perspectives*, 108(6):563, jun 2000.

[146] Joel Schwartz, Stacey E Alexeeff, Irina Mordukhovich, Alexandros Gryparis, Pantel Vokonas, Helen Suh, and Brent A Coull. Association between long-term exposure to traffic particles and blood pressure in the Veterans Administration Normative Aging Study. *Occupational and Environmental Medicine*, 69(6):422–427, 2012.

[147] Joel Schwartz, Marie-Abele Bind, and Petros Koutrakis. Estimating Causal Effects of Local Air Pollution on Daily Deaths: Effect of Low Levels. *Environmental Health Perspectives*, 125(1), may 2016.

[148] Joel Schwartz, Brent Coull, Francine Laden, and Louise Ryan. The effect of dose and timing of dose on the association between airborne particles and survival. *Environmental Health Perspectives*, 116(1):64–69, 2008.

[149] Joel Schwartz, Douglas W Dockery, and Lucas M Neas. Is Daily Mortality Associated Specifically with Fine Particles? *Journal of the Air & Waste Management Association*, 46(10):927–939, oct 1996.

[150] Joel D Schwartz. The distributed lag between air pollution and daily deaths. *Epidemiology*, 11(3):320–326, may 2000.

[151] Terence A R Seemungal, Gavin C Donaldson, Angshu Bhowmik, Donald J Jeffries, and Jadwiga A Wedzicha. Time Course and Recovery of Exacerbations in Patients with Chronic Obstructive Pulmonary Disease. *American Journal of Respiratory and Critical Care Medicine*, 161:1608–1613, 2000.

[152] Lianne Sheppard, Richard T Burnett, Adam A Szpiro, Sun Young Kim, Michael Jerrett, C. Arden Pope, and Bert Brunekreef. Confounding and exposure measurement error in air pollution epidemiology. *Air Quality, Atmosphere and Health*, 5(2):203–216, 2012.

[153] Liuhua Shi, Antonella Zanobetti, Itai Kloog, Brent A. Coull, Petros Koutrakis, Steven J. Melly, and Joel D. Schwartz. Low-concentration PM2.5 and mortality: Estimating acute and chronic effects in a population-based study. *Environmental Health Perspectives*, 124(1):46–52, 2016.

[154] Robert J. Shiller. A Distributed Lag Estimator Derived from Smoothness Priors. *Econometrica*, 41(4):775, jul 1973.

[155] Richard L Smith, Dan Spitzner, Yuntae Kim, and Montserrat Fuentes. Threshold dependence of mortality effects for fine and coarse particles in Phoenix, Arizona. *Journal of the Air & Waste Management Association (1995)*, 50(8):1367–1379, 2000.

[156] Adam A Szpiro and Christopher J Paciorek. Measurement error in two-stage analyses, with application to air pollution epidemiology. *Environmetrics*, 24(8):501–517, 2013.

[157] Adam A Szpiro, Lianne Sheppard, Sara D Adar, and Joel D Kaufman. Estimating acute air pollution health effects from cohort study data. *Biometrics*, 70(1):164–174, 2014.

[158] Adam A. Szpiro, Lianne Sheppard, and Thomas Lumley. Efficient measurement error correction with spatially misaligned data. *Biostatistics*, 12(4):610–623, 2011.

[159] Duncan C Thomas. *Statistical Methods in Environmental Epidemiology*. Oxford University Press, USA, 2009.

[160] US EPA. Air Quality System (AQS): www.epa.gov/aqs.

[161] US EPA. Integrated science assessment for particulate matter. Technical report, U.S. Environmental Protection Agency, Washington, DC, 2009.

[162] US EPA. Integrated Science Assessment for Ozone and Related Photochemical Oxidants Integrated Science Assessment for Ozone and Related Photochemical Oxidants. Technical report, US Environmental Protection Agency, Washington, DC, 2013.

[163] Aaron van Donkelaar, Randall V. Martin, Michael Brauer, Ralph Kahn, Robert Levy, Carolyn Verduzco, and Paul J. Villeneuve. Global estimates of ambient fine particulate matter concentrations from satellite-based aerosol optical depth: Development and application. *Environmental Health Perspectives*, 118(6):847–855, 2010.

[164] Valerie Viallefont, Adrian E. Raftery, and Sylvia Richardson. Variable selection and Bayesian model averaging in case-control studies. *Statistics in Medicine*, 20(21):3215–3230, 2001.

[165] Chi Wang, Francesca Dominici, Giovanni Parmigiani, and Corwin Matthew Zigler. Accounting for uncertainty in confounder and effect modifier selection when estimating average causal effects in generalized linear models. *Biometrics*, 71(3):654–665, sep 2015.

[166] Chi Wang, Giovanni Parmigiani, and Francesca Dominici. Bayesian effect estimation accounting for adjustment uncertainty. *Biometrics*, 68(3):661–71, sep 2012.

[167] Meng Wang, Bert Brunekreef, Ulrike Gehring, Adam Szpiro, Gerard Hoek, and Rob Beelen. A New Technique for Evaluating Land-use Regression Models and Their Impact on Health Effect Estimates. *Epidemiology*, 27(1):51–56, 2016.

[168] Meng Wang, Paul D Sampson, Jianlin Hu, Michael Kleeman, Joshua P Keller, Casey Olives, Adam A. Szpiro, Sverre Vedal, and Joel D Kaufman. Combining Land-Use Regression and Chemical Transport Modeling in a Spatiotemporal Geostatistical Model for Ozone and PM 2.5. *Environmental Science & Technology*, 50(10):5111–5118, may 2016.

[169] Yan Wang, Itai Kloog, Brent A Coull, Anna Kosheleva, Antonella Zanobetti, and Joel D Schwartz. Estimating Causal Effects of Long-Term PM2.5 Exposure on Mortality in New Jersey. *Environmental Health Perspectives*, 124(8), apr 2016.

[170] Joshua Warren, Montserrat Fuentes, Amy H Herring, and Peter H Langlois. Spatial-temporal modeling of the association between air pollution exposure and preterm birth: identifying critical windows of exposure. *Biometrics*, 68(4):1157–67, dec 2012.

[171] Leah J Welty, Roger D Peng, Scott L Zeger, and Francesca Dominici. Bayesian distributed lag models: estimating effects of particulate matter air pollution on daily mortality. *Biometrics*, 65(1):282–91, mar 2009.

[172] Jennifer Weuve, Robin C Puett, Joel Schwartz, Jeff D Yanosky, Francine Laden, and Francine Grodstein. Exposure to particulate air pollution and cognitive decline in older women. *Archives of Internal Medicine*, 172(3):219–227, feb 2012.

[173] WHO. Air quality guidelines: global update 2005: particulate matter, ozone, nitrogen dioxide, and sulfur dioxide. Technical report, World Health Organization, 2006.

[174] WHO. Health effects of particulate matter. Policy implications for countries in eastern Europe, Caucasus and central Asia. Technical report, World Health Organization, 2013.

[175] Ander Wilson, Ana G Rappold, Lucas M Neas, and Brian J Reich. Modeling the effect of temperature on ozone-related mortality. *The Annals of Applied Statistics*, 8(3):1728–1749, sep 2014.

[176] Ander Wilson and Brian J Reich. Confounder selection via penalized credible regions. *Biometrics*, 70(4):852–861, dec 2014.

[177] Ander Wilson, Corwin Matthew Zigler, Chirag J Patel, and Francesca Dominici. Model-averaged confounder adjustment for estimating multivariate exposure effects with linear regression. *Biometrics*, 74(3):1034–1044, sep 2018.

[178] Chit-Ming Wong, Nuntavarn Vichit-Vadakan, Haidong Kan, and Zhengmin Qian. Public Health and Air Pollution in Asia (PAPA): a multicity study of short-term effects of air pollution on mortality. *Environmental Health Perspectives*, 116(9):1195–1202, 2008.

[179] Shu Yang, Guido W Imbens, Zhanglin Cui, Douglas E Faries, and Zbigniew Kadziola. Propensity score matching and subclassification in observational studies with multi-level treatments. *Biometrics*, 4(2001):n/a–n/a, mar 2016.

[180] Jeff D Yanosky, Christopher J Paciorek, Francine Laden, Jaime E Hart, Robin C Puett, Duanping Liao, and Helen H Suh. Spatio-temporal modeling of particulate air pollution in the conterminous United States using geographic and meteorological predictors. *Environmental Health*, 13(1):63, 2014.

[181] Jeff D. Yanosky, Christopher J. Paciorek, and Helen H. Suh. Predicting chronic fine and coarse particulate exposures using spatiotemporal models for the northeastern and midwestern United States. *Environmental Health Perspectives*, 117(4):522–529, 2009.

[182] Ka Yee Yeung, Roger E. Bumgarner, and Adrian E. Raftery. Bayesian model averaging: Development of an improved multi-class, gene selection and classification tool for microarray data. *Bioinformatics*, 21(10):2394–2402, 2005.

[183] Michael T. Young, Matthew J. Bechle, Paul D. Sampson, Adam A. Szpiro, Julian D. Marshall, Lianne Sheppard, and Joel D. Kaufman. Satellite-Based NO 2 and Model Validation in a National Prediction Model Based on Universal Kriging and Land-Use Regression. *Environmental Science & Technology*, 50(7):3686–3694, apr 2016.

[184] Antonella Zanobetti, Matt P Wand, Joel Schwartz, and Louise M Ryan. Generalized additive distributed lag models: quantifying mortality displacement. *Biostatistics*, 1(3):279–292, 2000.

[185] Antonella Zanobetti, Elena Austin, Brent A Coull, Joel Schwartz, and Petros Koutrakis. Health effects of multi-pollutant profiles. *Environment International*, 71:13–19, 2014.

[186] Antonella Zanobetti, Marie S O'Neill, Carina J Gronlund, and Joel D Schwartz. Susceptibility to mortality in weather extremes: effect modification by personal and small-area characteristics. *Epidemiology*, 24(6):809–19, 2013.

[187] Antonella Zanobetti, Joel Schwartz, and Douglas W Dockery. Airborne Particles Are a Risk Factor for Hospital Admissions for Heart and Lung Disease. *Environmental Health Perspectives*, 108(11):1071, nov 2000.

[188] Antonella Zanobetti, Joel Schwartz, Evi Samoli, Alexandros Gryparis, Giota Touloumi, Richard Atkinson, Alain Le Tertre, Janos Bobros, Martin Celko, Ayana Goren, Bertil Forsberg, Paola Michelozzi, Daniel Rabczenko, Emiliano Aranguez Ruiz, and Klea Katsouyanni. The temporal pattern of mortality responses to air pollution: a multicity assessment of mortality displacement. *Epidemiology*, 13(1):87–93, 2002.

[189] Scott L. Zeger, Francesca Dominici, Aidan McDermott, and Jonathan M. Samet. Mortality in the medicare population and Chronic exposure to fine Particulate air pollution in urban centers (2000-2005). *Environmental Health Perspectives*, 116(12):1614–1619, 2008.

[190] Scott L Zeger, Duncan Thomas, Francesca Dominici, Jonathan M Samet, Joel Schwartz, Douglas Dockery, and Aaron Cohen. Exposure Measurement Error in Time-Series Studies of Air Pollution: Concepts and Consequences. *Environmental Health Perspectives*, 108(5):419, may 2000.

[191] Yeying Zhu, Donna L Coffman, and Debashis Ghosh. A Boosting Algorithm for Estimating Generalized Propensity Scores with Continuous Treatments. *Journal of Causal Inference*, 3(1):417–24.e5, jan 2015.

[192] Corwin M Zigler and Francesca Dominici. Point: clarifying policy evidence with potential-outcomes thinking–beyond exposure-response estimation in air pollution epidemiology. *American Journal of Epidemiology*, 180(12):1133–1140, 2014.

[193] Corwin M Zigler, Francesca Dominici, and Yun Wang. Estimating causal effects of air quality regulations using principal stratification for spatially correlated multivariate intermediate outcomes. *Biostatistics*, 13(2):289–302, apr 2012.

[194] Corwin Matthew Zigler, Kim Chanmin, Christine Choirat, John Barrett Hansen, Yun Wang, Lauren Hund, Jonathan Samet, Gary King, and Francesca Dominici. Causal Inference Methods for Estimating Long-Term Health Effects of Air Quality Regulations. Technical Report 187, Health Effects Institute, Boston, MA, 2016.

[195] Corwin Matthew Zigler and Francesca Dominici. Uncertainty in Propensity Score Estimation: Bayesian Methods for Variable Selection and Model-Averaged Causal Effects. *Journal of the American Statistical Association*, 109(505):95–107, jan 2014.

25

Connecting Exposure to Outcome: Exposure Assessment

Adam A. Szpiro

University of Washington, Seattle, WA

CONTENTS

25.1 Background and overview of this chapter

The objective in environmental epidemiology is to quantify the association between an environmental exposure and a health outcome measure, in a population under study, with the intention of generalizing to the population at large or to sub-groups of interest. The standard approach is to fit a regression model with health outcomes y_i and corresponding exposures x_i for subjects $i = 1, \ldots, n$ with additional health model covariates $\mathbf{z}_i = (z_{i1}, \ldots, z_{ip}) \in \mathbb{R}^p$, including an intercept. For example with a continuous outcome we might fit the linear model,

$$y_i = x_i\beta + \mathbf{z}_i\boldsymbol{\beta}_z + \epsilon_i, \tag{25.1}$$

where conditional on covariates the ϵ_i are assumed to satisfy $E(\epsilon_i) = 0$ and the target of inference is the health effect parameter β. As in all epidemiology studies, there are a number of complexities in estimating and interpreting β, including causality, model misspecification, time scale of action, and confounding. A distinguishing feature of environmental studies from the statistical perspective is the added challenge of exposure assessment in order to obtain a reasonable surrogate for x_i, which typically cannot be observed directly.

For concreteness, we focus in this chapter on air pollution cohort studies, where the exposure of interest is the long-term average pollutant concentration outside a subject's home. Even within air pollution epidemiology, there are many variants on this study design, and it must be acknowledged that our paradigm neglects important concerns such as subject mobility, time spent indoors, long-term averaging period, and outdoor-indoor infiltration. Many of these topics are discussed in the chapter by Lianne Sheppard elsewhere in this volume, while we focus on the interwoven statistical challenges involved in predicting exposure and utilizing predicted exposures in a health analysis.

Although in principle it may be possible to directly measure the outdoor exposure at every study subject's home, budgetary and logistical constraints typically make this impractical. Instead, the common approach is to use exposure monitoring data x_i^* from stations $j = 1, \ldots, n^*$ at different locations in the same region as the study subjects and then to employ spatial statistics to predict or estimate the unmeasured exposures of interest. The monitoring stations can be government regulatory sites, research monitors, or a mixture of these. The ancillary information that makes spatial prediction possible is accurate measures of geographic location (i.e., latitude and longitude) $\mathbf{s}_i$ and $\mathbf{s}_j^* \in \mathbb{R}^2$ for each subject and monitor, respectively, which can be obtained by a combination of global positioning system (GPS), mobile phone based geolocation technology, and address based geocoding. In addition, modern geograhic information systems (GIS) make it possible to calculate a large number (denoted here by r) of geographically referenced covariates, which we indicate by the assumed known mapping $\mathbf{R}(\mathbf{s}) : \mathbb{R}^2 \to \mathbb{R}^r$. We give concrete examples of geographic covariates later in this chapter. Other predictors of air pollution levels such as output from deterministic propagation models (e.g., the the Community Multi-scale Air Quality model (CMAQ) for regional sources or the California Line Dispersion Model (CALINE) for local sources) and remote satellite sensing also have important roles to play. For the purposes of this chapter, these additional sources of information can be treated analogously to GIS covariates (after suitable temporal averaging).

We will begin by providing a brief overview of spatial statistical methods that are commonly applied to exposure assessment for air pollution cohort studies. We will initially focus on the relatively elegant but idealized mathematical formulations of land-use-regression (LUR) and universal kriging (UK). We will then discuss three applied examples that highlight the range of approaches to model selection that go into the practical application of these methods. An unsurprising but unsettling common thread in all three examples is that the modeling and estimation strategy is entirely focused on optimizing out-of-sample prediction accuracy. This is reasonable since the goal is to predict unobserved exposures, but it is unsettling since these predictions are only intermediate values on the pathway to health effect estimates. Next, we switch gears and discuss what is known from a statistical perspective about the implications of using spatial statistics to predict unknown exposures in epidemiology regression analyses. That is, we describe statistical theory on the consequences for health effect inference of the measurement error resulting from using predicted rather than true exposures. We then circle back and consider the question of whether optimizing out-of-sample prediction accuracy is really the best way to select a model for exposure assessment. Recently published results confirm that maximizing prediction accuracy is not necessarily the best thing to do. We explain these findings in light of relevant measurement error theory and provide a novel example of an alternative approach to model selection that is specifically tailored to optimize health effect inference rather than exposure prediction accuracy. This area of statistical methodology is ripe for development.

25.2 Spatial statistics for exposure assessment

25.2.1 Land-Use Regression and Universal Kriging

In general terms, we now descrribe two commonly used spatial modeling strategies for air pollution exposure assessment. First, land use regression (LUR) is the term of art that describes fitting a linear regression with monitoring data as the outcome variables and corresponding GIS covariates as the explanatory variables, and then using predictions from this model with GIS covariates at subject locations to predict the unmeasured exposures. In the simplest form, without any attempt to reduce dimension of the set of GIS covariates, this is accomplished by fitting the model

$$\mathbf{X}^* = \mathbf{R}^* \boldsymbol{\gamma} + \boldsymbol{\eta}^* \tag{25.2}$$

by ordinary least squares (OLS), where $\mathbf{X}^*$ is the vector of n^* concentrations at monitor locations, $\mathbf{R}^*$ is the $n^* \times r$ matrix of GIS covariates at these locations, and $\boldsymbol{\eta}^*$ is treated as *i.i.d.*, corresponding to the assumption of spatial independence conditional on GIS covariates. If $\hat{\boldsymbol{\gamma}}$ is the OLS solution, then the set of predicted exposures is $\hat{\mathbf{w}} = \mathbf{R}\hat{\boldsymbol{\gamma}}$, where $\mathbf{R}$ is the $n \times r$ matrix of GIS covariates at subject locations, and the elements of $\hat{\mathbf{w}}$ are used in place of the unobserved exposures in (25.1).

Notice that LUR uses spatial data as encoded in GIS covariates in a regression framework to predict unobserved concentrations from monitoring data, but it does not explicitly account for the fact that air pollution levels as nearby locations are likely to be more similar to each other than at locations further apart in space. This phenomenon is typically described as residual spatial correlation, and a powerful way of taking advantage of it is universal kriging (UK), wherein we now, at least formally, assume

$$\begin{pmatrix} \mathbf{X} \\ \mathbf{X}^* \end{pmatrix} = \begin{pmatrix} \mathbf{R} \\ \mathbf{R}^* \end{pmatrix} \boldsymbol{\gamma} + \begin{pmatrix} \boldsymbol{\eta} \\ \boldsymbol{\eta}^* \end{pmatrix}, \tag{25.3}$$

with

$$\begin{pmatrix} \boldsymbol{\eta} \\ \boldsymbol{\eta}^* \end{pmatrix} \sim N\left(0, \boldsymbol{\Sigma}_{(\eta\eta^*)}(\boldsymbol{\theta})\right), \tag{25.4}$$

for a positive definite matrix function $\boldsymbol{\Sigma}_{(\eta\eta^*)}(\cdot)$ and unknown parameter $\boldsymbol{\theta}$. It is useful to introduce the decomposition

$$\boldsymbol{\Sigma}_{(\eta\eta^*)}(\cdot) = \begin{pmatrix} \boldsymbol{\Sigma}_{\eta}(\cdot) & \boldsymbol{\Sigma}_{\eta\eta^*}(\cdot) \\ \boldsymbol{\Sigma}_{\eta^*\eta}(\cdot) & \boldsymbol{\Sigma}_{\eta^*}(\cdot) \end{pmatrix}.$$

In UK models, $\boldsymbol{\theta}$ comprises the range, partial sill, and nugget parameters from a geostatistical model where correlation is purely a function of distance between observation locations. That is, UK typically assumes the correlation between two locations is invariant to absolute location (stationarity) and invariant to rotation (isotropy). Further details and specific forms of spatial correlation structure are summarized by Cressie [6] and (author?) [1].

We can estimate $\hat{\boldsymbol{\gamma}}$ and $\hat{\boldsymbol{\theta}}$ by maximum likelihood (ML), restricted maximum likelihood (REML), or another nonlinear optimization approach using only the monitoring data, and then predicted values at subject locations are given by the conditional expectation formula and then the predicted values at subject locations are given by the conditional expectation formula

$$\begin{aligned} \hat{\mathbf{w}} &= E(\mathbf{X}|\mathbf{X}^*; \hat{\boldsymbol{\gamma}}, \hat{\boldsymbol{\theta}}) \\ &= \mathbf{R}\hat{\boldsymbol{\gamma}} + \boldsymbol{\Sigma}_{\eta\eta^*}(\hat{\boldsymbol{\theta}}) \boldsymbol{\Sigma}_{\eta^*}^{-1}(\hat{\boldsymbol{\theta}})(\mathbf{X}^* - \mathbf{R}^*\hat{\boldsymbol{\gamma}}). \end{aligned}$$

The conditional expectation reduces to the same prediction equation as LUR if we estimate that there is no spatial correlation, but in general it allows predictions at subject locations near monitors to borrow information directly from the observed values at those monitors.

In practice, there are many variants on LUR and UK which are applied to air pollution data when the goal is exposure assessment for air pollution epidemiology studies. One important consideration is the need to deal with a very large number of GIS covariates compared to the number of monitoring locations, necesitataing either variable selection or dimension reduction. In the next several subsections, we summarize three specific implementations that have been published recently, with the objective of providing a view into the range of possibilities and also highlighting how much of the exposure modeling process involves decisions that fall outside the relatively straightforward probability models described above.

25.2.2 Example 1: Stepwise Variable Selection in LUR

The European Study of Cohorts for Air Pollution Effects (ESCAPE) project was funded by the Europen Union to investigate the long-term adverse effects on human health in Europe of exposure to air pollution. The primary pollutants of interest in ESCAPE were gaseous nitrogen dioxide (NO_2), total gaseous oxides of nitrogen (NO_x), and various classes of particulate matter (fine particles $PM_{2.5}$, $PM_{2.5}$ absorbance , and combined fine and coarse particles PM_{10}). We summarize the exposure modeling for NO_x, noting that similar approaches were used for other pollutants [2]. Measurenments of NO_x concentrations were taken in 20 European study areas (cities or regions within countries) between 2008 and 2012, at between 40 and 80 monitoring sites per region. Measurements were taken during three 14 day periods at different seasons, and after adjustment for temporal variation using a central site, these were combined to represent annual averages at each monitor. Locations of ESCAPE study subjects and pollutant concentration monitors were determined based on GPS and high quality maps.

Separate LUR models were fit for each study area. In each model approximately 120 GIS covariates were considered for inclusion in each model, using a selection algorithm summarized below. The availab le covariates included measures of land use (e.g., residential, commercial, port, urban, green) averaged over circular buffers ranging from 100 meters to 5,000 meters in radius, household and residential density averaged over similar buffers, measures of road length and traffic in buffers ranging from 25 meters to 1,000 meters, distances to major roads, and local altitude. A version of forward stepwise selection was used to choose GIS covariates for inclusion in each study area's LUR model, with an emphasis on optimzing out-of-sample prediction accuracy for area-specific models, but also with the requirement that the coefficient for each covariate in the final model was physically plausible (e.g., distance to a major road must have a negative coefficient). Initially univariate regression models were fit with each candidate GIS covariate as the independent variable, and the one with the highest within-sample R^2 and a slope in the interpretable direction was used as the starting model. Additional GIS covariates were added, using the criterion of the largest increase in within-sample R^2 subject to the condition that the increase was at least 1% and all covariates maintained an interpretable direction. Finally, variables were sequentially removed from the model if they had p-values greater thana 0.1. At the completion of variable selection, several model fit diagnostics were performed, sometimes resulting in GIS covariates being excluded from the final model. There was no specific attempt to fit models such as UK that would exploit spatial correlation, although the Moran's I statistics was calculated based on the final model fits and suggested small amounts of correlation. Finally, out-of-sample prediction accuracy was estimated for each region using leave-one-out cross validation (taking the selected model as fixed). The final LUR models included between 2 and 7 GIS covariates and had cross validated R^2 values ranging from 0.39 to 0.88 (in-sample R^2 values that ranged from 0.49 to 0.92).

25.2.3 Example 2: Distance Decay Variable Selection (ADDRESS) in LUR

An intriguing variant on forward stepwise selection was proposed by Su et al. [10] and was initially applied to gaseous NO_2, NO_x, and NO (nitrogen oxide) data collected in the Los Angeles region in 2006–2007 as part of a project funded by the California Air Resources Board (CARB) to assess the impact of outdoor air pollution on respiratory health in children. The monitoring for this study consisted of approximately 180 sites in a single study area, with two 14 day averages that were combined to approximate annual averages. Land-use and traffic related GIS covariates were calculated for buffer sizes ranging from 50 meters to 15,000 meters in intervals of 100 meters. Rather than treating all of these as potentially independent predictors of pollution and relying on an automatic procedure to make efficient use of them, Su et al. [10] devised "A Distance Decay Regression Selection Strategy" (ADDRESS), which modifies forward selection by limiting the number of buffer sizes for a given characteristic that can enter into the final model.

The essential idea of the ADDRESS algorithm is to consider all buffer sizes for a type GIS covariate together when determining which covariate to add to the LUR model during stepwise selection, with a semi-automated approach to choosing from among the buffers. As a first step, univariate correlations are computed for each GIS covariate with the measured pollutant levels, but rather than picking the covariate with the highest correlation these values are displayed as distance-decay plots for each class of covariate. Visual examination yields an optimal buffer size or "distance of influence" for each class of GIS covariates (e.g., road length, population density). A single covariate from among these best-in-class candidates is included in the model based on having the highest correlation and a p-value less than 0.05. This process is repeated, examining correlations with residuals from the interim model rather than raw concentration data, until no further covariates meet the criteria for inclusion.

The resulting model for NO_x in Los Angeles included 7 GIS covariates, all of which had coefficients with interpretable signs even though this was not a requirement in the ADDRESS algorithm. Optimal buffer sizes ranged from 100 meters to 11,000 meters. As in Beelen et al. [2] there was no specific attempt to exploit spatial correlation through UK or a similar procedure, but longitude and latitude were allowed to enter the model as linear terms. Out-of-sample prediction accuracy was estimated by holding out 16 monitors from the entire model selection procedure, resulting in an estimated R^2 of 0.91 (in-sample R^2 was 0.85). Typically we expect out-of-sample R^2 to be lower than in-sample due model fitting bias, but in this case the small size of the hold-out validation set available to approximate the true prediction error is likely responsible for the counter-intuitive result.

25.2.4 Example 3: Lasso Followed by Exhaustive Search Variable Selection in LUR and UK

As part of the MESA Air study of the association between air pollution exposure and subclinical cardiovascular disease, gaseous NO_2, NO_x, and NO_2 were monitored for 14 day periods at approximately 150 locations in the Los Angeles are, in three seasons during 2006 – 2007 [9] The primary motivation for collecting these data was to combine them with other research and regulatory monitoring data in a spatio-temporal model to predict long-term averages [12], but as in intermediate step Mercer et al. [9] developed season-specific LUR and UK models based on the 14 day snapshots. As in the first two examples, a large number of GIS covariates were available, including a range of buffer sizes from 50 meters to 15,000 meters for road length and land use categories.

Rather than stepwise variable selection as in the first two examples, Mercer et al. [9] employed a novel combination of the least absolute shrinkage and selection operator (lasso) [7] with limited exhaustive search of candidate covariates, according to the following algorithm. Lasso is a variant on OLS in a linear regression model that uses an L^1 penalty to shrink coefficients toward zero while forcing some of them to vanish, resulting in a more parsimonious model. Mercer et al. [9] use lasso as a pre-screening tool by tuning the penalty parameter λ to select a model with 15-20 GIS covariates out of the initial list of approximately 65. The penalized coefficient estimates are discarded since the ultimate goal is to fit a LUR model by OLS or a UK model by ML. Rather, the 15-20 covariates with non-zero coefficients in the lasso model are taken as candidate covariates in an exhaustive search algorithm that fits LUR or UK with all possible subsets and selects the best model by 10-fold cross validation. As suggested by Su et al. [10] in the ADDRESS algorithm discussed above, no more than a single buffer size for a given GIS covariate is included in candidate models in the exhaustive search.

Separate LUR and UK models were constructed for each of the summer, winter, and autumn datasets in Los Angeles. The UK models were slightly more parsimonious than the LUR models (6–7 covariates in LUR compared to 4–6 covariates in UK) since spatial smoothing in UK fills in gaps that would otherwise require inclusion of additional covariates in the regression. The cross-validated estimates of out-of-samples prediction accuracy were also somewhat higher for UK (R^2s of 0.74, 0.60, and 0.67 for LUR compared to 0.75, 0.72, and 0.74 for UK). We will revisit using lasso followed by exhaustive search for model selection at the end of this chapter, when we give an example of selecting the exposure model to optimize health effect inference, rather than the intermediate metric of prediction accuracy.

25.2.5 Example 4: Accurate exposure prediction does not necessarily improve health effect estimation

In the previous subsections, we outlined the mathematical underpinnings of LUR and UK as spatial statistical tools for exposure assessment when air pollution concentration data are spatially misaligned for cohort studies, meaning that concentration measures are available at monitor locations that are distinct from subject locations. We also considered three real-world examples of applications of these methods to air pollution data, illustrating the central role of model selection to reduce the effective dimensionality of the regression component of this model, given the fact that modern GIS systems can calculate essentially unlimited covariates at known monitor and subject locations.

Not surprisingly, all three variable selection algorithms we considered were designed to minimize out-of-sample prediction error, as estimated by either hold out validation or cross-validation. This is not necessarily ill-advised as it is fairly standard to design spatial prediction models with the goal of minimizing error at new locations. However, it also begs the question of whether minimizing prediction error is the most appropriate way to evaluate and choose between spatial models in this context, since this is not the ultimate objective. Instead, we will be combining the predicted concentrations with health data from a cohort in order to learn about the association between the true unmeasured exposure and that health outcome. That is, we are deriving these exposure predictions in order to learn about the relationship between true exposure and health outcomes at the population level.

We have recently shown by example that among a given class of exposure models, the one that maximizes out-of-sample prediction accuracy is not necessarily the one that leads to the best health effect estimates [13]. We will summarize the key results from this example in the remainder of this subsection. An important point to emphasize, however, is that the apparent paradox is not the result of poor estimation of out-of-sample prediction accuracy due to the limitations of cross-validation or holdout validation. Rather, since this is a simulation example, we are able to compare the true prediction accuracy of different candidate models, and we find that the one with the least error leads to health effect estimates that are more biased and more variable than an alternative model.

Suppose that the health model (25.1) holds with no additional covariates such that

$$y_i = \beta_0 + x_i\beta + \epsilon_i,$$

for $i = 1, \ldots, n$, with the ϵ_i i.i.d. $N(0, \sigma_\epsilon^2)$ and each $x_i = \sum_{k=1}^3 4R_{ki}$ where the R_{ki} are i.i.d. $N(0, 1)$. We observe the R_{ki} (which we conceptualize as GIS covariates), but not the x_i. We also observe monitoring data x_j^* for $j = 1, \ldots, n^*$ and corresponding R_{kj}^*, which similarly satisfy $x_j^* = \sum_{k=1}^3 4R_{kj}^*$. The difference in the monitoring data is that the third GIS covariate R_{3j} is less variable than it is in the subject population while the other two are similar. Specifically, we assume the R_{1j}^* and R_{2j}^* are i.i.d. $N(0, 1)$ whereas the R_{3j}^* are i.i.d. $N(0, 0.01)$. Knowing the form of the model for x_i and x_j^* as a regression on three covariates, but not the values of the coefficients, it would be most natural to estimate the coefficients using the available monitoring data, predict the unobserved x_i using the regression model, and plug these estimates into the health model to obtain $\hat{\beta}$ by OLS. We simulated this approach, setting $n = 10,000$, $n^* = 100$, $\sigma_\epsilon^2 = 25$, $\beta_0 = 1$, and $\beta = 2$, and calculated the out-of-sample prediction R^2 to be 0.73 and the bias and standard error of the resulting $\hat{\beta}$ to be -0.035 and 0.23, respectively. Computer code for these simulations is available as an online supplement to the paper by Szpiro et al. [13].

Suppose that we either do not know that the exposure model contains the third GIS covariate or that we choose to intentionally misspecify the exposure model by omitting this covariate from the regression. Given the relatively limited variability of R_{3j}^* it is plausible that a data analyst would at least consider taking this approach out of concern that the coefficient for this covariate will be poorly estimated. Even given the limited variability of R_{3j}^*, however, it turns out that excluding it from the model results in much less accurate exposure prediction, with an out-of-sample R^2 of only 0.50, so model selection methods like those reviewed in our examples from Section 25.2 would very reasonably guide an analyst back to a model that included all three covariates. However, the health effect estimates from the misspecified exposure model with only two GIS covariates are improved, with bias and standard error of 0.001 and 0.16, respectively, compared to the previously indicated -0.035 and 0.23 for the full exposure model. This seemingly paradoxical behavior goes away if n is much smaller. The remainder of this chapter is dedicated to reviewing measurement error theory that explains this example and to suggesting how that theory can be leveraged to devise novel variable selection methods that optimize health effect estimation rather than exposure prediction accuracy.

25.3 Measurement error for spatially misaligned data

25.3.1 Correctly specified LUR or UK exposure model

Recalling the notation from Section 25.2.1, the true unmeasured exposure for subject i is x_i, but what we have available to estimate β is the predicted value based on the conditional expectation formula

$$\hat{\mathbf{W}} = \mathbf{R}\hat{\gamma} + \Sigma_{\eta\eta^*}(\hat{\theta})\Sigma_{\eta^*}^{-1}(\hat{\theta})(\mathbf{X}^* - \mathbf{R}^*\hat{\gamma}),$$

where the estimated regression coefficients $\hat{\gamma}$ and covariance parameters $\hat{\theta}$ are obtained by fitting the UK model to the monitoring data by ML or REML. In the case of LUR the conditional expectation reduces to $\hat{\mathbf{W}} = \mathbf{R}\hat{\gamma}$. Writing $\hat{\mathbf{W}} = (\hat{w}_i)_{i=1}^n$ and using the $\hat{w}_i$ in place of x_i to estimate β in (25.1) introduces a form of covariate measurement error, which in general can introduce bias and affect the standard errors of $\hat{\beta}$. An unusual aspect of this measurement error, in contrast to the kind typically encountered in epidemiological studies [4], is that we created the error by means of the exposure modeling procedure applied to the available data, and it is not simply the result of a flawed measurement instrument. This does not mean we have the choice of not introducing error, but it does have two important implications.

First, in order to understand the measurement error's statistical properties, we should begin by studying the statistical properties of the spatial model that gave rise to predictions, rather than by assuming one or more standard measurement error models such as classical error that is independent of the x_i or Berkson error that is independent of the $\hat{w}_i$. Second, it opens the door to modifying the spatial prediction model to mitigate some of the impact of the measurement error on health effect estimation, a direction we will begin to explore later in this chapter.

We initially assume the UK exposure model given by (25.3) and (25.4) correctly characterizes the data-generating mechanism that gives rise to exposure and monitoring data. Under these assumptions, we follow the spatial measurement error framework developed by Szpiro et al. [14]. Since we assume that (25.3) and (25.4) hold in the data-generating mechanism, we begin analyzing the measurement error by assuming the parameters γ and θ were known without error. It would still be necessary to predict exposures at subject locations, but in this hypothetical scenario we would be able to calculate

$$\mathbf{W} = \mathbf{R}\gamma + \Sigma_{\eta\eta^*}(\theta)\Sigma_{\eta^*}^{-1}(\theta)(\mathbf{X}^* - \mathbf{R}^*\gamma),$$

and use the elements of $\mathbf{W} = (w_i)_{i=1}^n$ in place of the x_i in the health model. Although the w_i are derived by optimally smoothing the observed x_j^*, there is still measurement error in the sense that $w_i \neq x_i$. We call $u_{i,BL} = (x_i - w_i)$ the *Berkson-like* error because it represents that part of the true exposure that is missed, even under the optimal prediction approach. Similar to Berkson measurement error, this results in a surrogate exposure that is less variable than the unobserved true value, although it is distinct from Berkson error in that it does not share the same independence properties.

A second crucial point concerning Berkson-like error is that it does not induce bias in health effect estimates, at least for linear health models, another property shared by Berkson error [4]. To see this, we note that the health model can be written

$$y_i = w_i\beta + (\epsilon_i + \beta u_{i,BL}), \tag{25.5}$$

from which it follows that using w_i in place of x_i (treating $\beta u_{i,BL}$ as part of the unobserved residual) does not introduce bias as long as $E(u_{i,BL}|w_i) = 0$. This latter equality follows from $E(u_{i,BL}|w_i) = E(u_{i,BL}|\mathbf{X}^*)$ and the fact that

$$E\left(\mathbf{X} - \mathbf{R}\gamma + \Sigma_{\eta\eta^*}(\theta)\Sigma_{\eta^*}^{-1}(\theta)|\mathbf{X}^*\right) = 0$$

since θ and α hold in the true generating mechanism; see Szpiro et al. [14] for further details.

The other component of measurement error in the present scenario comes from finite sample variability in estimating the exposure model parameters θ and γ by $\hat{\theta}$ and $\hat{\gamma}$, respectively. Focusing on this aspect of error, we call $u_{i,CL} = (w_i - \hat{w}_i)$ the *classical-like* error. Like classical measurement error, this introduces variability into the surrogate exposure $\hat{w}_i$ that is not present in the unobserved true values x_i. Classical measurement error is independent of the y_i and is independent across observations, conditions that do not hold for classical-like error since all of the data are spatially correlated and errors in exposure model parameters induce shared variability across observations. Nonetheless, it is evident that the additional variability is uninformative for estimating β, leading to the intuition that it should introduce bias of the regression coefficient toward zero, corresponding to the null hypothesis of no association. Interestingly, this intuition turns out to be only partially correct. Simulations reported by Szpiro et al. [14] show evidence of bias, but it is relatively small compared to the standard errors for $\hat{\beta}$ and it is away from the null, in contrast to classical measurement error, which typically results in pronounced bias toward the null. Explaining this behavior theoretically remains an open question, but the next subsection provides some helpful intuition.

25.3.2 Incorrectly specified LUR or regression spline exposure model

The framework for analyzing measurement error from spatially misaligned data described in Section 25.3.1 is appealing in that it is mathematically elegant, and it highlights the important distinction between error that is intrinsic to the smoothing that is needed to make predictions at locations without data (Berkson-like error) and the error induced by uncertainty in estimating the smoothing parameters (classical-like error). There are, however, scientific and statistical motivations for reconsidering some of the underlying assumptions, the result of which has been development of the analytic framework described in this Section [3, 11].

The most obvious assumption in Section 25.3.1 is that the exposure model given by (25.3) and (25.4) correctly describes the data-generating mechanism. Of course, essentially all statistical models are wrong to some degree, so it should not be surprising that this model is at least partially incorrect. However, if we refer back to the practical applications in Examples 1-3, we see that the large number of geographic predictors and spatial smoothing models available necessitates model selection prior to arriving at (25.3) and (25.4), which suggests that model misspecification is non-negligible. Indeed, given the complexity of the underlying geography, meteorology, emissions, and atmospheric chemistry, it is difficult to imagine constructing a statistical model that comes near fully describing the structure of a sparsely observed air pollution surface.

Building on these general concerns about model misspecification, we focus specifically on the plausibility of the spatial random effect in (25.4). It is traditional to formulate spatial statistics in terms of a latent spatial random effect that leads to the universal kriging algorithm for spatial interpolation by way of the conditional expectation formula (25.5). However, the fact that this formulation leads to spatial smoother with favorable properties does not imply that it is a plausible description of the data-generating mechanism. Indeed, as we argued in Szpiro and Paciorek [11], when our interest is in long-term exposures within a specific country or geographic region, it makes much more sense to regard the unobserved latent spatial surface as fixed, and to focus on spatial variability in the monitor and subject locations rather than on spatial variability of the surface itself. Working in a frequentist setting, this means that our hypothetical repeated sampling framework should involve asking how our results would have been different if we had selected study subjects and exposure monitors at different locations, rather than how our results would have been different if the air pollution surface looked different than the one actually observed.

Motivated by the above considerations, we proposed the following alternate analytical framework for exposure modeling in Szpiro and Paciorek [11]. We believe this framework more faithfully represents the underlying sources of randomness and that it highlights some important statistical implications of measurement error that were not evident in Section 25.3.1. It is worth emphasizing at this stage that the distinction from Section 25.3.1 is not in the choice of exposure model. Rather, the distinction is in how we understand the statistical properties of the predictions generated by our chosen model, in light of assumptions about what is fixed and random in the data-generating mechanism. We focus in this section on pure LUR models in order to streamline the exposition. Note, however, that even without believing that the random effect in a kriging model is real (in the sense of corresponding to actual stochastic variability in the spatial air pollution pattern), an analyst may choose to utilize UK as a tool for spatial smoothing since it tends to perform well. Extending the theoretical results in this section to a UK exposure model remains an open question; see Bergen and Szpiro [3] for more discussion of this point and some related results for low-rank kriging spatial models [8].

As in Section 25.3.1, we simplify the exposition by omitting additional covariates (Szpiro and Paciorek [11] and Cefalu and Dominici [5] discuss this point further), so we are interested in the health model

$$y_i = x_i \beta + \epsilon_i,$$

where the y_i are observed for all subjects, but instead of the actual subject exposures we observe monitoring data, x_j^*, for $j = 1, \ldots, n^*$, at different locations $\mathbf{s}_j^*$. We now regard the spatial locations $\mathbf{s}_i$ and $\mathbf{s}_j^*$ of study subjects and monitors as realizations of spatial random variables with unknown densities $g(\mathbf{s})$ and $h(\mathbf{s})$, respectively, and corresponding distribution functions $G(\mathbf{s})$ and $H(\mathbf{s})$. Throughout, we assume the subject locations are chosen independently of the monitoring locations.

Conditional on the $\mathbf{s}_i$, we assume the x_i satisfy

$$x_i = \Phi(\mathbf{s}_i) + \eta_i,$$

with i.i.d. mean zero η_i. The function $\Phi(\mathbf{s})$ is a deterministic spatial surface that is potentially predictable by covariates and spatial smoothing, and the η_i represent variability between exposures for subjects at the very nearby physical locations. We assume an analogous model for the monitoring data at locations $\mathbf{s}_j^*$, with the same deterministic spatial field $\Phi(\mathbf{s}_j^*)$ and with instrument error represented by η_j^* having variance $\sigma_{\eta^*}^2$.

We let $\mathbf{R}(\mathbf{s})$ be a known function from $\mathbb{R}^2$ to $\mathbb{R}^r$ that incorporates r LUR covariates, termed the spatial basis. We do not assume the spatial basis is sufficiently rich to represent all of the structure in $\Phi(\mathbf{s})$, so we allow for misspecification in the sense that $\Phi(\mathbf{s}) \neq \mathbf{R}(\mathbf{s})\gamma$ for some $\mathbf{s} \in \mathbb{R}^2$, for any choice of γ. The practical implementation of LUR (after model selection) is to derive an estimate $\hat{\gamma}$ by ordinary least squares (OLS)

$$\hat{\gamma} = \arg\min{}_{\boldsymbol{\xi}} \sum_{j=1}^{n^*} \left(x_i^* - \mathbf{R}(\mathbf{s}_j^*)\boldsymbol{\xi} \right)^2. \tag{25.6}$$

and then use the estimated exposure, $\hat{w}_i = \hat{w}(\mathbf{s}_i) = \mathbf{R}(\mathbf{s}_i)\hat{\gamma}$, in place of x_i. Under standard regularity conditions [16], $\hat{\gamma}$ is asymptotically normal and converges a.s. to γ^* as $n^* \to \infty$, where γ^* is the solution to

$$\gamma^* = \arg\min{}_{\boldsymbol{\xi}} \int \left(\Phi(\mathbf{s}) - \mathbf{R}(\mathbf{s})\boldsymbol{\xi} \right)^2 dH(\mathbf{s}). \tag{25.7}$$

We are now in a position to define the intermediate quantity $w_i^* = w(\mathbf{s}_i) = \mathbf{R}(\mathbf{s}_i)\boldsymbol{\gamma}^*$ and decompose the measurement error

$$u_{i,BL}^* + u_{i,CL}^* = (x_i - w_i^*) + (w_i^* - \hat{w}_i).$$

The Berkson-like component, $u_{i,BL}^*$, is the information lost from smoothing even with un-limited monitoring data (a form of exposure model misspecification), and the classical-like component, $u_{i,CL}^*$, is variability that arises from estimating the parameters of the exposure model based on monitoring data at n^* locations. Notice that we have introduced the superscript asterisk to distinguish the measurement error decomposition here from the one in Section 25.3.1, since the intermediate quantities w_i and w_i^* are intrinsically different, even for the same exposure model, as they are defined based on different assumptions about the true data-generating mechanism.

In the framework of Section 25.3.1, Berkson-like error did not induce any bias in a linear health model. This is not always the case in the present analytic setting, but we can at least show that $\hat{\beta}$ is consistent for β as $n \to \infty$, if either the exposure model is correctly specified (i.e., $\Phi(\mathbf{s}) = \mathbf{R}(\mathbf{s})\boldsymbol{\gamma}$ for some $\mathbf{s} \in \mathbb{R}^2$, for some $\boldsymbol{\gamma}$) or if it holds that probability distribution of $\mathbf{R}(\mathbf{s})$ is the same if $\mathbf{s}$ is sampled from $G(\mathbf{s})$ or $H(\mathbf{s})$. A stronger but more interpretable version of the latter condition is that $g(\mathbf{s})$ and $h(\mathbf{s})$ are the same densities, meaning that monitor and subject locations are drawn from the same probability distribution. Under any of these assumptions, it is straightforward to show that Berkson-like error only inflates the SE of $\hat{\beta}$, and furthermore that this inflation vanishes at the same rate as the part of the SE due only to variability in sampling subjects (i.e., the part that does not involve measurement error) since the Berkson-like error behaves like part of the residual in a linear health model.

As in Section 25.3.1, the classical-like error induces bias and increases variability of $\hat{\beta}$, but now we are able to quantify both of these impacts asymptotically. The relevant asymptotic in the expressions below is that we consider the limit as $n^* \to \infty$, under the assumption that n is arbitrarily large. We emphasize this asymptotic regime by denoting $\hat{\beta}_{n,n^*}$ for the estimate of β with n study subjects and n^* monitors, and we write the asymptotic mean and variance as

$$\frac{1}{\beta}E(\hat{\beta}_{\infty,n^*} - \beta) \approx 2\frac{\int w(\mathbf{s}_1)w(\mathbf{s}_2)\text{Cov}\,(\hat{w}(\mathbf{s}_1), \hat{w}(\mathbf{s}_2))\,dG(\mathbf{s}_1)dG(\mathbf{s}_2)}{\left(\int w(\mathbf{s})^2 dG(\mathbf{s})\right)^2}$$

(25.8)

$$-\frac{\int \text{Var}\,(\hat{w}(\mathbf{s}))\,dG(\mathbf{s})}{\int w(\mathbf{s})^2 dG(\mathbf{s})} - \frac{\int w(\mathbf{s})E\,(\hat{w}(\mathbf{s}) - w(\mathbf{s}))\,dG(\mathbf{s})}{\int w(\mathbf{s})^2 dG(\mathbf{s})}$$

and

$$\frac{1}{\beta^2}\text{Var}(\hat{\beta}_{\infty,n^*}) \approx \frac{\int w(\mathbf{s}_1)w(\mathbf{s}_2)\text{Cov}\,(\hat{w}(\mathbf{s}_1), \hat{w}(\mathbf{s}_2))\,dG(\mathbf{s}_1)dG(\mathbf{s}_2)}{\left(\int w(\mathbf{s})^2 dG(\mathbf{s})\right)^2}.$$

(25.9)

These expressions hold under regularity assumptions and under conditions that guarantee no bias from Berkson-like error. Technical details, including the precise definitions we use for asymptotic mean and variance, are given by Szpiro and Paciorek [11]. Since $\hat{w}(\mathbf{s}) = \mathbf{R}(\mathbf{s})\hat{\boldsymbol{\gamma}}$, it is easy to see that these estimates of bias and variance from classical-like error can be calculated from the first and second moments of $\hat{\boldsymbol{\gamma}}$, which are in turn estimable from the data. It may be surprising to see a term corresponding to bias in $\hat{\boldsymbol{\gamma}}$ since this is the fit to a linear regression estimating $\boldsymbol{\gamma}$, but there can be non-zero bias in the case of a misspecified mean model with random sampling of the dependent and independent variables in a linear regression. Estimating this bias from the data requires a non-standard multinomial approximation to a von Mises expansion with the empirical process of monitor locations [11, 15].

We can make two interesting observations from (25.8) and (25.9). First, the asymptotic bias and variance from classical-like error are of the same order of magnitude, meaning that the bias can be regarded as asymptotically negligible. Nonetheless, in some examples it is still sizable and adversely affects inference, so correction methods that use (25.8) should be considered. The second observation is that the bias is a sum of positive and negative terms, corresponding to bias away from and toward the null, respectively. The positive term dominates precisely when there is high correlation between predictions across subject locations, which corresponds to having a low number of effective degrees of freedom in the exposure model. Translating this observation to a universal kriging spatial model, we speculate that the bias away from the null we mentioned in Section 25.3.1 is attributable to exposure surfaces that were estimated to be highly spatially correlated.

We also now have theoretical tools that enable us to interpret the seemingly paradoxical results in Example 4. The phenomenon at work is a tradeoff between Berkson-like and classical-like error. The correctly specified exposure model with all three predictors is subject to less Berkson-like error than the misspecified version with only two predictors, but this comes at the cost of increased classical-like error. Each version of the exposure model satisfies a set of assumptions that guarantees no bias from Berkson-like error (the model with three predictors is correctly specified, whereas the one with two predictors has the property that the probability distribution of $\mathbf{R}(\mathbf{s})$ is the same if $\mathbf{s}$ is sampled from $G(\mathbf{s})$ or $H(\mathbf{s})$), and the sample size for the health study is large enough ($n = 10,000$) to ensure that the variance from Berkson-like error is smaller than the contribution from classical-like error. Therefore, even though the overall prediction accuracy is better for the model with all three predictors, the bias and variance of $\hat{\beta}$ are smaller when we only use two predictors since there is less classical-like error. Indeed, the numerical results summarized in Example 4 are well approximated by asymptotic calculations based on (25.8) and (25.9).

25.4 Optimizing exposure modeling for health effect estimation rather than prediction accuracy

We have now seen that flexible spatial statistical methods combining LUR with smoothing through universal kriging can be used to solve the problem of spatial misalignment in air pollution cohort studies by predicting exposure at subject locations based on monitoring data at differnent locations. This process results in measurement error because the predicted exposures are different from the unobserved true values, and we have summarized two paradigms for analyzing the effects of this measurement error. Our first Examples 1-3 illustrate that practical applications of spatial prediction methods typically involve a fairly intensive model selection stage guided by an attempt to maximize out-of-sample prediction accuracy. On the other hand, Example 4 demonstrates that this is not always the best way to ensure optimal health effect inference. Furthermore, our decomposition of measurement error into Berkson-like and classical-like components and the asymptotic bias and variance expressions in Section 25.3.2 help to explain and quantify this apparent paradox.

Putting these pieces together leads naturally to the idea that we can specifically tailor our first stage exposure model to improve health effect inference rather than the intermediate goal of accurate exposure predcition, perhaps using expressions like (25.8) and (25.9). Bergen and Szpiro [3] showed that this approach can be successfully used to select the smoothing parameter in a low-rank kriging exposure model. We close this chapter by showing that we

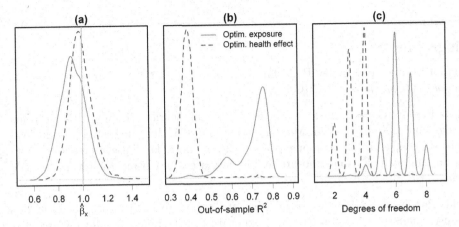

FIGURE 25.1
Exposure model selection and dimension reduction based on asymptotic estimates of the MSE of $\hat{\beta}$ is preferable to minimizing the error in exposure predictions. When the hybrid lasso/limited all-subset regression method optimizes an estimate of the MSE of $\hat{\beta}$ rather than exposure prediction accuracy: (a) $\hat{\beta}$ is less biased and less variable, (b) exposure predictions have more error, as quantified by out-of-sample R^2, and (c) the selected exposure model has fewer predictors.

can also improve on variable selection using a similar approach. What follows is a simulation study that compares different variants of the lasso followed by exhaustic search variable selection methods used by Mercer et al. [9] and summarized in our Example 3.

We generalize the simulation study from Szpiro et al. [13] by constructing a true exposure surface as a linear combination of eight independent covariates, each one of which is distributed normally with mean zero and unit standard deviation in the subject population. Half of the covariates are distributed have the same distribution at monitoring locations, and the other half are much less variable with standard deviations of 0.1 instead of 1.0. We select between the exposure model with all eight covariates and all possible misspecified versions that omit one or more of the covariates. The parameter values are $\sigma_\eta^2 = 16$, $\sigma_{(\eta^*)}^2 = 16$, $\sigma_\epsilon^2 = 6$, $n = 1000$, and $n^* = 100$. Computer code for this simulation is available from the publisher's website.

Following Mercer et al. [9], we prescreen the possible submodels using lasso penalization and then select between the reduced subset of candidate models by comparing their respective estimated mean squared error (MSE) for $\hat{\beta}$ or by comparing their estimated out-of-sample prediction accuracies calculated by cross-validation. That is, for a sequence of penalty parameters λ, we use lasso to derive estimates $\hat{\gamma}_\lambda$ with some elements equal to zero. For each λ, we then denote by $\mathbf{R}_\lambda(\mathbf{s})$ the candidate submodel comprised of only those predictors with non-zero coefficients in $\hat{\gamma}_\lambda$. We do not directly use the coefficients from these penalized models, but rather refit the unpenalized versions with restricted sets of predictors. Figure 25.1 summarizes the simulation results, demonstrating that we achieve improved estimation of the health effect parameter of interest β by using our approach to optimizing the estimated MSE of $\hat{\beta}$ compared to the more traditional approach of optimizing prediction accuracy for the exposure. In addition, we find that these improved results tend to result from selecting more parsimonious exposure models. Further development of these ideas and development of more general methods for optimizing exposure model selection is an area ripe for further research.

Bibliography

[1] Sudipto Banerjee, Bradley P Carlin, and Alan E Gelfand. *Hierarchical modeling and analysis for spatial data*. CRC Press, 2014.

[2] Rob Beelen, Gerard Hoek, Danielle Vienneau, Marloes Eeftens, Konstantina Dimakopoulou, Xanthi Pedeli, Ming-Yi Tsai, Nino Künzli, Tamara Schikowski, Alessandro Marcon, et al. Development of no 2 and no x land use regression models for estimating air pollution exposure in 36 study areas in europe–the escape project. *Atmospheric Environment*, 72:10–23, 2013.

[3] Silas Bergen and Adam A Szpiro. Mitigating the impact of measurement error when using penalized regression to model exposure in two-stage air pollution epidemiology studies. *Environmental and Ecological Statistics*, 22(3):601–631, 2015.

[4] Raymond J Carroll, David Ruppert, Leonard A Stefanski, and Ciprian M Crainiceanu. *Measurement error in nonlinear models: a modern perspective*. CRC press, 2006.

[5] Matthew Cefalu and Francesca Dominici. Does exposure prediction bias health effect estimation? the relationship between confounding adjustment and exposure prediction. *Epidemiology (Cambridge, Mass.)*, 25(4):583, 2014.

[6] Noel Cressie. *Statistics for spatial data*. John Wiley & Sons, 2015.

[7] Trevor Hastie, Robert Tibshirani, Jerome Friedman, and James Franklin. The elements of statistical learning: data mining, inference and prediction. *The Mathematical Intelligencer*, 27(2):83–85, 2005.

[8] EE Kammann and Matthew P Wand. Geoadditive models. *Journal of the Royal Statistical Society: Series C (Applied Statistics)*, 52(1):1–18, 2003.

[9] Laina D Mercer, Adam A Szpiro, Lianne Sheppard, Johan Lindström, Sara D Adar, Ryan W Allen, Edward L Avol, Assaf P Oron, Timothy Larson, L-J Sally Liu, et al. Comparing universal kriging and land-use regression for predicting concentrations of gaseous oxides of nitrogen (no x) for the multi-ethnic study of atherosclerosis and air pollution (mesa air). *Atmospheric Environment*, 45(26):4412–4420, 2011.

[10] Jason G Su, Michael Jerrett, Bernardo Beckerman, Michelle Wilhelm, Jo Kay Ghosh, and Beate Ritz. Predicting traffic-related air pollution in los angeles using a distance decay regression selection strategy. *Environmental research*, 109(6):657–670, 2009.

[11] Adam A Szpiro and Christopher J Paciorek. Measurement error in two-stage analyses, with application to air pollution epidemiology. *Environmetrics*, 24(8):501–517, 2013.

[12] Adam A Szpiro, Paul D Sampson, Lianne Sheppard, Thomas Lumley, Sara D Adar, and Joel D Kaufman. Predicting intra-urban variation in air pollution concentrations with complex spatio-temporal dependencies. *Environmetrics*, 21(6):606–631, 2010.

[13] Adam A Szpiro, Christopher J Paciorek, and Lianne Sheppard. Does more accurate exposure prediction necessarily improve health effect estimates? *Epidemiology (Cambridge, Mass.)*, 22(5):680, 2011.

[14] Adam A Szpiro, Lianne Sheppard, and Thomas Lumley. Efficient measurement error correction with spatially misaligned data. *Biostatistics*, 12(4):610–623, 2011.

[15] Aad W Van der Vaart. *Asymptotic statistics*, volume 3. Cambridge university press, 2000.

[16] Halbert White. A heteroskedasticity-consistent covariance matrix estimator and a direct test for heteroskedasticity. *Econometrica: Journal of the Econometric Society*, pages 817–838, 1980.

26

Environmental epidemiology study designs

Lianne Sheppard

University of Washington, Seattle, Washington

CONTENTS

26.1 Introduction

Exposure is the Achilles heel of environmental epidemiology because the validity of these studies depends upon the quality of the exposure estimate used in the analysis. Thus I start this chapter with a discussion of exposure.

Exposure assessment covers the entire process of estimating exposures, including the design of the exposure data collection, exposure measurement, and exposure modeling. The purposes of environmental exposure assessment fall into three broad classes: environmental surveillance, quantification of population exposure distributions for risk assessment, and inference about health effects in environmental epidemiology studies. This chapter focuses on using environmental exposures for inference about health effects in environmental epidemiology; the other two applications of environmental exposure assessment are beyond its scope. The ultimate application of the exposure assessment affects how the assessment should be designed, how and what exposures are measured, and the optimal approach to exposure modeling. While most chapters in this book focus exclusively on statistical methods for modeling environmental exposures, this is only one aspect of a good exposure assessment. Researchers want valid inference in their environmental epidemiology studies, i.e., the estimate of the target parameter of interest is unbiased and has proper confidence interval coverage. This is challenging because of exposure measurement error. Statisticians can make important contributions to exposure assessment design and exposure modeling in order to improve inference in environmental epidemiology.

Summarizing from an overview by Brunekreef and colleagues ([1] chapter 1), environmental epidemiology is a sub-discipline of epidemiology that focuses on understanding the determinants of environmental exposures on health outcomes. While epidemiology more broadly studies population distributions of disease and health, environmental epidemiology specifically targets the exposure-disease relationship for environmental exposures. These exposures are typically restricted to physical, chemical, and non-infectious biological substances that come from the general (as opposed to workplace) environment. Examples include noise, arsenic, and mold spores, respectively. There are many other environmental sources of exposure, such as dietary practices, the social environment, the workplace, and infectious agents; while often these are considered their own areas (i.e. nutritional, social, occupational, and infectious disease epidemiology), in some contexts these areas are considered to be part of environmental epidemiology. One of the ultimate goals of environmental epidemiology is to develop a knowledge base of the health effects of environmental exposures that can in turn influence policies to protect public health. For instance, as part of the scientific process mandated by the Clean Air Act, the Environmental Protection Agency reviews air pollution studies in order to determine whether current air pollution standards are adequate to protect public health with an adequate margin of safety.

A fundamental aspect of design of environmental epidemiology studies is the consideration of the important patterns and routes of exposure that need to be captured, and identification of the types and time scale(s) of their effect on the health outcome of interest. Researchers need to determine whether exposure occurs through ingestion, inhalation, and/or dermal absorption routes. They must consider how it varies in time and space as well as how personal behaviors impact this variation. They must determine the most plausible time scale(s) of the health effects. While the term "health effects" generically refers to the impact of an exposure on some aspect of health or bodily function, types of effects can vary considerably. For instance, health effects could be local or systemic, acute or chronic, reversible or irreversible. These distinctions will have impacts on the study design; in addition the time scale of the target health effect is a fundamentally important design consideration. Most environmental epidemiology studies can be divided into those that focus on acute (or short-term) effects, where health outcomes are triggered by exposures over the previous minutes, hours, or days, and those that focus on chronic (or long-term) effects where outcomes develop from exposures that accumulate over months, years, or decades. An example of an acute effect study is a case-crossover study to estimate the association between short-term fine particulate matter exposure and incidence of acute myocardial infarction[2] while an example of a long-term effect study is a cohort study such as the Multi-Ethnic Study of Atherosclerosis Air pollution study, MESA Air, which aimed to estimate the effect of long-term particulate matter exposure and progression of subclinical cardiovascular disease[3].

The ability to assess exposure is an essential feature of the success of environmental epidemiology studies. Exposure metrics need to be biologically relevant, including being well aligned with a biologically plausible exposure window. The exposure estimates should be sufficiently accurate and precise, and the study needs to have sufficient exposure contrasts that reflect the target population. The ultimate exposure assessment design goal for an epidemiologic study application is to ensure that inference about the target parameter of interest has sufficient power, is valid, and is generalizable beyond the study sample. To date there has been very little work on the design of environmental epidemiology studies that directly takes into account exposure assessment design and, when exposure will be obtained from a prediction model, how to select the exposure prediction model. Recently methods for exposure measurement error correction for statistical predictions of exposure in air pollution cohort studies have provided new insights into the importance of network design for inference [4]. Not only does the size of the monitoring network matter, but the placement of monitors has implications for the validity of the resulting health effect estimates. (See the discussion of cohort studies below for further elaboration.)

Two design challenges that are common to environmental epidemiology studies are the relatively low and typically ubiquitous exposures, and the need to estimate very small health effects. (Note that regardless of the size of effects, the ubiquity of many environmental exposures means that they have important public health impacts.) For instance, in time series studies of air pollution health effects relative rate estimates for interquartile range (IQR) change in exposure are often on the order of 1.01 to 1.05 (e.g. [5]) Time series studies have been able to estimate these small increases in risk by leveraging huge populations. Some environmental epidemiology studies, particularly famous historical examples such as John Snow's work on the relationship between drinking water exposure and the cholera epidemic in London, have benefitted from higher exposures and larger exposure contrasts that make it easier to detect health effects. Finally, environmental epidemiology shares many of the common challenges faced by all branches of epidemiology such as confounding, information, and selection bias, as well as exposure measurement error.

The remaining sections of this chapter are organized to consider study designs classified by time scales of effects. Studies that focus on short-term variation of exposures include time series, case-crossover, and panel studies. In my experience these are more commonly used in environmental epidemiology applications than in other branches of epidemiology. Studies that focus on long-term average exposures and chronic effects include cross-sectional, case-control, and cohort studies. All these studies are examples of analytic studies, i.e., studies that focus on estimating the relationship between exposures and health outcomes (i.e. responses). I will not consider descriptive epidemiology studies in this chapter.

26.2 Studies that focus on short-term variation of exposures and acute effects

Below I summarize the most pertinent aspects of acute effects designs. This broadens and updates the Dominici, Sheppard & Clyde [6] review of study designs for the health effects of air pollutants. Air pollution epidemiology is unique because there exist widespread time-varying ambient pollution monitoring data that can be leveraged to study acute health impacts. Most other environmental exposures do not have adequate data for the dominant route of exposure to support acute effects study designs.

26.2.1 Ecologic time series studies

Ecologic time series studies relate time-varying (e.g. daily average) estimates of exposure to time-varying numbers of events. These are common in the field of air pollution epidemiology to relate day-to-day variation in pollutants (such as particulate matter less than 2.5 microns in diameter ($PM_{2.5}$)) to day-to-day variation in total mortality counts. This is due to the easy availability of time series of pollutants and administrative data on morbidity and mortality outcomes. The simplest time series studies are for single sites; they use monitoring data from a single or an average of multiple monitors for one geographic area (such as a city) and daily counts of health outcomes for the same area. Multi-site studies combine estimates from multiple locations using meta-analysis methods. The National Morbidity, Mortality and Air Pollution Study (NMMAPS) is the best known multi-site time series study[7].

Analysis of time series studies uses a generalized additive model for the expected counts related to a linear predictor with an exponential link function, written for example as

$$E(Y_t) = \exp[\beta_0 + X_{t-\ell}\beta_X + S(\text{time}, \lambda_1) + S(\text{temp}, \lambda_2) + \text{DOW} \times \beta_{DOW}].$$

Data are typically on the daily time scale. A measure of exposure (e.g. the pollutant concentration, $X_{t-\ell}$) on the same or a lagged day (ℓ) is included; its coefficient β_X is the exposure effect parameter of interest. The exposure parameterization can be extended to include multiple distributed lags using a distributed lag model (e.g., [8]). Control for confounding is essential for valid interpretation of the exposure effect; the functions $S(\cdot)$ are included to control for temporal variation (time: annual and seasonal), and temperature trends (temp), respectively, while DOW is a set of indicator variables to control for day of week. Unlike most other analyses of count data, air pollution time series studies do not include an offset term to account for the population at risk. Because the population at risk varies slowly and not day to day, omission of this term is justified by the assumption that the confounder model is adequate to account for changes in the at-risk population over time. Typically analysts use parametric splines (e.g. natural cubic splines) or nonparametric smoothing splines to parameterize $S(\cdot)$. The smoothing parameter is λ; it is selected to control the degree of large-scale variation in time and temperature that is removed from the inference about the association between short-term change in air pollution and day-to-day variation of the health outcome. Use of a dependence model can be incorporated to address residual correlation, although in many applications the rich control for temporal confounding removes most of the residual dependence.

Many aspects of the analysis of time series studies need careful attention since the amount of pollutant variation used for inference is swamped by the variability of the confounders. Previous work in the air pollution epidemiology setting has shown that a fairly rich confounder adjustment model is needed to ensure the exposure effect estimate is not biased due to confounding, but that once this complexity has been achieved, further confounding control does not improve the inference[9]. Bhaskaran [10] provides a basic overview of time series regression studies in environmental epidemiology and tools for analysis of these studies.

Design of time series studies must consider the typical event counts, the day to day variation in the pollutant of interest, the number of days in the time series, and the anticipated effect size. Some studies have also combined estimates from multiple cities in order to provide more precise effect estimates, such as in the NMMAPS study[7]. Time series studies take advantage of the existence of long exposure time series from regulatory monitoring site measurements, making the assumption that these measurements represent population exposures. Within-city heterogeneity of pollutants is typically ignored in time series studies, although some authors have considered using modeled exposures to capture spatial variation in time series and evaluated their impacts in simulation studies [11, 12].

26.2.2   Case-crossover studies

The case-crossover design is an observational study analogue to the classic crossover study, an experimental design where subjects receive both intervention and control conditions in order to ascertain the average within-subject effect of the intervention. In the case-crossover study, outcomes occur at known times and in settings where recent exposure is believed to acutely affect the outcome. Only cases are used. The estimated health effect, the average within-subject effect of the exposure, is derived from within-subject contrasts of the exposure just prior to the outcome (called the "index" exposure) to exposures at comparison or "referent" times. The sets of index and referent times are called referent windows; referent times can be restricted to occur only before the event or can be defined as *a priori* strata such that referent times after the event are possible. Referent windows are typically restricted to be close in time to the event, such as within the same month. The key design feature for a case-crossover study is the choice of the referent selection approach. This is analogous to choosing the approach to matching in a case-control study. As I discuss below, some referent schemes give biased health effect estimates; the bias is called "overlap bias".

There are some ways that the analogy between a matched case-control and a case-crossover study can be misleading. [13] While the assumption of independence of exposures within strata in a case-control study is typically reasonable, temporal autocorrelation is often present in an exposure time series. In air pollution and other case-crossover studies with shared exposures, the exposures are known and two events at the same time will have similar or identical referents; in contrast, for matched case-control studies there is no between-stratum similarity in exposure. Furthermore, for shared exposures such as air pollution it is possible to quantify exposure after a subject is no longer at risk of the event. Finally, stratification for matched case-control studies is based on factors other than the response. In contrast, case-crossover strata are defined by the timing of the response. If not chosen properly, such that the strata are disjoint and non-overlapping, case-crossover study referent windows can be non-independent and the effect estimates subject to overlap bias.

The case-crossover design is well suited to the study of the health impacts of time-varying environmental exposures. It relies on the assumption that there is no systematic change over time in either the exposure or outcome within the referent window. As in their crossover study analogues, exposures must act quickly on the outcome and their effects should not persist so there is no carry-over effect of previous exposures on the outcome. The design was introduced by Maclure in an analysis of the effect of sexual activity on myocardial infarction[14]. In this and all case-crossover studies, it is essential that the exposure can be quantified at all times within the referent window. An important feature of the referent scheme design is that the referents should represent the expected exposure distribution at times when a case does not occur.

The case-crossover design has several important strengths relative to matched case-control studies. By relying only on data from cases, it is possible to estimate health effects in settings where non-cases are difficult or expensive to obtain. By focusing on within-subject contrasts it automatically controls for confounders that only vary between subjects. Time-dependent confounding can also be controlled by thoughtful referent selection design. For instance, if season is an important time-dependent confounder, referent windows can be restricted to fall completely within one season. It is worthwhile noting that the case-crossover design can be viewed as a stratified version of a time series study since both designs rely only on cases. A time series study controls for time-varying confounding through modeling while a case-crossover study uses matching. Furthermore, because its analysis is based on conditional logistic regression (as is described below), it is straightforward to assess individual-level effect modification in a case-crossover study.

In air pollution epidemiology many of the earliest case-crossover studies considered the impact of air pollutants on non-accidental mortality (see e.g. Janes et al's [15] compilation of studies published between 1999 and 2004). These papers were published before the operating characteristics of the referent design were well understood. A key feature of air pollution case-crossover studies is that the exposures are shared across members of the study population because, typically, they are derived from a single monitor or a daily average of measurements from multiple monitors. Shared exposure time series are fixed and known regardless of the health outcome or referent sampling scheme. Thus the health effect estimate needs to be unbiased for a single realization of the exposure series. This is more difficult to achieve than ensuring the estimate is unbiased on average over multiple realizations of the exposure series, the condition needed for unshared exposure series. Sexual activities and coffee drinking are examples of unshared exposure series.

Case-crossover studies are analyzed using conditional logistic regression. Typically this is framed as a model for the hazard where the subject-specific baseline hazard is constant over the entire referent window, leading to a model of the form $\lambda_i = \exp\{x_{it}\beta\}$ for individual

i at time t, baseline hazard λ_i, and predictors x_{it}. (The assumption of a constant baseline hazard is often reasonable over the short time periods considered in case-crossover studies.) As described by Lumley and Levy[13], for disjoint strata $S(t)$ defined *a priori*, this leads to the estimating function

$$U_i(\beta) = x_{it} - \sum_{t \in S(t)} \frac{x_{it}\exp\{x_{it}\beta\}}{\sum_{s \in S(s)} \exp\{x_{it}\beta\}}.$$

In order to ensure this estimating function is unbiased, it is essential to consider how $S(t)$ is defined. To avoid overlap bias it is most straightforward to ensure that the referent windows can be derived from strata that can be defined *a priori* such as with the time-stratified design (Figure 26.1.A: Each dash-color combination represents a distinct stratum. The timing of the index exposure determines the stratum. All other times within the stratum correspond to the referent exposures.). Referent sampling schemes that are subject to overlap bias typically are defined as a direct function of the timing of the outcome. In that case multiple referent windows overlap and cannot be written as disjoint partitions of time. Examples include the symmetric bi-directional (Figure 26.1.B) and restricted unidirectional (Figure 26.1.C) designs. These do not lead to valid estimating functions because the index time uniquely defines the referent window. Janes et al [15, 16] showed that the magnitude of the overlap bias is unpredictable and, while often small, can be important. Many of the referent selection strategies used in published studies, particularly the earliest studies, were subject to overlap bias.

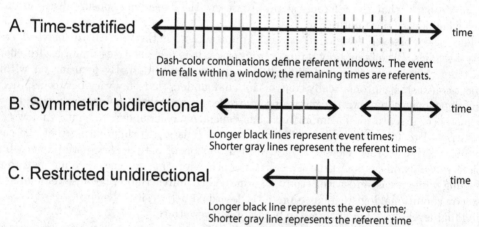

FIGURE 26.1
Referent Design Examples

Janes et al proposed a taxonomy of referent schemes according to their statistical properties. Designs that avoid overlap bias are localizable, meaning that there exists an unbiased estimating function that is restricted to the referent windows. The time-stratified referent design is one example of a localizable design (Figure 26.1.A). In a non-localizable referent design, unbiased estimating functions that are restricted to the referent windows do not exist. For instance, all referent designs where the referent window is a direct function of the timing of the event, such as the symmetric bidirectional (Figure 26.1.B) and restricted unidirectional designs (Figure 26.1.C), are examples of non-localizable referent designs. Localizable referent designs can further be divided into ignorable and non-ignorable groups. Ignorable designs allow the referent sampling scheme to be ignored in conducting the analysis. The time-stratified referent design is an ignorable design. In that case, conditional logistic regression analysis will be unbiased when applied using standard software implementations without further modification to account for the sampling design (such as incorporating an offset term). Non-ignorable designs require some attention to the referent design in the analysis. An example is the semi-symmetric bidirectional design where one referent is chosen at random from the days before and after the event, at a fixed lag from the event. At the beginning and end of the time series only one referent can be chosen; an offset is required to obtain the correct likelihood. [15]

Design of case-crossover studies must consider the number of events, the appropriate referent selection approach, the exposure contrast present within the referent windows, and the timeframe when the referent windows will avoid temporal confounding. When the exposures are shared, the design must also address the potential for overlap bias.

26.2.3 Panel studies

The panel study design focuses on understanding the association of short-term changes in a time-varying exposure on a health outcome. Panel studies are particularly useful in settings where it is feasible to quantify individual-level time-varying environmental exposures, and to assess their relationship to health outcomes that have sufficient variation in a small group. Panel studies are also particularly useful for the study of outcomes that are difficult to quantify on large populations as long as they vary sufficiently over time within an individuals. Biomarkers and clinical assessments such as lung function are good examples. In air pollution epidemiology, the panel study design has been used to estimate the association between total personal exposure and respiratory outcomes in people with asthma, such as lung function or asthma exacerbations. For instance, in research conducted in Seattle, several small panel studies used measured personal air pollutant exposures[17-21], while one that leveraged the pre-randomization diary data for a clinical study of children with asthma was considerably larger but its exposure assessment was much more limited because it did not have personal air pollution measurements[22]. A panel study that relies on community-level ambient monitoring data rather than personal exposure measurements is an example of a semi-individual design.[23, 24] From an exposure perspective, such panel studies have no more advantage than case-crossover or time series studies. However, from an outcome assessment perspective, the panel design may still be worthwhile even when only community-level exposure information is available.

Panel studies are a type of longitudinal study where the scientific emphasis is on short-term variations. Thus they often involve intensive daily measurements on all members of the panel over a relatively short time period of e.g., a few weeks. Otherwise, their models and parameter interpretations are similar to any longitudinal study where there are i individuals with both outcome and exposure information at t times. The most basic panel study model (omitting confounders) is the marginal model

$$g(E(Y_{it} \mid x_{it})) = \beta_0 + x_{it}\beta^M$$

where Y_{it} is the outcome, $g(\cdot)$ the link function, and x_{it} the exposue. This can be reframed as a conditional model by adding a subject-specific random effect, e.g.,

$$g(E(Y_{it} \mid x_{it}, i)) = \beta_0 + b_i + x_{it}\beta^C.$$

Some panel studies use transition models that condition on individual subjects' previous responses:

$$g(E(Y_{it} \mid x_{it}, Y_{it-1})) = \beta_0 + x_{it}\beta^T + Y_{it-1}\gamma.$$

In all cases the regression parameter β, whether it be β^M, β^C, or β^T, is the target parameter of interest, though its interpretation depends upon the model being fit. Analysis approaches for these models are well documented by Diggle et al [25]. In the context of air pollution epidemiology, Janes et al [26] describe these three models in some detail, provide parameter interpretations, and, through worked data analysis examples, compare and contrast the estimates for binary and continuous outcomes. They also present examples of exploratory analyses useful for panel (and longitudinal) studies.

Because panel studies are expensive, they often are conducted on relatively small populations. This makes it difficult to fit complex models or interrogate the data about its structure (e.g. for the dependence model). This feature can also magnify challenges such as handling of missing data and adequate confounding control. Design of panel studies should consider the adequacy of within-subject exposure contrasts.

26.3   Studies that focus on long-term average exposures and chronic health effects

26.3.1   Cohort studies

Cohort studies ascertain a population of individuals with diverse exposures and follow these over time for disease events. In contrast to designs discussed in the previous section, the cohort design is most useful for the study of long-term exposures and health outcomes that are impacted by this long-term exposure. In air pollution epidemiology examples of cohort studies include the Southern California Children's Health Study (CHS) [27-31], the Harvard Six Cities study[32], and the MESA Air study[3]. These studies assessed lung function growth, total mortality, and subclinical markers of cardiovascular disease, respectively. Cohort studies have also been used to assess carcinogenic effects of occupational exposures such as the effect of diesel exhaust on lung cancer in the Diesel Exposure to Miners study[33] and the incidence of nasal sinus cancer among Welsh nickel refiners[34-36].

For studies of longitudinal change of subclinical effects, as was the focus of the MESA Air study, longitudinal mixed models have been proposed that allow estimation of exposure effects for both cross-sectional exposures at baseline and for the impact of exposure on rate of change in the outcome. These analyses use a longitudinal mixed model of the form

$$Y_{it} = (\alpha_0 + a_i + x_{i0}\alpha_1 + z_{i0}\alpha_2) + [\beta_0 + b_i + x_{it}\beta_1 + w_{it}\beta_2] v_{it} + \{u_{it}\gamma\} + e_{it}$$

where for person i at their t^{th} follow-up visit, Y_{it} is the measured outcome; x_{i0} is the average exposure concentration in the appropriate time period prior to baseline; x_{it} is the long-term average exposure prior to exam t; z_{i0}, is a vector of covariates related to baseline outcome; w_{it} and u_{it} are vectors of covariates related to the current outcome; v_{it} is the time between the baseline and the tth follow-up visit; a_i, and b_i are the subject-specific random effects; and e_{it} is the measurement error associated with Y_{it}. The brackets distinguish the three components of the model: (predictors of the baseline outcome), [predictors of decline from baseline], and {time-varying transient (measurement-specific) effects on the outcome}. The model captures two parameterizations of the relationship between exposure and the outcome: $x_{i0}\alpha_1$ characterizes the cross-sectional association between exposures preceding the baseline exam and baseline response, while $x_{it}v_{it}\beta_1$ characterizes the outcome change as a function of long-term average exposure during follow-up. Often the rate of change parameter β_1 is the primary parameter of interest while the cross-sectional parameter α_1 is of secondary interest.

Berhane et al[37] and discussants[38] provided thoughtful discussion of statistical issues to consider in estimating long-term effects of air pollutants for continuous, binary, and time to event outcomes. Their multi-level model paradigm for estimating environmental exposure effects included three levels to separately address exposure contrasts within subjects across time, within communities across subjects, and between communities. They provided examples from the CHS, and their multi-level modeling paradigm allows important features of longitudinal change, such as age-specific lung function growth, to be incorporated into their models.

Bryan and Heagerty developed methods to estimate differences between groups in longitudinal rate of change for settings where there is non-linear change over time.[39] They assume that differences between covariate groups in rate of change are proportional. They provide methods for both marginal and subject-specific mixed models.

One of the important challenges with cohort studies of environmental (and occupational) exposures is the need to model exposure. All the studies cited above have used some form of predicted exposure, though in a few (specifically the Six Cities study and many of the early results from CHS), community-level measurements were used as a proxy for the exposure for all members of the community. While exposure assessment is a huge challenge in environmental epidemiology, there are many settings where some relevant aspect of exposure can be predicted. This could be as simple as an indicator variable for private vs. municipal water source as a proxy for exposure to arsenic and other water contaminants, or proxy exposure measures such as age and year of first employment, and time since first employment in the Welsh nickel refiners. Methods for correcting inference to account for spatial prediction of exposures have been proposed by Szpiro et al [4, 40]. Under a data-generating model for the exposure that treats the surface as fixed and the locations measured within it as random, Szpiro and Paciorek[4] note that two conditions are needed for valid inference: spatial compatibility – that the locations measured on the exposure surface represent the locations to be predicted in the cohort, and covariate compatibility – that the spatially varying confounders in the health analyses are also included in the exposure model. This work decomposes the measurement error induced into two components: a Berkson-like piece due to not being able to completely predict the underlying exposure surface, and a classical-like piece due to estimating the exposure model parameters. Correction for measurement error can be accomplished by applying a bootstrap.

26.3.2   Case-control and cross-sectional studies

Case-control and cross-sectional studies are not unique to environmental epidemiology so the following description is extremely brief.

Case-control study designs select individuals from a (hypothetical) source population based on their case vs. non-case status and then determine their exposure for analysis. They are more efficient than cohort studies because it is feasible to study rare diseases with many fewer non-cases included. However, their validity depends on how well the sampling represents the source population. Many cohort studies include nested case-control substudies; these allow more thorough assessment of confounding factors than is feasible in the entire cohort. For instance, in the Diesel Exposure in Miners study, the nested case-control study allowed detailed assessment of the confounding effect of smoking on the relationship between diesel exhaust exposure and lung cancer incidence; this was not possible in the larger cohort study[41].

Analysis of case-control study designs uses logistic regression where the case/control status is the outcome. While subject selection in this design is based on the disease outcome and the exposure is determined after sampling, under the condition that the case-control sampling represents the source population and is unrelated to the exposure, a standard application of logistic regression models can be applied for valid inference of exposure effects in case-control datasets[42].

Cross-sectional studies assess a cross-section of a population (or sample thereof) and its exposure at a specific time. Analysis, typically using a regression model, estimates the relationship between exposure contrasts and health outcomes. There are many reasons why cross-sectional studies are more limited than cohort studies. For instance, they are subject to length-biased sampling such that individuals who are healthy or with long-duration adverse health outcomes are overrepresented relative to individuals who leave the population due to short-duration adverse health outcomes. Furthermore, when exposures vary over time, cross-sectional study data and analyses don't allow analysts to sort out exposure effects due to heterogeneity between subjects from the impact of exposure on individuals over time.

26.4   Summary and Discussion

This chapter covered the most important analytic study designs for environmental epidemiology, with emphasis on designs that have received the most attention in this epidemiologic sub-discipline. The most common applications are in air pollution epidemiology because there exists considerable regulatory monitoring data that can be used in epidemiologic studies, in large part because the regulatory measurements inform individual and population exposures. A natural extension to the designs discussed in this chapter are pooled studies and meta-analyses, such as was done in the NMMAPS study; these methods are broadly applicable tools for combining information. They are useful for obtaining more precise estimates of exposure effects while properly accounting for heterogeneity between areas or studies.

Some early challenges in air pollution epidemiology have been solved, most notably, how to conduct referent selection in case-crossover studies[43], and the proper application of generalized additive models to ensure correct standard error calculations in time series studies[44, 45]. Many research challenges remain, including modeling of and inference for multiple concurrent environmental exposures, developing a better understanding the properties of health effect estimates when mismeasured exposures are used, and the importance of exposure assessment design in inference about health effects.

Recently efforts to better understand the health effects due to environmental exposures have adopted causal modeling methods in order to address a slightly different question than is common in air pollution epidemiology.[46] The common epidemiology question, "what is the association between the pollution exposure and a health outcome?" is reframed to determine the consequences of specific regulatory actions by asking "what is the causal effect of the regulatory actions on health outcomes?" Studies that address the latter question are "accountability" studies. By using causal modeling methods, researchers can formally estimate this causal effect from data by thinking about observational studies in a randomized study framework and considering how to ensure they approximate randomization. Using scientific understanding of exposure, one must define treatment and control conditions, and determine the counterfactual scenario, i.e., what would have happened had the treatment condition not occurred. A potential outcome is the outcome that would have happened if an individual had been randomized to the other condition. Many of the concerns with observational studies, most notably uncontrolled confounding, are addressed in a causal modeling framework by using tools such as propensity scores and principal stratification to approximate an experimental design for the study since randomization creates comparable groups on average and ensures there is no confounding with treatment.[47] Once comparable treatment and control groups have been created in this hypothetical experiment, analysis can proceed to estimate the average treatment effect.

Bibliography

[1] Baker DB, Nieuwenhuijsen MJ: Environmental epidemiology: Study methods and application: Oxford University Press; 2008.

[2] Sullivan J, Sheppard L, Schreuder A, Ishikawa N, Siscovick D, Kaufman J: Relation between short-term fine-particulate matter exposure and onset of myocardial infarction. Epidemiology 2005, 16(1):41-48.

[3] Kaufman JD, Adar SD, Barr RG, Budoff M, Burke GL, Curl CL, Daviglus ML, Roux AVD, Gassett AJ, Jacobs DR et al: Association between air pollution and coronary artery calcification within six metropolitan areas in the USA (the Multi-Ethnic Study of Atherosclerosis and Air Pollution): a longitudinal cohort study. Lancet 2016, 388(10045):696-704.

[4] Szpiro AA, Paciorek CJ: Measurement error in two-stage analyses, with application to air pollution epidemiology. Environmetrics 2013, 24(8):501-517.

[5] Sheppard L, Levy D, Norris G, Larson TV, Koenig JQ: Effects of ambient air pollution on nonelderly asthma hospital admissions in Seattle, Washington, 1987-1994. Epidemiology 1999, 10(1):23-30.

[6] Dominici F, Sheppard L, Clyde M: Health effects of air pollution: A statistical review. International Statistical Review 2003, 71(2):243-276.

[7] Dominici F, Samet JM, Zeger SL: Combining evidence on air pollution and daily mortality from the 20 largest US cities: a hierarchical modelling strategy. Journal of the Royal Statistical Society Series a-Statistics in Society 2000, 163:263-284.

[8] Almon S: The distributed lag between capital appropriations and expenditures. Econometrica: Journal of the Econometric Society 1965:178-196.

[9] Peng RD, Dominici F, Louis TA: Model choice in time series studies of air pollution and mortality. Journal of the Royal Statistical Society Series a-Statistics in Society 2006, 169:179-198.

[10] Bhaskaran K, Gasparrini A, Hajat S, Smeeth L, Armstrong B: Time series regression studies in environmental epidemiology. International Journal of Epidemiology 2013, 42(4):1187-1195.

[11] Goldman GT, Mulholland JA, Russell AG, Strickland MJ, Klein M, Waller LA, Tolbert PE: Impact of exposure measurement error in air pollution epidemiology: effect of error type in time-series studies. Environmental Health 2011, 10.

[12] Goldman GT, Mulholland JA, Russell AG, Gass K, Strickland MJ, Tolbert PE: Characterization of ambient air pollution measurement error in a time-series health study using a geostatistical simulation approach. Atmospheric Environment 2012, 57:101-108.

[13] Lumley T, Levy D: Bias in the case-crossover design: implications for studies of air pollution. Environmetrics 2000, 11(6):689-704.

[14] Maclure M: The case-crossover design - a method for studying transient effects on the risk of acute events. American Journal of Epidemiology 1991, 133(2):144-153.

[15] Janes H, Sheppard L, Lumley T: Case-crossover analyses of air pollution exposure data - Referent selection strategies and their implications for bias. Epidemiology 2005, 16(6):717-726.

[16] Janes H, Sheppard L, Lumley T: Overlap bias in the case-crossover design, with application to air pollution exposures. Statistics in Medicine 2005, 24(2):285-300.

[17] Allen RW, Mar T, Koenig J, Liu LJS, Gould T, Simpson C, Larson T: Changes in lung function and airway inflammation among asthmatic children residing in a woodsmoke-impacted urban area. Inhalation Toxicology 2008, 20(4):423-433.

[18] Koenig JQ, Mar TF, Allen RW, Jansen K, Lumley T, Sullivan JH, Trenga CA, Larson TV, Liu LJS: Pulmonary effects of indoor- and outdoor-generated particles in children with asthma. Environmental Health Perspectives 2005, 113(4):499-503.

[19] Liu LJS, Box M, Kalman D, Kaufman J, Koenig J, Larson T, Lumley T, Sheppard L, Wallace L: Exposure assessment of particulate matter for susceptible populations in Seattle. Environmental Health Perspectives 2003, 111(7):909-918.

[20] Mar TF, Jansen K, Shepherd K, Lumley T, Larson TV, Koenig JQ: Exhaled nitric oxide in children with asthma and short-term PM2.5 exposure in Seattle. Environmental Health Perspectives 2005, 113(12):1791-1794.

[21] Trenga CA, Sullivan JH, Schildcrout JS, Shepherd KP, Shapiro GG, Liu LJS, Kaufman JD, Koenig JQ: Effect of particulate air, pollution on lung function in adult and pediatric subjects in a Seattle panel study. Chest 2006, 129(6):1614-1622.

[22] Yu OC, Sheppard L, Lumley T, Koenig JQ, Shapiro GG: Effects of ambient air pollution on symptoms of asthma in Seattle-area children enrolled in the CAMP study. Environmental Health Perspectives 2000, 108(12):1209-1214.

[23] Kunzli N, Tager IB: The semi-individual study in air pollution epidemiology: A valid design as compared to ecologic studies. Environmental Health Perspectives 1997, 105(10):1078-1083.

[24] Sheppard L: Insights on bias and information in group-level studies. Biostatistics 2003, 4(2):265-278.

[25] Diggle P, Heagerty P, Liang K, Zeger S: Analysis of longitudinal data. Oxford: Statistical Science Series 2002.

[26] Janes H, Sheppard L, Shepherd K: Statistical Analysis of Air Pollution Panel Studies: An Illustration. Annals of Epidemiology 2008, 18(10):792-802.

[27] Gauderman WJ, Avol E, Gilliland F, Vora H, Thomas D, Berhane K, McConnell R, Kuenzli N, Lurmann F, Rappaport E et al: The effect of air pollution on lung development from 10 to 18 years of age. New England Journal of Medicine 2004, 351(11):1057-1067.

[28] Gauderman WJ, Vora H, McConnell R, Berhane K, Gilliland F, Thomas D, Lurmann F, Avol E, Kunzli N, Jerrett M et al: Effect of exposure to traffic on lung development from 10 to 18 years of age: a cohort study. Lancet 2007, 369(9561):571-577.

[29] Jerrett M, Shankardas K, Berhane K, Gauderman WJ, Kunzli N, Avol E, Gilliland F, Lurmann F, Molitor JN, Molitor JT et al: Traffic-related air pollution and asthma onset in children: A prospective cohort study with individual exposure measurement. Environmental Health Perspectives 2008, 116(10):1433-1438.

[30] McConnell R, Berhane K, Gilliland F, Molitor J, Thomas D, Lurmann F, Avol E, Gauderman WJ, Peters JM: Prospective study of air pollution and bronchitic symptoms in children with asthma. American Journal of Respiratory and Critical Care Medicine 2003, 168(7):790-797.

[31] Peters JM, McConnell R, Berhane K, Millstein J, Lurmann F, Gauderman J, Avol E, Gilliland F, Thomas D: Air pollution and the incidence, prevalence and severity of childhood asthma: Results from the Southern California Children's Health Study. Epidemiology 2002, 13(4):S132-S132.

[32] Dockery DW, Pope CA, Xu XP, Spengler JD, Ware JH, Fay ME, Ferris BG, Speizer FE: An association between air-pollution and mortality in six U.S. cities. New England Journal of Medicine 1993, 329(24):1753-1759.

[33] Attfield MD, Schleiff PL, Lubin JH, Blair A, Stewart PA, Vermeulen R, Coble JB, Silverman DT: The Diesel Exhaust in Miners Study: A Cohort Mortality Study With Emphasis on Lung Cancer. Jnci-Journal of the National Cancer Institute 2012, 104(11):869-883.

[34] Doll R, Morgan LG: cancers of lung and nasal sinuses in nickel workers. British Journal of Cancer 1970, 24(4):623-&.

[35] Doll R, Mathews JD, Morgan LG: cancers of lung and nasal sinuses in nickel workers - reassessment of period of risk. British Journal of Industrial Medicine 1977, 34(2):102-105.

[36] Kaldor J, Peto J, Easton D, Doll R, Hermon C, Morgan L: models for respiratory cancer in nickel refinery workers. Journal of the National Cancer Institute 1986, 77(4):841-848.

[37] Berhane K, Gauderman WJ, Stram DO, Thomas DC: Statistical issues in studies of the long-term effects of air pollution: The Southern California Children's Health Study. Statistical Science 2004, 19(3):414-434.

[38] Meiring W, Sheppard L, Wakefield JC, Le ND, Zidek JV, Berhane K, Gauderman WJ, Stram DO, Thomas DC: Statistical issues in studies of the long-term effects of air pollution: The southern california children's health study - comment. Statistical Science 2004, 19(3):434-449.

[39] Bryan M, Heagerty PJ: Direct regression models for longitudinal rates of change. Statistics in Medicine 2014, 33(12):2115-2136.

[40] Szpiro AA, Sheppard L, Lumley T: Efficient measurement error correction with spatially misaligned data. Biostatistics 2011, 12(4):610-623.

[41] Silverman DT, Samanic CM, Lubin JH, Blair AE, Stewart PA, Vermeulen R, Coble JB, Rothman N, Schleiff PL, Travis WD et al: The Diesel Exhaust in Miners Study: A Nested Case-Control Study of Lung Cancer and Diesel Exhaust. Journal of the National Cancer Institute 2012, 104(11):855-868.

[42] Breslow NE, Day N: Statistical methods in cancer research. Vol. 1, The analysis of case-control studies: IARC; 1981.

[43] Mittleman MA: Optimal referent selection strategies in case-crossover studies - A settled issue. Epidemiology 2005, 16(6):715-716.

[44] Dominici F, McDermott A, Hastie TJ: Improved semiparametric time series models of air pollution and mortality. Journal of the American Statistical Association 2004, 99(468):938-948.

[45] Lumley T, Sheppard L: Time series analyses of air pollution and health: Straining at gnats and swallowing camels? Epidemiology 2003, 14(1):13-14.

[46] Zigler CM, Dominici F: Point: Clarifying Policy Evidence With Potential-Outcomes Thinking-Beyond Exposure-Response Estimation in Air Pollution Epidemiology. American Journal of Epidemiology 2014, 180(12):1133-1140.

[47] Rubin DB: For objective causal inference, design trumps analysis. Annals of Applied Statistics 2008, 2(3):808-840.

Part IV

Topics in Climatology

27

Modeling and assessing climatic trends

Peter F. Craigmile

Department of Statistics, The Ohio State University, Columbus, Ohio, USA.

Peter Guttorp

Department of Statistics, University of Washington, Seattle, Washington, USA and Norwegian Computing Center, Oslo, Norway

CONTENTS

27.1 Introduction

Studies of climate change often involve the statistical analysis of climate variables (or their meteorological counterparts) indexed in space and time [98]. Usually these variables are observed directly from instrumental measurements (e.g., from thermometers for temperature; from rain gauges for precipitation), but it is becoming more common to observe data indirectly from satellite measurements or as the output of global or regional climate models. As the datasets become more complex, the statistical analyses can become more involved.

A common statistical problem is to model and assess trends in climate data [19, 21, 84, 90, 102]. Trends are often described by smooth changes in certain features of a stochastic process over longer scales. Traditionally trend refers to smooth changes in the mean of a process over time but, as we will illustrate, this definition can be extended to allow us to consider the smooth changes of other characteristics of a stochastic process, or can be extended to other dimensions of change, such as over space. Often it can help to think of smoothing in terms of its effect upon the low and high frequency components of the climate series $\{Y_t\}$; namely, smoothing emphasizes the low frequency components of the series ("the trend"), and reduces the effect of the high frequency component ("the noise").

In the latest IPCC report [46] the issue of trends is explained in Box 2.2. While they discuss both linear and nonlinear trend models, they chose to emphasize linear trends, arguing that "The linear trend fit is used in this chapter because it can be applied consistently to all the data sets, is relatively simple, transparent and easily comprehended, and is frequently used in the published research assessed here." (p.180)

In this chapter we mainly focus on the estimation and assessment of trend in the presence of dependence in time. Statistical methods that assume that the errors (after removal of the trend component) are independent and identically distributed are usually a poor assumption for climate data. Indeed assuming that the errors can be approximated by a simple time series model such as the autoregressive process of order one, AR(1), is also unlikely to represent the residual variation in the climate process. The choice of the temporal (and spatial) scale for the trend is difficult and not well defined. We present methods to estimate both linear and nonlinear trends, with associated uncertainty quantification. Trend estimation can be different from changepoint analysis (the estimation of breaks in a time series). (For a discussion of changepoints and trends see [36].) We discuss the estimation of trend in nontraditional settings, such as in the analysis of climate extremes, and point to future directions in the statistical assessment of climatic trends.

27.2   Two motivating examples

We motivate the estimation of trend in climate time series using two datasets of average annual temperature. Each series has a well-defined estimate of uncertainty. The first series we consider is the average temperature over the contiguous United States (US), while the other is a global temperature series.

27.2.1   US average temperature anomaly

[90] produced statistical estimates of US temperature anomalies from 1897–2008, using the US Historical Climatology Network data set version 2 [70], corrected for the fact that the time of day that measurements are made at can differ by site. Since US land temperature is a statistical estimate and not a direct measurement, it has a quantifiable standard error, with components coming from measurement error at individual stations, spatial dependence, natural variability, orographic effects, etc. [90] was mainly concerned with the statistical estimation of this standard error. Figure 27.1(a) shows a time series plot of the anomalies with respect to the 1961–1990 climatology (i.e., residuals from the average temperature over this time period). The shaded gray region denotes simultaneous 95% confidence intervals for this mean, calculated using the estimated standard errors with a Bonferroni correction [1]. Figure 27.1(b) shows that the standard errors decrease until 1975, and then increase again to the end of the record. The reason for this increase is the removal of stations mainly due to reduced funding.

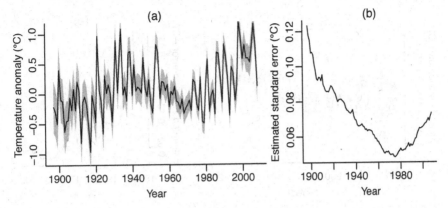

FIGURE 27.1
(a) US annual mean temperature anomalies with respect to the 1961–1990 climatology. The gray shaded region denotes simultaneous 95% confidence intervals for the mean, calculated using estimated standard errors from [90]; (b) A plot of the estimated standard errors for the mean by year.

27.2.2   Global temperature series

The Berkeley Earth project [84] uses isotropic geostatistical tools (kriging; see Chapter 5) to estimate the global mean temperature. The land data have been collected from 14 databases and almost 45,000 stations. One of the main differences between the Berkeley Earth approach and most other global approaches is that the former group does not attempt to "homogenize" stations [96]. If a measurement device is moved or replaced, it is considered a different station, rather than being "corrected." For a statistician, this seems to be a natural approach. The ocean data used in the Berkeley Earth series [83] come from the Hadley Center sea surface temperature data set HadSST3 [51, 52], modified by kriging of missing grid squares. This allows for a statistically justifiable estimate of global mean estimation uncertainty. Figure 27.2, formatted identically as Figure 27.1, demonstrates that the standard errors tend to be higher in the past. In particular, temperature reconstructions were very unreliable before about 1880.

That temperatures tend to increase with the years is obvious in the global temperature series, but possible trend effects are more nuanced for the US average temperatures. Both series exhibit dependence over time that need to be accounted for before we can assess the significance of possible trends. We also need to account for the uncertainty in each mean series, as measured by the time-varying standard errors.

27.3   Time series approaches

In this chapter we mainly focus on methods of estimation and assessment of trend for time series processes observed discretely in time. Let $T \subset \mathbb{Z}$ denote a set of possible time points. Suppose we observe a climate series $\{y_t : t \in T\}$ regularly sampled in time. Let $\{Y_t : t \in T\}$ denote the associated discrete-time time series process, the stochastic process that generated $\{y_t\}$. An *additive decomposition* for trend then assumes that

$$Y_t = \mu_t + \eta_t, \quad t \in T, \tag{27.1}$$

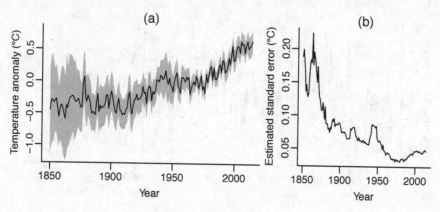

FIGURE 27.2

(a) Global annual mean temperature anomalies with respect to the 1961–1980 climatology. The gray shaded region denotes simultaneous 95% confidence intervals for the mean, using estimated standard errors from [84]; (b) A plot of the estimated standard errors for the mean by year.

where we refer $\{\mu_t : t \in T\}$ to be the *trend component* and $\{\eta_t : t \in T\}$ to be the *(irregular) noise component*, that captures everything in Y_t that is not captured by the trend. We could also extend the definition to include a seasonal component, $\{s_t : t \in T\}$ that repeats over time. (For a further discussion of statistical methods for estimating seasonal components in time series, see, e.g., [8, 12, 77].)

A multiplicative decomposition is often used to analyze data that are positively skewed or exhibit a mean-variance relationship. The *multiplicative decomposition* for trend posits that

$$Y_t = \mu_t \times \eta_t, \quad t \in T.$$

Clearly an additive decomposition (27.1) for a time series is not unique, as it may not be obvious for a given application what constitutes the trend $\{\mu_t\}$ and what is the noise $\{\eta_t\}$. While [50] encapsulates that "the essential idea of trend is that it shall be smooth", he does not indicate "how smooth". An estimate of trend is defined purely in terms of the statistical model or method that we use to estimate it. As with any additive statistical estimation procedure, anything that is not captured by the estimated trend, $\widehat{\mu}_t$ say, will appear in the estimate of the errors $\widehat{\eta}_t = y_t - \widehat{\mu}_t$. This has led to defining the trend in terms of how smooth the process is. For example: (i) letting $\mu_t = f(t)$ for some deterministic function f of time t, we assume a certain number of derivatives for f or by defining the trend in terms of linear combinations of known smooth functions [99, 106]; (ii) using certain functions such as wavelets we can define the trend in terms of averages over certain temporal scales [9, 11, 19].

There are cases for which the trend tends to be less smooth. For example, letting μ_t be related to known covariates, plus a possibly smooth function of time t, may lead to a rougher trends. Introducing an abrupt change in μ_t (e.g., using the broken stick model presented later in this chapter) or using stochastic models for μ_t often also produces rougher trends.

27.3.1   Candidate models for the noise

Different statistical models for the noise will influence our ability to estimate climatic trends. Thus, before we outline statistical methods for estimating and assessing trend, we discuss commonly used classes of time series models for the noise. A typical assumption is that the noise process $\{\eta_t\}$ is a mean zero stationary process. For a mean zero time series process, stationarity requires that the covariance of η_t and η_{t+h} depends only on the time lag h, and not on the time index t. This assumption simplifies estimation of the parameters driving the noise, and can be a reasonable assumption if the trend is able to capture the time-varying features of climate. For further details of stationary processes see Chapter 3.

As we argued at the start of the chapter, it is typically unreasonable to assume that the noise process is uncorrelated in time (also known as a *white noise process*) or independent and identically distributed in time (also known as an *IID process*). In climate studies an *autoregressive process of order one, AR(1),* also called a red-noise process, is commonly used to model the noise $\{\eta_t\}$. This model is defined by letting

$$\eta_t = \phi \eta_{t-1} + Z_t, \quad t \in T,$$

where $\{Z_t\}$ is a white noise or IID process. This process is stationary when the autoregressive parameter ϕ satisfies $|\phi| < 1$. More general dependence structures can be obtained by increasing the order of autoregression to yield an autoregressive process of order p, AR(p):

$$\eta_t = \sum_{j=1}^{p} \phi_j \eta_{t-j} + Z_t, \quad t \in T.$$

We can also filter the $\{Z_t\}$ process to obtain the ARMA(p, q) process, the autoregressive moving average process of orders p and q:

$$\eta_t = \sum_{j=1}^{p} \phi_j \eta_{t-j} + Z_t + \sum_{k=1}^{q} \theta_k Z_{t-k}, \quad t \in T.$$

Since a stationary ARMA model can be written as an infinite order AR model (that in practice would be truncated to finite order), some statisticians working in climate prefer to use lower order AR processes rather than ARMA processes (the parameters of AR processes can be easier to estimate than those of ARMA processes, for example).

In more recent years, long memory processes have been used as a model for noise in studies of climate [4, 19, 21, 48, 56]. The choice of this class of processes could be due to the fact these models capture the self-similar behavior of climate processes over long time scales [6], but also because stationary long memory processes exhibit slowly decaying auto-correlations. These slowly decaying correlations give the appearance of local deviations that we commonly see in climate series, but lead to greater uncertainty in the trend estimates.

For our two motivating datasets, we need to extend the class of models for the noise to include modulated stationary noise, with time dependent variance. A modulated noise process $\{\eta_t\}$ is defined by

$$\eta_t = \sigma_t \epsilon_t, \quad t \in T,$$

where $\{\sigma_t\}$ are the time-varying standard deviations and $\{\epsilon_t\}$ is a stationary process.

27.3.2 Linear trends

For some climate series it may be reasonable to assume that the mean level μ_t of the process at time t, is linearly associated with some covariate x_t of interest. Commonly this covariate x_t may be a linear function of time, but it could be another series such as carbon dioxide levels at time t (or some smoothed version of this series). A *simple linear regression/trend model* is then given by

$$\mu_t = \beta_0 + \beta_1 x_t, \quad t \in T. \tag{27.2}$$

A naive estimator of the intercept parameter β_0 and slope parameter β_1 is given by *ordinary least squares (OLS)*. For series of length N, The OLS estimates,

$$\arg\min{}_{\beta_0,\beta_1} \sum_{t=1}^{N} (y_t - \beta_0 - \beta_1 x_t)^2, \tag{27.3}$$

are

$$\widehat{\beta}_0 = \bar{y} - \widehat{\beta}_1 \bar{x} \quad \text{with} \quad \widehat{\beta}_1 = r\frac{s_y}{s_x},$$

where $\bar{y}$ is the sample mean of $\{y_t\}$, $\bar{x}$ is the sample mean of $\{x_t\}$, s_y is the sample standard dervation of $\{y_t\}$, s_x is the sample standard dervation of $\{x_t\}$, and r is the (Pearson) sample correlation between $\{x_t\}$ and $\{y_t\}$. When the noise $\{\eta_t\}$ is IID we have that the OLS estimates are the best linear unbiased estimates (BLUE) of β_0 and β_1 (they are unbiased estimates of the true parameters, and the variance of any linear combinations of the parameter estimates is smallest amongst all linear estimators). Since $\{\eta_t\}$ are usually not independent, a key question is how the OLS estimator behaves when $\{\eta_t\}$ is a dependent time series. When $\{\eta_t\}$ is a mean zero stationary process the OLS estimators are unbiased but are no longer the best when it comes to minimizing their variance among linear competitors. The same is true for modulated stationary processes.

When the variances of $\{\eta_t\}$ are changing with time, but the series are still independent, the best linear unbiased estimators come from *weighted least squares (WLS)*. The WLS estimates are

$$\arg\min{}_{\beta_0,\beta_1} \sum_{t=1}^{N} w_t(y_t - \beta_0 - \beta_1 x_t)^2, \tag{27.4}$$

where $w_t = 1/\text{var}(\eta_t)$ is the reciprocal of the variance of the noise at time t.

When the error terms are dependent, let $\mathbf{\Sigma} = \text{cov}(\boldsymbol{\eta})$ the covariance matrix for the noise, where $\boldsymbol{\eta}$ is the column vector of the η_t, meaning that $\mathbf{\Sigma}_{t,t'} = \text{cov}(\eta_t, \eta_{t'})$ for each t and t'. Letting $\boldsymbol{y}$ and $\boldsymbol{x}$ be the column vectors of the y_t and x_t, respectively, the *general least squares (GLS)* estimates,

$$\arg\min{}_{\beta_0,\beta_1} (\boldsymbol{y} - \beta_0\mathbf{1} - \beta_1\boldsymbol{x})^T \mathbf{\Sigma}^{-1} (\boldsymbol{y} - \beta_0\mathbf{1} - \beta_1\boldsymbol{x}) \tag{27.5}$$

(where $\mathbf{1}$ is a vector of ones), are the BLUEs of β_0 and β_1. OLS and WLS is a special case of GLS when $\mathbf{\Sigma}$ is a scaling of the identity matrix, or a diagonal matrix with diagonal entries $\text{var}(\eta_t)$, respectively.

Usually $\mathbf{\Sigma}$ is unknown. In that case, OLS is commonly used for exploratory purposes, even when we have correlated noise. Using the residuals calculated using via the OLS estimates,

$$\widehat{\eta}_t = y_t - \widehat{\beta}_0 - \widehat{\beta}_1 x_t,$$

we find candidate time series models. Using a given candidate model, we plug in the estimated $\mathbf{\Sigma}$ from the model in (27.5). Alternatively the time series parameters can be part of the minimization process [39], using OLS, followed by a number of steps of GLS, re-estimating the time series parameters after each OLS or GLS step.

If we are willing to assume a distribution for the noise process $\{\eta_t\}$, then we can estimate the parameters of the trend and time series process jointly using maximum likelihood (ML). For example, suppose that $\{\eta_t\}$ is a Gaussian process (i.e., the joint distribution of the noise at any collection of time indexes is normal). Let $\boldsymbol{\theta}$ denote the parameters characterizing the noise model, which drives $\boldsymbol{\Sigma}$. The likelihood function that needs to be maximized is

$$
\begin{aligned}
&L(\beta_0, \beta_1, \boldsymbol{\theta}) \\
&= (2\pi)^{-N/2} [\det(\boldsymbol{\Sigma})]^{-1/2} \exp\left\{ (\boldsymbol{y} - \beta_0 \mathbf{1} - \beta_1 \boldsymbol{x})^T \boldsymbol{\Sigma}^{-1} (\boldsymbol{y} - \beta_0 \mathbf{1} - \beta_1 \boldsymbol{x}) \right\}.
\end{aligned}
$$

It can be shown that the ML estimates of β_0 and β_1 are the GLS estimates substituting in the ML estimate of $\boldsymbol{\theta}$ into $\boldsymbol{\Sigma}$. Typically the ML estimate of $\boldsymbol{\theta}$ is not available in closed form for most time series models; it is obtained via numerical optimization.

Assuming the noise vector to be Gaussian with mean $\mathbf{0}$ (a vector of zeros) and covariance $\boldsymbol{\Sigma}$, the vector of the GLS estimators $\widehat{\boldsymbol{\beta}} = (\widehat{\beta}_0, \widehat{\beta}_1)^T$ of $\boldsymbol{\beta} = (\beta_0, \beta_1)^T$ have a bivariate normal sampling distribution:

$$
\widehat{\boldsymbol{\beta}} \sim \mathcal{N}_2\left(\boldsymbol{\beta}, \left[\boldsymbol{X}^T \boldsymbol{\Sigma}^{-1} \boldsymbol{X} \right]^{-1} \right), \tag{27.6}
$$

where here the design matrix $\boldsymbol{X}$ is a $N \times 2$ matrix with first column all ones and the second column being $\boldsymbol{x}$. This distributional assumption assumes that we have the correct model for both the trend and the noise. In practice when doing statistical inference for $\boldsymbol{\beta}$, we plug in an estimate of $\boldsymbol{\Sigma}$. The sampling distribution of the trend estimate, $\widehat{\boldsymbol{\mu}} = \boldsymbol{X}\widehat{\boldsymbol{\beta}}$ of $\boldsymbol{\mu} = (\mu_1, \dots, \mu_n)^T$ follows naturally [81]:

$$
\widehat{\boldsymbol{\mu}} \sim \mathcal{N}_n(\boldsymbol{\mu}, \boldsymbol{X}\mathrm{cov}(\widehat{\boldsymbol{\beta}})\boldsymbol{X}^T). \tag{27.7}
$$

The standard error for the fitted curve is then

$$
\mathrm{se}(\widehat{\mu}_t) = \sqrt{\boldsymbol{x}_t^T \mathrm{cov}(\widehat{\boldsymbol{\beta}})\boldsymbol{x}_t} \quad t \in T, \tag{27.8}
$$

where $\boldsymbol{x}_t$ is the tth row of $\boldsymbol{X}$. We can also derive a marginal sampling distribution for $\widehat{\mu}_t$ based on the t distribution when $\boldsymbol{\Sigma} = \sigma^2 \boldsymbol{R}$. In this case $\boldsymbol{R}$ is a positive definite matrix and σ^2 is some positive variance parameter that is estimated using OLS; multivariate inference follows from F-tests [81]. (The t and F distribution assumes that the parameters characterizing $\boldsymbol{R}$ are known.)

27.3.3 Nonlinear and nonparametric trends

Clearly there is no *a priori* reason why the trend component $\{\mu_t\}$ of the climate series needs to be linear in time or in some covariate. Again, introducing the covariate $\{x_t\}$ and considering it now to be defined as a linear function of time, a simple model that allows for the trend to be nonlinear is the *polynomial trend/regression model*. A polynomial trend model of degree p has functional form

$$
\mu_t = \beta_0 + \sum_{j=1}^{p} \beta_j x_t^j, \quad t \in T.
$$

As with the simple linear trend model the coefficients in the polynomial model $\{\beta_j\}$ can be estimated in a number of different ways (e.g., OLS, WLS, GLS, and ML). As this is a linear model [81], the properties of each estimator will again be driven by the statistical properties of the noise process, and these properties are essentially the same as for the simple linear trend model. Compared to (27.6), the GLS estimator now has a $(p + 1)$-variate normal sampling distribution for a Gaussian noise process, where the design matrix $\boldsymbol{X}$ has ith row

$$
\boldsymbol{x}_i = (1, x_i, x_i^2, \dots, x_i^p)^T, \quad i = 1, \dots, N.
$$

(27.7) and (27.8) also hold for this model. In choosing a polynomial trend model for data we need to select the degree of the polynomial, p. This is a common problem in both regression and time series analysis, where there is a need to balance the the smoothness of the trend that we observe in the climate series, with writing down a model that adequately captures the noise. We should be wary of overfitting (writing down a statistical model for the trend for which the estimated coefficients are highly uncertain). Overfitting leads to statistical models with poor predictive performance. Penalized regression and penalized likelihood methods can be applied to ensure that we do not fit models that are overly complicated and poor descriptors of the data. Demonstrating the idea of penalized ML methods, we will use the Akaike Information Criterion (AIC) in our case studies in Section 27.4 below. There are several other criteria, such as BIC, AICc, etc. [61].

There are others examples of parametric models for the trend beyond polynomial models. To see this, introducing the $p+1$ functions $\{f_j(x) : j = 0, \ldots, p\}$, defined as

$$f_j(x) = x^j, \quad j = 0, \ldots, p,$$

we see that the polynomial trend can be written in more a general notation:

$$\mu_t = \sum_{j=0}^{p} \beta_j f_j(x_t), \quad t \in T.$$

Replacing the polynomial functions by other smooth functions, lead to more generalized models for trend. Commonly used examples of so-called *basis functions* include sinusoids [8, 77], wavelets [9, 11, 19, 78], and splines [99, 101, 106]. Local regression methods based on kernels can also be used [20, 67]. Given the multitude of possible functions and methods that can be used to model trend, model selection and comparison becomes more problematic. We again need to be wary of overfitting; penalized regression models are popularly used for fitting trends constructed from a linear combination of basis functions. In the Bayesian paradigm there is extensive research into building priors for the regression parameters to enforce smoothness, while guarding against overfitting by enforcing sparseness [16, 42, 43, 76].

Compared to the case of linear trends, uncertainty quantication for nonlinear and non-parametric trend estimates are typically more involved and more computationally intensive to calculate. Depending on the paradigm (Bayesian or frequentist), Markov chain Monte Carlo (MCMC) [13, 38], or Monte Carlo and other resampling methods are commonly used for assessing the uncertainty or significance of trend estimates [15, 19, 33, 57].

A particularly simple and effective nonlinear model for trend is the *broken stick model*, which is a first order spline model with one knot. The broken stick model is useful when there is a qualitative change in the system at some time point. Sometimes this time point is known, sometimes only suspected. As with linear models, generalized least squares or maximum likelihood, for example, can be used to estimate the parameters in the model, while estimating the dependence in the errors. Given a change point time (or knot) τ we fit a straight line to $y_1, \ldots, y_\tau$ as a function of $x_1, \ldots, x_\tau$, and a connecting line to $y_{\tau+1}, \ldots, y_N$ as a function of $x_{\tau+1}, \ldots, x_N$. If desired, the parameter τ can also be estimated in the model. If a smoother version is desirable, a *bent cable model* [18] connects the two linear parts with a quadratic connector. The model is then

$$\mu_t = \beta_0 + \beta_1 t + \beta_2 \left[\frac{(t - \tau + \gamma)^2}{4\gamma} I\{|t - \tau| \leq \gamma\} + (t - \tau)I\{t > \tau + \gamma\} \right], \quad t \in T. \quad (27.9)$$

In other words, a straight line is fitted between indexes 1 and $\tau - \gamma$, a quadratic between $\tau - \gamma$ and $\tau + \gamma$, and another line between $\tau + \gamma$ and n, with the three segments connecting continuously. The model can be fit using least squares or via ML using the bentcableAR R package (`https://cran.r-project.org/web/packages/bentcableAR/`) – this package is able to estimate the changepoint and can calculate confidence intervals for the trend estimate. For a climate application, see [107].

27.3.4   Smoothing and filtering to estimate the trend

The *moving average (MA) filter* can be used to estimate the trend of a climate series $\{Y_t : t = 1, \ldots, n\}$ nonparametrically. The trend estimate from an MA(q) filter is defined by

$$\widehat{\mu}_t = \sum_{j=-q}^{q} \left[\frac{1}{2q+1}\right] Y_{t-j}, \quad t = q+1, \ldots, n-q$$

(care needs to be taken to define the trend estimate for $t < (q+1)$ and $t > (n-q)$). The value of q (a positive integer) controls the level of smoothing of the filter, while controlling the influence of the noise process upon the trend estimate (i.e., it manages the bias–variance tradeoff in estimating the trend: if q is too large the trend will be biased, while if q is too small the variability is too high). Any MA(q) filter provides an unbiased estimate of a linear or locally-linear trend (Chapter 1 of [12]).

Again for some positive integer q, a more general smoothing filter $\{a_j : j = -q, \ldots, q\}$ can be designed so that the trend estimate,

$$\widehat{\mu}_t = \sum_{j=-q}^{q} a_j Y_{t-j},$$

can unbiasedly estimate trends of a given degree of polynomial (e.g., Exercise 1.12 of [12]). For applications to climate series, see the review in [68].

Another common trend estimate is a cubic spline [89], which is a piecewise cubic function with differentiable connectors at a series of knots. A common choice of knot positions is to place them at equal quantiles of the x-variable, which is usually time for climate applications. The number of quantiles is called the degrees of freedom. Roughly speaking if there are N observations in the data set, there are N/df observations between each knot. Other common approaches in climate analyses use generalized additive models (GAM) ([47]; 2.SM.3) or singular spectrum analysis (SSA) [80].

With the additive decomposition for trend given by (27.1), uncertainty quantification for these trend estimates follow by assuming a model for the noise, and calculating the covariance of the filtered noise using the stationarity preserves a stationary result (See Chapter 3).

We need to be cautious with some choices of smoothing filter, because it is possible to introduce features in the estimated trend that do not exist in the original time series. Looking at the so-called spectral properties of filters can indicate issues that may occur with smoothing and filtering; see, e.g., [77] for more details of the spectral approach.

27.3.5   Removing or simplifying trend by differencing

Rather than estimating the trend component for a climate series, we may choose to remove or simplify it. For example the differencing operator, ∇, has the ability to remove trend by taking a polynomial of degree p and yielding a polynomial of degree $p-1$ (e.g., Exercise 1.10, p.42 of [12]). Formally for the series $\{Y_t\}$, the differencing operator ∇ is defined by

$$\nabla Y_t = Y_t - Y_{t-1}, \quad t \in T.$$

Now suppose that $\mu_t = \beta_0 + \beta_1 t$ is a linear trend in t. Then $\nabla \mu_t = \beta_1$ for all t. Thus differencing a linear trend yields a constant trend, with a trend equal to the slope of the original series. This demonstrates another way to estimate the slope for a simple linear trend model as defined by (27.2): we difference the climate series, and estimate the slope using the sample mean for the differencing time series. The statistical properties of this estimate of the slope may be involved, and are driven by the statistical properties of the differenced noise process, $\{\nabla \eta_t\}$.

27.3.6 Hierarchical and dynamic linear model decompositions for trend

Writing down statistical models defined in terms of differencing, allows for stochastic representations of trend. These stochastic representations are more commonly used in financial applications [97], but can also be used in climatic analyses. A simple example of a model defined using differencing is the autoregressive integrated moving average ARIMA(p, d, q) process that was defined in Chapter 3.

By noting that an ARIMA model can be written as a form of random walk model, we are drawn to define stochastic models for trend in terms of a hierarchical statistical model [32, 49]. Here follows a simple example. Again suppose $\{Y_t : t \in T\}$ is our time series process of interest and assume the additive decomposition of

$$Y_t = \mu_t + \eta_t, \quad t \in T,$$

where $\{\eta_t\}$ is the noise process. Now, rather than assuming that the trend μ_t is deterministic we assume a random walk model for μ_t:

$$\mu_t = \mu_{t-1} + Z_t, \quad t \in T,$$

where $\{Z_t\}$ is a white noise or IID process. For this model we can estimate the latent trend component process using the computationally efficient Kalman filtering algorithm [32, 64, 91]. Parameter estimation follows in the Gaussian process setting using ML. In the Bayesian context this is an example of a *Dynamic Linear Model (DLM)* ([45]; also see Chapter 4). This class of models can be related to the smoothing and spline methods discussed above [49, 65, 103].

27.4 Two case studies

We now return to the two data sets presented in Section 27.2. We will demonstrate that failure to investigate residual structure, or to make assumptions appropriate for the data, can yield incorrect conclusions. All our analyses were carried out using the R software package [76] – the R code is available from http://www.stat.osu.edu/~pfc/.

27.4.1 US annual temperatures

The standard approach in the climate literature is to fit a linear trend to a time series using ordinary least squares (27.3). Doing that for the US annual temperature anomalies yields an estimate of $0.55°C$ per century (see Table 27.1 for details of all the fits in this section). The slope is highly significantly different from zero, but that assumes the noise is IID and does not take into account the variability of the measurements, leading to an overfit of the more uncertain early measurements. In order to deal with this, we do a weighted

TABLE 27.1
Fits of linear trend slopes to US annual average temperature anomalies

Model	Slope per century	Standard error	Slope P-value	AIC	Ljung-Box P-value
OLS	0.55	0.12	<0.0001	116.0	0.000
WLS	0.48	0.14	0.0006	124.8	0.000
Shen et al.	0.57	0.17	0.0008	N/A	N/A
AR(1)	0.55	0.15	0.0004	110.6	0.066
Weighted AR(1)	0.47	0.18	0.0107	116.1	0.013
AR(4)	0.58	0.20	0.0038	109.3	0.593
Weighted AR(4)	0.53	0.26	0.0490	110.1	0.741
ARMA(3,1)	0.64	0.24	0.0092	107.4	0.814
Weighted ARMA(3,1)	0.60	0.32	0.0655	109.1	0.849

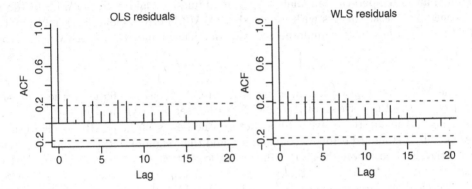

FIGURE 27.3
Sample autocorrelation functions for the residuals from the ordinary and weighted least squares linear trend fit to the US annual mean temperature anomalies.

least squares fit (27.4), with weights inversely proportional to the variance estimates. The resulting estimate is somewhat smaller (0.48°C per century) and somewhat more uncertain, although still highly significant (again assuming the noise is independent). We also remark that any estimate can be sensitive to choice of time range for the data (in this case, the years 1897–2008).

Looking at the sample autocorrelation function (ACF) of the residuals from the two least squares fits in Figure 27.3, we see clear evidence of dependence for both fits. Indeed, compared to the OLS fit, there is evidence that the dependence is stronger once we account for the change in variability over time using the WLS fit. The last column of Table 27.1, showing the Ljung-Box [66] P-value when testing that the residuals are IID (based on 10 lags of the ACF), demonstrates that there is significant dependence in the OLS and WLS residuals that needs to be accounted for.

The most common time series model in the climate literature is AR(1). To fit the two AR(1) models (unweighted: assuming a constant variance over time; weighted: assuming that the variance is time-varying) we use the generalized least squares estimator, given by (27.5). In essence, [90] use this approach to estimate slopes. The resulting slopes are almost the same as for the corresponding fits using the independence assumption, but the standard errors become 20-30% larger. The fitted slopes are still significantly different from zero

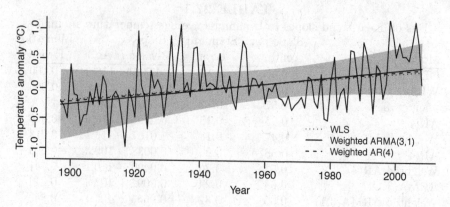

FIGURE 27.4
US annual mean temperature anomalies with trend lines overlaid from the WLS fit (dotted line), generalized least squares with weighted ARMA(3,1) errors (solid line), and weighted AR(4) errors (dashed line). Simultaneous 95% confidence bands for the ARMA(3,1) trend are shown in gray.

under the AR(1)-assumption for the noise structure. At the $\alpha = 0.05$ level, the Ljung-Box test rejects the null hypothesis of IID noise for either set of AR(1) residuals.

Sticking to the autoregressive error structure, the best such fit is AR(4). The resulting slopes as well as standard errors become larger, and the weighted AR(4) test for the significance of the slope parameter has a P-value of just lower than 0.05. Restricting to ARMA(p, q) models with $p \leq 10$ and $q \leq 10$, the best ARMA-type error structure is ARMA(3,1), which also yield the largest slopes. The weighted ARMA(3,1)-slope is not significantly different from zero. Figure 27.4 shows three of the fitted lines. All fits from these models pass the Ljung-Box IID noise test.

Table 27.1, in addition to showing that the statistical significance of the slope strongly depends on the assumptions made, indicates that unweighted ARMA(3,1) model has the best fit according to the AIC criterion. Choosing the model with the lowest AIC assumes that the errors are Gaussian and that the trend is linear. Since the weighted residuals are slightly less skewed than the unweighted residuals, we choose the weighted ARMA(3,1) or weighted AR(4) general least square fits to explain the dependence, which leads us to conclude that the linear trend is only weakly significant, once we account for the time-varying variance and dependence in the series.

27.4.2   Global annual mean temperature

When looking at the global temperature time series, we start by fitting a straight line using least squares. The linear fit is not particularly good, and the residuals (Figure 27.5) indicate that a quadratic fit might be better.

Using a similar approach to that in Section 27.4.1, we use a generalized least squares fit to a trend that is a quadratic function of time, with time series errors following an ARMA(4,1)-model and weights corresponding to the estimated observation variance. Figure 27.6 shows the fitted model. Again, the residuals pass various tests for IID noise. We calculate the standard error of the fitted curve from (27.8). A Bonferroni calculation shows that a simultaneous 95% confidence band is given by going up and down 3.6 standard errors from the fitted curve.

A broken stick model is suggested by the idea that at some point in the twentieth century

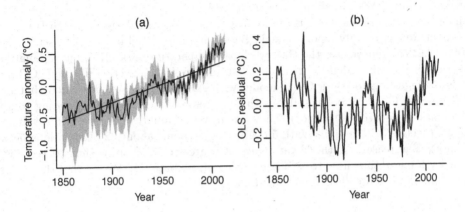

FIGURE 27.5
(a) Global annual mean temperature anomalies with respect to the 1951-1970 climatology.
The gray shaded region denotes simultaneous 95% confidence intervals for the mean, using
estimated standard errors from [84]. The trend line correspond to ordinary least squares.
(b) Residuals from the ordinary least squares fit.

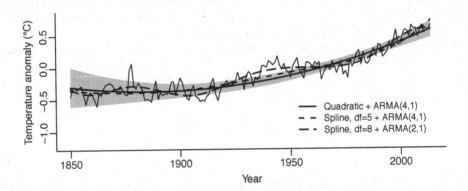

FIGURE 27.6
Global annual mean temperature anomalies with respect to the 1951-1970 climatology.
Different line types denote different estimated trends. The gray shaded regions denotes
simultaneous 95% confidence bands for the quadratic trend estimate, assuming ARMA(4,1)
errors and time-varying variances.

the human-generated greenhouse gases may have started to dominate the natural forcings of the climate system, thereby changing the rate of increase in global temperatures. The estimated broken stick model changes slope at 1909 (confidence interval 1894 – 1924). When fitting a bent cable model using AR(4) error structure to the global temperature series, the resulting curve is very similar to the quadratic fit. The quadratic part of the fit goes between 1863 and 2014; i.e. almost the entire series. Both the broken stick and bent cable models fall inside the quadratic simultaneous confidence band.

Since the global series is poorly fit by a straight line, the decision in the IPCC Fifth Assessment Report to only report straight lines is not supported by this data set. It should be noted that the report uses the Hadley global temperature series [71], but an analysis of that series yields nearly identical results to our analysis in this section. In the IPCC report there is a suggestion to use a GAM model with AR(1) error. We have tried that fit, as well as an SSA fit and a cubic spline with 8 degrees of freedom (corresponding to about 20 observations between each knot). They are all very similar. In Figure 27.6 we show the spline fit, together with a spline with 5 degrees of freedom (about 32 observations between each knot). The reason for the latter choice of degrees of freedom is that 30 years is the standard amount of time used to estimate a changing climate [41]. We see that the smoother of the spline fits is quite close to the quadratic fit.

27.5   Spatial and spatio-temporal trends

Statistical methods of trend estimation and assessment extend naturally to the spatial and spatio-temporal settings. (For a review of spatial and spatio-temporal models, see e.g., [28, 37, 60]; see [19] for an early discussion of spatial trend estimation.) Not all of the papers that appear in this section are climate examples, but the methodology can be applied to the analysis of climate variables observed over space and time.

An an example, suppose we wish to estimate trend for a point-referenced (also known as a geostatistical) spatial process $\{Z(s) : s \in D \subset \mathbb{R}^p\}$, based on n observations from the process, $z = (z(s_1), \ldots z(s_n))^T$. Assuming an additive decomposition for this process $\{Z(s)\}$ we have

$$Z(s) = \mu(s) + \eta(s), \quad s \in D,$$

where $\{\mu(s) : s \in D\}$ is the spatial trend component and $\{\eta(s) : s \in D\}$ is the spatial noise component. As in the time series case, we typically assume that the noise is a mean zero stationary process. Again, linear model representations are commonly used for the trend component:

$$\mu(s) = \sum_{j=0}^{p} \beta_j \, x_j(s),$$

where $\{x_j(s) : s \in D\}$ are $p+1$ spatial basis functions. The simplest case is often to include latitude and longitude as covariates. But traditionally, simple polynomials [28, 37], splines [27, 73, 87, 105], and other functions such as wavelets are used to model the trend in spatial settings [2, 27, 72]. Inference follows in the same way as for time series (e.g., via least squares, ML, penalized methods, and Bayesian methods).

Also, given that we can represent spline models (and indeed other basis models) using random spatial processes [54, 73], we could also assume that the trend component $\{\mu(s) :$

$s \in D\}$ is a random process that smoothly varies over space. In such a case we need to specify the model for the climate variables in such a way that the trend component captures the smooth (long-range) variation of the process over space, and the errors captures the noise, the short range variation of the process over space (cf [86]).

These ideas extend to modeling spatial trends for areal processes [7, 37], point processes (see e.g., [3, 10, 31, 58, 59] for temporal and spatial examples), and to river networks [74]. Spatio-temporal methods follow naturally [24, 28, 63, 75, 85, 100, 104].

27.6 Assessing climatic trends in other contexts

Up to this point we have considered the trend to be smooth changes over a mean over time and/or space. (Indeed the trend for a point process, as measured by the intensity function [31], can also be considered to be a mean number of events occurring in a specific time interval or spatial region.) More recently there has been an interest in modeling trends in other features of the distribution of climate; for example, trends in the variance [24], trends in the quantiles [17, 62, 82], and trends in extremes.

In the latter case, there are countless examples of using spatio-temporal models to assess trends in the extreme-value distribution of climate variables, usually in the location parameter of the generalized extreme value distribution for block maxima. Some examples of extreme value studies using different climate variables include: temperature [14, 25, 35, 55], paleoclimate proxies of temperature [64], and precipitation [22, 30, 88]. The use of hierarchical Bayesian modeling has revolutionized our ability to examine trends in extremes. For example, we can introduce a spatial process for the spatially-varying location parameter of the extreme value distribution to introduce some (albeit limited) smoothness over space. See [30] for a comprehensive review of the issues underlying the modeling of spatial extremes.

27.7 Discussion

In this chapter we have discussed the many issues underlying the modeling and assessment of climate trends. Defining trend to be smooth changes in time or space, we reviewed the idea that trend is ill-posed because we have not specified how smooth a trend we have. This requires us to introduce some prior belief about the trend (relative to the noise), via a suitable model or modeling framework for the climate variable of interest. For example, a Gaussian additive model with a specified component capturing the trend as the mean of the process may be reasonable for modeling long-term average temperatures, whereas for modeling precipitation block monthly maxima, the location parameter of a generalized extreme value distribution is more appropriate. We demonstrated that trends do not need to be linear, but once we allow for nonlinear trends, we need to be wary of overfitting to the climate data at hand.

Our modeling setup indicates that the measurement errors modulate a stationary, zero mean error process, i.e., that the natural variability does not change over time. This may be a strong assumption, but there is little evidence of changing climate variability in the literature.

Some authors have argued that the increasing temperature trend observed in many climate series is due to a long memory error structure, caused by the slow reaction to forcing in oceans as compared to atmosphere. This has led to a rich literature in researchers trying to purely describe the variation (and indeed the trends themselves) using only long memory processes (for a review, see, e.g., [56]). [84] demonstrated that a linear trend, after accounting for long memory errors, was statistically significant when estimated from the [44] global temperature series. (For other examples see [19, 24, 63, 94]; see [21] for further discussion of the long memory versus trend question.)

It is not possible to highlight all methods for assessing and modeling trend in this chapter. For example, tests for trend based on Kendall's tau statistics have some popularity in environmental and climate science [29, 34]. We also note the popularity of principal components (also known as empirical orthogonal components; see, e.g., [6, 95, 98]) as trend components in atmospheric and oceanic sciences. In addition, the dynamic linear models approach of [45] has been applied to power projection in Brazil by [64].

There are numerous future directions for statistical research into trend estimation. Using statistical learning methodologies, there is interest in applying different loss functions to the efficient, while robust, estimation of trend [53]. Further development of spatio-temporal methods for trend estimation is also needed in non-Gaussian and multivariate settings; a thought-provoking example in climate is the joint estimation of trends in the occurrence and intensity of precipitation amounts [23, 40].

Bibliography

[1] H. Abdi. Bonferroni and Šidák corrections for multiple comparisons. In N. J. Salkind, editor, *Encyclopedia of Measurement and Statistics*. Sage, Thousand Oaks, CA, 2007.

[2] G Álvarez and B Sansó. Bayesian wavelet regression for spatial estimation. *Journal of Data Science*, 6:219–229, 2008.

[3] A. Baddeley and R. Turner. Modelling spatial point patterns in R. In *Case studies in spatial point process modeling*, pages 23–74. Springer, New York, NY, 2006.

[4] L. Barboza, B. Li, M. P. Tingley, and F. G. Viens. Reconstructing past temperatures from natural proxies and estimated climate forcings using short-and long-memory models. *The Annals of Applied Statistics*, 8:1966–2001, 2014.

[5] J. Beran. *Statistics for Long Memory Processes*, volume 61 of *Monographs on Statistics and Applied Probability*. Chapman and Hall, New York, NY, 1994.

[6] L. M. Berliner, C. K. Wikle, and N. A. Cressie. Long-lead prediction of Pacific SSTs via Bayesian dynamic modeling. *Journal of Climate*, 13:3953–3968, 2000.

[7] J. Besag, J. York, and A. Mollié. Bayesian image restoration, with two applications in spatial statistics. *Annals of the Institute of Statistical Mathematics*, 43:1–20, 1991.

[8] P. Bloomfield. *Fourier Analysis of Time Series (Second Edition)*. Wiley-Interscience, New York, NY, 2000.

[9] D. R. Brillinger. Some river wavelets. *Enivronmetrics*, 5:211–220, 1994.

[10] D. R. Brillinger. Trend analysis: time series and point process problems. *Environmetrics*, 5:1–19, 1994.

[11] D. R. Brillinger. Some uses of cumulants in wavelet analysis. *Nonparametric Statistics*, 6:93–114, 1996.

[12] P. J. Brockwell and R. A. Davis. *Introduction to Time Series and Forecasting (Second Edition)*. Springer, New York, NY, 2002.

[13] S. Brooks, A. Gelman, G. L. Jones, and X.-L. Meng. *Handbook of MCMC*. Chapman and Hall/CRC, Boca Raton, FL, 2011.

[14] B. G. Brown and R. W. Katz. Regional analysis of temperature extremes: Spatial analog for climate change? *Journal of Climate*, 8:108–119, 1995.

[15] P. Cabilio, Y. Zhang, and X. Chen. Bootstrap rank tests for trend in time series. *Envronmetrics*, 24:537–549, 2013.

[16] C. M. Carvalho, N. G. Polson, and J. G. Scott. The horseshoe estimator for sparse signals. *Biometrika*, 97:465–480, 2010.

[17] R. Chandler and M. Scott. *Statistical Methods for Trend Detection and Analysis in the Environmental Sciences*. John Wiley & Sons, 2011.

[18] G. S. Chiu and R. A. Lockhart. Bent-cable regression with autoregressive noise. *Canadian Journal of Statistics*, 38:386–407, 2010.

[19] R. J. Chorley and P. Haggett. Trend-surface mapping in geographical research. *Transactions of the Institute of British Geographers*, 37:47–67, 1965.

[20] W. S. Cleveland, E. Grosse, and W. M. Shyu. Local regression models. In J. M. Chambers and T. J. Hastie, editors, *Statistical Models in S*, chapter 8. Wadsworth and Brooks/Cole, 1992.

[21] T. A Cohn and H. F. Lins. Nature's style: Naturally trendy. *Geophysical Research Letters*, 32:5, 2005.

[22] D. Cooley, D. Nychka, and P. Naveau. Bayesian spatial modeling of extreme precipitation return levels. *Journal of the American Statistical Association*, 102:824–840, 2007.

[23] P. S. P. Cowpertwait. A generalized spatial-temporal model of rainfall based on a clustered point process. In *Proceedings of the Royal Society of London A: Mathematical, Physical and Engineering Sciences*, volume 450, pages 163–175. The Royal Society, 1995.

[24] P. F. Craigmile and P. Guttorp. Space-time modelling of trends in temperature series. *Journal of Time Series Analysis*, 32:378–395, 2011.

[25] P. F. Craigmile and P. Guttorp. Can a regional climate model reproduce observed extreme temperatures? *Statistica*, 73:103, 2013.

[26] P. F. Craigmile, P. Guttorp, and D. B. Percival. Trend assessment in a long memory dependence model using the discrete wavelet transform. *Environmetrics*, 15:313–335, 2004.

[27] N. Cressie and G. Johannesson. Fixed rank kriging for very large spatial data sets. *Journal of the Royal Statistical Society, Series B (Statistical Methodology)*, 70:209–226, 2008.

[28] N. A. C. Cressie and C. K. Wikle. *Statistics for Spatio-Temporal Data*. Wiley, New York, NY, 2011.

[29] P. F. Darken, G. I. Holtzman, E. P. Smith, and C. E. Zipper. Detecting changes in trends in water quality using modified Kendall's tau. *Environmetrics*, 11:423–434, 2000.

[30] A. C. Davison, S. A. Padoan, and M. Ribatet. Statistical modeling of spatial extremes. *Statistical Science*, 27:161–186, 2012.

[31] P. J. Diggle. *Statistical Analysis of Spatial and Spatio-Temporal Point Patterns, Third Edition*. CRC Press, Boca Raton, FL, 2013.

[32] J. Durbin and S. J. Koopman. *Time Series Analysis by State Space Methods*. Oxford University Press, Oxford, United Kingdom, 2012.

[33] B. Efron and R. J. Tibshirani. *An Introduction to the Bootstrap*. CRC press, Boca Raton, FL, 1994.

[34] A. H. El-Shaarawi and S. P. Niculescu. On Kendall's tau as a test of trend in time series data. *Environmetrics*, 3:385–411, 1992.

[35] M. Fuentes, J. Henry, and B. Reich. Nonparametric spatial models for extremes: Application to extreme temperature data. *Extremes*, 16:75–101, 2013.

[36] C. Gallagher, R. Lund, and M. Robbins. Changepoint detection in climate time series with long-term trends. *Journal of Climate*, 26:4994–5006, 2013.

[37] A. Gelfand, P. Diggle, M. Fuentes, and P. Guttorp. *Handbook in Spatial Statistics*. Chapman and Hall, Boca Raton, FL, 2010.

[38] A. Gelman, J. Carlin, H. Stern, D. Dunson, A. Vehtari, and D. Rubin. *Bayesian Data Analysis, 3rd edition*. Chapman and Hall, Boca Raton, FL, 2013.

[39] H. Goldstein. Multilevel mixed linear model analysis using iterative generalized least squares. *Biometrika*, 78:43–56, 1986.

[40] P. Guttorp. Analysis of event based precipitation data with a view towards modeling. *Water Resources Research*, 24:35–44, 1988.

[41] P. Guttorp. Statistics and climate. *Annual Reviews of Statistics*, 1:87–101, 2014.

[42] C. Hans. Bayesian lasso regression. *Biometrika*, 96:835–845, 2009.

[43] C. Hans. Elastic net regression modeling with the orthant normal prior. *Journal of the American Statistical Association*, 106:1383–1393, 2011.

[44] J. E. Hansen and S. Lebedeff. Global trends of measured surface air temperature. *Journal of Geophysical Research*, 92:13345–13372, 1987.

[45] J. Harrison and M. West. *Bayesian Forecasting and Dynamic Models*. Springer, New York, NY, 1999.

[46] D. L. Hartmann, A. M. G. Klein-Tank, M. Rusticucci, L. V. Alexander, S. Brönnimann, Y. Charabi, F. J. Dentener, E. J. Dlugokencky, D. R. Easterling, A. Kaplan, B. J. Soden, P. W. Thorne, M. Wild, and P. M. Zhai. Observations: Atmosphere and surface. In T. F. Stocker, D. Qin, G.-K. Plattner, M. Tignor, S. K.

Allen, J. Boschung, A. Nauels, Y. Xia, V. Bex, and P. M. Midgley, editors, *Climate Change 2013: The Physical Science Basis. Contribution of Working Group I to the Fifth Assessment Report of the Intergovernmental Panel on Climate Change*, chapter 2, pages 159–254. Cambridge University Press, Cambridge, United Kingdom and New York, NY, USA, 2013.

[47] D. L. Hartmann, A. M. G. Klein Tank, M. Rusticucci, L. V. Alexander, S. Brönnimann, Y. Charabi, F. J. Dentener, E. J. Dlugokencky, D. R. Easterling, A. Kaplan, B. J. Soden, P. W. Thorne, M. Wild, and P. M. Zhai. Observations: Atmosphere and surface supplementary material. In T.F. Stocker, D. Qin, G.-K. Plattner, M. Tignor, S. K. Allen, J. Boschung, Y Nauels, A. Xia, V. Bex, and P.M. Midgley, editors, *Climate Change 2013: The Physical Science Basis. Contribution of Working Group I to the Fifth Assessment Report of the Intergovernmental Panel on Climate Change*. Cambridge University Press, Cambridge, United Kingdom and New York, NY, USA, 2013.

[48] S. Hussain and A. Elbergali. Fractional order estimation and testing, application to Swedish temperature data. *Environmetrics*, 10:339–349, 1999.

[49] R. Hyndman, A. B. Koehler, J. K. Ord, and R. D. Snyder. *Forecasting with Exponential Smoothing: the State Space Approach*. Springer Science and Business Media, Berlin, Germany, 2008.

[50] M. Kendall. *Time Series*. Charles Griffin, London, 1973.

[51] J. J. Kennedy, N. A. Rayner, R. O. Smith, M. Saunby, and D. E. Parker. Reassessing biases and other uncertainties in sea-surface temperature observations since 1850 part 1: measurement and sampling errors. *Journal of Geophysical Research*, 116(D14103), 2011.

[52] J. J. Kennedy, N. A. Rayner, R. O. Smith, M. Saunby, and D. E. Parker. Reassessing biases and other uncertainties in sea-surface temperature observations since 1850 part 2: biases and homogenisation. *Journal of Geophysical Research*, 116(D14104), 2011.

[53] S.-J. Kim, K. Koh, S. Boyd, and D. Gorinevsky. l_1 trend filtering. *SIAM review*, 51:339–360, 2009.

[54] W Kleiber and D.W. Nychka. Equivalent kriging. *Spatial Statistics*, 12:31–49, 2015.

[55] A. Kottas, Z. Wang, and A. Rodriguez. Spatial modeling for risk assessment of extreme values from environmental time series: a Bayesian nonparametric approach. *Environmetrics*, 23:649–662, 2012.

[56] D. Koutsoyiannis. Climate change, the Hurst phenomenon, and hydrologic statistics. *Hydrological Sciences Journal*, 48:3–24, 2003.

[57] S. N. Lahiri. *Resampling Methods for Dependent Data*. Springer, New York, NY, 2013.

[58] JF Lawless and K Thiagarajah. A point-process model incorporating renewals and time trends, with application to repairable systems. *Technometrics*, 38:131–138, 1996.

[59] A. Lawson. On tests for spatial trend in a non-homogeneous Poisson process. *Journal of Applied Statistics*, 15:225–234, 1988.

[60] N. D. Le and J. V. Zidek. *Statistical analysis of environmental space-time processes.* Springer, New York, 2006.

[61] H. Lee and S. K. Ghosh. Performance of information criteria for spatial models. *Journal of Statistical Computing and Simulation*, 79:93–106, 2009.

[62] K. Lee, H-J Baek, and C Cho. Analysis of changes in extreme temperatures using quantile regression. *Asia-Pacific Journal of Atmospheric Sciences*, 49:313–323, 2013.

[63] R. T. Lemos and B. Sansó. A spatio-temporal model for mean, anomaly, and trend fields of north Atlantic sea surface temperature. *Journal of the American Statistical Association*, 104:5–18, 2009.

[64] L. M. M. Lima, E. Popov, and P. Damien. Modeling and forecasting of Brazilian reservoir inflows via dynamic linear models. *International Journal of Forecasting*, 30:464–476, 2014.

[65] F. Lindgren and H. Rue. On the second-order random walk model for irregular locations. *Scandinavian Journal of Statistics*, 35:691–700, 2008.

[66] G. Ljung and G. Box. On a measure of lack of fit in time series models. *Biometrika*, 65:297–303, 1978.

[67] C. Loader. *Local Regression and Likelihood.* Springer, New York, NY, 1999.

[68] M. E. Mann. Smoothing of climate time series revisited. *Geophysical Research Letters*, 35, 2008.

[69] E. Mannshardt, P. F. Craigmile, and M. P. Tingley. Statistical modeling of extreme value behavior in North American tree-ring density series. *Climatic Change*, 117:843–858, 2013.

[70] M. J. Menne, C. N. Williams Jr., and R. S. Vose. The U.S. Historical Climatology Network monthly temperature data, version 2. *Bull. Amer. Meteor. Soc.*, 90:993–1007, 2009.

[71] C. P. Morice, J. J. Kennedy, N. A. Rayner, and P. D. Jones. Quantifying uncertainties in global and regional temperature change using an ensemble of observational estimates: The HadCRUT4 dataset. *Journal of Geophysical Research*, 117, 2012.

[72] D. Nychka, C. K. Wikle, and J. A. Royle. Multiresolution models for nonstationary spatial covariance functions. *Statistical Modelling*, 2:315–331, 2002.

[73] D. W. Nychka. Spatial-process estimates as smoothers. In M. G. Schimek, editor, *Smoothing and Regression. Approaches, Computation and Application*, pages 393–424. Wiley, New York, NY, 2000.

[74] D. O'Donnell, A. Rushworth, A. W. Bowman, E. M. Scott, and M. Hallard. Flexible regression models over river networks. *Journal of the Royal Statistical Society, Series C (Applied Statistics)*, 63:47–63, 2013.

[75] C. J. Paciorek and J. S. McLachlan. Mapping ancient forests: Bayesian inference for spatio-temporal trends in forest composition using the fossil pollen proxy record. *Journal of the American Statistical Association*, 104:608–622, 2009.

[76] T. Park and G. Casella. The Bayesian lasso. *Journal of the American Statistical Association*, 103:681–686, 2008.

[77] D. B. Percival and A. Walden. *Spectral Analysis for Physical Applications*. Cambridge University Press, Cambridge, United Kingdom, 1993.

[78] D. B. Percival and A Walden. *Wavelet Methods for Time Series Analysis*. Cambridge University Press, Cambridge, United Kingdom, 2000.

[79] R Core Team. *R: A Language and Environment for Statistical Computing*. R Foundation for Statistical Computing, Vienna, Austria, 2016.

[80] S. Rahmstorf. A semi-empirical approach to projecting future sea-level rise. *Science*, 315:368–370, 2007.

[81] N. Ravishanker and D. K. Dey. *A First Course in Linear Model Theory*. Chapman and Hall, 2001.

[82] B. J. Reich. Spatiotemporal quantile regression for detecting distributional changes in environmental processes. *Journal of the Royal Statistical Society: Series C (Applied Statistics)*, 61:535–553, 2012.

[83] R. Rohde and R. Muller. The average temperature of 2014. Results from Berkeley Earth. Technical report, Berkeley Earth, 2015.

[84] R. Rohde, R. A. Muller, R. Jacobsen, E. Muller, S. Perlmutter, A. Rosenfeld, J. Wurtele, D. Groom, and C. Wickham. A new estimate of the average earth surface land temperature spanning 1753 to 2011. *Geoinfor Geostat: An Overview*, 1, 2013.

[85] P. D. Sampson, A. A. Szpiro, L. Sheppard, J. Lindström, and J. D. Kaufman. Pragmatic estimation of a spatio-temporal air quality model with irregular monitoring data. *Atmospheric Environment*, 45:6593–6606, 2011.

[86] H. Sang and J. Z. Huang. A full scale approximation of covariance functions for large spatial data sets. *Journal of the Royal Statistical Society, Series B (Statistical Methodology)*, 74:111–132, 2011.

[87] L. M. Sangalli, J. O. Ramsay, and T. O. Ramsay. Spatial spline regression models. *Journal of the Royal Statistical Society, Series B (Statistical Methodology)*, 75:681–703, 2013.

[88] E. M. Schliep, D. Cooley, S. R. Sain, and J. A. Hoeting. A comparison study of extreme precipitation from six different regional climate models via spatial hierarchical modeling. *Extremes*, 13:219–239, 2009.

[89] J. F. Scinocca, D. B. Stephenson, T. C. Bailey, and J. Austin. Estimates of past and future ozone trends from multimodel simulations using a flexible smoothing spline methodology. *Journal of Geophysical Research: Atmospheres*, 115, 2010.

[90] S. S. P. Shen, S. K. Lee, and J. Lawrimore. Uncertainties, trends, and hottest and coldest years of U.S. surface air temperature since 1895: An update based on the USHCN V2 TOB data. *J. Climate*, 25:4185–4203, 2012.

[91] R. H. Shumway and D. S. Stoffer. An approach to time series smoothing and forecasting using the EM algorithm. *Journal of Time Series Analysis*, 3:253–264, 1982.

[92] R. H. Shumway and D. S. Stoffer. *Time series analysis and its applications: with R examples*. Springer, New York, NY, 2010.

[93] R. Smith. Long-range dependence and global warming. In V. Barnett and F. Turkman, editors, *Statistics for the Environment*, pages 141–161. John Wiley, Hoboken, NJ, 1993.

[94] D. B. Stephenson, V. Pavan, and R. Bojariu. Is the North Atlantic Oscillation a random walk? *International Journal of Climatology*, 20:1–18, 2000.

[95] H.J. Thiebaux and M.A. Pedder. *Spatial Objective Analysis*. Academic Press, Cambridge, MA, 1987.

[96] B. Trewin. Exposure, instrumentation, and observing practice effects on land temperature measurements. *WIREs Climate Change*, 1:490–506, 2010.

[97] R. S. Tsay. *Analysis of Financial Time Series (Third Edition)*. Wiley, New York, NY, 2010.

[98] H. von Storch and F. W. Zwiers. *Statistical analysis in climate research*. Cambridge University Press, Cambridge, United Kingdom, 2002.

[99] G. Wahba. *Spline Models for Observational Data*. SIAM (Society for Industrial and Applied Mathematics), Philadelphia, 1990.

[100] L. A. Waller, B. P. Carlin, H. Xia, and Gelfand A. E. Hierarchical spatio-temporal mapping of disease rates. *Journal of the American Statistical association*, 92:607–617, 1997.

[101] Y. Wang. Smoothing spline models with correlated random errors. *Journal of the American Statistical Association*, 93:341–348, 1998.

[102] G. C. Weatherhead, E. C.and Reinsel, G. C. Tiao, X-L Meng, D. Choi, W-K Cheang, T. Keller, J. DeLuisi, D. J. Wuebbles, and J. Kerr. Factors affecting the detection of trends: Statistical considerations and applications to environmental data. *Journal of Geophysical Research*, 103(D14):17–149, 1998.

[103] W. E. Wecker and C. F. Ansley. The signal extraction approach to nonlinear regression and spline smoothing. *Journal of the American Statistical Association*, 78:81–89, 1983.

[104] C. K. Wikle, R. F. Milliff, D. Nychka, and L. M. Berliner. Spatiotemporal hierarchical Bayesian modeling: Tropical ocean surface winds. *Journal of the American Statistical Association*, 96:382–397, 2001.

[105] S. N. Wood. Thin plate regression splines. *Journal of the Royal Statistical Society, Series B (Statistical Methodology)*, 65:95–114, 2003.

[106] S. N. Wood. *Generalized Additive Models: An Introduction with R*. CRC press, Boca Raton, FL, 2006.

[107] C. Xu, H. Liu, A. P. Williams, Y. Yin, and X. Wu. Trends toward an earlier peak of the growing season in Northern Hemisphere mid-latitudes. *Global Change Biology*, 22:2852–2860, 2016.

28

Climate Modelling

David B. Stephenson

Department of Mathematics, University of Exeter

CONTENTS

28.1 Aim

This chapter aims to provide a statistically-oriented introduction to climate modelling. The following section starts by discussing the concept of climate and presents several different statistical interpretations. Section 28.3 then describes climate processes and modelling, and the main classes of climate model. Design of climate change experiments and ensemble simulation is covered in Section 28.4 followed by a discussion in Section 28.5 of how to make inference about actual climate from data produced by climate model simulations.

Selected references are given to relevant literature rather than attempting to provide a comprehensive review.

28.2   What is climate?

Climate is a statistical concept. Broadly speaking, climate is a statistical summary of weather, and hence climate science provides interesting challenges for environmental statisticians. Many of the issues in climate science rely on statistical insight e.g. the recent debate about the early-2000s slowdown (the so-called *hiatus*) in global warming (Fyfe et al. 2016).

A simple operational definition is that climate is a statistical summary of a sample of either observed or simulated weather data. For example, the 30 year time means of past weather observations from 1961-90 used by the World Meteorological Organisation to summarise *normal* conditions at different locations on the planet[1]. To account for seasonal variations, long-term means are generally computed separately for different seasons e.g. December-February mean for boreal winter and July-August for boreal summer. Such measures provide simple estimates of the typical weather one might expect to observe in different seasons at specific locations. However, this approach is rather limited for summarising weather that is not independent and identically distributed in time due to the presence of trends (Hawkins and Sutton, 2016) and modes of climate variability such as the El Ninõ-Southern Oscillation.

A more powerful interpretation of climate is to consider it to be an *expectation* of weather rather than a sample mean[2]. Expectation here should be understood not simply as the expectation of a weather variable but the expectation of any function of such a variable, for example, the exceedance of the variable above a high threshold that can be used to estimate the probability of extreme events (see Cooley and Sain 2010). Interpretation in terms of expectation relies upon representing weather probabilistically as a random variable. For example, time series models can be developed to represent historical data (see Section 3 of Chapter [crossref Craigmile and Guttorp chapter]) and then expectations can be calculated from such models. So rather than arbitrary sample statistics, climate may be considered to be the set of parameters in a probability model capable of faithfully representing weather data. At a deeper level, one might also consider these parameters to be uncertain and so climate is then a latent process for representing weather. In this sense, climate is a model concept rather than something that truly exists and can be found in the real world.

One may wonder where the uncertainty comes from for representing weather as a random variable: our incomplete knowledge or fundamental indeterminism? A major source of aleatoric uncertainty arises due to the chaotic evolution of weather. Poincaré (1914) realised this when he wrote:

> ...*even if it were the case that the natural laws had no longer any secret for us, we could still only know the initial situation approximately. If that enabled us to predict the succeeding situation with the same approximation, that is all we require, and we should say that the phenomenon had been predicted, that it is governed by laws. But it is not always so - it may happen that small differences in the initial conditions produce very great ones in the final phenomena. A small error in the former will produce an enormous*

---

[1] The original meaning of the word "climate" was "latitude zone of the Earth" but later evolved to mean the typical weather of such zones.

[2] "Climate is what we expect, weather is what we get" - Robert Heinlein in "Time Enough for Love" 1973

error in the latter... The meteorologists see very well that the equilibrium is unstable, that a cyclone will be formed somewhere, but exactly where they are not in a position to say; a tenth of a degree more or less at any given point, and the cyclone will burst here and not there, and extend its ravages over districts it would otherwise have spared...

Weather variables behave similarly to pseudo-random number generators in that there is rapid loss of information about initial conditions (i.e. the seed of pseudo-random number generators). This gives rise to what appears to be almost nondeterministic behaviour, which can then be described stochastically. Alternatively, one can define the probability of weather by considering ergodic invariant measures of attractors of dynamical systems (e.g. Drótos et al. 2015), however, the existence and uniqueness of such entities is not easy to prove especially for non-stationary systems.

At a more fundamental level, the evolution of weather is not strictly deterministic because of thermodynamic and quantum mechanical sources of uncertainty. Since weather is described by thermodynamic variables such as temperature, which are only statistical summaries of air molecules, it lacks information on the position and momenta of all the constituent air molecules. Furthermore, such variables can never be known with perfect certainty even for individual molecules due to quantum effects c.f. Heisenberg's uncertainty principle. The proverbial butterfly that creates chaos by flapping its wings (Lorenz, 1995), is in a similar quantum mixture of dead and alive states as Schrödinger's cat (Gribbin, 2011), and so can only perturb the atmosphere once it is observed to be alive. Climate models ignore such sources of indeterminism yet still simulate chaotic sequences of weather that are highly uncertain due to imperfect knowledge in the initial conditions and/or the model equations.

In summary, climate is a statistical concept that has several different interpretations. If one considers climate to be a probabilistic representation of weather then it is is a model-dependent concept defined by how we represent our beliefs. Climate is not so much an intrinsic part of physical reality but a particular and non-unique way of describing it. To paraphrase de Finetti's statement about probability, *"climate does not exist"* (De Finetti, 2017).

28.3 Climate modelling

Climate models are mathematical models for understanding and predicting the behaviour of the climate system (McGuffie and Henderson-Sellers, 2013). They represent our knowledge and beliefs about how climate processes operate.

28.3.1 Climate processes

The climate system is truly complex. However, physical conservation laws provide firm principles for model development. For example, one expects energy, mass, and momentum to be locally conserved, which is the basis for the dynamical equations used to model the atmosphere and oceans, as explained in the historical accounts of Roulston and Norbury (2013) and Edwards (2011).

The Earth's climate system is driven by solar radiation from the Sun. About one third of the solar radiation is reflected back out to space, and so $240\ Wm^{-2}$ on average is absorbed primarily by the land and ocean surfaces. The Earth's surface then emits infra-red radiation back out to space. Gases in the atmosphere such as carbon dioxide, methane, water vapour etc. absorb infra-red radiation and so reduce the amount of outgoing infra-red radiation (*the greenhouse effect*). The tropics receive more solar radiation on average than the poles, which drives large atmospheric circulations, and in turn oceanic circulations, that both transport heat towards the high latitudes. Convective heating of the atmosphere and rotation of the Earth cause the atmospheric circulations to be fluid-dynamically unstable, which leads to the fascinating weather we experience on Earth.

To represent these processes realistically, it is necessary to model many coupled components that interact with one another in complicated ways. For example, the atmospheric winds drive the oceans, which in turn heat the atmosphere. Furthermore, these two fluid components interact with other components such as the cryosphere (e.g. sea ice) and the land surface including vegetation. In addition, processes on small spatial scales in the atmosphere and ocean are influenced by large-scale conditions but can also influence the larger scales e.g. storms can maintain contiental-scale high pressure systems.

28.3.2 Classes of climate model

Several different classes of climate model have been developed in the past 50 years. Climate models vary in dimensionality from simple conceptual models having only one variable (e.g. global mean temperature) to very high dimensional General Circulation Models having millions of variables required for resolving weather systems down to 10km horizontal resolution across the whole globe. The more sophisticated models explicitly represent atmospheric evolution and so can be thought of as pseudo-weather generators. The main classes of climate model commonly used in climate science are described below:

28.3.2.1 General Circulation Model (GCM)

GCMs (also sometimes referred to Global Climate Models) are augmented fluid dynamical models that aim to represent the general circulation of the global atmosphere and oceans. Atmospheric GCMs (AGCMs) evolved in the 1960s out of global Numerical Weather Prediction (NWP) models that were developed to make daily weather forecasts (Smagorinsky, 1983). The weather prediction models were extended to be able to do long climate simulations by adding simple representations of boundary components such as the land surface (e.g. crude hydrology models) and sea-ice. The development of AGCMs inspired the development of global ocean GCMs in the 1960s capable of describing the large-scale circulation in the world's oceans and then these were coupled to AGCMs to make so-called *coupled GCMs* (Manabe and Bryan, 1969). The fluid dynamics is modelled using a simplified form of the Navier-Stokes equations known as the *primitive equations* suitable for thin shells of fluid on a rotating sphere. This results in a coupled set of non-homogeneous 1st order partial differential equations having the general form:

$$\frac{\partial \mathbf{x}}{\partial t} = Q(\mathbf{x}, \nabla \mathbf{x}) + F(s, t) \tag{28.1}$$

where $\mathbf{x}(\mathbf{s}, \mathbf{t})$ is a vector field of prognostic variables defined over space-time, for example, temperature, horizontal wind velocity components, pressure, and humidity for the equations used to model the atmosphere in GCMs. The operator $Q(.)$ includes quadratic non-linearities caused by transport (advection) of mass, momentum, and energy (e.g. $du/dt = \partial_t u + u\partial_x u + v\partial_y u + w\partial_z u$). The general circulation is also a non-homogeneous forced-dissipative system where $F(.)$ represents external forcing and dissipation e.g. heating from radiation and the Earth surface, surface drag, etc. The solutions of this set of equations cannot be obtained analytically and so the equations have to be solved numerically. Numerical solution requires a finite representation and so these equations are usually approximated by ordinary finite difference equations involving values defined on a regular space-time lattice[3]. Current AGCMs typically have horizontal grid spacing of about 10-100km, time steps of around 10 minutes, and about 10-50 vertical levels. Coupled GCMs now typically produce around 2-10 years of simulated data per day when run on massively parallel supercomputers.

28.3.2.2 Regional Climate Model (RCM)

Because of their coarse horizontal resolution, GCMs misrepresent smaller scale features (e.g the Alps), which are known to influence on local weather and climate. One way to try to overcome this problem is to *downscale* global GCM output by using it to force a higher resolution climate model that covers only a specific region. This nested approach can be particularly useful for investigating more extreme weather such as storms and extreme local rainfall e.g. RCM projections of extreme weather over Europe (Beniston et al., 2007). Although providing more spatial detail in the simulated weather, downscaling neglects how this may feed back onto larger spatial scales represented by the GCM. Regional climate models generally have similar fluid equations to those of GCMs but often have more detailed surface parameterisations e.g. orography, vegetation, wind gusts, etc. For more details of the added value of such an approach see Feser et al. (2011).

28.3.2.3 Earth System Model (ESM)

In recent decades, much attention has been paid to adding explicit representations of bigeo-chemical processes to coupled GCMs. For example, rather than prescribing carbon dioxide concentrations in the atmosphere by adjusting $F(s,t)$, many models now calculate carbon dioxide concentrations based on how the carbon cycle might respond to carbon dioxide emis-sions. This involves modelling carbon sources and sinks in the biosphere and increases the number of prognostic variables. Other biogeochemical processes that are now represented include aerosols and sulphur cycle and ozone chemistry. Earth System Models are increas-ingly used in climate change projections - see Flato (2011) for a review. For computational speed, fast conceptual versions of ESMs have also been developed known as Earth system Models of Intermediate Complexity (EMIC), which have highly simplified representations of the atmosphere (see Claussen et al. 2002).

28.3.2.4 Low-order models

In addition to models having many variables, it is also useful to consider highly simplified conceptual models of climate with only a few (or even one) variables. For example, a simple heat balance equation can provide much insight into how global mean temperature $x(t)$ will respond to future changes in radiative forcing $F(t)$ (Gregory and Forster, 2008)

$$C\frac{dx}{dt} = -\lambda x + F(t) \tag{28.2}$$

where C is the heat capacity of the atmosphere and Earth surface and λ is the strength of the feedback in the system. Such models representing averages of variables over space have been widely used to improve process understanding e.g. single vertical column radiative-convective models of the atmosphere, box models for oceanography and carbon cycle, etc.

---

[3]some AGCMs use a truncated set of spherical harmonics rather than finite differences to calculate horizontal spatial derivatives.

28.3.2.5   Stochastic Climate Models

The models above are approximations to a continuum because they do not represent variations smaller than a predefined spatial resolution. One can argue that such truncation could and should be partially remedied by adding stochastic noise to the forcing to account for unrepresented sub-grid processes (Palmer and Williams, 2008). For example, a simple yet widely used stochastic climate model can be constructed by adding Gaussian white noise to the right hand side of Eqn. (28.2) i.e. $F(t) \rightarrow F(t) + \epsilon$ where $\epsilon \sim N(0, \sigma^2)$ (Hasselmann, 1976; and references therein). The noise term represents irregular weather variability and the resulting stochastic differential equation can easily be solved by spectral or finite difference methods. The resulting *red noise* process is similar to a first order autoregressive AR(1) process. This approach can easily be extended to most other low order conceptual models (Wilks, 2008). Another motivation for adding randomness to the forcing is to account for our epistemic uncertainty in the physical parameterisations. It is hoped that by making a climate model inherently stochastic, one may be able to reduce and even eliminate biases present in the deterministic model. However, unlike the conservation principles used to formulate the deterministic component, there is very little guiding theory on how to correctly specify the random components.

In summary, in the past 50 years, climate modellers have developed a rich hierarchy of models ranging from low order models to high resolution models having thousands of grid point variables. The high resolution high complexity models are the main tools used to make climate change predictions supported by understanding gained from the simpler models.

28.4   Design of Experiments for Climate Change

Design of climate model experiments is limited by the huge computational expense of running GCMs and by the diverse range of different applications they attempt to inform. Up until the mid-1990s, it was common to publish analysis of single simulations from individual climate models, some of which did not even include an annual cycle in radiation e.g. so-called *perpetual January* simulations. With such an approach, inference about the real world was based on implausible what-if assumptions such as "what would the actual climate do if it behaved like that of this model".

With increasing computing power, climate scientists are now able to explore a wider range of possibilities by running sets of simulations known as *ensembles* (Collins et al. 2012). Such ensemble data can be used to perform sensitivity analysis and uncertainty quantification. Various types of ensemble are used to sample the different sources of uncertainty. Ensemble experiments allow one to test the sensitivity of results to choice of initial conditions, model parameters for a given climate model, future emission scenarios, and choice of climate model. These types of ensemble are briefly described below - for more information see Flato et al. (2013).

Simulated data from such ensembles together with past observations provides the data from which one then attempts to infer the future behaviour of the single realization of the real climate system (Stephenson et al. 2012).

28.4.1 Initial condition ensembles (ICE)

Because of the chaotic nature of weather simulated by climate models, the time series produced by such models are highly sensitive to initial conditions i.e. after a short period of around two weeks, small differences in atmospheric initial conditions lead to very different realisations of simulated weather. This aleatoric uncertainty due to natural variation puts a fundamental limit on how accurately climate can be predicted and so needs to be quantified. Climate models are generally not ergodic, especially in the presence of forcings having long-run trends, and so it is necessary to do repeated simulations of climate models for each choice of model parameters. Such replication is computationally expensive and different modelling centres make different choices as to how perturb initial conditions (e.g. in the atmosphere and ocean components) and how many runs (replications) to make. With state-of-the-art climate models, typically from 1-10 simulations are made for each choice of model parameters (see Table 28.1).

28.4.2 Perturbed Physics Ensembles (PPE)

In addition to fundamental constants such as the radius of the Earth, climate models also have many tunable parameters that need to be prescribed. For example, the latest climate model used by the Met Office in the UK currently has around 50 tunable parameters in addition to parameters that can be estimated from measurements e.g. the spatial distribution of vegetation types, radiative absorption spectra of gases such as carbon dioxide, etc. (personal communication, David Sexton and Mat Collins). The simulated climate is sensitive in varying degrees to the choice of each these parameters and how they interact. Most of the parameters are usually chosen to be "best guess" values guided by past experience and scientific knowledge and then a small subset are adjusted (*tuned*) after running short simulations in order to minimise excessive biases in the simulations. Because of the high dimensionality of the parameter space and the computational expense of making simulations, it is not generally feasible to tune the parameters automatically unless one makes use of statistical emulators to interpolate the response behaviour in parameter space e.g. Williamson and Blaker (2014).

28.4.3 Multi-Model Ensembles (MME)

A further source of uncertainty arises from the structure of climate models being different. Climate modellers make choices about which equations to use and how to implement them numerically, which leads to models having different schemes and parameter sets. This gives rise to additional uncertainty not captured by PPEs (Yokohata et al., 2012).

Comparison of coupled GCMs has been greatly facilitated by Coupled Model Intercomparison Projects (CMIPs) organised by the World Climate Research Programme's Working Group on Coupled Modelling. The Lawrence Livermore National Laboratory collected simulations of the past, present and future climate in 2005-6 for phase 3 (CMIP3) and in 2011-13 for phase 5 (CMIP5). The vast amount of simulated data from more than 20 modelling centres is made freely available on online servers to allow analysis by scientists outside the modeling centres (36 terabytes for CMIP3 and 2 petabytes for CMIP5; personal communication, Karl Taylor). The resulting publications inform the climate assessment reports produced by the Intergovernmental Panel on Climate Change (IPCC). CMIP also provides an invaluable framework for coordinated design of common sets of experiments that modelling centres then perform with their models (Taylor et al. 2012).

TABLE 28.1

Climate change simulations made for the most recent phase 5 of the Coupled Model Intercomparison Project (http://cmip-pcmdi.llnl.gov/cmip5/). Subsets of the data can be visualised and downloaded from https://climexp.knmi.nl.

Modelling Centre	Country	Climate Model	Number of simulations				
			Historical	RCP2.6	RCP4.5	RCP6.0	RCP8.5
BCC	China	BCC-CSM1.1	3	1	1	1	0
		BCC-CSM1.1-m	3	1	1	1	0
CCCma	Canada	Can ESM2	5	5	5	0	5
CMCC	Italy	CMCC-CM	1	0	1	0	1
		CMCC-CMS	1	0	1	0	1
CNRM-CERFACS	France	CNRM-CM5	10	1	1	0	5
CSIRO-BOM	Australia	ACCESS1.0	1	0	1	0	1
		ACCESS1.3	3	0	1	0	1
CSIRO-QCCCE	Australia	CSIRO-Mk3.6.0	10	10	10	10	10
EC-EARTH	Europe	EC-EARTH	9	2	7	0	8
FIO	China	FIO-ESM	3	3	3	3	3
GCESS	China	BNU-ESM	1	1	1	0	1
INM	Russia	INM-CM4	6	4	4	1	4
IPSL	France	IPSL-CM5A-LR	6	4	4	1	4
		IPSL-CM5A-MR	3	1	1	1	1
		IPSL-CM5B-LR	1	0	1	0	1
LASG-CESS	China	FGOALS-g2	5	1	1	0	1
MIROC	Japan	MIROC5	5	3	3	3	3
		MIROC-ESM	3	1	1	1	1
		MIROC-ESM-CHEM	1	1	1	1	1
MOHC	UK	HadGEM2-AO	1	1	1	1	1
		HadGEM2-CC	1	0	1	0	1
		HadGEM2-ES	4	4	4	3	4
MPI-M	Germany	MPI-ESM-LR	3	3	3	0	3
		MPI-ESM-MR	3	1	3	0	1
MRI	Japan	MRI-CGCM3	3	1	1	1	1
NASA GISS	US	GISS-E2-H p2	5	1	5	1	1
		GISS-E2-H p3	6	1	5	1	1
		GISS-E2-H-CC p1	1	0	1	0	0
		GISS-E2-R p1	6	1	6	1	1
		GISS-E2-R p2	6	1	5	1	1
		GISS-E2-R p3	6	1	6	1	1
		GISS-E2-R-CC p1	1	0	1	0	0
NCAR	US	CCSM4	6	5	6	6	6
NCC	Norway	NorESM1-M	3	1	1	1	1
		NorESM1-ME	1	1	1	1	1
NIMR/KMA	Korea	HadGEM2-AO	1	1	1	1	1
NOAA GFDL	US	GFDL-CM3	5	1	1	1	1
		GFDL-ESM2G	3	1	1	1	1
		GFDL-ESM2M	1	1	1	1	1
NSF-DOE-NCAR	US	CESM1(BGC)	1	0	1	0	1
		CESM1(CAM5)	3	3	3	3	2

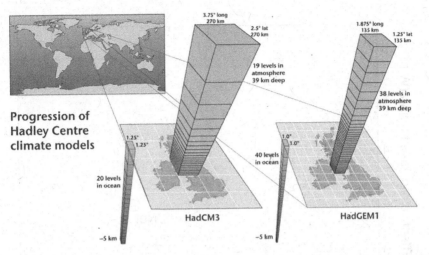

FIGURE 28.1
Schematic showing the grid cells used in recent GCMs developed by the UK Met Office.
(Reprinted with permission ©Crown Copyright Met Office).

28.4.4 Climate change projections

Table 28.1 summarises the climate change simulations made with 42 models from 22 modelling centres for the most recent CMIP5 experiment. Climate simulations were made using greenhouse gas concentrations in the 20th century (historical) and in four scenarios known as Representative Concentration Pathways (RCPs), which represent how greenhouse gas emissions may evolve in the future. The four RCPs, RCP2.6, RCP4.5, RCP6.0, and RCP8.5, are named after equivalent radiative forcing values in the year 2100 relative to pre-industrial values (see Fig. 28.2). RCP2.6 optimistically assumes that global annual greenhouse gas emissions peak between 2010-2020, with emissions declining substantially thereafter, whereas RCP8.5 assumes that emissions will continue to increase throughout the 21st century. Emissions peak around 2040 and 2080 for intermediate scenarios RCP4.5 and RCP6.0, respectively.

Modelling centres decide how many different initial condition simulations to make for each scenario, which leads to an unbalanced design as can be clearly noted in Table 28.1. The design of such experiments is complex because of many competing scientific and policy needs (see CMIP6 rationale in Eyring et al. 2016). However, it is worth noting that such designs have so far made very little use of statistical knowledge about design of experiments in order to obtain better balanced experiments e.g. optimal size of ensembles, balance between different scenarios, etc.

Figure 28.3 summarises the global mean surface temperature response from these CMIP5 simulations under the different scenarios. The mean of all the simulations clearly reveals warming for all the scenarios, with RCP6.0 and RCP8.5 leading to more than 2^0C additional warming by the end of the 21st century. These results provide important evidence for informing policy makers on climate change strategies. For more discussion and interpretation of these results see Collins et al. (2013).

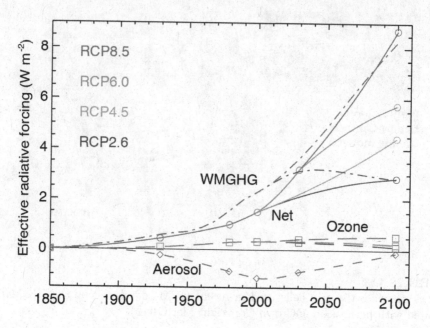

FIGURE 28.2
Effective global mean radiative forcing for the scenarios used in recent CMIP5 simulations: colours indicate the RCPs with red for RCP8.5, orange RCP6.0, light blue RCP4.5, anddark blue RCP2.6. (Reprinted with permission from Myhre et al. (2013)).

28.5   Real world inference from climate model data

It remains an important statistical challenge of how to best use climate model and observational data to infer the behaviour of the real climate system (Stephenson et al. 2012).

28.5.1   Current practice

The most common approach for estimating the expected climate change response is to calculate the multi-model mean of the MME i.e. the arithmetic mean of the individual model simulations e.g. Fig 28.3. This simple yet heuristic approach of one vote per model gives equal weight to each climate model irregardless of a) how many simulations each model has contributed, b) how interdependent the models are or c) how well each model has performed at simulating past climate (Sansom et al., 2013). No use is made of past observations unless past *performance metrics* have been used to screen out poorly performing models before taking the mean (e.g. Santer et al., 2009). This approach ignores model errors and provides only an estimate of the mean response with no credible estimate of how uncertain the individual response of the real climate system is likely to be.

28.5.2   Imperfect climate model and observational data

Despite models being able to capture many of the features of the climate system, climate scientists continue to spend much time striving to develop models that have smaller differ-

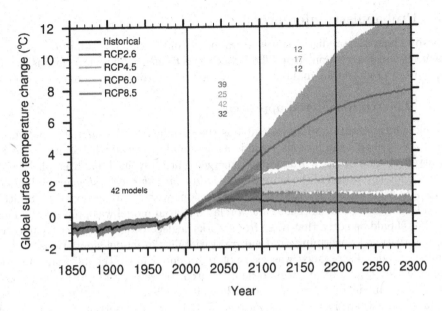

FIGURE 28.3
CMIP5 simulated global mean surface air temperature differences from 1986-2005 means for different RCP scenarios. The multi-model means are shown as solid lines for each RCP and ensemble spread is indicated by shading (1.64 standard deviation). Only one ensemble member is used from each model and numbers in the figure indicate the number of different models contributing to the different time periods. Discontinuities at 2100 are due to different numbers of models performing the extension runs beyond the 21st century and have no physical meaning. No ranges are given for the RCP6.0 projections beyond 2100 as only two models are available. (Reprinted with permission from Collins et al. (2013)).

ences to the observed behaviour of the climate system. Many of these biases are substantial compared to the climate change response and have resisted many years of model development e.g. double equatorial convergence zones, overly zonal storm tracks, too little monsoon rainfall over India, etc..

Rather than compare imperfect models with imperfect observations, it is useful to consider the concept of *model discrepancy* defined as the difference between the *actual climate* (i.e. that observable with perfect measurements) and climate model output. Model discrepancy can be considered to be analogous to *observation error* i.e. the difference between the actual climate and observations. In principle, by specifying statistical models for both model discrepancy and observation error, it is possible to then infer actual climate from simulated data from climate models and data from past observations.

However, unlike observation error, very little theory or guidance exists on how to specify model discrepancy. While model discrepancy is most likely independent of observation error, it is certainly not independent between different models due to the presence of common biases that are caused by various model components being related to one another (Knutti et al., 2013) Furthermore, model discrepancy is unlikely to be stationary in time and beliefs about how it may change have been shown to have substantial impact on predicted future climate (Buser et al. 2009; Ho et al. 2012).

28.5.3　Probabilistic inference

To go beyond heuristic predictions, it is necessary to make explicit assumptions about model discrepancies and observation error. The following sections briefly review various approaches that have been proposed for making statistical inference from multi-model ensembles.

28.5.3.1　The truth-centered approach

The simplest interpretation of an ensemble is the so-called *truth-centered* paradigm, which assumes that the model simulations are independent and identically distributed (i.i.d.) about the observed climate. Such a framework emerged naturally in Tebaldi et al. (2005) as a probabilistic interpretation of the heuristic *reliability ensemble averaging* method introduced by Giorgi and Mearns (2003). Multivariate extensions were then developed by Smith et al. (2009) and Tebaldi and Sansó (2009) and a related spatial model was proposed by Furrer et al. (2007). It should be noted that these frameworks neglected various sources of uncertainity such as observational uncertainty and natural variability in model simulations.

Common model discrepancy was also not represented until Tebaldi and Sansó (2009) proposed treating it as a fixed effect to be estimated. However, a more flexible approach is to treat common model discrepancy as a random effect. Chandler (2013) proposed a random effects framework where instead of being distributed around observed climate, model simulations are considered to be distributed about a latent variable (the ensemble consensus), which in turn is distributed about the actual climate.

28.5.3.2　The coexchangeable approach

Rather than make strong i.i.d. assumptions, Rougier et al. (2013) and Rougier and Goldstein (2014) used the concept of exchangeability to propose an alternative framework capable of representing common model discrepancy. The basic idea is to assume that a carefully chosen subset of model simulations are exchangeable with one another (i.e. statistically indistinguishable), and are also exchangeable with a linear transformation of the observations. Exchangeability is represented by introducing a random ensemble consensus variable, but in contrast to Chandler et al. (2013), the direction of conditioning is reversed - the actual climate is distributed about the ensemble consensus.

28.5.4　Summary

Despite progress in developing inferential frameworks, there is still much more to be done, such as how to prescribe discrepancy parameters, which are extremely difficult to elicit from climate scientists due to the lack of any guiding principles on common bias. The existing multi-model frameworks also neglect various sources of uncertainty, for example, how to make optimal use of perturbed physics ensembles to account for tuning of climate model parameters. The current frameworks also do not address *emergent constraints* i.e. the climate change response being dependent on the present day state of each model (Bracegirdle and Stephenson, 2012). Finally, the estimation methods are generally quite slow (e.g. MCMC) and so are not easy to apply to large spatial gridded climate data sets.

28.6 Concluding remarks

This chapter has provided an introduction to the concept of climate and how it is modelled. There is clearly a need for statistical modelling and interpretation in climate science. The very concept of climate is statistical yet is still not unambiguously defined. Better statistical interpretation could help avoid pitfalls associated with misleading absolute concepts such as "THE climate trend". Statisticians could also be usefully engaged in the development and testing of stochastic climate models, which require both physical and statistical expertise. Design of experiments and model tuning are other obvious areas that need more attention, for example, CMIP could produce more accurate projections if the number of simulations in past and future scenarios were more balanced. Finally, statisticians have a critical role to play in helping to make climate data relevant for decision-making, for example, in making reliable inference about future regional climate change.

Bibliography

[1] Beniston, M., D.B. Stephenson, O.B. Christensen, C.A.T. Ferro, C. Frei, S. Goyette, K. Halsnaes, T. Holt, K. Jylha, and B. Koffi B. 2007. Future extreme events in European climate: an exploration of regional climate model projections, *Climatic Change*, 81:71-95.

[2] Bracegirdle, T.J. and D.B. Stephenson 2012. More precise predictions of future polar winter warming estimated by multi-model ensemble regression. Climate Dynamics, 39, 2805-2821.

[3] Buser, C.M., Künsch, H.R., Lüthi, D., Wild, M., and C. Shär 2009. Bayesian multi-model projection of climate: bias assumptions and interannual variability, Climate Dynamics, 33, 849-868.

[4] Chandler, R.E. 2013. Exploiting strength, discounting weakness: combining information from multiple climate simulators. *Phil Trans R Soc A* 371(1991):20120388.

[5] Claussen, M., L.A. Mysak, A.J. Weaver, M. Crucifix, T. Fichefet, M.-F. Loutre, S.L. Weber, J. Alcamo, V.A. Alexeev, A. Berger, R. Calov, A. Ganopolski, H. Goosse, G. Lohmann, F. Lunkeit, I.I. Mokhov, V. Petoukhov, P. Stone, and Z. Wang. 2002: Earth System Models of Intermediate Complexity: Closing the Gap in the Spectrum of Climate System Models. *Climate Dynamics* 18:579-586.

[6] Collins, M., R.E. Chandler, P.M. Cox, J.M. Huthnance, J.C. Rougier, and D.B. Stephenson 2012. Quantifying future climate change, *Nature Climate Change*, 2, 403-409.

[7] Collins, M., R. Knutti, J. Arblaster, J.-L. Dufresne, T. Fichefet, P. Friedlingstein, X. Gao, W.J. Gutowski, T. Johns, G. Krinner, M. Shongwe, C. Tebaldi, A.J. Weaver and M. Wehner, 2013: Long-term Climate Change: Projections, Commitments and Irreversibility. *In: Climate Change 2013: The Physical Science Basis. Contribution of Working Group I to the Fifth Assessment Report of the Intergovernmental Panel on Climate Change* [Stocker, T.F., D. Qin, G.-K. Plattner, M. Tignor, S.K. Allen, J.

Boschung, A. Nauels, Y. Xia, V. Bex and P.M. Midgley (eds.)]. Cambridge University Press, Cambridge, United Kingdom and New York, NY, USA.

[8] Cooley, D., and S. R. Sain, 2010. Spatial hierarchical modeling of precipitation extremes from a regional climate model. Journal of Agricultural, Biological, and Environmental Statistics, 15, 381-402.

[9] De Finetti, 2017. Theory of Probability: A Critical Introductory Treatment. John Wiley and Sons, 596 pp.

[10] Drótos, G., T. Bódai, and T. Tél, 2015. Probabilistic Concepts in a Changing Climate: A Snapshot Attractor Picture. J. Climate, 28, 32753288.

[11] Edwards, P.N.. 2011: History of climate modeling. *WIREs Clim Change.* 2:128139.

[12] Eyring, V., Bony, S., Meehl, G. A., Senior, C. A., Stevens, B., Stouffer, R. J., and Taylor, K. E., 2016: Overview of the Coupled Model Intercomparison Project Phase 6 (CMIP6) experimental design and organization, Geosci. Model Dev., 9, 1937-1958.

[13] Feser, F., B. Rockel, H. von Storch, J. Winterfeldt, and M. Zahn 2011. Regional Climate Models Add Value to Global Model Data: A Review and Selected Examples. *Bulletin of the American Meteorological Society*, 92(9): 1181.

[14] Flato, G. M. 2011. Earth system models: an overview. *WIREs Clim Change*, 2:783800.

[15] Flato, G., J. Marotzke, B. Abiodun, P. Braconnot, S.C. Chou, W. Collins, P. Cox, F. Driouech, S. Emori, V. Eyring, C. Forest, P. Gleckler, E. Guilyardi, C. Jakob, V. Kattsov, C. Reason and M. Rummukainen, 2013: Evaluation of Climate Models. In: Climate Change 2013: The Physical Science Basis. Contribution of Working Group I to the Fifth Assessment Report of the Intergovernmental Panel on Climate Change [Stocker, T.F., D. Qin, G.-K. Plattner, M. Tignor, S.K. Allen, J. Boschung, A. Nauels, Y. Xia, V. Bex and P.M. Midgley (eds.)]. Cambridge University Press, Cambridge, United Kingdom and New York, NY, USA.

[16] Furrer, R., Sain, S. R., Nychka, D. W. and Meehl, G. A., 2007. Multivariate Bayesian analysis of atmosphere-ocean general circulation models, Environmental and Ecological Statistics 14(3), 249-266.

[17] Fyfe, J.C., G.A. Meehl, M.H. England, M.E. Mann, B.D. Santer, G.M. Flato, E. Hawkins, N.P. Gillett, S-P. Xie, Y. Kosaka, and N.C. Swart 2016. Making sense of the early-2000s warming slowdown. *Nature Climate Change*, 6:224-228.

[18] Giorgi, F. and Mearns, L. O., 2003. Probability of regional climate change based on the Reliability Ensemble Averaging (REA) method, Geophysical Research Letters 30(12), 1629.

[19] Gregory, J. M. and Forster, P. M., 2008. Transient climate response estimated from radiative forcing and observed temperature change. *J. Geophys. Res.*, 113:D23.

[20] Gribbin, J., 2011. In Search of Schr odinger's Cat: Quantum Physics And Reality. Random House Publishing Group.

[21] Hasselman, K., 1976. Stochastic climate variability. Tellus, 28:473-485.

[22] Hawkins, E. and Sutton, R., 2016. Connecting climate model projections of global temperature change with the real world, Bulletin of the American Meteorological Society, 97, 963.

[23] Ho, C.K., Stephenson D.B., Collins M., Ferro C.A.T., Brown S.J. 2012. Calibration strategies: a source of additional uncertainty in climate change projections, Bulletin of the American Meteorological Society, 93, 21-26.

[24] Knutti, R., D. Masson, and A. Gettelman, 2013. Climate model genealogy: Generation CMIP5 and how we got there, Geophys. Res. Lett., 40, 11941199.

[25] Lorenz, E.N., 1995. The Essence of Chaos. Taylor and Francis. 227 pp.

[26] Manabe, S. and K. Bryan 1969. Climate Calculations with a Combined Ocean-Atmosphere Model. *Journal of the Atmospheric Sciences*, 26:786-789.

[27] McGuffie, K. and A. Henderson-Sellers 2013. A Climate Modelling Primer. Wiley, 3rd Edition, 288 pp.

[28] Myhre, G., D. Shindell, F.-M. Bron, W. Collins, J. Fuglestvedt, J. Huang, D. Koch, J.-F. Lamarque, D. Lee, B. Mendoza, T. Nakajima, A. Robock, G. Stephens, T. Takemura and H. Zhang, 2013: Anthropogenic and Natural Radiative Forcing. In: Climate Change 2013: The Physical Science Basis. Contribution of Working Group I to the Fifth Assessment Report of the Intergovernmental Panel on Climate Change [Stocker, T.F., D. Qin, G.-K. Plattner, M. Tignor, S.K. Allen, J. Boschung, A. Nauels, Y. Xia, V. Bex and P.M. Midgley (eds.)]. Cambridge University Press, Cambridge, United Kingdom and New York, NY, USA.

[29] Poincaré, H. 1914. Science and Method. Translated by Francis Maitland 2007. Cosimo, New York.

[30] Palmer, T.N. and P. Williams (Editors) 2008. Stochastic Physics and Climate Modelling. Special issue of *Phil. Trans. Roy. Soc.*, 366:2419-2641.

[31] Rougier, J.C., M. Goldstein, and L. House 2013. Second-order exchangeability analysis for multi-model ensembles, *Journal of the American Statistical Association*, 108:852-863.

[32] Rougier, J.C. and M. Goldstein 2014. Climate Simulators and Climate Projections, *Annual Review of Statistics and Its Application* 1:103-123.

[33] Roulstone, I. and J. Norbury 2013. Invisible in the Storm: The Role of Mathematics. Princeton University Press, 346 pp.

[34] Sansom, P. G., D. B. Stephenson, C.A.T. Ferro, G. Zappa, and L.C. Shaffrey, 2013: Simple uncertainty frameworks for selecting weighting schemes and interpreting multi-model ensemble climate change experiments, *Journal of Climate*, 26:4017-4037.

[35] Santer, B., et al., 2009: Incorporating model quality information in climate change detection and attribution studies. *Proc. Natl. Acad. Sci.* 106:14778 14783.

[36] Smagorinsky, J. 1983:The Beginnings of Numerical Weather Prediction and General Circulation Modeling: Early Recollections. *Advances in Geophysics* 25:3-37.

[37] Smith, R. L., Tebaldi, C., Nychka, D. W. and Mearns, L. O., 2009. Bayesian Modeling of Uncertainty in Ensembles of Climate Models, Journal of the American Statistical Association, 104(485), 97-116.

[38] Stephenson, D.B., M. Collins, J.C. Rougier, and R.E. Chandler 2012: Statistical problems in the probabilistic prediction of climate change. *Environmetrics*. 23(5):364-372.

[39] Taylor, K.E., R.J. Stouffer, and G.A. Meehl 2012: An Overview of CMIP5 and the experiment design. *Bull. Amer. Meteor. Soc.*, 93:485-498.

[40] Tebaldi, C. and Sansó, B., 2009. Joint projections of temperature and precipitation change from multiple climate models: A hierarchical Bayesian approach, Journal of the Royal Statistical Society. Series A: Statistics in Society 172(1), 83-106.

[41] Tebaldi, C., Smith, R. L., Nychka, D. W. and Mearns, L. O., 2005. Quantifying uncertainty in projections of regional climate change: A Bayesian approach to the analysis of multimodel ensembles, Journal of Climate, 18(10), 1524-1540.

[42] Williamson, D. and A.T. Blaker 2014: Evolving Bayesian Emulators for Structured Chaotic Time Series, with application to large climate models. *SIAM Journal on Uncertainty Quantification* 2:1-28.

[43] Wilks, D.S. 2008: Effects of stochastic parametrization on conceptual climate models. *Phil. Trans. Roy. Soc.*, 366:2475-2488.

[44] Yokohata, T., J. Annan, M. Collins, C. Jackson, M. Tobis, M. Webb, and J. Hargreaves, 2012: Reliability of multi-model and structurally different single-model ensembles. *Clim. Dyn.* 39:599616.

29

Spatial Analysis in Climatology

Douglas Nychka

Colorado School of Mines, Golden, Colorado

Christopher K. Wikle

University of Missouri, Columbia, Missouri

CONTENTS

29.1   Introduction

An emerging view of the Earth's physical environment is as a system that encompasses the interactions and dynamics among many different components, including the atmosphere, the ocean, the cryopsphere, the land surface, and the external drivers from human activities. Climatology is concerned with multivariate interactions in the Earth system over a wide range of spatial and temporal scales. Here the focus is on the *distribution* of phenomena as opposed to specific events. To assist in understanding the components of such a system, researchers collect observations, explore relationships among variables, perform inference, and predict behavior across and between these scales of variability. Thus, by its very nature, climatology is a spatial science and closely dependent on maps (their creation, interaction and evolution). Indeed, one of the most important climate classification schemes, the so-called Köppen climate classification, was developed by Wladimir Köppen in 1884 to display world climate zones visually on a map. As climatology gradually developed from a qualitative geographical science to a more physically and statistically based science, the use of rigorous statistical methods to account for spatial information became more commonplace. Besides interpreting observations of the Earth system, spatial analysis is also important for interpreting the output of numerical simulations (e.g. climate models) and for transforming irregular data sets into more convenient data products. Thus, throughout this chapter we take a broad view of climate "data" that can include the results of deterministic models, derived parameters, and previous analyses. Another running theme throughout this chapter is that, although spatial statistical methods have the advantage of formal uncertainty quantification owing to their probabilistic foundation, there are instances in the climate sciences where such advanced methods are not used, either due to lack of demonstrated benefit, a focus on other issues (e.g. systematic biases), or ease of explanation to data users. It is also the case that many of the exploratory and descriptive methods that have traditionally been used in climate science and in meteorology and oceanography (see Chapter 33) have motivated the development of new formal statistical methods for spatial data.

In this chapter, we divide the spatial analysis techniques into three primary areas useful in climatology: (1) exploratory/descriptive analysis, and (2) the development of data products, and (3) spatio-temporal methods in climatology. The purpose of this overview is to introduce some of the key references, guidelines and concepts for the interested reader.

29.2　Exploratory/Descriptive Analysis

An overview of statistical methods used for exploratory and descriptive analysis in climatology can be found in both the climate science literature (e.g., [73] and [80]) and the statistics literature (e.g., [34] and [16]). These methods might loosely be classified as "moment-based," "spectral-based," and "eigen-decomposition-based." We outline some components of each of these below, but refer the interested reader to the aforementioned references for details. We also recommend Chapter 5 for background on spatial statistical methods, and to Chapter 33 for statistical methods in the closely related field of oceanography.

29.2.1　Moment-Based Methods

Following good data science, climatologists summarize their data in terms of empirical means and variances. In many cases, these summaries are calculated by averaging/marginalizing over time at data locations, and the results are plotted as maps (of course, they also consider summaries plotted in time by averaging over space). This is consistent with the now limited definition of climatology being the study of weather averaged over longer periods of time. For example, a working definition of climate for a location might be the 30 year average of a given atmospheric variable. Although a richer definition of climate includes consideration of the full multivariate distribution of variables, scrutiny of the first and second moments of key variables is still important.

As a way to reinforce some of the main points in this chapter we will use a moderate sized data example to illustrate specific methods. The NOAA Extended Reconstructed Sea Surface Temperature (SST) is a global, monthly analysis from 1854 to the present derived from the International Comprehensive Ocean Atmosphere Dataset (ICOADS). Where there are missing data, they are filled in by statistical methods and so this is technically a *data*

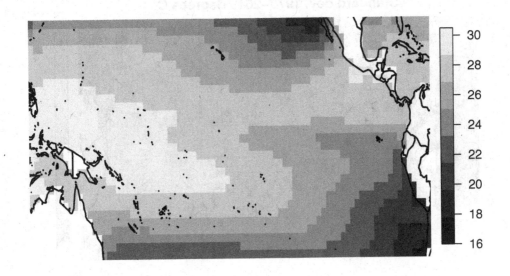

FIGURE 29.1
Pacific sea surface temperature mean field for the period 1970- 2015 in degrees C.

product (See Section 29.4). Here we focus on the Pacific Basin and on monthly SST over the years 1970 through 2015. The spatial domain is a rectangle of 2 degree grid cells that include the Equitorial Pacific Ocean ([-29, 29] latitude × [124, 290] longitude). Figure 29.1 shows the mean SST for this period and might be interpreted as an estimate of the climatological mean field for this part of the ocean. However, we note that this simple summary can mask long term changes in the mean temperatures over time (e.g. due to climate change) and also variability in the SST over space. Also, one should factor in the issue that these data are based on observations that may have potential biases and other errors. Figure 29.2 reports the standard deviation field for these data and the trend obtained by a least squares fit. Both of these fields help to give some insight into this space/time data and clearly emphasize the complex spatial aspects of the SST distribution. Although these kind of summaries are valuable for the exploration of the spatial structure, any formal testing or inference needs to specify a spatial statistical model and some choices will be presented in Section 29.4.2.

29.2.2 Spectral-Based Methods

Among the variations present in climatological data are cyclic and quasi-cyclic behaviors that are often tied to physical mechanisms. There has been historical interest in characterizing such behavior using spectral methods from time-series analysis (see Chapter 33). In particular, *spectral analysis* in time series identifies frequencies (or cycles) in the data record that exhibit increased variability. *Cross-spectral analysis* finds frequencies that exhibit high variability in two or more time series, and considers whether the variability at those frequencies occur at different time lags between the series. These methods have been extended to consider spatial variability as well and the two primary approaches that take into account spatial information are (1) spatial cross-spectral analysis, and (2) frequency-wavenumber analysis.

Standard dev. 1970–2015 degrees C

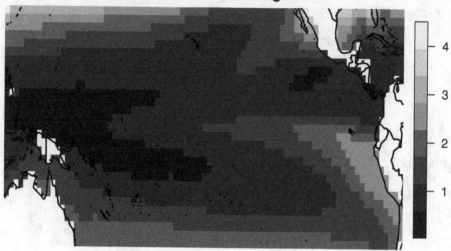

Trend 1970–2015 degrees C/year

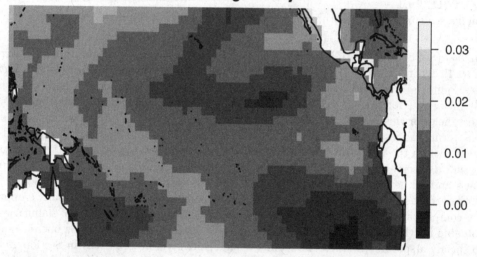

FIGURE 29.2
Pacific sea surface temperature standard deviation and linear trend fields for the period
1970- 2015 in degrees C.

Spatial cross-spectral analysis is simply the spectral analysis of multivariate time series
where the time series are drawn from two or more spatial locations. In particular, the analysis
of the strength of coherence and phase differences at different frequencies can be useful for
understanding how wave phenomena, which are endemic in the climate system, propagate.
For example, as summarized in [16] (Sections 3.5 and 5.2), [96] used bivariate cross-spectral
analysis to verify the existence and propagation behavior of atmospheric waves that had
been suggested by theory. In a significant feat of statistical exploration, R. Madden and P.
Julian [53] used such cross spectral analysis to discover the 30-60 day oscillation that is now

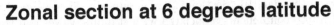

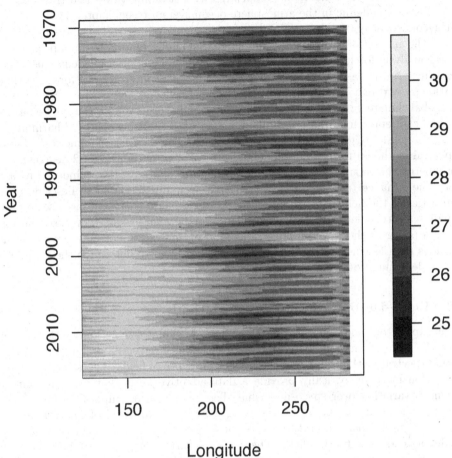

FIGURE 29.3
A Hovmöller diagram for monthly mean sea surface temperature and for the latitude band at 6 degrees transecting the Pacific Ocean basin.

named in their honor (the Madden-Julian Oscilation, MJO), and has proven to be one of the most influential climate oscillations in terms of its societal impacts.

Frequency-wavenumber analysis is typically performed with one-dimensional spatial data that varies through time (e.g., we might consider data along a given latitude). Figure 29.3 is an exploratory plot, known as as a Hovmöller diagram, for the SST field (in deg C) at the latitude line of 6 degrees North latitude and is a standard graphical tool in the geosciences to represent how a 1-d spatial process is changing over time. In this figure we see the seasonal variation at this latitude band and also the multi-year oscillation that is the El Niño/ La Niña process. In general, the goal is to understand the relative power or spectral density across a range of temporal frequencies and spatial wave numbers. The wavenumber is simply the number of wave (sinusoidal) components in a given spatial domain (i.e., cycles per unit distance); it is the spatial analog to frequency in the temporal domain (i.e., number of cycles per unit time). Thus, as the wavenumber increases, the spatial scale becomes smaller. As outlined in [16] (Section 5.2), one typically decomposes these frequency-wavenumber spectra into components associated with standing waves and traveling waves, both of which are endemic in the atmosphere-ocean system (see Chapter 33).

In the context of spatial data there is also a spectral decomposition that can be used to determine the amount of variation at different spatial scales. Similar to a temporal analysis, a spectral analysis for large, regular data fields can be computed efficiently using a fast Fourier transform. It has the disadvantage that it is not as interpretable for fields that are non-stationary and may also be sensitive to boundary effects.

For global climate data the spectral representation for processes on the sphere has a long history and can represent variation at different spatial scales using spherical harmonic basis functions ([44], [81], [67]). In fact this representation is also the basis of numerical algorithms (i.e., spectral methods) for climate modeling at modest resolutions (e.g. 2 degrees).

For spherical harmonics, basis functions at the same spatial scale are indexed by a wave number. The variance (or "energy") at a given wavenumber is used to describe physical characteristics of the field. For example, the relative kinetic energy in large scale features and small scale turbulence in the atmosphere can be characterized by how the energy decays as a power of the wave number. This kind of analysis is not only used to evaluate observations such as surface winds (e.g. [89], [46]) but also to determine the properties of climate model simulations (e.g. [24]).

29.2.3 Eigen-Decomposition-Based Methods

Of all the exploratory and descriptive methods used in climatology, arguably the most powerful are the suite of methods based in some sense on an eigen-decomposition. Although traditional spectral methods are effective for specific types of fields (e.g., stationary), eigen-decomposition methods typically provide a data-adaptive basis that can represent more heterogenous variation over space and time. These are spatio-temporal variants and extensions of some of the ordination methods presented in Chapter 11 of this volume. Eigen methods have an extensive literature so we focus here on the basic techniques to provide the reader a road map to this field. These are: empirical orthogonal functions (EOFs), spatio-temporal canonical correlation analysis (ST-CCA), and empirical principal oscillation patterns (POPs) or empirical normal modes (EMNs). For details on these and other eigen-decomposition methods, see [73], [34], [80], and [16].

To set the context for these methods we will represent the space/time process by the function $z(s, t)$ with s being the spatial locations. These are often in latitude/longitude coordinates, although care must be taken to ensure that the map projection is appropriate and that spatial covariance functions are valid in these coordinates (see Chapter 5). Here t denotes the time. In the simplest case, the observations are made at n spatial locations $\{s_i\}$ for $i = 1, \ldots, n$ and at T times $\{t_j\}$ for $j = 1, \ldots, T$. The observations are assumed to be complete and in this case are organized as an $n \times T$ matrix or table, Z. Rows index space and columns index time.

29.2.3.1 Empirical Orthogonal Functions (EOFs)

EOFs are essentially the climatologist's implementation of principal component analysis (PCA) (see Chapter 11 of this volume) but with the specific focus on interpreting space/time data matrices such as Z above. The motivation behind this decomposition is the theoretical representation of $z(s, t)$ as

$$z(s, t) = \sum_{\nu=1}^{\infty} \varphi_\nu(s) a_\nu(t), \tag{29.1}$$

where ϕ_ν are spatial basis functions and a_ν are coefficients for the basis functions at each time. The goal is to use a limited sum of terms that will approximate z well and be physically interpretable. To this end we construct the $n \times n$ empirical spatial covariance matrix, $\mathbf{C}_z$. Specifically, $\mathbf{C}_z$ is made up of elements $c_z(\boldsymbol{s}_i, \boldsymbol{s}_j)$, the sample covariance between locations i and j with the repeated observations over time dimension serving as replicates. The spectral decomposition theorem then gives

$$\mathbf{C}_z = \mathbf{\Phi}\mathbf{\Lambda}\mathbf{\Phi}',$$

where $\mathbf{\Phi}$ is the (orthonormal) matrix of eigenvectors $\mathbf{\Phi} = [\phi_1, \ldots, \phi_n]$. Here we interpret each of these column vectors as $\phi_k = (\phi_k(\mathbf{s}_1), \ldots, \phi_k(\mathbf{s}_n))'$ and $\mathbf{\Lambda}$ is a diagonal matrix of eigenvalues such that $\lambda_1 \geq \lambda_2 \geq \ldots \geq \lambda_n \geq 0$. The eigenvectors in this case, being indexed in space, can be represented as a spatial map – these maps are the *EOFs*. They are empirical because they are derived from the data, orthogonal due to the orthogonality of eigenvectors, and they can be interpreted as functions over space (evaluated on a discrete set of points.) The projection of the data onto the kth EOF map corresponds to the kth so-called principal component time series. Due to orthogonality, the projection is just

$$a_k(t_j) = \phi'_k \mathbf{z}_j, \tag{29.2}$$

where $\mathbf{z}_j$ is the jth column vector of $\mathbf{Z}$.

Another way to obtain this decomposition is to first center the data matrix so that each row has zero mean (subtract the spatial means from the columns of $\mathbf{Z}$). With the centered matrix given by $\mathbf{Z}^* = \mathbf{Z} - \bar{\mathbf{Z}}$ (where $\bar{\mathbf{Z}} = z\mathbf{1}'_T$ and z is the n-dimensional spatial mean vector with $\mathbf{1}_T$ the $T \times 1$ vector of ones), we form the (unique) singular value decomposition

$$\mathbf{Z}^* = \mathbf{\Phi}\mathbf{\Gamma}\mathbf{V}^t.$$

Here, $\mathbf{\Phi}$ is the same as above, $\mathbf{\Gamma}$ is a diagonal matrix of T singular values with $\mathbf{\Gamma}\mathbf{\Gamma}' = \mathbf{\Lambda}$, and V is a $T \times T$ orthogonal matrix. Setting $A = \mathbf{\Gamma}\mathbf{V}^t$ this matrix has the same entries as in (29.2) and it follows that $\mathbf{Z}^* = \mathbf{\Phi}\mathbf{A}$.

EOFs can be used for two purposes. One is to distill from the data some scientific understanding of important spatial patterns of variability. This is accomplished by plotting the spatially-indexed EOF eigenvectors as a map and noting the proportion of variance in the data accounted for by that map (e.g. $\lambda_k / \sum_i \lambda_i$). But, just as one has to be careful interpreting loadings in PCA, one has to be careful when interpreting the EOF patterns in terms of dynamical properties of the underlying climatological process (see [55]). Secondly, these EOFs can be thought of as spatial basis functions and can be used to reduce the dimension of the data. That is, similar to PCA, the leading EOFs are associated with higher variance patterns in the data, and it is often the case that just a few EOFs accommodate the vast majority of the variance in the dataset. Although these represent the optimal basis with respect to variation, they are not in general the optimal bases in terms of other components of the dynamical system ([55]).

Figure 29.4 illustrates the two leading EOFs for the SST data. Note that the first EOF places weight on the region of the Pacific close to South America and the second has more broadly distributed weight in the equatoral Pacific. Figure 29.5 gives the times series of coefficients for these two EOFs. The first suggests a strong annual cycle with some variation in amplitude. The second indicates some temporal structure on a longer time scale than a year and this is related to the ENSO phenomenon. The recipe for approximating the jth column of the data matrix using these two EOFs and the coefficients is

$$\mathbf{z}_j \approx \bar{z} + \phi_1 a_1(t_j) + \phi_2 a_2(t_j).$$

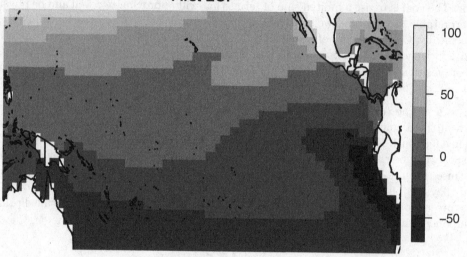

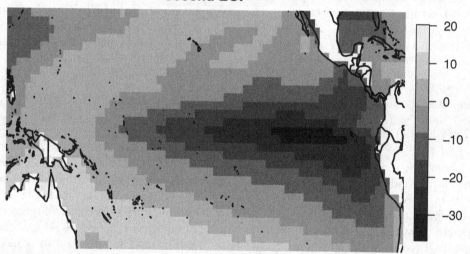

FIGURE 29.4
First and second EOFs computed from the the centered, monthly sea surface temperature
fields

This expression makes it clear that the two EOFs represent spatial patterns of variability
that are controlled at a particular time by the two coefficients, thus suggesting the need
for space-time dependence when building models for the elements of Z. Finally, we see
that from Figure 29.6 the fraction of variance explained by the leading EOFs increases
rapidly, with more than 85% of the variance explained by these two EOFs and more than
95% explained by 6 EOFs. This kind of reduction is not uncommon in climate data and

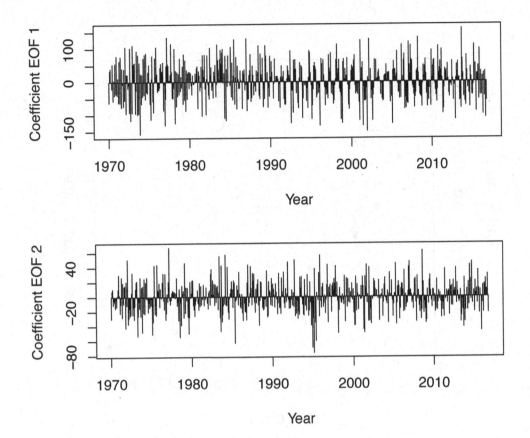

FIGURE 29.5
Coefficients for the first and second EOFs from Figure 29.4

indicates that much of the variation in the spatial fields over time can be reduced to several coefficients.

As described in [73] and [16] (Chapter 5), there are many extensions to the standard EOF analysis. Some of these include complex EOFs (used for data that have strong propagating spatial features), cyclostationary EOFs (used when one has periodicities in the data), and multivariate and extended EOFs (when one has multiple data sets or data at different time lags, respectively).

In addition, one can consider EOF analysis in time by the decomposition of the empirical temporal covariance matrix. In this case the eigenfunctions are time series and the projection of the data onto these basis functions are spatial fields (see [65] for an in-depth treatment). This switch of roles is easy to understand given the singular value decomposition of the space-time data matrix. The decomposition works for the matrix or its transpose. This version is useful for extracting low dimensional components of a time trend and examining how they may vary over space.

The connection of the EOF decomposition to random processes can be made in continuous space through a Karhunen-Loéve (KL) expansion. Consider $z(s,t)$ on some continuous spatial domain D with the expansion given in (29.1), where we assume $z(s,t)$ has zero mean and spatial covariance function given by $c_z(s,r) = E(z(s,t), z(r,t))$. The KL expan-

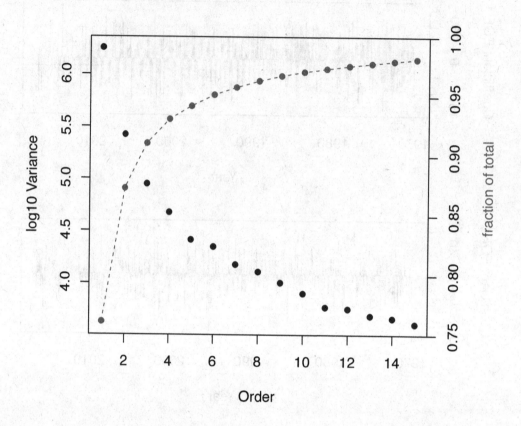

FIGURE 29.6
Variance explained by the EOF decomposition. Dashed line and points are the cumulative fraction of total variance explained as function of the number of EOFs used to reconstruct the original space-time data table (Z). Points indicate the contribution of each EOF to the total variance of the field.

sion then considers the decomposition of this covariance function as

$$c_z(s, r) = \sum_{\nu=1}^{\infty} \lambda_\nu \varphi_\nu(s) \varphi_\nu(r),$$

where $\{\varphi_\nu(\cdot) : \nu = 1, 2, \dots\}$ are eigenfunctions and $\{\lambda_\nu : \nu = 1, 2, \dots\}$ are the eigenvalues corresponding to the solution of the Fredholm integral equation given by

$$\int_D c_z(s, r) \varphi_\nu(s) ds = \lambda_\nu \varphi_\nu(r),$$

where the eigenfunctions are orthonormal. For details see [16] (Section 5.3). We can then project the process onto the eigenfunctions to get

$$a_\nu(t) = \int_D z(s, t) \varphi_\nu(s) ds, \quad t = 1, 2, \dots.$$

In this case, it can be shown that $E(a_i(t), a_j(t)) = \lambda_j$ when $i = j$ and 0 otherwise, and also that $var(a_1(t)) \geq var(a_2(t)) \geq \ldots$. Thus, the relevance of this theory is that it provides the strong random process-based motivation for EOF analysis. From a more practical perspective, it suggests that if one does not account for the spatial support associated with each observation when doing the empirical covariance matrix decomposition, one will not obtain EOFs that match the eigenfunctions from the KL expansion (e.g., [13]). This continuous space perspective also suggests how one might extend the empirically-derived basis functions to continuous space in a coherent manner (e.g., [60] and [8]).

29.2.3.2 Spatio-Temporal Canonical Correlation Analysis (ST-CCA)

Just as EOFs are the geophysical manifestation of PCA in statistics, one can also extend the canonical correlation analysis (CCA) methodology in multivariate statistics to the spatial context. Most often, this corresponds to two variables indexed at the same time but at potentially different sets of spatial locations. Although, it can also correspond to other cases, such as the case where one has a single spatio-temporal variable and seeks to optimize the correlation to a lagged value of that same variable (see [16], Chapter 5).

The basic idea in ST-CCA is that one makes two new sets of variables from linear combinations of the original variables such that the new variables are maximally correlated. That is, we take a linear combination of the spatial observations of one variable, say Z, at each time and then try to find the weights so that this new variable is maximally correlated with the spatial linear combination of another variable, say X. After one finds these weights, then one seeks weights for another set of new variables that are maximally correlated subject to being orthogonal to the first set. This process can be repeated until the complete cross-covariance matrix is decomposed into pairs of maximally correlated variables. It has long been known in statistics that these optimal weights are obtained through the eigen-decomposition of a cross-covariance or cross-correlation matrix. For example, as shown in [16] (Section 5.6), the left and right singular vectors of the singular value decomposition of the matrix product, $(\mathbf{C}_Z)^{-1/2}\mathbf{C}_{Z,X}(\mathbf{C}_X)^{-1/2}$, correspond to the aforementioned optimal weights, and the singular values of this decomposition correspond to the associated maximal correlations. In this case, $\mathbf{C}_Z$ and $\mathbf{C}_X$ refer to the empirical spatial covariance matrices for the Z and X data sets, respectively, and $\mathbf{C}_{Z,X}$ corresponds to the empirical cross-covariance matrix. As with EOFs, the associated singular vectors in this case are spatially indexed and one can represent them as maps. In addition, the new time-varying variables can be created by projecting the data sets onto these singular vectors. As described in [79] (Chapter 14), one often has to perform this analysis after first reducing the dimensionality, for example, by first projecting onto a relatively few EOFs.

29.2.3.3 Empirical Principal Oscillation Patterns (POPs)/Empirical Normal Modes (ENMs)

Climatological processes are inherently dynamical. That is, they are best described as spatial processes that evolve through time. Given a vector representation of the data, e.g. $\mathbf{z}_t = (z(\mathbf{s}_1; t), \ldots, z(\mathbf{s}_n; t))'$, a simple yet powerful dynamical representation of such data is given by a spatio-temporal vector autoregression model of order 1 (VAR(1)):

$$\mathbf{z}_t = \mathbf{M}\mathbf{z}_{t-1} + \boldsymbol{\eta}_t,$$

where it is assumed that the data have been centered (the time mean at each spatial location has been subtracted off). It is typically assumed that $\boldsymbol{\eta}_t$ is independent in time and normally distributed with zero mean and spatial variance/covariance matrix $\mathbf{Q}$. Furthermore, it is assumed that $\boldsymbol{\eta}_t$ is independent of $\mathbf{z}_{t-1}$ for all t (e.g., see Chapter 4 in this volume). Simple method of moments estimates of $\mathbf{M}$ and $\mathbf{Q}$ are given by $\hat{\mathbf{M}} = \mathbf{C}_Z^{(1)}(\mathbf{C}_Z^{(0)})^{-1}$ and $\hat{\mathbf{Q}} = \mathbf{C}_Z^{(0)} - \hat{\mathbf{M}}\mathbf{C}_Z^{(1)\prime}$, where $\mathbf{C}_Z^{(0)}$ and $\mathbf{C}_Z^{(1)}$ are the lag 0 and lag 1 (in time) empirical spatial covariance matrices, respectively (e.g. see [16] Section 3.4).

As discussed in [79] and summarized in [73] (Chapter 15) and [16] (Sections 3.2, 5.5), the linear dynamics of the process underlying the data can be deduced from the eigenvectors and eigenvalues of the empirical propagator matrix $\hat{\mathbf{M}}$. For example, the eigenvector with the maximum (in modulus) eigenvalue corresponds to the maximally growing initial condition of the system that yields the most variance at a given time. As with the other decomposition methods, the eigenvectors are again indexed in space in this case and can be visualized as spatial maps. In addition, if the eigenvector and associated eigenvalue are complex, then that particular empirical mode is associated with an oscillation. Although these decompositions can be used to aid in dimension reduction, most often their use corresponds to exploratory and diagnostic analyses. As with ST-CCA, in practice one often projects the original data into a lower-dimensional space (e.g. via EOFs) before doing the estimated POP/ENM analysis, especially in cases where there are large numbers of spatial data locations. For the SST data when $\hat{M}$ is found for the first 6 EOFs (having removed the annual cycle) $\hat{M}$ is close to a diagonal matrix. However, there is fairly high autocorrelation between successive months for these 6 coefficient series.

29.3   Figures

The evaluation of the spatial and temporal distribution of these and other simple measures of central tendency and variability are quite useful, but more interesting relationships are typically found by various measures of cross-dependence. These also help to quantify the spatial coherence of many climate fields. Summaries of the covariances and correlations among the variables in a field are ubiquitous in climate science. In the case of the SST example, one could consider the correlation between the monthly temperatures in different grid boxes, using the different values over time as (psuedo) replicates. Figure 29.7 is a field reporting the correlation of the highlighted grid box (centered at [-2,206] lat/lon) with the remaining grid boxes in this domain. Here we see higher correlations for grid boxes closer to this reference location and longer correlation scales in the east-west direction as opposed to the north-south direction. Finally, there also appears to be stronger correlations farther to the east when compared to similar distances to the west. These features make sense in terms of the dynamics of monthly SST fields and the El Niño/La Niña phenomenon. They also suggest that traditional spatial methods, such as the correlogram or variogram that rely on stationarity (see Chapter 5) may be misleading when applied to the SST field. This is because the spatial dependence does not appear to be a simple function of distance but must also take into account the direction. Of course any complete analysis of these fields would also consider correlations among different locations and investigate potential change in correlations over time.

It is important to distinguish the correlations among monthly mean SST with the estimated climatological mean field from Figure 29.7. For the mean field there is only one realization and so the dependence is more difficult to quantify. In this case one can use a variogram to explore how dependence in this mean field depends on distance of separation. Some statistical models for the spatial dependence are reviewed in Chapter 5.

Besides cross correlations among the values in a field it is useful to consider the correlations between spatial variables and other distinct climatological variables and at different

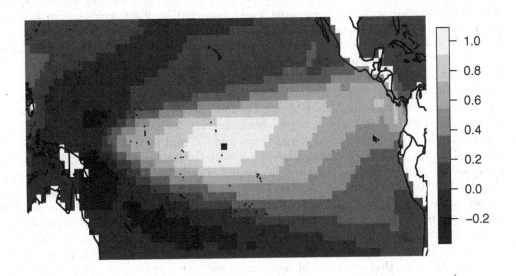

FIGURE 29.7
Sea surface monthly temperature correlations among grid boxes. The image plot indicates
the sample correlation over the period 1970- 2015 and based on monthly mean temperatures
between each grid box in this basin with the central grid box filled in black.

time lags. As an example, one might examine how the SST at a location in the central
Pacific ocean is related to surface air temperature in the Central US six months in advance.
Typically, this kind of analysis might construct a field over the Pacific ocean that indicates
the correlation of each ocean grid cell with the temperature in the Central US. When considered systematically, such analyses suggest predictive relationships that are inherent in the
atmosphere-ocean system. These are termed *teleconnections* to indicate a useful relationship, but at a distance where a direct physical and dynamical explanation is not possible.
An overview of such methods can be found in [73] (Chapter 12) and [16] (Chapter 5).

29.4 Data Products

Spatial statistics has played a significant role in the climate sciences in terms of the creation of *data products*. Historically, this was evident in the use of spatial methods for data
assimilation in terms of optimal prediction theory, specification of spatial dependence, and
state-space formulations of the problem. In addition, purely spatial statistical methods for
large data sets have been useful in recent years to produce complete spatial and spatio-
temporal datasets from irregular observations. We describe some of the methods below.

29.4.1 Data Assimilation

An important use of spatial and spatio-temporal statistical methods in climatology concerns
the construction of complete data products. For example, the creation of so-called *reanalysis*

data sets is a retrospective model-data fusion activity in which complete (in time and space) climate system state variables are created from a combination of available observations and mechanistic models (see Chapter 7). *Analysis* in weather prediction refers to the estimated meteorological fields as the result of combining a forecast with the current observational data. The *re-* refers to applying this process with the same geophysical forecast model throughout the time period of the data product. This is a type of *data assimilation* in which one typically considers very high-dimensional state-space model frameworks (observation model and process model; e.g., see Chapter 4 of this volume) and seeks an optimal estimate of the state of the system, accounting for both observational and model uncertainty (e.g., see [18], [6], [45] for climate-oriented overviews, and [76] and [57] for statistical overviews). Principal challenges in this framework have to do with the realistic characterization of model error and the shear volume of data and huge state dimension. Although there are rigorous statistical treatments of the problem (e.g., [90]), the operational climate centers have taken a more engineering approach to the problem. If the atmospheric model provides an accurate relationship between observed and unobserved variables, then the assimilation process can be used to infer these unobserved states. Thus, a clear benefit from reanalysis data products is the availability of variables on a regular grid that are not directly observed. It is important to realize, however, that estimates of these variables is an extrapolation and interpolation of the observations. If the physical models do not represent some variables well, such as precipitation at the surface, then reanalysis products in these cases can be biased and not accurate.

Perhaps the biggest disadvantage of most reanalysis efforts is that they give no quantification of the uncertainty ([63]) in the estimated fields. Thus, inference for patterns in the reanalysis is difficult. One exception is the recent work on a 20th century reanalysis that provides error estimates based on an ensemble Kalman filter used as the assimilation method [14].

Alternatives to data assimilation are data products that use observations directly and create regular spatial fields. These gridded data products for surface temperature, precipitation, and other variables, can be either global in extent, combining land and ocean observations, or for specific regions. Although many of these efforts use fairly simple averaging schemes to create gridded observations (e.g., HADCRU) others employ more sophisticated geostatistical methods following optimal statistical prediction approaches discussed in Chapter 5 of this volume. The *ClimateDataGuide* [22] is a web-based climate data resource maintained by the National Center for Atmospheric Research that provides expert guidance and an introduction to current climate data sets. In particular, this guide gives summary tables to help compare different products and links to individual sites for downloading. We note that some longstanding and highly cited work in this area are the climate fields from the UK Hadley Center, HadCRUT4 for land temperature and HadSST3 for ocean temperature. Here, no spatial interpolation is done as the estimates are weighted averages of gridded estimates. Uncertainty is quantified, however, through a large ensemble of fields ([56]).

The product from the NASA Goddard Institute of Space Studies (GISS) is also tied to weighted averages of observations within a grid box but is updated regularly and considers homogeneity adjustments to the station data such as urban warming ([37]). NOAA's Merged Land Ocean Surface Temperature Analysis ([80]) is also tied to averaging station data over grid boxes and homogenizes data to reduce biases. The work from the University of Delaware (UDEL) group ([94]) and the Berkeley Earth Surface Temperatures (BEST) ([69]) provide surface temperature analysis that are based on more sophisticated spatial methods with UDEL involving spatial weighting and BEST explicitly fitting a flexible covariance function and pursing optimal interpolation. For monthly and daily temperature and rainfall for the US, the PRISM ([20], [19], http://www.prism.oregonstate.edu) data products provide high resolution fields based on an algorithm adjusting for local terrain effects and spatial

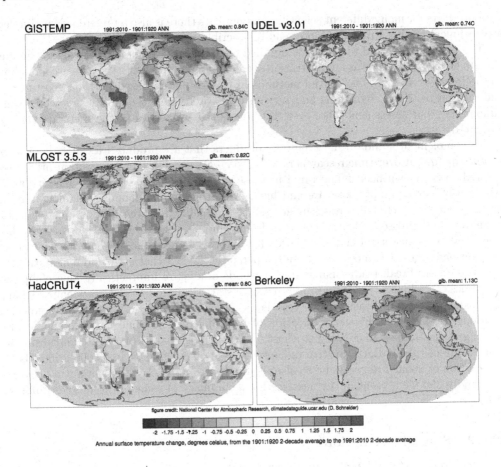

FIGURE 29.8

Comparison of temperature data products from the *Climate Data Guide*. Plotted are the the differences in average annual temperature between two time periods: 1901 -1920 and 1991- 2010.

smoothing. Similar to the problems of reanalysis products, however, many of these gridded data sets do not include rigorous standard errors or other measures of uncertainty, although some measures can be derived by cross-validation and also using climate model simulations. Figure 29.8 is reproduced from the *ClimateDataGuide* web site and gives an interesting comparison across several of these products for temperature. Some of these products mainly report the anomalies of the temperatures with respect to a fixed baseline period. Thus, it makes more sense to compare differences between two periods rather than absolute values at a specific time. Here what is reported is the difference in average annual temperature between two time periods: 1901 -1920 and 1991- 2010.

It may be surprising to a reader from outside climate science that more advanced and sophisticated spatial methods are not brought to bear on creating climate fields. There are several reasons, however, that support this choice. The focus on deriving regional and global temperature series typically finds a lack of sensitivity to the complexity of the spatial methods when climate information is aggregated up to large spatial scales. Also, it is well known that station observational data can have systematic biases of many kinds and often data analysis resources are concentrated on reducing these biases (known as homogenization) rather than pursing more complex statistical methodology. Finally, for widely distributed data products for many different scientific communities there is also a preference for simpler, less technical methods that are more readily described to a general audience.

The reader is referred to [41] for a detailed comparison of different methods for finding climate fields over Europe as an illustration of the different approaches that are used and how methods are evaluated. The spatial methods in this study are based both on geostatistics and also on heuristic algorithms and rely on cross-validation to quantify prediction skill. Similar to other assessments, however, the inferential properties of the methods are not evaluated.

To complement standard data analysis products that are produced for global coverage we anticipate that over time researchers will also create specialized analyses that are better tailored to their own needs. The rapid increase in computational resources for data analysis as well as community software and high level programming languages is now making spatial analysis of climate processes accessible to smaller research groups and students. Moreover the increase in spatial coverage from remotely sensed observations and the homogenization of historical climate records has improved the value of a spatial analysis of observations. These advances have been tempered by the fact that large and non-stationary spatial datasets break traditional spatial methods or require impractically long computation times. To address these issues, approximate statistical models and new computational algorithms have been developed for very large data sets and large numbers of prediction locations. Besides the obvious application of handling observational data we should also note a growing interest in creating stochastic/statistical emulators of climate model simulations. State-of-the-art Earth System models can easily have more than 50,000 grid cells (at a resolution of about 1 degree) and so also require these approaches. This literature is growing rapidly and we present in the next section a summary of some of the more established methods.

29.4.2 Spatial Prediction

As a way to compare different spatial models and methods we abstract the basic problem of estimating a climate field on an arbitrary set of points (typically a regular grid) from observations taken at irregular spatial locations. This is obviously not a new data analysis problem. In this section, however, we take a distinctly statistical approach where the spatial prediction methods also come with measures of prediction uncertainty. We also focus this discussion on methods to handle large spatial data sets as these are becoming the norm in climate analysis.

Following Chapter 5 we have a set of observations $\mathbf{z} \equiv (z(\mathbf{s}_1), \ldots, z(\mathbf{s}_n))'$ with the relationship

$$z(\mathbf{s}_i) = f(\mathbf{s}_i) + e_i, \tag{29.3}$$

where $f(\cdot)$ is the climate field of interest and $\{e_i\}$ reflect measurement errors or other departures from f. Typically these errors are assumed to be Gaussian, mean zero and uncorrelated and independent of f. When f is modeled as a Gaussian process with specific mean $(\mu(\mathbf{s}))$ and covariance function $(c(\mathbf{s}, \mathbf{s}'))$ then there is ample theory and software supporting spatial prediction. The statistical algorithms (e.g., equation 16 from Chapter 5) are dominated by linear algebra applied to the inverse of the covariance matrix for $\mathbf{z}$. For large spatial data sets this standard approach is not practical in terms of computer memory or computational time and alternatives will be presented below.

For a given spatial data set of m observations represented by the vector $\mathbf{z}$, one can show that spatial linear models such as (29.3) and the Kriging models given in Chapter 5 can be written in terms of a "smoother matrix", $\mathbf{S}$, i.e., $\hat{\mathbf{f}} = \mathbf{S}\mathbf{z}$, where $\hat{\mathbf{f}} = (\hat{f}(\mathbf{s}_1), \ldots, \hat{f}(\mathbf{s}_n))'$. The structure of $\mathbf{S}$ is typically such that the data locations that are closer to the prediction locations get more weight than those that are further away. The specific form of $\mathbf{S}$ depends on the nature of the covariance function chosen to represent the true process, f, and depends on the assumptions associated with the elements of (29.3). Thus the form of $\mathbf{S}$ is typically derived from the basic spatial statistics model for the underlying climate field rather than prescribing the weights in $\mathbf{S}$ directly. Conversely, a weighting scheme such as inverse weighted distance does not map to a simple spatial statistics model for f and the measurement error. The reader should also keep in mind that typically, $\mathbf{S}$ will involve several statistical parameters that need to be determined from data and the parameters may enter in a nonlinear form when specifying this smoothing matrix.

The simplest way to reduce the spatial computations are to use neighborhood ideas to limit the size of the problem. To this end there are many versions of local Kriging that essentially follow a spatial analysis for a window of spatial locations and then repeat this across the whole spatial domain ([33], [98], and [34]). A difficulty with these strategies is to combine spatial predictions across different windows and to estimate statistical parameters that may be the same across larger areas of the domain. We note, however, that such an approach is easy to parallelize and so can scale to large computing platforms. A related approach is to taper the covariance matrix itself, inducing many zero covariance values and then use sparse matrix methods for the statistical computations. Tapering has the advantage that it works with the entire spatial data and so finesses the issues of how to merge different local predictions. This suggests a simpler algorithm where sparse decompositions are just substituted for the dense ones. Both of these local approaches can be justified through the *screening effect* for spatial prediction. Predictions based on a large enough local neighborhood are not improved by adding more distant observations. The reason is that *conditional* on the local observations these additional observations do not add much new information.

Another approach to spatial prediction relies on an explicit representation of the field using basis functions. Following the presentation in Chapter 5, it is useful to represent a spatial process $\{f(\mathbf{s}) : \mathbf{s} \in D\}$ in terms of a generic expansion in basis functions:

$$f(\mathbf{s}) = \mu(\boldsymbol{s}) + \sum_{j=1}^{n_\alpha} \phi_j(\mathbf{s})\alpha_j + \eta(\boldsymbol{s}), \tag{29.4}$$

where $\mu(\boldsymbol{s})$ is a fixed and low dimensional function that depends on the location (e.g. a linear model or regression function), the $\{\phi_j(\mathbf{s})\}$ are basis functions, and $\{\alpha_j\}$ are random effects or equivalently random coefficients (see [84] for a comprehensive discussion). The final term, $\eta(\boldsymbol{s})$, is included as a localized spatial process. In some cases if the process described by the basis functions is rich enough that $\eta(\boldsymbol{s})$ can be set to zero. In general, for the three terms (mean, basis expansion, and residual term) in (29.4) to be identified, one must have information about the mean structure (or, covariates that help explain that mean), and the number and type of basis functions to include in the expansion. Care needs to be taken in formulating and interpreting this model if the mean function has similar spatial components to the basis functions. If this is the case then the contribution is confounded between these two parts of the model.

We also assume that the $\{\alpha_j\}$ are multivariate Gaussian, with a mean of zero and covariance matrix $\mathbf{C}_\alpha$. Note that this form can approximate the KL representation (see Section 29.2.3.1) by letting the sum tend to infinity, taking the basis functions as the eigenfunctions, and setting $\mathbf{C}_\alpha$ as a diagonal matrix consisting of the eigenvalues. However, in practice the basis functions are selected for computational efficiency and $\mathbf{C}_\alpha$ need not be diagonal. It is useful to represent this model in vector/matrix notation. Let $\boldsymbol{\Phi}$ be the matrix with elements $\boldsymbol{\Phi}_{i,j} = \phi_j(\boldsymbol{s}_i)$, and we define the following vectors in terms of their spatially-referenced elements, $\boldsymbol{e}_i = e(\boldsymbol{s}_i)$, $\boldsymbol{\eta}_i = \eta(\boldsymbol{s}_i)$, and $\boldsymbol{\mu}_i = \mu(\boldsymbol{s}_i)$. The observation model can be rewritten as

$$z = \mu + \Phi\alpha + \eta + e. \tag{29.5}$$

Depending on the assumptions one makes concerning the number of basis functions (n_α), the type of basis function, the distribution of the random effects, and the presence and/or nature of the error process, one can gain significant computational advantages when it comes to spatial prediction. The following is a list of some of the more effective methods and the associated assumptions relative to the models in (29.3) and (29.4).

- *Special Function Models:* It is traditional in the climate sciences to consider basis expansions based on known functions, particularly Fourier basis functions and certain classes of "special functions" (e.g., spherical harmonics, Hermite polynomials, etc.). The reason for this preference is because of the strong connection between such representations and the underlying solutions to simplified versions of the governing Navier-Stokes partial differential equations that control the weather and climate system. This also facilitates inference on many geophysical processes because of the inherent multi-scale nature of these basis functions. For example, it is common to think of the spatial scales associated with Fourier modes when investigating energy transfer and turbulent cascades.

  From a purely spatial prediction perspective, this basis function expansion can also be quite effective. For example, for predictions on a specified regular grid, one can use a complete Fourier expansion and exploit the computational efficiency of multi-resolution (discrete Fourier transform) algorithms as well as the inherent decorrelation induced on the random coefficients. That is, it is well known that for stationary spatial processes, if Φ consists of Fourier basis functions, then the elements of α are asymptotically independent, leading to a diagonal covariance specification on these random effects. In the cases where the covariance function is known (e.g., a Matèrn covariance function), one can specify the functional form of these variances analytically, conditioned on the parameters that control the covariance function (see [83] , [91] , [62]). In this case one considers a complete set of Fourier basis functions ($n = n_\alpha$) and $\eta \equiv 0$, and $\alpha \sim Gau(0, diag(\mathbf{d}))$. In cases where there is no theoretical form for the random effects dependence structure, one can still assume sparsity to facilitate computation (e.g. [59], who use multi-resolution wavelet bases). In other cases, one can use scientifically-motivated special functions in a low-rank specification as discussed below; e.g. see [90] , who use the theoretical basis functions associated with the shallow-water equations and β plane solution (see also Chapter 33).

- *Low-Rank Models:* A low-rank prediction model assumes an under-complete set of basis functions such that $n_\alpha \ll n$ in (29.4), with η accounting for left-over variation from the rank reduction. A classic example of a low-rank model in climatology is one based on the KL/EOF representation described previously (see Section 29.2.3.1). In that case, Φ corresponds to the first n_α spatial EOF basis functions, with the dependence structure in the η error term again being diagonal (e.g, see [13], [61], [8]). Such decompositions are often considered in spatio-temporal dynamical model applications, achieving significant dimension reduction (see [87]). As discussed below and in [84], other basis functions may be more appropriate depending on the problem at hand.

  In both the spatial and spatio-temporal case, the computational advantage of such methods typically comes from application of the Sherman-Morrison-Woodbury (SMW) matrix identity.[1] Because of the importance of this identity it is useful to give the details of its

---

[1]The SMW identity in its most basic form is just $(I_m + A'A)^{-1} = I_m - A'(I_n + AA')^{-1}A$ with A an $m \times n$ matrix and I_p the identity matrix of dimension p. This can be verified by simple linear algebra. The more general expression follows by rearrangement using the symmetric matrix square roots of $\mathbf{C}_\alpha$ and $\mathbf{V}$.

role. The implied smoother matrix $\mathbf{S}$ for prediction at the observation locations has the form:

$$\mathbf{S} = \mathbf{\Phi}(\mathbf{\Phi}\mathbf{C}_\alpha\mathbf{\Phi}' + \mathbf{V})^{-1}(\mathbf{\Phi}\mathbf{C}_\alpha\mathbf{\Phi}'),$$

where $\mathbf{V}$ is a matrix that contains the variance/covariance matrix of the residual process η plus the measurement error variance covariance matrix. (That is, $\mathbf{V}$ is the covariance matrix of $\eta + e$). When n is reasonably large, the inverse and determinant of $(\mathbf{\Phi}\mathbf{C}_\alpha\mathbf{\Phi}' + \mathbf{V})$ are computationally expensive to evaluate; the number of operations grows as $O(n^3)$. However, the SMW formula allows one to write this equivalently as

$$(\mathbf{\Phi}\mathbf{C}_\alpha\mathbf{\Phi}' + \mathbf{V})^{-1} = \mathbf{V}^{-1} - \mathbf{V}^{-1}\mathbf{\Phi}(\mathbf{\Phi}'\mathbf{V}^{-1}\mathbf{\Phi} + \mathbf{C}_\alpha^{-1})^{-1}\mathbf{\Phi}'\mathbf{V}^{-1}. \tag{29.6}$$

Thus, when $n_\alpha \ll n$ and $\mathbf{V}^{-1}$ is readily available, one has shifted the computational burden to an n_α-dimensional inverse which is much more tractable. The choice of structure for $\mathbf{\Phi}$, $\mathbf{V}$, and $\mathbf{C}_\alpha$ lead to different implementation strategies.

In principle, almost any basis set can be used in the reduced-rank setting. A popular approach that uses bisquare basis functions is the so-called "Fixed Rank Kriging" method ([15]). In addition, the so-called "predictive process" class of models ([66], [2], [27]) and "discrete kernel convolution models" (e.g., [3], [26]) can be thought of as reduced-rank models, but where the basis functions are induced by a specified covariance function on a discrete set of locations and a kernel function, respectively. In each of these approaches, the elements of the random effects vector (α) are spatially indexed (unlike the EOF and special function expansions). In order for reduced rank spatial methods to adequately accommodate small-scale spatial dependence, one typically should account for dependence in the residual process, η (e.g., [87], [7], and [71]).

- *Over-Complete Models:* Over-complete systems can be written as basis expansions where $n_\alpha > n$. This strategy assumes an efficient form for $\mathbf{C}_\alpha^{-1}$ (e.g., a sparse matrix) and uses the SMW formula to work with this matrix directly. Typically, these approaches make use of computationally favorable basis specifications. A popular and effective example is the stochastic partial differential equation (SPDE) approach of [52], that uses compactly supported basis functions implied by triangularizations of the spatial domain, and models the precision matrix $\mathbf{C}_\alpha^{-1}$ in terms of a Markov random field specification within a Laplace approximation framework. The LatticeKrig method of [58], uses a compactly supported, multi-resolution basis and considers dependence of the random effects through a Markov random field model.

- *Neighbor-Based Models:* Accepting the common situation that the spatial weight matrix $\mathbf{S}$ has effectively compact support in most cases, one can build spatial prediction models to accommodate such behavior explicitly. One such way is through the use of tapering, where one effectively truncates a spatial covariance function according to distance (e.g. [30] and [48]). Alternatively, one can consider an approximation of the likelihood by a product of conditional distributions that are given by the process at reference locations conditioned on relatively small neighbor sets (e.g., a composite likelihood for spatial data). This again implies a sparse precision matrix and leads to efficient calculation (e.g. [77], [75], [76], [21]).

29.4.3   Inference for spatial fields

The previous section focused on the models and estimates of climate fields based on spatial data. One advantage of pursuing a spatial analysis within a statistical framework is the opportunity to also generate measures of uncertainty for the spatial estimates. In this section we review some techniques for drawing inferences from predictions of a spatial field.

It is standard in geostatistics to generate prediction confidence intervals based on Kriging. In the simplest case these are based on the conditional variance formula displayed in Chapter 5 and assuming that the prediction error is mean zero and approximately normally distributed. Both of these assumptions should be scrutinized and a useful technique is cross-validation to assess the validity of the prediction intervals. For example, one would omit say 10% of the spatial data, fit the spatial model to the remaining 90% of the sample, and use this fit to predict the omitted portion. The degree to which the prediction intervals contain the omitted observations at the correct level of confidence is a useful way to judge the value of the intervals. It should be noted that in working with station and instrument data systematic biases can be present and these might be diagnosed as systematic errors in out-of-sample prediction.

This basic strategy for quantifying prediction uncertainty needs to be modified for large data sets. Exact prediction standard errors are computationally intensive to find and not feasible for a large number of spatial locations. An alternative is to use conditional simulation of the prediction error fields as a Monte Carlo method to evaluate the conditional variances. (see [58] for an example).

The standard frequentist approach for prediction uncertainty can be extended to include the uncertainty in the statistical model itself. For example, covariance and regression parameters are usually estimated and including their uncertainty provide a more complete view of the uncertainty in the prediction. In particular, prediction standard errors can be sensitive to the relative variances of the measurement error and the process marginal variance. An elegant solution is to use a Bayesian model for the spatial prediction that will naturally incorporate uncertainty from all the aspects of the model that are not known. The foundational work by M. Handcock and M. Stein [35] is a pioneering accomplishment before the advent of Markov chain Monte Carlo (MCMC) computational technology. Subsequent developments of Bayesian spatial methods have leveraged the more general research and computational algorithms for Bayesian hierarchical models (e.g., see the reviews in [1] and [16] (Chapter 4). Moreover, MCMC algorithms coupled with flexible models for the process and observations have resulted in practical software packages in R such as spBayes ([26]). These, and other approximate methods, remain an active area of research. Examples of such approximations include the use of likelihood approximations (e.g. [73], [95]), Laplace approximations (e.g. [70], [17], [52]), ensemble Kalman filters (e.g. [57], [47]), variational Bayesian methods [68], and nearest neighbor ideas [21].

For climate analysis and field comparisons one often wants to interpret the estimated field in terms of its shape, contours, or other features that include more than a single point. One useful kind of description is the exceedance region. For example, for a given threshold u one might want to estimate the exceedance set $E^+ = \{s : f(s) > u\}$. Inference for E^+ is more complicated than just using point-wise prediction intervals because one must take into account the dependence among spatial predictions. Some representative work in this area includes [97], [28], [74], and [7]. Here we summarize one approach used by J. French and S. Sain [29] to emphasize the distinction from point-wise intervals. Given an estimate for the field at s, $\hat{g}(s)$, consider the upper exceedance set:

$$A^+ = \{s : \frac{\hat{f}(s) - u}{\sigma(u)} > C\}, \tag{29.7}$$

where $\sigma(u)$ is an estimate of the point-wise prediction standard error and C is a constant to be determined. To obtain a $(1-\alpha)$ level frequentist inference one needs to choose C so that probability that E^+ is a subset of A^+ is $1-\alpha$. French and Sain propose to approximate the value for C by a Monte Carlo calculation applied to a fine grid of spatial prediction points. Briefly, one simulates a Gaussian random field f on a fine grid and from this generates synthetic data, based on the observation locations and measurement error. One computes the spatial prediction for this simulated case and finds the largest value for C, say C^*, such that $E \subseteq A^+$. This is possible because as a simulated field one has the the "true" f from the initial step. This algorithm is repeated to determine a distribution for C^* and the $(1 - \alpha)$ quantile of this distribution is now used to construct the confidence region for the original data, i.e. the set A^+.

This procedure can modified to consider the exceedance region *below* a threshold, $E^- = \{s : f(s) \leq u\}$. The result is a confidence set A^- such that $E^- \subseteq A^-$ at a specified probability. An important application of this lower bound is to use the *complement* of the lower bound confidence set to inform the upper exceedance region. Specifically, suppose that A^- is an α confidence set for E^-. Then we have that the complement of A^- is contained in E^+ with probability $(1 - \alpha)$. The net result is the sequence: $(A^-)^c \subseteq E^+ \subseteq A^+$ where both inclusions have $(1 - \alpha)$ confidence. Note that that this inclusion can bracket the "u" contour line of f if the coverage is adjusted to be simultaneous for both the lower and upper confidence sets. Figure 29.9 reproduces the exceedance inference from French and Sain's analysis ([29]) of a monthly precipitation field over Oregon (October 1998) and with u taken at $\sqrt{250}mm$. The confidence level is .9 and in this case we see relatively large uncertainty when comparing the nesting confidence sets to the total area of this region.

29.4.3.1   Spatial and Spatio-Temporal Field Comparison

It is common in the climate sciences to compare a spatial or spatio-temporal summary of model output or observations to another summary of observations or results from another model. Given this has also been of interest in numerical weather forecasting, many methods have been developed for more than 50 years. Here, we give a very brief summary of some of these methods used in the related areas of map comparison "significance" and prediction verification. This is by no means an exhaustive list, and we refer the interested reader to the excellent summaries in [73] (Chapter 18), [80] (Chapter 7), [16] (Section 5.7), and [43].

One can compare continuous response spatial or spatio-temporal fields that reside on a finite spatial domain (e.g. grid) with a simple summary measure such as *root mean squared error* (RMSE), which is just the square root of the average of point-wise squared differences. Similarly, if the magnitude of the fields is less important than the pattern, one can consider the *anomaly correlation coeffcient*, which is the average of the correlations at each grid point between the two fields relative to some common reference field (e.g. climatology). These single measures are useful but limited in the sense that the anomaly correlation coefficient takes into account linear association between the fields but not the bias, whereas the MSPE is a good measure of overall grid point accuracy, but doesn't take into account linear association.

More complicated measures to compare maps have been used in the climatological sciences. For example, [64] consider a decomposition of the Frobenius norm between two spatio-temporal fields (organized into matrices) in terms of the difference between the centroids of the fields, their variation, and patterns of evolution (see the review in [82]). One can also consider differences relative to spatial (and temporal) scales, e.g., in terms of EOFs or wavelet transforms ([9], [10] , and [31]). In the case where one is actually interested in a measure of "significance" between maps, the usual issues related to multiple testing must be addressed. That is, as the number of hypothesis tests that are considered simultaneously increases, the false positive or Type I error level (the error associated with incorrectly rejecting the null hypothesis if the null hypothesis is, in fact, true) is actually less (and potentially, much less) than the stated Type I error level. This problem also arises when

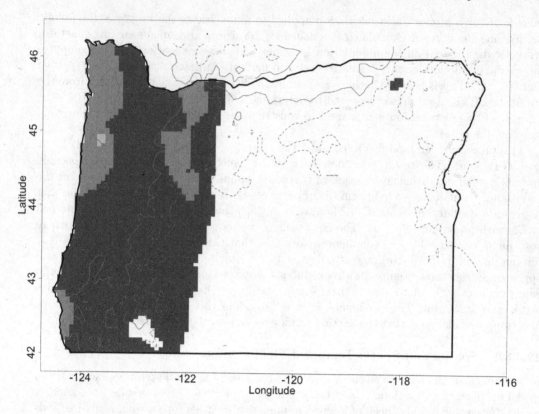

FIGURE 29.9

Exceedance regions for total monthly precipitation over Oregon. The 90 percent upper (blue) and lower (orange) exceedance sets based on the observations from October 1998 with the exceedance level set at 250 mm. For reference, the yellow intermediate region indicates where the spatial predicted values (i.e. the conditional mean) exceed this threshold.

considering multiple confidence intervals. Traditionally, one can use standard multiplicity corrections (e.g., Bonferroni) if the field sizes (i.e., number of tests or simultaneous intervals) are relatively small. Otherwise, to account for the higher number of comparisons typically seen in spatial map comparison problems, one can use Monte Carlo permutation procedures (see [25] as an example), or false discovery rate procedures (e.g., [72], see also the review in [93]). Note that rigorous inference for comparing fields is difficult because of the inherent spatial dependence commonly found in climate fields and so naive application of testing methods that assume independence are not valid.

In addition to methods that can compare across space, time, and scale, one needs to consider the issues associated with phase and orientation of "objects" (e.g. "features" such as fronts, storm cells, standing wave location, etc.). That is, even if these objects or features don't match on a grid point basis, the fact that they are in some sense "close" may be important. Examples that take into account such morphometric characteristics include [40], [23], and [54], see also the overview in [12]. These methods are related to the notion of developing "optimal fingerprints" to study the likely spatial signatures of climate change compared to internal climate variability (e.g. [38], [50], [5], [51], [36], and see also Chapter 35 of this volume).

In the context of probabilistic prediction comparisons, one's prediction is in the form of a probability distribution. Often for large models this distribution is expressed in an approximate form by a sample of fields, an ensemble, that represent a random sample from the probabilistic forecast distribution. In either case the interest is in evaluating the quality of that predictive distribution given some realization of a reference (e.g., true process) distribution. Such comparisons are best accomplished formally through the use of *scoring rules*, which assign scores to predictions that in some sense "reward" good forecasts and punish poor ones. These rules can either be *positively-oriented* or *negatively-oriented*, in which case they assign higher or lower scores to better predictions, respectively. Typically, to avoid potential inconsistencies, such scoring rules should be *strictly proper* (e.g., see [32], and the summary in [11]). In essence, a proper scoring rule maximizes (for positively-oriented scores) the expected score (i.e., expected value of the scoring rule over the distribution of potential realizations) relatively to any other predictive distribution. Such rules have the nice property that they make it difficult to "cheat" with prediction in the sense that moving away from the predictive distribution (the "true beliefs") is penalized.

29.4.4 Spatio-Temporal Prediction

Most climatological problems are inherently spatio-temporal, and the creation of data products outside the standard operational data assimilation framework require statistical models similar in spirit to the spatial models described in the previous section. That is, they typically rely on a latent random effects representation in which the random effects vary in time. A crucial difference in the spatio-temporal case, necessitated by the curse of dimensionality and justified by the etiology of the scientific processes, is that models for the associated random effects are dynamic in the sense that the random effects evolve through time. Low-rank models work well in this framework since much of the underlying dynamical variability resides on lower-dimensional manifolds. However, rather than just consider time-varying parameters, one has to make sure that parameterizations of the dynamic models in these settings can actually accommodate realistic process behavior (e.g., advection, diffusion, growth, decay, boundaries, etc.). This suggests that both linear and nonlinear models must be considered (e.g., [77]).

The "big data" nature of the spatio-temporal climatological prediction necessitates computational considerations. As summarized in [16] (Chapter 7) and [74], the biggest challenge for spatio-temporal modeling in environmental applications is accommodating the curse of dimensionality in terms of the process dimension and the number of parameters that must be estimated. As mentioned, the former is typically accounted for by reduced-rank representations and the latter is often accounted for through hierarchical Bayesian specifications. In particular, one can reduce the number of parameters in these models by hard-thresholding based on scientific knowledge. Some examples are banded but asymmetric structure in transition matrices to accommodate advection and diffusion, emulator-assisted prior specification (e.g., [38]), specification of parameters as processes at lower levels of model hierarchies, or regularization-based parameter estimation (e.g., [77]). Other types of approximations can be considered in the spatio-temporal case similar to the examples listed in Section 29.4.3. Additional discussion of spatio-temporal modeling relevant to climatology and applications such as long lead forecasting are included in the Oceanography chapter in this volume (Chapter 33).

29.5 Conclusion

In this chapter we have outlined some of the traditional and current uses of spatial and spatio-temporal statistical methods in climatology. The selected topics are not at all exhaustive, and other important uses of spatial and spatio-temporal statistics in climatology are included in this volume: climate change in Chapter 28, spatial extremes in Chapter 32, oceanography in Chapter 33, paleoclimate in Chapter 34, detection and attribution in Chapter 35, and health effects in Chapter 36. As stated in the introduction, the ubiquity of spatial and spatio-temporal methods in climate science is not surprising given that the climate system consists of the complex interactions of many different spatio-temporal processes across a wide variety of time and spatial scales.

29.6 Acknowledgments

Nychka was supported by the National Center for Atmospheric Research (NCAR) and NSF Grant 141785. NCAR is managed by the University Corporation for Atmospheric Research under the sponsorship of the NSF. Wikle was partially supported by the US National Science Foundation (NSF) and the U.S. Census Bureau under NSF grant SES-1132031, funded through the NSF-Census Research Network (NCRN) program.

Bibliography

[1] Sudipto Banerjee, Bradley P Carlin, and Alan E Gelfand. *Hierarchical modeling and analysis for spatial data*. Crc Press, 2014.

[2] Sudipto Banerjee, Alan E Gelfand, Andrew O Finley, and Huiyan Sang. Gaussian predictive process models for large spatial data sets. *Journal of the Royal Statistical Society: Series B (Statistical Methodology)*, 70(4):825–848, 2008.

[3] Ronald Paul Barry, M Jay, and Ver Hoef. Blackbox kriging: spatial prediction without specifying variogram models. *Journal of Agricultural, Biological, and Environmental Statistics*, pages 297–322, 1996.

[4] Andrew F Bennett. *Inverse modeling of the ocean and atmosphere*. Cambridge University Press, 2005.

[5] L Mark Berliner, Richard A Levine, and Dennis J Shea. Bayesian climate change assessment. *Journal of Climate*, 13(21):3805–3820, 2000.

[6] L Mark Berliner, Christopher K Wikle, and Noel Cressie. Long-lead prediction of pacific ssts via bayesian dynamic modeling. *Journal of climate*, 13(22):3953–3968, 2000.

[7] David Bolin and Finn Lindgren. Excursion and contour uncertainty regions for latent gaussian models. *Journal of the Royal Statistical Society: Series B (Statistical Methodology)*, 77(1):85–106, 2015.

[8] Jonathan R Bradley, Christopher K Wikle, and Scott H Holan. Regionalization of multiscale spatial processes by using a criterion for spatial aggregation error. *Journal of the Royal Statistical Society: Series B (Statistical Methodology)*, 2016.

[9] Grant Branstator, Andrew Mai, and David Baumhefner. Identification of highly predictable flow elements for spatial filtering of medium-and extended-range numerical forecasts. *Monthly weather review*, 121(6):1786–1802, 1993.

[10] William M Briggs and Richard A Levine. Wavelets and field forecast verification. *Monthly Weather Review*, 125(6):1329–1341, 1997.

[11] Jochen Broecker. Probability forecasts. *Forecast Verification: A Practitioner's Guide in Atmospheric Science, Second Edition*, pages 119–139, 2012.

[12] Barbara G Brown, Eric Gilleland, and Elizabeth E Ebert. Forecasts of spatial fields. *Forecast Verification: A Practitioner's Guide in Atmospheric Science, Second Edition*, pages 95–117, 2012.

[13] Ayala Cohen and Richard H Jones. Regression on a random field. *Journal of the American Statistical Association*, 64(328):1172–1182, 1969.

[14] Gilbert P Compo, Jeffrey S Whitaker, Prashant D Sardeshmukh, Nobuki Matsui, Robert J Allan, Xungang Yin, Byron E Gleason, Russell S Vose, Glenn Rutledge, Pierre Bessemoulin, et al. The twentieth century reanalysis project. *Quarterly Journal of the Royal Meteorological Society*, 137(654):1–28, 2011.

[15] Noel Cressie and Gardar Johannesson. Fixed rank kriging for very large spatial data sets. *Journal of the Royal Statistical Society: Series B (Statistical Methodology)*, 70(1):209–226, 2008.

[16] Noel Cressie and Christopher K Wikle. *Statistics for spatio-temporal data*. John Wiley & Sons, 2015.

[17] Botond Cseke and Tom Heskes. Approximate marginals in latent gaussian models. *Journal of Machine Learning Research*, 12(Feb):417–454, 2011.

[18] Roger Daley. Atmospheric data analysis, cambridge atmospheric and space science series. *Cambridge University Press*, 6966:25, 1991.

[19] Christopher Daly, Michael Halbleib, Joseph I Smith, Wayne P Gibson, Matthew K Doggett, George H Taylor, Jan Curtis, and Phillip P Pasteris. Physiographically sensitive mapping of climatological temperature and precipitation across the conterminous united states. *International journal of climatology*, 28(15):2031–2064, 2008.

[20] Christopher Daly, Ronald P Neilson, and Donald L Phillips. A statistical-topographic model for mapping climatological precipitation over mountainous terrain. *Journal of applied meteorology*, 33(2):140–158, 1994.

[21] Abhirup Datta, Sudipto Banerjee, Andrew O Finley, and Alan E Gelfand. Hierarchical nearest-neighbor gaussian process models for large geostatistical datasets. *Journal of the American Statistical Association*, 111(514):800–812, 2016.

[22] Clara Deser, A Dai, J Fasullo, J Hurrell, K Trenberth, D Shea, and D Schneider. Climatedataguide. https://climatedataguide.ucar.edu. Accessed: 2017-02-16.

[23] EE Ebert and JL McBride. Verification of precipitation in weather systems: Determination of systematic errors. *Journal of Hydrology*, 239(1):179–202, 2000.

[24] Jason P Evans, Fei Ji, Gab Abramowitz, and Marie Ekström. Optimally choosing small ensemble members to produce robust climate simulations. *Environmental Research Letters*, 8(4):044050, 2013.

[25] Johannes J Feddema, Keith W Oleson, Gordon B Bonan, Linda O Mearns, Lawrence E Buja, Gerald A Meehl, and Warren M Washington. The importance of land-cover change in simulating future climates. *Science*, 310(5754):1674–1678, 2005.

[26] Andrew O Finley, Sudipto Banerjee, and Bradley P Carlin. spbayes: an r package for univariate and multivariate hierarchical point-referenced spatial models. *Journal of Statistical Software*, 19(4):1, 2007.

[27] Andrew O Finley, Huiyan Sang, Sudipto Banerjee, and Alan E Gelfand. Improving the performance of predictive process modeling for large datasets. *Computational statistics & data analysis*, 53(8):2873–2884, 2009.

[28] Joshua P French. Confidence regions for the level curves of spatial data. *Environmetrics*, 25(7):498–512, 2014.

[29] Joshua P French, Stephan R Sain, et al. Spatio-temporal exceedance locations and confidence regions. *The Annals of Applied Statistics*, 7(3):1421–1449, 2013.

[30] Reinhard Furrer, Marc G Genton, and Douglas Nychka. Covariance tapering for interpolation of large spatial datasets. *Journal of Computational and Graphical Statistics*, 15(3):502–523, 2006.

[31] Eric Gilleland, David Ahijevych, Barbara G Brown, Barbara Casati, and Elizabeth E Ebert. Intercomparison of spatial forecast verification methods. *Weather and forecasting*, 24(5):1416–1430, 2009.

[32] Tilmann Gneiting and Adrian E Raftery. Strictly proper scoring rules, prediction, and estimation. *Journal of the American Statistical Association*, 102(477):359–378, 2007.

[33] Timothy C Haas. Kriging and automated variogram modeling within a moving window. *Atmospheric Environment. Part A. General Topics*, 24(7):1759–1769, 1990.

[34] Dorit M Hammerling, Anna M Michalak, and S Randolph Kawa. Mapping of co2 at high spatiotemporal resolution using satellite observations: Global distributions from oco-2. *Journal of Geophysical Research: Atmospheres*, 117(D6), 2012.

[35] Mark S Handcock and Michael L Stein. A bayesian analysis of kriging. *Technometrics*, 35(4):403–410, 1993.

[36] A Hannart, J Pearl, FEL Otto, P Naveau, and M Ghil. Causal counterfactual theory for the attribution of weather and climate-related events. *Bulletin of the American Meteorological Society*, 97(1):99–110, 2016.

[37] James Hansen, Reto Ruedy, Mki Sato, and Ken Lo. Global surface temperature change. *Reviews of Geophysics*, 48(4), 2010.

[38] Klaus Hasselmann. On the signal-to-noise problem in atmospheric response studies. *Meteorology of tropical oceans*, pages 251–259, 1979.

[39] David Higdon. A process-convolution approach to modelling temperatures in the north atlantic ocean. *Environmental and Ecological Statistics*, 5(2):173–190, 1998.

[40] Ross N Hoffman, Zheng Liu, Jean-Francois Louis, and Christopher Grassoti. Distortion representation of forecast errors. *Monthly Weather Review*, 123(9):2758–2770, 1995.

[41] Nynke Hofstra, Malcolm Haylock, Mark New, Phil Jones, and Christoph Frei. Comparison of six methods for the interpolation of daily, european climate data. *Journal of Geophysical Research: Atmospheres*, 113(D21), 2008.

[42] Ian Jolliffe. *Principal component analysis*. Wiley Online Library, 2002.

[43] Ian T Jolliffe and David B Stephenson. *Forecast verification: a practitioner's guide in atmospheric science*. John Wiley & Sons, 2012.

[44] Richard H Jones. Stochastic processes on a sphere. *The Annals of mathematical statistics*, 34(1):213–218, 1963.

[45] Eugenia Kalnay. *Atmospheric modeling, data assimilation and predictability*. Cambridge university press, 2003.

[46] Ioanna Karagali, Merete Badger, Andrea N Hahmann, Alfredo Peña, Charlotte B Hasager, and Anna Maria Semprevia. Spatial and temporal variability of winds in the northern european seas. *Renewable energy*, 57:200–210, 2013.

[47] Matthias Katzfuss, Jonathan R Stroud, and Christopher K Wikle. Understanding the ensemble kalman filter. *The American Statistician*, 70(4):350–357, 2016.

[48] Cari G Kaufman, Mark J Schervish, and Douglas W Nychka. Covariance tapering for likelihood-based estimation in large spatial data sets. *Journal of the American Statistical Association*, 103(484):1545–1555, 2008.

[49] WB Leeds, CK Wikle, J Fiechter, J Brown, and RF Milliff. Modeling 3-d spatio-temporal biogeochemical processes with a forest of 1-d statistical emulators. *Environmetrics*, 24(1):1–12, 2013.

[50] Stephen S Leroy. Detecting climate signals: Some bayesian aspects. *Journal of Climate*, 11(4):640–651, 1998.

[51] Nicholas Lewis. An objective bayesian improved approach for applying optimal fingerprint techniques to estimate climate sensitivity. *Journal of Climate*, 26(19):7414–7429, 2013.

[52] Finn Lindgren, Håvard Rue, and Johan Lindström. An explicit link between gaussian fields and gaussian markov random fields: the stochastic partial differential equation approach. *Journal of the Royal Statistical Society: Series B (Statistical Methodology)*, 73(4):423–498, 2011.

[53] Roland A Madden and Paul R Julian. Detection of a 40–50 day oscillation in the zonal wind in the tropical pacific. *Journal of the atmospheric sciences*, 28(5):702–708, 1971.

[54] Athanasios Christou Micheas, Neil I Fox, Steven A Lack, and Christopher K Wikle. Cell identification and verification of qpf ensembles using shape analysis techniques. *Journal of Hydrology*, 343(3):105–116, 2007.

[55] Adam H Monahan, John C Fyfe, Maarten HP Ambaum, David B Stephenson, and Gerald R North. Empirical orthogonal functions: The medium is the message. *Journal of Climate*, 22(24):6501–6514, 2009.

[56] Colin P Morice, John J Kennedy, Nick A Rayner, and Phil D Jones. Quantifying uncertainties in global and regional temperature change using an ensemble of observational estimates: The hadcrut4 data set. *Journal of Geophysical Research: Atmospheres*, 117(D8), 2012.

[57] Douglas Nychka and Jeffrey L Anderson. Data assimilation. *Handbook of Spatial Statistics, edited by: Gelfand, A., Diggle, P., Guttorp, P., and Fuentes, M., Chapman & Hall/CRC, New York*, 2010.

[58] Douglas Nychka, Soutir Bandyopadhyay, Dorit Hammerling, Finn Lindgren, and Stephan Sain. A multiresolution gaussian process model for the analysis of large spatial datasets. *Journal of Computational and Graphical Statistics*, 24(2):579–599, 2015.

[59] Douglas Nychka, Christopher Wikle, and J Andrew Royle. Multiresolution models for nonstationary spatial covariance functions. *Statistical Modelling*, 2(4):315–331, 2002.

[60] Ch Obled and JD Creutin. Some developments in the use of empirical orthogonal functions for mapping meteorological fields. *Journal of Climate and Applied meteorology*, 25(9):1189–1204, 1986.

[61] Ch Obled and JD Creutin. Some developments in the use of empirical orthogonal functions for mapping meteorological fields. *Journal of Climate and Applied meteorology*, 25(9):1189–1204, 1986.

[62] Christopher J Paciorek. Bayesian smoothing with gaussian processes using fourier basis functions in the spectralgp package. *Journal of statistical software*, 19(2):nihpa22751, 2007.

[63] Wendy S Parker. Reanalyses and observations: Whats the difference? *Bulletin of the American Meteorological Society*, 97(9):1565–1572, 2016.

[64] Rudolph W Preisendorfer and Tim P Barnett. Numerical model-reality intercomparison tests using small-sample statistics. *Journal of the Atmospheric Sciences*, 40(8):1884–1896, 1983.

[65] Rudolph W Preisendorfer and Curtis D Mobley. *Principal component analysis in meteorology and oceanography*, volume 17. Elsevier Science Ltd, 1988.

[66] Joaquin Quiñonero-Candela and Carl Edward Rasmussen. A unifying view of sparse approximate gaussian process regression. *Journal of Machine Learning Research*, 6(Dec):1939–1959, 2005.

[67] David A Randall. Spherical harmonics and related topics, 2014.

[68] Qian Ren, Sudipto Banerjee, Andrew O Finley, and James S Hodges. Variational bayesian methods for spatial data analysis. *Computational Statistics & Data Analysis*, 55(12):3197–3217, 2011.

[69] Robert Rohde, Richard A Muller, Robert Jacobsen, Elizabeth Muller, Saul Perlmutter, Arthur Rosenfeld, Jonathan Wurtele, Donald Groom, and Charlotte Wickham. A new estimate of the average earth surface land temperature spanning 1753 to 2011. *Geoinfor Geostat: An Overview*, 1(1):1–7, 2012.

[70] Håvard Rue, Sara Martino, and Nicolas Chopin. Approximate bayesian inference for latent gaussian models by using integrated nested laplace approximations. *Journal of the royal statistical society: Series b (statistical methodology)*, 71(2):319–392, 2009.

[71] Huiyan Sang and Jianhua Z Huang. A full scale approximation of covariance functions for large spatial data sets. *Journal of the Royal Statistical Society: Series B (Statistical Methodology)*, 74(1):111–132, 2012.

[72] Xiaotong Shen, Hsin-Cheng Huang, and Noel Cressie. Nonparametric hypothesis testing for a spatial signal. *Journal of the American Statistical Association*, 97(460):1122–1140, 2002.

[73] Scott A Sisson, Yanan Fan, and Mark M Tanaka. Sequential monte carlo without likelihoods. *Proceedings of the National Academy of Sciences*, 104(6):1760–1765, 2007.

[74] Max Sommerfeld, Stephen Sain, and Armin Schwartzman. Confidence regions for excursion sets in asymptotically gaussian random fields, with an application to climate. *arXiv preprint arXiv:1501.07000*, 2015.

[75] Michael L Stein, Zhiyi Chi, and Leah J Welty. Approximating likelihoods for large spatial data sets. *Journal of the Royal Statistical Society: Series B (Statistical Methodology)*, 66(2):275–296, 2004.

[76] Jonathan R. Stroud, Michael L. Stein, and Shaun Lysen. Bayesian and maximum likelihood estimation for gaussian processes on an incomplete lattice. *Journal of Computational and Graphical Statistics*, 0(ja):0–0, 0.

[77] Aldo V Vecchia. Estimation and model identification for continuous spatial processes. *Journal of the Royal Statistical Society. Series B (Methodological)*, pages 297–312, 1988.

[78] H Von Storch and FW Zwiers. *Statistical analysis in climatology*. Cambridge University Press, Cambridge, 1999.

[79] Hans von Storch, Gerd Bürger, Reiner Schnur, and Jin-Song von Storch. Principal oscillation patterns: A review. *Journal of Climate*, 8(3):377–400, 1995.

[80] Russell S Vose, Derek Arndt, Viva F Banzon, David R Easterling, Byron Gleason, Boyin Huang, Ed Kearns, Jay H Lawrimore, Matthew J Menne, Thomas C Peterson, et al. Noaa's merged land–ocean surface temperature analysis. *Bulletin of the American Meteorological Society*, 93(11):1677–1685, 2012.

[81] Warren M Washington and Claire L Parkinson. *An introduction to three-dimensional climate modeling*. University science books, 2005.

[82] TML Wigley and BD Santer. Statistical comparison of spatial fields in model validation, perturbation, and predictability experiments. *Journal of Geophysical Research: Atmospheres*, 95(D1):851–865, 1990.

[83] Christopher K Wikle. A kernel-based spectral model for non-gaussian spatio-temporal processes. *Statistical Modelling*, 2(4):299–314, 2002.

[84] Christopher K Wikle. Low-rank representations for spatial processes. In *Handbook of Spatial Statistics*, pages 107–118. CRC Press, 2010.

[85] Christopher K Wikle. Modern perspectives on statistics for spatio-temporal data. *Wiley Interdisciplinary Reviews: Computational Statistics*, 7(1):86–98, 2015.

[86] Christopher K Wikle and L Mark Berliner. A bayesian tutorial for data assimilation. *Physica D: Nonlinear Phenomena*, 230(1):1–16, 2007.

[87] Christopher K Wikle and Noel Cressie. A dimension-reduced approach to space-time kalman filtering. *Biometrika*, pages 815–829, 1999.

[88] Christopher K Wikle and Mevin B Hooten. A general science-based framework for dynamical spatio-temporal models. *Test*, 19(3):417–451, 2010.

[89] Christopher K Wikle, Ralph F Milliff, and William G Large. Surface wind variability on spatial scales from 1 to 1000 km observed during toga coare. *Journal of the atmospheric sciences*, 56(13):2222–2231, 1999.

[90] Christopher K Wikle, Ralph F Milliff, Doug Nychka, and L Mark Berliner. Spatiotemporal hierarchical bayesian modeling tropical ocean surface winds. *Journal of the American Statistical Association*, 96(454):382–397, 2001.

[91] Christopher K Wikle and J Andrew Royle. Dynamic design of ecological monitoring networks for non-gaussian spatio-temporal data. *Environmetrics*, 16(5):507–522, 2005.

[92] Daniel S Wilks. *Statistical methods in the atmospheric sciences*, volume 100. Academic press, 2011.

[93] DS Wilks. On "field significance" and the false discovery rate. *Journal of applied meteorology and climatology*, 45(9):1181–1189, 2006.

[94] Cort J Willmott and Scott M Robeson. Climatologically aided interpolation (cai) of terrestrial air temperature. *International Journal of Climatology*, 15(2):221–229, 1995.

[95] Simon N Wood. Statistical inference for noisy nonlinear ecological dynamic systems. *Nature*, 466(7310):1102–1104, 2010.

[96] Michio Yanai and Takio Maruyama. Stratospheric wave disturbances propagating over the equatorial pacific. *Journal of the Meteorological Society of Japan. Ser. II*, 44(5):291–294, 1966.

[97] Jian Zhang, Peter F Craigmile, and Noel Cressie. Loss function approaches to predict a spatial quantile and its exceedance region. *Technometrics*, 50(2):216–227, 2008.

[98] Zhengyuan Zhu and Yichao Wu. Estimation and prediction of a class of convolution-based spatial nonstationary models for large spatial data. *Journal of Computational and Graphical Statistics*, 19(1):74–95, 2010.

30

Assimilating Data into Models

Amarjit Budhiraja
University of North Carolina, Chapel Hill

Eric Friedlander
University of North Carolina, Chapel Hill

Colin Guider,
University of North Carolina, Chapel Hill

Christopher KRT Jones
University of North Carolina, Chapel Hill

John Maclean
University of North Carolina, Chapel Hill

CONTENTS

30.1 Introduction

Data abound in studies of the environment, and their abundance is increasing at an extraordinary rate. The instruments we use to observe the environment, be it the atmosphere, ocean, land or ice, are constantly improving in their accuracy and efficiency. Moreover, they follow the technological trend of becoming less expensive as time moves on. It is tempting then to think we will eventually enjoy enough observational data that a complete picture of the world around us will be available and constitute a virtual replica from which we can conclude both how the environment works, and how it will change. But more data does not necessarily mean more understanding. It is hard from raw data alone to conclude how different effects are related. Correlation may be detected in data, but to establish causation will usually involve more experimentation than passive data can provide. Needed is the ability to vary conditions and see what results, but these experiments may not be feasible in environmental applications. For instance: imagine wanting to see what would happen if the world's oceans were to increase by $5°C$ in globally averaged temperature. Luckily, we have models that allow us to experiment in such ways. These vary from stripped down models that focus on a small number of key effects in a particular environmental application to the few very large, physically inclusive and computationally expensive models, residing at dedicated centers around the world, that amount to computational replicas of the entire Earth system.

Scientific advances owe equally to models and data, and both will remain relevant and key to further understanding. Observations drive model development, and model development often drives data acquisition. It therefore is particularly prudent to have these two sides of the scientific coin work in concert. If we get information from data and from models, then should we not get the most complete picture from directly combining the informational content from each? This is a mathematical and statistical question: how to combine the output of model investigations and observational data. The area that is dedicated to studying and developing the best approaches to this issue is called *Data Assimilation (DA)*. The image intentionally conveyed by the term is that a model is running and producing output, while the data are brought in to update and/or amend the output as it progresses, thus the data are being assimilated into the model.

The main impetus for the development of DA came in the 1980s from our need to predict the weather. At that time, and based on decades of prior work, numerical models were beginning to improve in forecast skill and to drive an improvement in the skill of human forecasters ([18], [13] Ch. 1). From the seminal work of Lorenz two decades earlier [17], it was understood that an inherent sensitivity resided within the weather models that would lead to an inevitable discrepancy between model forecast and actual weather. The idea then arose (see [20]) that observational data could be used to correct the system state and keep the forecast on track. Thus was born the area of *Numerical Weather Prediction*

(NWP) which underpins the great success of modern-day weather forecasts, whether they be of tomorrow's precipitation or the track of a category 5 hurricane.

Data assimilation methods are then aimed at giving a mathematical framework for suitably updating the system state predicted by a model through the use of observations of the state collected from sensory instruments. The main thrust of decades of research has been to develop mathematically justified and computationally tractable methods for DA that make the best use of the model and observations. A key point is that both models and data contain errors, and confidence will reside in each to varying degrees. The key characteristic of a DA scheme will then be how it resolves the issue of balancing the anticipated model error versus that residing in the observational data.

30.1.1   The core of a DA scheme

The heart of any DA scheme lies in the step at which data are incorporated into our (mathematical) description of the system state. This description will come from applying some model to the situation under study. In the NWP example, it might be the prediction of the weather based on estimates from 6 hours previously. We take this description to be a vector $x \in \mathbb{R}^M$, which could be thought of as physical variables, such as temperature, pressure and wind speed, listed at all of the grid points of our numerical scheme. The dimension M may be large and this presents an issue that we will discuss below. But, for understanding the logic of the DA step, the size of M is not relevant.

We start with our current "guess" of the system state, this is the current (mathematical) description of the system state. We denote this x^b where the superscript b stands for *background*. The observations are denoted y^o and are also assumed to be a vector but of possibly quite different dimension. We shall set $y^o \in \mathbb{R}^m$. In general then, $m \neq M$ and often, particularly in geophysical applications, $m \ll M$, which reflects the fact that the number of observations is often far less than the number of variables involved in describing the state of the system. Relating the observations to the system is a function $h : \mathbb{R}^M \to \mathbb{R}^m$, called the observation operator.

The core step is then to produce a new system state from the knowledge of x^b and y^o, which is called the analysis state and denoted by x^a. So the DA scheme amounts to a way in which we make the following transition concrete:

$$\{x^b, y^o\} \to x^a. \tag{30.1}$$

This chapter gives an overview of several different ways to realize this move from input information, coming from both model and data, to the best-informed estimate of the system state.

30.1.2   Model and observations

There is potential for confusion when the term model is used in a DA context. We usually think of a model as a mathematical representation of a physical situation. Such a model might be given by a differential equation or statistical process. But in DA, the term model is used in a related but technically different way. When a differential equation for a spatially dependent system is solved, the underlying physical space is covered by a grid and the physical variables are evaluated at the nodes of that grid. The computation then is aimed at calculating these physical variables at the grid points based on the underlying physical laws, which are manifest in the differential equations. The computation then "solves" those equations numerically to evaluate the system state. But in this context, the system state is thought of as the set of physical variable values at the grid points. The "model" is then thought of as the computational process that updates this state vector, and no longer the original mathematical model.

The observation operator is a critical element of the DA process and much can hide behind it. First, imagine that observations are of physical variables at grid points. In this case, the observation operator would just be a projection onto the components of the state vector that are being measured, and we could consider the observation space $\mathbb{R}^m$ to be a subspace of the state space $\mathbb{R}^M$. But, of course, it is in general not so simple. The grid points are those determined by the numerical scheme being used in the model and there is no reason why observational measurements should be made at those points. We can, secondly, imagine the observations being of physical variables at points other than grid points. The corresponding "observational" values at observation locations can then be concluded by (linear) interpolation from model values nearby the observation location in question. In both of these cases, the observation operator would be linear, but the relationship between observations and physical state variables may demand a more complex, possibly nonlinear observation operator, if the observations are not of physical variables (e.g. temperature) but of variables entangled with them (e.g. radiance). In such cases the observation operator may involve a model in its own right.

Data assimilation methods usually do not attempt to 'invert' the observation operator to map a data point in the observation space to a point in the state space. In other words, we work with the observations in their own space and not the full state space. This is because there are typically many fewer observations than there are model variables, and so much less error is introduced by converting model variables (e.g. temperature) at model grid points into observed variables (e.g. satellite radiances) at observation locations, than vice versa. See for instance [13] for an elaboration of this point.

30.1.3   Challenges of DA

Numerical Weather Prediction is used to make weather forecasts, typically ranging from 6 hours to 10 days after the initial state. The model for NWP is based on the primitive equations: conservation laws applied to the atmosphere, to which some approximations or simplifications must be made. This results in a system of seven partial differential equations, pertaining to velocity, density, pressure, temperature, and humidity.

As described above, these equations are not solved exactly but instead numerically approximated at discrete points in space. The computational model is obtained by discretizing (or "gridding up") the spatial domain in km or degrees in the longitude x and latitude y, and in km or pressure for the height z. For example, NOAA's Global Forecasting System uses a resolution of $0.5°$ horizontally and 20 layers vertically from the surface of the earth. Thus, there are $720 \times 360 \times 20 = 5,184,000$ grid points. Between seven and dozens of variables must be specified at each of these grid points to initialize or compute with this model, so that the numerical model has a dimension of $\mathcal{O}(10^8 - 10^9)$, which is the size M of the state space in the previous paragraphs.

Observations (data) are available from weather balloons, planes, satellites, etc. These observations are heavily concentrated over land (particularly North America and Eurasia). Observations are usually not made at model grid points, and often are not of model variables but instead of related quantities. Such a situation leads to the type of complex observation operator mentioned in the previous subsection. A good example is that of satellite observations (these are often of radiances) where physical variables, such as temperature, have to be inferred from the direct observations. This inference is then encoded in the observation operator.

Although these observations may be plentiful and the resulting dimension m of the observation space may be high, they are likely to be many fewer than the number of dimensions in the state space, M. It may appear that m will catch up with M over time as we get more and more observations, but it is just as certain that M will increase accordingly as it largely depends on the current state of computing power.

The expense of the model also presents a direct obstacle to the implementation of DA methods. As we shall see, optimally assimilating data into the model state space may require that multiple instances of the model are run in order to form an *ensemble* of predictions, or may require that a cost function involving the model is minimized. In either case the computational cost of running the model is one of the key limiters on the applicability of DA.

30.1.4 Approaches to DA

There are three main approaches to DA, i.e., to realizing the analysis step (30.1). Each stems from a different historical tradition and we will explain them in that way here. But it should be noted, and will be seen throughout this chapter, that they are each very strongly related to each other. DA is often presented with one of the approaches as the primary focus depending on the background of the authors. Our goal in this overview is to describe all of these different approaches and hopefully convey the strengths of the various viewpoints.

30.1.4.1 Variational approach

This is based on the formulation of a cost function that suitably penalizes deviations of a candidate state vector, x, from the model forecast and from the observations. This method comes from Optimization theory. We start with a background state x^b which is a first guess at the state of the system and observations y^o. For each of these a covariance matrix expresses the confidence we have in them: $\mathbf{B}$ for the background and $\mathbf{R}$ for the observational data. Each piece of information is then weighted with the inverses of their respective covariance matrices and the cost function is the sum of two quadratic terms.

$$J(x) = \frac{1}{2}\left[(x - x^b)^T \mathbf{B}^{-1}(x - x^b) + (y^o - h(x))^T \mathbf{R}^{-1}(y^o - h(x))\right] . \qquad (30.2)$$

The analysis then corresponds to a state vector x^a that minimizes this cost function. If the observation operator is linear, this is the standard least-squares problem for which we have a unique minimizer. For nonlinear h a standard strategy is to linearize the observation operator in some appropriate fashion, which approximates the problem of minimizing (30.2) by a least-squares problem and provides a unique approximate minimizer. Note however that the cost function given by (30.2) in general may have multiple minima.

30.1.4.2 Kalman gain approach

This approach to DA produces the analysis x^a from a linear combination of the background and the observations. This can be expressed as

$$x^a = x^b + \mathbf{W}\left(y^o - h\left(x^b\right)\right), \qquad (30.3)$$

where $\mathbf{W} \in \mathbb{R}^{M \times m}$ is a suitable weight matrix, called the *gain matrix*. It is calculated from the same covariance matrices $\mathbf{B}$ and $\mathbf{R}$ that appear in the cost function of the variational approach. The ubiquitous form of the gain matrix in DA is the *Kalman gain*,

$$\mathbf{W} = \mathbf{B}\mathbf{H}^T(\mathbf{R} + \mathbf{H}\mathbf{B}\mathbf{H}^T)^{-1}, \qquad (30.4)$$

where $\mathbf{H}$ is a suitable linearization of h. This gain matrix was originally derived as part of the *Kalman-Bucy Filter* or *Kalman Filter*, which we will define later, and originated in control theory [11, 12]. However the Kalman gain appears in multiple other DA schemes, and we derive it here assuming that x^b and y^o are independent of each other and distributed normally with covariances given respectively by $\mathbf{B}$ and $\mathbf{R}$. In this simple context, as a DA scheme this is known as *Optimal Interpolation* (OI).

We posit a "true value of the system state" which we denote x^t (t should not be confused with time here!). The mean of the distributions of x^b and y^o are then taken to be x^t and $h(x^t)$ respectively. The OI scheme then requires that the Mean-Squared-Error (MSE) $E\left[(x^a - x^t)^T(x^a - x^t)\right]$ is minimized. Letting $\epsilon^a = x^a - x^t$ we have

$$\epsilon^a = x^a - x^t = x^b + \mathbf{W}\left(y^o - h\left(x^b\right)\right) - x^t$$
$$= (x^b - x^t) + \mathbf{W}y^o - \mathbf{W}h(x^b).$$

A key step is to linearize h around x^b which yields, writing $\epsilon^b = x^b - x^t$ and $\epsilon^o = y^o - h(x^t)$,

$$\epsilon^a \approx (x^b - x^t) + \mathbf{W}y^o - \mathbf{W}[h(x^t) + \mathbf{H}(x^b - x^t)]$$
$$= \epsilon^b + \mathbf{W}\epsilon^o - \mathbf{W}\mathbf{H}\epsilon^b = \mathbf{W}\epsilon^o + (I - \mathbf{W}\mathbf{H})\epsilon^b, \qquad (30.5)$$

where $\mathbf{H}$ is now defined explicitly to be the linearization of h at x^b. It can be shown using elementary matrix calculus that the Kalman gain (30.4) minimizes the MSE $E\left[(\epsilon^a)^T\epsilon^a\right]$, if it is expressed using (30.5).

To connect with the variational approach, we will see that if h is replaced by the linearization $\mathbf{H}$ in both (30.2) and (30.3) then the minimizer of the cost function $J(x)$ in (30.2) is given by x^a from (30.3) with $\mathbf{W}$ given by the Kalman gain (30.4).

30.1.4.3   Probabilistic approach

The key difference in the probabilistic or Bayesian approach to DA is that instead of updating point estimates of the state or $x^b \mapsto x^a$, it is full probability distributions that are instead updated. Each probability density function (pdf) is written as $p()$, and must be interpreted by looking at the variables it is conditioned upon. For example, $p(x)$ represents the likelihood function of the state with no knowledge of observations: this is usually called the *prior* or *forecast* distribution. The *posterior* or *analysis* distribution is found by conditioning the forecast distribution on the observation, forming $p(x|y^o)$, and the question of data assimilation is how to find, or approximate, this posterior density.

By an application of Bayes' formula, one can write the posterior density as

$$p(x|y^o) \propto p(y^o|x) \times p(x), \qquad (30.6)$$

where $p(y^o|x)$ is the conditional density, or *likelihood*, of the observation given that the state is x. In principle, this reduces the task of finding the posterior density into two tasks: find or approximate the prior or forecast density $p(x)$, and the likelihood $p(y^o|x)$.

The probabilistic approach can be related to both of the above approaches. First, it is directly related to the variational method as follows: Under a Gaussian assumption on the prior $p(x)$,

$$p(x) \propto \exp\left\{-(x - x^b)^T\mathbf{B}^{-1}(x - x^b)\right\},$$

and with the likelihood of the observations given by

$$p(y^o|x) \propto \exp\left\{-(y^o - h(x))^T\mathbf{R}^{-1}(y^o - h(x))\right\},$$

it is easy to write the Bayesian posterior in terms of the variational cost function (30.2), with

$$p(x|y^o) \propto \exp\left\{-J(x)\right\}. \qquad (30.7)$$

As in the variational section more can be said if the observation operator is linear, so that $h(x) = \mathbf{H}x$. In this case the likelihood $p(y^o|x)$ is Gaussian, and so the posterior $p(x|y^o)$ is Gaussian as well. The global minimizer of the cost function $J(x)$ is then the mode (and mean) of $p(x|y^o)$. The connection to the Kalman gain approach, as before, is that the mean of $p(x|y^o)$ is given by (30.3) with $\mathbf{W}$ from (30.4).

Note that (30.7) can be made into a general relationship between variational and Bayesian approaches for arbitrary prior and likelihood densities by writing the variational cost function $J(x)$ as a combination of the logarithms of these densities.

30.1.5 Perspectives on DA

The different ways of looking at a data assimilation problem are evident from the previous section. It is worth pointing out that some of the difference comes from the fundamentally different perspectives of statisticians and applied mathematicians. The applied mathematician will tend to look for a single answer to the question: "what is the best estimate of the system state?" The underlying view here is that there is some "truth" about the system out there and the task is to approximate that as closely as possible. This perspective is reflected in the formulation described above in Section 30.1.4.2. The same view underlies the variational method. Mathematically, this view encounters problems when the observation operator is nonlinear and the cost function (30.2) is not necessarily quadratic. The cost function may then have multiple local minima, and there may be no global minimum (although this is a non-generic situation). Efforts then to find a minimum can take the person implementing the DA scheme down blind alleys. Nevertheless, this view is powerful and compelling and has driven much of the historical development of DA.

The statistician will take a different view by seeking to determine how likely the system is to be in a certain state. The viewpoint here is that even if there is a "true state of the system," we cannot know it and can only assign probabilities to prospective states. The idea of DA, from this perspective, is then to combine all the information, coming from model runs and observations, to formulate the "best" probability density function for the state of the system. This is clearly encoded in Bayes' Formula (30.6). This shifts the problem from estimating a particular state to computing a probability distribution.

It might be expected that one or other of these viewpoints would have won out, but both remain important and influential. The statistician's approach, encoded in Bayes' Formula, is now taken as providing the framework for DA [16, 19] but the variational and Kalman gain methods remain dominant as operational problem solvers. The reason for this equivocation lies in the nature of the problems that DA is asked to address. As discussed above, the geophysical applications, such as weather and climate prediction, that have driven DA development have very high dimensional state spaces. The methods that have been developed for approximating the pdfs involved in Bayes' Formula, such as particle filters, work very well in low dimensions but do not scale to high dimensions. In particular, they do not require any assumptions of Gaussianity, nor any need to linearize the observation operator. This makes them very appealing, especially since they approximate the full pdf and not just the mean (or mode), but if they cannot deal with high dimensions, then other methods will still be needed.

To summarize this point, there is an inherent tension in current DA research between nonlinearity and dimension. The statistical methods do very well in dealing with nonlinearity in low dimensions, but some kind of linearization and the methods of Kalman Filtering and/or the variational approach are needed to deal with high dimensional problems.

30.1.6 Chapter Overview

In the following sections we elaborate on fundamental DA schemes from the variational, Kalman gain and Bayesian approaches. The notation used will reflect the approach taken, as described in the sections above, but may also change to reflect the dependence of the DA problem on time.

Section 30.2 is devoted to the variational approach introduced in 30.1.4.1. The notation of 'background' x^b is used throughout this section to refer to the model variables before the assimilation step. We introduced this notation in 30.1.4.1 and its use reflects a historical use in NWP. The 'background' was frequently a guess for the state produced from historical data, e.g. a climatological mean.

We describe the 3D-Var scheme in Section 30.2.1, that is concerned with estimating the state given observations at the same fixed time.

In Section 30.2.2 we present a second variational method, 4D-Var, that assimilates observations from multiple time instants simultaneously in order to estimate the state at an initial time, and consequently the notation for observations will change to include time subscripts. We comment that the problem of estimating the state given observations from a later time is known as *smoothing*, and that 4D-Var is the only smoothing method that we consider.

Section 30.3 is concerned with the Bayesian approach described in 30.1.4.3. This involves a switch in notation as we transfer from the applied mathematics perspective to the statistical perspective. In prior sections capitals typically denote matrices, or the large model dimension M. In this section, capitals will denote random variables. Moreover, all the methods described in this section will be *sequential*. This describes a particular recursive dependence of the DA problem on time, in which we would like to alternate between using the model to propagate the state forward in time, and using the observations at that time to improve or correct the forecast. Consequently the notation for both model and observations will include a time index.

In Section 30.3.2 we derive the celebrated Kalman Filter from the Bayesian approach. The Kalman Filter is a method of the form (30.3)–(30.4) in which we also track the confidence in the analysis. As described at the end of Section 30.1.4.3, there is a close connection between the deterministic and the Bayesian approaches, and in this section we present, in the context of the Kalman filter, a translation between the notation and perspectives of the Bayesian and the Kalman gain approach. In the same section we discuss extensions of the Kalman Filter suitable for (weakly) nonlinear models or observation operators, and show a connection between the Kalman Filter and 4D-Var schemes.

In Section 30.3.3 we describe the particle filter, another DA method that arises from the Bayesian formulation. This method does not need, nor use, linearity in the model or observations, nor Gaussianity in the prior or posterior pdf.

Section 30.4 is concerned with modifications of the above methods that are made in practice to mitigate the weak points of each method. This is an extensive topic and the focus of much research; we do not attempt a comprehensive review of the literature but instead present some of the key techniques.

30.2 Variational Methods

The two fundamental variational methods are usually referred to as 3D-Var and 4D-Var. As described above, 3D-Var is a method that finds a minimizer of the cost function (30.2), while

4D-Var is concerned with an extension of the 3D-Var problem in which the observations are taken at different time points and consequently minimizes a different, but related, cost function.

30.2.1 3D-Var

Consider the cost function (30.2), which seeks to balance the observation increment $y^o - h(x)$ with the deviation of the analysis from the background $x - x^b$, where the terms are weighted by the inverse of their respective covariance matrices, and which we rewrite here for convenience:

$$J(x) = \frac{1}{2}\left[(x - x^b)^T \mathbf{B}^{-1}(x - x^b) + (y^o - h(x))^T \mathbf{R}^{-1}(y^o - h(x))\right].$$

Thus the 'more uncertain' term is given 'lesser weight' in this cost. By linearizing h around x^b, and defining $d(x) = y^o - h(x)$, one obtains

$$J(x) \approx \frac{1}{2}\left[(x - x^b)^T \mathbf{B}^{-1}(x - x^b) + (y^o - h(x^b)\right.$$
$$\left. - \mathbf{H}_{x^b}(x - x^b))^T \mathbf{R}^{-1}(y^o - h(x^b) - \mathbf{H}_{x^b}(x - x^b))\right]$$
$$= \frac{1}{2}\left[(x - x^b)^T \mathbf{B}^{-1}(x - x^b) + (d(x^b)\right. \tag{30.8}$$
$$\left. - \mathbf{H}_{x^b}(x - x^b))^T \mathbf{R}^{-1}(d(x^b) - \mathbf{H}_{x^b}(x - x^b))\right]. \tag{30.9}$$

The analysis x^a is defined to be the minimizer of the above approximation of the cost function, which, using elementary vector calculus, yields

$$x^a = x^b + \mathbf{W}d(x^b), \tag{30.10}$$
$$\mathbf{W} = \left(\mathbf{B}^{-1} + \mathbf{H}_{x^b}^T \mathbf{R}^{-1}\mathbf{H}_{x^b}\right)^{-1} \mathbf{H}_{x^b}^T \mathbf{R}^{-1}. \tag{30.11}$$

The matrix $\mathbf{W}$ defined by (30.11) is same as the gain matrix defined in (30.4), so that the minimizer of (30.2) is the same as the analysis given by (30.3). This equivalence is a variant of the Sherman-Morrison-Woodbury identity, but can also be proved by an elegant construction of (30.3) in a block matrix equation ([13], Section 5.5.1).

In applications in NWP, the calculation of $(\mathbf{B}^{-1} + \mathbf{H}_{x^b}^T \mathbf{R}^{-1}\mathbf{H}_{x^b})^{-1}$ required to obtain $\mathbf{W}$ in either (30.4) or (30.11) is often too expensive to compute, as it requires inverting a matrix of size $\mathbb{R}^{M \times M}$, and we recall that $M \sim \mathcal{O}(10^9)$. This prevents the direct implementation of Kalman gain approaches to DA in NWP. The advantage of 3D-Var is that the variational description (30.2) allows one to avoid the exact calculation of (30.10), instead finding an approximation by minimization algorithms. Moreover such algorithms usually result in a more accurate estimates of x^a since they will not just use the linearization of the h function around the background x^b, but recursively linearise h around successive guesses for the analysis. Recall that in order to compute $\mathbf{W}$ in the OI scheme h was linearized around x^b, however in 3D-Var one can improve the point about which linearization takes place using an iterative method. This is discussed in the following section.

30.2.1.1 Incremental Method for 3D-Var

The 3D-Var analysis can be approximated without requiring the evaluation of (30.11). The algorithm consists, first, of an outer loop that creates an approximation of the cost function with the observation operator linearized around a guess for the analysis, and an inner loop in which the cost function (30.8) is minimized. Below is a step-by-step description of the algorithm.

1. Set $x^{(1)} := x^b$ and $i := 1$.

2. (Outer Loop) Define i-th approximation of the gradient

$$\nabla J(x) \approx \nabla J_i(x) = \mathbf{B}^{-1}(x - x^b) - \mathbf{H}^T_{x^{(i)}} \mathbf{R}^{-1}(d(x) - \mathbf{H}_{x^{(i)}}(x - x^b))$$

by linearizing h around $x^{(i)}$, the current best guess for x^a.

3. If the gradient is "close enough" to zero then accept $x^{(i)}$ as x^a. Namely, if $|\nabla J_i(x^{(i)})| < \epsilon$ where ϵ is some predefined threshold then stop and take the final analysis as $x^a = x^{(i)}$. Otherwise:

    4. (Inner Loop) Set $\eta^{(0)} = x^{(i)}$ and call a minimization subroutine of the form $\eta^{(j+1)} = \eta^{(j)} + \alpha_k f(\nabla J_i(\eta^{(j)}))$. Methods for choosing α_k and f are described elsewhere but a simple example is choosing a fixed $\alpha_k := \alpha$ and $f(x) = -x$.

    5. When a minimum, $\eta^{(J)}$, has been found, set $x^{(i+1)} = \eta^{(J)}$, increment i by 1 and repeat steps 2-5.

We remark that the above method requires that $\mathbf{B}^{-1}$ is known, and that the estimation of the background covariance matrix $\mathbf{B}$ is itself a major problem in NWP (see for instance [6]).

There exist a variety of computation techniques (PSAS, preconditioning, etc.) that the user can use to reduce the number of iterations needed for the above algorithm to converge. The improved accuracy of the algorithm stems from the ability to relinearize h in step 4. The algorithm in general will not converge without additional conditions on h, however experimental results have shown that the incremental method can achieve much more accurate estimates than OI without a significant increase in computational complexity.

30.2.2   4D-Var

Implicit in the previous variational methods has been that all components of the observation vector y^o were collected at the same time instant. In this section we give an overview of the 4D-Var scheme, that assimilates observations at multiple time instants by minimizing a global cost function.

Suppose we want to assimilate observations collected at times $\{t_n\}_{n=1}^N$ over some time interval. Denote these observations by $\{y_n\}_{n=1}^N$. The 4D-Var method minimizes a cost function analogous to (30.8), but that uses a numerical integrator to compare state estimates to observations at multiple time steps. We distinguish between two formulations of 4D-Var, strong and weak constraint.

30.2.2.1   Strong constraint 4D-Var

We now assume the numerical model to be perfect. This implies that given that the state at time t_0 is x, one can obtain a perfect state value $m_n(x)$ at every time instant t_n through a nonlinear model integrator m_n. The goal of the DA scheme is to find the initial condition that best matches the background x^b and all the observations. The background is viewed as some imprecise 'first guess' of the state at time instant t_0. The model in this formulation is a strong constraint because the cost function does not allow predicted state values to deviate from those obtained from a forward integration of the model.

The initial condition uniquely determines the model trajectory and thus all the analysis values. Once again we denote by $\mathbf{B}$ the background covariance matrix, which captures the uncertainty associated with the forecast at time t_0 and denote by $\mathbf{R}_n$ the observation covariance matrix at time t_n which quantifies the uncertainty associated with the observation at time t_n. The cost function is then given as follows

$$J(x) = \frac{1}{2}\left[(x - x^b)^T \mathbf{B}^{-1}(x - x^b) - \sum_{n=1}^{N}(y_n - h(m_n(x)))^T \mathbf{R}_n^{-1}(y_n - h(m_n(x)))\right]. \qquad (30.12)$$

Intuitively, by minimizing the above cost function, we are selecting the x at time t_0 that is not too far from the background and also not too far from the observations when x is "pushed forward" to the corresponding time and transformed through the observation function h.

Minimization of J is similar to the method used for 3D-Var but more approximations need to be made to deal with the nonlinearily of the forecast operator m_n. Most of the computational effort is in suitably approximating the gradient of J. In order to accomplish this we separate the components of J and write $J(x) = J^b(x) + J^o(x)$ where

$$J^b(x) = \frac{1}{2}\left[(x - x^b)^T \mathbf{B}^{-1}(x - x^b)\right]$$

and

$$J^o(x) = \frac{1}{2}\left[\sum_{n=1}^{N}(y_n - h(m_n(x)))^T \mathbf{R}_n^{-1}(y_n - h(m_n(x)))\right]. \qquad (30.13)$$

Note that $\nabla J = \nabla J^b + \nabla J^o$. Clearly

$$\nabla J^b(x) = \mathbf{B}^{-1}(x - x^b),$$

but due to the (typically nonlinear) model and observation operator, $\nabla J^o(x)$ cannot be so easily computed. Suppose starting from the forecast x^b at time t_0 the state $x_n^b = m_n(x^b)$ at time instant n is computed. The state update function m_n can be written as $m_{n-1,n} \circ m_{n-2,n-1} \cdots \circ m_{0,1}$ where $m_{j-1,j}$ is the function that propagates the state at time instant $j-1$ to j. We denote by $\mathbf{M}[t_{n-1}, t_n]$ the linearization of $m_{n-1,n}$ around x_{n-1}^b. Using these matrices the gradient ∇J^o can be approximated as

$$\nabla J^o(x) \approx \sum_{n=1}^{N}\left(\mathbf{H}_n \mathbf{M}[t_{n-1}, t_n]\cdots \mathbf{M}[t_1, t_2]\mathbf{M}[t_0, t_1]\right)^T d_n(x)$$

$$= \sum_{n=1}^{N}\mathbf{M}^T[t_1, t_0]\cdots \mathbf{M}^T[t_n, t_{n-1}]\mathbf{H}_n^T d_n(x)$$

where $d_n(x) = \mathbf{R}_n^{-1}y_n - h(m_n(x))$ and $\mathbf{H}_n$ is the gradient of h at $x_n^b = m_n(x^b)$. We used in the above that the adjoint of a matrix is its' transpose. The adjoint operator integrates backwards in time. Therefore, every iteration of the minimization algorithm requires two steps. First, the model must be integrated forward using a candidate analysis state x in order to calculate d_n. Then a backward pass must be made using the adjoint to calculate the gradient.

30.2.2.2 Incremental Method for 4D-Var

As with 3D-Var, one may find it impossible to calculate the minimizer of $J(x)$; we present here an approach that approximates the analysis, and does not use x_n^b as the only point of linearization but rather repeatedly generates linearization points through a recursive gradient descent algorithm. Specifically, the algorithm works as follows.

1. (Outer loop) Choose a starting point $x_0^{(1)}$ (normally x^b is chosen) and set $i = 1$.

2. Integrate the model forward to calculate

$$x_n^{(i)} = m_{n-1,n}(x_{n-1}^{(i)})$$

These values are used as points about which linearization is done to calculate $\mathbf{M}[t_{n+1}, t_n]$ and $\mathbf{H}_n$.

3. Using the $x_n^{(i)}$'s obtained from the forward integration calculate the weighted observation increments $d_n = \mathbf{R}_n^{-1}(h(x_n^{(i)}) - y_n)$.

4. Using these d_n's calculate the gradient

$$\nabla J_i\left(\{x_0^{(i)}, \dots, x_n^{(i)}\}\right) = \mathbf{B}^{-1}(x - x^b)$$

$$+ \sum_{n=1}^{N} \mathbf{M}^T[t_1, t_0] \cdots \mathbf{M}^T[t_n, t_{n-1}]\mathbf{H}_n^T d_n$$

5. If $\left|\nabla J_i\left(\{x_0^{(i)}, \dots, x_n^{(i)}\}\right)\right| < \epsilon$, where ϵ is a predetermined threshold, then accept each $x_n^{(i)}$ as x_n^a. Otherwise:

6. (Inner loop) Set $\eta^{(0)} = x_0^{(i)}$ and call a minimization subroutine of the form $\eta^{(j+1)} = \eta^{(j)} + \alpha_k \nabla J(\eta^{(j)})$.

7. When a minimum, $\eta^{(J)}$, has been found, set $x_0^{(i+1)} = \eta^{(J)}$, increment i by 1 and repeat steps 2-7.

30.2.2.3 Weak Constraint 4D-Var

The perfect model assumption can be relaxed and this is known as weak constraint 4D-Var. Instead of selecting an optimal initial condition, the algorithm produces a set of state values $\{x_0, x_1, \dots, x_N\}$ that approximate the state at each time from t_0 to t_N. The cost function (30.8) gains an extra term that measures the difference between the state estimate x_n and the push-forward of the model at the previous time step, $m_{n-1,n}(x_{n-1})$. Consequently the state estimates x_n are only approximately consistent with the model. The cost function is now written as

$$J(x_0, \dots, x_N) = \frac{1}{2}\bigg((x_0 - x^b)^T \mathbf{B}^{-1}(x_0 - x^b)$$

$$+ \sum_{n=1}^{N}(x_n - m_{n-1,n}(x_{n-1}))^T \mathbf{Q}^{-1}(x_n - m_{n-1,n}(x_{n-1}))$$

$$+ \sum_{n=1}^{N}(y_n - h(x_n))^T \mathbf{R}_n^{-1}(y_n - h(x_n))\bigg),$$

where $\mathbf{Q}$ is the model error covariance.

The above cost function also has a statistical interpretation. Suppose that $X_0 \sim \mathcal{N}(x^b, \mathbf{B})$ and

$$X_{n+1} = m_{n-1,n}(X_n) + \epsilon_n^b, \ Y_n = h(X_n) + \epsilon_n^o,$$

where $\epsilon_n^b \sim \mathcal{N}(0, \mathbf{Q})$, $\epsilon_n^o \sim \mathcal{N}(0, \mathbf{R}_n)$ are mutually independent (and independent of X_0) Normal random variables. Then $J(x_0, \dots, x_N)$ equals, up to a constant, the negative log of the conditional density of $(X_0, \dots X_N)$(evaluated at $(x_0, \dots, x_N)$) given $(Y_1, \dots Y_N) = (y_1, \dots y_N)$. Thus minimizing the cost function corresponds to maximizing this log-likelihood.

30.3 Bayesian formulation and sequential methods

We now reformulate the problem of Data Assimilation in statistical terms, changing the notation we have previously established in order to match that used by statisticians. We adopt the Bayesian approach, in which Bayes' law (30.6) transitions prior beliefs about the model variables into a posterior density via the likelihood function. The analogue of the prior density was previously the background state or first guess; one way to convert these deterministic quantities into a prior is to model the prior as Gaussian, with mean given by the background and variance given by the level of confidence in the background, that is $\mathbf{B}$ in (30.2), (30.4). We model the true state sequence and the observation sequence as a collection of random variables $\{(X_n, Y_n)\}_{n=0}^{N}$ that describe a two component (state, observation) Markov chain. Here X_n represents the true state of the system at time instant t_n and Y_n the observation collected at the same time instant. We will use lower case x_n, y_n to refer to realizations of these random variables. Note that an observation that has been made will always be a realization of the random variable, and so will always be written in lower case.

The underlying state process $\{X_n\}_{n=0}^{N}$ is unobservable, so all inference must be based on $\{Y_n\}_{n=0}^{N}$.

The use of time-dependent state and observation variables suggests one should employ a sequential Data Assimilation algorithm to recursively estimate the state given ever-updating observations. We now describe how this can be done using a Bayesian approach.

Let us describe the complete sets of state variables and observations from t_0 to t_n by $X_{0:n} = \{X_0, X_1, \ldots, X_n\}$, $Y_{1:n} = \{Y_1, \ldots, Y_n\}$, and similarly define the realizations of those random variables by $x_{0:n}$, and $y_{1:n}$. Then Bayes' rule says that

$$p(x_{0:n}|y_{1:n}) \propto p(y_{1:n}|x_{0:n})p(x_{0:n}) \,. \tag{30.14}$$

Analogously to Section 30.1.4.3, $p(x_{0:n}|y_{1:n})$ is the posterior density, $p(y_{1:n}|x_{0:n})$ is the likelihood of observing $y_{1:n}$, and $p(x_{0:n})$ is the prior or forecast density.

Working with (30.14) would require updating the joint distribution on $X_{0:n}$ in order to assimilate the observation y_n, for each n. One would also have to calculate the joint likelihood of every observation, $Y_{1:n}$. Rather than working with these joint densities, we now formulate the *filtering* problem, in which assumptions on the structure of the model and observations allow us to rewrite (30.14) so that we can recursively update the pdf for the state X_n at time t_n, instead of updating the joint density $X_{0:n}$. We then consider the Kalman Filter and Particle Filter, so-called because they solve the filtering problem.

30.3.1 Filtering

We assume that the distribution of Y_n given $\{Y_k\}_{k=0}^{n-1}$ and X_n depends only on X_n, and the distribution of X_n given X_{n-1} and $\{Y_k\}_{k=0}^{n-1}$ depends only on X_{n-1}. This dependency structure is shown in Figure 30.1. Given these assumptions, (30.14) can be rewritten as

$$p(x_n|y_{1:n}) \propto p(y_n|x_n)p(x_n|y_{1:n-1}), \tag{30.15}$$

where the forecast density is $p(x_n|y_{1:n-1})$ and the posterior density is $p(x_n|y_{1:n})$. The question of computing the posterior density $p(x_n|y_{1:n})$ is the filtering problem, and methods that accomplish this are called filtering methods or filters. To compute the forecast density $p(x_n|y_{1:n-1})$, one must integrate the posterior density $p(x_{n-1}|y_{1:n-1})$ from the previous

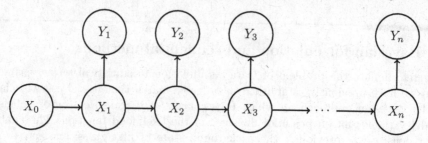

FIGURE 30.1
Dependency Structure for Nonlinear State Space Model

time step, as follows:

$$p(x_n|y_{1:n-1}) = \int p(x_n|x_{n-1})p(x_{n-1}|y_{1:n-1})dx_{n-1}. \tag{30.16}$$

This is the probabilistic analogue of a forecast model in the case of noiseless state dynamics. The posterior density is then given by (30.15). Note that point estimates of the analysis x_n^a can be given by the mean (or other measures of centrality such as the median or the mode) of the posterior density $p(x_n|y_{1:n})$.

In general, closed form expressions for the forecast and posterior distributions are not readily available for typical physical applications due to nonlinearity of state space models and observation functions. However there is one important setting where closed form expressions can be given, namely that of the classical *Kalman filter*.

30.3.2   The Kalman Filter

Suppose that the state-observation Markov process has the dependence structure in Section 30.3.1 and is described through the following linear model with Gaussian errors

$$X_n \sim \mathcal{N}\left(\mathbf{M}_n\, x_{n-1},\, \mathbf{Q}_n\right), \tag{30.17}$$

where $X_n,\, x_n \in \mathrm{R}^M$ and $\mathbf{M}_n,\, \mathbf{Q}_n \in \mathrm{R}^{M\times M}$, and suppose that observations are sampled independently and with Gaussian errors from the linear observation operator $\mathbf{H}_n$, with

$$Y_n \sim \mathcal{N}\left(\mathbf{H}_n\, x_n,\, \mathbf{R}_n\right), \tag{30.18}$$

where $Y_n \in \mathrm{R}^m$, $\mathbf{H}_n \in \mathrm{R}^{m\times M}$, and $\mathbf{R}_n \in \mathrm{R}^{m\times m}$. Namely, the conditional distribution of X_n given $(X_{0:n-1}, Y_{0:n-1})$ is Normal with conditional mean $\mathbf{M}_n X_{n-1}$ and conditional variance $\mathbf{Q}_n$ and conditional distribution of Y_n given $(X_{0:n}, Y_{0:n-1})$ is Normal with conditional mean $\mathbf{H}_n X_n$ and conditional variance $\mathbf{R}_n$. We assume without loss of generality that $X_0 \sim N(0, \mathbf{B})$.

It is easy to check that the linear, Gaussian form of the model and observations imply that the prior, likelihood and posterior pdf in (30.15) are all Gaussian. In this case, each distribution can be specified by giving its mean and covariance. We define the mean of the prior density $p(x_n|y_{1:n-1})$ to be $x_{n|n-1}$, and its covariance to be $\mathbf{P}_{n|n-1}$; similarly for the posterior density $p(x_n|y_{1:n})$, we define the mean to be $x_{n|n}$ and the covariance to be $\mathbf{P}_{n|n}$. This notation refers back to the Bayesian formulation of the filter; the first subscript gives the present time step, and the second subscript gives the time step of the last observation that we condition on. Using this notation in the model (30.17), we see that the forecast step in the analysis cycle consists of updating the mean and covariance

$$x_{n|n-1} = \mathbf{M}_n \, x_{n-1|n-1} \,, \tag{30.19}$$

$$\mathbf{P}_{n|n-1} = \mathbf{M}_n \, \mathbf{P}_{n-1|n-1} \, \mathbf{M}_n^T + \mathbf{Q}_n \,. \tag{30.20}$$

Before we construct the mean and covariance of the posterior density, let us pause to establish the connection between the variables used above and the Kalman gain approach of Section 30.1.4.3.

The mean of the prior density can be interpreted as the background x^b. In fact the only difference is that the notation used in this section allows for multiple assimilation steps, taken sequentially, while the notation of the background x^b in Section 30.1.4.2 assumes we are assimilating once at a fixed time. This has a standard name in the Kalman gain approach, one uses the term *forecast* for the sequential prediction of the model state. One can define $x_n^f := x_{n|n-1}$ and $\mathbf{B}_n := \mathbf{P}_{n|n-1}$ as time-varying analogues of the background x^b and confidence in the background, $\mathbf{B}$, respectively. Observe that $\mathbf{B}$ needed to be known a priori for OI and 3D-Var, while the Kalman Filter calculates the (optimal) value of $\mathbf{B}$ at each time step.

Returning to the derivation, one can use (30.17)–(30.20) and Bayes' rule (30.6) to determine that the posterior density is also Gaussian, with mean and covariance given by

$$x_{n|n} = \mathbf{P}_{n|n} \left(\mathbf{H}_n^T \, \mathbf{R}_t^{-1} \, y_n + \mathbf{P}_{n|n-1}^{-1} \, x_{n|n-1} \right) \,,$$

$$\mathbf{P}_{n|n} = \left(\mathbf{H}_n^T \, \mathbf{R}_t^{-1} \mathbf{H}_n + \mathbf{P}_{n|n-1}^{-1} \right)^{-1} \,.$$

These can be rewritten as

$$x_{n|n} = x_{n|n-1} + \mathbf{K}_n \left(y_n - \mathbf{H}_n x_{n|n-1} \right) \,, \tag{30.21}$$

$$\mathbf{P}_{n|n} = (I - \mathbf{K}_n \, \mathbf{H}_n) \, \mathbf{P}_{n|n-1} \,, \tag{30.22}$$

where the matrix $\mathbf{K}_n$ is given as

$$\mathbf{K}_n = \mathbf{P}_{n|n-1} \, \mathbf{H}_n^T \left(\mathbf{H}_n \, \mathbf{P}_{n|n-1} \, \mathbf{H}_n^T + \mathbf{R}_t \right)^{-1} \,. \tag{30.23}$$

Equations (30.19) – (30.23) fully describe the analysis cycle for the Kalman filter at time instant t.

We now complete the connection between the Bayesian formulation of the Kalman Filter presented here and the Kalman gain approach of Section 30.1.4.2. If we change notation as described above, replacing $x_{n|n-1}$ with the forecast x_n^f and $\mathbf{P}_{n|n-1}$ with the confidence $\mathbf{B}_n$, then (30.21) is clearly (30.3) (but with a linearized observation operator) and (30.23) is clearly the Kalman gain (30.4). Apart from the sequential nature of the forecast and of the confidence in the forecast, the only difference is that one now also obtains a confidence level in the analysis, $\mathbf{P}_{n|n}$. It is easy to check (see for instance [22]) that for the linear model (30.17) and data collection (30.18) the Kalman Filter provides the minimum MSE estimate of the true system state, by calculating the optimal confidence levels in the forecast and analysis at each time step.

30.3.2.1 Extensions

The simplicity and tractability of the Kalman filter has led to its use even for settings where the model conditions for its validity are not satisfied. One common setting is where the linear functions $\mathbf{M}_n x$ and $\mathbf{H}_n x$ are replaced by general nonlinear functions $m_n(x)$ and $h_n(x)$. In such a setting an adaptation of the Kalman filter, usually referred to as the *extended Kalman filter*, approximates the densities in the analysis cycle (30.15)–(30.16) by Gaussian approximations obtained by replacing $\mathbf{M}_n$ and $\mathbf{H}_n$ in (30.19) – (30.23) with the linearizations of m_n and h_n about appropriate points. We refer the reader to [10] for details.

Another approach that attempts to bypass the linearization of m_n is the so-called *ensemble Kalman filter*. Here one uses the full nonlinear state equation

$$x_n = m_n(x_{n-1}) + \tau_n \tag{30.24}$$

to simulate state values that are used to approximate the forecast density, and the nonlinear observation function h_n. More precisely, having obtained a Gaussian approximation for the posterior distribution at step $n-1$, one takes L samples from this distribution labeled as $\{X_{n-1}^i\}_{i=1}^L$. Using this and L samples $\{\tau_n^i\}$ of the noise in the state model, one uses the nonlinear state equation (30.24) to produce X_n^i according to

$$X_n^i = m_n(X_{n-1}^i) + \tau_n^i , \tag{30.25}$$

where $i = \{1, \cdots, L\}$. The samples approximate the forecast distribution at time n as a Gaussian density p_n^f with mean

$$x_{n|n-1} = \frac{1}{L} \sum_{i=1}^L X_n^i , \tag{30.26}$$

and covariance given by the covariance of the ensemble, with the (i, j)-th entry given by

$$\left(\mathbf{P}_{n|n-1}\right)_{ij} = \frac{1}{L-1} \left(X_n^i - x_{n|n-1}\right)\left(X_n^j - x_{n|n-1}\right)^T . \tag{30.27}$$

This forecast distribution is used to produce a Gaussian posterior density by linearizing the observation function h_n and either using a modified form of equations (30.21)–(30.22) in which the observations are perturbed slightly for each ensemble member [4], or using the class of Ensemble Square Root Filters. We refer the reader to [8] for details.

30.3.2.2   Equivalence of 4D-Var and KF

Recall that strong constraint 4D-Var, formulated in Section 30.2.2.1, assumes that the model is perfect and attempts to find the optimal initial condition $x^a(t_0)$ which is the best match to both the background x^b and a set of observations $\{y_0, y_1, \ldots, y_N\}$ at times $t_0, t_1, \ldots, t_N$. This initial condition then uniquely determines the analysis states $x^a(t_n)$, obtained by integrating $x^a(t_0)$ with the model.

In this Section we argue that the analysis state $x^a(t_N)$ obtained from 4D-Var at the final observation time exactly matches the mean of the final analysis state $x_{N|N}$ obtained from a Kalman filter with the same initial condition and observations, provided the model is perfect and linear and the observations are linear.

Recall that the Kalman Filter provides a recursive solution to the *filtering* problem, i.e. it computes $p(x(t_n)|y_0, \ldots, y_n)$, the conditional distribution of the state at the current time given all observations up to the current time. The analysis state $x^a(t_0)$ in 4D-Var computes the mode of $p(x(t_0)|y_0, \ldots, y_N)$, namely the conditional distribution of the state at the initial time given all future observations (and this is one example of a *smoothing* problem, where the state is conditioned on observations at a later time). We will denote the *pushforward* of this density under the n time step map M_n as $M_n * p(x(t_0)) = p(x(t_n))$ (cf. [16], Ch.1).

The equivalence of 4D-Var and the KF under the restrictive assumptions above can now be seen as follows. In the case of a perfect, deterministic model, the push-forward of the smoothing density, i.e. $M_N * p(x_0|y_0, \ldots, y_N)$, is the filtering density at the final observation time, $p(x(t_N)|y_0, \ldots, y_N)$. This is due to the fact that state dynamics is noiseless (see for example [16], Section 2.5). As can be seen from Section 30.3.2, in the case of a perfect linear model with a linear observation operator, the final analysis step $x^a(t_N)$ of the Kalman filter is the mode of the filtering distribution (note that for a Gaussian posterior, the mean and mode are identical). On the other hand, in the case of a perfect linear model with linear observation operator, the analysis $x^a(t_0)$ of 4D-Var is the mode of the smoothing distribution at the time t_0. From this and the linearity of M_n it follows that $x^a(t_N)$ is the mode of $p(x(t_N)|y_0, \ldots, y_N)$, proving the claimed equivalence. To be clear, this statement does not apply to the iterative approach to 4D-Var we described in Section 30.2.2, because that method approximates the analysis x^a. Instead, here $x^a(t_0)$ refers to the unique minimizer of (30.12), which by definition is the mode of the smoothing distribution.

30.3.3 Particle Filters

The methods discussed in all previous sections rely at some point on linearity of the observation operator and/or of the model, or have used a linearization approximation. In this section we give a brief overview of a collection of particle based schemes that are quite flexible, do not rely on linearization of the dynamics and are well suited for parallel computing architecture. These methods have a long history; we refer the reader to [7] for a comprehensive review of this area. The basic idea is to replace the high dimensional integrals in (30.15)-(30.16) with suitable Monte-Carlo sample averages. Instead of using probability densities to describe the distributions, here we will use discrete probability measures supported on finitely many points. These points and their weights will evolve in time to give the forecast measure Π_n^f and posterior measure Π_n at times t_n. One can compute the analysis state x_n^a by computing an integral with respect to the measure Π_n.

30.3.3.1 A basic particle filter

Suppose that Π_{n-1} is given as a discrete probability measure supported on points $x_{n-1}^1, \ldots x_{n-1}^L$ and with corresponding weights $p_{n-1}^1, \ldots p_{n-1}^L$. Here L represents the number of particles that are used to approximate the distribution Π_{n-1}. The two key steps in the analysis cycle are as follows:

Prediction step. Propagate each of the particles $x_{n-1}^i \mapsto \hat{x}_n^i$ using the nonlinear state dynamics (30.24). This requires simulating L noise random variables τ_n^i, $i = 1, \ldots L$. Given such random variables, $\hat{x}_n^i$ are defined as

$$\hat{x}_n^i = m_n(x_{n-1}^i) + \tau_n^i.$$

This gives the forecast probability distribution Π_n^f as a discrete probability measure concentrated on L points $\{\hat{x}_n^i\}_{i=1}^L$ with weights $\{p_{n-1}^i\}_{i=1}^L$.

Filtering step. Update the weights $\{p_{n-1}^i\}_{i=1}^L$ using the observation Y_n by setting $p_n^i = cp_{n-1}^i R(\hat{x}_t^i, Y_n)$, where

$$R(x, y) \doteq \exp\left\{-\frac{1}{2}(Y_n - h(\hat{x}_n^i))^T \mathbf{R}_t^{-1}(Y_n - h(\hat{x}_n^i))\right\}. \tag{30.28}$$

The posterior distribution Π_n is then defined as the discrete measure with support points $\{x_n^i = \hat{x}_n^i\}_{i=1}^L$ and weights $\{p_n^i\}$.

Although this scheme is easy to implement, it suffers from severe degeneracy, especially in high dimensions. The main difficulty is that after a few time steps all the weights tend to concentrate on a very few particles which drastically reduces the effective sample size. A common remedy for this paucity of significant particles is to occasionally re-sample in order to refresh the particle cloud.

30.3.3.2   Particle filter with resampling

The main idea here is to periodically resample with replacement from the discrete distribution Π_n to obtain a uniform distribution of weights. Of course, resampling adds extra noise to the approximating scheme so it is important not to resample too frequently. Fix a resampling lag parameter $\alpha \in \mathbb{N}$. This parameter specifies the number of time steps between successive resampling steps. Suppose that Π_{n-1} is given as a discrete probability measure supported on points $x_{n-1}^1, \ldots x_{n-1}^L$ and corresponding weights $p_{n-1}^1, \ldots p_{n-1}^L$. Suppose first that n/α is not an integer. In this case the posterior distribution Π_n is given exactly as before in Section 30.3.3.1. If n/α is an integer we further modify the above discrete probability measure Π_n as follows. Take a random sample of size L from the discrete distribution $\{(x_n^1, p_n^1), \ldots, (x_n^L, p_n^L)\}$. Relabel the new points as $(x_n^1, \ldots, x_n^L)$. The posterior distribution Π_n is then given as the discrete distribution: $\{(x_n^1, p_n^1), \ldots, (x_n^L, p_n^L)\}$ where all the p_n^i are set equal to $1/L$.

30.3.3.3   Variance reduction: Deterministic allocation and residual resampling

As noted earlier, one drawback of the above algorithm is that it unnecessarily introduces extra variability in the algorithm due to random sampling. This limitation motivates the study of various variance reduction schemes. We describe below one such commonly used scheme, that preserves the highest weighted particles, in numbers proportional to their weight, in a deterministic procedure. Another variance reduction scheme is given in the next subsection. Let α be as in the last subsection. The key difference here is that at a resampling step, i.e. when n/α is an integer, instead of random sampling with replacement from $(x_n^1, p_n^1), \ldots, (x_n^L, p_n^L)$ we do a partial deterministic allocation as follows. Let $k_i = \lfloor L p_n^i \rfloor$ and branch (i.e. duplicate) x_n^i into k_i particles. This procedure yields $\sum_{i=1}^L k_i \leq L$ particles. The remaining particles are resampled randomly, as follows. Set $L_r \doteq L - \sum_{i=1}^L k_i$ and $w_n^i = L p_n^i - k_i$. Now resample L_r particles using random sampling from the distribution $\{(x_n^1, c w_n^1), \ldots, (x_n^L, c w_n^L)\}$, where $c = 1 / \sum_{i=1}^L w_n^i$ is a normalizing constant. Combining the deterministic and stochastic resampled particles gives a total of L particles which are relabeled as $x_n^1, \ldots, x_n^L$. Finally, as before, the posterior distribution Π_n is given as the discrete distribution: $\{(x_n^1, p_n^1), \ldots, (x_n^L, p_n^L)\}$ where all the p_n^i are set equal to $1/L$.

30.3.3.4   Branching particle filter

In this scheme(cf. [5]) the number of particles is allowed to change at each time step. Let α be as before. The main steps are as follows. Suppose that Π_{n-1} is given as a discrete measure on L_{n-1} points x_{n-1}^i, $i = 1, \ldots, L_{n-1}$ with equal weights (namely $1/L_{n-1}$). Propagate each of the particles $x_{n-1}^i \mapsto \hat{x}_n^i$ using the nonlinear state dynamics (30.24) as in Section 30.3.3.1. The main difference from Sections 30.3.3.1 – 30.3.3.3 is that instead of reweighing the particles as in the above algorithms we "branch and kill particles". More precisely, if n/α is not an integer, the particle $\hat{x}_n^i$ branches into a random number (denoted by ζ_n^i) of particles. The ζ_n^i must be unbiased and satisfy an additional regularization condition (see [5]); one choice of the distribution of ζ_n^i is

$$P(\zeta_n^i = \lfloor \gamma_n^i \rfloor + 1) = \gamma_n^i - \lfloor \gamma_n^i \rfloor = 1 - P(\zeta_n^i = \lfloor \gamma_n^i \rfloor),$$

where

$$\gamma_n^i = \frac{L_{n-1}\beta_n^i}{\sum_{i=1}^{L_{n-1}} \beta_n^i}, \quad \beta_n^i = R(\hat{x}_n^i, Y_n).$$

This results in a total of say L_n particles denoted as $\{x_n^i\}_{i=1}^{L_n}$. The posterior distribution Π_n is the discrete distribution: $\{(x_n^1, p_n^1), \ldots, (x_n^{L_n}, p_n^{L_n})\}$ where all the p_n^i are set equal to $1/L_n$. In order to manage the explosion or decay of the number of particles L_n we add a resampling step to restore the number of particles to L at every α time steps, which is carried out as in Section 30.3.3.2 or 30.3.3.3.

30.3.3.5   Regularized particle filters

One common difficulty with particle filters is that they suffer from the lack of diversity among particles. This problem can be particularly severe in settings where the noise in state dynamics is degenerate. In order to treat this difficulty one usually considers regularized particle filters(cf. [7, Chapter 12]) which corresponds to replacing the sampling from the discrete distribution $\{(x_n^1, p_n^1), \ldots, (x_n^L, p_n^L)\}$ by that from an absolutely continuous approximation. The key idea is to use kernel density smoothers. The two basic versions of such regularized particle filters correspond to regularization at the prediction step and regularization at the filtering step. We only describe the latter, details of the first scheme can be found in [7, Chapter 12].

    A regularization kernel K is a symmetric probability function on $\mathbb{R}^k$ satisfying $\int_{\mathbb{R}^k} xK(x)dx = 0$ and $\int_{\mathbb{R}^k} \|x\|^2 K(x)dx < \infty$. For a smoothing parameter $\gamma \in (0,\infty)$, referred to as the bandwidth, denote $K_\gamma(x) \doteq \frac{1}{\gamma^k}K(x/\gamma)$. The two commonly used kernels are the Gaussian kernel and the Epanechnikov kernel (cf. [7]). The basic algorithm is as follows. We consider the regularization of the algorithm in Section 30.3.3.2. For simplicity we take the lag parameter $\alpha = 1$. Suppose that Π_{n-1} is given as a discrete measure on L points x_{n-1}^i, $i = 1, \ldots, L$ with equal weights $1/L$. Define $\Pi_{n-1}^f = \{(\hat{x}_n^1, 1/L), \ldots, (\hat{x}_n^L, 1/L)\}$ as in Section 30.3.3.1. Update the weights $\hat{p}_n^i = 1/L \mapsto p_n^i$ using the observation Y_n by setting $p_n^i \doteq c\hat{p}_n^i R(\hat{x}_n^i, Y_n)$, where c is the normalization constant. Draw $\{\tilde{x}_n^i, i = 1, ..., L\}$ from the discrete distribution $\{(x_n^1, p_n^1), \ldots, (x_n^L, p_n^L)\}$ and generate $\{\epsilon^i\}_{i=1}^L$, i.i.d. from the kernel K. Define (abusing notation) $x_n^i \doteq \tilde{x}_n^i + \gamma\epsilon^i$. The posterior distribution Π_n is then the discrete distribution: $\{(x_n^1, p_n^1), \ldots, (x_n^L, p_n^L)\}$ where all the p_n^i are set equal to $1/L$. Such a random jiggling of the particle locations can be particularly important if the stochastic dynamical system governing the state evolution is very sensitive to the initial condition.

30.4   Implementation of DA methods

This section is concerned with the following practical question: given a particular DA problem, consisting of a numerical model, which may be low or high dimensional, linear or weakly or strongly nonlinear, and data, how does one go about selecting a DA scheme, what are the common pitfalls of each DA scheme, and how are these pitfalls typically mitigated?

    A particular DA problem may be suitable for several DA schemes, or only one, or indeed none at all, and the suitability of a particular scheme may not be obvious from the approach it is formulated under. For instance Optimal Interpolation, which we developed in Section 30.1.4.2 as a simple implementation of the Kalman gain approach, was developed as a method for DA in Numerical Weather Prediction (NWP), and was only later associated with the Kalman gain approach, which originated in control theory. One might ask then why the Kalman Filter, that is the optimal scheme formulated under the Kalman gain approach, was never used in NWP, given that the linearization of the model and observation operator needed for the Kalman Filter are used in NWP. In fact, the Kalman Filter has never been used for NWP, and NWP centers adopted the variational scheme 3D-Var and some later moved to using 4D-Var.

This historical quirk in the development of DA schemes is due to the scaling of the computational cost of each scheme with the dimension of the model. As outlined in Section 30.1.3, numerical models for NWP may have a dimension of order 10^9. We mention in Section 30.2.1 that the direct computation of a Kalman gain type matrix is too expensive for a model of this high dimension, and this rules out for the time being the use of the Kalman Filter in NWP; the variational schemes 3D-Var and 4D-Var employ the incremental approximations of Sections 30.2.1.1 and 30.2.2.2 to avoid computing the gain matrix. The ensemble Kalman filter is, similarly, significantly computationally cheaper in the analysis step than directly evaluating the Kalman gain, though at the cost of requiring an ensemble of runs of the numerical model.

There is another large gap in performance between the Kalman gain approach and the Particle Filter. In particular, without exploiting a special structure in the model or advanced filtering schemes, use of the Particle Filter will require the model dimension to be at most 10–20. This restriction is also due to the computational cost of the Particle Filter; one finds that the number of particles required in the Particle Filter scales exponentially with the effective dimension of the model, and this exorbitant scaling in computational cost quickly becomes unreasonable.

The above paragraph may seem to imply that there is no advantage to the Kalman or Particle Filter. In practice, though, this is not the case. The (non)linearity of the model is another major factor in the selection of a DA scheme. A crucial point is that, while the schemes orginally developed under the variational or Kalman gain approach can be reformulated under the Bayesian framework, it is not possible to reformulate the particle filter under the variational or Kalman gain approach. For example, the Kalman Filter requires that the model and observations be linear with Gaussian errors, and so the posterior pdf is always Gaussian; the ensemble Kalman Filter does not require that the model be linear, but attempts to fit the best Gaussian posterior pdf to the forecast ensemble. By contrast, the particle filter does not require that the model be linear and does not require that the posterior pdf be Gaussian. In consequence, the ability of 3D-Var, 4D-Var, the ensemble Kalman Filter and Particle Filter to handle nonlinearity (or nonGaussianity) is roughly inverse to their computational cost.

One therefore must decide to include or eliminate DA schemes from consideration by comparing the model dimension, the nonlinearity of the model, and what is known or expected about the structure of the posterior pdf.

As a closing remark for this section, we note that the capabilities and in particular the drawbacks of each DA scheme are very much a topic for active research. For instance, recent developments in the ensemble Kalman Filter are competitive with 4D-Var in some situations, and the question of the relative drawbacks of each method is quite complicated; see [14, 23] for a review and numerical implementation. Meanwhile operational NWP centers have spearheaded the development of hybrid variational schemes. These schemes blend the fixed background covariance used in standard 3D-Var or 4D-Var with an evolving covariance matrix, estimated from an ensemble in a manner like the ensemble Kalman Filter, in a manner designed to yield the benefits of both schemes [2, 3, 15].

30.4.1 Common modifications for DA schemes

It is common in the Data Assimilation literature to employ some strategies in the application of a DA scheme to mitigate the known biases or flaws in that scheme. These strategies do not change the scheme to such an extent that the scheme is known under a different name, but are typically mentioned in concert with the scheme.

30.4.1.1 Localization

A problem known as 'spurious correlations' occurs when state variables are updated by any DA scheme using distant observations. For instance, the local weather in North Carolina should not affect the weather in California. However, it is sometimes the case that nonzero entries in the forecast covariance matrix will create a (spurious) correlation between distant sites, due to a coincidental pattern in the forecasts or to noise in the numerical model. The strategy employed to mitigate this problem is to *localize* the background or forecast covariance matrix (respectively $\mathbf{B}$ and $\mathbf{P}_{n|n-1}$), which ensures observations can only affect state variables within some selected distance. This can be effected by for instance taking the Schur product of the background or forecast covariance matrix with a matrix that is 0 on every entry that links distant locations, and 1 everywhere else, or by using some other cut-off function. For instance, if there is one state variable measured at locations $\{1, 2, 3\}$ and we want to localize so that only adjacent locations are updated by observations, then we would rewrite the forecast covariance matrix as

$$\mathbf{P}_{n|n-1} \mapsto \mathbf{P}_{n|n-1} \circ \begin{bmatrix} 1 & 1 & 0 \\ 1 & 1 & 1 \\ 0 & 1 & 1 \end{bmatrix},$$

where $\circ$ denotes element-wise multiplication.

30.4.1.2 Inflation

This section is concerned only with ensemble methods, particularly the ensemble Kalman Filter. One would like for the data assimilation problem to correctly account for the uncertainty or errors in the forecast, and this is done via the forecast covariance matrix. However, the forecast covariance matrix (30.27) is rank deficient, as the number L of ensemble members is typically much less than the dimension M of the numerical model, and consequently some directions in model space in which the forecast is uncertain may not be represented in the forecast covariance matrix. Furthermore, the forecast covariance matrix usually will not take into account the presence of model error, which is a feature in any application of DA. For these reasons, the ensemble Kalman Filter will typically under-estimate the forecast covariance matrix. The strategy adopted to remedy this defect is to artificially increase the forecast covariance. There are broadly two ways this may be done: multiplicative covariance inflation [1], which involves choosing a number $\delta > 0$ and replacing the forecast covariance matrix $\mathbf{P}_{n|n-1}$ according to

$$\mathbf{P}_{n|n-1} \mapsto (1 + \delta)\mathbf{P}_{n|n-1},$$

and additive inflation [9], in which one adds a small amount of variance along the diagonal, replacing

$$\mathbf{P}_{n|n-1} \mapsto \mathbf{P}_{n|n-1} + \delta \mathbf{I}.$$

Both methods have advantages. Multiplicative inflation preserves the rank and range of the forecast covariance matrix; that is, relationships between variables do not change. Additive inflation adds a little significance to each variable in the forecast, and so will prevent the covariance matrix from collapsing over a few dominant modes, but changes the relationships between variables. These inflation schemes can be tuned to relax the ensemble in a way that matches the prior, or the observations; see for instance [21]. Either or both methods can be combined with localization.

Bibliography

[1] Jeffrey L Anderson and Stephen L Anderson. A monte carlo implementation of the nonlinear filtering problem to produce ensemble assimilations and forecasts. *Monthly Weather Review*, 127(12):2741–2758, 1999.

[2] Mark Buehner, PL Houtekamer, Cecilien Charette, Herschel L Mitchell, and Bin He. Intercomparison of variational data assimilation and the ensemble kalman filter for global deterministic nwp. part i: Description and single-observation experiments. *Monthly Weather Review*, 138(5):1550–1566, 2010.

[3] Mark Buehner, PL Houtekamer, Cecilien Charette, Herschel L Mitchell, and Bin He. Intercomparison of variational data assimilation and the ensemble kalman filter for global deterministic nwp. part ii: One-month experiments with real observations. *Monthly Weather Review*, 138(5):1567–1586, 2010.

[4] Gerrit Burgers, Peter Jan van Leeuwen, and Geir Evensen. Analysis scheme in the ensemble kalman filter. *Monthly weather review*, 126(6):1719–1724, 1998.

[5] Dan Crisan, Pierre Del Moral, and Terry Lyons. *Discrete filtering using branching and interacting particle systems*. Citeseer, 1998.

[6] G. Descombes, T. Auligné, F. Vandenberghe, and D. M. Barker. Generalized Background Error covariance matrix model (GEN_BE v2.0). *Geoscientific Model Development Discussions*, 7:4291–4352, July 2014.

[7] Arnaud Doucet, Nando De Freitas, and Neil Gordon. An introduction to sequential monte carlo methods. In Arnaud Doucet, Nando De Freitas, and Neil Gordon, editors, *Sequential Monte Carlo methods in practice*, pages 3–14. Springer, 2001.

[8] Geir Evensen. *Data assimilation: the ensemble Kalman filter*. Springer Science & Business Media, 2009.

[9] Thomas M Hamill and Jeffrey S Whitaker. Accounting for the error due to unresolved scales in ensemble data assimilation: A comparison of different approaches. *Monthly weather review*, 133(11):3132–3147, 2005.

[10] Andrew H Jazwinski. *Stochastic processes and filtering theory*. Courier Corporation, 2007.

[11] Rudolph E Kalman and Richard S Bucy. New results in linear filtering and prediction theory. *Journal of basic engineering*, 83(1):95–108, 1961.

[12] Rudolph Emil Kalman. A new approach to linear filtering and prediction problems. *Journal of basic Engineering*, 82(1):35–45, 1960.

[13] Eugenia Kalnay. *Atmospheric modeling, data assimilation and predictability.* Cambridge university press, 2003.

[14] Eugenia Kalnay, Hong Li, Takemasa Miyoshi, Shu-Chih Yang, and Joaquim Ballabrera-Poy. 4-d-var or ensemble kalman filter? *Tellus A*, 59(5):758–773, 2007.

[15] David D Kuhl, Thomas E Rosmond, Craig H Bishop, Justin McLay, and Nancy L Baker. Comparison of hybrid ensemble/4dvar and 4dvar within the navdas-ar data assimilation framework. *Monthly Weather Review*, 141(8):2740–2758, 2013.

[16] Kody Law, Andrew Stuart, and Zygalakis Konstantinos. *Data Assimilation: A Mathematical Introduction*, volume 62. Springer Texts in Applied Mathematics, 2015.

[17] Edward N Lorenz. Deterministic nonperiodic flow. *Journal of the atmospheric sciences*, 20(2):130–141, 1963.

[18] Peter Lynch. *The emergence of numerical weather prediction: Richardson's dream.* Cambridge University Press, 2006.

[19] Sebastian Reich and Colin Cotter. *Probabilistic forecasting and Bayesian data assimilation.* Cambridge University Press, 2015.

[20] Ian Roulstone and John Norbury. *Invisible in the Storm: the role of mathematics in understanding weather.* Princeton University Press, 2013.

[21] Jeffrey S Whitaker and Thomas M Hamill. Evaluating methods to account for system errors in ensemble data assimilation. *Monthly Weather Review*, 140(9):3078–3089, 2012.

[22] Christopher K Wikle and L Mark Berliner. A bayesian tutorial for data assimilation. *Physica D: Nonlinear Phenomena*, 230(1):1–16, 2007.

[23] Shu-Chih Yang, Matteo Corazza, Alberto Carrassi, Eugenia Kalnay, and Takemasa Miyoshi. Comparison of local ensemble transform kalman filter, 3dvar, and 4dvar in a quasigeostrophic model. *Monthly Weather Review*, 137(2):693–709, 2009.

31

Spatial Extremes

Anthony C. Davison

École Polytechnique Fédérale de Lausanne, Switzerland

Raphaël Huser

King Abdallah University of Science and Technology, Thuawal, Saudi Arabia

Emeric Thibaud

École Polytechnique Fédérale de Lausanne, Switzerland

CONTENTS

31.1 Introduction

Climate change is perceptible in shifting patterns of rainfall, sea ice, temperatures and other phenomena. Although this affects entire distributions of observations, the largest impacts on humans and on the environment that sustains us are likely to be due to extreme events. The fact that such events take place over time and space has led to a surge of research on modelling complex extreme events over the past decade, with many developments in statistical theory and methods that have already influenced applications. The purpose of this chapter, which should be read in conjunction with Chapter 8, is to summarise these developments. Space limitations and its rapid development make it impossible to fully describe this area of research. Other recent summaries are [13], [18], [20] and [67].

Spatial analysis of extremes may be performed for a variety of reasons, including:

(a) the estimation of changes in extremes, e.g., increases in daily maximum air temperatures, by combining data from different sites while accounting for the dependence among them;

(b) the attribution of particular rare events, such as the 2003 European heatwave, to possible causes, such as human impacts on climate;

(c) the estimation of risk at a single important location, such as the site of a nuclear installation. Borrowing strength by including data from elsewhere will often reduce estimation uncertainty, particularly when the data available at the site itself are limited—which is often the case;

(d) the estimation of overall risk for single large events, such as hurricanes or floods, which may be key to assessing potential losses for insurance companies or for planning public security interventions. Likewise risk of crop failure due to drought is a crucial element of food security.

These different settings involve different emphases in modelling. Any spatial dependence in cases (a), (b) and (c) needs to be accounted for, but the details are not usually critical, whereas in case (d) the pattern of dependence will be a key feature. An individual farmer seeking insurance against loss of income will need the probability of disastrous weather at a specific spatial site, whereas failure of an important food crop due to a prolonged heatwave will depend on its spatial extent, so an overall risk assessment for public authorities must take into account the probability of simultaneous crop failure throughout a large region. In the former case a model providing accurate spatial interpolation to a single point will be adequate, whereas the second case will entail accurate modeling of a complex joint distribution, which is much harder and requires specialized models such as *max-stable processes*.

An obvious question is why specialized models are needed in such settings. The key issue is that, just as the univariate Gaussian distribution provides a good model for averages but a poor one for extremes, leading to inaccurate estimates of tail probabilities, joint tail probabilities may be badly mis-estimated by standard geostatistical models. The Gaussian distribution has no shape parameter analogous to the degrees of freedom in the Student t distribution, and thus cannot encompass different rates of tail probability decay. Moreover the rate of decay of Gaussian probabilities for joint events is determined by the correlation coefficient, and, in addition to being insufficiently flexible, this model implies that the variables become ever more independent as the events become rarer, which is not always true in applications. This motivates the study of special models particularly adapted to multivariate and spatial extremes.

We consider some quantity $Y(x)$, say, to be concrete, annual maximum rainfall, where the point x lies in a domain $\mathcal{X}$. Usually $x = (s, t)$ has a spatial component $s \in \mathcal{S}$ and a temporal component $t \in \mathcal{T}$, and for simplicity we suppose that $\mathcal{X} = \mathcal{S} \times \mathcal{T}$. We wish to model the properties of $\{Y(x) : x \in \mathcal{X}\}$, in order to estimate the probabilities of rare events, $\Pr\{Y(x) \in \mathcal{R}\}$, where the set $\mathcal{R}$ is extreme in some suitable sense. We might, for example, attempt to represent rainfall liable to lead to flooding at some point on a river by taking $\mathcal{R}$ to represent very large aggregate rainfall over a short period in a catchment area upstream.

Although the goal is to understand the properties of $Y(x)$ within $\mathcal{X}$, data are generally available only at a finite subset $\mathcal{X}' = \mathcal{S}' \times \mathcal{T}'$ of $\mathcal{X}$. One basic classification of such problems depends on $\mathcal{X}'$: if $\mathcal{S}'$ consists of a few sites, each with a long series of measurements at times $t \in \mathcal{T}'$, as might be the case with long-term temperature measurements, then the data may be 'time-rich but space-poor'; whereas if $\mathcal{S}'$ consists of a grid of thousands of points but the set $\mathcal{T}'$ is rather limited then the data may be 'space-rich but time-poor'. 'Space- and time-rich' data, such as five-minute radar observations of rainfall on a detailed spatial grid, in principle allow rich modelling of complex phenomena. In these three cases the observation times and sites are non-random, but in others they may be haphazard: for example, large forest fires appear at random points in space and time, and it is essential to model this in addition to the areas of the fires.

A second classification depends on whether extrapolation from $\mathcal{X}'$ to $\mathcal{X}$ is needed. The purpose of analysing temperature, rainfall or wind observations at measurement stations in $\mathcal{S}'$ is often to make predictions for the entire set $\mathcal{S}$, and this must be allowed by the model. The relationship between gridded data and the underlying phenomenon, e.g., between climate model rainfall reanalysis data and observed point rainfall, is often much less clear, and often in this setting it makes sense to take $\mathcal{S}' = \mathcal{S}$, so only temporal extrapolation may be required.

In Section 31.2 we set out a framework for complex extremes that generalizes the discussion in Chapter 8 and which allows a subsequent treatment of models both for maxima and for threshold exceedances. Section 31.3 describes some of prominent extremal models, and discusses how they may be extended to encompass an important phenomenon known as asymptotic independence. The following two sections concern inference: in Section 31.4 we sketch some ideas useful for initial data analysis and for the assessment of model fit, and in Section 31.5 we discuss the fitting of models by likelihood methods. Section 31.6 illustrates the earlier ideas and techniques in the context of Saudi Arabian rainfall in the region of Jeddah, and of extreme summer temperatures around the Spanish capital, Madrid. The chapter ends by outlining some topics that we have been unable to treat here.

31.2 Max-stable and related processes

31.2.1 Poisson process

Max-stable and related processes are the space- and space-time analogues of the extreme-value distributions arising in scalar and multivariate settings, and play a key role below. A general discussion is possible in terms of the Poisson process [51].

A Poisson process is a stochastic model for a set of random points $\mathcal{P}$ lying in a state space $\mathcal{E}$, and is defined by two properties of the random variables

$$N(\mathcal{A}) = |\{x : x \in \mathcal{P} \cap \mathcal{A}\}|, \quad \mathcal{A} \subset \mathcal{E},$$

that count how many points of $\mathcal{P}$ lie in a set $\mathcal{A}$: for any collection of disjoint subsets $\mathcal{A}_1, \ldots, \mathcal{A}_n \subset \mathcal{E}$, the $N(\mathcal{A}_1), \ldots, N(\mathcal{A}_n)$ are independent; and $N(\mathcal{A})$ has the Poisson distribution with mean $\mu(\mathcal{A})$, where μ is called the *mean measure* of the process. The measure μ must be *non-atomic*, i.e., $\mu(\{x\}) = 0$ for any singleton $x \in \mathcal{E}$, and moreover $\mu(\emptyset) = 0$, whereas $\mu(\mathcal{E})$ may be infinite, in which case $N(\mathcal{E})$ is infinite with probability one. For technical reasons, below we only consider sets $\mathcal{A}$ for which $\mu(\mathcal{A}) < \infty$.

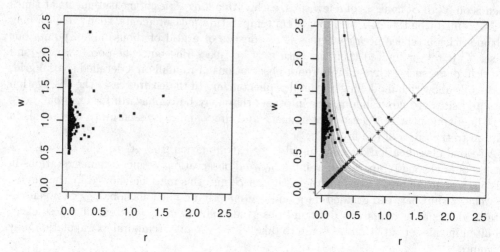

FIGURE 31.1

Poisson process example. Left panel: first 1000 points (r, w) of a Poisson process sequentially generated on $\mathbb{R}_+^2$ with intensity function (31.3). Right panel: mapping of the points shown in the left panel to $q = rw$, shown as $+$ on the diagonal, with the mapping function shown by the curved grey lines.

Consider, for example, a Poisson process with $\mathcal{E} = \mathbb{R}_+^2$, that generates points $x = (r, w)$ according to the mean measure

$$\mu\{(r, \infty) \times (w, \infty)\} = \frac{1}{r} \times \Phi(-\sigma^{-1}\log w - \sigma/2), \quad r, w > 0, \qquad (31.1)$$

where $\sigma > 0$ and Φ is the standard normal distribution function. This corresponds to setting $\mathcal{P} = \{(R_i, W_i) : i = 1, 2, \ldots\}$, where $R_1 > R_2 > \cdots > 0$ are generated sequentially by setting $R_i = (E_1 + \cdots + E_i)^{-1}$, with $E_i \stackrel{\text{iid}}{\sim} \exp(1)$, and $W_i = \exp(\sigma\varepsilon_i - \sigma^2/2)$, where $\varepsilon_i \stackrel{\text{iid}}{\sim} N(0, 1)$, independent of the E_i; note that $\mathrm{E}(W_i) = 1$. The first 1000 points of a realisation of such a process are shown in the left-hand panel in Figure 31.1; the full realisation would have an infinity of points at the left-hand edge of the panel, because $\mu\{(r, \infty) \times (0, \infty)\} = 1/r \to \infty$ as $r \to 0$. The mean measure has an *intensity function* $\dot{\mu}$ given by its derivative at the upper right corner of a rectangle $(r', r) \times (w', w)$, i.e.,

$$
\begin{aligned}
\dot{\mu}(r, w) &= \frac{\partial^2 \mu\{(r', r) \times (w', w)\}}{\partial r \partial w} \\
&= \frac{\partial^2}{\partial r \partial w} \{\mu(r', w') - \mu(r, w') - \mu(r', w) + \mu(r, w)\} \qquad (31.2) \\
&= \frac{1}{r^2} \times \frac{1}{\sigma w} \phi(-\sigma^{-1}\log w - \sigma/2), \quad r, w > 0, \qquad (31.3)
\end{aligned}
$$

where we have written $\mu(r, w) = \mu\{(r, \infty) \times (w, \infty)\}$ and so forth, and ϕ denotes the standard normal density function. Note that $\mu(\mathcal{A})$ equals the integral of $\dot{\mu}$ over $\mathcal{A}$.

A key role in constructing models for spatial extremes is played by the *mapping theorem*, which under mild conditions states that if a function g does not create atoms, then $\mathcal{P}^* = g(\mathcal{P})$ also follows a Poisson process. As a simple example we might take $g(r, w) = rw$, corresponding to setting $Q_i = R_i W_i$, which amounts to collapsing the points shown in the left-hand panel of Figure 31.1 onto the diagonal line shown in the right-hand panel. Clearly $\mu[\{(r, q/r) : r > 0\}] = 0$ for each $q > 0$, so this transformation does not create atoms. We obtain the mean measure of this new Poisson process by noting that $Q = RW > q$ if and only if $R > q/W$, and the corresponding set $\mathcal{A}_q = \{(r, w) : rw > q\}$ has measure

$$
\begin{aligned}
\mu(\mathcal{A}_q) &= \int_0^\infty \frac{1}{\sigma w} \phi(-\sigma^{-1}\log w - \sigma/2) \int_{r=q/w}^\infty \frac{1}{r^2} \, dr \, dw \\
&= \int_0^\infty \frac{1}{\sigma w} \phi(-\sigma^{-1}\log w - \sigma/2) \left[-\frac{1}{r}\right]_{q/w}^\infty dw \\
&= \frac{1}{q} E(W) = \frac{1}{q}, \quad q > 0.
\end{aligned}
\tag{31.4}
$$

Hence the process with points $Q_i = R_i W_i$ is also Poissonian, with the same mean measure as the R_i. Note that the calculation leading to (31.4) requires only that W is positive and satisfies $E(W) = 1$.

The restriction of $\mathcal{P}$ to a subset $\mathcal{E}'$ of $\mathcal{E}$ clearly also follows a Poisson process, with mean measure $\mu'(\mathcal{A}) = \mu(\mathcal{E}' \cap \mathcal{A})$. For example, if we let $\mathcal{E} = (0, \infty)$, consider $R_1, R_2, \ldots$ and let $\mathcal{E}' = (z', \infty)$ for some $z' > 0$, then we retain only those points R_i exceeding z'. As $\mu(\mathcal{E}') = 1/z'$ is finite, these R_i can be generated by first simulating a Poisson variable N' with mean $1/z'$, and if $N' = n$, simulating n independent variables on the interval (z', ∞) with survivor function z'/z; these Pareto variables have probability density function z'/z^2 $(z > z')$.

31.2.2 Classical results

To connect the above rather abstract discussion with classical results for maxima of a random sample $X_1, \ldots, X_n$, note for any real b_n and positive a_n that

$$
\frac{\max(X_1, \ldots, X_n) - b_n}{a_n} \leq y \quad \Leftrightarrow \quad N_n\{(y, \infty)\} = 0,
\tag{31.5}
$$

where $N_n(\mathcal{A}) = \sum_{j=1}^n I\{(X_j - b_n)/a_n \in \mathcal{A}\}$ for $\mathcal{A} \subset \mathbb{R}$; $N_n(\mathcal{A})$ has a binomial distribution. If the extremal types theorem applies to $X_1, \ldots, X_n$, then there exist sequences $\{b_n\}$ and $\{a_n\} > 0$ such that limiting probability of (31.5) is of generalized extreme-value (GEV) form, and this implies that the binomial random variables $N_n\{(y, \infty)\}$ satisfy

$$
\lim_{n \to \infty} \Pr\left[N_n\{(y, \infty)\} = 0\right] = \exp\left\{-\left(1 + \xi \frac{y - \eta}{\tau}\right)_+^{-1/\xi}\right\},
\tag{31.6}
$$

where $a_+ = \max(a, 0)$ and $\xi, \eta \in \mathbb{R}$ and $\tau > 0$ are respectively shape, location and scale parameters. Thus as $n \to \infty$ the point processes $\mathcal{P}_n = \{(X_j - b_n)/a_n : j = 1, \ldots, n\}$ converge to a limiting Poisson process $\mathcal{P}$ with mean measure $\mu\{(y, \infty)\} = \{1 + \xi(y - \eta)/\tau\}_+^{-1/\xi}$. The limiting generalized extreme-value distribution for the maximum corresponds to the probability that the set $(y, \infty) \cap \mathcal{P}$ contains no points, a *void probability* of $\mathcal{P}$. This distribution arises as a limit, and thus provides an approximation that should improve as n increases. Likewise the Poisson process approximation may be poor unless n is sufficiently large.

The process $\mathcal{P}$ can be transformed to have mean measure $1/z$ on $\mathbb{R}_+$ by setting $z = \{1 + \xi(y - \eta)/\tau\}_+^{1/\xi}$, and fitting (31.6) to sample maxima can be regarded as finding the η, τ and ξ that best achieve this. On this transformed scale the maximum has limiting distribution function $\exp(-1/z)$ $(z > 0)$; this, the standard Fréchet distribution, corresponds to (31.6) with $\eta = \tau = \xi = 1$.

The generalized extreme-value distribution is max-stable; in the case of independent standard Fréchet variables $Z, Z_1, \ldots, Z_n$, this means that

$$n^{-1} \max(Z_1, \ldots, Z_n) \overset{D}{=} Z, \quad n = 1, 2, \ldots, \tag{31.7}$$

where $\overset{D}{=}$ means 'has the same distribution as'; these equalities correspond to setting $b_n = 0$ and $a_n = n$ in (31.5).

The generalized Pareto distribution (GPD) emerges on noting that large values of the transformed original variables $Z_j = \{1 + \xi(X_j - \eta)/\tau\}_+^{1/\xi}$ approximately follow a Poisson process on $\mathbb{R}_+$ with mean measure $1/z$. Those points for which $X_j > u$, or equivalently $Z_j > \{1 + \xi(u - \eta)/\tau\}_+^{1/\xi}$, for some high threshold u, satisfy

$$
\begin{aligned}
\Pr(X_j > y \mid X_j > u) &= \frac{\Pr(X_j > y)}{\Pr(X_j > u)} \\
&\doteq \frac{\{1 + \xi(y - \eta)/\tau\}_+^{-1/\xi}}{\{1 + \xi(u - \eta)/\tau\}_+^{-1/\xi}} \\
&= \left(1 + \xi \frac{y - u}{\sigma_u}\right)_+^{-1/\xi}, \quad y > u,
\end{aligned}
$$

where $\sigma_u = \tau + \xi(u - \eta)$; this yields the generalized Pareto distribution, which is therefore the limiting model for threshold exceedances that corresponds to fitting (31.6) to maxima. This distribution is threshold-stable: it is easy to check that the distribution of each X_j, conditional on the event $X_j > u' > u$, is also generalized Pareto, with the same shape parameter ξ. Like the GEV and Poisson process, the GPD approximation stems from a limit and may be poor if the threshold u is too low.

As we shall now see, these results for scalar extremes, which are also derived in §§8.2.1–8.2.3, extend to multivariate and spatial settings.

31.2.3 Spectral representation

The Poisson process allows a general discussion of the classical extreme-value models in terms of a spectral representation, to be described below. This applies to the univariate setting, the multivariate setting, and to the functional setting, which we now describe.

Max-stability is the key property of the generalized extreme-value distribution that underpins its use for the estimation of rare event probabilities. On the standard Fréchet scale this may be expressed as (31.7), and the analogous equation for max-stable processes, which generalize the GEV to multivariate and functional settings, is

$$n^{-1} \max\{Z_1(x), \ldots, Z_n(x)\} \overset{D}{=} Z(x), \quad x \in \mathcal{X}, n = 1, 2, \ldots; \tag{31.8}$$

we require (31.8) to hold for the entire process $\{Z(x) : x \in \mathcal{X}\}$. When $\mathcal{X}$ is a finite set this corresponds to the max-stability of the multivariate extreme-value distributions.

To generalize our previous discussion, we extend the Poisson process (R, W) so that $\{W(x) : x \in \mathcal{X}\}$ is a random process taking values in a suitable space of functions, with $W(x) \geq 0$ and $\mathrm{E}\{W(x)\} = 1$ for each x. The topology on the function space, and the distribution of W, must be sufficiently rich that the measure of (R, W) remains non-atomic. Despite this added complexity, we continue to use the term points for (R, W) and $Q(x) = RW(x)$.

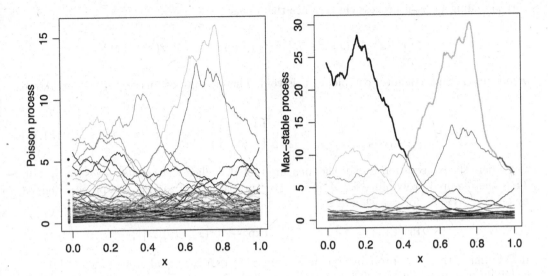

FIGURE 31.2

Construction of a max-stable process. Left panel: first 100 points of a Poisson process $\{(r_i, w_i) : i \in \mathbb{N}\}$, where the $w_i(x)$ are realizations of a log-Gaussian process on $(0, 1)$ and the r_i are shown at the left of the panel. Right panel: construction of the resulting realization of the max-stable process, $z(x) = \sup_i q_i(x)$, (heavy) as the pointwise supremum of individual processes $q_i(x) = r_i w_i(x)$. Most of the q_i are tiny because of the small associated r_i.

Any max-stable process may then be constructed through the *spectral representation* [24]

$$Z(x) = \sup_{i=1}^{\infty} Q_i(x), \quad x \in \mathcal{X}, \tag{31.9}$$

where $Q_i(x) = R_i W_i(x)$, with the R_i defined as before and the $W_i(x)$ independent replicates of $W(x)$ independent of the R_i. Then the point $Q_i(x)$ may be interpreted as the ith event, whose overall 'scale' and profile are respectively R_i and $W_i(x)$. The left-hand panel of Figure 31.2 shows an example in which $\mathcal{X} = (0, 1)$ and $W(x) = \exp\{\sigma\varepsilon(x) - \sigma^2/2\}$, where $\varepsilon(x)$ is a stationary Gaussian process with mean zero, unit variance and correlation function $\text{corr}\{\varepsilon(x), \varepsilon(x + h)\}$ $(x, x + h \in \mathcal{X})$; in the plot $\sigma = 1$ and the correlation function is $\exp(-h/\beta)$, with $\beta = 0.5$.

Since $W(x) \geq 0$ and $\text{E}\{W(x)\} = 1$ for each x, the mean measure for $Q(x) = RW(x)$ is (31.4), so $\text{Pr}\{Z(x) \leq z\} = \exp(-1/z)$ for each x; the max-stable process is then called *simple*.

It may be shown that any max-stable process may be constructed via (31.9), though the representation is not unique—recall that in deriving (31.4) we saw that the same mean measure, and hence the same Poisson process, would have been obtained for any positive random variable W with unit expectation. This non-uniqueness may be exploited to provide efficient simulation algorithms [27, 28, 57, 59, 81].

To obtain the joint distribution of $Z(x)$ for values of x lying in a subset $\mathcal{D}$ of $\mathcal{X}$, note that the event $\{Z(x) \leq z(x) : x \in \mathcal{D}\}$ occurs if and only if $R_i W_i(x) \leq z(x)$ for all $x \in \mathcal{D}$ and all i, and this is equivalent to $R_i \leq \inf_{x \in \mathcal{D}} z(x)/W_i(x)$ $(i = 1, 2, \ldots)$. In terms of the corresponding Poisson process this implies that the set

$$\{(r, w) : rw(x) \leq z(x), x \in \mathcal{D}\}^c = \left\{(r, w) : r > \inf_{x \in \mathcal{D}} z(x)/w(x)\right\},$$

where superscript c denotes complement, is void. The measure of this set may be expressed as

$$\int \nu(dw) \int_{\inf_{x \in \mathcal{D}}\{z(x)/w(x)\}} \frac{dr}{r^2} = \mathrm{E}\left[\sup_{x \in \mathcal{D}}\left\{\frac{W(x)}{z(x)}\right\}\right] = V\{z(x) : x \in \mathcal{D}\}, \qquad (31.10)$$

say, where the expectation is over the measure ν of W, called the *angular measure* by [14]. Thus the required probability for $Z(x)$ is a void probability of the Poisson process defined by $Q(x) = RW(x)$, for $x \in \mathcal{D}$, and so

$$\Pr\{Z(x) \leq z(x), x \in \mathcal{D}\} = \exp\left[-V\{z(x) : x \in \mathcal{D}\}\right]. \qquad (31.11)$$

If $\mathcal{D}$ is finite, then this probability equals expression (8.5); it equals the standard Fréchet distribution $\exp\{-1/z(x)\}$, $z(x) > 0$, for a singleton $\mathcal{D} = \{x\}$.

The exponent function V plays a central role in inference and modelling. Since $Z(x) \leq z(x)$ ($x \in \mathcal{D}$) if and only if all the $Q_i(x) = R_i W_i(x)$ fall into the set $\{(r, w) : rw(x) \leq z(x), x \in \mathcal{D}\}$, the measure of the Poisson process with points $Q_i(x)$ is determined by setting

$$\mu[\{q : q(x) \leq z(x), x \in \mathcal{D}\}^c] = V\{z(x) : x \in \mathcal{D}\};$$

thus (31.11) is the void probability of $\{q : q(x) \leq z(x), x \in \mathcal{D}\}^c$ for the Poisson process.

The case $\mathcal{D} = \{x_1, \ldots, x_D\}$ is particularly important, because in practice data are observed only on finite sets, and inference must therefore be based on observations of $Z(x_1), \ldots, Z(x_D)$. Writing $z(x_d) = z_d$ for simplicity, we let

$$V(z_1, \ldots, z_D) = \mu(\mathcal{A}_z), \qquad z_1, \ldots, z_D > 0,$$

where $\mathcal{A}_z$ is the complement of $[0, z_1] \times \cdots \times [0, z_D]$ in $\mathcal{E} = [0, \infty)^D \setminus \{0\}$; the origin cannot be included because the limiting Poisson process gives infinite measure for sets that contain it.

Expression (31.10) implies that V is homogeneous of order -1, i.e.,

$$V\{az(x) : x \in \mathcal{D}\} = a^{-1}V\{z(x) : x \in \mathcal{D}\}, \qquad a > 0.$$

The max-stability property (31.8) is easily established, because the event $n^{-1}\max\{(Z_1(x), \ldots, Z_n(x)\} \leq z(x)$ is equivalent to $Z_j(x) \leq nz(x)$ ($j = 1, \ldots, n$), and since the $Z_j(x)$ are independent, this occurs with probability

$$\begin{aligned}(\exp\left[-V\{nz(x) : x \in \mathcal{D}\}\right])^n &= \left(\exp\left[-n^{-1}V\{z(x) : x \in \mathcal{D}\}\right]\right)^n \\ &= \exp\left[-V\{z(x) : x \in \mathcal{D}\}\right].\end{aligned}$$

If we set $z(x) \equiv z$, then the homogeneity of V yields

$$\Pr\{Z(x) \leq z, x \in \mathcal{D}\} = \Pr\left\{\max_{x \in \mathcal{D}} Z(x) \leq z\right\} = \exp\left(-\theta_D/z\right), \qquad z > 0, \qquad (31.12)$$

where the *extremal coefficient* θ_D equals V evaluated with $z(x) \equiv 1$ ($x \in \mathcal{D}$). Hence

$\max_{x \in D} Z(x)$ has a Fréchet distribution; its parameter θ_D can be shown to lie between 1 and $|D|$. These bounds respectively correspond to perfect dependence and complete independence of the components of $Z(x)$. In particular, if $D = \{x_1, \ldots, x_D\}$, then $1 \leq \theta_D \leq D$, with smaller values indicating stronger dependence of the extremes. Like the variogram in equation (5.8), empirical estimates of θ_D are useful for exploratory purposes and for checking the adequacy of fitted models; see §31.4 and Figure 31.7.

31.2.4 Exceedances

Max-stable processes provide models for quantities such as annual maximum rainfall at a number of sites in a region, but in some applications it is preferable to model individual extreme events. In terms of the discussion above, this requires deciding which of the $Q_i(x)$ are to be regarded as extreme, and then basing inference on them, so we need to restrict the Poisson process $\{Q_i(x)\}$ to some suitable set $\mathcal{E}'$, for example by taking $\mathcal{E}' = \{q : \rho(q) > 1\}$ for some *risk functional* ρ [23, 81]. We might, for example, take

$$\rho_1(Q) = \sup_{x \in D} Q(x)/z(x), \quad \rho_2(Q) = \inf_{x \in D} Q(x)/z(x), \quad \rho_3(Q) = \int_D Q(x)/z(x)\, dx,$$

corresponding respectively to events that exceed a threshold function $z(x)$ at at least one site in D, those that exceed $z(x)$ everywhere in D and those for which the average of $Q(x)/z(x)$ over D is sufficiently large. Thus the type of rare event selected can be tailored to the problem, as in §31.6.2. Some care is needed in the choice of $\mathcal{E}'$, since likelihood inference involves the integral $\mu(\mathcal{E}')$, which must be both finite and capable of being computed rapidly. If the risk functional satisfies $\rho(aQ) = a\rho(Q)$ for $a > 0$, then $\rho(Q) > 1$ implies that $R\rho(W) > 1$, and then the argument leading to (31.4) can be extended to show that $\mu(\mathcal{E}') = \mathrm{E}\{\rho(W)\}$ depends only on the distribution of W.

31.3   Models

31.3.1   General

By determining their profiles, and thus their extent, their shape, their roughness, and their degree of spatial dependence, the random processes $W(x)$ play a key role in the construction of extreme events using (31.9). The $W(x)$ must be non-negative and satisfy $\mathrm{E}\{W(x)\} = 1$ ($x \in \mathcal{X}$), but many processes, both stationary and non-stationary, satisfy these minimal requirements. In applications it will usually be important to consider the scale of extreme events, their roughness, and their orientation, if any, and all of these may vary over the spatial domain $\mathcal{X}$ if it is large or very heterogeneous. For example, extreme rainfall events are typically smaller in size and more variable than major heat waves, and they may have a directionality given by particular weather patterns.

Below we describe some widely-used forms for $W(x)$ based on an underlying Gaussian or Student-t process; ideas from the extensive literature on Gaussian-based geostatistics may then be ported to the extremal setting. Anisotropic and non-stationary variograms can be used [4, 45].

Throughout the following discussion $\{\varepsilon(x)\}$ represents a zero-mean Gaussian process with variogram

$$\gamma(x_1, x_2) = \text{var}\{\varepsilon(x_1) - \varepsilon(x_2)\}, \quad x_1, x_2 \in \mathcal{X},$$

and $\{\varepsilon(x)\}$ may be either stationary or intrinsically stationary. In the stationary case $\text{var}\{\varepsilon(x)\} = \sigma^2$ is constant on $\mathcal{X}$, so the variogram is bounded,

$$0 \leq \gamma(x_1, x_2) = 2[\sigma^2 - \text{cov}\{\varepsilon(x_1), \varepsilon(x_2)\}] \leq 2\sigma^2,$$

and we may define a correlation function $c(x_1, x_2) = \text{corr}\{\varepsilon(x_1), \varepsilon(x_2)\}$. In the intrinsically stationary case the increments $\varepsilon(x_1) - \varepsilon(x_2)$ are stationary, and then the variogram may be unbounded. In this case we may define $\varepsilon'(x) = \varepsilon(x) - \varepsilon(x')$, where $x' \in \mathcal{X}$ is a fixed site, so that $\varepsilon'(x') = 0$ with probability one, and then it can be shown that

$$\text{cov}\{\varepsilon'(x_1), \varepsilon'(x_2)\} = \tfrac{1}{2}\{\gamma(x_1, x') + \gamma(x_2, x') - \gamma(x_1, x_2)\},$$

thereby expressing the covariance function of the process $\{\varepsilon'(x)\}$ in terms of the variogram of the original process.

In many applications the variogram is taken to be *stationary*, i.e., it depends only on $h = x_1 - x_2$, and *isotropic*, i.e., it depends only on the length $\|h\|$ of h. A common choice is the so-called *stable variogram*, $\gamma(h) = (\|h\|/\lambda)^\kappa$, with $\lambda > 0$ and $\kappa \in (0, 2]$, as variation in κ and λ yields max-stable processes with quite different roughnesses and scales for spatial dependence. The corresponding correlation function is $c(x_1, x_2) = \exp\{-\gamma(h)\}$.

Use of a stationary or intrinsically stationary process may impact the properties of the resulting extremal model: roughly speaking, models with bounded variance cannot represent independence of extremes at long distances, whereas those with unbounded variance can.

As mentioned in §5.2.1, new variograms and covariance functions can be constructed from existing ones. For example, [1] found that a linear combination of a variogram representing meteorological dependence and a covariance function representing flow-dependence along a river network gave a good spatial model for extreme river flows.

31.3.2   Brown–Resnick process

A simple idea is to set $W(x) = \exp[\varepsilon'(x) - \text{var}\{\varepsilon'(x)\}/2]$, where $\{\varepsilon'(x)\}$ is based on an intrinsically stationary Gaussian process. Clearly $W(x)$ is non-negative and has unit expectation throughout $\mathcal{X}$, and it turns out that (31.9) is a strictly stationary simple max-stable process, known as a Brown–Resnick process [6, 49, 50], whose distribution depends only on γ. Such processes are popular because a wide range of variograms can be employed with them, they are relatively easily simulated [28, 29, 57], and likelihoods for them have an explicit form; see §31.5.3.

If the definition of $W(x)$ is modified by replacing $\varepsilon'(x)$ by a stationary Gaussian process, then $\gamma(x_1, x_2)$ is bounded and the resulting max-stable process will be dependent even at very long ranges, which is often unrealistic in applications. Hence an intrinsically stationary process is often preferred.

The bivariate distribution function for such models may be written as

$$\Pr\{Z(x_1) \leq z_1, Z(x_2) \leq z_2\} = \exp\{-V(z_1, z_2)\}, \quad z_1, z_2 > 0, x_1, x_2 \in \mathcal{X},$$

with

$$V(z_1, z_2) = \frac{1}{z_1}\Phi\left\{\frac{a}{2} + \frac{1}{a}\log\left(\frac{z_2}{z_1}\right)\right\} + \frac{1}{z_2}\Phi\left\{\frac{a}{2} + \frac{1}{a}\log\left(\frac{z_1}{z_2}\right)\right\}, \tag{31.13}$$

where Φ is the standard normal cumulative distribution function and a is the positive square root of $\gamma(x_1, x_2)$. In this case the pairwise extremal coefficient, corresponding to taking $\mathcal{D} = \{x_1, x_2\}$, may be written as

$$\theta(x_1, x_2) = V(1, 1) = 2\Phi\left\{\gamma^{1/2}(x_1, x_2)/2\right\}, \quad x_1, x_2 \in \mathcal{X},$$

so small and large values of $\gamma(x_1, x_2)$ respectively correspond to strong and to weak dependence; as $\gamma(x_1, x_2) \to 0$ and $\gamma(x_1, x_2) \to \infty$, we see that

$$V(z_1, z_2) \to \max(1/z_1, 1/z_2), \quad V(z_1, z_2) \to 1/z_1 + 1/z_2,$$

corresponding respectively to complete dependence of $Z(x_1)$ and $Z(x_2)$ and to their independence.

The generalization of (31.13) to D variables involves the $(D-1)$-dimensional multivariate normal distribution function; see [56, Remark 2.5]. One drawback with using it is that repeated computation of this multivariate normal distribution as part of an iterative estimation algorithm can be costly, so in realistic cases one is constrained to $D \leq 50$ or so, at least for maximum likelihood estimation [23].

31.3.3 Extremal-t process

Extremal-t processes [60] arise as limits of renormalized maxima of elliptical processes, which include Student t processes. In this case $W(x) = m_\alpha \varepsilon(x)_+^\alpha$, where $\alpha > 0$, $m_\alpha = \pi^{1/2} 2^{1-\alpha/2}/\Gamma\{(\alpha+1)/2\}$ with $\Gamma(\cdot)$ the gamma function, and $\varepsilon(x)$ is a stationary Gaussian process with mean zero, unit variance and correlation function $\mathrm{corr}\{\varepsilon(x_1), \varepsilon(x_2)\} = c_\alpha(x_1, x_2)$. When $\alpha = 1$ this yields the so-called *Schlather process* [72]. Under certain conditions the extremal-t process converges to a Brown–Resnick process as $\alpha \to \infty$. For example, if $c_\alpha(x_1, x_2) = \exp\{-(2\alpha)^{-1}(\|x_2 - x_1\|/\lambda)^\kappa\}$ for some $\lambda > 0$ and $\kappa \in (0, 2]$, then the limiting Brown–Resnick process has $\gamma(x_1, x_2) = (\|x_2 - x_1\|/\lambda)^\kappa$. Hence an extremal-$t$ process will typically fit data at least as well as a Brown–Resnick process. Its bivariate exponent function $V(z_1, z_2)$ equals

$$\frac{1}{z_1} T_{\alpha+1}\left\{-\frac{c}{b} + \frac{1}{b}\left(\frac{z_2}{z_1}\right)^{1/\alpha}\right\} + \frac{1}{z_2} T_{\alpha+1}\left\{-\frac{c}{b} + \frac{1}{b}\left(\frac{z_1}{z_2}\right)^{1/\alpha}\right\}, \quad z_1, z_2 > 0, \qquad (31.14)$$

where $T_\nu(\cdot)$ is the cumulative distribution function of the Student t distribution with ν degrees of freedom, $c = c_\alpha(x_1, x_2)$ and $b^2 = (1 - c^2)/(\alpha + 1)$. It is straightforward to see that $\theta(x_1, x_2) = 2T_{\alpha+1}[\{(1-c)(1+\alpha)/(1+c)\}^{1/2}]$, which has upper bound $2T_{\alpha+1}\{(1+\alpha)^{1/2}\}$; thus independence can only be attained as $\alpha \to \infty$. The higher-order exponent functions are given by [56, Theorem 2.3]; see also [81].

31.3.4 Other models

A different class of max-stable models [65, 66, 76] is based on a hierarchical specification of the joint distribution of generalized extreme-value variables in terms of latent positive stable random variables that account for spatial dependence of the extremes; it can be viewed as a noisy version of the Smith model described below. One advantage of this formulation over those described above is that it allows Bayesian inference based on standard Markov chain Monte Carlo methods, and a second advantage is that the latent variable construction can itself have a hierarchical structure, though Monte Carlo algorithms for this model can be slow.

Max-stable processes may be modified in various ways. For example, skew-normal and skew-t processes can be used, rather than Gaussian or t processes, and these may be useful when asymmetries are present [2]. Another extension is multiplication by the indicator of a random compact set $\mathcal{B}$ independent of $W(\cdot)$, resulting in

$$W_{\mathcal{B}}(x) = W(x)I(x \in \mathcal{B})/\mathrm{E}(|\mathcal{B}|), \quad x \in \mathcal{X},$$

where the expectation $\mathrm{E}(|\mathcal{B}|)$ must be finite in order to ensure that $\mathrm{E}\{W_{\mathcal{B}}(x)\} \equiv 1$. If we write $\alpha(x_1, x_2) = \mathrm{Pr}(x_1, x_2 \in \mathcal{B})$, then it is straightforward to check that the corresponding exponent function is

$$\{1 - \alpha(x_1, x_2)\} \left(\frac{1}{z_1} + \frac{1}{z_2} \right) + \alpha(x_1, x_2)V(z_1, z_2), \quad z_1, z_2 > 0.$$

Thus if, for example, $\mathcal{B}$ is a disk of fixed diameter L centred at a point uniformly distributed on $\mathcal{X}$, then x_1 and x_2 cannot both lie in $\mathcal{B}$ when $\|x_1 - x_2\| > L$, and $Z(x_1)$ and $Z(x_2)$ must be independent, whereas if $x_1 \approx x_2$, then the exponent function will almost be that of the max-stable process based on W. [43] use this idea to model extreme hourly rainfall; their random sets are stylized representations of the space-time extent of rainfall cells, each containing an independent process $W(x)$.

Yet more processes may be constructed by letting $W(x) = f(x; Y)/f_y(Y)$, where Y has density f_y on some space $\mathcal{Y}$, and $\int_{\mathcal{Y}} f(x; y)\, dy = 1$ for each x. The simplest such model, the *Smith process* [77], has $\mathcal{Y} = \mathcal{X} = \mathbb{R}^k$, $f_y(y)$ an arbitrary density supported on $\mathbb{R}^k$ and $f(x; y) = \phi_k(x - y; \Omega)$, where $\phi_k(\cdot; \Omega)$ is the k-variate Gaussian density with covariance matrix Ω. The corresponding random functions $Q(x) = RW(x)$ consist of k-variate Gaussian densities centered at random points of $\mathbb{R}^k$, whose shape is determined by Ω and whose maximum height is determined by R and $|\Omega|$. These and similar processes obtained by using other standard probability density functions [25] are typically too smooth to be realistic in applications, and, though [85] proposed a rougher variant, we do not recommend them for use in practice.

31.3.5   Asymptotic independence

Most of the models described above presuppose that the transformed extremes of the phenomenon under investigation are *asymptotically dependent*, i.e.,

$$\lim_{z \to \infty} \mathrm{Pr}\{Z(x_2) > z \mid Z(x_1) > z\} = 2 - \theta(x_1, x_2) > 0, \quad x_1, x_2 \in \mathcal{X}, \qquad (31.15)$$

where $\theta(x_1, x_2) = V(1, 1)$ is the exponent function corresponding to $Z(x_1)$ and $Z(x_2)$. This implies that dependence between $Z(x_1)$ and $Z(x_2)$ persists even for very rare events, since the conditional probability of a rare event at x_2, given the occurrence of an equally rare event at x_1, remains non-zero as $z \to \infty$. An alternative, which arises if $V(z_1, z_z) = 1/z_1 + 1/z_2$, is that extremes are exactly independent, but this is not a useful model because data almost always show some degree of dependence. An intermediate possibility is needed, whereby the limit in (31.15) is zero, but the conditional probability decreases as $z \to \infty$. This is often plausible from physical arguments and is common in applications, of which it can be an important feature. Such *asymptotic independence* models are a topic of current research [e.g., 46, 48, 61, 85]. Earlier models that have this property are based on copulas [38, 70, 71]. Spatial applications of asymptotic independence models may also be found in [19] and [80].

31.4   Exploratory procedures

Simple nonparametric estimation procedures are valuable for initial analysis and for checking the validity of fitted models. Most are based on pairs of extremal observations, which offer direct insight into the dependence structure of the data, in analogy with quantities such as the variogram of classical geostatistics, which is defined in terms of the average squared difference between observations. Since extremal observations need not possess moments, a rank-based approach is preferable.

Suppose that observations $Z_1 \equiv Z(x_1)$ and $Z_2 \equiv Z(x_2)$ at sites $\mathcal{D} = \{x_1, x_2\} \in \mathcal{X}$ arise from a simple max-stable process. A natural measure of the dependence between Z_1 and Z_2 is the extremal coefficient $\theta_{\mathcal{D}}$ defined in (31.12); we have $\theta_{\mathcal{D}} \in [1, 2]$, with smaller values corresponding to stronger dependence, so we can expect $\theta_{\mathcal{D}}$ to increase as the distance $\|x_1 - x_2\|$ grows.

Estimation of $\theta_{\mathcal{D}}$ may be based on the F-madogram [15]

$$\psi_{\mathcal{D}} = \tfrac{1}{2}\mathrm{E}\left\{|F(Z_1) - F(Z_2)|\right\},$$

where $F(z) = \exp(-1/z)$ $(z > 0)$ is the standard Fréchet distribution. Since $|a - b| = 2\max(a, b) - a - b$, and the distribution functions of Z_1, Z_2 and $\max(Z_1, Z_2)$ are respectively $F(z)$, $F(z)$ and $\exp(-\theta_{\mathcal{D}}/z)$ $(z > 0)$, it is straightforward to check that $\psi_{\mathcal{D}} = \tfrac{1}{2}(\theta_{\mathcal{D}} - 1)/(\theta_{\mathcal{D}} + 1)$ and therefore that

$$\theta_{\mathcal{D}} = \frac{1 + 2\psi_{\mathcal{D}}}{1 - 2\psi_{\mathcal{D}}}.$$

An estimator $\widehat{\theta}_{\mathcal{D}}$ based on independent pairs $(Z_{1,1}, Z_{2,1}), \ldots, (Z_{1,n}, Z_{2,n})$ may be obtained by replacing $F(Z_1)$ by the rank estimator $R_{1,j}/(n + 1)$ $(j = 1, \ldots, n)$, where $R_{1,j}$ is the rank of $Z_{1,j}$ among $Z_{1,1}, \ldots, Z_{1,n}$, and likewise with $F(Z_2)$, and then replacing $\psi_{\mathcal{D}}$ by $\{2n(n + 1)\}^{-1}\sum_{j=1}^{n}|R_{1,j} - R_{2,j}|$, its empirical counterpart. If the underlying process is stationary, then $\theta_{\mathcal{D}}$ depends only on $h = x_1 - x_2$, so a plot of the $\widehat{\theta}_{\mathcal{D}}$ for all possible pairs of points $\mathcal{D}$ can be smoothed or binned and used for diagnostic or confirmatory purposes. As in the classical case, some care must be taken in interpreting such plots, since data from D underlying sites yield a cloud of $D(D - 1)/2$ correlated estimates $\widehat{\theta}_{\mathcal{D}}$; moreover these empirical estimates may be very variable. [54] extend the F-madogram to estimation of the function $V_{\mathcal{D}}(a, 1 - a)$ $(0 < a < 1)$.

The anticipated increase in $\theta_{\mathcal{D}}$ with $\|x_1 - x_2\|$ makes a natural analogy with the variogram, but since $\Pr(Z_2 > z \mid Z_1 > z) \sim 2 - \theta_{\mathcal{D}}$ as $z \to \infty$, the quantity $2 - \theta_{\mathcal{D}}$ may be interpreted as the probability of a large event at x_2 given a correspondingly large event at x_1. A plot of $2 - \widehat{\theta}_{\mathcal{D}}$ against distance is known as an *extremogram*, analogous to the correlogram of classical time series analysis, and has been proposed for use particularly with time series [17]; resampling can be used to gauge which probabilities are indistinguishable from zero. Cross-extremograms or extremes of two or more series can also be defined, but like cross-correlograms, they are vulnerable to spurious correlations for which allowance must be made.

The conditional probability $2 - \theta_{\mathcal{D}}$ has a natural interpretation for asymptotically dependent processes, but equals zero for asymptotically independent processes, whatever their rate of tail decay, so it is natural to seek a measure of asymptotic independence. See [11] and [52] for examples and further discussion.

As many standard models are based on underlying Gaussian processes, which are defined in terms of their first and second moments, standard fitting procedures will tend to match these to the data, either explicitly or implicitly. We might therefore expect theoretical and empirical estimates of pairwise quantities, such as the extremal coefficient $\theta(h)$ of two sites a distance h apart, to match well enough to make model failure more difficult to detect. This suggests using higher-order, in addition to pairwise, quantities for checking model fit. For example, if $\mathcal{D} = \{x_1, \ldots, x_D\}$ is a subset of sites, then it is easy to see that $\max_{x \in \mathcal{D}} Z(x)$ has the Fréchet distribution $\exp(-\theta_{\mathcal{D}}/z)$. Thus we might compare quantiles of the observed maxima for $\mathcal{D}$ with those from a fitted model, for sets $\mathcal{D}$ that are spatially close or spatially far-flung; if the fitted model is adequate, the empirical and simulated distributions should match. Similarly, empirical probabilities of certain spatial events involving more than two sites could be compared with estimates based on the model.

31.5 Inference

31.5.1 General

Although there is a growing literature on non-parametric inference for extremal processes [e.g., 32], parametric modelling is most used in applications. One reason for this is that it is usually more straightforward to interpret a parametric model, and a second is that a common goal of fitting such processes to data is the estimation of rare event probabilities. Typically this is performed by Monte Carlo simulation from one or more fitted models, which may be easier for a parametric model. Simulation from max-stable processes is discussed by [28] and references therein. [29] give an algorithm for simulation conditional on the observed values of the max-stable process at certain points.

Statistical estimation for extreme values is always subject to mis-specification bias, since the classes of models fitted are typically based on asymptotic arguments that do not apply to the data themselves. It can be hard to detect lack of fit, because extremal models tend to be rather flexible and power for goodness-of-fit assessment is limited owing to the small number of extreme events. Moreover the uncertainty surrounding extrapolation to events even rarer than those already observed is typically so large that bias is a secondary consideration. It is essential to assess the sensitivity of conclusions both to the choice of the model and to the degree of 'extremeness' of the data: this entails checking that the fitted models do not change greatly when the threshold or block size is varied over a reasonable range.

Below we focus on likelihood-based estimation, but other proposals include robust estimation based directly on the joint distribution function [89], approximate Bayesian computation [35] and score-based methods [23].

Most of the discussion above has presupposed that the data have been transformed to the standard Fréchet scale, but in practice this transformation should be built into the data analysis. For example, if annual maxima at points $\mathcal{D} = \{x_1, \ldots, x_D\} \subset \mathcal{X}$ are available, then it will often be reasonable to suppose that their marginal distributions are generalized extreme-value with location, scale and shape parameters $\eta(x)$, $\tau(x)$ and $\xi(x)$ that vary smoothly over $\mathcal{X}$ as functions of parameters φ. Often we express $\eta(x)$ and $\log \tau(x)$ as linear combinations of basis functions and explanatory variables such as altitude, but assume that $\xi(x)$ is constant, since variation in the shape parameter can be hard to detect and the value of ξ may be seen as an intrinsic aspect of the phenomenon being modelled. Since we can write

$$Z(x; \varphi) = [1 + \xi(x; \varphi)\{Y(x) - \eta(x; \varphi)\}/\tau(x; \varphi)]_+^{1/\xi(x; \varphi)},$$

where $Y(x)$ is the observed maximum at s and $Z(x; \varphi)$ has a standard Fréchet distribution, the joint probability density function of $Y(x_1), \ldots, Y(x_D)$ can be written in terms of the standardized variables

$$z(x_d; \varphi) = [1 + \xi(x_d; \varphi)\{y_d - \eta(x_d; \varphi)\}/\tau(x_d; \varphi)]_+^{1/\xi(x_d;\varphi)}, \quad d = 1, \ldots, D,$$

as

$$f_{Z(x_1),\ldots,Z(x_D)}\{z(x_1; \varphi), \ldots, z(x_D; \varphi); \vartheta\} \times \prod_{d=1}^{D} \frac{dz(x_d; \varphi)}{dy_d}, \tag{31.16}$$

where ϑ represents parameters in the joint distribution of the Zs and the second term is a Jacobian.

Maximum likelihood estimation will involve maximising a product of expressions of the form (31.16), for example with a term for each different year of observed maxima, over both φ and ϑ. Computational considerations may entail the estimation of φ separately from ϑ in a two-step procedure, though in principle this is undesirable because an overall measure allowing for the uncertainty of both $\widehat{\vartheta}$ and $\widehat{\varphi}$ is then awkward to obtain. In the applications below we found it necessary to use this two-step approach, but then obtained a combined assessment of uncertainty by applying it to bootstrap datasets obtained by resampled subsets of the original data that could be taken as independent. Initial values for φ can be obtained by maximising an independence likelihood, which uses the same smooth functions for spatial variation of the marginal parameters η, τ and ξ as in (31.16), but treats the Zs as independent standard Fréchet variables. Similar computations are needed for likelihoods based on copulas.

31.5.2 Likelihood inference for maxima

We now discuss the computation of the first term in (31.16) in the case of inference based on maxima. The key problem is that this involves D-fold differentiation of the joint cumulative distribution function (31.11), resulting in a combinatorial explosion. For example, with $D = 3$ we have

$$f(z_1, z_2, z_3; \vartheta) = \exp(-V)\left(-V_{123} + V_1 V_{23} + V_2 V_{13} + V_3 V_{12} - V_1 V_2 V_3\right), \tag{31.17}$$

using shorthand notation in which

$$V = V(z_1, z_2, z_3; \vartheta), \quad V_d = \frac{\partial V(z_1, z_2, z_3; \vartheta)}{\partial z_d}, \quad V_{d_1 d_2} = \frac{\partial^2 V(z_1, z_2, z_3; \vartheta)}{\partial z_{d_1} \partial z_{d_2}},$$

and so forth. The number of terms in the density is the number of partitions of the set $\mathcal{D}$; with $\mathcal{D} = \{x_1, x_2, x_3\}$ there are five such partitions, corresponding to the terms on the right-hand side of (31.17), but with $D = 10$ there are around 10^5 terms. Thus in realistic settings it is infeasible to compute the full likelihood, and a number of ways to avoid doing so have been proposed.

Composite likelihood inference [53, 82] uses simpler likelihood components. The most common is a pairwise likelihood, in which the components involve only pairs of observations, for example replacing (31.17) by

$$f(z_1, z_2; \vartheta) \times f(z_1, z_3; \vartheta) \times f(z_2, z_3; \vartheta). \tag{31.18}$$

Here only bivariate exponent functions such as (31.13) and (31.14) and their first and second derivatives are needed. For this approach to be useful, ϑ must be identifiable from the chosen marginal distributions, as is generally the case with Gaussian-based models. Although maximising a composite likelihood yields consistent estimators, $\widehat{\vartheta}_C$, say, measures of uncertainty and tools for model comparison must compensate for the re-use of observations in terms such as (31.18). For example, if z_1, z_2, z_3 were independent, then (31.18) would equal $\{f(z_1; \vartheta) f(z_2; \vartheta) f(z_3; \vartheta)\}^2$, i.e., the square of the correct likelihood contribution, and although $\widehat{\vartheta}_C$ would then equal the usual maximum likelihood estimate, standard errors based on the observed information matrix $\widehat{J}$ would be too small by a factor $\sqrt{2}$.

Standard errors for $\widehat{\vartheta}_C$ can be obtained by resampling or using a 'sandwich' variance matrix $\widehat{J}^{-1}\widehat{K}\widehat{J}^{-1}$. Suppose that the data fall into independent blocks indexed by $i \in \mathcal{I}$, and that the ith block consists of dependent observations $z_{i,d}$. In many cases $\mathcal{I}$ will correspond to data from different years or extreme events, assumed to be independent, whereas the data $z_{i,1}, \ldots, z_{i,D}$ for block i are likely to be dependent. Then with $\ell_{i;d',d}(\vartheta) = \log f(z_{i,d'}, z_{i,d}; \vartheta)$ and pairwise likelihood $\sum_i \sum_{d' < d} \ell_{i;d',d}(\vartheta)$, we have

$$\widehat{K} = \sum_{i \in \mathcal{I}} \sum_{d' < d} \frac{\partial \ell_{i;d',d}(\vartheta)}{\partial \vartheta} \frac{\partial \ell_{i;d',d}(\vartheta)}{\partial \vartheta^{\mathrm{T}}}\bigg|_{\vartheta = \widehat{\vartheta}_C}, \quad \widehat{J} = -\sum_{i \in \mathcal{I}} \sum_{d' < d} \frac{\partial^2 \ell_{i;d',d}(\vartheta)}{\partial \vartheta \partial \vartheta^{\mathrm{T}}}\bigg|_{\vartheta = \widehat{\vartheta}_C}.$$

Data are often missing from some blocks, and then the internal sum is taken over only those pairs available for that block. The matrix $\widehat{K}$ can be unstable, and then resampling of blocks to obtain standard errors is preferable, even though it can be slow.

[9], [39] and [42] investigate the potential increase in estimator variance due to use of composite rather than full likelihoods, and conclude that pairwise likelihoods will often provide an acceptable compromise between statistical efficiency and computational complexity. In some cases the computational burden can be reduced and the statistical properties of $\widehat{\vartheta}_C$ can be enhanced by weighting likelihood contributions $\ell_{i;d',d}(\vartheta)$, for example downweighting pairs that are spatially distant, since these might be expected to contribute less information.

Models are compared using the composite likelihood information criterion, CLIC $= 2\{\mathrm{tr}(\widehat{J}^{-1}\widehat{K}) - \widehat{\ell}_C\}$, where $\widehat{\ell}_C$ is the maximum log composite likelihood; low values of CLIC are preferred. The re-use of observations can lead to very large values of CLIC, and it may be useful to rescale it for comparability with standard information criteria such as AIC; in (31.18), for example, the log likelihood contribution should be halved, and if there were D maxima, it would be divided by $D - 1$. See [20] and [63] for more details.

A second solution to the explosion of terms in expressions such as (31.17) is due to [78] and stems from noticing that the terms in expressions such as the right-hand side of (31.17) correspond to different ways in which the extremes can occur: the first corresponds to a single event leading to all three maxima, the last to maxima from three separate events, and the others to the ways in which three maxima might arise from two separate events. Thus if it was known that z_1 arose from one event, whereas z_2 and z_3 occurred together, then the appropriate likelihood contribution would be $V_1 V_{23} \exp(-V)$ rather than the whole of (31.17). If the required derivatives of V are available then the resulting likelihood is computationally feasible even in high-dimensional cases, but the resulting maximum likelihood estimators depend on the choice of partition and can be badly biased if this choice is wrong [84]. In practice an empirical declustering rule is typically used to estimate the partition, and the approach does not seem to be very robust. It can be improved by estimating the partition in a more formal way, either using a stochastic EM algorithm [44] or, in a Bayesian context, by including the partition in a Monte Carlo sampling algorithm, and then integrating over it [79].

31.5.3 Likelihood inference for threshold exceedances

Analysis of individual events, typically those exceeding some suitable threshold, allows more detailed modelling and more precise inferences than analysis of maxima, and is particularly useful when assessing simultaneous risks. In principle the approach based on the Poisson process approximation using a parametric measure $\mu_\vartheta(\cdot)$ is straightforward: a set $\mathcal{E}'$ having finite measure is chosen to select rare events, and then estimation is based on a Poisson process likelihood. Suppose, for example, that events $q_1, \ldots, q_n$ have been observed at D points $x_1, \ldots, x_D$, so that $q_j \equiv \{q_j(x_1), \ldots, q_j(x_D)\}$, and that $\mathcal{E}' = \{q : \rho(q) > 1\}$ for some functional ρ. Then the selected data will be those q_j for which $\rho(q_j) > 1$, say $q'_1, \ldots, q'_{n'}$, and the likelihood is

$$L_{\text{Pois}}(\vartheta) = \exp\{-\mu_\vartheta(\mathcal{E}')\} \times \prod_{j=1}^{n'} \dot\mu_\vartheta(q'_j), \tag{31.19}$$

where $\dot\mu_\vartheta(q)$ denotes the Poisson process intensity evaluated at q. If $\rho(q) = \sum_{d=1}^{D} q(x_d)/u_d$ for some $u_1, \ldots, u_D > 0$, then $\mathcal{E}' = \{(r, w) : \sum_{d=1}^{D} rw(x_d)/u_d > 1\}$ and $\mu_\vartheta(\mathcal{E}')$ is a constant, so only the second term on the right-hand side of (31.19) is needed; it is known as the *spectral likelihood* [12, 34], though perhaps *angular likelihood* would be a better term. Its drawback is that the events q'_j thus chosen may have some very small components, making the extremal approximation questionable. An alternative ensuring that all events are far from the axes would be to take $\rho(q) = \min_d q(x_d)/u_d$, but this can greatly reduce the number of events available for estimation, especially if extremal dependence is weak. We therefore often take $\rho(q) = \max_d q(x_d)/u_d$, but use a *censored likelihood* to avoid including the exact values of small components, as we now describe. This uses less precise information and thus decreases the precision of our estimates, but it reduces the bias due to inclusion of non-extreme values.

Censored likelihood is most easily explained in the bivariate case shown in Figure 31.3, where $\mathbb{R}_+^2$ is partitioned into four regions $\mathcal{R}_{I_1, I_2}$, corresponding to the values of the indicators $I_1 = I(Q_1 > u_1)$ and $I_2 = I(Q_2 > u_2)$ of the events that the Q_d exceed thresholds u_d. If $I_1 = I_2 = 1$, i.e., both variables exceed the thresholds, then $q \in \mathcal{R}_{1,1}$, the extremal model is regarded as valid and the likelihood contribution from $q = (q_1, q_2)$ is $\dot\mu_\vartheta(q)$. If $I_1 = 1$, $I_2 = 0$, then $q \in \mathcal{R}_{1,0}$, so the precise value of q_1 is used but q_2 is left-censored at u_2, giving likelihood contribution $\int_0^{u_2} \dot\mu_\vartheta\{(q_1, y)\}\,dy$. Likewise when $I_1 = 0$, $I_2 = 1$, the likelihood contribution is $\int_0^{u_1} \dot\mu_\vartheta\{(x, q_2)\}\,dx$. If $I_1 = I_2 = 0$, then $q \in \mathcal{R}_{0,0}$, where the extremal model cannot be trusted, so we use a Poisson process likelihood with region $\mathcal{E}' = \mathbb{R}_+^2 \setminus \mathcal{R}_{0,0} = \mathcal{E}_u$, say, thus having censored contributions from $\mathcal{R}_{0,1}$ and $\mathcal{R}_{1,0}$ and uncensored contributions from $\mathcal{R}_{1,1}$. This likelihood involves computation of $\mu(\mathcal{E}_u) = V(u_1, u_2)$, where V is the exponent function corresponding to $W(x)$; see (31.11). One can check that $\dot\mu_\vartheta(q) = -\partial^2 V(q_1, q_2)/\partial q_1 \partial q_2$ and that the terms needed for the censored contributions have forms such as $-\partial V(u_1, q_2)/\partial q_2$, with obvious extensions to higher dimensions.

Two important examples of the computations above correspond to the Brown–Resnick and extremal-t processes, for which $\dot\mu_\theta$ is given in [86] and [81] respectively. The corresponding censored likelihoods require integrals of $\dot\mu_\theta$, convenient forms for which are also given in these papers.

31.6 Examples

31.6.1 Saudi Arabian rainfall

To illustrate some of the challenges arising when modeling spatial extremes, we analyze block maxima of Saudi Arabian rainfall data, collected by the Tropical Rainfall Measuring Mission from 1 January 1998 to 31 December 2014. Over this 17-year period, three-hourly rainfall satellite measurements (in mm/hr) were gathered over the globe at a resolution

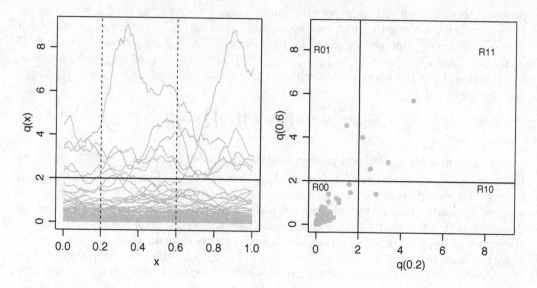

FIGURE 31.3

Censored likelihood inference for individual events $q_i(x)$. Left panel: 100 realizations $q_i(x)$ with threshold $u = 2$ shown by the horizontal line, and dashed vertical lines at $x_1 = 0.2$ and $x_2 = 0.6$. Right panel: pairs $(q_i(x_1), q_i(x_2))$ corresponding to the intersections of the dashed lines in the left-hand panel with the $q_i(x)$, and threshold $u = 2$ shown by the black lines. Points lying in $\mathcal{R}_{1,1}$ have $q_i(x_1), q_i(x_2) > u$, and are treated as uncensored, those lying in $\mathcal{R}_{0,1}$ have likelihood contributions corresponding to the event $q_i(x_1) \leq u$ but $q_i(x_2)$ known exactly, those in $\mathcal{R}_{0,0}$ correspond to $q_i(x_1), q_i(x_2) \leq u$, etc.

of 0.25°, corresponding to around 28km on the equator. Here we consider a tropical arid region consisting of 750 of these grid cells near the Red Sea in Saudi Arabia; see Figure 31.4. We used these rainfall data in our illustration because they are freely available online and provide fairly good spatio-temporal coverage with almost no missing values, although their spatial resolution is quite coarse and might not accurately represent the most localized rainfall events. Data with higher spatial resolution would clearly be desirable, but using them would increase the associated computational burden.

In the last decade, Jeddah, the second largest city in Saudi Arabia, has been hit several times by convective storms, with short but intense rainfall causing extensive flash-floods, damage and deaths [26, 88]. In order to distinguish independent storms and to reduce the number of zero observations, we compute the daily cumulative rainfall at each grid cell, and then extract the 17 annual daily rainfall maxima for each cell, yielding a 'space-rich but time-poor' dataset comprising a total of $17 \times 750 = 12750$ annual maxima—though spatial dependence means that the number of 'equivalent independent' maxima each year is lower than 750. Basic statistics for the daily rainfall totals are reported in Table 31.1, and clearly suggest that the distribution of positive rainfall intensities is highly right-skewed and heavy-tailed. Because the study region is arid, some grid cells experienced no rain in certain years, so six annual maxima, in different grid cells, are exactly equal to zero. We deal with this by censoring of low maxima, as described below. An alternative would be to take maxima over multiple years, but this could significantly reduce the precision of our estimates, due to the low number of temporal replicates.

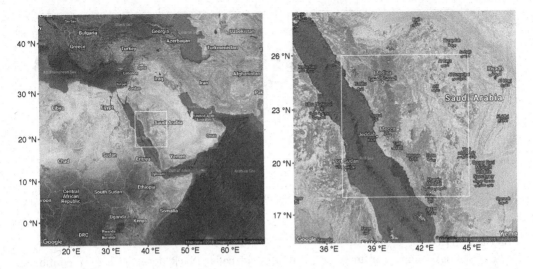

FIGURE 31.4
Satellite map and limits (white square) of the study region (left-hand panel), shown in detail in the right-hand panel for the Saudi Arabian rainfall data example.

TABLE 31.1
Basic statistics for the daily rainfall totals (measured in mm/day), combined over all grid cells in the study region in Western Saudi Arabia. The central moments (first row) and empirical quantiles (second row) concern positive rainfall intensities (i.e., non-zero observations).

Prop. wet days		Mean	Variance	Skewness	Kurtosis		
4.2%		5.2	56.2	3.9	29.8		

Minimum	25%	50%	75%	90%	95%	99%	Maximum
0.3	0.9	2.7	6.3	12.9	19.2	36.0	152.1

We assume that any annual rainfall maxima exceeding $u = 3\text{mm/day}$ follow the generalized extreme-value (GEV) distribution with spatially-varying location, scale and shape parameters, $\eta(s)$, $\tau(s)$ and $\xi(s)$, where $s \in \mathcal{S}'$ indicates a specific grid cell and $\mathcal{S}'$ is the collection of all grid cells in the study region. That is,

$$\Pr\{Z(s) \le z\} = \exp\left[-\left\{1 + \xi(s)\frac{z - \eta(s)}{\tau(s)}\right\}_+^{-1/\xi(s)}\right], \quad z > u = 3\text{mm/day},$$

where $a_+ = \max(0, a)$ and $Z(s)$ denotes an annual maximum observed at site $s \in \mathcal{S}'$. As there are only 17 temporal replicates, it is important to borrow strength across sites for accurate estimation of marginal parameters, but on a topographically and climatically diverse region of this size (880×880 km^2, see Figure 31.4), it is difficult to find simple relationships that well capture the spatial variation of the marginal parameters. We therefore adopt a local censored likelihood approach and, neglecting spatial dependence at this stage, we maximize the locally-weighted log likelihood [3, 8, 21]

$$\ell_{s_0}(\eta, \tau, \xi) = \sum_{i=1}^{17} \sum_{d \in \mathcal{N}(s_0)} \omega(\|s_d - s_0\|)\log g_u(z_{i,d}; \eta, \tau, \xi) \tag{31.20}$$

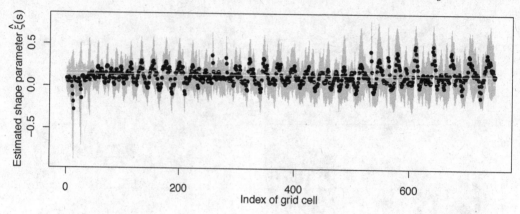

FIGURE 31.5
Estimated GEV shape parameter $\widehat{\xi}(s)$ at all grid cells (black dots) with 95% confidence intervals (grey segments), from a local likelihood fit using a biweight function with bandwidth $b = 80$km. Estimates are highly correlated across grid cells. The horizontal lines show $\widehat{\xi}$ (solid) when assumed to be constant over the study region, with 95% confidence intervals (dashed).

for each grid cell $s_0 \in \mathcal{S}'$; here $z_{i,d}$ is the ith observed annual maximum observed at the dth nearby station, $\mathcal{N}(s_0)$ is an index set corresponding to grid cells within a small neighborhood of s_0, $\omega(h)$ is a weight function that depends on the distance $h = \|s_d - s_0\| \geq 0$ between s_0 and s_d, and $g_u(z; \eta, \tau, \xi)$ is the censored GEV likelihood contribution, i.e., $g_u(z; \eta, \tau, \xi) = g(z; \eta, \tau, \xi)$ if $z > u$, and $g_u(z; \eta, \tau, \xi) = G(u; \eta, \tau, \xi)$ if $z \leq u$, where g and G respectively denote the GEV density and distribution functions. To obtain smooth marginal surfaces, we weight the log likelihood contributions by the biweight function $\omega(h) = \{1 - (h/b)^2\}_+^2$ with bandwidth $b > 0$. The choice of b entails a bias-variance tradeoff, with larger values producing smoother surfaces with less local detail. Some experimentation showed that taking $b = 80$km yields reasonable marginal fits, while greatly reducing parameter estimation uncertainty by increasing the number of observations used to estimate the margins. The uncertainty of our estimates was assessed using a non-parametric bootstrap, whereby we resampled the 17 years of data, in order to retain the spatial structure of the observations, but assuming that the years are independent replicates.

The estimated location parameter $\eta(s)$ and, to some extent, the scale parameter $\tau(s)$, are quite well-estimated, but the shape parameter estimates (corresponding standard errors) vary from -0.28 (0.30) to 0.49 (0.13), implying completely different tail behavior for the extremes, which is harder to interpret. The shape parameter is typically difficult to estimate, see Figure 31.5, and here it is plausible that there are too few years to obtain reliable results. To deal with this, we fix $\xi(s) \equiv \xi$ over the entire region $\mathcal{S}'$, and use a profile likelihood based on (31.20) to estimate ξ over the grid $\mathcal{G}_\xi = \{-0.50, -0.49, \ldots, 1.00\}$, i.e., $\widehat{\xi} = \arg\max_{\xi \in \mathcal{G}_\xi} \sum_{s \in \mathcal{S}} \ell_s(\widehat{\eta}_\xi(s), \widehat{\tau}_\xi(s), \xi)$, where $\widehat{\eta}_\xi(s)$ and $\widehat{\tau}_\xi(s)$ are the estimated location and scale parameters for fixed ξ. We obtain $\widehat{\xi} = 0.14$ (0.03), which implies that the rainfall distribution is slightly heavy-tailed. Figure 31.5 suggests that a constant shape parameter is reasonable for most grid cells, while being estimated with much lower uncertainty. The estimated location and scale parameters, $\widehat{\eta}_{\widehat{\xi}}(s)$ and $\widehat{\tau}_{\widehat{\xi}}(s)$, are displayed in Figure 31.6 and reveal interesting spatial patterns reflecting the topography of the study region. The middle and bottom panels of Figure 31.6 show the M-year return levels,

$$z_M(s) = G^{-1}\{1 - 1/M; \eta(s), \tau(s), \xi(s)\}, \quad s \in \mathcal{S}',$$

for $M = 10, 20, 50$ and 100 years, estimated at all grid points of the study region. The city of Jeddah appears to be at high risk, which corroborates empirical evidence, although our results must be interpreted with care, because of the large estimation uncertainty due to the heavy tails the relatively short time series, and the rather coarse spatial resolution.

After transforming the data to a common unit Fréchet scale, we used four stationary isotropic max-stable processes to assess the spatial dependence of extreme rainfall events over the entire study region. The models used were the Brown–Resnick process with variogram $\gamma(s_1, s_2) = (\|s_1 - s_2\|/\lambda)^\kappa$, $\lambda > 0$, $\kappa \in (0, 2]$, the Schlather process with powered exponential correlation function $c(s_1, s_2) = \exp\{-\gamma(s_1, s_2)\}$, the extremal-$t$ process with same correlation $c(s_1, s_2)$ and degrees of freedom $\alpha > 0$, and the Smith process with Gaussian density kernels defined through the diagonal covariance matrix $\Omega = \lambda^2 I_2$, which corresponds to the Brown–Resnick process with $\kappa = 2$. We also considered geometrically anisotropic models, obtained by replacing the Euclidean distance $\|s_1 - s_2\|$ in the models above by the Mahalanobis distance $h_\mathcal{M}$, where

$$h_\mathcal{M}^2 = (s_1 - s_2)^{\mathsf{T}} \begin{pmatrix} \cos(\theta) & -\sin(\theta) \\ \sin(\theta) & \cos(\theta) \end{pmatrix} \begin{pmatrix} 1 & 0 \\ 0 & a^2 \end{pmatrix} \begin{pmatrix} \cos(\theta) & -\sin(\theta) \\ \sin(\theta) & \cos(\theta) \end{pmatrix}^{\mathsf{T}} (s_1 - s_2),$$

with $a > 0$ and $\theta \in [-\pi, \pi]$. The parameter a reflects the degree of anisotropy, as it corresponds to the ratio of the principal axes of dependence contours, while θ is the angle with respect to the west-east direction; see [4]. All models were fitted by pairwise likelihood, using a random selection of 5100 pairs of sites less than 800km apart with pairwise distances being approximately uniform in $[25, 800]$km, and conditioning on the observed maxima exceeding the threshold $u = 3$mm/day. The pairwise conditional likelihood may be expressed in shorthand notation as

$$\ell(\vartheta) = \sum_{i=1}^{17} \sum_{d' < d} w_{d',d} I(z_{i,d'} > u'_d, z_{i,d} > u_d) \log\left\{\frac{\exp(-V)(V_1 V_2 - V_{12})}{p(u_{d'}, u_d)}\right\},$$

where V, V_1, V_2 and V_{12} are the bivariate exponent function and its partial and mixed derivatives of the corresponding max-stable model, evaluated at $(z_{i,d'}, z_{i,d})$, $u_{d'}$ and u_d denote the threshold $u = 3$mm/day transformed to the unit Fréchet scale, $p(u_{d'}, u_d) = 1 - \exp(-1/u_{d'}) - \exp(-1/u_d) + \exp\{-V(u_{d'}, u_d)\}$ is the probability that the threshold u is exceeded at both sites simultaneously, and $w_{d',d}$ is a binary weight determined by the selection of pairs. In less arid climates it would often be reasonable to assume that all the annual maxima were large enough for the extremal model to apply, in which case one would effectively set $u_d = u_{d'} = 0$ and $p(u_{d'}, u_d) = 1$ in the expression above, so that all the indicator functions equal unity.

Table 31.2 reports the parameter estimates and 95% confidence intervals obtained using a non-parametric bootstrap to resample years of data and thus reflect the overall uncertainty from estimating both the margins and the dependence structure. It also reports the composite likelihood information criterion, CLIC, rescaled to be comparable to the AIC in the independence case, which may be used to compare fitted models.

As the confidence interval for the anisotropy parameter a always includes unity, there is no strong evidence against isotropy. Overall, the CLIC values are only very slightly in favor of anisotropic models, except for the extremal-t model, for which the isotropic model seems to perform better. The CLIC suggests that the isotropic extremal-t model is the best, but the estimated degrees of freedom $\hat{\alpha}$ is quite unstable, which results in large and asymmetric

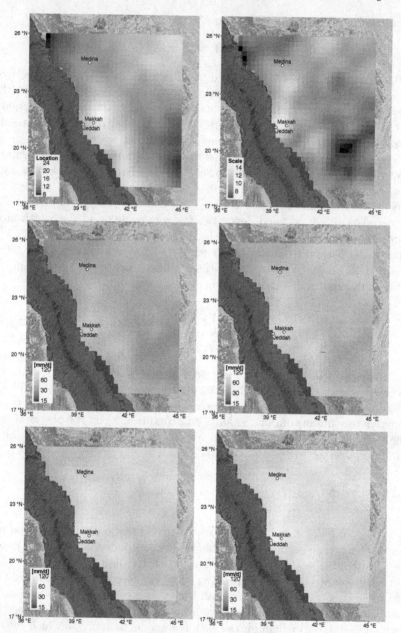

FIGURE 31.6

Saudi Arabian rainfall analysis. *Top*: location $\eta(s)$ (left) and scale $\tau(s)$ (right) parameters of the GEV distribution estimated from a local likelihood fit using a biweight function with bandwidth $b = 80$km. The estimated shape parameter, constant over the region, is $\widehat{\xi} = 0.14$. *Middle and bottom*: Estimated 10-year (middle left), 20-year (middle right) and 50-year (bottom left) and 100-year (bottom right) return levels (in mm/day) at each grid point of the study region, plotted on a common logarithmic scale.

TABLE 31.2

Estimated dependence parameters for all max-stable processes fitted to the Saudi Arabian rainfall annual maxima. Subscripts denote 95% confidence intervals obtained using a nonparametric bootstrap procedure. Results in the top rows are for isotropic models, while the bottom row are for models with geometric anisotropy. The last column reports the difference in composite likelihood information criterion with respect to the best model, rescaled to be comparable to the AIC in the independence case; lower values are better.

Isotropic max-stable models

Model	λ [km]	κ	α	a	θ	CLIC
Smith	$34_{[26,39]}$					124
Schl.	$44_{[34,53]}$	$1.46_{[1.19,1.84]}$				362
B.–R.	$13_{[8,16]}$	$0.71_{[0.52,0.94]}$				23
Ext.-t	$333_{[165,1357]}$	$0.90_{[0.63,1.13]}$	$5.9_{[3.9,13.1]}$			0

Anisotropic max-stable models

Model	λ [km]	κ	α	a	θ	CLIC
Smith	$31_{[28,37]}$			$0.82_{[0.72,1.33]}$	$0.19_{[-1.48,2.02]}$	119
Schl.	$42_{[31,54]}$	$1.47_{[1.20,1.82]}$		$0.89_{[0.70,1.28]}$	$0.23_{[-0.32,1.26]}$	362
B.–R.	$12_{[7,21]}$	$0.72_{[0.53,0.95]}$		$0.71_{[0.52,1.81]}$	$-0.12_{[-0.25,1.41]}$	21
Ext.-t	$424_{[176,1352]}$	$0.90_{[0.64,1.10]}$	$6.2_{[4.1,14.8]}$	$1.37_{[0.55,1.66]}$	$1.37_{[-0.30,1.42]}$	41

confidence intervals for α and the range λ. Therefore, the isotropic Brown–Resnick model, which provides a good balance of fit and parsimony, may be preferred. Its estimated shape parameter κ suggests that the fitted process is fairly rough, as might be expected, and the scale parameter λ seems to be both small and quite well-estimated. The Schlather model does not allow independence at long ranges, and gives the worst fit, followed by the Smith model, which is too smooth to be realistic. The relative small difference between the CLIC values for the Brown–Resnick and extremal-t model suggests that their fits are similar. Figure 31.7 plots the fitted bivariate extremal coefficients $\theta(s_1, s_2)$ as a function of distance $h = \|s_1 - s_2\|$ for all isotropic models, compared to the empirical counterpart, binned by distance class. The Brown–Resnick and extremal-t max-stable processes provide reasonable fits, the Smith model is too rigid to appropriately capture the decay of dependence with distance, and the Schlather model is unable to capture the long-range independence. These problems are common in applications, where these models are rarely better than the others.

To further explore how the fitted isotropic max-stable models capture higher-dimensional distributions, Table 31.3 compares empirical and fitted extremal coefficients $\theta_D \in [1, D]$ for sets of sites $\mathcal{D} = \{s_1, \ldots, s_D\}$ around Jeddah. All the fitted models tend to overestimate the strength of dependence, but the Brown–Resnick process provides the best fit and the Schlather model the worst fit, as one might expect from Figure 31.7. Surprisingly, the fit of the extremal-t model in high dimensions differs slightly from the Brown–Resnick model, despite their similar results for pairs of sites.

Figure 31.8 compares the map of annual rainfall maxima for 2009, a year with intense rainfall in Jeddah, with data simulated from the fitted isotropic Brown–Resnick model, using the exact simulation algorithm of [28]. Overall, the spatial patterns and forms of dependence observed in the simulations tend to agree with the data, although the 2009 annual maxima seem to be slightly smoother. To illustrate the ability of this modeling framework to assess spatial aggregated risk, we generate 10^5 independent simulations from the fitted isotropic

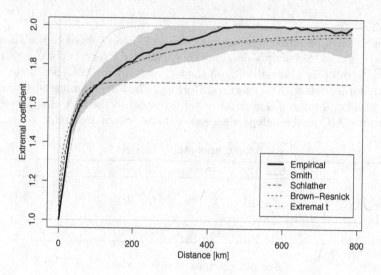

FIGURE 31.7
Saudi Arabian rainfall analysis. Empirical and fitted bivariate extremal coefficients $\theta(s_1, s_2)$, plotted as functions of distance $\|s_1 - s_2\|$, for the isotropic Smith, Schlather, Brown–Resnick and extremal-t models. The empirical extremal coefficients are binned by distance class, and the grey shaded area is a bootstrap 95% pointwise confidence band.

Brown–Resnick model, which we then use to compute the probability $p(v)$ that the annual maximum averaged over the 14 grid cells less than 50km from Jeddah or Makkah (the second and third largest cities in Saudi Arabia with about 4 and 2 million inhabitants, respectively) exceeds a given high threshold v, i.e., $p(v) = \Pr\{|\mathcal{S}^\star|^{-1} \sum_{s \in \mathcal{S}^\star} Z(s) > v\}$ with $\mathcal{S}^\star = \mathcal{N}(s_J) \cup \mathcal{N}(s_M) \subset \mathcal{S}'$ for some neighborhoods $\mathcal{N}(s_J)$ and $\mathcal{N}(s_M)$ of Jeddah and Makkah, respectively. We obtain $p(50\text{mm/day}) = 0.072$ (return period 13.9 years), $p(71.1\text{mm/day}) = 0.019$ (return period 53.6 years) and $p(100\text{mm/day}) = 0.0041$ (return period 245.1 years), where 71.1mm/day corresponds to the daily rainfall level observed on November 25, 2009, during the Jeddah floods, which caused 122 fatalities.

31.6.2 Spanish temperatures

To illustrate challenges in the modeling of threshold exceedances, we analyze extreme temperatures measured at 12 stations near Madrid in Spain, see Figure 31.9; five stations are located in the Madrid conurbation. The station elevations vary from 540m for Toledo to 1894m for Navacerrada, resulting in different climates over this region. Daily maximum temperatures for the years 1920–2016 were obtained from the European Climate Assessment and Dataset website. To avoid having to treat seasonality we focus on the months from June to September. Most of the time series have missing values; just one is complete, and only five stations have records before 1960. Missing values are typical of environmental applications based on observational data. This application is an example of modeling a 'time-rich but space-poor' dataset, as estimating extremal dependence based on 12 stations is challenging.

Space-time modeling of extreme temperatures must take into account two aspects: the space-time variability of marginal distributions (climate) and the space-time dependence of individual observations resulting from heatwaves over periods of a few days (weather). In this application we focus on modeling the spatial dependence at both the marginal and observation levels. We assume that the marginal distributions are constant over time

TABLE 31.3

Saudi Arabian rainfall analysis. Empirical and fitted extremal coefficients $\theta_D \in [1, D]$ for sets of sites $\mathcal{D} = \{s_1, \ldots, s_D\}$ around Jeddah. The fitted coefficients are for the four isotropic models and are calculated using the exact expression of the exponent function V in dimension D. Subscripts denote 95% bootstrap confidence intervals.

Region $\mathcal{D} = \{s_1, \ldots, s_D\}$	D	Empirical $\widehat{\theta}_D$	Smith	Schl.	B.–R.	Ext.-t
$[39, 40]°\mathrm{E} \times [21, 22]°\mathrm{N}$	14	$4.17_{[1.90, 6.44]}$	3.47	3.16	4.41	3.44
$[39, 41]°\mathrm{E} \times [21, 23]°\mathrm{N}$	62	$14.25_{[6.50, 22.00]}$	9.73	6.06	11.27	8.71
$[39, 42]°\mathrm{E} \times [21, 24]°\mathrm{N}$	142	$20.90_{[9.54, 32.26]}$	19.00	9.01	20.10	15.96

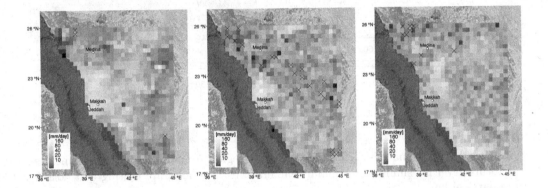

FIGURE 31.8

Saudi Arabian rainfall analysis. Annual rainfall maxima for 2009 (left), and two simulated maps (middle and right) based on the fitted isotropic Brown–Resnick max-stable model. $\times$ represents values below the threshold of $u = 3$mm.

and attempt to identify clusters of spatial extreme high temperatures to avoid modeling temporal dependence.

We identified extreme observations by retaining only observations exceeding thresholds taken to equal the 0.95 quantiles of each of the 12 series. These observations are grouped, due to the temporal dependence resulting from hot spells, and to avoid modeling this we identified clusters of extremes over space and time and reduced each cluster to one observation at each sites. Two extreme observations—possibly occurring at different site—were assumed to be in the same cluster if they occurred within five days of each other. Applying this procedure gave 311 separate spatial extreme events with observations at all 12 stations, which we used for our modeling. This is equivalent to selecting cluster extreme observations using the risk functional $\rho(q) = \max_{d=1}^D q(s_d)/u_d$ where the u_d are the thresholds at the $D = 12$ sites.

As with the rainfall example, the marginal and dependence structures were estimated separately. We first fitted the extreme observations at each site using the generalized Pareto distribution with spatially varying parameters, and then used the fitted marginal model to transform the data to the unit Fréchet scale and used the censored Poisson process approach to estimation.

For the marginal model we used an independence likelihood to fit generalized Pareto distributions with constant shape parameter and with a scale parameter that varies linearly with the station elevation. The estimated shape parameter was $\widehat{\xi} = -0.39_{[-0.43, -0.37]}$; the

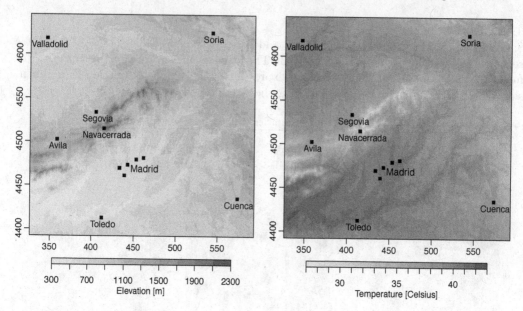

FIGURE 31.9
Spanish temperature analysis. Left panel: the study region around Madrid. Right panel: ten-year return level for summer maximum temperatures. The axes are easting and northing coordinates in km.

subscripts here and below are 95% confidence intervals obtained by bootstrapping the 311 multivariate extreme events. The estimated shape parameter for extreme temperatures is often negative, though here the uncertainty is particularly small. This model gave a reasonably good fit at the 12 sites of our dataset. The right-hand panel of Figure 31.9 shows the ten-year maximum summer temperature return levels given by this marginal model.

Extremal models for threshold exceedances presuppose that the data are asymptotically dependent and threshold-stable. To assess this we computed conditional exceedance probability curves, $\Pr\{F(Y_2) > u \mid F(Y_1) > u\}$, $u \in (0,1)$, for all pairs of stations. The left-hand panel of Figure 31.10 shows the conditional probability curve for one of the Madrid stations and Avila. This curve supports asymptotic dependence, as the estimate of (31.15) seems to be strictly positive. Moreover, taking into account the large uncertainty close to $u = 1$, these showed no evidence of non-stability above the 0.95 quantile, and thus cast no doubt on the suitability of the Poisson process model. Exploratory analysis shows that extremal dependence is strong and barely weakens with distance over this small region; see the right-hand panel of Figure 31.10. Heatwaves affect large regions simultaneously, resulting in strong spatial dependence.

A process $q(s)$ corresponding to a Brown–Resnick model with variogram $\gamma(s_1, s_2) = (\|s_1 - s_2\|/\lambda)^\kappa$, $\lambda > 0$, $\kappa \in (0,2]$, was fitted to the transformed data using the method described in Section 31.5.3. We used the censored full likelihood approach of [86] to obtain $\widehat{\lambda} = 139_{[49,303]}$ km and $\widehat{\kappa} = 0.17_{[0.13,0.21]}$. The large value and uncertainty for $\widehat{\lambda}$ is due to the strong dependence in extreme temperatures, which makes the likelihood flat. The small value of $\widehat{\kappa}$ is due to the small-scale variability in the data—here the fitted log-Gaussian random fields in the Poisson process model are non-differentiable. Figure 31.10 shows the fitted extremogram for our Brown–Resnick model, in terms of distances between any two sites. The model seems to underestimate the dependence somewhat. The empirical exceedance probability at very small distances is less than unity, which forces the value of $\widehat{\kappa}$ to be very small.

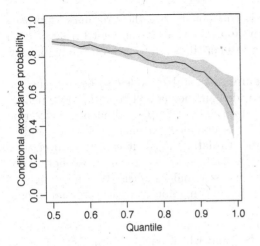 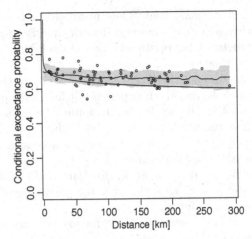

FIGURE 31.10

Spanish temperature analysis. Left panel: Conditional exceedance probability $\Pr\{F(Y_1) > u \mid F(Y_2) > u\}$ for Madrid and Avila, for $u \in (0.5, 1)$ with a 95% overall confidence region in grey. Right panel: Extremogram, see Section 31.4. The black dots are the empirical estimates obtained from each pair of stations. The rough curve is an estimate of the extremogram obtained by smoothing the black dots (the grey area represents a 95% overall confidence region for this curve). The smooth curve is obtained from the fitted Brown–Resnick model.

31.7 Discussion

Our discussion above has focused on spatial extensions of extremal models that are max-stable or threshold-stable, which are appropriate for the estimation of overall risk for single large events. Although well-justified by asymptotic arguments, the effort of fitting such models may be difficult to justify when the goal of analysis is the estimation of trends in extremes, the attribution of individual extreme events to possible causes, or the estimate of risk at a single site, and in any case it will often be valuable to compare and contrast inferences from different models. Numerous authors [e.g., 7, 10, 16, 36, 38, 62, 70, 71] have fitted hierarchical models in which time series of extremes at sites x_d are modelled using the generalized Pareto or GEV distributions, variation in whose parameters is modelled using Gaussian random fields. Markov chain Monte Carlo techniques, which allow the borrowing of strength between sites, are typically used to fit such models, but as these have asymptotic independent tail behaviour, their extremes may exhibit limited spatial structure [20]; whether this is critical will depend on the application. In other settings, such as attribution studies [e.g., 87, 91], fitting is performed using independence likelihoods or estimating functions. A comparison of the use of spatial models and independence likelihoods is provided by [90].

Fitting max-stable models to data can be difficult, for a variety of reasons. It can be hard to estimate both the marginal distributions and the dependence structure simultaneously, both because of the associated computational burden and because of tradeoffs between the two sets of parameters. If pairwise likelihood is used for fitting, estimation of any sort of complex surfaces for the marginal parameters is compromised by the use only of pairs of observations, even if the dependence structure can be estimated without too heavy a computational load. Thus in applications it is common to use the two-phase approach adopted in our examples. In principle Bayesian approaches allow the joint fitting of both structures, but existing Markov chain Monte Carlo algorithms can be awkward to implement and slow to converge [69].

Although the Smith model was one of the first max-stable models to be described in simple terms, in our experience it has never provided an adequate fit to data, so we believe that it should not be used in applications. Its realizations are simply too smooth to describe environmental data well. The Schlather model has bounded extremal coefficient, and in many cases this does not provide a good match to data, though it can do so over very limited spatial domains. Thus it too should usually be avoided. In any case both arise as limits of the more flexible Brown–Resnick and extremal-t models. One advantage of these more flexible classes is the possibility of using existing non-stationary and anisotopic covariance functions.

Poisson process models for exceedances are closely linked to Pareto processes, which have been investigated by [23], [30], [37] and [81] and which emerge from the former by conditioning.

Some problems involve multivariate spatial extremes—for example, simultaneous high values of different pollutants may be of interest in public health studies. The limited literature on this topic includes [40], [58] and [83].

An important issue in certain applications is downscaling, i.e., the relation between values of a process at particular sites and its time and/or space integrals; see [33] for a discussion of this in the setting of extremes. Another key issue can be estimation of the chance of extremes occurring simultaneously [31], which is closely related to the exponent function.

The use of max-stable processes as discussed in this chapter can be justified asymptotically for block maxima with increasing block size. Max-stability, however, strongly restricts the dependence structure, and in practice dependence is often seen to weaken as events become more extreme, a feature that max-stable and related threshold-stable models fail to capture. This can arise when the blocks are too small for the asymptotic regime to apply, e.g., when the data are asymptotically independent. Recent attempts to provide models that encompass this include [5, 46, 47, 48], but further research to develop 'penultimate' processes that provide adequate fits at sub-asymptotic regimes is needed. A related topic is the development of spatial models for jointly characterizing extreme and non-extreme data, i.e., in the tails and the bulk of the distribution. Some progress has been made in the univariate case [55, 64], but more research is needed to extend this to complex settings.

31.8   Computing

Several R packages allow the fitting of max-stable models. The most comprehensive is SpatialExtremes [68]. RandomFields [73, 74] is also useful for simulation of max-stable and related processes. The package hkevp [75] has several functions useful for fitting and using the latent variable models discussed in §31.3.4, and the function abba of extRemes [41] also provides experimental code for this model. mvPot [22] provides functions useful for multivariate peaks over threshold analysis, including a numerically efficient implementation of the multivariate Gaussian distribution function for use in extreme-value likelihoods.

Acknowledgement

We thank Luigi Lombardo for providing the Saudi rainfall dataset. The work was supported by the Swiss National Science Foundation, and completed while the first author was visiting the Institute for Mathematical Sciences, National University of Singapore. Support from the KAUST Supercomputing Laboratory and access to Shaheen is gratefully acknowledged.

Bibliography

[1] P. Asadi, A. C. Davison, and S. Engelke. Extremes on river networks. *Annals of Applied Statistics*, 9:2023–2050, 2015.

[2] B. Beranger, S. A. Padoan, and S. A. Sisson. Models for extremal dependence derived from skew-symmetric families. *Scandinavian Journal of Statistics*, 44:21–45, 2017.

[3] J. Blanchet and J.-D. Creutin. Co-occurrence of extreme daily rainfall in the French Mediterranean region. *Water Resources Research*, 53:9330–9349, 2017.

[4] J. Blanchet and A. C. Davison. Spatial modelling of extreme snow depth. *Annals of Applied Statistics*, 5:1699–1725, 2011.

[5] G. Bopp, B. A. Shaby, and R. Huser. A hierarchical max-infinitely divisible process for extreme areal precipitation over watersheds. arXiv preprint 1804.04588, 2018.

[6] B. M. Brown and S. I. Resnick. Extreme values of independent stochastic processes. *Journal of Applied Probability*, 14:732—739, 1977.

[7] E. Casson and S. G. Coles. Spatial regression models for extremes. *Extremes*, 1:449–468, 1999.

[8] D. Castro and R. Huser. Local likelihood estimation of complex tail dependence structures, applied to U.S. precipitation extremes. arXiv:1710.00875, 2018.

[9] S. Castruccio, R. Huser, and M. G. Genton. High-order composite likelihood inference for max-stable distributions and processes. *Journal of Computational and Graphical Statistics*, 25:1212–1229, 2016.

[10] S. G. Coles and E. Casson. Extreme value modelling of hurricane wind speeds. *Structural Safety*, 20:283–296, 1998.

[11] S. G. Coles, J. Heffernan, and J. A. Tawn. Dependence measures for extreme value analyses. *Extremes*, 2:339–365, 1999.

[12] S. G. Coles and J. A. Tawn. Modelling extreme multivariate events. *Journal of the Royal Statistical Society series B*, 53:377–392, 1991.

[13] D. Cooley, J. Cisewski, R. J. Erhardt, S. Jeon, E. Mannshardt, B. O. Omolo, and Y. Sun. A survey of spatial extremes: Measuring spatial dependence and modeling spatial effects. *REVSTAT—Statistical Journal*, 10:135–165, 2012.

[14] D. Cooley, B. Hunter, and R. L. Smith. Univariate and multivariate extremes for the environmental sciences. In A. E. Gelfand, M. Fuentes, J. A. Hoeting, and R. L. Smith, editors, *Handbook of Environmental and Ecological Statistics*. CRC Press, 2018.

[15] D. Cooley, P. Naveau, and P. Poncet. Variograms for max-stable random fields. In P. Bertail, P. Doukhan, and P. Soulier, editors, *Dependence in Probability and Statistics*, volume 187 of *Lecture Notes in Statistics*, pages 373–390. Springer, New York, 2006.

[16] D. Cooley, D. Nychka, and P. Naveau. Bayesian spatial modeling of extreme precipitation return levels. *Journal of the American Statistical Association*, 102:824–840, 2007.

[17] R. A. Davis and T. Mikosch. The extremogram: A correlogram for extreme events. *Bernoulli*, 15:977–1009, 2009.

[18] A. C. Davison and R. Huser. Statistics of extremes. *Annual Review of Statistics and its Application*, 2:203–235, 2015.

[19] A. C. Davison, R. Huser, and E. Thibaud. Geostatistics of dependent and asymptotically independent extremes. *Mathematical Geosciences*, 45:511–529, 2013.

[20] A. C. Davison, S. A. Padoan, and M. Ribatet. Statistical modelling of spatial extremes (with Discussion). *Statistical Science*, 27:161–186, 2012.

[21] A. C. Davison and N. I. Ramesh. Local likelihood smoothing of sample extremes. *Journal of the Royal Statistical Society series B*, 62:191–208, 2000.

[22] R. de Fondeville. *mvPot: Multivariate Peaks-over-Threshold Modelling for Spatial Extreme Events*, 2016. R package version 0.1.1.

[23] R. de Fondeville and A. C. Davison. High-dimensional peaks-over-threshold inference. *Biometrika*, 105:575–592, 2018.

[24] L. de Haan. A spectral representation for max-stable processes. *Annals of Probability*, 12:1194–1204, 1984.

[25] L. de Haan and T. T. Pereira. Spatial extremes: Models for the stationary case. *Annals of Statistics*, 34:146–168, 2006.

[26] L. Deng, M. F. McCabe, G. Stenchikov, J. P. Evans, and P. A. Kucera. Simulation of flash-flood-producing storm events in Saudi Arabia using the weather research and forecasting model. *Journal of Hydrometeorology*, 16:615–630, 2015.

[27] A. B. Dieker and T. Mikosch. Exact simulation of Brown–Resnick random fields at a finite number of locations. *Extremes*, 18:301–314, 2015.

[28] C. Dombry, S. Engelke, and M. Oesting. Exact simulation of max-stable processes. *Biometrika*, 103:303–317, 2016.

[29] C. Dombry, F. Éyi Minko, and M. Ribatet. Conditional simulation of max-stable processes. *Biometrika*, 100:111–124, 2013.

[30] C. Dombry and M. Ribatet. Functional regular variations, Pareto processes and peaks over threshold. *Statistics and its Interface*, 8:9–17, 2015.

[31] C. Dombry, M. Ribatet, and S. Stoev. Probabilities of concurrent extremes. *Journal of the American Statistical Association*, (DOI: 10.1080/01621459.2017.1356318), 2018.

[32] J. H. J. Einmahl, A. Kiriliouk, A. Krajina, and J. J. Segers. An M-estimator of spatial tail dependence. *Journal of the Royal Statistical Society, Series B*, 78:275–298, 2016.

[33] S. Engelke, R. de Fondeville, and M. Oesting. Extremal behavior of aggregated data with an application to downscaling. *Biometrika*, page to appear, 2019.

[34] S. Engelke, A. Malinowski, Z. Kabluchko, and M. Schlather. Estimation of Hüsler–Reiss distributions and Brown–Resnick processes. *Journal of the Royal Statistical Society, series B*, 77:239–265, 2015.

[35] R. J. Erhardt and R. L. Smith. Approximate Bayesian computing for spatial extremes. *Computational Statistics & Data Analysis*, 56:1468–1481, 2014.

[36] L. Fawcett and D. Walshaw. A hierarchical model for extreme wind speeds. *Applied Statistics,* 55:631–646, 2006.

[37] A. Ferreira and L. de Haan. The generalized Pareto process; with a view towards application and simulation. *Bernoulli*, 20:1717–1737, 2014.

[38] M. Fuentes, J. Henry, and B. Reich. Nonparametric spatial models for extremes: Application to extreme temperature data. *Extremes*, 16:75–101, 2013.

[39] M. G. Genton, Y. Ma, and H. Sang. On the likelihood function of Gaussian max-stable processes. *Biometrika*, 98:481–488, 2011.

[40] M. G. Genton, S. Padoan, and H. Sang. Multivariate max-stable spatial processes. *Biometrika*, 102:215–230, 2015.

[41] E. Gilleland and R. W. Katz. New software to analyze how extremes change over time. *Eos*, 92(2):13–14, 2011.

[42] R. Huser and A. C. Davison. Composite likelihood estimation for the Brown–Resnick process. *Biometrika*, 100:511–518, 2013.

[43] R. Huser and A. C. Davison. Space-time modelling of extreme events. *Journal of the Royal Statistical Society series B*, 76:439–461, 2014.

[44] R. Huser, C. Dombry, M. Ribatet, and M. G. Genton. Full likelihood inference for max-stable data. arXiv:1703.08665, Stat to Appear.

[45] R. Huser and M. G. Genton. Non-stationary dependence structures for spatial extremes. *Journal of Agricultural, Biological and Environmental Statistics*, 21:470–491, 2016.

[46] R. Huser, T. Opitz, and E. Thibaud. Bridging asymptotic independence and dependence in spatial extremes using Gaussian scale mixtures. *Spatial Statistics*, 21:166–186, 2017.

[47] R. Huser, T. Opitz, and E. Thibaud. Max-infinitely divisible models and inference for spatial extremes. arXiv preprint 1801:02946, 2018.

[48] R. Huser and J. Wadsworth. Modeling spatial processes with unknown extremal dependence class. *Journal of the American Statistical Association*, (DOI: 10.1080/01621459.2017.1411813), 2018.

[49] Z. Kabluchko. Extremes of independent Gaussian processes. *Extremes*, 14:285–310, 2011.

[50] Z. Kabluchko, M. Schlather, and L. de Haan. Stationary max-stable fields associated to negative definite functions. *Annals of Probability*, 37:2042–2065, 2009.

[51] J. F. C. Kingman. *Poisson Processes*. Clarendon Press, New York, 1993.

[52] A. W. Ledford and J. A. Tawn. Statistics for near independence in multivariate extreme values. *Biometrika*, 83:169–187, 1996.

[53] B. G. Lindsay. Composite likelihood methods. *Contemporary Mathematics*, 80:220–241, 1988.

[54] P. Naveau, A. Guillou, D. Cooley, and J. Diebolt. Modelling pairwise dependence of maxima in space. *Biometrika*, 96:1–17, 2009.

[55] P. Naveau, R. Huser, P. Ribereau, and A. Hannart. Modeling jointly low, moderate, and heavy rainfall intensities without a threshold selection. *Water Resources Research*, 52:2753–2769, 2016.

[56] A. K. Nikoloulopoulos, H. Joe, and H. Li. Extreme value properties of multivariate t copulas. *Extremes*, 12:129–148, 2009.

[57] M. Oesting, Z. Kabluchlo, and M. Schlather. Simulation of Brown–Resnick processes. *Extremes*, 15:89–107, 2011.

[58] M. Oesting, M. Schlather, and P. Friedrichs. Statistical post-processing of forecasts for extremes using bivariate Brown–Resnick processes with an application to wind gusts. *Extremes*, 20:309–332, 2017.

[59] M. Oesting, M. Schlather, and C. Zhou. Exact and fast simulation of max-stable processes on a compact set using the normalized spectral representation. *Bernoulli*, 24:1497–1530, 2018.

[60] T. Opitz. Extremal t processes: Elliptical domain of attraction and a spectral representation. *Journal of Multivariate Analysis*, 122:409–413, 2013.

[61] T. Opitz. Modeling asymptotically independent spatial extremes based on Laplace random fields. *Spatial Statistics*, 16:1–18, 2016.

[62] T. Opitz, R. Huser, H. Bakka, and H. Rue. INLA goes extreme: Bayesian tail regression for the estimation of high spatio-temporal quantiles. *Extremes*, 21:441–462, 2018.

[63] S. Padoan, M. Ribatet, and S. Sisson. Likelihood-based inference for max-stable processes. *Journal of the American Statistical Association*, 105:263–277, 2010.

[64] I. Papastathopoulos and J. A. Tawn. Extended generalised Pareto models for tail estimation. *Journal of Statistical Planning and Inference*, 143:131–143, 2013.

[65] B. J. Reich and B. A. Shaby. A hierarchical max-stable spatial model for extreme precipitation. *Annals of Applied Statistics*, 6:1430–1451, 2012.

[66] B. J. Reich, B. A. Shaby, and D. Cooley. A hierarchical model for serially-dependent extremes: A study of heat waves in the western US. *Journal of Agricultural, Biological and Environmental Statistics*, 19:119–135, 2014.

[67] M. Ribatet. Spatial extremes: Max-stable processes at work. *Journal de la Société Française de Statistique*, 154:156–177, 2013.

[68] M. Ribatet. *SpatialExtremes: Modelling Spatial Extremes*, 2015. R package version 2.0-2.

[69] M. Ribatet, D. Cooley, and A. C. Davison. Bayesian inference from composite likelihoods, with an application to spatial extremes. *Statistica Sinica*, 22:813–845, 2012.

[70] H. Sang and A. E. Gelfand. Hierarchical modeling for extreme values observed over space and time. *Environmental and Ecological Statistics*, 16:407–426, 2009.

[71] H. Sang and A. E. Gelfand. Continuous spatial process models for spatial extreme values. *Journal of Agricultural, Biological and Environmental Statistics*, 15:49–65, 2010.

[72] M. Schlather. Models for stationary max-stable random fields. *Extremes*, 5:33–44, 2002.

[73] M. Schlather, A. Malinowski, P. J. Menck, M. Oesting, and K. Strokorb. Analysis, simulation and prediction of multivariate random fields with package RandomFields. *Journal of Statistical Software*, 63:1–25, 2015.

[74] M. Schlather, A. Malinowski, M. Oesting, D. Boecker, K. Strokorb, S. Engelke, J. Martini, F. Ballani, Ol. Moreva, J. Auel, P. J. Menck, S. Gross, U. Ober, C. Berreth, K. Burmeister, J. Manitz, P. Ribeiro, R. Singleton, B. Pfaff, and R Core Team. *RandomFields: Simulation and Analysis of Random Fields*, 2017. R package version 3.1.50.

[75] Q. Sebille. *hkevp: Spatial Extreme Value Analysis with the Hierarchical Model of Reich and Shaby (2012)*, 2016. R package version 1.1.2.

[76] B. A. Shaby and B. J. Reich. Bayesian spatial extreme value analysis to assess the changing risk of concurrent high temperatures across large portions of European cropland. *Environmetrics*, 23:638–648, 2012.

[77] R. L. Smith. Max-stable processes and spatial extremes. Unpublished, 1990.

[78] A. G. Stephenson and J. A. Tawn. Exploiting occurrence times in likelihood inference for componentwise maxima. *Biometrika*, 92:213–227, 2005.

[79] E. Thibaud, J. Aalto, D. S. Cooley, A. C. Davison, and J. Heikkinen. Bayesian inference for the Brown–Resnick process, with an application to extreme low temperatures. *Annals of Applied Statistics*, 10:2303–2324, 2016.

[80] E. Thibaud, R. Mutzner, and A. C. Davison. Threshold modeling of extreme spatial rainfall. *Water Resources Research*, 49:4633–4644, 2013.

[81] E. Thibaud and T. Opitz. Efficient inference and simulation for elliptical Pareto processes. *Biometrika*, 102:855–870, 2015.

[82] C. Varin, N. Reid, and D. Firth. An overview of composite marginal likelihoods. *Statistica Sinica*, 21:5–42, 2011.

[83] S. Vettori, R. Huser, and M. G. Genton. Bayesian modeling of air pollution extremes using nested multivariate max-stable processes. arXiv preprint 1804.04588, 2018.

[84] J. L. Wadsworth. On the occurrence times of componentwise maxima and bias in likelihood inference for multivariate max-stable distributions. *Biometrika*, 102:705–711, 2015.

[85] J. L. Wadsworth and J. A. Tawn. Dependence modelling for spatial extremes. *Biometrika*, 99:253–272, 2012.

[86] J. L. Wadsworth and J. A. Tawn. Efficient inference for spatial extreme value processes associated to log-Gaussian random functions. *Biometrika*, 101:1–15, 2014.

[87] Z. Wang, Y. Jiang, H. Wan, J. Yan, and X. Zhang. Detection and attribution of changes in extreme temperatures at regional scale. *Journal of Climate*, 30:7035–7047, 2017.

[88] V. Yesubabu, C. Venkata Srinivas, S. Langodan, and I. Hoteit. Predicting extreme rainfall events over Jeddah, Saudi Arabia: Impact of data assimilation with conventional and satellite observations. *Quarterly Journal of the Royal Meteorological Society*, 142:327–348, 2016.

[89] R. Yuen and S. Stoev. CRPS M-estimation for max-stable models. *Extremes*, 17:387–410, 2014.

[90] F. Zheng, E. Thibaud, M. Leonard, and S. Westra. Assessing the performance of the independence method in modeling spatial extreme rainfall. *Water Resources Research*, 51:7744–7758, 2015.

[91] F. W. Zwiers, X. Zhang, and Y. Feng. Anthropogenic influence on long return period daily temperature extremes at regional scales. *Journal of Climate*, 24:881–892, 2011.

32

Statistics in Oceanography

Christopher K. Wikle

University of Missouri, Columbia

CONTENTS

32.1 Introduction

The world's oceans cover over 70 percent of the earth's surface. These oceans, commonly believed to have spawned life over 3 billion years ago, are an immensely complex system, with critical interactions across the primary systems of the planet – the atmosphere, biosphere, and crysosphere. One of the most interesting (and challenging) aspects of the ocean is that it operates over an enormous range of spatial and temporal scales, and its interactions and internal dynamics can be highly nonlinear – which makes characterizing its interaction with these various systems very challenging. These challenges have traditionally been further exacerbated by the lack of *in situ* observations of the ocean state and ocean biology - with primary observations coming from ships of opportunity, research cruises, tidal gauges, and moored buoys, which are limited in spatial and temporal coverage. However, in recent years there has been an enormous increase in oceanographic data coming from autonomous systems (e.g., Argo floats, buoyancy gliders, drifters, radio telemetry) and remotely sensed observations of the surface and near surface. Oceanography is in a unique situation of simultaneously having a wealth of information (for surface processes) and a dearth of information (for subsurface processes and lower trophic ecosystem components). Thus, the complexity of the system and its interactions, as well as challenges imposed by its observational record, make uncertainty quantification a fundamental component of oceanography. Not surprisingly, there has been a long history of using relatively advanced statistical methods in oceanography to help deal with the uncertainties in data and process knowledge for both inference and prediction.

Historically, the statistical methods used to analyze oceanographic data are very similar to the methods used in the atmospheric sciences. An introduction to statistics from this perspective can be found in the book by [68], with more advanced treatments in the classic books by [73] and [80]. Statistical treatments of this material can be found in [34] and Chapter 5 of [12]. Review papers with discussion at the interface of statistics and oceanography include [10] and [79]. In addition, as noted below, several chapters in this volume contain methods that are often used in oceanography.

The primary purpose of this chapter is to give a brief overview of some of the methods that have proven useful in the past, as well as some more modern methods that have shown promise in recent years. In cases where the methods are described elsewhere in this volume, details are omitted and the discussion is focused more on how these methods are, and have been, used in oceanography. More detailed description and examples are provided for the methods that are not covered in other chapters. The first section discusses standard descriptive multivariate methods. The next section discusses time series methods, with more focus given to the frequency-domain (spectral) approaches commonly used in oceanography. Spatio-temporal methods are then discussed briefly, setting up the more practical sections on data assimilation and long-lead forecasting that follow.

32.2   Descriptive Multivariate Methods

Most of the classical methods of multivariate statistics (e.g., principal component analysis (PCA), factor analysis, discrimination and classification, and cluster analysis) are used in oceanographic data analysis (see [80]). However, PCA (also known as empirical orthogonal function (EOF) analysis) and canonical correlation analysis (CCA) are perhaps the most commonly used, and are typically repurposed to accommodate spatio-temporal data. We discuss these in more detail below.

32.2.1   PCA and EOF Analysis

Principal component analysis is discussed in the "Ordination" chapter of this volume (Chapter 11). Further, as discussed in the "Spatial Analysis in Climatology" chapter (Chapter 30), PCA is typically referred to as empirical orthogonal function (EOF) analysis in the atmospheric and oceanic sciences when one has spatio-temporal data. Most often, EOFs are spatial maps associated with the eigenvectors of an empirical spatial covariance matrix, and the projection of the original data onto these EOFs gives time series that are referred to as "principal component time series."

Historically, practitioners tend to use EOF analysis in a descriptive fashion to characterize the spatial patterns associated with time series showing the greatest variability in a data set. To aid in interpreting the spatial patterns, practitioners often rotate the EOFs either through an orthonormal or oblique rotation, as with factor analysis in traditional multivariate statistics (e.g., see [34] for a discussion). It is important to exercise care when attempting to interpret EOF patterns (or their rotated versions) in terms of kinematic or dynamic components of the underlying process (e.g,. see [55]).

In addition to descriptive interpretation, EOFs can be useful as spatial basis functions for low-rank modeling of spatial or spatio-temporal processes (typically, for a fixed set of observation locations), as described in Chapter 30. Finally, as discussed briefly in Chapter 30, there are a large number of extensions to traditional EOF analysis that can be used to consider more complicated dependence structures in the data (see also [73] and [12]), and one can consider EOFs from the alternative time-space perspective, in which the eigenvectors of an empirical temporal covariance matrix are the EOF (time-series) patterns, and the data projected onto these time series are spatial fields.

32.2.2 Spatio-Temporal Canonical Correlation Analysis (ST-CCA)

Canonical correlation analysis (CCA) in its classical multivariate statistical usage is discussed in Chapter 11 of this volume. As discussed in Chapter 30, it is more often used in the atmospheric and oceanic sciences in the context of spatio-temporal data (i.e., space-time CCA, or ST-CCA). In this case, one typically is interested in two spatio-temporal fields, either two variables (e.g., sea surface temperature (SST) and sea surface height) with the same space and time observations, two contemporaneous variables at different locations (e.g., January SST in the western tropical Pacific ocean and January precipitation over the western U.S.), or variables at different (lagged) times (e.g., January SST in the Pacific and soil moisture in the U.S. corn belt in July).

As discussed in Chapter 30, the ST-CCA approach is used to find new variables via linear combinations of one field that are maximally correlated with a new variable defined to be a linear combination of the other field. Typically, the weights in these linear combinations are spatially indexed and the new variables are then time series (but, it can be done the other way as well). Again, this can be shown to correspond to the eigen-decomposition of a weighted spatial cross-covariance matrix (see Chapter 30). It is well known that there is a strong connection between CCA and multivariate multiple regression (e.g., [45]). Arguably, the ST-CCA methodology has seen the most success as a long-lead forecast methodology; e.g., where one seeks the prediction of some spatio-temporal quantity several months into the future, usually forced by SST (see Section 32.6 below for more discussion).

32.3 Time Series Methods

Oceanographic data arise from dynamical processes that impart serial dependencies. Thus, the use of statistical time series methods, such as those described in the "Time Series Methodology" and "Dynamic Models" chapters of this volume (i.e., Chapters 3 and 4), is wide-spread in oceanography (see also the discussion of climate trend detection in Chapter 28). In particular, the use of autoregressive time-domain processes, multivariate dynamic models, and frequency-domain (spectral) methods are used extensively. Some specific uses of these methods in oceanography are briefly described in the following subsections. A comprehensive overview of statistical methods for time series analysis can be found in [64].

32.3.1   Time-Domain Methods

Autoregressive moving average (ARMA) processes are discussed in Chapter 3 of this volume. Although there are certainly cases in oceanography where one might consider moving average processes, more focus has traditionally been placed on autoregressive (AR) models. This is because the simple first-order AR (i.e., AR(1)) model provides a good and scientifically plausible noise model for many real-world phenomena. That is, geophysical processes in the ocean (and atmosphere) are often best thought of as a superposition of interacting wave phenomena, and these are scaled (in space and time) such that the larger wave structures correspond to more energy (or variance) than smaller wave structures. Thus, the "background noise" is typically well-represented by a so-called "red noise" spectrum (see Section 32.3.2 below), which shows more spectral energy in lower frequencies and less energy in higher frequency phenomena. In addition, second-order AR processes (AR(2) processes) are very useful in oceanography because they can accommodate quasi-cyclical behavior, which is endemic in most atmospheric and oceanographic processes (see Chapter 3 of [12] for more detail).

One important use of AR noise processes in oceanography concerns hypothesis tests in the presence of serial dependence. In particular, consider the case of testing for differences in two means when there is serial dependence in the errors. Although such tests can be performed from likelihood perspectives in the linear mixed model context (e.g., see [49]), they are often performed in practice by simply adjusting the degrees of freedom of a traditional two sample test to to account for positive serial dependence (which is the usual case for oceanographic data). In this case, the effective number of observations is less than the actual number because they are not independent and the dependence is positive. Specifically, assuming AR(1) errors, the effective sample size can be shown to be $\tilde{n} \approx n\{1 - \widehat{\rho}(1)\}/\{1 + \widehat{\rho}(1)\}$, where n is the number of samples, and $\widehat{\rho}(1)$ is an estimate of the lag-1 autocorrelation coefficient (see Chapter 3 in this volume and Chapter 5 of [80]).

Time series in oceanography do not exist in isolation. It is therefore typically more realistic to consider multivariate methods. In particular, the vector autoregression (VAR) model can be quite useful (see [44] and [12], Chapter 3, for more details). Such models allow lagged values of the various time series to linearly influence future values of each of the time series, with an additive multivariate innovation noise process (typically, assumed to be independent in time, but dependent across variables). As described in Sections 32.4, 32.5, and 32.6 below, these models have been used quite successfully in the spatio-temporal context (i.e., where each series corresponds to a different spatial location) to perform data assimilation and long-lead forecasting. Typically, however, they are used in a state-space context (see Section 32.4 below and Chapter 4) in which there is a separate measurement equation accounting for observation error, and the latent underlying physical process is assumed to follow a VAR model. These models are often termed "dynamical models."

32.3.2   Frequency-Domain Methods

The frequency domain or spectral representation of time series has proven to be an effective data analysis methodology in oceanography due to the fact that the ocean is a fluid whose properties are well-described by the superposition of waves (not just on the ocean surface) across many scales of spatial and temporal variability. Frequency domain statistical methods are discussed in the books of [61], [73], [8], [64], [80], [12], and most advanced books on time series analysis. Here, we focus on three approaches that are used fairly extensively in oceanography: univariate and bivariate spectral analysis, and space-time spectral analysis. Given these topics are not covered elsewhere in this volume, more detail is provided here than in the other sections of this chapter (but, more complete discussion and details can be found in the aforementioned references).

32.3.2.1 Univariate Spectral Analysis

Although the theoretical aspects of spectral analysis are best considered from a continuous time perspective, its practical utility in oceanography is in the context of discrete time observations. Let $\{y_t : t = 1, \ldots, T\}$ be a time series (assumed to arise from a stationary process – see Chapter 3) and consider the expansion of y_t given by

$$y_t = a_0 + \sum_{k=1}^{T/2} \alpha(\omega_k)\phi_k(t), \tag{32.1}$$

where $\phi_k(t) \equiv \cos(\omega_k t) + i\sin(\omega_k t)$, $k = 1, \ldots, T/2$, are complete and orthogonal Fourier basis functions, $\omega_k \equiv 2\pi k/T$ is the frequency, and $\alpha(\omega_k) = a_k + ib_k$ are complex-valued expansion coefficients (note that $b_0 = b_{T/2} = 0$). For a simple example of how these basis functions are used in practice, see Chapter 3 of [12]. Importantly, (32.1) describes how variability in the time series is partitioned among cyclical components with amplitudes and phases given by $\sqrt{a_k^2 + b_k^2}$ and $\tan^{-1}(-b_k/a_k)$, respectively. The evaluation of the relative values of these amplitudes and phases is called *harmonic analysis*. To obtain these, one must estimate the $\{\alpha(\omega_k)\}$ coefficients, which can be done with the discrete Fourier transform

$$\alpha(\omega_k) = \sum_{t=1}^{T} y_t \phi_k^{-1}(t), \tag{32.2}$$

for $k = 1, \ldots, T/2$, where $\alpha(\omega_0) = a_0 \equiv \bar{y}$ (the mean of $\{y_t\}$), and $\phi_k^{-1}(t)$ are the inverse of the Fourier basis functions given in (32.1).

Spectral analysis considers the amplitudes as a function of frequency. First, we define the *periodogram* as the squared modulus of $\alpha(\omega_k)$:

$$\widehat{I}(\omega_k) \equiv \alpha(\omega_k)\bar{\alpha}(\omega_i) \equiv |\alpha(\omega_k)|^2, \tag{32.3}$$

where the overbar (–) corresponds to the complex conjugate. Thus, $\widehat{I}(\omega_k)$ corresponds to the amplitude associated with frequency ω_k. The periodogram is an estimator of the spectral density function, $f(\omega)$, which can be defined as

$$f(\omega) = \sum_{\tau=-\infty}^{\infty} \gamma_Y(\tau)\exp(-2\pi i\omega\tau), \quad -1/2 \leq \omega \leq 1/2, \tag{32.4}$$

where $\gamma_Y(\tau)$ is the autocovariance function (ACVF) associated with the process generating y_t (see Chapter 3).

Unfortunately, despite being asymptotically unbiased for stationary processes, the periodogram is not a good estimator of the spectrum because it is biased under finite samples and it is not a consistent estimator. Most estimation approaches then work to balance a tradeoff between variance and bias of the estimator (as is typically the case in statistical estimation). This is an important topic and a complete discussion is beyond the scope of this chapter (see the aforementioned references for details). Briefly, to address bias one first typically applies a *data taper* (or *data window*) to the data, in which observations are replaced by weighted observations (e.g., by $y_t w_t$), where the weights are some function that gives low (or zero) weights to the values near the ends of the series and higher weights to observations in the middle of the data record (e.g., see [8], [61]).

There are many alternatives to reduce variance in spectral estimation, most related in some way to applying a filter to the periodogram estimates. Perhaps the simplest approach is to *smooth the periodogram*. Although there are a number of kernel smoothers that can be used for this purpose, the most basic is just to average the periodogram estimates in some window of length $2P$ around a frequency of interest (sometimes called a "box filter"):

$$\widehat{f}(\omega_k) = \frac{1}{2P+1} \sum_{j=k-P}^{k+P} \widehat{I}(\omega_j). \tag{32.5}$$

Obviously, the ability to estimate at the lowest and highest frequencies are impacted by the choice of P, as is the tradeoff between variance and bias in the spectral estimate (see [9] for additional discussion and recommendations).

One can also approach the spectral estimation problem from a parametric perspective, in which a model (usually, an ARMA model) is estimated, and since its spectrum has a known analytical form in terms of those parameters, one can evaluate the spectrum based on the estimates (e.g., see [61]).

Example: Chapter 30 contains examples showing Pacific SSTs and their importance in climatological analysis. Here, we consider a time series of mean-centered monthly SST data (deg C) averaged in the Niño 3.4 region of the tropical Pacific (i.e., the region outlined by 5N-5S, 170W-120W) for the period from January 1970 - September 2016 (see the solid line in Figure 32.1). The power spectral density for this series, obtained by smoothing the periodogram (with $P = 6$) is shown in Figure 32.2. This plots shows several frequency regions that exhibit large spectral peaks. In particular, the annual and semiannual cycles are shown by the peaks centered at frequencies of 0.083 and 0.167 cycles/month, respectively (corresponding to cycles with periods of 12 months and 6 months, respectively). Perhaps the most interesting component of this estimate spectrum corresponds to the broad area of high spectral density between frequencies of about 0.015 and 0.035 cycles/month (or, periods between 66.7 and 28.5 months - i.e., 2.4 - 5.6 years or so). This corresponds to the known periods associated with El Niño/La Niña (ENSO) events.

32.3.2.2   Bivariate Spectral Analysis

One can gain additional insight in oceanographic data by considering the lead/lag behavior of two or more time series; e.g., whether one variable (say, surface wind) at some location is strongly related to another variable (say, SST) at some other location at some time lag. One can evaluate such dependencies by extending the notion of a stationary covariance function to the bivariate case, i.e., the so-called *cross-covariance function*. Given two time series, $\{y_t^{(i)}\}$ and $\{y_t^{(j)}\}$, generated by $Y_t^{(i)}$ and $Y_t^{(j)}$, respectively, define the stationary cross-covariance function as

$$\gamma_{ij}(\tau) \equiv \text{cov}(Y_t^{(i)}, Y_{t+\tau}^{(j)}), \tag{32.6}$$

for time lags $\tau = \{0, \pm 1, \pm 2, \ldots\}$ and note that $\gamma_{ij}(-\tau) = \gamma_{ji}(\tau)$.

Although the cross covariance function can be useful by itself, it is often the case in oceanographic and atmospheric processes that the lead/lag dependence between two series is driven by certain frequencies (i.e., wave modes). Thus, we consider the bivariate extension of spectral analysis. Under simple regularity conditions, the *cross-spectrum* between two series $Y_t^{(1)}$ and $Y_t^{(2)}$ is defined as:

$$f_{12}(\omega) = \sum_{\tau=-\infty}^{\infty} \gamma_{12}(\tau) \exp(-2\pi i \omega \tau), \quad -1/2 \leq \omega \leq 1/2 \tag{32.7}$$

$$\equiv c_{12}(\omega) - i\, q_{12}(\omega), \tag{32.8}$$

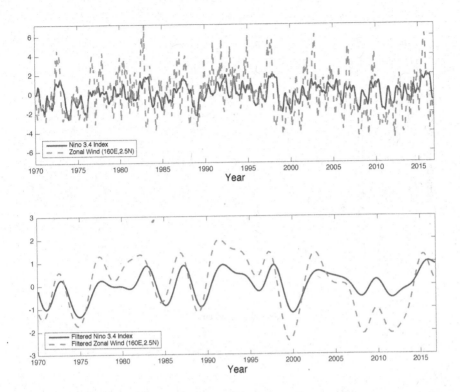

FIGURE 32.1
Top panel: Time series of mean-centered monthly sea surface temperature (SST) in deg C in the so called Niño 3.4 region of the tropical Pacific (i.e., the region outlined by 5N-5S, 170W-120W) for the period from January 1970 - September 2016 (solid line); corresponding time series for the zonal (east-west) component of the near surface wind (in units of m/s) at (160E, 2.5N) (dashed line). Bottom panel: low pass filtered versions of the same time series shown in the top panel.

where $f_{12}(\omega)$ is complex-valued and we call its real component, $c_{12}(\omega)$, the *co-spectrum*, and its imaginary component, $q_{12}(\omega)$, the *quadrature spectrum*. These can be manipulated to give useful features of the cross-spectrum. In particular, the *squared-coherence function* is given by:

$$\mathrm{coh}_{12}^2(\omega) = \frac{|f_{12}(\omega)|^2}{f_1(\omega)f_2(\omega)} = \frac{c_{12}^2(\omega)+q_{12}^2(\omega)}{f_1(\omega)f_2(\omega)}, \tag{32.9}$$

where $f_i(\omega)$ corresponds to the spectral density of $Y_t^{(i)}$, for $i = 1, 2$ given by equation (32.4). The squared coherence is analogous to the squared correlation coefficient and takes values between $0 \leq \mathrm{coh}_{12}^2(\omega) \leq 1$. It provides an indication of the square of the linear association between the components of the two processes at the frequency ω. We also define the *phase spectrum*:

$$\theta_{12}(\omega) = \tan^{-1}\left(\frac{-q_{12}(\omega)}{c_{12}(\omega)}\right), \tag{32.10}$$

which takes values between $-\pi$ and π. The phase spectrum is usually considered together

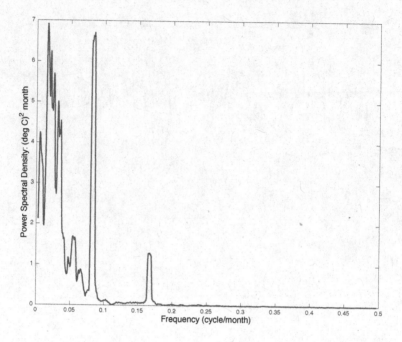

FIGURE 32.2
Power spectral density estimate (in units of $(deg\ C)^2\ month$) for the Niño 3.4 index shown in Figure 32.1. The smoothed periodogram estimate is shown, with $P = 6$.

with the squared coherence; that is, a constant phase that corresponds to a peak in the squared coherence indicates that one series leads/lags the other by the amount suggested by the phase (see the example below).

The calculation of the bivariate spectral estimates can be done through the bivariate extension of the periodogram. Let $\{\alpha_i(\omega_k)\}$ be the discrete Fourier transform (32.2) of the discrete time series $\{y_t^{(i)}\}$. The *cross-periodogram* estimate is then defined as:

$$\widehat{I}_{12}(\omega_k) \equiv \alpha_1(\omega_k)\overline{\alpha}_2(\omega_k), \tag{32.11}$$

where the overbar indicates the complex conjugate. The cross-periodogram estimate has the same issues (i.e., bias and consistency) as the univariate periodogram estimate discussed previously. As in the univariate case, one way to deal with these issues is by smoothing the estimated cross-periodogram across frequencies:

$$\widehat{f}_{12}(\omega_k) = \frac{1}{2P+1} \sum_{j=k-P}^{k+P} \widehat{I}_{12}(\omega_j). \tag{32.12}$$

Then, along with the estimates of the individual spectra (i.e., $\widehat{f}_1(\omega_k)$, $\widehat{f}_2(\omega_k)$ given by (32.5)), we can get estimates of the squared coherence, co-spectrum, quadrature spectrum, and phase spectrum by plugging these into their respective definitions. For details on the statistical properties of these estimators, see [61], [8], [64], or almost any book on advanced time series analysis.

Example: In addition to the Niño 3.4 SST index, the top panel of Figure 32.1 also shows (dashed line) a corresponding time series for the zonal (east-west) component of the near surface wind at (160E, 2.5N). As discussed in Section 32.5.1, there are strong connections between near surface winds and properties of the ocean circulation. Figure 32.3 shows the results of the bivariate spectral analysis between these two series. In particular, the top panel shows an estimate of the squared coherence function between the two time series obtained from smoothed periodogram and cross-periodogram estimates (with $P = 10$), and the bottom panel shows the associated estimate of the phase spectrum (in radians). Notice the very high coherence between the two series in the lower frequency range associated with the ENSO phenomena discussed in the univariate spectral example (Figure 32.2). In particular, consider the frequencies between 0.015 and 0.035 cycles/month (i.e., periods between 66.7 and 28.5 months). This range has high squared coherence estimates (around 0.84 on average) and the negative phase suggests that the Niño 3.4 series lags the zonal wind series by about 1/2 radians over this frequency range on average. This corresponds to the ENSO series lagging the wind series by about 3.6 months. This can be seen in the time series if we apply a low-pass filter as in the bottom panel of Figure 32.1, which clearly shows that the filtered zonal wind typically achieves its peaks and valleys just before the filtered ENSO series. This lead/lag relationship has proven useful in building models for the long-lead prediction of SST associated with ENSO (e.g., see [7] and Section 32.6 below).

32.3.2.3   Space-Time Spectral Analysis

As discussed briefly in Chapter 30, cross-spectral analysis can be extended to the case of one-dimensional spatial data that vary in time. Such an analysis is sometimes known as a *frequency-wavenumber analysis*. The purpose is to identify peaks in the spectral density estimates as a function of temporal frequency and spatial wavenumber (recall from Chapter 30 that the spatial wavenumber is the spatial analog to frequency in temporal spectral analysis and is just the number of sinusoidal components in a given spatial domain). With some fairly restrictive assumptions, one can decompose such a time-frequency spectrum into components associated with standing waves and traveling waves. This is important for fluid dynamical systems such as the atmosphere and ocean, which often exhibit such behavior. More details can be found in [12] and [73].

32.4   Spatio-Temporal Methods

The ocean system is inherently spatio-temporal. As such, many of the statistical models used to characterize oceanographic data and processes have been based on spatio-temporal statistical models (for recent overviews of spatio-temporal statistical methods, see [12] and [74]). In oceanography, most statistical models for spatio-temporal processes are dynamic models (see Chapter 4 of this volume), in which spatial processes evolve through time, following the etiology of the underlying process evolution. These models are typically state-space models, in which there is an observation or data model where the data are conditioned on a latent process, with biases and observational uncertainty accounted for in this stage. The latent state process is then assumed to evolve dynamically, typically via a Markov assumption, with spatially-dependent noise forcing.

Statistical spatio-temporal process models in oceanographic applications have typically been assumed to be linear, which is a reasonable assumption when one is focused on longer time scales and short time-scale nonlinear processes are approximated as stochastic forcing

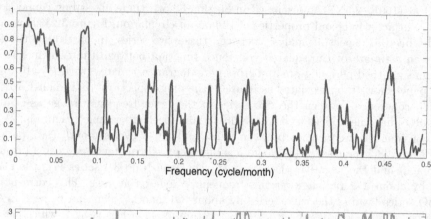

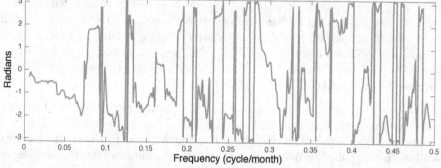

FIGURE 32.3

Cross-spectral analysis of the Niño 3.4 index and the zonal wind time series shown in Figure 32.1. Top panel: estimate of the squared coherence function between the two time series obtained from smoothed periodogram and cross-periodogram estimates with $P = 10$. Bottom panel: estimate of the phase spectrum (in radians) associated with the squared coherence estimates in the top panel.

in the linear evolution model. In some cases, specific parameterizations of these models can be motivated by considering the underlying dynamical mechanisms that govern the system (e.g., see [77] and Chapter 4 of this volume). More recently, fully nonlinear spatio-temporal statistical models have become more tractable for many modeling applications (see Section 32.6 below). The complex nature of these process models (nonlinearity and/or complex parameterizations) has led to implementation from a hierarchical perspective in which the parameters from the first two stages are assigned distributions, often with spatial or temporal structure. The parameters can also depend on exogenous covariates and various shrinkage mechanisms. The hierarchical versions of these models are most often considered from a Bayesian inferential paradigm.

Given that the dimension of the data sets and processes under consideration in oceanographic applications are fairly large, there is usually a need to reduce the process and/or parameter dimensionality when using spatio-temporal models. In the context of oceanographic and atmospheric examples, this is natural as there is a long history of thinking about these processes as "fluids" in which the spatial dimension can be expressed naturally in terms of expansions of wave modes. Chapter 30 in this volume gives a fairly extensive overview of spectral-based expansions (e.g., eigen-decomposition methods such as EOFs, science-based special functions, low-rank methods, over-complete methods, and neighbor-based models). Historically, most oceanographic spectral expansions in data assimilation and long-lead forecasting applications have been in terms of Fourier modes, science-based special functions, or EOFs. The example in Section 32.6 below gives a brief example of such methods for long-lead forecasting.

32.5   Methods for Data Assimilation

Data assimilation is a broad term that refers to the combination of observational data with *a priori* scientific information, usually mechanistic models (e.g., see [76]). This is a type of data fusion (see Chapter 7) and most of its development has come through the atmospheric and oceanic sciences (see Chapter 31 of this volume for details, and Chapter 30 for a discussion relative to spatial prediction).

Data assimilation is fundamentally important in oceanography because there are so few observations for most of the critical state process variables, that to obtain coherent and physically/biologically realistic fields, one must rely on the mechanistic models that pertain to the system (see [79] for a statistically-oriented description of the governing system of equations in oceanography). The dearth of observations for most of the state variables and an overabundance of observations from satellites, combined with the enormous range of time and spatial scales present in the system, make oceanographic data assimilation quite challenging. Several entire monographs have been devoted to the topic (e.g., see [21], [6]). As discussed in the aforementioned sources, the primary methods for data assimilation are based in optimal estimation theory, variational analysis, or Bayesian methodology.

As discussed above, the amount and quality of *in situ* data for ocean variables is somewhat limited and even if one is not performing a complete assimilation of the entire ocean system, there may be a need to construct complete spatial fields representing the overall climatological features of some component of the ocean state, as well as its uncertainty. Many of the statistical approaches to these problems have been conducted within a Bayesian hierarchical modeling framework (e.g., [26], [36], [39], [40]). Other challenges related to data assimilation in the ocean context that have required substantial statistical modeling effort are concerned with assimilating forcing fields (e.g., near surface winds) and biogeochemical variables (e.g., lower trophic ecosystem components). These are discussed briefly in the following subsections (see [79] for a more complete overview).

32.5.1   Assimilating Near Surface Ocean Winds

Near surface ocean winds are a critical component of the atmosphere/ocean interface as they are directly responsible for the transfer of momentum to the ocean and they modulate the exchanges of heat and fresh water to and from the upper ocean. Typically, the surface winds that arise from weather center global analysis products do not contain realistic energy content over oceanic regions. The advent of space-borne scatterometer instruments in the later decades of the 20th century provided the first high resolution wind observations over the oceans on daily timescales. However, these observations are "gappy" in the sense that they arise from polar orbiting instruments that have fairly limited spatial swaths on each orbit and typically cannot observe along the nadir angle. The goal of a statistical data assimilation is then to blend the complete, but energy-deficient, weather center analyses with the incomplete, yet energy-realistic, satellite surface wind observations in order to provide spatially complete wind fields at sub-daily intervals while managing the uncertainties associated with the different data sources and the blending procedure.

There have been a number of modeling applications focused on the near surface wind assimilation problem (see [54], [12], and [79] for recent overviews). These models typically include both types of data (analysis fields and satellite wind observations) in the data model, and the process model is motivated by a fairly simple, yet dynamically plausible, physical model (i.e., leading order terms and/or approximations of the primitive equations) with random parameters and some mechanism to ensure that the energy content of the predicted wind fields is scientifcally plausible. Uncertainty quantification is accomplished through the hierarchical Bayesian paradigm. Output from such models have been shown to be useful in the context of providing inputs to ocean forecasting systems (e.g., [62]).

32.5.2   Assimilating Lower Trophic Ecosystem Components

As summarized in [79], ocean biogeochemisty concerns the interaction between the ocean biology, chemistry and physical properties of the ocean. It is a very complex system and is essential for driving the ecology of the ocean food web. Analysis of this marine ecosystem is complicated by numerous sources of uncertainty in the observations, understanding of the underlying scientific process and the complexities that are involved in the numerous nonlinear interactions across variables and scale. The biological variables are poorly sampled and their observation errors are non-Gaussian.

The complexity of the lower trophic marine ecosystem assimilation problem has led to the development of innovative statistical methodology for parameter estimation and prediction (e.g., see the recent review in [63]). For example, [25] considered perhaps the first Markov chain Monte Carlo approach to the problem and adaptive Metropolis-Hastings algorithms were considered for a nonlinear marine ecosystem model by [24]. In addition, this problem has led to the development of novel sequential importance sampling, particle filter, ensemble Kalman filter, and hierarchical Bayesian algorithms (e.g., [25], [16], [65], [17], [56], [20], and [18]).

A challenging aspect of statistical data assimilation for ocean ecosystem processes is the efficient inclusion of the nonlinear governing deterministic equations that describe the complex interactions between the physical ocean and biological processes. In cases where where the model associated with these mechanistic equations is too computationally expensive to be included efficiently in a traditional statistical estimation algorithm, statistical surrogates (i.e., emulators) can be used. In essence, an emulator is simply a statistical model fit to the computer model. This surrogate model can then be used to predict the output of the mechanistic model under various inputs, parameter settings, and forcing (see the recent overview in [27]). In the context of the ocean biogeochemical processes, the statistical surrogate model is typically a nonlinear dynamical model. Recent examples in the context of ocean ecosystem data assimilation can be found in [31], [38], [37], and [78], and a broader overview can be found in [63].

It is important to note that Bayesian hierarchical models are also extensively used in statistical modeling of higher trophic levels of the ocean ecosystem. In particular, in fisheries management, stock-recruitment models are an important tool and have been considered from a Bayesian perspective for quite some time (e.g., [69], [28], [47], [15], [53], [29]). More recently, Bayesian methods have been used to consider distributions and movement of marine mammals (e.g., [72], [11], [48]).

32.6 Methods for Long-Lead Forecasting

Given that the ocean has components that evolve on fairly long time scales, there can be predictive skill over a time span of many months (i.e., so-called *long-lead forecasting*). Indeed, statistically derived empirical models can perform as well or better than deterministic models in this context. In this case, "long-lead" refers to time scales beyond those commonly predicted by meteorological forecast models. Although the time-scales considered are relatively long, various degrees of useful predictability are achievable due to the exploitation of the different time scales between sea and land processes, and the fact that these processes are closely coupled. These so-called *teleconnections* between the ocean and the atmosphere have long been exploited to build statistical models (e.g., [13] ,[1],[3], [75], [41]). A very brief summary of some of the common approaches is given in the following subsections.

32.6.1 Multivariate Methods

There is a long history in the oceanic and atmospheric sciences of using multivariate linear methods to produce long-lead forecasts. These include spatio-temporal CCA (e.g., see Section 32.2.2 above) as well as multivariate multiple regression (see [2], [73], [80]). It is often the case that these methods require a dimension reduction stage, typically accommodated by means of an eigen-decomposition (e.g., EOFs) as described above in Section 33.2.2.8. For the most part, these methods rely on traditional multivariate statistical procedures and the inherent spatial and temporal nature of the problems is not considered explicitly.

32.6.2 Autoregressive and State-Space Approaches

As discussed above, it is well-established that the ocean-atmosphere system is a nonlinear dynamical system consisting of many processes of different spatial and temporal scales. As discussed in Section 32.4, this system is often very well represented by linear evolution of the longer time-scale components of the process that are forced by Gaussian white noise (in time), where the forcing process is spatially-dependent and represents the short time-scale nonlinear effects (e.g., see the overview in [58]). Such first-order Markov models can be written in terms of linear continuous time or discrete time dynamical systems, although typically they must be projected into some lower-dimensional space. A natural long-lead forecasting application of such models considers the 6-12 month forecast of the ocean state. Examples in the climate literature include [59], [60], and [57], who, by projecting tropical Pacific SST anomalies onto their leading EOFs, show that first-order Markov process with Gaussian spatial noise can give quite successful long-lead forecasts relevant to the El Niño and La Niña (ENSO) prediction.

In general, reduced-order linear Markov models have proven quite effective for long-lead forecasting. However, it is well-known that such methods break down in the presence of significant nonlinear behavior (e.g., the ENSO phenomenon is known to exhibit nonlinearity; e.g. [30]). One can still use the low-order Markovian perspective to model such phenomena, with some modification. For example, [7] consider a reduced-order VAR(1) model but allow the transition operator and innovation covariance matrices to be regime-dependent, thereby inducing nonlinear behavior. Alternatively, [35] demonstrate the effectiveness of a quadratic nonlinear model for long-lead prediction of ENSO from a classical regression perspective, and [77] generalized this type of quadratic model in the fully Bayesian state-space framework, in which the process is "hidden" and parameters are random. The important part of these models is the quadratic interaction of the hidden process components. The downside to these models is that they require estimation of a large number of parameters. This can be mitigated by regularization and/or science-based parameterization (see the discussion in [74]).

32.6.3 Alternative Approaches

Traditionally, analog forecasting methods have proven effective for long-lead forecasting. First used as a heuristic method by operational weather forecasters in the 1920s, and re-popularized by [42], analog forecasting is a simple nonlinear forecasting method. In essence, this approach involves the identification of past trajectories of the state of a system (i.e., analogs) that are similar to the current state of the system (i.e., the initial condition). The current state of the system is then forecasted forward by assuming that the initial condition will evolve similarly to the trajectory of the identified past analogs. In the spatio-temporal context, one considers the trajectories of spatial fields (or their projections into lower-dimensional space) through time. The method has been shown to be effective for many climatological long-lead forecasting applications and climate downscaling (e.g., [4], [70], [82]). Traditional analog implementations are somewhat *ad hoc* and only recent have they been placed in a more statistically and mathematically rigorous framework (e.g., [14], [52], [81],[51]).

Traditional machine learning based models have also been used to perform long-lead forecasting (e.g., see the overview in [32]). For example, [67] and [66] use traditional feed-forward neural networks for long-lead forecasting of SST. The next generation of such models, so-called "deep learning" methods common in machine learning artificial intelligence applications, are also starting to be used in the climate forecasting framework (e.g., [23]). Although it is generally the case that feed-forward neural networks are not able to account adequately for the time dependence present in complex spatio-temporal processes, recurrent neural networks (RNNs), developed in engineering during the 1980s (e.g., see the overview in [43]), are designed to allow cycles and sequences in their hidden layers. These methods have not seen as much use for spatio-temporal prediction, but they have been used with great success in natural language processing, where word sequences (including their temporal dependence) are crucial. Traditionally, RNN models have been difficult to implement in spatio-temproal forecast problems due to the so-called vanishing gradient problem in the back-propagation algorithm used to estimate weights in the model's hidden layers. This has been mitigated to a great extent in the modern "deep learning" implementations.

Alternatively, one can minimize the number of weights that need to be learned using so-called echo state networks (ESNs; [33]) and liquid state machines (LSMs; [46]). These approaches, now known more generally as "reservoir computing" methods, consider hidden layers that have sparse connections and yet still allow for sequential interactions. A critical feature of reservoir models is that the connectivity and the weights for the hidden units are fixed and, perhaps surprisingly, randomly assigned. This is known as the reservoir, and it is typically of higher dimension than the input. The reservoir states are then mapped to the desired output. Importantly, only the weights at this mapping phase are estimated. Historically, "deep learning" algorithms are limited in oceanographic forecasting applications because they do not formally provide estimates of forecast uncertainty. However, this is an active area of research and there are examples of long-lead ocean forecasting that utilize these methods and provide uncertainty quantification (e.g., see [50]).

32.6.4 Long-Lead Forecast Example: Pacific SST

As discussed previously, long-lead forecasts of SSTs are important in oceanography and meteorology because of the importance of the ENSO phenomena and its impacts on weather patterns worldwide. It has consistently been shown to be a process for which statistical forecast models can do as well or better than deterministic models (e.g., [5], [71]). One of the "classic" statistical models in this context is a simple VAR(1) state-space model applied to a reduced dimensional state-vector of SSTs (e.g., see [59] and the overview in Chapter 9 of [12]). This is the model considered for the following example.

We consider SST anomalies for $m = 2,371$ (non land) locations on a 2 deg by 2 deg grid over the region from -30 S to 30 N latitude and 124 E to 290 E longitude (see the top panel of Figure 32.4) for the period from January 1970 through November 2015. In particular, we train the model on data up through May 2015 and produce a 6 month forecast (out-of-sample) for November 2015 (which corresponded to a fairly intense El Niño event). We consider a simple Bayesian hierarchical model implementation as given in the next subsection.

32.6.4.1   SST Forecast Model

Let the SST observations be given by the m-dimensional vectors, $\mathbf{Z}_t$, for $t = 1, \ldots, T$, where $T = 545$. Consider the following data and process models for $t = 1, \ldots, T$,

$$\mathbf{Z}_t = \boldsymbol{\Phi}\mathbf{Y}_t + \boldsymbol{\epsilon}_t, \tag{32.13}$$

$$\mathbf{Y}_{t+\tau} = \mathbf{M}\mathbf{Y}_t + \boldsymbol{\eta}_{t+\tau}, \tag{32.14}$$

where $\mathbf{Y}_t$ is a $p = 11$-dimensional latent process, $\boldsymbol{\Phi}$ is an $m \times p$ matrix of EOF basis functions, $\boldsymbol{\epsilon}_t \sim N(0, \sigma_\epsilon^2 \mathbf{I})$, $\mathbf{M}$ is a $p \times p$ transition matrix, and $\boldsymbol{\eta}_t \sim N(\mathbf{0}, \mathbf{Q})$, is the innovation noise process (for all t). In the example considered here, $\tau = 6$, corresponding to a 6 month lead-time forecast. Specifically, in this example, we are interested in a forecast of the filtered process $\mathbf{Y}_{T+\tau}^P \equiv \boldsymbol{\Phi}\mathbf{Y}_{T+\tau}$ given the data up to time T. In some cases, one might wish to factor in the truncation error associated with the basis expansion in the forecast (e.g., see [12] Chapter 9 for more discussion).

For simplicity, fairly naive priors are given for the hyperparameters in the this model, with the results quite insensitive to these choices. In particular,

$$\begin{aligned}
\text{vec}(\mathbf{M}) &\sim N(\text{vec}(0.8\,\mathbf{I}_p), 100\,\mathbf{I}_{p^2}), \\
\mathbf{Q}^{-1} &\sim W((\mathbf{S}_\eta\, d_\eta)^{-1}, d_\eta), \text{ with } \mathbf{S}_\eta = 100\,\mathbf{I}, d_\eta = p - 1, \\
\sigma_\epsilon^2 &\sim IG(q_\epsilon, r_\epsilon), \text{ with } q_\epsilon = 0.1, r_\epsilon = 100,
\end{aligned}$$

where $\text{vec}(\cdot)$ corresponds to a vectorization of the matrix argument, $W(\cdot, d_\eta)$ refers to a Wishart distribution with d_η degrees of freedom, and IG corresponds to an inverse gamma distribution. Because there are SST data for quite a long time period in this example, the results are not sensitive to the initial conditions, so rather than specify prior distributions for the initial conditions, we simply fix them at $\mathbf{Y}_t = \boldsymbol{\Phi}'\mathbf{Z}_t$; $t = 1, \ldots, \tau = 6$.

A simple Gibbs sampler was used to implement the Bayesian model (see Chapter 8 of [12] for the specific algorithm). The MCMC algorithm was run for 6,000 iterations, and the first 1,000 iterations were considered the "burn-in" period. The algorithm converges fairly quickly.

32.6.4.2   Results

Figure 32.4 shows the results for the long-lead prediction of November 2015 SST anomalies given SST observations through May 2015. The top panel shows the observed SST anomalies and the second panel shows the posterior mean forecast. Although the forecast clearly suggests an El Niño (warm) signature in the tropical Pacific ocean, it is substantially less intense than what was observed. In addition, the forecast uncertainties (pixelwise 2.5%-tiles

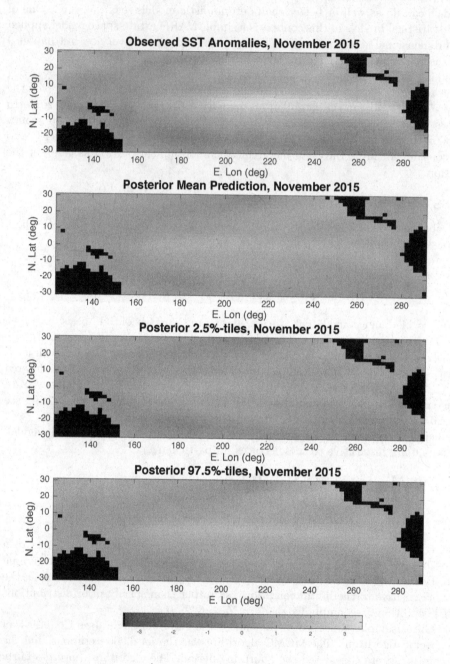

FIGURE 32.4
Sea surface temperature anomalies (deg C) for November 2015. Top panel: Observed anomalies; Second panel: posterior mean forecast of SST anomalies given data up to May 2015; Third panel: pixel-wise 2.5%-tiles from the posterior predictive distribution for November 2015; Fourth panel: pixel-wise 97.5%-tiles from the posterior predictive distribution for November 2015.

and 97.5%-tiles from the posterior predictive distribution, shown in the third and fourth panels of Figure 32.4, respectively) are also clearly suggesting a warm event in the forecasted SST anomalies, with the potential for a fairly intense El Niño. However, even in this case, the credible intervals do not include as an intense warm event as actually occurred. This is not uncommon, and it has been shown that nonlinear dynamical models can often do a better job at long lead forecasting of warm events (e.g., see [7], [35], [77], [12], [22], and references therein.)

32.7 Conclusion

Space limitations prevent the presentation of complete overviews of all of the various types of statistical models and methods that have been applied in oceanography. Rather, the purpose of this chapter is to highlight the broad nature of the methods that have been considered, especially those that are tied to the methods presented elsewhere in this volume. The literature review is not meant to be exhaustive, but simply meant to be a starting place for entry into the various subjects.

32.8 Acknowledgments

The writing of this chapter was partially supported by the US National Science Foundation (NSF) and the U.S. Census Bureau under NSF grant SES-1132031, funded through the NSF-Census Research Network (NCRN) program.

Bibliography

[1] T.P. Barnett. Statistical prediction of north american air temperatures from pacific predictors. *Monthly Weather Review*, 109(5):1021–1041, 1981.

[2] T.P. Barnett and K. Hasselmann. Techniques of linear prediction, with application to oceanic and atmospheric fields in the tropical pacific. *Reviews of Geophysics*, 17(5):949–968, 1979.

[3] T.P. Barnett and R. Preisendorfer. Origins and levels of monthly and seasonal forecast skill for united states surface air temperatures determined by canonical correlation analysis. *Monthly Weather Review*, 115:1825–1850, 1987.

[4] T.P. Barnett and R.W. Preisendorfer. Multifield analog prediction of short-term climate fluctuations using a climate state vector. *Journal of the Atmospheric Sciences*, 35(10):1771–1787, 1978.

[5] A.G. Barnston, Y. He, and M.H. Glantz. Predictive skill of statistical and dynamical climate models in sst forecasts during the 1997-98 el niño episode and the 1998 la niña onset. *Bulletin of the American Meteorological Society*, 80(2):217–243, 1999.

[6] A.F. Bennett. *Inverse modeling of the ocean and atmosphere.* Cambridge University Press, 2005.

[7] L.M. Berliner, C.K. Wikle, and N. Cressie. Long-lead prediction of pacific ssts via bayesian dynamic modeling. *Journal of Climate*, 13(22):3953–3968, 2000.

[8] P. Bloomfield. *Fourier analysis of time series: an introduction.* John Wiley & Sons, 2004.

[9] C. Chatfield. *The analysis of time series: an introduction.* CRC press, 2016.

[10] D.B. Chelton. Physical oceanography: a brief overview for statisticians. *Statistical Science*, pages 150–166, 1994.

[11] N. Cressie, C.A. Calder, J.S. Clark, J.M. Ver Hoef, and C.K. Wikle. Accounting for uncertainty in ecological analysis: the strengths and limitations of hierarchical statistical modeling. *Ecological Applications*, 19(3):553–570, 2009.

[12] N. Cressie and C.K. Wikle. *Statistics for spatio-temporal data.* John Wiley & Sons, 2011.

[13] R.E. Davis. Predictability of sea surface temperature and sea level pressure anomalies over the north pacific ocean. *Journal of Physical Oceanography*, 6(3):249–266, 1976.

[14] T. DelSole, L. Jia, and M. K. Tippett. Decadal prediction of observed and simulated sea surface temperatures. *Geophysical Research Letters*, 40(11):2773–2778, 2013.

[15] M.W. Dorn. Advice on west coast rockfish harvest rates from bayesian meta-analysis of stock-recruit relationships. *North American Journal of Fisheries Management*, 22(1):280–300, 2002.

[16] M. Dowd. A sequential monte carlo approach for marine ecological prediction. *Environmetrics*, 17(5):435–455, 2006.

[17] M. Dowd. Estimating parameters for a stochastic dynamic marine ecological system. *Environmetrics*, 22(4):501–515, 2011.

[18] M. Dowd, E. Jones, and J. Parslow. A statistical overview and perspectives on data assimilation for marine biogeochemical models. *Environmetrics*, 25(4):203–213, 2014.

[19] G. Evensen. Sequential data assimilation with a nonlinear quasi-geostrophic model using monte carlo methods to forecast error statistics. *Journal of Geophysical Research All Series-*, 99:10–10, 1994.

[20] J. Fiechter, R. Herbei, W. Leeds, J. Brown, R.F. Milliff, C.K. Wikle, A. Moore, and T. Powell. A bayesian parameter estimation method applied to a marine ecosystem model for the coastal gulf of alaska. *Ecological modelling*, 258:122–133, 2013.

[21] M. Ghil and P. Malanotte-Rizzoli. Data assimilation in meteorology and oceanography. *Advances in geophysics*, 33:141–266, 1991.

[22] D.W. Gladish and C.K. Wikle. Physically motivated scale interaction parameterization in reduced rank quadratic nonlinear dynamic spatio-temporal models. *Environmetrics*, 25(4):230–244, 2014.

[23] A. Grover, A. Kapoor, and E. Horvitz. A deep hybrid model for weather forecasting. In *Proceedings of the 21th ACM SIGKDD International Conference on Knowledge Discovery and Data Mining*, pages 379–386. ACM, 2015.

[24] H. Haario, E. Saksman, and J. Tamminen. An adaptive metropolis algorithm. *Bernoulli*, pages 223–242, 2001.

[25] R. Harmon and P. Challenor. A markov chain monte carlo method for estimation and assimilation into models. *Ecological modelling*, 101(1):41–59, 1997.

[26] D. Higdon. A process-convolution approach to modelling temperatures in the north atlantic ocean. *Environmental and Ecological Statistics*, 5(2):173–190, 1998.

[27] D. Higdon, J.D. McDonnell, N. Schunck, J. Sarich, and S.M. Wild. A bayesian approach for parameter estimation and prediction using a computationally intensive model. *Journal of Physics G: Nuclear and Particle Physics*, 42(3):034009, 2015.

[28] R. Hilborn, E.K. Pikitch, and M.K. McAllister. A bayesian estimation and decision analysis for an age-structured model using biomass survey data. *Fisheries Research*, 19(1):17–30, 1994.

[29] D. Hirst, G. Storvik, M. Aldrin, S. Aanes, and R.B. Huseby. Estimating catch-at-age by combining data from different sources. *Canadian Journal of Fisheries and Aquatic Sciences*, 62(6):1377–1385, 2005.

[30] Martin P Hoerling, Arun Kumar, and Min Zhong. El niño, la niña, and the nonlinearity of their teleconnections. *Journal of Climate*, 10(8):1769–1786, 1997.

[31] M.B. Hooten, W.B. Leeds, J. Fiechter, and C.K. Wikle. Assessing first-order emulator inference for physical parameters in nonlinear mechanistic models. *Journal of Agricultural, Biological, and Environmental Statistics*, 16(4):475–494, 2011.

[32] W.W. Hsieh. *Machine learning methods in the environmental sciences: Neural networks and kernels*. Cambridge university press, 2009.

[33] H. Jaeger. Echo state network. *Scholarpedia*, 2(9):2330, 2007.

[34] I. Jolliffe. *Principal component analysis*. Wiley Online Library, 2002.

[35] D. Kondrashov, S. Kravtsov, A.W. Robertson, and M. Ghil. A hierarchy of data-based enso models. *Journal of Climate*, 18(21):4425–4444, 2005.

[36] M. Lavine and S. Lozier. A markov random field spatio-temporal analysis of ocean temperature. *Environmental and Ecological Statistics*, 6(3):249–273, 1999.

[37] W.B. Leeds, C.K. Wikle, and J. Fiechter. Emulator-assisted reduced-rank ecological data assimilation for nonlinear multivariate dynamical spatio-temporal processes. *Statistical Methodology*, 17:126–138, 2014.

[38] W.B. Leeds, C.K. Wikle, J. Fiechter, J. Brown, and R.F. Milliff. Modeling 3-d spatio-temporal biogeochemical processes with a forest of 1-d statistical emulators. *Environmetrics*, 24(1):1–12, 2013.

[39] R.T. Lemos and B. Sansó. A spatio-temporal model for mean, anomaly, and trend fields of north atlantic sea surface temperature. *Journal of the American Statistical Association*, 104(485):5–18, 2009.

[40] R.T. Lemos, B. Sansó, and F.D. Santos. Hierarchical bayesian modelling of wind and sea surface temperature from the portuguese coast. *International Journal of Climatology*, 30(9):1423–1430, 2009.

[41] C.H.R. Lima and U. Lall. Hierarchical bayesian modeling of multisite daily rainfall occurrence: Rainy season onset, peak, and end. *Water Resources Research*, 45(7):W07422, 2009.

[42] E. N. Lorenz. Atmospheric predictability as revealed by naturally occurring analogues. *Journal of the Atmospheric Sciences*, 26(4):636–646, 1969.

[43] M. Lukoševičius and H. Jaeger. Reservoir computing approaches to recurrent neural network training. *Computer Science Review*, 3(3):127–149, 2009.

[44] H. Lütkepohl. *New introduction to multiple time series analysis*. Springer Science, 2005.

[45] J.G. Lutz and T.L. Eckert. The relationship between canonical correlation analysis and multivariate multiple regression. *Educational and Psychological Measurement*, 54(3):666–675, 1994.

[46] W. Maass, T. Natschläger, and H. Markram. Real-time computing without stable states: A new framework for neural computation based on perturbations. *Neural computation*, 14(11):2531–2560, 2002.

[47] M.K. McAllister and G.P. Kirkwood. Bayesian stock assessment: a review and example application using the logistic model. *ICES Journal of Marine Science: Journal du Conseil*, 55(6):1031–1060, 1998.

[48] B.T. McClintock, J.M. London, M.F. Cameron, and P. L. Boveng. Bridging the gaps in animal movement: hidden behaviors and ecological relationships revealed by integrated data streams. *Ecosphere*, 8(3), 2017.

[49] C.E. McCulloch and S.R. Searle. *Generalized, linear, and mixed models*. John Wiley and Sons, 2001.

[50] P. L. McDermott and C. K. Wikle. An ensemble quadratic echo state network for non-linear spatio-temporal forecasting. *Stat*, 6(1):315–330, 2017. sta4.160.

[51] Patrick L McDermott, Christopher K Wikle, and Joshua Millspaugh. A hierarchical spatiotemporal analog forecasting model for count data. *Ecology and evolution*, 8(1):790–800, 2018.

[52] P.L. McDermott and C.K. Wikle. A model-based approach for analog spatio-temporal dynamic forecasting. *Environmetrics*, 2015.

[53] C.G.J. Michielsens and M.K. McAllister. A bayesian hierarchical analysis of stock recruit data: quantifying structural and parameter uncertainties. *Canadian Journal of Fisheries and Aquatic Sciences*, 61(6):1032–1047, 2004.

[54] R.F. Milliff, A. Bonazzi, C.K. Wikle, N. Pinardi, and L. M. Berliner. Ocean ensemble forecasting. part i: Ensemble mediterranean winds from a bayesian hierarchical model. *Quarterly Journal of the Royal Meteorological Society*, 137(657):858–878, 2011.

[55] A.H. Monahan, J. C. Fyfe, M. H.P. Ambaum, D.B. Stephenson, and G.R. North. Empirical orthogonal functions: The medium is the message. *Journal of Climate*, 22(24):6501–6514, 2009.

[56] J. Parslow, N. Cressie, E. P. Campbell, E. Jones, and L. Murray. Bayesian learning and predictability in a stochastic nonlinear dynamical model. *Ecological Applications*, 23(4):679–698, 2013.

[57] C. Penland. A stochastic model of indopacific sea surface temperature anomalies. *Physica D: Nonlinear Phenomena*, 98(2-4):534–558, 1996.

[58] C. Penland. Stochastic linear models of nonlinear geosystems. In *Nonlinear dynamics in geosciences*, pages 485–515. Springer, 2007.

[59] C. Penland and T. Magorian. Prediction of niño 3 sea surface temperatures using linear inverse modeling. *Journal of Climate*, 6(6):1067–1076, 1993.

[60] C. Penland and P.D. Sardeshmukh. The optimal growth of tropical sea surface temperature anomalies. *Journal of climate*, 8(8):1999–2024, 1995.

[61] D.B. Percival and A.T. Walden. *Spectral analysis for physical applications*. Cambridge University Press, 1993.

[62] N. Pinardi, A. Bonazzi, S. Dobricic, R.F. Milliff, C.K. Wikle, and L. M. Berliner. Ocean ensemble forecasting. part ii: Mediterranean forecast system response. *Quarterly Journal of the Royal Meteorological Society*, 137(657):879–893, 2011.

[63] M. Schartau, P. Wallhead, U. Hemmings, J.and Löptien, I. Kriest, S. Krishna, B.A. Ward, TT Slawig, and A. Oschlies. Reviews and syntheses: Parameter identification in marine planktonic ecosystem modelling. *Biogeosciences (BG)*, 14(6):1647–1701, 2017.

[64] R.H. Shumway and D.S. Stoffer. *Time series analysis and its applications: with R examples*. Springer Science, 2010.

[65] J.R. Stroud, M.L. Stein, B.M. Lesht, D.J. Schwab, and D. Beletsky. An ensemble kalman filter and smoother for satellite data assimilation. *Journal of the American Statistical Association*, 105(491):978–990, 2010.

[66] B. Tang, W.W. Hsieh, A.H. Monahan, and F.T. Tangang. Skill comparisons between neural networks and canonical correlation analysis in predicting the equatorial pacific sea surface temperatures. *Journal of Climate*, 13(1):287–293, 2000.

[67] F.T. Tangang, W.W. Hsieh, and B. Tang. Forecasting regional sea surface temperatures in the tropical pacific by neural network models, with wind stress and sea level pressure as predictors. *Journal of Geophysical Research: Oceans*, 103(C4):7511–7522, 1998.

[68] H. J. Thiébaux. *Statistical Data Analysis for Ocean and Atmospheric Sciences*. Academic Press, 1994.

[69] G.G. Thompson. A bayesian approach to management advice when stock-recruitment parameters are uncertain. *Fishery Bulletin*, 90(3):561–573, 1992.

[70] H.M. Van den Dool. Searching for analogues, how long must we wait? *Tellus A*, 46(3):314–324, 1994.

[71] G.J. Van Oldenborgh, S.Y. Philip, and M. Collins. El niño in a changing climate: a multi-model study. *Ocean Science*, 1(2):81–95, 2005.

[72] J.M. Ver Hoef and J.K. Jansen. Space-time zero-inflated count models of harbor seals. *Environmetrics*, 18(7):697–712, 2007.

[73] H. Von Storch and F.W. Zwiers. *Statistical analysis in climatology*. Cambridge University Press, Cambridge, 1999.

[74] C.K. Wikle. Modern perspectives on statistics for spatio-temporal data. *Wiley Interdisciplinary Reviews: Computational Statistics*, 7(1):86–98, 2015.

[75] C.K. Wikle and C.J. Anderson. Climatological analysis of tornado report counts using a hierarchical bayesian spatiotemporal model. *J. Geophys. Res*, 108(D24):9005, 2003.

[76] C.K. Wikle and L.M. Berliner. A bayesian tutorial for data assimilation. *Physica D: Nonlinear Phenomena*, 230(1):1–16, 2007.

[77] C.K. Wikle and M.B. Hooten. A general science-based framework for dynamical spatio-temporal models. *Test*, 19(3):417–451, 2010.

[78] C.K. Wikle, W.B. Leeds, and M.B. Hooten. Models for ecological models: Ocean primary productivity. *Chance*, 29(2):23–30, 2016.

[79] C.K. Wikle, R.F. Milliff, R. Herbei, and W.B. Leeds. Modern statistical methods in oceanography: A hierarchical perspective. *Statistical Science*, pages 466–486, 2013.

[80] D.S. Wilks. *Statistical methods in the atmospheric sciences*. Academic Press, 2011.

[81] Z. Zhao and D. Giannakis. Analog forecasting with dynamics-adapted kernels. *Nonlinearity*, 29(9):2888–2939, 2016.

[82] E. Zorita and H. Von Storch. The analog method as a simple statistical downscaling technique: comparison with more complicated methods. *Journal of Climate*, 12(8):2474–2489, 1999.

33

Paleoclimate reconstruction: looking backwards to look forward

Peter F. Craigmile

The Ohio State University, Columbus, Ohio

Murali Haran

The Pennsylvania State University

Bo Li

University of Illinois at Urbana–Champaign

Elizabeth Mannshardt

North Carolina State University, Raleigh

Bala Rajaratnam

University of California, Davis

Martin Tingley

The Pennsylvania State University

CONTENTS

33.1 Introduction

Paleoclimatology is the study of past climate. While in rare cases we may be able to study past climate from "directly observed" historical record, typically we infer past climate indirectly through the use of proxies. These proxies, such as tree rings, ice cores, corals, and pollen, have climate-sensitive characteristics that can be measured (e.g., tree ring density; the ratio of dissolved oxygen isotopes in ice cores and coral; the signature and relative abundance of pollen species). By associating these climate sensitive measurements to observed instrumental records of climate such as temperature and precipitation, we are able to predict or hindcast past climate along with, hopefully, an associated measure of uncertainty for our predictions. This task is called *paleoclimate reconstruction*. For example, tree ring densities of certain species of trees in the upper latitudes, suitably normalized to remove growth effects, are approximately linearly associated with temperature. By modeling the relationship between tree ring density and observed temperature, we can produce predictions of past temperature.

In paleoclimate reconstruction problems, climate scientists differentiate between two types of prediction or hindcast problems: index reconstructions (IR) and climate field reconstruction (CFR) [47]. In IR we predict climate on the basis of index series (typically averages over larger spatial scales such as Northern hemispheric or El Ñino Southern Oscillation (ENSO) reconstructions; see [2] for an example). Some versions of IR are also known as "composite plus scale" (CPS) [47]. IR seeks to produce hindcasts in time, and is a simpler statistical problem compared to CFR, which seeks to produce hindcasts in both space and time. Thus, as with many other studies carried out in climate, paleoclimate reconstruction is a spatio-temporal statistical problem. This problem is challenging because the spatial and temporal domains of the proxy and instrumental records may not overlap. Instrumental records are only available more recently in time, but have reasonably good spatial and temporal coverage. Proxy measurements are expensive to collect and are typically observed sparsely in space, and proxy records typically have longer but more uncertain temporal coverage. The paleoclimate reconstruction often involves multiple proxies. Since different proxies are sensitive to climate in varied ways, the relationships between different proxy measurements and instrumental records need to be modeled distinctly and carefully. If multiple climate variables are to be reconstructed simultaneously the paleoclimate reconstruction can become a multivariate problem. For example, the reconstruction is a bivariate problem if we attempt to reconstruct temperature and precipitation together.

In this chapter we review the statistical methodologies underlying paleoclimate reconstruction, providing some commentary on existing and outstanding problems for statisticians. We also discuss more recent research that blends the study of paleoclimate with climate models, investigating how researchers should compare reconstructions to climate model output. We close with a discussion of further ideas and research in statistical paleoclimatology.

33.2 Paleoclimate reconstruction: looking backwards

Informally, a paleoclimatic reconstruction produces predictions or hindcasts (and measures of uncertainty) of an underlying climate process, given direct observations of the climate system and indirect observations of climate-sensitive paleo proxies.

Formally, we introduce paleoclimate reconstruction in terms of a hierarchical Bayesian model (HBM) that link these direct and indirect measurements to some underlying climate process of interest. (Section 33.2.2 contains many references to Bayesian and non-Bayesian statistical methods used for paleoclimate reconstruction.) We introduce a simplified description of a framework introduced by [91]. (Also see [43] for a review of other methods of reconstruction.) For an IR we are interested in predicting a climate process series $\{C_t : t \in T\}$, where T is the time domain of interest (typically $T \subset \mathbb{Z}$ for a discrete-time climate series, but it is also possible that $T \subset \mathbb{R}$ for a continuous-time series, which can be useful for modeling time-uncertain proxy data). In the context of CFR, our climate process is $\{C(s,t) : s \in D, t \in T\}$ for some spatial domain D and temporal domain T. Commonly the spatial domain of interest is some region $D \subset \mathbb{R}^2$. Without loss of generality we will discuss the IR case in the definition of the HBM – with the addition of a spatial index for the models that we define below the CFR case follows similarly.

Now we introduce processes that capture the proxies and instrumental records. With one proxy and one instrumental record, let $\{P_t : t \in T\}$ denote the proxy process and $\{Z_t : t \in T\}$ the associated instrumental process. For example with $\{C_t\}$ being the average latent temperature series over a given spatial domain, $\{P_t\}$ could be the average tree ring density over that domain, and $\{Z_t\}$ the observed average instrumental temperature, again over the same domain.

An HBM for paleoclimate reconstruction then requires the following model components to be defined:

1. The model for the instrumental process, $f(\{Z_t\}|\{C_t\}, \boldsymbol{\theta}_Z)$. This model is conditioned on the underlying climate process and depends on a vector of parameters $\boldsymbol{\theta}_Z$. In the simplest setting we could imagine that Z_t is a noisy version of C_t for each $t \in T$: Assuming independence over time, with normal measurement errors, we have

$$Z_t \sim \mathcal{N}(C_t, \sigma_Z^2), \quad t \in T,$$

where σ_Z^2 is the measurement error. The likelihood is then

$$f(\{Z_t\}|\{C_t\}, \boldsymbol{\theta}_P) = \prod_t d\mathcal{N}(Z_t; C_t, \sigma_Z^2),$$

where $d\mathcal{N}(z; \mu, \sigma^2)$ is the probability density function (pdf) of a $\mathcal{N}(\mu, \sigma^2)$ random variable (RV) evaluated at z. We can see that $\boldsymbol{\theta}_Z = (\sigma_Z^2)^T$ for this simpler model. Note that the product in the likelihood for the instrumental process is over time points for which we have instrumental data (typically "recent" in time); the time points for which we will typically have proxy data will be different ("in the past" and "recent"). We discuss extensions below.

2. The model for the proxy process, $f(\{P_t\}|\{C_t\}, \boldsymbol{\theta}_P)$. This likelihood is again conditioned on the underlying climate process and depends on a vector of parameters $\boldsymbol{\theta}_P$. Writing down the likelihood for the proxy process is more involved than the model for the instrumental process as we need to specify how the proxy is related to the climate process. In paleoclimatology the functional relationship between

the proxy and climate process is called the *forward model* [23, 24, 91]. In statistical paleoclimatology we often write the forward model as a distribution that disentangles the deterministic and random components of the proxy, the latter of which can simply be measurement errors. A simple example used for tree ring density is to assume that after accounting for measurement error, there is a linear relationship between tree ring density and temperature; in this case, assuming conditional independence we suppose

$$P_t \sim \mathcal{N}(\beta_0 + \beta_1 C_t, \sigma_P^2), \quad t \in T,$$

for regression parameters β_0 and β_1 and measurement error variance σ_P^2. The likelihood in this case is

$$f(\{P_t\}|\{C_t\}, \boldsymbol{\theta}_P) = \prod_t d\mathcal{N}(P_t; \beta_0 + \beta_1 C_t, \sigma_P^2),$$

with $\boldsymbol{\theta}_P = (\beta_0, \beta_1, \sigma_P^2)^T$. Again we discuss more general cases below.

3. The prior distribution for the climate process, $\pi(\{C_t\}|\boldsymbol{\theta}_C)$, which depends on another vector of parameters $\boldsymbol{\theta}_C$ (we discuss the prior distributions of the other model parameters below). In the IR case, typically a prior distribution is provided that captures the smoothness of the climate process in time (e.g., an autoregressive time series model [5, 54], or uses covariate information from climate forcings). Further details are discussed below.

To complete the model, in addition to prior of the climate process we also need specific prior distributions, $\pi(\boldsymbol{\theta})$, for the remaining parameters $\boldsymbol{\theta} = (\boldsymbol{\theta}_Z, \boldsymbol{\theta}_P, \boldsymbol{\theta}_C)^T$. Informative priors are commonly used in the specification of this part of the model [5, 54, 92].

Given the specification of the likelihood for the instrumental and proxy processes, and the prior distributions for the climate process and other model parameters, the posterior distribution $\pi(\{C_t\}, \boldsymbol{\theta}|\{Z_t\}, \{P_t\})$ satisfies

$$\pi(\{C_t\}, \boldsymbol{\theta}|\{Z_t\}, \{P_t\})$$
$$\propto \ f(\{Z_t\}|\{C_t\}, \boldsymbol{\theta}_Z) \, f(\{P_t\}|\{C_t\}, \boldsymbol{\theta}_P) \, \pi(\{C_t\}|\boldsymbol{\theta}_C) \, \pi(\boldsymbol{\theta}).$$

In most situations the posterior distribution is not available in a closed-form expression and Markov chain Monte Carlo is used to provide a sample-based approximation to the posterior distribution [5, 54, 92].

For an IR, the posterior distribution of climate typically has reduced variance for more recent time periods, and the uncertainty increases as we go back in time. This reflects the fact that we have more proxy and instrument records in recent time periods. The imprint of climate from proxy records through the forward model, along with the prior distribution of the climate process, drives the posterior distribution of climate as we move further and further into the past. For the CFR, spatio-temporal reconstructions have more uncertainty over space and time, due to the sparsity of instrumental and proxy records. Clearly, the selection of meaningful and accurate prior models for the climate process is key to producing more informative paleoclimate reconstructions. For such purposes, information from climate model runs can be used to form a prior that can compensate for the sparsity of data.

To further illustrate this HBM framework we next work through a detailed example of a paleoclimate reconstruction.

33.2.1 A multiproxy, multiforcing IR reconstruction

The power of a hierarchical Bayesian model (HBM) lies in its flexibility in modeling the complex structure or relationship among multiple data sets. The paleoclimate reconstruction usually involves data from different sources and thus the HBM naturally provides an efficient and feasible framework for integrating paleodata of various characteristics. As an example for an index reconstruction, suppose we have different types of proxies, $P_1, \ldots, P_m$. Each proxy records the climate via a different mechanism and the statistical model should respect each proxy's characteristics. For example, if the proxy responds to climate variation through a very slow evolving process such as the heating diffusion into the earth and rocks, then the proxy only relates to the very low frequency variation of the climate change. On the contrary, if the proxy responds quickly to annual or even seasonal variation of climate, then the proxy mainly relates to the high frequency components of climate change. For these reasons, we model the proxy data in the hierarchy as below:

$$
\begin{aligned}
P_{1t} &= \beta_{10} + \beta_{11} g_1(\{C_t\}) + \epsilon_{1t}, \\
\vdots\ &= \qquad \vdots \\
P_{mt} &= \beta_{m0} + \beta_{m1} g_m(\{C_t\}) + \epsilon_{mt},
\end{aligned}
\tag{33.1}
$$

where $g_i(\cdot)$ for each $i = 1, \ldots, m$ are transformation functions that link the climate and proxies. These transformation functions can either be linear or nonlinear, depending on how the proxy reflects the climate evolution. The error terms $\{\epsilon_{it}\}$, $i = 1, \ldots, m$, are due to both the measurement errors in proxies and the imperfect linear relationship between the proxy and the transformed climate. The instrumental climate can contain a substantial amount of noise relative to the true climate. Therefore it is more appropriate to explicitly model those measurement errors in instrumental climate in order to take all major errors into account in the reconstruction:

$$
Z_t = \beta_0 + \beta_1 C_t + \zeta_t,
$$

where $\{\zeta_t\}$ may be modeled as white noise or a dependent time series process.

After modeling the likelihood of the observations, we now focus on the prior for climate, $\{C_t\}$. Depending on the purpose of the reconstruction, we can either introduce the climate forcings into the prior or simply model the climate as a stochastic process. Suppose including forcings is appropriate, and using the solar irradiance S_t, volcanism V_t, and greenhouse gases G_t as examples for forcings, the prior for climate can take the form of

$$
C_t = \alpha_0 + \alpha_1 S_t + \alpha_2 V_t + \alpha_3 G_t + \eta_t,
\tag{33.2}
$$

where $\{\eta_t\}$ are the errors in temperature that cannot be described by the variability in forcings. When it is not appropriate to include forcings, then we can simply model the prior as

$$
C_t = \eta_t.
\tag{33.3}
$$

The models for $\{\epsilon_{it}\}$ in (33.1), and $\{\eta_t\}$ in (33.2) or (33.3) are discussed in Section 33.2.2.3.

33.2.2 Statistical issues in paleoclimate reconstruction

33.2.2.1 Incorporating climate forcings in the HBM

The variabilities in different proxies are driven by the climate evolution and thus they serve as the noise-contaminated reflection of climate change. Forcings, on the contrary, are the internal drivers of the climate evolution, and any variation in forcings will eventually lead to fluctuation in the climate system. The major forcings include solar irradiance, volcanism, greenhouse gases, and aerosols, the first two of which are natural forcings, while the latter are anthropogenic forcings. We can also include the effect of orbital forcings [22] upon climate.

Since the forcings determine the large scale climate variation, it is very natural to employ them to improve the reconstruction. The four major forcings are estimated via different techniques. For example, the solar irradiance is derived from measurements of fluctuations of Berium ^{10}Be production rates which is modulated by solar magnetic variability, and the volcanic series is based on synthesis of individual ice cores and in some cases on historical records of large eruptions. Both [54] and [5] demonstrated that including forcings can make the temperature reconstruction better calibrated and more sharp. Although forcings play a very important role, when the primary interest is to use paleoclimate reconstructions to verify climate models, it is more appropriate to perform the reconstruction with no forcings included. Otherwise the modeling is circular in the sense that the forcings are also major inputs for the algorithm built in climate models.

33.2.2.2   Handling proxies separately

There is a large variety of climate proxies, including, measurements on tree rings, pollen assemblage, ice core, corals, boreholes, and speleothems [47, 65, 71]. The various proxies differ in their temporal resolutions. Tree rings and some corals provide an observation each year, and are accurately dated by layer counting. Pollen assemblage retain a smoothed record of climate variation due both to the persistency properties of mature plants [10] and the low-resolution dating of the lake core sediments from which pollen assemblages are extracted. Speleothem and borehole depth profiles behave as low pass filters that only retain the long term trends of climate.

The noise structures of different proxies are also characterized differently. For example, the error terms in the tree ring–climate relationship largely reflect that trees are imperfect recorders of local climate. With pollen data, however, an important source of uncertainty is due to the substantial dating errors in the radiocarbon technology. Another difference among different proxies is their relationship to climate variation. The relationship between pollen and temperature is sigmoidal, and therefore a logistic model has been developed to capture this nonlinear relationship.

Clearly, it is not optimal to treat all different proxies in the same way when trying to extract climate signal from them. [54] proposed a Bayesian framework to integrate proxies of different characteristics into one single coherent climate reconstruction. They studied the added value of tree ring, pollen, and borehole proxies representing, respectively, high, medium, and low frequency components of the climate signal. If forcings are not included in the model, they found that moderate to low frequency proxies need to be added to compensate for the information loss.

There is a growing literature on the development of scientifically-motivated forward models that describe how proxies record variations in the climate system [11, 23, 24, 97]. Hierarchical statistical modeling, combined with Bayesian inference, provides a framework for specifying and fitting a rich variety of models that are in line with current scientific understanding of the proxies.

33.2.2.3   Modeling temporal dependence

Climate reconstructions, either with direct or indirect regression models (see a discussion for these two frameworks in [4]), or with Bayesian hierarchical models, generally treat errors as white noise or short memory time series model such as autoregressive (AR) models with order 1 or 2 [5, 54, 90, 92, 93, 94, 95, 99, 100]. [70] and [82] demonstrate the choice of AR order could differ by the type of proxy. [36] use models based on Brownian motion and [75] use a stochastic volatility normal-inverse Gaussian model. [53] use principal components analysis (PCA) (also known as empirical orthogonal functions, EOF) in an IR context. Also worth noting is the method by [31] which models the temporal dependence nonparametrically.

There is evidence [19, 44, 45, 84] that the climate dynamics is governed by a long-memory stochastic process (see, e.g., [6] for a definition), which raises an interesting question on whether long-memory error models would be helpful with the reconstruction or uncertainty quantification. A further question is which error model is most appropriate for modeling the relationship between proxy and temperature and between temperature and forcings. Under the Bayesian modeling framework, [5] investigated this question by using both the land-surface temperature and the combined land and ocean temperature data sets with the [59] proxy network. [5] reveal that when forcings are included in the reconstruction model then reconstructions are robust to the choice of different time series models for the error (there is a suggestion that a long memory model for the errors improve the accuracy of the uncertainty quantification).

33.2.2.4   Modeling spatial dependence and spatio-temporal reconstructions

Due to spatially wide-spread data availability, many paleoclimate reconstruction efforts have primarily focused on reconstructing Northern hemispheric average temperature anomalies. Such reconstructions correspond to hindcasting a single time-series; for examples (not comprehensive) see [21, 46, 52, 55, 60, 61] for an overview and discussion in [47]. [73] consider IR for each continent. There has also been considerable effort to undertake paleoclimate reconstructions spatio-temporally for particular regions, hemispheres, or for the entire globe. These entail reconstructing a random field.

Climate field reconstructions (CFR) are inherently more difficult as there is an immediate need to specify (either directly or indirectly) a spatial model to relate temperature and proxies to other temperature and proxies, which are located at different spatial and temporal points. The problem is exacerbated by the fact that the number of grid points on even a coarse spatial grid by far exceed the number of years of proxy data. Hence there is a need to obtain a parsimonious spatial representation to overcome the dimensionality issue. A commonly used approach is to use a covariance function to model the relationships in space and time [92, 93]. [92] introduce the BARCAST method that, in addition to being a multiproxy reconstruction method, introduces a spatio-temporal model for the climate process, $\{C(s,t)\}$ (BARCAST stands for "a Bayesian Algorithm for Reconstructing Climate Anomalies in Space and Time"). The authors suppose that climate process is Gaussian with a constant mean. The covariance structure for the climate process is separable – AR(1) in time, and exponential and isotropic in space. Applications of this method include development of an imputation method to improve the estimation of climate anomalies from time series with missing data [90], a 400 year reconstruction of high northern latitude temperatures with a focus on recent extremes [94], an analysis of the divergence between temperature and tree ring records [95], and a multiproxy reconstruction of European temperatures [58]. The exponential covariance function and the assumption of separability in time and space implicit in the BARCAST methodology can be relaxed further. Section 33.2.2 discusses many of the recent state-of-the-art methods in multiproxy spatial paleoclimate reconstructions. These methods have explored more sophisticated spatial modeling approaches.

Another approach is to use regularization approaches to deal with the dimensionality issue as in Reg-EM [81]. Here the ill-conditioning of the covariance of the field is overcome by using a ridge-type penalty. [83] uses canonical correlation analysis (CCA) to undertake dimensionality reduction. Earlier work on CFR is also seen in [60] that uses inverse ordinary least squares regression in combination with principal component analysis, and [62] and [77] that apply regularized expectation maximization using truncated total least squares. Recently, [31] uses Markov random fields (MRF) to model the covariance of the spatial field. The MRF approach leverages the conditional independences in space in a natural way.

MRF models may often be parameterized with fewer parameters as compared to specifying a model for the covariance matrix directly. [56] and [57] investigated methods to compare different climate fields including CFRs.

There is also a rich literature on space-for-time substitution reconstruction methods [35, 75, 78, 88]. Assuming ergodicity of the relevant processes, the modern spatial distribution of climates and proxies is used to calibrate the proxy-climate relationship that is used for predictions. These models do not model the spatial dependence directly, but instead they use the heterogenous proxy observed in recent time periods to describe the relationship between proxies and climate in the past.

33.2.2.5 Missing values and data augmentation

The very nature of temperature and proxy measurements gives rise to records that are often missing in both space and time. Moreover, the lengths of proxy records are quite different for the various classes of proxies. Discarding proxy time series with missing values is not a viable option as doing so can lead to a loss of valuable information. Thus paleoclimate reconstructions on real data (versus reconstructions on simulated pseudoproxy data) have to contend with missing values. Perhaps the most popular approach to deal with missing values is to use the EM algorithm, and its variant, Reg-EM [81]. Reg-EM casts the entire reconstruction problem as an imputation problem where missing values in the proxies/temperature are treated similarly, regardless of whether they appear in the calibration period or in the hindcast period. More specifically, for each year [81] uses the multivariate normal distribution to jointly model both the temperature and proxy records on a spatial grid. Reg-EM uses the EM algorithm to estimate the mean and covariance of the joint distribution of the temperature and proxies. Various regularizations can be used in sample-starved settings in order to obtain stable parameter estimates (Reg-EM is an example of this). The parameter estimates are then used to impute the missing values in the spatial field. Imputation of missing values in the hindcast period are then considered to form the reconstruction. The advantage of the EM algorithm approach is that it is rooted in established statistical theory.

33.2.2.6 Temporal uncertainty

As discussed above, the dating of proxies can be an important source of uncertainty in paleoclimate reconstructions, especially when observing long, but time-uncertain, proxy records. Significant development has been made to develop statistical models that account for time-uncertain proxies – examples include [1, 3, 33, 35, 75].

In the context of HBMs, [100] describe how dating uncertainty, for layer counted proxies such as tree rings, can be incorporated into a hierarchical framework for space-time reconstruction. The key insight is that the inferred spatial covariance of the climate variable(s) imposes a constraint on the population of chronologies associated with the time-uncertain proxies. [100] assume that, for each time-uncertain proxy, there exists an ensemble of age models that are a priori equally likely. These probabilities are then updated based on the current draw of the climate variables. In this way, simultaneous inference on both the chronologies and the space-time climate process can be achieved. The same concept can be extend to proxies dated using radiocarbon or other isotopes. Indeed, common chronology models, such as the Bayesian accumulation (BACON) model of [8], could be imbedded with the hierarchical model, as described in [100].

33.2.2.7 Non-Gaussian paleoclimate reconstruction

Gaussian spatial and temporal models are often used for characterizing proxy measurements such as tree ring width and density, and chemical measurements such as isotope ratios. This is certainly common for hindcasting past temperatures. In more recent years there has been a growing need to work with random variables that are not continuous, or, if continuous, do not follow a Gaussian distribution.

Pollen counts or their relative abundance is an example of a proxy measurement that is discretely valued. For example, [72] consider the reconstruction of temperature based on counts of pollen from some indicator taxon. They employ a binomial distribution to model the counts, associating larger probabilities of abundance with warmer temperatures. (Also see [72] for a thorough review of pollen-based reconstruction methods.) Multivariate versions of pollen-based reconstructions include using Dirichlet-multinomial [36], zero-inflated extensions [78], and compound Poisson-Gamma processes [75].

Non-Gaussian continuous climate variables occur in paleoclimate studies of precipitation and accumulation [12, 68, 97] and the analysis of climate extremes. For example, in the latter case, [64] investigate spatially varying trends and dependencies in the parameters characterizing the distribution of extremes of a proxy dataset through the development of an HBM that implements spatially varying coefficients. This directly reconstructs extremal climate behavior by linking extremes in the proxy record to extremes in the instrumental record, rather than modeling mean climate behavior. See also, e.g., [49], [69], and [18] for other applications of extreme value theory to modeling climate proxies.

33.2.2.8   Multivariate reconstructions

An especially challenging problem is to reconstruct several climate variables at once on the basis on proxy measurements. In the HBM construction defined above we need the forward model to be a function of more than one climate variable. Statistical inference is challenging as there is often identifiability or a lack of Bayesian learning if we try to infer about multiple climate variables jointly on the basis of a single proxy. To tackle this issue we need to assume informative priors on the climate variables, including the relationship between them. Even so, the computations are non-trivial. In the context of tree ring widths, see [97]; these ideas are further explored in [96]. Another solution is to use more than one proxy in the reconstruction, each with a different relationship with the climate variables.

The space-for-time substitution methodology provides an alternative strategy to construct multiple climate variables simultaneously. For example [36] reconstruct the mean temperature of the coldest month as well as the growing degree days above 5°C in prehistoric times at a location in Ireland from fossilized pollen.

There is also a large literature focused on combining numerous climate variables into a single index, such as the Palmer Drought Severity Index [74], which are then reconstructed from climate-sensitive proxies [15, 16, 17].

33.2.3   Ideas and good practices

Many of the papers discussed here, reconstructing a single climate time series within our HBM framework on the basis of instrumental, proxy, and possibly forcing series. While there are challenges in specifying the forward model and the other time series models in the HBM, the statistical methodology required to producing reconstructions (i.e., sampling from one time series given some other time series) is straightforward. Computations are typically not particularly challenging in this fairly rich information-in-time scenario. For instance, our research group has been able to carry out computations on personal laptops, with code for a fairly routine Markov chain Monte Carlo algorithm implemented in the language R [76].

As we move to CFRs, the challenges increase. We now need to build spatio-temporal models that accurately capture the dependencies we observe in the data and the climate variables of interest. Proxy records can be sparse in space, but rich in time. Instrumental records are often data products, and not raw measurements. Instrumental datasets are often massive. Spatio-temporal models for climate variables are not easy to specify. The increasing complexity of specifying and fitting the models, in combination with a greater need for more prior information about model components, leads to more demanding statistical inference. For instance, the computations would typically require specialized techniques such as sparse matrix algorithms or reduced-rank approaches for high-dimensional spatial or spatio-temporal data sets. But, this should not deter us from using the HBM approach for CFR. Understanding a complicated model via a number of conditional distributions, while fully propagating uncertainty quantification is exactly what HBMs excel at, and is what a successful paleoclimate reconstruction requires.

A key challenge moving forward is to understand how a climate process behaves spatio-temporally. In the next section, we discuss the role that climate models can play in helping us to understand climate, and their relationship to paleoclimate reconstructions.

33.3 Climate models and paleoclimate

Climate models are mathematical models used to simulate various aspects of and interactions between climate system processes, and can include the oceans, atmosphere, and the land surface. Climate models play a central role in modern climate science, and are a critical tool for projecting the future climate under various emission scenarios; see, for example, the reports of the Intergovernmental Panel on Climate Change (IPCC) ([14, 27, 51, 86]). Climate models and climate reconstructions from proxies are two different, if overlapping, sources of information on past and projected future climate. For instance, paleoreconstructions may be used to discriminate between competing climate models. Paleoreconstructions may also be used to assess or calibrate (infer the parameters of) climate models that are used to make projections about future climate (see Section 33.3.3.1). Alternatively, climate models may also be used to improve paleoreconstructions (see Section 33.3.2.1). Hence there are interesting challenges and opportunities in studying the interaction between these sources of information. Here we outline some basic questions at the intersection of the analysis of paleoclimate reconstructions and climate models, along with a discussion of some existing work in this area and avenues for future research.

33.3.1 Climate model assessment using paleoclimate reconstructions

There is a growing body of literature focused on comparing climate model output to late Holocene paleoclimate reconstructions; recent reviews include [66], [65], and [79]. As discussed in [89], a wide range of tools, of varying levels of mathematical sophistication, have been brought to bear on the problem. These include qualitative comparisons [50, 63], fuzzy logic [32], selection from an ensemble of climate simulations based on distance metrics [30], data assimilation (e.g., [29] and Chapter 35), and ideas from detection and attribution [37, 38].

Statistically rigorous approaches for studying climate models via paleoclimate reconstructions have recently been considered in [89] and in [67], the latter building upon a series of papers [39, 40, 87]. Both papers address the same question: how can the paleoclimate reconstructions be used to discriminate between different climate models and different forcing scenarios?

To fix ideas, consider a study of global temperatures from k different global circulation models (GCMs) for a single volcanic forcing scenario. Both the k model outputs and an ensemble of paleoclimate reconstructions of the same global temperature time series are available to learn about the forced and unforced components of the temperature series. The ensemble of paleoclimate reconstructions is necessary to account for the uncertainty in the reconstructions – using a single "best estimate" reconstruction would ignore potentially large uncertainties in our reconstruction. Assuming a Bayesian reconstruction of the time series, for instance produced using BARCAST [92], it is straightforward to simply use the time series draws from the posterior distribution; such an ensemble is a byproduct of a standard Markov chain Monte Carlo (MCMC) simulation-based approach to Bayesian inference.

A starting assumption is that the paleoclimate reconstructions provide an approximation to the true temperature series. The ensemble accounts for uncertainties about the true temperature. Of interest is the following question: How well does the climate model-simulated temperature series "predict" (i.e., capture) the behavior of the true series? A simple approach to answer this questions would be to regress the paleoclimate reconstructed series as the response with the climate model simulation as predictor. Such an approach, however, essentially treats each model-forcing combination separately and does not allow for a study of the relationships between the different components of the reconstructions and the model. In particular, what is often of interest is the component of the "true" (latent) series that may be attributable to a forcing. Hence, a useful framework would explicitly model the forced and unforced components of the climate.

An HBM approach makes it easy to specify common latent climate processes and relate them to different sets of data. For instance, this permits a model specification as follows: model how the paleoclimate reconstruction relates to the latent temperature series, which in turn is modeled as the sum of the forced and unforced components of the temperature series. In other words we assume that the latent temperature series is made up of a component due to the forcings plus a component due to the natural variability of the climate system. Similarly, the climate model-simulated series may be decomposed into forced and unforced components. Then, to relate the reconstructed and simulated temperatures, a linear regression model may be specified between their respective forced components, with the forced component of paleoclimate reconstruction as the predictor and the forced component of the model output as the response. The residual from this regression contains some information on how the forced components differ.

The unforced components for the paleoclimate reconstructions and the climate model simulations are assumed to arise from the same independent generating process. In this approach, information about the distribution of the natural variability of the climate system comes from control runs (climate model simulations that are run without forcings). More generally, because there are k different GCM runs, it is possible to tease apart the forced components from the unforced components. The methodology outlined above involves repeating the statistical analysis separately for each set of forcing combinations. As a result of such an analysis, for instance, [89] are able to show that of the three volcanic forcing scenarios they consider, one results in superior agreement with the paleoclimate reconstruction. Superiority here is measured in a number of different ways: examining the slope parameter and variance of the discrepancy in the regression model, and comparing the fraction of the variability in the forced component of the simulation that can be explained by the regression relationship.

Paleoreconstructions may be useful in principle for distinguishing among competing climate models. One caveat is that current available proxy-based reconstruction may not be informative enough to discriminate among different models. For instance, [89] find that comparisons of the spatial average of a 600 year High Northern Latitude temperature reconstruction to suites of last millennium climate simulations from the GISS 2E and CSIRO models suggest that the proxy-based reconstructions are able to discriminate only between the crudest features of the simulations within each ensemble.

33.3.2   Statistical issues

33.3.2.1   Considering and embracing the lack of independence

A subtle but potentially problematic issue is that paleoclimate reconstructions are often used to calibrate climate models and therefore the paleoclimate reconstructions used to validate the climate model are not, strictly speaking, an independent source of information. Because the calibration process used is typically informal or unknown, it is difficult to envision a rigorous way of adjusting for the fact that we may be using the paleoclimate reconstructions twice. Furthermore, as mentioned earlier some climate reconstructions have also employed the forcings in GCMs in order to improve the quality of the reconstruction [5, 54]. Although the forcings are used in very different fashion in the reconstruction procedure and the GCM formulation, such reconstructions should not be used to validate the forced response in models due to the circularity issue.

Efforts to mathematically or statistically formalize links between climate models and paleoclimate reconstructions remain relatively sparse. In HBMs involving relationships between these latent variables, there is the possibility of identifiability and confounding issues affecting the interpretability of the parameters. This issue becomes even more pronounced when data are spatially or temporally dependent (cf. [41]).

HBMs can reflect that paleoclimate proxies, instrumental data, and climate models contain different (even if overlapping) types of information about the climate. For example, proxies are imperfect recorders of local-scale climate variability, whereas climate models generally represent climate averaged over spatial areas defined by the model grid. Furthermore, the climate models are tuned to the instrumental record. Any statistically rigorous joint analysis of reconstructed and modeled climate must account for the potential spatial and temporal misalignment between the two sources of information.

33.3.2.2   Refining model components

Standard statistical modeling must be considered to ensure that the assumptions underpinning the statistical model relating paleoclimate proxies, instrumental data, and climate models to climate are reasonable. For instance, linear regression relationships may not be adequate for capturing how the two time series are related. Furthermore, there may be important lags between the two time series that are currently unaccounted for in the models such as those presented in [89].

Normal error structures may not always be appropriate, for example when a volcanic eruption impacts temperatures. It is important to account for dependence in the model appropriately. Dependent error terms are also useful in adjusting for misspecification in the model. For example, if there are trends in the residuals because the linear regression was inadequate, the dependent errors can pick them up. This is a positive when it comes to model fitting, but can be a negative when it comes to interpreting the model parameters relating the different processes of interest.

As in paleoclimate reconstruction, the more information about the underlying climate process we can provide, the better our inferences about the climate models become. [89] use information from control runs from a climate model to learn about the time series properties of temperature. Further research into statistical models for the climate process, in the framework introduced by [89], is warranted.

33.3.3   Research directions

33.3.3.1   Paleoclimate-based climate model calibration

The calibration of a model, such as a climate model, involves inferring appropriate parameter settings of the model based on observations or reconstructed observations of the physical process being modeled. Rigorous statistical calibration summarizes the information about parameters in the form of probability distributions, which in turn may be used to generate future projections from the model, while also incorporating parametric uncertainty. Going far back in time allows for the study of the impact of climatic events or periods in the Earth's climate history where conditions were very different from conditions in the past few centuries. For example, [80] estimate climate sensitivity, an important parameter in climate models, by calibrating their model using paleoclimate reconstructions from the Last Glacial Maximum.

Given the relatively short period of time when instrumental records are available, paleoclimate reconstructions may be very valuable in that they allow researchers to examine the behavior of the models over much longer time periods than otherwise possible. For instance, [13] show how using paleoclimate reconstructions that inform the very long term behavior of the West Antarctic ice sheet result in sharper projections about the ice sheet's future state. In particular, unrealistic simulations with overshoots in past ice retreat and projected future regrowth are eliminated. When calibration is done entirely based on modern ice sheet observations, the uncertainties related to future ice sheet projections are larger allowing for a small but non-negligible probability of no contributions (or negative contributions) to sea level rise from the ice sheet. By incorporating paleoclimate data, this possibility is virtually eliminated.

33.3.3.2   Making climate projections using paleoclimate reconstructions

Suppose we are interested in making projections based on an ensemble of models, and the goal is to form a multi-model mean of future projections, with weights based on their past agreement with the proxies. A probabilistic approach to averaging across multiple models is Bayesian model averaging (BMA) [42], which essentially provides weights based on the probability of each model given the observations. For example, [7] use historic space-time resolved temperature data to derive model weights for an ensemble of GCMs, while also incorporating the space-time dependence in the observations. This approach derives model weights, which allows for the assessment of the relative skill of the models in terms of how well they match observations. Projections of future temperatures may therefore be obtained by taking a weighted average of the GCM projections. The projections therefore incorporate historical data, and also characterize uncertainties due to space-time dependence and other sources of variation. Paleoclimate reconstructions may have much to contribute to work with multi-model ensembles as well, by incorporating paleoclimate information from far greater time ranges than is possible from instrumental records; cf [9, 34, 79] and the references therein. Instrumental records are available for mostly about 170 years, whereas paleoclimate records can span several centuries. Thus paleoclimate records have the potential to assess broad qualitative aspects of climate models, if not even more nuanced local phenomenon.

33.3.3.3   Paleoclimate reconstructions using climate models

When making weather forecasts, it is common to combine information from numerical weather forecasts based on physical models with observational data. The general idea of combining these two sources of information is often referred to as *data assimilation* (See Chapter 35 for a general discussion of this topic). The resulting forecasts can then, in principle, include both the knowledge of the physical system as well as empirical information based on recent observations. There is a long history of methods for combining information from dynamic models and observations, with the Kalman filter [48] and many extensions, notably the ensemble Kalman filter [25, 26], being among the most well known. Roughly speaking, these approaches involve sequentially updating forecasts, conditional on the physical model and the observations. As the error structures become more complicated and the modeling approach becomes more flexible, more sophisticated methods like the ensemble Kalman filter and variants are useful, as well as Bayesian approaches to combining information [28]. More recently, some of these ideas have entered paleoclimate reconstructions (cf. [85]), where information from the sparse and noisy paleoclimate reconstructions is augmented with more spatially and temporally resolved climate model output, which in turn can provide more local (in space and time) reconstructions of past climate.

33.4   Discussion: looking forward

In this chapter we have discussed the important idea that in paleoclimate studies and climate studies more generally we can use statistical models to relate instrumental records, data from paleoclimate proxies, and climate model output to climate to learn about climate processes of interest. Using statistical models we are able to capture both the certainty and uncertainty inherent in relating each data source to climate. Through careful model assessment, we can understand how each data source contributes to our knowledge of climate.

Given the incomplete information about climate that are embedded in proxy records, it is worth asking what we cannot learn about climate using the statistical methods that we have discussed in this chapter. Part of the statistical difficulty in learning about the latent climate process is related to the sparseness of the measurements, but also in the fact that the measurement support for each data source can be different; to our knowledge there has been little research into methods of change-of-support applied to paleoclimate – see [91], Section 33.5.2 for a further discussion of the problem and references.

Throughout, we have reinforced the notion that successful studies tend to involve "good" information about the various processes of interest and their interaction. This requires a solid understanding of not only climate science but also the science embodied by paleoclimate proxies; for example, the study of how the interaction of trees with their environment relate to climate [96, 97, 98]. When there is a lack of knowledge about a relationship, we can build statistical models that can carefully mimic features inherent in the data [11, 20, 35, 54]. The inclusion of important covariates and climate forcings in particular are key to building reliable models for climate.

Paleoclimate reconstructions are central to climate science just as understanding our history is crucial for studying our present and future. Statistical paleoclimatology has allowed researchers to *look back* to learn about past climate. There are exciting opportunities to develop statistical methodologies to look back and also *look forward*, combining information from instruments, proxies, and climate model output to learn about the future of our climate. Given the importance of incorporating both scientific and statistical insights, the future of this research involves sustained interdisciplinary collaborations between climate scientists and statisticians.

Acknowledgments

PFC is supported in part by the US National Science Foundation (NSF) under grants NSF-DMS-1407604 and NSF-SES-1424481, and the National Cancer Institute of the National Institutes of Health under Award Number R21CA212308. The content is solely the responsibility of the authors and does not necessarily represent the official views of the National Institutes of Health. MH was partially supported by NSF-DMS-1418090 and Network for Sustainable Climate Risk Management (SCRiM) under NSF cooperative agreement GEO-1240507. BL was partially supported by NSF-DPP-1418339 and NSF-AGS -1602845. BR was supported in part by the US NSF under grants DMS-CMG-1025465, AGS-1003823, DMS-1106642, and DMS-CAREER-1352656, and by the US Air Force Office of Scientific Research grant award FA9550-13-1-0043.

Bibliography

[1] H A. Björn , R. H. Shumway, D., and K. L. Verosub. Linear and nonlinear alignment of time series with applications to varve. *Environmetrics*, 19:409–427, 2008.

[2] C. M. Ammann, M. G. Genton, and B. Li. Technical Note: Correcting for signal attenuation from noisy proxy data in climate reconstructions. *Climate of the Past*, 6:273–279, 2010.

[3] K. J. Anchukaitis and J. E. Tierney. Identifying coherent spatiotemporal modes in time-uncertain proxy paleoclimate records. *Climate Dynamics*, 41:1291–1306, 2013.

[4] M. Auffhammer, B. Li, B. Wright, and S.-J. Yoo. Specification and estimation of the transfer function in dendroclimatological reconstructions. *Environmental and Ecological Statistics*, 22:105–126, 2015.

[5] L. Barboza, B. Li, M. P. Tingley, and F. G. Viens. Reconstructing past temperatures from natural proxies and estimated climate forcings using short-and long-memory models. *The Annals of Applied Statistics*, 8:1966–2001, 2014.

[6] J. Beran. *Statistics for Long Memory Processes*, Chapman and Hall, New York, 1994.

[7] K. S. Bhat, M. Haran, A. Terando, and K. Keller. Climate projections using bayesian model averaging and space–time dependence. *Journal of Agricultural, Biological, and Environmental Statistics*, 16:606–628, 2011.

[8] M. Blaauw and J. A. Christen. Flexible paleoclimate age-depth models using an autoregressive gamma process. *Bayesian Analysis*, 6:457–474, 2011.

[9] P. Braconnot, S. P. Harrison, M. Kageyama, P. J. Bartlein, V. Masson-Delmotte, A. Abe-Ouchi, B. Otto-Bliesner, and Y. Zhao. Evaluation of climate models using palaeoclimatic data. *Nature Climate Change*, 2:417–424, 2012.

[10] K. J. Brown, J. S. Clark, E. C. Grimm, J. J. Donovan, P. G. Mueller, B. C. S. Hansen, and I. Stefanova. Fire cycles in North American interior grasslands and their relation to prairie drought. *Proceedings of the National Academy of Sciences of the United States of America*, 102:8865–8870, 2005.

[11] J. Brynjarsdóttir and L. M. Berliner. Bayesian hierarchical modeling for paleoclimate reconstruction from geothermal data. *The Annals of Applied Statistics*, 5:1328–1359, 2011.

[12] C. A. Calder, P. F. Craigmile, and E. Mosley-Thompson. Spatial variation in the infuence of the North Atlantic oscillation on precipitation across Greenland. *Journal of Geophysical Research: Atmospheres*, 113(D6), 2008.

[13] W. Chang, M. Haran, P. Applegate, and D. Pollard. Calibrating an ice sheet model using high-dimensional binary spatial data. *Journal of the American Statistical Association*, 111:57–72, 2016.

[14] M. Collins, R. Knutti, J. Arblaster, J.-L. Dufresne, T. Fichefet, P. Friedlingstein, X. Gao, W.J. Gutowski, T. Johns, G. Krinner, M. Shongwe, C. Tebaldi, A.J. Weaver, and M. Wehner. Long-term climate change: Projections, commitments and irreversibility. In T.F. Stocker, D. Qin, G.-K. Plattner, M. Tignor, S.K. Allen, J. Boschung, A. Nauels, Y. Xia, V. Bex, and P.M. Midgley, editors, *Climate Change 2013: The Physical Science Basis. Contribution of Working Group I to the Fifth Assessment Report of the Intergovernmental Panel on Climate Change*, pages 1029–1136, Cambridge, United Kingdom and New York, NY, USA, 2013. Cambridge University Press.

[15] E. R. Cook, D. M. Meko, D. W. Stahle, and M. K. Cleaveland Drought reconstructions for the continental united states. *Journal of Climate*, 12:1145–1162, 1999.

[16] E.R. Cook, R. Seager, M.A. Cane, and D.W. Stahle. North American drought: Reconstructions, causes, and consequences. *Earth Science Reviews*, 81:93–134, 2007.

[17] E.R. Cook, C.A. Woodhouse, C.M. Eakin, D.M. Meko, and D.W. Stahle. Long-Term Aridity Changes in the Western United States, 2004.

[18] D. Cooley, Naveau P., and Jomelli V. A Bayesian hierarchical extreme value model for lichenometry. *Environmetrics*, 17:555–574, 2006.

[19] P. F. Craigmile, P. Guttorp, and D. B. Percival. Trend assessment in a long memory dependence model using the discrete wavelet transform. *Environmetrics*, 15:313–335, 2004.

[20] P. F. Craigmile, M. P. Tingley, and J. Yin. Paleoclimate reconstruction using statistical non-linear forward models. In *Proceedings of the 59th ISI World Statistics Congress, Hong Kong*, 2013.

[21] J. Esper, F.H. Schweingruber, and M. Winiger. 1300 years of climatic history for Western Central Asia inferred from tree-rings. *The Holocene*, 12:267–277, 2002.

[22] D. L. Evans, H. J. Freeland, J. D. Hays, J. Imbrie, and N. J. Shackleton. Variations in the earth's orbit: Pacemaker of the ice ages? *Science*, 198:528–530, 1977.

[23] M. N. Evans, B. K. Reichert, A. Kaplan, K. J. Anchukaitis, E. A. Vaganov, M. K. Hughes, and M. A. Cane. A forward modeling approach to paleoclimatic interpretation of tree-ring data. *Journal of Geophysical Research*, 111:3008, 2006.

[24] M.N. Evans, S.E. Tolwinski-Ward, D.M. Thompson, and K.J. Anchukaitis. Applications of proxy system modeling in high resolution paleoclimatology. *Quaternary Science Reviews*, 76:16–28, 2013.

[25] G. Evensen. Sequential data assimilation with a nonlinear quasi-geostrophic model using Monte Carlo methods to forecast error statistics. *Journal of Geophysical Research: Oceans*, 99(C5):10143–10162, 1994.

[26] G. Evensen. The ensemble Kalman filter: Theoretical formulation and practical implementation. *Ocean dynamics*, 53:343–367, 2003.

[27] G. Flato, J. Marotzke, B. Abiodun, P. Braconnot, S.C. Chou, W. Collins, P. Cox, F. Driouech, S. Emori, V. Eyring, C. Forest, P. Gleckler, E. Guilyardi, C. Jakob, V. Kattsov, C. Reason, and M. Rummukainen. Evaluation of climate models. In T.F. Stocker, D. Qin, G.-K. Plattner, M. Tignor, S.K. Allen, J. Boschung, A. Nauels, Y. Xia, V. Bex, and P.M. Midgley, editors, *Climate Change 2013: The Physical Science Basis. Contribution of Working Group I to the Fifth Assessment Report of the Intergovernmental Panel on Climate Change*, pages 741–866, Cambridge, United Kingdom and New York, NY, USA, 2013. Cambridge University Press.

[28] M. Fuentes and A. E. Raftery. Model evaluation and spatial interpolation by Bayesian combination of observations with outputs from numerical models. *Biometrics*, 61:36–45, 2005.

[29] H. Goosse, E. Crespin, A. de Montety, M. E. Mann, H. Renssen, and A. Timmermann. Reconstructing surface temperature changes over the past 600 years using climate model simulations with data assimilation. *Journal of Geophysical Research: Atmospheres*, 115(D9), 2010.

[30] H. Goosse, H. Renssen, A. Timmermann, R. S. Bradley, and M. E. Mann. Using paleoclimate proxy-data to select optimal realisations in an ensemble of simulations of the climate of the past millennium. *Climate Dynamics*, 27:165–184, 2006.

[31] D. Guillot, B. Rajaratnam, and Emile-Geay J. Statistical paleoclimate reconstructions via Markov random fields. *Annals of Applied Statistics*, 9:324–352, 2015.

[32] J. Guiot, J.-J. Boreux, P. Braconnot, and F. Torre. Data-model comparison using fuzzy logic in paleoclimatology. *Climate dynamics*, 15:569–581, 1999.

[33] E. Haam and P. Huybers. A test for the presence of covariance between time-uncertain series of data with application to the Dongge Cave speleothem and atmospheric radiocarbon records. *Paleoceanography*, 25:PA2209, 2010.

[34] J. C. Hargreaves and J. D. Annan. On the importance of paleoclimate modelling for improving predictions of future climate change. *Climate of the Past*, 5:803–814, 2009.

[35] J. Haslett, A. Parnell, and M. Salter-Townsend. Modelling temporal uncertainty in palaeoclimate reconstructions. In *Proceedings of the 21st International Workshop on Statistical Modelling*, pages 26–37, 2006.

[36] J. Haslett, M. Whiley, S. Bhattacharya, M. Salter-Townshend, S. P. Wilson, J. R. M. Allen, B. Huntley, and F. J. G. Mitchell. Bayesian palaeoclimate reconstruction. *Journal of the Royal Statistical Society: Series A (Statistics in Society)*, 169:395–438, 2006.

[37] G. Hegerl, J. Luterbacher, F. González-Rouco, S. F. B. Tett, T. Crowley, and E. Xoplaki. Influence of human and natural forcing on European seasonal temperatures. *Nature Geoscience*, 4:99–103, 2011.

[38] G.C. Hegerl, T.J. Crowley, M. Allen, W.T. Hyde, H.N. Pollack, J. Smerdon, and E. Zorita. Detection of human influence on a new, validated 1500-year temperature reconstruction. *Journal of Climate*, 20:650–666, 2007.

[39] A. Hind and A. Moberg. Past millennial solar forcing magnitude. *Climate Dynamics*, 41:2527–2537, 2013.

[40] A. Hind, A. Moberg, and R. Sundberg. Statistical framework for evaluation of climate model simulations by use of climate proxy data from the last millennium – Part 2: A pseudo-proxy study addressing the amplitude of solar forcing. *Climate of the Past*, 8:1355–1365, 2012.

[41] J. S. Hodges and B. J. Reich. Adding spatially-correlated errors can mess up the fixed effect you love. *The American Statistician*, 64:325–334, 2010.

[42] J. A. Hoeting, D. Madigan, A. E. Raftery, and C. T. Volinsky. Bayesian model averaging: a tutorial. *Statistical Science*, 14:382–417, 1999.

[43] M. K, Hughes and C. M. Ammann. The future of the past – an earth system framework for high resolution paleoclimatology: editorial essay. *Climatic Change*, 94:247–259, 2009.

[44] P. Huybers and W. Curry. Links between annual, Milankovitch and continuum temperature variability. *Nature*, 441:329–332, 2006.

[45] J. Imbers, A. Lopez, C. Huntingford, and M. Allen. Sensitivity of climate change detection and attribution to the characterization of internal climate variability. *Journal of Climate*, 27:3477–3491, 2014.

[46] L. Janson and B. Rajaratnam. A methodology for robust multiproxy paleoclimate reconstructions and modeling of temperature conditional quantiles. *Journal of the American Statistical Association*, 109:63–77, 2014.

[47] P. D. Jones, K. R. Briffa, T. J. Osborn, J. M. Lough, T. D. van Ommen, B. M. Vinther, J. Luterbacher, E. R. Wahl, F. W. Zwiers, M. E. Mann High-resolution palaeoclimatology of the last millennium: a review of current status and future prospects. *The Holocene*, 19:3–49, 2009.

[48] R. E. Kalman. A new approach to linear filtering and prediction problems. *Journal of basic Engineering*, 82:35–45, 1960.

[49] R.W. Katz, G.S. Brush, and M.B. Parlange. Statistics of extremes: Modeling ecological disturbances. *Ecology*, 86:1124–1134, 2005.

[50] D.S. Kaufman, D.P. Schneider, N.P. McKay, C.M. Ammann, R.S. Bradley, K.R. Briffa, G.H. Miller, B.L. Otto-Bliesner, J.T. Overpeck, B.M. Vinther, et al. Recent warming reverses long-term arctic cooling. *Science*, 325:1236, 2009.

[51] B. Kirtman, S.B. Power, J.A. Adedoyin, G.J. Boer, R. Bojariu, I. Camilloni, F.J. Doblas-Reyes, A.M. Fiore, M. Kimoto, G.A. Meehl, M. Prather, A. Sarr, C. Schar, R. Sutton, G.J. vanOldenborgh, G. Vecchi, and H.J. Wang. Near-term climate change: Projections and predictability. In T.F. Stocker, D. Qin, G.-K. Plattner, M. Tignor, S.K. Allen, J. Boschung, A. Nauels, Y. Xia, V. Bex, and P.M. Midgley, editors, *Climate Change 2013: The Physical Science Basis. Contribution of Working Group I to the Fifth Assessment Report of the Intergovernmental Panel on Climate Change*, pages 953–1028, Cambridge, United Kingdom and New York, NY, USA, 2013. Cambridge University Press.

[52] T.C.K. Lee, F.W. Zwiers, and M. Tsao. Evaluation of proxy-based millennial reconstruction methods. *Climate Dynamics*, 31:263–281, 2008.

[53] B. Li, D.W. Nychka, and C.M. Ammann. The 'hockey stick' and the 1990s: a statistical perspective on reconstructing hemispheric temperatures. *Tellus A*, 59:591–598, 2007.

[54] B. Li, D.W. Nychka, and C.M. Ammann. The value of multi-proxy reconstruction of past climate. *Journal of the American Statistical Association*, 105:883–911, 2010.

[55] B. Li, D. W. Nychka, and C. M. Ammann. The "hockey stick" and the 1990s: A statistical perspective on reconstructing hemispheric temperatures. *Tellus A*, 59:591–598, 2007.

[56] B. Li and J. E Smerdon. Defining spatial comparison metrics for evaluation of paleoclimatic field reconstructions of the common era. *Environmetrics*, 23:394–406, 2012.

[57] B. Li, X. Zhang, and J. E. Smerdon. Comparison between spatio-temporal random processes and application to climate model data. *Environmetrics*, 27: 267–279, 2016.

[58] J. Luterbacher, J. P. Werner, J. E. Smerdon, L. Fernandez-Donado, F. J. Gonzalez-Rouco, D. Barriopedro, F. C. Ljungqvist, U. Büntgen, E. Zorita, S. Wagner. European summer temperatures since roman times. *Environmental Research Letters*, 11:024001, 2016.

[59] M. E. Mann, Z. Zhang, M. K. Hughes, R. S. Bradley, S. K. Miller, S. Rutherford, and F. Ni. Proxy-based reconstructions of hemispheric and global surface temperature variations over the past two millennia. *Proceedings of the National Academy of Sciences*, 105:13252–13257, 2008.

[60] M. E. Mann, R. S. Bradley, and M. K. Hughes. Global-scale temperature patterns and climate forcing over the past six centuries. *Nature*, 392:779–787, 1998.

[61] M.E. Mann and P.D. Jones. Global surface temperatures over the past two millennia. *Geophysical Research Letters*, 30:1820, 2003.

[62] M.E. Mann, S. Rutherford, E. Wahl, and C. Ammann. Robustness of proxy-based climate field reconstruction methods. *Journal of Geophysical Research*, 112:D12109, 2007.

[63] M.E. Mann, Z. Zhang, S. Rutherford, R.S. Bradley, M.K. Hughes, D. Shindell, C.M. Ammann, G. Faluvegi, and F. Ni. Global signatures and dynamical origins of the little ice age and medieval climate anomaly. *Science*, 326:1256–1260, 2009.

[64] E. Mannshardt, P. F. Craigmile, and M. P. Tingley. Statistical modeling of extreme value behavior in North American tree-ring density series. *Climatic Change*, 117:843–858, 2013.

[65] V. Masson-Delmotte, M. Schulz, A. Abe-Ouchi, J. Beer, A. Ganopolski, J.F. GonzalezRouco, E. Jansen, K. Lambeck, J. Luterbacher, T. Naish, T. Osborn, B. Otto-Bliesner, T. Quinn, R. Ramesh, M. Rojas, X. Shao, and A. Timmermann. Information from paleoclimate archives. In T.F. Stocker, D. Qin, G.-K. Plattner, M. Tignor, S.K. Allen, J. Boschung, A. Nauels, Y. Xia, V. Bex, and P.M. Midgley, editors, *Climate Change 2013: The Physical Science Basis. Contribution of Working Group I to the Fifth Assessment Report of the Intergovernmental Panel on Climate Change*, pages 383–464, Cambridge, United Kingdom and New York, NY, USA, 2013. Cambridge University Press.

[66] A. Moberg. Ccomparisons of simulated and observed northern hemisphere temperature variations during the past millennium - selected lessons learned and problems encountered. *Tellus B*, 65, 2013.

[67] A. Moberg, R. Sundberg, H. Grudd, and A. Hind. Statistical framework for evaluation of climate model simulations by use of climate proxy data from the last millennium – Part 3: Practical considerations, relaxed assumptions, and using tree-ring data to address the amplitude of solar forcing. *Clim. Past*, 11:425–448, 2015.

[68] E. Mosley-Thompson, C. R. Readinger, P. Craigmile, L. G. Thompson, and C. A. Calder. Regional sensitivity of greenland precipitation to NAO variability. *Geophysical Research Letters*, 32(L24707), 2005.

[69] P. Naveau and C.M. Ammann. Statistical distributions of ice core sulfate from climatically relevant volcanic eruptions. *Geophysical Research Letters*, 32(L05711), 2005.

[70] A. Neumaier and T. Schneider. Estimation of parameters and eigenmodes of multivariate autoregressive models. *ACM Transactions on Mathematical Software (TOMS)*, 27:27–57, 2001.

[71] NRC. *Surface Temperature Reconstructions for the Last 2000 Years*. The National Academies Press, Washington, D.C., 2006.

[72] C. Ohlwein and E. R. Wahl. Review of probabilistic pollen-climate transfer methods. *Quaternary Science Reviews*, 31:17–29, 2012.

[73] PAGES 2k Consortium. Continental-scale temperature variability during the past two millennia. *Nature Geoscience*, 6:339–346, 2013.

[74] W. C. Palmer. *Meteorological drought*, volume 30. US Department of Commerce, Weather Bureau Washington, DC, USA, 1965.

[75] A. C. Parnell, J. Sweeney, T. K. Doan, M. Salter-Townshend, J. R. M. Allen, B. Huntley, and J. Haslett. Bayesian inference for palaeoclimate with time uncertainty and stochastic volatility. *Journal of the Royal Statistical Society*, Series C, 64:115–138, 2014.

[76] R Core Team. *R: A Language and Environment for Statistical Computing*. R Foundation for Statistical Computing, Vienna, Austria, 2016.

[77] S. Rutherford, M. E. Mann, T. L. Delworth, and R. J. Stouer. Climate field reconstruction under stationary and nonstationary forcing. *Journal of Climate*, 16:462–479, 2003.

[78] M. Salter-Townshend and J. Haslett. Fast inversion of a flexible regression model for multivariate pollen counts data. *Environmetrics*, 23:595–605, 2012.

[79] G. A. Schmidt, J. D. Annan, P. J. Bartlein, B. I. Cook, E. Guilyardi, J. C. Hargreaves, S. P. Harrison, M. Kageyama, A. N. LeGrande, and B. Konecky. Using palaeo-climate comparisons to constrain future projections in CMIP5. *Climate of the Past*, 10:221–250, 2014.

[80] A. Schmittner, N. M. Urban, J. D. Shakun, N. M. Mahowald, P. U. Clark, P.J. Bartlein, A. C. Mix, and A. Rosell-Melé. Climate sensitivity estimated from temperature reconstructions of the last glacial maximum. *Science*, 334:1385–1388, 2011.

[81] T. Schneider. Analysis of incomplete climate data: estimation of mean values and covariance matrices and imputation of missing values.. *Journal of Climate*, 14:853–871, 2001.

[82] T Schneider and A Neumaier. Algorithm 808: ARfit—a Matlab package for the estimation of parameters and eigenmodes of multivariate autoregressive models. *ACM Transactions on Mathematical Software (TOMS)*, 27:58–65, 2001.

[83] J. E. Smerdon, A. Kaplan, and D. Chang. On the origin of the standardization sensitivity in regem climate field reconstructions. *Journal of Climate*, 21:6710–6723, 2008.

[84] R. Smith. Long-range dependence and global warming. In V. Barnett and F. Turkman, editors, *Statistics for the Environment*, pages 141–161. John Wiley, Hoboken, NJ, 1993.

[85] N. J. Steiger, G. J. Hakim, E. J. Steig, D. S. Battisti, and G. H. Roe. Assimilation of time-averaged pseudoproxies for climate reconstruction. *Journal of Climate*, 27:426–441, 2014.

[86] T.F. Stocker, D. Qin, G.-K. Plattner, L.V. Alexander, S.K. Allen, N.L. Bindoff, F.-M. Breon, J.A. Church, U. Cubasch, S. Emori, P. Forster, P. Friedlingstein, N. Gillett, J.M. Gregory, D.L. Hartmann, E. Jansen, B. Kirtman, R. Knutti, K. KrishnaKumar, P. Lemke, J. Marotzke, V. Masson-Delmotte, G.A. Meehl, I.I. Mokhov, S. Piao, V. Ramaswamy, D. Randall, M. Rhein, M. Rojas, C. Sabine, D. Shindell, L.D. Talley, D.G. Vaughan, and S.-P. Xie. Technical summary. In T.F. Stocker, D. Qin, G.-K. Plattner, M. Tignor, S.K. Allen, J. Boschung, A. Nauels, Y. Xia, V. Bex, and P.M. Midgley, editors, *Climate Change 2013: The Physical Science Basis. Contribution of Working Group I to the Fifth Assessment Report of the Intergovernmental Panel on Climate Change*, page 33115, Cambridge, United Kingdom and New York, NY, USA, 2013. Cambridge University Press.

[87] R. Sundberg, A. Moberg, and A. Hind. Statistical framework for evaluation of climate model simulations by use of climate proxy data from the last millennium – Part 1: Theory. *Climate of the Past*, 8:1339–1353, 2012.

[88] J. E. Tierney and M. P. Tingley. A Bayesian, spatially-varying calibration model for the TEX 86 proxy. *Geochimica et Cosmochimica Acta*, 127:83–106, 2014.

[89] M. Tingley, P. F. Craigmile, M. Haran, B. Li, E. Mannshardt, and B. Rajaratnam. On discriminating between GCM forcing configurations using Bayesian reconstructions of Late-Holocene temperatures. *Journal of Climate*, 28:8264–8281, 2015.

[90] M. P. Tingley. A Bayesian ANOVA Scheme for Calculating Climate Anomalies, with Applications to the Instrumental Temperature Record. *Journal of Climate*, 25:777–791, 2012. Code is available at ftp://ftp.ncdc.noaa.gov/pub/data/paleo/softlib/anova.

[91] M. P. Tingley, P. F. Craigmile, M. Haran, B. Li, E. Mannshardt, and B. Rajaratnam. Piecing together the past: statistical insights into paleoclimatic reconstructions. *Quaternary Science Reviews*, 35:1–22, 2012.

[92] M. P. Tingley and P. Huybers. A Bayesian algorithm for reconstructing climate anomalies in space and time. Part I: Development and applications to paleoclimate reconstruction problems. *Journal of Climate*, 23:2759–2781, 2010.

[93] M. P. Tingley and P. Huybers. A Bayesian algorithm for reconstructing climate anomalies in space and time. Part II: Comparison with the regularized expectation-maximization algorithm. *Journal of Climate*, 23:2782–2800, 2010.

[94] M. P. Tingley and P. Huybers. Recent temperature extremes at high northern latitudes unprecedented in the past 600 years. *Nature*, 496:201–205, 2013.

[95] M. P. Tingley, A. R. Stine, and P. Huybers. Temperature reconstructions from tree-ring densities overestimate volcanic cooling. *Geophysical Research Letters*, 41:7838–7845, 2014.

[96] J. Tipton, M. Hooten, N. Pederson, M. Tingley, and D. Bishop. Reconstruction of late holocene climate based on tree growth and mechanistic hierarchical models. *Environmetrics*, 27:42–54, 2016.

[97] S. E. Tolwinski-Ward, M. P. Tingley, M. N. Evans, M. K. Hughes, and D. W. Nychka. Probabilistic reconstructions of local temperature and soil moisture from tree-ring data with potentially time-varying climatic response. *Climate Dynamics*, 44:791–806, 2015.

[98] S. E. Tolwinski-Ward, M. E. Evans, M. K. Hughes, and K. J. Anchukaitis. An efficient forward model of the climate controls on interannual variation in tree-ring width. *Climate Dynamics*, 36:2419–2439, 2011.

[99] J. P. Werner, J. Luterbacher, and J. E. Smerdon. A Pseudoproxy Evaluation of Bayesian Hierarchical Modeling and Canonical Correlation Analysis for Climate Field Reconstructions over Europe. *Journal of Climate*, 26:851–867, 2013.

[100] J. P. Werner and M. P. Tingley. Technical note: Probabilistically constraining proxy age–depth models within a Bayesian hierarchical reconstruction model. *Climate of the Past*, 11:533–545, 2015.

34

Climate Change Detection and Attribution

Dorit Hammerling

University Corporation for Atmospheric Research

Matthias Katzfuss

Texas A&M University at College Station

Richard Smith

University of North Carolina, Chapel Hill

CONTENTS

34.1 Introduction

Climate-change detection and attribution is an important area in the climate sciences, specifically in the study of climate change. Statements such as "It is extremely likely that human influence has been the dominant cause of the observed warming since the mid-20th century" are frequently found in the Assessment Reports of the Intergovernmental Panel on Climate Change (IPCC). These types of statements are largely based on to the synthesis of results from detection and attribution studies [6]. Broadly speaking, the goal of climate-change detection and attribution methods is to differentiate if observed changes in variables quantifying weather (e.g., temperature or rainfall amounts) are consistent with processes internal to the climate system or are evidence for a change in climate due to so-called external forcings [24]. External forcings are often categorized into natural and anthropogenic (human-caused) forcings, where solar and volcanic activity are examples of natural forcings and increased greenhouse gas emissions and land use change are examples of anthropogenic forcings. Figure 34.1 shows a typical example of a detection and attribution study for long-term temperature change. In this example, natural forcings alone can not explain the observed temperature-change, but a combination of human-caused and natural forcings can.

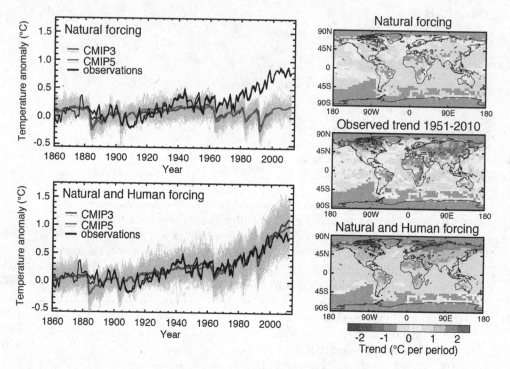

FIGURE 34.1

Example of a detection and attribution study, reproduced from FAQ 10.1, Figure 34.1 IPCC 2013: The Physical Science Basis. Time series of global and annual-averaged surface temperature change from 1860 to 2010. The top left panel shows results from two ensemble of climate models driven with just natural forcings, shown as thin blue and yellow lines; ensemble average temperature changes are thick blue and red lines. Three different observed estimates are shown as black lines. The lower left panel shows simulations by the same models, but driven with both natural forcing and human-induced changes in greenhouse gases and aerosols. (Right) Spatial patterns of local surface temperature trends from 1951 to 2010. The upper panel shows the pattern of trends from a large ensemble of Coupled Model Intercomparison Project Phase 5 (CMIP5) simulations driven with just natural forcings. The bottom panel shows trends from a corresponding ensemble of simulations driven with natural + human forcings. The middle panel shows the pattern of observed trends from the Hadley Centre/Climatic Research Unit gridded surface temperature data set 4 (HadCRUT4) during this period.

One key challenge is that (for planet Earth) we can only observe a single realization of climate over space and time. This fact makes it intrinsically difficult to detect changes and to attribute them to specific forcings without further constraining information. This is where climate models play on important role, as they can be used to test the evolution of pathways under different forcings scenarios. The prevailing paradigm is that the climate system is a chaotic system, meaning that minute initial condition changes can lead to varying outcomes, and our observed climate is one specific realization of that system. The variability associated with the chaotic nature of the system, referred to as internal variability, is typically estimated from control runs, which are are climate model runs without any external forcings. [34] provide a more detailed conceptual overview of the problem of climate change detection and attribution from a statistical point of view.

Climate model output needs to be calibrated to agree with observed weather, in that the climate models might be biased or be scaled differently than the observations. For example, this means that it is very difficult to directly compare the global average temperature today to that of, say, 100 years ago, and be able to attribute the increase in temperature to specific forcing scenarios obtained from climate models. What is used (in place of absolute changes) are the patterns of changes in the climate in response to a given forcing, which are referred to as fingerprints [35]. This way, the focus is less on whether an increase in global average temperatures is more consistent with the observed temperatures, but whether observed changes are greater in a specific region than in another, i.e. how well the patterns of change match.

The most commonly employed framework to address this problem is linear regression, where the observed change is the response variable and a linear combination of the patterns corresponding to the specific external forcing scenarios (obtained from climate models) are the explanatory variables. The inferential goal is the determination of the regression coefficients associated with the different forcings. Their estimated values and uncertainty ranges establish if a change has been detected and to which combination of scenarios it can be attributed.

A different area, discussed in Section 34.4, is extreme event attribution. The focus of extreme event attribution is on assessing specific events such as, for example, an extreme flood. The main goal is to determine if anthropogenic influences have changed the probability of occurrence or magnitude for this particular event. A concept commonly used within this framework is the Fraction of Attributable Risk (FAR), which is defined as $FAR = (p_1 - p_0)/p_1$, where p_1 is the probability of an extreme event with anthropogenic forcings, and p_0 without. [56] and [41] provide recent reviews and we provide a more statistically focussed review here.

Another area, which we will not discuss any further, is the fact that distributions can change in many ways. The simplest, and most commonly considered case, is a change in the mean. But changes in other characteristics of climate, such as the variance, the magnitude or frequency of extreme values, and even changes in the dependence structure over time (e.g., higher likelihood of droughts due to an extended period of no rain) are important and of interest. Here, we will only discuss work focused on changes in the mean.

34.2   Statistical model description

Regression-based climate-change detection and attribution can be viewed as a multivariate spatial or spatio-temporal regression problem, where we express an observed signal as a linear combination of different forcings scenarios. For global studies, the observations and forced responses are typically available as gridded quantities, which are often further aggregated to coarser grids (e.g., to a $2.5° \times 2.5°$ grid, resulting in $144 \times 72 = 10,368$ grid cells or to a $5° \times 5°$ grid, resulting in $72 \times 36 = 2,592$ grid cells). Observations and corresponding forced responses are often averaged in time, e.g. decadal averages, or expressed as estimated slope coefficients from a simple linear regression in each grid cell. Estimating slope coefficients is a straightforward way to smooth out short-term climate variability, which is overlaid on the longer-term trend we are trying to detect. For example, the quantity describing the observations could be slope coefficient estimates based on 30 years of temperature or rainfall observations.

Hence, let $\mathbf{y} = (y_1, \ldots, y_n)'$ be a vector of the true quantity describing the observations at the n grid cells, and the vectors $\mathbf{x}_1, \ldots, \mathbf{x}_J$ the analogous (true) quantities that would have occurred under the J different forcing scenarios. Let $\mathbf{X} = (\mathbf{x}_1, \ldots, \mathbf{x}_J)$, and define the $n \times n$ covariance matrix characterizing the internal climate variability (without any forcing) as $\mathbf{C}$. We can then write the commonly assumed linear regression model in the form of a conditional distribution,

$$\mathbf{y}|\mathbf{X}, \boldsymbol{\beta}, \mathbf{C} \sim \mathcal{N}_n \Big(\sum_{j=1}^{J} \beta_j \mathbf{x}_j, \mathbf{C} \Big), \tag{34.1}$$

where $\mathcal{N}_n$ denotes an n-variate normal or Gaussian distribution.

Within this context, climate change detection is viewed as testing whether each of the β_j is equal to zero or not. Assuming that $\mathbf{x}_1$ corresponds to the anthropogenic forcing, the conclusion that $\beta_1 \neq 0$ implies that human-caused climate change (with regard to the specific observed quantity as defined) has been detected. Attribution extends this framework by testing if the β_j are equal to unity, under the assumption that the mean responses for each forcing have been removed and that the responses are additive [e.g., 49]. Under the assumption of normality, the maximum-likelihood estimate for $\boldsymbol{\beta}$ is identical to the generalized-least-squares estimate, which is the solution approach typically pursued within the climate science community.

On first glance, this problem seems trivial. The challenge is, however, that in practice, $\mathbf{y}$, $\mathbf{X}$, $\boldsymbol{\beta}$, and $\mathbf{C}$ are all unknown. With the exception of $\boldsymbol{\beta}$, all these unknown quantities are high-dimensional, and our means to learn more about them are rather limited and modeling choices have to be made. Modern understanding further acknowledges that the observations are unknown as well, and can only be observed with measurements errors. Reconstructed and observational data sets are nowadays often provided in the form of ensembles, which allows for the estimation of an observational error covariance matrix to be incorporated in the modeling procedure. The following section describes the development of solution approaches leading up to the most recent formulation using Bayesian hierarchical modeling.

34.3   Methodological development

Climate-change detection and attribution methods have been developed by a variety of groups in the climate science and, to a lesser degree, in the statistics community and notations have varied accordingly. In this section, we apply the notation used in the corresponding original literature.

34.3.1   The beginning: Hasselmann's method and its enhancements

The original name for detection and attribution was *the optimal fingerprint method* [22, 25]. The idea was that human-caused greenhouse gas emissions would not only result in increased temperature overall, but would exhibit distinctive patterns or fingerprints, and with careful analysis, these could be detected in the observational signal. In contrast, other possible causes of warming, such as increases in solar output, would result in quite different patterns. If we were able to detect a pattern that was closer to that associated with greenhouse gases than with changes in solar output, that could be taken as evidence that greenhouse gases, rather than solar variations, were the cause of the changes we saw. The patterns to be expected were taken from climate models in which the different possible forcing factors could be separated into different model runs.

As originally formulated by Hasselmann [21], the setting was as follows:

1. The overall signal (for example, the change over time in a modeled temperature field) is represented by an n-dimensional vector $\overline{\Phi}$;

2. The estimated change from observation data is written Φ;

3. We assume $\Phi - \overline{\Phi} \sim \mathcal{N}(0, \mathbf{C})$ (multivariate normal with mean 0 and covariance matrix $\mathbf{C}$);

4. $\mathbf{C}$ estimated from data but *treated as known*;

5. The null hypothesis $H_0 : \overline{\Phi} = 0$ is tested using a χ^2 test.

As Hasselmann showed through a detailed example, this formulation is too simple without further structure on the signal. For example, the amplitude of signal required to be detected at a given level of significance increases with the dimension of the signal itself. Therefore, it is desirable to make use of further information on the anticipated form of the signal.

To bring this idea into the analysis, Hasselmann assumed we could write the signal as a linear combination of individual signals (the fingerprints). Therefore, we write $\overline{\Phi} = B\overline{\Psi}$ where B is a $n \times p$ matrix of known basis functions (interpreted as a p-dimensional "signal"). A revised estimate $\tilde{\Phi}$ is chosen to minimize $|\tilde{\Phi} - \Phi|^2$. This in turn is used to construct a revised χ^2 test statistic. A key part of the method is *expansion in principal components*. In the climate literature, principal components are known as Empirical Orthogonal Functions or EOFs. Hasselmann anticipated that it might in practice be necessary to restrict to a small number of leading EOFs (he suggested between 5 and 20).

The initial paper of Hasselmann was followed by a number of extensions and ramifications in the 1990s, e.g. [22, 23]. North and Stevens [39] presented a particularly simple derivation of the main results using linear algebra and the elementary theory of linear models. Even at this time, however, it was also implicit that a reduction in dimension (for example, restricting the signal to the leading EOFs) was needed to make the method applicable in practice.

The method started to influence the broader climate community with a series of papers in the mid-1990s applying these ideas to large climate datasets, see in particular [25, 52]. For example, Hegerl *et al.* [25] used a guessed greenhouse gas signal from a climate model, information about natural climate variability derived from control runs, and global near-surface temperature temperature observations. The null hypothesis, that changes in observed temperatures could be explained by natural variability, was rejected with a p-value < 0.05. However, they acknowledged considerable uncertainty about natural variability and did not take into account signals from other forcing factors such as solar variation.

The parallel paper by Santer and co-authors [52] focussed on the vertical structure of temperatures through the atmosphere. One particular issue here is the contrast between warming of the troposphere and cooling of the stratosphere, a pattern that one would expect to be particularly indicative of greenhouse gas warming, whereas other conceivable sources of atmospheric warming, for example if solar radiation were generally increasing, would not lead to such a characteristic vertical pattern of temperature changes. In this paper, they enhanced their conclusions by incorporating other signals besides greenhouse gases (they included stratospheric ozone in their model, as well as sulfate aerosols — small particles in the atmosphere, generally caused by human industrial processes, that have the effect of cooling the atmosphere and thereby partially mitigating the greenhouse gas effect). They also compared results from two climate models to examine the sensitivity of their results to model-dependent uncertainties, and, like [25], used control runs from climate models to assess natural variability, a key step in formulating a statistical significance test. They concluded "it is likely that this trend is partially due to human activities, though many uncertainties remain, particularly relating to estimates of natural variability."

34.3.2 The method comes to maturity: Reformulation as a regression problem; random effects and total least squares

Levine and Berliner [34] showed how Hasselmann's equations could be reformulated as a linear regression problem. The same formulation was proposed independently by Allen and Tett [3] with an observational signal $\mathbf{y}$ regressed on a finite number of covariates $\mathbf{x}_1, ..., \mathbf{x}_J$ representing J signals (for example, greenhouse gases, sulfate aerosols, solar variation, volcanoes) that were supposed to be derived from a climate model. Allen and Stott [2, 4] made the important extension of treating the signals themselves as random quantities, while a paper by Huntingford *et al* [29] showed how to extend the methodology to multiple climate models. The methodology defined in these papers is at the core of many present-day detection and attribution studies. The next part of our review therefore develops the methodology in these papers in some detail, though we refer to the original papers for full details.

First we outline the paper [3]. The model assumed by them was of the form

$$\mathbf{y} = \mathbf{X}\beta + \mathbf{u} \tag{34.2}$$

where

- $\mathbf{y}$ is vector of observations ($\ell \times 1$, where ℓ is the number of grid cells — several thousand in a typical climate model);

- $\mathbf{X}$ is matrix of J response patterns ($\ell \times J$ — here J is typically small, for example 4 if the response patterns correspond to greenhouse gases, sulfate aerosols, solar variability and the effects of volcanic eruptions — the first two of these are referred to as *anthropogenic forcing factors* and the latter two as *natural forcing factors*;

- $\mathbf{u}$ is "climate noise", assumed normal with mean 0 and covariance matrix $\mathbf{C}$;

- We assume there exists a normalizing matrix $\mathbf{P}$ such that $\mathbf{P}\mathbf{C}\mathbf{P}^T = \mathbf{I}$, $\mathbf{C}^{-1} = \mathbf{P}^T\mathbf{P}$.

Then the model (34.2) may be rewritten

$$\mathbf{P}\mathbf{y} = \mathbf{P}\mathbf{X}\beta + \mathbf{P}\mathbf{u} \tag{34.3}$$

where noise $\mathbf{P}\mathbf{u}$ has covariance matrix $\mathbf{I}$.

The Gauss-Markov Theorem implies that the optimal estimator in (34.3) is

$$\tilde{\beta} = (\mathbf{X}^T\mathbf{P}^T\mathbf{P}\mathbf{X})^{-1}\mathbf{X}^T\mathbf{P}^T\mathbf{P}\mathbf{y}$$

with covariance matrix

$$V(\tilde{\beta}) = (\mathbf{X}^T\mathbf{C}^{-1}\mathbf{X})^{-1}.$$

A confidence ellipsoid for β may be derived from the distributional relationship

$$(\tilde{\beta} - \beta)^T(\mathbf{X}^T\mathbf{C}^{-1}\mathbf{X})^{-1}(\tilde{\beta} - \beta) \sim \chi_J^2.$$

The main difficulty in this elegant construction stems from the dimension of the sampling space. Evidently we need an estimate of the covariance matrix $\mathbf{C}$, and the established method [25, 52] is to use control runs of the climate model, but in its general form $\mathbf{C}$ is an $\ell \times \ell$ matrix, and in the best-case scenario we would not have more than about 2,000 years of control-run simulations from which to estimate $\mathbf{C}$. We could have a have a vector of n independent "noise" simulations $\mathbf{y}_N$ and then estimate $\widehat{\mathbf{C}} = \frac{1}{n}Y_N Y_N^T$, but typically $n << \ell$ so $\widehat{\mathbf{C}}$ is singular.

To resolve this issue, in practice the following steps are followed:

- Restrict to κ EOFs with largest variance (equivalent to replacing $\mathbf{P}$ by $\mathbf{P}^\kappa$, consisting of the κ eigenvectors of $\mathbf{C}$ with largest eigenvalues);

- The set of control runs is split into two, one part being used to estimate $\mathbf{C}$ and the other part to estimate β and the associated variance estimates and tests of significance;

- These estimates lead to an estimate $\tilde{V}(\tilde{\beta})$ with ν degrees of freedom, where ν corresponds to the *effective sample size* of the control runs. The concept of effective sample size is closely related to the problem of testing for the mean or a trend in an autocorrelated process, which is covered in detail in Chapter 27 of this volume. Allen and Tett referred to the paper by Zwiers and von Storch [60] which has been widely cited in the climate literature.

The end result of these manipulations is a formal test statistic for the significance of β,

$$(\tilde{\beta} - \beta)^T \tilde{V}(\tilde{\beta})^{-1}(\tilde{\beta} - \beta) \sim J F_{J,\nu}$$

which is readily adapted to testing just a subset of the components of β or some linear combination of those components.

The final methodological development of Allen and Tett was a procedure for testing the fit of the statistical model. Define

$$\tilde{\mathbf{u}} = \mathbf{y} - \mathbf{X}\tilde{\beta}.$$

Then

$$r^2 = \tilde{\mathbf{u}}^T \mathbf{C}^{-1} \tilde{\mathbf{u}} \sim \chi^2_{\kappa - J}.$$

With independent control runs

$$\tilde{\mathbf{u}}^T \widehat{\mathbf{C}}^{-1} \tilde{\mathbf{u}} \sim (\kappa - J) F_{\kappa - J,\nu} \text{ approximately.}$$

This can be used as a diagnostic on the model fit and also to guide the choice of κ.

34.3.3 Accounting for noise in model-simulated responses: the total least squares algorithm

The next major methodological development was due to Allen and Stott [2, 4]. They recognized that a flaw in model (34.2) was that it treated the components of the $\mathbf{X}$ matrix as known, whereas in practice, these components (the outputs of a climate model with different forcing factors) are subject to their own random errors due to the internal variability of the climate system. As a first approximation, these random errors should have distributions similar to those of the control runs, so it would be reasonable to assume they have the same means and covariances.

Allen and Stott [4] rewrote (34.2) as follows. First, note that the term $\mathbf{X}\beta$ in (34.2) may also be written $\sum_{j=1}^{J} \mathbf{x}_j \beta_j$ where $\mathbf{x}_j$ is the model-generated signal. Second, assume each observed $\mathbf{x}_j$ is a perturbation of some "true signal" and can therefore be rewritten $\mathbf{x}_j - \mathbf{u}_j$ where $\mathbf{x}_j$ is the true signal and $\mathbf{u}_j$ a random error. This leads to the model

$$\mathbf{y} = \sum_{j=1}^{J} (\mathbf{x}_j - \mathbf{u}_j)\beta_j + \mathbf{u}_0 \tag{34.4}$$

where $\mathbf{u}_0, \mathbf{u}_1, ..., \mathbf{u}_J$ are assumed to be random errors with a common distribution, $\mathcal{N}(0, \mathbf{C})$ in the typical case that we are considering.

To fit the model (34.4), an appropriate algorithm is not ordinary least squares (OLS), but *Total Least Squares* (TLS), which we discuss next.

According to (34.4), the distribution of $\mathbf{y}$ is normal with mean $\sum_j \mathbf{x}_j \beta_j$ and covariance $(1 + \sum_j \beta_j^2)\mathbf{C}$ so the likelihood function is proportional to

$$|\mathbf{C}|^{-1/2}(1 + \sum_j \beta_j^2)^{-\ell/2} \exp\left\{ -\frac{1}{2} \frac{(\mathbf{y} - \sum_j \mathbf{x}_j \beta_j)^T \mathbf{C}^{-1}(\mathbf{y} - \sum_j \mathbf{x}_j \beta_j)}{(1 + \sum_j \beta_j^2)} \right\}$$

Therefore, one possible estimator of $\boldsymbol{\beta}$, assuming $\mathbf{C}$ known, would chose $\beta_1, ..., \beta_J$ to minimize

$$\frac{(\mathbf{y} - \sum_j \mathbf{x}_j \beta_j)^T \mathbf{C}^{-1}(\mathbf{y} - \sum_j \mathbf{x}_j \beta_j)}{(1 + \sum_j \beta_j^2)} + \ell \log(1 + \sum_i \beta_j^2) \tag{34.5}$$

However, there is a practical difficulty with including the second term in (34.5), which arises from the determinant part of the multivariate normal density: it depends critically on the dimension of the signal ℓ, and as we have already seen, in practice the estimation is carried out in only a low-dimensional subset of the true sampling space.

In fact, the model (34.4) and the estimator (34.5) are special cases of the general *errors in variables* (EIV) regression formulation due to Gleser [14], who proposed a *generalized least squares* algorithm minimizing only the quadratic exponent part of the likelihood function, ignoring that part that arises from the determinant. One argument made by Gleser to support this procedure was that it was less dependent on the errors having an exact multivariate normal distribution.

Applied in this context, Gleser's formulation would choose $\beta_1, ..., \beta_J$ to minimize just the first term of (34.5), in other words

$$Q = \frac{(\mathbf{y} - \sum_j \mathbf{x}_j \beta_j)^T \mathbf{C}^{-1}(\mathbf{y} - \sum_j \mathbf{x}_j \beta_j)}{(1 + \sum_j \beta_j^2)}. \tag{34.6}$$

The solution of (34.6) is the TLS estimator of $\boldsymbol{\beta}$.

Allen and Stott [4] discussed several variants of TLS, but noted that "the differences [among different approaches] are likely to be much less important than the impact of neglecting response-pattern noise altogether." In the simple case of a single regressor $\mathbf{x}$ the formula amounts to minimizing the sum of squares of perpendicular distances from the data points to the best-fit line, instead of the the sum of squares of vertical distances which is the standard OLS procedure. In this form, the method apparently originated in a paper of Adcock [1].

To see that minimizing (34.6) is in fact equivalent to the Allen-Stott solution, we note the following. After applying a pre-whitening operator, Allen and Stott sought constants $v_0, v_1, ..., v_J$ to minimize $(v_0 \mathbf{y} - \sum_{j=1}^{J} v_j \mathbf{x}_j)^T \mathbf{C}^{-1}(v_0 \mathbf{y} - \sum_{j=1}^{J} v_j \mathbf{x}_j)$ subject to the constraint

$\sum_{j=0}^{J} v_j^2 = 1$, and then defined $\beta_j = v_j/v_0$ for $j = 1, ..., J$. However, the two are the same for the following reason. Fix v_0 and write $v_j = \beta_j v_0$ for $j = 1, ..., J$. Then Allen and Stott minimized $v_0^2(\mathbf{y} - \sum_j \beta_j \mathbf{x}_j)^T \mathbf{C}^{-1}(\mathbf{y} - \sum_j \beta_j \mathbf{x}_j)$ subject to the constraint $1 = \sum_{j=0}^{J} v_j^2 = v_0^2(1 + \sum_{j=1}^{J} \beta_j^2)$ which implies $v_0^2 = 1/(1 + \sum_{j=1}^{J} \beta_j^2)$, reducing to (34.6).

34.3.4 Combining multiple climate models

A further extension of this framework was introduced by Huntingford and co-authors [29] to allow for the possibility of multiple climate models. The key assumption here is that, in addition to the internal noise variability between successive runs of any given model, there is also an "inter-model variability" between the output of one model and another. The covariance matrix for the inter-model variability of signal j (denoted $\mathbf{G}_j$ in the discussion to follow) is assumed to be different from the covariance of the internal variability, and is therefore estimated from the model runs. The result is a model that depends on multiple random components but which may also be estimated by *errors in variables* (EIV) methodology, extending the TLS concept.

In more detail, Huntingford *et al.* extended model (34.4) into

$$\mathbf{y} = \sum_{j=1}^{J} (\overline{\mathbf{x}}_j - \mathbf{u}_j - \mathbf{v}_j)\beta_j + \mathbf{u}_0 \tag{34.7}$$

where $\overline{\mathbf{x}}_j$ is the mean over all M climate models and $\mathbf{v}_j$ is an additional noise term that represents the variability among models around $\overline{\mathbf{x}}_j$. In effect, $\mathbf{v}_j$ is treated as an additional random effect with mean $\mathbf{0}$ and a covariance matrix which we write here as $\mathbf{G}_j$ (different for each forcing variable j). Huntingford *et al.* proposed a specific algorithm to estimate $\mathbf{G}_j$ from ensembles of the individual model runs.

The $\mathbf{u}_j$ terms in (34.7) are again assumed to be dominated by the internal variability component of the noise; however, since in this case it is explicit that climate model runs are averaged — we assume a total of M climate models, where the mth climate model has K_m ensemble members, but the model averages $\overline{\mathbf{x}}_j$ are assumed to be the *unweighted* averages of the M model averages — the natural assumption is to assume $\mathbf{u}_j \sim \mathcal{N}[\mathbf{0}, \kappa \mathbf{C}]$ where $\kappa = M^{-2} \sum_{m=1}^{M} K_m^{-1}$.

This is an instance of the general EIV algorithm where coefficients $\beta_1, ..., \beta_J$ and denoised values $\mathbf{y}^*, \mathbf{x}_1^*, ..., \mathbf{x}_J^*$ are chosen to minimize

$$Q^* = (\mathbf{y} - \mathbf{y}^*)^T \mathbf{C}^{-1}(\mathbf{y} - \mathbf{y}^*) + \sum_{j=1}^{J} (\mathbf{x}_j - \mathbf{x}_j^*)^T (\mathbf{G}_j + \mathbf{C})^{-1}(\mathbf{x}_j - \mathbf{x}_j^*) \tag{34.8}$$

subject to the constraint

$$\mathbf{y}^* = \sum_{j=1}^{J} \mathbf{x}_j^* \beta_j. \tag{34.9}$$

Huntingford *et al.* [29] cited a paper by Nounou *et al.* [40] for the algorithm used to solve (34.8). Hannart *et al.* [19] pointed out that the method of [40] does not actually solve the correct version of the EIV problem; instead, they proposed an alternative method due to Schaffrin and Wieser [53].

In practice, the whole optimization takes place in the space of PCs of the internal variability and restricted to the first r components, where r is relatively small; thus, $\mathbf{C}$ in (34.8) may be replaced by $\mathbf{I}_r$ and the $\mathbf{G}_j$'s are $r \times r$ sample covariance matrices based on the departures of the individual model runs from the inter-model average.

34.3.5   Recent advances

Recent years have seen a number of new technical developments, especially regarding the estimation of the matrix $\mathbf{C}$, or incorporating the uncertainty of $\mathbf{C}$ into general inference statements about detection and attribution.

It was noted already in Section 34.3.2 that the sample estimate of the matrix $\mathbf{C}$ is typically singular, due to the limited number of control runs available. The traditional approach to this is to restrict attention to κ EOFs, which is equivalent to a low-rank approximation to the covariance matrix. As an alternative, [48] considered a shrinkage estimator of the form

$$\alpha \widehat{\mathbf{C}} + \gamma \mathbf{I}, \tag{34.10}$$

where $\widehat{\mathbf{C}}$ is the sample covariance matrix and $\mathbf{I}$ denotes the identity matrix. This is known as the *Ledoit-Wolf estimator* and is one of the earliest examples of how to improve the properties of a high-dimensional covariance matrix by regularization [33].

Specifically, [48] used the Ledoit-Wolf estimator to construct an optimal test for detection, and showed it is more powerful than the standard test based on restriction to leading EOFs. They also argued the method was more efficient in the sense that the estimator could be based on small samples of model runs, avoiding the need for long control runs.

[49] extended this approach to attribution. Recall from Section 34.3.2 that the conventional approach to attribution uses two independent estimates of $\mathbf{C}$, one for the initial prewhitening and the second for estimation of β and associated covariance estimates and tests. [49] also used two independent estimates, with the Ledoit-Wolf estimator being used for prewhitening but then a regular covariance matrix estimator (without regularization) for the estimation and testing part of the procedure. The latter was primarily motivated by trying to keep the resulting statistical tests and confidence limits relatively simple though ultimately they still recommended Monte Carlo procedures. The methods were worked out both for the Ordinary Least Squares case (Section 34.3.2) and for Total Least Squares (Section 34.3.3).

An entirely different formulation of the detection and attribution problem was given in [50]. Noting that the conventional approach treats the shape of the responses $\mathbf{x}_j$ as known but the magnitudes β_j as unknown, they questioned why that should be a natural assumption, and whether it was more logical to test whether both the shape and magnitude of the climate model response were correct. However, they continued to recognize that both the observations and the climate model responses are subject to error. They also preserved the key assumption of *additivity* that has been a feature of every approach discussed in this chapter. With those points in mind they proposed relating observed $Y, X_1, ..., X_J$ and "true" $Y^*, X_1^*, ..., X_J^*$ by the equations

$$Y^* = \sum_{j=1}^{J} X_j^*,$$

$$Y = Y^* + \epsilon_Y, \ \epsilon_Y \sim \mathcal{N}(0, \Sigma_Y),$$

$$X_j = X_j^* + \epsilon_{X_j}, \ \epsilon_{X_j} \sim \mathcal{N}(0, \Sigma_{X_j}), \ j = 1, ..., J,$$

where an additional twist is they assume the covariance matrices Σ_Y and Σ_{X_j} to be known. This assumption appears to have been made largely to permit exact distributional calculations, and they acknowledged that in practice it would be necessary to use a plug-in approach with empirical covariance estimates. Thus, while these developments may well lead to more satisfactory estimation and testing procedures, the full accounting for covariance matrix uncertainty in a context that also allows for climate model variability is still an open problem.

A different approach to these issues was started by Hannart [18], who proposed a hierarchical regression approach, which accounts for uncertainty in the climate covariance matrix by making inference on the covariance matrix and on β in a single statistical model. Specifically, the climate covariance matrix $\mathbf{C}$ was assumed to follow an inverse-Wishart prior and subsequently integrated out. In contrast to (34.10), this allowed shrinkage toward target matrices other than the identity, for example covariance matrices that account for spatial dependence. By carrying out the integral with respect to $\mathbf{C}$ analytically, the resulting inference procedure is computationally feasible even without pre-reduction of the dimension of the data.

Katzfuss and co-workers [31] considered an empirical Bayesian hierarchical framework in the context of regression-based detection and attribution. The Bayesian hierarchical formulation ensures that all uncertainties represented by the model are propagated to inference on the regression coefficients of interest. Returning to the traditional expansion of $\mathbf{C}$ in terms of EOFs, their model used a Bayesian model averaging approach to probabilistically infer the optimal number κ of EOFs, instead of choosing a fixed truncation value as in previous approaches. In addition, their model took into account that not only $\mathbf{X}$ but also the observations $\mathbf{y}$ are typically not precisely known. More precisely, they accounted for uncertainty in $\mathbf{y}$ due to a finite number of incomplete and noisy measurements as represented by an ensemble of observations. Their Bayesian hierarchical model was fitted using an efficient Markov chain Monte Carlo (MCMC) procedure that also integrated out analytically all high-dimensional quantities.

Another interesting issue is the treatment of the unknown true mean forcing signals $\mathbf{X}$. The standard practice in most recent approaches [e.g., 18] is to profile (i.e., maximize) out the signals. In contrast, [31] integrated out $\mathbf{X}$ under the assumption of an improper uniform prior. Eliminating unknown nuisance quantities via integration is the standard procedure in Bayesian inference. In the statistics literature on (the simpler) errors-in-variables regression [e.g., 8, 38], it has been noticed that maximization (called a functional approach) can ignore uncertainty and lead to inconsistent estimation of variance parameters. A preliminary simulation study in a simplified detection-and-attribution setting indicated that, for small sample size, integrated likelihood can be conservative with lower power, while the profile likelihood can lead to false positives, but more comprehensive simulations are needed. Also explored should be the effect of an informative spatial or spatio-temporal prior on the unknown signals $\mathbf{X}$, as opposed to the uniform prior used in [18, 31].

34.4 Attribution of extreme events

34.4.1 Introduction

A different kind of question related to detection and attribution concerns the attribution of extreme events. As a concrete example, the extremely active hurricane season in the summer of 2017 led to widespread devastation across the Caribbean, Puerto Rico and the southern United States. A natural question is to what extent such events may be considered to have been "caused by" climate change, where one must be extremely careful about what exactly is meant by "caused by". As an example, as if the time of writing of this chapter, three papers have been published analyzing the influence of anthropogenic climate change on the extreme precipitations produced by Hurricane Harvey at the end of August 2017 [12, 42, 51]. However, the 2017 hurricane season is only one of numerous instances of extreme weather events in recent years where questions have naturally arisen about the influence of anthropogenic climate change. In response, an extensive literature has grown up.

The subject is usually considered to have begun with the paper of Stott, Stone and Allen [55] which analyzed the European heatwave of 2003. This heatwave produced temperatures more than 10°C above the seasonal norm for several consecutive days across much of Central Europe and, by some estimates, was responsible for as many as 70,000 excess deaths. The paper [55] argued that the probability of such an event was increased by a factor of 4 (with a 90% lower confidence bound of 2) compared with a hypothetical counterfactual world without greenhouse-gas warming. Other papers analyzing the 2003 heatwave such as [5, 30, 54] supported the claim of a strong anthropogenic influence on this event. Later papers such as [20, 26, 43] generally supported the anthropogenic influence on a variety of extreme events though using a wide range of methodologies. However, not every paper in this field conveyed the same message. For example Dole and co-authors [10] argued that the 2010 heatwave that badly affected western Russia was most likely a natural event associated with a blocking pattern in the atmosphere, though they did not address the possibility that the frequency of blocking patterns could itself be increasing as a result of global warming. Hoerling and co-authors [27] made similar arguments in discussing the 2011 Texas drought/heatwave, noting that "the principal factor contributing to the heat wave magnitude was a severe rainfall deficit during antecedent and concurrent seasons related to anomalous sea surface temperatures ... that included a La Niña event" while the human-induced contribution to the probability of a new temperature was much smaller.

The wide variety of methods being used for these assessments, as well as occasional disputes over the results, led the National Academy of Sciences to commission a review of the whole field. Their report [41] appeared in 2016. Our own review follows some of the structure of the National Academy report, though necessarily with much condensation, and focuses specifically on the statistical issues these questions raise.

34.4.2   Framing the question

There are so many different ways of defining the problem that the National Academy report [41] devoted a whole chapter to the "framing question". We follow their approach here, and focus on the specifically statistical issues that they raise.

Many researchers beginning with [55] have used the "fraction of attributable risk" as the primary quantity of interest. Given a specific extreme event, let p_1 be the probability of that event under a scenario that includes all known forcing factors that influence climate, and p_0 the counterfactual probability of the same event under natural forcings only (including random internal variation). Climate models are needed here because p_0 can only be estimated from a model; typically, parallel runs of a climate model (or several climate models) are used to that p_0 and p_1 can be estimated in a way that makes comparisons possible.

The *Fraction of Attributable Risk* (FAR) is defined to be

$$FAR = 1 - \frac{p_0}{p_1}. \tag{34.11}$$

The reason for the name is that, in the typical case where $p_1 > p_0$, we can partition the total probability of the event (p_1) into two components, p_0 for the natural contribution and $p_1 - p_0$ for the anthropogenic contribution. Thus (34.11) represents the fraction that may be "attributed" to the human influence.

However, FAR is not the only measure used. Another is the risk ratio

$$RR = \frac{p_1}{p_0} \tag{34.12}$$

which, although equivalent to (34.11) in the sense that either formula can be transformed into the other, in some respects has a more natural interpretation — for example, the RR represents the proportion by which insurance claims from extreme weather events would be expected to rise in a world subject to anthropogenic forcings compared with one that is not. Another argument is that RR corresponds to statements of risk that are common in medical research, such as "smoking increases the probability of lung cancer by a factor of X", page 34 of [41].

Our own preference is in favor of RR, and this is reinforced by several arguments made in [41]. For example:

1. FAR can be misleading when it is very close to 1 — for example, FAR=0.99 might not seem much different from FAR=0.999 but the latter represents a ten times greater risk ratio;

2. The "fractional" interpretation of FAR only makes sense when it is between 0 ad 1 — indeed, FAR is often truncated at 0, especially in confidence intervals — but there is actually nothing pathological about the possibility that $p_0 > p_1$. Indeed, [41] make the argument that there would probably be many more reported instances with $p_0 > p_1$ were it not for the implicit bias that such events are unlikely to be observed. It is better to put things on an even footing and treat the cases $RR > 1$ and $RR < 1$ as equally interesting and important, at least until the weight of evidence suggests to the contrary;

3. Tests and confidence intervals — an inherent issue with estimating extreme event probabilities is that they are very uncertain, so confidence intervals (or Bayesian credible intervals) tend to be very wide. In the language of hypothesis testing, a formal test of the null hypothesis $p_0 = p_1$ may well result in acceptance of that hypothesis. The Academy report [41] cautions *against* concluding "there is no effect" in these cases. To quote the report (p. 35), "Failure to reject the null hypothesis of no effect should not be regarded as evidence in favor of these being no effect." Although one could make a similar assertion with respect to FAR, the issue is more clear-cut when framed in terms of RR.

In recent years, there have been a number of attempts to reformulate detection and attribution in the language of causality research. A particularly influential paper was by Hannart and co-authors [17].

According to the classical theory of Hume [28], quoted by [17], "We may define a cause to be an object followed by another, where, if the first object had not been, the second never had existed." In the language of events, an event X may be said to cause an event Y if Y cannot occur in the absence of X, or in other words, $Y \implies X$. This immediately suggests some probabilistic relationships. [17] review the modern theory of causal inference including the use of graphical relationships to represent causality in system of interacting variables.

Suppose we have (0,1)-valued random variables X and Y where in the present context $X = 1$ is associated with the presence of an anthropogenic effect and $Y = 1$ with the occurrence of an extreme climate event. We may also define Y_x to be the value Y would take if X were fixed at x. In a world of perfect causality where $Y = 1$ if and only if $X = 1$, we would have $Y_0 = 0$, $Y_1 = 1$.

In this context, [17] following [46], defines

$$
\begin{aligned}
\mathrm{PN} &= \Pr\{Y_0 = 1 \mid Y = 1, X = 1\}, \\
\mathrm{PS} &= \Pr\{Y_1 = 1 \mid Y = 0, X = 0\}, \\
\mathrm{PNS} &= \Pr\{Y_0 = 0, Y_1 = 1\}.
\end{aligned}
$$

Here PN is referred to as "the probability of necessary causation", PS as "the probability of sufficient causation", and PNS as "the probability of necessary and sufficient causation". This subdivision into "necessary" and "sufficient" causation is the main new feature of their approach.

Under two additional assumptions, *monotonicity* and *exogeneity* (of X) they show that the above expressions reduce to

$$PN = \max\left\{1 - \frac{p_0}{p_1}, 0\right\},$$

$$PS = \max\left\{1 - \frac{1 - p_1}{1 - p_0}, 0\right\},$$

$$PNS = \max\{p_1 - p_0, 0\},$$

Here, monotonicity is the property that $Y_1 \geq Y_0$ with probability 1, while exogeneity of X essentially means that X is external to the system, in other words, not changed by any of the other variables being observed.

Thus when $p_1 \geq p_0$, PN corresponds exactly to the FAR. PS and PNS are new measures which do not seem to have been used previously in the extreme event attribution context.

[17] goes on to consider the implications of these definitions for the European heatwave of 2003. Assuming the same climate variables and probability calculations as [55], for which p_0 was estimated to be $\frac{1}{1000}$ and p_1 to be $\frac{1}{250}$, the probability of necessary causation is 0.75, equal to the FAR discussed earlier. However, PS and PNS are both of the order of 0.003, implying very low evidence for sufficient causation.

However, they also consider other interpretations of the data for which the distinction between PN and PS is less clear-cut. The calculations of [55] were based on temperature anomalies with respect to 1961–1990 averages exceeding the threshold of 1.6°C (the second largest summer-mean anomaly in the dataset). However, had the threshold been set substantially lower, both p_0 and p_1 would be larger and hence so would be both PS and PNS. For thresholds in the lower tail of the distribution, we find PS close to 1; as stated by [17], "anthropogenic CO_2 emissions are virtually certainly a sufficient cause, and virtually certainly not a necessary cause, of the fact that the summer of 2003 was not unusually cold." They also point out that when referred to a much longer time period than one year, even if we (unrealistically) assume stationarity in time, an event of the form "the threshold will be exceeded at least once in the next hundred years" (rather than for one specific year) will lead to much larger p_0 and p_1 and therefore a higher probability for sufficient causation as well as necessary causation.

In summary, the use of causal inference methods in climate research is still a new field but it is growing. Ebert-Uphoff and Deng [11] used Bayesian networks to examine the causal relationships among four large-scale climate circulation indexes and, very recently, Hannart and Naveau [16] have proposed an ambitious reformulation of tradition detection and attribution theory in the language of causal inference. We expect to see much more research of this nature in the next few years.

34.4.3　Other "Framing" Issues

Here, we review more briefly several other framing issues discussed by [41]

1. **Choice of climate variable.** The action of the 2003 European heatwave focussed on a few days in early August on a region of western Europe that included most of France, parts of Germany, Switzerland and northern Italy, but did not extend as far west as the Iberian peninsula or into Scandinavia or Eastern Europe. Yet [55] took as their climate variable of interest the annual summer (June,

July, August) temperature means over a large geographical area (30°N to 50°N latitude, 10°W to 40°E longitude). There were a few motivations for defining an event over a substantially larger spatial and temporal scale than the one observed. For example, one concern was selection bias, discussed further below — focussing on a climatic variable that had very locally extreme behavior in the observational record would attract the criticism that the event had been specifically selected for this reason, whereas by choosing a more generic spatial and temporal scale, the focus was more on the appearance of extreme events in general than this one particular event. However, another reason was that [55] recognized that an extreme event attribution analysis was unlikely to be successful if the considered variables were not well represented in climate models.

Subsequent extreme event analyses have generally focussed on smaller spatial regions, and sometimes smaller time windows as well, than [55], but the general principle remains, that it is better to expand the spatial and temporal coverage to reduce concerns about selection bias and to focus on variables that are well represented in climate models. [41] noted some other considerations: (a) different physical variables, e.g. some analyses of the California drought found no anthropogenic effect if precipitation levels were considered on their own but did find an effect when low precipitation was combined with high temperature: (b) where attribution analysis is based primarily on observations, such as comparing recent records with those of an earlier period, it is important that the observations should be of high quality and consistently measured over the whole time period; (c) robustness of results — "a robust attribution analysis would show that results are qualitatively similar across a range of event definitions, acknowledging that quantitative results are expected to differ somewhat because of difference of definition."

2. **Changes in frequency or changes in magnitude?** The discussion of p_0 and p_1 assumed that the interest is in changes of frequency for an event of fixed magnitude, but one can ask a parallel question with frequency and magnitude interchanged. For example, given a historical estimate of the 100-year return value[1] for a given climatic variable, how would that estimate change if the underlying climate conditions changed? That question, for example, underlies the production of flood risk maps by the Federal Emergency Management Agency (FEMA). [51] give examples of both types of calculations.

3. **Conditioning.** One of the most contentious issues in this field in recent years has been the desirability of *conditioning* as a vital component of the analysis. Trenberth and colleagues [57] argued that when an extreme event depends on the presence of some feature of large-scale atmospheric circulation, it may be problematic to try to attribute the circulation event itself to anthropogenic effects, but, conditional on the appearance of the circulation pattern, other variables that affect the development of an extreme event, such as SST, may be much more clearly attributable to human influence. Examples that they gave included Superstorm Sandy, that caused widespread flooding over New York and New Jersey in 2012, and the Colorado floods of 2013. Therefore, they suggested, the entire analysis should be conducted conditionally on the presence of whatever large-scale circulation feature initiated the event of interest.

With reference to the Colorado floods, a specific example of this kind of analysis was given by Pall and co-workers [44]. The flooding events in question

---

[1]that value which is exceeded in any given year with probability 0.01

happened during September 2013, and caused over \$2 billion damage and nine fatalities. However, the authors noted, "the unusual hydrometeorology of the event...challenges standard frameworks [for attribution]... because they typically struggle to simulate and connect the large-scale meteorology associated with local weather processes." Consequently, these authors developed an approach that was part statistical, part dynamical based on simulations of the local weather conditioned on observed synoptic-scale meteorology. The key meteorological point seems to be that warmer air holds more moisture (the Clausius-Clapeyron relationship) and therefore exacerbates the magnitude of precipitation events within a developing weather system. The authors looked at this from both a "frequency" and "magnitude" point of view, concluding that the magnitude of the extreme event was increased by 30% as a result of anthropogenic climate change, or conversely, the probability of an event of the given magnitude, conditional on the synoptic weather pattern, was increased by a factor of 1.3 compared with what might have been expected in a pre-industrial world.

For the analysis of [44], it appears that an attempt to take into account the probability of the triggering synoptic-scale event would not have been successful because there was no reasonable basis for determining how this event could have been changed by the anthropogenic influence. In other cases, however, there may be a choice: do we base the analysis solely on the conditional probabilities or do we also take into account the probabilities of the conditioning event? As described by [41], the choice lies between considering

$$\frac{\Pr_f\{E \mid N\}}{\Pr_c\{E \mid N\}} \tag{34.13}$$

or

$$\frac{\Pr_f\{E \mid N\}}{\Pr_c\{E \mid N\}} \times \frac{\Pr_f\{N\}}{\Pr_c\{N\}} = \frac{\Pr_f\{E \cap N\}}{\Pr_c\{E \cap N\}}. \tag{34.14}$$

Here E is the event of interest, N is the conditioning event (which may be El Niño, hence the choice of initial) and $\Pr_f$ and $\Pr_c$ denote probabilities in the observed (factual) and counterfactual worlds. The controversy, such as it is, appears to hinge on the question of whether it is preferable to base the inferences on (34.13) in place of (34.14).

In fact, there is a sound statistical argument for conditioning based on R.A. Fisher's theory of conditional inference and the related concept of *ancillary statistics*. We do not attempt a detailed review here, since there is a extensive and large literature, but we refer to [13] for a relatively recent review.

The key point of this theory is that conditional inference is always indicated when the distribution of the conditioning variable N is independent of the quantity being estimated, which in this case, is the influence of anthropogenic climate change on the probability of the event E. Such a variable N is known as an ancillary statistic. In meteorological terms, if the event N is not affected by climate change, then it is valid to use (34.13) in place of (34.14).

As things stand, this may seem a trivial conclusion, because if N is not influenced by climate change, the second factor on the left side of (34.14) will be 1 and there is no distinction between (34.13) and (34.14). However, there is also an extensive theory of approximate or "local" ancillarity and the overall conclusion is that conditioning is still appropriate in this case [7].

In practice, the more realistic difficulty is that E and N are physical measurements, not random variables satisfying some theoretical distribution, and there may simply not be enough information to determine whether either or both are different under the factual and the counterfactual scenarios. Under such circumstances, arguing conditionally on N seems a logical way to proceed.

4. **Selection bias.** The final "framing" issue we discuss here is that of selection bias — the idea that the selection of an event to study may itself bias the conclusions drawn from it. As noted by [41], selection bias may take various forms, the most pervasive being occurrence bias, "bias from studying only events that occur." As noted already, a partial solution may be to define the climate events of interest on a sufficiently large temporal and spatial scale that extreme local fluctuations do not bias the results. [41] conclude that "selection bias [is] almost inevitable in event attribution applied to individual events" but caution that "Such selection biases interfere with the ability to draw general conclusions about anthropogenic influence on extreme events collectively."

A possible direction for future methodological development is to adjust statistical event attribution techniques to account explicitly for their tendency to focus on local spatial and/or temporal extremes, in similar manner to the use of scan statistics in spatial epidemiology, e.g. [32]. In short, we might adjust the extreme event probabilities to allow for a selection bias effect, but then proceed as in earlier analyses with the comparisons of scenarios that do or do not include the anthropogenic component.

34.4.4 Statistical methods

In this section we do not pursue further the various type of framing issues but assume the problem is essentially the basic one that motivated this whole section: given the interest in a specific event E, and the possibility of estimating the event E from parallel runs of either a climate model (under "all forcings" versus "natural forcings" scenarios) or an observational dataset (under "present-day" versus "pre-industrial" conditions), how would we actually estimate the probabilities p_1 and p_0, the respective probabilities under those two scenarios? These probabilities may then be used to estimate the FAR, the RR, or various other measures of interest. We briefly review the main methods for estimation p_1 and p_0.

1. **Methods based on normal distributions.** The simplest methods assume the underlying variables are normally distributed. This assumption was common in the early days of the subject [5, 30, 54] and still surfaces from time to time [47]. In general, we don't recommend normal-theory approaches because, even when the overall distribution is close to normal, as is usually the case with temperatures, deviations from normality in the tails of the distribution can cause serious biases to the estimated probabilities.

2. **Adapting conventional detection and attribution theory.** For example, Min and co-authors [37] fitted the generalized extreme value (GEV) distribution to extremes in spatially averaged precipitation variables and then used a probability integral transform (if G is the CDF fitted to a random variable Y and if Φ denotes standard normal CDF, then $\Phi^{-1}(G(Y))$ is standard normal) to transform the variables to normality. On the resulting normal scale, they then applied conventional detection and attribution theory to determine the statistical significance of the anthropogenic component and to compute estimates and confidence intervals for the various regression coefficients. Similar techniques have been used

in other papers such as [59, 61]. Compared with other techniques considered in this section, these methods do not lead to numerical estimates of p_1 and p_0 and hence FAR, RR etc., but in principle they could be: use the fitted GEV distribution to estimate p_1 and then the same transformation applied to the detection and attribution model without the anthropogenic component to estimate p_0, but this would be a decidedly roundabout method of estimating p_0! Overall, this method seems less well suited to studying the attribution of a single event than to understand the anthropogenic influence on extreme events generally, but that remains an important consideration for climate research.

3. **Methods based on counting exceedances in model simulations.** A nonparametric method for estimating p_1 and p_0 is simply to count the number of exceedances of the threshold of interest in parallel all-forcings and natural-forcings model runs. Tests and confidence intervals may then be based on standard statistical theory for binomial distributions. This method has the considerable advantage of simplicity, when it is applicable; but against that, nonparametric methods cannot be used at all when they involve extrapolating beyond the range of the climate model data. In practice it seems to be used in two situations: (i) when the supposed extreme event is not actually very extreme at all, at least when conditioned on suitable large-scale variables [45], (ii) when the analyst has available a fast-running model for which generating a large ensemble of model runs is not a problem [12]. The "climateprediction.net" experiment (https://www.climateprediction.net/) is a citizen science project to generate very large ensembles of climate model runs through volunteers running climate simulation programs on their laptops, but the emphasis of extreme event attribution analysis in recent years has switched towards trying to get very fast "operational time" results, and the time taken to collect large ensembles through a distributed network would appear to impede that.

4. **Methods based on extreme value theory.** These methods appear, perhaps surprisingly, to be state of the art in this field at the time of writing. For a review of different methods of (univariate) extreme value analysis, see Chapter 8 of this volume. The methods may be divided into two broad categories, (i) block maxima methods, where the Generalized Extreme Value (GEV) distribution is fitted to maxima over blocks of fixed time length (usually the time length of a block is taken to be one year, in which case it is also called the annual maxima method), or (ii) threshold exceedance methods, in the most common form of which the Generalized Pareto Distribution (GPD) is fitted to the exceedances over a fixed threshold. The original paper of [55] used the GPD to characterize the tail distribution of European summer temperatures, but recent applications have more frequently used the GEV applied to annual maxima. A significant issue with these methods is how to characterize uncertainty of the extreme-event probabilities calculated from the fitted distributions — both frequentist (bootstrap) and Bayesian methods have been applied but there seems to be no universal agreement at this time. We do not review these methods further here because Chapter 8 gives a full account.

34.4.5   Application to precipitation data from Hurricane Harvey

We now return to our earlier discussion of Hurricane Harvey, and specifically the extreme precipitations produced by that event in the region surrounding Houston, by reviewing the three papers that we are aware of that have discussed that event.

1. **The World Weather Attribution Project.** This project is a consortium of researchers in The Netherlands, U.K. and U.S.A. who aim to produce "rapid attributions" of extreme weather events, sometimes within a week of the event in question. Two recent examples are [58] for the August 2016 Louisiana floods and [42] for the flooding associated with Hurricane Harvey. Both studies used the same statistical methods and covered essentially the same geographical region, so they used many of the same meteorological variables as well. The essential idea is to compile several datasets, both observational and model-based, that use different data sources, different spatial resolutions and (with the models) different forcing factors, including historical runs based on all known forcing factors, pre-industrial control runs and "static" experiments with forcings fixed at levels corresponding to various points in time (e.g. 1860, 1940, 1990, 2015) in order to compare equilibrium climate behavior under different forcing scenarios.

   The basic method is to fit generalized extreme value distributions (see Section 34.4.6 below) with adjustment for a covariate which they typically take as global mean temperature for a given year. In the notation of (34.15) below, the model of [58] allows both $\log \eta_t$ and $\log \tau_t$ to be linearly dependent on a global temperature variable T'. The GEV is fitted to each of the observational and model datasets, omitting the extreme event that stimulated the study, and the results compared to evaluate consistency across observations-based and model-based analyses. Typically, p_1 and p_0 are evaluated by setting T' to be its value in the present day and its value at some historical date, e.g. 2017 versus 1900 in [42]. Standard errors and uncertainty bounds are evaluated largely by bootstrapping, with a spatial block bootstrap recommended in spatially aggregated datasets to reduce bias due to spatial dependence.

   For three-day maxima from 85 stations in the Gulf Coast region, the estimated return value associated with the 2017 event is around 9,000 years, and a risk ratio of four compared with the corresponding estimate for 1900. Somewhat lower return level estimates (of the order 200–800 years) are obtained for events based on the spatial maximum over a region than for events at a specific location, but similar risk ratios of the order of 4–6 compared with 1900. In all cases, the range of uncertainties associated with these estimates is very wide. Estimates based on model runs typically show somewhat lower risk ratios but narrower confidence limits, which the authors interpret as evidence of an anthropogenic effect.

2. **Risser and Wehner.** These authors [51] provided an alternative extreme value analysis of precipitations during August 2017. They computed seven-day maximum precipitation values from raingauges, aggregated over two regions near Houston: a small and large region of approximately 33,000 km^2 and 105,000 km^2 respectively. The GEV was fitted to annual maximum data from 1950–2016 using a model similar in structure to (34.15) and (34.16) below, but with two covariates: the Niño 3.4 index as a measure of El Niño activity, which the authors identify as a natural variation, and annual global CO_2 measurements. They estimated return values for the 2017 event of the order of 30–100 years (larger for the large region than the small region) and also risk ratios of the order 4–8 (with lower confidence bounds of the order 1.5–4).

   For this paper, a direct "attribution" statement was not possible because climate model runs were not used, but the authors argued that an equivalent if somewhat weaker statement could be made in the language of Granger causality [15]. They argued, in effect, that confidence intervals for risk ratios with lower bounds > 1 are consistent with Granger causality at a "likely" level of uncertainty (66%

confidence) though not at a "very likely" level (90% confidence). In this sense, the results provide a valid attribution statement.

3. **Emanuel.** The third author [12] to have examined Hurricane Harvey precipitations took a completely different approach, based on a model for directly simulating hurricanes and their associated storm rainfalls. The model relies on 'global climate model data to generate the large-scale state of the system and then randomly perturbs that state to create hurricane-like disturbances. Thus, the approach is still dependent on global-scale models for the large-scale variables, but improves significantly on global-scale models in its resolution of specific hurricane events.

For storm totals at the single point of Houston, Texas, this approach suggests a return value in excess of 2,000 years, though with huge uncertainty as very few of the model runs get close to that level of precipitation. A side comment here is that although this was based on the "counting exceedances" approach (see Section 34.4.4), an extreme value theory approach might also be productive in reducing the uncertainty of estimation at the very end of the observed range of data. A second calculation based on total rainfall over the state of Texas suggests an annual exceedance probability of around 1% in 1981–2000, increasing to around 18% in 2081–2100 under the representation concentration pathway 8.5 scenario (sometimes called the "business as usual" scenario, since it assumes no significant slowing down in the rate of emissions of greenhouse gases).

This paper was the only one of the three to extrapolate future probabilities of a Harvey-type event, but if the 18% estimate is realistic, for the probability that a Harvey-sized event will occur somewhere in the Gulf region in any given year, that is a disturbing conclusion.

34.4.6   An example

The following example is more limited than the three studies just cited [12, 42, 51] because it only considers precipitation from one station (Houston Hobby Airport) but it nevertheless shows that many of the same effects that they cite as evidence of anthropogenic influence are also present here. In common with [51], seven-day precipitation totals were calculated each year for the hurricane season from July–November. We also calculated annual average sea surface temperatures (SST) for the entire Gulf of Mexico, computed from July of the year preceding the precipitation year through June of the same year; this time window was considered most likely to influence the following hurricane season. Figure 34.2 illustrates the data. Three points are immediately apparent. First, the 7-day precipitation total associated with Hurricane Harvey is by far the largest in history, more than twice the second-largest value. Second, SSTs have also increased steadily since the 1970s, and the Gulf of Mexico SST mean for 2016-17 was the largest in history for that variable. Third, based on the straight line fit in (c), there is some evidence that 7-day precipitation maxima and SSTs are correlated, though the statistical significance of that is hard to judge from the plot.

A more formal analysis of the latter point may be based on the Generalized Extreme Value (GEV) distribution for annual maxima: see Chapters 8 and 31 of the present volume for a detailed discussion of extreme value theory and the GEV distribution in particular.

If Y_t denotes the maximum precipitation value for year t, we assume Y_t follows the GEV distribution in the form

$$\Pr\{Y_t \le y\} = \exp\left[-\left\{1 + \xi\left(\frac{y - \eta_t}{\tau_t}\right)\right\}_+^{-1/\xi}\right], \qquad (34.15)$$

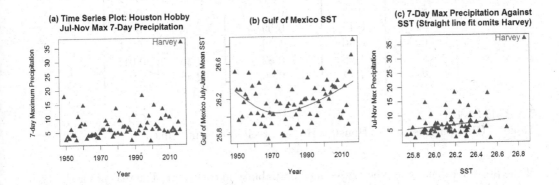

FIGURE 34.2

Precipitation in Houston and Gulf of Mexico SST. (a) Maximum 7-day precipitation total from Houston Hobby airport, computed from July–November each year. (b) Gulf of Mexico mean July–June sea surface temperature, each year from 1948-49 through 2016-17. The fitted trend curve is based on a spline with 4 DF and is consistent with overall Northern hemisphere temperature trends during this time period. (c) Plot of maximum 7-day precipitation against Gulf mean SST, with a fitted straight line omitting the 2017 outlier. Public data sources: Daily precipitation from the Global Historical Climatological Network (National Centers for Environmental Information, U.S.A.; monthly sea surface temperatures from HadISST (U.K. Meteorological Office.)

where the subscript $+$ denotes positive part, η_t and τ_t are allowed to vary with year and, in accordance with common practice in this field, the shape parameter ξ is treated as a constant. Recall that [51] used two covariates, the Niño 3.4 index and annual global CO_2 means. Here, we are assuming that Gulf of Mexico SST will include any El Niño effect that influences precipitation, but to be consistent with [51], we also included annual global CO_2 means from the RCP database (https://tntcat.iiasa.ac.at/RcpDb).

The following models are considered: each of η_t and $\log \tau_t$ is a linear function of up to two covariates, where the covariates considered are SST_t (Gulf of Mexico annual mean SST in year t) and $CO2_t$ (global mean CO_2 in year t). For numerical stability, SST_t is expressed as the deviation from 26°C and $CO2_t$ is replaced by $0.01(CO2_t - 350)$. This gives 16 possible models of which the Akaike Information Criterion chooses the following:

$$
\begin{aligned}
\eta_t &= \theta_1 + \theta_4 SST_t + \theta_5 CO2_t, \\
\log \tau_t &= \theta_2 + \theta_6 SST_t, \\
\xi &= \theta_3.
\end{aligned}
\tag{34.16}
$$

The fitted parameters are given in Table 34.1.

Next, this model is used to calculate exceedance probabilities in various years corresponding to the observed 2017 value due to Hurricane Harvey. First, we smoothed the SST values, using the same smoothing spline as in Figure 34.2(b). The reason for smoothing is that we are interested in long-term climatic effects, not individual-year fluctuations, and smoothing the SSTs seems a logical way to achieve that. Second, the model defined by (34.15) and (34.16) was refitted using Bayesian methods, assuming a flat prior. The reason for this is to allow the uncertainty of the estimates to be expressed in terms of posterior distributions. The three curves in Figure 34.3(a) represent the 17th, 50th and 83rd percentiles

Parameter	Estimate	Standard error	t-statistic	p-value
θ_1	4.70	0.29	16.22	0.00
θ_2	0.56	0.13	4.25	0.00
θ_3	0.15	0.09	1.64	0.10
θ_4	3.06	1.49	2.06	0.04
θ_5	1.95	0.82	2.36	0.018
θ_6	1.24	0.50	2.48	0.013

TABLE 34.1
Table of GEV parameters for Houston Hobby precipitation maxima.

of the posterior density for the exceedance probability in each year. The reason for the 17th and 83rd percentiles is that the posterior probability between them is 0.66; according to the Intergovernmental Panel on Climate Change uncertainty guidelines [36], it is *likely* that the true value lies between these bounds[2]. For the specific occurrence in 2017, the calculation shows a posterior median exceedance probability of 0.0019 (return value 525 years) with a *likely* range from 0.00022 to 0.00685 (return values 145 to 4472 years).

Climate model data have been downloaded from the CMIP5 model archive and used to calculate annual SST means over the Gulf of Mexico. These are available under three scenarios: (a) historical all-forcings data up to 2005 or 2012; (b) historical natural-forcings data up to 2005 or 2012; (c) future forcings data under the RCP 8.5 scenario, often called the "business as usual" scenario because it does not presume any significant effort to slow down greenhouse gas emissions. All model runs have been converted to anomalies and where natural-forcings data ended before 2017, we simply assumed the last available value (for 2005 or 2012) was also valid up to 2017. We combined the all-forcings and RCP 8.5 data to obtain a continuous record of data from 1949 up to 2080 which was taken as the end-year for this assessment. This exercise was repeated for four climate models; where multiple ensembles were available from the same model, we averaged over ensembles.

The model Gulf of Mexico SSTs do not follow the observational data very closely so, in order to use the regression model fitted previously to observational SSTs, we proceed as follows. The observational SSTs for 1949–2017 are regressed on two covariates: first, the difference between historical-forcings and natural-forcings climate model runs, and, second, the natural-forcings climate model runs on their own. The two components together are then used to define the "all forcings" signal and the second component on its own is used to define the "natural forcings" signal. Both components are represented via smoothing splines to give a smooth signal. This exercise is repeated for each of the four climate models and also with all four models averaged to give the curves in Figure 34.3(b). A curious feature of these curves, which we are not able to fully explain, is that even the natural-forcings curves seem to show an upwards trend towards the end of the series.

This exercise was repeated to obtain future projections of Gulf of Mexico SST up to 2080; see Figure 34.3(c). Since there are no natural-forcings projections over this time period, only the RCP 8.5 values are shown.

We now repeat the calculation of the probability of a Harvey-sized event under the circumstances, (a) for 2017 under all forcings, (b) for 2017 under natural forcings, (c) for 2080 under RCP 8.5. The calculation is repeated for all four climate models and for the average over the four models; we used the same posterior density output as before to obtain Bayesian posterior curves. Finally, we took the ratio of (a) to (b) (relative risk for 2017

---

[2]The stronger terms *very likely* and *virtually certain* are used for events with probability at least 0.9 and 0.99, respectively.

under the all-forcungs and natural-forcings scenario), and the ratio of (c) to (a) (relative risk for a Harvey-sized event in 2080 compared with 2017). The results are in Table 34.2.

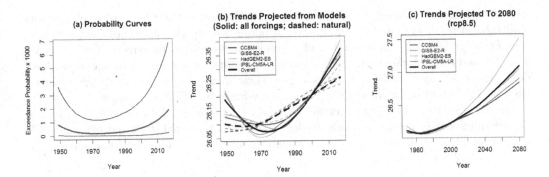

FIGURE 34.3
Probability Curves and SST Projections. (a) Projected probability (red curve) and 66% confidence bounds (green curves) for the probability of a Harvey-sized event at Houston Hobby airport, 1949-2017. (b) Projected SSTs in the Gulf of Mexico under all-forcings (solid curves) and natural-forcings (dashed curves) for four climate models, and all four models averaged. (c) Projected SSTs through 2080, under the RCP 8.5 scenario for four climate models, and all four models averaged.

Model	Present			Future		
	Lower	Mid	Upper	Lower	Mid	Upper
CCSM4	1.5	2.0	3.2	9.0	26.2	133
GISS-E2-R	1.8	2.5	4.8	13.5	43.5	244
HadGEM2-ES	1.6	2.1	3.5	23.6	73.3	415
IPSL-CM5A-LR	1.5	2.0	3.3	10.8	33.8	186
Combined	1.7	2.4	4.4	'14.3	46.0	254

TABLE 34.2
Relative risks. The columns labelled "Present" refer to relative risks for the 2017 event under an all-forcings scenario versus a natural-forcings scenario, computed under four climate models and with all four models combined. Lower, mid and upper bounds correspond to the 17th, 50th and 83rd percentiles of the posterior distribution. The columns labelled "Future" are relative risks for such an event in 2080 against 2017; same conventions regarding climate models and percentiles.

For the combined-model results, the relative risk of the Harvey precipitation under all-forcings versus natural-forcings scenarios is estimated as 2.4, "likely" between 1.7 and 4.4. For all five sets of model results in Table 34.2, the lower bound exceeds 1, proving that it's "likely" that anthropogenic conditions affected Harvey. This is consistent with the earlier results reported by [12, 42, 51].

For the relative risks of a Harvey-sized event in 2080 against 2017, the posterior means range from 26 to 73, with "likely" bounds ranging from 9 to 415. Evidently, the uncertainty range for future projections is very wide. Recalling that Emanuel [12] obtained an estimated relative risk of 18 by complete different methods, there seems to be some agreement that a drastic rise in the frequency of this type of event is to be expected.

Further details of these results will be developed elsewhere.

34.4.7   Another approach

Diffenbaugh and co-authors [9] also sought to quantify the increase of extreme event probabilities as a result of global warming, though taking a more global view of the problem in computing probabilities for a number of extreme events related to extreme temperatures, droughts and extreme rain events. They compared results obtained using both observational data and climate models. A particular feature of their approach was the use of some standard statistical goodness of fit procedures (Kolmogorov-Smirnov and Anderson-Darling tests) to assess the agreement between observational and climate-model data after first correcting for the difference in means between pre-industrial climate model data and detrended observations. They recommend rejecting any climate model which fails the Anderson-Darling test at a p-value of 0.05.

34.5   Summary and open questions

Climate change detection and attribution refers to a set of statistical tools to relate observed changes to external forcings, specifically to anthropogenic influences. While this issue can be viewed in different ways, the most commonly applied framework is linear regression. The problem formulation per se seems straight forward, but the challenges lie in the high dimensionality of the problem and the large number of unknown quantities in the context of limited observations. Current methods differ in their complexity of the problem formulation and what assumptions are being made to reduce the dimensionality of the problem. Most methods implemented so far are of frequentist nature and Bayesian implementations have only recently appeared on the scene.

While many of the approaches discussed address some of the methodological challenges, there is of yet no model framework to address them all comprehensively. For example, most current frameworks assume the different model runs to be independent realizations from a common random quantity. This viewpoint is justifiable in cases where all model runs come from the same climate model or all come from different climate models, but less so if we have multiple, and potentially unequal numbers, of replicates from multiple climate models. In this case a formulation explicitly accounting for inter- and intra-model variability, as considered by [29], is needed. An analogous issue exists with control runs coming from different models. Having a way to use them jointly would drastically increase the amount of information available to estimate internal variability. The assumed covariance structure of observations is also relatively simple in current methods [e.g., 31], if observational uncertainty is considered at all. With the advent of observational products now routinely being provided as ensembles rather than a single data set, which used to render data-driven observational covariance estimation practically impossible, more complex covariance structures can be envisioned. Other directions include joint inference on multiple properties, e.g. different temperature layers in the atmosphere, and incorporating non-linear interactions.

The alternative field of extreme event attribution is still rapidly growing and would seem to offer excellent opportunities for involvement by statisticians. For example, although the standard univariate methods of extreme value theory are becoming standard in this field, none of the references cited in this chapter has made any use of bi/multivariate or spatial extreme value theory, though there is extensive statistical theory in both cases as documented in Chapters 8 and 31 of this volume. Therefore, there are many possibilities for extensions of the methods as they currently exist.

34.6 Acknowledgments

This work was partly supported by NSF grants DMS-1127914 and DMS-1638521 to the Statistical and Applied Mathematical Sciences Institute, and NSF grant DMS-1106862 to the Research Network on Statistics in the Atmosphere and Ocean Sciences (STATMOS). In addition, Katzfuss' research was partially supported by NSF grants DMS-1521676 and DMS-1654083 and Smith's by NSF grant DMS-1242957.

Bibliography

[1] R.J. Adcock. A problem in least squares. *The Analyst*, 5:53–54, 1981.

[2] M.R. Allen, P.A. Stott, and G.S. Jones. Estimating signal amplitudes in optimal fingerprinting, part II: application to a general circulation model. *Climate Dynamics*, 21:493–500, 2003.

[3] M.R. Allen and S.F.B. Tett. Checking for model consistency in optimal fingerprinting. *Climate Dynamics*, 15:419–434, 1999.

[4] Myles R. Allen and Peter A. Stott. Estimating signal amplitudes in optimal fingerprinting, part I: theory. *Climate Dynamics*, 21:477–491, November 2003.

[5] M. Beniston and H.F. Diaz. The 2003 heatwave as an example of summers in a greenhouse climate? Observations and climate model simulations for Basel, Switzerland. *Global and Planetary Change*, 44:73–81, 2004.

[6] N.L. Bindoff, P.A. Stott, K.M. AchutaRao, M.R. Allen, N. Gillett, D. Gutzler, K. Hansingo, G. Hegerl, Y. Hu, S. Jain, I.I. Mokhov, J. Overland, J. Perlwitz, R. Sebbari, and X. Zhang. *Detection and Attribution of Climate Change: from Global to Regional*, book section 10, pages 867–952. Cambridge University Press, Cambridge, United Kingdom and New York, NY, USA, 2013.

[7] D.R. Cox. Local ancillarity. *Biometrika*, 67:279–286, 1981.

[8] A. M. Cruddas, N. Reid, and D. R. Cox. A time series illustration of approximate conditional likelihood. *Biometrika*, 76(2):231–237, 1989.

[9] N.S. Diffenbaugh, D. Singh, J.S. Mankin, D.E. Horton, D.L. Swain, D. Touma, A. Charland, Y. Liu, M. Haugen, M. Tsiang, and B. Rajaratnam. Quantifying the influence of global warming on unprecented extreme climate events. *PNAS*, 114(19):4881–4886, 2017.

[10] R. Dole, M. Hoerling, J. Perlwitz, J. Eischeid, P. Pegion, T. Zhang, X.-W. Quan, T. Xu, and D. Murray. Was there a basis for anticipating the 2010 Russian heat wave? *Geophysical Research Letters*, 38:L06702, doi:10.1029/2010GL046582, 2011.

[11] Imme Ebert-Uphoff and Yi Deng. Causal discovery for climate research using graphical models. *Journal of Climate*, 25:5648–5665, 2012.

[12] K. Emanuel. Aseessing the present and future probability of Hurricane Harvey's rainfall. *PNAS*, 114(48):12681–12684, 2017.

[13] M. Ghosh, N. Reid, and D.A.S. Fraser. Ancillary statistics: A review. *Statistica Sinica*, 20:1309–1332, 2010.

[14] L.J. Gleser. Estimation in a multivariate "errors in variables" regression model: Large sample results. *Annals of Statistics*, 9:24–44, 1981.

[15] C.W. Granger. Investigating causal relations by econometric models and cross-spectral methods. *Econometrica*, 37(3):424–438, 1969.

[16] A. Hannart and P. Naveau. Probabilities of causation of climate changes. *Journal of Climate*, doi:10.1175/JCLI-D-17-0304.1, in press, 2018.

[17] A. Hannart, J. Pearl, F.E.L. Otto, and M. Ghil. Causal counterfactual theory for the attribution of weather and climate-related events. *BAMS*, 97:99–110, 2016.

[18] Alexis Hannart. Integrated optimal fingerprinting: Method description and illustration. *Journal of Climate*, 29(6):1977–1998, 2016.

[19] Alexis Hannart, Aurélien Ribes, and Philippe Naveau. Optimal fingerprinting under multiple sources of uncertainty. *Geophysical Research Letters*, 41(4):1261–1268, 2014.

[20] J. Hansen, M. Sato, and R. Ruedy. Perception of climate change. *PNAS PLUS*, pages https://protect–us.mimecast.com/s/ K0T4CW6w8Df5zOmN0u65gIKy?domain=pnas.org, 2012.

[21] K Hasselmann. On the signal-to-noise problem in atmospheric response studies. In D.B. Shaw, editor, *Meteorology of Tropical Oceans*, pages 251–259. Royal Meteorological Society, 1979.

[22] K Hasselmann. Optimal fingerprints for the detection of time-dependent climate change. *Journal of Climate*, 6(10):1957–1971, 1993.

[23] K. Hasselmann. Multi-pattern fingerprint method for detection and attribution of climate change. *Climate Dynamics*, 13:601–611, 1997.

[24] Gabriele Hegerl and Francis Zwiers. Use of models in detection and attribution of climate change. *Wiley Interdisciplinary Reviews: Climate Change*, 2(4):570–591, 2011.

[25] Gabriele C Hegerl, Hans von Storch, Klaus Hasselmann, Benjamin D Santer, Ulrich Cubasch, and Philip D Jones. Detecting greenhouse-gas-induced climate change with an optimal fingerprint method. *Journal of Climate*, 9(10):2281–2306, 1996.

[26] M. Hoerling, J. Eischeid, X. Quan, and T. Xu. Explaining the record US warmth of 2006. *Geophysical Research Letters*, 34:L17704, doi:10.1029/2007GL030643, 2007.

[27] M. Hoerling, A. Kumar, R. Dole, J. Nielsen-Gammon, J. Eischeid, J. Perlwitz, X.-W. Quan, T. Zhang, P. Pegion, and M. Chen. Anatomy of an extreme event. *Preprint*, pages https://protect–us.mimecast.com/s/ 6Glr-CXD7MEUXB2okmi6OqbCz?domain=esrl.noaa.gov, 2012.

[28] David Hume. *An Enquiry Concerning Human Understanding*. Dover, 2004.

[29] Chris Huntingford, Peter A. Stott, Myles R. Allen, and F. Hugo Lambert. Incorporating model uncertainty into attribution of observed temperature change. *Geophysical Research Letters*, 33(5):L05710, 2006.

[30] C.C. Jaeger, J. Krause, A. Haas, R. Klein, and K. Hasselmann. A method for computing the fraction of attributable risk related to climate damages. *Risk Analysis*, 28(4):815–823, 2008.

[31] Matthias Katzfuss, Dorit Hammerling, and Richard L. Smith. A Bayesian hierarchical model for climate change detection and attribution. *Geophysical Research Letters*, 44(11):5720–5728, 2017. 2017GL073688.

[32] M. Kulldorf. A spatial scan statistic. *Communications in Statistics — Theory and Methods*, 26(6):1481–1496, 1997.

[33] O. Ledoit and M. Wolf. A well-conditioned estimator for large-dimensional covariance matrices. *Journal of Multivariate Analysis*, 88(2):365–411, 2004.

[34] Richard A Levine and L. Mark Berliner. Statistical principles for climate change studies. *Journal of Climate*, 12:564–574, 1999.

[35] F. C. Lott, P. A. Stott, D. M. Mitchell, N. Christidis, N. P. Gillett, L. Haimberger, J. Perlwitz, and P. W. Thorne. Models versus radiosondes in the free atmosphere: A new detection and attribution analysis of temperature. *Journal of Geophysical Research: Atmospheres*, 118(6):2609–2619, 2013.

[36] M.D. Mastrandrea, C.B. Field, T.F. Stocker, O. Edenhofer, K.L. Ebi, D.J. Frame, H. Held, E. Kriegler, K.J. Mach, P.R. Matschoss, G.-K. Plattner, G.W. Yohe, and F.W. Zwiers. Guidance note for lead authors of the ipcc fifth assessment report on consistent treatment of uncertainties. *Intergovernmental Panel on Climate Change*, www.ipcc.ch, 2010.

[37] S.-K. Min, X. Zhang, F.W. Zwiers, and G.C. Hegerl. Human contribution to more-intense precipitation extremes. *Nature*, 470:378–381, 2011.

[38] J. Neyman and Elizabeth L. Scott. Consistent estimates based on partially consistent observations. *Econometrica*, 16(1):1–32, 1948.

[39] G. North and M. Stevens. Detecting climate signals in the surface temperature record. *Journal of Climate*, 11(4), 1998.

[40] Mohamed N. Nounou, Bhavik R. Bakshi, Prem K. Goel, and Xiaotong Shen. Process modeling by Bayesian latent variable regression. *AIChE Journal*, 48(8):1775–1793, August 2002.

[41] National Academy of Sciences. *Attribution of Extreme Weather Events in the Context of Climate Change*. National Academies Press, Washington, D.C., 2016.

[42] G.J. van Oldenborgh, K. van der Wiel, A. Sebastian, R. Singh, J. Arrighi, F. Otto, K. Haustein, S. Li, G. Vecchi, and H. Cullen. Attribution of extreme rainfall from Hurricane Harvey, August 2017. *Environmental Research Letters*, 12:124009, 2017.

[43] P. Pall, T. Aina, D.A. Stone, P.A. Stott, T. Nozawa, A.G.J. Hilberts, Lohmann D., and M.R. Allen. Anthropogenic greenhouse gas contribution to flood risk in England and Wales in autumn 2000. *Nature*, 470:382–386, 2011.

[44] P. Pall, C.M. Patricola, M.F. Wehner, D.A. Stone, C.J. Paciorek, and W.D. Collins. Diagnosing conditional anthropogenic contributions to heavy Colorado rainfall in September 2013. *Weather and Climate Extremes*, 1706.03388, 2018.

[45] Pardeep Pall, Tolu Aina, Dáithí A Stone, Peter A. Stott, Toru Nozawa, Arno G J Hilberts, Dag Lohmann, and Myles R. Allen. Anthropogenic greenhouse gas contribution to flood risk in England and Wales in autumn 2000. *Nature*, 470:382–386, March 2011.

[46] Judea Pearl. *Causality: Models, Reasoning and Inference (second edition)*. Cambridge University Press, 2009.

[47] S. Rahmstorf and D. Coumou. Increase of extreme events in a warming world. *PNAS*, 108(44):17905–17909, 2011.

[48] Aurélien Ribes, Jean-Marc Azais, and Serge Planton. Adaptation of the optimal fingerprint method for climate change detection using a well-conditioned covariance matrix estimate. *Climate Dynamics*, 33:707–722, May 2009.

[49] Aurélien Ribes, Serge Planton, and Laurent Terray. Application of regularised optimal fingerprinting to attribution. Part I: method, properties and idealised analysis. *Climate Dynamics*, 41(11-12):2817–2836, April 2013.

[50] Aurélien Ribes, Francis W. Zwiers, Jean-Marc Azais, and Philippe Naveau. A new statistical approach to climate change detection and attribution. *Climate Dynamics*, 48(1):367–386, Jan 2017.

[51] M.D. Risser and M.F. Wehner. Attributable human-induced changes in the likelihood and magnitude of the observed extreme precipitation during Hurricane Harvey. *Geophysical Research Letters*, 44:1000–1000, 2013.

[52] B.D. Santer, K.E. Taylor, T.M.L. Wigley, T.C. Johns, P.D. Jones, D.J. Karoly, J.F.B. Mitchell, A.H. Oort, J.E. Penner, V. Ramaswamy, M.D. Schwarzkopf, R.J. Stouffer, and S. Tett. A search for human influences on the threman structure of the atmopshere. *Nature*, 382:39–46, 1996.

[53] B. Schaffrin and A. Wieser. On weighted total least-squares adjustment for linear regression. *J. Geod*, 82:415–421, 2008.

[54] C. Schär, P.L. Vidale, D. Lüthi, C. Frei, C. Häberli, M. Liniger, and C. Appenzeller. The role of increasing temperature variability in European summer heatwaves. *Nature*, 427:332–336, 2004.

[55] P.A. Stott, D.A. Stone, and M.R. Allen. Human contribution to the European heatwave of 2003. *Nature*, 432:610–614, 2004.

[56] Peter A. Stott, Nikolaos Christidis, Friederike E. L. Otto, Ying Sun, Jean-Paul Vanderlinden, Geert Jan van Oldenborgh, Robert Vautard, Hans von Storch, Peter Walton, Pascal Yiou, and Francis W. Zwiers. Attribution of extreme weather and climate-related events. *Wiley Interdisciplinary Reviews: Climate Change*, 7(1):23–41, 2016.

[57] K.E. Trenberth, J.T. Fasullo, and T.G. Shepherd. Attribution of climate extreme events. *Nature Climate Change*, 5(8):725–730, 2015.

[58] K. van der Wiel, S.B. Kapnick, G.J. von Oldenborgh, K. Whan, P. Sjoukje, G.A. Vecchi, R.K. Singh, J. Arrighi, and H. Cullen. Rapid attribution of the august 2016 flood-inducing extreme precipitation in south louisana to climate change. *Hydrology and Earth System Sciences*, 21:897–921, 2017.

[59] X. Zhang, J.F. Wang, F.W. Zwiers, and S.-K. Hegerl, G.C.and Min. Attributing intensification of precipitation extremes to human influence. *Geophysical Research Letters*, 40(19):5252–5257, 2013.

[60] F.W. Zwiers and H. von Storch. Taking serial correlation into account in tests of the mean. *Journal of Climate*, 8:336–351, 1995.

[61] F.W. Zwiers, X. Zhang, and Y. Feng. Anthropogenic influence on long return period daily temperature extremes at regional scales. *Journal of Climate*, 24(3):881–892, 2011.

35

Health risks of climate variability and change

Kristie L. Ebi
University of Washington, Seattle

David Hondula
Arizona State University Tempe, Arizona

Patrick Kinney
Boston University School of Public Health

Andrew Monaghan
University Corporation for Atmospheric Research

Cory W. Morin
University of Washington, Seattle

Nick Ogden
Public Health Agency of Canada, Saint-Hyacinthe, Québec, Canada

Marco Springmann
Oxford University, Oxford, UK

CONTENTS

35.1   Introduction

Climate change is happening, is largely human induced, and will have serious consequences for human health (IPCC 2014). The health consequences of climate variability and change are diverse, potentially affecting the burden of a wide range of health outcomes, including illnesses and deaths related to cardiovascular and respiratory ailments, infectious and vector-borne disease, malnutrition, and poor water quality. Changing weather patterns can affect the magnitude and pattern of morbidity and mortality from extreme weather and climate events as well as from changing concentrations of ozone, particulate matter, and aeroallergens (Smith et al. 2014). Changing weather patterns and climatic shifts also may create environmental conditions that facilitate alterations in the geographic range, seasonality, and incidence of climate-sensitive infectious diseases in some regions, such as the spread of malaria into highland areas in parts of Sub-Saharan Africa (Smith et al. 2014). Changes in water availability and agricultural productivity could affect undernutrition, particularly in parts of Africa and Asia (Lloyd et al. 2011). While climate change may benefit some health outcomes in certain regions, the overall balance will be detrimental for health and well-being, especially in low- and low-middle-income countries that experience higher burdens of climate-sensitive health outcomes (Smith et al. 2014).

The pathways between climate change and health outcomes are often complex and indirect, making attribution challenging. Climate change may not be the most important driver of climate-sensitive health outcomes over the next few decades, but could be significant past mid-century. It is a stress multiplier, putting pressure on vulnerable systems, populations, and regions. Efforts to detect the extent to which changing weather patterns are affecting current health burdens and to attribute some of that change to climate change are only just beginning (Ebi et al. 2017). For example, cases of heat-related mortality and Lyme disease in Canada, and Vibrio emergence in the northern Baltic were attributed to recent climate change.

Understanding the health risks of climate change entails multiple unique challenges, including the magnitude, pattern, and rate of climate change over smaller spatial scales are inherently uncertain. Also, weather patterns will continue to change until at least mid-century no matter the extent to which greenhouse gas emissions (drivers of climate change) are reduced in the short term. Further, the magnitude and pattern of climate-sensitive health risks past mid-century will be determined largely by the extent to which emissions are reduced over coming decades and the extent to which health systems and society at large are strengthened to manage current impacts and prepare for projected risks (IPCC 2014). In other words, significant reductions in greenhouse gas emissions (mitigation) in the next few years will be critical to preventing more severe climate change later in the century, but will have limited benefits with respect to changing weather patterns in the short term.

Reducing and managing health risks over the next few decades will require modifying health systems to prepare for, cope with, and recover from health consequences of climate variability and change; these changes are part of what is termed adaptation (Ebi 2009). Adaptation will be required throughout the century, with the extent of mitigation a key determinant of the ability of health systems to manage risks projected later in the century (Smith et al. 2014). No matter the success of adaptation and mitigation, residual risks from climate change will burden health systems, particularly in low- and middle-income countries (LMICs) (Ebi and del Barrio 2016).

35.2   Framework of the health risks of climate variability and change

The magnitude and pattern of the health risks from climate change arise from the characteristics of the hazards created by changing weather patterns, the extent of exposure of human and natural systems to the hazard, the susceptibility of those systems to harm, and their ability to cope with and recover from exposure (IPCC 2012; Steinbruner et al. 2013). A climate-related hazard can alter vulnerability to future events by changing one or more of these factors. For example, experiencing a hazard can alter future exposure to similar hazards when coastal barriers are constructed to reduce the risk of flooding associated with storm surges. Or, individuals and communities can be made more or less susceptible by affecting access to and/or effective functioning of healthcare facilities or the proportion of the population vulnerable to an event.

From the perspective of the health sector, *vulnerability* is viewed as the summation of all risk and protective factors that determine whether an individual or subpopulation experiences adverse health outcomes from exposure (Balbus and Malina 2009). One element of vulnerability is *sensitivity*, which is an individual or subpopulation's increased responsiveness to an exposure, often for biological reasons such as the presence of a chronic disease. The poor, children, pregnant women, individuals with chronic medical conditions, and individuals with mobility and/or cognitive constraints are at increased risk of adverse health outcomes during an extreme event (Balbus and Malina 2009). In addition, the social determinants of health influence vulnerability. These include access to health care services, access to and quality of education, availability of resources, transport options, social capacity, and social norms and culture.

Environmental, social, economic, and governance/institutional factors influence vulnerability to climate-related hazards (Cardona et al. 2012). Environmental dimensions include physical variables (e.g. location-specific context for human-environment interactions); geography, location, and place; and settlement patterns and development trajectories. Social and behavioral dimensions include demographic variables (education, human health and well-being); cultural variables; and institutions and governance. Crosscutting factors include relevant and accessible science and technology. In the health sector, important factors include the health of the population and the status of health systems (e.g. ability of healthcare facilitates, laboratories, and other parts of the health system to manage an extreme event).

35.3   Quantifying the associations between weather/climate and health

Exploring data on i) cases of human disease; ii) human sensitivity to disease risks; and iii) the level of risk to which humans are exposed can provide useful predictive relationships of the effects of weather and climate on health. The latter can be subdivided into the level of risk in the environment and the frequency and intensity with which humans are exposed to this risk. The strengths and weaknesses of these data are described and some recommendations made for working with them, considering the types and quality of data may have greater or lesser value for different analysis objectives.

35.3.1   Cases of adverse health outcomes

Human case data are often amongst the most difficult to obtain and use. Privacy concerns mean that data access is often difficult. In some jurisdictions, privacy laws severely limit the utility of human case data by limiting the spatio-temporal precision of data on cases and provision of information about affected patients; these may act as confounders on environmental data (Ogden et al. 2015). Further, while human cases have the advantage of being collected in some sort of systematic way during surveillance in many countries for many health outcomes, considerable caution must be used in employing them as outcome variables in statistical models.

- Case definitions vary over time (e.g. Bacon et al. 2008) and these have to be carefully accounted for as they can alter the accuracy of case detection. As a consequence, the numbers of reported cases can be misinterpreted as increasing or decreasing in incidence associated with weather or climate in medium-term and long-term longitudinal analyses.

- The accuracy and timeliness of diagnostic tests or modeling algorithms also impact the validity of comparisons of human case incidence and environmental variables. In Africa, many cases of infection are attributed in malaria when other infectious diseases are more likely the cause, even though they may be reported as malaria (Ghai et al 2016).

- There are inherent difficulties in using case data for attributing changes in environmental variables with changes in incidence. The main difficulties are with linking the place and time of exposure to environmental variables with disease case data. Temporally, illnesses associated with air quality require accumulated exposures to manifest. It is difficult to appropriately link these accumulated exposures to detectable disease outcomes (which themselves are often not specifically caused by the exposure). In most cases, this requires large sample sizes and long time series that increase uncertainties about linking exposure to individual cases (Ebi et al. 2017; Stieb et al. 2005). Vectorborne (mosquito-borne or tick-borne) infectious diseases are for the most part acquired by the single bite of an infected arthropod. Many infectious diseases, including West Nile virus or other arboviruses (Rudolph et al 2014), have relatively short latent periods (days) between infection and onset of clinical manifestations and likely diagnosis. However, others, such as Lyme disease, can have relatively long latent periods between infection and likely diagnosis (weeks to months) meaning that precision in the location and time of infection becomes difficult even if some information on location of infection is obtainable (Wormser et al. 2006; Ogden et al. 2015).

- Surveillance systems themselves may reduce the utility of data for attribution to possible environmental causes. This may occur simply because the system itself or the method of reporting actually prevents the direct association of human cases with environmental exposures. For example, under national surveillance in the U.S. for infectious diseases, the geo-locator for human cases is the county of residence, which may be very different from the county where disease exposure occurred. Changes in the number of cases can be mistakenly associated with environmental conditions rather than at the site of acquisition. For Lyme disease, this has led to changes in incidence in U.S. states being attributed to climate change when part of the data are from states in the southern and central U.S., such as Texas and Colorado, where most or all Lyme disease cases would be travel acquired (Tuite et al 2013; Forrester et al 2016)

35.3.2 Human sensitivity to disease risk

Population subgroups are differentially sensitive to disease risk (Balbus and Malina 2009). For example, older adults, children, individuals with certain chronic diseases such as diabetes, and individuals taking certain medications such as beta-blockers, and individuals who are less physically fit for their age, are more sensitive to exposure to high ambient temperatures. Data on these particularly vulnerable groups are infrequently available, reducing the ability to quantify exposure-response relationships at the level at which health systems can develop the most effective policies and programs.

35.3.3 Exposure risk

It is clear that estimation, wherever possible, of risk values in the environment is far more immediate and spatiotemporally precise than measuring health outcomes in the human population. This is because these are much easier to precisely measure in the environment, be that air quality, water quality, or the level of risk from vectorborne disease estimated by the abundance of host-seeking infected vectors, than it is to measure human disease outcomes (Gertler et al. 2014; Ogden et al. 2014; Højris et al. 2016; Schneider et al. 2016). There is often much more confidence, therefore, in analyses that involve measurement of the levels of disease risk in the environment as the outcome when exploring the effects of climatic explanatory variables in spatial and time-series analyses. An ongoing issue, however, is the need to quantify (and not merely identify) the relationship between levels of causes of risk in the environment, and incidence of human disease. This is often difficult for all of the reasons identified in the human case data section (e.g. Hinckley et al. 2016).

35.3.4 Analyzing associations between exposures to weather/climate variability and adverse health outcomes

Robust epidemiology and statistical analysis are needed to associate and measure the effects of environmental factors/changes such as climate and climate change with disease outcomes or measures of risk in the environment, while eliminating possible sources of bias and confounding. There is a robust and growing literature base describing and quantifying associations between weather, climate variability, climate, and adverse health outcomes. These analyses typically focus on event-based analyses (e.g. morbidity and mortality associated with a heatwave or other extreme weather or climate event); time-series analyses to estimate short-term associations between climate-sensitive health outcomes and environmental variables; and climate and/or climate variability via analyses of long time series. All analyses use standard statistical approaches that are described in other chapters. The only difference is how the exposure is conceptualized and quantified, as discussed in the following sections. All analyses should account for possible confounding factors associated with both the exposure and the health outcome.

Time series analysis has become a commonly and widely applied statistical toolset for quantifying exposure-response functions linking temperature to adverse health outcomes (e.g. Egondi et al. 2012; see chapter on time series in this volume). Most often the interest is in understanding how temporal variations in weather variables influence health outcomes, which leads to estimating exposure-response functions using regression models. Regression assumes that the outcome measures are independent. However, with time series data, outcomes tend to be correlated in time. Thus, an important feature of time series regression models is the ability to remove or control for serial correlation in the data. Other issues with time series analyses include spatial autocorrelation, such as the level of population immunity, seasonal variations within the time series, and appropriate lag times (Imai and

Hashizume 2015). Aggregation of data on exposures, risk levels, and health outcomes are significant issues. Choosing the right time window is crucial to aligning exposures and health outcomes in biologically meaningful ways.

While time series analysis methods could be applied to climate-related hazards besides temperature (e.g. storms, floods), to date most research focused on temperature effects. Therefore, we focus on the application of several statistical methods for analyzing time series of temperature data and health outcomes, especially mortality. These methods have widespread application because of the ready availability of multi-year records of daily temperature measurements and daily death counts in many places in the world. Another reason is the ease of analysis using freely available software for generalized linear, additive, and/or nonlinear modeling, particularly in R. In contrast to event-based analyses that depend on the occurrence of truly extreme events, time series analysis can reveal and quantify small but important functional dependencies of health outcomes on short-term variations in temperature that might be difficult to discern in the data. Time series analyses typically include possible lagged and/or accumulated effects. Both hot and cold temperatures are associated with excess deaths, leading to U-or J-shaped relationships, necessitating non-linear statistical models to capture the full range of impacts.

35.3.5   Spatial analyses

Geography, location, and place are important determinants of climate-sensitive health risks. All components of the risk management framework—hazard, exposure, and vulnerability—vary spatially at scales ranging from people and households to continents and the entire globe (IPCC 2012). Climate-health researchers are interested in understanding and measuring the spatial patterns of each of these components to identify places and populations that may be at elevated risk of adverse health events, at present and projected into the future, and to understand the drivers of observed spatial variability.

Two major thrusts of spatial analysis recently emerged in the climate-health community. The first leverages theoretical notions of vulnerability, susceptibility, hazard, and exposure, to produce maps of *expected* patterns of adverse health impacts associated with climate-sensitive hazards. The second leverages spatially explicit health outcome data to relate *observed* patterns to spatially varying determinants of risk. Spatial variation in mortality and/or morbidity records with known geographic identifiers (e.g. address, latitude and longitude coordinates, census tract, postal code) can be related to risk factors anticipated to contribute to the variation in the outcome data. The risk factors used in these types of analyses are the same as those considered in vulnerability mapping (section 35.3.6), but here those variables serve as independent variables for statistical models in which mortality or morbidity records are the dependent variables, or serve as the basis for derived dependent variables such as a relative risk of mortality or morbidity. The methods used for these types of assessments largely draw from the fields of health geography, spatial ecology, and spatial epidemiology. A wide range of dependent variables and statistical models have been used—a decision often based on the temporal resolution at which health data are available or have been aggregated.

Aggregation methods include summation or averaging of the number of cases of a particular outcome over a given time span (e.g. during/after a specific hazard event, an entire season/year, multiple years) and classification of outcome occurrence (e.g. a binary yes/no dichotomization). The choice in aggregation scheme determines the types of statistical approaches available for measuring spatial associations, assuming all necessary modeling assumptions are met. Binary indicators, for example, are well-suited for logistic regression, whereas a continuous indicator of incidence rate or relative risk makes it possible to use ordinary least squares linear regression or non-linear regression models. Many studies adopt an ecological approach in which mortality or morbidity records are aggregated to continuous spatial units like census tracts or postal codes.

The modifiable area unit problem (MAUP) is a challenge known to geographers and spatial statisticians by which the results of a spatial analysis are in part dependent on the manner in which the boundaries are drawn between units, irrespective of spatial scale (Fotheringham and Wong 1991; Arsenault et al. 2013). Other studies use area-level effect estimates as the dependent variable for second-stage modeling or directly include spatial variables in time series and case-crossover designs as effect measure modifiers (e.g. Gronlund et al. 2016). Data mining techniques such as classification and regression trees and neural networks can be used to understand spatial variation in climate-sensitive health outcomes, which rely on fewer assumptions about the underlying structure of the data and can help model unknown but complex, non-linear associations and interactions between multiple independent variables and the health outcome.

Co-production of spatial risk assessment strategies with end-users is important, because evidence suggests they have not yet realized their full potential in practice (Wolf et al. 2015). These analyses are intended to provide decision support for planners, preparedness coordinators, and policymakers addressing short-term preparedness and response needs as well as longer-term, systemic drivers of risk in complex socio-ecological systems. Environmental justice concerns are especially well suited for spatial analysis because historical development and planning practices have influenced the places where more vulnerable individuals live around the world, often to the detriment of their health (Bolin et al. 2013).

35.3.6 Vulnerability mapping

Vulnerability maps are useful for identifying where the societal impacts—including those related to health—of climate-sensitive hazards are likely to be most severe. Approaches to vulnerability mapping vary widely for different research and planning applications (see Preston et al. 2011 for a thorough review). A consistent theme in vulnerability mapping studies is the interest in understanding the spatial distribution of known or anticipated health risk factors. For example, health agencies interested in protecting populations most vulnerable to extreme heat in their jurisdiction might produce at the census tract scale a map of the population density of elderly people and a map of median income. A more comprehensive approach involves the creation of a vulnerability index that aggregates many different risk factors into one quantitative score. These indices are often generated using dimension reduction techniques to reduce collinearity such as principal components analysis (PCA), in which the orthogonal (transformed) components are summed to produce the overall score (e.g. Cutter et al. 2003, Reid et al. 2009, Wolf and McGregor 2013). Once scores are created for each geographic region, the mapping scheme and/or classification of risk (e.g. "high vulnerability") can inform developing and deploying decision-support tools.

Vulnerability mapping tends to require several subjective decisions. The weighting of different principal components early in the analyses will influence the final vulnerability score. Many researchers assume that equal weighting of components is appropriate in the absence of information that would justify an alternative choice. At the beginning of the process of vulnerability mapping, variable selection itself may have a significant influence on the final product. Sensitivity analysis performed at each stage of the process can help reveal the impact of each subjective decision on the assessment and encourage deliberation regarding the rationale for particular choices.

Data used in vulnerability maps for demographic and socioeconomic attributes are commonly sourced from publicly available census records. Some researchers expand their analysis to include factors related to landscape characteristics, including land cover/land use and vegetation coverage. Others include factors related to infrastructure, including air conditioning availability, building age, and building materials. Selection of the appropriate spatial scale for vulnerability mapping requires careful consideration of uncertainty in the underlying data sets. In the U.S., for example, the margin of error in census block-level American Community Survey estimates of socioeconomic and demographic characteristics can dwarf contrasts of interest between geographic units (e.g. Bazuin and Fraser 2013, Spielman et al. 2014).

A subset of vulnerability mapping efforts combines information about underlying social and infrastructure systems with information about the anticipated severity of particular hazards to produce *hybrid hazard-vulnerability maps*. As a socially-oriented vulnerability analysis includes no explicit indicators of hazard severity, the addition of hazard or exposure information can help guide resources and planning efforts to communities more likely to be impacted and face disproportionate adverse health effects and challenges in recovery. Examples of the types of data sources used to map hazard severity include projected storm surge inundation zones in coastal regions, floodplain zones near major rivers and washes, and temperature observations across a metropolitan region representing the urban heat island effect (e.g. Emrich and Cutter 2011). Several methods can be used to produce these hybrid maps. One method employs a bivariate mapping scheme in which vulnerability information is plotted using one color/intensity gradient (e.g. red to blue), and the hazard information is plotted using a different gradient (e.g. dark to light). A second method integrates the hazard data itself into the vulnerability index calculation (via principal components analysis or other information) as an additional input variable or set of input variables. This method often necessitates reclassification of the hazard variable into the geographic units for which socioeconomic and demographic data are available (e.g. Harlan et al. 2013).

35.3.7   Cautions and considerations

As true in all disciplines, using statistical ecological approaches to quantify spatial patterns and associations, climate-health researchers employing these techniques must give careful considerations to issues associated with scale, boundaries, spatial autocorrelation, and interpretation of results. Scale and boundary issues include resolving the units at which the analysis can be conducted based on inconsistency in data availability from different sources as well as the scale at which robust effect estimates can be drawn. Spatial autocorrelation is a critical consideration. A fundamental assumption of traditional regression methods is that the cases of independent variables are independent from one another. When dealing with spatial data, this assumption is frequently not met, as observations located more geographically proximate to one another are more likely to be related. Statistical models relating spatial independent variables to spatial dependent variables will frequently require additional terms accounting for the spatial structure of the data (Anselin and Bera 1998) (for applications within the climate context, see Uejio et al. 2011, Hondula et al. 2015). Finally, when reporting the modeled associations from ecological analyses, care must be given to the fact that the associations were derived for the units at which the analysis were conducted—aggregated units or groups, not individuals. Drawing erroneous conclusions about individuals based on group-level analysis is a common misinterpretation known as the ecological fallacy (e.g. Duneier 2006).

A major goal of vulnerability and ecological analyses is coordinating and improving mechanisms, resource management, and response preparedness. However, neither the vulnerability mapping approach nor the outcomes mapping/modeling approach provides perfect information for guiding intervention measures aiming at reducing health threats associated with climate-sensitive hazards. Both approaches face challenges related to scale, sample size, and the quality and availability of necessary input data. In the case of vulnerability mapping, it is difficult to anticipate the extent to which the variables included in index calculations correctly depict true susceptibility to hazards. The variables typically included in such analyses are often surrogates for indicators of risk for which data are not continuously available at the scale of interest. For example, access to backup electrical power from a generator or other source may be protective against adverse health effects associated with a wide variety of hazards, but data about backup generators are not routinely or widely collected. The assumption in vulnerability mapping is that the important resources and mechanisms that will ultimately determine health risks are sufficiently represented via surrogate variables that are routinely collected and easily accessible.

Conversely, maps and models of historically observed adverse health outcomes associated with climate-sensitive hazards have their own limitations for preparedness and response efforts. First and foremost, there is no guarantee that the communities and populations most significantly impacted by prior hazards will be the most impacted by a future event. Spatial analyses based on data sets with long periods of record (e.g. a decade or more) might not be sensitive to ongoing patterns of social and infrastructure change in certain locations. The health data used in such analyses only provide information about one part of the entirety of the exposure that leads to adverse health outcomes. Analyses based on residential address information, for example, do not necessarily align with the locations at which dangerous exposure occurs. Furthermore, most analyses focus on only one or a small set of adverse health outcomes, and thus do not consider the full range of health impacts from mortality through illness and discomfort. Finally, these analyses are only able to reveal associations between different variables and provide no direct evidence of causation, which could lead to under-informed or misguided intervention measures if important drivers of risk are unknown to the analyst.

Direct comparisons between vulnerability and outcomes maps and models can yield insights to improve the techniques used for both. Perhaps more importantly, such comparisons can help facilitate researchers and practitioners identifying the important causal mechanisms associated with spatial variability in adverse health outcomes. One lesson learned is that vulnerability assessments should be tailored to local contexts and settings to the greatest extent possible. The ideal "recipe" for vulnerability mapping likely differs from place to place, and may also differ from hazard to hazard. In an evaluation of a heat vulnerability index in the western United States, Reid et al. (2012) found state-to-state differences in the association between the index and indicators of mortality. Reid et al. (2012) also noted that in certain geographies, the vulnerability index was an equally informative predictor of spatial patterns in health outcomes on days associated with normal and hot weather. In the Phoenix, Arizona metropolitan area, Chuang et al. (2015) reported that a generic heat vulnerability index only identified approximately half of the census tracts with elevated incidence of heat-related hospitalizations. Another lesson is that known vulnerability and risk factors are imperfect representations of the person- and household-level data that could provide more comprehensive information about susceptibility, risk perception, and coping strategies (Harlan et al. 2013).

In the cases where vulnerability and outcomes maps and models misalign, a valuable research opportunity may await the ambitious and creative climate-health researcher. The "off diagonals" of a conventional two-by-two contingency table (i.e. high vulnerability and low outcome incidence or low vulnerability and high outcome incidence) may be particularly informative for future exploration. In the case of communities with low vulnerability and high outcome incidence, unknown variables are leading to higher risk than anticipated or health outcomes are being incorrectly classified. On the other hand, in the case of communities with high vulnerability and low outcome incidence, unknown variables are leading to lower risk than anticipated or health outcomes are being undercounted. In these communities, it is possible that the unknown variables are effective adaptation strategies that are not well represented in conventional analyses. Uncovering these adaptation strategies and disseminating them more broadly could help other communities prepare for climate-sensitive hazards. The best means for uncovering the hidden determinants of climate-sensitive health risks involves methods better suited for building qualitative understandings drawn from sociology and anthropology. A healthy dose of "shoe leather" epidemiology (e.g. engagement in data collection; Duneier 2004), alongside continuing advancements in numerical and statistical techniques, would propel the community of climate-health researchers and practitioners forward in their efforts to reduce preventable causes of morbidity and mortality associated with changing weather patterns, sea level rise, and ocean acidification from climate change.

35.4   Modeling future risks of climate-sensitive health outcomes

Two basic modeling approaches are used to project the future geographic distribution and burden of climate sensitive health outcomes; these approaches can reach different conclusions. The first step is to quantify the relationships between environmental and health variables. Statistically-based approaches (often called empirical) generally use regression or other model-based methods to establish relationships between a health outcome and one or more meteorological and non-meteorological (e.g. socioeconomic, demographic) variables thought to be key drivers of the outcome. Process-based (biological or mechanistic) approaches are deterministic models that generally simulate the impact of weather variables under different scenarios on the health outcome(s) of interest by solving equations describing the instantaneous state of a [disease] system at discrete time steps (e.g. hourly, daily or weekly). Whichever approach is used, projected changes in meteorological variables under different scenarios of climate change are then used to estimate the future distribution and burden of the health outcome. Challenges to process-based modeling include poorly constrained values for parameters in the equations, and the possibility that there are unknown processes that are not described by the model equations.

Quantifying health risks using statistical models generally requires long time series of meteorological variables and some proxy of disease burden or risk (e.g. case data or presence and/or abundance data for a disease vector). Other quasi-static non-meteorological variables (e.g. census data) may also be needed. The datasets need to match in terms of spatial and temporal resolution. This can be difficult, especially in low-income and remote areas where health and meteorological data are unreliable or not collected. In such instances data often need to be aggregated to broader spatial scales or longer temporal scales to ensure robust results, with the tradeoff being that the model simulates outcomes at lower resolution.

Disease burden data are most often the limiting factor when conducting climate and heath analysis because of the geographic scale of the environmental data: remotely sensed and model-based meteorological datasets have global or near-global coverage and although they have ever increasing spatial and temporal resolution, the resolution of the environmental and health data often don't match. The most common approach to modeling climate sensitive health outcomes is using some form of regression analysis including multivariate, logistic, and Poisson; however, other methods are also used (e.g. wavelet analysis, eigenanalysis).

Depending on the chosen time scale and the nature of the relationship, there may be a lag between a given meteorological/climate event and the health outcome. For example, precipitation is often associated with increases in mosquito-transmitted diseases, like dengue fever, by increasing the availability of reproductive habitat. However, there will be a delay between the precipitation and the time it takes for new mosquitoes to develop, take an infectious blood meal, complete the extrinsic incubation period, and infect a new host through a subsequent blood meal. This process can take several weeks and therefore must be accounted for in the regression analysis.

Challenges to empirical modeling include the lack of validated absence data for many locations and projecting non-meteorological variables more than a few years into the future. Further, because the current magnitude and pattern of health burdens are based on past development patterns, the degree to which these relationships are predictive of future occurrence is unclear.

In process-based models, the model equations are solved at discrete time steps (e.g. hours, days, weeks) and thus the state of the system evolves through time. In contrast to empirical models that require little knowledge of a system to develop, process-based models often require detailed knowledge about the system and thus are generally applied to comparatively well-understood systems. Process-based models can be useful for understanding and predicting complex systems that cannot be fully depicted with linear relationships. Vectorborne disease systems are an example, where environmental variables can influence the vector, pathogen, and human population very differently causing feedbacks and nonlinear dynamics. For most environmental health applications, process-based models are driven with meteorological data, although not all models are data driven. Equations are used to describe population (host, vector, reservoir), pathogen life cycle (development, survival, dissemination), and/or transmission dynamics.

Once relationships between weather and other environmental variables and health outcomes are quantified, the next step is to project how the weather / climate variables could change with climate change.

35.4.1 Spatiotemporal issues and other considerations when using climate projections

A number of issues must be considered when applying meteorological fields from climate change projections to process-based or empirically based models to project climate-sensitive outcomes for future periods. These can broadly be characterized as spatial limitations, temporal limitations, and other factors.

Spatial limitations for future climate projections are related to the relatively coarse resolution of global climate models (GCMs) compared to the scales of relevance for human health outcomes. GCMs are complex models that simulate the atmosphere, oceans, sea ice and land surface across the globe and are used to project climate 10s to 100s of years into the future. Because of their complexity and the long periods that they simulate, GCMs must compromise their spatial resolution (the spacing of the model "mesh" that covers Earth) to efficiently run on supercomputers. The most recent generation of GCM simulations, which informed the Fifth Assessment Report of the Intergovernmental Panel on Climate Change (IPCC 2014; IPCC 2013), had a spatial resolution on the order of 100-200 km, depending on the GCM. Such a coarse resolution may not be adequate for accurately depicting climatic gradients near mountains or coastlines. Thus, for applications that use GCM results to assess the impacts of climate change on a given sector (e.g. health, agriculture), the projected meteorological fields are often downscaled to finer spatial resolution using statistical or dynamical approaches. In statistical downscaling, empirical relationships are established between a coarse GCM and observation-based dataset(s) from the recent past. These relationships, which are able to account for some biases in GCMs, are then used to downscale the present and future GCM data to some desired resolution such as 1-2 km (e.g. Wilby et al. 1998). In dynamical downscaling, a regional climate model (RCM) is run to simulate climate variability at finer spatial resolutions over a limited area, with initial and boundary conditions provided by the coarser GCM (e.g. Mearns et al. 2009). Importantly, while statistical and dynamical downscaling techniques allow more realistic climatic gradients to be resolved, they are subject to their own sources of uncertainty due to the downscaling methods. Because most of the climate data used for projecting future health outcomes are derived from downscaled GCM data (e.g. the popular Worldclim dataset of Hijmans et al. 2005), users should be aware that the data may have biases that vary in space and by season. For example, temperatures may be a few degrees higher than they should be during springtime. Fortunately, in addition to the availability of future data, GCM outputs are also available over a historical (baseline) period that represents the recent past. As a best practice, users of downscaled GCM data for health applications should first validate the variables of interest (temperature, precipitation, etc.) for the baseline period versus weather observations taken during that period. This resulting information can then be used to either bias correct the GCM data, or to establish bounds of uncertainty for the future projections.

Temporal limitations are generally less of a constraint than spatial limitations, but nonetheless can pose challenges. The main temporal limitation that users of GCM projections encounter is the low temporal resolution at which the data are available. Until recently, many GCM datasets, particularly if they were downscaled, are only available as monthly averages. Having access to monthly data will not likely be a problem if the GCM data are driving empirical models of health outcomes because they usually employ monthly, seasonal, or annual data. However, process-based models often operate with daily, weekly, or hourly time steps, and thus require GCM inputs with higher temporal resolution. Some downscaled GCM datasets have begun to address the need for higher temporal resolution (NASA 2016). There are also methods for temporally downscaling monthly GCM climate fields to daily time steps, for example by using stochastic weather generators (Wilby et al. 1998). A second potential limitation is that some downscaled GCM datasets may not be available for the desired future periods of interest. For example, the future projections from Worldclim (Hijmans et al. 2005) are available for two future periods: 2041-2060 and 2061-2080; users wishing to examine near-future (2021-2040) or far-future (2081-2100) must find alternative datasets or create a custom dataset. A third aspect for consideration is the need to employ historical and future "time slices" for examining climate change impacts, rather than examining only one or a few years. The reason for doing so is to avoid spurious results due to natural variability that occurs on annual timescales (i.e., one very cold or warm year that may be an outlier) or longer timescales (e.g. El Nino events) and isolate the signal of anthropogenic climate change that operates on decadal and longer timescales. At a minimum, a time slice should be 10 years long, and ideally should be 20 or more years. While one does not necessarily need to average all years of a time slice together (particularly if they are interested in outliers such as the warmest or coldest years), it is important to work with time slices to examine the impacts of climate change in a statistically robust manner.

Other factors must also be considered when modeling climate-sensitive health outcomes. First, the choice of GCM(s) will impact results: there are approximately 20 or so GCMs used by different modeling centers worldwide. While the GCMs have much in common in terms of the basic equations they employ, they each describe climate system processes slightly differently, which leads to different results. As a best practice, when using GCM data to model climate sensitive health outcomes, at least several different GCMs should be used to account for GCM uncertainty. Fields from multiple GCMs (e.g. temperature, precipitation) can either be averaged together before being used to drive a health outcome model, or the health outcome model can be driven by each GCM dataset, and the subsequent 'ensemble' of health outcomes can be assessed. Second, GCMs are run for different future emissions scenarios (see Section 1.6), and the chosen scenario(s) will impact results: the pace at which humans will continue to emit climate altering pollutants (carbon dioxide, methane, aerosols, etc.) in the future is unknown. Humanity may continue emitting at a rapid pace, or may implement mitigation measures to reduce emissions; how these potential emissions pathways play out will have profound impacts on the magnitude of future climate change, and thus must be considered by those who are modeling climate-sensitive health outcomes. Users can do this by employing GCM data from two or more different emissions scenarios. The most recent generation of GCM simulations that informed the IPCC Fifth Assessment Report (IPCC 2013) employed four emissions scenarios, known as Representative Concentration Pathways (RCPs; Moss et al. 2010). The RCPs span a range of possible future emissions scenarios, including high (RCP2.6 and RCP4.5), moderate (RCP6.0), and low (RCP8.5) mitigation of emissions. Third, the choice of a baseline period will impact results: all future projections of climate change or climate sensitive health outcomes are referenced to a baseline historical period for which we already know the outcome. Baseline periods can vary widely. For example, a user might compare projected changes for 2061-2080 versus the 1960-1990 period, or the 1990-2005 period. While the choice of baseline period is often limited to the available GCM data, or by the temporal extent of historical health outcome data, users should be aware that future changes will be sensitive to the chosen baseline. In particular, when comparing results among multiple studies projecting the same health outcome, keep in mind that changes for the same future period may be different because the baseline period differs.

35.4.2   Projections using statistical models

Several steps may be involved in the projections, depending on the goals of the analysis. Once robust relationships are established and highly correlated explanatory variables are properly accounted for using PCA or another method, they can be used in model construction. During this process, the dataset is usually divided into training and testing datasets. The training dataset is used to construct the best-fit model. The model is then evaluated by driving it with the explanatory variables in the test dataset and comparing the results with the health outcome in the test dataset. This will help determine the robustness of the model and aid in quantifying accuracy, but it can be difficult when the health outcome data are limited.

Finally, an accurate and robust model can be used for evaluating risk under future climate regimes using projected climate change data as input. However, it is important to recognize the limitations of this form of analysis. Future climate regimes may be significantly different from the conditions on which the model was constructed, thus forcing the model to extrapolate to make future projections of climate-related health burdens. This can be problematic given some of the nonlinear relationships between environmental variables and health outcomes. Empirical models can also have diminishing accuracy outside of the region for which they were developed, and therefore extrapolating results to different regions (particularly those with greatly different climatic or socioeconomic conditions) can be problematic.

Despite these limitations, empirical models can be very useful. While a general understanding is necessary, extensive knowledge about a particular climate-health outcome system may not be required. Simple models can be useful for exploratory analyses. Under appropriate assumptions and applications, results from broadly-utilized and well-understood methods can be easy to interpret. Because they are built using observational data, they generally have good short-term accuracy, and they may be more accurate than process-based models when applied to the specific region over which they were developed. Consequently, empirical models are best used for identifying environmental-health outcome relationships, especially in subject areas where knowledge is limited, and can be implemented to investigate recent and near future (when environmental conditions have not significantly changed) health risks with respect to environmental drivers.

35.4.3   Projections using process-based models

Process-based models use equations representing known biophysical processes to simulate the mechanisms through which environmental variables can influence health outcomes. Although the models often do not necessarily require health outcome data to build the model (an advantage), outcome data is often needed for model evaluation and parameterization. Parameterization is the process of determining the "constants", or parameters, in the equations that define the process-based model. Some parameter values are not known or well-studied, but adjusting their values can have considerable impact on the model results. Therefore, multiple simulations are generally performed to represent a range (i.e. probability distribution) of possible parameter values (Morin et al. 2015).

Important applications of process-based models include intervention analysis, outbreak investigation, and projections of future risk (as exemplified above). Because the models are built using mechanistic relationships (as opposed to empirical relationships), they are better able to project health outcomes under climate change conditions. However, their potential sensitivity to possibly unknown or changing parameter values should always be considered when performing any form of risk analysis; this is best accomplished by performing multiple simulations under a range of potential conditions and analyzing and reporting the characteristics of the results.

For example, Morin et al. (2015) performed 10,000 simulations of dengue fever cases in San Juan, Puerto Rico and found that the values of two parameters – the quantity and type of water habitats – had significant impacts on simulated cases. Furthermore, the parameter values for the best-fit simulations differed between years and could not be treated as constants. Therefore, the accuracy of process-based simulations can sometimes be questionable without the availability of health outcome data to evaluate and parameterize the model.

Process-based models can be used for a variety of applications because they are built based on the underlying mechanisms through which the system operates instead of empirical associations. For example, Springmann et al. (2016) linked a detailed agricultural modelling framework, the International Model for Policy Analysis of Agricultural Commodities and Trade (IMPACT), to a comparative risk assessment of changes in fruit and vegetable consumption, red-meat consumption, and body weight for deaths from coronary heart disease, stroke, cancer, and an aggregate of other causes. The change in the number of deaths due to climate-related changes in weight and diets were projected for the combination of four greenhouse gas emissions and three socio-economic pathways for 2050. The model projected climate change would lead to per-capita reductions of 3.2% ($\pm$ 0.4%), 4.0% ($\pm$ 0.7%), and 0.7% ($\pm$ 0.1) in global food availability, fruit and vegetable consumption, and red-meat consumption, respectively. Those changes were associated with 529,000 climate-related deaths globally (95% confidence interval (CI): 314,000-736,000), representing a 28% (95% CI: 26-33%) reduction in the number of deaths that would be avoided due to changes in dietary and weight-related risk factors between 2010 and 2050. Twice as many climate-related deaths were associated with reductions in fruit and vegetable consumption than with climate-related increases in the prevalence of underweight, and most climate-related deaths were projected to occur in South and East Asia.

35.4.4   Other considerations

One challenge for statisticians and epidemiologists in projecting the future health risks of climate change is how to estimate exposure-response relationships when temperatures are higher than at present (Rocklov and Ebi 2012). Similar concerns arise when studying health outcomes, such as waterborne diseases, associated with extreme precipitation events. Considerable scholarship on low dose extrapolation has led to agreement on methods and best practices. With climate change altering weather variables and their variability beyond historical trends, the highest exposure values in a particular region are expected to be higher in future decades; under some pathways of greenhouse gas emissions, these values could be much higher than current experience (IPCC 2013). Modelers of the health impacts of ambient temperatures are thus making assumptions about human responses associated with exposures outside the range of their data that significantly affect the magnitude of projected future risks. Improved approaches, such as kernel methods, to estimating the health impacts in the tails of temperature distributions could be a helpful. A further complexity is how to incorporate the extent of adaptation over coming decades. Rocklov and Ebi (2012) present an example suggesting that assuming a linear association continues into the future may result in projections that over- or under-estimate the likely extent of risk. Additional research is needed that explores plausible shapes of future exposure-response relationships, incorporating different assumptions of the magnitude and pattern of climate change and of the timing and effectiveness of adaptation (Ebi et al. 2016).

35.4.5 Scenarios

The climate change research community established a scenario process to facilitate integrated analyses of future climate impacts, vulnerabilities, adaptation, and mitigation, including policy trade-offs and synergies (Riahi et al. 2016). The process is organized around a matrix architecture, with scenarios formed from combinations of (1) pathways of greenhouse gas emissions across this century (RCPs) and (2) pathways describing alternative socioeconomic development (termed Shared Socioeconomic Pathways or SSPs) (O'Neill et al. 2015; van Vuuren et al. 2015). The RCPs are used by earth system modelers to project changes in weather and climate patterns, including sea level rise (IPCC 2013). The SSPs describe development pathways that will lead to different degrees of vulnerability and burdens of climate-sensitive health outcomes over time, before considering climate change (Ebi 2014). The five SSP's are sustainability, regional rivalry, inequality, fossil-fueled development, and a middle-of-the-road development.

Independent of climate change, socioeconomic factors will alter future burdens of climate-sensitive health outcomes and the status of public health and health care infrastructure, making it easier or more difficult to prepare for and effectively manage the health risks of a changing climate (Ebi 2014). Understanding these patterns can be used to inform policy-makers about the extent to which climate change could affect the geographic range, seasonality, and incidence of climate-sensitive health outcomes under different assumptions of future socioeconomic development; about the extent to which adaptation and mitigation policies could avoid those health risks and increase the capacity of the health sector to prepare for, cope with, and recover from the hazards associated with a changing climate, and what these policies could cost; and about how the balance of adaptation and mitigation could alter health burdens over time (Ebi 2014). Considering different socioeconomic futures provides more credible analyses of the range of possible future health risks associated with climate change, promoting understanding of where best to place scarce human and financial resources to manage risks.

For example, in a world aiming for sustainable development (SSP1), population health improves significantly, with increased emphasis on enhancing public health and health care functions that, in turn, increase the capacity to prepare for, respond to, cope with, and recover from climate-related health risks, before considering any impacts of climate change. Coordinated, worldwide efforts through international institutions and non-governmental organizations to achieve sustainable development goals increase access to safe water, improved sanitation, medical care, education, and other factors in underserved populations.

Life expectancies increase in low-income countries with decreasing burdens of key causes of childhood mortality. However, as more children survive to adulthood, burdens of non-communicable diseases increase.

In comparison, in a world characterized by regional rivalry (SSP3), there would be increasing mortality from climate-related health outcomes and possibly falling life expectancy in low-income countries because of increased childhood mortality, although some sub-regions enjoy better health. Low- and middle-income countries experience a double burden of infectious and chronic climate-related health outcomes. Climate-related infectious diseases increase in high-income countries with reduced funding for surveillance and monitoring programs, limited investment in research and technology development, less effective institutions for ensuring food and water safety, and poor access to health care. Large regions of the world are food and water insecure.

Projections of the health risks of climate change are incorporating the RCPs and SSPs. For example, Monaghan et al. (2016) investigated how alternate emissions and socioeconomic pathways could modulate future human exposure to *Ae. aegypti*, the mosquito that transmits the viruses that cause dengue and chikungunya, two of the most important vector-borne viral diseases globally. Occurrence patterns for *Ae. aegypti* for 2061-2080 were mapped globally using empirically downscaled air temperature and precipitation projections from the Community Earth System Model, for RCPs 4.5 and 8.5. Population growth was quantified using gridded global population projections consistent with the SSP storylines SSP3 (Regional Rivalry) and SSP5 (Fossil-fueled Development). The annual average number of people exposed globally to *Ae. aegypti* was 3.8 billion in the 1950-2000 reference period; this number was projected to statistically significantly increase by 8-12% by 2061-2080 if only climate change was considered, and by 127-134% (SSP3) and 59-65% (SSP5) if climate and population change were considered (lower and upper values of each range are for RCP4.5 and RCP8.5 respectively). Regionally, Australia, Europe and North America were projected to have the largest percentage increases in human exposure to *Ae. aegypti* resulting from climate change alone.

35.5 Conclusions

As detailed in this chapter, the unique nature of climate change as a changing "exposure" adds complexity to approaches to quantifying current associations between weather / climate variables and health outcomes, including time series analyses (e.g. temperature-mortality associations) and spatial regression / ecological analyses (e.g. vulnerability mapping). Further, to be effective in informing policies and interventions to reduce the burdens of climate-sensitive health outcomes, the results of these analyses also need to take into consideration local factors that could affect the magnitude and pattern of the associations. The use of empirical and process-based models of environmentally driven health outcomes over short time scales may have significant value for decision-makers by moving from a culture of surveillance and monitoring to prediction and prevention. Collaborating with end users (policy and decision makers) before, during, and after statistical analyses is critical for creating intuitive, effective, and sustainable models for reducing negative climate-sensitive-outcomes.

The complexities are far greater when modeling future risks of climate-sensitive health outcomes. Consideration is needed not just of changes in weather patterns due to climate change, but also changes in exposure and vulnerability associated with different socio-economic development pathways. Because the future is inherently uncertain, scenarios are used to explore and evaluate the range and character of risks possibly associated with future climate change and development pathways.

With policy- and decision-makers wanting a better understanding of current and projected climate-related health risks, statistical methods will likely evolve to take the unique nature of climate change as an exposure into account. Strengthening the evidence base, including by improving health system and vector surveillance to provide a more robust baseline from which to measure changes in health outcomes, and providing more nuanced projections of future risks can help communities and nations more effectively prepare for an uncertain but undoubtedly different future.

Bibliography

[1] Anselin L, Bera AK. Spatial dependence in linear regression models with an introduction to spatial econometrics. Statistics Textbooks and Monographs. 1998 Feb 3;155:237-90.

[2] Arsenault J, Michel P, Berke O, Ravel A, Gosselin P. How to choose geographical units in ecological studies: proposal and application to campylobacteriosis. Spat Spatiotemporal Epidemiol. 2013 Dec;7:11-24.

[3] Bacon RM, Kugeler KJ, Mead PS; Centers for Disease Control and Prevention (CDC). Surveillance for Lyme disease–United States, 1992-2006. MMWR Surveill Summ. 2008 Oct 3;57(10):1-9.

[4] Balbus JM, Malina C. 2009. Identifying Vulnerable Subpopulations for Climate Change Health Effects in the United States. *J Occup Environ Med* 51 (1): 33–37.

[5] Bazuin JT, Fraser JC. How the ACS gets it wrong: The story of the American Community Survey and a small, inner city neighborhood. Applied Geography. 2013 Dec 31;45:292-302.

[6] Cardona OD, Ordaz MG, Reinoso E, Yamin L, Barbat B, et al. 2012. CAPRA: Comprehensive Approach to Probabilistic Risk Assessment; International Initiative for Risk Management Effectiveness. A: World Conference on Earthquake Engineering. Paper prepared for the 15th World Conference on Earthquake Engineering, Lisbon.

[7] Descloux E, Mangeas M, Menkes CE, Lengaigne M, et al (2012) Climate-based models for understanding and forecasting dengue epidemics. PLoS Negl Trop Dis 6:e1470.

[8] Ebi KL. 2009. Public health responses to the risks of climate variability and change in the United States. *J Occup Environ Med* 51:4-12

[9] Ebi KL. 2014. Health in the new scenarios for climate change research. *Int J Environ Res Public Health*. 10, 1-x manuscripts; doi:10.3390/ijerph100x000x

[10] Ebi KL, Otmani del Barrio M (2016) Lessons learned on health adaptation to climate variability and change: experiences across low- and middle-income countries. *Environ Health Perspect* (in press).

[11] Ebi KL, Ziska LH, Yohe GW. 2016. The shape of impacts to come: lessons and opportunities for adaptation from uneven increases in global and regional temperatures. Climatic Change https://link.springer.com/article/10.1007/s10584-016-1816-9.

[12] Egondi, T., Kyobutungi, C., Kovats, S., Muindi, K., Ettarh, R., & Rocklöv, J. (2012). Time-series analysis of weather and mortality patterns in Nairobi's informal settlements. Global Health Action, 5.

[13] Focks DA, Haile DG, Daniels E, Mount GA (1993) Dynamic life table model for *Aedes aegypti* (L.) (Diptera, Culicidae)–analysis of the literature and model development. J Med Entomol 30:1003–1017.

[14] Forrester JD, Brett M, Matthias J, Stanek D, Springs CB, Marsden-Haug N, Oltean H, Baker JS, Kugeler KJ, Mead PS, Hinckley A. Epidemiology of Lyme disease in low-incidence states. Ticks Tick Borne Dis. 2015 Sep;6(6):721-3.

[15] Ghai RR, Thurber MI, El Bakry A, Chapman CA, Goldberg TL. Multi-method assessment of patients with febrile illness reveals over-diagnosis of malaria in rural Uganda. Malar J. 2016 Sep 7;15:460.

[16] Gertler AW, Moshe D, Rudich Y. Urban PM source apportionment mapping using microscopic chemical imaging. Sci Total Environ. 2014 Aug 1;488-489:456-60.

[17] Hinckley AF, Meek JI, Ray JA, Niesobecki SA, Connally NP, Feldman KA, Jones EH, Backenson PB, White JL, Lukacik G, Kay AB, Miranda WP, Mead PS. Effectiveness of Residential Acaricides to Prevent Lyme and Other Tick-borne Diseases in Humans. J Infect Dis. 2016 Jul 15;214(2):182-8.

[18] Højris B, Christensen SC, Albrechtsen HJ, Smith C, Dahlqvist M. A novel, optical, on-line bacteria sensor for monitoring drinking water quality. Sci Rep. 2016 Apr 4;6:23935.

[19] Hosking J, Campbell-Lendrum D (2012) How well does climate change and human health research match the demands of policymakers? A scoping review Environ Health Perspect 120: 1076-1082.

[20] Imai C, Hashizume M. A systematic review of methodology: time series regression analysis for environmental factors and infectious diseases. Trop Med Health. 2015 Mar;43(1):1-9.

[21] Intergovernmental Panel on Climate Change (IPCC). 2014. Climate Change 2014: Synthesis Report, Summary for Policymakers. Pp 32. IPCC, Geneva, Switzerland. https://www.ipcc.ch/pdf/assessment-report/ar5/syr/AR5_SYR_FINAL_SPM.pdf.

[22] IPCC. 2013: Summary for Policymakers. In: Climate Change 2013: The Physical Science Basis. Contribution of Working Group I to the Fifth Assessment Report of the Intergovernmental Panel on Climate Change [Stocker, T.F., D. Qin, G.-K. Plattner, M. Tignor, S.K. Allen, J. Boschung, A. Nauels, Y. Xia, V. Bex and P.M. Midgley (eds.)]. Cambridge University Press, Cambridge, United Kingdom and New York, NY, USA.

[23] Lloyd SJ, Kovats RS, Chalabi Z. 2011. Climate Change, Crop Yields, and Undernutrition: Development of a Model to Quantify the Impact of Climate Scenarios on Child Undernutrition. *Environmental Health Perspectives* 119 (12): 1817–23.

[24] Mearns, L. O., W. J. Gutowski, R. Jones, L.-Y. Leung, S. McGinnis, A. Nunes, and Y. Qian, 2009: A regional climate change assessment program for North America. *EOS Transactions AGU*, **90**, 311.

[25] Monaghan AJ, Sampson KM, Steinhoff DF, Ernst KC, Ebi KL, et al. 2016. The potential impacts of 21st century climatic and population changes on human exposure to the virus vector mosquito Aedes aegypti. Climatic Change DOI 10.1007/s10584-016-1679-0.

[26] Morin CW, Monaghan AJ, Hayden MJ, Barrera R, Ernst K. 2015. Meteorologically Driven Simulations of Dengue Epidemics in San Juan, PR. PloS Negl Trop Dis 9(8):e0004002.

[27] Moss, R. H., J. A. Edmonds, and K. A. Hibbard, 2010: The next generation of scenarios for climate change research and assessment. *Nature,* **463,** 747–756, doi:10.1038/nature08823.

[28] NASA, 2016: NEX Global Daily Downscaled Climate Projections. Accessed September 29, 2016 at: https://nex.nasa.gov/nex/projects/1356/.

[29] Ogden NH, Koffi JK, Lindsay LR, Fleming S., Mombourquette DC, Sanford C, et al. Surveillance for Lyme disease in Canada, 2009-2012. Can Comm Dis Rep 2015;41: 132-45.

[30] O'Neill BC, Kriegler E, Ebi KL, Kemp-Benedict E, Riahi K, Rothman D, et al. 2016. The roads ahead: narratives for shared socioeconomic pathways describing world futures in the 21st century. *Global Environmental Change* doi:10.1016/j.gloenvcha.2015.01.004

[31] Riahi K, Van Vuuren DP, Kriegler E, Edmonds J, O'Neill B, Fujimori S, t al. 2016. The Shared socioeconomic pathways and their energy, land use, and greenhouse gas emissions implications: an overview, *Global Environmental Change,* http://dx.doi.org/10.1016/j.gloenvcha.2015.01.004

[32] Rocklov J, Ebi KL. 2012. High dose extrapolation in climate change projections of heat-related mortality. *JABES* doi:10.1007/s13253-012-0104-z

[33] Rudolph KE, Lessler J, Moloney RM, Kmush B, Cummings DA. Incubation periods of mosquito-borne viral infections: a systematic review. Am J Trop Med Hyg. 2014 May;90(5):882-91.

[34] Schneider J, Valentini A, Dejean T, Montarsi F, Taberlet P, Glaizot O, Fumagalli L. Detection of Invasive Mosquito Vectors Using Environmental DNA (eDNA) from Water Samples. PLoS One. 2016 Sep 14;11(9):e0162493.

[35] Shi Y, Liu X, Kok S-Y, Rajarethinam J, et al (2016) Three-month-real-time dengue forecast models: An early warning system for outbreak alerts and policy decision support in Singapore. Environ Health Perspect 124:1369-1375.

[36] Siraj AS, Santos-Vega M, Bouma MJ, Yadeta D, Carrascal DR, Pascual M (2014) Altitudinal changes in malaria incidence in highlands of Ethiopia and Columbia. Science 343: 1154-1158.

[37] Smith KR, Woodward A, Campbell-Lendrum D, Chadee DD, Honda Y, et al. 2014. Human Health: Impacts, Adaptation, and Co-Benefits. In *Climate Change 2014: Impacts, Adaptation, and Vulnerability; Part A: Global and Sectoral Aspects.* Contribution of Working Group II to the Fifth Assessment Report of the Intergovernmental Panel on Climate Change, edited by C. B. Field, V. R. Barros, D. J. Dokken, K. J. Mach, M. D. Mastrandrea, et al. pp 709–54. New York: Cambridge University Press.

[38] Spielman SE, Folch D, Nagle N. Patterns and causes of uncertainty in the American Community Survey. Applied Geography. 2014 Jan 31;46:147-57.

[39] Springmann M, Godfray HCJ, Rayner M, Scarborough. 2016. Analysis and valuation of the health and climate change co-benefits of dietary change. PNAS 113:4146-4151.

[40] Steinbruner JD, Stern PC, Husbands JL. 2013. *Climate and Social Stress: Implications for Security Analysis; National Climate Assessment.* Washington, DC: National Research Council.

[41] Stieb DM, Doiron MS, Blagden P, Burnett RT. Estimating the public health burden attributable to air pollution: an illustration using the development of an alternative air quality index. J Toxicol Environ Health A. 2005 Jul 9-23;68(13-14):1275-88.

[42] Tuite AR, Greer AL, Fisman DN. Effect of latitude on the rate of change in incidence of Lyme disease in the United States. CMAJ Open. 2013 Apr 16;1(1):E43-7.

[43] Wilby, R. L., T. M. L. Wigley, D. Conway, P. D. Jones, B. C. Hewitson, J. Main, and D. S. Wilks, 1998: Statistical downscaling of general circulation model output: A comparison of methods. *Water Resources Research,* **34**, 2995-3008.

[44] Hijmans, R.J., S.E. Cameron, J.L. Parra, P.G. Jones and A. Jarvis, 2005. Very high resolution interpolated climate surfaces for global land areas. International Journal of Climatology 25: 1965-1978.

[45] Wormser GP, Dattwyler RJ, Shapiro ED, Halperin JJ, Steere AC, Klempner MS, et al. The clinical assessment, treatment, and prevention of Lyme disease, human granulocytic anaplasmosis, and babesiosis: clinical practice guidelines by the Infectious Diseases Society of America. Clin Infect Dis 2006;43: 1089-134.

[46] van Vuuren DP, Kriegler E, O'Neill BC, Ebi KL, Riahl K, Carter TR, Edmonds J, Hallegatte S, Kram T, Mathur R, Winkler H 2014. A new scenario framework for climate change research: scenario matrix architecture. *Climatic Change* 122:373-386. doi:10.1007/s10584-013-0906-1

[47] Benmarhnia, T., Deguen, S., Kaufman, J. S., & Smargiassi, A. (2015). Review article: Vulnerability to heat-related mortality: A systematic review, meta-analysis, and meta-regression analysis. *Epidemiology, 26*(6), 781-793.

[48] Bolin, B., Barreto, J. D., Hegmon, M., Meierotto, L., & York, A. (2013). Double exposure in the sunbelt: the sociospatial distribution of vulnerability in Phoenix, Arizona. In *Urbanization and Sustainability* (pp. 159-178). Springer Netherlands.

[49] Cutter, S. L., Boruff, B. J., & Shirley, W. L. (2003). Social vulnerability to environmental hazards. *Social science quarterly, 84*(2), 242-261.

[50] Duneier, M. (2004). Scrutinizing the heat: On ethnic myths and the importance of shoe leather. *Contemporary Sociology, 33*(2), 139-150.

[51] Duneier, M. (2006). Ethnography, the ecological fallacy, and the 1995 Chicago heat wave. *American Sociological Review, 71*(4), 679-688.

[52] Emrich, C. T., & Cutter, S. L. (2011). Social vulnerability to climate-sensitive hazards in the southern United States. *Weather, Climate, and Society, 3*(3), 193-208.

[53] Fotheringham, A. S., & Wong, D. W. (1991). The modifiable areal unit problem in multivariate statistical analysis. *Environment and planning A, 23*(7), 1025-1044.

[54] Gronlund, C. J., Zanobetti, A., Wellenius, G. A., Schwartz, J. D., & O'Neill, M. S. (2016). Vulnerability to renal, heat and respiratory hospitalizations during extreme heat among US elderly. *Climatic Change*, 1-15.

[55] Harlan, S. L., Declet-Barreto, J. H., Stefanov, W. L., & Petitti, D. B. (2013). Neighborhood Effects on Heat Deaths: Social and Environmental Predictors of Vulnerability in Maricopa County, Arizona. *Environ Health Perspect, 121*, 197-204.

[56] Heaton, M. J., Sain, S. R., Greasby, T. A., Uejio, C. K., Hayden, M. H., Monaghan, A. J., ... & Wilhelmi, O. V. (2014). Characterizing urban vulnerability to heat stress using a spatially varying coefficient model. *Spatial and spatio-temporal epidemiology, 8*, 23-33.

[57] Hondula, D. M., Davis, R. E., Saha, M. V., Wegner, C. R., & Veazey, L. M. (2015). Geographic dimensions of heat-related mortality in seven US cities. *Environmental research, 138*, 439-452.

[58] Preston, B. L., Yuen, E. J., & Westaway, R. M. (2011). Putting vulnerability to climate change on the map: a review of approaches, benefits, and risks. *Sustainability Science, 6*(2), 177-202.

[59] Reid, C. E., O'Neill, M. S., Gronlund, C. J., Brines, S. J., Brown, D. G., Diez-Roux, A. V., & Schwartz, J. (2009). Mapping community determinants of heat vulnerability. *Environmental health perspectives, 117*(11), 1730.

[60] Reid, C. E., Mann, J. K., Alfasso, R., English, P. B., King, G. C., Lincoln, R. A., ... & Woods, B. (2012). Evaluation of a heat vulnerability index on abnormally hot days: an environmental public health tracking study. *Environmental health perspectives, 120*(5), 715.

[61] Uejio, C. K., Wilhelmi, O. V., Golden, J. S., Mills, D. M., Gulino, S. P., & Samenow, J. P. (2011). Intra-urban societal vulnerability to extreme heat: the role of heat exposure and the built environment, socioeconomics, and neighborhood stability. *Health & Place, 17*(2), 498-507.

[62] Wolf, T., Chuang, W. C., & McGregor, G. (2015). On the science-policy bridge: do spatial heat vulnerability assessment studies influence policy?. *International journal of environmental research and public health, 12*(10), 13321-13349.

[63] Wolf, T., & McGregor, G. (2013). The development of a heat wave vulnerability index for London, United Kingdom. *Weather and Climate Extremes, 1*, 59-68.

Index

Printed in the United States
by Baker & Taylor Publisher Services